Lecture Notes in Computer Science 2695

Edited by G. Goos, J. Hartmanis, and J. van Leeuwen

Springer
Berlin
Heidelberg
New York
Barcelona
Hong Kong
London
Milan
Paris
Tokyo

Lewis D. Griffin   Martin Lillholm (Eds.)

Scale Space Methods in Computer Vision

4th International Conference, Scale Space 2003
Isle of Skye, UK, June 10-12, 2003
Proceedings

 Springer

Series Editors

Gerhard Goos, Karlsruhe University, Germany
Juris Hartmanis, Cornell University, NY, USA
Jan van Leeuwen, Utrecht University, The Netherlands

Volume Editors

Lewis D. Griffin
Imaging Sciences
5th Floor Thomas Guy House, Guy's Campus, London, SE1 9RT, UK
E-mail: Lewis.Griffin@kcl.ac.uk

Martin Lillholm
IT University of Copenhagen
Glentevej 67-69, Copenhagen NV, Denmark
E-mail: grumse@it-c.dk

Cataloging-in-Publication Data applied for

A catalog record for this book is available from the Library of Congress

Bibliographic information published by Die Deutsche Bibliothek
Die Deutsche Bibliothek lists this publication in the Deutsche Nationalbibliografie;
detailed bibliographic data is available in the Internet at <http://dnb.ddb.de>.

CR Subject Classification (1998): I.4, I.3.5, I.5, I.2.10, G.1.2

ISSN 0302-9743
ISBN 978-3-540-40368-5 Springer-Verlag Berlin Heidelberg New York

Springer-Verlag Berlin Heidelberg New York
a member of BertelsmannSpringer Science+Business Media GmbH

http://www.springer.de

© Springer-Verlag Berlin Heidelberg 2003

Typesetting: Camera-ready by author, data conversion by Olgun Computergrafik
Printed on acid-free paper   SPIN 10927977   06/3142   5 4 3 2 1 0

Preface

Following previous conferences in Utrecht (1997), Corfu (1999) and Vancouver (2001), this year's Scale Space Conference took place on the Isle of Skye, Scotland, UK. The conference was an independent stand-alone affair following two previous successful meetings as a satellite. This attests to the still-growing interest in the field 6 years after its first dedicated conference and 10 years after the first US-NSF/EC-ESPRIT sponsored workshop.

Scale Space 2003 received 101 abstract submissions of which 88 were submitted as full high-quality papers; this was an increase of 47% over the previous meeting! Fifty-six reviewers completed 288 reviews, resulting in a mean of 3.3 reviews. Reviewing was administered by Richard van de Stadt's excellent Cyber-Chair software, which, through its expertise and bidding mechanisms, ensured a tight fit between papers and reviewers. The results of the reviewing were the primary basis of the Program Board's decision to accept 56 papers (64% of submissions), of which 31 (35% of submissions) were accepted as oral presentations.

We were delighted that three experts in fields related to Scale Space accepted our invitation to join the conference. Aapo Hyvärinen reviewed his and others' work on ICA and images, Mark Georgeson presented psychophysical data on features and explained it using Scale Space style models, and Markus van Almsick described his implementation of Scale Space symbolics and numerics as a Mathematica add-on package.

I would like to express my appreciation to the authors of the submitted papers, and to the members of the program committee, who provided timely and significant reviews. Thanks also to the program board and to Martin Lillholm (assistant chair) for their essential work. Thanks to the British Machine Vision Association for underwriting the conference. Thanks to Carlotta Graham and the other staff at Sabhal Mór Ostaig, the college on Skye where the conference was held. Special thanks to Richard van de Stadt, the creator of the excellent CyberChair package used to administrate the submission, reviewing and proceedings preparation.

London, April 2003   Lewis Griffin

Organization

Scale Space 2003 was organized by the Division of Imaging, School of Medicine, King's College London together with the IT University of Copenhagen, Denmark. The conference was supported by the British Machine Vision Association.

General Board

Olivier Faugeras (INRIA Sophia Antipolis, France)
Lewis D. Griffin (King's College London, UK)
Tony Lindeberg (KTH, Sweden)
Bart ter Haar Romeny (Eindhoven, The Netherlands)

Program Board

Rein van den Boomgaard (Amsterdam, The Netherlands)
Lewis D. Griffin (King's College London, UK)
Joachim Weickert (Saarland, Germany)
Mads Nielsen (IT-U Copenhagen, Denmark)

Program Committee

Miguel Alemán-Flores	Universidad de Las Plamas de Gran Canaria, Spain
Luis Álvarez	Universidad de Las Plamas de Gran Canaria, Spain
Simon Arridge	University College London, UK
Imiya Atsushi	Chiba University, UK
Danny Barash	New York Unversity, USA
Philipp Batchelor	King's College London, UK
Michel Bister	Multimedia University, Malaysia
Rein van den Boomgaard	University of Amsterdam, The Netherlands
Freddy Bruckstein	Technion IIT, Israel
Frederic Cao	INRIA Rennes, France
Laurent Cohen	Université Paris Dauphine, France
James Crowley	INRIA Rhônes Alpes, France
Rachid Deriche	INRIA Sophia Antipolis, France
François Dibos	Université Paris-Dauphine, France

Invited Talks

Mark Georgeson (Aston University, UK)
"Explaining Psychophysical Data on Feature Marking and Blur Matching Using a Scale Space Model of Receptive Fields"

Aapo Hyvärinen (Helsinki University of Technology, Finland)
"Independent Component Analysis and Related Models of the Statistical Structure of Natural Images: Implications for Visual Coding"

Markus van Almsick (Eindhoven University of Technology, The Netherlands)
"Efficient Coding of Scale-Space Applications with Mathematica"

Table of Contents

Temporal Scale Spaces

Shape

Motion & Stereo

Poster Session 1

Poster Session 2

On Manifolds in Gaussian Scale Space

Arjan Kuijper

Image group, IT-University of Copenhagen
Glentevej 67, DK-2400 Copenhagen, Denmark

Abstract. In an ordinary 2D image the critical points and the isophotes through the saddle points provide sufficient information for classifying the image into distinct regions belonging to the extrema (i.e. a collection of bright and dark blobs), together with their nesting due to the saddle isophotes. For scale space images, obtained by convolution of the image with a Gaussian filter at a continuous range of widths for the Gaussian, things are more complicated. Here only scale space saddle points occur. They are related to spatial saddle points and spatial extrema and can thus provide a scale space based segmentation and hierarchy. However, a spatial extremum can be related to multiple scale space saddles. The key to solve this ambiguity is the investigation of both the scale space saddles and the iso-intensity manifolds (the extension of isophotes in scale space) through them. I will describe the different situations that one can encounter in this investigation, which scale space saddles are relevant, give examples and show the difference between selecting the relevant and the non-relevant ("void") scale space saddles.

1 Introduction

In image analysis most operators intrinsically have some size, or scale. Inevitably, if one is not aware of this fact, the wrong scale can be used. One way to avoid this risk is by taking into account all possible, reasonable, scales. The basic operator for just observing an image - an array of discrete numbers - under some almost trivial assumptions like linearity, isotropy, no preferred direction or scale, and separability of the kernel then becomes a *Gaussian filter*, as argued by Florack et al. [10].

The gist of a so-called *Gaussian scale space* originated from Koenderink [16] and Witkin [29] with as starting point the non-enhancement of local extrema of the signal (Witkin) or image (Koenderink) as the scale increases. However, some twenty years earlier scale space was already introduced in Japanese literature by Ijimia [14].

Although several books on scale space, e.g. [8,12,23,27,28], and much literature have appeared and biannually scale space conferences [13,15,25] are held, the exploration of the so-called *deep structure* of Gaussian scale space – i.e. the image at all scales simultaneously – is still premature. Mostly scale space is used as a tool, while the tool itself is not investigated.

In this paper I elaborate on previous work [17,18,19,20,21,22] in which results on deep structure were presented. In section 2 I will summarise the main results

L.D. Griffin and M. Lillholm (Eds.): Scale-Space 2003, LNCS 2695, pp. 1–16, 2003.

of this work leading to a unique uncommitted topological segmentation of the image. This is based on the hierarchical structure implicitly present in the scale space image and the movement of the critical points under blurring (increasing scale). Of particular interest is the combination of *iso-intensity manifolds* and *saddle points*, both in the *scale space image*. They were mentioned by Koenderink [16] and Griffin and Colchester [11]. The extraction of iso-manifolds in linear scale space for feature tracking has been reported by Fidrich [5,6,7]. She used intersections of two manifolds to determine curves representing differential invariants. Eberly discussed the geometrical properties of general scale spaces and appropriate metrics for them [4]. One result is that using a space with an Euclidean metric leads to the linear Gaussian scale space. It appears that the iso-intensity manifolds through the scale space saddles provide necessary and sufficient information for the hierarchy and segmentation [17,18,19,22].

However, sometimes also so-called *void* scale space saddles with their corresponding manifolds occurred [17,21]. In section 3 I will present a classification of the iso-intensity manifolds as they can be found in the scale space image. It will be explained why "void" scale space saddles are *not* of interest for the hierarchy and segmentation.

Section 4 shows some applications and provides "visual" insight of the theory presented. Conclusions will be drawn in section 5.

2   The Deep Structure of Gaussian Scale Space

In view of the ample literature on linear scale space, only a brief introduction is given. The interested reader is referred to the literature mentioned in the introduction and references therein. We note that there are also non-linear scale spaces [12] and that besides a Gaussian scale space also other linear scale spaces exist [3]. In the remainder of this paper "scale space" is used as shorthand notation for "Gaussian scale space".

2.1   Gaussian Scale Space

Let $L(\mathbf{x})$ be an image with $\mathbf{x}$ an n-dimensional spatial variable (point) and L the intensity measured at the point. The *scale space image* $L(\mathbf{x};t)$ is defined as the convolution of L with a Gaussian:

$$L(\mathbf{x};t) = \int_{\mathbb{R}^n} \frac{1}{\sqrt{4\pi t}^n} e^{-\frac{|\mathbf{x}-\mathbf{y}|^2}{4t}} L(\mathbf{y}) \, d\mathbf{y}$$

As one can verify, the Gaussian scale space image satisfies the diffusion equation: $\partial_t L(\mathbf{x};t) = \Delta L(\mathbf{x};t)$ and $\lim_{t\downarrow 0} L(\mathbf{x};t) = L(\mathbf{x})$.

The *critical points* of an image are those points with zero gradient: $\nabla_{\mathbf{x}} L(\mathbf{x}) = 0$. Their type is determined by the *Hessian matrix*, containing all mixed second order derivatives: $H(\mathbf{x}) = \nabla^T \nabla L(\mathbf{x})$. If all eigenvalues have the same sign the point is an *extremum*, a *minimum* if all are positive and a *maximum* if all are negative. If the eigenvalues are both positive and negative, the point is a *saddle point.* If an eigenvalue is zero (and consequently $\det H = 0$), the point is degenerate and called *catastrophe point* [26] . For images this is non-generic, since it demands an extra requirement.

In scale space the critical points lie on a *critical curve*: $\nabla_{\mathbf{x}} L(\mathbf{x}; t) = 0$. The critical curve consists of (spatial) extrema and saddle points as well as their connection points, the catastrophe points. At these points *generically* exactly one extremum and one saddle point are annihilated or created at increasing scale. For a thorough treatment of the non-trivial transfer of catastrophe theory [26] into scale space, see e.g. Damon [1,2]. It roughly boils down to the fact that scale can be regarded as an extra parameter enabling the most simple catastrophe demanding only one parameter. So generically, critical curves don't intersect. The part of the critical curve containing extrema is called *extremum branch*, the one containing saddle *saddle branch*.

Under some conditions of the initial image like non-negativeness of the intensity and presence of a Jordan curve around all extrema [24], the image at some large scale contains only one extremum. As a consequence, all but one critical curves exist on a finite interval of scales. Most of them start and end in the initial image, even with "wiggles" when the curve increases, decreases and increases again due to the presence of a creation event [17,20]. It should be noted that also closed loops can occur [17,20].

2.2   Critical Points in Scale Space

At *Scale space critical points* $\nabla_{\mathbf{x}} L(\mathbf{x}; t) = 0$ and $\partial_t L(\mathbf{x}; t) = 0$. The first requirement implies that scale space critical points are located on a critical curve. The second requirement equals $\Delta L(\mathbf{x}; t) = 0$ due to the diffusion equation. Consequently, it implies a zero trace of the (spatial) Hessian matrix, and thus both positive and negative eigenvalues, and scale space critical points *must* be scale space saddles [22]. Note that for signals ($n = 1$) catastrophe points coincide with scale space saddles. For images of dimension larger than one this is non-generic.

From the definition of a scale space saddle it follows directly that at a critical curve the intensity has a local extremum at the scale space saddle: it will change from increasing ($\partial_t L(\mathbf{x}; t) > 0$) to decreasing ($\partial_t L(\mathbf{x}; t) < 0$) or vice versa. Consequently, scale space critical points can be found easily when the critical curve and its intensities are known. It thus makes sense to visualise the intensity of the critical curve as a function of scale. The extremum branches are monotonically decreasing (maximum) or increasing (minimum). Saddle curves wiggle if scale space saddles are present. At catastrophe points two curves come together with the same non-zero sign of $\partial_t L$ [11]. This is clear from the fact that when parameterising the critical curve in this whole, the catastrophe point is a regular point [9].

2.3   Iso-intensity Manifolds

Iso-intensity manifolds in scale space that encapsulate extrema always vanish at some scale. This is due to the property of "non-enhancement of local extrema",

Fig. 1. Iso-intensity manifolds and critical curves in scale space. From left to right the intensity increases. a) Two manifolds with one top. Each one intersects an extremum branch. b) They touch at the scale space saddle. c) One iso-intensity manifold remains with two tops, at each extremum branch one. d) The extremum manifold intersects only the left extremum branch. e) The nesting of the manifolds.

"maximum principle", or "non-creation of new level lines" [16,23]. Note that this does not forbid creations of critical points. Consequently, iso-intensity manifolds in $(\mathbf{x}; t)$-space form closed "realms" (Koenderink [16]). They can be visualised as a mountain landscape [21] (a surface in a 3D scale space image, etc.), in which one or multiple tops are present. The number of tops depends on the number of spatial extrema on the manifold. This is visualized in Figure 1, where four different iso-intensity manifolds are shown, together with the critical curves. Each top of a manifold intersects an different extremum branch.

2.4   Scale Space Hierarchy

As Figure 1 may indicate, the nesting of the iso-intensity manifolds impose a hierarchical structure on the original image. Starting with the intensity of one extremum at the initial image, one finds a nesting of iso-intensity manifolds around this extremum (cf. the nesting of isophotes around an extremum in 2D).

At some intensity the manifold touches another manifold emerging from another extremum, cf. Figure 1b. This occurs at a scale space saddle, corresponds via the saddle branch and its critical curve to one of the extrema. So unique segments in scale space can be assigned to an extremum and the nesting of them results in the topological hierarchy tree, with scale space segments as branches and information of the scale space saddle and the catastrophe point as node (cf. Figure 1, where the right segment is related to the annihilating extremum). It enables an topological segmentation, since to each extremum a region is assigned [17,18,19].

In [18] the authors reported the existence of multiple scale space saddles on a saddle branch. This obviously results in multiple iso-intensity manifolds and extra nodes in the hierarchy tree. Questions rise in how far this is redundant or essential information. It also may influence the positions of the nodes in the hierarchy tree. In the following section I will answer these questions by classifying the manifolds that may be present. The key to solve this potential problem is given by the concept of *void scale space saddle*. This subset of scale space saddles is defined as those scale space saddles that do *not* connect two distinct iso-intensity manifolds.

3   Classification of Manifolds

In this section the different types of manifolds in Gaussian scale space are given, as far as they are relevant for the hierarchy as described in the previous section. All manifolds in scale space contain either one or zero scale space saddles - by genericity, just as an isophote contains either one or zero spatial saddles. However, as will be clear in the remainder of this section, not all scale space saddles lie on iso-intensity manifolds that give rise to distinct segments in scale space. The segments are related to extrema, which are in turn related to extrema branches and consequently saddle branches. Therefore a classification is give based on the number of scale space saddles that a saddle branch contains.

In this classification we will only address critical curves containing a saddle branch that annihilates with a maximum. The saddle - minimum variant follows analogously. Since we are especially interested in the behaviour of the iso-intensity manifolds through scale space saddles in relation with the hierarchical structure and induced topological segmentation, the combination annihilation - scale space saddle is taken. In this context creations are not relevant, since they are only protuberances of the critical curve [17,20]. In the examples shown one can consider the first scale taken as either the initial scale or the creation scale. This makes no difference for understanding.

In the examples sets of two connected plots are given. The first plot shows the intensities of the saddle and maximum branch as a function of increasing scale. Consequently, the intensity of the maximum decreases monotonically: all its eigenvalues are negative and so their sum, equalling the trace of the Hessian and the scale derivative. The saddle branch is allowed to have multiple local extrema: the scale space saddles. The second plot visualises the iso-intensity manifolds in the 2D $(x;t)$-space, reducing them thus to isophotes. The remaining dimensions can be visualised by imagining this as a cross-section through a mountain landscape containing bridges at scale space saddles, cf. Figure 1.

Special attention must be paid to non-generic events. At a catastrophe points, $\det H = 0$, so one of the eigenvalues λ_i equals zero. The case that more eigenvalues are zero is non-generic. The same holds for a combination of them being zero. So generically, $\operatorname{tr} H \neq 0$ at a catastrophe point. For two dimensional images this implies that at a catastrophe point one eigenvalue is non-zero. In higher dimensional images, for example 3D, not only the case $\lambda_1 = \lambda_2 = \lambda_3 = 0$, but also cases like $\lambda_1 = -\lambda_3$ and $\lambda_2 = 0$ at a catastrophe point are *not* generic in 3D.

However, dealing with discrete images and finite precision, one may encounter non-genericities [21]. In this case the catastrophe point and a scale space saddle cannot be distinguished. In the following classification this can be visualised by coinciding the catastrophe point with the last scale space saddle before annihilation.

3.1   Zero Scale Space Saddles

If the saddle branch does not contain any scale space saddles from the initial scale until the annihilation, the intensities of both the saddle and the maximum

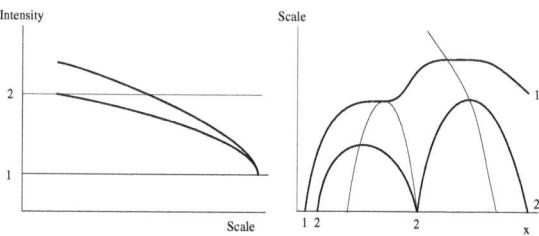

Fig. 2. a: Intensities of the maximum (top) and saddle branch. No scale space saddle are present. b: Critical curves (thin) and Iso-intensity manifolds (thick) in the plane of the dominant spatial variable and scale. Intensity 1 at the catastrophe forms a manifold through the catastrophe point. Intensity 2 through the saddle point at the initial image separates two distinct manifolds.

are decreasing, see Figure 2a. The largest value on the saddle branch is thus the value of the spatial saddle in the initial image (intensity 2). At this point two distinct manifolds intersect as shown in Figure 2b. At intensity 1, the one of the annihilation, there is only one manifold.

Example 1. Let $L(x, y; t) = \frac{1}{6}x^3 + xt - 3y^2 - 6t$ for $t \geq -10$. Then the critical curve follows from $\frac{1}{2}x^2 + t = 0$ and $-6y = 0$ and is given by $(x, y; t) = (\pm\sqrt{-2t}, 0; t)$ for $-10 \leq t \leq 0$. The intensity of the critical curve is $L_{\pm}(x(t), 0; t) = -6t \pm \frac{2}{3}t\sqrt{-2t}$, for $0 \geq t \geq -10$. The determinant of the Hessian reads $-6x$, so the origin is the catastrophe point. For a potential scale space saddle $\partial_t L = 0$, yielding $x - 6 = 0$, i.e. $t = -18$, which is outside the scale domain. The intensities of the branches are both decreasing: $\partial_t L_{\pm}(x(t), 0; t) = -6 \pm \sqrt{-2t}$, which is negative in the scale domain.

3.2   One Scale Space Saddle

If the saddle branch contains one scale space saddle, its intensity at it is larger than that at the catastrophe, see Figure 3a. The intensity of the saddle increases – passing intensity 1 – until it reaches the scale space saddle (intensity 2), and then decreases until its annihilation point (intensity 1).

The values on the saddle branch attained between the scale space saddle and the catastrophe point are also attained at some smaller scale. This is visible in Figure 3 where for intensities between 1 and 2, the manifold has three intersections with the critical curve: two with the saddle branch and one with the maximum.

Example 2. Take $L(x, y; t) = x^3 + 6xt - y^2 - 2t$. The critical curve follows the path $(x, y; t) = (\pm\sqrt{-2t}, 0; t)$ for $t \leq 0$. Since the determinant of the Hessian is $-12x$ and its trace $6x - 2$, the maximum and the saddle annihilate at the origin, while the saddle branch (positive x values) contains a scale space saddle at $(x, y; t) = (1/3, 0; -1/18)$. The intensities on the critical curve are given by

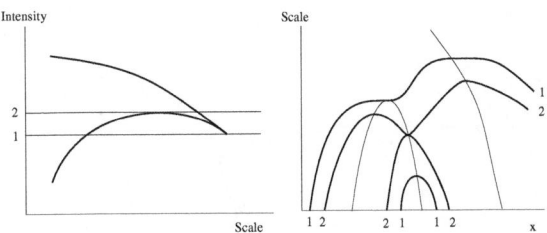

Fig. 3. a: Intensities of the maximum (top) and saddle branch. b: Critical curves (thin) and Iso-intensity manifolds (thick) in the plane of the dominant spatial variable and scale. At the scale space saddle two distinct manifolds intersect.

$L_+(x(t), 0; t) = 4t\sqrt{-2t} - 2t$ and $L_-(x(t), 0; t) = -4t\sqrt{-2t} - 2t$. One can verify that L_+ is decreasing strict monotonically, as it corresponds to the maximum. The saddle branch, L_-, contains a local maximum at $t = -1/18$.

3.3   Two Scale Space Saddles

Recalling Figure 3, a saddle branch with an extra scale space saddle can occur in two cases: The intensity of the second scale space saddle is either larger than the catastrophe point (case A), or smaller (case B).

Case A. The first case is visualised in Figure 4. Since the number of scale space saddles is even, the intensity of the saddle in the initial image (intensity 4) is taken to derive two distinct manifolds. Now the scale space saddle with intensity 3 intersects the saddle branch also at a smaller scale: it is the part of a manifold that intersects itself. The same analogy holds for the scale space saddle with intensity 2. It is part of the manifold through both branches that is wrapped around the manifold with intensity 3. Consequently, both scale space saddles are void.

Case B. In the second case, the intensity of the catastrophe point lies between those of the scale space saddles, see figure 4. Again the intensity of the saddle in the initial image (intensity 4) yields two distinct manifolds and the scale space saddles do not connect distinct manifolds. Only the one with intensity 3 is related to a manifold through the extremum branch. The other one is not related to the critical curve. Again, both scale space saddles are void.

3.4   More Scale Space Saddles

This classification can be infinitely extended as one can imagine. However, inserting more scale space saddles basically implies inserting more local extrema

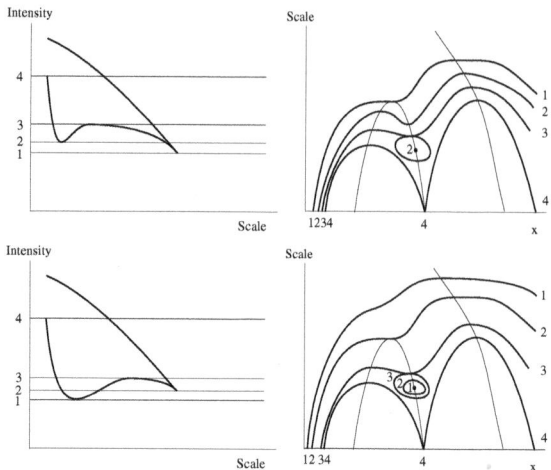

Fig. 4. a (Case A) and c (Case B): Intensities of the maximum (top) and saddle branch. b (Case A) and d (Case B): Critical curves (thin) and Iso-intensity manifolds (thick) in the plane of the dominant spatial variable and scale. The two manifolds through the saddle at the initial image are distinct. The scale space saddles each form the self-intersecting locations of one single manifold.

on a saddle branch and thus inserting more self-intersecting locations of an iso-intensity manifold and thus inserting void scale space saddles. This holds unless, of course, when the intensity of a newly inserted scale space saddle is the local maximum on the saddle branch – still regarding maximum-saddle annihilations. In case of minimum-saddle annihilations "highest" becomes "lowest", and so on. The classification thus follows the following rules:

1. If the highest value of the saddle branch is attained at the initial image, the iso-intensity manifold through this saddle contains two distinct parts. If the branch contains scale space saddles, then at these points manifolds intersect themselves and the scale space saddles are void.
2. If the highest value is attained at a scale space saddle, the manifold through it contains two distinct parts. If more scale space saddles are present, the manifolds through them are not distinct and the scale space saddles are void.

4   Examples and Applications

4.1   Artificial Image

In the artificial image shown in Figure 5a, five extrema are present of which 4 maxima and one minimum. Its critical curves in scale space (existing of 113 calculated samples that are logarithmically taken) are visualised in Figure 5b. Clearly, four (dark-grey) saddle branches are visible. One extremum branch remains.

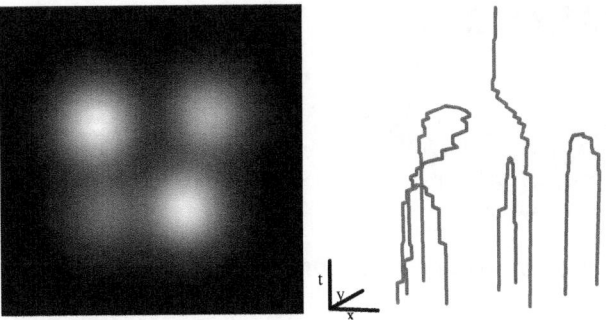

Fig. 5. a: Artificial blob image. b: Its critical curves containing extrema (bright) and saddle (dark) branches.

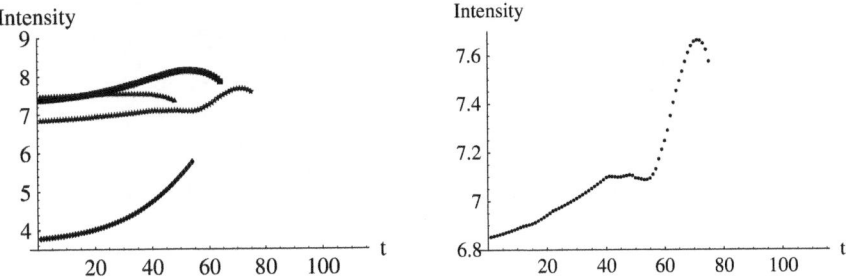

Fig. 6. a: Intensities of the saddle branches. b: The saddle branch with multiple scale space saddles around intensity 7.1.

The intensities of the saddle branches of the critical curves are plotted in Figure 6a. One can see that the saddle branch with the lowest intensities is monotonically increasing. It annihilates with the minimum, so for segmentation the lowest value on the saddle branch, i.e. the one of the saddle at the initial image, should be taken. The two upper branches both contain one scale space saddle. Of the last saddle branch a close-up is shown in Figure 5b. Around the value of approximately 7.1 four scale space saddles can be seen in the scale sample range 40 to 60. Obviously they are void, and the scale space saddle with maximal intensity is located at approximately 7.66, at scale sample 71.

4.2   Zero Scale Space Saddles

The minimum-saddle branch doesn't contain scale space saddles. The iso-intensity manifold in scale space of the value of the saddle at the initial image is shown in Figure 7a. Here two distinct manifolds intersecting each other at the saddle are visible. In contrast, the iso-intensity manifold with the value of the saddle at the catastrophe, Figure 7b, shows only one manifold.

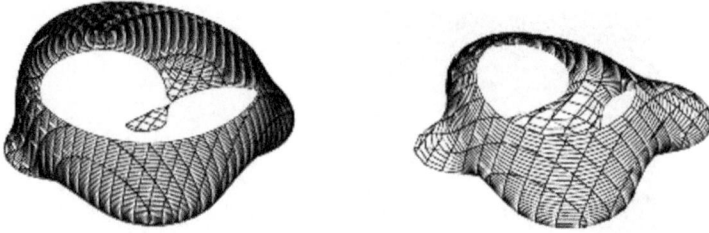

Fig. 7. a: Correct iso-intensity manifold for the minimum-saddle pair. b: The wrong one.

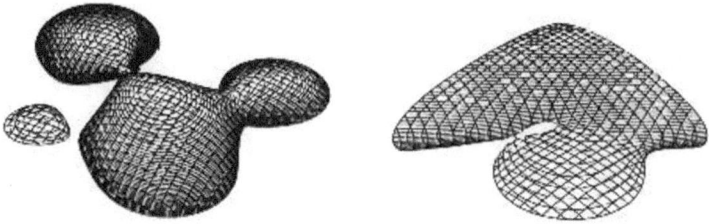

Fig. 8. a: Correct iso-intensity manifold for the maximum-saddle pair with multiple scale space saddles. b: Close-up around a iso-intensity manifold through a void scale space saddle.

4.3   Multiple Scale Space Saddles

The iso-intensity manifolds through the correct scale space saddle, in the case of multiple scale space saddles on the saddle branch, is shown in Figure 8. Again, two distinct manifolds (one top-left and one at the right) intersecting each other are clearly visible. A third manifold is isolated and not related to the scale space saddle. Note that the iso-intensity manifolds for the two branches with one scale space saddle on the saddle branch are alike.

If one of the void saddles is chosen, the iso-intensity manifold indeed exists of one single manifold that intersects itself leaving room for a hole, as shown in Figure 8b.

The global situation is visible in Figure 9. It appears that the manifold also encapsulates the extremum bottom-left, in contrast with the correct iso-intensity. As one can see from these images, the extremum top-left in the original image, Figure 5a, is related by the scale space saddle to the two extrema on the right. However, due to the blurring process the extremum bottom-left exhibits a disturbing influence. One may also think of the saddle being attracted firstly by the extremum bottom-left, but taking a final annihilation relation with the extremum top-left[1]. This is also visible in the behaviour of the critical curves, Figure 5b, where this saddle branch takes a kind of detour before annihilation in contrast to the other saddle branches.

---

[1] One may also imagine the saddle as being rejected by the extremum bottom-left, since it preferred another saddle to annihilate with.

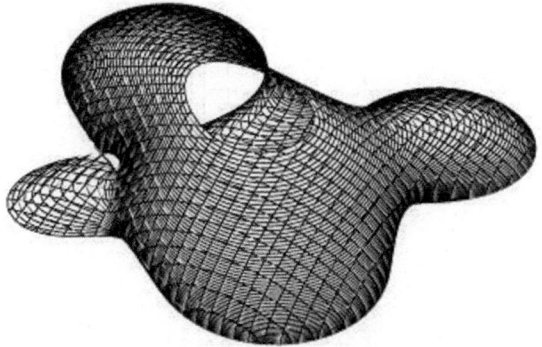

Fig. 9. False iso-intensity manifold for the maximum-saddle pair with multiple scale space saddles. The intensity belonging to a void scale space saddle is chosen.

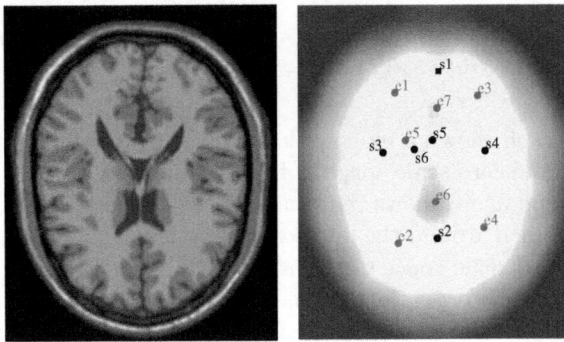

Fig. 10. a) MR image. b) Image on scale 8.37, with the labelled spatial critical points: seven extrema and six saddles.

4.4 MR Image

Secondly the MR image of Figure 10a is taken. Since it contains many critical points, the scale space image starting at scale 8.37 is used. This scale is chosen for visualisation purposes, since it yields a reasonable number of critical points to handle with. The first image is visualised in Figure 10b, together with its seven extrema and six saddle points. A scale space containing 87 scales was built and critical paths were calculated.

The intensities of some of them are shown in Figure 11.

The various types of scale space saddle point occurrences are present in this sequence.

Clearly the first graph shows non-generic behaviour since the annihilation co-incides with the scale space saddle. Furthermore two void scale space saddles are present but both not affect the proposed segmentation since they are both larger than the intensity of the annihilation (in this case it is a minimum annihilating).

The second graph shows the "standard" situation when the intensity of the scale space saddle is larger than the annihilation intensity.

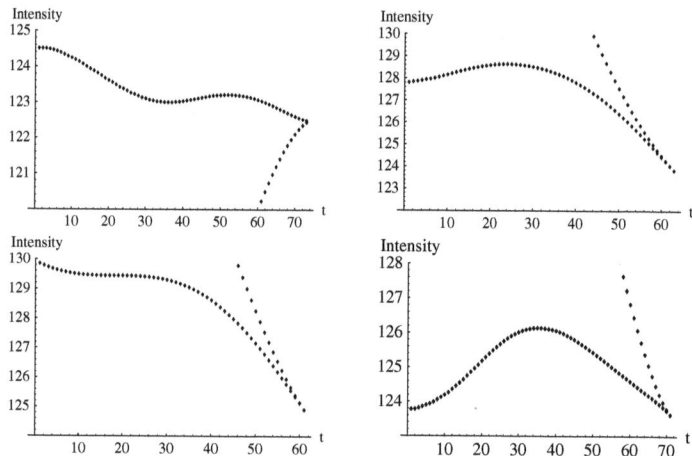

Fig. 11. Intensities of the critical curves combing the critical points. From left to right, top to bottom: a) e6-s1 b) e1-s3 c) e3-s4 d) e2-s2.

The third graph shown a monotonically decreasing saddle intensity and thus corresponds to the zero scale space saddle situation.

The last graph again shows the "standard" situation, although the situation of the saddle branch around the first scales may give rise to the assumption that there also a (void) scale space saddle is located.

The segments corresponding to the intensities of the correct values of the scale space saddles (and initial saddle for the third curve) are shown in the top row of Figure 12.

To show that one should be careful, in the bottom row the segments corresponding to false intensity values are shown: The initial saddle for the minimum (left) and annihilation intensities for the three maxima (second to forth). For the minimum inside the brain, background is selected for this false intensity. As can be seen in Figure 12, bottom left, the connection from the inner part to the exterior is made via a saddle connecting the parts related to the left and right maxima's, cf. the second and third image in the top row of Figure 12. Just as in the previous section a void saddle is chosen. The consequence for this in the original image is shown in Figure 13. To the left the correct segment is shown, while the white segment of Figure 13b is clearly connected to the background

The regions belonging to the maxima regions become at least twice as large. Here the problematic – and wrong – situation occurs that two regions are found that encapsulate the same number of extrema, viz. the second and the last image of the bottom row of Figure 12. This obviously violates the uniqueness of the hierarchy.

5   Conclusion and Discussion

In this paper we investigated the structure of iso-intensity manifolds through scale space saddles in scale space. They can be classified into two essentially

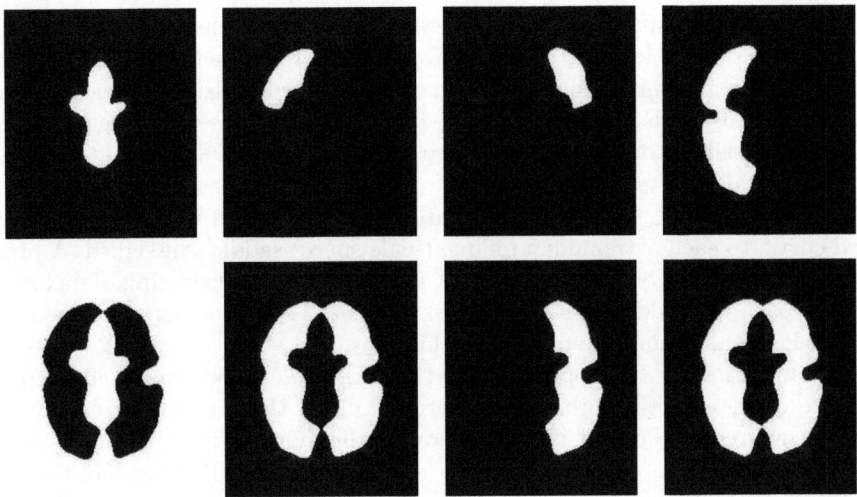

Fig. 12. Shown in white are the segments of Figure 10b determined by the critical curves of Figure 11. They correspond to, from left to right, the extrema e6, e1,e3, and e2. The top row depicts the correct selection of the intensity values, while the bottom row gives an example of a selection of wrong values. Note that the segment belonging to e2 *does* encapsulate e1 (top row), but *does not* encapsulate e3, cf. [17,19,22].

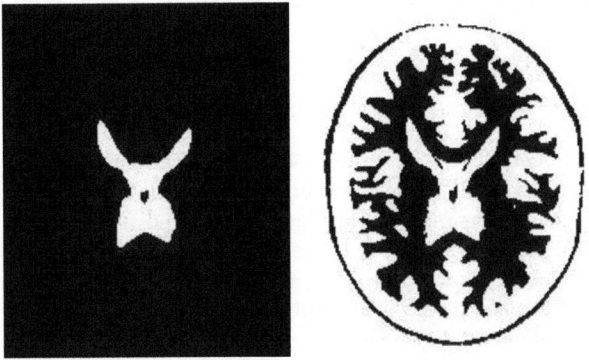

Fig. 13. Segment of extremum e6 of Figure 10a – the unblurred MR image – according to the a) correct, and b) wrong intensity value, cf. Figure 12 top left and bottom left, respectively.

different cases. The first set consists of manifolds that have only one point in common, the scale space saddle. Each manifold yields a pair of manifolds with the same intensity if the scale space saddle is removed. The second set is formed by the manifolds that self-intersect. If the scale space saddle is removed from such a manifold, still one manifold remains. The scale space saddle in case is defined as *void*.

For the scale space hierarchy (tree) and segmentation it is essential that scale space saddles yielding manifolds from the first set are taken. Then for each extremum in the initial image a unique region can be assigned, together with the region to which it hierarchically belongs. In some cases - if the number of scale space saddles on the saddle branch is even - the spatial saddle at the initial scale must also be taken into account.

Since each scale space saddle is related to an extremum by means of the critical curve, to each extremum a unique (scale space) saddle is assigned. A procedure to select the correct (scale space) saddle is given: considering a maximum-saddle pair, the saddle with the highest intensity is to be assigned to the maximum. For a minimum the opposite holds.

Examples show the appearance of the various described manifolds and relevance of the selection procedure. More details of the algorithm to derive the hierarchy tree, the use of it and several applications, can be found elsewhere [17,19,22]. In case of multiple scale space saddles on a saddle branch of the critical curve, the procedure as described above was used, albeit without justification.

Acknowledgements

The author would like to thank Luc Florack of Eindhoven University of Technology, the Netherlands, for interesting discussions leading to this research and Senter-IOP grant IBV99006 for funding.

References

1. J. Damon. Local Morse theory for solutions to the heat equation and Gaussian blurring. *Journal of Differential Equations*, 115(2):386–401, 1995.
2. J. Damon. Local Morse theory for Gaussian blurred functions. In *Sporring et al. [27]*, pages 147–162, 1997.
3. R. Duits, L. M. J. Florack, J. de Graaf, and B. M. ter Haar Romeny. On the axioms of scale space theory, 2001. submitted.
4. D. Eberly. A differential geometric approach to anisotropic diffusion. In *ter Haar Romeny, [12]*, pages 371–392, 1994.
5. M. Fidrich. Iso-surface extraction in 4-D with applications related to scale space. In *Proceedings of DGCI'96, 6th International Conference on Discrete Geometry for Computer Imagery (Lyon, France, November 1996)*, pages 257–268, 1996. Lecture Notes in Computer Science 1176.
6. M. Fidrich. Following feature lines across scale. In *ter Haar Romeny et al. [13]*, pages 140–151, 1997.
7. M. Fidrich. Iso-surface extraction in n-D applied to tracking feature curves across scale. *Image and Vision Computing*, 16(8):545–556, 1998.
8. L. M. J. Florack. *Image Structure*, volume 10 of *Computational Imaging and Vision Series*. Kluwer Academic Publishers, Dordrecht, The Netherlands, 1997.
9. L. M. J. Florack and A. Kuijper. The topological structure of scale-space images. *Journal of Mathematical Imaging and Vision*, 12(1):65–80, February 2000.

10. L. M. J. Florack, B. M. ter Haar Romeny, J. J. Koenderink, and M. A. Viergever. Linear scale-space. *Journal of Mathematical Imaging and Vision*, 4(4):325–351, 1994.

11. L. D. Griffin and A. Colchester. Superficial and deep structure in linear diffusion scale space: Isophotes, critical points and separatrices. *Image and Vision Computing*, 13(7):543–557, September 1995.

12. B. M. ter Haar Romeny, editor. *Geometry-Driven Diffusion in Computer Vision*, volume 1 of *Computational Imaging and Vision Series*. Kluwer Academic Publishers, Dordrecht, 1994.

13. B. M. ter Haar Romeny, L. M. J. Florack, J. J. Koenderink, and M. A. Viergever, editors. *Scale-Space Theory in Computer Vision: Proceedings of the First International Conference, Scale-Space'97, Utrecht, The Netherlands*, volume 1252 of *Lecture Notes in Computer Science*. Springer-Verlag, Berlin, July 1997.

14. T. Iijima. Basic theory of pattern normalization (for the case of a typical one dimensional pattern). *Bulletin of the Electrotechnical Laboratory*, 26:368–388, 1962. (in Japanese).

15. M. Kerckhove, editor. *Scale-Space and Morphology in Computer Vision*, volume 2106 of *Lecture Notes in Computer Science*. Springer -Verlag, Berlin Heidelberg, 2001.

16. J. J. Koenderink. The structure of images. *Biological Cybernetics*, 50:363–370, 1984.

17. A. Kuijper. *The Deep Structure of Gaussian Scale Space Images*. PhD thesis, Utrecht University, 2002.

18. A. Kuijper and L. M. J. Florack. Hierarchical pre-segmentation without prior knowledge. In *Proceedings of the 8th International Conference on Computer Vision (Vancouver, Canada, July 9–12, 2001)*, pages 487–493, 2001.

19. A. Kuijper and L. M. J. Florack. Logical filtering in scale space. Technical Report UU-CS-2002-018, Department of Computer Science, Utrecht University, 2002. Accepted for publication as "The hierarchical structure of images" in IEEE Transactions on Image Processing.

20. A. Kuijper and L. M. J. Florack. The relevance of non-generic events in scale space. In *Proceedings of the 7th European Conference on Computer Vision (Copenhagen, Denmark, May 28–31, 2002)*, pages 190–204, 2002.

21. A. Kuijper and L. M. J. Florack. Understanding and modeling the evolution of critical points under Gaussian blurring. In *Proceedings of the 7th European Conference on Computer Vision (Copenhagen, Denmark, May 28–31, 2002)*, pages 143–157, 2002.

22. A. Kuijper, L. M. J. Florack, and M. A. Viergever. Scale space hierarchy. *Journal of Mathematical Imaging and Vision*, 18(2):169–189, April 2003.

23. T. Lindeberg. *Scale-Space Theory in Computer Vision*. The Kluwer International Series in Engineering and Computer Science. Kluwer Academic Publishers, 1994.

24. M. Loog, J. J. Duistermaat, and L. M. J. Florack. On the behavior of spatial critical points under Gaussian blurring, a folklore theorem and scale-space constraints. In *Kerckhove [15]*, pages 183–192, 2001.

25. M. Nielsen, P. Johansen, O. Fogh Olsen, and J. Weickert, editors. *Scale-Space Theories in Computer Vision*, volume 1682 of *Lecture Notes in Computer Science*. Springer -Verlag, Berlin Heidelberg, 1999.

26. T. Poston and I. N. Stewart. *Catastrophe Theory and its Applications*. Pitman, London, 1978.

27. J. Sporring, M. Nielsen, L. M. J. Florack, and P. Johansen, editors. *Gaussian Scale-Space Theory*, volume 8 of *Computational Imaging and Vision Series*. Kluwer Academic Publishers, Dordrecht, second edition, 1997.
28. J. A. Weickert. *Anisotropic Diffusion in Image Processing*. Teubner, Stuttgart, 1998.
29. A. P. Witkin. Scale-space filtering. In *Proceedings of the Eighth International Joint Conference on Artificial Intelligence*, pages 1019–1022, 1983.

Many-to-Many Matching
of Scale-Space Feature Hierarchies
Using Metric Embedding

M. Fatih Demirci[1], Ali Shokoufandeh[1], Yakov Keselman[2],
Sven Dickinson[3], and Lars Bretzner[4]

[1] Department of Computer Science, Drexel University
Philadelphia, PA 19104, USA
{mdemirci,ashokouf}@mcs.drexel.edu
[2] School of Computer Science, Telecommunications and Information Systems
DePaul University, Chicago, IL 60604, USA
ykeselman@cs.depaul.edu
[3] Department of Computer Science, University of Toronto
Toronto, Ontario, Canada M5S 3G4
sven@cs.toronto.edu
[4] Computational Vision and Active Perception Laboatory
Department Of Numerical Analysis and Computer Science
KTH, Stockholm, Sweden
bretzner@nada.kth.se

Abstract. Scale-space feature hierarchies can be conveniently represented as graphs, in which edges are directed from coarser features to finer features. Consequently, feature matching (or view-based object matching) can be formulated as graph matching. Most approaches to graph matching assume a one-to-one correspondence between nodes (features) which, due to noise, scale discretization, and feature extraction errors, is overly restrictive. In general, a subset of features in one hierarchy, representing an abstraction of those features, may best match a subset of features in another. We present a framework for the many-to-many matching of multi-scale feature hierarchies, in which features and their relations are captured in a vertex-labeled, edge-weighted graph. The matching algorithm is based on a metric-tree representation of labeled graphs and their low-distortion metric embedding into normed vector spaces. This two-step transformation reduces the many-to-many graph matching problem to that of computing a distribution-based distance measure between two such embeddings. To compute the distance between two sets of embedded, weighted vectors, we use the Earth Mover's Distance under transformation. To demonstrate the approach, we target the domain of multi-scale, qualitative shape description, in which an image is decomposed into a set of blobs and ridges with automatic scale selection. We conduct an extensive set of view-based matching trials, and compare the results favorably to matching under a one-to-one assumption.

L.D. Griffin and M. Lillholm (Eds.): Scale-Space 2003, LNCS 2695, pp. 17–32, 2003.

1   Introduction

The problem of object recognition is often formulated as that of matching configurations of image features to configurations of model features. Such configurations are often represented as vertex-labeled graphs, whose nodes represent image features (or their abstractions), and whose edges represent relations (or constraints) between the features. For scale-space structures, represented as hierarchical graphs, relations can represent both parent/child relations as well as sibling relation. To match two graph representations (hierarchical or otherwise) means to establish correspondences between their nodes. To evaluate the quality of a match, one defines an overall distance measure, whose value depends on both node and edge similarity.

Due to the importance of the recognition problem (reformulated in terms of graph matching), there has been a growing interest in developing efficient algorithms for matching vertex-labeled graphs. Previous work on graph matching (see Section 2) has typically focused on the problem of finding a one-to-one correspondence between the vertices of two graphs. However, the assumption of one-to-one correspondence is a very restrictive one, for it assumes that the primitive features (nodes) in the two graphs agree in their level of abstraction. In scale-space (hierarchical) structures, this restrictive assumption takes the form of assuming that corresponding features exist at the same level. Unfortunately, there are a variety of conditions that may lead to graphs that represent visually similar image feature configurations yet do not contain a single one-to-one node correspondence. For example, due to noise or segmentation errors, a single feature (node) in one graph may map to a collection of broken features (nodes) in another graph. Or, due to scale differences, a single, coarse-grained feature in one graph may map to a collection of fine-grained features in another graph. In general, we seek not a one-to-one correspondence between image features (nodes), but rather a many-to-many correspondence.

Several existing approaches to the problem of many-to-many graph matching suffer from computational inefficiency and/or from inability to handle small perturbations in graph structure. This paper seeks a solution to this problem while addressing drawbacks of existing approaches. Drawing on recently-developed techniques from the domain of low-distortion graph embedding, we have explored an *efficient* method for mapping a graph's structure to a set of vectors in a low-dimensional space. This mapping not only simplifies the original graph representation, but it *retains important information* about both local (neighborhood) as well as global graph structure. Moreover, the mapping is *stable* with respect to noise in the graph structure.

The above embedding is applicable only to undirected graphs, in which a metric (undirected) distance can be defined between every pair of nodes. Although scale-space structures may contain undirected edges, information is mostly encoded by hierarchical, non-metric relations, such as parent/child relations. We accommodate these constraints by moving this information into the nodes as feature distributions over the values of incident, oriented edges. Although pulling the oriented edge information into the node would seem to weaken the representation, it's important to note that the resulting node encodes contextual (or neighborhood) information about its relations to adjacent nodes in the graph.

Armed with a low-dimensional, robust vector representation of an input graph's structure, many-to-many graph matching can now be reduced to the much simpler problem of matching weighted distributions of points in a normed vector space, using a *distribution-based* similarity measure. We consider one such similarity measure, known as the Earth Mover's Distance, and show that the many-to-many vector mapping that realizes the minimum Earth Mover's Distance *corresponds to* the desired many-to-many matching between nodes of the original graphs. The result is a more efficient and more stable approach to many-to-many graph matching that, in fact, includes the special case of one-to-one graph matching. To illustrate the approach, we apply it to the problem of view-based object recognition, in which views are represented as graphs.

2   Related Work

The problem of object recognition is often reformulated as that of matching feature graphs. Several researchers have developed algorithms that find one-to-one correspondences between graph nodes. Shapiro and Haralick [20] proposed a matching algorithm based on comparing weighted primitives (weighted attributes and weighted relation tuples) using a normalized distance for each primitive property that is inexactly matched. Pellilo et al. [16] devised a quadratic programming framework for matching association graphs using a maximal clique reformulation, while Gold and Rangarajan [8] used graduated assignment for matching graphs derived from feature points and image curves. Siddiqi et al. combined a bipartite matching framework with a spectral decomposition of graph structure to match shock graphs [22], while Shokoufandeh et al. [21] extended this framework to directed acyclic graphs that arise in multi-scale image representations. Hancock and his colleagues have also proposed numerous frameworks for graph matching, including [14].

The problem of many-to-many graph matching has also been studied, most often in the context of edit-distance (see, e.g., [13,19]). In such a setting, one seeks a minimal set of re-labelings, additions, deletions, merges, and splits of nodes and edges that transform one graph into another. However, the edit-distance approach has its drawbacks: 1) it is computationally expensive (polynomial time algorithms are available only for trees); 2) the method in its current form does not accommodate edge weights; and 3) the cost of an editing operation often fails to reflect the underlying visual information (for example, the visual similarity of a contour and its corresponding broken fragments should not be penalized by the high cost of merging the many fragments). In the context of line and segment matching, Beveridge and Riseman [4] addressed this problem via exhaustive local search. Although their method found good matches reliably and efficiently (due to their choice of the objective function and a small neighborhood size), it is unclear how this can be generalized to other types of feature graphs and objective functions.

In contrast to advances in solving matching problems on specially-structured graphs, such as trees or directed acyclic graphs, there has been much less progress in solving the problem of many-to-many matching in general graphs. In a novel generalization of Scott and Longuet [18], Kosinov and Caelli [11] showed how inexact graph matching can be solved using the re-normalization of projections of vertices into the eigenspaces of graphs along with a form of relational clustering. Our framework differs from their approach in that: (1) it can handle information encoded in a graph's nodes, which is desirable in many vision applications; (2) it does not require an ad hoc clustering step; and (3) it provides a well-bounded, low-distortion metric representation of graph structure. In relation to low-distortion metric representations of graphs, Indyk [10] provides a comprehensive survey of recent advances and applications of low-distortion graph embedding. For recent results related to the properties of low-distortion tree embedding, see [1,15].

3   Metric Embedding of Graphs

During the last decade, low-distortion embedding has become recognized as a very powerful tool for designing efficient algorithms. In low-distortion embedding of metric spaces into normed spaces, we consider mappings $f : \mathcal{A} \to \mathcal{B}$, where $\mathcal{A}$ is a set of points in the original metric space, with distance function $\mathcal{D}(.,.)$, $\mathcal{B}$ is a set of points in the (host) d-dimensional normed space $||.||_k$, and for any pair $p, q \in \mathcal{A}$ we have

$$\frac{1}{c}\mathcal{D}(p,q) \leq ||f(p) - f(q)||_k \leq \mathcal{D}(p,q) \tag{1}$$

for a certain parameter c, known as the *distortion*. Intuitively, such an embedding will enable us to reduce problems defined over *difficult* metric spaces, $(\mathcal{A}, \mathcal{D})$, to problems over *easier* normed spaces, $(\mathcal{B}, ||.||_k)$. As can be observed from Equation 1, the closer c is to 1, the better the target set $\mathcal{B}$ mimics the original set $\mathcal{A}$. Consequently, the distortion parameter c is a critical characteristic of embedding, f.

Perhaps the most fundamental existence result in computational embedding is due to Bourgain [5]:

Lemma 1. *Any finite metric space $(\mathcal{A}, \mathcal{D})$ can be embedded into a finite normed space $||.||_2$ of dimension at most $\log |\mathcal{A}|$ with distortion $O(\log |\mathcal{A}|)$.*

This result is important since even an exponential[1] matching algorithm in the normed space may be tractable. However, $O(\log |\mathcal{A}|)$ distortion is too high; we seek an embedding with a much lower distortion.

3.1   Low-Distortion Embedding

Our interest in low-distortion embedding is motivated by its ability to transform the problem of many-to-many matching in finite graphs to geometrical

---

[1] in the dimension of the target space.

problems in low-dimensional vector spaces. Specifically, let $G_1 = (\mathcal{A}_1, E_1, \mathcal{D}_1)$, $G_2 = (\mathcal{A}_2, E_2, \mathcal{D}_2)$ denote two graphs on vertex sets $\mathcal{A}_1$ and $\mathcal{A}_2$, edge sets E_1 and E_2, under distance metrics $\mathcal{D}_1$ and $\mathcal{D}_2$, respectively ($\mathcal{D}_i$ represents the distances between all pairs of nodes in G_i). Ideally, we seek a single embedding that can map each graph to the same vector space, in which the two embeddings can be directly compared. However, in general, this is not possible without introducing unacceptable distortion.

We will therefore tackle the problem in two steps. First, we will seek low-distortion embeddings f_i that map sets $\mathcal{A}_i$ to normed spaces $(\mathcal{B}_i, ||.||_k)$, $i \in \{1, 2\}$. Next, we will align the normed spaces, so that the embeddings can be directly compared. Using these mappings, the problem of many-to-many vertex matching between G_1 and G_2 is therefore reduced to that of computing a mapping $\mathcal{M}$ between subsets of $\mathcal{B}_1$ and $\mathcal{B}_2$.

In practice, the robustness and efficiency of mapping $\mathcal{M}$ will depend on several parameters, such as the magnitudes of distortion of the $\mathcal{D}_i$'s under the embeddings, f_i's, the computational complexity of applying the embeddings, f_i's, the efficiency of computing the actual correspondences (including alignment) between subsets of $\mathcal{B}_1$ and $\mathcal{B}_2$, and the quality of the computed correspondence. The latter issue will be addressed in Section 5.

The problem of low-distortion embedding has a long history for the case of planar graphs, in general, and trees, in particular. More formally, the most desired embedding is the subject of the following conjecture:

Conjecture 1. [9] Let $G = (\mathcal{A}, E)$ be a planar graph, and let $M = (\mathcal{A}, \mathcal{D})$ be the shortest-path metric for the graph G. Then there is an embedding of M into $||.||_p$ with $O(1)$ distortion.

This conjecture has only been proven for the case in which G is a tree. Although the existence of such a distortion-free embedding under $||.||_k$-norms was established in [12], no deterministic construction was provided. One such deterministic construction was given by Matoušek [15], suggesting that if we could somehow map our graphs into trees, with small distortion, we could adopt Matoušek's framework.

3.2 Tree Metric of a Distance Function

Before we can proceed with Matoušek's embedding, we must choose a suitable *tree metric* for our graphs. Let $G = (\mathcal{A}, E)$ denote an edge-weighted graph with real edge weights $\mathcal{W}(e)$, $e \in E$. We will say that $\mathcal{D}$ is a metric for G if, for any three vertices $u, v, w \in \mathcal{A}$, $\mathcal{D}(u, v) = \mathcal{D}(v, u) \geq 0$, $\mathcal{D}(u, u) = 0$, and $\mathcal{D}(u, v) \leq \mathcal{D}(u, w) + \mathcal{D}(w, v)$. In general, there are many ways to define metric distances on a weighted graph. The best-known metric is the shortest-path metric $\delta(.,.)$, i.e., $\mathcal{D}(u, v) = \delta(u, v)$, the shortest path distance between u and v for all $u, v \in \mathcal{A}$. We will say that the edge weighted tree $\mathfrak{T} = \mathfrak{T}_G(V', E')$ is a tree metric for G, with respect to distance function $\mathcal{D}$, if for any pair of vertices u, v, the length of the unique path between them in $\mathfrak{T}$ is equal to $\mathcal{D}(u, v)$. The problem of approximating (or fitting) an $n \times n$ distance matrix $\mathcal{D}$ by a tree metric $\mathfrak{T}$ is known as the *Numerical Taxonomy* problem.

The Numerical Taxonomy problem is closely related to that of constructing an *additive metric* distance, i.e., a metric distance D that satisfies the *4-point* condition, $\mathcal{D}[x, y] + \mathcal{D}[z, w] \leq \max\{\mathcal{D}[x, z] + \mathcal{D}[y, w], \mathcal{D}[x, w] + \mathcal{D}[y, z]\}\ \forall x, y, z, w$. A stronger version of the 4-point condition is the *ultra-metric* condition. A metric D is an ultra-metric if, for all points $x, y, z,$, $D[x, y] \leq \max\{D[x, z], D[y, z]\}$. Observe that an ultra-metric is a type of tree metric defined on rooted trees, where the distance to the root is the same for all leaves in the tree. This is a critical property in the construction of metric trees for distance functions.

The following Theorem relates the existence of a tree metric to the 4-point condition:

Theorem 1. (see [6]) *A metric $\mathcal{D}$ is additive if and only if it is a tree metric.*

In fact, if there is a tree metric $\mathfrak{T}$ coinciding exactly with $\mathcal{D}$, it is unique and constructible in linear time [24]. If $\mathcal{D}$ is not an additive metric there might be no tree metric $\mathfrak{T}$ that exactly coincides with $\mathcal{D}$. In this case, we can approximate $\mathcal{D}$ under norms, such as $||.||_k$, $k \geq 1$. That is, we want to find a tree metric $\mathfrak{T}$ minimizing $||\mathfrak{T} - \mathcal{D}||_k$.

In the event that G is not a tree, we will use an approximation framework proposed by Agarwala et al. [1]. They consider the approximate numerical taxonomy problem for additive metrics under the $||.||_\infty$ norm. The construction of a tree metric $\mathfrak{T}$ in their algorithm is achieved by transforming the general tree metric problem to that of ultra-metrics. Their algorithm will generate an approximation tree metric $\mathfrak{T}$, to an optimal additive metric under $||.||_\infty$ in time $O(n^2)$ (see [1] for details). It should be noted that this construction does not maintain the vertex set of G invariant, i.e., $V(G) \subseteq V'(\mathfrak{T})$. We will have to make sure that in the embedding process (see Section 3.4), the extra vertices generated during the metric tree construction are eliminated.

An example of the embedding applied to a multi-scale blob decomposition [21] is shown in Figure 1. The gesture image (a) consists of 5 regions (the topmost region is not shown in the image). The complete graph in (b) captures the Euclidean distance between the centroids of the regions, while (c) is the metric tree representation of the multi-scale decomposition (with additional vertices).

3.3   Path Partition of a Graph

The construction of the embedding depends on the notion of a path partition of a graph. In this subsection, we introduce the path partition, and then use it in the next subsection to construct the embedding. Given a weighted graph $G = (V, E)$ with metric distance $\mathcal{D}(.,.)$, let $\mathfrak{T} = (V', \mathfrak{E})$ denote a metric tree representation of G, whose vertex distances are approximately $\mathcal{D}(.,.)$. In the event that G is a tree, $\mathfrak{T} = G$; otherwise $\mathfrak{T}$ is the metric tree of G. To construct the embedding, we will assume that $\mathfrak{T}$ is a rooted tree. It will be clear from the construction that the choice of the root does not affect the distortion of the embedding.

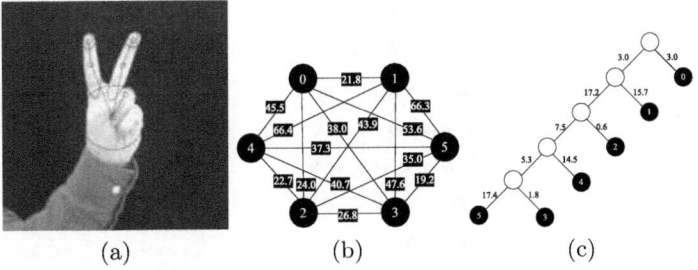

(a)   (b)   (c)

Fig. 1. Metric tree embedding of Euclidean distances for blob and ridge decomposition.

The dimensionality of the embedding of $\mathfrak{T}$ depends on the caterpillar dimension, denoted by cdim($\mathfrak{T}$), and is recursively defined as follows [15]. If $\mathfrak{T}$ consists of a single vertex, we set cdim($\mathfrak{T}$) = 0. For a tree $\mathfrak{T}$ with at least 2 vertices, cdim($\mathfrak{T}$) $\leq k + 1$ if there exist paths $P_1, \ldots, P_r$ beginning at the root and otherwise pairwise disjoint, such that each component $\mathfrak{T}_j$ of $\mathfrak{T} - \mathfrak{E}(P_1) - \mathfrak{E}(P_2) - \cdots - \mathfrak{E}(P_r)$ satisfies cdim($\mathfrak{T}_j$) $\leq k$. Here, $\mathfrak{T} - \mathfrak{E}(P_1) - \mathfrak{E}(P_2) - \cdots - \mathfrak{E}(P_r)$ denotes the tree $\mathfrak{T}$ with the edges of the P_i's removed, and the components $\mathfrak{T}_j$ are rooted at the single vertex lying on some P_i. The caterpillar dimension can be determined in linear time for a rooted tree $\mathfrak{T}$, and it is known that cdim($\mathfrak{T}$) $\leq \log(|V'|)$ (see [15]).

The construction of vectors $f(v)$, for $v \in V$, depends on the notion of a *path partition* of $\mathfrak{T}$. The path partition $\mathfrak{P}$ of $\mathfrak{T}$ is empty if $\mathfrak{P}$ is single vertex; otherwise $\mathfrak{P}$ consists of some paths $P_1, \ldots, P_r$ as in the definition of cdim($\mathfrak{T}$), plus the union of path partitions of the components of $\mathfrak{T} - \mathfrak{E}(P_1) - \mathfrak{E}(P_2) - \cdots - \mathfrak{E}(P_r)$. The paths $P_1, \ldots, P_r$ have level 1, and the paths of level $k \geq 2$ are the paths of level $k - 1$ in the corresponding path partitions of the components of $\mathfrak{T} - \mathfrak{E}(P_1) - \mathfrak{E}(P_2) - \cdots - \mathfrak{E}(P_r)$. Note that the paths in a path partition are edge-disjoint, and their union covers the edge-set of $\mathfrak{T}$.

To illustrate these concepts, consider the tree shown in Figure 2. The three darkened paths from the root represent the three level 1 paths. Following the removal of the level 1 paths, we are left with 6 connected components that, in turn, induce seven level 2 paths, shown with lightened edges[2]. Following the removal of the seven level 2 paths, we are left with an empty graph. Hence, the caterpillar dimension (cdim($\mathfrak{T}$)) is 2. It is easy to see that the path partition $\mathfrak{P}$ can be constructed using a modified depth-first search in $O(|V'|)$ time.

3.4   Construction of the Embedding

Given a path partition $\mathfrak{P}$ of $\mathfrak{T}$, we will use m to denote the number of levels in $\mathfrak{P}$, and let $P(v)$ represent the unique path between the root and a vertex $v \in V$. The first segment of $P(v)$ of weight l_1 follows some path P^1 of level 1 in $\mathfrak{P}$, the

---

[2] Note that the third node from the root in the middle level 1 branch is the root of a tree-component consisting of five nodes that will generate two level 2 paths.

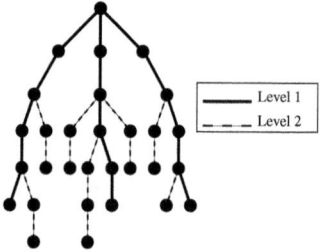

Fig. 2. Path partition of a tree.

second segment of weight l_2 follows a path P^2 of level 2, and the last segment of weight l_α follows a path P^α of level $\alpha \leq m$. The sequences $\langle P^1, \ldots, P^\alpha \rangle$ and $\langle l_1, \ldots, l_\alpha \rangle$ will be referred to as the *decomposition sequence* and the *weight sequence* of $P(v)$, respectively.

To define the embedding $f : V \to \mathcal{B}$ under $||.||_2$, we let the relevant coordinates in $\mathcal{B}$ be indexed by the paths in $\mathfrak{P}$. The vector $f(v)$, $v \in V$, has non-zero coordinates corresponding to the paths in the decomposition sequence of $P(v)$. Returning to Figure 2, the vector $f(v)$ will have 10 components (defined by three level 1 paths and seven level 2 paths). Furthermore, every vector $f(v)$ will have at most two non-zero components. Consider, for example, the lowest leaf node in the middle branch. Its path to the root will traverse three level 2 edges corresponding to the fourth level 2 path, as well as three level 1 edges corresponding to the second level 1 path.

Such embedding functions have become fairly standard in the metric space representation of weighted graphs [15]. In fact, Matoušek [15] has proven that setting the i-th coordinate of $f(v)$, corresponding to path P^k, $1 \leq k \leq \alpha$, in decomposition sequence $\langle P^1, \ldots, P^\alpha \rangle$, to

$$f(v)_i = \sqrt{l_k \left[l_k + \sum_{j=1}^{\alpha} \max\left(0, l_j - l_k/2m\right) \right]}$$

will result in a small distortion of at most $\sqrt{\log \log |V'|}$. It should be mentioned that although the choice of path decomposition $\mathfrak{P}$ is not unique, the resulting embeddings are isomorphic up to a transformation. Computationally, constructions of $\mathfrak{T}$, $\mathfrak{P}$, and $\mathcal{B}$ are all linear in terms of $|V|$ and $|\mathfrak{E}|$. It should be noted that although the tree $\mathfrak{T}$ has been defined based on set V', we have only embedded the vertices of G (set V).

4   Encoding Scale-Space Features

The distance metric defined on the graph structure is based on the undirected edge weights. While the above embedding has preserved the distance metric, it has failed to preserve any oriented relations, such as the hierarchical relations common to scale-space structures. This is due to the fact that oriented relations do not satisfy the symmetry property of a metric. We can retain this important information in our embedding by moving it into the nodes as node attributes, a technique used in the encoding of directed topological structure in [22], directed geometric structure in [21], and shape context in [3]. Encoding in a node the attributes of the oriented edges incident to the node requires computing distributions on the attributes and assigning them to the node. For example, a node with a single parent at a coarser scale and two children at a finer scale might encode a relative scale distribution (histogram) as a node attribute. The resulting attribute provides a contextual signature for the node which will be used by the matcher (Section 5) to reduce matching ambiguity.

Specifically, let $G = (V, E)$ be a graph to be recognized. For every pair of vertices, we let $R_{u,v}$ denote the attribute vector associated with the pair (u, v). The entries of each such vector represent the set of oriented relations R between u, v. For a vertex $u \in V$, we let $N(u)$ denote the set of vertices $v \in V$ adjacent to u. For a relation $p \in R$, we will denote $\mathcal{P}(u, p)$ as the set of values for relation p between u and all vertices in $N(u)$, i.e., $\mathcal{P}(u, p)$ corresponds to entry p of vector $R_{u,v}$ for $v \in N(u)$. Feature vector $\mathcal{P}_u$ for point u is the set of all $\mathcal{P}(u, p)$'s for $p \in R$. Observe that every entry $\mathcal{P}(u, p)$ of vector $\mathcal{P}_u$ can be considered as a local distribution (*histogram*) of feature p in the neighborhood $N(u)$ of u. We adopt the method of [21], in which the distance function for two such vectors $\mathcal{P}_u$ and $\mathcal{P}_p$ is computed through a weighted combination of Hausdorff distances between $\mathcal{P}(u, p)$ and $\mathcal{P}(u', p)$ for all values of p.

5   Distribution-Based Many-to-Many Matching

By embedding vertex-labeled graphs into normed spaces, we have reduced the problem of many-to-many matching of graphs to that of many-to-many matching of weighted distributions of points in normed spaces. However, before we can match two point distributions, we must map them into the same normed space. This involves reducing the dimension of the higher-dimensional distribution and transforming one of the distributions with respect to the other. Given a pair of weighted distributions in the same normed space, the Earth Mover's Distance (EMD) framework [17] is then applied to find an optimal match between the distributions. The EMD approach computes the minimum amount of work (defined in terms of displacements of the masses associated with points) it takes to transform one distribution into another.

5.1   Embedding Point Distributions in the Same Normed Space

Embeddings produced by the graph embedding algorithm can be of different dimensions and are defined only up to a distance-preserving transformation (a translated and rotated version of a graph embedding will also be a graph embedding). Therefore, in order to apply the EMD framework, we must first perform a "registration" step, whose objective is to project the two distributions into the same normed space. The resulting transformation is expected to minimize the initial EMD between the distributions.

Our transformation is based on Principal Components Analysis (PCA). Namely, the projection of the original vectors onto the subspace spanned by the first K right singular vectors of the covariance matrix retains the maximum information about the original vectors among all projections onto subspaces of dimension K. Hence, projecting the two distributions onto the first K right singular vectors of their covariance matrices will equalize their dimensions while losing minimal information. Specifically, assuming that K is the minimum of the two dimensions, we define embeddings $P_x(x_i) = W_x^T(x_i - \mu_x)/\sigma_x$ and $P_y(y_i) = W_y^T(y_i - \mu_y)/\sigma_y$ as follows:

$$\mu_x \leftarrow (\textstyle\sum_i w_i x_i)/\textstyle\sum_i w_i$$
$$\mu_y \leftarrow (\textstyle\sum_i w_i y_i)/\textstyle\sum_i w_i$$
$$\sigma_x^2 \leftarrow (\textstyle\sum_i w_i \|x_i - \mu_x\|)/\textstyle\sum_i w_i$$
$$\sigma_y^2 \leftarrow (\textstyle\sum_i w_i \|y_i - \mu_y\|)/\textstyle\sum_i w_i$$
$$\Sigma_{xx} \leftarrow (\textstyle\sum_i w_i(x_i - \mu_x)(x_i - \mu_x)^T)/\textstyle\sum_i w_i$$
$$\Sigma_{xx} = U_x D_x V_x^T \text{ is the SVD of } \Sigma_{xx}$$
$$W_x \leftarrow \text{first } K \text{ columns of } V_x$$
$$\Sigma_{yy} \leftarrow (\textstyle\sum_i w_i(y_i - \mu_y)(y_i - \mu_y)^T)/\textstyle\sum_i w_i$$
$$\Sigma_{yy} = U_y D_y V_y^T \text{ is the SVD of } \Sigma_{yy}$$
$$W_y \leftarrow \text{first } K \text{ columns of } V_y$$

5.2   The Earth Mover's Distance

The Earth Mover's Distance (EMD) [17,7] is designed to evaluate dissimilarity between two multi-dimensional distributions in some feature space. The EMD approach assumes that a distance measure between single features, called the *ground distance*, is given. The EMD then "lifts" this distance from individual features to full distributions. Moreover, if the weights of the distributions are the same, and the ground distance is a metric, EMD induces a metric distance [17]. However, the main advantage of using EMD lies in the fact that it subsumes many histogram distances and permits partial matches in a natural way. This important property allows the similarity measure to deal with uneven clusters and noisy data sets.

Computing the EMD is based on a solution to the well-known *transportation problem* [2], whose optimal value determines the minimum amount of "work" required to transform one distribution into the other. More formally, let $P = \{(p_1, w_{p_1}), \ldots, (p_m, w_{p_m})\}$ be the first distribution with m points, and let $Q = \{(q_1, w_{q_1}), \ldots, (q_n, w_{q_n})\}$ be the second distribution with n points. Let $D = [d_{ij}]$ be the ground distance matrix, where d_{ij} is the ground distance between points p_i and q_j. Our objective is to find a flow matrix $F = [f_{ij}]$, with f_{ij} being the flow between points p_i and q_j, that minimizes the overall cost:

$$\text{Work}(P, Q, F) = \sum_{i=1}^{m} \sum_{j=1}^{n} f_{ij} d_{ij}$$

subject to the following list of constraints:

$$f_{ij} \geq 0, \ 1 \leq i \leq m, \ 1 \leq j \leq n$$
$$\sum_{j=1}^{n} f_{ij} \leq w_{p_i}, \ 1 \leq i \leq m$$
$$\sum_{i=1}^{m} f_{ij} \leq w_{q_j}, \ 1 \leq j \leq n$$
$$\sum_{i=1}^{m} \sum_{j=1}^{n} f_{ij} = \min \left(\sum_{i=1}^{m} w_{p_i}, \sum_{j=1}^{n} w_{q_j} \right)$$

The optimal value of the objective function, Work(P, Q, F), defines the Earth Mover's Distance between the two distributions.

The above formulation assumes that the two distributions have been aligned. However, recall that a translated and rotated version of a graph embedding will also be a graph embedding. To accommodate pairs of distributions that are "not rigidly embedded", Cohen and Guibas [7] extended the definition of EMD, originally applicable to pairs of fixed sets of points, to allow one of the sets to undergo a transformation. Assuming that a transformation $T \in \mathcal{T}$ is applied to the second distribution, distances d_{ij}^{T} are defined as $d_{ij}^{T} = d(p_i, T(q_j))$, and the objective function becomes Work$(P, Q, F, T) = \sum_{i=1}^{m} \sum_{j=1}^{n} f_{ij} d_{ij}^{T}$. The minimal value of the objective function defines the Earth Mover's Distance between the two distributions that are allowed to undergo a transformation from $\mathcal{T}$.

Cohen and Guibas [7] also suggested an iterative process (which they call **FT**, short for "an optimal **F**low and an optimal **T**ransformation") that achieves a local minimum of the objective function. Starting with an initial transformation $T^{(0)} \in \mathcal{T}$ from a given $T^{(k)} \in \mathcal{T}$, they compute the optimal flow $F = F^{(k)}$ that minimizes the objective function, Work$(P, T^{(k)}(Q), F)$, and from a given optimal flow, $F^{(k)}$, they compute an optimal transformation, $T = T^{(k+1)}, \in \mathcal{T}$ that minimizes the objective function, Work$(P, T(Q), F^{(k)})$. The iterative process stops when the improvement in the objective function value falls below a threshold. The resulting optimal pair (F, T) depends on the initial transformation $T^{(0)}$. Starting the iteration from several initial transformations increases the chances of obtaining a global minimum.

5.3 Choosing an Appropriate Transformation

For our application, the set $\mathcal{T}$ of allowable transformations consists of only those transformations that preserve distances. Therefore, we use a weighted version of the Least Squares Estimation algorithm [23] to compute an optimal distance-preserving transformation given a flow between the distributions. Specifically, given a set of pairings $\{(x_i, y_i, w_i)\}$ (the flow of weight w_i is sent from point x_i to point y_i), we define the transformation $T(x) = cRx + t$ in accordance with [23] as follows:

$$\mu_x \leftarrow (\textstyle\sum_i w_i x_i) / \textstyle\sum_i w_i$$
$$\mu_y \leftarrow (\textstyle\sum_i w_i y_i) / \textstyle\sum_i w_i$$
$$\sigma_x^2 \leftarrow (\textstyle\sum_i w_i \|x_i - \mu_x\|) / \textstyle\sum_i w_i$$
$$\sigma_y^2 \leftarrow (\textstyle\sum_i w_i \|y_i - \mu_y\|) / \textstyle\sum_i w_i$$
$$\Sigma_{xy} \leftarrow (\textstyle\sum_i w_i (y_i - \mu_y)(x_i - \mu_x)^T) / \textstyle\sum_i w_i$$
$$R \leftarrow UV^T, \text{ where } UDV^T \text{ is the SVD of } \Sigma_{xy}$$
$$c \leftarrow \sigma_y / \sigma_x$$
$$t \leftarrow \mu_y - cR\mu_x$$

The original proof of optimality of the transformation [23] is easily adapted to the weighted case. Namely, assuming that the flows from the x_i's to the y_i's are integer, each weighted pairing $\{(x_i, y_i, w_i)\}$ is replaced by w_i unweighted pairings $\{(x_i^j, y_i^j)\}$, which makes the original proof applicable. Collecting appropriate terms, we get weighted versions of the original equations. Fractional flows are reduced to integer flows by multiplying all fractions by their least common denominator.

5.4   The Final Algorithm

Our algorithm for many-to-many matching is a combination of the previous procedures. Specifically, given two vertex-labeled graphs G_1 and G_2, we first find isometric embeddings of the graphs into low-dimensional normed spaces, obtaining two weighted distributions. We then "register" one distribution with respect to the other so as to minimize the (original) EMD between them. We then apply the FT iteration of the transformation version of the EMD framework [7] to minimize the (extended) EMD. The pairing of points minimizing the EMD corresponds to a weighted many-to-many pairing of nodes. We summarize our approach in Algorithm 1.

Algorithm 1 Many-to-many graph matching.

1: Compute the metric tree $\mathfrak{T}_i$ corresponding to G_i (see Section 3.2).
2: Construct low-distortion embeddings $f_i(\mathfrak{T}_i)$ of $\mathfrak{T}_i$ into $(\mathcal{B}_i, ||.||_2)$ according to Section 3.4.
3: Compute low-distortion embeddings $\mathcal{E}_i = P_i(f_i(\mathfrak{T}_i))$ into $(\mathcal{B}, ||.||_2)$ according to Section 5.1.
4: Compute the EMD between $\mathcal{E}_i$'s by applying the FT iteration (Section 5.2), computing the optimal transformation T according to Section 5.3.
5: Interpret the resulting optimal flow between $\mathcal{E}_i$'s as a many-to-many vertex matching between G_i's.

---

6   Experiments

We tested the many-to-many matching algorithm on the COIL-20 database of Columbia University consisting of 72 views per object. A representative view of each object is shown in Figure 3(a). The multi-scale blob decomposition is then computed for each view using the algorithms described in [21] (and illustrated in Figure 1). For the experiments, we compute the tree metric corresponding to the complete edge-weighted graph defined on the regions of the scale-space decomposition of the view. The edge weights are computed as a function of the distances between the centroids of the regions in the scale-space representation. Next, each tree will be embedded into a normed space with low distortion. This procedure results in a database of weighted point-sets, each representing an embedded graph.

To test the matching algorithm on the resulting database, we removed 36 (of the 72) representative views of each object (every other view) and used these as

Fig. 3. Columbia University Image Library (COIL-20) database.

queries to the remaining view database (the other 36 views for each of the 20 objects). We then computed the distance between each "query" view and each of the remaining database views. Ideally, for any given query view i of object j, $v_{i,j}$, the matching algorithm should return either $v_{i+1,j}$ or $v_{i-1,j}$ as the closest view. We will classify this as a correct matching. Figure 4 presents a subset of the matching experiments for Object 9 of the COIL-20 database, with a correct matching in almost all cases.

The results of the experiment is presented in Figure 5, with darker points representing the closer matches. Based on the overall matching statistics, we observed that in all but 10.74% of the experiments, the closest match selected by our algorithm was a neighboring view. This is clearly evident from the darker diagonal entries of Figure 5. Among the mismatches, the closest view belonged to the same object in 80% of the cases. It should be noted that these results can be considered worst case for two reasons. First, the original 72 view per object sampling resolution was tuned for an eigenimage approach. Given the high similarity among neighboring views, it could be argued that our matching criterion is overly harsh, and that perhaps a measure of "viewpoint distance", i.e., "how many views away was the closest match" would be less severe. In any case, we anticipate that with fewer samples per object, neighboring views would be more dissimilar, and our matching results would improve. Second, and perhaps more importantly, many of the objects are symmetric, and if a query neighbor has an identical view elsewhere on the object, that view might be chosen (with equal distance) and scored as an error. Many of the objects in the database are rotationally symmetric, yielding identical views from each viewpoint and likely errors.

Both the embedding and matching procedures can accommodate occlusion. This is due to the fact that the path partitions for unoccluded portions of the graph are unaffected by occlusion. During the projection step, the projections of unoccluded nodes will also be unaffected by occlusion. Finally, the matching procedure is an iterative process driven by flow optimization which, in turn, depends only on local features, and is thereby unaffected by occlusion. We are in the process of conducting occlusion experiments and expect to report them shortly.

7   Conclusions and Future Work

We have presented a computationally efficient approach to many-to-many matching of multi-scale feature representations in terms of blobs and ridges. The ap-

Query	Model											
	2.70	7.05	6.54	7.78	10.37	5.89	13.41	12.30	20.34	13.90	19.60	19.53
	4.14	5.56	5.18	7.98	10.30	4.24	12.34	11.23	20.01	12.24	18.40	17.73
	6.39	**2.34**	2.68	4.17	5.97	5.94	17.03	15.87	25.28	16.74	22.99	22.17
	6.07	4.04	4.04	**3.17**	4.44	6.64	17.26	15.92	25.82	17.17	23.53	22.80
	7.27	6.39	6.55	5.31	**3.88**	8.26	18.20	16.88	26.74	17.85	24.57	23.81
	5.31	4.20	5.25	5.67	**3.21**	5.63	17.08	15.86	25.20	17.07	23.49	22.79
	9.61	11.65	11.21	13.81	16.00	**6.80**	7.07	8.20	14.92	9.05	13.74	14.65
	13.64	15.32	14.85	17.35	19.28	11.80	**2.69**	3.70	14.20	6.75	10.93	12.19
	14.34	16.03	15.23	17.92	19.90	12.21	5.28	**3.54**	14.61	4.61	8.96	10.33
	13.50	14.90	14.41	17.39	19.32	11.44	6.56	**4.13**	15.00	5.25	8.97	9.98
	17.16	18.97	18.34	21.28	23.11	15.70	7.95	7.85	13.52	**4.23**	4.73	6.17
	20.53	22.30	21.25	24.17	26.18	18.77	11.46	11.59	14.48	7.14	3.02	**2.75**
	20.19	20.90	19.92	22.89	24.87	18.18	12.19	12.27	14.91	7.94	6.53	**3.24**

Fig. 4. Sample matching results for object 9 of the COIL-20 database, in which rows and columns can be interleaved to form the set of sequential views. The diagonal and next lower diagonal therefore represent the neighboring views of the query (row). Only one query, entry (10,8), was incorrectly matched.

proach is based on a combination of metric tree representation and low-distortion embedding of graphs to normed spaces with a distribution-based similarity measure. The matching framework presented attempts to exploit both topological and geometrical properties of multi-scale feature hierarchies. Due to the strengths of the two components, our approach is able to establish robust, many-to-many correspondences in the presence of noise. Preliminary matching experiments on the COIL-20 database demonstrate that our method performs very well subject to view sampling constraints. Our work on matching of multi-scale features is in its preliminary stages, and we plan to extend its scope in several directions: 1) to study the viewpoint invariance of the multi-scale blob decomposition within our many-to-many matching framework; 2) to study the initial conditions of the FT iteration to improve matching results; 3) to exploit the

Fig. 5. The matching results for the COIL-20 database. The rows represent the query views (36 views per object), and the columns representing model views (36 views per object). Each row represents the matching results for a query view against the whole database. The intensity of entries represents the quality of the matching, with black representing maximum similarity between the views and white minimum similarity.

possibility of using embedded vector representations as signatures for indexing purposes; and 4) to conduct occlusion and noise sensitivity experiments.

Acknowledgment

Ali Shokoufandeh acknowledges the partial support provided by a grant from National Science Foundation (NSF) ITR-DM-0219176. The work of Yakov Keselman is supported in part by the NSF grant No. 0125068. Sven Dickinson acknowledges the support of Natural Science and Engineering Research Council of Canada.

References

1. R. Agarwala, V. Bafna, M. Farach, M. Paterson, and M. Thorup. On the approximability of numerical taxonomy (fitting distances by tree metrics). *SIAM Journal on Computing*, 28(2):1073–1085, 1999.
2. R. K. Ahuja, T. L. Magnanti, and J. B. Orlin. *Network Flows: Theory, Algorithms, and Applications*, pages 4–7. Prentice Hall, Englewood Cliffs, New Jersey, 1993.
3. S. Belongie, J. Malik, and J. Puzicha. Shape matching and object recognition using shape contexts. *IEEE PAMI*, 24(4):509–522, April 2002.
4. R. Beveridge and E. M. Riseman. How easy is matching 2D line models using local search? *IEEE Transactions on Pattern Analysis and Machine Intelligence*, 19(6):564–579, June 1997.
5. J. Bourgain. On Lipschitz embedding of finite metric spaces into Hilbert space. *Israel Journal of Mathematics*, 52:46–52, 1985.
6. P. Buneman. The recovery of trees from measures of dissimilarity. In F. Hodson, D. Kendall, and P. Tautu, editors, *Mathematics in the Archaeological and Historical Sciences*, pages 387–395. Edinburgh University Press, Edinburgh, 1971.
7. S. D. Cohen and L. J. Guibas. The earth mover's distance under transformation sets. In *Proceedings, 7th International Conference on Computer Vision*, pages 1076–1083, Kerkyra, Greece, 1999.

8. Steven Gold and Anand Rangarajan. A graduated assignment algorithm for graph matching. *IEEE Transactions on Pattern Analysis and Machine Intelligence*, 18(4):377–388, 1996.

9. A. Gupta, I. Newman, Y. Rabinovich, and A. Sinclair. Cuts, trees and l_1 embeddings. *Proceedings of Symposium on Foundations of Computer Scince*, 1999.

10. P. Indyk. Algorithmic aspects of geometric embeddings. In *Proceedings, 42nd Annual Symposium on Foundations of Computer Science*, 2001.

11. S. Kosinov and T. Caelli. Inexact multisubgraph matching using graph eigenspace and clustering models. In *Proceedings of SSPR/SPR*, volume 2396, pages 133–142. Springer, 2002.

12. N. Linial, E. London, and Y. Rabinovich. The geometry of graphs and some of its algorithmic applications. *Proceedings of 35th Annual Symposium on Foundations of Computer Scince*, pages 557–591, 1994.

13. T.-L. Liu and D. Geiger. Approximate tree matching and shape similarity. In *Proceedings, 7th International Conference on Computer Vision*, pages 456–462, Kerkyra, Greece, 1999.

14. B. Luo and E.R.Hancock. Structural matching using the em algorithm and singular value decomposition. *IEEE Transactions on Pattern Analysis and Machine Intelligence*, 23:1120–1136, 2001.

15. J. Matoušek. On embedding trees into uniformly convex Banach spaces. *Israel Journal of Mathematics*, 237:221–237, 1999.

16. M. Pelillo, K. Siddiqi, and S. Zucker. Matching hierarchical structures using association graphs. *IEEE Transactions on Pattern Analysis and Machine Intelligence*, 21(11):1105–1120, November 1999.

17. Y. Rubner, C. Tomasi, and L. J. Guibas. The earth mover's distance as a metric for image retrieval. *International Journal of Computer Vision*, 40(2):99–121, 2000.

18. G. Scott and H. Longuet-Higgins. An algorithm for associating the features of two patterns. *Proceedings of Royal Society of London*, B244:21–26, 1991.

19. T. Sebastian, P. Klein, and B. Kimia. Recognition of shapes by editing shock graphs. In *IEEE International Conference on Computer Vision*, pages 755–762, 2001.

20. L. G. Shapiro and R. M. Haralick. Structural descriptions and inexact matching. *IEEE Transactions on Pattern Analysis and Machine Intelligence*, 3:504–519, 1981.

21. A. Shokoufandeh, S.J. Dickinson, C. Jönsson, L. Bretzner, and T. Lindeberg. On the representation and matching of qualitative shape at multiple scales. In *Proceedings, 7th European Conference on Computer Vision*, volume 3, pages 759–775, 2002.

22. K. Siddiqi, A. Shokoufandeh, S. Dickinson, and S. Zucker. Shock graphs and shape matching. *International Journal of Computer Vision*, 30:1–24, 1999.

23. S. Umeyama. Least-squares estimation of transformation parameters between two point patterns. *IEEE Transactions on Pattern Analysis and Machine Intelligence*, 13(4):376–380, April 1991.

24. M. S. Waterman, T. F. Smith, M. Singh, and W. A. Beyer. Additive evolutionary trees. *J. Theor. Biol.*, 64:199–213, 1977.

Content Based Image Retrieval
Using Multiscale Top Points
A Feasibility Study

Frans Kanters, Bram Platel, Luc Florack, and Bart M. ter Haar Romeny

Eindhoven University of Technology
Den Dolech 2, Postbus 513
5600 MB Eindhoven, The Netherlands
{F.M.W.Kanters,B.Platel,L.M.J.Florack,B.M.terHaarRomeny}@tue.nl
http://www.bmi2.bmt.tue.nl/image-analysis/

Abstract. A feasibility study for a new method for content based image retrieval is presented. First, an image representation using multiscale top points is introduced. This representation is validated using a minimal variance reconstruction algorithm. The image retrieval problem can now be translated into comparing distances between point sets. For this purpose the proportional transportation distance (PTD) is used. A method is proposed using multiscale top points and their reconstruction coefficients in the PTD to define these distances between images. We present some experiments with promising results on a database with face images.

1 Introduction

In this paper a feasibility study is presented for using multiscale top points in a content based image retrieval system. The goal of such a system is to find, given an image, the closest matches to that image in a large image database, looking at the image content. Many of such systems exist but IBM's Query by Image Content (QBIC)[12] is probably the best-known one. The user has to specify a number of parameters prior to the searching, so much labor is still needed here. Some examples of more automated systems are Virage [4] and VisualSeek [17] which are based on texture and material structure. Color histogram based systems are also commonly used [18,19]. These systems, however, do not take into account the spatial distribution of the features used. Some examples of systems which do use spatial information are systems which use segmented image regions [11,1]. In the following sections we propose a completely new approach based on an image representation using multiscale top points.

2 Image Representation Using Multiscale Top Points

2.1 Introduction

Sometimes it is useful to represent an image in a different way than by it's pixel values. A "good" representation should contain all the information desired for

L.D. Griffin and M. Lillholm (Eds.): Scale-Space 2003, LNCS 2695, pp. 33–43, 2003.

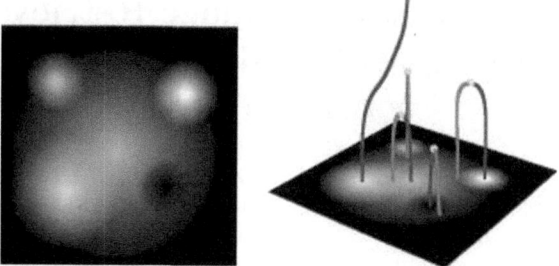

Fig. 1. Simple example top points. Left: original image, right: 3D view of critical paths and top points.

the application and can thus be different for each application. In this chapter we propose an image representation using multiscale top points and some local derivatives in those points. We will verify if the representation contains information about the original image by making a reconstruction from only these top points and some of their local N-jet properties.

2.2   Critical Points, Critical Paths and Top Points

In the Gaussian scale space of a 2D image, a number of points with special differential characteristics can be found [3,2,10,6,9]. In this paper we define the following points:

Definition 1. *Spatial critical points are points where the spatial gradient is zero. For 2D images these points are maxima, minima or saddles.*

Definition 2. *Critical paths are the paths that spatial critical points follow in scale.*

Definition 3. *Top points are critical points where the Hessian degenerates (det H=0). For generic 2D images, these points are annihilations or creations of saddles with maxima or minima.*

where the Hessian of a 2D image f is given by:

$$Hf = \nabla^T \nabla f = \begin{pmatrix} \partial_x^2 f & \partial_x \partial_y f \\ \partial_y \partial_x f & \partial_y^2 f \end{pmatrix} \tag{1}$$

Figure 1 shows a simple image with the critical paths and top points. Note that according to Loog et al. [10] under certain weak conditions always one critical path from a maximum or minimum will be left at the highest scale.

In 1D, the top points form a good representation of the signal according to [5]. However, in 2D it is still not proven that using top points results in a good representation of an image. In the next section we will try to reconstruct the original image using it's top points and the local N-jet properties of the image at those points.

2.3 Minimal Variance Reconstructions from Multi Scale Points

We present a reconstruction algorithm that preserves top points in the reconstruction and is as smooth as possible. It is based on the minimal variance reconstruction algorithm by Nielsen and Lillholm [13]. We have two constraints: the reconstruction must have the same local N-jet properties at the top points and the variance must be minimal. A quadratic functional is chosen for simplicity of implementation. When using sufficiently many constraints the precise form of the functional does not significantly influence the reconstruction. More information about the reconstruction algorithm can be found in [7]. Given a set of filters ϕ_i we thus have to minimize:

$$S[\hat{f}] \stackrel{def}{=} \frac{1}{2} \| \hat{f} \|_{L^2}^2 + \sum_i \lambda_i \langle f - \hat{f} \mid \phi_i \rangle \tag{2}$$

Where $\langle . \mid . \rangle$ is an inner product in a real Hilbert space[1]. The first part satisfies the minimal variance constraint, the second part makes sure that the features are preserved. Using the functional derivative we obtain:

$$\frac{\delta S[\hat{f}]}{\delta \hat{f}} \stackrel{def}{=} \hat{f} - \sum_i \lambda_i \, \phi_i \tag{3}$$

we can determine the unique solution of $\frac{\delta S[\hat{f}]}{\delta \hat{f}} = 0$:

$$\hat{f} = \sum_i \lambda_i \, \phi_i \tag{4}$$

which is the $\hat{f}$ that minimizes the variance if the coefficients λ_i are calculated by substitution of (4) in:

$$\langle f - \hat{f} \mid \phi_j \rangle = 0 \tag{5}$$

Now we define the following filters:

$$\phi_{i,\nu_1 \ldots \nu_k}(x, y) = \sqrt{2t_i}^k \nabla_{\nu_1 \ldots \nu_k} \phi(x - x_i, y - y_i, t_i) \tag{6}$$

with ϕ the standard Gaussian at scale t centered at the origin, thus

$$\phi(x, y, t) = \frac{1}{4\pi t} e^{-\frac{x^2 + y^2}{4t}} \tag{7}$$

Using (6) in (4), we obtain our reconstruction formula for top points, using derivatives up to the second order. Note that the summation convention for repeated indices applies for all spatial indices $\mu = x, y$ and $\rho = x, y$. The reconstruction formula becomes:

$$\hat{f}(x, y) = \sum_{i=1}^{N} a_i \phi_i(x, y) + b_i^{\mu} \phi_{i,\mu}(x, y) + c_i^{\mu\rho} \phi_{i,\mu\rho}(x, y) \tag{8}$$

subject to the constraints:

---

[1] Or bilinear functional on the product space of fiducial filters (containing at least the Gaussian family) and its topological dual (i.e. the raw images).

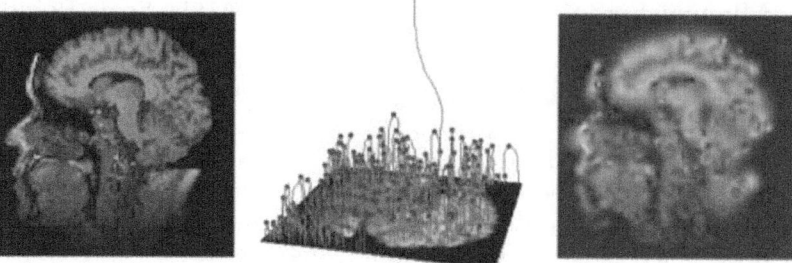

Fig. 2. Reconstruction from top points of mrbrain.tif. Left: original image, Center: 3D view of critical paths and top points, Right: reconstruction from top points, using all features with order $k \leq 2$.

$$\langle f - \hat{f} \mid \phi_i \rangle = 0, \ \langle f - \hat{f} \mid \phi_{i,\mu} \rangle = 0 \quad \text{and} \quad \langle f - \hat{f} \mid \phi_{i,\mu\rho} \rangle = 0 \qquad (9)$$

The coefficients a_i, b_i^x, b_i^y, c_i^{xx}, c_i^{xy} and c_i^{yy} can be calculated from the constraints using simple linear algebra. Each coefficient determines how much the corresponding derivative of the Gaussian is used for the reconstructed image at point i. Later we will use these coefficients for our image retrieval system.

Given a number of (top) points and their local derivatives up to the second order we can make a reconstruction of the image using the reconstruction formula (8). Figure 2 shows the result of a reconstruction using only top points and some local derivatives. Note that the reconstruction is not a perfect one, but for our purpose it is only important to show that the information carried within the top points is sufficiently rich.

3   Content Based Image Retrieval Using Multiscale Top Points

3.1   Introduction

In this section we try to use the top point representation for image matching in a database. How much two images "look" alike is now translated into how "close" two point sets are to each other. Of course, a lot of different distance measures can be used here. In each top point we can calculate some properties (e.g. derivatives, reconstruction coefficients) which can be taken into account for the distance. The number of top points will generally be different for each image. For the distance measurement we chose to use the Proportional Transportation Distance (PTD) as introduced by P. Giannopoulos and R. Veltkamp [14], which is based on the Earth Movers Distance (EMD) [20]. The PTD is a pseudo metric which can be used on weighted point sets. The advantage of the PTD lies in the fact that it holds the triangle inequality which can be used to speed up the retrieval process.

3.2   Proportional Transportation Distance (PTD)

First, we define weighted point sets.

Definition 4. *Let* $A = \{a_1, a_2, \ldots, a_m\}$ *be a weighted point set such that* $a_i = (x_i, w_i), i = 1, \ldots, m$, *where* $x_i \in \mathbb{R}^k$ *with* $w_i \in \mathbb{R}^+ \cup \{0\}$ *being its corresponding weight. Let also* $W = \sum_{i=1}^{m} w_i$ *be the total weight of set* A.

Definition 5. *Let* $B = \{b_1, b_2, \ldots, b_n\}$ *be a weighted point set such that* $b_j = (x_j, u_j), j = 1, \ldots, n$, *where* $x_j \in \mathbb{R}^k$ *with* $u_j \in \mathbb{R}^+ \cup \{0\}$ *being its corresponding weight. Let also* $U = \sum_{j=1}^{n} u_j$ *be the total weight of set* B.

The PTD can then be defined as follows:

Definition 6. *Let* A *and* B *be two weighted point sets and* d_{ij} *a ground distance between point* a_i *and* b_j. *The set of all feasible flows* $\mathcal{F} = [f_{ij}]$ *from* A *to* B, *is now defined by the following constraints:*

(i) $f_{ij} \geq 0, i = 1, \ldots, m, j = 1, \ldots, n$
(ii) $\sum_{j=1}^{n} f_{ij} = w_i, i = 1, \ldots, m$
(iii) $\sum_{i=1}^{m} f_{ij} = \frac{u_j W}{U}, j = 1, \ldots, n$
(iv) $\sum_{i=1}^{m} \sum_{j=1}^{n} f_{ij} = W$

then the PTD can be defined as:

$$PTD(A, B) = \frac{min_{F \in \mathcal{F}} \sum_{i=1}^{m} \sum_{j=1}^{n} f_{ij} d_{ij}}{W}$$

The PTD can be seen as the minimum amount of work needed to transform A into a new set A' that resembles B. In particular, we redistribute A's total weight from the position of its points to the position of B's points leaving the old percentages of weights in B the same. For calculation of the PTD we used a fast implementation of the Earth Movers Distance (EMD) which is publicly available [20,15]. In our case we used normalized weights such that $U = W = 1$.

There are two major parameters with which we can experiment. The first one is the ground distance d_{ij}. All top points have the following properties: $p_i = (x_i, y_i, t_i, a_i, b_i^x, b_i^y, c_i^{xx}, c_i^{xy}, c_i^{yy})$, with $a_i, b_i^x, b_i^y, c_i^{xx}, c_i^{xy}, c_i^{yy}$ the reconstruction coefficients of (8). One possible ground distance is the Euclidian distance for only the x and y coordinates, discarding all other features. Another possibility is to include the reconstruction coefficients of each point in the ground distance. The second tuneable parameter is the weight w_i for each point. We can use equal weights for all points but we can also use different weights for each point, for example by using the reconstruction coefficients of (8) in the weights. In the next section a number of different settings for these parameters are used for content based image retrieval experiments.

4   Experimental Results

In the previous section we described a method to calculate a distance between two weighted point sets, the PTD. We can use this PTD for content based image retrieval in a large database. In the experiments described in this section, we used a subset of the Face database from the Olivetti Research Laboratory, made by Samaria and Harter [16]. The subset consists of 200 images of faces from 20 people (10 images each, with different deviations such as pose, glasses, distortion). From every image of this set, the top points and the reconstruction coefficients were calculated. For the content based image retrieval experiment we used the first image of each person as a query and looked at the 9 images with the smallest PTD to this image, using different parameters.

4.1   Experiment 1

In our first experiment the ground distance is defined as:

$$d_{ij} = \sqrt{(x_i - x_j)^2 + (y_i - y_j)^2 + (t_i - t_j)^2} \tag{10}$$

and the weights are equally distributed, thus:

$$w_i = \frac{1}{m} \tag{11}$$

$$u_j = \frac{1}{n} \tag{12}$$

While the ground distance is obviously ad hoc and the weights do not contain any information, the results are surprising. As can be seen in Fig. 3, the images in the database with the smallest PTD are mostly from the same person. Note that in the second row, the results seem to be independent of the glasses. Also note that the third row is one of the worst query results from the database for this experiment (in this case, the hair line seems to have a strong influence). Somehow, the *structure* of the faces is more important than the pose of the person for the PTD algorithm.

4.2   Experiment 2

In our second experiment the ground distance is defined as in (10), but the weights are different. Our first try was to take the mean of the absolute values of the reconstruction coefficients (admittedly somewhat an ad hoc choice):

$$w_i = \frac{|a_i| + |b_i^x| + |b_i^y| + |c_i^{xx}| + |c_i^{xy}| + |c_i^{yy}|}{6} \tag{13}$$

$$u_j = \frac{|a_j| + |b_j^x| + |b_j^y| + |c_j^{xx}| + |c_j^{xy}| + |c_j^{yy}|}{6} \tag{14}$$

The results of this first try can be seen in Fig. 4. As can be seen, the results are absolutely not close to the results from the first experiment. It seems

Fig. 3. Experiment 1: $d_{ij} = \sqrt{(x_i - x_j)^2 + (y_i - y_j)^2 + (t_i - t_j)^2}$ and weights are equally distributed. Leftmost images are queries, neighboring images are closest matches, with increasing PTD.

that some points have a much too large weight. Looking at the range of the reconstruction coefficients, the problem becomes more clear. Sometimes the reconstruction coefficients can be as high as 10^6, resulting in very strong points, where distance is not very important anymore. As a second try we take the log of the weights from (13):

$$w_i = \log\left(\frac{|a_i| + |b_i^x| + |b_i^y| + |c_i^{xx}| + |c_i^{xy}| + |c_i^{yy}|}{6}\right) \qquad (15)$$

$$u_j = \log\left(\frac{|a_j| + |b_j^x| + |b_j^y| + |c_j^{xx}| + |c_j^{xy}| + |c_j^{yy}|}{6}\right) \qquad (16)$$

The results are shown in Fig. 5. The results are much better than with (13-14), but still not as good as with equally distributed weights. It is clear that one has to think more carefully about the way in which the incommensurable features are to be incorporated, instead of just averaging the features as in (13-14).

4.3   Experiment 3

For image retrieval, it is desired to be independent of translations and rotations of the query image. When looking at the top point structure of an image, one question is how much information is contained in the position of the top points. As a final experiment we tried a ground distance which is translation invariant, because the x and y coordinates are left out. Here things are becoming mildly interesting. We propose the ground distance as:

$$d_{ij} = \sqrt{(t_i - t_j)^2 + (a_i - a_j)^2 + (b_i^\nu - b_j^\nu)^2 + (c_i^{\mu\rho} - c_j^{\mu\rho})^2} \qquad (17)$$

Fig. 4. Experiment 2a: $d_{ij} = \sqrt{(x_i - x_j)^2 + (y_i - y_j)^2 + (t_i - t_j)^2}$ and mean of absolute reconstruction coefficients as weights. Leftmost images are queries, neighboring images are closest matches, with increasing PTD.

Fig. 5. Experiment 2b: $d_{ij} = \sqrt{(x_i - x_j)^2 + (y_i - y_j)^2 + (t_i - t_j)^2}$ and Log of mean of absolute reconstruction coefficients as weights. Leftmost images are queries, neighboring images are closest matches, with increasing PTD.

for all spatial indices ν, μ, ρ. The weights are taken the same as in (15). The results for this experiment are shown in Fig. 6. It is very surprising to see that the results are better without the position of the top points in the ground distance. Still there are some queries which give problems, as can be seen in the fourth row of Fig. 6. The big advantage with this ground distance is the invariance regarding translation.

Note that for all described methods, some odd results can show up. For example the fifth image on the third row of Fig. 6 appears to be very different

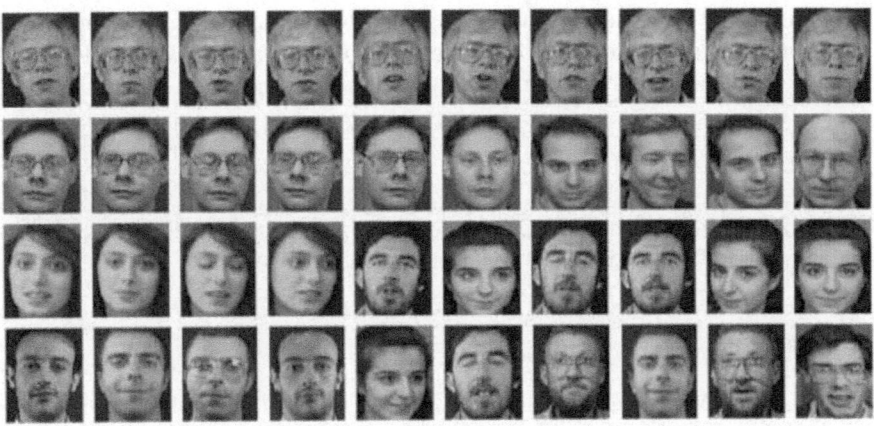

Fig. 6. Experiment 3a: $d_{ij} = \sqrt{(t_i - t_j)^2 + (a_i - a_j)^2 + (b_i^\nu - b_j^\nu)^2 + (c_i^{\mu\rho} - c_j^{\mu\rho})^2}$ and Log of mean of absolute reconstruction coefficients as weights. Leftmost images are queries, neighboring images are closest matches, with increasing PTD.

from the rest, but for the algorithm it is closer to the query than the sixth image. These results can not be explained at this moment and have to be further investigated.

5   Conclusions and Discussion

In this paper we presented a method for content based image retrieval using a multiscale top points representation of an image. The first experiments on a face database of 200 images are promising, but a lot of further research must be done.

There are two main parameters in the PTD that must be further investigated, the ground distance and the weights of the points. In our experiments we made some ad hoc choices for the ground distance. Better results will be possible if other ground distances are used. It is possible to obtain translation, rotation and scale invariance by choosing the correct ground distance. For the weights of the points in the point sets holds the same; we only tested a few rather ad hoc weight values. The importance of a single top point for the reconstruction of the image should be expressed in the weight for that top point. In one test we used the reconstruction coefficients to calculate the weights, but it might be easier to use the local derivatives in the top points instead.

With this representation it might also be possible to search for substructures in an image. If for example the top point representation of a face is known, it might be possible to search for a subset of top points in a complete scene of different faces.

At this moment, no *topological* information from the top points is used. However, it is possible to make a graph from the top points which does contain topological information. The point matching then becomes graph matching, which is a well known research area. It might be that content based image retrieval using such graphs is more robust than using point sets.

Acknowledgements

This work is part of the DSSCV project supported by the IST Programme of the European Union (IST-2001-35443).

References

1. C. Carson, M. Thomas, S. Belongie, J. M. Hellerstein, and J. Malik. Blobworld: A system for region-based image indexing and retrieval. In *Third International Conference on Visual Information Systems*. Springer, 1999.
2. L. Florack and A. Kuijper. The topological structure of scale-space images. *Journal of Mathematical Imaging and Vision*, 12(1):65–79, February 2000.
3. L. D. Griffin and A. C. F. Colchester. Superficial and deep structure in linear diffusion scale space: Isophotes, critical points and separatrices. *Image and Vision Computing*, 13(7):543–557, September 1995.
4. Amarnath Gupta and Ramesh Jain. Visual information retrieval. *Communications of the ACM*, 40(5):70–79, 1997.
5. P. Johansen, S. Skelboe, K. Grue, and J. D. Andersen. Representing signals by their top points in scale-space. In *Proceedings of the 8th International Conference on Pattern Recognition (Paris, France, October 1986)*, pages 215–217. IEEE Computer Society Press, 1986.
6. S. N. Kalitzin, B. M. ter Haar Romeny, A. H. Salden, P. F. M. Nacken, and M. A. Viergever. Topological numbers and singularities in scalar images: Scale-space evolution properties. *Journal of Mathematical Imaging and Vision*, 9(3), November 1998.
7. F.M.W. Kanters, L.M.J. Florack, B. Platel and B.M. ter Haar Romeny. Image reconstruction from multiscale critical points. Elsewhere in these proceedings.
8. M. Kerckhove, editor. *Scale-Space and Morphology in Computer Vision: Proceedings of the Third International Conference, Scale-Space 2001, Vancouver, Canada*, volume 2106 of *Lecture Notes in Computer Science*. Springer-Verlag, Berlin, July 2001.
9. A. Kuijper and L.M.J. Florack. The application of catastrophe theory to image analysis. Submitted to Image and Vision Computing.
10. M. Loog, J. J. Duistermaat, and L. M. J. Florack. On the behavior of spatial critical points under Gaussian blurring. a folklore theorem and scale-space constraints. In Kerckhove [8], pages 183–192.
11. Wei-Ying Ma and B. S. Manjunath. Netra: A toolbox for navigating large image databases. *Multimedia Systems*, 7(3):184–198, 1999.
12. M.Flickner, H.Sawhney, et.al. Query by image and video content: the qbic system. *IEEE Computer*, 28(9):23–32, 1995.
13. M. Nielsen and M. Lillholm. What do features tell about images? In Kerckhove [8], pages 39–50.
14. R. Veltkamp P. Giannopoulos. A pseudo-metric for weighted point sets. In *ECCV 2002, LNCS 2352*, pages 715–730. Springer, 2002.

15. Y. Rubner. Code for earth movers distance (emd).
    http://vision.stanford.edu/~rubner/emd/default.htm.
16. F. Samaria and A. Harter. Parameterisation of a stochastic model for human face
    identification, 1994.
17. J. Smith and S. Chang. Single color extraction and image query, 1995.
18. Markus Stricker and Michael Swain. The capacity and the sensitivity of color
    histogram indexing. Technical Report TR-94-05, University of Chicago, 3 1994.
19. M. Swain and D. Ballard. Color indexing. *International Journal on Computer
    Vision*, 7(1):11–32, 1991.
20. L.J. Guibas Y. Rubner, C. Tomasi. A metric for distributions with applications to
    image databases. In *IEEE International Conference on Computer Vision, Bombay,
    India*, pages 59–66, 1998.

Feature Coding with a Statistically Independent Cortical Representation

Roberto Valerio[1], Rafael Navarro[1], Bart M. ter Haar Romeny[2], and Luc Florack[2]

[1] Instituto de Óptica "Daza de Valdés" - CSIC, Madrid, Spain, 28006
{r.valerio,r.navarro}@io.cfmac.csic.es
[2] Faculteit Biomedische Technologie - Technische Universiteit Eindhoven
Eindhoven, The Netherlands, 5600 MB
{B.M.terHaarRomeny,L.M.J.Florack}@tue.nl

Abstract. Current models of primary visual cortex (V1) include a linear filtering stage followed by a gain control mechanism that explains some of the nonlinear behavior of neurons. The nonlinear stage has been modeled as a divisive normalization in which each input linear response is squared and then divided by a weighted sum of squared linear responses in a certain neighborhood. In this communication, we show that such a scheme permits an efficient coding of natural image features. In our case, the linear stage is implemented as a four-level Daubechies decomposition, and the nonlinear normalization parameters are determined from the statistics of natural images under the hypothesis that sensory systems are adapted to signals to which they are exposed. In particular, we fix the weights of the divisive normalization to the mutual information of the corresponding pair of linear coefficients. This nonlinear process extracts significant, statistically independent, visual events in the image.

1   Introduction

Traditional linear models of the responses of V1 neurons in primate visual cortex [1, 2] were highly attractive because of their simplicity, but they failed to explain essential nonlinear features of neural responses. In recent years, various authors have shown that the nonlinear behavior of V1 neurons can be modeled by including a gain control stage, known as *divisive normalization* [3, 4, 5, 6], after the linear filtering step. In this nonlinear stage the input linear responses are squared and then divided by a weighted sum of squared neighboring responses in space, orientation and scale, plus a regularizing constant.

The hypothesis that sensory systems are adapted to the signals to which they are exposed implies that parameters of the divisive normalization are related to the statistics of natural images. More specifically, the efficient coding hypothesis [7] states that an efficient group of neurons should be able to encode as much information as possible, or in other words, that all their responses should be statistically independent. Different versions of this hypothesis were formulated by Barlow [8], who proposed that early sensory neurons remove statistical redundancy in the input signal, and by other authors [9, 10, 11, 12, 13].

Statistically-derived divisive normalization models have shown their utility to characterize the nonlinear response properties of typical neurons in sensory systems

L.D. Griffin and M. Lillholm (Eds.): Scale-Space 2003, LNCS 2695, pp. 44–56, 2003.

[14, 15, 16]. In addition, Simoncelli and Schwartz [14] showed that statistically-derived divisive normalization reduces pairwise statistical dependences between output responses. It is well known that linear methods cannot eliminate higher-order statistical dependencies [e.g. 14, 17, 18]. Empirical results by Simoncelli and Schwartz [14] suggest that statistically-derived divisive normalization can significantly reduce higher order statistical dependencies between adjacent responses, which further supports the efficient coding hypothesis. In a recent paper, Hoyer and Hyvärinen [19] have used an alternative framework to show how the responses of V1 complex cells could be sparsely represented by a higher-order neural layer, which leads to contour coding and end-stopped receptive fields.

In this communication, we use a statistically-derived divisive normalization model of the nonlinear information processing in V1 cells to obtain almost statistically-independent and sparse descriptors of natural images, which code relevant features in images. The main novelty of this scheme resides in the way of fixing the weights of the divisive normalization: each weight is fixed to the mutual information of the corresponding pair of linear coefficients. This is computationally more efficient.

2   Methods

2.1   Model of the Response of V1 Neurons

Following the current models of primary visual cortex (V1), the model proposed here consists of a linear decomposition followed by a nonlinear divisive normalization stage.

2.1.1   Linear Stage
The linear stage is a four-level orthogonal wavelet decomposition based on Daubechies filters of order 4 (db8), that is with 4 vanishing moments and 8 coefficients [20]. The basis functions of this linear transform are localized in space, spatial frequency and orientation. This gives rise to 12 subbands (horizontal, vertical and diagonal for each of the 4 scales) plus an additional low-pass channel. Multi-scale and multi-orientation orthogonal wavelet transforms like this are very popular for image representation.

The marginal statistics of the resulting coefficients are known to be highly kurtotic and therefore non-Gaussian [e.g. 21, 22], and the joint statistics exhibit nonlinear dependencies. The left panel in Fig. 1 shows a typical orthogonal wavelet conditional histogram of a natural image. We can see that coefficients are decorrelated, since the expected value of the ordinate is approximately zero independently of the abscissa and therefore the covariance is close to zero as well. This makes an important difference between orthogonal and non-orthogonal linear transforms, since in the non-orthogonal cases "close" coefficients are correlated and the expected value of the ordinate is not zero but varies linearly with the abscissa. This is what we can see in the right panel in Fig. 1, which corresponds to a non-orthogonal Gabor pyramid [23].

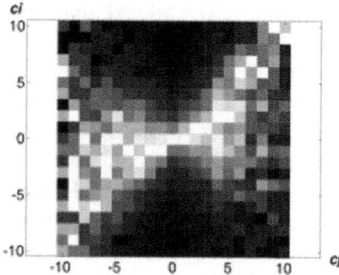 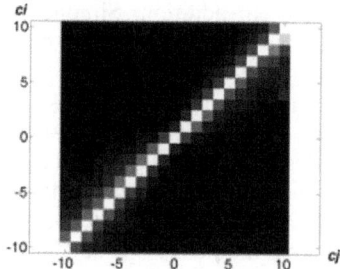

Fig. 1. Conditional histograms of two neighbor coefficients (one is the vertical neighbor of the other) in the lowest (finest) scale vertical subband of the Daubechies (left), and Gabor (right) pyramid of the "Einstein" standard test image.

On the other hand, the "bowtie" shape of the left histogram reveals that coefficients are not statistically independent. This shape suggests that the variance of a coefficient depends on the value of the neighboring coefficient. The dependence is such that the variance of the ordinate scales with the squared value of the abscissa.

All pairs of coefficients taken either in space, frequency or orientation always show this type of dependence [14, 17, 18], while the strength varies depending on the specific pair chosen. Intuitively, dependence will be stronger for coefficients that are closer, while it will decrease with distance along any axis. The form of the histograms is robust across a wide range of images, and different pairs of coefficients [15, 24]. In addition, this is a property of natural images but is not of the particular basis functions chosen [15].

Several distributions have been proposed in the context of V1 models to describe the conditional statistics of the coefficients obtained by projecting natural images onto an orthogonal linear basis [e.g. 15, 16]. Here we will consider a Gaussian model. Other closely related models have been proposed out of this context. Among them are the Gaussian Scale Mixture (GSM) models [e.g. 25, 26], which model a group of nearby wavelet coefficients as the product of two independent random variables: a scalar random variable and a zero-mean Gaussian random vector. Although GSM models permit, in theory, to get statistically independent coefficients through a nonlinear normalization, these models are not considered here because: first, this normalization procedure differs from the classical divisive normalization in V1 models and second, the Gaussian-distributed output coefficients that one would obtain, do not seem, intuitively, to be a proper model of the sparse responses of V1 neurons (for a discussion on sparse coding in V1 see [e.g. 27, 28]).

Assuming a Gaussian model, the conditional probability $p(c_i | \{c_j\})$ of an orthogonal wavelet coefficient c_i of a natural image, given the other wavelet coefficients $\{c_j\}$ ($j \neq i$), can be modeled as a zero-mean Gaussian density with a variance that depends on a linear combination of the squared coefficients $\{c_j^2\}$ ($j \neq i$) plus a constant a_i^2, $(a_i^2 + \sum_{j \neq i} b_{ij} c_j^2)$ [16]:

$$p(c_i \,|\, \{c_j\}) = \frac{1}{\sqrt{2\pi(a_i^2 + \sum_{j\neq i} b_{ij} c_j^2)}} \exp\left\{-\frac{c_i^2}{2(a_i^2 + \sum_{j\neq i} b_{ij} c_j^2)}\right\} \tag{1}$$

In the model, a_i^2 and $\{b_{ij}\}$ ($i \neq j$) are free parameters and can be determined by maximum-likelihood (ML) estimation. Operating with Eq. 1 we obtain the following ML equation:

$$\{a_i^2, b_{ij}\} = \arg\min_{\{a_i^2, b_{ij}\}} \mathbb{E}\left\{\frac{c_i^2}{a_i^2 + \sum_{j\neq i} b_{ij} c_j^2} + \log(a_i^2 + \sum_{j\neq i} b_{ij} c_j^2)\right\} \tag{2}$$

where $\mathbb{E}$ denotes expected value. In practice, we can compute $\mathbb{E}$ for each subband, averaging over all spatial positions of a set of natural images.

2.1.2   Nonlinear Stage

The nonlinear stage consists basically of a divisive normalization, in which the responses of the previous linear filtering stage, c_i, are squared and then divided by a weighted sum of squared neighboring responses in space, orientation and scale, $\{c_j^2\}$, plus a constant, d_i^2:

$$r_i = \frac{c_i^2}{d_i^2 + \sum_j e_{ij} c_j^2} \tag{3}$$

Parameters d_i^2 and $\{e_{ij}\}$ in Eq. 3 can be determined from the statistics of natural images if we accept the hypothesis that sensory systems are adapted to the statistical properties of the signals to which they are exposed [7, 8]. We will refer as optimal divisive normalization to the one defined by the values of the parameters (constant d_i^2 and weights $\{e_{ij}\}$) that yields the minimum mutual information, or equivalently minimizes statistical dependence, between normalized responses for a set of natural images.

For a general formulation, we start from the definition of the mutual information, also called Kullback-Leibler divergence [29] of the normalized responses r_i:

$$MI(r_1, r_2, ..., r_n) = \int_0^{+\infty} ... \int_0^{+\infty} p(r_1, r_2, ..., r_n) \log\left(\frac{p(r_1, r_2, ..., r_n)}{p(r_1) \cdot p(r_2) ... p(r_n)}\right) dr_1 dr_2 ... dr_n \tag{4}$$

If we use the change of variable theorem [30] to express the conditional probability density of the normalized responses $p(r_1, r_2, ..., r_n)$ in terms of that of the linear inputs $p(c_1, c_2, ..., c_n)$, and change the integration variables we get:

$$MI(r_1, r_2, ..., r_n) = \tag{5}$$

$$= \int_{-\infty}^{+\infty} ... \int_{-\infty}^{+\infty} p(c_1, c_2, ..., c_n) \log\left(\frac{|c_1 \cdot c_2 ... c_n| \cdot p(c_1, c_2, ..., c_n)}{r_1 \cdot r_2 ... r_n \cdot p(r_1) \cdot p(r_2) ... p(r_n) \cdot |\det[\mathbf{Id} - \mathbf{E} \cdot \mathbf{R}]|}\right) dc_1 dc_2 ... dc_n$$

where **Id** denotes the identity matrix, **E** the matrix of weights $\{e_{ij}\}$, and **R** the diagonal matrix of the normalized responses $\{r_i\}$.

And therefore the general expression for the set of parameters that minimizes the mutual information is:

$$\{d_i^2, e_{ij}\} = \arg \min_{\{d_i^2, e_{ij}\}} MI(r_1, r_2, ..., r_n) = \tag{6}$$

$$= \arg \max_{\{d_i^2, e_{ij}\}} \int_{-\infty}^{+\infty} ... \int_{-\infty}^{+\infty} p(c_1, c_2, ..., c_n) \log\left(r_1 \cdot r_2 \, ... \, r_n \cdot p(r_1) \cdot p(r_2) \, ... \, p(r_n) \cdot \left| \det[\mathbf{Id} - \mathbf{E} \cdot \mathbf{R}] \right|\right) dc_1 dc_2 \, ... \, dc_n$$

In addition to the nice feature of giving statistical independent coefficients, the described model has another important property, namely it is invertible. Note that the linear stage is obviously invertible. Regarding the nonlinear stage (except for the signs, which need to be stored), we can reconstruct the squared input linear coefficients c_i^2 from the normalized responses r_i and the parameters of the divisive normalization, d_i^2 and $\{e_{ij}\}$, by simply operating in Eq. 3. Since both stages, linear and nonlinear, of the nonlinear image representation scheme are invertible (if we keep the signs of the linear coefficients), it is possible to recover the input image from its nonlinear decomposition.

2   Results

In order to illustrate the features of the cortical code resulting from the above model we have undertaken some experiments with several standard test images.

First, to model the conditional statistics of the orthogonal wavelet coefficients of an image, we used the Gaussian model in Eq. 1. We have studied the strength of the dependences between coefficients in space, frequency and orientation, finding that to a first approximation, most of the statistical dependence is concentrated within a neighborhood formed by adjacent coefficients. Thus, we considered in Eq. 1 the 299 coefficients $\{c_j\}$ $(j \neq i)$ adjacent to c_i along the four dimensions (a 5x5 square box in the 2D space for each of the 12 subbands). An alternative way to determine which coefficients to include in the conditioning set, is to obtain the set of intraband and interband coefficients that minimizes the Kullback-Leibler divergence between the real conditional density of the wavelet coefficients and the statistical model of this density [24].

Once we have chosen the conditioning set, the Gaussian model non-negative parameters, a_i^2 and $\{b_{ij}\}$ $(i \neq j)$, are adjusted in the following way, instead of using Eq. 2, for the sake of computational efficiency: a_i^2 is fixed to a small value (0.1) and $\{b_{ij}\}$ $(i \neq j)$ are substituted by the mutual information of the corresponding pair of coefficients, c_i and c_j, calculated independently for each subband of the wavelet pyramid over all the coefficients of the input natural image. Note that this choice of parameters makes sense since a_i^2 is basically a regularizing constant to avoid dividing by zero, and the stronger the statistical dependence between two given coefficients, the greater is the corresponding b_{ij} parameter. As an example, Fig. 2 shows the resulting values of the $\{b_{ij}\}$ $(i \neq j)$ parameters for the lowest (finest) scale vertical subband of the "Lena" image. In general, $\{b_{ij}\}$ $(i \neq j)$ values are much less than one and highest values correspond to those neighboring coefficients in space, orientation and scale that have strongest statistical dependency with a certain coefficient.

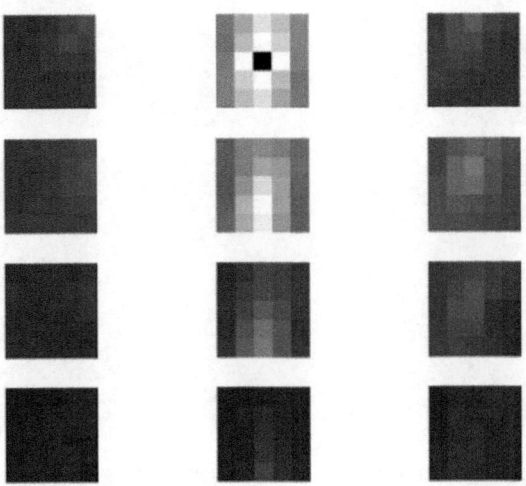

Fig. 2. $\{b_{ij}\}$ $(i \neq j)$ parameters for the lowest (finest) scale vertical subband of the "Lena" image. The 4 scales are vertically arranged (the lowest scale on top), and horizontally the 3 orientations (horizontal, vertical and diagonal on the left, in the middle and on the right respectively).

On the other hand, two considerations must be taken into account in order to efficiently attain an optimal divisive normalization that minimizes statistical dependence between output responses for natural images, that is, in order to efficiently solve Eq. 6. First, the mutual information of the normalized responses, $MI(r_1, r_2, ..., r_n)$, (Eq. 4) does not depend on the value of the normalization parameters e_{ii} (it can be demonstrated that $\frac{\partial MI(r_1, r_2, ..., r_n)}{\partial e_{ii}} = 0$, no matter the statistics of the linear inputs), so that these parameters can be fixed to any value (zero, say, for the sake of simplicity).

Second, the choice of parameters $d_i^2 = a_i^2$, $e_{ii} = 0$ and $e_{ij} = b_{ij}$, where a_i^2 and b_{ij} are the parameters of the Gaussian model (precisely the choice considered by Schwartz and Simoncelli in [16]) is an approximated solution of Eq. 6, easy to be computed, as it has been empirically demonstrated [16]. It is important to note that if we particularize the marginal probability of the normalized responses, $p(r_i)$, to that case, we obtain a highly kurtotic probability density function, which implies that the resulting code is sparse:

$$p(r_i) = \frac{1}{\sqrt{2\pi \cdot r_i}} \cdot \exp\left\{-\frac{r_i}{2}\right\} \tag{7}$$

When we apply the model described above, we obtain representations similar to that of Fig. 3.

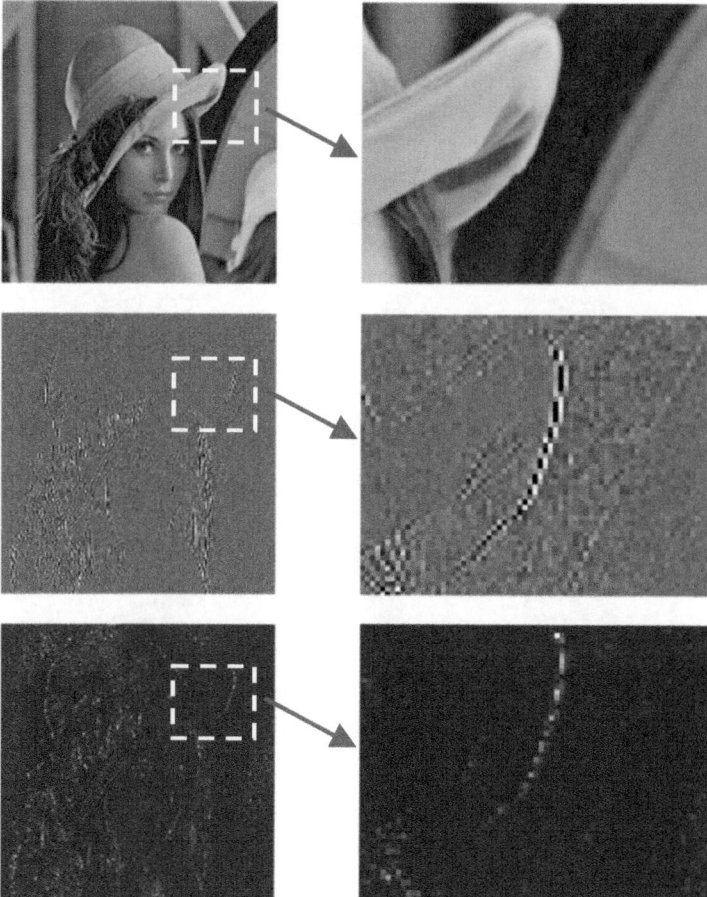

Fig. 3. Lowest (finest) scale vertical subband of the Daubechies (middle row) and the nonlinear divisive normalization (bottom row) decomposition of the "Lena" image.

Intuitively, the nonlinear transform has the effect of randomizing the image representation in order to reduce statistical dependencies between coefficients belonging to the same structural feature, or in other words, the effect of the divisive normalization is to choose which coefficients are most effective for describing a given image structure, similarly to sparsification in those models based on an overcomplete and non-orthogonal basis set (see [27]).

Fig. 4 represents three conditional histograms of two adjacent samples in space and illustrates the progressive statistical independence achieved by successive application of the linear (wavelet) and nonlinear (divisive normalization) transforms. The upper panel corresponds to the original image so that p_i and p_j are adjacent pixels (p_j is the right down neighbor pixel of p_i). The nearly 1 slope indicates the strong correlation between coefficients. The central panel shows the conditional histogram of the wavelet coefficients c_i and c_j. This linear transform cannot remove higher-order statistical dependencies, as suggested by the "bowtie" shape of the histogram. The bottom panel in Fig. 4 shows the conditional histogram between two adjacent output responses (r_i and r_j). After normalization, output statistical dependencies are practically removed since the resulting conditional histogram is basically independent on the value of the abscissa. Fig. 4 shows also a numerical evaluation of statistical dependence in terms of mutual information (Kullback-Leibler divergence). Consistently with Fig. 4, the mutual information, MI, is high, about 1.5, in the image domain. MI is much lower in the wavelet domain, after removing linear correlations. The divisive normalization further decreases MI to very low values close to zero.

Table 1 illustrates the "robustness" of the model in the sense that results do not depend much neither on the parameters values, nor on the particular input natural image. Specifically, Table 1 shows some numerical measures of statistical dependence in terms of mutual information for 5 standard test images ("Boats", "Elaine", "Goldhill", "Peppers" and "Sailboat"). As we can see, even if we use the parameters values previously calculated for the "Lena" image, the normalized responses are more statistically independent than the corresponding wavelet coefficients for any of the 5 images.

Table 1. Mutual information between two neighboring wavelet coefficients c_i and c_j (c_j is the right up neighboring coefficient of c_i), and between the corresponding normalized responses r_i and r_j (using the parameters values of the "Lena" image), of 5 standard test images (see Fig. 5). The considered subband is always the lowest (finest) scale vertical one.

	c_i , c_j	r_i , r_j
"Boats"	0.18	0.06
"Elaine"	0.05	0.04
"Goldhill"	0.10	0.03
"Peppers"	0.10	0.05
"Sailboat"	0.13	0.05

The effect of divisive normalization on marginal statistics is illustrated in Fig. 6. As we can see, the marginal density of the nonlinear responses, $p(r_i)$, is much more kurtotic, or in other words more peaked at zero, than the marginal density of the linear inputs, $p(c_i)$. In addition, we can observe that if we multiply the nonlinear responses by a certain constant k, the resulting marginal density $p(r_i' = k \cdot r_i)$ closely fits the expression in Eq. 7, which is consistent with the assumption that $MI(c_i, c_j) \approx k \cdot b_{ij}$ ($i \neq j$) implicitly used in the model implementation.

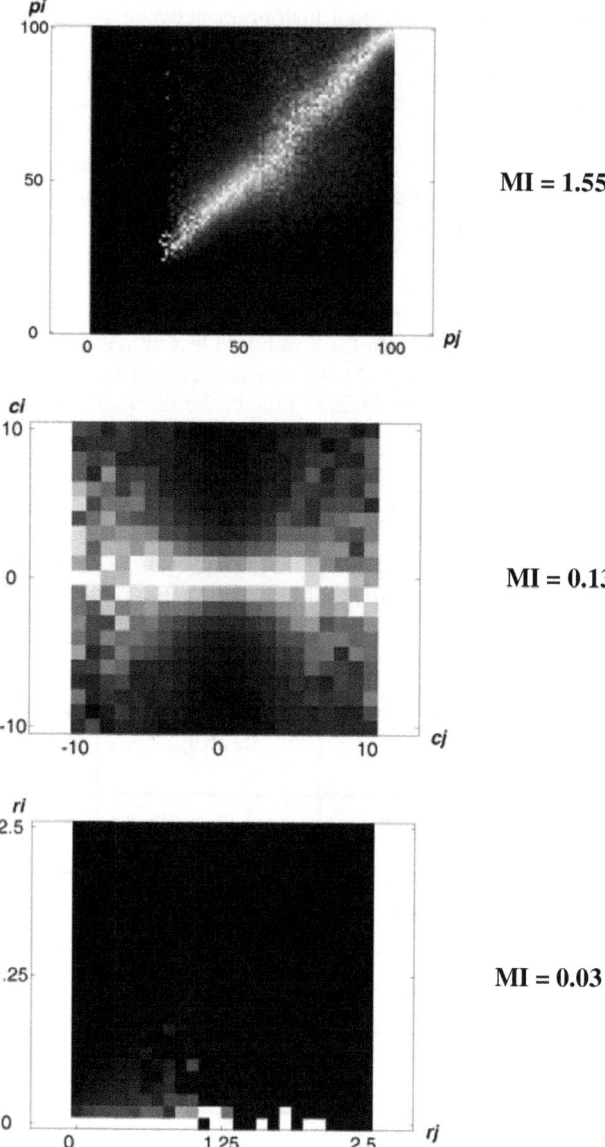

Fig. 4. Conditional histograms and mutual information (MI) values of two neighboring pixels p_i and p_j (p_j is the right up neighboring pixel of p_i), wavelet coefficients c_i and c_j, and nonlinear responses r_i and r_j of the "Lena" image. The considered subband is the lowest (finest) scale vertical one. Mutual information has been calculated from 200 bin joint histograms in the interval (-100 , 100) of the corresponding random variables after fixing their standard deviation (related to the mode) to 5 in order to compare the results.

Fig. 5. Standard test images used in the experiments. From top to bottom and left to right: "Boats", "Einstein", "Elaine", "Goldhill", "Lena", "Peppers" and "Sailboat".

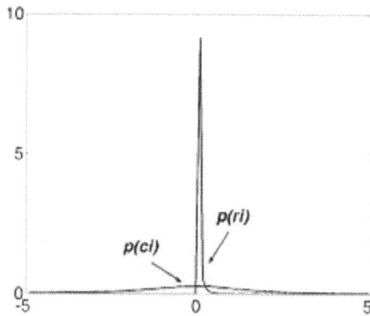 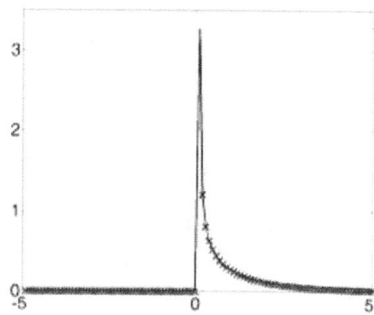

Fig. 6. Left panel: Marginal density functions of wavelet coefficients (c_i) and of nonlinear responses (r_i) in the lowest (finest) scale vertical subband of the "Lena" image. Right panel: Density function of the nonlinear responses in the lowest (finest) scale vertical subband after multiplying them by a certain constant k. With x's it is represented

$$p(r_i) = \frac{1}{\sqrt{2\pi \cdot r_i}} \cdot \exp\left\{-\frac{r_i}{2}\right\}$$

3   Conclusions

We have used a statistically-derived model of the nonlinear information processing in V1 neurons to illustrate some features of the cortical code. The key principle is the efficient coding hypothesis, that is, that the resulting responses should be as statistically independent as possible.

One of the main contributions is the method for obtaining the parameters values of the model nonlinear step. Basically, d_i^2 is fixed to a small value (0.1) and $\{e_{ij}\}$ ($i \neq j$) are substituted by the mutual information of the corresponding pair of coefficients, c_i and c_j. This method is computationally efficient and thus permits to consider a large set of neighboring coefficients.

Numerical results constitute the empirical proof for both the theoretical formulation and the validity of the numerical procedures, and suggest that the statistically-independent and sparse cortical code is a very efficient representation of the most frequent features of natural images.

The model is robust in the sense that results do not depend much on the linear orthogonal decomposition, the model of the conditional statistics of the linear coefficients, the training set of natural images, the computing errors (for example in estimating the parameters), or even the particular input natural image.

Finally, this nonlinear scheme of image representation, which has better statistical properties than the popular orthogonal wavelet transforms, is potentially useful for many image analysis and processing applications, such as restoration, synthesis, fusion, coding and compression, registration, etc., because of its invertibility and the great importance of statistical independence in these applications. Similar schemes [18, 31] have already been used very successfully in image analysis and processing applications.

Acknowledgments

This research was supported by the Spanish Commission for Research and Technology (CICYT) under grant DPI2002-04370-C02-02. Roberto Valerio was supported by a Madrid Education Council and Social European Fund Scholarship for Training of Research Personnel, and by a City Hall of Madrid Scholarship for Researchers and Artists in the Residencia de Estudiantes.

References

1. D. Hubel, and T. Wiesel, "Receptive fields, binocular interaction, and functional architecture in the cat's visual cortex", *J. Physiol (Lond)*, 160, pp. 106-154, 1962.
2. J. A. Movshon, I. D. Thompson, and D. J. Tholhurst, "Spatial summation in the receptive fields of simple cells in the cat's striate cortex", *J. Physiol (Lond)*, 283, pp. 53-77, 1978.
3. A. B. Bonds, "Role of inhibition in the specification of orientation selectivity of cells in the cat striate cortex", *Visual Neuroscience*, 2, pp. 41-55, 1989.
4. W. S. Geisler, and D. G. Albrecht, "Cortical neurons: Isolation of contrast gain control", *Vision Research*, 8, pp. 1409-1410, 1992.
5. D. J. Heeger, "Normalization of cell responses in cat striate cortex", *Visual Neuroscience*, 9, pp. 181-198, 1992.
6. M. Carandini, D. J. Heeger, and J. A. Movshon, "Linearity and normalization in simple cells of the macaque primary visual cortex", *J. Neuroscience*, 17, pp. 8621-8644, 1997.
7. F. Attneave, "Some informational aspects of visual perception", *Psych. Rev*, 61, pp. 183-193, 1954.
8. H. B. Barlow, "Possible principles underlying the transformation of sensory messages", *Sensory Communication*, p. 217, MIT Press, 1961.
9. S. B. Laughlin, "A simple coding procedure enhances a neuron's information capacity", *Z. Naturforsch*, 36C, pp. 910-912, 1981.
10. J. J. Atick, "Could information theory provide an ecological theory of sensory processing?", *Netw. Comput. Neural Syst.*, 3, pp. 213-251, 1992.
11. J. H. van Hateren, "A theory of maximizing sensory information", *Biol. Cybern.*, 68, pp. 23-29, 1992.
12. D. J. Field, "What is the goal of sensory coding?", *Neural Comput.*, 6, pp. 559-601, 1994.
13. F. Rieke, D. A. Bodnar, and W. Bialek, "Naturalistic stimuli increase the rate and efficiency of information transmission by primary auditory afferents", *Proc. R. Soc. London B*, 262, pp. 259-265, 1995.
14. E. P. Simoncelli, and O. Schwartz, "Modeling Surround Suppression in V1 Neurons with a Statistically-Derived Normalization Model", *Advances in Neural Information Processing Systems*, 11, pp. 153-159, 1999.
15. M. J. Wainwright, O. Schwartz, and E. P. Simoncelli, "Natural image statistics and divisive normalization: modeling nonlinearities and adaptation in cortical neurons", *Statistical Theories of the Brain*, eds. R. Rao, B. Olshausen, and M. Lewicki, MIT Press, 2001.
16. O. Schwartz, and E. P. Simoncelli "Natural signal statistics and sensory gain control", *Nature neuroscience*, 4(8), pp. 819-825, 2001.
17. B. Wegmann, and C. Zetzsche, "Statistical dependence between orientation filter outputs used in an human vision based image code". *Proc. SPIE Vis. Commun. Image Processing*, pp. 1360, 909-922, Soc. Photo-Opt. Instrum. Eng, Lausanne, Switzerland, 1990.

18. E. P. Simoncelli, "Statistical Models for Images: Compression, Restoration and Synthesis", *Asilomar Conf. Signals, Systems, Comput.*, pp. 673-679, IEEE Comput. Soc, Los Alamitos, CA, 1997.
19. P. O. Hoyer, and A. Hyvärinen, "A multi-layer sparse coding network learns contour coding from natural images", *Vision Research*, 42(12), pp. 1593-1605, 2002.
20. I. Daubechies, *Ten Lectures on Wavelets*, CBMS-NSF Lecture Notes nr. 61, SIAM, 1992.
21. D. J. Field, "Relations between the statistics of natural images and the response properties of cortical cells", *J. Opt. Soc. Am. A*, 4(12), pp. 2379-2394, 1987.
22. S. G. Mallat, "A theory for multiresolution signal decomposition: The wavelet representation", *IEEE Pat. Anal. Mach. Intell.*, 11, pp. 674-693, 1989.
23. O. Nestares, R. Navarro, J. Portilla, and A. Tabernero, "Efficient Spatial-Domain Implementation of a Multiscale Image Representation Based on Gabor Functions", *Journal of Electronic Imaging*, 7(1), pp. 166-173, 1998.
24. R. W. Buccigrossi, and E. P. Simoncelli, "Image compression via joint statistical characterization in the wavelet domain" *IEEE Transactions on Image Processing*, 8(12), pp. 1688-1701, 1999.
25. M. J. Wainwright, and E. P. Simoncelli, "Scale Mixtures of Gaussians and the Statistics of Natural Images", *Advances in Neural Information Processing Systems*, 12, pp. 855-861, 2000.
26. M. J. Wainwright, E. P. Simoncelli, and A. S. Willsky, "Random Cascades on Wavelet Trees and Their Use in Modeling and Analyzing Natural Imagery", *Applied and Computational Harmonic Analysis*, 11(1), pp. 89-123.
27. B. A. Olshausen, and D. J. Field, "Sparse coding with an overcomplete basis set: A strategy employed by V1?", *Vision Research*, 37, pp. 3311-3325, 1997.
28. B. A. Olshausen, "Sparse Codes and Spikes", *Statistical Theories of the Brain*, eds. R. Rao, B. Olshausen, and M. Lewicki, pp. 257-272, MIT Press, 2002.
29. S. Kullback, and R. A. Leibler, "On information and sufficiency", *The Annals of Mathematical Statistics*, 22, pp. 79-86, 1951.
30. A. Papoulis, *Probability, random variables and stochastic processes* (3rd ed.), McGraw-Hill, Inc, Singapore, 1991.
31. J.Malo, F. Ferri, R. Navarro, and R. Valerio, "Perceptually and Statistically Decorrelated Features for Image Representation: Application to Transform Coding". *Proceedings of the 15TH International Conference on Pattern Recognition*, 3, pp. 242-245, IEEE Computer Society, Barcelona, Spain, 2000.

Scale-Space Image Analysis
Based on Hermite Polynomials Theory

Sherif Makram-Ebeid and Benoit Mory

MediSys Research Group, Philips France,
51 rue Carnot, B.P. 301, F-92156 SURESNES Cedex, France
{sherif.makram-ebeid,benoit.mory}@philips.com

Abstract. The Hermite transform allows to locally approximate an image by a linear combination of polynomials. For a given scale σ and position $\underline{\xi}$, the polynomial coefficients are closely related to the differential jet (set of partial derivatives of the blurred image) for the same scale and position. By making use of a classical formula due to Mehler (late 19th century), we establish a linear relationship linking the differential jets at two different scales σ and positions $\underline{\xi}$ involving Hermite polynomials. Pattern registration and matching applications are suggested.

We introduce a Gaussian windowed correlation function $K(\underline{v})$ for locally matching two images. When taking the mutual translation parameter $\underline{v}$ as independent variable, we express the Hermite coefficients of $K(\underline{v})$ in terms of the Hermite coefficients of the two images being matched. This new result bears similarity with the Wiener-Khinchin theorem which links the Fourier transform of the conventional (flat-windowed) correlation function with the Fourier spectra of the images being correlated. Compared to the conventional correlation function, ours is more suited for matching localized image features.

The mathematical tools we propose are shown to have attractive computational features. Numerical simulations using synthetic 1D and 2D test patterns demonstrate the advantages of our proposals for signal and image matching in terms of accuracy and low algorithm complexity.

1 Introduction

One of the key issues in signal and image processing is to identify appropriate data representations to suit a given range of applications. The widest spread such representation is the Fourier transform which is quite suited, for example, to capture the oscillatory behaviour of signals and noise [2,15]. When local features or patterns have to be analysed or processed, Gaussian windowed Fourier or Gabor transforms are preferred [10,11]. A large family of wavelet transforms aim at fulfilling similar needs [4,12].

In the present article, we study orthogonal polynomial representations that are suited to represent signal and image data within a Gaussian window. Such a window is unique in the sense that it is optimal with respect to scale-space axioms and is dimensionally separable [5,7]. Furthermore, as engineers and physicists know, it offers the best simultaneous frequency-space localization compromise [1].

L.D. Griffin and M. Lillholm (Eds.): Scale-Space 2003, LNCS 2695, pp. 57–71, 2003.

For a fixed Gaussian window, the corresponding Hermite polynomial representation is strongly related to the biologically motivated Gabor representation [11]. It has advantages over Gabor wavelets because it is more convenient for handling rotations, translations and scale changes as will be shown in the next sections. As with Gabor or with other wavelets transforms, systematic analysis or processing of a complete signal or image requires an array of overlapping windows [10,11,14]. In general, this should be preferably done for a range of different scales (*i.e.* different sizes of Gaussian windows in our case) within a multi-resolution pyramid. A very powerful tool for dealing with such window arrays is the wavelet frames theory [4,11,12,14]. In this article, we concentrate on the local aspects that constitute essential building blocks of a multiple window multi-resolution pyramid that we wish to develop in the future.

1.1   Orthogonal Polynomials for Signal and Image Processing

We consider signals and images as scalar valued functions of one or several variables. We start here by dealing with one dimensional signals or, equivalently, functions of one real variable. Orthogonal polynomials provide a convenient tool for locally approximating such functions [6,17]. Thus, one approximates a function $f(x)$ as a linear combination of polynomials $p_i(x)$ having degrees i going from 0 to infinity. The approximation is performed by a least mean square technique that consists in minimizing the weighted quadratic error Q defined by

$$Q = \int_{-\infty}^{+\infty} w(x) \left(f(x) - \sum_{i=0}^{\infty} a_i p_i(x) \right)^2 dx \qquad (1)$$

where a_i are the polynomial coefficients and where $w(x)$ is a non-negative weight function. Regions of x for which $w(x)$ is large are those for which one wishes the approximation to be accurate even when keeping a small number of terms. The coefficients a_i are those which minimize the quadratic error Q. For a polynomial basis which is orthogonal for the weight function $w(x)$, they are given by

$$a_i = f_i/c_i \text{ where } f_i = \int_{-\infty}^{+\infty} w(x)f(x)p_i(x)\, dx$$
$$\text{and } c_i = \int_{-\infty}^{+\infty} w(x)\left(p_i(x)\right)^2 dx. \qquad (2)$$

The weight function $w(x)$ defines L_2 metrics in the vector-space of functions $f(x)$ for which any distinct pair of polynomials in the basis are orthogonal. A direct consequence of this is that one can define a scalar product between any two functions $f(x)$ and $g(x)$ through Parseval's theorem [1]

$$\langle f, g \rangle_w \equiv \int_{-\infty}^{+\infty} w(x)f(x)g(x)\, dx = \sum_{i=0}^{\infty} \frac{f_i g_i}{c_i}$$
$$\text{where } g_i = \int_{-\infty}^{+\infty} w(x)g(x)p_i(x)\, dx. \qquad (3)$$

The weighted scalar product $\langle f, g \rangle_w$ can therefore be expressed either as an integral in x-space or else in terms of polynomial transform coefficients f_i and g_i. The discrete sum in the rightmost member of Eq. 3 defines an equivalent transfom-space (index i-labelled) scalar product; its corresponding metrics is referred to, in this article, as the l_2-metrics. The inverse of the normalization constants c_i introduce weights in the l_2-metrics playing a role similar to that of $w(x)$ in the x-space L_2-metrics corresponding to the scalar product $\langle f, g \rangle_w$.

1.2 Hermite Transform
and Related Scale-Space Differential Structure

As mentioned earlier, the Gaussian window is a natural choice for image processing. For one dimension, we use for $w(x)$ a normalized window of arbitrary size σ and centroid ξ

$$ w(\sigma, \xi; x) = \frac{1}{\sqrt{2\pi\sigma^2}} \exp\left(-\frac{(x - \xi)^2}{2\sigma^2} \right). \tag{4} $$

The corresponding orthogonal basis is that of Hermite polynomials [17] $H_i\left((x - \xi)/\sigma\sqrt{2}\right)$ for which the normalization constants c_i defined in Eq. 2 is given by $c_i = 2^i i!$. The Hermite transform of an arbitrary function $f(x)$ is given by the coefficients f_i defined in Eq. 2 with the above special weight function and with $p_i(x)$ replaced by the above Hermite polynomials. Hermite coefficients f_i can also be mathematically written in terms of derivatives of the function $F = w \otimes f$ obtained by convolving function $f(x)$ with a normalized Gaussian kernel of size σ. By using the basic properties of Hermite polynomials [17], it is a simple exercise to show that [9]

$$ f_i(\sigma, \xi) = 2^{i/2} \sigma^i \frac{d^i F}{dx^i}\bigg|_{x=\xi}. \tag{5} $$

A particular class of functions f is that of polynomials. In this case, the blurred function F is also a polynomial and the resulting Hermite expansion will include a finite number of non-vanishing terms. The results obtained in this article can then be considered as exact tools to manipulate polynomial approximations. A similar remark extends to the multidimensional case (see section 3 below).

If the independent variable x is expressed in units of $\sigma\sqrt{2}$, the pre-factor $2^{i/2}\sigma^i$ in Eq. 5 can be omitted. Eq. 5 will be used later to derive other results. It can be used, in practice, to obtain low-order Hermite coefficients f_i with $i < 4$ by numerically deriving a blurred version $F(x)$ of $f(x)$. However, this numerical method cannot be extended to compute higher order coefficients because of the occurrence of severe rounding errors. Computation of the set of coefficients f_i, from Eq. 2, requires the knowledge of the non-blurred function $f(x)$ for x-values at which the window function $w(x)$ is not negligibly small. With this reservation in mind, Eq. 5 shows that the Hermite transform is directly related to the derivatives of $F(x)$ which define the differential structure of function $f(x)$ at scale σ.

2   Basic Properties of Hermite Transform

In this main section, we study the basic properties of the Hermite transform that can be useful for signal and image processing. We make use of orthogonal polynomials theory which is well established [17] and which has been extended to any number of dimensions [6].

2.1   Dimensional Separability and Its Computational Advantages

A d-dimensional image is defined here as a real-valued function $f(\underline{x})$ of the d-dimensional position vector $\underline{x} = (x_1, x_2, .., x_d)^T$. As for the 1D case, one can locally analyse $f(\underline{x})$ within an isotropic Gaussian window of size σ (representing scale) around any window centre $\underline{\xi} = (\xi_1, \xi_2, \ldots, \xi_d)^T$. The isotropic d-dimensional window can be considered as a product of one-dimensional windows. Likewise, the orthogonal polynomial basis is made up of products of 1D Hermite polynomials, which are shown to form a complete set [6]. Each one of those polynomials is identified by a multi-index or d-dimensional vector index $I = (i_1, i_2, .., i_d)$ where all component indices i_k are non-negative integers. The corresponding d-dimensional Hermite transform coefficient f_I can now be written as

$$f_I(\sigma, \underline{\xi}) = \frac{1}{(2\pi\sigma^2)^{d/2}} \int_{\underline{x} \in \mathbb{R}^d} \exp\left(-\frac{\|\underline{x} - \underline{\xi}\|^2}{2\sigma^2}\right) P_I(\underline{x}) f(\underline{x}) \, d\underline{x}$$

$$\text{where } P_I(\underline{x}) = \prod_{k=1}^{d} H_{i_k}\left(\frac{x_k - \xi_k}{\sqrt{2\sigma^2}}\right)$$

(6)

and the corresponding polynomial approximation for $f(\underline{x})$ in the window is

$$f(\underline{x}) \cong \sum_I f_I(\sigma, \underline{\xi}) P_I(\underline{x}) / c_I$$

(7)

where c_I is the normalization constant which for a multi-index $I = (i_1, i_2, .., i_d)$ is expressed as a product of 1D normalization factors. The summation in Eq. 7 extends over all multi-indices I one wishes to use in the approximation, all integer component indices i_k are constrained to be non-negative.

Eq. 6 and Eq. 7 define the d-dimensional transform pair for going from direct x-space representation to transform $I-$space representation and vice-versa. Numerical computation algorithms can take good advantage of the dimensional factorisation that is inherent in these equations. To avoid burdening the notation, we limit ourselves here to the 2D case but our observations are readily generalized to any number of dimensions d. For the 2D case, the multi-indices I are written as pairs of non-negative integers $I = (i, j)$. A change of variable is used so as to bring the centre of the Gaussian window to the origin $\underline{\xi} = (0, 0)^T$, the coordinate units are selected so that $\sigma\sqrt{2} = 1$ and the space independent positional variable is written as $(x, y)^T$. Eq. 6 can now be reformulated as

$$f_{i,j} = \frac{1}{\pi} \int_{\underline{x} \in \mathbb{R}^2} e^{-x^2 - y^2} H_i(x) H_j(y) \, f(x,y) \, dx \, dy$$

$$= \frac{1}{\sqrt{\pi}} \int_{y=-\infty}^{+\infty} e^{-y^2} H_j(y) f_i(y) \, dy \qquad (8)$$

$$\text{where } f_i(y) = \frac{1}{\sqrt{\pi}} \int_{x=-\infty}^{+\infty} e^{-x^2} H_i(x) f(x,y) \, dx.$$

In other words, one may first take the 1D Hermite transform $f_i(y)$ along the x-axis for each of the y-coordinates and then take the 1D Hermite transform along the y-axis for each value of i to get the set of coefficients $f_{i,j}$. A symmetric procedure allows getting the inverse transform in a similar manner. To estimate the algorithmic complexity of those procedures, assume that numerical integration along the x- or y-axes make use of L discretization points and that the i and j indices take values in the interval $[0, M-1]$. The number of operations needed for each of the transform and its inverse is of the order of $L^2 M + LM^2$. So, if N stands for the largest of L and M, the computation load is of the order of $2N^3$ and for any number of dimensions d it can be easily seen to be of the order of $d \times N^{d+1}$. If the orthogonal basis was not dimensionally separable, the corresponding computational load would have been of the order of N^{2d} which is much larger. The larger the image dimension, the more significant is the saving.

Another useful consequence of Eq. 8 is that $f_{0,j}$ can be interpreted as the 1D Hermite transform of the function $f_0(y)$ obtained by Gaussian smoothing $f(x,y)$ in the x-direction around point $(0,y)^T$.

2.2 Dealing with Rotation in 2D

In two or more dimensions, the Hermite transform is particularly suited for dealing with rotations [9,14]. This is because the basis polynomial set behave like tensors [16] under isometric transformations (pure rotations and/or mirror symmetries) provided the window centroid is preserved. As in the previous section, we bring the centre of the Gaussian window to the origin $\underline{\xi} = (0,0)^T$ and the coordinate units are selected so that $\sigma\sqrt{2} = 1$. There are $(p+1)$ basis polynomials $P_{i,j}(x,y) = H_i(x) H_j(y)$ for which the total order $(i+j)$ has a fixed value p, they can be written in the form $H_i(x) H_{p-i}(y)$ with $i = 0, 1, \ldots, p$. In the transformed referential those order p polynomials can be expressed linearly in terms of order p polynomials in the original referential by means of a square $(p+1)^2$ matrix $A^{(p)}$. Converting the set of degree p monomials $x^i y^j$ (with $p = i + j$) requires exactly the same matrix. The same holds for converting all $f_{i,p-i}$ coefficients of total order p. Thus, if the coordinate axes are rotated by an angle θ so that any $(x,y)^T$ vector is changed to $(x',y')^T$ according to the law:

$$\begin{pmatrix} x' \\ y' \end{pmatrix} = \begin{pmatrix} \cos\theta & \sin\theta \\ -\sin\theta & \cos\theta \end{pmatrix} \begin{pmatrix} x \\ y \end{pmatrix}, \qquad (9)$$

then, in the new referential, any monomial $(x')^i (y')^{p-i}$ (with $0 < i < p$) can be expressed as

$$(x')^i(y')^{p-i} = (x.\cos\theta + y.\sin\theta)^i(-x.\sin\theta + y.\cos\theta)^{p-i} \equiv \sum_{j=0}^{j=p} A_{i,j}^{(p)} x^j y^{p-j}. \quad (10)$$

The coefficients of matrix $A^{(p)}$ are evaluated by expanding of the middle expression of Eq. 10 in terms of the monomials $(x)^j(y)^{p-j}$. As already stated above, we can deduce how Hermite coefficients are converted under rotation namely

$$f'_{i,p-i} = \sum_{j=0}^{j=p} A_{i,j}^{(p)} f_{j,p-j}. \quad (11)$$

The matrix coefficients take a particularly simple form when $i = 0$ and when $i = p$ (top and bottom lines of $A^{(p)}$) since they can then be expressed (using Eq. 10) from the binomial theorem

$$A_{0,j}^{(p)} = (-1)^j \binom{p}{j} \sin^j\theta.\cos^{p-j}\theta \quad \text{and} \quad A_{p,j}^{(p)} = \binom{p}{j} \cos^j\theta.\sin^{p-j}\theta \quad (12)$$

$$\text{where} \quad \binom{p}{j} = \frac{p!}{(p-j)!j!} = \binom{p}{p-j} \quad \text{are binomial coefficients.} \quad (13)$$

If only rotations have to be dealt with, it may be convenient to make use of the polar Gauss-Laguerre transform [8,9,14] in place of the Cartesian Hermite transform. The two kinds of transform are strongly related. Going from one to the other is a simple matter [14]. However, it is convenient to start from the computationally efficient (dimensionally separable) Cartesian form to deduce the polar form when needed.

2.3   Mehler Formula and Its Applications to Scale-Space

To deal with Hermite domain translations, scale changes and blurring, we suggest using a formula due to F.G. Mehler (late 19th Century, see [17], p. 380). This formula is very valuable when studying the convergence of Hermite polynomial expansions. Mehler's formula reads

$$(1 - t^2)^{-1/2} \exp\left(\frac{2txy - t^2(x^2 + y^2)}{(1-t^2)}\right) = \sum_{i=0}^{\infty} \frac{t^i}{2^i i!} H_i(x) H_i(y) \quad (14)$$

for any value of t in the range $-1 < t < 1$. The proof of this key formula is straightforward using 2D Fourier transform [18]. By multiplying both sides of Eq. 14 by $e^{-x^2} f(\xi + \sigma\sqrt{2}x)/\sqrt{\pi}$ where f is an arbitrary 1D function and integrating both sides from $-\infty$ to $+\infty$, one gets

$$\frac{1}{\sqrt{\pi(1-t^2)}} \int_{x=-\infty}^{+\infty} e^{-\frac{(ty-x)^2}{1-t^2}} f(\xi + \sigma\sqrt{2}x)\,dx = \sum_{i=0}^{\infty} \frac{f_i(\sigma\,;\,\xi)}{2^i i!} t^i H_i(y) \quad (15)$$

where $f_i(\sigma\,;\xi) = \frac{1}{\sqrt{2\pi\sigma^2}}\int_{-\infty}^{+\infty} e^{-\frac{(x-\xi)^2}{2\sigma^2}} H_i\left(\frac{x-\xi}{\sigma\sqrt{2}}\right) f(x)\,dx$ is the Hermite coefficient of f for a Gaussian window centred at $x = \xi$ and of size σ. With change of variables, this result can be rewritten as

$$F(\sigma\sqrt{1-t^2};\xi+tv) = \sum_{i=0}^{\infty} \frac{(\sigma t)^i}{2^{i/2}i!} H_i\left(\frac{v}{\sigma\sqrt{2}}\right) F^{(i)}(\sigma;\xi)$$

$$= \sum_{i=0}^{\infty} \frac{t^i f_i(\sigma;\xi)}{2^i i!} H_i\left(\frac{v}{\sigma\sqrt{2}}\right) \qquad (16)$$

where $F(\gamma;\eta)$ stands for the signal f Gaussian smoothed with kernel $\sigma = \gamma$ and evaluated at $x = \eta$ and $F^{(i)}(\gamma;\eta)$ is the corresponding i^{th} derivative. By taking t close to unity, one reconstructs the signal without blurring.

Mehler Formula Related Signal Processing LEMMA: *Eq. 16 implies that for $|t| < 1$, the Hermite coefficient of order i is attenuated by t^i; the resulting Hermite expansion is a version of signal f blurred with a Gaussian kernel of size $\sigma\sqrt{1-t^2}$ and then zoomed by a factor $1/t$ about the window centre.*

An equivalent formulation was derived by Martens [13] for deblurring isotropic Gaussian blur in images.

New Analytical Expression for 1D Scale-Space Transformations

Florack *et al* [7] have analyzed the influence of simultaneous translation and scale reduction excursions on scale-space local jets. They have studied the constraints on such excursions that ensure convergence of their Taylor expansions. Mehler's formula allows us to more specifically put the corresponding results in an explicit and compact analytical form provided appropriate pairs of "conjugate zooming and blurring operators" are applied. To do so, we generalize Eq. 16 by replacing signal f by its jth derivative and replacing the values of F and of its derivatives by their corresponding Hermite coefficients using Eq. 5 to get

$$f_j(\sigma\alpha\,;\xi+\beta v) = \alpha^j\left(\sum_{i=0}^{\infty} \frac{\beta^i}{2^i i!} H_i\left(\frac{v}{\sigma\sqrt{2}}\right) f_{i+j}(\sigma;\xi)\right) \qquad (17)$$

where $\alpha = \sqrt{1-t^2}$ and $\beta = t$ so that $\alpha^2 + \beta^2 = 1$. This allows to express Hermite coefficients for a Gaussian window translated to $x = \xi + \beta v$ and referred to a window of size $\sigma\alpha$ in terms of the Hermite coefficients for an original window centred at $x = \xi$ and of size σ. If one is interested in polynomial approximations not exceeding a maximum degree n, all coefficients of order larger than n can be replaced by zeros and the above relations are expressed as a linear relation between the $(n+1)$-dimensional Hermite coefficient vectors for the two windows. The matrix incurred in this transformation is an upper triangular one where any row j is obtained from the first one $(j = 0)$ by a shift and a scalar multiplication by α^j.

New 1D l_2-Matching Based on a Further Use of Mehler Formula

Take two signals f and g for which Hermite coefficients are known in an original window centred at $x = \xi$ and of size σ. Eq. 17 can be used to compare scaled and translated versions of the two signals in Hermite coefficient domain. To do so, we operate a scale and translation change on f and g with parameters $\alpha = \alpha_i$, $\beta = \beta_i$ and $v = v_i$ with $i = 1$ for f and $i = 2$ for g and with constraints that $\alpha_i^2 + \beta_i^2 = 1$. One can, furthermore, smooth and zoom the two functions before comparing them. The corresponding Hermite approximations to order n are

$$F(\xi + \beta_1 v_1 + \alpha_1 tu) \cong \sum_{i=0}^{n} \frac{t^i f_i(\sigma\alpha_1 \,;\xi + \beta_1 v_1)}{2^i i!} H_i\left(\frac{u}{\sigma\sqrt{2}}\right) \qquad (18)$$

and

$$G(\xi + \beta_2 v_2 + \alpha_2 tu) \cong \sum_{i=0}^{n} \frac{t^i g_i(\sigma\alpha_2 \,;\xi + \beta_2 v_2)}{2^i i!} H_i\left(\frac{u}{\sigma\sqrt{2}}\right) \qquad (19)$$

where F and G are obtained from the functions f and g by convolving with Gaussian low-pass kernels of size $\sigma\alpha_1\sqrt{1 - t^2}$ and $\sigma\alpha_2\sqrt{1 - t^2}$ respectively. Parseval theorem allows writing the L_2 (x-space) difference norm of the difference of blurred functions in terms of a modified index-labelled l_2 difference norm:

$$\frac{1}{\sqrt{\pi}} \int_{u=-\infty}^{+\infty} \left(e^{-\frac{u^2}{2}} F(\xi + \beta_1 v_1 + \alpha_1 tu) - e^{-\frac{u^2}{2}} G(\xi + \beta_2 v_2 + \alpha_2 tu)\right)^2 du$$

$$\cong \sum_{i=0}^{n} \frac{t^{2i}}{2^i i!} \left(f_i(\sigma\alpha_1 \,;\xi + \beta_1 v_1) - g_i(\sigma\alpha_2 \,;\xi + \beta_2 v_2)\right)^2. \qquad (20)$$

2.4   New Hermite-Domain 1D Windowed Cross-Correlation

In the previous section, we have proposed a method to evaluate the mean square difference of two signals for matching purposes. We now investigate the correlation function for two signals versus relative translation and scale parameters. We attempt to do this in such a way that the two signals being correlated play symmetrical roles. Furthermore, we build the correlation function in a hierarchical fashion (coarse to fine) using the smallest possible number of Hermite coefficients of each signal. To simplify notation, we assume that the Gaussian window is centred at the ordinates origin and that the ordinates unit is chosen so that $\sqrt{2\sigma^2} = 1$.

Take two 1D functions $f(x)$ and $g(x)$ with Hermite transforms f_i and g_i. Rather than predicting how each of these two coefficient sets are modified with translation or scale change, we study instead the behaviour of the external product $h(x, y) = f(x).g(y)$ of these two functions. We thus generate a 2D function out of the two 1D functions. The 2D Hermite coefficients of this new 2D function are given by $h_{i,j} = f_i g_j$. Let us now perform a $(-\theta)$ rotation from the (x, y) referential to the rotated (u, v) so that

$$\begin{pmatrix} u \\ v \end{pmatrix} = \begin{pmatrix} \cos\theta & \sin\theta \\ -\sin\theta & \cos\theta \end{pmatrix} \begin{pmatrix} x \\ y \end{pmatrix} \quad \text{and hence} \quad \begin{pmatrix} x \\ y \end{pmatrix} = \begin{pmatrix} \cos\theta & -\sin\theta \\ \sin\theta & \cos\theta \end{pmatrix} \begin{pmatrix} u \\ v \end{pmatrix}. \qquad (21)$$

Eq. 12 tells us how to use the above $h_{i,j}$ coefficients in the (x, y) referential in order to compute the coefficients $h'_{0,p}$ in the (u, v) referential with the first index equal to zero $(i = 0)$ and the second index equal to an arbitrary non-negative integer $(j = p)$. From the remark in the last paragraph of section 2.1, one sees that these $h'_{0,p}$ coefficients are just the Hermite coefficients of the 1D function $K(v)$ obtained by Gaussian smoothing in the u direction, i.e.

$$K(v) = \frac{1}{\sqrt{\pi}} \int_{-\infty}^{+\infty} e^{-u^2} f(\alpha.u - \beta.v) g(\beta.u + \alpha.v) du \qquad (22)$$

where $\alpha = \cos\theta$ and $\beta = \sin\theta$. In other words, the Hermite coefficients K_p of $K(v)$ are given by $h'_{0,p}$ which from Eq. 12 gives

$$K_p = h'_{0,p} = \sum_{j=0}^{j=p} \left((-1)^j \binom{p}{j} \sin^j \theta \cos^{p-j} \theta \right) \cdot f_j.g_{p-j}. \qquad (23)$$

The fact that $K(v)$ is a windowed cross-correlation of the two functions f and g is particularly clear if one chooses $\theta = \pi/4$ which, with a change of independent variable in the integrand of Eq. 23, yields

$$K\left(\frac{\tau}{\sqrt{2}}\right) = \sqrt{\frac{2}{\pi}} \int_{-\infty}^{+\infty} e^{-2u^2} f\left(u - \frac{\tau}{2}\right) g\left(u + \frac{\tau}{2}\right) du. \qquad (24)$$

The special case for $\theta = \pi/4$ and with identical f and g functions defines a "windowed auto-correlation". More generally, one may remark that the independent variable u in the integral of Eq. 23 is multiplied by a factor $\cos\theta$ in the argument of f and by $\sin\theta$ in the argument of g. The independent variable v parameterises a translation of the argument of f relative to that of g. For a fixed translation parameter, by setting

$$\cos\theta = \alpha = 1 / \sqrt{1 + \zeta^2} \quad \text{and} \quad \sin\theta = \beta = \zeta / \sqrt{1 + \zeta^2} \qquad (25)$$

the above rotation matrix equations relating (x, y) to (u, v) can be reinterpreted by eliminating the parameter u; they are equivalent to an affine transformation from argument x of f to argument y of g which can be written as

$$y = \zeta x + \alpha^{-1} v, \text{ with inverse } x = \zeta^{-1} y - \beta^{-1} v \qquad (26)$$

where ζ is a relative scaling parameter and $\alpha^{-1} v$ (or $-\beta^{-1} v$) as relative translations.

Advantages of Our Hermite-Domain Windowed Correlation

Eqs. 23-26 provide new original features relative to earlier proposed correlation functions. They allow to deal with translation for any given scale change when computing the Gaussian windowed cross-correlation (Eq. 22). Furthermore, the Hermite transform of the cross-correlation function can be deduced explicitly from the Hermite coefficients of the two functions f and g. When searching for a best match, our approach allows very substantial computational savings by working in transform space and concentrating on noise-robust low-order coefficients. Den Brinker [3] have earlier reported related Laguerre transform domain correlation results. However, the Ref.[3] Laguerre counterpart of Eq. 23 is a less convenient infinite series (in place of the finite sum in Eq. 23). In addition to its more compact form, our Hermite domain correlation can readily extend to several dimensions by virtue of dimensional separability.

3   Dealing with any Number of Image Dimensions

The dimensional separability of the Hermite transform greatly facilitates dealing with any number of dimensions. A particularly useful feature is the relation with the differential structure of an image. The result expressed by Eq. 5 is readily generalized for a d-dimensional image with an isotropic Gaussian window of size σ around a centre ξ. The Hermite coefficient for a multi-index $I = (i_1, i_2, \ldots, i_d)$ of rank $n = \sum_{k=1}^{d} i_k$ is given by:

$$f_I(\sigma, \underline{\xi}) = 2^{n/2} \sigma^n \left. \frac{\partial^n F}{\partial x_1^{i_1} \partial x_2^{i_2} \ldots \partial x_d^{i_d}} \right|_{\underline{x}=\underline{\xi}} . \tag{27}$$

3.1   Reducing and Extending the Number of Transform Dimensions

In section 2.4, we have generated a 2D function by taking the external product of two 1D functions. This proved useful for computing the localized cross-correlation and its Hermite transform. This procedure can be extended to any number of dimensions. Take a function $f(\underline{u})$ defined for m-dimensional vector arguments $\underline{u}$ and a function $g(\underline{v})$ defined for n-dimensional positional vectors $\underline{v}$, their product $f(\underline{u}).g(\underline{v})$ can be treated as a function of the $m + n$-dimensional vector $\underline{x} = (\underline{u}, \underline{v}) = (u_1, \ldots, u_m, v_1, \ldots, v_n)$. It is not hard to see that the Hermite coefficient of this product function for the $m + n$-dimensional multi-index $I = (i_1, i_2, .., i_{m+n})$ is equal to $f_{(i_1, \ldots, i_m)} g_{(i_{m+1}, \ldots, i_{m+n})}$.

The inverse situation is also of interest. In section 2.1, for example, it was shown that $f_{0,j}$ are the 1D Hermite coefficients of $f_0(y)$ obtained by Gaussian smoothing a 2D function $f(x, y)$ along the x-axis. This was instrumental in deriving the new Hermite domain correlation result of Eq. 23.

3.2   Generic Referential Change Readily Handled
in Hermite Transform Domain Using our Results

The results obtained in sections 2.3 to 2.4 readily extend to any number of dimensions through the dimensional separability properties of the Hermite transform.

Take two d-dimensional images $f(\underline{x})$ and $g(\underline{y})$. Assume that for any pair of co-ordinates x_k and y_k of the positional vectors $\underline{x}$ and $\underline{y}$ are transformed into u_k and v_k through a unitary transformation defined by

$$\begin{pmatrix} x_k \\ y_k \end{pmatrix} = \begin{pmatrix} \alpha_k & -\beta_k \\ \beta_k & \alpha_k \end{pmatrix} \begin{pmatrix} u_k \\ v_k \end{pmatrix} \quad \text{where } \alpha_k^2 + \beta_k^2 = 1. \tag{28}$$

The results of sections 2.3 to 2.4 directly apply to each of the dimensions. The corresponding partition-matrix relation between the two $(2 \times d)$-dimensional vectors $(\underline{x}, \underline{y})^T$ and $(\underline{u}, \underline{v})^T$ can be expressed as

$$\begin{pmatrix} \underline{x} \\ \underline{y} \end{pmatrix} = \begin{pmatrix} A & -B \\ B & A \end{pmatrix} \begin{pmatrix} \underline{u} \\ \underline{v} \end{pmatrix} \tag{29}$$

where A and B are $d \times d$ diagonal matrices with diagonal elements given by α_k and β_k respectively. If $\underline{v}$ is considered to be the translation parameter, the corresponding coordinate transformation is expressed by

$$\underline{y} = Z\underline{x} + A^{-1}\underline{v}, \text{ with inverse } \underline{x} = Z^{-1}\underline{y} - B^{-1}\underline{v} \tag{30}$$

with $Z = BA^{-1} = A^{-1}B$ a diagonal $d \times d$ matrix with diagonal elements given by $\zeta_k = \beta_k/\alpha_k$ and matrix Z^{-1} is its inverse with diagonal elements given by $1/\zeta_k = \alpha_k/\beta_k$.

This, however, is not the most general transformation class which can be conveniently dealt with in Hermite transform domain. As already mentioned before, pure rotations are easily handled in view of the tensor properties of the Hermite coefficients. One can apply rotation operators R_1 and R_2 in $\underline{x}$ and in $\underline{y}$ respectively around their window centres so that the generic form of transformation that can conveniently be handled is of the form:

$$\underline{y} = R_2^T Z R_1 \underline{x} + R_2^T A^{-1}\underline{v}, \text{ with inverse } \underline{x} = R_1^T Z^{-1} R_2 \underline{y} - R_1^T B^{-1}\underline{v}. \tag{31}$$

3.3 Extending our l_2 Norm and Correlation Results to Several Dimensions

Generalizing Eq. 16 to several variables is straightforward. It is easily shown, for example, that multiplying each Hermite coefficient by a factor t^n where n is the coefficient's rank (sum of the indices i_k), the resulting Hermite expansion is a version of f blurred with an isotropic Gaussian kernel of size $\sigma\sqrt{1 - t^2}$ and zoomed about the window centre by a factor of $1/t$. The other results of sections 2.3 and 2.4 extend to several dimensions by making use of the dimensional separability of the Hermite transform. The generic class of referential changes of section 3.2 can be easily dealt with in this manner.

4 Numerical Simulations

Numerical simulations are performed to illustrate the results of the previous sections using 1D and 2D synthetic data as seen in Figures 1–4. The legends of

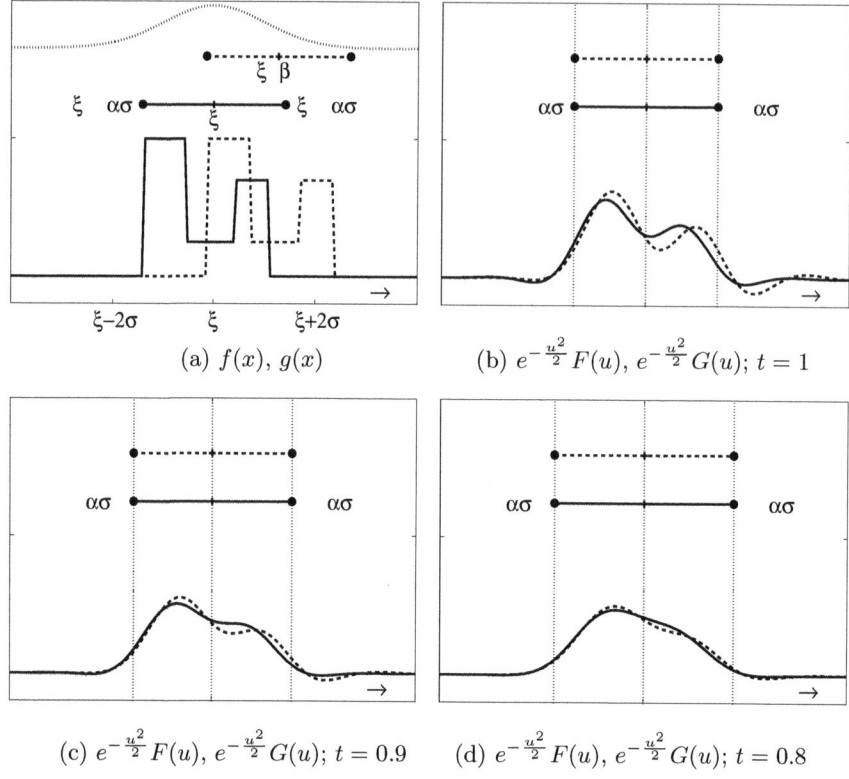

(a) $f(x)$, $g(x)$

(b) $e^{-\frac{u^2}{2}} F(u)$, $e^{-\frac{u^2}{2}} G(u)$; $t = 1$

(c) $e^{-\frac{u^2}{2}} F(u)$, $e^{-\frac{u^2}{2}} G(u)$; $t = 0.9$

(d) $e^{-\frac{u^2}{2}} F(u)$, $e^{-\frac{u^2}{2}} G(u)$; $t = 0.8$

Fig. 1. l_2-matching in Hermite domain is equivalent to L_2-matching in space domain: (a) Gaussian window (above), solid curve for signal f, dashed curve for signal g. (b-d) blurred and shifted signals F (solid curves) and G (dashed curve) reconstructed with different values of t. The relationship between l_2 matching in Hermite domain and L_2 matching in space domain is expressed by Eqs. 18,19,20. Referring to those equations, dashed lines (resp. solid lines) represent signal f (resp. g) and its blurred version F (resp. G) multiplied by $e^{-u^2/2}$; α and β are set to $1/\sqrt{2}$ and the translation parameters are set to compensate for the relative shift of the two signals ($v_1 = 0$ and $\beta_2 v_2 = 1.25\sigma$). The Hermite reconstruction to order 12 (b) with $t = 1$ shows an oscillatory reconstruction error (Gibbs phenomenon) for signal g occurring near the right extreme of the Gaussian window. Using smaller values of t as shown in (c) and (d) attenuates the faster varying higher order polynomials (right hand side of Eq. 20). This is the same as blurring and zooming the signals, thus reducing Gibbs oscillations and bringing them to higher values of u where the $e^{-u^2/2}$ factor is small. However, the reduction in Gibbs oscillations is accompanied by a blurring of the patterns and a loss of their discriminating high frequency features.

the figures provide comments and discussions. In the future, we intend to perform systematic benchmarking studies to compare our pattern matching approach with state of the art techniques for a real medical image test basis.

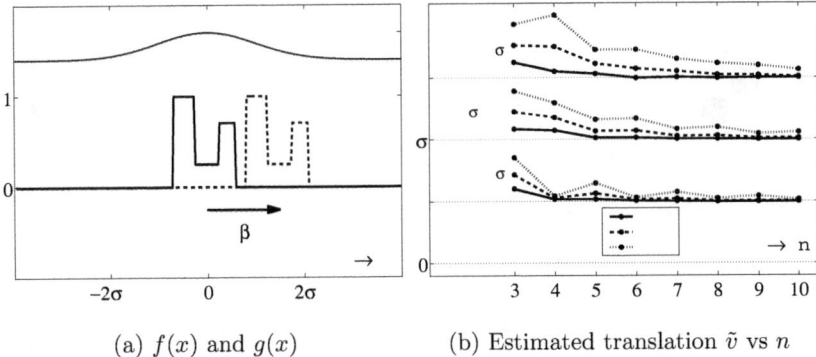

(a) $f(x)$ and $g(x)$ (b) Estimated translation $\tilde{v}$ vs n

Fig. 2. Translation estimation using the Hermite domain windowed correlation of section 2.4. In (a) above: Gaussian window, below: solid curve for signal f, dashed curve for the shifted signal g. In (b) estimated translations $\tilde{v}$ versus maximum Hermite polynomial order n for different values of $0 < t < 1$. Parameter t defines a rank-dependent attenuation of Hermite coefficients of the correlation function (equivalently smoothing and zooming according to section 2.3). Relative translation $\tilde{v}$ is estimated from the peak of the Hermite reconstructed smoothed-zoomed correlation function $K_s(tv)$ using Eqs. 23,25,16 with $\zeta = 1$. Accurate translation estimation is obtained with $t = 0.5$ (full curves) with $n > 4$ and for translations up to 1.5σ. Larger values of t yield slower and more oscillatory convergence to the exact translation with n increasing.

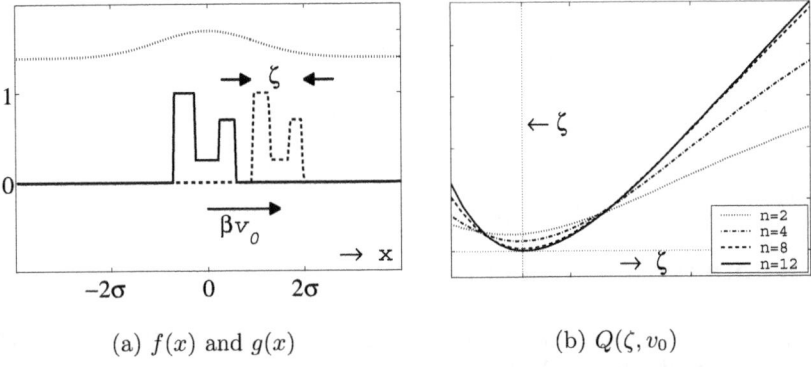

(a) $f(x)$ and $g(x)$ (b) $Q(\zeta, v_0)$

Fig. 3. Relative scale change estimation between two artificially scaled and translated signals. (a) Gaussian window (above), solid curve for signal f, dashed curve for signal g, (b) l_2 matching error $Q(\zeta, \tilde{v})$ versus relative scale ζ. For each scale ζ ranging from 0.5 to 2, the best translation $\tilde{v}$ is obtained from the peak of the correlation function in Hermite domain (as in Fig. 2). The corresponding l_2 matching error $Q(\zeta, \tilde{v})$ is then estimated from the right hand side of Eq. 20 with $t = 0.5$. The different curves correspond to different maximum Hermite polynomial orders. The minimum error is located at $\tilde{\zeta} \approx \zeta_0$ which corresponds to the true (synthetic data) value for approximation order n larger than 4. Even for $n = 2$ the minimum error is not too far ($\tilde{\zeta} = 0.7$).

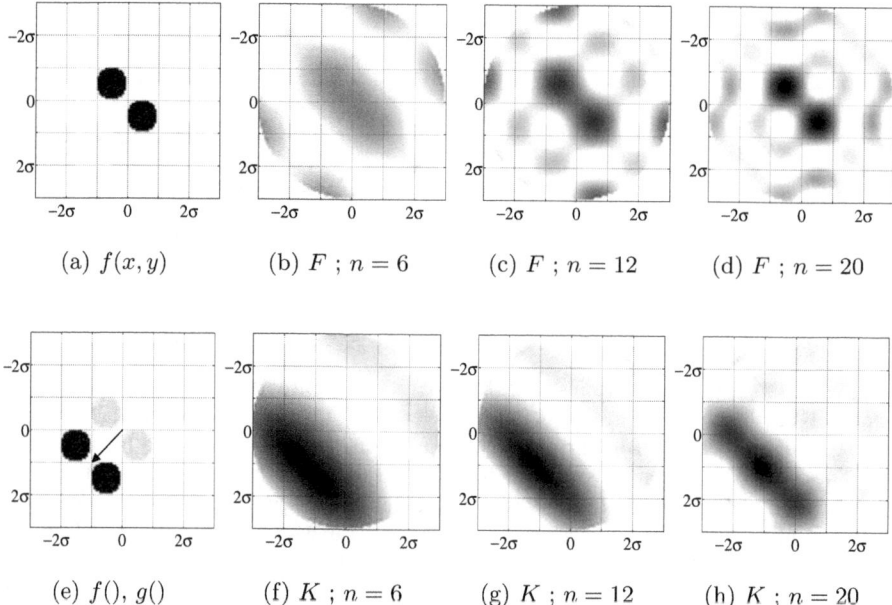

(a) $f(x,y)$ (b) F ; $n = 6$ (c) F ; $n = 12$ (d) F ; $n = 20$

(e) $f()$, $g()$ (f) K ; $n = 6$ (g) K ; $n = 12$ (h) K ; $n = 20$

Fig. 4. Illustration with a simple 2D pattern. The test pattern (a) is made out of two disjoint disks. (b), (c) and (d) show reconstructed images with $n = 6, 12$ and 20 respectively. In (e), the gray pattern $f()$ is a copy of (a) and, the pattern $g()$ in black is the same as $f()$ shifted by $\underline{v}_0 = (-\sigma, \sigma)^T$. (f), (g) and (h) show the 2D reconstructed correlation function $K(\underline{v})$ for the two versions of the pattern for $n = 6, 12$ and 20 respectively. In this experiment, no rank-dependent attenuation is applied when reconstructing $K(\underline{v})$ from its inverse Hermite transform (*i.e.* t was set to 1). The main correlation peak is located at the correct translation value in all cases. For $n = 20$ (g), two secondary peaks are observed corresponding to translations for which only one of the 2 disks are matched. All patterns are reconstructed within a radius 3σ about the window centre and set to zero beyond this radius.

5   Summary and Conclusions

In this article, we suggest new links between scale-space and Hermite transform theory. A formula due to Mehler is found to have interesting implications for Hermite transform pattern manipulations. In scale-space language, this formula tells us that attenuating rank n coefficients of a local jet by a factor of t^n (with $|t| < 1$) is equivalent to blurring and zooming the image about the window centre.

Based on this formula, we derive a new result allowing to analytically express the effect of scaling and translation in Hermite domain. The differential jet (or equivalently Hermite coefficients vector) can be used for pattern matching. The corresponding l_2 metrics is shown to be related to Gaussian weighted L_2 metrics in space domain. This allows the comparison, in Hermite domain, of mutually rotated, shifted and scaled pattern pairs. Comparing blurred versions of the patterns is shown to be equivalent to a modification of the l_2 metrics.

A new weighted cross-correlation method is also proposed. It plays a role similar to that of the Wiener-Khinchin theorem that deals with flat-windowed correlation in Fourier domain. Our result allows the evaluation of Gaussian windowed correlation in Hermite domain.

By using our techniques, computational and robustness advantages can be expected. Several applications may be envisaged for registration, motion estimation and pattern matching. The windowed correlation result may, furthermore, be applied to code and characterize textures in images.

References

1. Arfken, G.: Mathematical Methods for Physicists, 3rd ed. Orlando, FL: Academic Press, (1985)
2. Bracewell, R.: The Fourier Transform and Its Applications, 3rd ed. New-York: McGraw-Hill, (1999)
3. Den Brinker, A.C.: Calculation of the Local Cross-Correlation Function on the Basis of the Laguerre Transform, IEEE Trans. Signal Processing, SP-41, (1993) 1980–1982
4. Daubechies, I.: Ten Lectures on Wavelets, Society for Industrial and Applied Mathematics, Philadelphia (1992)
5. Duits, R., Florack L.M.J., Ter Haar Romeny B.M., De Graaf J.: On the axioms of scale-space theory, Proc. 4th IASTED Internat. Conf. on Signal and Image Processing (SIP 2002), Kauai, Hawaii, August (2002) 12–14
6. Dunkl C.F., Xu, Y.: Orthogonal Polynomials of Several Variables, Encyclopaedia of Mathematics and its Applications, 81, Cambridge University Press (2001)
7. Florack, L., Ter Haar Romeny, B., Viergever, M., Koenderink, J. : The Gaussian Scale-Space Paradigm and the Multiscale Local Jet, International Journal of Computer Vision, 18, (1996) 61–75
8. Jacovitti, G., Neri, A.: Multiresolution Circular Harmonic Decomposition, IEEE Trans. on Signal Processing, vol. 48, (2000) 3242–3247
9. Koenderink, J.J., Van Doorn, A.: Generic Neighbourhood Operators, IEEE Trans. on Signal Pattern Analysis and Machine Intelligence, PAMI-14, (1992) 597–605
10. Kruger, V., Sommer, G.: Gabor Wavelet Networks for Efficient Head-Pose Estimation, Image and Vision Computing, vol. 20, (2002) 665–672
11. Lee, T.S.: Image Representation unsing 2D Gabor wavelets, IEEE Transaction on Pattern Analysis And Machine Intelligence, PAMI-18 (1996) 959–971
12. Mallat, S.: A Wavelet Tour of Signal Processing, Academic Press (1998)
13. Martens, J.-B.: Adaptive Image Processing by Means of Polynomial Transform, Human Vision, Visual Processing and Digital Displays, III, (B.E. Rogowitz, Ed.), Proc. SPIE vol. 1666, (1992) 276–287
14. Martens, J.-B.: Local Orientation Analysis in Images by Means of the Hermite Transform, IEEE Trans. on Image Processing, vol. 6, (1997) 1103–1116
15. Schröder, H., Blume, H.: One- and Multidimensional Signal Processing- Algorithms and Applications in Image Processing, John Wiley & Sons Ltd (2000)
16. Spain, B.: Tensor Calculus, Oliver and Boyd Publishing, Edinburgh, New-York, 3rd Ed. (1960)
17. Szegö, G.: Orthogonal Polynomials, American Mathematical Society Colloquium Publications, vol. XXIII, Providence Rhode Island (1939), 3rd ed.(1967)
18. Watson, G.N.: Journal of the London Mathematical Society, vol. 8, (1933) 194–199

A Complete System of Measurement Invariants for Abelian Lie Transformation Groups

Yaron Gvili[1] and Nir Sochen[2]

[1] Department of Computer Sciences
[2] Department of Applied Mathematics
University of Tel-Aviv, Ramat-Aviv, Tel-Aviv 69978, Israel
{gvili,sochen}@post.tau.ac.il

Abstract. We present a complete system of functionally independent invariants for Abelian Lie transformation groups acting on an image. The invariants are based on measurements, given by inner product of predesigned functions and the image. We build on steerable filters and adopt a Lie theoretical approach that is applicable to any dimensionality. A complete characterization of Lie measurement invariants of a general irreducible component of the group, termed block invariants, is provided. We show that invariants for the entire group can be taken as the union of the invariants of its components. The system is completed by deriving invariants between components of the group, termed cross invariants.

1 Introduction

The problem of invariance to transformations has been studied extensively in the past. There are many different types of invariants in image processing and computer vision. We will present below invariants which are obtained from dense measurements of an image. For the fascinating subject of differential invariants the interested reader is referred to the excellent book by Olver [7]. Differential invariants for higher codimensions were recently derived as well [11]. Another direction of research is moment invariants. These invariants, to linear transformations, including translation, scaling and rotation, have been introduced by Hu [3] as early as 40 years ago. In this remarkable work the theories of algebraic invariants and moment invariants have been connected. Keren [4] applied algebraic invariants to models which are symbolically defined by implicit polynomials. A fundamental problem with moments is that they vanish for symmetric or antisymmetric images and hence information is lost. To correct this Palaniappan et al [8] proposed improved moment invariants. In this method acentric moments satisfying properties similar to the regular moments are used and their displacement depends on regular moments of the image. The invariants are constructed from these new moments. Zitová and Flusser [8] extended the scope of moment invariance to convolution, rotation, scaling, translation and contrast changes. Invariant distance metrics have also been a focus of research. Werman and Weinshall [12] proposed an invariants distance metric for 2D point sets. Simard et al [10] introduced the locally invariant tangent distance metric using a Lie theoretical approach. Semi-differential invariants were presented and studied in [6]. Kernel invariants were introduced by Segman et al [9] who showed a method to

L.D. Griffin and M. Lillholm (Eds.): Scale-Space 2003, LNCS 2695, pp. 72–85, 2003.

estimate the transformation between two images, considering an Abelian Lie transformation group model for the deformation between the images, by applying frequency domain techniques to the canonical coordinates of the group. Many uses of Lie groups, Lie algebras and their representations are found in the excellent book by Lenz [5]. The seminal work by Teo and Hel-Or [2] reformulated the problem of steerability, estimation and invariance of features on a unified Lie algebraic ground. It offers the basic differential equations which are satisfied by the measurement invariants as well. However, explicit expressions for the complete system of invariants, derived from these equations, have never been derived.

We present here, for the first time, a complete system of invariants[1]. The system of invariants we develop is based exclusively on inner products with a positive definite weighting function w, defined by $\langle \phi, \psi \rangle_w \doteq \int \phi \psi w$. The invariance is with respect to deformations modelled by an Abelian Lie transformation group. The restriction to a Lie group transformation is not a serious limitation, as most deformations considered in computer vision and pattern recognition, e.g. translation, rotation, scaling, projection and many more are Lie transformations groups.

The rest of this paper is organized as follows. In section 2 we describe the framework due to Teo and Hel-Or [2]. In section 3 we analyze irreducible components of Abelian Lie transformation groups by considering their conjugate generator. In section 4 we develop the complete system of differential invariants. In section 5 we discuss experimental results. We conclude in section 6.

2 Steerability and Equivariance

We consider conjugate generators of Lie transformation groups and their connection to the action of the group on measurements.

2.1 Lie Transformation Groups and Generators

A Lie transformation group $G(\tau)$ is a set of transformations in a k dimensional parameter domain P parameterized by $\tau \doteq (\tau_1, \ldots, \tau_k)$ with a continuous group structure. Recall that a group structure implies the following. For any $T(\alpha), T(\beta), T(\gamma) \in G(\tau)$

- Closure: The composition $T(\alpha\beta) \doteq T(\alpha)T(\beta)$ belongs to $G(\tau)$,
- Existence of the identity: There exists the identity transformation $I \in G(\tau)$ such that $IT(\alpha) \equiv T(\alpha)I \equiv T(\alpha)$,
- Existence of an inverse: There exists an inverse transformation $T(\alpha)^{-1} \in G(\tau)$ such that $T(\alpha)T(\alpha)^{-1} \equiv T(\alpha)^{-1}T(\alpha) \equiv I$,
- Associativity: $(T(\alpha)T(\beta))T(\gamma) \equiv T(\alpha)(T(\beta)T(\gamma)) \doteq T(\alpha\beta\gamma)$.

Note that commutativity is not implied.

In the following, we will consider only the two dimensional case. The extension to higher dimensions is straightforward. We denote $I(x, y)$ as the image and analyze the following Lie transformation groups.

---

[1] Partial results were submitted to CVPR2003

- Uniform brightness scaling: $G_\sigma(\sigma)$ is the group of transformations of the form $\mathcal{T}(\sigma)I(x,y) \doteq e^\sigma I(x,y)$,
- y translation and y scaling: $G_{\mu_y,\sigma_y}(\tau_y,\sigma_y)$ is the group of transformations of the form $\mathcal{T}(\tau_y,\sigma_y)I(x,y) \doteq I(x, e^{-\sigma_y}y - \tau_y)$,
- Rotation and uniform scaling: $G_{\mu_\theta,\sigma_r}(\tau_\theta,\sigma_r)$ is the group of transformations of the form $\mathcal{T}(\tau_\theta,\sigma_r)I(x,y) \doteq I(e^{-\sigma_r}x', e^{-\sigma_r}y')$ where $x' \doteq x\cos\tau_\theta - y\sin\tau_\theta$ and $y' \doteq x\sin\tau_\theta + y\cos\tau_\theta$ are the rotated coordinates and r,θ are the polar coordinates.

Of fundamental importance to the analysis of Lie groups is the studying of the infinitesimal action of the group with respect to each of its parameters about the identity. The differential operators defined by

$$\mathcal{L}_i \doteq \frac{d}{d\tau_i}\mathcal{T}(\tau)\Big|_{\tau=0} \equiv \frac{\partial x}{\partial \tau_i}\frac{\partial}{\partial x} + \frac{\partial y}{\partial \tau_i}\frac{\partial}{\partial y} + \frac{\partial}{\partial \tau_i}\Big|_{\tau=0}, \quad i = 1,\ldots,k \qquad (1)$$

are called generators of the group. We follow the convention that the identity transformation corresponds to $\tau = 0$ for all group parameterizations hereafter. By equation (1) we have

$$
\begin{aligned}
G_\sigma &\rightsquigarrow & \mathcal{L}_\sigma &\equiv \mathcal{I} & & & (2)\\
G_{\mu_y,\sigma_y} &\rightsquigarrow & \mathcal{L}_{\mu_y} &\equiv -\partial_y, & \mathcal{L}_{\sigma_y} &\equiv -y\partial_y \\
G_{\mu_\theta,\sigma_r} &\rightsquigarrow & \mathcal{L}_{\mu_\theta} &\equiv x\partial_y - y\partial_x \equiv -\partial_\theta, & \mathcal{L}_{\sigma_r} &\equiv -x\partial_x - y\partial_y \equiv -r\partial_r.
\end{aligned}
$$

The set of elements $\{\tau_i\mathcal{L}_i\}$ is called the *tangent space* - a linear vector space. The commutator, or the Lie bracket, defines a multiplication rule for the this space by $[\mathcal{L}_1,\mathcal{L}_2] \doteq \mathcal{L}_1\mathcal{L}_2 - \mathcal{L}_2\mathcal{L}_1$. The resulting algebra is called the Lie algebra associated with the Lie group. Each element of the group may be generated then from an element of the tangent space (i.e. the associated Lie algebra) using the *exponential map* $\mathcal{T}(\tau_i) \equiv e^{\tau_i\mathcal{L}_i}$ and $\mathcal{T}(\tau) \equiv e^{\tau_1\mathcal{L}_1+\cdots+\tau_k\mathcal{L}_k}$. The vanishing of the commutator for each pair of generators of a given group is equivalent to the commutativity, or Abelianity, of the group. It is easy to verify that $[\mathcal{L}_{\mu_y},\mathcal{L}_{\sigma_y}] \neq 0$ and hence G_{μ_y,σ_y} is non-commutative, but $[\mathcal{L}_{\mu_\theta},\mathcal{L}_{\sigma_r}] \equiv 0$ and hence G_{μ_θ,σ_r} is commutative. The action of an Abelian group is separable by $\mathcal{T}(\tau) \equiv e^{\tau_1\mathcal{L}_1}\cdots e^{\tau_k\mathcal{L}_k}$ irrespective of the order of multiplication. We will hereafter exploit the separability of Abelian groups and treat them as compositions of single parameter groups each generated by a single generator of the group.

2.2 Action on Measurements

Let $\mathbf{b} \doteq (b_0,\ldots,b_n)^t$ be a vector of measuring functions applied to the image $I(x,y)$. We denote the resulting vector of measurements, also called *features*, by

$$(f_0,\ldots,f_n)^t \doteq \mathbf{f} \doteq \langle \mathbf{b}, I\rangle \equiv \iint \mathbf{b}I\,dx\,dy. \qquad (3)$$

We define the *transformed features* as those resulting from applying a transformation $\mathcal{T}(\tau)$ to the image and then taking measurements, denoted

$$(\mathcal{T}(\tau)f_0,\ldots,\mathcal{T}(\tau)f_n)^t \doteq (f_0(\tau),\ldots,f_n(\tau))^t \doteq \mathbf{f}(\tau) \equiv \iint \mathbf{b}\mathcal{T}(\tau)I(x,y)\,dx\,dy. \qquad (4)$$

Note that by our group parameterization convention $\mathbf{f} \doteq \mathbf{f}(0)$. It is possible to select measuring functions for which the transformed features may be computed exactly from the original features. For example, take G_{μ_x} as the group of x translations defined by $\mathcal{T}(\tau_x)I(x,y) \doteq I(x - \tau_x, y)$ and choose $\mathbf{b} \equiv (e^x, e^x + xe^x)$. Then

$$f_0(\tau_x) \equiv \iint e^x I(x - \tau_x, y)dxdy \equiv \iint e^{u+\tau_x} I(u, y)dudy \tag{5}$$

$$\equiv e^{\tau_x} \iint e^u I(u, y)dudy \equiv e^{\tau_x} f_0$$

and

$$f_1(\tau_x) \equiv \iint (e^x + xe^x)I(x - \tau_x, y)dxdy \tag{6}$$

$$\equiv \iint (e^{u+\tau_x} + (u + \tau_x)e^{u+\tau_x})I(u, y)dudy$$

$$\equiv e^{\tau_x} \iint (\tau_x e^u + e^u + ue^u)I(u, y)dudy$$

$$\equiv e^{\tau_x}(\tau_x f_0 + f_1).$$

Hence $\mathbf{f}(\tau_x)$ are linear in $\mathbf{f}$ in a manner solely dependent on τ_x. This property is fundamental and we formalize it next.

The set of all images attainable from the image $I(x,y)$ by applying a transformation from the group is called the *image orbit* and denoted $O(G(\tau), I) \doteq \{\mathcal{T}(\tau)I | \tau \in P\}$. Similarly, the set of all features corresponding to the image orbit is called *feature orbit* and denoted $O(G(\tau), \mathbf{f}) \doteq \{\langle \mathbf{b}, \mathcal{T}(\tau)I \rangle | \tau \in P\}$. The span of the feature orbit is called a *feature space*.

Definition 1. *A feature space, derived from the measuring functions* $\mathbf{B}$*, is called equivariant under* $G(\tau)$ *if* $\mathbf{f}(\tau) \equiv A(\tau)\mathbf{f}$ *where* A *is a matrix solely dependent on* τ*. In this case the latter relation is termed interpolation equation and* $\mathbf{b}$ *are called equivariant measuring functions [2].*

We also refer to the functions $\mathbf{b}$ as *steerable filters* [1] due to the possibility to steer, or transform, the filters outputs directly instead of steering the measuring functions and filtering again. This property is very useful to applications and saves valuable computation time. Since equivariance is a linear property then without loss of generality we will hereafter assume $\mathbf{b}$ are linearly independent, i.e. a basis. For G_{μ_x} we have

$$\begin{pmatrix} f_0(\tau_x) \\ f_1(\tau_x) \end{pmatrix} \equiv e^{\tau_x} \begin{pmatrix} 1 & 0 \\ \tau_x & 1 \end{pmatrix} \begin{pmatrix} f_0 \\ f_1 \end{pmatrix} \quad \Rightarrow \quad A_{\mu_x}(\tau_x) \equiv e^{\tau_x} \begin{pmatrix} 1 & 0 \\ \tau_x & 1 \end{pmatrix}. \tag{7}$$

Teo and Hel-Or [2] demonstrated that any generator $\mathcal{L}$ has a *conjugate generator* $\overline{\mathcal{L}}$ satisfying $\langle \phi, \mathcal{L}\psi \rangle \equiv \langle \overline{\mathcal{L}}\phi, \psi \rangle$. Since a Lie transformation group $G(\tau)$ is generated by its generators using the exponential map, the conjugate generators generate an isomorphic *conjugate group* $\overline{G}(\tau)$ satisfying $\langle \phi, \mathcal{T}(\tau)\psi \rangle \equiv \langle \overline{\mathcal{T}}(\tau)\phi, \psi \rangle$. This useful property of measurements allows us to treat the action of the group on an unknown image as equivalent to an action on a known measuring function. It comes as no surprise that conjugate generators directly connect equivariance of a feature space with respect to the group to its equivariance with respect to the components of the group, and we formalize this in the following Theorem.

Theorem 1. *A feature space* F, *derived from the measuring functions* **b**, *is equivariant with respect to the group* $G(\tau)$ *if and only if* $\overline{T}(\tau)\mathbf{b} \equiv A(\tau)\mathbf{b}$ *or equivalently if and only if* $\overline{\mathcal{L}}_i\mathbf{b} \equiv B_i\mathbf{b}$ *for some matrices* B_i *for* $i = 1, \ldots, k$. *In the latter case* $A(\tau) \equiv e^{\tau_k B_k + \cdots + \tau_1 B_1}$. *The matrices* B_i *are called the representation matrices,* A *is called the interpolation matrix and* F *is called an equivariant measuring space (EMS) [2].*

For G_{μ_x} we have

$$B_{\mu_x} \equiv \begin{pmatrix} 1 & 0 \\ 1 & 1 \end{pmatrix}, \quad A_{\mu_x}(\tau_x) \equiv e^{\tau_x B_{\mu_x}} \equiv e^{\tau_x} \begin{pmatrix} 1 & 0 \\ \tau_x & 1 \end{pmatrix}. \tag{8}$$

An EMS is a linear vector space. This is evident from the linearity of the interpolation equation. Let $\mathbf{b}_1, \mathbf{b}_2$ be EMS bases and P be a regular matrix P then using a similar argument we find that $P\mathbf{b}_1$, the direct sum $\mathbf{b}_1 \oplus \mathbf{b}_2$ and the Kronecker product $\mathbf{b}_1 \otimes \mathbf{b}_2$ also span EMSs. This leads to the following fundamental Theorem [2].

Theorem 2. *Let* $\mathbf{b}_1, \mathbf{b}_2$ *be two equivariant measuring spaces bases of size* n *with respect to the same single-parameter Lie transformation group* G_τ *with conjugate generator* $\overline{\mathcal{L}}$, *and let* $\overline{\mathcal{L}}\mathbf{b}_1 = B_1\mathbf{b}_1, \overline{\mathcal{L}}\mathbf{b}_2 = B_2\mathbf{b}_2$ *for some matrices* B_1, B_2. *Then* $\mathbf{b}_1 = P\mathbf{b}_2 \iff B_1 = PB_2P^{-1}$ *for any* $n \times n$ *non-singular matrix* P.

Hence *similar* matrices correspond to the same feature space, and the representation matrices may be categorized by their *Jordan form*. We will therefore hereafter assume that without loss of generality B is in *block Jordan form* i.e. B is an $m \times m$ matrix with eigenvalue λ of multiplicity m along its main diagonal, 1s along its lower secondary diagonal and zero otherwise. In the above example, B_{μ_x} is in block Jordan form. If otherwise B has more than one Jordan block then it corresponds to a direct sum of spans, each associated with a Jordan block representation matrix of B for which the assumption holds.

3   Conjugate Generators

In this section we analyze conjugate generators, formulate standard bases for their EMSs and demonstrate results for standard Lie transformation groups.

3.1   Deriving Conjugate Generators

To derive conjugate generators we need only two rules [2]. The *multiplicative rule* is $\langle \phi, c\psi \rangle \equiv \langle c\phi, \psi \rangle$ for any function c and the *derivative rule* for x is

$$\langle \phi, \partial_x \psi \rangle \equiv \iint \phi(\partial_x \psi) dx dy \tag{9}$$

$$\equiv \int \phi\psi dy \,|_{-\infty}^{\infty} - \iint (\partial_x \phi)\psi dx dy$$

$$\equiv \langle -\partial_x \phi, \psi \rangle.$$

Table 1. Conjugate Generators

Transformation	Integration Domain	Operator	Generator	Conjugate Generator
brightness scaling	$(-\infty, \infty)$	$e^\tau I(x)$	$\mathcal{I}$	$\mathcal{I}$
x-translation	$(-\infty, \infty)$	$I(x - \tau)$	$-\partial_x$	∂_x
x-scaling	$(-\infty, \infty)$	$I(e^{-\tau} x)$	$-x\partial_x$	$\mathcal{I} + x\partial_x$
x-projective	$(-\infty, \infty)$	$I(\frac{x}{1+\tau x})$	$-x^2\partial_x$	$2x + x^2\partial_x$
rotation	$[-\pi, \pi]$	$I(\theta - \tau)$	$-\partial_\theta$	∂_θ
uniform scaling	$(-\infty, \infty)$	$I(e^{-\tau} r)$	$-r\partial_r$	$2\mathcal{I} + r\partial_r$

This rule applies similarly to y. Note that the vanishing of the boundary term is by assumption that the image is bounded. In case of a compact transformation the required assumption is that image is periodic with respect to the generator, which is generally the case, as well as the basis. Considering variables different than the integration variables x and y, one must account for the Jacobian of the variables change. For example, the derivative rule for r becomes

$$\langle \phi, \partial_r \psi \rangle \equiv \iint \phi(\partial_r \psi) r \, dr \, d\theta \tag{10}$$

$$\equiv \int \phi\psi r \, dr \, d\theta \, |_{-\infty}^{\infty} - \iint (\partial_r \phi r)\psi \, dr \, d\theta$$

$$\equiv 0 + \iint ((\partial_r \phi)r + \phi)\psi \, dr \, d\theta$$

$$\equiv \langle -(\partial_r + r^{-1})\phi, \psi \rangle.$$

Another way to handle generators of parameters other than x and y is to express them in x and y before applying the rules. Table 1 summarizes the conjugate generators of common transformations derived by applying these rules. Note that the results for the y transformations are similar to those of x.

3.2 Deriving Standard Bases

To determine a basis of a feature spaces with respect to a generator, termed *fundamental feature space*, we need to solve the equivariance equation. Solutions of the equivariance equation $\bar{\mathcal{L}}\mathbf{b}(x) = B\mathbf{b}(x)$ for $\mathbf{b}(x)$ yield an EMS basis with respect to the generator $\mathcal{L}$. They may be obtained using simple ODE techniques and are of the form $\mathbf{b}(x) = M(x)\mathbf{b}(0)$, where $M(x)$ is a matrix solely dependent on x, and with arbitrary $\mathbf{b}(0)$ as initial condition. For G_{μ_x} we have

$$M_{\mu_x}(x) \equiv e^x \begin{pmatrix} 1 & 0 \\ x & 1 \end{pmatrix} \tag{11}$$

The corresponding row space of M is used as the *standard basis*. For G_{μ_x} this is exactly $\mathbf{b} \equiv (e^x, e^x + xe^x)$. Solutions and their corresponding row spaces, i.e. the standard bases, are summarized in table 2. All integration are $dxdy$ with a regular inner product.

Table 2. Equivariance solutions and row space for B in block Jordan form with eigenvalue λ and multiplicity m

Transformation	Jacobian	Solution	Constraints	Standard Basis
brightness scaling	1	$\mathbf{1}_1$	$m = 1, \lambda = 1$	*any*
x-translation	1	e^{Bx}	$m \in \mathbb{N}, \lambda \in \mathbb{C}$	$\overrightarrow{\sum_{n=0}^{m-1} \frac{1}{n!} x^n e^{\lambda x}}$
x-scaling	e^{τ}	x^{B-I}	$m \in \mathbb{N}, \lambda \in \mathbb{C}$	$\overrightarrow{\sum_{n=0}^{m-1} \frac{1}{n!} \ln^n(x) x^{\lambda-1}}$
x-projective	$(-1 + \tau x)^{-2}$	$x^{-2} e^{-x^{-1}B}$	$m \in \mathbb{N}, \lambda \in \mathbb{C}$	$\overrightarrow{\sum_{n=0}^{m-1} \frac{(-1)^n}{n!} x^{-n-2} e^{-\lambda x^{-1}}}$
rotation	1	$e^{B\theta}$	$m = 1, \lambda \in i\mathbb{Z}$	$e^{\lambda\theta}$
uniform scaling	$e^{2\tau}$	r^{B-2I}	$m \in \mathbb{N}, \lambda \in \mathbb{C}$	$\overrightarrow{\sum_{n=0}^{m-1} \frac{1}{n!} \ln^n(r) r^{\lambda-2}}$

The Jacobian of the transformation, defined by $\det \partial x_i / \partial x'_j$ where x_i are the original variables and x'_j are the transformed variables, is also listed in table 2 for reference. $\overrightarrow{\sum}$ is the cumulative sum vector, e.g. $\overrightarrow{\sum_{i=1}^{n}} a_i \equiv (a_1, a_1 + a_2, \ldots, a_1 + \cdots + a_n)$. If a w weighted inner product is given simply multiply the bases by w^{-1}. The results stay the same since $\langle \phi, \psi \rangle \equiv \langle w^{-1}\phi, \psi \rangle_w$. No singularity is introduced since the weighting function w is positive definite by definition.

A fundamental issue arises here. Two isomorphic Lie algebras, or tangent spaces, are not guaranteed to yield the same solution. This may happen due to global constraints, i.e. when the corresponding Lie transformation groups are not isomorphic. For example the fundamental feature spaces of compact and non-compact single parameter groups are different. The integration domain of a non-compact group is invariant under the variables change of the transformation involved, i.e. it remains $(-\infty, \infty)$, but not so for a compact group. The solution for a compact group must be periodic with respect to the group parameter. This is the case for rotation in table 2 where the periodicity implies that the multiplicity must be 1 and the eigenvalue must be an integer imaginary number. Also note that further constraints apply to prevent the basis functions from introducing singularities.

Theorem 1 shows that a transformed feature of the standard basis, e.g. $\mathbf{f}(\tau_x)$ with respect to translation, is connected to the original features $\mathbf{f}$ via e.g.

$$f_i(\tau_x) = e^{\lambda \tau_x} \sum_{n=0}^{i} \frac{1}{n!} \tau_x^n f_{i-n}(0). \tag{12}$$

This is the *standard interpolation equation*. Other transformations have similar connections. One may verify that our example A_{μ_x} adheres to this equation.

4   Measurements Invariants

In this section we derive measurements invariants for B in block Jordan form and then extend to B in general Jordan form.

4.1 Block Invariants

An invariant $h(\mathbf{f})$ is a function of the features, which is a constant in the feature space when restricted to the orbit of the group. As was shown above it is enough to check that the invariant vanishes under the action of the conjugate generator:

$$\overline{\mathcal{L}}h(\mathbf{f}) \equiv (\mathbf{Bf})^t \nabla h(\mathbf{f}) = 0 \tag{13}$$

where $\nabla h(\mathbf{f})$ stands for $(\partial_{f_1}, \dots, \partial_{f_n})^t h(\mathbf{f})$. Trivially, any function of the invariants is also an invariant and we are interested in non-trivial and functionally independent maximal set of invariants. For G_{μ_x} we have

$$f_0 \partial_{f_0} h + (f_0 + f_1) \partial_{f_1} h = 0. \tag{14}$$

We term these as *block invariants*, since B is in block Jordan form. Generally, this PDE is difficult to solve, as the general technique of characteristic ODE for solving it involves $m-1$ repeated integrations for $m-1$ invariants. However, under our standard condition on B this PDE is tractable. The characteristic ODE for G_{μ_x} is

$$\frac{df_0}{f_0} = \frac{df_1}{f_0 + f_1}. \tag{15}$$

In general, this is a list of terms for $f_0, \dots, f_{m-1}$, all equal. The method we applied to solve this is to equate each term to the left most term. At each step $i = 1, \dots, m-1$ an invariant h_i is solved, f_i is expressed as a function of f_0 and $h_1, \dots, h_i$ and substituted into the next equation. This is possible due to the block Jordan form of B and produces simple expressions since $h_1, \dots, h_i$ are constant. For G_{μ_x} we have

$$h_1 \equiv \ln f_0 - \frac{f_1}{f_0} \quad \Rightarrow \quad f_1 \equiv f_0 \ln f_0 - f_0 h_1. \tag{16}$$

For the general case, it can be verified that the solutions, for $\lambda = 0$, i.e. $B_{i,j} \equiv \delta_{i,j+1}$, are

$$h_1 \equiv f_0,$$

and

$$h_i \equiv \frac{f_0^2}{i!} \left(\frac{f_1}{f_0} \right)^i - \sum_{n=1}^{i-2} \frac{h_{i-n}}{n!} \left(\frac{f_1}{f_0} \right)^n - f_0 f_i$$

for $i > 1$, \tag{17}

and the solutions for $\lambda \neq 0$, i.e. $B_{i,j} \equiv \lambda \delta_{i,j} + \delta_{i,j+1}$, are

$$h_i \equiv \frac{\lambda}{i!} \left(\frac{\ln f_0}{\lambda} \right)^i - \sum_{n=1}^{i-1} \frac{h_{i-n}}{n!} \left(\frac{\ln f_0}{\lambda} \right)^n - \frac{\lambda f_i}{f_0}. \tag{18}$$

4.2 Cross Invariants

Previously, we assumed B is in block Jordan form, i.e. it has a single block, and we analyzed the block for its standard basis functions and invariants. We now lift this assumption and derive further invariants. For B with s Jordan blocks we take the block invariants of each block separately as invariants. This is justified by correspondence of each block to a separate equivariant feature space. Indeed each feature space is closed under the linear action of the conjugate generator described by the separate representation block of B. Therefore the feature space corresponding to B is a direct sum of the feature spaces corresponding to its Jordan blocks. However, it is clear that additional invariants exist using features of more than one block. We term these as *cross invariants*. The method we applied to derive the invariants from equation (13) is to take the term in the characteristic ODE with the lowest degree for each block of the pair and equate them.

Let λ, Λ be eigenvalues of a pair of blocks and denote $f_0, \ldots, f_m$ and $F_0, \ldots, F_M$ the features of the corresponding blocks. If $\lambda, \Lambda \neq 0$ then we have

$$\frac{df_0}{\lambda f_0} \equiv \frac{dF_0}{\Lambda F_0} \quad \Rightarrow \quad h \equiv \Lambda \ln f_0 - \lambda \ln F_0, \tag{19}$$

and if $\lambda = 0$ then the we have

$$\frac{df_1}{\lambda f_0} \equiv \frac{dF_0}{\Lambda F_0} \quad \Rightarrow \quad h \equiv \ln F_0 - \frac{f_1}{f_0}. \tag{20}$$

It should be clear that the set of cross invariants corresponding to all possible pair of blocks is redundant. It suffices to take $s - 1$ functionally independent invariants. For example, if we denote $\lambda_1, \ldots, \lambda_s$ the eigenvalues of B then the set of cross invariants corresponding to the pairs (λ_1, λ_i) with $i = 2, \ldots, s$ is functionally independent and maximal. To see this notice that we equate s characteristic ODE terms for the cross invariants which correspond to one PDE constraint over s differentials.

4.3 A Complete System of Invariants

We have shown above that any EMS of an Abelian Lie transformation group has a standard basis for which the representation matrix B is in Jordan form. The block invariants and cross invariants derived are together a functionally independent and maximal set. This can be seen through simple counting of degrees of freedom. Assume that we have s blocks where the ith block has degree m_i, i.e. there are m_i features it corresponds to. For the ith block take $m_i - 1$ functionally independent block invariants. In addition take $s - 1$ functionally independent cross invariants as described above. It should be clear that the block and cross invariants together are functionally independent. This totals $s - 1 + \sum_{i=1}^{s}(m_i - 1) = -1 + \sum_{i=1}^{s} m_i$ functionally independent invariants for $\sum_{i=1}^{s} m_i$ features. Furthermore, we note that any other basis for the EMS is linearly connected to the standard basis and thus any invariant defined on the features of this measuring basis is functionally dependent on our standard invariants. We therefore conclude that this system of invariants is complete for Abelian Lie transformation groups.

5 Results

We have used slices of the images D2,D18,D23,D88,D91 from the Brodatz texture database of size 128×128 displayed in Figure 1 to test the theory. Deformations of

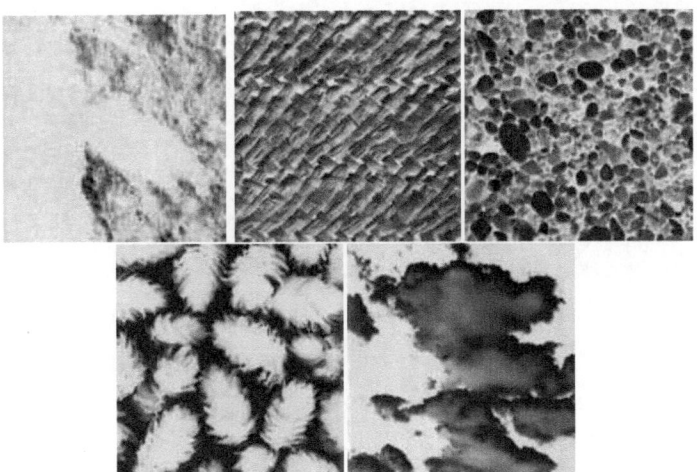

Fig. 1. The slices of the images D2,D18,D23,D88,D91 (left to right) used in the experiments

different transformation groups have been applied to the images. We have experimented with the x-,y-translation, x-,y-scaling, rotation and uniform scaling transformations and simulated them using a bilinear method. Measurements in the form of a regular inner product for selected standard bases have been taken and the corresponding invariants have been calculated. We have used one block of multiplicity $m = 8$ for non-compact transformations when testing features and block invariants. We have used 8 blocks of multiplicity $m = 1$ for compact and non-compact groups when testing cross invariants.

Equation (12) was verified against the features of the multiplicity 8 blocks as displayed in Figure 2. All relative errors were small and independent of the degree of the basis function or the amount of transformation. We assume that the errors are due to the interpolation.

We verified that the invariants are constant for deformations of an image and are different for different images, as demonstrated in Figure 3. Figures 4, 5 and 6 show the average relative errors of the block and cross invariants. All relative errors are below 6%, and most of them are orders of magnitude less.

6 Conclusions

We have presented a complete system of measurement invariants to deformations modelled by a Lie transformation group. The invariants are functions of standard features. These features are given as measurements, integral inner product of functions and the image, using a standard basis derived from the transformation group. The system of invariants is applicable to any Abelian Lie transformation group acting on the image, and the image space may be of any dimensionality.

We have shown that an Abelian Lie transformation group is separable. Using Theorem 2 all feature spaces that are equivariant with respect to a given generator have been categorized and standard bases for these spaces have been derived. Inner product with a

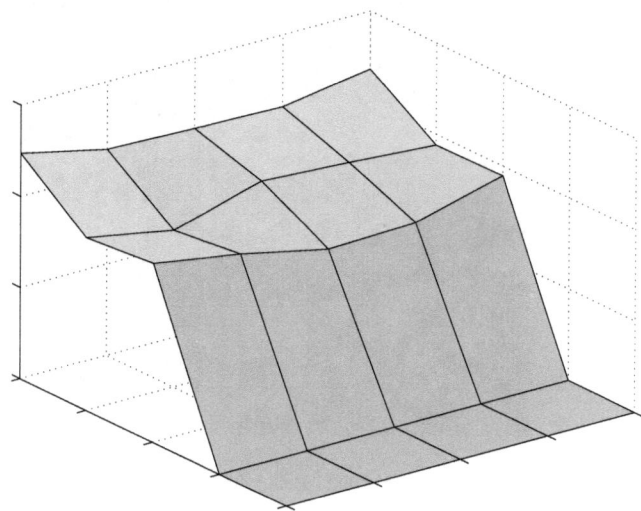

Fig. 2. Average relative error of features by equation (12). The images D2,D18,D23,D88,D91 are labelled (1)–(5), and the transformations are labelled: (1) x-translation with $\lambda = 1$, (2) y-translation with $\lambda = 0$, (3) x-scaling with $\lambda = 3$, (4) y-scaling with $\lambda = 3$, (5) uniform scaling with $\lambda = 4$

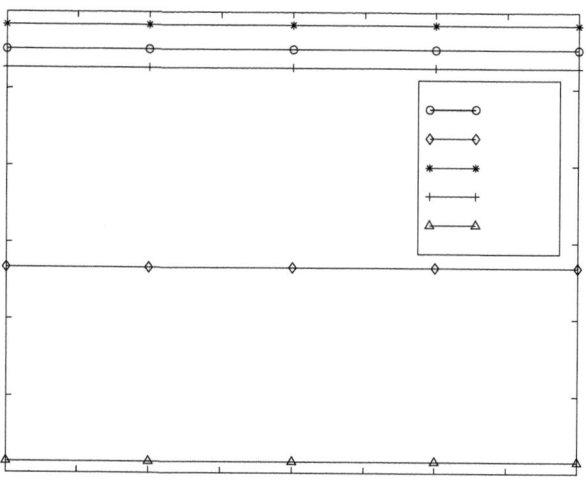

Fig. 3. The value of degree 0 invariant for D2,D18,D23,D88,D91 under x-translation with various parameterizations. The error is orders of magnitude less than this similarity

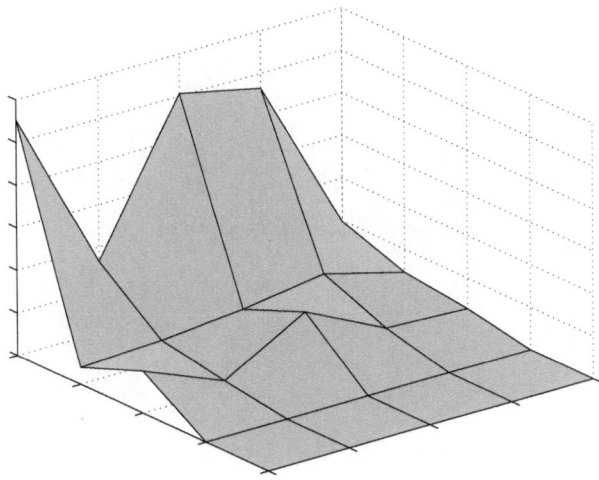

Fig. 4. Average relative error of block invariants by equation (17), (18). The images D2,D18,D23,D88,D91 are labelled (1)–(5), and the transformations are labelled: (1) x-translation with $\lambda = 1$, (2) y-translation with $\lambda = 0$, (3) x-scaling with $\lambda = 3$, (4) y-scaling with $\lambda = 3$, (5) uniform scaling with $\lambda = 4$

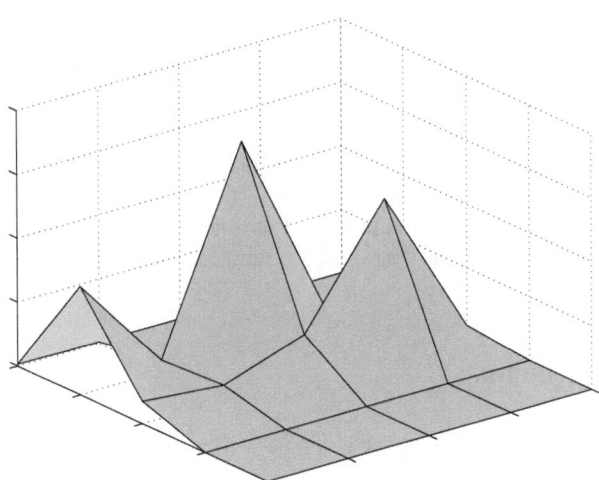

Fig. 5. Average relative error of cross invariants by equation (19). The images D2,D18,D23,D88,D91 are labelled (1)–(5), and the transformations are labelled: (1) x-translation (2) y-translation (3) x-scaling (4) y-scaling (5) rotation with $\lambda = 1, \ldots, 7$ and $\Lambda = 2, \ldots, 8$

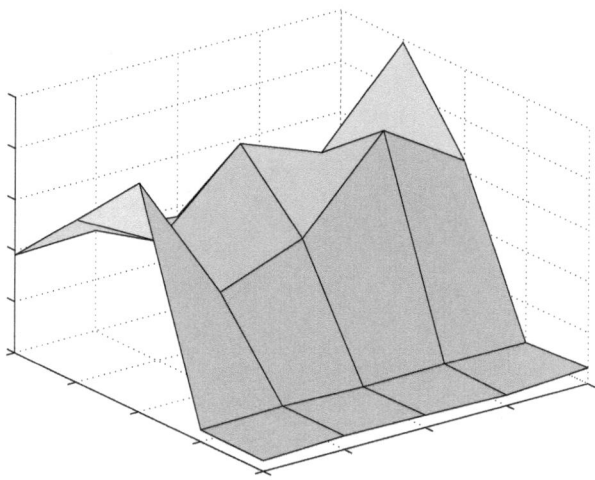

Fig. 6. Average relative error of cross invariants by equation (20). The images D2,D18,D23,D88,D91 are labelled (1)–(5), and the transformations are labelled: (1) x-translation (2) y-translation (3) x-scaling (4) y-scaling (5) uniform scaling with $\lambda = 0$ and $\Lambda = 1, \ldots, 6$

weighting function w may be accommodated by multiplying the standard bases by w^{-1}. The simple structure of the PDE derived from the Jordan form of the representation with respect to the standard basis has allowed us to arrive at the block invariants. The system of invariants has been completed with the cross invariants, functions of features from different blocks. The separability of Abelian groups has allowed us to take the direct sum of standard bases as a basis for the entire group, and the union of the corresponding invariants as invariants for the entire group. The results apply to any dimensionality by the use of a Lie theoretical approach. The relation to differential invariants i.e. the commutation relation with the semi-group generated in linear and non-linear scale-space theories is under current study.

Acknowledgements

The authors wish to thank Yacov Hel-Or for helpful discussions in the course of this work.

This research has been supported in part by the Israel Academy of Science, Tel-Aviv University fund, the Adams Center and the Israeli Ministry of Science.

References

1. W. T. Freeman and E. H. Adelson. The design and use of steerable filters. *IEEE Transactions on Pattern Analysis and Machine Intelligence*, 13(9):891–906, September 1991.

2. Y. Hel-Or and P. Teo. A common framework for steerability, motion estimation and invariant feature detection. Technical Report STAN-CS-TN-96-28, Stanford University, January 1996.

3. M. K. Hu. Visual pattern recognition by moment invariants. *IRE Transactions on Information Theory*, pages 179–189, February 1962.

4. D. Keren. Using symbolic computation to find algebraic invariants. *IEEE Transactions on Pattern Analysis and Machine Intelligence*, 16(11):1143–1149, November 1994.

5. R. Lenz. *Group Theoretical Methods in Image Processing*, volume 413 of *Lecture Notes in Computer Science*. Springer-Verlag, 1990.

6. T. Moons, E. Pauwels, L. Van Gool and A. Oosterlinck. Foundations of semi-differential invariants. *International Journal of Computer Vision*, (14):25–47, 1995.

7. P. J. Olver. *Equivalence, Invariants and symmetry*. Cambridge University Press, 1995.

8. M. A. Rodrigues. *Invariants for Pattern Recognition and Classification*, volume 42 of *Machine Perception and Artificial Intelligence*. World Scientific, 2000.

9. J. Segman, J. Rubinstein and Y.Y. Zeevi. The Canonical Coordinates Method for Pattern Deformation: Theoretical and Computational Considerations. *IEEE Transactions on Pattern Analysis and Machine Intelligence*, 14(12):1171–1183, December 1992.

10. P. Y. Simard, Y. A. Le Cun, J. S. Denker and B. Victorri. Transformation Invariance in Pattern Recognition - Tangent Distance and Tangent Propagation. Neural Networks: Tricks of the Trade, pp.239-27, 1996.

11. N. Sochen, On Affine Invariance in the Beltrami Framework for Vision and tangent propagation. In Proceedings of Variational and Level Set Methods '01, Vancouver, Canada, July 2001

12. M. Werman and D. Weinshall. Similarity and Affine Invariant Distances Between 2d Point Sets. *IEEE Transactions on Pattern Analysis and Machine Intelligence*, 17(8):810–814, August 1995.

Equivalence Results
for TV Diffusion and TV Regularisation

Thomas Brox[1], Martin Welk[1], Gabriele Steidl[2], and Joachim Weickert[1]

[1] Mathematical Image Analysis Group
Faculty of Mathematics and Computer Science, Building 27
Saarland University, 66041 Saarbrücken, Germany
{brox,welk,weickert}@mia.uni-saarland.de
http://www.mia.uni-saarland.de

[2] Faculty of Mathematics and Computer Science, D7, 27
University of Mannheim, 68131 Mannheim, Germany
steidl@math.uni-mannheim.de
http://www.kiwi.math.uni-mannheim.de

Abstract. It has been stressed that regularisation methods and diffusion processes approximate each other. In this paper we identify a situation where both processes are even identical: the space-discrete 1-D case
of total variation (TV) denoising. This equivalence is proved by deriving
identical analytical solutions for both processes. The temporal evolution
confirms that space-discrete TV methods implement a region merging
strategy with finite extinction time. Between two merging events, only
extremal segments move. Their speed is inversely proportional to their
size. Our results stress the distinguished nature of TV denoising. Furthermore, they enable a mutual transfer of all theoretical and algorithmic
achievements between both techniques.

1 Introduction

In the present paper we are concerned with two successful signal and image
restoration methods: diffusion filters and regularisation methods. Both techniques serve the same denoising purpose and both methods can be formulated
in terms of partial differential equations (PDEs). This has triggered several researchers to investigate connections between both paradigms.

In order to review their results, let us start with a brief description of 1-D
diffusion filtering. We consider a noisy signal as some function $f : [a, b] \rightarrow \mathbb{R}$.
The basic idea behind nonlinear diffusion filtering is to obtain a family $u(x, t)$ of
filtered versions of the signal $f(x)$ as the solution of a suitable diffusion process
with $f(x)$ as initial condition and homogeneous Neumann boundary conditions
[19]:

$$u_t = (g(u_x^2)\, u_x)_x \quad \text{on } (a, b) \times (0, \infty),$$

$$u(x, 0) = f(x) \quad \text{for all } x \in [a, b],$$

$$u_x(a, t) = u_x(b, t) = 0 \quad \text{for all } t \in (0, \infty),$$

(1)

L.D. Griffin and M. Lillholm (Eds.): Scale-Space 2003, LNCS 2695, pp. 86–100, 2003.

where subscripts denote partial derivatives, and larger diffusion times t correspond to more simplified signal representations.

Regularisation methods constitute an alternative to diffusion filters. Here the basic idea is to look for the minimiser u of the energy functional

$$E(u; \alpha, f) := \int_a^b \left((u - f)^2 + \alpha \, \Psi(u_x^2) \right) dx. \tag{2}$$

The first term of this functional encourages similarity between the original signal $f(x)$ and its filtered version $u(x)$, while the second term penalises deviations from smoothness. The increasing function Ψ is called *penaliser (regulariser)*, and the nonnegative *regularisation parameter* α serves as smoothness weight: larger values correspond to a more pronounced filtering.

As is explained in detail in [23], there are strong relations between regularisation methods and diffusion filters (see also [17,26]): A minimiser of (2) satisfies necessarily the Euler–Lagrange equation

$$\frac{u - f}{\alpha} = (\Psi'(u_x^2) \, u_x)_x,$$

with homogeneous Neumann boundary conditions. This equation may be regarded as a fully implicit time discretisation of the diffusion equation (1) with diffusivity $g(u_x^2) = \Psi'(u_x^2)$, initial value $f(x)$, and stopping time $t = \alpha$. Thus, one would expect that the minimiser of (2) *approximates* the diffusion filter (1), but is not identical to it. In [21,23] this relation has been used to establish scale-space properties for regularisation methods that resemble results for diffusion filters in [28].

While the before mentioned situation is an *approximation* only, one may be interested in results where one can establish *equivalence* between a diffusion filter and a regularisation method.

Nielsen et al. [18] have shown that the solution of the linear diffusion filter [14,29]

$$u_t = u_{xx}$$
$$u(x, 0) = f(x)$$

at time $t = \alpha$ may be regarded as the exact minimiser of an energy functional with an infinite number of penalising terms of arbitrarily high order:

$$E(u; \alpha, f) = \int_{\mathbb{R}} \left[(u - f)^2 + \sum_{k=1}^{\infty} \frac{\alpha^k}{k!} \left(\frac{d^k u}{dx^k} \right)^2 \right] dx.$$

An equivalent result has also been obtained earlier by Yuille and Grzywacz in the context of visual motion perception [30,31].

Another linear PDE-based filter is given by the pseudodifferential equation [10,9]

$$u_t = -\sqrt{-\frac{\partial^2}{\partial x^2}}\, u$$
$$u(x,0) = f(x).$$

Duits et al. [9] have shown that this so-called *Poisson scale-space* may be regarded as the exact minimiser of

$$E(u; \alpha, f) = \int_{\mathbb{R}} \left[(f - u)^2 + \sum_{k=1}^{\infty} \frac{\alpha^k}{k!} \left(\left(-\frac{d^2}{dx^2} \right)^{k/4} u \right)^2 \right] dx.$$

This discussion illustrates that only in the *linear* case, people have been able to derive energy functionals that are equivalent to the evolution equation. Unfortunately, these functionals are relatively complicated, since they involve an infinite number of regularising terms. This gives rise to the question if it is possible to derive also equivalence results in the *nonlinear* setting. Moreover, it would be nice if these energy functionals had simple regularisers that do not involve a large number of high-order derivatives.

We will address these problems in the present paper. To keep things as simple as possible, we will focus on the spatially discrete 1-D case. Surprisingly, it turns out that there exists a nonlinear framework in which there is an equivalence between diffusion filtering and regularisation that has a significantly simpler structure than equivalences in the linear case. This framework is given by so-called total variation (TV) denoising methods [22,2]. By deriving analytical solutions that are identical for TV diffusion and TV regularisation, we prove the equivalence of both paradigms.

Our paper is organised as follows. Section 2 gives an introduction to the continuous formulations of TV diffusion and TV regularisation. In Section 3 we shall derive the analytical solution for space-discrete TV diffusion, and in Section 4 we use the same proof structure to find an identical analytical solution for discrete TV regularisation. The paper will be concluded with a summary in Section 5.

Related Work. Space-discrete nonlinear TV diffusion creates a dynamical system with a discontinuous right hand side. Systems of this type – but with different force functions – have been proposed by Pollak et al. [20] for solving segmentation problems. Analytical results for some convex regularisation problems applied to specific test signals have been presented by Li [15]. Strong [25] derived analytical results in the case of continuous TV regularisation methods with step functions as initialisations. Equivalent results have been obtained by Mammen and van de Geer [16] for the taut-string algorithm in statistics; see also [13]. Our results are in accordance with these findings, but our proof shows that they can be derived in a different way: The structure of our proof is in complete analogy with the proof for the TV diffusion case. The results in the present paper can also be extended to investigate situations in which TV denoising methods are equivalent to wavelet shrinkage techniques. This is investigated in [24].

2   Continuous TV Diffusion and TV Regularisation

2.1   TV Diffusion

One-dimensional TV diffusion is a nonlinear diffusion filter that uses the unbounded diffusivity $g(u_x^2) = 1/|u_x|$. Hence it is based on the equation

$$u_t = \left(\frac{u_x}{|u_x|} \right)_x$$

This equation has been considered by Andreu et al. [2] under the name *total variation flow*. It requires no additional parameters (besides t), it is well-posed [2,4,11], it preserves the shape of some objects [4], and it leads to constant signals in finite time [3]. A numerical algorithm based on level sets has been proposed in [8].

2.2   TV Regularisation

TV regularisation has been proposed in its unconstrained form by Rudin, Osher and Fatemi [22], and in its constrained form by Acar and Vogel [1]. It uses the penaliser $\Psi(u_x^2) = 2|u_x|$. Hence, the constrained form minimises

$$E(u; \alpha, f) := \int_a^b \left((u - f)^2 + 2\alpha |u_x| \right) dx$$

This regularisation strategy is well-known for its good denoising capabilities and its tendency to create blocky, segmentation-like results. Well-posedness results have been established in [5]. A number of numerical schemes have been proposed including primal-dual methods [6], nonlinear Jacobi algorithms [7], and multigrid strategies [27].

3   Analytical Solution for Space-Discrete TV Diffusion

Let us now consider a space-discrete formulation of TV diffusion. We assume that the spatial grid size is 1 and that $f = (f_0, \ldots, f_{N-1})$ denotes a discrete version of $f(x)$ with N pixels. This leads to the following dynamical system:

$$\left. \begin{aligned} \dot{u}_0 &= \mathrm{sgn}(u_1 - u_0), \\ \dot{u}_i &= \mathrm{sgn}(u_{i+1} - u_i) - \mathrm{sgn}(u_i - u_{i-1}) \qquad (i = 1, \ldots, N-2), \\ \dot{u}_{N-1} &= -\mathrm{sgn}(u_{N-1} - u_{N-2}), \\ u(0) &= f. \end{aligned} \right\} \qquad (3)$$

In the following, we further set $u_{-1} := u_0$ and $u_N := u_{N-1}$, which may be regarded as a discretisation of the homogeneous Neumann boundary conditions. Since the right-hand side of this system is discontinuous, we need a more detailed specification of when a system of functions is said to satisfy these differential equations (cf. also [12]). A vector-valued function u is said to fulfil the system (3) over the time interval $[0, T]$ if the following holds true:

(I) u is an absolutely continuous vector-valued function which satisfies (3) almost everywhere, where sgn is defined by $\operatorname{sgn} w := 1$ if $w > 0$, $\operatorname{sgn} w := -1$ if $w < 0$, and may take any value in $[-1, 1]$ if $w = 0$.

(II) If $\dot{u}_i(t)$ and $\dot{u}_{i+1}(t)$ exist for the same t, and $u_{i+1}(t) = u_i(t)$ holds, then the expression $\operatorname{sgn}(u_{i+1}(t) - u_i(t))$ occurring in both the right-hand sides for $\dot{u}_i(t)$ and $\dot{u}_{i+1}(t)$ must take the same value in both equations.

Under these conditions we obtain the following result:

Proposition 1. (Properties of Space-Discrete TV Diffusion)
The system (3) has a unique solution $u(t)$ in the sense of (I) and (II). This solution has the following properties:

(i) (Finite Extinction Time)
There exists a finite time $T \geq 0$ such that for all $t \geq T$ the signal becomes constant:

$$u_i(t) = \frac{1}{N} \sum_{k=0}^{N-1} f_k \qquad \text{for all } i = 0, \dots, N-1.$$

(ii) (Finite Number of Merging Events)
There exists a finite sequence $0 = t_0 < t_1 < \dots < t_{n-1} < t_n = T$ such that the interval $[0, T)$ splits into sub-intervals $[t_j, t_{j+1})$ with the property that for all $i = 0, \dots, N - 2$ either $u_i(t) = u_{i+1}(t)$ or $u_i(t) \neq u_{i+1}(t)$ throughout $[t_j, t_{j+1})$. The absolute difference between neighbouring pixels does not become larger for increasing $t \in [t_j, t_{j+1})$.

(iii) (Analytical Solution)
In each of the sub-intervals $[t_j, t_{j+1})$ constant regions of $u(t)$ evolve linearly: For a fixed index i let us consider a constant region given by

$$u_{i-l+1} = \dots = u_i = u_{i+1} = \dots = u_{i+r} \qquad (l \geq 1, r \geq 0) \qquad (4)$$

and

$$u_{i-l} \neq u_{i-l+1} \text{ if } i - l \geq 0, \qquad u_{i+r} \neq u_{i+r+1} \text{ if } i + r \leq N - 1$$

for all $t \in [t_j, t_{j+1})$. We call (4) a region of size $m_{i,t_j} = l + r$. For $t \in [t_j, t_{j+1})$ let $\Delta t = t - t_j$. Then $u_i(t)$ is given by

$$u_i(t) = u_i(t_j) + \mu_{i,t_j} \frac{2\Delta t}{m_{i,t_j}},$$

where μ_{i,t_j} reflects the relation between the region containing u_i and its neighbouring regions. It is given as follows:
For inner regions (i.e. $i - l \geq 0$ and $i + r \leq N - 1$) we have

$$\mu_{i,t_j} = \begin{cases} 0 \text{ if } (u_{i-l}, u_i, u_{i+r+1}) \text{ is strictly monotonous,} \\ 1 \text{ if } u_i \text{ is minimal in } (u_{i-l}, u_i, u_{i+r+1}), \\ -1 \text{ if } u_i \text{ is maximal in } (u_{i-l}, u_i, u_{i+r+1}) \end{cases} \tag{5}$$

and in the boundary case ($i - l + 1 = 0$ or $i + r = N - 1$), the evolution is half as fast:

$$\mu_{i,t_j} = \begin{cases} 0 \text{ if } m = N, \\ \frac{1}{2} \text{ if } u_i \text{ is minimal in } (u_{i-l}, u_i, u_{i+r+1}), \\ -\frac{1}{2} \text{ if } u_i \text{ is maximal in } (u_{i-l}, u_i, u_{i+r+1}). \end{cases} \tag{6}$$

Proof.

Let u be a solution of (3). We show that u is uniquely determined and satisfies the rules (i)–(iii). Our proof proceeds in four steps.

1. If $\dot{u}(t)$ exists at a fixed time t and $u_i(t)$ lies at this time in some region

$$u_{i-l+1}(t) = \ldots = u_i(t) = \ldots = u_{i+r}(t) \qquad (l \geq 1, r \geq 0),$$

$$u_{i-l}(t) \neq u_{i-l+1}(t) \text{ if } i - l \geq 0, \quad u_{i+r}(t) \neq u_{i+r+1}(t) \text{ if } i + r \leq N - 1$$

of size $m_{i,t}$, then it follows by (3) and (II) in the non-boundary case $i - l \geq 0$ and $i + r \leq N - 1$ that

$$u_i(t) = \frac{1}{m_{i,t}} \sum_{k=-l+1}^{r} u_{i+k}(t),$$

and therefore

$$\begin{aligned} \dot{u}_i(t) &= \frac{1}{m_{i,t}} \sum_{k=-l+1}^{r} \dot{u}_{i+k}(t) \\ &= \frac{1}{m_{i,t}} \left(\text{sgn}\left(u_{i+r+1}(t) - u_i(t)\right) - \text{sgn}\left(u_i(t) - u_{i-l}(t)\right) \right) \\ &= \mu_{i,t} \frac{2}{m_{i,t}}, \end{aligned} \tag{7}$$

where $\mu_{i,t}$ describes the relation between the region containing u_i and its neighbours at time t as in (5). In the boundary case $i - l + 1 = 0$ or $i + r = N - 1$ we follow the same lines and obtain (7) with $\mu_{i,t}$ defined by (6).

2. Let $\dot{u}(t)$ exist in some small interval (τ_0, τ_1) and assume that $u_i(t) \neq u_{i+1}(t)$ for some $i \in \{0, \ldots, N-2\}$ and all $t \in (\tau_0, \tau_1)$. By continuity of u we may assume that $u_i(t) < u_{i+1}(t)$ throughout (τ_0, τ_1). The opposite case $u_i(t) > u_{i+1}(t)$ can be handled in the same way. Then we obtain by (7) and definition of $\mu_{i,t}$ for all $t \in (\tau_0, \tau_1)$ that

$$\dot{u}_i(t) \geq 0 \quad \text{if} \quad i - l \geq 0, \tag{8}$$
$$\dot{u}_i(t) > 0 \quad \text{if} \quad i - l + 1 = 0, \tag{9}$$
$$\dot{u}_{i+1}(t) \leq 0 \quad \text{if} \quad i + r \leq N - 2, \tag{10}$$
$$\dot{u}_{i+1}(t) < 0 \quad \text{if} \quad i + r = N - 1. \tag{11}$$

Set $w(t) := u_{i+1}(t) - u_i(t)$. Then the mean value theorem yields

$$w(\tau_1) - w(\tau_0) = (\tau_1 - \tau_0)\, \dot{w}(t^*)$$

for some $t^* \in (\tau_0, \tau_1)$ and we get by (8)–(11) that

$$w(\tau_1) - w(\tau_0) \leq 0$$

with strict inequality in the boundary case. Consequently, the difference between pixels cannot become larger in the considered interval. In particular, by continuity of u, pixels cannot be split. Once merged they stay merged.

3. Now we start at time $t_0 = 0$. Let t_1 be the largest time such that $\dot{u}(t)$ exists and no merging of regions appears in $(0, t_1)$. Then, for all $i \in \{0, \ldots, N-1\}$, a function u_i is in the same region with the same relations to its neighbouring regions throughout $[0, t_1)$. Thus, we conclude by (7) that

$$\dot{u}_i(t) = \mu_{i,0}\, \frac{2}{m_{i,0}} \qquad (t \in (0, t_1))$$

and consequently

$$u_i(t) = \mu_{i,0}\, \frac{2t}{m_{i,0}} + C_{i,0}$$
$$= f_i + \mu_{i,0}\, \frac{2t}{m_{i,0}} \qquad (t \in [0, t_1]),$$

where the last equality follows by continuity of u_i if t approaches 0.

4. We are now in the position to analyse the entire chain of merging events successively.

   Next we consider the largest interval (t_1, t_2) without merging events in the same way, where we take the initial setting $u(t_1)$ instead of f into account. Then we obtain

$$u_i(t) = \mu_{i,t_1}\, \frac{2t}{m_{i,t_1}} + C_{i,t_1},$$

   where $u_i(t_1) = \mu_{i,t_1} \frac{2t_1}{m_{i,t_1}} + C_{i,t_1}$ by continuity of u_i. Consequently

$$u_i(t) = u_i(t_1) + \mu_{i,t_1}\, \frac{2(t - t_1)}{m_{i,t_1}}.$$

Now we can continue in the same way by considering $[t_2, t_3)$ and so on. Since we have only a finite number N of pixels and some of these pixels merge at the points t_j the process stops after a finite number of n steps with output

$$u_i(t_n) = \frac{1}{N} \sum_{k=0}^{N-1} f_k$$

for all $i = 0, \ldots, N - 1$.

Conversely, it is easy to check that a function u with (i)–(iii) is a solution of the system (3). This completes the proof of the proposition. □

4   Analytical Solution for Discrete TV Regularisation

Next we will prove that discrete TV regularisation satisfies the same rules as space-discrete TV diffusion. For given initial data $f = (f_0, \ldots, f_{N-1})$ discrete TV regularisation consists in constructing the minimiser

$$u(\alpha) = \min_u E(u; \alpha, f) \qquad (12)$$

of the functional

$$E(u; \alpha, f) = \sum_{i=0}^{N-1} \left((u_i - f_i)^2 + 2\alpha \, |u_{i+1} - u_i| \right), \qquad (13)$$

where we suppose again Neumann boundary conditions $u_{-1} = u_0$ and $u_N = u_{N-1}$.

For a fixed regularisation parameter $\alpha \geq 0$, the minimiser of (13) is uniquely determined since $E(u; \alpha, f)$ is strictly convex in $u_0, \ldots, u_{N-1}$. Furthermore, $E(u, \alpha; f)$ is a continuous function in $u_0, \ldots, u_{N-1}, \alpha$. Consequently, $u(\alpha)$ is a (componentwise) continuous function in α.

The following proposition implies together with Proposition 1 the equivalence of space-discrete TV diffusion and discrete TV regularisation, if the diffusion time t is identical to the regularisation parameter α.

Proposition 2. (Properties of Discrete TV Regularisation)
The function $u(\alpha)$ in (12) is uniquely determined by the following rules:

(i) (Finite Extinction Parameter)
There exists a finite $A \geq 0$ such that for all $\alpha \geq A$ the signal becomes constant:

$$u_i(\alpha) = \frac{1}{N} \sum_{k=0}^{N-1} f_k \qquad \text{for all } i = 0, \ldots, N - 1.$$

(ii) (Finite Number of Merging Events)
  There exists a finite sequence $0 = a_0 < a_1 < \ldots < a_{n-1} < a_n = A$ such that the interval $[0, A)$ splits into sub-intervals $[a_j, a_{j+1})$ with the property that for all $i = 0, \ldots, N - 2$ either $u_i(\alpha) = u_{i+1}(\alpha)$ or $u_i(\alpha) \neq u_{i+1}(\alpha)$ throughout $[a_j, a_{j+1})$. The absolute difference between neighbouring pixels does not become larger for increasing $\alpha \in [a_j, a_{j+1})$.

(iii) (Analytical Solution)
  In each of the sub-intervals $[a_j, a_{j+1})$ constant regions of $u(\alpha)$ evolve linearly:
  For a fixed index i let us consider a constant region given by

$$u_{i-l+1} = \ldots = u_i = u_{i+1} = \ldots = u_{i+r} \qquad (l \geq 1, r \geq 0) \qquad (14)$$

  and

$$u_{i-l} \neq u_{i-l+1} \text{ if } i - l \geq 0, \qquad u_{i+r} \neq u_{i+r+1} \text{ if } i + r \leq N - 2 \qquad (15)$$

  for all $\alpha \in [a_j, a_{j+1})$. We call (14) a region of size $m_{i,a_j} = l + r$. For $\alpha \in [a_j, a_{j+1})$ let $\triangle\alpha = \alpha - a_j$.
  Then $u_i(\alpha)$ is given by

$$u_i(\alpha) = u_i(a_j) + \mu_{i,a_j} \frac{2\triangle\alpha}{m_{i,a_j}},$$

  where μ_{i,a_j} reflects the relation between the region containing u_i and its neighbouring regions. It is given as follows:
  For inner regions (i.e. $i - l \geq 0$ and $i + r \leq N - 2$) we have

$$\mu_{i,a_j} = \begin{cases} 0 \text{ if } (u_{i-l}, u_i, u_{i+r+1}) \text{ is strictly monotonous,} \\ 1 \text{ if } u_i \text{ is minimal in } (u_{i-l}, u_i, u_{i+r+1}), \\ -1 \text{ if } u_i \text{ is maximal in } (u_{i-l}, u_i, u_{i+r+1}) \end{cases} \qquad (16)$$

  and in the boundary case ($i - l + 1 = 0$ or $i + r = N - 1$), the evolution is half as fast:

$$\mu_{i,a_j} = \begin{cases} 0 \text{ if } m = N, \\ \frac{1}{2} \text{ if } u_i \text{ is minimal in } (u_{i-l}, u_i, u_{i+r+1}), \\ -\frac{1}{2} \text{ if } u_i \text{ is maximal in } (u_{i-l}, u_i, u_{i+r+1}). \end{cases} \qquad (17)$$

Proof:

Again our proof proceeds in four steps. It has a similar structure as the proof of Proposition 1.

1. Let us first verify the solution $u(\alpha)$ of (12) for an arbitrary but fixed $\alpha > 0$. If $u_i(\alpha)$ is contained in some region of size $m_{i,\alpha}$ with (14), (15), then, in case

$i - l \geq 0$ and $i + r \leq N - 2$, we have that $u(\alpha)$ can be obtained as minimiser of

$$E(u_0, \ldots, u_{i-l}, u_i, u_{i+r+1}, \ldots, u_{N-1}; \alpha, f)$$
$$= \sum_{k=-l+1}^{r} (u_i - f_{i+k})^2 + 2\alpha \left(|u_i - u_{i-l}| + |u_{i+r+1} - u_i| \right)$$
$$+ F(u_0, \ldots, u_{i-l}, u_{i+r+1}, \ldots, u_{N-1})$$

with some function F independent of u_i. By (14), (15) the partial derivative of E with respect to u_i exists and is given by

$$\frac{\partial E}{\partial u_i} = 2 \sum_{k=-l+1}^{r} (u_i - f_{i+k}) - 4\alpha \mu_{i,\alpha}.$$

Here $\mu_{i,\alpha}$ describes the relation between the region containing u_i and its neighbours for the regularisation parameter α as in (16). Setting the partial derivative to zero, we obtain that

$$u_i(\alpha) = \frac{1}{m_{i,\alpha}} \sum_{k=-l+1}^{r} f_{i+k} + \mu_{i,\alpha} \frac{2\alpha}{m_{i,\alpha}}. \tag{18}$$

In the boundary case $i - l + 1 = 0$ or $i + r = N - 1$ we follow the same lines and obtain (18) with $\mu_{i,\alpha}$ defined by (17).

2. Next we show that initially merged pixels will not be split for any α in a small interval $[0, a_1]$.
For $\alpha = 0$ we have that $u(0) = f$. Let $f_i = u_i(0)$ be contained in some region of the form

$$f_{i-l_0+1} = \ldots = f_i = f_{i+1} = \ldots = f_{i+r_0} \qquad (l_0, r_0 \geq 1)$$

and

$$f_{i-l_0} \neq f_{i-l_0+1} \text{ if } i - l_0 \geq 0, \quad f_{i+r_0} \neq f_{i+r_0+1} \text{ if } i + r_0 \leq N - 2.$$

By continuity of $u(\alpha)$ we can choose $\alpha_1 > 0$ so that $u_i(\alpha) \neq u_{i-l_0}(\alpha)$ and $u_{i+1}(\alpha) \neq u_{i+r_0}(\alpha)$ throughout $[0, \alpha_1)$. Assume that there exists $\alpha \in (0, \alpha_1)$ so that $u_i(\alpha) \neq u_{i+1}(\alpha)$, where we may assume that

$$u_i(\alpha) < u_{i+1}(\alpha). \tag{19}$$

The opposite case $u_i(\alpha) > u_{i+1}(\alpha)$ can be handled in the same way. Note that at time α more pixels than u_i and u_{i+1} may be separated. However, we have by (18) with some $1 \leq l \leq l_0$ and some $1 \leq r \leq r_0$ that

$$u_i(\alpha) = \frac{1}{l} \sum_{k=-l+1}^{0} f_{i+k} + \mu_{i,\alpha} \frac{2\alpha}{l} = f_i + \mu_{i,\alpha} \frac{2\alpha}{l},$$
$$u_{i+1}(\alpha) = \frac{1}{r} \sum_{k=1}^{r} f_{i+k} + \mu_{i+1,\alpha} \frac{2\alpha}{r} = f_i + \mu_{i+1,\alpha} \frac{2\alpha}{r},$$

where we see by (19) and (16), (17) that $\mu_{i,\alpha} \geq 0$ and $\mu_{i+1,\alpha} \leq 0$. Thus, $u_i(\alpha) \geq u_{i+1}(\alpha)$ which contradicts (19). Consequently $u_i(\alpha) = u_{i+1}(\alpha)$ throughout $[0, \alpha_1)$, i.e., the pixels of our initial region stay merged.

Let $a_1 > 0$ denote the largest number such that no merging of regions appears in $[0, a_1)$. Then we have for all $i = 0, \ldots, N - 1$ and all $\alpha \in [0, a_1)$ that $\mu_{i,\alpha} = \mu_{i,0}$ and regarding that $u(\alpha)$ is continuous that

$$u_i(\alpha) = f_i + \mu_{i,0} \frac{2\alpha}{m_{i,0}} \qquad (\alpha \in [0, a_1]). \tag{20}$$

3. Now we show that the absolute difference between neighbouring regions cannot become larger with increasing $\alpha \in [0, a_1)$.

   Without loss of generality let for some fixed index i

$$u_{i-l+1} = \ldots = u_i < u_{i+1} = \ldots = u_{i+r} \qquad (l, r \geq 1)$$

and

$$u_{i-l} \neq u_{i-l+1} \text{ if } i - l \geq 0, \quad u_{i+r} \neq u_{i+r+1} \text{ if } i + r \leq N - 2.$$

We consider the non-boundary case $i - l \geq 0$ and $i + r \leq N - 2$ first. By (20) we obtain for $\alpha + \delta \in [0, a_1)$, $\delta > 0$ that

$$d_i(\alpha) = u_{i+1}(\alpha) - u_i(\alpha) = f_{i+1} - f_i + 2\alpha \left(\frac{\mu_{i+1,0}}{r} - \frac{\mu_{i,0}}{l} \right),$$

$$d_i(\alpha + \delta) = u_{i+1}(\alpha + \delta) - u_i(\alpha + \delta) = f_{i+1} - f_i + 2(\alpha + \delta) \left(\frac{\mu_{i+1,0}}{r} - \frac{\mu_{i,0}}{l} \right)$$

and consequently

$$d_i(\alpha + \delta) - d_i(\alpha) = 2\delta \left(\frac{\mu_{i+1,0}}{r} - \frac{\mu_{i,0}}{l} \right).$$

By (16) it follows that

$$\frac{\mu_{i+1,0}}{r} - \frac{\mu_{i,0}}{l} = \begin{cases} 0 & \text{if } u_{i-l} < u_i < u_{i+1} < u_{i+r+1}, \\ -\frac{1}{r} & \text{if } u_{i-l} < u_i \text{ and } u_{i+1} > u_{i+r+1}, \\ -\frac{1}{l} & \text{if } u_{i-l} > u_i \text{ and } u_{i+1} < u_{i+r+1}, \\ -\frac{1}{r} - \frac{1}{l} & \text{if } u_{i-l} > u_i \text{ and } u_{i+1} > u_{i+r+1} \end{cases}$$

which yields the desired property $d_i(\alpha) \geq d_i(\alpha + \delta)$.

In case of boundary regions we follow the same lines but replace (16) by (17). Then we see that the absolute difference between neighbouring regions becomes smaller with increasing $\alpha \in [0, a_1)$.

4. We are now in the position to analyse the entire chain of merging events successively.

   For $\alpha > a_1$ and $\triangle\alpha = \alpha - a_1$, we consider

$$\tilde{u}_i(\triangle\alpha) = \min_u E(u; \triangle\alpha, u(a_1)).$$

We can repeat the same considerations as in Part 2 of the proof but with initial setting $u(a_1)$ instead of f. It follows that there exists a_2 such that for all $i = 0, \ldots, N - 2$ either $\tilde{u}_i(\triangle\alpha) = \tilde{u}_{i+1}(\triangle\alpha)$ or $\tilde{u}_i(\triangle\alpha) \neq \tilde{u}_{i+1}(\triangle\alpha)$ throughout $[a_1, a_2)$, where the absolute difference between neighbouring pixels does not become larger for increasing $\triangle\alpha$. Further, we obtain by (20) and (18) that

$$\tilde{u}_i(\triangle\alpha) = u_i(a_1) + \mu_{i,a_1}\frac{2\triangle\alpha}{m_{i,a_1}}$$

$$= \frac{1}{m_{i,a_1}} \sum_{j \in R_{i,a_1}} f_j + \mu_{i,a_1}\frac{2a_1}{m_{i,a_1}} + \mu_{i,a_1}\frac{2\triangle\alpha}{m_{i,a_1}},$$

where $R_{i,\alpha} = \{j : u_j(\alpha) \text{ is in the region of } u_i(\alpha)\}$ while m_{i,a_1} denotes the size of the region containing $u_i(a_1)$ and μ_{i,a_1} reflects the relation between the region containing $u_i(a_1)$ and its neighbouring regions. Since the relations between regions do not change for $\triangle\alpha \in [0, a_2 - a_1)$ we can rewrite $\tilde{u}_i(\triangle\alpha)$ as

$$\tilde{u}_i(\triangle\alpha) = \frac{1}{m_{i,a_1+\triangle\alpha}} \sum_{j \in R_{i,a_1+\triangle\alpha}} f_j + \mu_{i,a_1+\triangle\alpha}\frac{2(a_1 + \triangle\alpha)}{m_{i,a_1+\triangle\alpha}}$$

$$= \frac{1}{m_{i,\alpha}} \sum_{j \in R_{i,\alpha}} f_j + \mu_{i,\alpha}\frac{2\alpha}{m_{i,\alpha}}.$$

On the other hand, we have by (18) that

$$u_i(\alpha) = \frac{1}{m_{i,\alpha}} \sum_{j \in R_{i,\alpha}} f_j + \mu_{i,\alpha}\frac{2\alpha}{m_{i,\alpha}}.$$

Thus, $u_i(\alpha) = \tilde{u}_i(\triangle\alpha)$.

Now we can continue in the same way by considering $[a_2, a_3)$ and so on. Since we have only a finite number N of pixels and some of these pixels merge at the points a_j the process stops after a finite number of n steps with output $u(a_n)$ which by (18) reads as

$$u_i(a_n) = \frac{1}{N} \sum_{k=0}^{N-1} f_k$$

for all $i = 0, \ldots, N - 1$. This completes the proof. $\square$

5   Conclusions

In this article we have seen that in the 1-D case, space discrete TV diffusion and discrete TV regularisation are identical, if we identify the diffusion time with the regularisation parameter. Given the relatively complicated relations between the linear Gaussian and Poisson scale-spaces and their corresponding regularisation methods, these results may seem to be of surprising simplicity. However, they are natural consequences from the simple structure of the scale-space evolutions of both TV processes: The evolution can be regarded as a sequence of region merging events. Between two mergings, only extremal segments are allowed to move. Their velocity is proportional to the inverse of the pixel number, and it can be guaranteed that all segments merge within a finite extinction time. These properties are more transparent than those of most other discrete scale-space evolutions and put space-discrete TV denoising in an extraordinary position: It is not only a nonlinear scale-space that preserves discontinuities, it also does not require any additional parameters, it implements a multiscale segmentation, and – last but not least – it is equivalent to its corresponding regularisation method. We conjecture that the latter property also holds for the continuous TV diffusion and regularisation process. Moreover, we are investigating if it can be extended to the higher dimensional situation. If this is the case, the door will be opened for a direct transfer between the results of two previously separated worlds: a parabolic scale-space world and an elliptic regularisation world.

Acknowledgements

The research in this paper is partly funded by the projects WE 2602/1-1 and WE 2602/2-1 of the *Deutsche Forschungsgemeinschaft (DFG)*. This is gratefully acknowledged.

References

1. R. Acar and C. R. Vogel. Analysis of bounded variation penalty methods for ill–posed problems. *Inverse Problems*, 10:1217–1229, 1994.
2. F. Andreu, C. Ballester, V. Caselles, and J. M. Mazón. Minimizing total variation flow. *Differential and Integral Equations*, 14(3):321–360, Mar. 2001.
3. F. Andreu, V. Caselles, J. I. Diaz, and J. M. Mazón. Qualitative properties of the total variation flow. *Journal of Functional Analysis*, 188(2):516–547, Feb. 2002.
4. G. Bellettini, V. Caselles, and M. Novaga. The total variation flow in R^N. *Journal of Differential Equations*, 184(2):475–525, 2002.
5. A. Chambolle and P.-L. Lions. Image recovery via total variation minimization and related problems. *Numerische Mathematik*, 76:167–188, 1997.
6. T. F. Chan, G. H. Golub, and P. Mulet. A nonlinear primal–dual method for total-variation based image restoration. In M.-O. Berger, R. Deriche, I. Herlin, J. Jaffré, and J.-M. Morel, editors, *ICAOS '96: Images, Wavelets and PDEs*, volume 219 of *Lecture Notes in Control and Information Sciences*, pages 241–252. Springer, London, 1996.
7. T. F. Chan, S. Osher, and J. Shen. The digital TV filter and nonlinear denoising. *IEEE Transactions on Image Processing*, 10(2):231–241, Feb. 2001.
8. F. Dibos and G. Koepfler. Global total variation minimization. *SIAM Journal on Numerical Analysis*, 37(2):646–664, 2000.

9. R. Duits, L. Florack, J. de Graaf, and B. ter Haar Romeny. On the axioms of scale space theory. Technical Report 2002-01, Dept. of Biomedical Engineering, TU Eindhoven, The Netherlands, Jan. 2002.

10. M. Felsberg and G. Sommer. Scale-adaptive filtering derived from the Laplace equation. In B. Radig and S. Florczyk, editors, *Pattern Recognition*, volume 2032 of *Lecture Notes in Computer Science*, pages 95–106. Springer, Berlin, 2001.

11. X. Feng and A. Prohl. Analysis of total variation flow and its finite element approximations. Technical Report 1864, Institute of Mathematics and its Applications, University of Minnesota, Minneapolis, MN, July 2002. Submitted to Communications on Pure and Applied Mathematics.

12. A. F. Filippov. *Differential Equations with Discontinuous Righthand Sides*. Kluwer, Dordrecht, 1988.

13. W. Hinterberger, M. Hintermüller, K. Kunisch, M. von Oehsen, and O. Scherzer. Tube methods for BV regularization. Technical Report 6, Department of Computer Science, University of Innsbruck, Austria, Dec. 2002.

14. T. Iijima. Basic theory on normalization of pattern (in case of typical one-dimensional pattern). *Bulletin of the Electrotechnical Laboratory*, 26:368–388, 1962. In Japanese.

15. S. Z. Li. Close-form solution and parameter selection for convex minimization-based edge-preserving smoothing. *IEEE Transactions on Pattern Analysis and Machine Intelligence*, 20(9):916–932, Sept. 1998.

16. E. Mammen and S. van de Geer. Locally adaptive regression splines. *Annals of Statistics*, 25(1):387–413, 1997.

17. J.-M. Morel and S. Solimini. *Variational Methods in Image Segmentation*. Birkhäuser, Basel, 1994.

18. M. Nielsen, L. Florack, and R. Deriche. Regularization, scale-space and edge detection filters. *Journal of Mathematical Imaging and Vision*, 7:291–307, 1997.

19. P. Perona and J. Malik. Scale space and edge detection using anisotropic diffusion. *IEEE Transactions on Pattern Analysis and Machine Intelligence*, 12:629–639, 1990.

20. I. Pollak, A. S. Willsky, and H. Krim. Image segmentation and edge enhancement with stabilized inverse diffusion equations. *IEEE Transactions on Image Processing*, 9(2):256–266, Feb. 2000.

21. E. Radmoser, O. Scherzer, and J. Weickert. Scale-space properties of nonstationary iterative regularization methods. *Journal of Visual Communication and Image Representation*, 11(2):96–114, June 2000.

22. L. I. Rudin, S. Osher, and E. Fatemi. Nonlinear total variation based noise removal algorithms. *Physica D*, 60:259–268, 1992.

23. O. Scherzer and J. Weickert. Relations between regularization and diffusion filtering. *Journal of Mathematical Imaging and Vision*, 12(1):43–63, Feb. 2000.

24. G. Steidl, J. Weickert, T. Brox, P. Mrázek, and M. Welk. On the equivalence of soft wavelet shrinkage, total variation diffusion, total variation regularization, and SIDEs. Technical Report 26, Series SPP-1114, Department of Mathematics, University of Bremen, Germany, Feb. 2003.

25. D. M. Strong. *Adaptive Total Variation Minimizing Image Restoration*. PhD thesis, Department of Mathematics, University of California, Los Angeles, CA, 1997.

26. D. M. Strong and T. F. Chan. Relation of regularization parameter and scale in total variation based image denoising. Technical Report CAM-96-7, Department of Mathematics, University of California at Los Angeles, CA, U.S.A., 1996.

27. C. R. Vogel. *Computational Methods for Inverse Problems*. SIAM, Philadelphia, 2002.

28. J. Weickert. *Anisotropic Diffusion in Image Processing*. Teubner, Stuttgart, 1998.
29. J. Weickert, S. Ishikawa, and A. Imiya. Linear scale-space has first been proposed in Japan. *Journal of Mathematical Imaging and Vision*, 10(3):237–252, May 1999.
30. A. L. Yuille and N. M. Grzywacz. A computational theory for the perception of coherent visual motion. *Nature*, 333:71–74, 1988.
31. A. L. Yuille and N. M. Grzywacz. The motion coherence theory. In *Proc. Second International Conference on Computer Vision*, pages 344–354, Washington, DC, Dec. 1988. IEEE Computer Society Press.

Correspondences between Wavelet Shrinkage and Nonlinear Diffusion[*]

Pavel Mrázek[1], Joachim Weickert[1], and Gabriele Steidl[2]

[1] Mathematical Image Analysis Group
Faculty of Mathematics and Computer Science, Building 27
Saarland University, 66123 Saarbrücken, Germany
{mrazek,weickert}@mia.uni-saarland.de
http://www.mia.uni-saarland.de

[2] Faculty of Mathematics and Computer Science, D7, 27
University of Mannheim, 68131 Mannheim, Germany
steidl@math.uni-mannheim.de
http://www.kiwi.math.uni-mannheim.de

Abstract. We study the connections between discrete one-dimensional schemes for nonlinear diffusion and shift-invariant Haar wavelet shrinkage. We show that one step of (stabilised) explicit discretisation of nonlinear diffusion can be expressed in terms of wavelet shrinkage on a single spatial level. This equivalence allows a fruitful exchange of ideas between the two fields. In this paper we derive new wavelet shrinkage functions from existing diffusivity functions, and identify some previously used shrinkage functions as corresponding to well known diffusivities. We demonstrate experimentally that some of the diffusion-inspired shrinkage functions are among the best for translation-invariant multiscale wavelet shrinkage denoising.

1 Introduction

We consider a classical task of signal denoising: create an estimate u of an original signal z from its noisy measurement f, where

$$f = z + n,$$

and n denotes an additive noise function. Various methods have been proposed to remove the noise from z without sacrificing important structures such as edges, including rank-order filtering, mathematical morphology, stochastic methods, adaptive smoothing, wavelet techniques, partial differential equations (PDEs) and variational methods. Although these method classes serve the same purpose,

[*] This joint research was supported by the project *Relations between nonlinear filters in digital image processing* within the DFG–Schwerpunktprogramm 1114: *Mathematical methods for time series analysis and digital image processing*. This is gratefully acknowledged.

L.D. Griffin and M. Lillholm (Eds.): Scale-Space 2003, LNCS 2695, pp. 101–116, 2003.
© Springer-Verlag Berlin Heidelberg 2003

relatively few publications examine their similarities and differences, in order to transfer results from one of these classes to the others, or to design hybrid methods that combine the advantages of different classes. The present paper is a contribution in this direction, where we concentrate on two of these methods, namely nonlinear diffusion techniques and wavelet shrinkage.

Nonlinear diffusion creates a family of restored signals $u(t)$ by starting from the noisy signal f, and evolving it locally according to a process described by a nonlinear partial differential equation. This process is controlled by a diffusivity function g of the signal gradient. Typically, $g(s)$ is a nonnegative, nonincreasing function of the gradient magnitude, approaching zero as $s \to \infty$. This setting leads to the effect that smoothing of u proceeds faster in homogeneous regions (where the gradient is small, caused possibly by noise), and discontinuities (large gradient, hopefully corresponding to important features of the underlying signal) tend to be preserved. Depending on the choice of the diffusivity function g, a single nonlinear diffusion equation may cover a variety of nonlinear filters, including the original nonlinear diffusion of Perona and Malik [27] and its regularised variants [8,31], total variation (TV) diffusion [2], or balanced forward-backward (BFB) diffusion [21]. When applied to discrete data $\mathbf{f} = (f_i)_{i=0}^{N-1}$, the nonlinear diffusion filter creates a series of smoothed signals $\mathbf{u}^k := \mathbf{u}(k\tau)$ iteratively, starting from the noisy signal, $\mathbf{u}^0 = \mathbf{f}$.

Wavelet transforms express the signal in terms of wavelet coefficients, describing the signal variation at different scales. If the wavelet basis is chosen properly, a signal will be generally described by only a few significant wavelet coefficients, while moderate white Gaussian noise pollutes all the wavelet coefficients by a small amount. Signal denoising by wavelet shrinkage [13,14] starts from this assumption, and creates a smoothed version of the processed signal by the following three-step procedure:

1. *Analysis:* transform the noisy data f to the wavelet coefficients d_i^j, representing the signal at various scales j and positions i.
2. *Shrinkage:* apply a shrinkage function S_θ to the wavelet coefficients d_i^j, thus reducing the relative importance of small coefficients.
3. *Synthesis:* reconstruct a denoised version u of f from the shrunken wavelet coefficients.

The shrinkage parameter θ is chosen with respect to the amount of noise in the input signal. In general, the denoised solution u is obtained from f using a single step of this multiscale procedure, i.e. the method is applied noniteratively. The specific choice of the wavelets and the shrinkage functions allows a large variability of wavelet shrinkage methods.

In the present paper, we show equivalence between a single iteration of a 1-D explicit scheme for nonlinear diffusion on one side, and translation-invariant wavelet shrinkage with a single level of Haar wavelet decomposition on the other. This equivalence is obtained by constructing an appropriate shrinkage function S_θ to an existing diffusivity g, and vice versa.

Having asserted the equivalence between wavelet shrinkage and nonlinear diffusion for this special situation, it remains to be seen whether this connection brings any advantages in more general settings. We demonstrate numerically that the shrinkage functions derived from diffusivities are able to provide some of the best results when used for classical (i.e. multi-level, one step) translation-invariant wavelet shrinkage.

This paper is organised as follows. Section 2 presents nonlinear diffusion and develops its explicit discretisation in 1-D. Section 3 provides a brief introduction into translation-invariant Haar wavelet shrinkage. The connections between the two procedures are exploited in Section 4 to establish the conditions on diffusivity and shrinkage functions under which the two methods (restricted to one-step / one-scale) are equivalent. Some newly created shrinkage function are then tested experimentally, and compared to previously used ones. The paper is concluded with a summary in Section 6.

Related Work. Analysing the relations between regularisation methods and *continuous* wavelet shrinkage of functions, Chambolle *et al.* [5] showed that one may interpret wavelet shrinkage of functions as regularisation processes in suitable Besov spaces. In the case of Haar wavelets, Cohen *et al.* [9] showed that this approximates total variation regularisation. Later on, Chambolle and Lucier [6] considered iterated translation-invariant wavelet shrinkage and interpreted it as a nonlinear scale-space that differs from other scale-spaces by the fact that it is not given in terms of PDEs.

Regarding the relations between wavelet shrinkage denoising of *discrete* signals and nonlinear diffusion, not much research has been done so far. A recent paper by Coifman and Sowa [12] proposes TV diminishing flows that act along the direction of Haar wavelets. Recent work in which the authors are involved [29,30] investigates conditions under which equivalence between wavelet shrinkage of discrete signals, space-discrete TV diffusion or regularisation, and SIDEs (stabilised inverse diffusion equations) holds true.

Some recently proposed hybrid methods are based on combining wavelet shrinkage and TV regularisation methods [1,28]. Durand and Froment [15] proposed to address the problem of pseudo-Gibbs artifacts in wavelet denoising by replacing the thresholded wavelet coefficients by coefficients that minimise the total variation. Their method is also close in spirit to approaches by Chan and Zhou [7] who postprocessed images obtained from wavelet shrinkage by a TV-like regularisation technique. Coifman and Sowa [11] used functional minimisation with wavelet constraints for postprocessing signals that have been degraded by wavelet thresholding or quantisation. Candes and Guo [4] also presented related work, in which they combined ridgelets and curvelets with TV minimisation strategies. Recently, Malgouyres [23,24] proposed a hybrid method that uses both wavelet packets and TV approaches. His experiments showed that it may restore textured regions without introducing visible ringing artifacts.

This discussion shows that the previous papers typically focus on TV-based denoising techniques on the PDE side. Moreover, most of them present a continuous analysis rather than a discrete one. Our paper differs from previous work in this field by the fact that we do not restrict ourselves to a single diffusivity or shrinkage function, but introduce and analyse a general connection between a discrete diffusion scheme and Haar wavelet shrinkage. To this end, we investigate a large number of diffusivities and shrinkage functions.

2 Nonlinear Diffusion

2.1 Basic Concept

The basic idea behind nonlinear diffusion filtering [27] is to obtain a family $u(x,t)$ of filtered versions of the signal $f(x)$ as the solution of a suitable diffusion process

$$u_t = (g(|u_x|) u_x)_x \qquad (1)$$

with f as initial condition:

$$u(x,0) = f(x).$$

Here subscripts denote partial derivatives, and the diffusion time t is a simplification parameter: larger values correspond to stronger filtering.

The diffusivity $g(|u_x|)$ is a nonnegative function that controls the amount of diffusion. Usually, it is decreasing in $|u_x|$. This ensures that strong edges are less blurred by the diffusion filter than noise and low-contrast details. Depending on the choice of the diffusivity function, equation (1) covers a variety of filters. Here are some of the previously employed diffusivity functions:

A. Linear diffusivity [19]: $g(|x|) = 1,$

B. Charbonnier diffusivity [8]: $g(|x|) = \dfrac{1}{\sqrt{1 + \frac{|x|^2}{\lambda^2}}},$

C. Perona–Malik diffusivity [27]: $g(|x|) = \dfrac{1}{1 + \frac{|x|^2}{\lambda^2}},$

D. Weickert diffusivity [31]: $g(|x|) = \begin{cases} 1 & |x| = 0, \\ 1 - \exp\left(\frac{-3.31488}{(|x|/\lambda)^8}\right) & |x| > 0, \end{cases}$

E. TV diffusivity [2]: $g(|x|) = \dfrac{1}{|x|},$

F. BFB diffusivity: [21] $g(|x|) = \dfrac{1}{|x|^2}.$

Note that the diffusivities A–D are bounded from above by 1, while the diffusivities E and F are unbounded. In order to avoid theoretical and numerical difficulties, it is common to replace the latter ones by regularisations that make them bounded: e.g. one may use $g(|x|) = 1/\sqrt{\epsilon^2 + |x|^2}$ instead of the TV diffusivity.

Well-posedness results are available for the diffusivities A, B and E, since they lead to forward parabolic processes. For the diffusivities C, D and F, which may lead to backward parabolic equations, well-posedness questions are open in the continuous setting [22,20], while a space-discretisation seems to lead to well-posed processes [32].

2.2   Explicit Discretisation Scheme

When applied to discrete signals, the partial differential equation (1) has to be discretised. In this paper, we focus on explicit finite difference schemes. Substituting the spatial partial derivatives in (1) by finite differences (with the assumption of unit distance between neighboring pixels), and employing explicit discretisation in time, an explicit 1-D scheme for nonlinear diffusion can be written in the form

$$\frac{u_i^{k+1} - u_i^k}{\tau} = g(|u_{i+1}^k - u_i^k|)(u_{i+1}^k - u_i^k) - g(|u_i^k - u_{i-1}^k|)(u_i^k - u_{i-1}^k),$$

where τ is the time step size and the upper index k denotes the approximate solution at time $k\tau$. Separating the unknown u_i^{k+1} on one side, we obtain

$$u_i^{k+1} = u_i^k - \tau g(|u_i^k - u_{i+1}^k|)(u_i^k - u_{i+1}^k) + \tau g(|u_{i-1}^k - u_i^k|)(u_{i-1}^k - u_i^k). \quad (2)$$

The initial condition reads $u_i^0 = f_i$ for all i.

3   Wavelet Shrinkage

3.1   Basic Concept

The discrete wavelet transform represents a one-dimensional signal f in terms of shifted versions of a dilated lowpass scaling function φ, and shifted and dilated versions of a bandpass wavelet function ψ. In case of orthonormal wavelets, this gives

$$f = \sum_{i \in \mathbb{Z}} \langle f, \varphi_i^n \rangle \varphi_i^n + \sum_{j=-\infty}^{n} \sum_{i \in \mathbb{Z}} \langle f, \psi_i^j \rangle \psi_i^j, \quad (3)$$

where $\psi_i^j(s) := 2^{-j/2}\psi(2^{-j}s - i)$ and where $\langle \cdot, \cdot \rangle$ denotes the inner product in $L_2(\mathbb{R})$. If the measurement f is corrupted by moderate white Gaussian noise, then this noise is contained to a small amount in all wavelet coefficients $\langle f, \psi_i^j \rangle$, while the original signal is in general determined by a few significant wavelet coefficients [25]. Therefore, wavelet shrinkage attempts to eliminate noise from the wavelet coefficients by the following three-step procedure:

1. *Analysis*: transform the noisy data f to the wavelet coefficients $d_i^j = \langle f, \psi_i^j \rangle$ and scaling function coefficients $c_i^n = \langle f, \varphi_i^n \rangle$ according to (3).
2. *Shrinkage*: apply a shrinkage function S_θ with a threshold parameter θ to the wavelet coefficients, i.e., $S_\theta(d_i^j) = S_\theta(\langle f, \psi_i^j \rangle)$.
3. *Synthesis*: reconstruct the denoised version u of f from the shrunken wavelet coefficients:

$$u := \sum_{i \in \mathbb{Z}} \langle f, \varphi_i^n \rangle \varphi_i^n + \sum_{j=-\infty}^{n} \sum_{i \in \mathbb{Z}} S_\theta(\langle f, \psi_i^j \rangle) \psi_i^j.$$

In this paper we restrict our attention to Haar wavelets, well suited for piecewise constant signals with discontinuities. The Haar wavelet and scaling functions are given respectively by

$$\psi(x) = \mathbf{1}_{[0,\frac{1}{2})} - \mathbf{1}_{[\frac{1}{2},1)}, \tag{4}$$

$$\phi(x) = \mathbf{1}_{[0,1)} \tag{5}$$

where $\mathbf{1}_{[a,b)}$ denotes the characteristic function, equal to 1 on $[a,b)$ and zero everywhere else. Using the so-called "two-scale relation" of the wavelet and its scaling function, the coefficients c_i^j and d_i^j at higher level j can be computed from the coefficients c_i^{j-1} at lower level $j-1$ and conversely:

$$c_i^j = \frac{c_{2i}^{j-1} + c_{2i+1}^{j-1}}{\sqrt{2}}, \qquad d_i^j = \frac{c_{2i}^{j-1} - c_{2i+1}^{j-1}}{\sqrt{2}}, \tag{6}$$

and

$$c_{2i}^{j-1} = \frac{c_i^j + d_i^j}{\sqrt{2}}, \qquad c_{2i+1}^{j-1} = \frac{c_i^j - d_i^j}{\sqrt{2}}. \tag{7}$$

This results in a fast algorithm for the analysis step and synthesis step. Various shrinkage functions leading to qualitatively different denoised functions u were considered in literature, e.g.,

A. Linear shrinkage: $\qquad S(x) = \lambda x \qquad (\lambda \in [0,1])$,

B. Soft shrinkage [13]: $\qquad S_\theta(x) = \begin{cases} 0 & |x| \le \theta, \\ x - \theta\, \mathrm{sgn}(x) & |x| > \theta, \end{cases}$

C. Garrote shrinkage [16]: $\quad S_\theta(x) = \begin{cases} 0 & |x| \le \theta, \\ x - \frac{\theta^2}{x} & |x| > \theta, \end{cases}$

D. Firm shrinkage [17]: $\qquad S_{\theta_1,\theta_2}(x) = \begin{cases} 0 & |x| \le \theta_1, \\ \mathrm{sgn}(x)\frac{\theta_2(|x|-\theta_1)}{\theta_2-\theta_1} & \theta_1 < |x| \le \theta_2, \\ x & \theta_2 < |x|, \end{cases}$

E. Hard shrinkage [25]: $\qquad S_\theta(x) = \begin{cases} 0 & |x| \le \theta, \\ x & |x| > \theta. \end{cases}$

3.2   Discrete Translation-Invariant Scheme

In practice one deals with discrete signals $\mathbf{f} = (f_i)_{i=0}^{N-1}$, where, for simplicity, N is a power of 2. Then Haar wavelet shrinkage starts by setting $c_i^0 = f_i$ and proceeds by analysis (6), shrinkage, and synthesis (7). Let us just consider a *single* wavelet decomposition level, i.e., we set $n = 1$. Then, using the convention that $c_i = c_i^1$ and $d_i = d_i^1$, we can drop the superscripts $j = 0$ and $j = 1$. By (6) and (7), Haar wavelet shrinkage on one level produces the signal $\mathbf{u}^+ = (u_i^+)_{i=0}^{N-1}$ with coefficients

$$u_{2i}^+ = \frac{c_i + S_\theta(d_i)}{\sqrt{2}} = \frac{f_{2i} + f_{2i+1}}{2} + \frac{1}{\sqrt{2}} S_\theta\left(\frac{f_{2i} - f_{2i+1}}{\sqrt{2}}\right), \tag{8}$$

$$u_{2i+1}^+ = \frac{c_i - S_\theta(d_i)}{\sqrt{2}} = \frac{f_{2i} + f_{2i+1}}{2} - \frac{1}{\sqrt{2}} S_\theta\left(\frac{f_{2i} - f_{2i+1}}{\sqrt{2}}\right). \tag{9}$$

Note that the single Haar wavelet shrinkage step (8)–(9) decouples the input signal into successive pixel pairs: the pixel at position $2i-1$ has no direct connection to its neighbour at position $2i$, and the procedure is not invariant to translation of the input signal. To overcome this problem, Coifman and Donoho [10] introduced the so-called *cycle spinning*: the input signal is shifted, denoised using wavelet shrinkage, shifted back, and the results of all such shifts are averaged. This procedure is equivalent to thresholding of nondecimated wavelet coefficients which can be implemented efficiently using the *algorithme à trous* [18]. For our single decomposition level, we need only one additional shift to acquire translation invariance. The shifted Haar wavelet shrinkage yields the signal $\mathbf{u}^- = (u_i^-)_{i=0}^{N-1}$ with coefficients

$$u_{2i-1}^- = \frac{f_{2i-1} + f_{2i}}{2} + \frac{1}{\sqrt{2}} S_\theta\left(\frac{f_{2i-1} - f_{2i}}{\sqrt{2}}\right),$$

$$u_{2i}^- = \frac{f_{2i-1} + f_{2i}}{2} - \frac{1}{\sqrt{2}} S_\theta\left(\frac{f_{2i-1} - f_{2i}}{\sqrt{2}}\right).$$

Averaging the shifted results, one cycle of shift-invariant Haar wavelet shrinkage can be summarised into

$$u_i = \frac{u_i^- + u_i^+}{2}$$
$$= \frac{f_{i-1} + 2f_i + f_{i+1}}{4} + \frac{1}{2\sqrt{2}} S_\theta\left(\frac{f_i - f_{i+1}}{\sqrt{2}}\right) - \frac{1}{2\sqrt{2}} S_\theta\left(\frac{f_{i-1} - f_i}{\sqrt{2}}\right). \tag{10}$$

4   Correspondence of Diffusivities and Shrinkage Functions

4.1   Basic Considerations

In order to derive the relation between the explicit diffusion scheme and translation-invariant Haar wavelet shrinkage, we rewrite the first iteration step in (2) using the initial condition $u_i^0 = f_i$ and the simplified notation $u_i^1 = u_i$ as

$$u_i = \frac{f_{i-1} + 2f_i + f_{i+1}}{4} + \frac{f_i - f_{i+1}}{4} - \frac{f_{i-1} - f_i}{4}$$
$$- \tau g(|f_i - f_{i+1}|)(f_i - f_{i+1}) + \tau g(|f_{i-1} - f_i|)(f_{i-1} - f_i)$$
$$= \frac{f_{i-1} + 2f_i + f_{i+1}}{4}$$
$$+ (f_i - f_{i+1})\left(\frac{1}{4} - \tau g(|f_i - f_{i+1}|)\right)$$
$$- (f_{i-1} - f_i)\left(\frac{1}{4} - \tau g(|f_{i-1} - f_i|)\right). \tag{11}$$

This coincides with (10) if and only if

$$\frac{1}{2\sqrt{2}} S_\theta\left(\frac{x}{\sqrt{2}}\right) = x\left(\frac{1}{4} - \tau g(|x|)\right). \tag{12}$$

Equation (12) relates the shrinkage function S_θ of wavelet denoising to the diffusivity g of nonlinear diffusion. Provided that relation (12) holds true, a single step of wavelet shrinkage is equivalent to a single step of explicitly discretised nonlinear diffusion. The following two formulas are derived from (12) and can be used to obtain a shrinkage function S_θ from a diffusivity g, or vice versa.

$$S_\theta(x) = x\left(1 - 4\tau g(|\sqrt{2}x|)\right), \tag{13}$$

$$g(|x|) = \frac{1}{4\tau} - \frac{\sqrt{2}}{4\tau x} S_\theta\left(\frac{x}{\sqrt{2}}\right). \tag{14}$$

4.2   From Diffusivities to Shrinkage Functions

Let us now investigate equation (13) in detail. The examples from Section 3.1 show that typical shrinkage functions from the literature satisfy

$$S(x) \geq 0 \quad \text{for} \quad x > 0, \tag{15}$$

$$S(x) \leq 0 \quad \text{for} \quad x < 0. \tag{16}$$

One can show that these conditions are responsible for ensuring certain stability properties (so-called sign stability) of the shrinkage process. We can now specify the time step size τ in (13) such that these two conditions are always satisfied for bounded diffusivities. In Section 2.1 we have seen that the diffusivities A–D are bounded by 1. In order to ensure that the corresponding shrinkage functions satisfy (15)–(16), the time step size has to fulfil $\tau \leq 0.25$.

We observe that the linear diffusivity corresponds to the linear shrinkage function

$$S(x) = (1 - 4\tau)x.$$

Nonlinear shrinkage functions such as soft, garrote, firm and hard shrinkage satisfy $S'(0) = 0$, since the goal was to set small wavelet coefficients to zero. In order to derive shrinkage functions that correspond to the bounded nonlinear diffusivities B–D and satisfy $S'(0) = 0$ as well, let us now fix $\tau := 0.25$. Then we obtain the following novel shrinkage functions:

- The Charbonnier diffusivity corresponds to the shrinkage function

$$S_\lambda(x) = x\left(1 - \sqrt{\frac{\lambda^2}{\lambda^2 + 2x^2}}\right).$$

- The Perona–Malik diffusivity leads to

$$S_\lambda(x) = \frac{2x^3}{2x^2 + \lambda^2}.$$

– The Weickert diffusivity gives

$$S_\lambda(x) = \begin{cases} 0 & x = 0, \\ x \exp\left(-\frac{0.20718\,\lambda^8}{x^8}\right) & x \neq 0. \end{cases}$$

Figure 1 illustrates these bounded diffusivities and their shrinkage functions.

4.3 From Shrinkage Functions to Diffusivities

Having derived shrinkage functions from nonlinear diffusivities, let us now derive diffusivities from frequently used shrinkage functions. To this end, all we have to do is to plug in the specific shrinkage function into (14).
In the case of soft shrinkage, this gives the diffusivity

$$g(|x|) = \begin{cases} \frac{1}{4\tau} & |x| \leq \theta\sqrt{2}, \\ \frac{\sqrt{2}\theta}{4\tau|x|} & |x| > \theta\sqrt{2}. \end{cases}$$

If we select the time step size τ such that $\theta = 2\sqrt{2}\tau$, we obtain a stabilised TV diffusivity:

$$g(|x|) = \begin{cases} \frac{1}{4\tau} & |x| \leq 4\tau, \\ \frac{1}{|x|} & |x| > 4\tau. \end{cases}$$

In the same way one can show that garrote shrinkage leads to a stabilised BFB diffusivity for $\theta = \sqrt{2}\tau$:

$$g(|x|) = \begin{cases} \frac{1}{4\tau} & |x| \leq 2\sqrt{\tau}, \\ \frac{1}{|x|^2} & |x| > 2\sqrt{\tau}. \end{cases}$$

Firm shrinkage yields a diffusivity that degenerates to 0 for sufficiently large gradients:

$$g(|x|) = \begin{cases} \frac{1}{4\tau} & |x| \leq \sqrt{2}\theta_1, \\ \frac{\theta_1}{4\tau(\theta_2-\theta_1)}\left(\frac{\sqrt{2}\theta_2}{|x|} - 1\right) & \sqrt{2}\theta_1 < |x| \leq \sqrt{2}\theta_2, \\ 0 & |x| > \sqrt{2}\theta_2. \end{cases}$$

Such diffusivities have been considered in [3], where they have been motivated using priors from robust statistics.
Another diffusivity that degenerates to 0 can be derived from hard shrinkage:

$$g(|x|) = \begin{cases} \frac{1}{4\tau} & |x| \leq \sqrt{2}\theta, \\ 0 & |x| > \sqrt{2}\theta. \end{cases}$$

All diffusivities in this subsection are depicted in Figure 2.

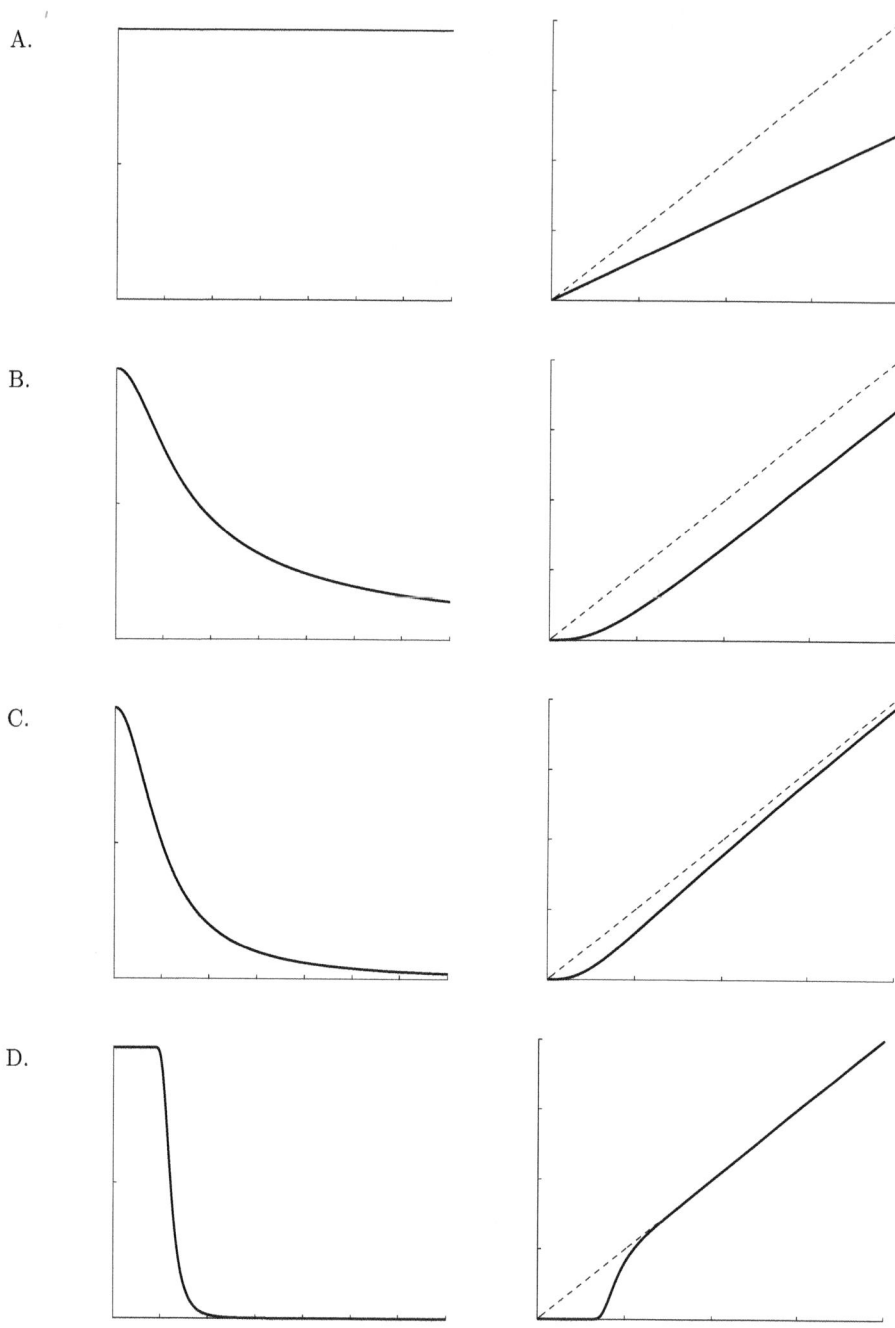

Fig. 1. Diffusivity functions (left), corresponding shrinkage functions (right).
A. Linear diffusion. B. Charbonnier diffusivity. C. Perona-Malik diffusivity. D. Weickert diffusivity. The functions are plotted for $\tau = 0.1$ (linear diffusion), and $\tau = 0.25$, $\lambda = 1$ (all others).

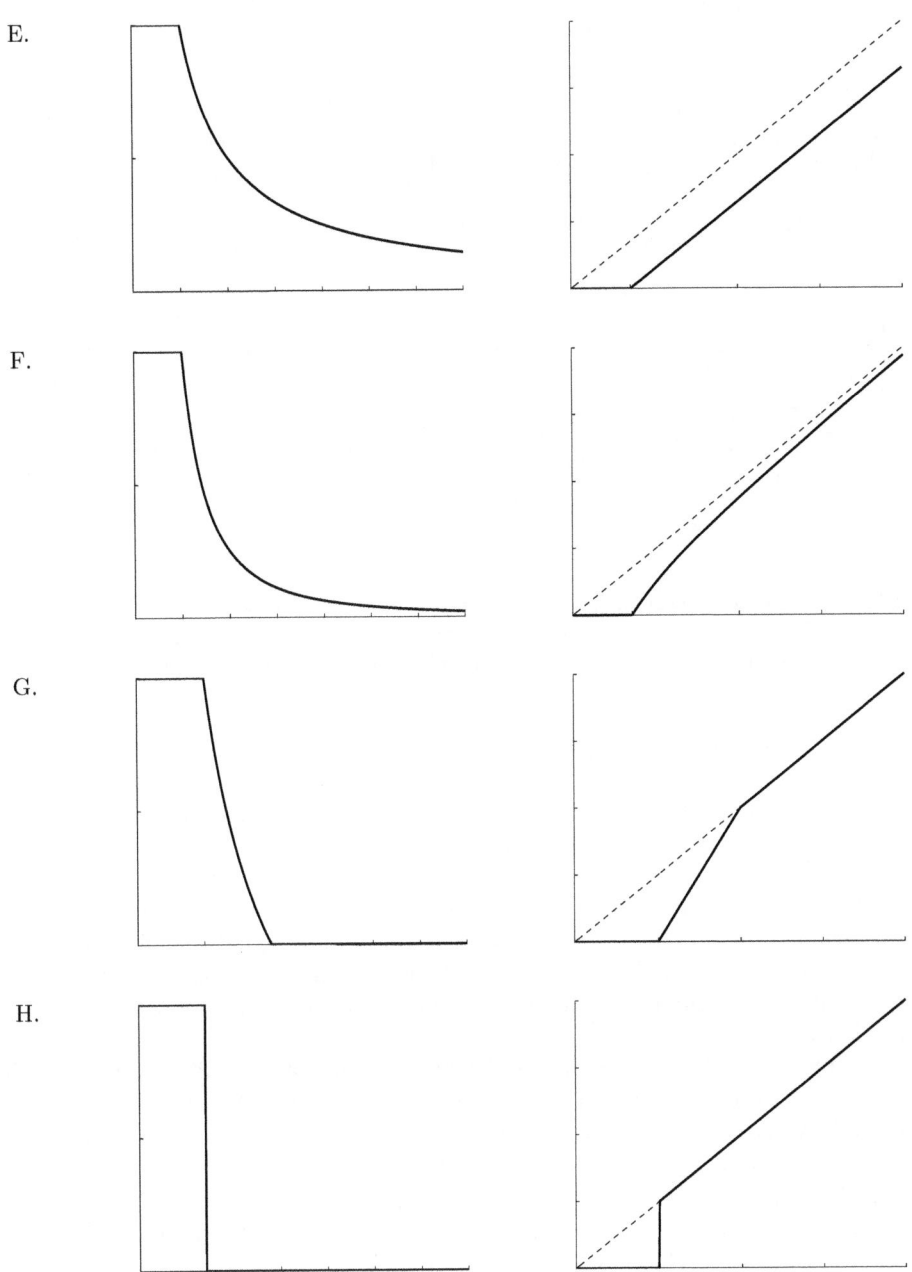

Fig. 2. Diffusivity functions (left), corresponding shrinkage functions (right).
E. TV flow and soft shrinkage. F. Balanced forward-backward (BFB) diffusivity and garrote shrinkage. G. Firm shrinkage. H. Hard shrinkage.
The functions are plotted for $\tau = 0.25$ (which corresponds to $\theta = \tau 2\sqrt{2}$ for soft shrinkage, and to $\theta = \sqrt{2\tau}$ for garrote; the other use $\theta = \theta_1 = 1$, $\theta_2 = 2$).

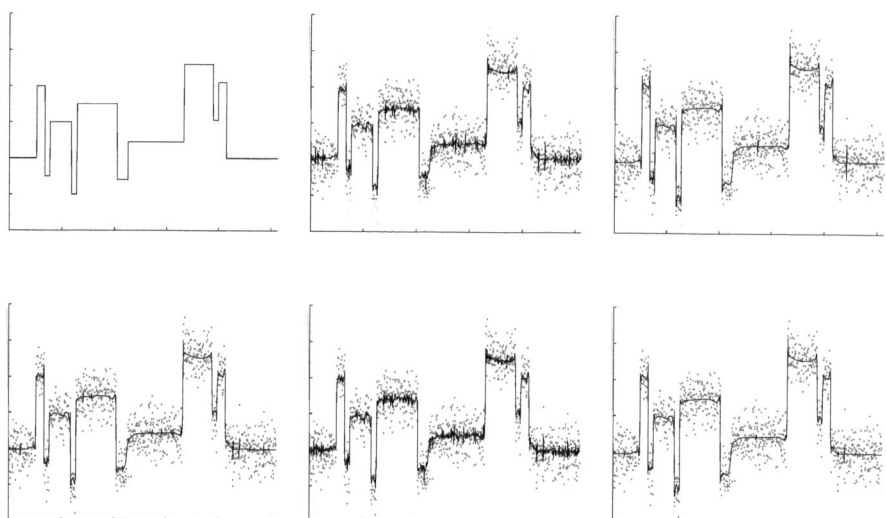

Fig. 3. Example of multiscale translation-invariant Haar wavelet denoising. Normal noise of SNR=8 was added to the ideal signal, and different shrinkage functions have been applied. The noisy signal is represented by dots, reconstructed signal by solid line.

5   Denoising Experiment

To test the applicability of the newly derived shrinkage functions from Subsection 4.2, we perform experiments with signal-denoising using the shift-invariant multiscale Haar wavelet transform from Section 3. The input signal *blocks*, one of the standard signals in wavelet denoising, mimics a scan line through a 2-D image depicting an object with several edges [14]. The signal is shown in Fig. 3. The same figure then shows examples of the results of multiscale Haar wavelet denoising when combined with several shrinkage functions introduced in previous sections.

Table 1 and Fig. 4 present additional experimental results obtained with the *blocks* data. Here we performed a series of experiments with several levels of additive zero-mean Gaussian noise in the input signal. The noise varies between SNR=1 and SNR=32, where the signal-to-noise ratio (SNR) is defined by $\text{SNR} = 20\log_{10}\frac{|z-\bar{z}|_2}{|n|_2}$, with z standing for the ideal signal with mean $\bar{z}$, and n representing noise. The noise is generated five times for each SNR level. Then we used multiscale wavelet denoising with various shrinkage functions, and searched for the optimal solution that can be obtained with this method. By optimal we mean the solution maximising the signal-to-noise ratio in the filtered signal.

Table 1 summarises the average optimal SNR after filtering obtained with different shrinkage functions; Fig. 4 presents the same information graphically, together with the standard deviation of the results. We observe that for all noise levels, the best signal-to-noise ratio is obtained by those shrinkage functions

Table 1. Numerical results (measured by mean signal-to-noise ratio in the filtered signal) of wavelet denoising for the *blocks* data of length 1024. Each column represents a given level of noise in the input image; each row contains the results for one shrinkage function.

SNR$_{in}$ Shrinkage method	1	2	4	8	16	32
Linear	3.6	4.2	5.5	8.7	16.1	32.0
Soft (TV)	10.0	10.8	12.6	16.2	24.0	39.9
Perona-Malik	9.9	10.8	12.7	16.8	25.8	44.6
Weickert	12.7	13.7	15.9	20.3	29.4	46.3
Garrote (BFB)	11.8	12.8	14.9	19.3	28.7	46.2
Firm	12.6	13.6	15.8	20.2	29.3	46.3
Hard	12.7	13.8	15.9	20.4	29.3	46.3

which put small wavelet coefficients to zero and keep larger coefficients almost unaffected. The functions with these properties include hard shrinkage, firm shrinkage and – to some extent – the garrote shrinkage on the wavelet side. Of the diffusion origin, the experimentally best shrinkage functions correspond to Weickert diffusivity, stabilised BFB diffusivity (which is equivalent to garrote shrinkage), and Perona-Malik diffusivity. Interestingly, these are diffusivities with nonmonotone flux functions that allow even contrast enhancement.

The second group of shrinkage functions decreases even large wavelet coefficients by a constant (or almost constant) value; the functions with this behaviour include soft shrinkage, TV flow corresponding to it, and Charbonnier diffusivity. It seems that this strategy is less successful numerically. These diffusivities lead to monotonically increasing flux functions and well-posed diffusion filters.

As a group of its own, the denoising performance of linear diffusion (or its shrinkage function) is far worse than that of the nonlinear methods.

6   Conclusions

We have analysed correspondences between explicit one-dimensional schemes for nonlinear diffusion and discrete translation-invariant Haar wavelet shrinkage. We have shown that if we restrict the methods to one discrete step and a single spatial level, the two approaches can be made equivalent, if suitable diffusivities or shrinkage functions are chosen.

This connection between nonlinear diffusion and wavelet shrinkage opens the gate for a fruitful exchange of ideas between the two worlds. In this paper, we derived new wavelet shrinkage functions from frequently used nonlinear diffusivities; vice versa, we showed that soft and garrote shrinkage may be regarded as stabilised TV or BFB diffusion, respectively. We experienced that the novel shrinkage functions corresponding to rapidly decreasing diffusivities are competitive with the best previously known shrinkage methods when applied to signal denoising with multiscale wavelet procedures.

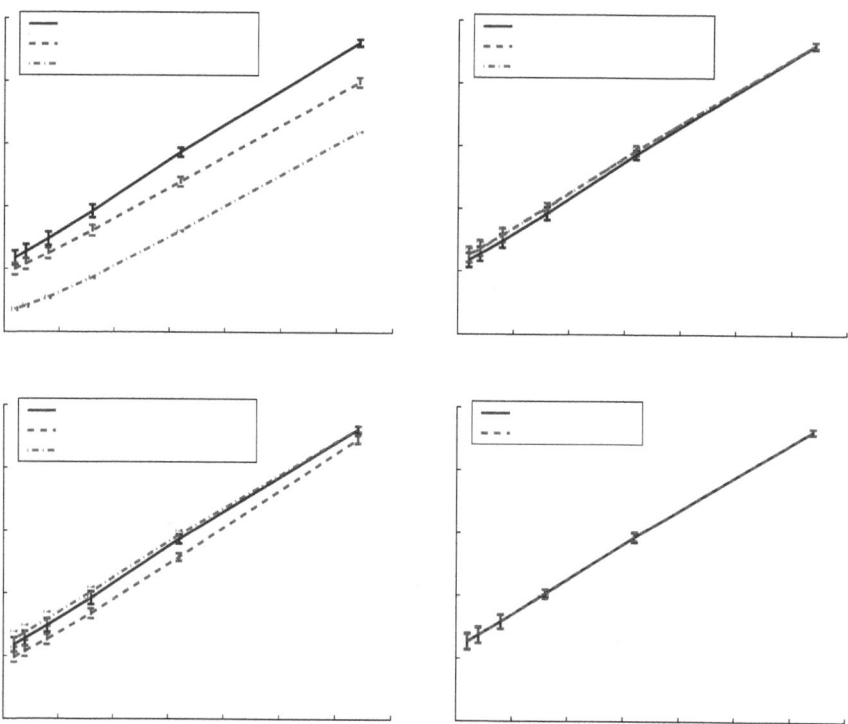

Fig. 4. Comparing the optimal denoising performance of shift-invariant multiscale wavelet shrinkage with various shrinkage functions. SNR of the filtered signal is plotted against SNR of the input; the higher the graph, the better the result. The input signal was *blocks*, length 1024.
Top left: garrote shrinkage (BFB diffusivity), soft shrinkage (TV flow) and linear diffusion. Top right: garrote (BFB), hard and firm shrinkages. Bottom left: garrote (BFB), Perona-Malik and Weickert functions. Bottom right: best from either world, hard shrinkage and Weickert diffusivity give comparable results.

The results in this paper can be extended in several directions. One can study iterated multi-scale wavelet shrinkage as a hybrid method combining the efficiency of multi-scale wavelet shrinkage with the quality of iterated diffusion filtering [26]. This hybrid method may be also explained as nonlinear diffusion applied to the Laplacian pyramid of the signal [29,30]. In our ongoing work, we are considering other wavelet bases and the two-dimensional case.

References

1. R. Acar and C. R. Vogel. Analysis of bounded variation penalty methods for ill–posed problems. *Inverse Problems*, 10:1217–1229, 1994.

2. F. Andreu, C. Ballester, V. Caselles, and J. M. Mazón. Minimizing total variation flow. *Differential and Integral Equations*, 14(3):321–360, March 2001.
3. M. J. Black, G. Sapiro, D. H. Marimont, and D. Heeger. Robust anisotropic diffusion. *IEEE Transactions on Image Processing*, 7(3):421–432, March 1998.
4. E. J. Candés and F. Guo. New multiscale transforms, minimum total variation synthesis: Applications to edge-preserving image reconstruction. *Signal Processing*, 82(11):1519–1543, 2002.
5. A. Chambolle, R. A. DeVore, N. Lee, and B. L. Lucier. Nonlinear wavelet image processing: variational problems, compression, and noise removal through wavelet shrinkage. *IEEE Transactions on Image Processing*, 7(3):319–335, March 1998.
6. A. Chambolle and B. L. Lucier. Interpreting translationally-invariant wavelet shrinkage as a new image smoothing scale space. *IEEE Transactions on Image Processing*, 10(7):993–1000, 2001.
7. T. F. Chan and H. M. Zhou. Total variation improved wavelet thresholding in image compression. In *Proc. Seventh International Conference on Image Processing*, Vancouver, Canada, September 2000.
8. P. Charbonnier, L. Blanc-Féraud, G. Aubert, and M. Barlaud. Two deterministic half-quadratic regularization algorithms for computed imaging. In *Proc. 1994 IEEE International Conference on Image Processing*, volume 2, pages 168–172, Austin, TX, November 1994. IEEE Computer Society Press.
9. A. Cohen, R. DeVore, P. Petrushev, and H. Xu. Nonlinear approximation and the space $BV(R^2)$. *Americal Journal of Mathematics*, 121:587–628, 1999.
10. R. R. Coifman and D. Donoho. Translation invariant denoising. In A. Antoine and G. Oppenheim, editors, *Wavelets in Statistics*, pages 125–150. Springer, New York, 1995.
11. R. R. Coifman and A. Sowa. Combining the calculus of variations and wavelets for image enhancement. *Applied and Computational Harmonic Analysis*, 9(1):1–18, July 2000.
12. R. R. Coifman and A. Sowa. New methods of controlled total variation reduction for digital functions. *SIAM Journal on Numerical Analysis*, 39(2):480–498, 2001.
13. D. L. Donoho. De-noising by soft thresholding. *IEEE Transactions on Information Theory*, 41:613–627, 1995.
14. D. L. Donoho and I. M. Johnstone. Ideal spatial adaptation by wavelet shrinkage. *Biometrica*, 81(3):425–455, 1994.
15. S. Durand and J. Froment. Reconstruction of wavelet coefficients using total-variation minimization. Technical Report 2001–18, Centre de Mathématiques et de Leurs Applications, ENS de Cachan, France, 2001.
16. H.-Y. Gao. Wavelet shrinkage denoising using the non-negative garrote. *Journal of Computational and Graphical Statistics*, 7(4):469–488, 1998.
17. H.-Y. Gao and A. G. Bruce. WaveShrink with firm shrinkage. *Statistica Sinica*, 7:855–874, 1997.
18. M. Holschneider, R. Kronland-Martinet, J. Morlet, and Ph. Tchamitchian. A real-time algorithm for signal analysis with the help of the wavelet transform. In J.M. Combes, A. Grossman, and Ph. Tchamitchian, editors, *Wavelets: Time-Frequency Methods and Phase Space*, pages 286–297. Springer-Verlag, 1987.
19. T. Iijima. Basic theory on normalization of pattern (in case of typical one-dimensional pattern). *Bulletin of the Electrotechnical Laboratory*, 26:368–388, 1962. In Japanese.
20. B. Kawohl and N. Kutev. Maximum and comparison principle for one-dimensional anisotropic diffusion. *Mathematische Annalen*, 311:107–123, 1998.

21. S. L. Keeling and R. Stollberger. Nonlinear anisotropic diffusion filters for wide range edge sharpening. *Inverse Problems*, 18:175–190, January 2002.
22. M. Kijima. *Markov Processes and Stochastic Modeling*. Chapman and Hall, New York, 1997.
23. F. Malgouyres. Combining total variation and wavelet packet approaches for image deblurring. In *Proc. First IEEE Workshop on Variational and Level Set Methods in Computer Vision*, pages 57–64, Vancouver, Canada, July 2001. IEEE Computer Society Press.
24. F. Malgouyres. Mathematical analysis of a model which combines total variation and wavelet for image restoration. *Inverse Problems*, 2(1):1–10, 2002.
25. S. Mallat. *A Wavelet Tour of Signal Processing*. Academic Press, San Diego, second edition, 1999.
26. Pavel Mrázek, Joachim Weickert, Gabriele Steidl, and Martin Welk. On iterations and scales of nonlinear filters. In O. Drbohlav, editor, *Computer Vision Winter Workshop 2003*, pages 61–66. Czech Pattern Recognition Society, 2003.
27. P. Perona and J. Malik. Scale space and edge detection using anisotropic diffusion. *IEEE Transactions on Pattern Analysis and Machine Intelligence*, 12:629–639, 1990.
28. L. I. Rudin, S. Osher, and E. Fatemi. Nonlinear total variation based noise removal algorithms. *Physica D*, 60:259–268, 1992.
29. G. Steidl and J. Weickert. Relations between soft wavelet shrinkage and total variation denoising. In L. Van Gool, editor, *Pattern Recognition*, volume 2449 of *Lecture Notes in Computer Science*, pages 198–205. Springer, Berlin, 2002.
30. G. Steidl, J. Weickert, T. Brox, P. Mrázek, and M. Welk. On the equivalence of soft wavelet shrinkage, total variation diffusion, total variation regularization, and SIDEs. Technical report, Series SPP-1114, Department of Mathematics, University of Bremen, Germany, 2003.
31. J. Weickert. *Anisotropic Diffusion in Image Processing*. Teubner, Stuttgart, 1998.
32. J. Weickert and B. Benhamouda. A semidiscrete nonlinear scale-space theory and its relation to the Perona–Malik paradox. In F. Solina, W. G. Kropatsch, R. Klette, and R. Bajcsy, editors, *Advances in Computer Vision*, pages 1–10. Springer, Wien, 1997.

Approximating Non-linear Diffusion

Erik Dam, Ole Fogh Olsen, and Mads Nielsen

IT University of Copenhagen
Glentevej 67
2400 Copenhagen NV
DK – Denmark
{erikdam,fogh,malte}@itu.dk
http://www.itu.dk/image

Abstract. We assess the feasibility of approximating non-linear diffusion processes with simple local Gaussian filters. The purpose of doing
this is twofold. Firstly, the theoretical implications are by themselves interesting. Secondly, a successful method would reduce the need for computationally expensive implementations of non-linear diffusion schemes.
We evaluate using isotropic and affine Gaussian filters for the task of
approximating the local diffusion for a number of non-linear diffusion
schemes. The approximations are firstly explored using an information
theoretical approach and secondly evaluated based on their performance
on a multi-scale segmentation application.
The results show that while the approximations do not perform quite
as well as the original non-linear scheme, the decrease in performance is
acceptable for the evaluated task. Furthermore, the affine approximations
perform significantly better than the isotropic.

1 Introduction

Non-linear diffusion have proven extremely useful in numerous applications:
noise-reduction, enhancement, restoration, and multi-scale segmentation [12,14,11,2].
This success is due to the ability to incorporate prior task-specific knowledge into
the diffusion process — typically by tuning parameters to the task at hand.

However, non-linear diffusion does also introduce a couple of basic problems.
Firstly, the parameters for the non-linear schemes have to be determined in some
more or less well-founded manner. We do not address this.

Secondly, the non-linear schemes are expensive in terms of computational
complexity. For applications that use non-linear diffusion for noise-reduction in
a pre-processing step, this is not problematic due to relatively short diffusion
times. For more demanding applications the use of non-linear diffusion is often
un-feasible — e.g. high resolution medical 3D scans will often impose severe
computational time problems.

One way to attack this problem of computational complexity is to introduce
sophisticated numerical implementations. One such example is the AOS scheme
for anisotropic diffusion [14].

L.D. Griffin and M. Lillholm (Eds.): Scale-Space 2003, LNCS 2695, pp. 117–131, 2003.

This paper aims at an alternative solution where the non-linear diffusion schemes such as anisotropic diffusion schemes are replaced by simpler schemes based on local Gaussian filters. Since the goal is to determine the "best" local Gaussian filter this has a certain scale-selection flavour and could be inspired by the methods that do this by maximization of scale-invariant expressions [8] or MDL minimization [4]. However, we aim at replacing a given non-linear diffusion scheme and therefore the scale-selection mechanism must have the desired diffusion time as a parameter.

We do not present such a scale-selection method — this is a fundamental feasibility study. However, we illustrate how such a method should perform. But the main focus is on the evaluation of the performance of local Gaussian filters that approximate a given non-linear diffusion scheme optimally. So we answer the question: *If we design the perfect scale-selection method for approximating non-linear diffusion, will it be good enough?*

The first section of this paper explains how we approximate non-linear diffusion schemes. This is done by extracting the local filters of the non-linear scheme and then approximating them by isotropic and affine Gaussian filters satisfying maximal entropy constraints. As a secondary result, these approximations allow us to quantify how non-linear the schemes are (i.e. how different from linear Gaussian diffusion they are).

Having defined the approximating filters we turn to the evaluation that we perform in two steps. First we explore the performance of the approximating diffusion schemes by evaluating them on random points in natural images from the Van Hateren collection [13].

Secondly we evaluate the approximating filters in a multi-scale segmentation setting where non-linear diffusion schemes have been shown to offer superior performance compared to linear diffusion.

2   Approximating Non-linear Diffusion

In order to approximate the non-linear diffusion schemes we first perform the desired non-linear scheme implemented through a simple explicit iterative scheme. This is done with an augmented implementation that records the actual local filters that are implicitly used in the diffusion.

These recorded local non-parametric filters are approximated with simple, parametric Gaussian filters based on information-theoretic criteria.

2.1   The Diffusion Echo

Most non-linear diffusion schemes have no explicit expression for (or representation of) the local filter that determines the diffusion in each point (i.e. pixel or voxel, the following works for arbitrary spatial dimension).

However, a simple method exists for obtaining these local filters. For each pixel an auxiliary image is created with the value 1 in this pixel and zero elsewhere. This is the discrete equivalent of the impulse function. The diffusion that is performed on the actual image is then performed in parallel on each auxiliary image. Thereby the impulse responses for the diffusion process are acquired. They determine where the "mass" in a pixel flows to during diffusion. The local diffusion filter for a pixel is then collected by picking the flow from each impulse response that flows to the specific pixel. The recorded impulse responses and local filters are called the *Diffusion Echo* [3].

The downside of this simple method is the computational complexity and the memory requirements. Independent of the efficiency of the underlying implementation of the diffusion scheme, the augmented method becomes at least $O(P^2)$ (where P is the number of pixels in the image) in the straightforward implementation. This can be lowered to $O(P \times F)$ (where F is an upper limit of the number of pixels in a local filter not being zero) if the extent of the impulse response can be limited. Nevertheless, this is still a quite restrictive complexity.

However, for some purposes only the moments of the filters are necessary and not the actual filters. Many implementations of non-linear diffusion schemes are based on iterative schemes where a local stencil is used to form a weighted average of some neighbourhood for each pixel/voxel. The specific scheme defines the local stencil.

In this case it is simple to record the moments (specifically mean and variance) of the local filter directly without recording the local impulse response. This is done by using the local stencil to average the moments from the previous iteration for each iteration step. This process adds no computational complexity to the diffusion method.

2.2   Maximum Entropy Approximation Filters

As described above we summarise the diffusion process with a local diffusion filter in each point of the domain. The convolution with the filter and the original image in that specific point will give exactly the same result as the diffusion process in the point. The filter values depend on the chosen diffusion scheme (including choice of parameters — e.g. iterations) and the original image.

We want to approximate this filter. A straightforward approach is to approximate the first few moments, say the mean and the variance. In order to select among the filters with the same mean and the variance we choose the filter with maximum entropy when the filter is viewed as a distribution.

A maximum entropy solution is a least committed appoach in the sense that it treats all locations as equally as possible under the given restrictions (here a specified mean and variance). In the case of no restrictions, a maximum entropy solution will result in a uniform distribution. In this way we avoid to introduce a bias towards a specific (unknown) purpose in our approximation.

The maximum entropy solution given a specific mean and variance is in the continuous case a Gaussian distribution. To stay in the continuous domain we calculate the variance of the diffusion echo filter by modelling it as a piecewise constant function on a continuous domain. When we apply the resulting approximating filter we also model the image as a piecewise constant function on a continuous domain.

By following Jaynes' maximum entropy principle [6] we minimize the Kullback-Leibler divergence [7] between the diffusion echo filter and the approximating filter.

2.3   Illustrating Diffusion Approximations

We briefly present the non-linear diffusion schemes that we investigate in the evaluation sections. The approximation scheme is illustrated by comparing the result with the use of the original non-linear schemes.

Weickert defines the anisotropic diffusion equation [14] by the PDE in equation 1. Here the diffusion on the image U is defined by the eigenvalues λ_1 and λ_2 for the diffusion tensor D where the eigenvectors are defined such that $\bar{v}_1 \parallel \nabla U_\sigma$, $\bar{v}_2 \perp \nabla U_\sigma$ (here ∇U_σ is the gradient evaluated at regularization scale σ). The diffusivity function w is used to determine these eigenvalues.

$$\frac{\partial U(t, \boldsymbol{x})}{\partial t} = div(\ D(\nabla U_\sigma)\ \nabla U\ ) \quad \text{where} \quad U(0, \boldsymbol{x}) = I(\boldsymbol{x}) \qquad (1)$$

$$\lambda_1 = w(m, \lambda, |\nabla U_\sigma|^2)$$

$$\lambda_2 = \theta + (1 - \theta)\ \lambda_1$$

$$w(m, \lambda, |\nabla U_\sigma|^2) = \begin{cases} 1 & |\nabla U_\sigma| = 0 \\ 1 - exp\left(\frac{-C_m}{\left(\frac{|\nabla U_\sigma|^2}{\lambda}\right)^m}\right) & |\nabla U_\sigma| > 0 \end{cases}$$

The parameter θ determines the degree of anisotropy and m the aggressiveness with which the edges are preserved (where edges are defined by the soft threshold λ, and C_m is calculated from m such that w is increasing for $|\nabla U_\sigma|^2 < \lambda$ and decreasing for $|\nabla U_\sigma|^2 > \lambda$). Weickert's edge enhancing diffusion scheme [14] is a special case (EED, $\theta = 1$ and $m = 4$), the regularized Perona-Malik scheme [10,2] can be approximated (RPM, $\theta = 0$ and $m = 0.75$), and for $\lambda \to \infty$ GAN becomes linear Gaussian diffusion (LIN).

For the illustrations we use the simple image in figure 1 where the point of interest is the center point. From a diffusion approximation point of view, this is a relatively challenging point since it is located just inside a corner.

Figures 1 and 2 show that the approximating filters do a reasonably good job for this example point — the quantitative differences are obvious, but the qualitative appearance of the diffused images are quite similar.

3   Information Theoretical Evaluation

The diffusion echo filters from the diffusion processes are approximated by a Gaussian filter. We evaluate both the use of isotropic and affine Gaussian filters for the approximations. The expectation is that in some cases the affine can give a better approximation than the isotropic. More advanced filters could be chosen, but that would defeat the purpose of making simple, approximating schemes.

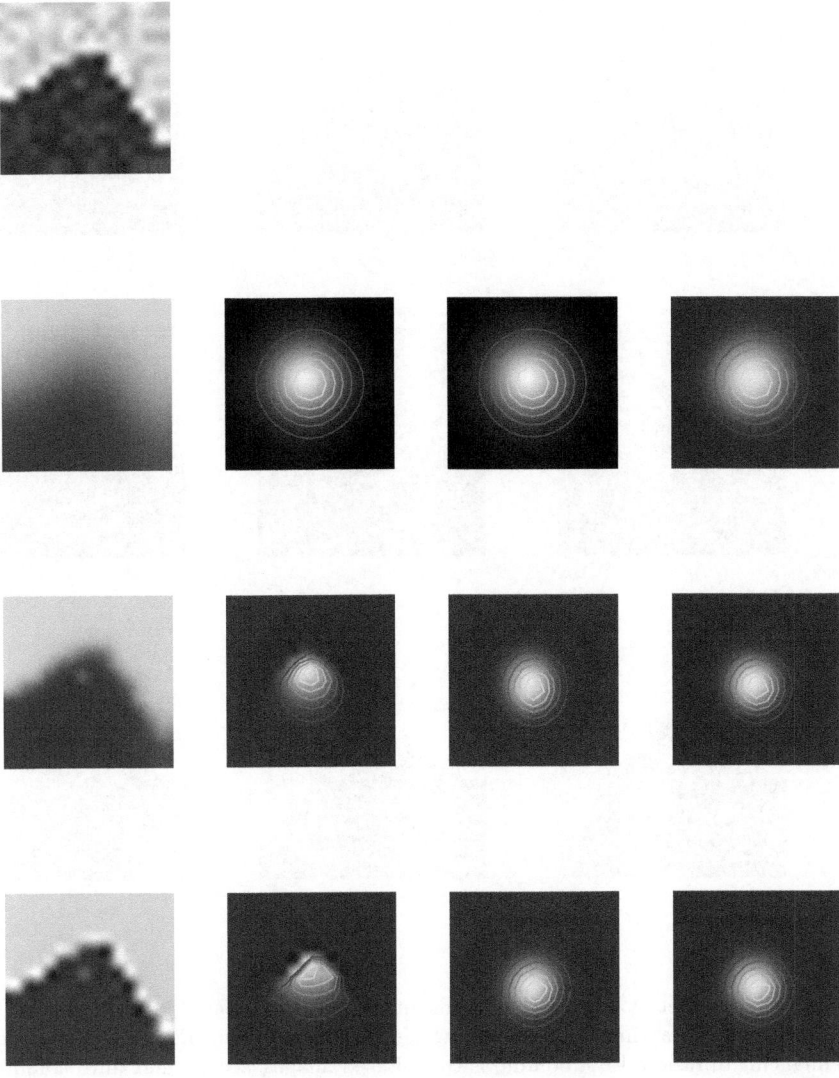

Fig. 1. Comparison of the actual diffusion echo filters with the approximated filters. The left column shows the original image and the result of applying linear Gaussian, Perona-Malik, and edge enhancing diffusion. The red dot is the center point where the filters are approximated. The second column shows the diffusion echo filters for each diffusion scheme. Third and fourth columns shows affine and isotropic approximating filters, respectively. The overlaid contours in warm colours are iso-curves. The entropy and KL measures show that EED is clearly more non-Gaussian than RPM, and that there is a surprisingly small difference between the isotropic and affine approximations. Contact the authors for color versions of the figures.

Fig. 2. The effect of applying the diffusion echo filter compared to the result of applying the approximating filters. The rows are linear Gaussian, Perona-Malik, and edge enhancing diffusion. First column is the diffused images, second and third columns are the approximated diffused images. Even though the quantitative differences between applying the non-linear filters and the approximated filters are evident, the qualitative appearence of the resulting images are quite similar.

The diffusion echo filter depends on the diffusion process as well as the original image. The influence of the original image is addressed by 10,000 repetitions of randomly selecting a point in a randomly selected image from the Van Hateren database of natural images (actually approximately half of the Van Hateren images are discarded due to poor image quality [5]). Around each selected point a neigbourhood of $N \times N$ pixels is selected and the diffusion is done on this subimage. The neighbourhood size N is defined such that a linear Gaussian filter at the center has three standard deviations inside the subimage. The diffusion time t gives the standard deviation by $2t = \sigma^2$ — we use 100 iterations with time step 0.2 so $N = 39$.

In order to assess the correspondence between the diffusion echo filter and the approximating filter we compare both the filters themselves and their effect on the image.

The filters can be perceived as distributions that govern the flow of mass in the image. Thereby a natural measure of difference is the Kullback-Leibler divergence (also known as the relative entropy) [7].

When measuring the difference in the effect of the diffusion on the images we measure the resulting intensity difference. The local intensity difference is the difference between intensities in the point of interest of the diffused image and the intensity resulting from convolving the original image with the approximating filter in the center of the subimage.

The figures 3 and 4 display the Kullback-Leibler divergence and the local intensity difference respectively with histograms over the 10,000 samples. Each subfigure consists of the histogram and a smaller figure with the same data but view different scaling and bin distribution.

All the large plots in the subfigures have the same scaling and the y-axis (the counts) has been scaled logarithmic. The smaller inserted are scaled according to the range of the data for that specific histogram and with linear y-axis.

In Figure 3 the mean of the histrogram gives us the mean Kullback-Leibler divergence between the approximation and the diffusion process. Pairs of mean KL divergence for (affine, isotropic) approximations for linear, RPM, and EED are respectively (0.0006, 0.0006), (0.0542, 0.0663), and (0.2898, 0.3566).

First, we see that linear and Perona-Malik diffusion can be approximated quite well. In the linear case that is trivial. If the linear diffusion process was a perfect approximation to convolving with a Gaussian the difference would be zero but due to the numeric limitation in discretisation that is not the case. The KL divergence of 0.0006 can been interpreted as the level of precision. In the case of edge enhancing diffusion the difference is quite high but as can be seen from the histogram a lot of cases can be approximated well but some approximations are very poor which on average gives a large difference.

Secondly, the numbers offer an ordering of the diffusion processes with EED as having behavior furthest away from linear. As expected the aggressive aniso-tropic EED scheme is more non-Gaussian than Perona-Malik.

Finally the affine approximations are evidently significantly better than the isotropic.

In figure 4 the local intensity difference are presented, hence this plot assesses the actual outcome of applying the approximations compared to the original diffusion. The local intensity differences should be compared to the range of the image (which is zero to one due to normalization).

Due to the symmetry in the formulation a mean of zero is expected and confirmed within the precision.

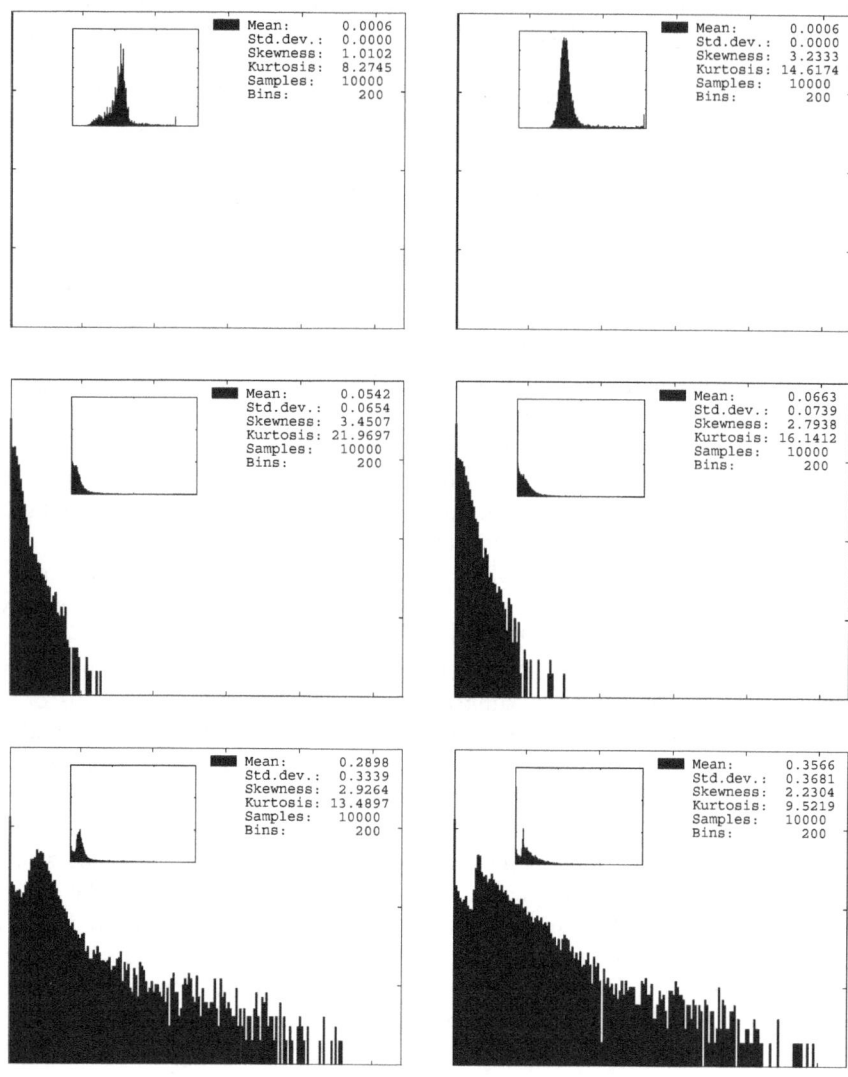

Fig. 3. Measuring Kullback-Leibler divergence between diffusion echo filters and approximations. The columns show the affine and isotropic approximations respectively. The rows show Linear, RPM and EED. The small plots show same data — with different axis, bins and scaling. Note that the large plots are scaled logarithmically.

Pairs of standard deviations for (affine, isotropic) approximations for linear, RPM, and EED are respectively (0.0002, 0.0002), (0.0067, 0.0096), and (0.0087, 0.0141) This supports the trend established in figure 3 which again indicates that the good approximations in the information theoretical sense actually gives good approximations of the diffusion processes.

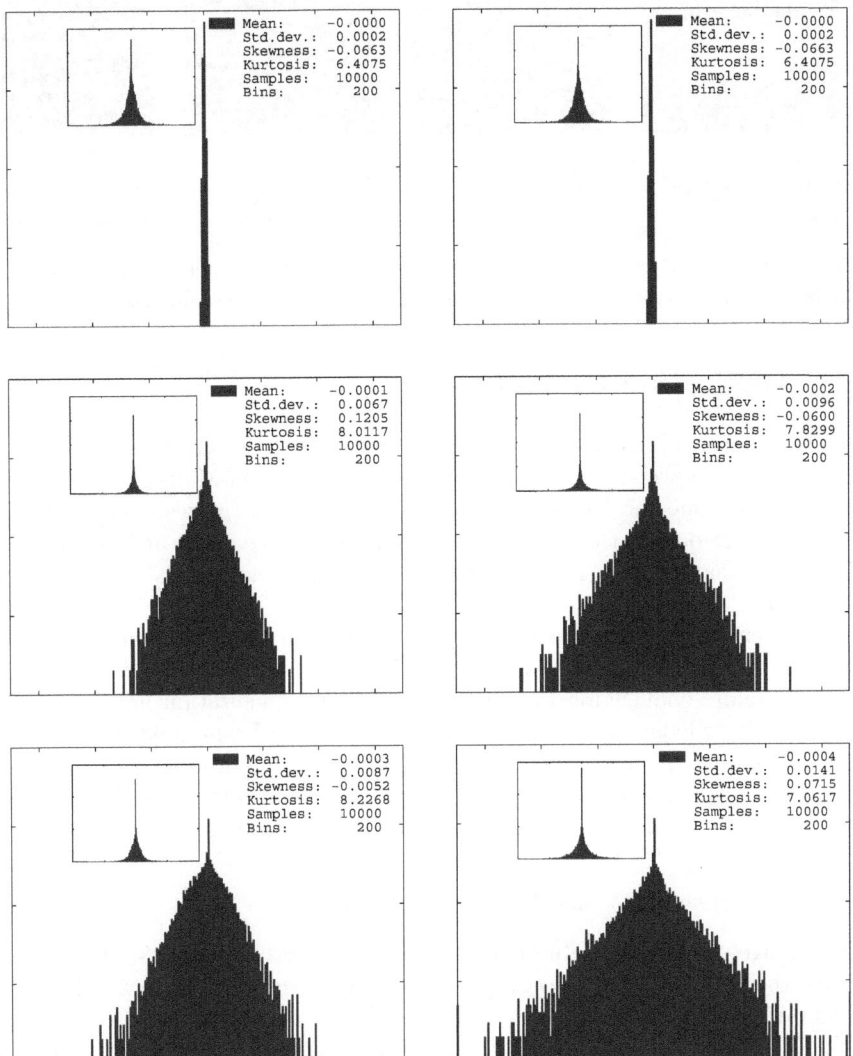

Fig. 4. Comparing local intensity difference between the diffusion process and the approximated diffusion process. Like figure 3 the columns are isotropic and affine approximations and the rows are Linear, RPM and EED.

The standard deviations again support that the affine approximations give significantly better results than the isotropic as expected.

It should be noted that the evaluation above is specific to the chosen diffusion schemes and the choice of parameters for these. Especially the choice of diffusion time could influence the results.

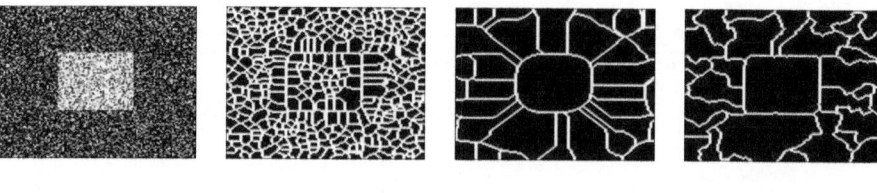

| Original example image | Watersheds at localisation scale | Watersheds at detection scale | Detection linked to localisation scale |

Fig. 5. Linking of watershed segments across scale. The segments at localisation scale is linked down to the localisation scale and thereby get fine scale precision.

4   Application Evaluation

The evaluation on natural images above showed that the approximations perform quite well — especially for the less non-Gaussian Perona-Malik scheme. However, the lack of a specific task makes the interpretation of the results slightly vague. As a counter-part to that we evaluate the performance of the approximated non-linear diffusion schemes in an interactive multi-scale watershed segmentation method.

The chosen task is segmentation of white matter tissue from brain scans. This task is chosen since the geometry of white matter tissue is quite complicated and therefore challenging for a non-committed segmentation method that uses no prior knowledge on the intensity distributions of brain matter.

The scans and corresponding ground truth segmentations are obtained from the *BrainWeb* site [1] (we use 9 slices with 10 slices between each from a simulated T1 MR brain scan, intensity non-uniformity level 20%, noise level 9%).

4.1   Multi-scale Watershed Segmentation

For images with reasonable contrast across object boundaries, the use of the watershed transformation on a gradient magnitude image is a well-known segmentation method.

The structures that are outlined by this partitioning are defined with respect to the scale at which the gradient is calculated. Different scales are therefore needed to locate objects of different sizes.

Linking of the segments across scale combines the simplification at the detection scale with the fine scale precision at the localisation scale (see figure 5). The segmentation method presented in [9] uses these localised segments as building blocks for the segmentation. The user can shift the detection scale and thereby select building blocks appropriate for sculpting the desired objects.

4.2   Non-linear Diffusion in MSWS

The original MSWS (multi-scale watershed segmentation) method relies on linear Gaussian scale-space to simplify the image. This simplification determines

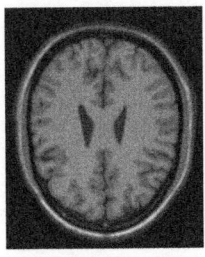 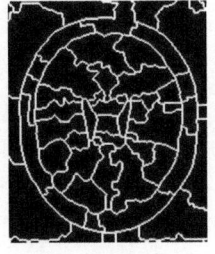 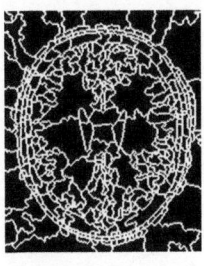

| Slice from BrainWeb scan | Groundtruth white matter | Segments from linear diffusion | Segments from GAN diffusion |

Fig. 6. Linear diffusion makes the building blocks merge across brain structures before reasonably large segments are formed. Therefore smaller building blocks must be selected in order to segment the white matter tissue. The use of non-linear diffusion allows the building blocks to grow within the tissue boundaries. The practical implication is that where 18 action are required to select 70% of the white matter using building block created using linear diffusion, only 5 are necessary with GAN diffusion. For 80% the action counts are 39 and 11, respectively.

how the watershed segments group into gradually larger building blocks corresponding to image structures at a given scale. In [2] the use of non-linear diffusion in MSWS was explored. Figure 6 illustrates how the building blocks are better suited to the application at hand.

In [2] the use of non-linear diffusion was also evaluated. The evaluation is based on a count of the minimal number of selections and deselections of building blocks in the segmentation. The parameters for the non-linear schemes were determined such that this count was minimized.

The simplest way to present the evaluation results is to normalize the performance (of the building blocks resulting from a non-linear scheme) with respect to the performance of linear diffusion. A good descriptor is then the ratio of actions required compared to linear diffusion. Table 1 shows that for each action used on building blocks resulting from linear diffusion, it is on average only necessary to use 0.42 or 0.32 when the building blocks resulted from Perona-Malik or GAN (where the optimal parameters give very little anisotropy, 0.06, and relatively low aggressiveness at 1.2).

We have chosen to disregard the edge enhancing diffusion scheme, EED, for this evaluation since it does not perform very well [2], and therefore is less interesting to approximate. The parameters for GAN given above produce a diffusion scheme that intuitively is in-between RPM and EED with a qualitative behaviour closer to RPM.

4.3 Approximating Non-linear Diffusion in MSWS

In section 2.1 we presented how to approximate the non-linear diffusion scheme using local isotropic and affine Gaussian filters. The approximation method can

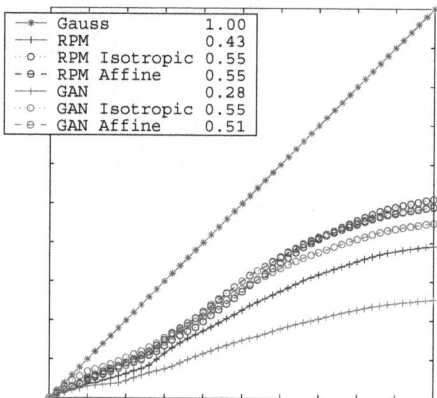

Fig. 7. The diffusion schemes are evaluated on the image in figure 6 on the number of actions needed to reach a given segmentation quality where the performance of linear diffusion building blocks is used as normalization. The graph shows the actions needed to reach the quality reached by a given number of actions using linear diffusion.

be used directly to generate filters approximating the non-linear diffusion in the segmentation method. For each scale we approximate the diffusion using the mean and variances recorded for each pixel. The approximating Gaussian filters are then applied and the gradient magnitude calculated. From there the MSWS method is exactly as described above.

The performance of the approximating filters can then be evaluated in the same manner as the non-linear schemes were evaluated in [2]. Figure 7 shows the performance of the approximating filters on the image in figure 6.

Table 1 shows how the filters perform on average on the entire test set. As stated above, for each action used on building blocks resulting from linear diffusion, it is on average only necessary to use 0.42 or 0.32 when the building blocks resulted from Perona-Malik or GAN. The approximating schemes perform significantly worse than the original non-linear schemes (they use between 27% and 73% more actions) but still maintain the main part of the advantage compared to linear diffusion (the ratios are between 0.50 and 0.65). Also worth noticing is that the affine approximations perform significantly better than the isotropic.

As a rule of thumb, it can be stated that for each 6 actions necessary to reach a given quality using linear diffusion, then only 2 actions are required with GAN diffusion, and 3 actions when the approximations of GAN are used.

4.4 Multi-scale Scale Selection

In the following we investigate the use of approximating filters in multi-scale watershed segmentation a bit further. Due to the simpler parameterisation of isotropic Gaussian filters we restrict ourselves to those for the analysis.

Table 1. The performance of the approximated non-linear diffusion scheme evaluated by the usability of the resulting multi-scale watershed segmentation building blocks. All schemes are compared using the performance of linear diffusion building blocks as yardstick in the first column. In the last column the approximating schemes are compared to the schemes they are approximating — so the approximating schemes use between 27% and 73% more actions than ideally.

Scheme	Ratio	Std. Dev.	Ratio compared to approx. scheme
Linear Gaussian	1.00	0.00	
Regularized Perona-Malik	0.42	0.15	
RPM approximated Isotropic	0.65	0.14	1.55
RPM approximated Affine	0.53	0.13	1.27
Generalized Anisotropic Nonlinear	0.32	0.07	
GAN approximated Isotropic	0.56	0.08	1.73
GAN approximated Affine	0.50	0.10	1.55

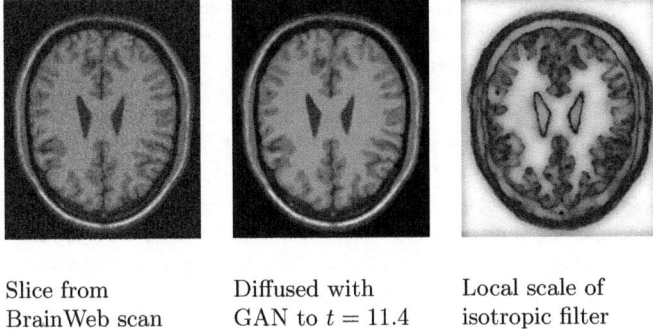

Slice from	Diffused with	Local scale of
BrainWeb scan	GAN to $t = 11.4$	isotropic filter

Fig. 8. The local scale as determined by the approximating isotropic filter. In areas far from a edge (as defined by the parameters for GAN) full diffusion is allowed. Around edges the diffusion is restricted, thus giving a lower local scale for the diffusion.

Implicit Scale Selection

At a given scale, in a given pixel the isotropic approximation filter implicitly determines a local scale. This local scale is illustrated in figure 8.

Implicit Multi-scale Scale Selection

In a standard multi-scale setting based on linear diffusion, each scale level is a hyper-plane in scale-space with a constant scale value. For the approximating isotropic filters, the local scale is selected at each point — thereby the scale levels become hyper-surfaces. These surfaces are bounded upwards in scale by the maximal scale the non-linear diffusion can reach with the given diffusion time (e. g. GAN diffusion becomes linear diffusion when used on a constant starting image — it then reaches this maximal scale in all points). At points where the diffusion is restricted (e.g. due to edges in the image) the hyper-surface will drop down to lower scales.

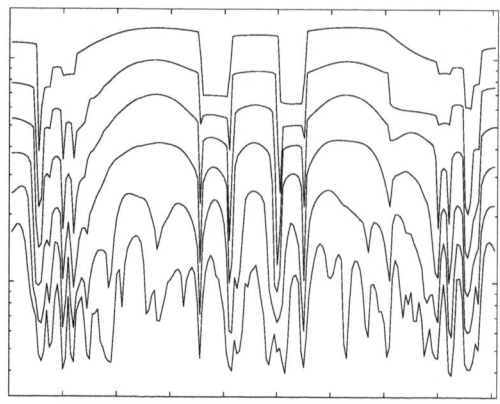

Fig. 9. This shows the local scale selected on a line across the ventricles in the image in figure 6 for 6 scale levels. The crossings of the ventricles are evident: at low scale levels we see drops in local scale at each boundary — at high scale levels there is a single nose-dive across the entire ridge.

This local scale selection is illustrated in the multi-scale setting in figure 9. It is this behaviour that a scale selection method would have to approximate in order to implement the approximating diffusion schemes directly.

Note that the scale level surfaces does not intersect. The local scale is monotonically increasing as a function of the diffusion time — provided the diffusion scheme meet a suitable causality principle.

5   Conclusion

In this paper we evaluate the feasibility of approximating non-linear diffusion schemes with isotropic and affine Gaussian filters. The basis are the local diffusion filters from the non-linear schemes that are extracted from the diffusion echo. These filters are compared to the approximating Gaussian filters.

The approximating filters are evaluated on natural images. Both the measures that compare the filters directly (Kullback-Leibler divergence) and the measure that compare the resulting diffused images (intensity difference) show that especially Perona-Malik diffusion can be approximated quite well. Furthermore, the affine approximation shows significantly better performance than the simpler isotropic Gaussian filters.

Secondly, the approximating filters are evaluated on their performance in a multi-scale segmentation method. This evaluation offers more concrete results due to the specific task. The results show that for each 6 actions necessary to reach a given segmentation quality using linear diffusion, then only 2 actions are required with GAN diffusion, and 3 actions when the approximations of GAN are used. While the approximations perform significantly worse than the original non-linear scheme, the major part of the advantage compared to linear diffusion is retained. Again, this evaluation showed that the affine filters approximate the non-linear schemes significantly better.

It should be noted that the approximating processes have not been optimised for solving the task but optimised to mimic the diffusion. Hence it is possible that another choice from the same class of filters would give a better performance in the segmentation task.

So, do we need the original non-linear diffusion schemes? The somewhat predictable answer is: If optimal performance is needed then yes — but if a relatively small decrease in performance is acceptable then no.

The obvious and necessary direction for future work is to establish methods for determining the parameters for the approximating filters directly from the desired diffusion process parameters and the local image structure.

References

1. D.L. Collins, A.P. Zijdenbos, V. Kollokian, J.G. Sled, N.J. Kabani, C.J. Holmes, & A.C. Evans. *Design and Construction of a Realistic Digital Brain Phantom.* IEEE Transactions on Medical Imaging, 17(3):463–468, June 1998. http://www.bic.mni.mcgill.ca/brainweb/.
2. Erik Dam & Mads Nielsen. *Non-Linear Diffusion for Interactive Multi-scale Watershed Segmentation.* In *Medical Image Computing and Computer-Assisted Intervention — MICCAI 2000,* volume 1935 of Lecture Notes in Computer Science, pages 216—225. Springer, October 2000.
3. Erik Dam & Mads Nielsen. *Exploring Non-Linear Diffusion: The Diffusion Echo.* In Michael Kerckhove, editor, *Scale-Space Theories in Computer Vision,* Lecture Notes in Computer Science. Springer, 2001.
4. G. Gomez, J.L. Marroquin, & L.E. Sucar. *Probabilistic Estimation of Local Scale.* In *Proc. of the ICPR,* 2000.
5. Griffin, Lillholm and Nielsen *Natural Image Profiles are most likely to be Step Edges.* Vision Research, submitted.
6. E.T.Jaynes. *Probability Theory: The Logic of Science.* http://omega.albany.edu:8008/JaynesBook.
7. S. Kullback & R. A. Leibler. *On Information Theory and Sufficiency.* Annals of Mathematical Statistics, Volume 22, 1951.
8. Tony Lindeberg. *Scale-Space Theory in Computer Vision.* Kluwer Academic Publishers, 1994.
9. Ole Fogh Olsen. *Multi-Scale Watershed Segmentation.* In Jon Sporring, Mads Nielsen, Luc Florack, & Peter Johansen, editors, *Gaussian Scale-Space Theory,* pages 191–200. Kluwer, 1997.
10. Pietro Perona and Jitendra Malik. Scale-space and edge detection using anisotropic diffusion. *IEEE PAMI,* 12(7):629 – 639, July 1990.
11. Guillermo Sapiro. *Geometric Partial Differential Equations and Image Analysis.* Cambridge, 2001.
12. Bart M. ter Haar Romeny, editor. *Geometry-Driven Diffusion in Computer Vision.* Kluwer Academic Publishers, 1994.
13. van Hateren and van der Schaaf. *Independent component filters of natural images compared with simple cells in primary visual cortex.* Proceedings of the Royal Society of London Series B - Biological Sciences, 265 (1394), 359-366, 1998.
14. Joachim Weickert. *Anisotropic Diffusion in Image Processing.* B. G. Teubner, Stuttgart, 1998.

A Generalized Discrete Scale-Space Formulation for 2-D and 3-D Signals

Ji-Young Lim and H. Siegfried Stiehl

Universität Hamburg, Fachbereich Informatik, Arbeitsbereich Kognitive Systeme
Vogt-Kölln-Str. 30, 22527 Hamburg, Germany
{lim,stiehl}@informatik.uni-hamburg.de
http://kogs-www.informatik.uni-hamburg.de/~lim

Abstract. This paper addresses the issue of a higher dimensional discrete scale-space (DSS) formulation. The continuous linear scale-space theory provides a unique framework for visual front-end processes. In practice, a higher dimensional DSS formulation is necessary since higher dimensional discrete signals must be dealt with. In this paper, first we examine the approximation fidelity of the commonly used sampled Gaussian. Second, we propose a generalized DSS formulation for 2-D and 3-D signals. The DSS theory has been presented at first by Lindeberg. While his 1-D DSS formulation is complete, the formulation as related to the extension to higher dimensions has not been fully derived. Furthermore, we investigate the properties of our derived DSS kernels and present the results of a validation study with respect to both smoothing and differentiation performance.

1 Introduction

It is theoretically proven ([1], [15]) that the isotropic Gaussian kernel is the unique kernel to generate the linear scale-space for continuous signals. Furthermore, the Gaussian is the only real-valued convolution kernel which gives the minimum uncertainty of the bandwidth-duration product ([2]) and it satisfies the necessary conditions required for being a lowpass filter. However, given the fact that the Gaussian kernel is defined in the continuous and infinite spatial domain, in practice we have to cope with bounded discrete signals and consequently a discrete Gaussian with compact support is required. A sampled Gaussian (SG) kernel is commonly used in practice, where the problem lies in the accuracy and validity of a SG kernel approximating the continuous Gaussian kernel. Two limitations of the Gaussian kernel were remarked in [13]; i) information loss caused by the unavoidable Gaussian truncation and ii) the prohibitive processing time due to the mask size. Also, it was shown in [5] that there exists a trade-off scale of the Gaussian kernel below which frequency filtering in the Fourier domain yields more accurate results than spatial filtering at the accompanying cost of computational load. In Sect. 2, we analyze the problems behind the SG kernel used as a convolution kernel, where we consider how to measure the fidelity of the approximation of the SG kernel with respect to the continuous Gaussian.

L.D. Griffin and M. Lillholm (Eds.): Scale-Space 2003, LNCS 2695, pp. 132–147, 2003.

The DSS theory presented at first by Lindeberg [11] is closely linked to the continuous scale-space theory through the discretization of the linear diffusion equation. In his work, the 1-D DSS formulation is well derived and complete, whereas the proposed higher dimensional DSS formulation has left open important questions. Motivated by this, in Sect. 3 we propose a generalized higher dimensional DSS formulation through a clear theoretical derivation that improves upon Lindeberg's higher dimensional DSS formulation. We investigate the properties of our derived DSS kernels in Sect. 4. Moreover, Sect. 5 presents the results of a validation study of the DSS kernel through which we analyze its performance with respect to both smoothing and differentiation.

2   Analysis of the Sampled Gaussian Kernel

According to the sampling theorem (see e.g. [2], [3]), it is theoretically possible to recover the full range of original function values with full accuracy given the condition that the function is "band-limited". Fig. 1 illustrates the sampling process of the Gaussian in the frequency domain: The spectrum of the Fourier transformed continuous Gaussian kernel ($F_G(\omega; t)$; t is the scale parameter) becomes replicated by sampling in the spatial domain through the Shah function, which corresponds to $F_G^{\#}(\omega; t)$. Since the Gaussian is not perfectly band-limited, when its spectrum is repeated, high-frequency components are overlapping. This effect is the so-called *aliasing*. Owing to the aliasing effect, the contribution of high-frequency components is superimposed on low-frequency components. In Fig. 1(c), the solid line results from the aliasing effect. Using the rectangle function for windowing, one aperiodic spectrum ($\hat{F}_G(\omega; t)$) can be cut off, and we call the filled area of both lobes of $\hat{F}_G(\omega; t)$ in Fig. 1(d) the *high-frequency tail*.

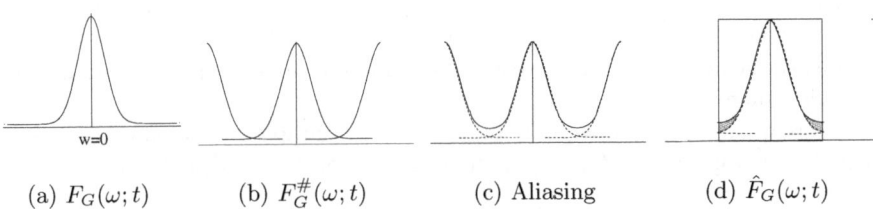

(a) $F_G(\omega; t)$ (b) $F_G^{\#}(\omega; t)$ (c) Aliasing   (d) $\hat{F}_G(\omega; t)$

Fig. 1. Sampling in the frequency domain: (a) The Fourier transformed continuous Gaussian, (b) replication occurs in the frequency domain by the sampling in the spatial domain, (c) high-frequency components influence on low frequencies around $|\omega| = \frac{\omega_0}{2}$, and (d) the cutoff spectrum contains the high-frequency tail caused by the aliasing.

For a given sampling period, we can derive numerically $\hat{F}_G(\omega; t)$, from which the amount of the high-frequency tail can be calculated. We fix here the sampling period to one (i.e. $T = 1$, or $\omega_0 = 2\pi$) for the reason that the input signal with which the Gaussian kernel is convolved is in general the intensity function of a digital image with an inter-pixel distance of 1. Since

$$\int_{-\infty}^{\infty} \hat{F}_G(\omega; t)\, d\omega = \int_{-\infty}^{\infty} F_G(\omega; t)\, d\omega, \tag{1}$$

holds (see [8, Eq. 3, Sect. 4] for the detailed derivation), which means that the amount of the high-frequency tail corresponds to the total amount of the contribution of high-frequency components influenced on low-frequency components owing to the aliasing effect, one can conclude that the smaller the amount of the high frequency tail is the better the sampling result is. Provided that the amount of the high-frequency tail is zero (i.e. no aliasing occurs), for example, one can fully reconstruct the continuous Gaussian from the sampled Gaussian. Accordingly, based on the fact derived in (1), the amount of the high-frequency tail denoted by $F_{G_{\mathrm{HFT}}}$ is calculated as

$$F_{G_{\mathrm{HFT}}} = 2\int_{\pi}^{\infty} F_G(\omega; t)\, d\omega = \sqrt{\frac{2\pi}{t}} \left(1 - \mathrm{erf}\left(\pi\sqrt{\frac{t}{2}}\right)\right),$$

where $\mathrm{erf}(x) = \frac{2}{\sqrt{\pi}} \int_0^x e^{\xi^2}\, d\xi$. $F_{G_{\mathrm{HFT}}}$ is a monotonously decreasing function of the scale parameter t. Since $F_{G_{\mathrm{HFT}}}$ is expressed by $\mathrm{erf}(x)$, an approximated value can be given only.

Sampling the higher dimensional isotropic Gaussian kernel is analogous to that of the 1-D Gaussian based on the separability property. Therefore, we refrain from describing here in detail the sampling process of higher dimensional Gaussian kernels (see [8, Sect. 4.3] for details). For a given sampling period $T = 1$, the amount of the high-frequency tail in both 2-D and 3-D can be calculated as

$$^2F_{G_{\mathrm{HFT}}} = (F_{G_{\mathrm{HFT}}})^2 \quad \text{and} \quad ^3F_{G_{\mathrm{HFT}}} = (F_{G_{\mathrm{HFT}}})^3,$$

which can be further generalized for an N-D Gaussian kernel to $(F_{G_{\mathrm{HFT}}})^N$. Fig. 2 depicts the high-frequency tail of the sampled Gaussian kernel in 1-D, 2-D, and 3-D in dependence on the scale parameter, from which one can easily recognize that as the scale parameter decreases the amount of high-frequency tail increases in each dimension.

Consequently, it can be generally said that a sampled Gaussian with a small scale is not appropriate for approximating the continuous Gaussian.

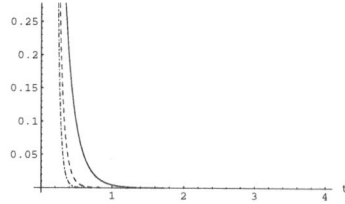

Fig. 2. The high-frequency tail of the sampled Gaussian kernel in 1-D, 2-D, and 3-D ($T = 1$). The solid curve, the dashed curve, and the dotted curve represent $F_{G_{\mathrm{HFT}}}$, $^2F_{G_{\mathrm{HFT}}}$, and $^3F_{G_{\mathrm{HFT}}}$, respectively.

3   DSS Formulation

Several open problems of Lindeberg's higher dimensional DSS formulation can be identified: i) The rationale behind the setting of coefficients for the point operator (i.e. $\frac{1}{2}$ for $\nabla^2_{\times 2}$ in 2-D as well as $\frac{1}{4}$ for both ∇^2_{+3} and $\nabla^2_{\times 3}$ in 3-D) has not been explained, ii) in the 2-D DSS formulation the answer to the question of how to determine the definite parameter γ for solving the semi-discretized diffusion equation was left open, and iii) the 3-D DSS formulation was not considered at all, which, however, is necessary e.g. for 3-D medical image analysis. Therefore, it is indispensable to contribute a generalized higher dimensional DSS formulation.

3.1   Preliminaries

The Neighborhood Connectivity. For a given point $p \in \mathbb{Z}^k$, we define its neighborhood as

$$^kN(p) = \{\xi \in \mathbb{Z}^k : (\|p - \xi\| \leq \sqrt{k}) \wedge (\xi \neq p)\}, \tag{2}$$

for $k \geq 1$. The neighborhood is classified according to the distance between a given point and its neighbors. That is, $^2N(p)$ consists of $^2N_1(p)$ and $^2N_{\sqrt{2}}(p)$, $^3N(p)$ comprises $^3N_1(p)$, $^3N_{\sqrt{2}}(p)$, and $^3N_{\sqrt{3}}(p)$, and $^NN(p)$ has $^NN_1(p)$, $^NN_{\sqrt{2}}(p)$, $\ldots$, $^NN_{\sqrt{N}}(p)$ (the number of elements of $^NN_{\sqrt{k}}(p)$ is equivalent to $\binom{N}{k}2^k$).

The Laplacian of the Higher Dimensional DSS Kernel. On the basis of numerical differentiation, the second-order derivative of $f(x)$ is approximated by

$$f(x + 1) - 2f(x) + f(x - 1) \approx \frac{\partial^2 f(x)}{\partial x^2}.$$

According to (2), for a given discrete signal $f : \mathbb{Z}^N \to \mathbb{R}$ its scale-space representation generated by the convolution with $T : \mathbb{Z}^N \times \mathbb{R}_+ \to \mathbb{R}$ (which we call the *higher dimensional DSS kernel*) satisfies in 2-D

$$\partial_t L(x, y; t) = a_1 \nabla^2_{2N_1} L(x, y; t) + a_2 \nabla^2_{2N_{\sqrt{2}}} L(x, y; t), \tag{3}$$

and in 3-D

$$\partial_t L(x, y, z; t) = a_1 \nabla^2_{3N_1} L(x, y, z; t) + a_2 \nabla^2_{3N_{\sqrt{2}}} L(x, y, z; t) + a_3 \nabla^2_{3N_{\sqrt{3}}} L(x, y, z; t), \tag{4}$$

for some constants $a_1 \geq 0$, $a_2 \geq 0$, and $a_3 \geq 0$.

3.2   2-D DSS Formulation

Equation (3) can be expressed as a normalized form given by

$$\partial_t L(x, y; t) = \frac{1}{2} \nabla^2 L(x, y; t)$$
$$= \frac{1}{2} \left((1 - \gamma) \nabla^2_{2N_1} L(x, y; t) + \lambda \gamma \nabla^2_{2N_{\sqrt{2}}} L(x, y; t) \right) \tag{5}$$

for $\gamma \in [0, 1]$ and $\lambda \in (0, 1)$. Note that in [9, Sect. 3.2], for the definition of the Laplacian of $^2N_{\sqrt{2}}$, we intended to steer the ratio of rotational symmetry between the Laplacians of $^2N_1$ and $^2N_{\sqrt{2}}$ only through parameter γ, and thus we avoided setting an additional ambiguous (or unexplained) coefficient. However, this does not give rise to a correct result, i.e. the 2-D DSS kernel does not satisfy the semi-group property. On the other hand, one cannot find any proof that the unexplained coefficient in the definition of $\nabla^2_{\times^2}$ given by Lindeberg is proper. Therefore, in a more generalized way, we set this coefficient as a variable λ and determine its proper value through the following theoretical derivation.

Equation (5) can be further discretized with the scale step $\triangle t$ as

$$L^{k+1}_{x,y} = L^k_{x,y} + \triangle t\left(\partial_t L^k_{x,y}\right) = L^k_{x,y} + \triangle t\frac{1}{2}\left((1-\gamma)\nabla^2_{2N_1}L + \lambda\gamma\nabla^2_{2N_{\sqrt{2}}}L\right),$$

where subscripts x and y denote the spatial coordinates and superscript k represents the iteration index. This discretization corresponds to the iteration with the 2-D discrete iteration kernel given by

$$T_{\triangle t} = \begin{pmatrix} \frac{\lambda\triangle t}{2}\gamma & \frac{\triangle t}{2}(1-\gamma) & \frac{\lambda\triangle t}{2}\gamma \\ \frac{\triangle t}{2}(1-\gamma) & 1-2\triangle t(1-\gamma+\lambda\gamma) & \frac{\triangle t}{2}(1-\gamma) \\ \frac{\lambda\triangle t}{2}\gamma & \frac{\triangle t}{2}(1-\gamma) & \frac{\lambda\triangle t}{2}\gamma \end{pmatrix}. \tag{6}$$

The generating function describing one iteration given in (6) corresponds to

$$^2\varphi_{step}(z, \chi) = (1 - 2\triangle t(1-\gamma+\lambda\gamma)) + \frac{\triangle t(1-\gamma)}{2}A + \frac{\lambda\triangle t\gamma}{2}B,$$

where

$$A = z^{-1} + z + \chi^{-1} + \chi \quad \text{and} \quad B = z^{-1}\chi^{-1} + z^{-1}\chi + z\chi^{-1} + z\chi,$$

and we obtain the generating function describing the composed transformation $(\triangle t = \frac{t}{n})$ as

$$^2\varphi_{composed,n}(z, \chi) = \left(1 + \frac{t}{n}\left(-2(1-\gamma+\lambda\gamma) + \frac{(1-\gamma)}{2}A + \frac{\lambda\gamma}{2}B\right)\right)^n.$$

Based on the fact that $\lim_{n\to\infty}(1+\frac{\alpha_n}{n})^n = e^\alpha$ if $\lim_{n\to\infty}\alpha_n = \alpha$, the generating function of the kernel describing the transformation from the original signal to the representation at a certain scale t is given by

$$^2\varphi_T(z, \chi) = \sum_{(m,n)\in\mathbb{Z}^2} T(m, n; t)z^m\chi^n = e^{t\left(-2(1-\gamma+\lambda\gamma)+\frac{(1-\gamma)}{2}A+\frac{\lambda\gamma}{2}B\right)}.$$

Its Fourier transform is derived by replacing the complex variables z and χ with e^{-iu} and e^{-iv} as

$$\mathcal{F}\left(^2\varphi_T(z, \chi)\right) = {}^2\psi_T(e^{-iu}, e^{-iv}) = {}^2\psi_T(\cos u - i\sin u, \cos v - i\sin v)$$
$$= e^{t(-2(1-\gamma+\lambda\gamma)+(1-\gamma)(\cos u+\cos v)+\lambda\gamma 2\cos u\cos v)},$$

which can be transformed into polar coordinates given a fixed value of radius r and an angular variable ϕ such that $u = r\cos\phi$ and $v = r\sin\phi$. It follows

$$^2\psi_T(r,\phi) = e^{(t\cdot k(r,\phi))},$$

where

$$
\begin{aligned}
k(r,\phi) = &-2(1 - \gamma + \lambda\gamma) + (1 - \gamma)(\cos(r\cos\phi) + \cos(r\sin\phi)) \\
&+ \lambda\gamma 2\cos(r\cos\phi)\cos(r\sin\phi).
\end{aligned}
\tag{7}
$$

Now, we determine the value of γ of $k(r,\phi)$ in (7) which gives the smallest angular variation of ϕ for a fixed value r. For examining the ϕ-dependency of γ from $k(r,\phi)$, we expand the MacLaurin series of $k(r,\phi)$ with respect to r

$$
k(r,\phi) = -\frac{1 - \gamma + 2\lambda\gamma}{2}r^2 + \frac{3 + 3\gamma(4\lambda - 1) + (1 - \gamma - 4\lambda\gamma)\cos 4\phi}{96}r^4 + O(r^6),
$$

where the smallest angular variation is achieved when $\gamma = \frac{1}{1+4\lambda}$. That is to say, $\gamma = \frac{1}{1+4\lambda}$ yields the least possible rotational asymmetry for the 2-D DSS kernel. In other words, maximal isotropy of T is guaranteed even for the discrete case. Substituting $\gamma = \frac{1}{1+4\lambda}$ for the 2-D iteration kernel of (6) yields

$$
T_{\Delta t} =
\begin{pmatrix}
\frac{\lambda}{2(1+4\lambda)}\Delta t & \frac{2\lambda}{1+4\lambda}\Delta t & \frac{\lambda}{2(1+4\lambda)}\Delta t \\
\frac{2\lambda}{1+4\lambda}\Delta t & 1 - \frac{10\lambda}{1+4\lambda}\Delta t & \frac{2\lambda}{1+4\lambda}\Delta t \\
\frac{\lambda}{2(1+4\lambda)}\Delta t & \frac{2\lambda}{1+4\lambda}\Delta t & \frac{\lambda}{2(1+4\lambda)}\Delta t
\end{pmatrix},
\tag{8}
$$

where $\Delta t > 0$. This iteration kernel is symmetric and normalized.

In order to constrain computational cost when the number of iterations increases, it would be favorable to apply a separable iteration kernel since the higher the dimension is the more efficient separable filters are ([6]). Therefore, we assume the 2-D iteration kernel given in (8) to be separable such that it should be constructed by convolution of the 1-D kernel with itself given by

$$
\begin{pmatrix} a & 1 - 2a & a \end{pmatrix} *
\begin{pmatrix} a \\ 1 - 2a \\ a \end{pmatrix} =
\begin{pmatrix}
\frac{\lambda}{2(1+4\lambda)}\Delta t & \frac{2\lambda}{1+4\lambda}\Delta t & \frac{\lambda}{2(1+4\lambda)}\Delta t \\
\frac{2\lambda}{1+4\lambda}\Delta t & 1 - \frac{10\lambda}{1+4\lambda}\Delta t & \frac{2\lambda}{1+4\lambda}\Delta t \\
\frac{\lambda}{2(1+4\lambda)}\Delta t & \frac{2\lambda}{1+4\lambda}\Delta t & \frac{\lambda}{2(1+4\lambda)}\Delta t
\end{pmatrix},
$$

for $0 < a \le \frac{1}{4}$, from which we obtain

$$
a = \frac{1}{6} \quad \text{and} \quad \Delta t = \frac{2}{9} + \frac{1}{18\lambda}.
$$

Besides, in order to satisfy the semi-group property, Δt should correspond to the variance of the 2-D discrete iteration kernel. This means that $\Delta t = 2a$ must hold. Finally, we have

$$
a = \frac{1}{6}, \quad \lambda = \frac{1}{2}, \quad \gamma = \frac{1}{3}, \quad \text{and} \quad \Delta t = \frac{1}{3}.
$$

It is noticeable that our theoretically derived value of λ is equal to that defined by Lindeberg [11], where, however, no formal explanation was given.

Consequently, the 2-D separable iteration kernel for the rotationally least asymmetric 2-D DSS kernel satisfying the semi-group property is given by

$$T_{\triangle t} = \begin{pmatrix} \frac{1}{6} & \frac{2}{3} & \frac{1}{6} \end{pmatrix} * \begin{pmatrix} \frac{1}{6} \\ \frac{2}{3} \\ \frac{1}{6} \end{pmatrix} = \begin{pmatrix} \frac{1}{36} & \frac{1}{9} & \frac{1}{36} \\ \frac{1}{9} & \frac{4}{9} & \frac{1}{9} \\ \frac{1}{36} & \frac{1}{9} & \frac{1}{36} \end{pmatrix}, \tag{9}$$

where $\triangle t = \frac{1}{3}$.

3.3   3-D DSS Formulation

Equation (4) can be expressed as a normalized form given by

$$\partial_t L = \frac{1}{2}\left((1-\gamma_1-\gamma_2)\nabla^2_{3N_1}L + \lambda_1\gamma_1\nabla^2_{3N_{\sqrt{2}}}L + \lambda_2\gamma_2\nabla^2_{3N_{\sqrt{3}}}L\right) \tag{10}$$

for $\gamma_1, \gamma_2 \in [0,1]$ and $\lambda_1, \lambda_2 \in (0,1)$. Following the line of thought of the 2-D case, in defining the Laplacians of $^3N_{\sqrt{2}}$ and of $^3N_{\sqrt{3}}$ in [9, Sect. 3.3], we intended to steer the ratio of rotational symmetry between the Laplacians of $^3N_1$, $^3N_{\sqrt{2}}$, and $^3N_{\sqrt{3}}$ only through parameters γ_1 and γ_2. Therefore, we did not define any a priori coefficients for the Laplacians of $^3N_{\sqrt{2}}$ and of $^3N_{\sqrt{3}}$, whereas Lindeberg set them both to $\frac{1}{4}$. However, our first account in [9, Sect. 3.3] did not give rise to a proper result (the semi-group property is not satisfied). On the other hand, it is neither evident nor proven why those parameters were set to $\frac{1}{4}$ by Lindeberg. As a consequence, we again approach this problem in a more generalized way by defining the coefficients of the Laplacians of $^3N_{\sqrt{2}}$ and of $^3N_{\sqrt{3}}$ as λ_1 and λ_2.

Equation (10) can be discretized with the scale step $\triangle t$ given by

$$L^{k+1}_{x,y,z} = L^k_{x,y,z} + \triangle t\left(\partial_t L^k_{x,y,z}\right)$$
$$= L^k_{x,y,z} + \triangle t\frac{1}{2}\left((1-\gamma_1-\gamma_2)\nabla^2_{3N_1}L + \lambda_1\gamma_1\nabla^2_{3N_{\sqrt{2}}}L + \lambda_2\gamma_2\nabla^2_{3N_{\sqrt{3}}}L\right),$$

where the parameters γ_1 and γ_2 play the role of preserving the rotational symmetry of the 3-D DSS kernel. Similarly to the 2-D case, based on the assumptions that i) the 3-D DSS kernel is rotationally least asymmetric, ii) the 3-D iteration kernel is separable, and iii) the 3-D DSS kernel satisfies the semi-group property, we determine the parameter values as follows (see [10, Sect. 3.2] for the detailed derivation):

$$\gamma_1 = \frac{4}{9}, \quad \gamma_2 = \frac{1}{9}, \quad \lambda_1 = \frac{1}{4}, \quad \lambda_2 = \frac{1}{4}, \quad a = \frac{1}{6}, \quad \text{and} \quad \triangle t = \frac{1}{3}.$$

As a consequence, the 3-D separable iteration kernel for the rotationally least asymmetric 3-D DSS kernel satisfying the semi-group property is given by

$$T_{\triangle t} = \begin{pmatrix} \frac{1}{6} & \frac{2}{3} & \frac{1}{6} \end{pmatrix}_x * \begin{pmatrix} \frac{1}{6} & \frac{2}{3} & \frac{1}{6} \end{pmatrix}_y * \begin{pmatrix} \frac{1}{6} & \frac{2}{3} & \frac{1}{6} \end{pmatrix}_z = \begin{pmatrix} \frac{1}{216} & \frac{1}{54} & \frac{1}{216} \\ \frac{1}{54} & \frac{1}{27} & \frac{1}{54} \\ \frac{1}{216} & \frac{1}{54} & \frac{1}{216} \end{pmatrix}_{z\pm1}, \begin{pmatrix} \frac{1}{54} & \frac{2}{27} & \frac{1}{54} \\ \frac{2}{27} & \frac{8}{27} & \frac{2}{27} \\ \frac{1}{54} & \frac{2}{27} & \frac{1}{54} \end{pmatrix}_z,$$

where $\triangle t = \frac{1}{3}$.

4 Properties of the DSS Kernels

4.1 Smoothing Kernel

The 1-D DSS kernel is given by

$$T\left(x; \frac{k}{3}\right) = *^k \left(\begin{array}{ccc} \frac{1}{6} & \frac{2}{3} & \frac{1}{6} \end{array}\right), \tag{11}$$

where $*^k$ is denoted as k-times self-convolution and $\frac{k}{3}$ corresponds to the variance. The coefficients of the 1-D DSS kernel generated by self-convolution given in (11) can be easily calculated using the z-transform of the given DSS kernel based on the property that convolution of sequences corresponds to multiplication of their z-transforms, from which

$$T\left(x; \frac{k}{3}\right) \circ\!\!-\!\!\bullet \left(\frac{1}{6}z^{-1} + \frac{2}{3} + \frac{1}{6}z\right)^k$$

follows. The 1-D DSS kernels are normalized to 1 for any k, and their implementation is simple and fast.

The higher dimensional DSS kernel is separable. For example, the smallest 2-D DSS kernel is given by

$$T\left(x, y; \frac{1}{3}\right) = T\left(x; \frac{1}{3}\right) * T\left(y; \frac{1}{3}\right) = \left(\begin{array}{ccc} \frac{1}{6} & \frac{2}{3} & \frac{1}{6} \end{array}\right)_x * \left(\begin{array}{ccc} \frac{1}{6} & \frac{2}{3} & \frac{1}{6} \end{array}\right)_y,$$

the coefficients of which can be easily calculated using its z-transform

$$T\left(x; \frac{1}{3}\right) * T\left(y; \frac{1}{3}\right) \circ\!\!-\!\!\bullet \left(\frac{1}{6}z^{-1} + \frac{2}{3} + \frac{1}{6}z\right) \cdot \left(\frac{1}{6}\omega^{-1} + \frac{2}{3} + \frac{1}{6}\omega\right)$$

$$\bullet\!\!-\!\!\circ \left(\begin{array}{ccc} \frac{1}{36} & \frac{1}{9} & \frac{1}{36} \\ \frac{1}{9} & \frac{4}{9} & \frac{1}{9} \\ \frac{1}{36} & \frac{1}{9} & \frac{1}{36} \end{array}\right).$$

The higher dimensional DSS kernel with larger variance, analogously to 1-D, can be derived through self-convolution given by

$$T\left(x, y; \frac{k}{3}\right) = *^k T\left(x, y; \frac{1}{3}\right) \circ\!\!-\!\!\bullet \left(\frac{1}{6}z^{-1} + \frac{2}{3} + \frac{1}{6}z\right)^k \cdot \left(\frac{1}{6}\omega^{-1} + \frac{2}{3} + \frac{1}{6}\omega\right)^k,$$

where k denotes the number of self-convolution.

4.2 Differencing Kernel

In order to apply the DSS kernel to images for the purpose of feature extraction, it is necessary to derive derivative operators. In contrast to the continuous case in which any nth-order derivatives of the Gaussian can be defined at any scale, it is not as simple to define derivative operators in the discrete case. By introducing the terminology "differencing operator" denoted by $\triangle$, we here discriminate the discrete derivative from the continuous derivative.

Based on the principles of numerical differentiation, one can approximate the first-order derivative by the difference quotient, for which there exist two formulae according to the number of points involved in the differencing. One is the two-point difference formula denoted by $\triangle_{even}$

$$f_{\triangle_{even}}(x) = \frac{f(x) - f(x - h)}{h} = f(x) - f(x - 1) \quad (h = 1),$$

while the other is the three-point difference formula denoted by $\triangle_{odd}$

$$f_{\triangle_{odd}}(x) = \frac{f(x + h) - f(x - h)}{2h} = \frac{f(x + 1) - f(x - 1)}{2} \quad (h = 1).$$

Based on these two formulae, we thoroughly derive two types of the 1-D DSS first-order differencing operator using the z-transform. The z-transform of $\triangle_{even}$ is given by

$$f_{\triangle_{even}}(x) = f(x) - f(x - 1) \circ\!\!-\!\!\bullet F(z) \cdot (1 - z),$$

where $F(z)$ corresponds to the z-transform of $f(x)$. The DSS first-order differencing kernel follows through application of $\triangle_{even}$ as

$$T_{\triangle_{even}}\left(x; \frac{k}{3}\right) \circ\!\!-\!\!\bullet \left(\frac{1}{6}z^{-1} + \frac{2}{3} + \frac{1}{6}z\right)^k (1 - z) \bullet\!\!-\!\!\circ *^k \left\{\frac{1}{6} \frac{2}{6} \frac{1}{6}\right\} * \{1 \; -1\}.$$

Analogously to $\triangle_{even}$, the z-transform of $\triangle_{odd}$ is given by

$$f_{\triangle_{odd}}(x) = \frac{f(x + 1) - f(x - 1)}{2} \circ\!\!-\!\!\bullet F(z) \cdot \left(\frac{1}{2}z^{-1} - \frac{1}{2}z\right),$$

and the DSS first-order differencing kernel follows from application of $\triangle_{odd}$

$$T_{\triangle_{odd}}\left(x; \frac{k}{3}\right) \circ\!\!-\!\!\bullet \left(\frac{1}{6}z^{-1} + \frac{2}{3} + \frac{1}{6}z\right)^k \left(\frac{1}{2}z^{-1} - \frac{1}{2}z\right)$$
$$\bullet\!\!-\!\!\circ *^k \left\{\frac{1}{6} \frac{2}{3} \frac{1}{6}\right\} * \left\{\frac{1}{2} \; 0 \; \frac{-1}{2}\right\}.$$

For a given higher dimensional DSS kernel, the differencing kernel through application of $\triangle_{x_\alpha}$ is derived from

$$T_{\triangle_{x_\alpha}}(\overrightarrow{x}; \cdot) = T(x_1; \cdot) * T(x_2; \cdot) * \cdots * T_{\triangle}(x_\alpha; \cdot) * \cdots * T(x_N; \cdot).$$

For example, applying $\triangle_{even,x}$ to $T(x, y; \frac{1}{3})$ results in

$$T_{\triangle_{even,x}}\left(x, y; \frac{1}{3}\right) = \left(\frac{1}{6} \; \frac{1}{2} \; \frac{-1}{2} \; \frac{-1}{6}\right)_x * \left(\frac{1}{6} \; \frac{2}{3} \; \frac{1}{6}\right)_y = \begin{pmatrix} \frac{1}{36} & \frac{1}{12} & \frac{-1}{12} & \frac{-1}{36} \\ \frac{1}{9} & \frac{1}{3} & \frac{-1}{3} & \frac{-1}{9} \\ \frac{1}{36} & \frac{1}{12} & \frac{-1}{12} & \frac{-1}{36} \end{pmatrix},$$

while applying $\triangle_{odd,y}$ to $T(x, y; \frac{1}{3})$ yields

$$T_{\triangle_{odd,y}}\left(x,y;\frac{1}{3}\right) = \left(\tfrac{1}{6} \ \tfrac{2}{3} \ \tfrac{1}{6}\right)_x * \left(\tfrac{1}{12} \ \tfrac{1}{3} \ 0 \ \tfrac{-1}{3} \ \tfrac{-1}{12}\right)_y = \begin{pmatrix} \frac{1}{72} & \frac{1}{18} & \frac{1}{72} \\ \frac{1}{18} & \frac{2}{9} & \frac{1}{18} \\ 0 & 0 & 0 \\ \frac{-1}{18} & \frac{-2}{9} & \frac{-1}{18} \\ \frac{-1}{72} & \frac{-1}{18} & \frac{-1}{72} \end{pmatrix}.$$

Normalization. Let us assume that i) $f(x)$ is a scale-space kernel, ii) $f(x)$ should be sufficiently smooth such that n-th order derivatives can be taken, iii) $f(x)$ is normalized such that $\int f(x)\,dx = 1$, and additionally iv) $f(x)$ is essentially compact, meaning that the kernel and all of its derivatives vanish sufficiently fast when $|x|$ goes to infinity. Provided that these assumptions are satisfied, $f(x)$ simply follows by partial integration, i.e.

$$\int \frac{(-1)^n}{n!} x^n f^{(n)}(x)\,dx = 1,$$

where n corresponds to the order of the derivative (see [4]). According to this rule, in the case of the first-order derivative (i.e. $n = 1$),

$$\int -x f'(x)\,dx = 1 \tag{12}$$

must hold. It is evident that the $T_{\triangle_{even}}$ and $T_{\triangle_{odd}}$ satisfy the normalization requirement given in (12), and thus these two kernels are normalized DSS differencing kernels.

Variance. For a given $f : \mathbb{R} \to \mathbb{R}$, its variance derived by the second central moment (by assuming its existence) can be written as

$$\mathrm{Var}(f(x)) = \int_{-\infty}^{\infty} x^2 |f(x)|\,dx, \tag{13}$$

in order to measure a dispersion of $f(x)$. According to (13), the variance of (even- and odd-number-sized) $T_{\triangle}(x;t)$ can be calculated, from which one can find that $\mathrm{Var}(T_{\triangle_{even}}(x;t))$ nicely equals $\mathrm{Var}(T_{\triangle_{odd}}(x;t))$ for any t even though the two kernels are different in local support and shape.

Integration. For a given z-transformed DSS kernel $T(z)$ (i.e. $T(z) \bullet\!\!-\!\!\circ T(x)$), $T_{\triangle}(z)$ is derived from multiplying $T(z)$ with $\triangle_z$, i.e.

$$T_{\triangle}(z) \cdot \triangle_z^{-1} = T(z) \cdot \triangle_z \cdot \triangle_z^{-1} = T(z).$$

Introducing the symbol $\blacktriangle$ that stands for $\triangle^{-1}$, we denote $\blacktriangle$ as the "discrete integration operator" (due to the duality of differentiation and integration). Multiplication of $\blacktriangle_z$ corresponds to convolution of $\blacktriangle$ in the spatial domain. Since there exist two types of $\triangle_z$, there are correspondingly two types of $\blacktriangle_z$, i.e. $\blacktriangle_{even,z}$ and $\blacktriangle_{odd,z}$.

$\blacktriangle_{even,z}$ is given by

$$\blacktriangle_{\text{even},z} = (\triangle_{\text{even},z})^{-1} = (1-z)^{-1},$$

where the inverse z-transform of $\blacktriangle_{\text{even},z}$ is derived as

$$\blacktriangle_{\text{even},z} = \frac{1}{1-z} \bullet\!\!-\!\!\circ \mathcal{H}(x) = \{\cdots 0\ 0\ 0\ \underset{x=0}{1}\ 1\ 1\cdots\}.$$

Integration of the DSS kernel through application of $\blacktriangle_{\text{even}}$ yields

$$T_{\blacktriangle_{\text{even}}}\left(x;\frac{k}{3}\right) \circ\!\!-\!\!\bullet T\left(z;\frac{k}{3}\right)\cdot\blacktriangle_{\text{even},z},$$

where, e.g. in the case of $k=1$,

$$T_{\blacktriangle_{\text{even}}}\left(x;\frac{1}{3}\right) \circ\!\!-\!\!\bullet \frac{1}{6}\left(-1+\frac{1}{z}+\frac{6}{1-z}\right) \bullet\!\!-\!\!\circ -\frac{1}{6}\delta(x)+\frac{1}{6}\delta(x+1)+\mathcal{H}(x),$$

i.e. $T_{\blacktriangle_{\text{even}}}(x;\frac{1}{3}) = \{\cdots 0\ \frac{1}{6}\ \frac{5}{6}\ 1\ \cdots\}$. Analogously, $\blacktriangle_{\text{odd},z}$ is given by

$$\blacktriangle_{\text{odd},z} = (\triangle_{\text{odd},z})^{-1} = \left(\frac{1}{2}z^{-1}-\frac{1}{2}z\right)^{-1} = \frac{2z}{1-z^2},$$

where the inverse z-transform of $\blacktriangle_{\text{odd},z}$ is derived as

$$\blacktriangle_{\text{odd},z} = \left(\frac{1}{1-z}-\frac{1}{1+z}\right) \bullet\!\!-\!\!\circ \mathcal{H}(x)-(-1)^x\mathcal{H}(x)$$
$$= \{\cdots 0\ 0\ 0\ \underset{x=0}{0}\ 2\ 0\ 2\cdots\}.$$

Integration of the DSS kernel through application of $\blacktriangle_{\text{odd}}$ is given by

$$T_{\blacktriangle_{\text{odd}}}\left(x;\frac{k}{3}\right) \circ\!\!-\!\!\bullet T\left(z;\frac{k}{3}\right)\cdot\blacktriangle_{\text{odd},z},$$

where, e.g. in the case of $k=1$,

$$T_{\blacktriangle_{\text{odd}},z}\left(x;\frac{1}{3}\right) \circ\!\!-\!\!\bullet \frac{1}{3}\left(-1+\frac{3}{1-z}+\frac{-1}{1+z}\right)$$
$$\bullet\!\!-\!\!\circ -\frac{1}{3}\delta(x)+\mathcal{H}(x)-\frac{(-1)^x}{3}\mathcal{H}(x),$$

i.e. $T_{\blacktriangle_{\text{odd}}}\left(x;\frac{1}{3}\right)=\{\cdots 0\ \frac{1}{3}\ \{\frac{4}{3}\ \frac{2}{3}\}\ \cdots\}$, where $\{\frac{4}{3}\ \frac{2}{3}\}$ denotes $\{\cdots\ \frac{4}{3}\ \frac{2}{3}\ \frac{4}{3}\ \frac{2}{3}\ \cdots\}$.

As a consequence, when one executes discrete integration and discrete differentiation simultaneously, it must be considered that $\blacktriangle_{\text{even}}$ ($\blacktriangle_{\text{odd}}$) is necessarily paired with $\triangle_{\text{even}}$ ($\triangle_{\text{odd}}$). Otherwise, one cannot expect a correct result since

$$T_{\blacktriangle_{\text{even}}}(z;\cdot)\cdot\triangle_{\text{odd},z} \neq T_{\blacktriangle_{\text{odd}}}(z;\cdot)\cdot\triangle_{\text{even},z} \neq T(z;\cdot).$$

5   Validation of the DSS Kernels

In this section, we validate the derived DSS kernel in comparison to the SG kernel in order to characterize its performance with respect to both smoothing and differentiation. As important performance criteria, we consider the accuracy of approximation, the fulfillment of the non-enhancement requirement (see [11]), and the accuracy of edge extraction. The criterion of accuracy of edge extraction is further divided into rotational invariance and steadiness from adjacency.

5.1   Accuracy of Approximation

We intend to measure how accurately a discrete convolution approximates a continuous convolution. To this end, we consider a continuous constant function $f(x) = c$ with $c \in \mathbb{R}$. Theoretically, convolution of a constant function with the normalized Gaussian kernel should result in the constant function again:

$$L(x;t) = f(x) * G(x;t) = c * G(x;t) = c. \qquad (14)$$

In practice, (14) is implemented by

$$L_d(x;t) = f_d(x) * G_d(x;t) = \{\cdots c \ c \ c \cdots\} * G_d(x;t),$$

where $f_d(x)$ is the discrete constant signal ($c = 100$ in the experiment), $G_d(x;t)$ is a discrete kernel. The approximation error of discrete convolution is given by

$$\bar{\xi}(t) = \frac{1}{(2n_l + 1)} \sum_{l=-n_l}^{n_l} \xi_l(t),$$

where $2n_l + 1$ corresponds to the number of coefficients of $L_d(x; \cdot)$ and $\xi_l(t) = |L_d(x_l;t) - L(x_l;t)|$.

The results of approximation accuracy are given in Fig. 3: $\bar{\xi}_T(t)$ consistently gives zero as t gradually increases, which implies that the discrete convolution with $T(x;t)$ accurately approximates the continuous convolution for any t, whereas $\bar{\xi}_{SG}(t)$ inconsistently varies over t and even attains a maximum at a small t. This unsatisfactory experimental result of the SG kernel is connected to the analytical result of the SG kernel given in Sect. 2. Consequently, it can be said that $T(x;t)$ is superior to $SG(x;t)$ with respect to approximation of discrete convolution when t gets smaller.

5.2   Fulfillment of the Non-enhancement Requirement

According to the prerequisites for the DSS formulation proposed by Lindeberg [11], a higher dimensional DSS kernel is assumed to obey the non-enhancement requirement. In order to examine the fulfillment of the non-enhancement requirement of the DSS kernel as well as of the SG kernel, we provide a synthetic image that has two local maxima and two local minima (one local extremum has

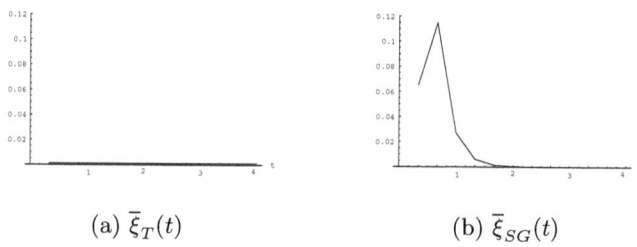

(a) $\overline{\xi}_T(t)$ (b) $\overline{\xi}_{SG}(t)$

Fig. 3. Accuracy of approximation of the DSS kernel compared with the SG kernel.

a high intensity contrast, whereas the other has a low intensity contrast). For a given synthetic image, using additive Gaussian noise we control the level of noise. We generate the scale-space representation through convolution with both the DSS kernel and the SG kernel. Then, we observe whether the local extrema of each scale-space representation are not enhanced, i.e. whether the intensity value of the local maxima (minima) does not increase (decrease) as the scale parameter gradually increases.

The experimental results show that the intensity values of the local maxima (minima) in the scale-space representation generated by both DSS kernel and the SG kernel do not increase (decrease) as the scale parameter increases. The level of noise and the intensity contrast can influence the shape of convergence of local extrema, however, they do not affect the principal non-enhancement behavior of the local extrema. Based on this result, as a consequence, it can be said that the DSS kernel as well as the SG kernel fulfill the non-enhancement requirement.

5.3 Accuracy of Edge Extraction

We attempt to examine the accuracy of edge extraction (based on the non-maximum suppression method subsequent to the gradient magnitude) using the DSS differencing kernels in comparison to using the SG differencing kernel. For evaluating the result of edge extraction, we use a synthetic image in order to identify easily its edge image (we call it the *edge atlas*). The accuracy of $T_{\triangle_{\text{even}}}$, $T_{\triangle_{\text{odd}}}$, and $SG_{\triangle}$ for edge extraction is assessed by measuring the error of extracted edges based on the edge atlas (see e.g. [12]). In concrete terms, we denote $P_{l,\text{edge-atlas}}$ and $P_{l,\text{extracted-edge}}$ as the edge loci of the edge atlas and those of the extracted edge image, respectively, where n_l corresponds to the total number of edge loci. We measure the error of edge extraction in global terms by

$$\overline{\psi} = \frac{1}{n_l} \sum_{l=1}^{n_l} \psi_l, \tag{15}$$

where $\psi_l = |P_{l,\text{edge-atlas}} - P_{l,\text{extracted-edge}}|$.

Rotational Invariance. The derived 2-D and 3-D DSS kernels are proven to be rotationally least asymmetric. We are now interested in the question whether

$T_{\triangle_{\mathrm{even}}}$ and $T_{\triangle_{\mathrm{odd}}}$ as well as $SG_{\triangle}$ used for edge extraction are rotationally invariant. We examine how consistent the edge extraction result of each discrete differencing kernel is under gradual rotation of an edge line. For this, we provide a series of ten synthetic images as shown in Fig. 4, where a straight edge line gradually rotates.

Fig. 4. A series of ten synthetic images $I_1 \cdots I_{10}$: A straight edge line gradually rotates.

The experimental results are given in Fig. 5, from which one can notice that i) for a given image edge extraction using $T_{\triangle_{\mathrm{even}}}$ is less accurate than using $T_{\triangle_{\mathrm{odd}}}$ and $SG_{\triangle}$, ii) the edge extraction error using $T_{\triangle_{\mathrm{odd}}}$ is almost identical with that using $SG_{\triangle}$, and iii) the accuracy result of edge extraction using $T_{\triangle_{\mathrm{even}}}$ from I_1 to I_{10} is rather inconsistent compared with that using $T_{\triangle_{\mathrm{odd}}}$ and $SG_{\triangle}$, which definitely appears in the case of the noiseless images. This shows that $T_{\triangle_{\mathrm{even}}}$ is inferior to both $T_{\triangle_{\mathrm{odd}}}$ and $SG_{\triangle}$ with respect to rotational invariance for edge extraction.

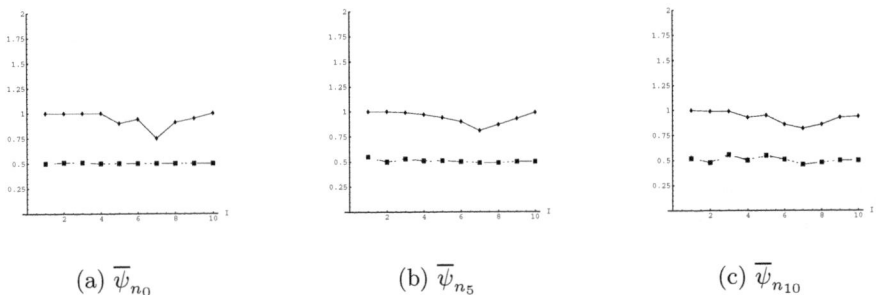

(a) $\overline{\psi}_{n_0}$ (b) $\overline{\psi}_{n_5}$ (c) $\overline{\psi}_{n_{10}}$

Fig. 5. Error of edge extraction applied to the images given in Fig. 4. n_0, n_5, and n_{10} denote that the images are noiseless, weakly noisy, and strongly noisy, respectively. ($\underline{\quad\blacklozenge\quad}$: $\overline{\psi}_{T_{\triangle_{\mathrm{even}}}}$, $\cdots\star\cdots$: $\overline{\psi}_{T_{\triangle_{\mathrm{odd}}}}$, $\cdot-\blacksquare-\cdot$: $\overline{\psi}_{SG_{\triangle}}$). Note that $\star$ and $\blacksquare$ almost overlap.

Steadiness from Adjacency. For the purpose of examining how steadily each discrete differencing kernel applied to an image (that contains closely adjacent edge structures) performs for edge extraction, we provide four synthetic images as shown in Fig. 6. Similarly to the case of rotational invariance, we apply $T_{\triangle_{\mathrm{even}}}$, $T_{\triangle_{\mathrm{odd}}}$, and $SG_{\triangle}$ to a given synthetic image and follow the procedure for evaluation of the edge extraction.

Fig. 7 illustrates the edge extraction results of three different types of the first-order differencing kernel applied to the weakly noisy images. From Fig. 7, one can observe that i) for a given image $\overline{\psi}_{T_{\triangle_{\mathrm{even}}}}$ is much larger than $\overline{\psi}_{T_{\triangle_{\mathrm{odd}}}}$

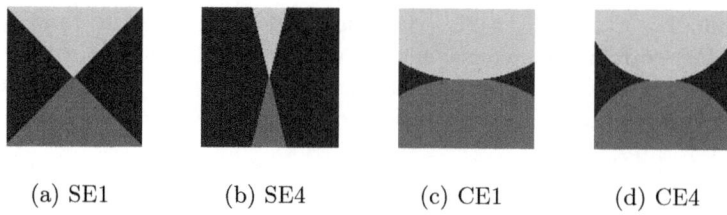

(a) SE1   (b) SE4   (c) CE1   (d) CE4

Fig. 6. Four synthetic images having closely adjacent edge structures.

and $\overline{\psi}_{SG_\triangle}$, ii) the values of $\overline{\psi}_{T_{\triangle_{odd}}}$ are similar to those of $\overline{\psi}_{SG_\triangle}$ on the whole, and iii) regardless of the type of the used discrete differencing kernel, SE1 is the steadiest image type of the adjacent edge structure in edge extraction (i.e. $\overline{\psi}_{SE1}$ is the smallest) and CE4 is the second steadiest one. Consequently, regarding the accuracy of edge extraction with respect to steadiness from adjacency, $T_{\triangle_{odd}}$ and $SG_\triangle$ are superior to $T_{\triangle_{even}}$ in general.

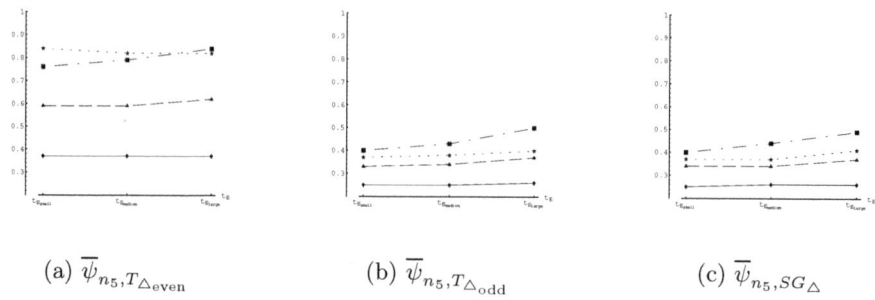

(a) $\overline{\psi}_{n_5, T_{\triangle_{even}}}$ (b) $\overline{\psi}_{n_5, T_{\triangle_{odd}}}$ (c) $\overline{\psi}_{n_5, SG_\triangle}$

Fig. 7. Error of edge extraction applied to the images given in Fig. 6. n_5 denotes that the images are weakly noisy (——◆—— : SE1, ···★··· : SE4, ·—■—· : CE1, ——▲—— : CE4).

6 Conclusion

In this paper, we addressed the issue of a higher dimensional DSS formulation. For the purpose of analyzing the problem associated with the commonly used sampled Gaussian for approximating the continuous Gaussian, we derived the approximation error caused by the sampling as a function of the scale parameter. This analysis explicates that in general a sampled Gaussian with a small scale is not appropriate for approximating the continuous Gaussian. By developing the generalized 2-D and 3-D DSS formulation through a theoretical derivation, we made a step forward in investigating the problem of how to correctly approach the higher dimensional DSS theory. Furthermore, we investigated the properties of the derived DSS kernels and carried out a validation study with respect to both smoothing and differentiation performance. Our investigation as well as the experimental results of the validation study show that the derived DSS kernel does not only match the performance of the SG kernel but also clearly exhibits superior performance with respect to both smoothing and differentiation.

Future work will include the investigation of the performance of higher-order differencing operators for feature extraction in higher dimensions. Also, a validation study with respect to the derived 3-D DSS kernel has to be carried out. Furthermore, it would be interesting to compare the derived DSS kernels with the SG kernel as well as e.g. Florack's scheme shown in [5] with respect to the cost of computational load.

Acknowledgement

The financial support by DAAD(German Academic Exchange Service) to the first author is greatly acknowledged.

References

1. J. Babaud, A. P. Witkin, M. Baudin, and R. O. Duda. Uniqueness of the Gaussian Kernel for Scale-Space Filtering. *IEEE Trans. on Pattern Analysis and Machine Intelligence*, 8(1):26-33, 1986.
2. R. N. Bracewell. *The Fourier Transform and Its Applications*. McGraw-Hill, 3rd edition, 2000.
3. N. Fliege. *Systemtheorie*. Teubner, 1991.
4. L. M. J. Florack, *Image Structure*, Kluwer Academic Publishers, 1997.
5. L. M. J. Florack. A Spatio-Frequency Trade-Off Scale for Scale-Space Filtering. *IEEE Trans. on Pattern Analysis and Machine Intelligence*, 22(9):1050-1055, 2000.
6. B. Jähne. *Digitale Bildverarbeitung*. Springer, 4. völlig neubearbeitete Auflage, 1997.
7. J. J. Koenderink. The Structure of Images. *Biological Cybernetics*, 50:363-370, 1984.
8. J. Y. Lim. On the Role of the Gaussian Kernel in Edge Detection and Scale-Space Methods. Technical Report FBI-HH-B-230/01, Fachbereich Informatik, Universität Hamburg, Germany, 2001.
9. J. Y. Lim. On the Discrete Scale-Space Formulation. Technical Report FBI-HH-B-231/01, Fachbereich Informatik, Universität Hamburg, Germany, 2001.
10. J. Y. Lim. The Supplemented Discrete Scale-Space Formulation. Technical Report FBI-HH-B-312/02, Fachbereich Informatik, Universität Hamburg, Germany, 2002.
11. T. Lindeberg. Scale-Space for Discrete Signals. *IEEE Trans. on Pattern Analysis and Machine Intelligence*, 12(3):234-264, 1990.
12. D. W. Paglieroni. A Unified Distance Transform Algorithm and Architecture. *Machine Vision and Applications*, 5(1):47-55, 1992.
13. L. Remaki and M. Cheriet. KCS-New Kernel Family with Compact Support in Scale-Space: Formulation and Impact. *IEEE Trans. on Image Processing*, 9(6):970-981, 2000.
14. J. Sporring, M. Nielsen, L. M. J. Florack, and P. Johansen. *Gaussian Scale-Space Theory*. Kluwer Academic Publishers, 1997.
15. L. Wu and Z. Xie. Scaling Theorems for Zero-Crossings. *IEEE Trans. on Pattern Analysis and Machine Intelligence*, 12(1):46-54, 1990.
16. A. P. Witkin. Scale-Space Filtering. In *Proc. of 8th Int. Joint Conf. Artificial Intelligence, Karlsruhe*, 1019-1021, 1983.

Real-Time Scale Selection
in Hybrid Multi-scale Representations

Tony Lindeberg and Lars Bretzner

Computational Vision and Active Perception Laboratory (CVAP)
Department of Numerical Analysis and Computer Science
KTH, SE-100 44 Stockholm, Sweden

Abstract. Local scale information extracted from visual data in a bottom-up manner constitutes an important cue for a large number of visual tasks. This article presents a framework for how the computation of such scale descriptors can be performed in real time on a standard computer.

The proposed scale selection framework is expressed within a novel type of multi-scale representation, referred to as hybrid multi-scale representation, which aims at integrating and providing variable trade-offs between the relative advantages of pyramids and scale-space representation, in terms of computational efficiency and computational accuracy. Starting from binomial scale-space kernels of different widths, we describe a family pyramid representations, in which the regular pyramid concept and the regular scale-space representation constitute limiting cases. In particular, the steepness of the pyramid as well as the sampling density in the scale direction can be varied.

It is shown how the definition of γ-normalized derivative operators underlying the automatic scale selection mechanism can be transferred from a regular scale-space to a hybrid pyramid, and two alternative definitions are studied in detail, referred to as variance normalization and l_p-normalization. The computational accuracy of these two schemes is evaluated, and it is shown how the choice of sub-sampling rate provides a trade-off between the computational efficiency and the accuracy of the scale descriptors. Experimental evaluations are presented for both synthetic and real data. In a simplified form, this scale selection mechanism has been running for two years, in a real-time computer vision system.

1 Introduction

Recent works have shown how the notion of automatic scale selection constitutes an essential complement to traditional scale-space representation. While a scale-space representation provides a well-founded framework to represent image structures at different scales, the scale-space representation by itself contains no explicit information about what scales are relevant for further processing.

For addressing the problem of choosing interesting scale levels from image data, a number of different approaches have been developed in the literature (see the review in section 2). If one aims at real-time performance, however, a

L.D. Griffin and M. Lillholm (Eds.): Scale-Space 2003, LNCS 2695, pp. 148–163, 2003.

common problem of most present approaches for automatic scale selection, is computational efficiency. Since scale selection is performed by either minimizing or maximizing feature measures over scales, the algorithms involve explicit search over scales. The purpose of this article is to show how these problems can be remedied for a class of scale selection methods based on normalized derivatives, and how real-time performance can be obtained on a standard PC.

2 Related Work

An early approach to scale selection focused on the detection of blob-like image features and scale levels were selected from local maxima over scales of a normalized measure of blob strength (Lindeberg 1993a). Later, this idea was generalized to a wide class of differential image features, by selecting scale levels from local maxima over scales of differential invariants expressed in terms of normalized derivatives (Lindeberg 1993b, Lindeberg 1994). This principle has been applied to various problems relating to the detection of image features (Lindeberg 1998b, Lindeberg 1998a, Chomat et al. 2000, Almansa & Lindeberg 2000, Pedersen & Nielsen 2000, Nielsen & Lillholm 2001, Kadir & Brady 2001). In particular, and motivated by the observation that single-scale ridge detection may be highly sensitive to the choice of scale level, special emphasis has been on the detection of ridges for medical image analysis (Pizer et al. 1994, Eberly et al. 1994, Koller et al. 1995, Lorenz et al. 1997, Sato et al. 1998, Staal et al. 1999, Frangi et al. 1999, Majer 2001). Moreover, for the purpose of obtaining zoom invariant image features for further processing, scale selection mechanisms have proven highly useful for interest point detection (Mikolajczyk & Schmid 2002) with applications to object recognition (Lowe 1999, Hall et al. 2000) and tracking (Bretzner & Lindeberg 1998, Laptev & Lindeberg 2001). Other approaches for scale selection have also been presented from the behaviour of entropy measures or error measures over scales (Jägersand 1995, Elder & Zucker 1996, Niessen & Maas 1996, Yacoob & Davis 1997, Lindeberg 1998c, Sporring & Weickert 1999, Pedersen & Nielsen 2001, Comaniciu et al. 2001, Hadjidemetriou et al. 2002).

3 Hybrid Pyramid Representation

Both pyramids (Burt & Adelson 1983, Crowley & Parker 1984, Jähne 1995, Simoncelli & Freeman 1995) and scale-space representations (Witkin 1983, Koenderink 1984, Lindeberg 1994, Florack 1997) have been developed from the idea of representing images at multiple scales in such a way that the resulting representation can be used as input to a large number of visual processes. Computationally, however, these concepts have their relative advantages and disadvantages.

A pyramid representation is highly efficient in the sense that it leads to a rapidly decreasing image size, while a scale-space representation successively becomes more redundant as the scale parameter increases. The highly discretized nature of a pyramid can, however, lead to algorithmic problems at coarse scales, while in scale-space representation the task of operating on the data will be successively simplified at coarser scales.

When processing data at a coarse scale in a scale-space representation, it thus seems natural that a certain amount of subsampling can be performed without affecting the performance too seriously. On the other hand, one could also consider decomposing the smoothing operation in a pyramid into a set of smoothing stages, so as to obtain a denser sampling along the scale direction. In this way, we obtain an *oversampled pyramid*, characterized by the fact that not every smoothing step is followed by a subsampling operation.

The goal of this section is to describe a general class of multi-scale representations, which comprises both regular pyramids, oversampled pyramids and scale-space representation as special cases. Due to space limitations, however, the presentation will sometimes be somewhat condensed. For a more extensive description, see (Lindeberg and Bretzner 2003).

3.1 Reduction Operators

Following (Burt & Adelson 1983, Crowley & Parker 1984), let us describe the the transformation between two adjacent scale levels in a pyramid by a reduction operator. For simplicity, let us assume that the pyramid is separable and that the size N of the smoothing filter is odd. Then, the transformation from the representation $L^{(i)}$ at the current scale level i, to the representation $L^{(i+1)}$ at the next coarser level $i+1$ is for some set of filter coefficients $c\colon \mathbb{Z} \to \mathbb{R}$ given by

$$L^{(i+1)} = \text{REDUCECYCLE}(L^{(i)}) \tag{1}$$

$$L^{(i+1)}(x) = \sum_{n=-(N-1)/2}^{(N-1)/2} c(n)\, L^{(i)}(sx - n). \tag{2}$$

Next, let us assume that the smoothing operation can be decomposed into several smoothing steps:

$$\begin{aligned} \text{REDUCECYCLE} := \ &\text{SUBSAMPLE} \\ &\text{SMOOTH}^+ \end{aligned} \tag{3}$$

where the notation OP^+ means that several operators of the form OP may occur. REDUCECYCLE is thus composed of one or more smoothing operations followed by a subsampling. The subsampling operation is here defined by

$$S = \text{SUBSAMPLE}(L;\ s) \tag{4}$$

$$S(x) = L(sx) \qquad (s \in \mathbb{Z}_+). \tag{5}$$

(where we usually choose $s = 2$) and each smoothing step according to

$$S = \text{SMOOTH}(L) \tag{6}$$

$$S(x) = \sum_{n=-N}^{N} c(n)\, L(x - n). \tag{7}$$

BIN3REDUCECYCLE := SUBSAMPLE   BIN5REDUCECYCLE := SUBSAMPLE
                  BIN3KERNEL   BIN5KERNEL

BIN3REDUCE6CYCLE := SUBSAMPLE
                   BIN3KERNEL   BIN5REDUCE3CYCLE := SUBSAMPLE
                   BIN3KERNEL   BIN5KERNEL
                   BIN3KERNEL   BIN5KERNEL
                   BIN3KERNEL   BIN5KERNEL
                   BIN3KERNEL
                   BIN3KERNEL

Fig. 1. Examples of regular and oversampled pyramids as generated using the notation for hybrid multi-scale representations defined in (3)–(12). By applying these reduction cycles repeatedly, we obtain pyramids that will be referred to as BIN3PYRAMID, BIN5PYRAMID, BIN3(6)PYRAMID and BIN5(3)PYRAMID, respectively.

For simplicity, let us assume that the coefficients of the smoothing operation originate from a discretization of the diffusion operator repeated K times

$$\text{SMOOTH}(L) = \text{DELTASMOOTH}(L;\ \Delta t, K) = [\text{DELTASMOOTH}(L;\ \Delta t, 1)]^K \quad (8)$$

where in one dimension the $\text{DELTASMOOTH}(L;\ \Delta t, 1)$ operator corresponds to convolution with a binomial diffusion filter of the following form

$$T = \text{DELTASMOOTH}(L;\ \Delta t, 1) \quad (9)$$

$$T(x) = \frac{\Delta t}{2} L(x - 1) + (1 - \Delta t) L(x) + \frac{\Delta t}{2} L(x + 1) \quad (10)$$

Thus, we can construct kernels such as the binomial three-kernel

$$\text{BIN3KERNEL} = \text{DELTASMOOTH}(\cdot;\ \tfrac{1}{2}, 1) = (\frac{1}{4},\ \frac{1}{2},\ \frac{1}{4}) \quad (11)$$

and the binomial five-kernel

$$\text{BIN5KERNEL} = \text{DELTASMOOTH}(\cdot;\ \tfrac{1}{2}, 2) = (\frac{1}{16},\ \frac{4}{16},\ \frac{6}{16},\ \frac{4}{16},\ \frac{1}{16}). \quad (12)$$

Moreover, we can define different types of oversampled pyramid representations as illustrated in figure 1. To index the levels in such a hybrid representations, we shall henceforth use the index $i \in [1 \ldots I]$ for the subsampling levels and the index $j \in [1 \ldots J]$ within each subsampling level.

3.2   Equivalent Convolution and Derivative Approximation Kernels

Since the representation at each level is constructed from a set of repeated smoothing and subsampling operations, which are all linear operations, the composed operation can equivalently be modeled as the result of applying one kernel $C^{(i,j)}$, termed *equivalent convolution kernel*, to the original image, followed by a pure subsampling step. If we define a dual operator to the REDUCECYCLE operator according to

$$\textsc{ExpandCycle} := \textsc{Smooth}^{+}$$
$$\textsc{Enlarge}$$

where the $\textsc{Enlarge}$ operation enlarges any D-dimensional image by a factor s

$$E = \textsc{Enlarge}(L) \tag{13}$$

$$E(x) = \begin{cases} s^{D}L(x/s) & \text{if all indices in } x \text{ are multiples of } s \\ 0 & \text{if any index in } x \text{ is not a multiple of } s \end{cases} \tag{14}$$

the equivalent convolution kernel corresponding to level (i,j) can be written

$$C^{(i,j)} = \textsc{ExpandAll}(\delta^{(i,j)}) \tag{15}$$

where $\delta^{(i,j)}$ is a discrete delta function at level (i,j) and $\textsc{ExpandAll}$ denotes the $\textsc{ExpandCycle}$ operators corresponding to the set of all the $\textsc{ReduceCycle}$ operators used for reaching this level. Similarly derivative approximations are computed by taking the grid spacing h at the current into explicit account

$$\partial_{x^{r}} \approx \mathcal{D}_{x^{r}} = \frac{1}{h^{|r|}} \, \delta_{x^{r}}, \tag{16}$$

at any level with resolution h in the pyramid, the corresponding *equivalent derivative approximation kernel* is given by

$$C^{(i,j)}_{x^{r}} = \textsc{ExpandAll}(\delta^{(i,j)}_{x^{r}}) \tag{17}$$

where higher dimensional difference approximations $\delta_{x^{r}} = \delta_{x_{1}^{r_{1}}} \delta_{x_{2}^{r_{2}}} .. \delta_{x_{D}^{r_{D}}}$ are expressed in terms of the one-dimensional rth order difference operator

$$\delta_{x^{r}} = \begin{cases} (\delta_{xx})^{r/2} & \text{if } r \text{ is even} \\ \delta_{x} \, \delta_{x^{r-1}} & \text{if } r \text{ is odd} \end{cases} \tag{18}$$

and δ_{x} and δ_{xx} denote the first-order symmetric difference operators with computational molecules $(-\frac{1}{2}, 0, \frac{1}{2})$ and $(1, -2, 1)$, respectively (see figure 2).

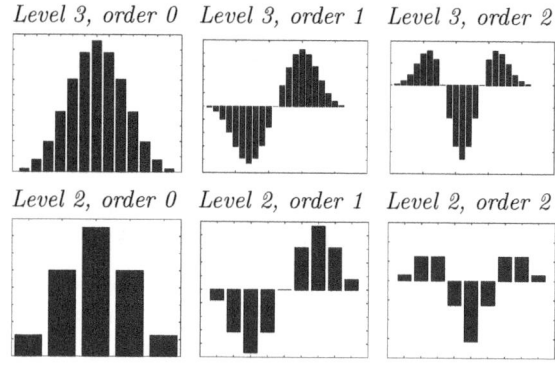

Fig. 2. Examples of equivalent convolution kernels and equivalent derivative approximation kernels for the $\textsc{Bin5Pyramid}$ derived from the $\textsc{Bin5ReduceCycle}$ in figure 1.

Table 1. Scale values for the different levels of two pure and one oversampled pyramid as defined in figure 1.

BIN3PYRAMID	BIN5PYRAMID	BIN5(3)PYRAMID		
0.0	0.0	0.0	1.0	2.0
0.5	1.0	3.0	7.0	11.0
2.5	5.0	15.0	31.0	47.0
10.5	21.0	63.0	127.0	191.0
42.5	85.0	255.0	511.0	767.0
170.5	341.0	1023.0	2047.0	3071.0

3.3   Measuring the Scale Parameter and the Subsampling Rate

For measuring the scale parameter $t_{(i,j)}$ at any level (i,j) in a hybrid pyramid, we will start from *the covariance matrix of the equivalent convolution kernel*:

$$t_{(i,j)} = (\det V(C^{(i,j)}))^{1/D} = (\det V(\text{EXPAND}(\delta^{(i,j)})))^{1/D} \qquad (19)$$

where $V(C)$ represents the spatial covariance matrix of a kernel C and D is the dimension of the signal. At coarser levels of resolution with grid spacing $h \in \mathbb{Z}_+$, the operator DELTASMOOTH(L; $\Delta t, K$) in (8) corresponds to scale values at levels k and $k+1$ that are related according to $t^{(i,j+1)} - t^{(i,j)} = K h^2 \Delta t$.

Table 1 shows the scale values for each level computed in this way for some of the pure and oversampled pyramids defined in figure 1.

Then, to describe how the grid spacing h depends on the scale parameter t in a hybrid pyramid let us introduce a *subsampling factor* ρ from the relation

$$h_{max} = \rho\sigma = \rho\sqrt{t} \qquad (20)$$

where for reasons of computational efficiency we define the actual grid spacing as the maximum power of two that does not exceed this upper bound

$$h(t,\rho) = \begin{cases} \max_{h'=2^{i-1}: \, i \in \mathbb{Z}_+ \backslash \{0\}} h' : h' < h_{max}(t,\rho) & \text{if } h_{max} \geq 1 \\ 1 & \text{otherwise} \end{cases} \qquad (21)$$

Thus, a subsampling factor of $\rho = 0$ corresponds to preserving the original resolution at all levels of scales, while increasing values of ρ correspond to higher degrees of subsampling at coarser scales.

In this context, self-similarity over scales (implying that $h \leq \rho\sqrt{t}$ holds with equality at the lowest pyramid level for any amount of subsampling) is obtained only if we precede the computation of the pyramid by a certain amount of pre-smoothing. If the total amount of smoothing in the composed SMOOTH$^+$ stage between two sub-sampling stages in (3) corresponds to a variance of $h^2 \Delta t_{cycle}$, where for hybrid pyramids generated according to (8) and (9) we have $h^2 \Delta t_{cycle} = h^2 J K \Delta t$, then it can be shown that the requirement of self-similarity over scales implies that the pre-smoothing t_{start} (i.e the scale of the first level) and the subsampling factor ρ must be given by

$$\rho = \sqrt{\frac{3}{\Delta t_{cycle}}}, \quad t_{start} = \frac{\Delta t_{cycle}}{3} \qquad (22)$$

Table 2. The subsampling rate ρ and the amount of pre-smoothing t_{start} for a few self-similar pyramids.

Pyramid	$t^{(i,j+1)} - t^{(i,j)}$	Levels	ρ	t_{start}	d_{mean}
BIN3PYRAMID	$h^2/2$	1	$\sqrt{6}$	1/6	2
BIN5PYRAMID	h^2	1	$\sqrt{3}$	1/3	2
BIN5(3)PYRAMID	h^2	3	1	1	2/3
BIN5(6)PYRAMID	h^2	6	$1/\sqrt{2}$	2	1/3

Table 2 shows values of ρ and t_{start} computed in this way for a few pyramids. In addition, this table also lists a measure of the average sampling density in the scale direction defined as $d_{mean} = (\tau(t^{(i+1,1)}) - \tau(t^{(i,1)}))/J$ where $\tau(t) = \log_2(t)$.

4   Scale Selection in Hybrid Multi-scale Representation

Our next goal is to express a scale selection mechanism within a hybrid pyramid representation. In previous works, it has been shown that a powerful principle for automatic scale selection consists of selecting interesting scale levels from the scales at which (possibly non-linear) combinations of γ-*normalized derivatives*

$$\partial_{\xi_i} = t^{\gamma/2} \partial_{x_i}, \tag{23}$$

assume local maxima over scales (see section 2). Intuitively, this corresponds to selecting scale levels at which the normalized feature response is locally strongest.

General Scale Invariance Property. A basic property of this scale selection method is as follows: If $\mathcal{D}(L)$ is a homogeneous differential expression, and if a local maximum of a signal f is detected at scale t_{locmax}, then under a rescaling of f by a factor s, this local maximum over scale is transferred to the scale level $s^2 t_{locmax}$.

Interpretation in Terms of L_p-Norms. With respect to the computation of derivatives of the scale-space representation, it can be shown that γ-normalization corresponds to normalizing the corresponding γ-normalized Gaussian derivative operators $g_{\xi^m}(\cdot;\ t) = t^{m\gamma/2} g_{x^m}(\cdot;\ t)$ to constant L_p-norms

$$\|g_{\xi^m}(\cdot;\ t)\|_p = \left(\int_{x \in \mathbb{R}^D} |g_{\xi^m}(\cdot;\ t)|^p dx \right)^{1/p} \tag{24}$$

over scales, where the parameter p in the L_p-norm is related to the parameter γ in the γ-normalized derivative concept according to

$$p = \frac{1}{1 + \frac{m}{D}(1 - \gamma)}, \tag{25}$$

where m is the order of differentiation and D denotes the dimension of the signal. Specifically, $\gamma = 1$ corresponds to $p = 1$ and thus to L_1-normalization of all the Gaussian derivative kernels.

4.1   Defining Normalized Derivatives with Spatial Subsampling

For transferring this notion of γ-normalized derivatives from a scale-space representation to a hybrid pyramid, our next goal is to define normalization parameters γ_r such that normalized derivative approximations can be written:

$$\mathcal{D}_{x^r,norm} = \gamma_r\,\mathcal{D}_{x^r}. \tag{26}$$

Here, two approaches will be considered and evaluated:

- *variance-based normalization:* multiplying the equivalent derivative approximation kernel (17) at any level in the pyramid by the variance (19) of the equivalent convolution kernel at the corresponding level

$$\gamma_{r,var} = \left(t^{(i,j)}\right)^{|r|/2} = \left(\det(V(C^{(i,j)}))^{1/D}\right)^{|r|/2} \tag{27}$$

- l_p-*normalization:* requiring the l_p-norm of the normalized equivalent derivative approximation kernel to be equal to the L_p-norm of the corresponding Gaussian derivative operator $\partial_{\xi^r} g(x;\ t)$

$$\gamma_{r,l_1}\|C_{x^r}^{(i,j)}\|_p = \|\partial_{\xi^r} g(x;\ t)\|_p \tag{28}$$

Experiments: Scale-Space Signatures for Gaussian Blobs. For a rotationally symmetric Gaussian blob with variance t_0 in two dimensions $f(x,y) = g(x,y;\ t_0)$ it can be shown that the evolution over scales of the γ-normalized Laplacian response at the center of the blob is in the case when $\gamma = 1$ given by

$$(\nabla^2_{norm}L)(0,0;\ t) = t\,(\partial_{xx} + \partial_{yy})L(0,0;\ t) = -\frac{t(t_0 + 2t)}{\pi(t_0 + t)^3} \tag{29}$$

and there is a unique maximum over scales in $-(\nabla^2_{norm}L)(0,0;\ t)$ at $t = t_0$.

Figure 3 shows a few examples of such scale-space signatures computed for Gaussian blobs of different sizes, using a separable BIN3(6)PYRAMID with an initial pre-smoothing stage. As can be seen from these graphs, l_p-normalization (stars) gives a closer approximation of the continuous behaviour (the solid curve) than variance-based normalization (crosses). Moreover, for variance-based normalization there are a number of "kinks" in the graph at the scales where subsamplings occur. In these respects, l_p-normalization has clear advantages compared to variance-based normalization.

4.2   Detecting Scale-Space Maxima

A method for complementary scale selection and detection of interest points consists of simultaneously maximizing differential entities over both space and scale. If $\mathcal{D}_{space}L$ denotes the differential entity used for spatial selection and if

$t_0 = 10$ $t_0 = 30$ $t_0 = 100$

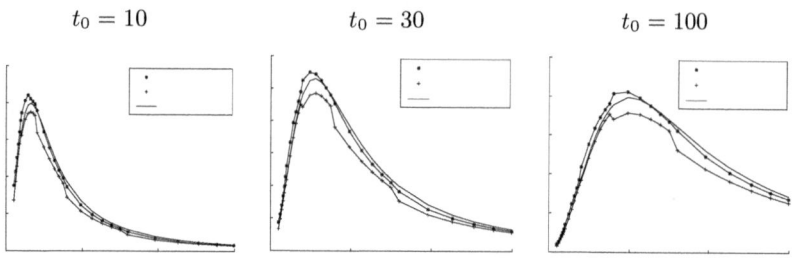

Fig. 3. Scale-space signatures of the (negative) normalized Laplacian response for rotationally symmetric Gaussian blobs with variances $t_0 = 10$, $t_0 = 30$ and $t_0 = 100$, respectively, computed using a separable BIN3(6)PYRAMID in two dimensions using l_p-normalization (stars) and variance-based normalization (crosses). For reference, the corresponding continuous behaviour is shown as well (solid curve).

$\mathcal{D}_{scale,norm}L$ is the γ-normalized differential entity used for scale selection, such *interest points with automatic scale selection* can be characterized by

$$\begin{cases} \nabla(\mathcal{D}_{space}L) = 0 \\ \mathcal{H}(\mathcal{D}_{space}L) \text{ negative definite} \\ \partial_t(\mathcal{D}_{scale,norm}L) = 0 \\ \partial_{tt}(\mathcal{D}_{scale,norm}L) \leq 0 \end{cases} \tag{30}$$

where $\mathcal{H}(\mathcal{D}_{space}L)$ denotes the Hessian of $\mathcal{D}_{space}L$. In the special case when $\mathcal{D}_{space}L = \mathcal{D}_{scale,norm}L$ such points are referred to as *scale-space maxima* of $\mathcal{D}_{scale,norm}L$. Our next goal is to investigate how the performance of a blob detector with automatic scale selection depends on the choice of normalization method as well as the subsampling rate ρ in the pyramid.

    To quantify the difference between these two normalization approaches, 1000 Gaussian images were generated containing one blob each with random variance between $t_0 = 10$ and $t_0 = 100$ and at a random position within a central 128×128 window in the image. The global maximum over scales of the normalized Laplacian response in the hybrid pyramid representation was detected, and a quadratic interpolation over scales was performed to estimate the scale $\hat{t}$ of the peak in the scale-space signature. The relative error in the estimate was computed

$$\varepsilon_n = \log_2\left(\frac{\hat{t}_n}{t_{0,n}}\right) \tag{31}$$

and the performance was measured in terms of the following descriptors

$$\varepsilon_{mean} = \frac{1}{N}\sum_{n=1}^{N} \varepsilon_n, \qquad \varepsilon_{spread} = \sqrt{\frac{\sum_{n=1}^{N} \varepsilon_n^2}{N}} \tag{32}$$

where N is the number of blobs. These error measures were then transformed into relative error factors measured in dimension length $\sigma = \sqrt{t}$ according to

$$r_{mean} = \sqrt{2^{\varepsilon_{mean}}}, \qquad r_{spread} = \sqrt{2^{\varepsilon_{spread}}} \tag{33}$$

Table 3. Performance of the scale selection method when performing simultaneous spatial and scale selection based on scale-space maxima of the normalized Laplacian response using different types of hybrid multi-scale representations and either l_p-normalization or variance-based normalization.

Pyramid type	l_p-normalization		variance-based	
	r_{mean}	r_{spread}	r_{mean}	r_{spread}
BIN3PYRAMID	0.65	1.61	0.62	1.70
BIN5PYRAMID	0.78	1.34	0.77	1.36
BIN3(6)PYRAMID	0.93	1.11	0.93	1.15
BIN5(3)PYRAMID	0.93	1.12	0.92	1.15
BIN3(12)PYRAMID	0.96	1.08	0.95	1.13
BIN5(6)PYRAMID	0.94	1.10	0.94	1.13

Table 4. Measures of the spatial localization error when performing simultaneous spatial and scale selection based on scale-space maxima of the normalized Laplacian response using different types of hybrid multi-scale representations and either l_p-normalization or variance-based normalization.

Pyramid type	l_p-normalization		variance-based	
	δ	δ_{rel}	δ	δ_{rel}
BIN3PYRAMID	1.86	0.32	1.76	0.29
BIN5PYRAMID	1.21	0.21	1.21	0.21
BIN3(6)PYRAMID	0.18	0.03	0.05	0.01
BIN5(3)PYRAMID	0.19	0.03	0.07	0.01
BIN3(12)PYRAMID	0.05	0.01	0.03	0.00
BIN5(6)PYRAMID	0.05	0.01	0.02	0.00

where the ideal case corresponds to $r_{mean} = 1$ and $r_{spread} = 1$. In addition, the absolute error in the estimated position $(\hat{x}, \hat{y})$ was measured as $\delta = \sqrt{(\hat{x} - x_0)^2 + (\hat{y} - y_0)^2}$ and a relative error measure in relation to the scale level $\sigma_0 = \sqrt{t_0}$ was defined as $\delta_{rel} = \delta/\sigma_0$. This procedure was repeated for different types of separable two-dimensional pyramids as shown in tables 3–4.

As can be seen from the results, there is a substantial variation in the accuracy of the estimate local maximum over scales depending on the type of pyramid — the oversampled BIN3(6)PYRAMID and the BIN5(3)PYRAMID perform significantly better than the regular BIN3PYRAMID and the BIN5PYRAMID, and further improvement is obtained if we increase the amount of oversampling by using a BIN3(12)PYRAMID or a BIN5(6)PYRAMID. In all of these cases, l_p-normalization leads to better performance measures than variance-based normalization. For this reason, we will henceforth prefer l_p-normalization.

Concerning the spatial localization error, we can see how the error decreases as we increase the degree of oversampling in the hybrid pyramid, by decreasing ρ and h_{max}. For the BIN3(6)PYRAMID, the BIN5(3)PYRAMID, the BIN3(12)PYRAMID and the BIN5(6)PYRAMID, the average error in all cases corresponds to a fraction of a pixel, and true sub-pixel accuracy is obtained for these data.

Table 5. Performance of the scale selection method when adding extended coarser scale level search and triquadratic interpolation to the previously developed method for performing simultaneous spatial and scale selection based on scale-space maxima of the normalized Laplacian response (see table 3). The numerical values show the mean r_{mean} and the spread r_{spread} of the relative error according to (31) for 1000 Gaussian blobs with random variances between $t_0 = 10$ and $t_0 = 100$.

Pyramid type	l_p-normalization		variance-normalization	
	r_{mean}	r_{spread}	r_{mean}	r_{spread}
BIN5PYRAMID	1.196	1.250	1.182	1.239
BIN5(3)PYRAMID	1.006	1.032	0.999	1.180
BIN5(6)PYRAMID	0.996	1.019	0.999	1.082

4.3 Post-processing the Scale-Space Maxima from a Hybrid Pyramid

While the previous results show that scale-space maxima can be detected in a hybrid pyramid using conceptually very clean operations, there is a minor complication with the previous approach. From the quantitative measure r_{mean} shown in table 3, it can be seen that there is a certain bias in the scale selection procedure that leads to an average underestimate of the scale estimate by 4 to 7 % for the sample types of oversampled hybrid pyramid representations that have been evaluated here.

When analysing the image data in more detail, it can be observed that a major reason for this scale bias is due to the detection of local maxima when translational invariance has been violated by the subsampling step. If the position of the original blob is far away from the closest grid point at the scale levels around the scale level t_0 at which it would be detected without spatial subsampling, the magnitude of the normalized Laplacian at the available grid points at the desired scale level $t_k \approx t_0$ may be significantly smaller than they would have been without spatial subsampling. As a result of this, the values of the normalized Laplacian at lower scale levels may be higher (since the grid sampling there is denser), which in turn means that a lower scale level is selected than in the ideal case without spatial subsampling.

To reduce this problem, an additional post-processing stage is applied: If a scale-space maximum is detected at a scale level where the next coarser scale level is at lower resolution, then a computation of image values at (one level of) finer resolution is initiated in a spatial 3×3 neighbourhood around the scale space maximum at this pyramid level. If the magnitude of the normalized differential entity is greater at this scale, then the scale-space maximum is translated to this nearest coarser scale level. Moreover, a tri-quadratic interpolation is performed in a $3 \times 3 \times 3$ neighbourhood in space and scale to estimate the position and the scale of the scale-space maximum with subpixel accuracy.

Table 5 shows the results obtained by adding these two post-processing stages to the previously methodology. As can be seen from a comparison with table 3, for the BIN5(3)PYRAMID and the BIN5(6)PYRAMID the average bias in the scale estimate is reduced by basically one order of magnitude, from 6–7 % to

Table 6. Computation times (in ms) for blob detection in different hybrid pyramids with and without the additional post-processing stage for scale localization. The timings have been performed on a 2.4 GHz DELL PC with a Pentium 4 processor.

Pyramid type	ρ	500 blobs		1000 blobs	
		det	det+loc	det	det+loc
BIN5PYRAMID	1.73	16	32	17	45
BIN5(2)PYRAMID	1.22	23	51	25	79
BIN5(3)PYRAMID	1.00	39	66	43	97
BIN5(4)PYRAMID	0.87	55	89	63	127
BIN5(5)PYRAMID	0.77	72	105	81	153
BIN5(6)PYRAMID	0.71	88	121	101	173

Table 7. The spatial and scale localization errors for different subsampling factors ρ using l_p-normalization. The experiments were performed on 1000 Gaussian blobs with random position and random variances between 10 and 100.

Pyramid type	ρ	δ (pixels)	r_{spread}
BIN5PYRAMID	1.73	1.72	1.250
BIN5(2)PYRAMID	1.22	0.52	1.050
BIN5(3)PYRAMID	1.00	0.29	1.032
BIN5(4)PYRAMID	0.87	0.18	1.022
BIN5(5)PYRAMID	0.77	0.12	1.022
BIN5(6)PYRAMID	0.71	0.11	1.019

0.4–0.6 %. Moreover, the measure r_{spread} of the spread in the scale values is reduced from 10–12 % to 1–3 %.

5   Trade-off: Computational Efficiency vs. Accuracy

From the experiments on blob detection with automatic scale selection, we have seen how decreasing the value of ρ improves the accuracy of the results. On the other hand, increasing ρ improves the computational efficiency, since fewer grid points are computed. Thus, the hybrid pyramid concept allows us to obtain different trade-offs between computational efficiency vs. accuracy by varying ρ.

To quantify this trade-off, we started out by measuring the computational efficiency in the following way: For a given image size of 384*288 pixels, a threshold on the magnitude of the blob response was determined such that around 500 blobs would be detected between $t_{min} = 4$ and $t_{max} = 2000$ in a BIN5(6)PYRAMID. Keeping this threshold fixed, blobs were then detected using the BIN5PYRAMID, BIN5(2)PYRAMID, ... BIN5(5)PYRAMID. A similar experiment was performed using a lower threshold on the blob response, determined in such a way that about 1000 blobs would be obtained in the BIN5(6)PYRAMID. Table 6 shows the computation time for detecting scale-space extrema in this way, with and without using the additional localization stage described in section 4.3. To allow for comparison, a denser estimation of the scale and localization errors for Gaussian blob detection was also performed for the same types of pyramids and using the methodology described in section 4.2 — see table 7.

scale localization error vs. time   *spatial localization error vs. time*

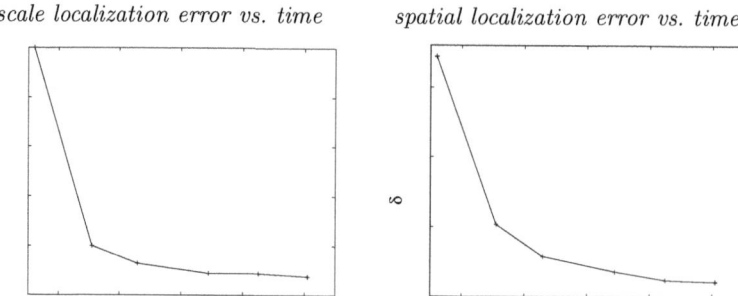

Fig. 4. Trade-offs between the localization error (vertical axis) and the computation time (horizontal axis) for hybrid pyramids with different values of ρ: (left) scale localization error, (right) spatial localization error.

If we regard these measures as representative indicators of the computational effort and the computational accuracy in the scale estimates, we thus obtain the following trade-off curves for how ρ affects r_{spread} and the computation time:

6   Stability of the Scale Descriptors

In addition to the abovementioned quantitative experiments on synthetic data with ground truth, it is of particular interest to investigate the stability of the scale descriptors on real-world images. To investigate this, we performed the following experiment: An image sequence was taken for a set of uniformly spaced distances to an object. In each image, blob detection was performed by detecting scale-space extrema of the normalized Laplacian response in a Bin5(6)-pyramid using l_p-normalization. Five scale-space maxima were selected manually in the first frame, and these features were matched over time as illustrated in figure 5.

For each one of these five features, a straight line of the form $\frac{1}{\sqrt{t}} = A\tau + B$ was fit to the data (with τ denoting time), and the time to collision was estimated by extrapolating the line to $\tau \to \infty$ (see figure 6). Here, the mean value of the five different estimates of the time to collision was 14.89 time units and the standard deviation 0.30 time units. Considering that these estimates are based on measurements at single points in scale-space, the results show how scale descriptors computed from a hybrid multi-scale representation can be stable enough to be used as a visual cue in its own right.

7   Summary and Discussion

We have presented a general framework for defining subsampled multi-scale representations in such a way that the theory comprises both traditional pyramid representations and discrete scale-space as limiting cases. Regular pyramids arise as a special case when we have only one scale level between any pair of successive subsampling stages (*i.e.* a reduction cycle with $J = 1$), while a regular discrete

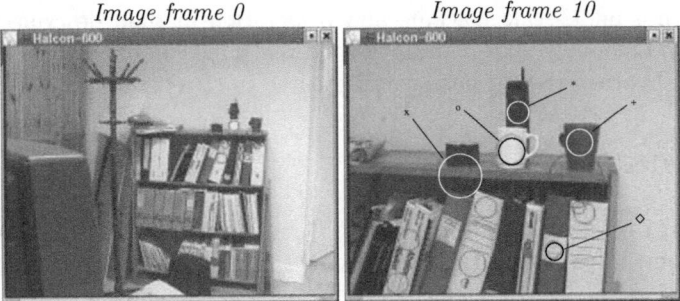

Fig. 5. Two out of eleven images in an image sequence used for testing the stability of the scale descriptors over time. In each image, a set of detected image features is indicated, out of which a subset has been matched over time and been used for measuring variations in scale levels over time. In the last image, five scale-space maxima used for scale measurements have been marked by corresponding symbols used in figure 6.

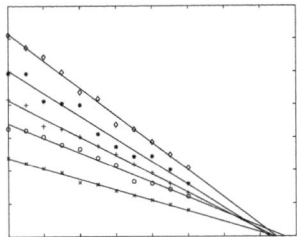

Fig. 6. Graph showing the variation over time of $1/\sqrt{t}$ for five image features matched over time as a camera approaches an object with uniform velocity.

scale-space representation is obtained as the limiting case if we let the scale increment Δt in the diffusion smoothing operator tend to zero, while keeping the product of $J\Delta t$ constant and equal to the maximum scale level t_{max} that needs to be accessed. Since this family of multi-scale representations provides a way to express different trade-offs between the relative advantages of pyramids and scale-space representation, we refer to it as hybrid multi-scale representations.

Then, we presented a theory for how scale selection mechanisms based on the maximization over scales of γ-normalized derivatives can be expressed within this family of subsampled multi-scale representations. Two ways of defining normalized derivatives in the presence of spatial subsampling have been studied, and it has been shown that the approach referred to as l_p-normalization performs significantly better than the possibly more straightforward approach of variance-based normalization. Specifically, we have quantified how the steepness of a hybrid representation, parameterized by the subsampling rate ρ, allows us to obtain different trade-offs between computational accuracy as enabled by dense sampling and computational efficiency as promoted by sparse sampling.

We have also shown how the scale descriptors computed from a hybrid multi-scale representation are stable enough to be used as a cue in its own right. Combined with a multi-scale tracking and recognition method described elsewhere (Laptev & Lindeberg 2001), an integrated real-time computer vision based on a simplified hybrid pyramid has been presented in (Bretzner et al. 2002).

Acknowledgments

The support from the Swedish Research Council, the Royal Swedish Academy of Sciences and the Knut and Alice Wallenberg Foundation is gratefully acknowledged. We would like to thank Pascal Grostabussiat and Jani Niemenmaa for their earlier contributions to this project. Efficient implementations of scale-selection are also being developed by James Crowley (personal communication) and David Lowe (personal communication).

References

Almansa, A. & Lindeberg, T. (2000), 'Fingerprint enhancement by shape adaptation of scale-space operators with automatic scale-selection', *IEEE Transactions on Image Processing* 9(12), 2027–2042.

Bretzner, L., Laptev, I. & Lindeberg, T. (2002), Hand-gesture recognition using multi-scale colour features, hierarchical features and particle filtering, *Face and Gesture'02*, 63–74.

Bretzner, L. & Lindeberg, T. (1998), 'Feature tracking with automatic selection of spatial scales', *Computer Vision and Image Understanding* 71(3), 385–392.

Burt, P. J. & Adelson, E. H. (1983), 'The Laplacian pyramid as a compact image code', *IEEE Trans. Comm.* 9:4, 532–540.

Chomat, O., de Verdiere, V., Hall, D. & Crowley, J. (2000), Local scale selection for Gaussian based description techniques, *ECCV'00*, Springer LNCS 1842, 117–133.

Comaniciu, D., Ramesh, V. & Meer, P. (2001), The variable bandwidth mean shift and data-driven scale selection, *ICCV'01*, 438–445.

Crowley, J. L. & Parker, A. C. (1984), 'A representation for shape based on peaks and ridges in the Difference of Low-Pass Transform', *IEEE-PAMI* 6(2), 156–170.

Eberly, D., Gardner, R., Morse, B., Pizer, S. & Scharlach, C. (1994), 'Ridges for image analysis', *J. Math. Im. Vis.* 4(4), 353–373.

Elder, J. H. & Zucker, S. W. (1996), Local scale control for edge detection and blur estimation, *in* 'ECCV'96', 57–69..

Florack, L. M. J. (1997), *Image Structure*, Kluwer, Netherlands.

Frangi, A. F., Niessen, W. J., Hoogeveen, R. M., van Walsum, T. & Viergever, M. A. (1999), Quantitation of vessel morphology from 3D MRI, *'MICCAI*, 358–367.

Hadjidemetriou, E., Grossberg, M. D. & Nayar, S. K. (2002), Resolution selection using generalized entropies of multiresolution histograms, *ECCV'02*, Springer LNCS 2350, 220–235.

Hall, D., de Verdiere, V. & Crowley, J. (2000), Object recognition using coloured receptive fields, *ECCV'00*, Springer LNCS 1842, 164–177.

Jägersand, M. (1995), Saliency maps and attention selection in scale and spatial coordinates: An information theoretic approach, *ICCV'95*, 195–202.

Jähne, B. (1995), *Digital Image Processing*, Springer-Verlag.

Kadir, T. & Brady, M. (2001), 'Saliency, scale and image description', *IJCV* 45, 83–105.

Koenderink, J. J. (1984), 'The structure of images', *Biol. Cyb.* 50, 363–370.

Koller, T. M., Gerig, G., Szèkely, G. & Dettwiler, D. (1995), Multiscale detection of curvilinear structures in 2-D and 3-D image data, *ICCV'95*, 864–869.

Laptev, I. & Lindeberg, T. (2001), Tracking of multi-state hand models using particle filtering and a hierarchy of multi-scale image features, *Scale-Space'01*, Springer LNCS 2106, 63–74.

Lindeberg, T. (1993a), 'Detecting salient blob-like image structures and their scales with a scale-space primal sketch: A method for focus-of-attention', *IJCV* 11(3), 283–318.

Lindeberg, T. (1993b), On scale selection for differential operators, *SCIA'93*, 857–866.

Lindeberg, T. (1994), *Scale-Space Theory in Computer Vision*, Kluwer, Netherlands.

Lindeberg, T. (1998a), 'Edge detection and ridge detection with automatic scale selection', *IJCV* 30(2), 117–154.

Lindeberg, T. (1998b), 'Feature detection with automatic scale selection', *IJCV* 30(2), 77–116.

Lindeberg, T. (1998c), 'A scale selection principle for estimating image deformations', *Image and Vision Computing* 16(14), 961–977.

Lindeberg, T. & Bretzner, L (2003), Real-time scale selection in hybrid multi-scale representations, Technical report, KTH, Stockholm, Sweden.

Lorenz, C., Carlsen, I.-C., Buzug, T. M., Fassnacht, C. & Weese, J. (1997), Multi-scale line segmentation with automatic estimation of width contrast and tangential direction in 2D and 3D medical images, *CVRMed-MRCAS'97*, Springer LNCS 1205, 233–242.

Lowe, D. (1999), Object recognition from local scale-invariant features, *ICCV'99*, 1150–1157.

Majer, P. (2001), The influence of the γ-parameter on feature detection with automatic scale selection, *Scale-Space'01*, Springer LNCS 2106, 245–254.

Mikolajczyk, K. & Schmid, C. (2002), An affine invariant interest point detector, *ECCV'02*, Springer LNCS 2350, 128–142.

Nielsen, M. & Lillholm, M. (2001), What do features tell about images, *Scale-Space'01*, Springer LNCS 2106, 39–50.

Niessen, W. & Maas, R. (1996), Optic flow and stereo, *in* J. Sporring et al (eds) *Gaussian Scale-Space Theory*, Kluwer.

Pedersen, K. S. & Nielsen, M. (2000), 'The Hausdorff dimension and scale-space normalisation of natural images', *J. Visual Com. and Im. Repr.* 11(2), 266–277.

Pedersen, K. S. & Nielsen, M. (2001), Computing optic flow by scale-space integration of normal flow, *Scale-Space'01*, Springer LNCS 2106, 14–25.

Pizer, S. M., Burbeck, C. A., Coggins, J. M., Fritsch, D. S. & Morse, B. S. (1994), 'Object shape before boundary shape: Scale-space medial axis', *J. Math. Im. Vis.* 4, 303–313.

Sato, Y., Nakajima, S., Shiraga, N., Atsumi, H., Yoshida, S., Koller, T., Gerig, G. & Kikinis, R. (1998), '3D multi-scale line filter for segmentation and visualization of curvilinear structures in medical images', *Medical Image Analysis* 2(2), 143–168.

Simoncelli, E. P. & Freeman, W. T. (1995), The steerable pyramid: A flexible architecture for multi-scale derivative computation, *ICIP'95*, 444–447.

Sporring, J. & Weickert, J. A. (1999), 'Information measures in scale-spaces', *IEEE-IT* 45(3), 1051–1058.

Staal, J., Kalitzin, S., ter Haar Romeny, B. & Viergever, M. (1999), Detection of critical structures in scale-space, *Scale-Space'99*, Springer LNCS 1682, 105–116.

Witkin, A. P. (1983), Scale-space filtering, *8th IJCAI*, pp. 1019–1022.

Yacoob, Y. & Davis, L. S. (1997), Estimating image motion using temporal multi-scale models of flow and acceleration. In: *Motion-Based Recognition*, Kluwer.

A Scale Space for Contour Registration Using Minimal Surfaces

Christopher V. Alvino and Anthony J. Yezzi Jr.[*]

Georgia Institute of Technology, School of Electrical and Computer Engineering,
Atlanta, GA 30332-0250, USA
{alvino,ayezzi}@ece.gatech.edu

Abstract. Previously, we presented a method for contour registration using minimal surfaces. This method involves embedding each of two unregistered two-dimensional contours into two parallel planes separated in three-dimensional space. The minimal surface is then computed between the two contours via mean curvature flow. We then evolve the rigid registration of one of the two contours which in turn changes the minimal surface. Mean curvature flow of the surface and evolution of the curve registration both support a consistent energy functional, i.e., area of the connecting surface. We review the implementation details and show an example registration.

In this paper we concentrate on developing this method as a registration scale space. The separation of the two contour planes serves as a scale space parameter, larger separations producing coarser registrations. At the finest scale, which occurs as the separation distance approaches zero, this registration method is identical to minimizing the set-symmetric difference between the interiors of the contours. Thus, this method can be viewed as a geometric generalization of set-symmetric difference registration. We explain the scale space properties of this registration method theoretically and experimentally. Through examples we show how at increasingly coarser scales, our method overcomes increasingly coarser local minima apparent in set-symmetric difference registration. In addition, we present sufficient conditions for existence of the minimal surface connecting two contours. This condition yields an upper bound for the separation distance between two contours and gives an estimate for the coarsest registration scale.

1 Introduction

1.1 Proposed Work

Contour registration has been established as a fundamental problem in computer vision and medical imaging; the literature is rich in techniques for registration [2,8]. In a previously submitted paper, we have introduced a method for registration of contours using minimal surfaces, that arises from developing a natural

---

[*] This work was supported by NSF grant CCR-0133736 and NIH grant R01-HLS0004-01A1.

L.D. Griffin and M. Lillholm (Eds.): Scale-Space 2003, LNCS 2695, pp. 164–179, 2003.

geometric generalization of the set symmetric difference between the interiors of two sets. In this paper, we will elaborate more of the scale space properties of the proposed method.

We develop a novel method for rigid registration of two dimensional contours that relies on connecting two contours, separated in three dimensional, by a minimal surface. The novel contribution of this paper is not the computation of the minimial surfaces, but rather the evolution of the registration as presribed by the family of minimal surfaces (the problem of computer minimal surfaces we performed in a way described by Chopp in [3]). Once the minimal surface is computed between the two contours, the registration of the top contour is evolved rigidly in an attempt to minimize the surface area of the connecting surface. We will explain the method and its implementation as well as stating its connection to set symmetric difference. We show an example that presents the method as a scale space, with the separation distance between the two contours the scale space parameter.

1.2   Past Work in Registration

A significant amount of work has been done in image registration. Reviews of registration methods can be found in [2,8]. Point based techniques, such as those used by West et. al. and by Zhang [12,13] attempt to match feature points by finding optimal registrations. Sebastian, Klein and Kimia have developed a method based on shape outlines, that aligns based on a measure of similarity between the intrinsic properties of the curve such as curvature and length [9]. Feldmar and Ayache discuss rigid and affine registration of free-form surfaces [4]. Hansen and Morse develop a method for multiscale registration using scale trace correlation [6].

The technique proposed in this paper, can be considered a transformation technique, where a one contour is gradually transformed to match the second via a connecting minimal surface.

2   Method

2.1   Initialization

In the initialization, we will embed two, two dimensional contours into three dimensional space and create a surface between them. We will also define some terminology in this section that will be used throughout the paper.

Let vector functions $\mathbf{C} : [0, 1] \rightarrow \Re^2$ and $\hat{\mathbf{C}} : [0, 1] \rightarrow \Re^2$ explicitly represent two simple closed contours which we wish to register. We represent these contours implicitly as the 0 level sets of the functions $I : \Re^2 \rightarrow \Re$ and $\hat{I} : \Re^2 \rightarrow \Re$, such that $I(\mathbf{C}(s)) = \hat{I}(\hat{\mathbf{C}}(s)) = 0$ for $s \in [0, 1]$. We will construct the level set functions such that $I < 0$ and $\hat{I} < 0$ in the interior of the contours and $I > 0$ and $\hat{I} > 0$ in the exterior of the contours.

We then embed the functions I and $\hat{I}$ into three dimensional space, and thus embed the contours. Let $\Phi_0 : \Re^2 \times [0, z_{\mathrm{M}}] \rightarrow \Re$ be the three dimensional level set function in which the contours will be embedded. The embedding is represented by setting $\Phi_0(x, y, 0) = I(x, y)$ and $\Phi_0(x, y, z_{\mathrm{M}}) = \hat{I}(x, y)$. The variable $z_{\mathrm{M}} > 0$ has special significance, representing the distance between the embedded contours, will be called the *separation distance*.

To create a surface between the two contours, with the contours as its boundary, we will interpolate between the contours, initializing the function so that,

$$\Phi_0(x, y, z) = (1 - z)I(x, y) + z\hat{I}(x, y), \tag{1}$$

for $z \in (0, z_{\mathrm{M}})$. The surface, S, is now said to exist at the zero level set of this function Φ_0, that is, $S = \Phi_0^{-1}(0)$.

We will call the planes in which the contours are embedded the *top* and *bottom contour planes* at $z = z_{\mathrm{M}}$ and at $z = 0$ respectively. Similarly, the *middle contour plane* will refer to the slice of the function Φ_0 when $z = z_{\mathrm{M}}/2$ and the *middle contour* will refer to the zero level set of Φ_0 when $z = z_{\mathrm{M}}/2$.

2.2   Evolution to Minimal Surface

Finding the minimal surface between these two contours is equivalent to finding the surface with constant zero mean curvature that has the contours as its boundary. The minimal surface that connects the two contours exists when z_{M} is sufficiently small. A more detailed discussion of the sufficient condition for existence is discussed in the appendix of this paper.

The initial surfaces described in Section 2.1 will not, except in certain degenerate cases, have the property of being minimal surfaces. Throughout this paper it will be necessary to obtain the surface of minimal area that connects the two fixed contours. For this purpose we have chosen to employ an evolution technique explained in [3]. The idea is that by evolving the surface in the direction of the surface normal with speed proportional to the mean curvature, H, we ensure the fastest possible decrease in surface area. Therefore, we wish to evolve the level sets of the function that represents the surface in this fashion.

Consider the function $\Phi : \Re^2 \times [0, z_{\mathrm{M}}] \times [0, \infty) \to \Re$, whose first three arguments are the spatial variables x, y, and z and whose last argument is the temporal variable t. The temporal variable t will be used to state the surface area minimization evolution as a partial differential equation. The initial value of Φ is set to the initialization in Section 2.1, that is, $\Phi(x, y, z, 0) = \Phi_0(x, y, z)$.

When we evolve $\Phi(x, y, z, t)$ according to the non-linear partial differential equation,

$$\frac{\partial \Phi}{\partial t} = H\|\nabla\Phi\|, \tag{2}$$

to guarantee that the surface area of each level set is descreasing as quickly as possible. In order to not keep the contours fixed throughout this evolution, we do not evolve the function Φ at $z = 0$ or at $z = z_{\mathrm{M}}$, but we do evolve at every value of z in between.

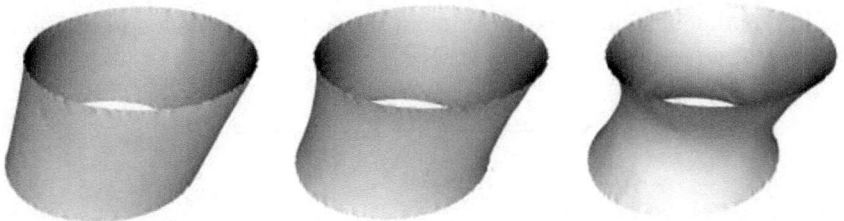

Fig. 1. Oblique view of three different stages of surface minimizing evolution for shifted circular contours. Surface created from initialization (left), surface after 125 seconds of surface evolution (center), and final minimal surface (right).

The mean curvature, H, of the surface is given by a function of derivatives of Φ,

$$H = \frac{\left(\begin{array}{c} \Phi_{xx}\left(\Phi_y^2+\Phi_z^2\right)+\Phi_{yy}\left(\Phi_x^2+\Phi_z^2\right)+\Phi_{zz}\left(\Phi_x^2+\Phi_y^2\right) \\ -2\Phi_{xy}\Phi_x\Phi_y - 2\Phi_{yz}\Phi_y\Phi_z - 2\Phi_{xz}\Phi_x\Phi_z \end{array} \right)}{2\left(\Phi_x^2 + \Phi_y^2 + \Phi_z^2\right)^{3/2}}, \tag{3}$$

as explained in [3].

Evolving the function Φ by Eq. 2 ensures that the zero level set approaches the minimal surface that connects the two contours. Fig. 1 shows the surface connecting two circular contours at various stages of this area minimizing evolution.

2.3   Gradient Evolution for Rigid Registration

The previous two sections described the initialization of a minimal surface evolution between two contours. This section describes how we rigidly register the top contour using the minimal surface. Since we wish to obtain a two dimensional rigid registration, we evolve the top contour rigidly within the top contour plane to ensure the most rapid decrease in area of the minimal surface, while holding the bottom contour fixed. This will be done using two projections.

Unconstrained Top Contour Evolution. Assume, temporarily, that the top contour were not constrained to remain within the top contour plane. If the goal of the evolution was to minimize surface area, the contour at a given point would move in a direction downward along the surface, orthogonal to both the tangent vector to the contour, $\mathbf{T}$, and the surface normal, $\mathbf{N}$.

More explicitly, the outward unit normal to the surface is $\mathbf{N} = (N_x, N_y, N_z) = \nabla\Phi$. The normal vector to the contour $\mathbf{n}$ is the projection of $\mathbf{N}$ onto the plane of the top contour. It is given by $\mathbf{n} = (N_x, N_y)/\sqrt{N_x^2 + N_y^2}$. The tangent vector to the contour is in the plane of the top contour and is a 90 degree rotation of $\mathbf{n}$, i.e., $\mathbf{T} = (N_y, -N_x)/\sqrt{N_x^2 + N_y^2}$. The vector, $\mathbf{V}$, that is orthogonal to both $\mathbf{T}$

and $\mathbf{N}$, and points along the direction of the surface, can be found by computing the vector cross-product of $\mathbf{T}$ with $\mathbf{N}$,

$$\mathbf{V} = \frac{\left(-N_x N_z, -N_y N_z, N_x^2 + N_y^2\right)}{\sqrt{N_x^2 + N_y^2}}. \tag{4}$$

Projection to Top Contour Plane. The vector $\mathbf{V}$ shows the direction in which the contour would move in 3D space to minimize surface area most rapidly. Constraining the contour to remain in the top contour plane yields the explicit contour evolution, $(\mathbf{V} \cdot \mathbf{n})\mathbf{n} = -N_z \mathbf{n}$.

Projection to Rigid Registration. By projecting the explicit contour evolution $-N_z \mathbf{n}$, onto the group of rigid motions, we ensure that the top contour's shape and size do not change.

Define a rigid motion, $g : \Re^2 \rightarrow \Re^2$, by,

$$g(\hat{\mathbf{C}}) = R(\hat{\mathbf{C}} - \mathbf{m}) + T + \mathbf{m}, \tag{5}$$

where R is the rotation matrix that characterizes rotation around the point $\mathbf{m}$. R is parameterized by rotation angle θ. T is the translation vector parameterized by T^x and T^y, the x and y components of the translation.

We can convert the explicit evolution of the contour, $\mathbf{C}_t$, into an evolution of the rigid motion parameters, θ, T^x, and T^y. This is done by evolving the rigid group parameters to maximize the inner product

$$J = \int_0^L \left(-N_z \mathbf{n} \cdot \mathbf{n}\right) \left(\mathbf{g} \cdot \mathbf{n}\right) ds, \tag{6}$$

where ds is the arc length element and L is the length of the contour.

In this fashion we obtain a gradient evolution for the rigid group parameters that ensures motion of the contour in the direction that minimizes surface area. They are,

$$\theta_t = J_\theta = \int_0^L \left(-N_z \mathbf{n} \cdot \mathbf{n}\right) \left(\mathbf{g}_\theta \cdot \mathbf{n}\right) ds = -\int_0^L N_z \left(\mathbf{g}_\theta \cdot \mathbf{n}\right) ds \tag{7}$$

$$(T^x)_t = J_{T^x} = \int_0^L \left(-N_z \mathbf{n} \cdot \mathbf{n}\right) \left(\mathbf{g}_{T^x} \cdot \mathbf{n}\right) ds = -\int_0^L N_z \frac{N_x}{\sqrt{N_x^2 + N_y^2}} ds \tag{8}$$

$$(T^y)_t = J_{T^y} = \int_0^L \left(-N_z \mathbf{n} \cdot \mathbf{n}\right) \left(\mathbf{g}_{T^y} \cdot \mathbf{n}\right) ds = -\int_0^L N_z \frac{N_y}{\sqrt{N_x^2 + N_y^2}} ds, \tag{9}$$

where,

$$(\mathbf{g}_\theta \cdot \mathbf{n}) = \frac{1}{\sqrt{N_x^2 + N_y^2}} \begin{bmatrix} -\sin\theta(x - m_x) - \cos\theta(y - m_y) \\ \cos\theta(x - m_x) - \sin\theta(y - m_y) \end{bmatrix} \cdot \begin{bmatrix} N_x \\ N_y \end{bmatrix}. \tag{10}$$

Fig. 2. An example of rigid registration. Initial surface between two unregistered rectangles (left), minimal surface between unregistered rectangles (second from left), surface during rigid registration (second from right), and surface after rigid registration (right).

Evolution of the rigid registration parameters in this fashion will change the top contour. In the proposed registration method, we evolve the registration of the top contour by equations (7), (8), and (9) while ensuring that the surface continues to stay minimal by equation (2). Continued evolution of the top contour in this fashion, while ensuring that the connecting surface remains minimal, produces a rigid registration of the contour $\hat{\mathbf{C}}$, summarizing our proposed method.

In Fig. 2 we show an example of the alignment of two misaligned rectangles while constraining the registration to consist of only rotations and translations, as described in this section. The correct registration of these two shapes is a 30 degree rotation followed by a 10 pixel translation in the x direction and a 20 pixel translation in the y direction. The proposed minimal surface method produces a registration of 29.3 degrees followed by an x translation of 9.8 pixels and a y translation of 19.6 pixels, accurate to within half of a pixel.

2.4   Motivation

In this section we will motivate the proposed method by explaining its connection to set symmetric difference registration. In fact, we will show that set symmetric difference registration is a special case of the proposed method. The cost functional in set symmetric difference registration is, as the name suggests, the mismatch between the interiors of the two contours. We will explain how this mismatch generalizes to surface area, thus motivating surface area as the new cost functional. In addition, we will explain how the proposed method eliminates local minima that occur when registering by minimizing SSD, thus justifying the proposed method as a useful generalization. We show an example of the proposed method avoiding a registration local minimum.

Set Symmetric Difference. Let A and B be sets that are the interiors of two contours. The SSD of A and B is, $A \cup B - A \cap B$, that is, all points in the union that are not in the intersection. The area of the SSD gives a measure for how poorly two contours are aligned. Perfect alignment of two contours yields an area of 0. Very poorly aligned contours yield very high SSD areas. These

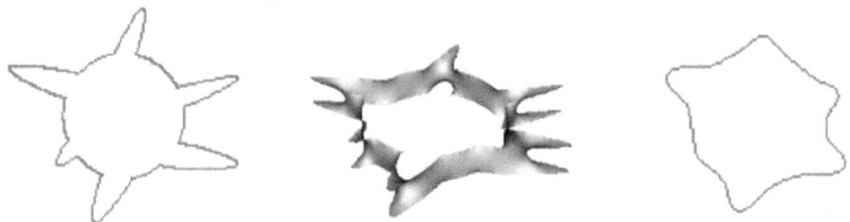

Fig. 3. An illustration of the smoothing effect of minimal surface on the intermediate contours. Original contour (left), minimal surface connecting this contour with replica of itself (center), middle contour of minimal surface (right).

properties make it a natural cost functional to measure the alignment between two contours.

The minimal surface method proposed in this paper is a generalization of SSD registration. This can be seen by imagining the set symmetric difference to comprise the area of a surface as the two contours are pulled apart in space.

Therefore, SSD registration can be viewed as a special case of the proposed method, when the separation distance, $z_M = 0$, in which the area of the connecting surface simplifies to the area of the SSD.

Elimination of Local Minima. Evolving to minimize the area of the SSD will produce usable results when the interiors of the contours are convex. However, certain contours with nonconvex interiors develop local minima when set symmetric difference registration is used. This happens when the registration must pass through a region of higher SSD area in order to reach the minimum SSD area. Since the evolution method will not pass through this region of higher SSD area, the resulting registration is a suboptimal solution where SSD area is locally minimized but not globally minimized. We will refer to this as a local minimum.

Minimal surfaces, having constant zero mean curvature, are smooth. In addition, as the separation distance z_M increases, the intermediate contours between $z = 0$ and $z = z_M$ become smoother. The surface thus acts as an interface between the two contours that eliminates some of the fine scale structure of the original contours. Fig. 3 shows this smoothing effect. On the left is the original contour that has six protrusions of varying length surrounding a circular center. In the middle is the minimal surface connecting this contour with an identical replica of itself, each of the two contours embedded in their respective contour planes. The right contour is the middle contour for this surface, that is, the zero level set of $\Phi(x, y, z_M/2)$. Notice that, in the middle contour, much of the fine scale structure was eliminated by the smoothing of the mean curvature flow.

As the separation distance, z_M, is increased, the middle contour of the surface becomes more circular and the interface between the two contours becomes more coarse, i.e., the fine scale structure that can produce local minima is eliminated. As a result, the registration evolution may pass through regions in the registration space that would have produced higher surface area for low values

Fig. 4. Contour and registration local minimum. "E"-shaped contour (left), minimal surface connecting contour and translated replica of itself (center), and registration local minimum (right).

Fig. 5. Registration overcoming local minimum by increasing separation distance to 25. Minimal surface connecting contour and translated replica (left), registration overcoming local minimum (center), and final correct registration of top contour (right).

of the separation distance. Figs. 4 and 5 illustrate this concept. Fig. 4 shows an "E"-shaped contour which is of dimension 140 voxels by 90 voxels. In the center, it shows the minimal surface connecting this contour with a translated replica of itself, when the separation distance is $z_M = 12$ voxels. Note that the surface connects the legs of the "E". On the right is the final registration of this contour, which results in an incorrect alignment. The contour does not get to the correct misaligning the legs of the "E" would produce a surface with much higher area, thus making the incorrect alignment on the right a local minimum. Note that SSD registration behaves in a similar fashion, reaching a local minimum in the registration space.

Fig. 5 shows the proposed method when the separation between the two contours is $z_M = 25$. On the left is the minimal surface between the contour and its translated replica. Notice in this case how the minimal surface does not directly connect the legs of the "E", but instead smoothly connects by curving inward, removing the fine scale legs for contours in between the top and bottom. This smooth, inward curving connection allows the registration to pass through the position that was unattainable when the separation distance was too low due to the existence of the fine scale features. The registration passing through this intermediate region is shown in the center of Fig. 5. On the right is the final correct registration, showing how the method overcame the local minimum.

3   Scale Space

3.1   Scale Space Parameter

The natural scale space parameter for this registration method is the separation distance between the two contour planes. As this height becomes smaller, the proposed method behaves more like set symmetric difference registration, and thus the finest scale information in the contour is represented. As this height becomes larger, less fine scale information about the contour is considered. This method acts as a viable scale space because the connection between the contours, namely the surface, becomes smoother as the distance between the contour planes increases.

3.2 Extreme Cases

The two extreme cases of the scale parameter help us understand this method more clearly. As the separation distance approaches zero, this method behaves more closely to set-symmetric difference registration. When the separation distance is too small, the minimal surface closely resembles the shape of the original contours, and the method behaves similarly to SSD registration.

When the separation distance is very large (larger than 1.325 times the exscribed circle), the minimal surface that connects the two contours will fail to exist. Each pair of contours has a critical distance at which the connecting surface exists but won't exist for higher separation distances. This critical distance is the coarsest possible scale at which registration using the proposed method can be performed. When the scale parameter is at its critical value, we expect a low number of registration local minima. In the next section we experimentally support these claims using an example with structure at varying scales. When the separation distance, z_M is greater than $1.325r_{ex}$, where r_{ex} is the radius of the exscribed circle, the minimal surface will not exist for any contours. When the separation distance is less than $1.325r_{in}$, where r_{in} is the radius of the inscribed circle, a connecting minimal surface is guaranteed to exist by the maximum principle of minimal surfaces. This will be further discussed in the appendix. Thus the region of interest for the separation distance is in the range from 0 to $1.325r_{in}$; in this range, registration can be performed with the proposed method.

3.3 Local Minima Decreasing

In this section we will show an example that has multiple local minima at different scale levels that will serve as experimental support that this method is a scale space in the separation distance parameter.

Fig. 6 shows the example contour, that has a multiscale structure. The finest scale features of the contours are the 8 long extended thin protrusions. These thin protrusions create local minima while registering with set symmetric difference or when the separation distance is very low. The intermediate scale features of the contour are the four stems from which the thin protrusions extend. They will create local minima in the presented method when the separation distance is small or medium, but not always when the separation distance is large. The coarsest scale feature of the contour is the circular middle that would remain after several iterations of curvature flow on the contour. A few example of the local minima produced by this method are displayed in Fig. 7.

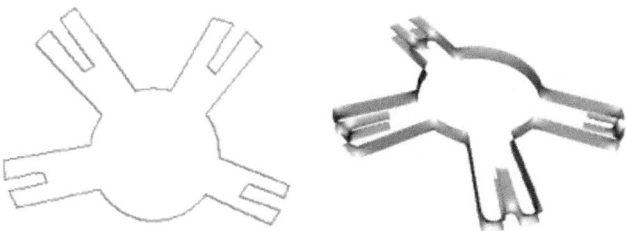

Fig. 6. Contour with features at multiple scales (left) and minimal surface resulting from correct registration of contour with a separation distance of 10 (right).

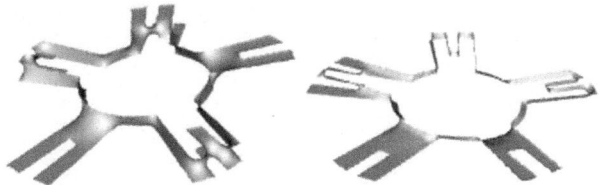

Fig. 7. Two local minima for contour with multiple scale features. The method can not correct for incorrect initial alignment due to local minima. The separation distances for the surfaces 5 and 10 for the left and right surfaces respectively.

Fig. 8 shows a plot of the number of local minima as a function of the separation distance between the two contours for selected separation distances. At the highest separation distances, there are the fewest number of local minima and at the lowest separation distances there are the highest number of local minima. Thus is it experimentally verified that the presented method is a reasonable scale space method.

4   Implementation

In this section we explain appropriate implementation details of the proposed method. First, we will discuss the level set representation of the surface and its details. The level set representation and mean curvature flow of the surface is nearly identical to the method of Chopp in [3], with noted exceptions. Then we will discuss the implementation of the partial differential equations. Finally, we will explain the simultaneous evolution of the registration and minimal surface.

4.1   Level Set Representation of Surface

As explained in Section 2, we implicitly represent the contours and the surface that connects the contours as the level set of the function $\Phi(x, y, z, t)$. To create the minimal surface, we evolve the Φ function with mean curvature flow as stated in Equation (2). Using level sets to represent the curve and surface implicitly

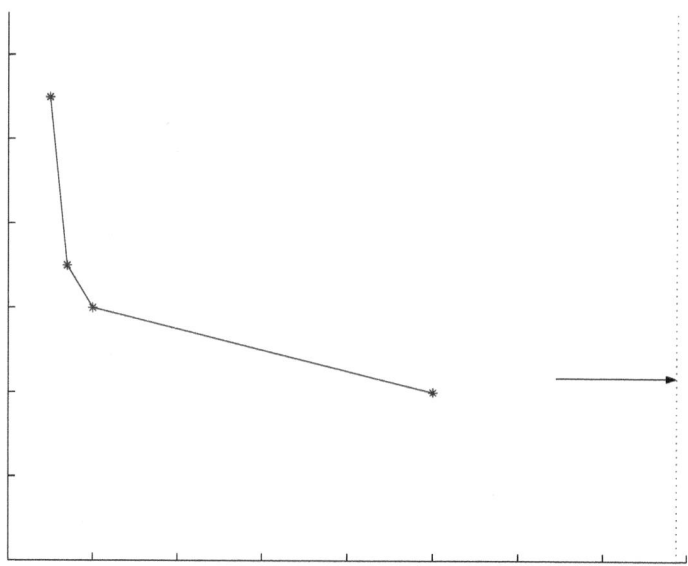

Fig. 8. Number of local minima as a function of the separation distance between the two contours for the contour in Fig. 6. On the right is shown the pinchoff distance, i.e., the separation distance at which the minimal surface fails to exist.

allows us to handle natural topology changes of the surface that occur during mean curvature flow [3].

We represent the interior of the surface as negative level set values and the exterior of the surface as positive level set values. This enables us to get the outward surface normal vector when we compute the gradient, $\nabla\Phi$, at any point on the surface.

Numerical problems can develop if we simply evolve the function Φ. The reason for these numerical problems is that surrounding level sets to the surface can bunch up or spread out. We are primarily interested in what happens to the surface, i.e, the zero level set. Therefore, after each evolution step of the function Φ, in order to avoid these numerical problems, we recalculate the values surrounding the surface in order to maintain equidistance level sets. To do this, we replace the values of the function at points that surround the surface by a value representing the distance to the nearest surface point. This keeps the level sets equally spaced and ensures a numerically well-behaved flow. We then multiply all points interior to the surface by -1, so that the sign convention of the level set function is preserved. This technique, as seen in [3], is known as the signed distance transform.

Although we use the function Φ to represent the surface, we do not store Φ on the entire grid. Since we are only interested in the evolution of the surface and not of the entire Φ function, we only store the values of Φ on a narrow band near the zero level set. Only storing values on this narrow band reduces computational cost significantly [10].

4.2 Freezing of Top and Bottom Contours

When we evolve the surface with Equation (2) in order to minimize surface area, we want to keep the top and bottom contours constant. Unlike the method proposed in [3], we freeze the top and bottom contours by simply not evolving the Φ function when $z = 0$ or when $z = z_M$. We still recompute the level set functions on the top and bottom contours with the signed distance transform, as mentioned above, to keep the level sets regularly spaced. Although Chopp's method, enforcing chains where the fixed contours lie, is more general, we do not need such generality since the contours are always in the top and bottom planes.

4.3 PDE Implementation

Spatial Derivatives. Implementation of this work relies on calculation of spatial derivatives on the discrete grids. On the interior of the three dimensional region, we estimate spatial derivatives using the central difference approximation. On the boundaries of the three dimensional region we use the one-sided difference approximation.

Mean Curvature Flow. The temporal derivatives in Equation (2) is implemented with Euler's method. Euler's method is equivalent to estimating the time derivative by its one-sided finite difference approximation, that is,

$$\Phi_t = (\Phi(x, y, z, t + \Delta t) - \Phi(x, y, z, t)) / \Delta t. \tag{11}$$

Substituting Φ_t with the evolution in Equation (2) and rearranging, we obtain the update equation for Φ,

$$\Phi(x, y, z, t + \Delta t) = \Phi(x, y, z, t) + \Delta t \, H(x, y, z, t) \| \nabla \Phi(x, y, z, t) \|. \tag{12}$$

Note that the Δt parameter is known as the time step, and is chosen as large as possible, as long as the flow will remain stable and well behaved. Also note that the implementation of Equation (12) is spatially discrete, that is, the mean curvature is computed with discrete spatial derivatives as mentioned above.

We evolve the surface with mean curvature flow, i.e., the flow that minimizes surface area, until the surface area decreases slowly enough to be considered nearly minimal. We stop the mean curvature flow when the difference in surface area between two successive time steps of Equation (12) is less than 0.01. In this manner, we ensure that the surface connecting the two contours is nearly minimal.

Registration Flow. The proposed rigid registration method relies on the surface already being minimal. Once the surface is minimal, as described in the previous section, we evolve the registration parameters by Equations (7), (8), and (9). We evolve these parameters with the forward Euler method, obtaining,

$$\theta(n+1) = \theta(n) + \eta J_\theta \tag{13}$$
$$T^x(n+1) = T^x(n) + \eta J_{T^x} \tag{14}$$
$$T^y(n+1) = T^y(n) + \eta J_{T^y}. \tag{15}$$

Each time we update the rigid registration parameters, we are moving the top contour to reflect this movement. If we move the top contour too much, the contour will no longer be connected to the surface. This is undesirable in our proposed method. Therefore, η is often scaled to prevent the contour becoming disconnected from the surface.

Note that here we use the letter n to denote the time step and not t as in Equation (12). This is because of the nesting of the algorithm as will be described in the next section.

Simultaneous Evolution Algorithm. The algorithm we used in our proposed method is, to first initialize by embedding the contours within their respective contour planes, then to initialize and minimize the connecting surface, then to modify the registration parameters while ensuring that the surface remains minimal. The structure of this algorithm is,

1. Embed contours in respective contour planes
2. Initialize surface
3. Evolve surface with mean curvature flow until minimized
4. Evolve registration parameters once
5. Go to step 3 if registration is not converged.

Step 3 of this algorithm involves evolving with mean curvature flow as per Equation (12). This evolutions happens separately from the evolutions of the registration parameters, and thus they are denoted with time variable t. Each instance of step 3 requires multiple iterations of Equation (12). Step 4 of the algorithm happens when the surface is minimal and only occurs once per surface minimization. By nesting the surface minimization within the evolution of the registration parameters, we achieve the desired effect of always having a minimal surface guiding the motion of the registration parameters.

5   Conclusion

We have presented a novel method for registration of two dimensional images that not only a geometric extension of set symmetric difference registration, but is also a scale space. The method rigidly registers contours by evolving both the registration of a contour and the surface that connects them by minimizing a unified energy functional, i.e., surface area. We have explained the method in theory and its implementation and shown a number of examples. One particular example showed how the number of local minima decreased with increasing scale parameter. We have also identified a sufficient condition for existence of the minimal surface that connects two contours.

6 Appendix

We will explain a sufficient condition for existence of the minimal surface that connects two contours. The first section will describe the condition for existence of the catenoid. The second section will explain the condition for existence of the minimal surface connecting two contours using the maximum principle of minimal surfaces.

6.1 Existence of Catenoid

Parameterization. The catenoid is the only surface of revolution with the topology of a cylinder that is a minimal surface [5]. The parametric equation of the catenoid is,

$$(x(u,v), y(u,v), z(u,v)) = (p\cosh(v/p)\cos(u), p\cosh(v/p)\sin(u), v) , \quad (16)$$

where $u \in [0, 2\pi)$ and $v \in [-z_H/2, z_H/2]$ are the surfaces parameters and p is the radius of the middle contour. Note that in the appendix we use the convention that the top contour of the catenoid is at $z = z_H/2$ and the bottom contour of the catenoid is at $z = -z_H/2$. Although this does not follow the convention of the paper, it will prove useful in the analysis. It can be shown that this surface has constant zero mean curvature, although we will not show it here.

By setting $u = 0$ and $v = z_H/2$, which represents the top of the catenoid, we get that, $(r, 0, 0) = (p\cosh(z/2p), 0, 0)$, where r is the radius of the circular contour and the top and bottom of the catenoid. This shows that $r = p\cosh(z/2p)$, giving a relation between the radius of the top and bottom circular contours, r, the height of the catenoid, z, and the inner radius of the catenoid which occurs at $z = 0$, p.

Solving this function for positive z results in, $z = 2p\ln\left(\frac{r+\sqrt{r^2-p^2}}{p}\right)$.

When Does It Collapse? By finding out, for what value of z the current catenoid with top radius r fails to exist, we are essentially answering our title question. The exact z where the catenoids fail to exist are when $\frac{\partial r(p,z)}{\partial p} = 0$ for a given r, for this is where the bottom of the constant z curves intersect with the horizontal fixed r lines. This analysis proceeds as follows,

$$r(p, z) = p\cosh\left(\frac{z}{2p}\right) \text{ and } \frac{\partial r}{\partial p} = \cosh\left(\frac{z}{2p}\right) - \frac{z}{2p}\sinh\left(\frac{z}{2p}\right) . \quad (17)$$

Setting this partial derivative equal to zero will give us the appropriate value of p where z is maximum and the catenoid still exists. I will call these critical values p' and z'. Substituting for z and $\frac{r}{p'}$ for $\cosh(z'/2p')$ yields,

$$0 = r - \ln\left(\frac{r + \sqrt{r^2 - p'^2}}{p'}\right)\sqrt{r^2 - p'^2}. \tag{18}$$

By finding, for a given r, the p' that satisfies Eq. (18), we will have found the critical p' below which catenoids fail to exist. That p' will produce the largest possible catenoid height, z'.

An explicit function to get p' from r is, $p'(r) = \alpha r$, where α is the zero of the function,

$$f(x) = 1 - \ln\left(\frac{1 + \sqrt{1 - x^2}}{x}\right)\sqrt{1 - x^2}. \tag{19}$$

It can be seen that $p'(r)$ satisfies Eq. (18). by substitution into Equation (18) if α is a zero of $f(x)$. The zero of $f(x)$ is found numerically to be, $\alpha \approx 0.5524$.

It follows that, $z'(p'(r), r) = 2\alpha \ln\left(\frac{1 + \sqrt{1 - \alpha^2}}{\alpha}\right) r = \beta r$ where,

$$\beta = 2\alpha \ln\left(\frac{1 + \sqrt{1 - \alpha^2}}{\alpha}\right) \approx 1.325. \tag{20}$$

These results show that the critical p' and z' for a given r are given by, $p' = \alpha r$ and $z' = \beta r$. That is, the critical inner radius, p', and critical height, z', of a catenoid of fixed top radius are each directly proportional to the fixed top radius. Therefore, a catenoid with radius r will exist for heights less than or equal to $z = 1.325r$

6.2   Existence of Minimal Surface for Arbitrary Closed Contours

Arbirtary contours that are not concentric circles yield surfaces that are not catenoids and typically have no closed form parameterization. Consider an arbitrary closed contour, C, and its identical, aligned replica, C', lying directly below in, separated in respective contour planes by a distance z_M. In each plane, the contour lies between its inscribed circle and its exscribed circle with radii r_{in} and r_{ex} respectively. If we choose the separation distance so that $z_M < 1.325r_{in}$, a catenoid exists that connects the inscribed circles, as proven in Section 6.1. Since $r_{in} \leq r_{ex}$, and thus, $z_M < 1.325r_{in} \leq 1.325r_{ex}$. It follows that the catenoid of radius r_{ex} exists for this chosen z_M. It then follows from the maximum principle of minimal surfaces, that the minimal surface with contours C and C' as its boundary will exist and will completely contain the catenoid connecting the inscribed circles, that is, the minimal surface will exist for the chosen separation distance $z_M < 1.325r_{in}$.

This argument can be extended to show existence for the minimal surface connecting the top contour, D, and bottom contour E. Consider the inscribed circle lying within the intersection of the interior of the two contours. A catenoid with this radius, r_b, will be completely contained in the minimal surface connecting D and E and therefore by a similar argument as above, the minimal surface connecting D and E will exist when the separation distance is less than $1.325r_b$.

References

1. K. A. Brakke, *The Motional of a Surface by its Mean Curvature*, Princeton University Press, 1978.
2. L. Brown, "A survey of image registration techniques," *ACM Computing Surveys*, vol. 24, no. 4, pp. 325–376, 1992.
3. D. L. Chopp, "Computing minimal surfaces via level set curvature flow," *Journal of Computational Physics*, vol. 106, pp. 77–91, 1993.
4. J. Feldmar and N. Ayache, "Rigid, affine and local affine registration of free-form surfaces," *IJCV*, vol. 18, no. 2, pp.99-119, 1996.
5. D. Hoffman and W. H Meeks, "Minimal surfaces based on the catenoid," *Amer. Math. Monthly*, vol. 97, no. 8, pp. 702-730, 1990.
6. B. Hansen and B. Morse, "Multiscale Registration Using Scale Trace Correlation," *CVPR*. 1999.
7. C. Isenberg, *The Science of Soap Films and Soap Bubbles*, Dover Publications Inc., 1978.
8. H. Lester and S. R. Arridge "A survey of hierarchical non-linear medical image registration," *Pattern Recognition*, vol. 32, pp 129–149, 1999.
9. T. B. Sebastian, P. N. Klein and B. B. Kimia, "Alignment-based Recognition of Shape Outlines," *IWVF 2001* pp. 606–618, 2001.
10. J. A. Sethian, *Level Set Methods and Fast Marching Methods*, Cambridge University Press, 1999.
11. H. D. Tagare, D. O'Shea, and A. Rangarajan, "A geometric criterion for shape based non-rigid correspondece", *Fifth Intl. Conf. on Computer Vision (ICCV)*, pp. 434–439, 1995.
12. J. B. West, J. M. Fitzpatrick, and P. G. Batchelor. "Point-Based Registration under a Similarity Transform," *Proceedings SPIE*, 2001.
13. Z. Zhang, "Iterative point matching for registration of free-form curves and surfaces", *IJCV*, vol. 13, no. 2, pp. 147–176, 1994.

The Extrema Edges

Pablo Andrés Arbeláez and Laurent D. Cohen

CEREMADE, UMR CNRS 7534 Université Paris Dauphine
Place du maréchal de Lattre de Tassigny
75775 Paris cedex 16, France
{arbelaez,cohen}@ceremade.dauphine.fr

Abstract. We present a new approach to model edges in monochrome images. The method is divided in two parts: the localization of possible edge points and their valuation. The first part is based on the theory of minimal paths, where the selection of an energy and a set of sources determines a partition of the domain. Then, the valuation is obtained by the creation of a contrast driven hierarchy of partitions. The method uses only the original image and supplies a set of closed contours that preserve semantically important characteristics of edges.

1   Introduction

The presence of sharp discontinuities in the image intensity seems to play a fundamental role for the interpretation of visual information in humans. Therefore, edge detection has been a very active field of research since the early days of computer vision. Originally, edge detection techniques were motivated by the generalization to the plane of signal processing methods and the adaptation of regular analysis tools to the discrete domain. Thus, differentiation appeared as the natural operation to address the problem. Many estimations of the image derivatives and models for the edges have been proposed in the last decades. Examples include the zero crossings of the Laplacian [22], the maxima in the gradient direction [4] and the crest lines of the gradient's modulus. However, in spite of their diversity, the strategy in many edge detection methods consists in a differential approach and the use of local image information to measure the relevance of the edge points [30,10].

The classical approach to address this issue in the context of mathematical morphology is the characterization of edges as the watershed lines of the gradient's modulus [2,31]. Among the reasons for the large popularity of this method one can cite its intuitive definition, efficient algorithms for its implementation and the fact that the watersheds supply a set of closed edges. In the regular framework, the watersheds were defined as the skeleton by influence zones of a determined distance function [27]. These ideas inspired a construction of the watersheds using curve evolution [21].

The proposed approach to model the edges in an image follows the opposite direction. Our starting point is the theory of minimal paths, described in Section 2, where a partition of the domain is determined by the choice of an

L.D. Griffin and M. Lillholm (Eds.): Scale-Space 2003, LNCS 2695, pp. 180–195, 2003.

energy and a set of sources. In Section 3, we introduce an energy called the *path variation*, a generalization of the one dimensional total variation for functions of two variables. This energy preserves the geometric structure of the function and allows to work directly on the original image. In Section 4, the choice of the intensity extrema as sources provides a piecewise constant simplification of the image, whose discontinuities are designated as the *extrema edges* of the image. Finally, in Section 5, we consider the valuation of the extrema edges using global image information. For this purpose, a family of nested partitions, guided by a notion of contrast, is constructed.

2   Minimal Paths and Energy Partitions

This introductory section presents the mathematical framework for the rest of the paper. Basic definitions are recalled and the notations settled.

Let $\Omega \subset \mathbb{R}^2$ be a compact connected domain in the plane and $x, y \in \Omega$ two points. A *path* from x to y designates a continuous function $\gamma : [a, b] \to \Omega$ such that $\gamma(a) = x$ and $\gamma(b) = y$. The image of γ is then a curve in Ω. If $\gamma \in \mathcal{C}^1([0, L])$ and we consider an arc-length parametrization of γ (i.e. $\|\dot{\gamma}(s)\| = 1$, $\forall s \in [0, L]$), then L represents the Euclidean length of the path and its image is a rectifiable simple curve. The set of paths from x to y is noted by Γ_{xy} and the set of paths in Ω is noted by Γ_Ω.

Definition 1. *The **surface of minimal action**, or **energy**, of a potential function $P : \Omega \times \mathcal{S}^1 \to \mathbb{R}^+$ with respect to a source point $x_0 \in \Omega$, evaluated at x, is defined as*

$$E_0(x) = \inf_{\gamma \in \Gamma_{x_0 x}} \int_0^L P(\gamma(s), \dot{\gamma}(s)) \, ds \ .$$

When P depends only on the position $\gamma(s)$ and is strictly positive, the computation of the energy can be performed using Sethian's *Fast Marching* method [33], as detailed in [6].

The surface of minimal action with respect to a set of sources $S = \{x_i\}_{i \in J}$ is defined as the minimal individual energy:

$$E_S(x) = \inf_{i \in J} E_i(x) \ .$$

In the presence of multiple sources, a valuable information is provided by the interaction in the domain of a source x_i with the other elements of S, which is expressed through its *influence zone*:

$$Z_i = \{x \in \Omega \,|\, E_i(x) < E_j(x), \forall j \in J\} \ .$$

Thus, the influence zone, or briefly the *zone*, is a connected subset of the domain, completely determined by the energy and the rest of the sources. Their union is noted by:

$$Z(E, S) = \bigcup_{i \in J} Z_i \ .$$

The *medial set* is defined as the complementary set of $Z(E, S)$:

$$M(E, S) = \{x \in \Omega \mid \exists\, i, j \in J,\, i \neq j : E_S(x) = E_i(x) = E_j(x)\} .$$

Therefore, the selection of an energy and a set of sources defines an *energy partition* $\Pi(E, S)$ of the domain:

$$\Pi(E, S) = Z(E, S) \bigcup M(E, S) .$$

Energy minimizing paths have been used to address several problems in the field of computer vision, where the potential is generally defined as a function of the image. Examples include the global minimum for active contour models [6], shape from shading [17], continuous scale morphology [18], virtual endoscopy [8] and perceptual grouping [5].

3   The Path Variation

In the usual approach for the application of minimal paths to image analysis, a large part of the problem consists in the design of a relevant potential for a specific situation and type of images. However, we adopt a different perspective and use the notions of the previous section for the study of a particular energy, whose definition depends only on geometric properties of the image.

3.1   Definition

For functions of one real variable, the variation is a functional with known properties [14,29]. It was introduced by Jordan [16] as follows:
Let $f : [0, L] \to \mathbb{R}$ be a function, $\sigma = \{s_0, ..., s_n\}$ a finite partition of $[0, L]$ such that $0 = s_0 < s_1 < ... < s_n = L$ and Φ the set of such partitions.
The *variation*, or *total variation*, of f is defined as

$$v(f) = \sup_{\sigma \in \Phi} \sum_{i=1}^{n} |f(s_i) - f(s_{i-1})| .$$

If $f \in C^1([0, L])$, then the variation can be expressed as:

$$v(f) = \int_0^L |f'(s)|\, ds . \tag{1}$$

The path variation is a generalization of the total variation for two variable functions:

Definition 2. *The **path variation** of a function* $u : \Omega \subset \mathbb{R}^2 \to \mathbb{R}$ *with respect to a source point* $x_0 \in \Omega$, *evaluated at* x, *is defined as*

$$V_0(u)(x) = \inf_{\gamma \in \Gamma_{x_0 x}} v(u \circ \gamma) .$$

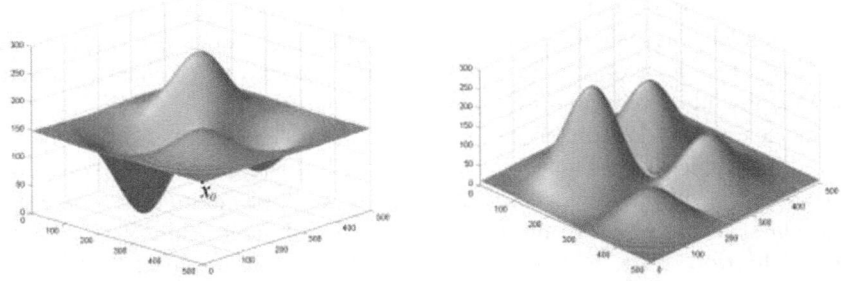

Fig. 1. Simple example: graphs of u and $V_0(u)$.

Thus, the path variation between two points is given by the minimal variation of the function on all the paths that join them.

Definition 3. *The space of functions of* **bounded path variation** *of Ω, noted by $BPV(\Omega)$ is defined by*

$$BPV(\Omega) = \{u : \Omega \to \mathbb{R} \mid \forall x_0, x \in \Omega, \exists \hat{\gamma} \in \Gamma_{x_0 x} : V_0(u)(x) = v(u \circ \hat{\gamma}) < \infty\} .$$

In the sequel, we suppose that u has bounded path variation.

If u is a continuously differentiable function, then (1) allows to reformulate $V_0(u)$ as

$$V_0(u)(x) = \inf_{\gamma \in \Gamma_{x_0 x}} \int_0^L |D_\gamma u(\gamma(s))| \, ds . \tag{2}$$

Hence, $V_0(u)$ may be seen as a surface of minimal action for the potential $P = |D_{\hat{\gamma}} u|$, the absolute value of the directional derivative of u in the tangent direction of the path.

The intuitive interpretation of the path variation is illustrated in Fig. 1: consider a particle moving along the graph of the function depicted on the left and starting at the source x_0. Then, as shown on the right, the value of $V_0(u)$ evaluated at x represents the minimal sum of ascents and descents to be travelled to reach the point x.

The path variation expresses the same notion as the concept of *linear variation*, introduced in [19], though in a formulation without paths, as a part of a geometric theory for functions of two variables .

The *component* of u containing x, noted by K_x, designates the maximal connected subset of Ω such that $u(y) = u(x)$, $\forall y \in K_x$. The importance of the components for the path variation is given by the following proposition, whose proof is an immediate consequence of Def. 2.

Proposition 1. *The path variation acts on the components of the function:*

$$\forall x, y \in \Omega, \ K_x = K_y \Rightarrow \forall x_0, \ V_0(u)(x) = V_0(u)(y) .$$

Therefore, each element of an energy partition induced by the path variation is a union of components of the function. Thus, the operator that associates $\Pi(V(u), S)$ to a set of sources S is connected [32] and its application simplifies the image while preserving its geometrical structure.

In the discrete domain, the component structure of the function can be represented in a region adjacency graph. Hence, with this approach, the computation of the path variation is reduced to finding the path of minimal cost on a graph. This classical problem can be solved using a greedy algorithm [9,20]. For a discrete definition of the path variation and implementation details, the reader is referred to [1].

3.2   Path Variation and Image Distance

In the context of mathematical morphology, the surface of minimal action associated to the potential $P = \|\nabla u\|$, given by the formula:

$$W_0(u)(x) = \inf_{\gamma \in \Gamma_{x_0 x}} \int_0^L \|\nabla u(\gamma(s))\| \, ds$$

was used to define the watershed transform in the continuous domain [26,23]. If, as for the class of Morse functions, u has only isolated critical points, then W_0 induces a distance transform on Ω, called the *image distance* [26] or the *topographic distance* [23].

The relation between W and V in the regular framework is expressed by the following property:

Proposition 2. *If u is a Morse image, $u \in BPV(\Omega)$ and $x_0 \in \Omega$, then*

$$|u(x) - u(x_0)| \leq V_0(u)(x) \leq W_0(u)(x), \ \forall x \in \Omega \ .$$

In particular, if x and x_0 belong to a line of steepest slope for u, then

$$|u(x) - u(x_0)| = V_0(u)(x) = W_0(u)(x) \ ,$$

The proof of this proposition [1] follows from simple calculus and the fact that, by definition, $|D_{\gamma}(u)| = \|\nabla u\|$ when $\dot{\gamma}$ is parallel to the gradient.

The behavior of these two energies can be compared using the test image shown on the right column of Fig. 2 and given by the simple formula $u(x) = c\|x - x_0\|$. The set of sources in this case is $S = \{x_0, x_1\}$, where x_0 is the upper left and x_1 the lower right corners of the domain. The central column shows the effect of the path variation: as a consequence of Prop. 1, u and $V_S(u)$ have in this example the same components and only their level is modified. The medial set $M(V(u), S)$, shown on black, is the component whose level is the average of the sources' levels. On the right column, we can observe that, since $\|\nabla u\|$ is constant, $W_S(u)$ is proportional to the Euclidean distance to the closest

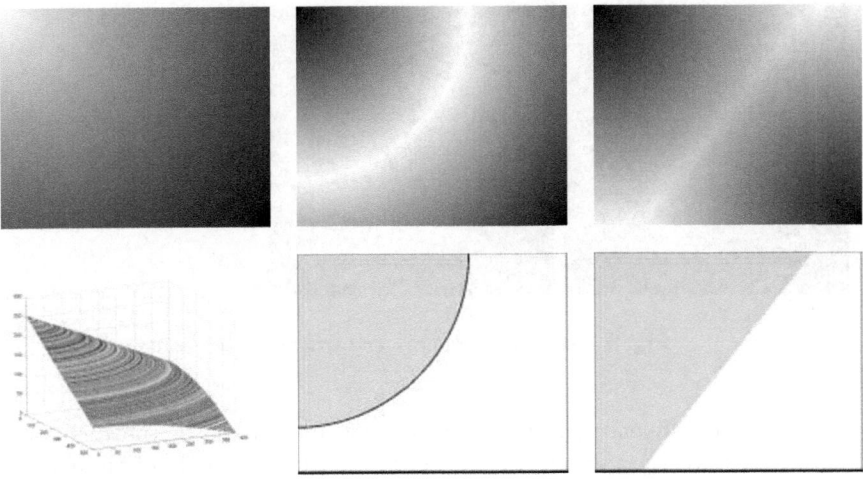

Fig. 2. From left to right. Top: u, $V_S(u)$ and $W_S(u)$. Bottom: Graph of u, energy partitions $\Pi(V(u), S)$ and $\Pi(W(u), S)$.

source and $M(W(u), S)$ corresponds to the medial line between the sources; however, in this example, the medial set falls in the intergrid space. Note that any other function for which $\|\nabla u\|$ is constant would produce the same partition $\Pi(W(u), S)$. This example illustrates how $\Pi(V(u), S)$, the partition induced by V, preserves the image structure better than $\Pi(W(u), S)$.

4   The Extrema Edges

4.1   The Extrema Partition

Surfaces of minimal action are often appropriated for a local level of analysis in the image. This is due to the fact that Def. 1 is based on an integration along the paths. Consequently, this type of energies may lose their meaning when the zones become too large. Besides, replacing a source $x_i \in S$ by another point $x'_i \in Z_i$ usually modifies the resulting energy partition.

Therefore, in order to construct an energy partition based on the path variation, the set of sources must be selected with care. Firstly, they should be physically representative of the image content. Secondly, each significant feature should contain at least one of them. Since they satisfy these conditions, the intensity extrema appear as natural candidates for the sources.

Definition 4. *The **extrema partition** of an image $u : \Omega \to \mathbf{R}$ is defined as $\Pi(V(u), ext(u))$, the energy partition induced by the path variation and the set of extremal components of u.*

Thus, Prop. 1 implies that the elements of the extrema partition are unions of components of u. By definition, they can be divided in two types: on the

Fig. 3. Test image, energy end extrema partition.

one hand, the influence zones of the extrema, interpreted as the *atoms* or *elemental zones* of the image; on the other hand, the elements of the medial set $M(V(u), ext(u))$ are designated as *boundary components* of the atoms.

Figure 3 illustrates our approach on a simple regular case. The function u, on the left, is a Gaussian blob, where the only extremal components are the center and the border of the squared domain. The image on the middle shows the energy $V_{ext(u)}(u)$, rescaled by a factor of 2 for better visualization. On the right, the extrema partition $\Pi(V(u), ext(u))$ is composed by two elemental zones and a circular boundary component, fragmented by the quantization.

4.2   Definition of the Extrema Edges

The effects of the extrema partition on smooth functions, suggest the use of the boundary components to model the edges in the image. Nevertheless, in practice, digital images are subsampled on a discrete grid. Consequently, as noted in the previous examples, important parts of the medial set may fall in the intergrid space. An alternative to surround this problem is to consider an energy partition composed only by zones. Thus, the elements of the medial set that would fall exactly in the grid are assigned to one of their neighboring influence zones.

Then, an approximation of the image can be constructed by the assignation of a *model* to represent each influence zone. The model is determined by the distribution of the image values; simple models are the mean or median value in the zone, or the level at the source. When the model is constant, the valuation of each zone by its model produces a piecewise constant approximation of the image, referred in the sequel as a *mosaic image*. The mosaic corresponding to the extrema partition will be called the *extrema mosaic* of u. Generally, on real images, the intensity at the extremum represents accurately the atom's levels. Hence, unless stated differently, this model was chosen.

The choice of the path variation as the energy and the spatial distribution of the intensity extrema provide a compromise between content conservation and simplification in the extrema mosaic. Perceptually, the effects of this piecewise constant approximation of the image can be better appreciated when the ratio between the number of components in the original image and the number of

Fig. 4. Image and extrema mosaic.

atoms is high. Figure 4 shows an example where this ratio is 68. On the left, we can observe the original image and, on the right, its extrema mosaic. This image illustrates well three important properties of our approach. First, a contrast enhancement in the butterfly's wings, mainly due to the choice of the zone model. Second, a reduction of the blur in the background, caused by the absorption of blurred contours and transition zones by neighboring atoms. Last, but not least, note how the boundaries of the atoms model accurately the contour information and, particularly, semantically important characteristics of edges such as corners and junctions. Therefore, they constitute a sound set of closed curves to search for edges in the image:

Definition 5. *The **extrema edges** of an image are defined as the discontinuities of its extrema mosaic.*

4.3 Extrema Edges and Watersheds

In mathematical morphology, the edges in an image u are usually modelled as the watershed lines of its gradient's modulus, $g = \|\nabla u\|$ [2,31]. Their construction can then be obtained by a *flooding* process: a gradient image, seen as a topographical surface, is pierced at its regional minima and progressively immersed in water. The water floods uniformly the valleys, or catchment basins of the minima, and, at the points where two lakes meet, a dam is built. When the surface is totally immersed, the union of the dams forms the watershed lines. This interpretation of the watershed transform inspired efficient algorithms for its implementation [34] and allowed the formalization of the watersheds in the continuous domain as the skeleton by influence zones of the image distance [26]. Furthermore, it suggested the interpretation of the minima of g as the dual concept of edges: the sources. In our notation, starting at a source $x_0 \in \Omega$, this energy can be written as

$$\widehat{W}_0(g) = W_0(g) + g(x_0) \ . \tag{3}$$

Thus, the energy associated to the segmentation by watersheds of an image u can be expressed as

$$\widehat{W}_{min(g)}(g) = \inf_{m_i \in min(g)} \widehat{W}_i(g) \ ,$$

where $min(g)$ denotes the set of regional minima of g. This continuous formulation motivated the implementation of the watersheds using the Fast Marching method [21].

Therefore, $\Pi(\widehat{W}(g), min(g))$, the energy partition associated to the watershed transform, has the following interpretation: the medial set $M(\widehat{W}(g), min(g))$ corresponds to the watershed lines of g and represent the edges in u. Besides, $Z(\widehat{W}(g), min(g))$, the zones of the minima, coincide with the lakes, or catchment basins of the topographical surface.

If we use V instead of W in (3), we obtain the following result, whose proof is based on Prop. 1 and 2 and the fact that, for Morse images, each catchment basin corresponds to the set of lines of steepest slope ending at its minimum [26].

Proposition 3. *If g is a Morse image and $g \in BPV(\Omega)$, then*

$$M(\widehat{V}(g), min(g)) = \bigcup_{x \in M(\widehat{W}(g), min(g))} K_x$$

Thus, the medial set of $\widehat{V}$ coincides with the set of components of the watershed lines. Hence, in the continuous domain, the use of V on the gradient generally produces edges thicker than the watersheds.

In practice, as happens for the boundary components, the watersheds are usually fragmented in real images. Therefore, in order to compare the extrema edges and the watersheds, we used their corresponding mosaics. Indeed, the construction of both mosaics depends on the same factors: the digital connectivity, the gray level on the zones and the rule of assignation for the medial set. However, the fundamental difference is that the former is defined in the original image, while the latter is built on the modulus of its gradient. Consequently, the watershed lines depend also on the choice of a discrete approximation of the gradient. Moreover, a smoothing step is usually performed by most gradient operators in order to well pose differentiation [30,10]. Since the smoothing implies a loss of information in the image content, the watersheds suffer from limited resolution in certain cases. These problems cannot be neglected in fields where the precision of the extracted features is an essential issue, as in medical image analysis.

Figure 5 depicts the mosaics associated to the different models of edges presented. The first row shows the original image, a detail of the *cameraman*, and the extrema mosaic. The second row depicts, on the left, the watershed mosaic

Fig. 5. Above: original image and extrema mosaic. Below: mosaics of the energy partitions $\Pi(\widehat{W}(g), min(g))$ and $\Pi(\widehat{V}(g), min(g))$.

constructed on the morphological gradient and, on the right, the mosaic corresponding to the choice of $\widehat{V}$ as the energy and the gradient's minima as sources. For all the cases 8-connectivity was used, the zone model was the source's level and the points in the medial set were assigned to the first source to reach them. As a consequence of the spatial distribution of the sources and their large number, all the methods preserve the main features in the scene, such as the silhouette of the man. However, the extrema mosaic enhances perceptually important details like the mouth or the inner parts of the camera that are lost in the mosaics of the second row. The loss of information is due to the absence of regional minima inside those features and, even if the result may be improved by changing the type of gradient operator or the connectivity, the problem is intrinsic to the use of the gradient image. Finally, since Prop. 3 implies that the partitions $\Pi(\widehat{W}(g), min(g))$ and $\Pi(\widehat{V}(g), min(g))$ differ mainly in their medial set, the two mosaics in the second row are almost identical.

5   Valuation of the Extrema Edges

Once a set of candidates for the edge points has been determined, the next problem is the integration of this local information into meaningful curves. In this section, we propose to construct a contrast driven hierarchy of partitions to provide global image information for the valuation of the extrema edges.

The idea of progressively merging regions of an initial partition has been used for a long time to address image segmentation problems [3,15,7,13,25]. In general, this type of methods can be implemented efficiently using a region adjacency graph (RAG), as described in [35,11].

A RAG is an undirected graph where the nodes correspond to connected regions of the domain. The links encode the vicinity relation and are weighted by a *dissimilarity* measure. The dissimilarity δ is a function defined for every couple of neighboring regions. It takes values in an interval $I = [0, \Lambda]$, referred in the sequel as the set of *indices* or *scales*.

Then, removing the links of the RAG for increasing values of the dissimilarity and merging the corresponding regions produces a family of nested partitions, or hierarchy, $\{\mathcal{P}_\lambda\}_{\lambda \in I}$, where every region in $\mathcal{P}_\mu$ is a disjoint union of regions in $\mathcal{P}_\lambda$, for $\mu \geq \lambda$. Therefore, in this context, the selection of the initial partition and the dissimilarity measure determines the resulting hierarchy.

The watershed flooding provides a classical example of hierarchical segmentation: the gradient's modulus is again flooded from its minima but, instead of building a dam at the meeting points, the lakes merge. Increasing levels of water produce coarser partitions and the resulting hierarchy is known as the *dynamics* [12]. In terms of a region merging process, the initial partition is composed by the watershed mosaic and the dissimilarity is defined as the height of the saddle point between two adjacent lakes, i.e. the minimal value of the gradient in the common border of the regions [24].

Since our purpose was to construct a contrast driven hierarchy, the dissimilarity was measured on the initial partition and only boundary information was taken into account. Thus, we considered a local dissimilarity: the absolute value of the gray level difference of neighboring regions on the initial partition. Then, the dissimilarity was defined as a function of the local dissimilarities' distribution in the common boundary of the regions. For the examples presented in this paper, the dissimilarity was the average of the local dissimilarities. The resulting hierarchy is noted by $\mathcal{H}$.

Figure 6 illustrates the application of $\mathcal{H}$. The left image displays the initial partition, the extrema mosaic of the *cameraman*. On the right, we can observe the segmentation corresponding to the scale $\lambda = 54$. Note how $\mathcal{H}$ expresses the perceived contrast in the image; at the scale presented, only contrasted regions remain in the segmentation, regardless of their size.

In order to measure the relevance of the extrema edges, the notion of *saliency image* of a hierarchy presents a particular interest:

The *saliency* of a pixel, with respect to a hierarchy of partitions $\{\mathcal{P}_\lambda\}_{\lambda \in I}$, is defined as the highest index λ for which the pixel belongs to a boundary of $\mathcal{P}_\lambda$. The valuation of each pixel by its saliency determines a *saliency image*.

Fig. 6. Extrema mosaic and segmentation for the scale $\lambda = 54$.

Fig. 7. Valuated extrema edges and threshold for $\lambda = 54$.

The saliency image provides a compact description of the hierarchy: a threshold λ in this image supplies the set of boundaries of the corresponding partition $\mathcal{P}_\lambda$. Thus, the usefulness of the saliency image is determined by the hierarchy. The saliency image of the dynamics hierarchy was used in [28] to valuate the watersheds.

Definition 6. *The **valuated extrema edges** of an image u correspond to the saliency image associated to the hierarchy $\mathcal{H}$, when the initial partition is the extrema mosaic of u.*

The left image of Fig. 7 shows the valuated extrema edges of the *cameraman*, while the right image displays the threshold corresponding to the scale $\lambda = 54$.

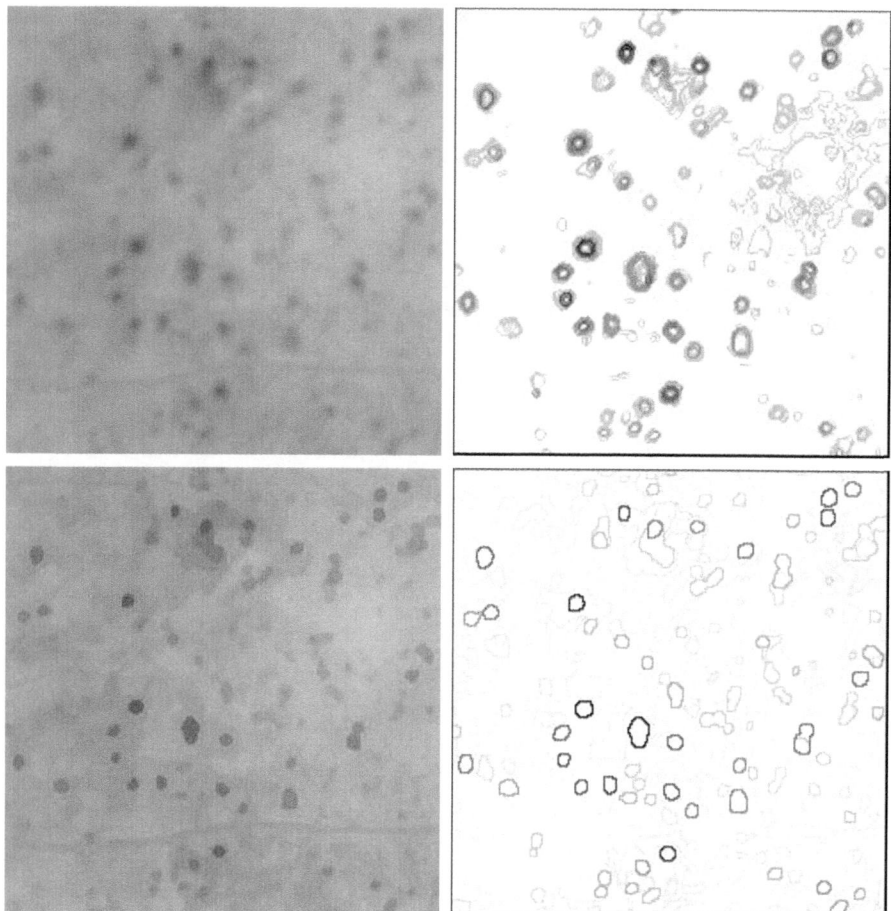

Fig. 8. Row 1: original image and saliency image. Row 2: extrema mosaic and valuated extrema edges.

The main properties of our edge model may be summarized as follows. First, the valuation is obtained using global contrast information and a simple threshold in the valuated extrema edges determines a set of meaningful closed curves. Second, the location of edges is conserved through the scales. Last, but not least, the use of the extrema mosaic preserves the geometric structure of the original image and enhances the semantically important characteristics of edges.

Note that $\mathcal{H}$ can also be applied directly to the original image; however, the use of the extrema mosaic generally improves the quality of the edges obtained. Figure 8 shows an application to medical image analysis where the use of the extrema mosaic as the initial partition is a crucial issue. The goal was to detect a pathology called the *drusen* - the dark spots - in images of the eye fundus, as the one shown on the top left. The variations in the background's intensity in retinal angiographies as well as the absence of abrupt discontinuities in the drusen boundaries make their extraction a difficult problem with classical edge detection methods. The top right image shows the *saliency image* associated to $\mathcal{H}$ when the initial partition is the original image. The image was rescaled for better visualization, but the scale Λ at which a single region remains is only 6. Since the transitions in the original image are smooth, the saliency image produces blurred edges. In contrast, the second row depicts the application of the extrema edges. On the left, we can observe the extrema mosaic, where the drusen can be clearly distinguished from the background. The right image depicts the valuated extrema edges, where $\Lambda = 58$. Note how the method provides the location and the shape of the drusen with precision. Furthermore, their saliency may be used to evaluate the magnitude of the disease.

6   Conclusion and Perspectives

We presented a new approach to model edges in the image. The method is divided in two parts. First, a set of possible edge points, the extrema edges, is defined and then a measure of saliency is assigned to every point in this set. The extrema edges are defined as the discontinuities of the mosaic image associated to the energy partition $\Pi(V(u), ext(u))$. Their valuation is obtained using global information through a family of nested partitions guided by a notion of contrast. The method uses only the original image to construct a contour map called the *valuated extrema edges*. A threshold in this image provides a set of closed curves where semantically important characteristics of edges are preserved.

Finally, this paper focused on monochrome images in order to emphasize the mathematical formulation of our approach and to establish a comparison with the continuous watershed transform. Nevertheless, a straightforward application to color images can be done by considering only their lightness channel. Alternatively, we are presently working on the generalization of our approach to vector-valued images.

References

1. P. A. Arbeláez. The path variation. Technical report, CEREMADE, 2003.
2. S. Beucher and F. Meyer. The morphological approach to segmentation: The watershed transformation. *Mathematical Morphology in Image Processing*, pages 433–481, 1992.
3. C. R. Brice and C. L. Fenema. Scene analysis using regions. *Artificial Intelligence*, 1:205–226, 1970.
4. J. Canny. A computational approach to edge detection. *IEEE Transactions on Pattern Analysis and Machine Intelligence*, 8(6):679–698, November 1986.
5. L. D. Cohen. Multiple contour finding and perceptual grouping using minimal paths. *Journal of Mathematical Imaging and Vision*, 14(3):225–236, 2001.
6. L. D. Cohen and R. Kimmel. Global minimum for active contour models: A minimal path approach. *International Journal of Computer Vision*, 24(1):57–78, August 1997.

7. L. D. Cohen, L. Vinet, P. Sander, and A. Gagalowicz. Hierarchical region based stereo matching. In *Proc. 6th Scandinavian Conference on Image Analysis*, 1989.
8. T. Deschamps and L. D. Cohen. Fast extraction of minimal paths in 3d images and applications to virtual endoscopy. *Medical Image Analysis*, 5(4):281–299, 2001.
9. E. W. Dijkstra. A note on two problems in connection with graphs. *Numerische Mathematic*, 1:269–271, 1959.
10. D. A. Forsyth and J. Ponce. *Computer Vision: A Modern Approach*. Prentice Hall, 2003.
11. L. Garrido, P. Salembier, and D. Garcia. Extensive operators in partition lattices for image sequence analysis. *Signal Processing*, 66(2):157–180, April 1998. Special Issue on Video Sequence Segmentation.
12. M. Grimaud. New measure of contrast: Dynamics. In *Image Algebra and Morphological Processing III*, SPIE, San Diego, CA., 1992.
13. R. Haralick and L. Shapiro. *Computer and Robot Vision*, volume I. Adison Wesley, 1992.
14. E. Hewitt and K. Stromberg. *Real and Abstract Analysis*. Springer Verlag, 1969.
15. S. L. Horowitz and T. Pavlidis. Picture segmentation by a directed split-and-merge procedure. In *Proceedings of the Second International Joint Conference on Pattern Recognition*, pages 424–433, 1974.
16. C. Jordan. Sur la série de fourier. *C. R. Acad. Sci. Paris Sér. I Math.*, 92(5):228–230, 1881.
17. R. Kimmel and A. M. Bruckstein. Global shape from shading. *CVIU*, 62(3):360–369, 1995.
18. R. Kimmel, N. Kiryati, and A. M. Bruckstein. Distance maps and weighted distance transforms. *Journal of Mathematical Imaging and Vision*, 6:223–233, May 1996. Special Issue on Topology and Geometry in Computer Vision.
19. A. S. Kronrod. On functions of two variables. *Uspehi Mathematical Sciences*, 5(35), 1950. In Russian.
20. R. Kruse and A. Ryba. *Data structures and program design in C++*. Prentice Hall, New York, 1999.
21. P. Maragos and M. A. Butt. Curve evolution, differential morphology and distance transforms applied to multiscale and eikonal problems. *Fundamenta Informaticae*, 41:91–129, 2000.
22. D. Marr and E. Hildreth. Theory of edge detection. In *Proc. of Royal Sociery of London*, volume B-207, pages 187–217, 1980.
23. F. Meyer. Topographic distances and watershed lines. *Signal Processing*, 38:113–125, 1994.
24. F. Meyer. Hierarchies of partitions and morphological segmentation. In Michael Kerckhove, editor, *Scale Space and Morphology in Computer Vision*, pages 161–182, 2001.
25. J. M. Morel and S. Solimini. *Variational Methods in Image Segmentation*. Birkhauser, 1995.
26. L. Najman. *Morphologie Mathématique: de la Segmentation d'Images à l'Analyse Multivoque*. PhD thesis, Université Paris Dauphine, 1994.
27. L. Najman and M. Schmitt. Watershed of a continuous function. *Signal Processing*, 38(1):99–112, July 1994.
28. L. Najman and M. Schmitt. Geodesic saliency of watershed contours and hierarchical segmentation. *IEEE Transactions on Pattern Analysis and Machine Intelligence*, 18(12):1163–1173, 1996.
29. I. P. Natansson. *Theory of Functions of a Real Variable*. Frederick Ungar Publishing, New York, 1964.

30. J. R. Parker. *Algorithms for Image Processing and Computer Vision*. John Wiley and Sons, 1997.

31. M. Schmitt and J. Mattioli. *Morphologie Mathématique*. Masson, 1994.

32. J. Serra and P. Salembier. Connected operators and pyramids. In SPIE, editor, *Image Algebra and Mathematical Morphology*, volume 2030, pages 65–76, San Diego CA., July 1993.

33. J. A. Sethian. *Level Set Methods and Fast Marching Methods*. Cambridge University Press, Cambridge, UK, 2 edition, 1999.

34. L. Vincent and P. Soille. Watersheds in digital spaces: an efficient algorithm based on immersion simulations. *PAMI*, 13(6):583–598, 1990.

35. T. Vlachos and A. G. Constantinides. Graph-theoretical approach to colour picture segmentation and contour classification. In *IEE Proc. Vision, Image and Sig. Proc.*, volume 140, pages 36–45, February 1993.

The Maximum Principle for Beltrami Color Flow

Lorina Dascal and Nir Sochen

Department of Applied Mathematics
University of Tel-Aviv, Ramat-Aviv, Tel-Aviv 69978, Israel
{lorina,sochen}@post.tau.ac.il

Abstract. We study, in this work, the maximum principle for the Beltrami color flow and the stability of the flow's numerical approximation by finite difference schemes. We discuss, in the continuous case, the theoretical properties of this system and prove the maximum principle in the strong and the weak formulations. In the discrete case, all the second order explicit schemes, that are currently used, violate, in general, the maximum principle. For these schemes we give a theoretical stability proof, accompanied by several numerical examples.

Keywords: Maximum Principle, Beltrami Framework, Parabolic PDE's, Finite difference schemes.

1 Introduction

The scale-space approach in low-level vision originated from several ideas: One point of view emphasized the multi-scale nature of images. Important features are to be found in all scale levels and one should use all scales in order to gain an understanding of the captured scene. The second, and somewhat related point of view, looked for a procedure of gradual simplification of the image. In each step the image should be a simplified version of the preceding step. New features, in a given step, that cannot be explained as a simplification of the previous step should not be present. This "causality" principle, introduced by Koenderink [11], leads directly to the maximum principle in the one-dimensional case.

Many generalizations of the scale-space linear flow exist. They are based on anisotropic and/or inhomogeneous diffusion flows. In order to serve as a basis for a multi scale analysis they should be "causal" flows. It is well known that the Perona-Malik continuous flow, for example, does not satisfy the maximum principle and cannot serve as a basis for a scale-space analysis. However, the closely related Partial Differential Equation (PDE) introduced by Catté, et al [2] has this property as does the discrete Perona-Malik flow [18].

There are several possible definitions of causality in higher dimensions. We restrict ourselves in this article to the study of the *maximum principle* feature for the color flow. We assume hereafter that the maximum principle is a necessary condition for causality. We treat in this paper the Beltrami flow for color images and show that it satisfies the maximum (minimum) principle in both the strong and the weak formulations.

L.D. Griffin and M. Lillholm (Eds.): Scale-Space 2003, LNCS 2695, pp. 196–208, 2003.

We follow the duality approach of Florack [6] and treat an image as a tempered distribution. This approach follows our intuition that by taking two pictures of the scene with two different devices we have two representations of "the same thing" and not two completely different objects. The duality approach describes the sensor space as a functional space. The data that we usually process, which result from the interaction of the physical/optical data and the sensor, are modelled as an inner product of the sensor function and the "true image". Under this approach, the set of images is equivalent to the set of linear functionals on the sensor functional space (i.e. distributions). It is natural from this point of view to study the flow equations on the image space directly. In order to do that we have to define the equations in the weak sense. Another reason to study weak solutions is noise. Since noise is a non-continuous and non-differentiable function in all its points, the corrupted initial image which is a sum of the "true image" and the noise, is non-continuous and non-differentiable as well. One needs, then, to resort to the weak formulation in order to be able to *define* the PDE based denoising algorithm.

The strong formulation is presented here for two reasons. The first and obvious reason is that "it is there". Since we can prove it, then it fits naturally to the rest of this study. The second reason is more technical. It is used as an approximation tool in the proof of the maximum principle for weak solutions. We will prove below that if a weak solution exists then it must satisfy the maximum (minimum) principle.

In the last part of this paper we study the maximum principle for the various discrete schemes by which the differential equation is approximated. We show the fact that the various derivatives are approximated to a given order is not enough to guarantee the maximum principle. Many common numerical schemes are proved *to violate* the maximum principle. We present a proof, though, of their stability along with examples that clearly demonstrate their failure to obey the maximum principle.

The paper is organized as follows: In section 2 we review the Beltrami Framework. In section 3 we deal with the continuous formulation of the maximum principle. We present the maximum principle theorem for the strong solution of the parabolic quasilinear system that characterizes the Beltrami color flow. In section 4 we introduce the weak (distributional) solution for this system and present the maximum principle in a weak formulation. In section 5 we discuss the properties of the second-order central differences scheme, which in general violates the maximum principle. For this scheme we give a theoretical stability proof. In section 6 we present numerical results. We summarize and conclude in Section 7.

2   The Beltrami Framework

Let us briefly review the Beltrami framework for non-linear diffusion in computer vision [13, 14, 7].

We represent an image and other local features as embedding maps of a Riemannian manifold in a higher dimensional space. The simplest example is a gray-level image which is represented as a 2D surface embedded in $\mathbb{R}^3$. We denote the map by $X : \Sigma \to \mathbb{R}^3$. Where Σ is a two-dimensional surface, and we denote the local coordinates on it by (σ^1, σ^2). The map U is given in general by $(U^1(\sigma^1, \sigma^2), U^2(\sigma^1, \sigma^2), U^3(\sigma^1, \sigma^2))$. In our example we represent it as follows $(U^1 = \sigma^1, U^2 = \sigma^2, U^3 = I(\sigma^1, \sigma^2))$. We choose on this surface a Riemannian structure, namely, a metric. The metric is a positive definite and a symmetric 2-tensor that may be defined through the local distance measurements:

$$ds^2 = g_{11}(d\sigma^1)^2 + 2g_{12}d\sigma^1 d\sigma^2 + g_{22}(d\sigma^2)^2.$$

The canonical choice of coordinates in image processing is the cartesian. For such choice, that we follow in the rest of the paper we identify $\sigma^1 = x^1$ and $\sigma^2 = x^2$. We use below the Einstein summation convention in which the above equation reads $ds^2 = g_{ij}dx^i dx^j$ where repeated indices are summed over. We denote the elements of the inverse of the metric by superscripts $g^{ij} = (g^{-1})_{ij}$.

Once the image is defined as an embedding mapping of Riemannian manifolds it is natural to look for a measure on this space of embedding maps.

2.1   Polyakov Action: A Measure on the Space of Embedding Maps

Denote by (Σ, g) the image manifold and its metric and by (M, h) the space-feature manifold and its metric, then the functional $S[U]$ attaches a real number to a map $U : \Sigma \to M$:

$$S[U^a, g_{ij}, h_{ab}] = \int dV \langle \vec{\nabla} U^a, \vec{\nabla} U^b \rangle_g h_{ab}$$

where dV is a volume element and $\langle \nabla R, \nabla B \rangle_g = g^{ij} \partial_{x_i} R \partial_{x_j} B$. This functional, for $m = 2$ and $h_{ab} = \delta_{ab}$, was first proposed by Polyakov [12] in the context of high energy physics, and the theory known as *string theory*.

Let us formulate the Polyakov action in matrix form: (Σ, G) is the image manifold and its metric as before. Similarly, (M, H) is the spatial-feature manifold and its metric. Define

$$A^{ab} = (\vec{\nabla} U^a)^t G^{-1} \vec{\nabla} U^b$$

The map $U : \Sigma \to M$ has a weight

$$S[U, G, H] = \int d^m \sigma \sqrt{g} \mathrm{Tr}(AH),$$

where m is the dimension of Σ and $g = \det(G)$.

Using standard methods in the calculus of variations the Euler-Lagrange equations with respect to the embedding (assuming Euclidean embedding space) are (see [13] for explicit derivation):

$$-\frac{1}{2\sqrt{g}} h^{ab} \frac{\delta S}{\delta U^b} = \frac{1}{\sqrt{g}} \partial_{x_i}(\sqrt{g} g^{ij} \partial_{x_j} U^a).$$

Or in matricial form

$$-\frac{1}{2\sqrt{g}}h^{ab}\frac{\delta S}{\delta U^b} = \underbrace{\frac{1}{\sqrt{g}}\mathrm{div}\left(\sqrt{g}G^{-1}\nabla U^a\right)}_{\Delta_g U^a}. \tag{1}$$

The extension for non-Euclidean embedding space is treated in [14, 15, 8].

The elements of the induced metric for color images are:

$$g_{ij} = \delta_{ij} + \beta^2 \sum_{a=1}^{3} U^a_{x_i} U^a_{x_j}, \tag{2}$$

where $\beta > 0$ is the ratio between the spatial and color distances. Note that this metric is different from the Di Zenzo matrix [20] (which is not a metric since it is not positive definite). A generalization of DiZenzo's gradient for color images has been investigated in [19] by constructing an anisotropic vector-valued diffusion model with a common tensor-valued structure descriptor.

The value of parameter β present in the elements of the metric g_{ij} is very important and determines the nature of the flow. In the limit $\beta \to 0$, for example, the flow degenerates to the decoupled channel by channel linear diffusion flow. In the other limit $\beta \to \infty$ we get a new non-linear flow. The gray-value analogue of this limit is the Total Variation flow of [9] (see details in [14]). Our proof of the extremum property is independent of the value of β though.

Since (g_{ij}) is positive definite, $g \equiv \det(g_{ij}) > 0$ for all σ^i. This factor is the simplest one that does not change the minimization solution while giving a reparameterization invariant expression. The operator that is acting on U^a is the natural generalization of the Laplacian from flat spaces to manifolds, is called the Laplace-Beltrami operator and is denoted by Δ_g.

The non-linear diffusion or scale-space equation emerges as a gradient descent minimization:

$$U^a_t = \frac{\partial}{\partial t}U^a = -\frac{1}{2\sqrt{g}}h^{ab}\frac{\delta S}{\delta U^b} = \Delta_g U^a. \tag{3}$$

The mathematical properties of this system, together with the initial value, given by the original noisy image and with Neumann boundary condition, are studied in the rest of the paper with an emphasis on the maximum principle.

3   Extremum Principle for Functional Solutions

We establish, in this subsection, the maximum principle for the strong solution of the initial boundary-value problem which characterizes the Beltrami color flow. We refer to the term strong solutions when we talk about solutions which are functions with some smoothness criteria that we detail below. Let us first introduce few notations: We denote the image domain by Ω. It is a bounded open domain in $\mathbb{R}^2$. We denote by $\partial\Omega$ the boundary of Ω. We define the space-time cylinder $Q_T = \Omega \times (0, T)$, and denote its lateral surface by $S_T = \{(x, t)|x \in \partial\Omega, t \in (0, T)\}$. We define also the parabolic boundary by the union of the bottom and the lateral boundaries of the cylinder $\Gamma_T = \Omega \bigcup S_T$.

The PDE is the gradient descent equation for the Polyakov action as was described in the previous section. We carry out explicitly the result of applying the derivation operator Div. The result is a sum of two terms: The first term results from applying the derivative on $\sqrt{g}G^{-1}$, and the second comes from applying the div on the gradient $div(\nabla U^a) = \Delta U^a$ which is the Laplacian. Remember that the metric, and consequently its inverse and its determinant, depends on first order derivatives. The application of the Div operator on it give rise to second order derivatives of the different channels as well. Rearranging the right hand side of Eq. (3) according to the second order derivatives and the coefficients thereof we arrive to the following coupled system of PDEs:

$$U_t^a = (F_b^a)^{ij} U_{x_i x_j}^b, \quad (x,t) \in Q_T , \tag{4}$$

where $a, b = 1, 2, 3$ are indices in color space, $i, j = 1, 2$ are spatial indices and summation is applied on all repeated indices. Note that (F_b^a) are nine 2x2 matrices. Denote by $H^a = U_{x_i x_j}^a$ the Hessian of U^a. This system of PDEs can be written in terms of a trace in spatial domain as

$$U_t^a = \text{Trace}\left(F_b^a H^b\right) , \quad (x,t) \in Q_T , \tag{5}$$

where, as before, the repeated b index implies a summation over the color indices. The initial and boundary conditions are

$$U^a(x,0) = U_0^a(x), \quad x \in \Omega \tag{6}$$

$$(F_b^a)\vec{\nabla} U^b \cdot \vec{n}\Big|_{S_T} = 0 \quad \text{(no summation on } b \text{ here)}, \tag{7}$$

where $\vec{n}$ is the outer normal to $\partial\Omega$ and the dot product denotes, as usual, the Euclidian scalar product on $\mathbb{R}^2$.

Lemma 1. *The nine 2x2 matrices (F_b^a) are symmetric, positive definite, and their elements $(F_b^a)^{ij}$ are rational functions of the first derivatives of the different channels. These matrix elements are, moreover, uniformly bounded functions on Q_T.*

Proof. The proof is by direct calculation. One finds for example:

$$\left(F_1^2\right)^{11} = -R_x G_x \frac{g_{22}^2}{g^2} + (R_x G_y + R_y G_x)\frac{g_{12}g_{22}}{g^2} - \frac{R_y G_y}{g}\left(1 + \frac{g_{12}^2}{g}\right)$$

$$\left(F_1^2\right)^{12} = \left(F_1^2\right)^{21} = \frac{R_x G_y + R_y G_x}{g} - \frac{R_x G_y + R_y G_x}{g^2}g_{11}g_{22} -$$

$$\frac{R_x G_y + R_y G_x}{g^2}g_{12}^2 + 2\frac{R_x G_x g_{22} + R_y G_y g_{11}}{g^2}g_{12}^2$$

$$\left(F_3^2\right)^{22} = -R_y G_y \frac{g_{11}^2}{g^2} + (R_x G_y + R_y G_x)\frac{g_{11}g_{12}}{g^2} - \frac{R_x G_x}{g}\left(1 + \frac{g_{12}^2}{g}\right) . \tag{8}$$

(here R, G, B denote the three components of the color vector $\vec{U}$)

These are rational functions of the first derivatives. The diagonal elements are strictly positive (by direct check) and the negativity of the discriminant implies the positive definiteness of this matrix. One can verify by direct check that the coefficients are bounded functions of the first derivatives values. Other matrices are checked along the same lines. □

Next we state the maximum principle for strong solutions of the coupled system of PDEs Eq. (4).

Theorem 1. *Let* $\overrightarrow{U_0} \in C^2(\Omega)$ *with bounded second derivatives. A solution* $\overrightarrow{U} \in C^{2,1}(\bar{Q}_T)$ *satisfies, then, the following maximum principle:*

$$1) \quad \max_{\bar{Q}_T} U^a = \max_{\Omega} U_0^a$$

$$2) \quad \max_{\bar{Q}_T} \sum_{a=1}^{3} U^a = \max_{\Omega} \sum_{a=1}^{3} U_0^a. \tag{9}$$

Proof. The proof make use of Lemma 1. We describe here the main steps of the proof. The details can be found in [3]. Note that assertion 1) does not imply, in principle assertion 2). The proof is based on the observation that the off diagonal matrices F_b^a with $a \neq b$ can be written as $U_{x_i}^a$ times a bounded function. It follows that if the maximum is attained in the interior of the cylinder then the first derivatives vanish at that point while the Hessian is negative definite. It implies the negativity of the right side of Eq. (4) while the left side is zero by the maximality of the function at that point. This excludes the interior points. The lateral boundary is shown to be excluded as well by using the Neumann boundary conditions. The upper boundary of the cylinder Q_T is a little more complicated and we leave the details to our technical report [3]. □

This theorem, besides its own value, serves as a basic approximation tool in the proof of the extremum principle for weak solutions of the non-linear system of the color Beltrami flow. We assume in the meantime that weak solutions from a proper space (which will be described below) exist and prove that if they exist, they obey the extremum principle.

4   Extremum Principle for Distributional Solutions

There are two reasons that convinced us to look at weak solutions for the, fairly complicated and highly non-linear, system of the color Beltrami partial differential equations. The first reason is the fact that this is a denoising algorithm. It means that the original image is corrupted with noise. One usually assumes an additive Gaussian or uniform noise. Since the noise is non differential function in ALL its points then one is not allowed to assume that the original noisy image is continuous, let alone differentiable. We consider, therefore, the initial corrupted image to belong to L^∞, i.e. for each color component $U_0^a \in L^\infty(Q_T)$. In order to define a PDE based denoising process, let alone solving it, we have to work with weak solutions. The second reason is that from the duality viewpoint of Florack [6] images should be considered as linear functionals on the sensor functional space. This space of linear functionals is the dual space, composed of distributions in the sense of Laurent Schwartz [10].

Let us introduce the following notations: We use the following scalar product on Q_T:

$$(u, v) = \int_{Q_T} (uv + u_{x_k} v_{x_k} + u_{x_k x_j} v_{x_k x_j}) dx\, dt$$

This scalar products defines naturally a norm which we denote by $||u|| = \sqrt{(u, u)}$. The Sobolev space of functions, with finite norm, over the cylinder Q_T is defined by $W_2^{2,0}(Q_T) = \{u : Q_T \to \mathbb{R}|\ ||u|| < \infty\}$. We will omit from now on the Q_T notation. We write, for example, the above functional space as $W_2^{2,0}$ and remember that all the functions in it are defined over the Q_T cylinder. More generally, the Sobolev space $W_r^{p,q}$ is the space of functions, for which the L_r norm of their first generalized p spatial derivatives and q time derivatives, is finite.

We are now in a position to define weak (generalized) solutions:

Definition 1. *A generalized (weak) solution of the system Eq. (4), with boundary and initial conditions Eqs. (7) and (6), is a vector function $\overrightarrow{U}$ whose components $U^a \in W_2^{2,0}(Q_T)$ for $a = 1, 2, 3$ and such that for any vector function $\overrightarrow{\eta}$ whose components $(\eta^a) \in W_2^{2,1}(Q_T)$, the following integral equations hold for almost all $t \in [0, T]$:*

$$\int_\Omega U^a \eta^a |_{t=0}^{t=T}\, dx - \int_\Omega U^a \eta_t^a\, dx\, dt = - \int_{Q_T} \left(\eta^a F_b^{a\ ij}\right)_{x_i} U_{x_j}^b. \tag{10}$$

For such weak solutions the following maximum principle holds:

Theorem 2. *Assume $\overrightarrow{U}_0 \in L^\infty(Q_T)$. For a weak (distributional) solution $\overrightarrow{U} \in W_2^{2,0}(Q_T)$ of the system $(4), (6), (7)$ we have for almost all $(x, t) \in Q_T$:*

$$\operatorname*{ess\,inf}_{\Omega} |U_0^a(x)| \le |U^a(x, t)| \le \operatorname*{ess\,sup}_{\Omega} |U_0^a(x)|, \tag{11}$$

$$\operatorname*{ess\,inf}_{\Omega} |\sum_{a=1}^{3} U_0^a(x)| \le |\sum_{a=1}^{3} U^a(x, t)| \le \operatorname*{ess\,sup}_{\Omega} |\sum_{a=1}^{3} U_0^a(x)|. \tag{12}$$

Remark that by definition :
$$\operatorname*{ess\,sup}_{x \in X} f(x) = \inf_{A_0 \subset S_0} (\sup_{X - S_0} f(x)), \quad where \quad S_0 = \{S \subset X | \mu(S) = 0\}$$

Proof. The proof is based on Sobolev Embedding Theorem [4] (by which the space $W_2^{2,0}(Q_T)$ is embedded in $C(\bar{Q}_T)$) and on a Density Theorem [5] (which asserts that the space $W_2^{2,0}(Q_T) \bigcap C^\infty(\bar{Q}_T)$ is dense in the space $W_2^{2,0}(Q_T)$).

By the density theorem we can approximate the solution $\overrightarrow{U}$ by smooth vector functions $\overrightarrow{U}_k \in W_2^{2,0}(Q_T) \bigcap C^\infty(\bar{Q}_T)$ such that $||U_k^a - U^a||_{W_2^{2,0}(Q_T)} \to 0$ as $k \to \infty$. The initial data $\overrightarrow{U}_0$ can also be approximated by smooth functions

$\overrightarrow{\Phi_k} \in C^\infty(\Omega)$, such that $\Phi_k^a \to U_0^a$ uniformly as $k \to \infty$. Now for each k consider the boundary value problem (4),(6),(7) corresponding to the vector function $\overrightarrow{U_k}$. For enough smooth solution of this problem, based on Theorem 9 we have the maximum principle: $|U_k^a| \leq \sup_\Omega |\Phi_k^a|$. Furthermore, by the Embedding Theorem there exists a constant C such that

$$|U^a| \leq ess \sup_{Q_T} |U_k^a - U^a| + ess \sup_{Q_T} |U_k^a|$$

$$\leq C \cdot ess \sup_{Q_T} \|U^a - U_k^a\|_{W_2^{2,0}(Q_T)} + ess \sup_\Omega |\Phi_k^a|$$

$$\leq C \cdot ess \sup_{Q_T} \|U^a - U_k^a\|_{W_2^{2,0}(Q_T)} + ess \sup_\Omega |U_0^a| + ess \sup_\Omega |U_0^a - \Phi_k^a|.$$

Letting $k \to \infty$, we obtain the maximum principle for the weak solutions from $W_2^{2,0}$:

$$|U^a| \leq ess \sup_\Omega |U_0^a|. \tag{13}$$

In a similar way one obtains (12). $\square$

5   The Discrete Maximum Principle and Stability

In this section we show that the commonly used central difference second order explicit schemes violate, in general, the discrete maximum principle. We give nevertheless, for these schemes, a theoretical proof of stability.

We work on a rectangular grid

$$x_i = i\Delta x, \quad y_j = j\Delta y, \quad t_m = m\Delta t,$$

$$i, j = 0, 1, 2, ...M; \quad m = 0, 1, 2, ...[T/\Delta t]$$

The spatial units are normalized such that $\Delta x = \Delta y = 1$. The approximate solution $(R_{ij}^m, G_{ij}^m, B_{ij}^m)$ samples the functions:

$$R_{ij}^m \equiv U^1(i\Delta x, j\Delta y, m\Delta t),$$

$$G_{ij}^m \equiv U^2(i\Delta x, j\Delta y, m\Delta t),$$

$$B_{ij}^m \equiv U^3(i\Delta x, j\Delta y, m\Delta t),$$

On the boundary we impose the Neumann boundary condition. This corresponds to a prolongation by reflection of the image across the boundary.

We replace the second spatial derivatives and the first time derivative by central difference and forward difference respectively. Based on (3), the first equation of the system (4),(6),(7) can be written in the form

$$U_t^1 = \frac{1}{\sqrt{g}} \text{Div}(D\nabla U^1). \tag{14}$$

The diffusion matrix is written here as

$$D = \begin{pmatrix} a & b \\ b & c \end{pmatrix}$$

where the coefficients are given in terms of the image metric: $a = g_{22}/\sqrt{g}$; $c = g_{11}/\sqrt{g}$; $b = -g_{12}/\sqrt{g}$. With this notation,thus, equation (14) is written as

$$U_t^1 = \frac{1}{\sqrt{g}}((aU_x^1 + bU_y^1)_x + (bU_x^1 + cU_y^1)_y) . \tag{15}$$

We approximate Eq. (15) by the following central difference explicit scheme:

$$R_{ij}^{m+1} = R_{ij}^m + \beta \Delta t O_{ij}(R^m, G^m, B^m) \tag{16}$$

where $O_{ij}(R^m, G^m, B^m)$ is the discrete version of the right side of Eq. (4) and is given explicitly, in the central difference framework, by

$$
\begin{aligned}
O_{ij} = \\
\frac{1}{\sqrt{g}}\Big[a_{i+\frac{1}{2},j}^m (R_{i+1,j}^m - R_{i,j}^m) - a_{i-\frac{1}{2},j}^m (R_{i,j}^m - R_{i-1,j}^m) \\
+ c_{i,j+\frac{1}{2}}^m (R_{i,j+1}^m - R_{i,j}^m) - c_{i,j-\frac{1}{2}}^m (R_{i,j}^m - R_{i,j-1}^m) \\
+ \frac{1}{4} b_{i,j+1}^m (R_{i+1,j+1}^m - R_{i-1,j+1}^m) - \frac{1}{4} b_{i,j-1}^m (R_{i+1,j-1}^m - R_{i-1,j-1}^m) \\
+ \frac{1}{4} b_{i+1,j}^m (R_{i+1,j+1}^m - R_{i+1,j-1}^m) - \frac{1}{4} b_{i-1,j}^m (R_{i-1,j+1}^m - R_{i-1,j-1}^m) \Big] , \tag{17}
\end{aligned}
$$

where the half indices are obtained by linear interpolation. The equations for the two other color components are discretized in the same manner. This scheme is stable under CFL-like bound requirements of the step time. The stability, as well as the lack of extremum principle property, can be learned from the following theorem.

Theorem 3. *If Δt satisfies the condition:*

$$\Delta t \leq \frac{1}{8\beta \max_{i,j}\{\frac{a_{i+\frac{1}{2},j}}{\sqrt{g_{i,j}}}, \frac{a_{i-\frac{1}{2},j}}{\sqrt{g_{i,j}}}, \frac{c_{i,j+\frac{1}{2}}}{\sqrt{g_{i,j}}}, \frac{c_{i,j-\frac{1}{2}}}{\sqrt{g_{i,j}}}\}}, \tag{18}$$

then the solution satisfies:

$$|R_{i,j}^m| \leq e^{\alpha t m} \max_{i,j} |R_{i,j}^0|,$$
$$|G_{i,j}^m| \leq e^{\alpha t m} \max_{i,j} |G_{i,j}^0|,$$
$$|B_{i,j}^m| \leq e^{\alpha t m} \max_{i,j} |B_{i,j}^0|,$$

$$where \quad \alpha = 2\beta \max_{ij} \frac{|b_{ij}^m|}{\sqrt{g_{ij}^m}} \leq 2\beta.$$

Proof. See details in [3] □.

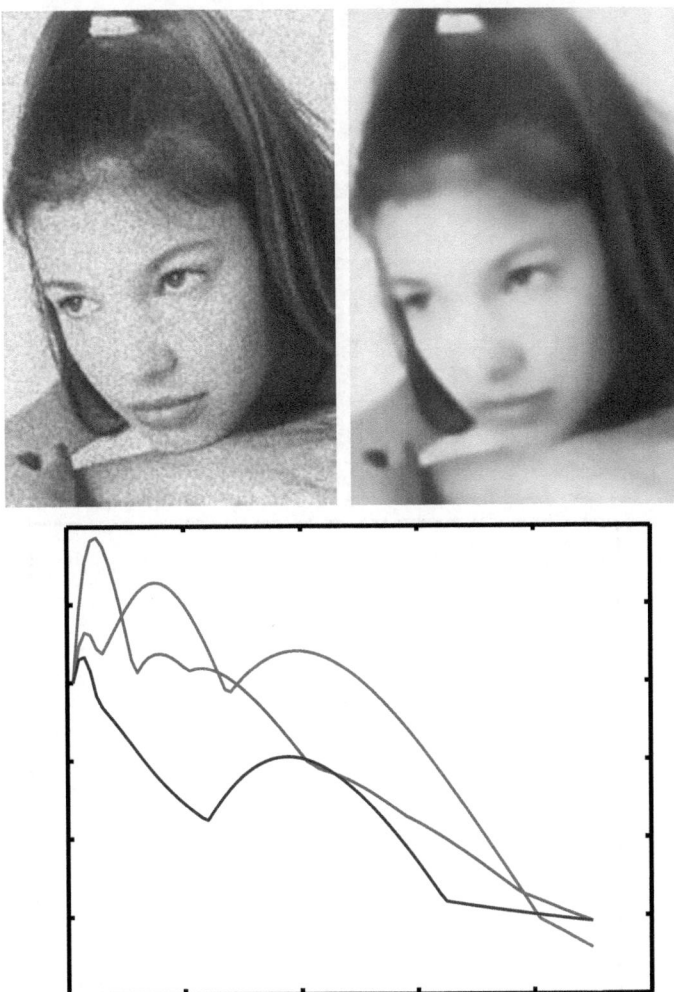

Fig. 1. Top-left: Noisy Camila image. Top-right: Result of the Beltrami flow after 90 iteration. Bottom: Plot of maximum of each of the channels versus number of iterations.Parameters: $\beta^2 = 100$, first scheme: $\Delta t = 0.0091$.

The inequalities in Theorem 3 show that the numerical solution is bounded in each iteration by the maximum value of the initial image multiplied by a factor. It guarantees that the flow does not blow up in finite time and ensures its stability. At the same time it is clear from the positivity of β that the maximum principle can possibly be violated. One can actually see it in practice (see Fig. 1).

The reason for this discrepancy between the continuous and discrete setting is the fact that this second order approximation is not a non-negative one. Indeed, the mixed derivatives in eq. (15) can create negative weights in certain pixels.

Fig. 2. Top: Noisy Iguana image. Middle: Denoised Iguana by the Beltrami flow with after 60 iterations. Bottom: Plot of maximum of each of the channels versus number of iterations. Parameters: $\beta^2=80$, $\Delta t = 0.0091$.

One can easily show that a scheme which is based on a nonnegative discretization does satisfy the discrete maximum principle. Based on this result, the problem of proving the discrete maximum principle boils down to the problem of finding a nonnegative second order difference approximation. In [18], Weickert proposed a way for building a nonnegative scheme. The non-negativity of his proposed scheme depends, though, on the condition number of the diffusion tensor D. Only in pixels were the condition number is smaller than $3 + 2\sqrt{2}$ the weights are non-negative. This limits the application of the scheme since in many images the condition number is typically higher than this limit in many pixels.

6   Details of the Implementation and Results

We present in this Section results that represent the numerical behavior of the above described numerical scheme. The initial data is given in three channels r, g and b in the range 0 to 256. We first transfer the images to the more perceptually adaptive coordinates $R = \log(1+r)$, $G = \log(1+g)$, $B = \log(1+b)$. The dynamic range for these variables is 0 to 8 and these adaptive coordinates do not limit the generality of our analysis. In the two examples presented below we corrupt an image with random noise and denoise it using the scheme mentioned above.

In the implementation, the parameters β and Δt were chosen to satisfy the stability condition Eq. (18).

Figure 2 demonstrates that the violation of the maximum principle does not obligatory occur in the central difference scheme. In this figure the numerical scheme respects the maximum principle.

Remark that for a too large time step one gets a violation of the maximum principle in the Iguana image as well. Yet this violation gets smaller and smaller until the maximum principle is satisfied as the time step becomes smaller and smaller. This is NOT the case for the Camila image where the break up of the maximal principle is stable and does not depend on the time step. It is not clear for us what are the special characteristic of an image that make it to respect or not the discrete maximum principle.

7   Concluding Remarks

We have studied in this paper the extremum principle condition for the color image Beltrami flow. This is done in the context of the possibility to build a scale-space from the solution to this complicated non-linear coupled system of PDEs. We adapted in this paper the duality paradigm of Florack and we regarded "true images" as tempered distributions. We investigate therefore, besides the strong solutions, the generalized (weak) solutions. We proved that both the strong and the weak solutions satisfy the extremum principle which is a necessary condition for causality, and therefore, for the construction of well-defined scale-space.

We addressed also the problem of numerically construct a well-defined scale-space. It is shown that, in contrast to the continuous case, the extremum principle is not automatically guaranteed. We prove that the central difference scheme *does not* guaranty the satisfaction of the extremum principle. It is important to note that we studied many variants of central and/or forward-backward schemes. They all shared similar behavior and the detailed description of these and other methods will be found in future work.

Acknowledgments

This research has been supported in part by the Israel Academy of Science, Tel-Aviv University fund, the Adams Center and the Israeli Ministry of Science.

References

1. L. Alvarez, P.L. Lions, J.M. Morel,"Image selective Smoothing and edge detection by Nonlinear Difusion.II", *Siam Journal on Numerical Analysis*, Vol.29, Issue (1992), 845-866.
2. F. Catte, P. L. Lions, J. M. Morel and T. Coll, "Image selective smoothing and edge detection by nonlinear diffusion", SIAM J. Num. Anal., vol. 29, no. 1, pp. 182-193, 1992.
3. L. Dascal and N. Sochen, "The Maximum principle for Beltrami color flow", Technical Report Tel-Aviv University, Israel, in preparation.
4. D.Gilbarg, N.Trudinger, "Elliptic partial differential equations of second order",Springer-Verlag, 1977.
5. L.Evans, "Partial differential equations",Berkeley Mathematics 1994.
6. L. Florack, "Duality principles in Image processing and Analysis", IVCNZ 1998.
7. R. Kimmel and R. Malladi and N. Sochen, "Images as Embedding Maps and Minimal Surfaces: Movies, Color, Texture, and Volumetric Medical Images", International Journal of Computer Vision 39(2) (2000) 111-129.
8. R. Kimmel and N. Sochen, "Orientation Diffusion or How to comb a Porcupine", *Journal of Visual Communication and Image Representation* 13:238-248, 2001.
9. L. Rudin, S. Osher and E. Fatemi, " Non Linear Total Variation Based Noise Removal Algorithms", *Physica D 60 (1992) 259-268.*
10. L. Schwartz, "Functional analysis", Courant Institute of Mathematical Sciences, 1964.
11. J.J. Koenderink, "The structure of image", Biol. Cybern., Vol.50, 36-370, 1984.
12. A. M. Polyakov, "Quantum geometry of bosonic strings", *Physics Letters*, 103B (1981) 207-210.
13. N. Sochen and R. Kimmel and R. Malladi, "From high energy physics to low level vision", Report, LBNL, UC Berkeley, LBNL 39243, August, Presented in ONR workshop, UCLA, Sept. 5 1996.
14. N. Sochen and R. Kimmel and R. Malladi, "A general framework for low level vision", *IEEE Trans. on Image Processing*, 7 (1998) 310-318.
15. N. Sochen and Y. Y. Zeevi, "Representation of colored images by manifolds embedded in higher dimensional non-Euclidean space", Proc. IEEE ICIP'98, Chicago, 1998.
16. N. Sochen and Y. Y. Zeevi, "Representation of images by surfaces embedded in higher dimensional non-Euclidean space", 4th International Conference on Mathematical Methods for Curves and Surfaces, Lillehamer, Norway, July 1997.
17. R. Kimmel and N. Sochen. "Orientation Diffusion or How to Comb a Porcupine ? ", Special issue on PDEs in Image Processing, Computer Vision, and Computer Graphics, Journal of Visual Communication and Image Representation. In press.
18. J. Weickert, " Anisotropic Diffusion in Image processing", Teubner Stuttgart, 1998.
19. J.Weickert,"Coherence-enhancing diffusion of color images", Image and Vision Computing, Vol.17, 201-212, 1999.
20. S. Di Zenzo ," A note on the gradient of a multiimage", Computer Vision, Graphics and Image Processing, 33:116-125, 1986.

The Monogenic Scale Space
on a Bounded Domain and Its Applications

Michael Felsberg[1,*], Remco Duits[2], and Luc Florack[2]

[1] Computer Vision Laboratory, Linköping University
S-58183 Linköping, Sweden
mfe@isy.liu.se
http://www.isy.liu.se/~mfe
[2] Eindhoven University of Technology, Department of Biomedical Engineering
P.O. Box 513, NL-5600 MB Eindhoven, The Netherlands
{R.Duits,L.M.J.Florack}@tue.nl
http://www.bmi2.bmt.tue.nl/image-analysis/

Abstract. In this paper we present a method to implement the monogenic scale space on a bounded domain and show some applications. The monogenic scale space is a vector valued scale space based on the Poisson scale space, which establishes a sophisticated alternative to the Gaussian scale space. The features of the monogenic scale space, including local amplitude, local phase, local orientation, local frequency, and phase congruency, are much easier to interpret in terms of image features evolving through scale than in the Gaussian case. Furthermore, applying results from harmonic analysis, relations between the features are obtained which improve the understanding of image analysis. As applications, we present a very simple but still accurate approach to image reconstruction from local amplitude and local phase and a method for extracting the evolution of lines and edges through scale.

1 Introduction

In the recent past, scale space theory has become a quite large field of research. Although there have been several attempts to base the theory upon a well defined set of axiomatics [1,2,3], the whole variety of approaches more or less boils down to Gaussian scale space, i.e., the solution of the heat equation. Since the link between image processing and the heat conduction is a priori not obvious, some authors motivated using the Gaussian kernel by the point spread function of a defocused optical system. From a practical point of view, a reasonable motivation for using the heat equation is the causality requirement [4].

1.1 Scale Spaces beyond the Gaussian Case

As already stated in [2], most scale space axiomatics do not necessarily lead to the Gaussian scale space as a unique solution. The authors proposed a class

---

* This work has been supported by DFG Grant FE 583/1-1.

L.D. Griffin and M. Lillholm (Eds.): Scale-Space 2003, LNCS 2695, pp. 209–224, 2003.

of so-called α-*scale spaces* which corresponds to a family of smoothing kernels, forming a continuous transition between the identity operator ($\alpha = 0$) and the Gaussian kernel ($\alpha = 1$). A particular interesting case in this context is the *Poisson scale space* ($\alpha = 1/2$), since it is the solution of the *Laplace equation* [5]. Hence, by computing the Poisson scale space of an nD signal, an $(n + 1)$D *harmonic function* is obtained. Harmonic functions have especially nice properties from a mathematical point of view, e.g., the maximum principle, infinite differentiability, and a *minimization of the gradient energy* [6].

Until recently it was commonly accepted that the Gaussian scale space was the only α-scale space with an infinitesimal generator, i.e., it was the only one describable in terms of an evolution equation. In [3] it has been shown that every α-scale space correspond to a (pseudo) differential equation based on a *fractional power of the negative Laplace operator*. Furthermore, the authors have verified the validity of an over-complete set of axioms[1]. The causality requirement, however, has to be relaxed slightly [7] in order to remain valid for the non-Gaussian scale spaces.

Besides the nice mathematical properties of the Poisson scale space, it also seems to be preferable for a number of other reasons. First, in contrast to most other α-scale space kernels, the analytic formulation of the Poisson kernel is known [5]. Second, the uncertainty of the Poisson kernel is known to be close to the optimum [8]. Finally, the Poisson kernel is the point spread function of an idealized defocused optical system (not the Gaussian!) [7].

1.2   Why Using a Vector Valued Scale Space?

A central topic in the analysis of linear scale space is the interpretation of points with special topological properties, e.g., top-points, annihilation points, saddle points [9,10,11]. Although it can be shown that these special points contain basically all information from the image, it is nearly impossible to relate the occurrences of special points to certain classical image features or interpretations in the image plane.

The same 'semantic gap' occurs in context of the Poisson scale space. However, due to its relation to harmonic analysis, the Poisson scale space can easily be extended to a vector valued scale space, the *monogenic scale space* [7,12][2]. The basic idea is to lift the whole problem by one level of differentiation, yielding the Poisson scale space as one component of a vector valued scale space, an $(n + 1)$D vector field (remember that the scale space of an nD signal lives in an $(n + 1)$D space). The other n components are the *harmonic conjugates* of the Poisson scale space [8][3].

---

[1] This set of axioms also includes the existence of an infinitesimal generator.

[2] Monogenic is just another term for Clifford analytic.

[3] In [3] the authors mention that the square root of the negative Laplace operator imply a Clifford algebra. Actually, a Clifford algebra formulation is helpful to derive fundamental properties of the monogenic scale space. Accordingly, in [8] the author makes use of advanced algebras.

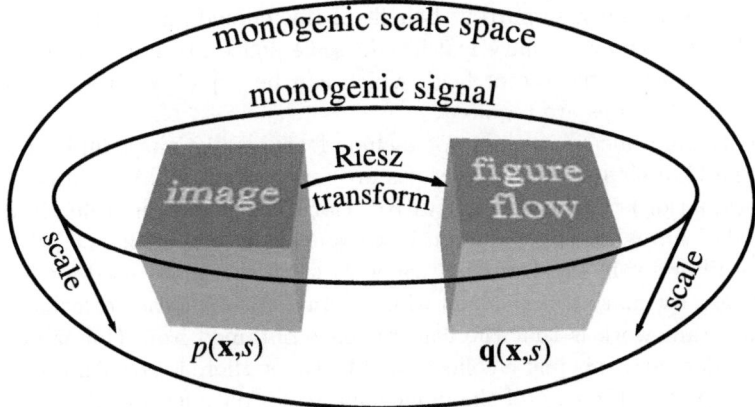

Fig. 1. Relations between Riesz transform, monogenic signal, Poisson scale space, and monogenic scale space. The Riesz transform of an image at an arbitrary scale yields the corresponding figure flow. The image and its figure flow form the monogenic signal. The monogenic signals at all scales build up the monogenic scale space. Alternatively, the Poisson scale space is obtained from the original image by Poisson filtering. Computing the harmonic conjugate yields the conjugate Poisson scale space (the figure flow at all scales). The Poisson scale space and its harmonic conjugate form the monogenic scale space.

For each scale, the smoothed signal and the corresponding part of the other n components are related by the *Riesz transform*. Combining a signal with its Riesz transform (the *figure flow* as we will see later) yields a generalization of the analytic signal, the *monogenic signal* [13], see fig. 1. Hence, the monogenic scale space can either be seen as a vector extension of the Poisson scale space $p(\mathbf{x}, s)$ or as the Poisson scale space of a vector valued signal (the monogenic signal), fig. 1.

As a generalization of the analytic signal, the monogenic signal provides the signal features *local amplitude* and *local phase*. Additionally, it also contains information about the *local orientation*. These three local features are pointwise orthogonal, which means that they represent independent information. The local amplitude represents the local intensity or dynamics, the local phase describes the local symmetry or grey value transition (even/line, odd/edge), and the local orientation describes the direction of highest signal variance. Obviously, all three signal properties can vary independently.

From the three features of the monogenic signal, one can derive secondary features by considering (combinations of) their spatial derivatives, e.g., local frequency (or scale), local curvature, and local extrema. However, using higher order derivatives reduces the stability and the robustness of the feature estimation. Considering the monogenic scale space instead of the monogenic signal at one scale solely, provides us with an additional dimension for the feature analysis.

This embedding of features into scale space improves the stability and robustness significantly and allows even to find previously undiscovered relationships, e.g., between edge detection by maximum detection and by phase congruency

[7]. The possibly most important benefit of this embedding is the fundamental result that the three primary features in scale space are not independent but form again a harmonic vector field, which can be exploited in all cases where second order features are required.

Traditional scale space does not include features like phase and orientation, unless the Gaussian smoothing kernels are replaced with Gabor filters (for the phase evaluation) or Gaussian derivatives (for the orientation evaluation). However, Gabor filters and Gaussian derivatives are not compatible, i.e., there is no common filter set in the Gaussian framework which can be used for both at a time, phase estimation and orientation estimation[4]. Another drawback of the Gaussian framework is that the feature space obtained from Gabor filters and Gaussian derivatives is just a collection of features; there is no such fundamental relationship as in the case of the monogenic scale space features.

2   The Monogenic Scale Space

In this section we give a short introduction to the monogenic scale space and its phase model for 2D signals. For more detailed explanations on the monogenic signal, the Riesz transform, and the phase model, refer to [13] or [8].

In order to make a clear distinction between the scale dimension and the spatial dimensions, we keep these dimensions separated, i.e., vectors $\mathbf{x} = (x, y)^T$ live in the spatial domain (2D space), and Δp is the Laplacian of p in the spatial domain. As a result, the 3D Laplace equation splits into two parts, according to $\Delta_{\mathbf{x},s} p = \Delta p + p_{ss} = 0$, where s indicates the scale coordinate.

2.1   Extending the Poisson Scale Space

We mentioned in the introduction that the monogenic scale space is a vector valued extension of the Poisson scale space and that this extension is obtained by lifting the level of differentiation. This becomes possible since the (partial) derivatives of harmonic functions are harmonic again. The original equation defining the Poisson scale space is of *Dirichlet kind*, i.e., the function values on the boundary are given by the initial image $f(\mathbf{x})$:

$$\Delta p = -p_{ss} \quad \text{for } s > 0 \quad \text{and} \quad p(\mathbf{x}, s)\Big|_{s=0} = f(\mathbf{x}) \ . \tag{1}$$

Replacing the Dirichlet boundary condition with a *Neumann boundary condition*, i.e., the values of the *normal derivatives* at the boundary are given, yields

$$\Delta \varphi = -\varphi_{ss} \quad \text{for } s > 0 \quad \text{and} \quad \varphi_s(\mathbf{x}, s)\Big|_{s=0} = f(\mathbf{x}) \ . \tag{2}$$

Since φ_s is harmonic and since $p = \varphi_s$ for $s = 0$, the uniqueness of the solution of (1) implies that $p = \varphi_s$ for all $s > 0$.

---

[4] Note that the Gaussian kernel and its derivatives are not in a quadrature relation and that differently oriented Gabor filters are not orthogonal, i.e., they are not suitable for orientation analysis.

The benefit from changing the boundary condition is that we can also consider the gradient field of φ in the image plane: $\mathbf{q} = \nabla\varphi$. The components of $\mathbf{q} = (q_1, q_2)^T$ are harmonic; they are called the *harmonic conjugates* of p. The 3D vector field $(q_1, q_2, p)^T$ is the 3D gradient field of the scalar field φ, hence its curl is zero everywhere for $s > 0$. Expressing the zero curl in components yields

$$\partial_x q_2 - \partial_y q_1 = 0 \tag{3}$$

$$\partial_x p - \partial_s q_1 = 0 \tag{4}$$

$$\partial_y p - \partial_s q_2 = 0 \ . \tag{5}$$

Furthermore, φ is harmonic which implies that the divergence of $(q_1, q_2, p)^T$ is also zero:

$$\nabla \cdot \mathbf{q} + p_s = 0 \ . \tag{6}$$

The latter equation can be considered as the continuity equation of Poisson scale space, which means that the figure flow (or image flow) is given by $\mathbf{q}(\mathbf{x}, s)$ (see also [3], eq. (9)).

The system of equations (3–6) with boundary condition $p(\mathbf{x}, 0) = f(\mathbf{x})$ can be considered as the PDE formulation of the monogenic scale space. At a first glance, finding a direct solution to this system seems to be rather complicated. However, the PDE (1) is part of the solution which simplifies solving the full system. Applying Fourier transform techniques, the solution for p reads

$$\mathcal{F}_2\{p(\cdot, s)\} = \mathcal{F}_2\{f\} \exp(-2\pi|\mathbf{u}|s) \tag{7}$$

where $\mathcal{F}_2$ denotes the 2D Fourier transform with the kernel $\exp(-i2\pi\mathbf{u} \cdot \mathbf{x})$ and $\mathbf{u} = (u, v)^T$ is the frequency vector. In the spatial domain, the equivalent convolution formulation reads

$$p(\mathbf{x}, s) = (f * P_s)(\mathbf{x}) \quad \text{where} \quad P_s(\mathbf{x}) = \frac{s}{2\pi(|\mathbf{x}|^2 + s^2)^{3/2}} \tag{8}$$

is the Poisson kernel. Using the derivative theorem of the Fourier transform, the harmonic conjugates of p are obtained as

$$\mathcal{F}_2\{\mathbf{q}(\cdot, s)\} = \mathcal{F}_2\{f\}\frac{-i\mathbf{u}}{|\mathbf{u}|} \exp(-2\pi|\mathbf{u}|s) \ , \tag{9}$$

which corresponds to the convolution

$$\mathbf{q}(\mathbf{x}, s) = (f * \mathbf{Q}_s)(\mathbf{x}) \quad \text{where} \quad \mathbf{Q}_s(\mathbf{x}) = \frac{\mathbf{x}}{2\pi(|\mathbf{x}|^2 + s^2)^{3/2}} \tag{10}$$

is the conjugate Poisson kernel (a vector function). Both results (8) and (10) can also be obtained using the Greens function. Letting s tend to zero in (10) yields an integral which can be evaluated in a principal value sense. The resulting transform is the *Riesz transform* with the kernel

$$\mathbf{h}_2(\mathbf{x}) = \frac{\mathbf{x}}{2\pi|\mathbf{x}|^3} \tag{11}$$

and the frequency response

 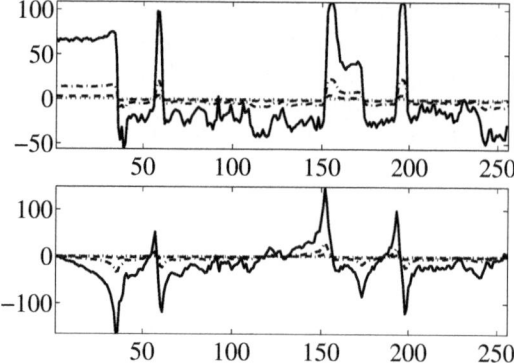

Fig. 2. Illustration of the 1D case: The slice cut of the image (left) yields the Poisson scale space in the upper right corner, plotted for scales: 0 (solid line), 0.5 (dash-dotted line), 1 (dashed line), and 1.5 (dotted line). The harmonic conjugate is plotted in the lower right corner for the same scales.

$$\mathcal{F}_2\{\mathbf{h}_2\} = -i\frac{\mathbf{u}}{|\mathbf{u}|} \ . \tag{12}$$

The Riesz transform is the 2D generalization of the Hilbert transform which preserves the most properties of the latter [13].

In order to illustrate how the components of the monogenic scale space evolve through scale, we have run an experiment on 1D data (a slice cut of an image, see fig. 2). The corresponding Poisson scale space and its harmonic conjugate (i.e., the Hilbert transform) are given for four different scales.

2.2 The Features of the Monogenic Scale Space

The aim of this section is to define first order features of the monogenic scale space, namely local amplitude, local orientation, and local phase. A reasonable requirement for these definitions is to be consistent with established definitions in the literature. Under the assumption of an *intrinsically 1D* neighborhood, i.e., a neighborhood where the signal varies only in one direction, it has been shown in [13] that the local amplitude and the local phase of the monogenic signal are consistent with the amplitude and the phase of the corresponding 1D analytic signal. Furthermore, it has been shown that the local orientation of the monogenic signal equals the orientation of the intrinsically 1D image patch.

The definitions of the monogenic scale space features are basically the same as for the monogenic signal, i.e., they are also consistent with the established definitions. The local amplitude is given by the magnitude of the 3D vector field. For most applications, however, it is more convenient to work with the *local attenuation*, the logarithm of the local amplitude:

$$A = \log(\sqrt{|\mathbf{q}|^2 + p^2}) = \frac{1}{2}\log(|\mathbf{q}|^2 + p^2) \ . \tag{13}$$

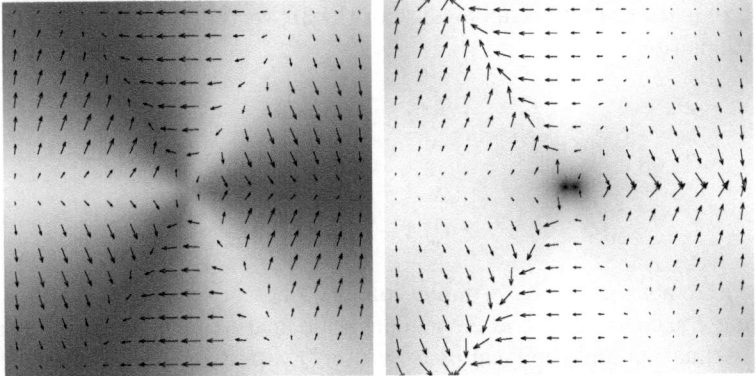

Fig. 3. Example for local features of the monogenic scale space at scale $s = 5$. Left: blurred Siemens star with three periods (grey value image) and corresponding Riesz transform (figure flow). Right: local attenuation (grey value image) and local phase vector field.

The local orientation and the local phase are best represented in a combined form, the *local phase vector*:

$$\mathbf{r} = \frac{\mathbf{q}}{|\mathbf{q}|} \tan^{-1} \left(\frac{|\mathbf{q}|}{p} \right) \qquad (14)$$

where $\tan^{-1}$ takes its values in the interval $[0, \pi)$. Whenever an explicit representation of orientation or phase is needed, the local orientation can be extracted from $\mathbf{r}$ as the orientation of the latter, and the local phase can be extracted by projecting $\mathbf{r}$ onto the local orientation[5]. The local features attenuation and phase vector are illustrated in fig. 3.

From the local features attenuation and phase vector, a couple of second order features can be extracted by means of derivative operators. Before we head for these features, we want to establish another fundamental relationship. In 1D the attenuation and the phase response of a minimum phase filter are related by the Hilbert transform [14]. The minimum phase property assures that there are no zeroes in the right halfplane, such that $\partial_z \log(f(z)) = f(z)^{-1} \partial_z f(z)$. Hence, if $f(z)$ is analytic in the right halfplane, $\log(f(z))$ is also analytic, and therefore, phase and attenuation form a Hilbert pair for every line $\text{Re}\{z\}$ fixed. Considering the analytic signal, the attenuation-phase principle can be transferred to the spatial domain: the *local* attenuation and the *local* phase form a Hilbert pair for any scale if the analytic signal is non-zero for any scale.

Under certain circumstances this equality generalizes to 2D. For local signal parts with an intrinsic dimension of one the local attenuation and the local phase vector of the monogenic signal form a Riesz triplet [7]. This means in particular

---

[5] Note that the orientation-direction ambiguity does not allow to extract 'the' correct sign of the local phase [8].

that the equations (3–6) remain valid for the 3D vector consisting of phase vector and attenuation:

$$\partial_x r_2 - \partial_y r_1 = 0 \tag{15}$$

$$\nabla \cdot \mathbf{r} + A_s = 0 \tag{16}$$

$$\partial_x A - \partial_s r_1 = 0 \tag{17}$$

$$\partial_y A - \partial_s r_2 = 0 \tag{18}$$

The requirement of intrinsic dimensionality one can easily be justified by the following consideration. The monogenic scale space can be computed from the local attenuation and the local phase vector by applying a generalized exponential function[6]:

$$(\mathbf{q}, p) = \exp(A) \left(\frac{\mathbf{r}}{|\mathbf{r}|} \sin |\mathbf{r}|, \cos |\mathbf{r}| \right) . \tag{19}$$

If the signal is intrinsically 2D, the derivative of the normalized phase vector does not vanish, which introduces some additional terms compared to the complex case. The deeper algebraic reason is the existence of commutator terms in the Campbell-Hausdorff formula [15].

Each of the four equations in (15–18) has a fundamental impact on the understanding of higher order image features. Consider a local image patch with intrinsic dimension one.

1. From (15) it follows that the local frequency is given by the divergence of the phase vector: $\omega = \nabla \cdot \mathbf{r}$.
2. From (17) and (18) it follows that the local amplitude maxima are points of phase congruency.
3. From (16) it follows that the local frequency is identical to the scale derivative of the local attenuation.

Each of these points need to be commented in order to understand their full consequences. The local frequency is normally computed by an implicit directional derivative of the local phase. This requires the local orientation to be known in advance and a steerable filter approximation of quadrature filters. Hence, the computation of local frequency is significantly simplified using the monogenic scale space. Note that the local frequency is reciprocal to the local scale of a signal (see third point).

Phase congruency is a method for edge and line detection based on the following idea [16]. At edges or lines, the local phase is independent of the local frequency in a wide range of scales. The original idea in [16] was to compare the local phase for a set of different scales. Much more elegant and also more efficient is it to calculate the *scale derivative* of the phase and to test it for zero crossings [8]. In case of the monogenic scale space, the latter method is equivalent

---

[6] This 'generalized exponential function' is simply the ordinary exponential series of a spinor [8].

to a maximum detection of the local amplitude in the main orientation[7], i.e., at local patches with intrinsic dimension one, the two contradictory approaches of maximum detection and phase congruency are actually equivalent.

It is well known that the isotropic local frequency can be computed from a ratio of log-normal bandpass filters, see e.g. [17]. However, it is not obvious if this is just a coincidence or if there is a more fundamental relationship. As it can be seen above, the latter is the case: isotropic local frequency is identical to the scale derivative of the local attenuation. The derivative operator can be considered as a infinitesimal difference, i.e., a ratio of local amplitudes.

The second order features discussed above were considered in the context of local patches with intrinsic dimension one. At points of high curvature, corners, and junctions, (15–18) are not fulfilled in general. In particular this means that the phase vector field might contain rotational components at those points. Furthermore, phase congruency and detection of local maxima behave differently; actually phase congruency is superior [7].

The surely most interesting extensions of the so far described features probably need higher order derivatives than first order, e.g., curvature. However, the deeper investigation of intrinsically 2D points is not covered by this paper. We rather focus on efficient and accurate implementations for the monogenic scale space on finite domains.

3   Implementation on a Bounded Domain

The solution for the monogenic scale space based on the convolution operators (8) and (10) implicitly assumes that the spatial domain has infinite extent. In practice, this is obviously not true. Applying the convolution operators to finite image planes requires assumptions about the signal outside the image plane. Some popular methods are either extensions with zeroes or periodic extensions. The latter is always used if the operators are applied in the Fourier domain. More advanced methods making use of e.g. normalized convolution [17] cannot be used in combination with odd filter kernels.

3.1   Eigenfunction Implementation for the Poisson Scale Space

A remaining popular method is the one assuming reflective boundaries for the convolution. This implicitly corresponds to the symmetric extension of the image at the boundaries. In context of a PDE formulation, this extension assures zero Neumann boundary conditions [12]. From a common perspective of scale space theory, the average grey value invariance is an often required property. This is fulfilled if we have zero Neumann boundary conditions since no intensity can leave or enter the image plane.

Hence, we extend (1) with conditions for the image boundary $\partial\Omega$ according to

---

[7] Note that parabolic points of the amplitude are also parabolic points of the attenuation. Furthermore, a zero crossing is only reported if the slope of the phase derivative is non-zero, corresponding to a non-zero main curvature of the attenuation.

$$\mathbf{n} \cdot \nabla p(\mathbf{x}, s)\big|_{\mathbf{x} \in \partial \Omega} = 0 \qquad (20)$$

where $\mathbf{n}$ is the normal of the boundary. In case of a rectangular domain, the horizontal derivatives at the left and the right border and the vertical derivatives at the bottom and the top border must vanish.

An elegant way to solve the PDE with extended boundary conditions is to make use of eigenfunctions. In case of a rectangular domain (for the sake of simplicity we assume $(x, y)^T \in [0, 1] \times [0, 1]$), the eigenfunctions are cosine oscillations [12], eq. (18): $l_{\mathbf{m}}(\mathbf{x}) = 2\cos(\pi m x)\cos(\pi n y)$ where $\mathbf{m} = (m, n)^T$. Hence, the Poisson scale space can be computed by applying a cosine transform to the image, multiplying the resulting coefficients $C_{\mathbf{m}}$ with the corresponding eigenvalues $\exp(-\pi|\mathbf{m}|s)$, and by transforming the image back to the spatial domain.

Note that the cosine transform only needs to be evaluated at discrete frequencies since the symmetric extensions (see fig. 4, left) of the image imply a periodic extension of the signal with doubled period. The required cosine transform is then simply the 2D Fourier transform of the extended symmetric signal (divided by four), since the latter is symmetric with respect to both coordinate axes, and hence, all sine-terms of the Fourier kernel are zero.

In the case of *sampled* signals on a bounded domain, the number of discrete frequencies is finite and the eigenfunctions $l_{\mathbf{m}}(\mathbf{x})$ boil down to discrete functions $l_{\mathbf{mx}}$. Since the reflection axes are located at $x = 0$ and $y = 0$ and since the sampling points are located at the centers of the pixels, we obtain the discrete functions $l_{\mathbf{mx}} = l_{\mathbf{m}}(\mathbf{x}/k)$ with $\mathbf{x} \in \{\frac{1}{2}, \frac{3}{2}, \ldots, k - \frac{1}{2}\} \times \{\frac{1}{2}, \frac{3}{2}, \ldots, k - \frac{1}{2}\}$ and $\mathbf{m} \in \{0, 1, \ldots, k-1\} \times \{0, 1, \ldots, k-1\}$ if k indicates the image height and width in pixels[8].

These functions $l_{\mathbf{mx}}$ are the basis functions of the *2D discrete cosine transform* (DCT2) which is a concatenation of 1D DCTs (see [18] for the definition of the DCT1). The DCT1 of length k can either be computed by applying the FFT1 of length $2k$ to a symmetrically extended signal (like one row/column in fig. 4, left) [18], or by reordering the samples and applying an FFT1 of length k [19].

3.2   Eigenfunction Implementation for the Monogenic Scale Space

Having an implementation for the Poisson scale space, we are still missing for a method to obtain its harmonic conjugates. These are easily obtained by the following method. Since we know that

$$p(\mathbf{x}, s) = \sum_{m=0}^{\infty} \sum_{n=0}^{\infty} C_{\mathbf{m}} l_{\mathbf{m}}(\mathbf{x}) \exp(-\pi|\mathbf{m}|s) = \sum_{m=0}^{\infty} \sum_{n=0}^{\infty} C_{\mathbf{m}}^p l_{\mathbf{m}}(\mathbf{x})$$

we can compute the corresponding harmonic potential $\varphi(\mathbf{x}, s)$ by

---

[8] Without loss of generality, we assume quadratic domains. The generalization to non-quadratic domains is straightforward.

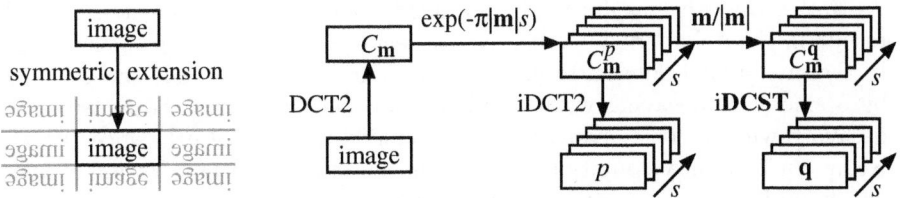

Fig. 4. Left: symmetric extension of the image. Right: algorithm for computing the monogenic scale space. The image is transformed using the DCT2. Multiplying with the eigenvalues and applying the inverse DCT2 results in the Poisson scale space p. Multiplying with $\mathbf{m}/|\mathbf{m}|$ and applying the inverse **DCST** yields the corresponding figure flow $\mathbf{q}$.

$$\varphi(\mathbf{x}, s) = \sum_{m=0}^{\infty} \sum_{n=0}^{\infty} \frac{C_\mathbf{m}}{-\pi|\mathbf{m}|} l_\mathbf{m}(\mathbf{x}) \exp(-\pi|\mathbf{m}|s) + C_{\mathrm{DC}}$$

such that $p = \varphi_s$. The harmonic conjugates of p are then obtained by

$$\mathbf{q}(\mathbf{x}, s) = \nabla\varphi(\mathbf{x}, s) = \sum_{m=0}^{\infty} \sum_{n=0}^{\infty} \frac{C_\mathbf{m}^p}{|\mathbf{m}|} \begin{pmatrix} m\sin(\pi m x)\cos(\pi n y) \\ n\cos(\pi m x)\sin(\pi n y) \end{pmatrix} . \qquad (21)$$

Similar as for the Poisson scale space implementation, we switch to discrete signals with $\mathbf{x} \in \{\frac{1}{2}, \frac{3}{2}, \ldots, k - \frac{1}{2}\} \times \{\frac{1}{2}, \frac{3}{2}, \ldots, k - \frac{1}{2}\}$, i.e., we obtain the basis functions

$$l'_{\mathbf{mx}} = \begin{pmatrix} \sin(\pi m x/k)\cos(\pi n y/k) \\ \cos(\pi m x/k)\sin(\pi n y/k) \end{pmatrix} . \qquad (22)$$

Note that the sine functions $\sin(\pi m x/k)$ only form a basis, if $m \in \{1, 2, \ldots, k\}$ (accordingly for $\sin(\pi n y/k)$)[9]. However, since $C_{(k,n)}$ and $C_{(m,k)}$ are zero for all m, n, it is sufficient to compute (22) for $\mathbf{m} \in \{0, 1, \ldots, k-1\} \times \{0, 1, \ldots, k-1\}$.

The basis functions (22) yield a vector valued transformation called **DCST**. Its two components are the two possible combinations of a DCT1 and a 1D discrete sine transform (DST1)[10]. The DST1 can either be computed by applying an FFT1 to the anti-symmetrically extended signal of length $2k$ or by reordering the samples and an FFT1 of length k.

In our algorithm (see fig. 4, right), the figure flow is obtained by multiplying the DCT spectrum of the Poisson scale space with $\mathbf{m}/|\mathbf{m}|$ and applying the inverse **DCST** component-wise (see (21)), i.e.,

$$\mathbf{iDCST}\{C_\mathbf{m}^q\} = \begin{pmatrix} \mathrm{iDST1}_m\{\mathrm{iDCT1}_n\{m/|\mathbf{m}|\, C_\mathbf{m}^p\}\} \\ \mathrm{iDCT1}_m\{\mathrm{iDST1}_n\{n/|\mathbf{m}|\, C_\mathbf{m}^p\}\} \end{pmatrix} .$$

---

[9] This follows from computing the discrete Fourier transform of the anti-symmetrically extended signal (size $2k$). In this case, the spectral component at the frequency k does not vanish, but the component at frequency 0 does.

[10] According to [19] the DST1 consists of sine functions with period $2k + 2$ (instead of $2k$) which is not appropriate in our case. Although both transforms are called DST, they differ and should not be mixed up.

Notice that the results in the bounded domain case can deviate from the results in the unbounded domain case. Consider for instance a plane sine-wave. In the unbounded domain case, the corresponding Poisson scale space is given by a sine-wave decaying with s and the harmonic conjugate is a cosine-wave decaying with s. This changes in the bounded domain case[11], since the zero Neumann boundary condition requires the sine to be continued symmetrically, which implies that at scales $s > 0$ we do not have a pure sine any more.

Furthermore, the harmonic conjugate component perpendicular to the boundary must vanish at the boundary for $s > 0$. This can easily be concluded from (4) and (5). Requiring $\partial_x p = 0$ ($\partial_y p = 0$) at the vertical (horizontal) boundaries yields that $\partial_s q_1 = 0$ ($\partial_s q_2 = 0$) and since for $s \to \infty$ $q_1 = q_2 \equiv 0$, we obtain $q_1 = 0$ ($q_2 = 0$). Taking the limit $s \to 0$, the perpendicular components at boundary at scale 0 also vanish. Hence, the figure flow of a sine-wave cannot be a pure cosine. Although this result is totally correct with respect to (4,5) and the zero Neumann boundary condition, it might not be the desired output for certain applications.

4   Examples and Applications

In this section we give some examples for the results obtained from the implementation described above, and sketch a sophisticated application of the introduced framework.

4.1   Visualization and Reconstruction from Local Features

The best way to visualize the monogenic scale space is of course an image sequence descending through scale. On paper, we must restrict to a single slice of the scale space or to isosurface plots. One example for a slice of the monogenic scale space is given in fig. 3. Typically, it is a good choice to represent the Poisson scale space as a grey value image and the Riesz transform (as its figure flow) as a 2D vector field. Similar we do for the polar representation (attenuation and phase vector).

A nice application of attenuation and phase vector forming a harmonic field (15–18) is the reconstruction from local attenuation and local phase (see e.g. [20]). Whereas methods based on traditional approaches like, e.g., Gabor filters, are quite complicated and computationally expensive, the reconstruction becomes nearly trivial for the monogenic signal.

The reconstruction from local attenuation works as follows. The local phase vector field at any scale is approximated by the Riesz transform of the local attenuation: $\mathbf{r}(\mathbf{x}, s) \approx (\mathbf{h}_2 * A(\cdot, s))(\mathbf{x})$. Plugging this into (19) for scale $s = 0$ yields

$$f(\mathbf{x}) = p(\mathbf{x}, 0) \approx \exp(A(\mathbf{x}, 0)) \cos(|(\mathbf{h}_2 * A(\cdot, s))(\mathbf{x})|) \ . \tag{23}$$

---

[11] The inverse remains unchanged for the bounded domain case: a cosine becomes a decaying cosine and its harmonic conjugate is a decaying sine.

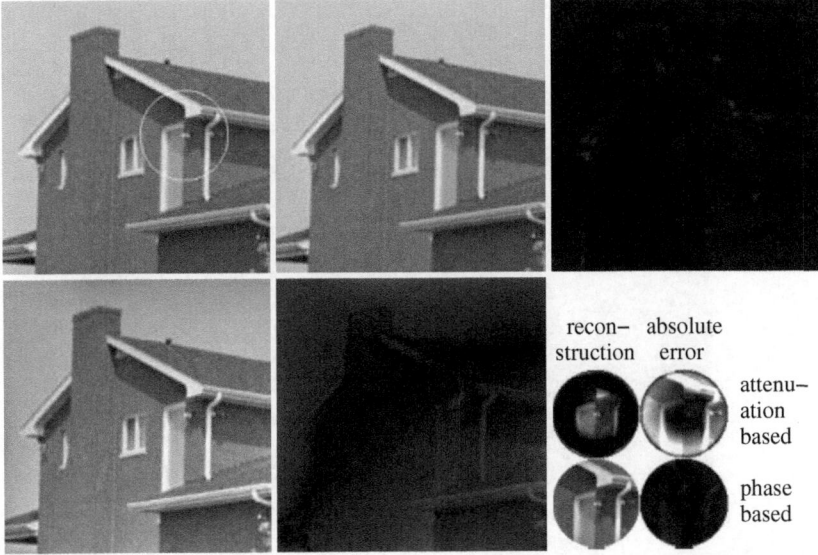

Fig. 5. Example for the reconstruction from the local attenuation and the local phase vector field. Upper row from left to right: original image (the circle indicates the area of local reconstruction), image reconstruction from attenuation, and corresponding absolute error (in the same scale). Bottom row from left to right: image reconstruction from phase vector field, corresponding absolute error (in the same scale), and local reconstructions.

The reconstruction based on the local phase vector field is slightly more complicated. The (DC-free) local attenuation at any scale is approximated by the (negative) Riesz transform of the phase vector field[12] at the same scale: $A(\mathbf{x}, s) - \sum_{\mathbf{x}} A(\mathbf{x}, s) \approx -(\mathbf{h}_2 * \mathbf{r}(\cdot, s))(\mathbf{x})$. The DC component of the attenuation is lost, since the Riesz transform of the phase vector field is by definition DC free. The DC component of the attenuation corresponds to a global factor of the image dynamics.

Plugging the approximated attenuation into (19) for scale $s = 0$ yields

$$f(\mathbf{x}) \approx \exp\left(\sum_{\mathbf{x}} A(\mathbf{x}, 0)\right) \exp(-(\mathbf{h}_2 * \mathbf{r}(\cdot, 0))(\mathbf{x})) \cos(|\mathbf{r}(\mathbf{x}, 0)|) + c_{\mathrm{DC}} \ , \quad (24)$$

where c_{DC} is a further DC correction term corresponding to a grey value shift.. The two correction terms (the DC component of the attenuation and the final DC correction) are required, since the absolute grey level and the dynamics information of the image are obviously lost in the reconstruction.

In fig. 5 examples for both reconstruction methods can be found, where the grey value dynamics of the phase-reconstructed images are rescaled and shifted

---

[12] This convolution between two vector functions yields a scalar function, i.e., is interpreted in terms of a scalar product.

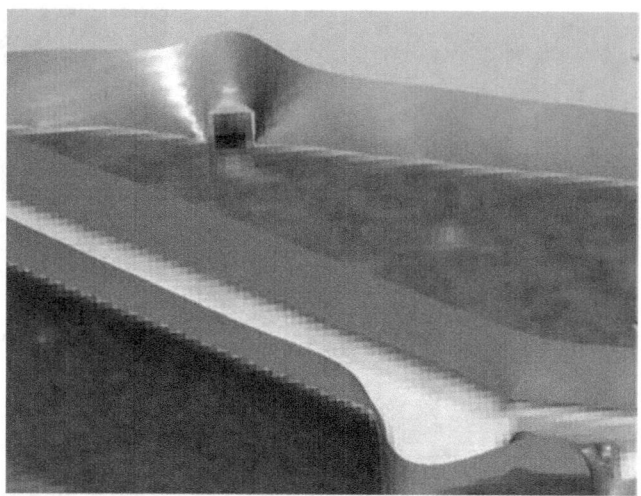

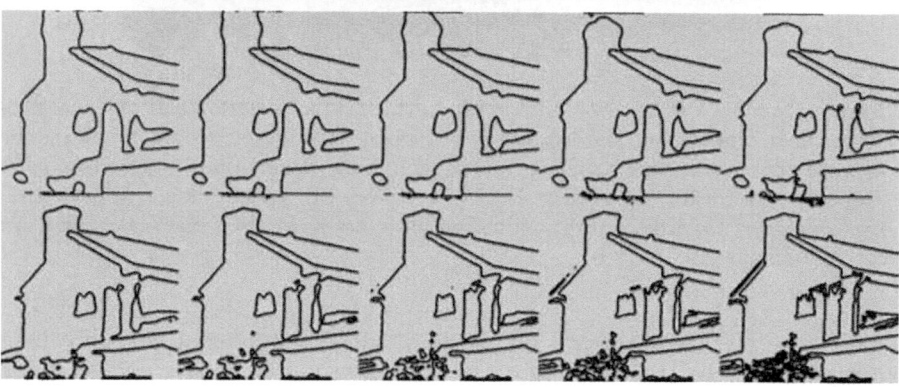

Fig. 6. Example for the phase congruency surface in scale space. The original image can be found in fig. 5. The upper part of this figure shows a detail of the 3D phase congruency surface with the corresponding part of the image at the ground plane. The lower part shows a sequence of slice cuts through the surface at different scales (range 20–2).

according to the discussion above. For the global reconstruction, the method based on the local attenuation obviously works better than the one based on the local phase vector field. The loss of the DC and dynamics information also occurs in a semi-local context, which results in a too dark reconstruction in the upper left corner and a too bright reconstruction in the lower right corner, see fig. 5 bottom left. Furthermore, the dynamics in the lower right corner is over-estimated.

For the local reconstruction however, the phase based method gives a very accurate result, see fig. 5 lower right, whereas the attenuation based reconstruction becomes very poor close to the border of the region. Hence, if local reconstructions are required, the phase based method is clearly preferable. Furthermore, the local phase vector field seems to represent the image structure more locally than the attenuation does.

4.2 Phase Congruency in Scale Space

As stated above, phase congruency is a method for edge and line detection. Using the scale derivative of the phase instead of finite differences, edges and lines can be detected for every single scale. Now, the decisive advantage of the monogenic scale space is that the phase congruency contours in each scale are connected to the contours at the neighbored scales. Hence, contours can be tracked from coarse to fine to avoid problems of disconnected edges and noise. Furthermore, such a scheme can directly be combined with methods for hierarchical processing.

Last but not least, one can apply methods from differential geometry to the phase congruency surfaces in order to extract certain features. For instance, instead of estimating the local 2D curvature from the 3D curvature of isophotes in Gaussian scale space, one can compute the local 2D curvature from the 3D curvature of surfaces which *directly represent an edge / line evolution through scale space*. To make this point clearer, consider the example in fig. 6. One can clearly see, how the isosurface fits into the small corner for decreasing scale. From this point of view, the slice cuts of the isosurface can be considered to be a primitive version (the contour is piecewise linear) of a diffusion snake. However, it should be easy to extend the contour finding to higher order functions, resulting in a new approach to diffusion snakes.

5   Conclusion

We have presented a new framework for image processing in scale space. The main idea is to extend the Poisson scale space to a vector valued scale space which is a harmonic vector field. Exploiting the nice mathematical properties of the monogenic scale space, new insight into the analysis of images was found. In order to make the new approach applicable to real data, an efficient algorithm has been presented, which allows to compute the monogenic scale space on a bounded domain based on standard algorithms like DCT and DST. As a result, the computation of each slice of the monogenic scale space, i.e., the monogenic signal at a certain scale, takes just as long as three FFT2. With about the same computational effort, images can be reconstructed if solely the attenuation or phase vector field is available. The high computational speed also allows to consider the evolution of phase congruency through scale, which appears to be a promising approach to image analysis for the future.

References

1. Weickert, J., Ishikawa, S., Imiya, A.: Linear scale-space has first been proposed in Japan. Mathematical Imaging and Vision 10 (1999) 237–252

2. Pauwels, E.J., Van Gool, L.J., Fiddelaers, P., Moons, T.: An extended class of scale-invariant and recursive scale space filters. IEEE Transactions on Pattern Analysis and Machine Intelligence 17 (1995) 691–701
3. Duits, R., Florack, L.M.J., de Graaf, J., ter Haar Romeny, B.M.: On the axioms of scale space theory. Journal of Mathematical Imaging and Vision (2003) to appear.
4. Koenderink, J.J.: The structure of images. Biological Cybernetics 50 (1984) 363–370
5. Felsberg, M., Sommer, G.: Scale adaptive filtering derived from the Laplace equation. In Radig, B., Florczyk, S., eds.: 23. DAGM Symposium Mustererkennung, München. Volume 2191 of Lecture Notes in Computer Science., Springer, Heidelberg (2001) 124–131
6. Silvester, P.P., Ferrari, R.L.: Finite Elements for Electical Engineers. Cambridge University Press (1983)
7. Felsberg, M., Sommer, G.: The monogenic scale-space: A unifying approach to phase-based image processing in scale-space. Journal of Mathematical Imaging and Vision (2003) to appear.
8. Felsberg, M.: Low-Level Image Processing with the Structure Multivector. PhD thesis, Institute of Computer Science and Applied Mathematics, Christian-Albrechts-University of Kiel (2002) TR no. 0203, available at http://www.informatik.uni-kiel.de/reports/2002/0203.html.
9. Koenderink, J.J.: A hitherto unnoticed singularity of scale-space. IEEE Transactions on Pattern Analysis and Machine Intelligence 11 (1989) 1222–1224
10. Damon, J.: Local Morse theory for solutions of the heat equation and gaussian blurring. Journal of Differential Equations 115 (1995) 368–401
11. Florack, L., Kuijper, A.: The topological structure of scale-space images. Journal of Mathematical Imaging and Vision 12 (2000) 65–79
12. Duits, R., Felsberg, M., Florack, L.M.J.: Scale space theory on a bounded domain. In: Scale Space Conference. (2003) accepted.
13. Felsberg, M., Sommer, G.: The monogenic signal. IEEE Transactions on Signal Processing 49 (2001) 3136–3144
14. Papoulis, A.: The Fourier Integral and its Applications. McGraw-Hill, New York (1962)
15. Hein, W.: Struktur- und Darstellungstheorie der klassischen Gruppen. Springer, Berlin (1990)
16. Kovesi, P.: Image features from phase information. Videre: Journal of Computer Vision Research 1 (1999)
17. Granlund, G.H., Knutsson, H.: Signal Processing for Computer Vision. Kluwer Academic Publishers, Dordrecht (1995)
18. Ahmed, N., Natarajan, T., Rao, K.R.: Discrete cosine transform. IEEE Transactions on Computers (1974) 90–93
19. Jain, A.K.: Fundamentals of Digital Image Processing. Prentice-Hall (1989)
20. Behar, J., Porat, M., Zeevi, Y.Y.: Image reconstruction from localized phase. IEEE Transactions on Signal Processing 40 (1992) 736–743

Using the Complex Ginzburg–Landau Equation for Digital Inpainting in 2D and 3D

Harald Grossauer and Otmar Scherzer

Department of Computer Science
University of Innsbruck
Technikerstr. 25
A–6020 Innsbruck, Austria
{harald.grossauer,otmar.scherzer}@uibk.ac.at
http://informatik.uibk.ac.at/infmath/

Abstract. Recently, several different approaches for digital inpainting have been proposed in the literature. We give a review and introduce a novel approach based on the complex Ginzburg–Landau equation. The use of this equation is motivated by some of its remarkable analytical properties. While common inpainting technology is especially designed for restorations of two dimensional image data, the Ginzburg–Landau equation can straight forwardly be applied to restore higher dimensional data, which has applications in frame interpolation, improving sparsely sampled volumetric data and to fill in fragmentary surfaces. The latter application is of importance in architectural heritage preservation. We discuss a stable and efficient scheme for the numerical solution of the Ginzburg–Landau equation and present some numerical experiments. We compare the performance of our algorithm with other well established methods for inpainting.

Keywords: Ginzburg–Landau equation, inpainting, diffusion filtering, non–linear partial differential equations, variational problems

1   Introduction

Inpainting is the process of restoration of missing image data; it is typically done by artists. *Digital Inpainting* is performed by computers requiring the user only to mark areas to be inpainted in a digitized image. Digital inpainting has several applications in photography [1], such as scratch removal or retouching. Combining inpainting algorithms with scratch detection algorithms (see e.g. [2] and the references therein) allows to almost automatically restore large sets of degraded images or even complete movies. A difficulty associated with digital inpainting is to set up a measure of visual sensitivity towards defects which can be used in computer code. An attempt in this direction is the *perceptually based physical error metric* introduced by Ramasubramanian, Pattanaik and Greenberg [3]. Today, the common opinion is that the human perceptual system is more sensitive to edges than to texture and most sensitive to junctions; see Caselles, Coll and Morel [4] as a paradigm of this statement in the computer science and

L.D. Griffin and M. Lillholm (Eds.): Scale-Space 2003, LNCS 2695, pp. 225–236, 2003.

mathematical literature. As a consequence a good inpainting algorithm should connect corresponding edges and extrapolate textures smoothly.

In the following we survey some recently proposed inpainting methods based on *level line strategies, partial differential equations (PDEs)* and *variational methods*. These methods are most relevant for a comparison with our work. Other topics related to image inpainting such as *texture synthesis* with statistical methods (see e.g. [5,6] and references therein) and image interpolation with *sampling methods* (see e.g. [7,8]) are not considered in this paper.

- An inpainting algorithm based on level lines has been proposed by Masnou and Morel [9,10]. It consists of several stages:
  1. All T–junctions — that are points where level lines hit the boundary of the inpainting domain — are tabulated.
  2. A table of pairs of compatible T–junctions is generated. Two T–junctions are *compatible*, if their associated level lines belong to the same grey level intensity and have the same orientation.
  3. From the set of candidates of level lines connecting compatible T–junctions, the one having the lowest *total generalized elastica energy*

  $$\sum \int_{L_{i,j}} (\alpha + \beta |\kappa|^p) ds$$

  is selected. Here $L_{i,j}$ denotes the level line connecting the T–junctions with the indices i and j, and the sum is with respect to all level lines. For convex inpainting domains these are just straight lines.

  The algorithm is computationally expensive: a triangulation of the inpainting domain has to be calculated and an optimal set of level lines out of all possible connections has to be found. Combination with a dynamic programming approach and sorting out inadmissible connections at an early stage keeps runtime complexity relatively low. The implementation presented in [9] does not allow inpainting domains with holes (doughnut shaped inpainting domains).
- Ballester et.al. [11] have proposed a variational method for inpainting. They derive a system of coupled PDEs to extrapolate grey level values and the gradient direction vector field smoothly into the inpainting domain. The system of PDEs is solved using level sets of the image intensity function. This makes the numerical results depend on implementational details, in particular the order in which the level sets are processed (cf. figure 6).
- Bertalmio et.al. [12] introduced an algorithm which imitates the work of manual inpainting of artists. The process is heuristically designed such that an "image smoothness measure" (in their case the image Laplacian) is constantly propagated along the level line direction into the inpainting domain. Numerically, their algorithm corresponds to an explicit finite difference scheme for the partial differential equation

  $$\frac{\partial u}{\partial t} = |\nabla u| \left(\nabla \Delta u \cdot \frac{\nabla^\perp u}{|\nabla^\perp u|} \right).$$

Here $\nabla u = \left(\frac{\partial u}{\partial x_1}, \frac{\partial u}{\partial x_2}\right)$ denotes the gradient, $\nabla^\perp u = \left(-\frac{\partial u}{\partial x_2}, \frac{\partial u}{\partial x_1}\right)$ denotes the vector orthogonal to the gradient, and $|\cdot|$ denotes the Euclidean distance. To stabilize the explicit scheme it is combined with a nonlinear diffusion filtering technique for u in the inpainting domain. Bertalmio, Bertozzi and Sapiro [13] have extended this work and embedded the algorithm into a framework of the Navier–Stokes equations.

– Chan and Shen [14,15] use TV–inpainting and solve the differential equation

$$\frac{\partial u}{\partial t} = \nabla \cdot \left(\frac{\nabla u}{|\nabla u|}\right) + \lambda_e \left(u^0 - u\right) \tag{1}$$

up to a stationary point. Here λ_e is an a–priori defined positive function which is zero on the domain to be inpainted. Outside the inpainting domain this equation denoises the image and thus makes the algorithm robust to noise. The inpainted image is composed from the stationary function in the inpainting domain and by the initial image u^0 outside.

Since this TV–inpainting fails to connect long thin structures Chan and Shen [15] propose to replace the diffusion term in (1) by $\nabla \cdot \left(\frac{g(|\kappa|)}{|\nabla u|} \nabla u\right)$, where κ denotes the *curvature* of u. This results in an inhomogeneous third order partial differential equation with forcing term $\lambda_e \left(u^0 - u\right)$.

In [16] Chan and Shen develop an inpainting algorithm based on connecting appropriate level lines by Euler elastica curves. Formulated as a fourth order PDE in terms of the image function u, this algorithm turns out to be a generalization of TV–inpainting and Bertalmio's algorithm, containing both of them as special cases if the involved parameters are set appropriately.

– Esedoglu and Shen [17] combine the Euler elastica functional and the Mumford–Shah functional to derive a fourth order PDE which concurrently does inpainting, denoising and segmentation of the image.

– Oliveira et.al. [18] propose an inpainting algorithm optimized for speed. It consists in repeatedly convolving an arbitrary continuation of u^0 in the inpainting domain with a filter mask. The inpainted image is composed from the convolved image in the inpainting domain and by u^0 outside. This solution behaves very similar to the solution of the linear diffusion equation on the inpainting domain with Dirichlet boundary conditions u^0. As it is well known, linear diffusion tends to blur edges, unless the user manually supplies additional information, e.g. an a–priori segmentation of the inpainting domain.

2   The Ginzburg–Landau Equation

2.1   Motivation

The Ginzburg–Landau equation was originally developed by Ginzburg and Landau [19] to phenomenologically describe phase transitions in superconductors near their critical temperature. The equation has proven to be useful in several distinct areas besides superconduction. It is used to model some types of chemical reactions like the famous Belousov–Zhabotinsky reaction, to describe boundary layers in multi–phase systems, and to describe the development of patterns and shocks in non–equilibrium systems (see [20,21,22] and references therein).

Solutions of the real valued Ginzburg–Landau equation develop homogeneous areas, which are separated by *phase transition regions*, that are interfaces of minimal area, see the comments following equation (6). In image processing homogeneous areas correspond to domains of constant grey value intensities, and phase transitions to edges. Thus the quoted properties make the real valued Ginzburg–Landau equation a reasonable method for high quality inpainting of binary images.

2.2   Physical Foundations of the Ginzburg–Landau Equation

Investigating the thermodynamic potential of superconductors Ginzburg and Landau derived the following approximation for the corresponding energy functional, depending on the *order function* $u : \Omega \to \mathbb{C}$:

$$F(u, \nabla u) := \frac{1}{2} \int_\Omega \underbrace{|-i\nabla u|^2}_{\text{kinetic term}} + \underbrace{\alpha |u|^2 + \frac{\beta}{2} |u|^4}_{\text{potential term}} \tag{2}$$

where α and β are physical constants. For a nontrivial minimizer to exist $\alpha < 0$ and $\beta > 0$ is necessary. The factor '$-i$' in the kinetic term is a holdover from quantum mechanics and is not essential. The state of minimal energy satisfies the Euler equation of $F(u, \nabla u)$:

$$\frac{\delta F}{\delta u} := \Delta u + \frac{1}{\varepsilon^2} \left(1 - |u|^2\right) u = 0 . \tag{3}$$

The equation has been rescaled such that the minima of the potential term function are attained at the sphere $|u| = 1$, which corresponds to the choice $\alpha = -\frac{1}{\varepsilon^2}$ and $\beta = -\alpha$.

In physics ε is called the *coherence length* and correlates to the width of the transition region, that is the width of the transient separating different phases. This can be highlighted in the case when the order function u is one–dimensional and real valued, $u : \mathbb{R} \to \mathbb{R}$, satisfying the boundary condition $\lim_{x \to \pm\infty} u(x) = \pm 1$. In this case an analytical formula for the solution of (3) can be given:

$$u(x) = \frac{e^{\frac{\sqrt{2}}{\varepsilon} x} - 1}{e^{\frac{\sqrt{2}}{\varepsilon} x} + 1} . \tag{4}$$

In figure 1 we have plotted $u(x)$ corresponding to different values of ε. It can be realized that the width between the phases ± 1 is approximately 4ε. For $\varepsilon \searrow 0$, $u(x)$ approaches the Heaviside function.

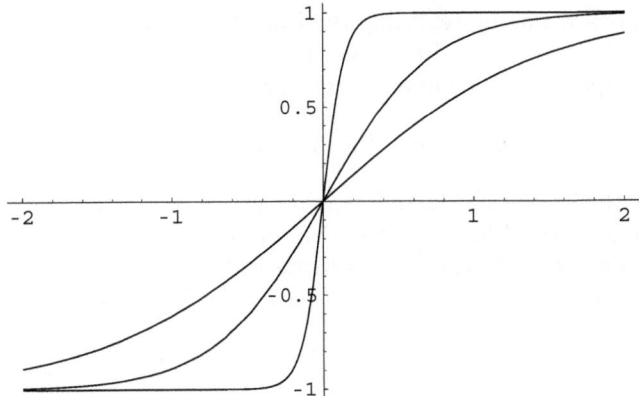

Fig. 1. Solutions of the one–dimensional real valued Ginzburg–Landau equation for $\varepsilon = 1, 0.5, 0.1$

2.3 Algorithm

In the following we describe the use of the Ginzburg–Landau equation for image inpainting of *grey* valued images. To this end we solve the *complex* Ginzburg–Landau equation on the inpainting domain with appropriate boundary data related to the input data.

Preparations. Let D be the domain of the image, usually a rectangular subset of $\mathbb{R}^2$. The *inpainting domain* is denoted by Ω; we assume that it is an open subset of D.

Let $\overline{u}^0 : D \to [-1, 1]$ be the grey–value intensity of an image scaled to the interval $[-1, 1]$; the values -1 and 1 correspond to pure white and black, respectively. The function $\overline{u}^0$ is identified with the real part of a complex valued function $u^0 : D \to \mathbb{C}$ by selecting the imaginary part as

$$\Im(u^0) = \sqrt{1 - \left(\Re(\overline{u}^0)\right)^2}.$$

such that $|u^0(x)| = 1$ for all $x \in D$.

A complex valued solution u of (3) will still have an absolute value of 1 almost everywhere but our inpainting (the real part of the solution) may contain any value from the interval $[-1, 1]$. Note that the complex Ginzburg–Landau equation can be considered a system of partial differential equations. Substituting back the solution for the imaginary component into the equation for the real component, we obtain a fourth order equation for the real part.

A common approach to inpaint color images, respectively vector valued images, is to inpaint the (color) components separately. In real world images the color components are typically not independent and a separation approach may lead to artifacts, like spurious colors or rainbow effects. The structure of equation (3) can formally be generalized to vector valued functions $u : D \to \mathbb{C}^n$ in a straight forward way by replacing the Euclidean distance in (3) by an appropriate norm $\| \cdot \|$ on $\mathbb{C}^n$. For RGB color images we found the maximum norm of the RGB–components to be most appropriate:

$$\|u(x)\| := \max\{|u^1(x)|, |u^2(x)|, |u^3(x)|\} \, .$$

Note that with this setting the differential equation (3), where $| \cdot |$ is replaced by $\| \cdot \|$, cannot be derived from a variational principle.

Now the problem of inpainting consists in finding $u : \Omega \to \mathbb{C}$ (resp. $\mathbb{C}^3$ for color images) which satisfies equation (3) and the Dirichlet boundary condition $u|_{\partial\Omega} = u^0|_{\partial\Omega}$.

Solving the Equation. To find a solution of equation (3) with Dirichlet boundary condition numerically we use a relaxation method (i.e. a steepest descent method) and solve the differential equation

$$\frac{\partial u}{\partial t} = \Delta u + \frac{1}{\varepsilon^2} \left(1 - |u|^2\right) u \tag{5}$$

up to a stationary point in time.

The reaction–diffusion type equation (5) with real valued u is a variant of the Allen–Cahn equation

$$u_t = \Delta u + \psi'(u). \tag{6}$$

According to [23] equation (6) is called Ginzburg–Landau equation if u is vector valued or complex and the potential $\psi(u)$ has a stable minimum for $|u| = 1$, which is true in our case. In [23] equation (5) with real valued scalar u is used to approximate mean curvature motion. For every ε there exists a unique bounded solution, which — in the limit $\varepsilon \searrow 0$ — consists of sets where either $u = +1$ or $u = -1$ and the interface moves according to mean curvature motion. A lot of mathematical theory about this matter is available, see [23] and references therein. Much less is known though when u is complex or vector valued. A comprehensive study for $u : \mathbb{R}^2 \to \mathbb{C}$ is given in [24].

To numerically integrate equation (5) we use an explicit, forward in time, finite difference scheme (see e.g. [25]). While irregular inpainting domains (which appear in text and scratch removal applications, cf. figures 2 and 3) can easily be handled with explicit schemes, the matrix equations to be solved with *implicit* schemes do not reveal regular structures, are difficult to set up numerically, and cannot be solved efficiently.

In the case of rectangular inpainting domains we have been able to compare the explicit finite difference method with an *implicit nonlinearity lagging* scheme [26], as well as semi–implicit techniques, such as Peaceman–Rachford [26], or a semi–implicit Fourier–spectral method [27], and found that for $\varepsilon \ll 1$ these methods did not give any obvious advantage concerning stability and computation time. This observation is in accordance with [28] where it is argued (though not rigorously proven) that even for implicit schemes a timestep restriction $\delta t < \mathcal{O}(\varepsilon^2)$ is necessary.

Fig. 2. The painting "Holy Family" from Michelangelo with scratches (*top left*). The scratches have been inpainted with the plain Ginzburg–Landau algorithm (*bottom left*). The picture shows the result of the same algorithm interleaved with some steps of coherence enhancing diffusion (*bottom right*). Detailed views of the red framed parts are compared (*top right*)

Fig. 3. Lenna inpainting

Discretizing (5) in space and time, the explicit finite difference method has the following form:

$$u_{i,j}^{t+1} = u_{i,j}^t + \delta t \cdot \left(\Delta u_{i,j}^t + \frac{1}{\varepsilon^2} \left(1 - \|u_{i,j}^t\|^2 \right) u_{i,j}^t \right) \tag{7}$$

where

$$\Delta u_{i,j} = u_{i-1,j} + u_{i+1,j} + u_{i,j-1} + u_{i,j+1} - 4u_{i,j} \ .$$

Here $u_{i,j}$ denotes the color vector intensity at the pixel point (i,j). As initial value we set $u^0|_\Omega = 0$. Equation (7) has been rescaled to get rid of the influence of space discretization. According to what was said before we have to choose a timestep $\delta t < \varepsilon^2$, which in particular shows that the time steps have to be extremely small for $\varepsilon \ll 1$, which is required for high contrast inpainting purposes. The iteration process (7) is stopped at time $\bar{t}$ if

$$\max_{i,j}\{|u_{i,j}^{\bar{t}} - u_{i,j}^{\bar{t}-1}|\}$$

drops below a certain threshold.

Postprocessing. The solution of the Ginzburg–Landau equation reveals high contrast in the inpainting domain, which makes it particularly suited for inpainting purposes. However, the level lines of the solution of the Ginzburg–Landau equation at the boundary of the inpainting domain might look kinky. In general this cannot be considered a bad habit as figure 4 shows, but in certain pictures (for instance in artistic drawings) this may look disturbing. For such applications we suggest to apply a couple of coherence enhancing diffusion steps (see e.g. [29]) to steer the direction of inpainting. In figure 2 we have inpainted a scratched digitized painting by Michelangelo. The inpainting with the Ginzburg–Landau equation is of high contrast, but the level lines look kinky as the enlargements of details in figure 2 show. The kinks can be smoothed via coherence enhancing diffusion.

2.4   Three Dimensional Inpainting

The Ginzburg–Landau equation can be generalized to any number of space dimension. Thus in particular it can be applied to inpaint three dimensional grey valued image intensity functions $u : \mathbb{R}^3 \to \mathbb{R}$; for example allowing optical improvement of sparsely sampled data or frame interpolation (treating a movie as a stack of images). The generalization for inpainting of three dimensional vector valued data (e.g. color images) is straight forward.

Inspired by the work of Davis et. al. [30] we also applied our algorithm for completion, respectively continuation, of 2D–surfaces, which are represented as zero level sets of auxiliary functions $u : \mathbb{R}^3 \to \mathbb{R}$. A typical settings for the auxiliary function is the signed distance function to the surface or by setting $u = \pm 1$ on opposite sides of the surface. Then the inpainting algorithm is applied to a volume containing the missing parts of the surface. The missing surface

Fig. 4. A corner of the cube was manually cut out (*left*). The completion attained from the linear diffusion equation resembles a membrane stretched across the edges (*middle*). With the Ginzburg–Landau equation a perfect corner is achieved (*right*)

is completed by the zero level set of the solution. We view an example of our algorithm applied to a synthetic three dimensional object in figure 4. The missing corner of the cube was manually cut out. The middle picture shows the result of surface completion using the linear diffusion equation for inpainting. As expected the resulting surface looks like a membrane stretched over the existing edges. In contrast the Ginzburg–Landau inpainting algorithm develops a hard corner. It depends on the individual application whether one prefers smooth and soft surfaces or hard edges and corners. By tuning the value of ε Ginzburg–Landau inpainting allows the user to choose any degree of "smoothness".

A more realistic application is shown in figure 5 where a hole in the left cheekbone has been filled. Using this kind of processing for medical data is quite dangerous but could be useful for refinement of data obtained in heritage recording projects [31].

3   Results and Conclusion

Results of our inpainting algorithm for text removal are presented in figure 3. In the Lenna picture 28779 pixels (which is about 11% of the image) have been overdrawn with white color. The inpainting was finished within a few seconds on a 1.5 GHz Pentium 4 PC running unoptimized C++ code under Linux. On a first look no visual deficiencies can be seen in the inpainted image. Closer examination reveals some fringes and kinks, most notable at the bottom edge of the hat. Combination with a couple of coherence enhancing diffusion steps (cf. subsection "Postprocessing" in section 2.3) can reduce these disturbing effects.

Further numerical experiments have shown that best results are obtained for *locally small* inpainting areas which makes our approach well suited for removing

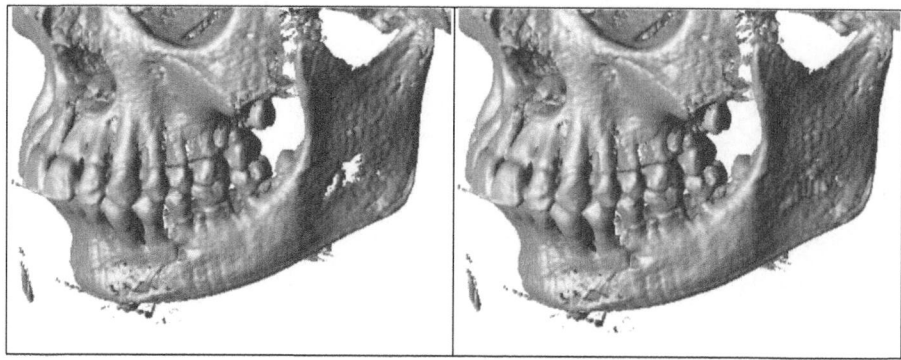

Fig. 5. The hole in the cheekbone has been filled

cracks or superimposed texts. By locally small we mean that the Hausdorff distance between Ω and $\partial\Omega$

$$d(\Omega) = \sup_{x \in \Omega} \inf_{y \in \partial\Omega} (|x - y|)$$

is smaller than the typical size of the image structures in the surrounding of Ω. This is intuitive since the intensity information can only be extended "reliably" from a given pixel into a small neighborhood. The inpaintings produced with the algorithms outlined in this paper differ by the contrast. Some reveal blurry inpainting away from the boundary like the Gaussian heat flow. The approach of Masnou & Morel produces high contrast inpaintings with straight or polygonal level lines, thus revealing artifical (desired or not) kinks. Our approach produces high contrast inpaintings as well. Using the parameter $\varepsilon > 0$ we are able to compromise between blurry and high contrast models, which can be used to weaken the visibility of unwanted kinks.

Our approach, as well as the inpainting algorithms outlined in section 1, is not applicable to inpaint textured regions. In such cases texture synthesis or combined texture–synthesis/inpainting algorithms are more appropriate, see [32] and references therein. See also [33] for a short comparison between local inpainting and texture synthesis approaches.

To juxtapose the results of the Ginzburg–Landau equation we discuss a frequently cited example in the area of inpainting: how should the noisy area in the first image of figure 6 be filled? There are many reasonable inpaintings since it is a synthetic image allowing no intuitive interpretation. Level line based algorithms are usually designed to establish short and smooth connections. Depending on the concrete numerical implementation the level set method automatically selects either the third or the fourth picture — the level set formulation itself does not have the information which inpainting to choose. From a mathematical point of view the Ginzburg–Landau inpainting (second picture) is most appropriate to retain the symmetry of the initial image.

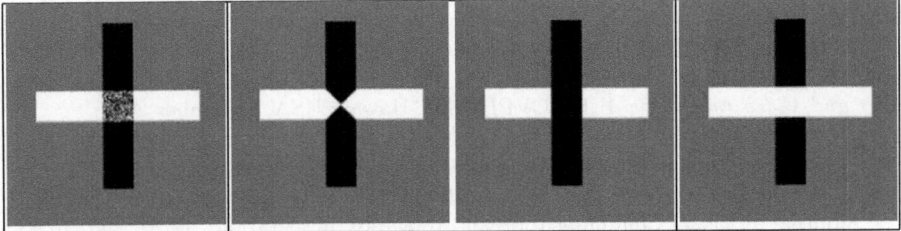

Fig. 6. Ambiguous example: (*first picture*) the noisy area should be inpainted, (*second picture*) inpainting via the Ginzburg–Landau algorithm, (*third and fourth picture*) inpainting via level set algorithm

Acknowledgement

This work is supported by the Austrian Science Fund (FWF), grant Y–123 INF.

References

1. *Lückenfüller und Farbmischer, Bildkorrekturverfahren: beim Menschen gelernt*, c't–Magazin für Computer Technik, 24/2002, Heise Zeitschriften Verlag, Hannover, p.190
2. L. Joyeux, O. Buisson, B. Besserer, S. Boukir, *Detection and removal of line scratches in motion picture films*, Proceedings of CVPR'99, IEEE Int. Conf. on Computer Vision and Pattern Recognition, Fort Collins, Colorado, USA, June 1999
3. M. Ramasubramanian, S. Pattanaik, D. Greenberg, *A Perceptually Based Physical Error Metric for Realistic Image Synthesis*, Proceedings of SIGGRAPH 99. In Computer Graphics Proceedings, Annual Conference Series, 1999, ACM SIGGRAPH, p.73–82
4. V. Caselles, B. Coll, J. Morel, *A Kanizsa programme*, Progress in Nonlinear Differential Equations and their Applications, Vol. 25, p.35–55, 1996
5. A.A. Efros and T.K. Leung, *Texture synthesis by non–parametric sampling*, Proceedings of the Seventh International Conference on Computer Vision, Corfu, Greece, 1999
6. H. Igehy, L. Pereira, *Image Replacement through Texture Synthesis*, Proceedings of the 1997 IEEE International Conference on Image Processing
7. M. Unser, *Sampling–50 years after Shannon*, Proceedings of the IEEE, vol. 88, no. 4, p.569–587, April 2000
8. P. Thenaz, T. Blu, M. Unser, *Handbook of Medical Imaging, Processing and Analysis*, I.N. Bankman, Ed., Academic Press, San Diego CA, USA, p.393–420, 2000
9. S. Masnou, *Disocclusion: A Variational Approach Using Level Lines*, IEEE Transactions on Signal Processing, 11(2), February 2002, p.68–76
10. S. Masnou, J.-M. Morel, *Level Lines based Disocclusion*, Proceedings of the 1998 IEEE International Conference on Image Processing, p.259–263
11. C. Ballester, M. Bertalmio, V. Caselles, G. Sapiro, J. Verdera, *Filling–In by Joint Interpolation of Vector Fields and Gray Levels*, IEEE Transactions on Signal Processing, 10(8), August 2001, p.1200–1211

12. M. Bertalmio, G. Sapiro, C. Ballester, V. Caselles, *Image inpainting*, Computer Graphics, SIGGRAPH 2000, July 2000
13. M. Bertalmio, A. Bertozzi, G. Sapiro, *Navier–Stokes, Fluid Dynamics, and Image and Video Inpainting*, IEEE CVPR 2001, Hawaii, USA, December 2001
14. T. Chan, J. Shen, *Mathematical Models for Local Nontexture Inpaintings*, SIAM Journal of Applied Mathematics, 62(3), 2002, p.1019–1043
15. T. Chan, J. Shen, *Non–Texture Inpainting by Curvature–Driven Diffusions (CDD)*, Journal of Visual Communication and Image Representation , 12(4), 2001, p.436–449
16. T. Chan, S. Kang, J. Shen, *Euler's Elastica and Curvature Based Inpainting*, SIAM Journal of Applied Mathematics, 63(2), pp. 564–592, 2002
17. S. Esedoglu, J. Shen, *Digital Inpainting Based on the Mumford–Shah–Euler Image Model*, European Journal of Applied Mathematics, 13, pp. 353-370, 2002
18. M. Oliveira, B. Bowen, R. McKenna, Y. Chang, *Fast Digital Inpainting*, Proceedings of the International Conference on Visualization, Imaging and Image Processing (VIIP 2001), Marbella, Spain, pp. 261–266
19. L. Landau, V. Ginzburg, *On the Theory of Superconductivity*, Journal of Experimental and Theoretical Physics (USSR), 20 (1950), p.1064
20. M. Ipsen, P. Sorensen, *Finite Wavelength Instabilities in a Slow Mode Coupled Complex Ginzburg–Landau Equation*, Physical Review Letters, Vol. 84/11, p.2389, 2000
21. M. van Hecke, E. de Wit, W. van Saarloos, *Coherent and Incoherent Drifting Pulse Dynamics in a Complex Ginzburg–Landau Equation*, Physical Review Letters, Vol. 75/21, p.3830, 1995
22. T. Bohr, G. Huber, E. Ott, *The structure of spiral–domain patterns and shocks in the 2D complex Ginzburg–Landau equation*, Physica D, Vol. 106, p.95–112, 1997
23. L. Ambrosio, N. Dancer, *Calculus of Variations and Partial Differential Equations*, Springer Verlag
24. F. Bethuel, H. Brezis, F. Hélein, *Ginzburg–Landau Vortices*. In "Progress in Nonlinear Differential Equations and Their Applications", 13 (1994), Birkhäuser
25. W. Press, B. Flannery, S. Teukolsky, W. Vetterling, *Numerical Recipes*, Cambridge University Press
26. J. Thomas, *Numerical Partial Differential Equations: Finite Difference Methods*, Texts in Applied Mathematics, Vol. 22
27. L. Chen, J. Shen, *Applications of semi–implicit Fourier–spectral methods to phase field equations*, Computer Physics Communications, Vol. 108, p.147–158, 1998
28. R. H. Nochetto, M. Paolini, C. Verdi, *A Dynamic Mesh Algorithm for Curvature Dependent Evolving Interfaces*, Journal of Computational Physics, 123, 1996, p.296–310
29. J. Weickert, *Anisotropic Diffusion in Image Processing*, B.G.Teubner, Stuttgart, 1998
30. J. Davis, S. Marschner, M. Garr, M. Levoy, *Filling holes in complex surfaces using volumetric diffusion*, to appear in First International Symposium on 3D Data Processing, Visualization, and Transmission Padua, Italy, June 19–21, 2002
31. J. Taylor, *Demonstration of Canadian 3D Technology for Heritage Recording in China*, Proceedings of the Italy–Canada 2001 Workshop on 3D Digital Imaging and Modeling Applications, Padova, Italy, April 3–4, 2001
32. M. Bertalmio, L. Vese, G. Sapiro, S. Osher, *Simultaneous Structure and Texture Image Inpainting*, submitted
33. http://www.people.fas.harvard.edu/~hchong/Spring2002/cs276r/

Least Squares and Robust Estimation
of Local Image Structure

Rein van den Boomgaard[1] and Joost van de Weijer[1]

Intelligent Sensory Information Systems
Computer Science Department
University of Amsterdam
The Netherlands
{rein,joostw}@science.uva.nl

Abstract. Linear scale space methodology uses Gaussian probes at scale s to observe the differential structure. In observing the differential image structure through the Gaussian derivative probes at scale s we implicitly construct the Taylor series expansion of the smoothed image. The Gaussian facet model, as a generalization of the classic Haralick facet model, constructs a polynomial approximation of the unsmoothed image. The measured differential structure therefore is closer to the 'real' structure then the differential structure measured using Gaussian derivatives.

At the points in an image where the differential structure changes abruptly (because of discontinuities in the imaging conditions, e.g. a material change, or a depth discontinuity) both the Gaussian derivatives and the Gaussian facet model diffuse the information from both sides of the discontinuity (smoothing across the edge).

Robust estimators that are classically meant to deal with statistical outliers can also be used to deal with these 'mixed model distributions'. In this paper we introduce the robust estimators of local image structure. Starting with the Gaussian facet model model where we replace the quadratic error norm with a robust (Gaussian) error norm leads to a robust Gaussian facet model.

We will show examples of using the robust differential structure estimators for luminance and color images, for zero and higher order differential structure. Furthermore we look at a 'robustified' structure tensor that forms the basis of robust orientation estimation.

1 Introduction

Linear scale-space theory of vision not only refers to the introduction of an explicit scale-parameter, it also refers to the use of differential operators to study the local structure of images. The classical way to observe the local differential image structure is to consider all Gaussian derivatives at scale s up to order N. Basically what we do is construct the Taylor series expansion of the *smoothed* image (i.e. the image observed at scale s). The Taylor polynomial thus is an approximation of the smoothed image and not of the original image.

L.D. Griffin and M. Lillholm (Eds.): Scale-Space 2003, LNCS 2695, pp. 237–254, 2003.

Instead of constructing a polynomial local model of the smoothed image we can equally well construct a polynomial approximation of the unsmoothed image. Our starting point is the *image facet* model as introduced by Haralick et. al. [1]. His facet model takes a polynomial function and fits it to the data observed in a small neighborhood in the image using a linear least squares estimation procedure. The image derivatives then can be calculated as the derivatives of the fitted analytical function.

Farnebäck [2] generalizes the Haralick facet model to incorporate spatial weights in order to express the relative importance of the image samples in estimating the parameters of the polynomial function. In the classic Haralick facet model all points in the local neighborhood are considered equally important.

The obvious choice for the spatial weighting kernel is of course the Gaussian kernel. Due to the fact that the derivatives of the Gaussian function are given by a polynomial (determined by the order of differentiation) times the Gaussian function itself, the coefficients in the polynomial function turn out to be a linear combination of the Gaussian derivatives.

The least squares estimation procedure considers all points in a local neighborhood, even in the situation where the local neighborhood is on the boundary of two regions in an image. Each of the regions on either side of the boundary may well be approximated with a low-order polynomial model. The regions can be so different that their union cannot be accurately described using the same low order polynomial model. The estimation procedure then compromises between the two regions: the edge will be smoothed.

In Section 2 we generalize the Gaussian facet model to deal with those multi-model situations. Instead of using a linear least squares estimation procedure we will use a robust estimation technique. A robust estimation technique will only consider the data points from one of the regions and will disregard the data from the other region as being statistical outliers. Robust estimation of local image structure is pioneered by Besl [3]. Our work (see also [4]) differs from the work of Besl in that we consider Gaussian aperture instead of 'crisp' neighborhoods in which the polynomial function is fitted. Furthermore we introduce a fixed point iteration procedure to find the robust estimate.

In Section 3 we derive iterative robust estimators of local image structure. The work presented in this papers is a generalization of the work presented in earlier work [4–6]. We will give some examples ranging from a simple zero order Gaussian facet model to a first order facet model for color images.

In Section 4 we describe a robust estimator for a derived image quantity: the local orientation (see also [6]). To that end we consider the often used orientation estimator based on a eigen analysis of the structure tensor. Robust estimation of the orientation turns out to be quite similar, the structure tensor is replaced with a 'robustified' version in which only the points are considered that closely fit the model (i.e. the points that are not outliers).

2   Least Squares Estimation of Local Image Structure

Locally around a point $\mathbf{x}$ we can approximate the image function f with a linear combination of basis functions ϕ_i, $i = 1, \ldots, K$:

$$\hat{f} = a_1\phi_1 + \cdots + a_K\phi_K$$

We can rewrite this as $\hat{f} = \Phi\mathbf{a}$ where $\Phi = (\phi_1\ \phi_2 \cdots \phi_K)$ and $\mathbf{a} = (a_1\ a_2 \cdots a_K)^\mathsf{T}$. The least squares estimator minimizes the difference ϵ of the image f and the approximation $\hat{f}$:

$$\epsilon(\mathbf{x}) = \int_{\mathbb{R}^d} \left(f(\mathbf{x} + \mathbf{y}) - \hat{f}(\mathbf{y}) \right)^2 W(\mathbf{y}) d\mathbf{y}$$

where W is the aperture function defining the locality of the model fitting. Note that the optimal fitting function $\hat{f}$ differs from position to position in the image plane. We thus have that $\hat{f}(\mathbf{y}) = \Phi(\mathbf{y})\mathbf{a}(\mathbf{x})$, i.e. $\hat{f}(\mathbf{y}) = a_1(\mathbf{x})\phi_1(\mathbf{y}) + \cdots + a_K(\mathbf{x})\phi_K(\mathbf{y})$.

The optimal parameter vector $\mathbf{a}$ is found by projecting the function f onto the subspace spanned by the basis functions in Φ. In this function space the inner product is given by:

$$f^\mathsf{T} g \equiv \langle f, g \rangle_W = \int_{\mathbb{R}^d} f(\mathbf{x})\, g(\mathbf{x})\, W(\mathbf{x})\, d\mathbf{x}$$

The inner product of functions f and g will also be denoted as $f^\mathsf{T} g$.

To derive the optimal parameter vector $\mathbf{a}$ we take the derivative of the error ϵ with respect to the parameter vector $\mathbf{a}$, set it equal to zero and solve for $\mathbf{a}$. We first write ϵ in terms of the inner product

$$\epsilon(\mathbf{x}) = (f_{-\mathbf{x}} - \Phi\mathbf{a})^\mathsf{T}(f_{-\mathbf{x}} - \Phi\mathbf{a})$$

where we have introduced a translated image $f_{-\mathbf{x}}(\mathbf{y}) = f(\mathbf{x} + \mathbf{y})$. The integral is now 'hidden' in the inner product of two functions. This can be rewritten as:

$$\epsilon(\mathbf{x}) = f_{-\mathbf{x}}^\mathsf{T} f_{-\mathbf{x}} - 2\mathbf{a}^\mathsf{T}\Phi^\mathsf{T} f + \mathbf{a}^\mathsf{T}\Phi^\mathsf{T}\Phi\mathbf{a}$$

Taking the derivative of ϵ with respect to $\mathbf{a}$ and setting this equal to 0 and solving for $\mathbf{a}$ we obtain:

$$\mathbf{a} = (\Phi^\mathsf{T}\Phi)^{-1}\Phi^\mathsf{T} f_{-\mathbf{x}} = \tilde{\Phi}^\mathsf{T} f_{-\mathbf{x}}$$

where $\tilde{\Phi} = \Phi(\Phi^\mathsf{T}\Phi)^{-1}$ is the *dual basis*. The functions in the dual basis, $\tilde{\Phi} = (\tilde{\phi}_1 \cdots \tilde{\phi}_K)$, are the functions such that the inner product $\tilde{\phi}_i^\mathsf{T} f_{-\mathbf{x}}$ equals the coefficient a_i in the approximation $\hat{f} = a_1\phi_1 + \cdots + a_K\phi_K$. The dual basis functions, multiplied with the aperture function, thus are the correlation kernels needed to calculate the coefficients in the polynomial image approximation.

The classic Haralick facet model uses a uniform weight function $W(\mathbf{x}) = 1$ for $\|\mathbf{x}\|_\infty \leq s$ and $W(\mathbf{x}) = 0$ elsewhere, i.e. a 'crisp' neighborhood within an axis aligned square of size $2s \times 2s$.

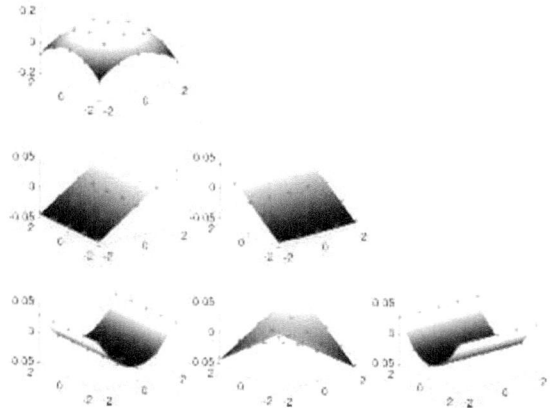

Fig. 1. The Haralick Facet Model. From left to right, top to bottom the dual basis functions are plotted. The shaded functions are the dual basis functions within a 2nd order facet model, the (red) stars correspond with the discrete dual functions. The neighborhood was taken to be of size 5×5. The scale s for the analytical kernel was set at $s = 2.42$. This value is the value to make the difference between the discrete and analytical facet models minimal. For larger neighborhoods $N \times N$ the correspondence becomes better and the analytical scale approaches $(N-1)/2 + \frac{1}{2}$.

For the second order polynomial basis:

$$\Phi = \left(1 \ x \ y \ \tfrac{1}{2}x^2 \ xy \ \tfrac{1}{2}y^2 \right)$$

the dual basis is

$$\tilde{\Phi} = \left(\frac{7}{8\,s^2} - \frac{15\,x^2}{16\,s^4} - \frac{15\,y^2}{16\,s^4} \ \frac{3\,x}{4\,s^4} \ \frac{3\,y}{4\,s^4} \ \frac{-15}{8\,s^4} + \frac{45\,x^2}{8\,s^6} \ \frac{9\,x\,y}{4\,s^6} \ \frac{-15}{8\,s^4} + \frac{45\,y^2}{8\,s^6} \right)$$

The dual basis functions are depicted in Fig. 1. The first dual basis function (multiplied with the aperture function) is the correlation kernel needed to calculate the coefficient of the constant basis function in the approximation of the local image patch. Observe that in the Haralick facet model, the first dual basis function is not everywhere positive. Fig. 1 also shows the discrete dual basis functions, these follow from a formulation of the facet model in a discrete image space as can be found in the work of Haralick.

Within a scale-space context the most natural choice is to start with a polynomial basis and a Gaussian aperture function $W = G^s$ where G^s is the Gaussian function at scale s. Again starting with the second order polynomial basis the dual basis is a different one due to the difference in the inner product (as a consequence of a different aperture function):

$$\tilde{\Phi} = \left(2 - \frac{x^2}{2\,s^2} - \frac{y^2}{2\,s^2} \ \frac{x}{s^2} \ \frac{y}{s^2} \ -s^{-2} + \frac{x^2}{s^4} \ \frac{x\,y}{s^4} \ -s^{-2} + \frac{y^2}{s^4} \right)$$

Again, a dual basis function, multiplied with the—Gaussian—aperture function is the correlation kernel needed to calculate the corresponding coefficient in the polynomial approximation of the local image patch. For the zero order coefficient the correlation kernel is a Gaussian function multiplied with a parabola: $(2 - \frac{x^2}{2\,s^2} - \frac{y^2}{2\,s^2})\,G^s(x,y)$. Again we see that the zero order coefficient in the polynomial image approximation requires a kernel with negative values.

The derivatives of the Gaussian function are equal to a polynomial function (a Hermite polynomial depending on the derivative taken) times the Gaussian function, we may write the correlation kernels associated with the dual basis functions in the Gaussian facet model as a linear combination of Gaussian derivatives. It is not hard to prove that the zero order coefficient in the second order Gaussian facet model is found by convolving the image f with the kernel:

$$G^s - \tfrac{1}{2}s^2\left(G^s_{xx} + G^s_{yy}\right)$$

Now we easily recognize where the negative values in the kernel come from. The term G^s is the Gaussian scale-space smoothing term. The term $-\tfrac{1}{2}s^2\left(G^s_{xx} + G^s_{yy}\right)$ is a well-known sharpening term: subtracting the Laplacian from the smoothed image, sharpens the image. The sharpening term is due to the fact that the Gaussian facet model approximates the original image, not the smoothed image.

Fig. 2. Zero-order coefficient in the Gaussian Facet Model. On the first row, from left to right: the original image, and the zero order coefficients in the Gaussian facet model of order 0,2,4 and 6. On the second row the convolution kernel is shown that, convoluted with the original image, results in the image above it.

It turns out that this observation is true for higher order facet models as well. For a 4th order Gaussian facet model, the kernel to calculate the zero order coefficient is:

$$G^s - \tfrac{1}{2}s^s\left(G^s_{xx} + G^s_{yy}\right) + \tfrac{1}{8}s^4\left(G^s_{xxxx} + 2\,G^s_{xxyy} + G^s_{yyyy}\right)$$

In Fig. 2 the kernels to calculate the zero order coefficient in the Gaussian facet model of orders 0, 2, 4 and 6 are depicted together with the convoluted images. Apparently the N-jet of an image observed at scale s encodes details of size less then s, i.e. from the N-jet observed at scale s a lot of detail can be reconstructed.

3   Robust Estimation of Local Image Structure

Consider again the error of the Gaussian weighted least squares approximation:

$$\epsilon(\mathbf{x}) = \int_{\mathbb{R}^d} \left(f(\mathbf{x} + \mathbf{y}) - \hat{f}(\mathbf{y}) \right)^2 G^s(\mathbf{y}) d\mathbf{y}$$

It is well known that this error definition is not well suited for those situations were we have outliers in our measurements. In the image processing context statistical outliers are not so frequently occurring. The effect that makes least squares estimates questionable is that when collecting measurements from a neighborhood in an image these are often not well modeled using a simple (facet) model. For instance we may model local image luminance quite well with a second order polynomial model *but not near edges* where we switch from one model instantiation to another. Such multi-model situations are abundant in computer vision applications and are most often due to the nature of the imaging process where we see abrupt changes going from one object to another object.

Multi-modality can be incorporated into sophisticated estimation procedures where we not only estimate (multi-)model parameters but also the geometry that separates the different regions (one for each model). One of the oldest examples is perhaps Hueckels edge detector [7] in which a local image patch is described with two regions separated by a straight boundary. The detector estimates this boundary and the parameters of the luminance distributions on each side of the edge.

In this paper we take a less principled approach. Instead of a multi-model approach we stick to a simpler one-model approach where we use a statistical *robust estimator* that allows us to consider part of the measurements from the local neighborhood to belong to the model we are interested in and disregard all other measurements as being 'outliers' and therefore not relevant in estimating the model parameters.

The crux of a robust estimation procedure is to rewrite the above error measure as:

$$\epsilon(\mathbf{x}) = \int_{\mathbb{R}^d} \rho(f(\mathbf{x} + \mathbf{y}) - \hat{f}(\mathbf{y})) \, G^s(\mathbf{y}) d\mathbf{y}$$

where ρ is the error norm. The choice $\rho(e) = e^2$ leads to the least squares estimator. Evidently measurements that are outliers to the 'true' model are weighted heavily in the total error measure. Reducing the influence of the large errors leads to *robust error norms*.

Writing $f_{-\mathbf{x}}(\mathbf{y}) = f(\mathbf{x} + \mathbf{y})$ and using the local linear model $\hat{f}(\mathbf{y}) = \Phi(\mathbf{y})\mathbf{a}(\mathbf{x})$ we obtain:

$$\epsilon(\mathbf{x}) = \int_{\mathbb{R}^d} \rho(f_{-\mathbf{x}} - \Phi\mathbf{a}(\mathbf{x})) \, G^s \, d\mathbf{y}$$

We omitted the spatial argument $\mathbf{y}$ for ease of notation. In this paper the 'Gaussian error norm' is chosen:

$$\rho(e) = 1 - \exp\left(-\frac{e^2}{2m^2}\right)$$

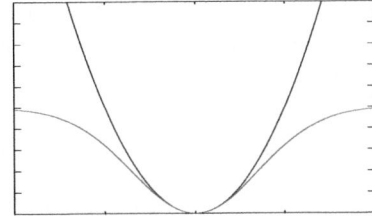

Fig. 3. Quadratic versus (robust) Gaussian error norm. The Gaussian error norm is of 'scale' $m = 0.7$.

The scale m in the error norm will be called the *model scale* to contrast it with the spatial scale s that is used in the spatial aperture function G^s. In Fig. 3 the error norm is sketched. Compared to the quadratic error norm this norm is 'clamped' at value 1. For $e \gg m$ the exact value of the error is not important any more. Gross outliers are therefore not given the weight to influence the estimation greatly.

The optimal model parameters are found by calculating the derivative of the error measure and setting this equal to zero:

$$\frac{\partial \epsilon}{\partial \mathbf{a}} = \frac{\partial}{\partial \mathbf{a}} \int_{\mathbb{R}^d} \rho(f_{-\mathbf{x}} - \Phi\mathbf{a}(\mathbf{x}))\, G^s \, dy$$

$$= \frac{\partial}{\partial \mathbf{a}} \int_{\mathbb{R}^d} \left(1 - \exp\left(\frac{(f_{-\mathbf{x}} - \Phi\mathbf{a}(\mathbf{x}))^2}{2\, m^2}\right)\right) G^S \, dy$$

$$= -2 \int_{\mathbb{R}^d} (f_{-\mathbf{x}} - \Phi\mathbf{a}(\mathbf{x}))\, \Phi \exp\left(\frac{(f_{-\mathbf{x}} - \Phi\mathbf{a}(\mathbf{x}))^2}{2\, m^2}\right) G^s \, dy$$

Setting this derivative equal to zero and rewriting terms we obtain:

$$\int_{\mathbb{R}^d} f_{-\mathbf{x}}\, \Phi \exp\left(\frac{(f_{-\mathbf{x}} - \Phi\mathbf{a}(\mathbf{x}))^2}{2\, m^2}\right) G^s \, dy =$$
$$\int_{\mathbb{R}^d} \Phi\mathbf{a}(\mathbf{x})\, \Phi \exp\left(\frac{(f_{-\mathbf{x}} - \Phi\mathbf{a}(\mathbf{x}))^2}{2\, m^2}\right) G^s \, dy \quad (1)$$

This can be rewritten as:

$$\int_{\mathbb{R}^d} f_{-\mathbf{x}}\, \Phi G^m \left(f_{-\mathbf{x}} - \Phi\mathbf{a}(\mathbf{x})\right) G^s \, dy = \int_{\mathbb{R}^d} \Phi\mathbf{a}(\mathbf{x})\, \Phi G^m \left(f_{-\mathbf{x}} - \Phi\mathbf{a}(\mathbf{x})\right) G^s \, dy \quad (2)$$

where G^m is the Gaussian function at scale m. This Gaussian function weighs the model distance, whereas the Gaussian function G^s weighs the spatial distance.

We define the operator Γ:

$$(\Gamma^m g)(\mathbf{y}) = G^m (f_{-\mathbf{x}}(\mathbf{y}) - \Phi(\mathbf{y})\mathbf{a}(\mathbf{x})))\, g(\mathbf{y})$$

i.e. the point wise multiplication of the function g with the model weight function. Now Γ^m acts as a diagonal (matrix) operator in the function space. Using the vectorial notation of the inner product we can write:

$$\Phi^{\mathsf{T}} \Gamma^m f_{-\mathbf{x}} = \Phi^{\mathsf{T}} \Gamma^m \Phi \mathbf{a}$$

This looks like a familiar weighted linear least squares equation that can be solved for the value of **a**. It is not, because Γ^m is dependent on **a**. Solving for **a** can be done using an *iterated weighted least squares* procedure:

$$\mathbf{a}^{i+1} = \left(\Phi^\mathsf{T}\Gamma(\mathbf{a}^i)\Phi\right)^{-1}\Phi^\mathsf{T}\Gamma(\mathbf{a}^i)f_{-\mathbf{x}} \tag{3}$$

Some examples of these robust estimators may clarify matters. In the next subsection we consider the most simple of all local structure models: a locally constant model. The resulting image operator turns out to be an iterated version of the bilateral filter introduced by Tomasi and Manduchi [8].

3.1   Zero-Order Image Structure

Consider a locally constant image model with only one basis function:

$$\Phi = (1)$$

i.e. the constant function. Eq.(3) then reduces to:

$$a_0^{i+1}(\mathbf{x}) = \frac{\int_{\mathbb{R}^d} f(\mathbf{x}+\mathbf{y})\, G^m(f(\mathbf{x}+\mathbf{y}) - a_0^i(\mathbf{x}))\, G^s(\mathbf{y})\, d\mathbf{y}}{\int_{\mathbb{R}^d} G^m(f(\mathbf{x}+\mathbf{y}) - a_0^i(\mathbf{x}))\, G^s(\mathbf{y})\, d\mathbf{y}}$$

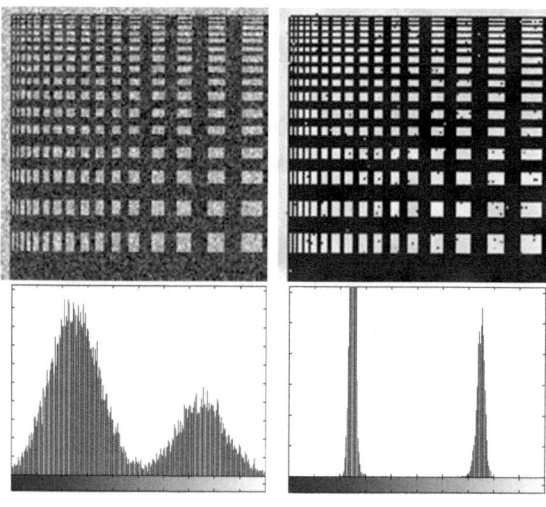

Fig. 4. Robust Estimation of Local Image Structure. On the first row a test image with noise added on the left and the result of the robust estimator based on a zero-order facet model. On the second row the histograms of the images above are depicted. Observe that the robust estimator is capable of finding the modes of both the distributions.

This is an iterated version of the bilateral filter as introduced by Tomasi and Manduchi [8]. It is also related to the filters introduced by Smith et. al. [9]. The bilateral filter thus implements one iteration of a robust estimator with initial value $a_0^0 = f$.

In previous papers [4, 5] we have analyzed robust estimation of the zero order local image structure. Some observations made are:

- The robust estimator finds the local mode in the local luminance histogram which is smoothed with a Gaussian kernel of scale m. The local mode that is found is the local maximum in the smoothed histogram that is closest to the initial value.
- Bilateral filtering implements one iteration of the robust estimator. From mean shift analysis we know that the first step in a mean shift algorithm is a large one in the direction of the optimal value. This explains the impressive results on the bilateral filter in reducing the noise while preserving the structure of images.
- The choice of an initial estimate is very important. We have found good results using the result of a linear least squares estimate as the initial estimate. In certain situations however the amount of smoothing induced by the least squares estimator sets the robust estimator at a wrong starting point leading to a local maximum in the histogram that does not correspond with the structure that we are interested in. This situation is often occurring in case the area of the structure of interest is less then the area of the 'background' (e.g. document images where there is more paper then ink visible). In such cases the image itself can be used as an initial estimate of the zero order local structure.

Fig. 5. Robust Estimation and Non-linear diffusion. On the left the original image of a flower. In the middle the robust estimation of the zero order local structure and on the right the result of iteratively applying one iteration of the robust estimator, each time using the image data from the previous iteration (this procedure is very much like a non-linear diffusion process).

- The results of robust estimation of local image structure bear great resemblance to the results of non-linear diffusion. The theoretical link between robust estimation and non-linear diffusion techniques has been reported before (see [10]). The main difference with the robust estimator technique described here is that in each iteration of a non-linear diffusion algorithm the image data resulting from the previous iteration is used. In the robust estimator described here we stick to the original image data and only update the parameter to be estimated. Fig. 5 shows the differences between these two procedures.

3.2 Higher-Order Image Structure

For the image in Fig. 4 the assumption of local constant image model is a correct assumption, for most natural images such a model is an oversimplification though. Then it is better to use a higher order model for the local image structure. We start with a simple first order model for 1D functions. The local basis is:

$$\Phi = \begin{pmatrix} 1 & x \end{pmatrix}$$

This leads to the matrix $\Phi^\mathsf{T} \Gamma^m \Phi$:

$$\begin{pmatrix} \int_{\mathbb{R}} G^m(f(x+y) - a_0^i - a_1^i y)G^s(y)dy & \int_{\mathbb{R}} y\, G^m(f(x+y) - a_0^i - a_1^i y)\, G^s(y)\, dy \\ \int_{\mathbb{R}} y\, G^m(f(x+y) - a_0^i - a_1^i y)\, G^s(y)\, dy & \int_{\mathbb{R}} y^2\, G^m(f(x+y) - a_0^i - a_1^i y)\, G^s(y)\, dy \end{pmatrix}$$

and vector $\Phi^\mathsf{T} \Gamma^m f_{-x}$:

$$\begin{pmatrix} \int_{\mathbb{R}} f(x+y)\, G^m(f(x+y) - a_0^i - a_1^i y)\, G^s(y)\, dy \\ \int_{\mathbb{R}} y\, f(x+y)\, G^m(f(x+y) - a_0^i - a_1^i y)\, G^s(y)\, dy \end{pmatrix}$$

The robust estimator of the local linear model is given by Eq.(3). Fig. 6 shows a univariate 'saw-tooth' signal corrupted with additive noise. Also shown are the robust estimates based on a zero order facet model and the robust estimate based on a first order facet model. It is obvious that a robust estimator based on a local constant model is not capable of reconstructing the saw tooth signal from the noisy observations. Using a local first order model leads to a far better reconstruction.

The first order robust facet model is easily generalized to 2D functions:

$$\Phi = \begin{pmatrix} \phi_{(00)} & \phi_{(10)} & \phi_{(01)} \end{pmatrix}$$
$$= \begin{pmatrix} 1 & x_1 & x_2 \end{pmatrix}$$

This leads to the matrix $\Phi^\mathsf{T} \Gamma^m \Phi$:

$$\begin{pmatrix} \int_{\mathbb{R}^2} G^m G^s dy & \int_{\mathbb{R}^2} y_1 G^m G^s dy & \int_{\mathbb{R}^2} y_2^2 G^m G^s dy \\ \int_{\mathbb{R}^2} y_1 G^m G^s dy & \int_{\mathbb{R}^2} y_1^2 G^m G^s dy & \int_{\mathbb{R}^2} y_1 y_2 G^m G^s dy \\ \int_{\mathbb{R}^2} y_2 G^m G^s dy & \int_{\mathbb{R}^2} y_1 y_2 G^m G^s dy & \int_{\mathbb{R}^2} y_2^2 G^m G^s dy \end{pmatrix}$$

to simplify the notation we have omitted the arguments of the functions in the integrand. For the G^m-function the argument is the model error $f(\mathbf{x}+\mathbf{y}) - a_{00} -$

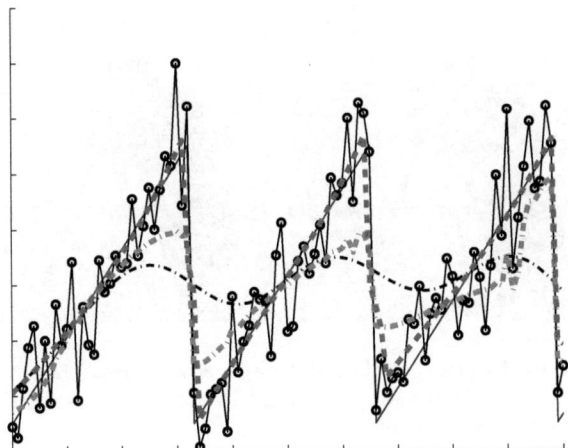

Fig. 6. Robust Estimation of Local Structure in 1D functions. A 'sawtooth' function with added noise is shown together with the Gaussian linear least squares estimate, i.e. the Gaussian smoothing (the thin 'sinusoidal' line), the robust estimate based on a zero order facet model (the dashed-dotted line) and the robust estimate based on a first order model (the thick dashed line). The spatial scale is 9 and the tonal (model) scale is 0.1. The number of iterations used is 10.

$a_{10}y_1 - a_{01}y_2$. The vector $\Phi^T \Gamma^m f_{-\mathbf{x}}$ equals

$$\begin{pmatrix} \int_{\mathbb{R}^2} f(\mathbf{x}+\mathbf{y})G^m(f(\mathbf{x}+\mathbf{y}) - a_{00} - a_{10}y_1 - a_{01}y_2)G^s(\mathbf{y})d\mathbf{y} \\ \int_{\mathbb{R}^2} y_1 f(\mathbf{x}+\mathbf{y})G^m(f(\mathbf{x}+\mathbf{y}) - a_{00} - a_{10}y_1 - a_{01}y_2)G^s(\mathbf{y})d\mathbf{y} \\ \int_{\mathbb{R}^2} y_2 f(\mathbf{x}+\mathbf{y})G^m(f(\mathbf{x}+\mathbf{y}) - a_{00} - a_{10}y_1 - a_{01}y_2)G^s(\mathbf{y})d\mathbf{y} \end{pmatrix}$$

Eq.(3) then can be used to calculate the new estimate of the optimal parameter vector $\mathbf{a}^{i+1}$.

In Fig. 7 the robust estimation of the zero order coefficient based on a first order facet model is shown. For this image the difference with a zero order facet model estimation can only be observed in regions of slowly varying luminance (like in the background).

3.3   Color Image Structure

In this section we generalize the robust facet models for scalar images to models for vectorial images. The analysis is done for color images but is valid for all vectorial images.

A color image $\mathbf{f} = (f^1 \; f^2 \; f^3)$ at any position $\mathbf{x}$ has three color components $f^1(\mathbf{x})$, $f^2(\mathbf{x})$ and $f^3(\mathbf{x})$. The local model for a color image using a basis

$$\Phi = (\phi_1 \; \phi_2 \; \cdots \; \phi_K)$$

Fig. 7. Robust Estimation of Local Image Structure. On the left the cameraman image with noise added and on the right the robust estimation of the zero order
coefficient in a first order facet model.

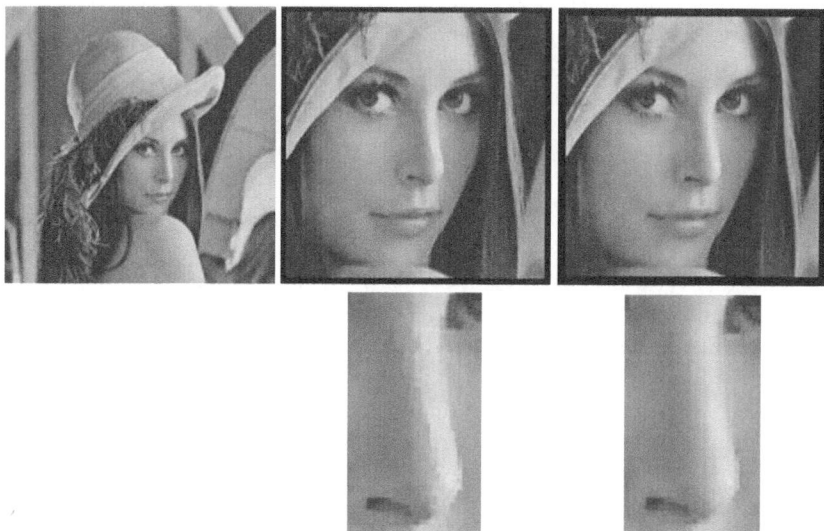

Fig. 8. Robust Estimation of Local Structure in Color Images. On the first
row from left to right: the 'Lena' image with some noise added to it, the zero-order
facet model based robust estimator of the color values and the robust estimator based
on a first order based facet model. On the second row we show a detail from the image
above.

is chosen as:

$$\hat{\mathbf{f}}(\mathbf{x} + \mathbf{y}) = \varPhi \mathbf{A} = \varPhi \left(\mathbf{a}_1 \ \mathbf{a}_2 \ \mathbf{a}_3 \right)$$

where $\mathbf{A} = \left(\mathbf{a}_1 \ \mathbf{a}_2 \ \mathbf{a}_3 \right)$ is the $K \times 3$ parameter matrix. Each of the columns $\mathbf{a}_i$ is the parameter vector in the approximation $\hat{f}_i = \varPhi \mathbf{a}_i$ of the i-th color component. Each of the color components is thus approximated as a linear combination of K basis functions. The model error is now written as:

$$\epsilon(\mathbf{x}) = \int_{\mathbb{R}^d} \rho \left((f_{-\mathbf{x}}^1 - \varPhi \mathbf{a}_1)^2 + (f_{-\mathbf{x}}^2 - \varPhi \mathbf{a}_2)^2 + (f_{-\mathbf{x}}^3 - \varPhi \mathbf{a}_3)^2 \right) G^s(\mathbf{y}) d\mathbf{y}$$

It is not hard to prove that in this case

$$\frac{\partial \epsilon}{\partial \mathbf{A}} = 0 \Longleftrightarrow \varPhi^\mathsf{T} \varGamma^m \mathbf{f} = \varPhi^\mathsf{T} \varGamma^m \varPhi \mathbf{A}$$

where $\varGamma^m$ is the 'diagonal' operator that multiplies a function point wise with the function: $G^m \left((f_{-\mathbf{x}}^1 - \varPhi \mathbf{a}_1)^2 + (f_{-\mathbf{x}}^2 - \varPhi \mathbf{a}_2)^2 + (f_{-\mathbf{x}}^3 - \varPhi \mathbf{a}_3)^2 \right)$. As $\varGamma^m$ is dependent on the parameter matrix $\mathbf{A}$ we arrive at a iterated weighted least squares estimator:

$$\mathbf{A}^{i+1} = (\varPhi^\mathsf{T} \varGamma^m(\mathbf{A}^i) \varPhi)^{-1} \varPhi^\mathsf{T} \varGamma^m(\mathbf{A}^i) \mathbf{f}$$

The estimation of the robust facet model for color images is thus almost the same as for scalar images. The three color components are dealt with independently, only the error weights operator $\varGamma^m$ is dependent on all three color components.

In Fig. 8 the robust estimators are shown that are based on a zero order facet model and on a first order facet model. Especially in the nose-region the first order model based robust estimator performs better then the zero order model based robust estimator.

4 Robust Estimation of Orientation

In the previous sections we have considered local image models for the image values (grey value and color). In this section we look at robust estimation of the orientation of image structures.

Oriented patterns are found in many imaging applications, e.g. in fingerprint analysis, and in geo-physical analysis of soil layers. The classical technique to estimate the orientation of the texture is to look at the set of luminance gradient vectors in a local neighborhood. In an image patch showing a stripe pattern in only one orientation we can clearly distinguish the orientation as the line cluster in gradient space perpendicular to the stripes (see Fig. 9(a-b)). A straightforward eigenvector analysis of the covariance matrix will reveal the orientation of the texture. The covariance matrix of the gradient vectors in an image neighborhood is often used to estimate the local orientation [11–14]

In case the local neighborhood is taken from the border of two differently oriented patterns (see Fig. 9) an eigenvector analysis of the covariance matrix will mix both orientations resulting in a 'smoothing' of the orientation estimation.

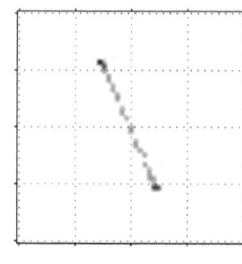
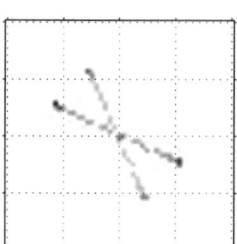

Fig. 9.   Histograms of gradient vector space. In (a) an image (64 × 64) is shown with in (b) the histogram of all gradient vectors (where darker shades indicate that those gradient vectors occur often in the image. In (c) a composition of two differently oriented patterns is shown with corresponding histogram in (d).

In case the regions showing different textures are of sufficient size it is possible to use a post-processing step to sharpen the smoothed orientation measurements. A classical way of doing so is the Kuwahara-Nagao operator [15–17]. At a certain position in an image this operator searches for a nearby neighborhood where the (orientation) response is more homogeneous then it is at the border. That response is then used at the point of interest. In this way the neighborhoods are not allowed to cross the borders of the textured regions. In [18] we have shown that the classic Kuwahara-Nagao operator can be interpreted as a 'macroscopic' version of a PDE image evolution that combines linear diffusion (smoothing) with morphological sharpening.

Again consider the texture in Fig. 9(a). The histogram of the gradient vectors in this texture patch is shown in Fig. 9(b). Let $\mathbf{v}$ be the true orientation vector of the patch, i.e. the vector perpendicular to the stripes. In an ideal image patch every gradient vector should be parallel to the orientation $\mathbf{v}$. In practice they will not be parallel. The error of a gradient vector $\mathbf{g}(\mathbf{y})$ observed in a point $\mathbf{y}$ with respect to the orientation $\mathbf{v}(\mathbf{x})$ of an image patch centered at location $\mathbf{x}$ is defined as:

$$e(\mathbf{x}, \mathbf{y}) = \|\mathbf{g}(\mathbf{y}) - (\mathbf{g}(\mathbf{y})^{\mathsf{T}}\mathbf{v}(\mathbf{x}))\mathbf{v}(\mathbf{x})\|$$

The difference $\mathbf{g}(\mathbf{y}) - (\mathbf{g}(\mathbf{y})^{\mathsf{T}}\mathbf{v}(\mathbf{x}))\mathbf{v}(\mathbf{x})$ is the projection of $\mathbf{g}$ on the normal to $\mathbf{v}$. The error $e(\mathbf{x}, \mathbf{y})$ thus measures the perpendicular distance from the gradient vector $\mathbf{g}(\mathbf{y})$ to the orientation vector $\mathbf{v}(\mathbf{x})$. Integrating the squared error over all positions $\mathbf{y}$ using a soft Gaussian aperture for the neighborhood definition we define the total error:

$$\epsilon(\mathbf{x}) = \int_{\Omega} e^2(\mathbf{x}, \mathbf{y})G^s(\mathbf{x} - \mathbf{y})dy \tag{4}$$

The error measure can be rewritten as:

$$\epsilon = \int_{\Omega} \mathbf{g}^T \mathbf{g} G^s d\mathbf{y} - \int_{\Omega} \mathbf{v}^T (\mathbf{g}\mathbf{g}^T) \mathbf{v} G^s d\mathbf{y}.$$

where we have omitted the arguments of the functions. Minimizing the error thus is equivalent with maximizing:

$$\int_{\Omega} \mathbf{v}^T (\mathbf{g}\mathbf{g}^T) \mathbf{v} G^s d\mathbf{y},$$

subject to the constraint that $\mathbf{v}^T\mathbf{v} = 1$. Note that $\mathbf{v}$ is not dependent on $\mathbf{y}$ so that we have to maximize:

$$\mathbf{v}^T \left(\int_{\Omega} (\mathbf{g}\mathbf{g}^T) G^s d\mathbf{y} \right) \mathbf{v} = \mathbf{v}^T \mu^s \mathbf{v}$$

where μ^s is the *structure tensor*.

Using the method of Lagrange multipliers to maximize $\mathbf{v}^T \mu^s \mathbf{v}$ subject to the constraint that $\mathbf{v}^T\mathbf{v} = 1$, we need to find an extremum of

$$\lambda(1 - \mathbf{v}^T\mathbf{v}) + \mathbf{v}^T \mu^s \mathbf{v}.$$

Differentiating with respect to $\mathbf{v}$ (remember that $d\mathbf{v}^T A\mathbf{v}/d\mathbf{v} = 2A\mathbf{v}$ in case $A = A^T$) and setting the derivative equal to zero results in:

$$\mu^s \mathbf{v} = \lambda \mathbf{v}. \tag{5}$$

The 'best' orientation thus is an eigenvector of the structure tensor. Substitution in the quadratic form then shows that we need the eigenvector corresponding to the largest eigenvalue.

The least squares orientation estimation works well in case all gradients in the ensemble of vectors in an image neighborhood all belong to the same oriented pattern. In case the image patch shows two oriented patterns the least squares estimate will mix the two orientations and give a wrong result.

A robust estimator is constructed by introducing the Gaussian error norm once again:

$$\epsilon(\mathbf{x}) = \int_{\Omega} \rho(e(\mathbf{x}, \mathbf{y})) G^s(\mathbf{x} - \mathbf{y}) d\mathbf{y} \tag{6}$$

In a robust estimator large deviations from the model are not taken into account very heavily. In our application large deviations from the model are probably due to the mixing of two different linear textures (see Fig. 9(c-d)).

The error, Eq.(6), can now be rewritten as (we will omit the spatial arguments):

$$\epsilon = \int_{\Omega} \rho \left(\sqrt{\mathbf{g}^T \mathbf{g} - \mathbf{v}^T (\mathbf{g}\mathbf{g}^T) \mathbf{v}} \right) G^s d\mathbf{y}.$$

Again we use a Lagrange multiplier to minimize subject to the constraint that $\mathbf{v}^T\mathbf{v} = 1$:

$$\frac{d}{d\mathbf{v}} \left(\lambda(1 - \mathbf{v}^T\mathbf{v}) + \int_{\Omega} \rho \left(\sqrt{\mathbf{g}^T \mathbf{g} - \mathbf{v}^T (\mathbf{g}\mathbf{g}^T) \mathbf{v}} \right) G^s d\mathbf{y} \right) = 0.$$

This leads to:

$$\eta(\mathbf{v})\mathbf{v} = \lambda\mathbf{v} \qquad (7)$$

where

$$\eta(\mathbf{v}) = \int_\Omega \mathbf{g}\mathbf{g}^\mathsf{T} G^m (\mathbf{g}^\mathsf{T}\mathbf{g} - \mathbf{v}^\mathsf{T}(\mathbf{g}\mathbf{g}^\mathsf{T})\mathbf{v}) G^s \, d\mathbf{y}.$$

The big difference with the least squares estimator is that now the matrix η is dependent on $\mathbf{v}$ (and on $\mathbf{x}$ as well). Note that η can be called a 'robustified' structure tensor in which the contribution of each gradient vector is weighted not only by its distance to the center point of the neighborhood, but also weighted according to its 'distance' to the orientation model. Weickert et. al. [19] also introduce a non linear version of the structure tensor that is close in spirit to the robust structure tensor η.

We propose the following *fixed point* iteration scheme to find a solution. Let $\mathbf{v}^i$ be the orientation vector estimate after i iterations. The estimate is then updated as the eigenvector $\mathbf{v}^{i+1}$ of the matrix $\eta(\mathbf{v}^i)$ corresponding to the largest eigenvalue, i.e. we solve:

$$\eta(\mathbf{v}^i)\mathbf{v}^{i+1} = \lambda\mathbf{v}^{i+1}$$

The proposed scheme is a generalization of the well-known fixed point scheme (also called *functional iteration*) to find a solution of the equation $v = F(v)$.

Note that the iterative scheme does not necessarily lead to the *global* minimum of the error. In fact often we are not even interested in that global minimum. Consider for instance the situation of a point in region A (with orientation α_1) that is surrounded by many points in region B (with orientation β). It is not to difficult to imagine a situation where the points of region B outnumber those in region A. Nevertheless we would like our algorithm to find the orientation α whereas the global minimum would correspond with orientation β. Because our algorithm starts in the initial orientation estimate and then finds the local minimum nearest to the starting point we hopefully end up in the desired *local* minimum: orientation α.

The choice for an initial estimate of the orientation vector is thus crucial in a robust estimator in case we have an image patch showing two (or more) striped patterns.

In Fig. 10 and Fig. 11 the robust estimation of orientation for a simple test image (without noise and with noise). For the robust estimation in both cases we have used the orientation in location $\mathbf{x}$ that resulted from the least squares estimator as the initial orientation vector in that point. In both cases only 5 iterations are used.

From both the noise free and the noise corrupted texture images it is evident that the robust estimation performs much better at the border of the textured regions.

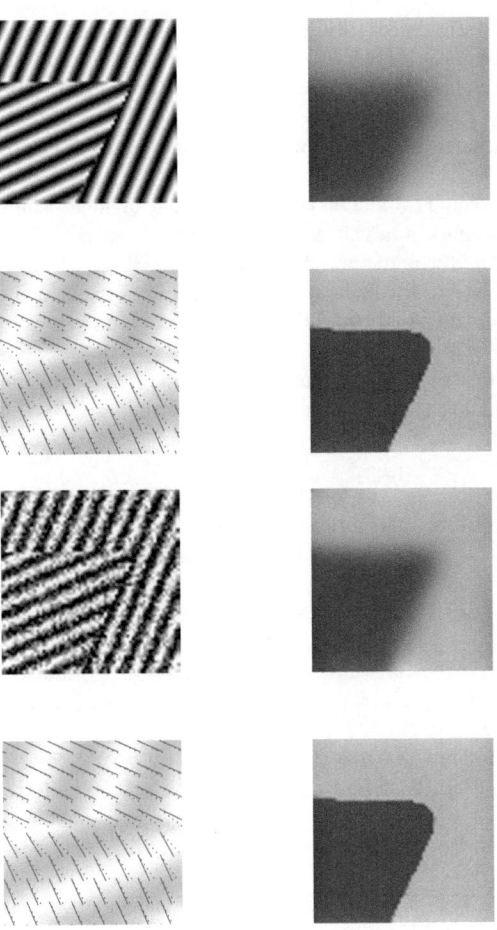

Fig. 10. Least Squares versus Robust Orientation Estimation. In (a) a generated noise free image is shown. The texture is made out of two regions each differently oriented. In (b) the orientation field $\alpha = \arctan(v_2/v_1)$ is shown that results from the least squares estimate. In (d) the orientation field is shown resulting from the robust estimation. In (c) a *detail* of the orientation vector fields for both the least squares estimation (dotted lines) and the robust estimation (solid lines) are shown.

Fig. 11. Least Squares versus Robust Orientation Estimation. In (a) a generated image with some noise added is shown. The texture is made out of two regions each differently oriented. In (b) the orientation field $\alpha = \arctan(v_2/v_1)$ is shown that results from the least squares estimate. In (d) the orientation field is shown resulting from the robust estimation. In (c) a *detail* of the orientation vector fields for both the least squares estimation (dotted lines) and the robust estimation (solid lines) are shown.

5   Conclusions

In this paper we have described robust estimators of local image structure. Our starting point was the Gaussian facet model, using a Gaussian soft aperture function to define the local neighborhood.

Using the Gaussian error norm resulted in an iterative procedure for the robust estimator that is essentially a fixed point iteration. The advantage is that very few iterations are needed in most cases (the maximum number of iterations used in the examples in this paper ranges from 5 to 10).

Through the Gaussian facet model we are able to link the robust estimation of local image structure with classical linear scale-space techniques.

References

1. Haralick, R., andT.J. Laffey, L.W.: The topographic primal sketch. IJPR (1983)

2. Farnebäck, G.: Spatial Domain Methods for orientation and velocity measurements. PhD thesis, Linköping University (1999)
3. Besl, P., Birch, J., Watson, L.: Robust window operators. Machine Vision and Applications **2** (1989) 179–191
4. van de Weijer, J., van den Boomgaard, R.: Local mode filtering. In: CVPR01. (2001) II:428–43
5. van den Boomgaard, R., van de Weijer, J.: On the equivalence of local-mode finding, robust estimation and mean-shift analysis as used in early vision tasks. In: International Conference on Pattern Recognition. (2002) 30927–30930
6. van den Boomgaard, R., van de Weijer, J.: Robust estimation of orientation for texture analysis. In: The 2nd International workshop on texture analysis and synthesis, Copenhagen (2002) 135—138
7. Heuckel, M.: An operator which locates edges in digital pictures. J. Association for computing machinery **18** (1971) 113–125
8. Bilateral filtering for gray and color images. In: International Conference of Computer Vision, ICCV 98. (1998)
9. Smith, S., Brady, J.: Susan: A new approach to low-level image processing. International Journal of Computer Vision **23** (1997) 45—78
10. Black, M., Sapiro, G., Marimont, D., Heeger, D.: Robust anisotropic diffusion. IEEE Trans. Image Processing **7** (1998) 421–432
11. Kass, M., Witkin, A.: Analyzing oriented patterns. Computer Graphics and Image Processing **37** (1987) 363–385
12. Bigun, J., Granlund, G., Wiklund, J.: Multidimensional orientation estimation with application to texture analysis and optical flow. IEEE Transactions on Pattern Analysis and Machine Intelligence **13** (1991) 775–789
13. Lindeberg, T., Garding, J.: Shape from texture from a multi-scale perspective. Technical Report ISRN KTH/NA/P–93/03–SE, Royal Institution of Technology, Stockholm (1993)
14. Weickert, J.: Anisotropic Diffusion in Image Processing. Tuebner Verlag, Stuttgart (1997)
15. Kuwahara, M., Hachimura, K., Eiho, S., Kinoshita, M.: Processing of ri-angiocardiographic images. In Preston, K., Onoe, M., eds.: Digital Processing of Biomedical Images. (1976) 187–202
16. Nagao, M., Matsuyama, T.: Edge preserving smoothing. Computer Graphics and Image Processing **9** (1979) 394–407
17. Bakker, P., van Vliet, L., Verbeek, P.: Edge preserving orientation adaptive filtering. In Boasson, M., Kaandorp, J., Tonino, J., Vosselman, M., eds.: ASCI'99, Proc. 5th Annual Conference of the Advanced School for Computing and Imaging. (1999) 207–213
18. van den Boomgaard, R.: The Kuwahara-Nagao operator decomposed in terms of a linear smoothing and a morphological sharpening. In Talbot, H., Beare, R., eds.: Mathematical Morphology, Proceedings of the 6th International Symposium on Mathematical Morphology, Sydney, Australia, CSIRO Publishing (2002) 283–292
19. Weickert, J., Brox, T.: Diffusion and regularization of vector- and matrix valued images. In: Inverse Problems, Image Analysis and Medical Imaging. Volume 313. (2002) 252—268

Regularity Classes for Locally Orderless Images

Luc Florack and Remco Duits

Eindhoven University of Technology
Department of Biomedical Engineering
P.O. Box 512, NL-5600 MB Eindhoven, The Netherlands

Abstract. Gaussian scale space permits one to compute image derivatives. The limitation to some finite order is not inherent in the paradigm itself (Gaussian blurred functions are always smooth), but is caused by the interplay of (at least) two external factors. One is the ratio of the Gaussian scale parameter versus the atomic scale that limits physically or perceptually meaningful sizes (e.g. pixel size, or in general any scale at which the image is "locally orderless"). The second factor involved is the fiducial level of tolerance. Together these factors conspire to determine a maximal order beyond which differential structure becomes meaningless. Thus they give rise to the notion of regularity classes for images akin to the conceptual C^k-classification pertaining to mathematical functions. We study the relationship between the maximal differential order k, the ratio of inner scale to atomic scale, and the prescribed tolerance level, and draw several conclusions that are of practical interest when considering image derivatives.

1 Introduction

Our goal is to establish an operational definition of regularity classes for images akin to the conceptual C^k-classification pertaining to mathematical functions. This is a nontrivial problem due to resolution limitations inherent in recordings of physical observables, as a result of which differential structure is ill-defined in the classical sense. Besides, classical differentiation is ill-posed [6], as a consequence of which it makes little sense to consider discrete approximations. Distribution theory provides a fundamental and at the same time practical solution for both problems [11], as it redefines derivatives in an intrinsically well-posed and operationally well-defined way through the use of a dual space of test functions (basically linear filters). Scale space theory is a particular instance whereby the test functions are taken to be normalised Gaussians and their derivatives for reasons amply discussed in the literature (see *e.g.* the seminal work by Koenderink [7,9] or any of the existing books on the subject [3,5,4,10,12]).

Still, despite the fact that scale space theory guarantees well-posedness *in principle*, it turns out impossible *in practice* to differentiate digital images beyond a certain order without running into difficulties. It is intuitively clear that the relevant parameter in this case is the resolution of the raw image relative to that at which local structure is resolved using the scale space paradigm. If this

L.D. Griffin and M. Lillholm (Eds.): Scale-Space 2003, LNCS 2695, pp. 255–265, 2003.

ratio is large, one expects differential structure to be well-defined, at least up to some order. Likewise it is reasonable to expect that for a given ratio differential structure will cease to be meaningful beyond a certain order. In this article the trade-off between raw image resolution, differentiation scale, differential order, and tolerance will be scrutinised.

In brief, we want to be able to say up to which order we may differentiate an image given a tolerance level and a fiducial ratio of differentiation scale (width of the Gaussian derivative filters) and intrinsic scale (graininess of the raw image). Alternatively, we may want to know which (minimal) scale we should take in order to justify a certain differential order while not violating a prescribed tolerance[1].

2   Theory

2.1   Ansatz

The approach is as follows. The raw image is embedded into a scale space representation according to the Gaussian scale space paradigm. Theoretically this enables us to extract derivatives of arbitrary order (for any fixed scale a scale space representation is an analytical function), but in practice there is an upper limit due to finite measurement tolerance and numerical effects induced by discretisation and quantisation. This limit on differential order can be revealed by considering the effect of spatially confined but otherwise arbitrary spatial deformations on differential structure. For fixed deformation scale (the range of confinement of the local deformations) one expects that beyond a certain order the induced distortion of derivatives will exceed a prescribed tolerance level. This puts a fundamental upper bound on differential order.

The spatial perturbations generically capture degrading factors such as discretisation effects, spatial noise correlations and correlations induced by the point spread function, and other fine-scale deformation artifacts. In view of genericity we decline from making specific assumptions about the details of such perturbations. In particular we do not want to commit ourselves to a particular grid.

A similar question for the case of dynamic perturbations without spatial deformations (additive noise) has been studied previously by Blom [1] in a statistical setting. Spatial and dynamic perturbations are in fact confounded.

2.2   Raw Image Scale

Consider a scalar image[2] $f : \mathbb{R}^n \to \mathbb{R}$. We will refer to it as the *raw image*. The details of its physical format are irrelevant here, except for the fact that it is of *finite resolution*. Let us adopt a linear model[3] and assume that

---

[1] Limitations due to finite scope will be ignored in this paper.

[2] We disregard limitations due to finite scope. Filter truncation will impose additional constraints on differentiability.

[3] Linearisation can be achieved via suitable reparametrisation of f_0 and f within some physical range.

$$f(x) = (f_0 \star \psi)(x) \overset{\text{def}}{=} \int d\xi\, f_0(\xi)\, \psi(x + \xi)\,,$$

for some nonnegative "naked" source distribution $f_0 \in \mathcal{S}'(\mathbb{R}^n)$ and nonnegative point spread function (PSF) $\psi \in \mathcal{S}(\mathbb{R}^n)$. Here, $\mathcal{S}(\mathbb{R}^n)$ is the Schwartz' space of smooth functions of rapid decay, and $\mathcal{S}'(\mathbb{R}^n)$ its topological dual, *i.e.* the space of tempered distributions [11]. The appropriateness of this model can be justified by virtue of the huge number of degrees of freedom contained in $\mathcal{S}'(\mathbb{R}^n)$ (note *e.g.* that it contains all $L^1(\mathbb{R}^n)$-functions, as well as all derivatives of point distributions of the Dirac type). Thus $\mathcal{S}'(\mathbb{R}^n)$ is certainly large enough to accomodate all physical source fields of interest. The PSF ψ has a "centre of gravity" as well as a finite scale. Assuming the centre of gravity is at the origin and the PSF is normalised to unit weight, its scale can be defined as

$$\delta \overset{\text{def}}{=} \left(\int d\xi\, \|\xi\|^2\, \psi(\xi) \right)^{\frac{1}{2}}.$$

The essence is that raw images

- are bounded, and
- are hardly affected by spatial variations confined to a ball of radius δ.

In fact we have $f \in L^\infty(\mathbb{R}^n) \cap C^\omega(\mathbb{R}^n)$, i.e. bounded and analytical (the latter, by the way, turns out irrelevant in subsequent considerations, since this pertains to the operationally void notion of differentiability in classical sense). By a suitable scaling we may assume that $\|f\|_{L^\infty} = 1$. The raw image scale δ is limited from below by pixel scale, but may potentially be much larger. Its defining property is that within a δ-ball there is total spatial disorder, although histogram information is reliable [8]. Put differently, we may apply any (arbitrarily interpolated) volume preserving diffeomorphism that is spatially confined to a δ-ball without essentially altering information content. This reflects the lack of spatial information at scales below δ. We shall henceforth refer to δ as the (spatial) *confinement scale*. In the next section we investigate the implications of a nonzero confinement scale for differentiability.

2.3 Distortions Induced by Spatial Deformations

To study the effect of deformations on differential structure, let us introduce the deformation map $\mathcal{F} : \mathbb{R}^n \to \mathbb{R}^n$, collectively capturing various sources of spatial noise. That is, instead of the "unperturbed" raw image $f \in L^\infty(\mathbb{R}^n) \cap C^\omega(\mathbb{R}^n)$ we consider a collection of perturbed images $\mathcal{F}^*f \equiv f \circ \mathcal{F}$. The ensemble of image instantiations $\mathcal{F}^*f$ for all diffeomorphisms[4] $\mathcal{F}$ stratified according to their distance to the identity map captures Koenderink's notion of "locally orderless images" [8]. We call a given diffeomorphism δ-*admissible* if $\|\mathcal{F}(x) - x\|_{L^\infty} = \max_{\mu=1,\ldots,n} \|\mathcal{F}^\mu(x) - x^\mu\|_{L^\infty} \leq \delta$ for some confinement scale $\delta > 0$. We will

---

[4] A diffeomorphism is a C^1-isomorphism with a C^1-inverse.

henceforth consider δ-admissible diffeomorphisms only. Thus δ-admissibility entails that $\mathcal{F}$ must be a diffeomorphism which looks very much like the identity map when resolved at inner scales $\sigma \gg \delta$, but is unconstrained below scale δ.

In classical sense, a δ-admissible choice of $\mathcal{F}$, regardless of how small δ is, completely destroys any differential information in the image f (ill-posedness [6]). We proceed by investigating the case of well-posed, finite-scale differentiation in the context of Gaussian scale space theory. Here one encounters integrals of the type

$$F[\phi] = \int dx\, f(x)\, \phi(x)\,,$$

representing a zeroth order image sample of f at the origin at a scale σ corresponding to the width of the normalised Gaussian filter ϕ. For any other point one simply shifts the filter accordingly, and the above integral becomes a correlation, or convolution. Without loss of generality we may set the width of the Gaussian equal to $\sigma \equiv 1$ (in each direction). The diffeomorphism $\mathcal{F}$ induces a change in the above functional, which can be represented in two equivalent ways using substitution of variables:

$$\mathcal{F}^*F[\phi] = \int dx\, \mathcal{F}^*f(x)\, \phi(x) \stackrel{\text{def}}{=} \int dy\, f(y)\, \mathcal{F}_*\phi(y) = F[\mathcal{F}_*\phi]\,,$$

with

$$\mathcal{F}_*\phi(y) = \det D\mathcal{F}^{\text{inv}}(y)\, \phi \circ \mathcal{F}^{\text{inv}}(y)\,,$$

in which $D\mathcal{F}^{\text{inv}}(y)$ is the Jacobian matrix of the transformation. In other words, we can model the effect of a spatial perturbation either at the level of the raw image or at the level of the filter representation (duality, cf. the monograph for details [3]). The latter will be our point of departure below.

More generally, if $\mathcal{D}_{\mu_1 \ldots \mu_k}$ denotes the k-th order partial derivative operator with respect to $x^{\mu_1} \ldots x^{\mu_k}$, then we have

$$\mathcal{D}_{\mu_1 \ldots \mu_k} \mathcal{F}^*F[\phi] = (-1)^k \int dx\, \mathcal{F}^*f(x)\, \phi_{\mu_1 \ldots \mu_k}(x) = (-1)^k \int dy\, f(y)\, \mathcal{F}_*\phi_{\mu_1 \ldots \mu_k}(y).$$

with $\phi_{\mu_1 \ldots \mu_k} \equiv \mathcal{D}_{\mu_1 \ldots \mu_k} \phi$ and $\mathcal{F}_*\phi_{\mu_1 \ldots \mu_k}(y) = \det D\mathcal{F}^{\text{inv}}(y)\, \phi_{\mu_1 \ldots \mu_k} \circ \mathcal{F}^{\text{inv}}(y)$. Although the perturbative effect of $\mathcal{F}$ on the derivatives will be seen to increase with order, it is not nearly as bad as in the case of classical differentiation (which re-emerges in the physically void limit $\sigma \downarrow 0$ by the "correspondence principle").

To reveal the effect of spatial noise, let us write $\mathcal{F}(x) = x + \delta\, \Xi(x)$, in which $\delta > 0$ is the confinement scale of the local deformation, and $\Xi \in C^1(\mathbb{R}^n \to \mathbb{R}^n)$, the space of uniformly bounded and (for the moment) continuously differentiable vector fields [2], with norm given by

$$\|\Xi\|_{L^\infty} \stackrel{\text{def}}{=} \max_{\mu=1,\ldots,n}\ \sup_{x \in \mathbb{R}^n} |\Xi^\mu(x)|\,.$$

The inverse automorphism is approximately $\mathcal{F}^{\text{inv}}(x) = x - \delta\, \Xi(x) + \mathcal{O}(\delta^2)$, and thus, omitting higher order terms in δ,

$$\det D\mathcal{F}^{\text{inv}}(x) \approx 1 - \delta\, \text{tr}\, \mathcal{D}\, \Xi(x) = 1 - \delta\, \text{div}\, \Xi(x)\,.$$

By slick choice of δ we may assume that $\|\Xi\|_{L^\infty} = 1$. We wish to compare perturbed and unperturbed results.

Lemma 1. *Write $\mathcal{F}(x) = x + \delta\,\Xi(x)$ and define*

$$\delta_{\mathcal{F}}\mathcal{D}_{\mu_1\dots\mu_k}F[\phi] \overset{\text{def}}{=} \mathcal{D}_{\mu_1\dots\mu_k}\mathcal{F}^*F[\phi] - \mathcal{D}_{\mu_1\dots\mu_k}F[\phi] = F[\mathcal{F}_*\phi_{\mu_1\dots\mu_k}] - F[\phi_{\mu_1\dots\mu_k}]\,,$$

Then we have

$$\delta_{\mathcal{F}}\mathcal{D}_{\mu_1\dots\mu_k}F[\phi] = -\delta \int dx\, f(x)\,\mathcal{L}_\Xi\,\phi_{\mu_1\dots\mu_k}(x) + \mathcal{O}(\delta^2)\,, \tag{1}$$

in which $\mathcal{L}_\Xi\psi = \psi\operatorname{div}\Xi + \nabla\psi\cdot\Xi$.

Proof. The result readily follows by noticing that

$$\det\mathcal{D}\mathcal{F}^{\text{inv}} = 1 - \delta\operatorname{div}\Xi + \mathcal{O}(\delta^2)\,,$$
$$\phi_{\mu_1\dots\mu_k}\circ\mathcal{F}^{\text{inv}} = \phi_{\mu_1\dots\mu_k} - \delta\,\phi_{\mu_1\dots\mu_k\mu}\Xi^\mu + \mathcal{O}(\delta^2)\,.$$

Thus there are two contributions to the overall deformation:

$$\delta_{\mathcal{F}}\mathcal{D}_{\mu_1\dots\mu_k}F[\phi] = \delta_1\mathcal{D}_{\mu_1\dots\mu_k}F[\phi] + \delta_2\mathcal{D}_{\mu_1\dots\mu_k}F[\phi]\,,$$

with

$$\delta_1\mathcal{D}_{\mu_1\dots\mu_k}F[\phi] = -\delta \int dx\, f(x)\,\phi_{\mu_1\dots\mu_k}(x)\operatorname{div}\Xi(x)\,, \tag{2}$$

$$\delta_2\mathcal{D}_{\mu_1\dots\mu_k}F[\phi] = -\delta \int dx\, f(x)\,\phi_{\mu_1\dots\mu_k\mu}(x)\,\Xi^\mu(x)\,. \tag{3}$$

This completes the proof. □

Note that Eq. (2) is a volumetric error that vanishes identically for volume preserving diffeomorphisms regardless of the input. For the moment we will retain the divergence term. Eq. (3) is a displacement error and is generically nonzero.

Let us consider a normalised Gaussian filter ϕ of unit scale,

$$\phi(x) \overset{\text{def}}{=} \frac{1}{\sqrt{2\pi}^n} \exp\left(-\frac{1}{2}x^2\right)\,,$$

and write it as $\phi(x) = \psi^2(x)$. Then we may rewrite Eqs. (2–3) using the definitions of the parabolic cylinder functions $p_{\mu_1\dots\mu_k}(x)$—in which one of the factors $\psi(x)$ is absorbed—cf. Appendix A, as follows:

$$\delta_1\mathcal{D}_{\mu_1\dots\mu_k}F[\phi] = -\delta\,\frac{(-1)^k}{(2\pi)^{n/4}} \int dx\, f(x)\,\psi(x)\,p_{\mu_1\dots\mu_k}(x)\operatorname{div}\Xi(x)\,, \tag{4}$$

$$\delta_2\mathcal{D}_{\mu_1\dots\mu_k}F[\phi] = -\delta\,\frac{(-1)^{k+1}}{(2\pi)^{n/4}} \int dx\, f(x)\,\psi(x)\,p_{\mu_1\dots\mu_k\mu}(x)\,\Xi^\mu(x)\,. \tag{5}$$

Here we have used the decomposition

$$\phi_{\mu_1\ldots\mu_k}(x) = \frac{(-1)^k}{(2\pi)^{n/4}}\, \psi(x)\, p_{\mu_1\ldots\mu_k}(x)\,.$$

As a result we can make the following, sharp estimates for their respective upper bounds:

$$|\delta_1 \mathcal{D}_{\mu_1\ldots\mu_k} F[\phi]| \leq \frac{\delta}{(2\pi)^{n/4}}\, \|f\|_{L^\infty}\, \|\phi\|_{L^1}\, \|p_{\mu_1\ldots\mu_k}\|_{L^2}\, \|\mathrm{div}\,\varXi\|_{L^\infty}\,, \tag{6}$$

$$|\delta_2 \mathcal{D}_{\mu_1\ldots\mu_k} F[\phi]| \leq \frac{\delta}{(2\pi)^{n/4}}\, \|f\|_{L^\infty}\, \|\phi\|_{L^1}\, \max_{\mu=1,\ldots,n}\, \|p_{\mu_1\ldots\mu_k\mu}\|_{L^2}\, \|\varXi\|_{L^\infty}\,. \tag{7}$$

In the particular case of volume preserving diffeomorphisms the divergence term cancels and we may in fact relax the regularity assumption on the generating vector field $\varXi$ accordingly. Using $\|\psi\|_{L^2} = \sqrt{\|\phi\|_{L^1}} = 1$, $\|f\|_{L^\infty} = 1$, and $\|\varXi\|_{L^\infty} = 1$, and assuming $\mathrm{div}\,\varXi = 0$ we obtain

$$|\delta_{\mathcal{F}} \mathcal{D}_{\mu_1\ldots\mu_k} F[\phi]| \leq \frac{\delta}{(2\pi)^{n/4}}\, \max_{\mu=1,\ldots,n}\, \|p_{\mu_1\ldots\mu_k\mu}\|_{L^2}\,.$$

The analogue in terms of multi-indices, which is somewhat more convenient for our purpose, is

$$|\delta_{\mathcal{F}} \mathcal{D}_{\mathbf{k}} F[\phi]| \leq \frac{\delta}{(2\pi)^{n/4}}\, \max_{\mu=1,\ldots,n}\, \|p_{\mathbf{k}+\mathbf{e}_\mu}\|_{L^2}\,,$$

in which $\mathbf{k} = (k_1,\ldots,k_n)$ is an n-dimensional multi-index of order $\|\mathbf{k}\| = k_1 + \ldots + k_n = k$ and $\mathbf{e}_\mu$ is the n-dimensional multi-index with entry 1 at μ-th position, and zeros elsewhere. The norm of the parabolic cylinder functions on the right hand side follows from the orthogonality condition

$$\int dx\, p_{\mathbf{k}}(x)\, p_{\mathbf{l}}(x) = \delta_{\mathbf{k}\mathbf{l}}\, \sqrt{2\pi}^{\,n}\, \mathbf{k}!$$

with the usual multi-index conventions

$$\delta_{\mathbf{k}\mathbf{l}} \overset{\mathrm{def}}{=} \delta_{k_1\ell_1}\ldots\delta_{k_n\ell_n} \quad \text{and} \quad \mathbf{k}! \overset{\mathrm{def}}{=} k_1!\ldots k_n!\,.$$

See Appendix A for details. Maximizing over all multi-indices of order k, using the fact that

$$\max_{\mathbf{k},|\mathbf{k}|=k}\, \max_{\mu=1,\ldots,n}\, (\mathbf{k}+\mathbf{e}_\mu)! = (k+1)!\,,$$

we finally arrive at

$$\max_{\mathbf{k},|\mathbf{k}|=k}\, |\delta_{\mathcal{F}} \mathcal{D}_{\mathbf{k}} F[\phi]| \leq \delta\, \sqrt{(k+1)!}\,.$$

Reintroducing spatial scale σ and dynamic units $\|f\|_{L^\infty}$ leads to

$$\max_{\mathbf{k},|\mathbf{k}|=k}\, |\sigma^k\, \delta_{\mathcal{F}} \mathcal{D}_{\mathbf{k}} F[\phi]| \leq \frac{\delta}{\sigma}\, \sqrt{(k+1)!}\, \|f\|_{L^\infty}\,.$$

Note that we now have naturally scaled derivatives $\sigma^k\, \mathcal{D}_{\mathbf{k}}$ on the left hand side.

This spatial noise estimate allows us to solve for a nonnegative integer value of $k \geq 0$ such that the right hand side remains below a prescribed tolerance bound, $\epsilon M_k \|f\|_{L^\infty}$, say, with

$$M_k \stackrel{\text{def}}{=} \sup_{f, \|f\|_{L^\infty}=1} \max_{\mathbf{k}, |\mathbf{k}|=k} \left| \sigma^k \mathcal{D}_{\mathbf{k}} F[\phi] \right| . \tag{8}$$

This k-dependent tolerance bound scales proportionally to the order of magnitude of k-th order derivatives. So

$$\frac{\delta}{\sigma} \sqrt{(k+1)!} \leq \epsilon M_k . \tag{9}$$

To determine M_k consider the following estimate:

$$\left| \sigma^k \mathcal{D}_{\mathbf{k}} F[\phi] \right| \leq \|f\|_{L^\infty} \left\| \sigma^k \mathcal{D}_{\mathbf{k}} \phi \right\|_{L^1} .$$

Again using the decomposition of Gaussian derivatives into parabolic cylinder functions and a Gaussian envelope ψ with $\|\psi\|_{L^2} = 1$ as before, setting $\sigma = 1$ (note that the L^1-norm of the scaled derivative filter does not depend on σ),

$$\mathcal{D}_{\mathbf{k}} \phi(x) = \frac{(-1)^k}{(2\pi)^{n/4}} \, \psi(x) \, p_{\mathbf{k}}(x) ,$$

we arrive at

$$\left| \sigma^k \mathcal{D}_{\mathbf{k}} F[\phi] \right| \leq \frac{1}{(2\pi)^{n/4}} \|f\|_{L^\infty} \|p_{\mathbf{k}}\|_{L^2} \|\psi\|_{L^2} = \sqrt{k!} \, \|f\|_{L^\infty} .$$

Thus we can set

$$M_k = \sqrt{k!} \tag{10}$$

in Eq. (8).

Let us now define $\sigma_{\min}$ as the one for which equality holds in Eq. (9), using the result of Eqs. (8) and (10):

$$\sigma_{\min} \stackrel{\text{def}}{=} \frac{\delta}{\epsilon} \sqrt{k+1} . \tag{11}$$

For each desired order this allows us to pick the minimally required scale given our tolerance criterion ϵ. Note that if we would insist on zero tolerance we would have to go to the hypothetical limit $\sigma_{\min} \to \infty$ (or return to the unrealistic mathematical assumption that $\delta = 0$).

Similarly, we may define the maximum admissible order of differentiation:

$$k_{\max} \stackrel{\text{def}}{=} \text{Entier} \left[\left(\epsilon \frac{\sigma_{\min}}{\delta} \right)^2 - 1 \right] \quad \text{subject to the condition} \quad k_{\max} \geq 0 , \tag{12}$$

i.e. the largest nonnegative integer that does not exceed the value of k for which Eqs. (9–10) hold. This gives us the upper bound on differentiability, again given the tolerance criterion ϵ. For $\epsilon = 0$ the defining equation for $k_{\max}$ has no solution,

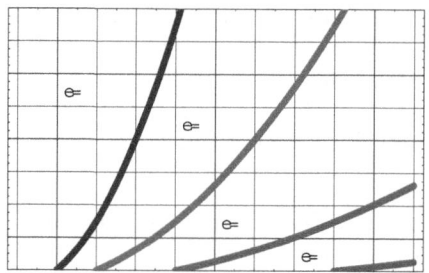

Fig. 1. Plot of the maximal (interpolated) order k_{max} (vertical axis) versus the scale ratio σ_{min}/δ (horizontal axis) for four different tolerance levels, $\epsilon = 0.125, 0.250, 0.500, 1.000$.

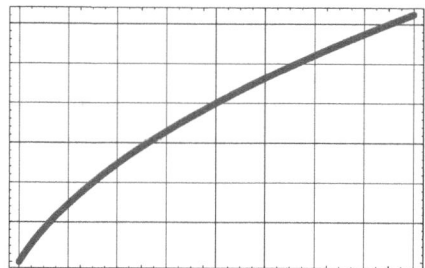

Fig. 2. Plot of the minimal scale required for extracting k-th order derivatives with a given tolerance relative to the scale needed for 0-th order computation (vertical axis) as a function of differential order (horizontal axis).

as expected. For $\sigma_{min}/\delta < 1/\epsilon$ we cannot even extract zeroth order samples! See Figure 1.

Yet another interpretation of the established result is that if we want to extract a k-th order derivative with the same tolerance as that characterising the zeroth order samples at some fiducial inner scale σ_0 we need to take a larger inner scale $\sigma_k \geq \sigma_0$, viz.

$$\sigma_k = \sqrt{k+1}\,\sigma_0 . \tag{13}$$

See Figure 2.

Finally, if we fix the minimal inner scale and compare the induced tolerances for k-th order derivatives relative to zeroth order we obtain

$$\epsilon_k = \sqrt{k+1}\,\epsilon_0 . \tag{14}$$

Thus the induced error indeed increases with order.

3   Conclusion

We have investigated the trade-off between various factors that affect the accuracy of differentiation using the Gaussian scale space paradigm. The factors of interest here are the spatial confinement scale δ, expressing the extent of local neighbourhoods within which the raw image is "locally orderless" (with pixel scale as a special and limiting instance), the inner scale σ of the Gaussian filter, the level of tolerance ϵ, which measures the relative amplitude of additive noise, and, of course, the order k of differentiation.

The main conclusion can be captured by a $\sqrt{k+1}$-law, expressing the relative degradation of output accuracy with differential order k for fixed confinement scale and inner scale. Alternatively, insisting on a fixed (maximally admissible) tolerance level and given a fixed confinement scale it expresses the relative increase of inner scale (compared to zeroth order scale) needed to maintain the same accuracy as for the zeroth order data at any higher order k. Yet another interpretation of this law is that it prescribes a minimal inner scale once we have fixed the admissible tolerance level for a given confinement scale as a function of differential order k.

In any case there exists an upper bound on differential order in realistic scenarios where we neither have zero measurement tolerance nor a vanishing confinement scale. In fact we have arrived at an operationally well-defined C^k-classification of images in the context of the Gaussian scale space paradigm, which is essentially different from the conceptual but operationally void C^k-classification used in the context of classical differentiation.

A   Hermite Polynomials and Parabolic Cylinder Functions

Definition 1 (Hermite Polynomials). *The Hermite polynomials in n dimensions are defined as follows:*

$$\mathrm{He}_{\mu_1\ldots\mu_k}(x) \overset{\text{def}}{=} (-1)^k \exp\left(\frac{1}{2}x^2\right) \frac{\partial^k}{\partial x^{\mu_1}\ldots\partial x^{\mu_k}} \exp\left(-\frac{1}{2}x^2\right).$$

Alternatively, in multi-index notation, with $\mathbf{k} = (k_1,\ldots,k_n)$ and $|\mathbf{k}| = k_1+\ldots+k_n$:

$$\mathrm{He}_{\mathbf{k}}(x) \overset{\text{def}}{=} (-1)^{|\mathbf{k}|} \exp\left(\frac{1}{2}x^2\right) \frac{\partial^{|\mathbf{k}|}}{\partial x^{\mathbf{k}}} \exp\left(-\frac{1}{2}x^2\right).$$

Gaussian derivatives can be expressed in terms of Hermite polynomials and the zeroth order Gaussian:

$$\phi_{\mu_1\ldots\mu_k}(x) = (-1)^k \phi(x) \mathrm{He}_{\mu_1\ldots\mu_k}(x).$$

Alternatively,

$$\phi_{\mathbf{k}}(x) = (-1)^{|\mathbf{k}|} \phi(x) \mathrm{He}_{\mathbf{k}}(x).$$

Definition 2 (Parabolic Cylinder Functions). *The parabolic cylinder functions in n dimensions are defined as follows:*

$$p_{\mu_1\ldots\mu_k}(x) \overset{\text{def}}{=} (-1)^k \exp\left(\frac{1}{4}x^2\right) \frac{\partial^k}{\partial x^{\mu_1}\ldots\partial x^{\mu_k}} \exp\left(-\frac{1}{2}x^2\right),$$

i.e.

$$p_{\mu_1\ldots\mu_k}(x) = \exp\left(-\frac{1}{4}x^2\right) \text{He}_{\mu_1\ldots\mu_k}(x).$$

Alternatively, in multi-index notation:

$$p_{\mathbf{k}}(x) \overset{\text{def}}{=} (-1)^{|\mathbf{k}|} \exp\left(\frac{1}{4}x^2\right) \frac{\partial^{|\mathbf{k}|}}{\partial x^{\mathbf{k}}} \exp\left(-\frac{1}{2}x^2\right)$$

or

$$p_{\mathbf{k}}(x) = \exp\left(-\frac{1}{4}x^2\right) \text{He}_{\mathbf{k}}(x).$$

These functions are orthogonal. In 1D we have:

$$\int dx\, p_k(x)\, p_\ell(x) = \delta_{k\ell}\, \sqrt{2\pi}\, k!.$$

In general,

$$\int dx\, p_{\mathbf{k}}(x)\, p_{\mathbf{l}}(x) = \delta_{\mathbf{kl}}\, \sqrt{2\pi}^{\,n}\, \mathbf{k}!.$$

with the conventions

$$\delta_{\mathbf{kl}} \overset{\text{def}}{=} \delta_{k_1\ell_1}\ldots\delta_{k_n\ell_n},$$
$$\mathbf{k}! \overset{\text{def}}{=} k_1!\ldots k_n!.$$

References

1. J. Blom, B. M. ter Haar Romeny, A. Bel, and J. J. Koenderink. Spatial derivatives and the propagation of noise in Gaussian scale-space. *Journal of Visual Communication and Image Representation*, 4(1):1–13, March 1993.
2. Y. Choquet-Bruhat, C. DeWitt-Morette, and M. Dillard-Bleick. *Analysis, Manifolds, and Physics. Part I: Basics.* Elsevier Science Publishers B.V. (North-Holland), Amsterdam, 1991.
3. L. M. J. Florack. *Image Structure*, volume 10 of *Computational Imaging and Vision Series*. Kluwer Academic Publishers, Dordrecht, The Netherlands, 1997.
4. B. M. ter Haar Romeny. *Front-End Vision and Multi-Scale Image Analysis*. Kluwer Academic Publishers, Dordrecht, The Netherlands, 2003. To appear.
5. B. M. ter Haar Romeny, L. M. J. Florack, J. J. Koenderink, and M. A. Viergever, editors. *Scale-Space Theory in Computer Vision: Proceedings of the First International Conference, Scale-Space'97, Utrecht, The Netherlands*, volume 1252 of *Lecture Notes in Computer Science*. Springer-Verlag, Berlin, July 1997.

6. J. Hadamard. Sur les problèmes aux dérivées partielles et leur signification physique. *Bul. Univ. Princeton*, 13:49–62, 1902.
7. J. J. Koenderink. The structure of images. *Biological Cybernetics*, 50:363–370, 1984.
8. J. J. Koenderink and A. J. van Doorn. The structure of locally orderless images. *International Journal of Computer Vision*, 31(2/3):159–168, April 1999.
9. J. J. Koenderink and A. J. van Doorn. Receptive field families. *Biological Cybernetics*, 63:291–298, 1990.
10. T. Lindeberg. *Scale-Space Theory in Computer Vision*. The Kluwer International Series in Engineering and Computer Science. Kluwer Academic Publishers, Dordrecht, The Netherlands, 1994.
11. L. Schwartz. *Théorie des Distributions*. Publications de l'Institut Mathématique de l'Université de Strasbourg. Hermann, Paris, second edition, 1966.
12. J. Sporring, M. Nielsen, L. M. J. Florack, and P. Johansen, editors. *Gaussian Scale-Space Theory*, volume 8 of *Computational Imaging and Vision Series*. Kluwer Academic Publishers, Dordrecht, The Netherlands, 1997.

Mode Estimation Using
Pessimistic Scale Space Tracking

Lewis D. Griffin[1] and Martin Lillholm[2]

[1] Radiological Sciences
5th Floor Thomas Guy House
Guys Campus, London SE1 9RT, UK
lewis.griffin@kcl.ac.uk

[2] IT University of Copenhagen
Glentevej 67-69, DK 2400, Copenhagen, Denmark
grumse@it-c.dk

Abstract. Estimation of the mode of a distribution over $\mathbb{R}^n$ from discrete samples is introduced and three methods for its solution are developed and evaluated. The first solution is based on Fréchet's definition of central tendencies. We show that algorithms based on this approach have only limited success due to the non-differentiability of the Fréchet measures. The second solution is based on tracking maxima through a Scale Space built from the samples. We show that this is more accurate than the Fréchet approach, but that tracking to very fine scales is unwarranted and undesirable. For our third method we analyze the reliability of the information across scale using an exact bootstrap analysis. This leads to a modified version of the Scale Space approach where unreliable information is downgraded (pessimistically) so that tracking into such regions does not occur. This modification improves the accuracy of mode estimation. We conclude with demonstrations on high-dimensional real and synthetic data, which confirm the technique's accuracy and utility.

1   Introduction

For a distribution D over a domain $\mathbb{R}^n$ the mode, like the mean and median, is simple to define[1], it is the element of the domain that maximizes $D(\vec{x})$; but unlike the mean and median, estimation of the mode from samples generated by the distribution is difficult. It is this estimation problem that concerns us here. In the remainder of the introduction we review previous approaches to this problem, then in sections 2-4 we describe and evaluate three (progressively better) approaches. In sections 5-6 we present results of our third technique, which we call *Pessimistic Scale Space Tracking*, when it is successfully applied to high-dimensional mode estimation problems. In section 7 we draw conclusions.

---

[1] For some non-generic distributions, this definition fails to define a unique mode but this possible complication will be ignored in the remainder.

L.D. Griffin and M. Lillholm (Eds.): Scale-Space 2003, LNCS 2695, pp. 266–280, 2003.
© Springer-Verlag Berlin Heidelberg 2003

1.1   Previous Methods for Mode Estimation

Mode estimation is straightforward if one has a prior expectation of the form of the density – one simply finds the MAP estimate of the density and reads off the mode. A more flexible version of this is the Gaussian mixture method (Everitt & Hand, 1981) that models a distribution as a sum of Gaussians of varying center and width. Again, the model is fitted to make the observed data as likely as possible and then the mode of the Gaussian mixture can be calculated by hill climbing. The power of this technique is in applications where the densities can be expected to be multi-modal; it is not a universal panacea for density estimation.

If no parametric form for the density is available then the naïve way to proceed is to bin the data, plot a histogram and read off its mode. Obvious problems with this are a shift-dependency on the bin counts and a dependency on the bin width. The shift-dependency problem was solved by kernel methods (Parzen, 1962) where the discrete data (modeled as a collection of delta functions) is convolved by a smoothing kernel (see figure 1). As with filtering images (Griffin, 1995), much discussion of the optimal kernel shapes followed Parzen's work but eventually the Gaussian was agreed on and this approach was popularized (B. W. Silverman, 1986).

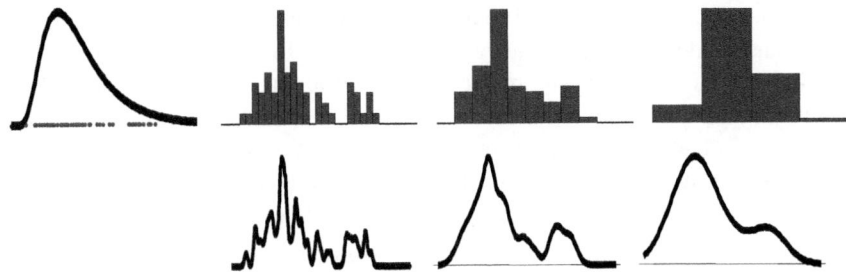

Fig. 1. Shows at top-left a log-normal distribution ($\mu = \sigma = 1$) and 64 random samples from the distribution. This density and the samples are used in sections 2-4. All three histograms (top) are of these samples, but they have different bin widths. The kernel density estimates (bottom) are created using Gaussian windows of scale equated with the corresponding histogram bin widths.

To solve the question of the optimal bin width, Parzen proposed that the optimal kernel should minimize the L^2 difference between the estimated and true densities. He further showed that this difference depends on (i) the kernel shape & width and (ii) the L^2-norm of the true density's 2nd derivative; but since we don't know the true density this fails to solve the problem. Numerous methods (Wand & Jones, 1995) have been proposed for finding a globally optimal kernel width, such as (i) to use a standard form for the true density (e.g. Gaussian (Fukunga, 1990)), (ii) 'plug-in methods' that estimate the density 2nd derivative by assuming normality of its 4th derivative (and so on), (iii) iterative plug-in (Scott, Tapia & Thompson, 1977). More recently, methods for estimating locally-optimal kernel width are being developed, including: (i) use of a 'pilot estimate' of the density obtained by filtering the data (raising the problem of what kernel width to use) (Hazelton, 1999), and (ii) iteratively estimating the density, thus bandwidths, thus density, etc. (Katkovnik & Shmulevich, 2002). The case of multivariate date is rarely tackled, an exception being (Gasser, Hall & Presnell, 1998); in order to use a pilot estimate technique, they make the strong assumption that the unknown density is the product of its marginal densities i.e. they assume independence between the different dimensions of the space.

All the methods reviewed above are either: (i) derived by asymptotic analysis, (ii) use heuristics, or (iii) make strong assumptions about the true density. Very often more than one of these criticisms applies.

2   Mode Estimation Using the Fréchet Definition

Our first method of mode estimation is based on Fréchet's definition of central location (Fréchet, 1948, Griffin, 1997). Given a distribution $D(\vec{x})$: $\mathbb{R}^n \rightarrow \mathbb{R}$ he defines a family of central locations $\vec{\mu}_{r \geq 0}$ as the minimizers of $\left(|\vec{x}|^r * D \right)^{\frac{1}{r}}$ (we will refer to this as Fréchet's measure). For $r = 2$ one obtains the mean, for $r = 1$ the median and in the limit as $r \rightarrow 0$ the mode. The definition is elegant in that it produces the mean, median and mode in a single framework that applies to a distribution of any dimension, but there is a concealed subtlety in the definition of the mode of which one should be aware. The subtlety is that in general there will be multiple local minimizers for $0 \leq r < 1$ and these minimizers change their locations continuously with r. To use the Fréchet definition operationally to define the mode, one must make an additional assumption or requirement that the entire family of central locations $\vec{\mu}_r$ should be continuous with respect to r. Using this assumption, one proceeds by first locating the median $\vec{\mu}_1$, then tracking down from $r = 1$ to $r = 0$ following a continuous path of local minimizers of Fréchet's measure as one does so – eventually at $r = 0$ one arrives at the mode.

Fréchet's definition is normally used with explicitly given distributions, but we wondered if it has use as an estimator. Our best effort to this end is shown in figure 2, where results of estimating the mode of a log-normal distribution from samples are presented. The contour plot shows Fréchet's measure (logged and negated for clarity) on which is superimposed the path that we have tracked as r was reduced. Although the result is clearly close to the mode of the distribution from which the samples came (*cf.* figure 1) this is partly good fortune. If one looks closely, one notices that for $r < 1$ there is almost no lateral movement of the tracked minimizer and because of this at r near 0, the global minimizer of Fréchet's measure is not found.

The explanation for the failure of tracking is not difficult to discover and is illustrated by the panels in the center and bottom right of figure 2. These show the Fréchet measure for $r = 0.37$ which corresponds to the level marked by the horizontal ticks adjacent to the contour plot. In the expanded view of the measure one can clearly see

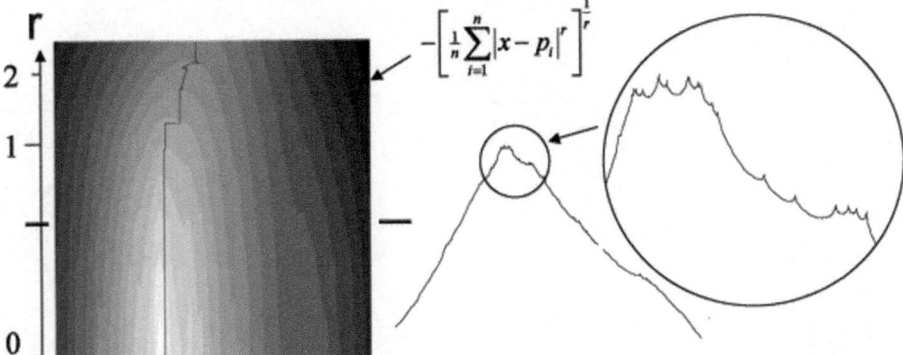

$$-\left[\frac{1}{n}\sum_{i=1}^{n}|x-p_i|^r\right]^{\frac{1}{r}}$$

Fig. 2. Shows an approach to mode estimation using Fréchet's definition of central tendency. The variable x ranges across the domain of the distribution and is the horizontal axis of all 3 plots. The p_i are the samples from the distribution, r is the parameter that indexes the family of central locations. See text (section 2) for more details.

its non-differentiable nature, which is why tracking has failed — gradient ascent algorithms find it difficult to stay on such singularly narrow ridges.

3   Mode Estimation Using Scale Space Tracking (SST)

Our second method is based on the more familiar kernel density approach to mode estimation (using Gaussians as our kernels), but rather than searching for a globally or locally optimal kernel widths we follow previous authors (Marchette & Wegman, 1997, Minnotte, Marchette & Wegman, 1998, Minonotte & Scott, 1993, B.W. Silverman, 1981) in considering the full continuum of widths (or scales). These authors have noted that one may track modes as they vary continuously in position and annihilate with anti-modes as scale is increased from fine-to-coarse. Indeed much of early Scale Space work was duplicated (or possibly preceded) in the field of non-parametric density estimation, including the observation that in 1-D, modes are never created as scale increases (B.W. Silverman, 1981). While these authors have concentrated on using the pattern of modes over scale (the 'mode tree') to understand and visualize multi-modality, one may also use it as a numerical method of mode estimation. In 1-D the idea is as follows: at sufficiently coarse scale there is exactly one mode; since modes are never created with increasing scale, this mode may be tracked through decreasing scale to one of the original sample points; and this sample point may be taken as an estimate of the mode of the true distribution[2]. An example of this is shown in figure 3.

---

[2] We are unaware of any proof that in two or more dimensions the equivalent result — that the final mode that exists at coarse scale can be traced back to a sample point at zero scale — holds. This does not have any impact on our algorithms though.

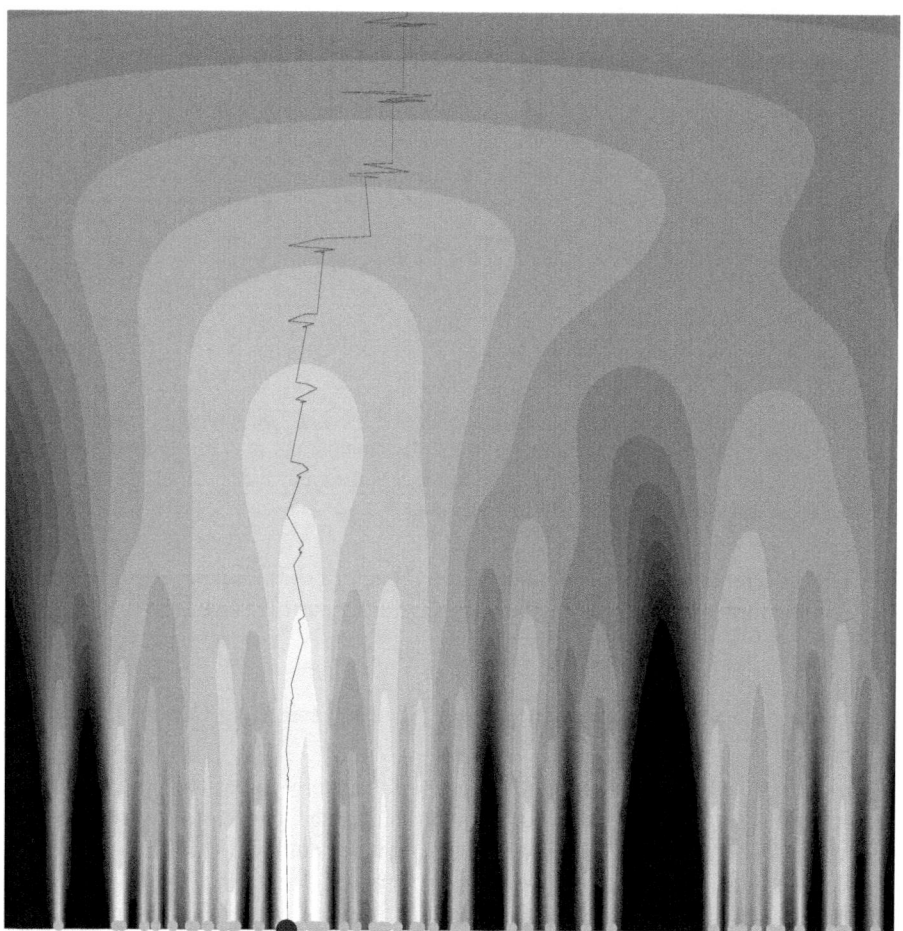

Fig. 3. Shows an approach to mode estimation using Scale Space Tracking (SST). The horizontal axis is the domain of the density from which the sample points (same as figure 1) along the bottom have been generated. The vertical axis is log scale. The jagged polyline is the path of the mode tracking and its lower endpoint is the estimate of the mode that results.

At this point we will introduce formalism for the general case of SST in a D-dimensional space. Given n sample points $\vec{p}_1, \dots, \vec{p}_n$ in D-dimensional space, define the number of sample points in a Gaussian bin, centred at $\vec{x}$ and of scale s, to be:

$w(\vec{x}, s) = \sum_{i=1}^{n} e^{-\frac{\|\vec{p}_i - \vec{x}\|^2}{4s}}$. The volume of such a bin is $v(s) = (4\pi s)^{D/2}$, so the kernel-

estimated density is $d(\vec{x}, s) = \dfrac{w(\vec{x}, s)}{v(s)}$. Figure 3 shows the logarithm (for improved

visualization) of such a density created from the 64 sample points introduced in figure 1.

To use this density for mode estimation, we start tracking at $\vec{x}^* = \left(\mu_1 \left[\{ p_{11}, ..., p_{n1} \} \right], ..., \mu_1 \left[\{ p_{1D}, ..., p_{nD} \} \right] \right)^T$ which is calculated by, separately for each dimension, taking the median of the sample point values; and $s^* = \left(\mu_2 \left[\{ iqr \left[\{ p_{11}, ..., p_{1n} \} \right], ..., iqr \left[\{ p_{D1}, ..., p_{Dn} \} \right] \} \right] \right)^2$ which is the square of the mean (over dimensions) of the inter-quartile range of the sample points. We use these robust measures to produce a starting point rather than the more obvious measures based on the sample mean and variance, because we often deal with highly kurtosed data (e.g. in section 6) for which mean and variance are very unstable measures.

To track from our starting point we could simply use a gradient ascent method to move through $\vec{x}$ and s, but this is too uncontrolled. Consider figure 3. The starting point is that at the top of the figure. If we gradient ascend the density function shown in the figure, we will immediately rush towards finer scales where higher kernel-estimated densities may be found. This charge to finer scales happens so quickly that the algorithm fails to track the path of the mode that exists at all scales. To prevent this it is not sufficient simply to remap (for example, logarithmically) the scale axis. Rather, we must carefully control the rate of reduction of s as is done in graduated non-convexity methods (Blake & Zisserman, 1987). We use a reduction rate $\bullet = 0.5$. First we track from s^* to γs^*, then from γs^* to $\gamma^2 s^*$ and so on. This is visible in figure 3 in which the intervals are equal size as the vertical axis is log scale. In **each** interval we reparameterize scale logarithmically, for example for the first interval we gradient ascend on the parameters $\vec{x}$ and t_1 where $s = \gamma s^* + e^{t_1}$ and t_1 is initialized to $t_1 = \ln \left(s^* \right) + \ln \left(1 - \gamma \right)$. We terminate the scheme when s is a small fraction of its starting value.

Three details of our implementation are worth mentioning. First, we gradient ascend the log-density rather than the density. This is forced on us when we are dealing with high-dimensional problems (e.g. $D=64$ in section 6) because the unlogged bin volume $v(s) = (4\pi s)^{D/2}$ is so large that it causes numerical overflow. The second detail is that we use Polak-Ribiere conjugate-gradient ascent (Press, Teukolsky, Vetterling & Flannery, 2002). This requires explicit gradient calculations but these are fast, simple and stable when using Gaussian kernels. The final detail is that when calculating the density and its derivatives we ignore points that are more than $8\sqrt{2s}$ from $\vec{x}$ as the Gaussian bin is of negligible weight at this distance. A consequence of this is that the ascent runs faster and faster as s reduces, so for computation time, it matters not if we track to unnecessarily fine scales.

We have evaluated the performance of this scheme when estimating the mode of the log-normal distribution in figure 1. For each sample size in the range $n = 2^2, ..., 2^{10}$ we generated that number of samples and used scale-space mode-tracking to estimate the mode. This was repeated 10^4 times for each sample size to get an accurate figure for the root-mean-squared estimation error. For comparison we performed an identical experiment using the Fréchet scheme of the previous section.

The results are shown in figure 4 where it can be seen that Fréchet is superior up to sample sizes of 90, but SST is superior for larger samples. We hypothesize that this is because for small sample sizes the Scale Space method is biased towards the mean, while the Fréchet method is biased towards the median.

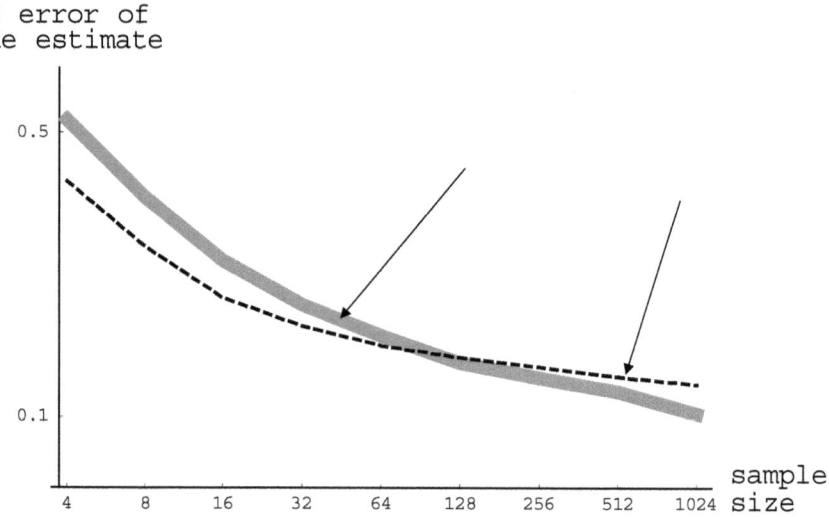

Fig. 4. Compares the performance on two approaches to mode estimation.

4   Mode Estimation
Using Pessimistic Scale Space Tracking (PSST)

The Scale Space tracking of section 3 has a flaw that became very apparent when we used it on high dimensional problems. Inevitably the mode-estimate that the method returns will be one of the original samples and for a very high dimensional problem it is tremendously unlikely that any one of the samples will correspond to the mode of the true distribution. This seems to limit performance. To fix this it seems desirable to prevent the tracking reaching very fine scales, but how is one to define very fine? We have already noted the profusion of literature on selecting the optimal kernel width in the Parzen density estimation approach. Figure 5 gives a clue to an alternative approach.

The most notable aspect of the histograms of figure 5 is the increase in the number of modes with smaller bin width, but there is another interesting aspect — the bin heights become noticeably quantized for small bin width, because they have so few samples. Having very few samples in a bin means that its height is an unreliable estimate of the true density. We can quantify this unreliability through a thought experiment. Suppose that we repeatedly run a bootstrap process of sampling-with-replacement our data to make synthetic data (Efron & Tibshirani, 1993). Say we have

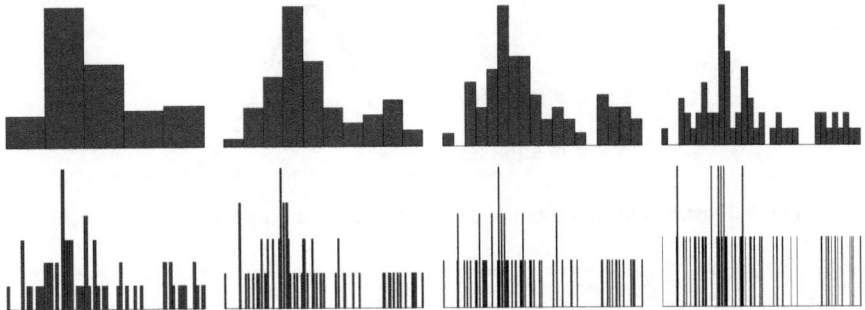

Fig. 5. Histograms of the 64 samples introduced in figure 1 with different bin widths.

a hard-edged bin containing m samples out of a total of n. If we bootstrap the data, the number of samples that we will see in the bin will be binomially distributed with mean m and standard deviation $\sqrt{m(n-m)}$. Our idea is to apply this same analysis to fuzzy Gaussian bins/kernels rather than hard-edged bins.

We continue the formalism introduced in the previous section, but now consider a fixed Gaussian kernel centred at $\vec{x}$ and of scale s. On the original data, this bin 'sees' $w(\vec{x},s)$ samples. Now consider a single point $\vec{q}$ picked (during bootstrap) uniformly from the $\vec{p}_i$. This random process induces a random bin-count process w_1. The mean and variance of this process will be:

$$\mu\left[w_1\left(\vec{x},s\right)\right]=\frac{1}{n}\sum_{i=1}^{n}e^{-\frac{\|\vec{p}_i-\vec{x}\|^2}{4s}}=\frac{1}{n}w\left(\vec{x},s\right)$$

$$V\left[w_1\left(\vec{x},s\right)\right]=\frac{1}{n}\sum_{i=1}^{n}\left(e^{-\frac{\|\vec{p}_i-\vec{x}\|^2}{4s}}\right)^2-\left(\mu\left[w_1\left(\vec{x},s\right)\right]\right)^2=\frac{1}{n}w\left(\vec{x},\tfrac{s}{2}\right)-\frac{1}{n^2}w\left(\vec{x},s\right)^2$$

Now consider n points $\vec{q}_1,\dots,\vec{q}_n$ generated by a full bootstrap process from the $\vec{p}_i$. Because each $\vec{q}_i$ is independent and bootstrap is sampling with replacement, this will induce a random bin count process w_n with mean and variance:

$$\mu\left[w_n\left(\vec{x},s\right)\right]=n\mu\left[w_1\left(\vec{x},s\right)\right]=w\left(\vec{x},s\right)$$

$$V\left[w_n\left(\vec{x},s\right)\right]=nV\left[w_1\left(\vec{x},s\right)\right]=w\left(\vec{x},\tfrac{s}{2}\right)-\tfrac{1}{n}w\left(\vec{x},s\right)^2$$

Note how the variance formula is a function of the bin weights at scale s and $\tfrac{s}{2}$. We illustrate this in figure 6 where the black kernel is of scale s and the grey kernel of scale $\tfrac{s}{2}$. This makes clear that the bootstrap bin-count process for Gaussian bins, by depending on two measurements of the data, is more geometrically sensitive than it is for hard-edged histogram bins, which depends on only one measurement. We also note that it is the nice property of the Gaussian that when squared it gives another Gaussian, which leads to this simple formula for the variance.

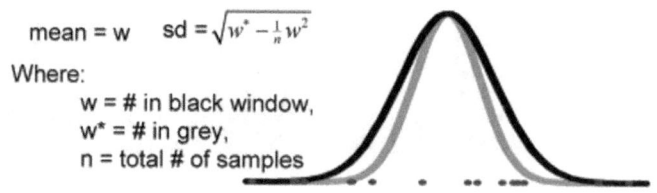

$$\text{mean} = w \qquad \text{sd} = \sqrt{w^* - \tfrac{1}{n}w^2}$$

Where:

w = # in black window,
w* = # in grey,
n = total # of samples

Fig. 6. Illustrates the bin count of the black aperture during a bootstrap process.

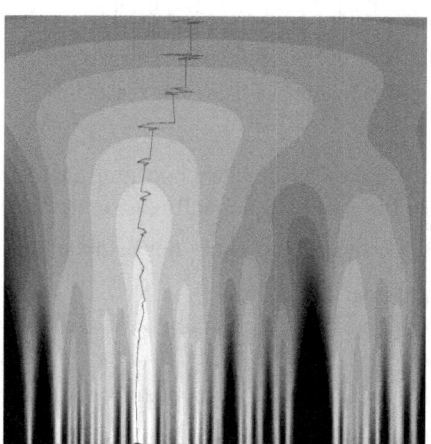
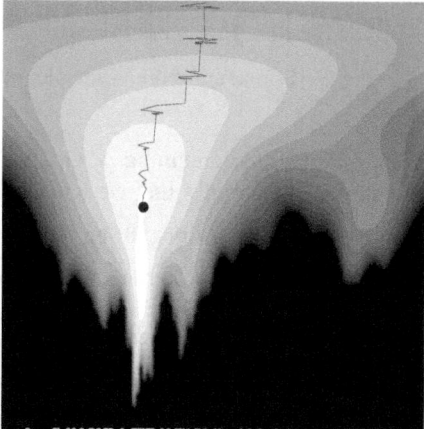

Fig. 7. Shows the standard Scale Space of the samples at the left (same as figure 3) and the 'Pessimistic Scale Space' on the right, where 2.23 sds of pessimism have been used. The jagged polylines show tracking in these Scale Spaces, and the dark discs mark the resulting mode estimates.

At this point in the development of our method we take a step that though plausible has no rigorous justification. To prevent tracking into bins that have high density but small numbers, we reduce the number seen by a bin by some number of standard deviations, where the standard deviation comes from the preceding bootstrap analysis. The 'logic' here is that if we reduce by (say) 1.645 sds we get approximately the bin count that would be exceeded on 95% of bootstrap replications. The 'approximately' arises because we make the assumption that the bin counts would be normally distributed across bootstrap replications. Figure 7 shows how the Scale Space changes with 2.23 sds of downgrading. We call such a downgraded Scale Space a 'Pessimistic Scale Space' as it is in a sense sceptical about the number of samples in a bin.

We are able to track in Pessimistic Scale Space in a very similar way to standard Scale Space. The following changes should be noted:

(1) rather use $d(\vec{x},s) = w(\vec{x},s)v(s)^{-1}$, we use

$$\breve{d}(\vec{x},s) = \left\lfloor w(\vec{x},s) - k\sqrt{w(\vec{x},\sfrac{s}{2}) - \tfrac{1}{n}w(\vec{x},s)^2} \right\rfloor v(s)^{-1}, \text{ where } k \text{ is the number of}$$

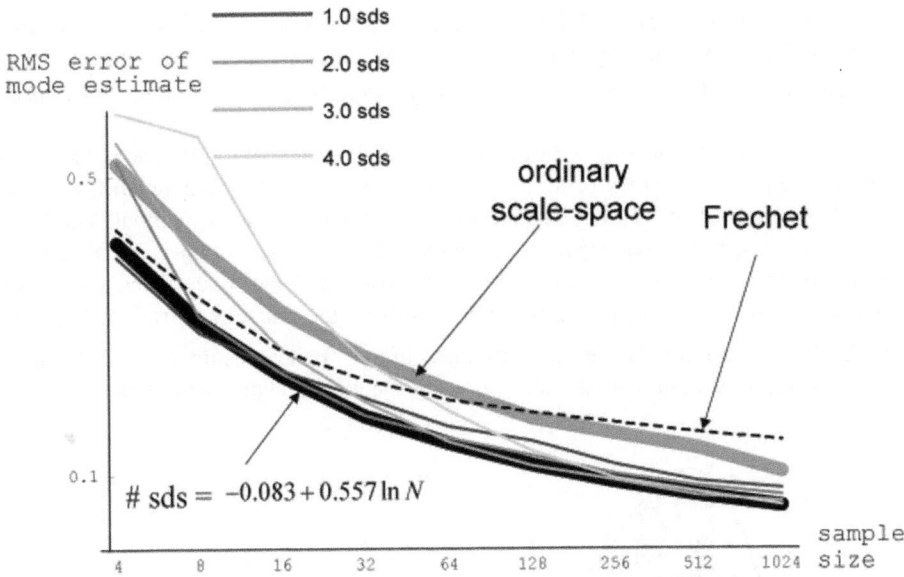

Fig. 8. Compares all three mode-estimation techniques presented. The coloured curves are from PSST with a fixed number of sds of pessimism. The black curve, which is approximately their envelope, is with a scheme where the number of sds depends on the sample size.

sds of pessimism, and the operator $\lfloor \ \rfloor$ returns the argument unchanged if it is positive and 0 otherwise

(2) similarly to before, we actually ascend $\ln\left(1+\breve{d}\right)$ rather than $\breve{d}$

(3) explicitly computed gradients are still used but of $\ln\left(1+\breve{d}\right)$ rather than $\ln\left(d\right)$. The computation of these is more complicated but not excessively.

(4) We stop the process when tracking is no longer decreasing the scale i.e. when we reach a maximum relative to $\bar{x}$ and s.

We have evaluated PSST using the methodology we used to compare the Fréchet and standard SST approaches. The results are shown in figure 8. Because we are unclear about how many sds of pessimism (k in the equation for $\breve{d}$ above) to use, we have tried $k=1, 2, 3$ and 4. What the figure shows is that the optimal k increases with the size of sample. For samples of $n \leq 12$ $k=1$ is best, for $12 < n \leq 50$ $k=2$ is best, for $50 < n \leq 800$ $k=3$ and for $n > 800$ $k=4$. Using this result and others not shown be have fitted a function to give the optimal k for n samples. It is $k_{opt} = -0.083 + 0.557 \ln n$. Figure 8 shows the accuracy of mode estimation using this scheme and it can be seen that it is certainly always better than ordinary Scale Space or Fréchet tracking, and generally better than any of the pessimistic schemes with fixed k. We will return to this point in the conclusion, but we note now that fitting of our k_{opt} formula has been done for a single 1-D distribution so it may not be general.

Nevertheless it is the method that we will use for the high-dimensional mode estimation problems presented in sections 5 and 6. We also note that the value $k = 2.23$ used to generate figure 7 is the optimal value (according to our formula) for 64 samples.

We conclude this section with figure 9, which is designed to give the reader a visual handle on the accuracy of mode estimation by PSST. We have generated 10^4 sets of 1024 samples from the log-normal distribution that we use throughout. The figure shows a histogram of one such set of samples. For each set we have calculated estimates of the mean and median in the usual way, and the mode using pessimistic tracking with 3.78 sds of pessimism which is the value that comes from our k_{opt} formula. We have found the 95% confidence limits of these estimates and displayed them in figure 9, along with the true values of the mean, median and mode.

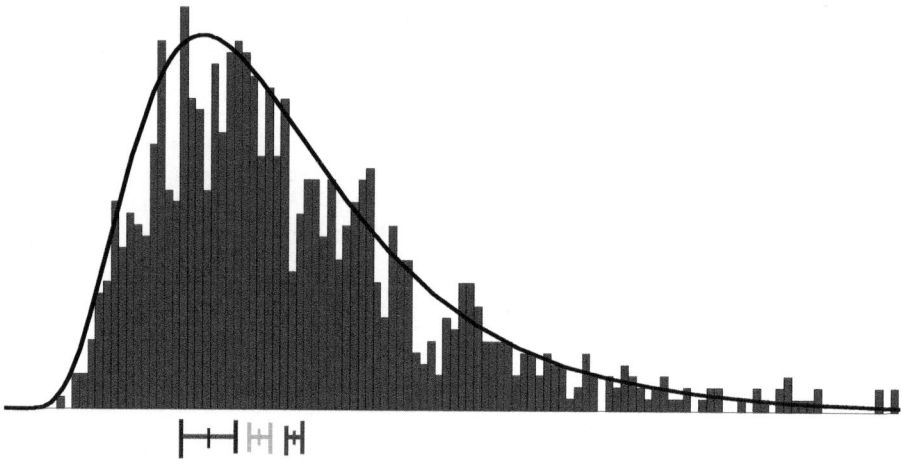

Fig. 9. A **typical** histogram of 1024 samples from the distribution shown by the black curve. Below this are 95% confidence intervals for mean estimation (right), median (middle) and mode (left) based on this number of samples i.e. not for this particular set of samples but for sets of samples of this size. The central ticks of each interval mark the true values. The mode estimation method used is pessimistic tracking with 3.78 sds of pessimism.

5   Results on Handwriting Data

Our first example of high dimensional mode estimation uses handwriting data (Alimoglu & Alpaydin, 1996). The data consists of poly-line representations of the digits 0-9, captured using a stylus, writing within a box on a pressure sensitive tablet. For each digit there are 1100 instances written by 44 subjects (i.e. 25 per subject). The data was captured at high temporal resolution but has been sub-sampled down to 8 points equally spaced along the stylus tip trajectory, so each sample is a point in $\mathbb{R}^{16}$

Estimating the mode by building hard-binned histograms is out of the question for such data, as even if we coarsely quantized each dimension into 8 ranges, the histogram would have 2^{48} bins. Even if one adopted some efficient coding scheme to avoid memory problems, this approach could not succeed, as the data is so sparse: consider that $1.54913^{16} = 1100$.

We have used the PSST method described in section 4, to estimate modes for each of the digits. For this we used $k=3.82$ sds of pessimism as suggested by our fitted formula $k_{opt} = -0.083 + 0.557 \ln n$. Each mode estimation computation took less than 1sec on a 1.1GHz PC. We show results in figure 10, along with a depiction of the raw data. We also show for comparison the mean digits (computed by straightforward averaging) and the dimension-wise-modal digits. The dimension-modal digits are computed by calculating the mode, using PSST, of each of the 16 dimensions separately.

Several points are notable about these results:

1. Several of the mean digits (particular 0, 5 & 8) are smaller than the raw digits; this is because of the sensitivity of the mean to outliers.
2. Most of the dimension-wise modal digits have five or more of their eight points on the border of the square; this is because each digits has been translated and scaled so that it has two of its points on the border.
3. By informal inspection, for all ten digits, the modal digit is either the best example of the digit or equal best with the mean or dimension-wise modal.

While we have no way of knowing what the true modes of these data are, our estimates are certainly plausible.

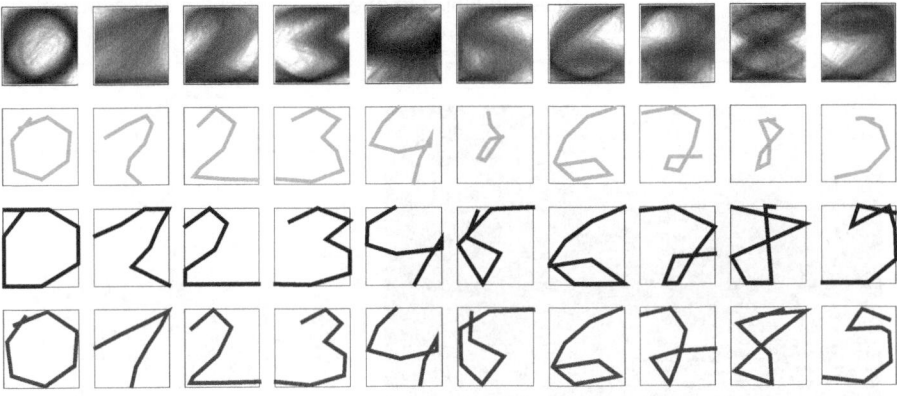

Fig. 10. On the top row the 1100 poly-lines for each digit have been combined. The 2nd row shows the mean digits, the 3rd row the dimension-wise-modal digits and the 4th row shows the estimated modal digits.

6   Results on Image Profiles

Our final example involves finding the mode of a distribution over the space of 1-D functions. We represent these functions with 64 samples so that this is a problem of finding the mode of a distribution over $\mathbb{R}^{64}$. We are interested in this problem as it is part of a program of research aimed at discovering feature classes by investigating natural image statistics (Tagliati & Griffin, 2001).

We will consider 1-D profiles extracted at random position and orientation from 3 classes of images – Gaussian noise, Brownian noise (Pedersen, submitted) and natural images (van Hateren & van der Schaaf, 1998) – examples of which are shown in figure 11. The figure also shows how we process these samples: we measure them with 0th and 1st order Gaussians ($\sigma = 7$) and then affinely scale them so that they then measure 0 and 1 with these two filters i.e. we bring them into the same metamery class (Koenderink, 1993) for these two filters. Although our primary interest is in the natural images we also use the noise images since we can prove what the mode is in these two cases: for Gaussian images the mode is a Gaussian first derivative, and for Brownian images the mode is an error function.

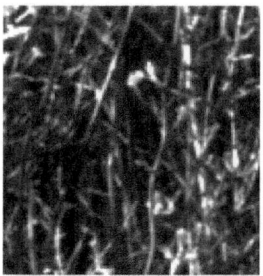  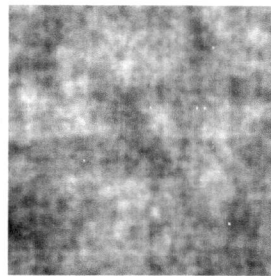

Gaussian Noise Images   Brownian Noise Images

 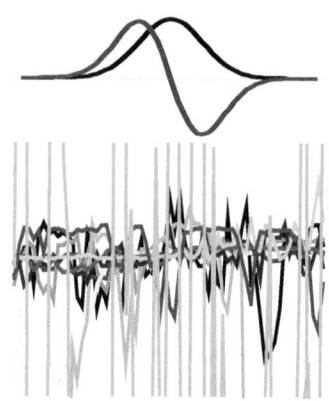

Fig. 11. Shows at top examples from the three classes of image that we use in the experiment of section 6. The panel at bottom left shows how we extract profiles from these images, and at right the filters that we measure profiles with and how 10 typical profiles look after scaling to bring them into the same metamery class for these filters.

We collect 2.4×10^6 of each class of scaled profiles (each of which is represented as a 64-dimensional vector) and use PSST to estimate the mode of each class. Each mode-estimation takes approximately 40mins on a 1.1GHz PC. We repeat this four times (with new profiles each time) for each class of images so that we can calculate confidence intervals on our estimated modes. The results are shown in figure 12. The figure shows that within the limits of the confidence intervals the estimated modes of the two classes of noise image are correct i.e. that the mode of the Gaussian image profiles is a Gaussian 1st derivative and the mode of the Brownian image profiles is an error function. The mode of the natural image profiles is close to being a step edge; the correctness of the mode estimates of the noise images makes us confident of the correctness of this result.

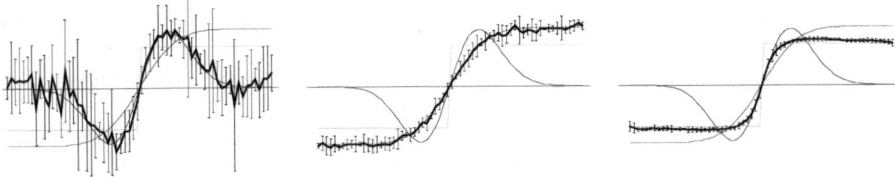

Fig. 12. Estimated modes (solid with error bars) based on 2.4×10^6 samples. Four independent estimates of each mode were calculated to derive the confidence intervals. Also shown in each plot is (i) the true mode of the Gaussian noise images – a Gaussian 1st derivative of the same scale as the filters that define the metamery class, (ii) the true mode of the Brownian noise images – the integral of a Gaussian of the same scale as the filters that define the metamery class, and (iii) a centred step edge within the metamery class – the mode of natural image profiles is similar to this curve.

7   Conclusions

We have presented three methods of mode estimation, but recommend the third – *Pessimistic Scale Space Tracking* – as the most effective. We have quantified its performance on a test 1-D log-normal distribution which will allow ready comparison with any methods proposed in the future. We have also demonstrated with the Gaussian and Brownian noise image profiles in section 6 that it recovers even high-dimensional modes correctly.

We are aware of two weak points in the derivation of our algorithm. The first is that there should be a better explanation of why pessimistic discounting should work at all. The second concerns our fitted equation that provides the optimum number of sds of pessimism to use as a function of the sample size. This equation is unjustified and based only on experiments with 1-D distributions and sample sizes up to 2^{10}. However the fact that the method is successful on 64-D distributions with very large sample sizes (2.4×10^6) suggests that it is not too bad a guess.

References

Alimoglu, F., & Alpaydin, E. (1996). Methods of combining multiple classifiers based on different representations for pen-based handwriting recognition. *5th Turkish AI & ANN Symposium* (Istanbul.

Blake, A., & Zisserman, A. (1987). Visual Reconstruction. (MIT Press.

Efron, B., & Tibshirani, R. (1993). An Introduction to the Bootstrap. (New York, London: Chapman and Hall.

Everitt, B.S., & Hand, D.J. (1981). Finite Mixture Distributions. (London: Chapman Hall.

Fréchet, M. (1948). Les elements aléatoires de nature quelconque dans un espace distancié. *Annales de l'Institut Henri Poincaré, X*, 215-308.

Fukunga, K. (1990). Statistical Pattern Recognition. (New York: Academic Press.

Gasser, T., Hall, P., & Presnell, B. (1998). Nonparametric estimation of the mode of a distribution of random curves. *Journal of the Royal Statistical Society Series B-Statistical Methodology, 60*, 681-691.

Griffin, L.D. (1995). Descriptions of Image Structure. (London: PhD thesis, University of London.

Griffin, L.D. (1997). Scale-imprecision space. *Image and Vision Computing, 15* (5), 369-398.

Hazelton, M.L. (1999). An optimal local bandwidth selector for kernel density estimation. *Journal of Statistical Planning and Inference, 77* (1), 37-50.

Katkovnik, V., & Shmulevich, I. (2002). Kernel density estimation with adaptive varying window size. *Pattern Recognition Letters, 23* (14), 1641-1648.

Koenderink, J.J. (1993). What is a feature? *Journal of Intelligent Systems, 3* (1), 49-82.

Marchette, D.J., & Wegman, E.J. (1997). The filtered mode tree. *Journal of Computational and Graphical Statistics, 6* (2), 143-159.

Minnotte, M.C., Marchette, D.J., & Wegman, E.J. (1998). The bumpy road to the mode forest. *Journal of Computational and Graphical Statistics, 7* (2), 239-251.

Minonotte, M.C., & Scott, D.W. (1993). The Mode Tree: A Tool for visualization of nonparametric density features. *J. Comp. & Graph. Statist., 2*, 51-68.

Parzen, E. (1962). On estimation of probability density function and mode. *Annals of Mathematical Statistics, 33*, 520-531.

Pedersen, K.S. (submitted). Properties of Brownian Image Models in Scale-Space. In: L.D. Griffin (Ed.) *Proc. Scale Space 2003* (Springer.

Press, W.H., Teukolsky, S.A., Vetterling, W.T., & Flannery, B.P. (2002). Numerical Recipes in C++. (Cambridge: Cambridge University Press.

Scott, D.W., Tapia, R.A., & Thompson, J.R. (1977). Kernel density estimation revisited. *Nonlinear Analysis, Theory, Methof & Application, 1*, 339-372.

Silverman, B.W. (1981). Using Kernel density Estimates to investigate Multimodality. *J. Roy. Statist. Soc. B, 43*, 97-99.

Silverman, B.W. (1986). Density Estimation for Statistics and Data Analysis. (London: Chapman Hall.

Tagliati, E., & Griffin, L.D. (2001). Features in Scale Space: Progress on the 2D 2nd Order Jet. In: M. Kerckhove (Ed.) *LNCS*, 2106 (pp. 51-62): Springer.

van Hateren, J.H., & van der Schaaf, A. (1998). Independent component filters of natural images compared with simple cells in primary visual cortex. *Proceedings of the Royal Society of London Series B-Biological Sciences, 265* (1394), 359-366.

Wand, M.P., & Jones, M.C. (1995). Kernel Smoothing. (London: Chapman Hall.

Properties of Brownian Image Models
in Scale-Space

Kim S. Pedersen

DIKU, University of Copenhagen
Universitetsparken 1
DK-2100 Copenhagen, Denmark

Abstract. In this paper it is argued that the Brownian image model is
the least committed, scale invariant, statistical image model which describes the second order statistics of natural images. Various properties
of three different types of Gaussian image models (white noise, Brownian
and fractional Brownian images) will be discussed in relation to linear
scale-space theory, and it will be shown empirically that the second order statistics of natural images mapped into jet space may, within some
scale interval, be modeled by the Brownian image model. This is consistent with the $1/f^2$ power spectrum law that apparently governs natural
images. Furthermore, the distribution of Brownian images mapped into
jet space is Gaussian and an analytical expression can be derived for
the covariance matrix of Brownian images in jet space. This matrix is
also a good approximation of the covariance matrix of natural images in
jet space. The consequence of these results is that the Brownian image
model can be used as a least committed model of the covariance structure
of the distribution of natural images.

1 Introduction

Bayesian statistical methods are becoming increasingly more popular in the fields
of image analysis and computer vision. In order to apply such methods one need
to introduce probability models of images. Such models might be derived by
empirical studies of the class of images under concern. An image model should
at least not contradict empirical findings. Several authors have introduced probability models of images, e.g. [2,7,17,18], where the goal is to model the generic
statistics of natural images. That is, to model the statistical properties that are
common among images of naturally occurring scenes. Such models have been
used as priors in the solution of various image analysis problems, e.g. Nielsen
and Lillholm [19] have used various simple probability models, such as the white
noise and Brownian models, in conjunction with image reconstruction. Pedersen and Nielsen [22] have used the fractional Brownian model as a basis for an
analysis of scale normalization of differential invariants.

This paper discusses Gaussian image models, with a focus on their properties
in linear scale-space. The class of Gaussian image models, to which the white
noise, Brownian and fractional Brownian models belong, is interesting because

L.D. Griffin and M. Lillholm (Eds.): Scale-Space 2003, LNCS 2695, pp. 281–296, 2003.

of its fundamental character and simplicity. E.g. the Brownian image model is a scale invariant stationary Gaussian stochastic function (or field), and as such it is the maximum entropy solution constrained only by the mean and covariance structure. In this paper the statistical properties of these image models and what properties they have in common with natural images will be discussed. Especially, it will be shown through an empirical study that the Brownian image model captures the second order statistics of natural images mapped into the so-called jet space. Natural images have been shown empirically to have a scale invariant covariance structure through the $1/f^2$ law of their ensemble power spectrum [3], and the empirical findings of this paper are consistent with this law. Empirical findings also show that higher order natural image statistics are scale invariant, e.g. [21].

Images of natural scenes have complex dependencies among pixels due to local geometric structures, such as edges, which is not modeled by Brownian images. Nevertheless, it is the opinion of the author that the Brownian image model can be useful as a simple probabilistic model which might serve as a basis for other models capturing the higher order statistics of natural images.

The work presented in this paper has some similarities with the work by Blom ct al. [1] on the second order statistics of scale-space derivatives of spatially correlated noise. Properties of Gaussian white noise in scale-space has also been studied by Majer [14].

In the continuous setting, an image $f(\boldsymbol{x}) : \mathbb{R}^2 \to \mathbb{R}$ can be viewed as a realization of a continuous stochastic function on the plane. In this case, a statistical image model is given by a probability distribution on an infinite dimensional vector space of functions (see e.g. [12]). In this paper only gray value images are discussed and it will be assumed for simplicity that images may take values in the full range of $\mathbb{R}$, ignoring that in a digital computer images are represented as discrete arrays of pixels and pixel values are represented by a discrete set of gray values in some interval. The discussion of this paper also holds in the discrete case.

An image is a set of measurements of the light that goes through the camera optics and hits the image plane. Each number in the image array quantifies the amount of light that hit the corresponding spot on the photographic film or the CCD chip in your digital camera. The light that hits this spot does not come from a single point in the physical world, but is a weighted integral of light originating ideally from every point in the physical world. The optics of the camera has a certain response function that dictates the weighting of the light. Therefore, an image should not be considered as a function in the normal sense of the word, but as an entity constructed through a set of response or test functions. Florack [4] and Mumford et al. [18] have argued independently for this view, and both propose that images should not be considered as functions, but rather as generalized functions (see e.g. [6]).

A *generalized function* on a subset of the plane $\Omega \subset \mathbb{R}^2$ is a linear functional $F[\phi] : \mathcal{D}(\Omega) \to \mathbb{R}$ that maps test functions $\phi(\boldsymbol{x}) \in \mathcal{D}(\Omega) \subset C_c^\infty(\Omega)$ onto $\mathbb{R}$, where $\mathcal{D}(\Omega)$ is a vector space of C^∞ functions on Ω with compact support.

Let the vector space of generalized functions on Ω be denoted as $\mathcal{D}'(\Omega)$. If we substitute the requirement of compact support for $\phi(\boldsymbol{x}) \in \mathcal{D}(\Omega)$ with the requirement that $\phi(\boldsymbol{x})$ must be rapidly decreasing[1] we get the so-called tempered generalized functions. Let $\mathcal{S}(\Omega)$ denote the vector space of test functions $\phi(\boldsymbol{x})$ that are rapidly decreasing. Linear functionals on $\mathcal{S}(\Omega)$ are called *tempered generalized functions* and the set of these functions is denoted as $\mathcal{S}'(\Omega)$. Notice that $\mathcal{D}(\Omega) \subset \mathcal{S}(\Omega) \subset \mathcal{S}'(\Omega) \subset \mathcal{D}'(\Omega)$.

I assume that naturally occurring images belong to the space of tempered generalized functions $\mathcal{S}'(\Omega)$, $\Omega \subset \mathbb{R}^2$. I therefore seek a probability distribution or rather a probability measure on $\mathcal{S}'(\Omega)$.

Florack [4] has pointed out the connection between linear scale-space theory [10] and tempered generalized functions. In scale-space theory the test functions are the scale-space aperture function $\psi(\boldsymbol{x}; s)$ and its partial derivatives $\partial_{x^n y^m} \psi(\boldsymbol{x}; s)$, which all decrease rapidly and therefore belong to $\mathcal{S}(\Omega)$. The linear scale-space aperture function is

$$\psi(\boldsymbol{x}; s) \equiv \frac{1}{2\pi s^2} \exp\left(-\frac{(x^2 + y^2)}{2s^2}\right)$$

where $\boldsymbol{x} = (x, y)^T$ and $s \geq 0$ denotes the scale parameter. The scale-space of an image $f(\boldsymbol{x})$, which is assumed to be a locally L_1-function[2], is a tempered generalized function given by

$$L(\boldsymbol{x}; s) \equiv \int_\Omega f(\boldsymbol{x}')\psi(\boldsymbol{x} - \boldsymbol{x}'; s)\, d\boldsymbol{x}' = f(\boldsymbol{x}) * \psi(\boldsymbol{x}; s)$$

with $L(\boldsymbol{x}; s = 0) \equiv f(\boldsymbol{x})$. That is, $L(\cdot; s)$ belongs to the space of tempered generalized functions $\mathcal{S}'(\Omega)$.

By introducing generalized functions the concept of differentiation of otherwise non-differentiable functions is made sensible (see e.g. [6]). In stochastic analysis the concept of generalized functions is often used for solving the problem of differentiability of realizations of stochastic functions (see e.g. [16]). By using the linear scale-space representation we can define derivatives of images $f(\boldsymbol{x})$ by applying the differential operator $\partial_{x^n y^m}$ to the test function $\psi(\boldsymbol{x}; s)$

$$L_{x^n y^m}(\boldsymbol{x}; s) = f(\boldsymbol{x}) * \partial_{x^n y^m} \psi(\boldsymbol{x}; s) \ .$$

The scale-space partial derivative $L_{x^n y^m}(\boldsymbol{x}; s)$ is a tempered generalized function belonging to $\mathcal{S}'(\Omega)$.

The layout of the paper is as follows: Sec. 2 starts out by discussing the Gaussian white noise image model, before going on to the Brownian image model and its generalization — the fractional Brownian image model — in Sec. 3. In Sec. 4 empirical results are shown that indicates that Brownian images model the covariance structure of natural images. Finally, Sec. 5 gives some concluding remarks.

---

[1] That is, all derivatives must decrease faster than polynomials.
[2] A function $f(\boldsymbol{x})$ is a locally L_1-function if $\int_K |f|\, d\boldsymbol{x} < \infty$ for all compact sets $K \subset \Omega$.

2   The Gaussian Model

In order to define the continuous Gaussian image model[3], one has to define a Gaussian probability measure on an infinite dimensional vector space of functions. Here I am interested in the space of tempered generalized functions $S'(\mathbb{R}^2)$. The distributions of stochastic vectors $(f_1, \ldots, f_N)$ obtained by sampling a stochastic function $f(\boldsymbol{x}) \in S'(\mathbb{R}^2)$ at different sets of $\boldsymbol{x}_1, \ldots, \boldsymbol{x}_N \in \mathbb{R}^2$ are called the *finite distributions* of the stochastic function $f(\boldsymbol{x})$. A *stochastic function* $f(\boldsymbol{x}) \in S'(\mathbb{R}^2)$ *is said to be Gaussian if all its finite distributions are Gaussian distributed stochastic vectors.* That is, if the stochastic vector $\boldsymbol{F} = (f(\boldsymbol{x}_1), \ldots, f(\boldsymbol{x}_N))^T$ has a probability density of the form

$$p(\boldsymbol{F}) = \frac{1}{Z} \exp\left(-(\boldsymbol{F} - \overline{\boldsymbol{F}})^T \mathbf{C}^{-1} (\boldsymbol{F} - \overline{\boldsymbol{F}})\right)$$

for all sets of $\boldsymbol{x}_1, \ldots, \boldsymbol{x}_N \in \mathbb{R}^2$ with a mean vector $\overline{\boldsymbol{F}}$ and covariance matrix $\mathbf{C}$. The normalization factor Z ensures that $\int p(\boldsymbol{F}) \, d\boldsymbol{x} = 1$. In general, all linear functionals of a Gaussian distributed stochastic function are also Gaussian distributed. Specifically, generalized functions $F[\phi] = \langle f, \phi \rangle = \int f(x)\phi(x) \, dx$ (for any test functions $\phi(x) \in S(\mathbb{R}^2)$) of a Gaussian distributed stochastic function $f(x)$ are also Gaussian distributed stochastic functions. *Hence the scale-space and scale-space derivatives of a Gaussian stochastic function are themselves Gaussian distributed stochastic functions.*

The *mean function* $\bar{f} \in S'(\mathbb{R}^2)$ of a probability measure p on $S'(\mathbb{R}^2)$ is defined such that the expectation $E[\cdot]$ of all continuous linear functionals $h : S'(\mathbb{R}^2) \to \mathbb{R}$ is

$$E[h] = h(\bar{f}) \ .$$

The function $C(\boldsymbol{x}, \boldsymbol{y})$ is the *covariance function* of the probability measure p on $S'(\mathbb{R}^2)$, if for all continuous linear functionals $h_1(f) = \int f(\boldsymbol{x})\phi_1(\boldsymbol{x}) \, d\boldsymbol{x}$ and $h_2(f) = \int f(\boldsymbol{x})\phi_2(\boldsymbol{x}) \, d\boldsymbol{x}$, we have

$$E[(h_1 - h_1(\bar{f})) \cdot (h_2 - h_2(\bar{f}))] = \iint \phi_1(\boldsymbol{x})\phi_2(\boldsymbol{y})C(\boldsymbol{x}, \boldsymbol{y}) \, d\boldsymbol{x} \, d\boldsymbol{y} \ .$$

A Gaussian probability measure is fully parameterized by its mean function $\bar{f}$ and covariance function $C(\boldsymbol{x}, \boldsymbol{y})$ [12].

Gaussian white noise on $S'(\mathbb{R}^2)$ is defined as a Gaussian stochastic function $\nu \in S'(\mathbb{R}^2)$ with zero mean $\overline{\nu}(\boldsymbol{x}) = 0$ and covariance given by

$$E[\langle \nu, \phi_1 \rangle \langle \nu, \phi_2 \rangle] = \sigma^2 \int \phi_1(\boldsymbol{x})\phi_2(\boldsymbol{x}) \, d\boldsymbol{x} \ .$$

See Fig. 1 (Top left) for a realization of the Gaussian white noise model. This is a generalization of the discrete independently identically distributed (i.i.d.) Gaussian image model where each pixel value is an independent Gaussian stochastic

---

[3] Lifshits [12] provides some insight on Gaussian stochastic functions in general.

variable, since if ϕ_1 and ϕ_2 have disjoint support, we have $E[\langle\nu,\phi_1\rangle\langle\nu,\phi_2\rangle] = 0$ and any disjoint open sets are independently distributed. In linear scale-space the Gaussian white noise $W(\boldsymbol{x};s) = \nu(\boldsymbol{x}) * \psi(\boldsymbol{x};s)$ has the covariance

$$C_W(\boldsymbol{x},\boldsymbol{y};s) = E[W(\boldsymbol{x};s)W(\boldsymbol{y};s)] = \frac{\sigma^2}{4\pi s^2}\exp\left(-\frac{(\boldsymbol{x}-\boldsymbol{y})^2}{4s^2}\right) . \tag{1}$$

Thus scale-space smoothing changes the covariance of white noise and introduces spatial dependencies. Remember that by definition $W(\boldsymbol{x};s=0) \equiv \nu(\boldsymbol{x})$, hence for scale $s = 0$ the spatial covariance of $W(\boldsymbol{x};s)$ reduces to $E[\nu(\boldsymbol{x})\nu(\boldsymbol{y})] = \sigma^2\delta(\boldsymbol{x}-\boldsymbol{y})$, where $\delta(\cdot)$ is the Dirac delta function[4].

A Gaussian stochastic function $f(\boldsymbol{x})$ is stationary if it has the same mean value $E[f(\boldsymbol{x})] = a \in \mathbb{R}$ for all $\boldsymbol{x} \in \mathbb{R}^2$, and its covariance $E[(f(\boldsymbol{x}) - E[f(\boldsymbol{x})]) \cdot (f(\boldsymbol{y}) - E[f(\boldsymbol{y})])]$ depends only on the separation $\boldsymbol{x} - \boldsymbol{y}$ for all $\boldsymbol{x},\boldsymbol{y} \in \mathbb{R}^2$, i.e. it does not depend on position. The covariance function of Gaussian white noise $\nu(\boldsymbol{x})$ may be written as $C(\boldsymbol{x}-\boldsymbol{y}) = \sigma^2\delta(\boldsymbol{x}-\boldsymbol{y})$ and its mean is $\overline{\nu}(\boldsymbol{x}) = 0$, hence white noise is stationary. Gaussian white noise in scale-space $W(\boldsymbol{x};s)$ defined by the covariance function of Eq. (1) is clearly also a stationary stochastic function.

Sample functions of Gaussian white noise $\nu(\boldsymbol{x})$ are not differentiable. The power spectrum[5] of a zero mean white noise image is flat, i.e. constant,

$$S(\boldsymbol{\xi}) = \int C(\boldsymbol{x})e^{-i\boldsymbol{\xi}\cdot\boldsymbol{x}}\,d\boldsymbol{x} = \sigma^2\int\delta(\boldsymbol{x})e^{-i\boldsymbol{\xi}\cdot\boldsymbol{x}}\,d\boldsymbol{x} = \sigma^2 .$$

This means that realizations of white noise have a uniform energy distribution across all frequencies and have energy in infinite frequencies, which implies that realizations of the white noise image model are not differentiable. Nevertheless, by introducing the concept of generalized functions and linear scale-space we may define derivatives of realizations of the white noise image model in a sensible way. By using the linear scale-space representation one can compute partial derivatives of $\nu(\boldsymbol{x})$ by convolution with partial derivatives of the scale-space aperture function $\psi_{x^n y^m}(\boldsymbol{x};s)$. All linear functionals of white noise $\nu(\boldsymbol{x})$ must be Gaussian, hence all partial scale-space derivatives $W_{x^n y^m}(\boldsymbol{x};s)$ must be zero mean Gaussian stochastic functions. This follows immediately from the definition of the Gaussian distribution on the space of tempered generalized functions $\mathcal{S}'(\mathbb{R}^2)$.

Any directional derivative $\partial W(\boldsymbol{x};s)/\partial\boldsymbol{m}\big|_{P_0}$ in the direction $\boldsymbol{m}$, $\|\boldsymbol{m}\| = 1$, at the point P_0 is a Gaussian stochastic variable, since $\partial W(\boldsymbol{x};s)/\partial\boldsymbol{m}\big|_{P_0}$ is linear in the two Gaussian variables $W_x = W_x(P_0;s)$ and $W_y = W_y(P_0;s)$

$$\left.\frac{\partial W(\boldsymbol{x};s)}{\partial\boldsymbol{m}}\right|_{(x,y)=P_0} = W_x(P_0;s)\cos(\theta) + W_y(P_0;s)\sin(\theta) \tag{2}$$

---

[4] The Dirac delta function $\delta(x)$ is a generalized function defined such that $\int\delta(x)\,dx = 1$ and $\langle\delta(x),\phi(x)\rangle = \phi(0)$. It may be used as the limit, $\lim_{s\to 0}\psi(\boldsymbol{x};s) \equiv \delta(\boldsymbol{x})$.

[5] The power spectrum of a stationary stochastic function is usually defined as the Fourier transform of the autocorrelation function. In the case of zero mean stochastic functions the autocorrelation reduces to the spatial covariance.

where θ is the angle between the positive x-axis and the directional vector $\boldsymbol{m}$. The distribution of directional derivatives of a Gaussian function is Gaussian and stationary. From Eq. (2) one can also conclude that W_x and W_y are independently distributed, since when $\boldsymbol{m}$ is parallel to the x-axis we have $\theta = 0$ and $\partial W(\boldsymbol{x};s)/\partial \boldsymbol{m}\big|_{P_0} = W_x(P_0;s)$, similarly for the y-axis case. This can also be seen in Eq. (5). Furthermore, $p(W_x)$ and $p(W_y)$ are identically Gaussian distributed. This implies that the joint distribution of (W_x, W_y) is an isotropic Gaussian distribution $p(W_x, W_y)$ with density $p(W_x, W_y) = \frac{1}{2\pi\sigma^2}\exp\left(-\frac{1}{2\sigma^2}(W_x^2 + W_y^2)\right)$.

The gradient magnitude $\|\nabla W\|$ of an image sampled from a Gaussian stochastic function must be Rayleigh distributed. To see this, let us compute the probability of the gradient magnitude $\|\nabla W\|$ having a value in the annulus in the (W_x, W_y)-plane given by the inner r_1 and outer r_2 radii. Since the joint distribution of (W_x, W_y) at the point P_0 is an isotropic Gaussian, we can make a change of variables to polar coordinates $(r = \|\nabla W\|, \theta)$ and write $p(W_x, W_y)$ as $p(r,\theta) = \frac{1}{2\pi\sigma^2}\exp\left(-\frac{r^2}{2\sigma^2}\right)$. The probability of a value in the annulus is therefore

$$P(\|\nabla W\| \in [r_1; r_2]) = \int_{r_1}^{r_2}\int_0^{2\pi} p(r,\theta)r\,d\theta\,dr = \int_{r_1}^{r_2} \frac{r}{\sigma^2}\exp\left(-\frac{r^2}{2\sigma^2}\right)dr\ .$$

Thus the probability density of the gradient magnitude is

$$p(\|\nabla W\|) = \frac{\|\nabla W\|}{\sigma^2}\exp\left(-\frac{\|\nabla W\|^2}{2\sigma^2}\right)\ , \tag{3}$$

which is exactly the density function of the Rayleigh distribution.

Differentiation of stochastic functions introduces spatial dependencies. For partial derivatives of Gaussian white noise in scale-space we may write the spatial covariance function as

$$E[W_{x^n y^m}(\boldsymbol{x}_1;s)W_{x^n y^m}(\boldsymbol{x}_2;s)]$$

$$= \sigma^2\int_{\mathrm{I\!R}^2} \partial_{x^n y^m}\psi(\boldsymbol{x}_0;s)\Big|_{\boldsymbol{x}_0=\boldsymbol{x}'-\boldsymbol{x}_1}\ \partial_{x^n y^m}\psi(\boldsymbol{x}_0;s)\Big|_{\boldsymbol{x}_0=\boldsymbol{x}'-\boldsymbol{x}_2}\ d\boldsymbol{x}'\ .$$

It is also possible to say something about the dependency and correlation between partial derivatives of different order. The k-jet space of images is the space of k-jets of functions on the plane $\mathrm{I\!R}^2 \mapsto \mathrm{I\!R}$ (see e.g. [11]). The k-jet of a scale-space image $L(\boldsymbol{x};s)$ is a map of functions into a subset of $\mathrm{I\!R}^N$,

$$j^k L(\boldsymbol{x};s) \equiv (L(\boldsymbol{x};s), L_x(\boldsymbol{x};s), L_y(\boldsymbol{x};s), \ldots, L_{x^n y^m}(\boldsymbol{x};s))^T \subset \mathrm{I\!R}^N \tag{4}$$

where $n + m = k$ and $N = (2+k)!/(2k!)$. The k-jet map of $L(\cdot;s)$ evaluated at a point $\boldsymbol{x}_0$ is a point in $\mathrm{I\!R}^N$, $j^k L(\boldsymbol{x}_0;s) \in \mathrm{I\!R}^N$. The distribution of Gaussian white noise in k-jet space is Gaussian, since scale-space derivatives of white noise are also Gaussian distributed. The covariance matrix of white noise in jet space is (see derivation in appendix A)

$$E[\langle W_{x^{n_1} y^{m_1}}(\boldsymbol{x};s), W_{x^{n_2} y^{m_2}}(\boldsymbol{x};s)\rangle]$$

$$= (-1)^{\frac{(n+m)}{2}+n_2+m_2}\frac{\sigma^2}{\pi s^{n+m+2}}\frac{n!m!}{2^{n+m+2}(n/2)!(m/2)!} \tag{5}$$

whenever both $n = n_1 + n_2$ and $m = m_1 + m_2$ are even integers, otherwise $E[\langle W_{x^{n_1} y^{m_1}}(\boldsymbol{x}; s), W_{x^{n_2} y^{m_2}}(\boldsymbol{x}; s)\rangle] = 0$. We see that the derivatives $W_{x^{n_1} y^{m_1}}(\boldsymbol{x}; s)$ and $W_{x^{n_2} y^{m_2}}(\boldsymbol{x}; s)$ are correlated whenever n and m are both even.

The Gaussian image model is not scale invariant, which can be seen from the fact that in the covariance matrix in k-jet space, Eq. (5), there is a scale dependency, $1/s^{n+m+2}$, which we can not get rid of by the standard scale normalization of derivatives [5], i.e. by multiplying each derivative $W_{x^n y^m}$ with s^{n+m}. In fact, the only way to scale normalize derivatives of white noise $W_{x^n y^m}$ is to multiply with s^{n+m+1}.

The Gaussian white noise image model is a poor model of natural images, because several of its properties differ from the empirical findings for natural images. First of all pixel intensity values of natural images are not Gaussian [3]. Secondly, pixel differences and general filter responses are not Gaussian distributed and in fact have a distribution that may be approximated by either the generalized Laplacian distribution [8,15] or the so-called Bessel K forms proposed by Srivastava et al. [25]. But most importantly, natural images exhibit scale invariance of the covariance structure [3,24], which as just mentioned is not the case for the Gaussian white noise image model. I will now discuss an extension of the Gaussian image model — the Brownian image model — which has the property of scale invariance.

3   The Brownian Model

Brownian motion was first described by the botanist R. Brown in 1828 as the random movement of pollen in water, i.e. a 1D path in $\mathbb{R}^3$. Around 1905–1908 Einstein and Bachelier developed a mathematical theory of Brownian motion. In 1923, Wiener proposed a model of the Brownian motion in the form of a stochastic function on $\mathbb{R}$. Here I will discuss the Brownian stochastic function on $\mathbb{R}^2$ or the *Brownian image* and its generalization, the *fractional Brownian image*. The Brownian image is a scale invariant stationary zero mean Gaussian stochastic function. The fractional Brownian image is stationary but not in general scale invariant, instead it exhibits the property of self similarity [16].

The Brownian image may be defined in several ways (see e.g. [12,16,23]), but I will only give the definition through the fractional Brownian image as proposed by Pentland [23]. The image $\beta_H(\boldsymbol{x}) : \mathbb{R}^2 \to \mathbb{R}$ is a *fractional Brownian image* if for all $\boldsymbol{x}, \Delta\boldsymbol{x} \in \mathbb{R}^2$

$$P\left(\frac{\beta_H(\boldsymbol{x} + \Delta\boldsymbol{x}) - \beta_H(\boldsymbol{x})}{\|\Delta\boldsymbol{x}\|^H} < y\right) = F(y)$$

where $F(y)$ is the cumulative probability distribution function of the increments. The parameter $0 < H \le 1$ controls the "roughness" of the image. In this paper I will concentrate on the case where $F(y)$ is the cumulative distribution function of a zero mean Gaussian distribution. Hence the Gaussian fractional Brownian image, or simply the *fractional Brownian image*, is a zero mean Gaussian stochastic function on $\mathbb{R}^2$. When $H = 1/2$, the fractional Brownian image $\beta_H(\boldsymbol{x})$ is called the Gaussian Brownian image $\beta(\boldsymbol{x}) = \beta_{1/2}(\boldsymbol{x})$ or simply the *Brownian image*. Interestingly, the Brownian image can be interpreted as the integral of white noise (in the sense of stochastic integrals, see e.g. [20]).

The power spectrum of the fractional Brownian image is $S(\boldsymbol{\xi}) = \sigma_0^2/\|\boldsymbol{\xi}\|^\alpha$ where $\alpha = 2H + 1$. For Brownian images ($H = 1/2$) this reduces to $S(\boldsymbol{\xi}) = \sigma_0^2/\|\boldsymbol{\xi}\|^2$, which means that the Brownian image has equal energy at equal frequency octaves. This property implies scale invariance of the covariance structure of the Brownian image model. This in turn implies that Brownian images have a scale invariant distribution because Brownian images are Gaussian, thence are completely parameterized by its mean and covariance. Field [3] has shown that broad ensembles of natural images have the same power spectrum as Brownian images, $S(\boldsymbol{\xi}) = \sigma_0^2/\|\boldsymbol{\xi}\|^2$, thus giving evidence of scale invariance of the second order statistics of natural images. The study by Ruderman and Bialek [24] shows that for certain classes of images, such as images of a forest, the power spectrum has the same form as for the fractional Brownian images, i.e. $S(\boldsymbol{\xi}) = \sigma_0^2/\|\boldsymbol{\xi}\|^\alpha$, $\alpha \neq 2$. This indicates that scale invariance might not hold for narrow ensembles of natural images, instead the second order statistics is self similar.

A sample of the periodic fractional Brownian image can be constructed by sampling zero mean i.i.d. Gaussian distributed Fourier coefficients $\boldsymbol{\xi}$ in the frequency domain with variance $\sigma_0^2/\|\boldsymbol{\xi}\|^\alpha$ and enforcing conjugate symmetry. The constant σ_0^2 can be thought of as a global variance offset. The inverse Fourier transformation yields a realization of the periodic fractional Brownian image (see Fig. 1).

The Gaussian fractional Brownian image inherits the properties of the zero mean Gaussian stochastic function described in the previous section. All partial scale-space derivatives of the fractional Brownian image are Gaussian stochastic functions. Following a similar argument as for the Gaussian white noise model we get that the gradient magnitude of the fractional Brownian image is Rayleigh distributed and given by Eq. (3). The fractional Brownian image has stationary Gaussian distributed increments [12], $\beta_H(\boldsymbol{x}) - \beta_H(\boldsymbol{y})$. Furthermore, the fractional Brownian image has self similar increments [16]. Brownian motion on $\mathbb{R}$ has independent increments, but this does not generalize to higher dimensions [12].

The fractional Brownian image is also a stationary stochastic function and from Wiener-Khinchin relations[6], we have that the power spectrum of a fractional Brownian image is the Fourier transform of the spatial covariance function $C(\boldsymbol{x}, \boldsymbol{y})$. The inverse Fourier transform of the power spectrum $S(\boldsymbol{\xi}) = \sigma_0^2/\|\boldsymbol{\xi}\|^\alpha$ yields a spatial covariance function $C(\boldsymbol{x}, \boldsymbol{y})$, which is obviously not proportional to the Dirac delta function. This leads to the conclusion that fractional Brownian images have spatial dependencies, contrary to white noise.

---

[6] The Wiener-Khinchin relations between the power spectrum $S(\boldsymbol{\xi})$ and the auto-correlation function $E[f(\boldsymbol{x})f(\boldsymbol{x} + \boldsymbol{y})]$ of a stationary stochastic function $f(\boldsymbol{x})$ are $S(\boldsymbol{\xi}) = \iint E[f(\boldsymbol{x})f(\boldsymbol{x} + \boldsymbol{y})]e^{-i\boldsymbol{\xi}\cdot\boldsymbol{y}}\, d\boldsymbol{y}$ and $E[f(\boldsymbol{x})f(\boldsymbol{x} + \boldsymbol{y})] = \frac{1}{(2\pi)^2}\iint S(\boldsymbol{\xi})e^{i\boldsymbol{\xi}\cdot\boldsymbol{y}}\, d\boldsymbol{\xi}$. For zero mean stochastic functions the autocorrelation function reduces to the spatial covariance function.

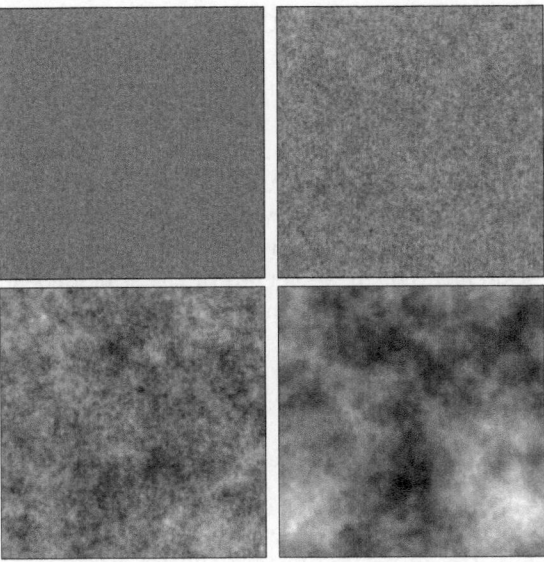

Fig. 1. (Top left) A sample from the Gaussian white noise image model. (Top right and bottom row) Samples from the fractional Brownian image model $\alpha = 1, 2, 3$.

As for all Gaussian stochastic functions the Brownian image $\beta(\boldsymbol{x})$ is fully described by specifying its mean and covariance functions. By definition the Brownian image has mean zero $\overline{\beta}(\boldsymbol{x}) = 0$. The covariance function of the Brownian image is defined in several ways in the literature [12,18]. Mumford and Gidas [18] argues that under the assumptions of stationarity and scale invariance the spatial covariance function of the Brownian image $\beta(\boldsymbol{x})$ must be

$$C(\boldsymbol{x}, \boldsymbol{y}) = \sigma_0^2 \log\left(\frac{1}{\|\boldsymbol{x} - \boldsymbol{y}\|}\right) .$$

For Brownian images in scale-space, $B(\boldsymbol{x}; s) = \beta(\boldsymbol{x}) * \psi(\boldsymbol{x}; s)$, we have the spatial covariance $C_B(\boldsymbol{x}, \boldsymbol{y}; s) = -\sigma_0^2 \int_{\mathbb{R}^2} \psi(\boldsymbol{x}' - \boldsymbol{x}; s)\psi(\boldsymbol{x}' - \boldsymbol{y}; s) \log\left(\|\boldsymbol{x} - \boldsymbol{y}\|\right) d\boldsymbol{x}'$.

All partial derivatives of a fractional Brownian image are zero mean Gaussian distributed, hence the distribution of such images mapped into k-jet space, Eq. (4), is zero mean Gaussian. The covariance matrix for the fractional Brownian image, $B_H(\boldsymbol{x}; s) = \beta_H(\boldsymbol{x}) * \psi(\boldsymbol{x}; s)$, in k-jet space can be calculated analytically (see appendix A for the derivation)

$$E\left[\langle B_{H,x^{n_1}y^{m_1}}(\boldsymbol{x}; s), B_{H,x^{n_2}y^{m_2}}(\boldsymbol{x}; s)\rangle\right] =$$
$$(-1)^{\frac{n+m}{2}+n_2+m_2} \frac{\sigma_0^2(n-1)!!(m-1)!!}{4\pi s^{n+m+2-\alpha}(n+m)!!} \Gamma\left(\frac{n+m-\alpha}{2}+1\right) \quad (6)$$

whenever both $n = n_1 + n_2$ and $m = m_1 + m_2$ are even integers, otherwise $E\left[\langle B_{H,x^{n_1}y^{m_1}}(\boldsymbol{x}; s), B_{H,x^{n_2}y^{m_2}}(\boldsymbol{x}; s)\rangle\right] = 0$. Double factorial is defined as $n!! =$

$n(n-2)(n-4)\cdots$. As was the case for Gaussian white noise, Eq. (5), we see that partial derivatives of fractional Brownian images are correlated whenever n and m are even integers. Derivatives of the fractional Brownian image have a scale dependency, $1/s^{n+m+2-\alpha}$, that we can not get rid of by scale normalization of the derivatives based on dimensional analysis [5], i.e. by scaling the derivatives by s^{n+m}. In fact the partial derivatives $B_{x^n y^m}(\boldsymbol{x}; s)$ of a fractional Brownian image must be scaled by $s^{n+m+1-\alpha/2}$, which is the scale normalization proposed by Pedersen and Nielsen [22]. Derivatives of Brownian images ($\alpha = 2$) on the other hand are scale invariant if we get rid of the scale dependency $1/s^{n+m}$ by scale normalization based on dimensional analysis. Lindeberg [13] made similar observations in conjunction with scale selection.

4   Covariance Structure of Natural Images in Jet Space

Interestingly, it turns out that the covariance structure of Brownian images in k-jet space, within some scale interval, is a good model of the covariance structure of natural images mapped into k-jet space. To be more precise, the eigenvalues of the covariance matrix of natural images and Brownian images in 3-jet space are equivalent up to a multiplicative constant within a scale interval. Fig. 3 shows graphs of eigenvalues of estimated covariance matrices in 3-jet space for four classes of images; white noise, fractional Brownian images ($\alpha = 1, 2, 3$), natural images (see Fig. 2), and logarithm of power spectrum of natural images. The jet space used here is defined by the map $j^3 L(\boldsymbol{x}; s)$, Eq. (4), where the zeroth order term $L(\boldsymbol{x}; s)$ is discarded. Furthermore, each derivative is scale normalized by $s^{n+m} L_{x^n y^m}(\boldsymbol{x}; s)$ and the scale normalized jet map $j^3 L(\boldsymbol{x}; s)$ is applied to the log-images $\log(f(x) + 1)$. The scale-space jets are sampled at different scales s and positions $\boldsymbol{x}$. Each data set has zero mean and the covariance matrices were estimated from these data sets. It can be seen from Fig. 3 that, not surprisingly, Brownian images ($\alpha = 2$) have constant eigenvalues across scale, contrary to fractional Brownian images and white noise, which have increasing and decreasing eigenvalues across scale due to these image models' lack of scale invariance. Natural images have a plateau of constant eigenvalues in the scale interval between approximately $s = 4$ ($\log(4) = 1.39$) and $s = 32$ ($\log(32) = 3.47$).

As a control experiment I have included images of logarithm of power spectra of natural images. These images have a statistics that is very different from both natural and Gaussian images and can therefore be used to evaluate whether the results of constant eigenvalues for natural images are an artifact of the method. It can be seen from Fig. 3 that log-power spectrum images do not have constant eigenvalues across scale.

Besides having constant eigenvalues in a scale interval, natural images seem to have the same relative interrelation among the eigenvalues as Brownian images. This can be seen from Fig. 4 (Left), which shows the graph of the natural image eigenvalues divided by the corresponding eigenvalues of the theoretical covariance matrix, Eq. (6), for Brownian images. Fig. 4 (Right) shows the mean value and standard deviation of the constant factor between eigenvalues across

Fig. 2. Two typical images from the van Hateren natural stimuli collection [26] used in the experiments.

scales. We see that in the scale interval $4 \leq s \leq 32$ the factor is nearly constant across eigenvalues and scale. The eigenvectors of the covariance matrix in k-jet space are the same up to a change of sign for all types of images and do not depend on the scale, since the covariance matrix is build by inner products of partial derivatives of the scale-space aperture (see Fig. 5 and appendix A).

In the scale interval $4 \leq s \leq 32$ natural images apparently have a scale invariant covariance structure similar to Brownian images. There might be several reasons why we do not have true scale invariance, i.e. invariance across all scales, for the eigenvalues of the covariance matrix of natural images. An image obviously has an inner and outer scale, bounded by the physical constraints of the camera. The lower bound is given by the pixel size, but the actual inner scale might be larger than this lower bound since the inner scale is equivalent to the size of the smallest discernable objects in the image. Similarly, the outer scale is often smaller than the physical size of the image and is given by the scale of the largest discernable objects. The actual inner and outer scales reflect the tendency of the photographer to take pictures of discernable objects, hence introducing a bias in represented object scales. In both Fig. 3 (Bottom left) and Fig. 4 we see that for small scales ($s < 4$) the eigenvalues increase with scale. This might be due to a lack of sufficient small scale structure in the van Hateren image database. That is, $s < 4$ might be below the average inner scale of van Hateren images. Similarly, at large scales ($s > 32$) the eigenvalues increase again, which might be due to the scale being larger than the average outer scale of van Hateren images. This increase might also be caused by a problem with saturation of the CCD elements in the camera, which gives large areas of white pixels hence introducing false large scale objects. The second order structure of images at scales $s < 4$ might be due to noise from the camera and the estimates of the image derivatives as well as motion blur. According to Blom et al. [1] the effect of noise on scale-space derivatives diminish as the scale increase, which leads to the conclusion that the scale invariance of the covariance for $4 \leq s \leq 32$ accounts for some large scale structural property of natural images.

The images in the van Hateren database depict a broad range of natural scenes, from scenes with several different large scale objects (Fig. 2 (left)) to

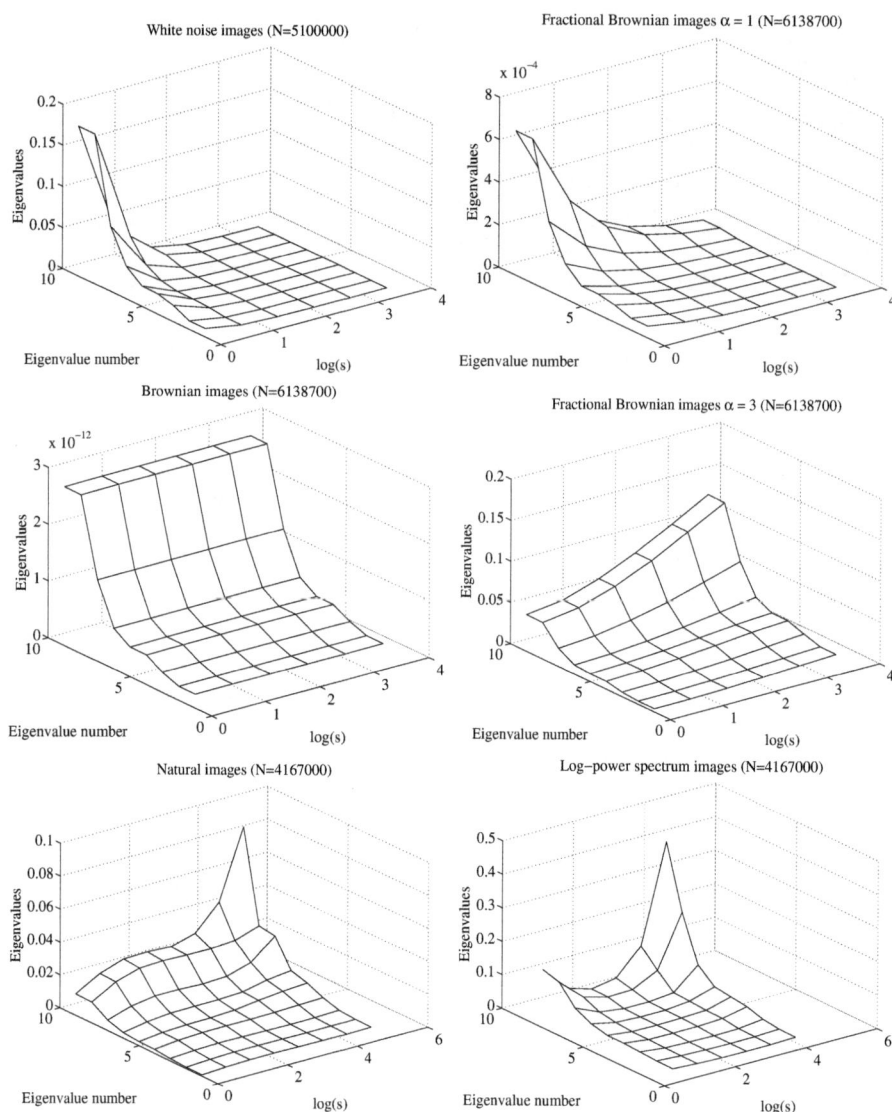

Fig. 3. Eigenvalues across scale s of an empirically estimated covariance matrix $E\left[\langle (s^{n_1+m_1}L_{x^{n_1}y^{m_1}}), (s^{n_2+m_2}L_{x^{n_2}y^{m_2}})\rangle\right]$ of images mapped into 3-jet space ($n_1 + m_1 \leq 3$, $n_2 + m_2 \leq 3$ and omitting the zeroth order term) by scale-normalized scale-space derivatives. (Top row) eigenvalues for white noise and fractional Brownian images ($\alpha = 1$). (Middle row) eigenvalues for Brownian ($\alpha = 2$) and fractional Brownian images ($\alpha = 3$). (Bottom row) eigenvalues for natural images and images of log-power spectrum of natural images. The size of each data set is given by the number N in parenthesis.

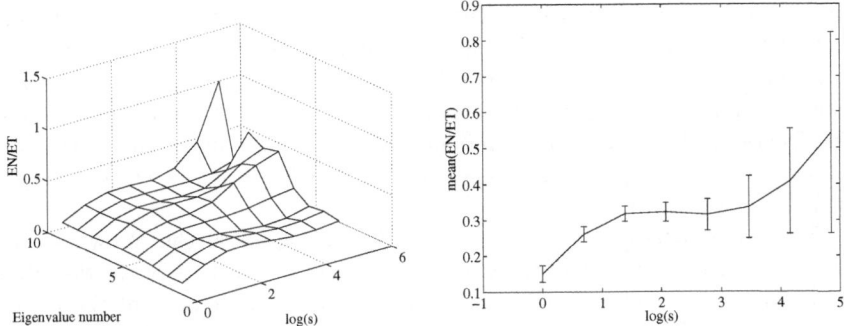

Fig. 4. (Left) Eigenvalues of natural images $EN(s)$ at different scales s divided by the eigenvalues of the theoretical covariance matrix for Brownian images ET, EN/ET. (Right) mean value, mean(EN/ET), and standard deviation (error bars) of the 9 eigenvalue factors EN/ET at each scale s.

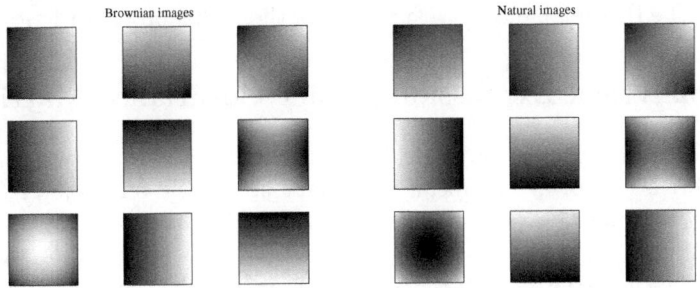

Fig. 5. (Left) Eigen-images of the theoretical covariance matrix, Eq. (6), of Brownian images in 3-jet space. (Right) Eigen-images of the empirically estimated covariance matrix of natural images in 3-jet space ($s = 8$). The eigen-images are ordered with the image for the largest eigenvalue in the top left corner and decreasing eigenvalues in the reading direction. Each eigen-image is generated by drawing the Taylor series corresponding to the 3-jet eigenvector within one scale unit.

fractal texture-like scenes (Fig. 2 (right)). This means that the results reported in this section holds for a very broad class of images that might be considered as representative of what we would call generic images.

5   Summary and Conclusion

In this paper I have discussed which statistical properties of natural images may be modeled by Gaussian image models, including the white noise, Brownian and the fractional Brownian image models. The conclusion is that the Gaussian white noise model is a poor model of the statistics of natural images. On the other hand the Gaussian Brownian image model seems to capture the scale invariant covariance structure of a broad class of natural images. Empirical results indicate that the covariance structure of natural images in 3-jet space corresponds to that of Brownian images within a scale interval. The connection between the covariance matrices for Brownian and natural images in jet space comes from the scale invariance property of the second order statistics of both types of images. The fractional Brownian image model might be a reasonable choice for narrow classes of images, e.g. images of a forest, but I have not given any empirical evidence of this claim.

The consequence of these results is that we can use the Brownian image model as a least committed model of the covariance structure of the distribution of natural images. The model is least committed in the sense that the Gaussian distribution is the maximum entropy solution given only constraints on the mean and covariance structure (see Jaynes [9]). Natural images have more complex dependencies among pixel values than Brownian images, since natural images are more structured, e.g. consists of local structures such as edges and blobs. Nevertheless, the Brownian image model can be used as a first step toward the creation of a probabilistic model of background images. Obviously we need to combine the Brownian model with models of the higher order statistics of images.

A   Analytical Expression for Jet Covariance Matrix

I will derive the expectation value $E[\cdot]$ of the inner product of two partial scale-space derivatives of a zero mean stationary stochastic function f, $L_{x^{n_1}y^{m_1}} = \partial_{x^{n_1}y^{m_1}}\psi * f$ and $L_{x^{n_2}y^{m_2}} = \partial_{x^{n_2}y^{m_2}}\psi * f$. Let the inner product be defined as $\langle L_{x^{n_1}y^{m_1}}, L_{x^{n_2}y^{m_2}} \rangle \equiv \iint L_{x^{n_1}y^{m_1}} \cdot L_{x^{n_2}y^{m_2}} \, dx \, dy$. The expectation value can be simplified by going into Fourier space using Plancherel's theorem, $\int_{\mathbb{R}^d} f(x)\overline{g(x)}dx = \frac{1}{(2\pi)^d} \int_{\mathbb{R}^d} \hat{f}(\xi)\overline{\hat{g}(\xi)}d\xi$,

$$E\left[\langle L_{x^{n_1}y^{m_1}}, L_{x^{n_2}y^{m_2}} \rangle\right]$$
$$= \frac{1}{(2\pi)^2} \iint (-1)^{n_2+m_2} i^{n+m} \xi^n \eta^m e^{-s^2(\xi^2+\eta^2)} E\left[|\hat{f}(\xi,\eta)|^2\right] d\xi \, d\eta \quad (7)$$

where $n = n_1 + n_2$ and $m = m_1 + m_2$.

Covariance Matrix of White Noise in Jet Space

Let the stochastic function $\nu(x,y)$ be a white noise image with variance σ^2 and power spectrum $E[|\hat{\nu}(\xi,\eta)|^2] = \sigma^2$. The expectation value in Eq. (7) is therefore,

$$E\left[\langle W_{x^{n_1}y^{m_1}}, W_{x^{n_2}y^{m_2}} \rangle\right]$$
$$= (-1)^{n_2+m_2} i^{n+m} \frac{\sigma^2}{(2\pi)^2} \int_{-\infty}^{\infty} \xi^n e^{-s^2\xi^2} \, d\xi \int_{-\infty}^{\infty} \eta^m e^{-s^2\eta^2} \, d\eta \ .$$

If we now use that

$$\int_{-\infty}^{\infty} \xi^n e^{-s^2\xi^2} \, d\xi = \begin{cases} \frac{1}{s^{1+n}}\Gamma\left(\frac{1+n}{2}\right) & \text{if } n \text{ even} \\ 0 & \text{otherwise} \end{cases}$$

and $\Gamma\left(\frac{1+n}{2}\right) = \frac{n!}{2^n (n/2)!}\sqrt{\pi}$ we get that

$$E\left[\langle W_{x^{n_1}y^{m_1}}, W_{x^{n_2}y^{m_2}}\rangle\right] = (-1)^{\frac{n+m}{2}+n_2+m_2}\frac{\sigma^2}{\pi s^{2+n+m}}\frac{n!m!}{2^{n+m+2}(n/2)!(m/2)!}$$

whenever both n and m are even integers, otherwise $E\left[\langle W_{x^{n_1}y^{m_1}}, W_{x^{n_2}y^{m_2}}\rangle\right]$ $= 0$.

Covariance Matrix of Fractional Brownian Images in Jet Space[7]

Let the stochastic function $\beta_H(x,y)$ be a fractional Brownian image, i.e. the power spectrum of $\beta_H(x,y)$ is $E[|\hat{\beta}_H(\xi,\eta)|^2] = \sigma_0^2/\|\boldsymbol{\xi}\|^\alpha$. Hence the expectation value of Eq. (7) can be written as

$$E\left[\langle B_{H,x^{n_1}y^{m_1}}, B_{H,x^{n_2}y^{m_2}}\rangle\right]$$
$$= \frac{1}{(2\pi)^2}\iint (-1)^{n_2+m_2}i^{n+m}\xi^n\eta^m e^{-s^2(\xi^2+\eta^2)}\frac{\sigma_0^2}{\|\boldsymbol{\xi}\|^\alpha}\,d\xi\,d\eta\ .$$

If we introduce polar coordinates $(\|\boldsymbol{\xi}\|, \theta)$ instead of (ξ, η) we get

$$\frac{\sigma_0^2}{(2\pi)^2}\iint\frac{(-1)^{n_2+m_2}i^{n+m}(\|\boldsymbol{\xi}\|\cos\theta)^n(\|\boldsymbol{\xi}\|\sin\theta)^m}{\|\boldsymbol{\xi}\|^\alpha}e^{-s^2\|\boldsymbol{\xi}\|^2}\|\boldsymbol{\xi}\|\,d\|\boldsymbol{\xi}\|\,d\theta$$
$$= (-1)^{n_2+m_2}i^{n+m}\frac{\sigma_0^2}{(2\pi)^2}\int_0^\infty \|\boldsymbol{\xi}\|^{n+m+1-\alpha}e^{-s^2\|\boldsymbol{\xi}\|^2}\,d\|\boldsymbol{\xi}\|$$
$$\cdot\int_0^{2\pi}(\cos\theta)^n(\sin\theta)^m\,d\theta\ .$$

Notice that

$$\int_0^{2\pi}(\cos\theta)^n(\sin\theta)^m\,d\theta = \begin{cases} 0, \text{if } n \text{ or } m \text{ is odd} \\ \frac{[(n-1)(n-3)\cdots 1]\cdot[(m-1)(m-3)\cdots 1]}{(n+m)(n+m-2)\cdots 2}2\pi \end{cases}$$

and

$$\int_0^\infty r^{n+m+1-\alpha}e^{-s^2 r^2}\,dr = \frac{1}{2s^{n+m+2-\alpha}}\Gamma\left(\frac{n+m-\alpha}{2}+1\right)\ .$$

We can now write the expectation value of interest as

$$E\left[\langle B_{H,x^{n_1}y^{m_1}}, B_{H,x^{n_2}y^{m_2}}\rangle\right]$$
$$= (-1)^{\frac{n+m}{2}+n_2+m_2}\frac{\sigma_0^2(n-1)!!(m-1)!!}{4\pi s^{n+m+2-\alpha}(n+m)!!}\Gamma\left(\frac{n+m-\alpha}{2}+1\right)$$

whenever both n and m are even integers, otherwise $E\left[\langle B_{H,x^{n_1}y^{m_1}}, B_{H,x^{n_2}y^{m_2}}\rangle\right]$ $= 0$.

---

[7] This derivation was developed in discussion with D. Mumford.

References

1. J. Blom, B.M. ter Haar Romeny, A. Bel, and J.J. Koenderink. Spatial derivatives and the propagation of noise in Gaussian scale space. *J. Vis. Comm. & Im. Repr.*, 4(1):1–13, 1993.
2. Z. Chi. Construction of stationary self-similar generalized fields by random wavelet expansion. *Probability Theory and Related Fields*, 121(2), 269–300 2001.
3. D. J. Field. Relations between the statistics of natural images and the response properties of cortical cells. *J. Optic. Soc. of Am.*, 4(12):2379–2394, 1987.
4. L. Florack. *Image Structure*. Kluwer Academic Publishers, 1997.
5. L. M. Florack, B. M. ter Haar Romeny, J. J. Koenderink, and M. A. Viergever. Linear scale-space. *Journal of Math. Imaging and Vision*, 4(4):325–351, 1994.
6. G. Friedlander and M. Joshi. *Introduction to The Theory of Distributions*. Cambridge University Press, 2nd edition, 1998.
7. U. Grenander and A. Srivastava. Probability models for clutter in natural images. *IEEE Trans. on Pattern Analysis and Machine Intelligence*, 23(4):424–429, 2001.
8. J. Huang and D. Mumford. Statistics of natural images and models. In *Proc. of IEEE Conf. on Computer Vision and Pattern Recognition*, 1999.
9. E. T. Jaynes. Information theory and statistical mechanics. *Physical review*, 106(4):620–630, 1957.
10. J. J. Koenderink. The structure of images. *Biol. Cybern.*, 50:363–370, 1984.
11. J. J. Koenderink and A. J. van Doorn. Representation of local geometry in the visual system. *Biological Cybernetics*, 55:367–375, 1987.
12. M. A. Lifshits. *Gaussian Random Functions*. Kluwer Academic Publishers, 1995.
13. T. Lindeberg. Feature detection with automatic scale selection. *International Journal of Computer Vision*, 30(2):79–116, November 1998.
14. P. Majer. Self-similarity of noise in scale-space. In *Proc. of Scale-Space'99*, LNCS 1682, pages 423–428. Springer Verlag, 1999.
15. S. Mallat. A theory for multiresolution signal decomposition: The wavelet representation. *IEEE Trans. on PAMI*, 11(7):674–693, July 1989.
16. B. B. Mandelbrot and J. W. van Ness. Fractional Brownian motions, fractional noises and applications. *SIAM Review*, 10(4):422–437, October 1968.
17. D. Mumford. The statistical description of visual signals. In *ICIAM'95*, 1996.
18. D. Mumford and B. Gidas. Stochastic models for generic images. *Quarterly of Applied Mathematics*, 59(11):85–111, March 2001.
19. M. Nielsen and M. Lillholm. What do features tell about images? In *Proc. of Scale-Space'01*, LNCS 2106, pages 39–50. Springer, 2001.
20. B. Øksendal. *Stochastic Differential Equations*. Springer, 5 edition, 2000.
21. K. S. Pedersen and A. B. Lee. Toward a full probability model of edges in natural images. In *Proc. of 7th ECCV*, LNCS 2350, pages 328–342. Springer Verlag, 2002.
22. K. S. Pedersen and M. Nielsen. The Hausdorff dimension and scale-space normalisation of natural images. *J. Vis. Comm. & Im. Rep.*, 11(2):266 – 277, 2000.
23. A. P. Pentland. Fractal-based description of natural scenes. *IEEE Trans. on Pattern Analysis and Machine Intelligence*, 6(6):661–674, November 1984.
24. D. L. Ruderman and W. Bialek. Statistics of natural images: Scaling in the woods. *Physical Review Letters*, 73(6):814–817, August 1994.
25. A. Srivastava, X. Liu, and U. Grenander. Universal analytical forms for modeling image probabilities. *IEEE Trans. on PAMI*, 24(9):1200–1214, September 2002.
26. J. H. van Hateren and A. van der Schaaf. Independent component filters of natural images compared with simple cells in primary visual cortex. *Proc. R. Soc. Lond. Series B*, 265:359 – 366, 1998.

Image Decomposition Application to SAR Images

Jean-François Aujol[1,2,*], Gilles Aubert[1],
Laure Blanc-Féraud[2], and Antonin Chambolle[3]

[1] Laboratoire J.A.Dieudonné, UMR CNRS 6621
Université de Nice Sophia-Antipolis, Parc Valrose, 06108 Nice Cedex 2, France
{aujol,gaubert}@ath.unice.fr
[2] ARIANA, projet commun INRIA/UNSA/CNRS
INRIA Sophia Antipolis, 2004, route des Lucioles, BP93
06902, Sophia Antipolis, Cedex, France
{Jean-Francois.Aujol,Laure.Blanc_Feraud}@sophia.inria.fr
[3] CEREMADE, CNRS UMR 7534
Université Paris IX - Dauphine, Place du Maréchal De Lattre De Tassigny
75775 Paris Cedex 16, France
antonin.chambolle@ceremade.dauphine.fr

Abstract. We construct an algorithm to split an image into a sum
$u + v$ of a bounded variation component and a component containing
the textures and the noise. This decomposition is inspired from arecent
work of Y. Meyer. We find this decomposition by minimizing a convex
functional which depends on the two variables u and v, alternatively in
each variable. Each minimization is based on a projection algorithm to
minimize the total variation. We carry out the mathematical study of our
method. We present some numerical results. In particular, we show how
the u component can be used in nontextured SAR image restoration.

Keywords: Total variation minimization, BV, texture, classification,
restoration, SAR images, speckle.

1 Introduction

1.1 Preliminaries

Image restoration is one of the major goals of image processing. A classical
approach consists in considering that an image f can be decomposed into two
components $u + v$. The first component u is well-structured, and has a simple
geometric description: it models the homogeneous objects which are present in
the image. The second component v contains both textures and noise. An ideal
model would split an image into three components $u + v + w$, where v should
contain the textures of the original image, and w the noise.

In Section 1, we begin by recalling some models proposed in the literature.
Then our model is introduced in Section 2. We give a powerful algorithm to compute the image decomposition we want to get. We carry out the mathematical

---

* partially supported by the GdR-PRC ISIS

L.D. Griffin and M. Lillholm (Eds.): Scale-Space 2003, LNCS 2695, pp. 297–312, 2003.

study of our model in Section 3. We then show some experimental results. In Section 4, we give an application to SAR images, the u component being a way to carry out efficient restoration.

1.2   Related Works

Rudin-Osher-Fatemi's model: Images are often assumed to be in BV, the space of functions with bounded variation (even if it is known that such an assumption is too restrictive [1]). We recall here the definition of BV:

Definition 1. $BV(\Omega)$ is the subspace of functions $u \in L^1(\Omega)$ such that the following quantity is finite:

$$J(u) = \sup \left\{ \int_\Omega u(x) \mathrm{div}\,(\xi(x)) dx / \xi \in C_c^1(\Omega; \mathbb{R}^2), \|\xi\|_{L^\infty(\Omega)} \leq 1 \right\} \quad (1)$$

where $C_c^1(\Omega; \mathbb{R}^2)$ is the space of functions in $C^1(\Omega; \mathbb{R}^2)$ with compact support in Ω. $BV(\Omega)$ endowed with the norm $\|u\|_{BV} = \|u\|_{L^1} + J(u)$ is a Banach space.

If $u \in BV(\Omega)$, the distributional derivative Du is a bounded Radon measure and (1) corresponds to the total variation $|Du|(\Omega)$.

In [2], the authors decompose an image f into a component u belonging to $BV(\Omega)$ and a component v in $L^2(\Omega)$. In this model v is supposed to be the noise. In such an approach, they minimize (see [2]):

$$\inf_{(u,v)\in BV(\Omega)\times L^2(\Omega)/f=u+v} \left(J(u) + \frac{1}{2\lambda}\|v\|_{L^2(\Omega)}^2 \right) \quad (2)$$

In practice, they try to compute a numerical solution of the Euler-Lagrange equation associated to (2). The mathematical study of (2) has been done in [4].

Meyer's model: In [3], Y. Meyer points out some limitations of the model developed in [2]. He proposes a variant which he believes is more adapted:

$$\inf_{(u,v)\in BV(\mathbb{R}^2)\times G(\mathbb{R}^2)/f=u+v} (J(u) + \lambda\|v\|_G) \quad (3)$$

The Banach space $G(\mathbb{R}^2)$ contains signals with large oscillations, and thus in particular textures and noise. We give here the definition of $G(\mathbb{R}^2)$.

Definition 2. $G(\mathbb{R}^2)$ is the Banach space composed of the distributions f which can be written

$$f = \partial_x g_1 + \partial_y g_2 = \mathrm{div}\,(g) \quad (4)$$

with g_1 and g_2 in $L^\infty(\mathbb{R}^2)$. On G, the following norm is defined:

$$\|v\|_G = \inf \left\{ \|g\|_{L^\infty(\mathbb{R}^2)} = \operatorname*{ess\,sup}_{x\in\mathbb{R}^2} |g(x)| / v = \mathrm{div}\,(g),\ g = (g_1, g_2), \right.$$
$$\left. g_1 \in L^\infty(\mathbb{R}^2), g_2 \in L^\infty(\mathbb{R}^2), |g(x)| = \sqrt{|g_1|^2 + |g_2|^2}(x) \right\} \quad (5)$$

In the space G, very oscillating functions have a small norm (see [3]) (and large oscillations are linked with textures and noises).

Vese-Osher's model: L. Vese and S. Osher have first proposed an approach for the resolution of Meyer's program. They have studied the problem (see [5]):

$$\inf_{(u,v)\in BV(\Omega)\times G(\Omega)} \left(\int |Du| + \lambda\|f - u - v\|_2^2 + \mu\|v\|_{G(\Omega)} \right) \tag{6}$$

where Ω is a bounded open set. To compute their solution, they replace the term $\|v\|_{G(\Omega)}$ by $\|\sqrt{g_1^2 + g_2^2}\|_p$ (where $v = \operatorname{div}(g_1, g_2)$). It approximates (6) when p goes to $+\infty$. For numerical reasons, the authors use the value $p = 1$ and they claim they did not see any visual difference when they used larger values for p. Then they formally derive the Euler-Lagrange equations. They report good numerical results.

These two authors, together with A. Solé, have proposed another approach to this problem in [6], where they propose a more direct algorithm in the case $\lambda = +\infty$ and $p = 2$.

2   Our Approach

In this section we introduce our model.It is inspired from the formulation of [5]. We first present it in the continuous setting. Then we propose a discretization, and provide a mathematical study and an algorithm for the discretized model.

2.1   Presentation

We propose to solve the problem:

$$\inf_{(u,v)\in BV(\Omega)\times G_\mu(\Omega)} \left(J(u) + \frac{1}{2\lambda}\|f - u - v\|_{L^2(\Omega)}^2 \right) \tag{7}$$

where

$$G_\mu(\Omega) = \{v \in G(\Omega) / \|v\|_G \leq \mu\} \tag{8}$$

We recall that $\|v\|_G$ is defined by (5). The parameter μ plays the same role as the one in problem (6). The larger μ is, the more v contains information, and therefore the more u is averaged. The smaller λ is, the smaller the L^2 norm of the residual $f - u - v$ is. We will render more precisely the link of our model with Meyer's one later. Let us introduce the following functional defined on $BV(\Omega) \times G(\Omega)$:

$$F(u,v) = \begin{cases} J(u) + \frac{1}{2\lambda}\|f - u - v\|_{L^2(\Omega)}^2 & \text{if } v \in G_\mu(\Omega) \\ +\infty & \text{if } v \in G(\Omega)\backslash G_\mu(\Omega) \end{cases} \tag{9}$$

$F(u,v)$ is finite if and only if (u,v) belongs to $BV(\Omega) \times G_\mu(\Omega)$. Problem (7) can thus be written:

$$\inf_{(u,v)\in BV(\Omega)\times G(\Omega)} F(u,v) \tag{10}$$

2.2   Discretization

We study (10) in the discrete case. We take here the same notations as in [7]. The image is a two dimensional array of size $N \times N$. We denote by X the Euclidean space $\mathbb{R}^{N \times N}$, and $Y = X \times X$. The space X will be endowed with the scalar product $(u, v)_X = \sum_{1 \le i, j \le N} u_{i,j} v_{i,j}$ and the norm $\|u\|_X = \sqrt{(u, u)_X}$. In Y, we use the Euclidean scalar product $(p, q)_Y = \sum_{1 \le i, j \le N} p^1_{i,j} q^1_{i,j} + p^2_{i,j} q^2_{i,j}$ with $p = (p^1, p^2)$ and $q = (q^1, q^2)$ in Y. To define a discrete total variation, we introduce a discrete version of the gradient operator. If $u \in X$, the gradient ∇u is a vector in Y given by: $(\nabla u)_{i,j} = ((\nabla u)^1_{i,j}, (\nabla u)^2_{i,j})$. with

$$(\nabla u)^1_{i,j} = \begin{cases} u_{i+1,j} - u_{i,j} & \text{if } i < N \\ 0 & \text{if } i = N \end{cases} \quad \text{and} \quad (\nabla u)^2_{i,j} = \begin{cases} u_{i,j+1} - u_{i,j} & \text{if } j < N \\ 0 & \text{if } j = N \end{cases}$$

The discrete total variation of u is then defined by:

$$J(u) = \sum_{1 \le i, j \le N} |(\nabla u)_{i,j}| \tag{11}$$

We also introduce a discrete version of the divergence operator. We define it by analogy with the continuous setting by $\mathrm{div} = -\nabla^*$ where ∇^* is the adjoint of ∇: that is, for every $p \in Y$ and $u \in X$, $(-\mathrm{div}\, p, u)_X = (p, \nabla u)_Y$. It is easy to check that:

$$(\mathrm{div}\,(p))_{i,j} = \begin{cases} p^1_{i,j} - p^1_{i-1,j} & \text{if } 1 < i < N \\ p^1_{i,j} & \text{if } i = 1 \\ -p^1_{i-1,j} & \text{if } i = N \end{cases} + \begin{cases} p^2_{i,j} - p^2_{i,j-1} & \text{if } 1 < j < N \\ p^2_{i,j} & \text{if } j = 1 \\ -p^2_{i,j-1} & \text{if } j = N \end{cases} \tag{12}$$

We are now in position to introduce the discrete version of the space G.

Definition 3.

$$G^d = \{v \in X \ / \ \exists g \in Y \text{ such that } v = \mathrm{div}\,(g)\} \tag{13}$$

and if $v \in G^d$:

$$\|v\|_{G^d} = \inf \{\|g\|_\infty \ / \ v = \mathrm{div}\,(g),$$

$$g = (g^1, g^2) \in Y, |g_{i,j}| = \sqrt{(g^1_{i,j})^2 + (g^2_{i,j})^2}\} \tag{14}$$

where $\|g\|_\infty = \max_{i,j} |g_{i,j}|$.

Moreover, we will denote:

$$G^d_\mu = \{v \in G^d \ / \ \|v\|_{G^d} \le \mu\} \tag{15}$$

We notice that

$$J(u) = \sup_{v \in G^d_1} (v, u)_X \tag{16}$$

and

$$\|v\|_{G^d} = \sup_{u \in X, J(u) \le 1} (u, v)_X \tag{17}$$

Proposition 1. *The space G^d identifies with the following subspace:*

$$X_0 = \{v \in X \ / \ \sum_{i,j} v_{i,j} = 0\} \tag{18}$$

Proof: Choose $v \in G^d$. There exists $g \in Y$ such that: $v = \mathrm{div}\,(g)$. But $\sum_{i,j}(\mathrm{div}\,g)_{i,j}$
$= (-\nabla^* g, 1)_Y = (g, \nabla 1)_X = 0$ i.e. $v \in X_0$. Hence $G^d \subset X_0$.

Conversely, let $v \in X_0$. Since the kernel of ∇ is the constant images, i.e. the vectors $x \in X$ such that $x_{i,j} = x_{i',j'}$ for all i, j, i', j', it is clear that a discrete Poincaré inequality holds: $\|x - \frac{1}{N^2}\sum_{i,j} x_{i,j}\|_X \leq c\|\nabla x\|_Y$. Hence one shows easily that the problem $\min_{x \in X} A(x)$, with $A(x) = \|\nabla x\|_Y^2 + 2(x, v)_X$, has a solution. This solution satisfies $A'(x) = 0$, that is, $-2\mathrm{div}\,(\nabla x) + 2v = 0$. Hence $v = \mathrm{div}\,(\nabla x) \in G^d$, and we conclude that $X_0 \subset G^d$.

∎

The discretized functional associated to (9), defined on $X \times X$, is given by:

$$F(u,v) = \begin{cases} J(u) + \frac{1}{2\lambda}\|f - u - v\|_X^2 & \text{if } v \in G_\mu^d \\ +\infty & \text{if } v \in X \backslash G_\mu^d \end{cases} \tag{19}$$

The problem we want to solve is:

$$\inf_{(u,v) \in X \times X} F(u,v) \tag{20}$$

2.3   Total Variation Minimization as a Project

Introduction: We recall that the Legendre-Fenchel transform of J is:

$$J^*(v) = \sup_u \left((u,v)_X - J(u)\right) \tag{21}$$

Since here J defined by (1) is homogeneous of degree one (i.e. $J(\lambda u) = \lambda J(u) \ \forall u$ and $\lambda > 0$), it is then standard (see [8]) that J^* is the indicator function of some closed convex set, which turns out to be the set G_1^d defined by (15):

$$J^*(v) = \chi_{G_1^d}(v) = \begin{cases} 0 & \text{if } v \in G_1^d \\ +\infty & \text{otherwise} \end{cases} \tag{22}$$

This can be checked out easily (see [7] for details). In [7], A. Chambolle proposes a nonlinear projection algorithm to minimize the total variation. The problem is:

$$\inf_{u \in X} \left(J(u) + \frac{1}{2\lambda}\|f - u\|_X^2\right) \tag{23}$$

The following result is shown:

Proposition 2. *The solution of (23) is given by:*

$$u = f - P_{G_\lambda^d}(f) \tag{24}$$

where P is the orthogonal projector on G_λ^d (defined by (15)).

Algorithm: [7] gives an algorithm to compute $P_{G_\lambda^d}(f)$. It indeed amounts to finding:

$$\min\left\{\|\lambda\operatorname{div}(p) - f\|_X^2 \ / \ p \in Y \ , \ |p_{i,j}| \le 1 \ \forall i,j = 1,\dots,N\right\} \tag{25}$$

This problem can be solved by a fixed point method:

$$p^0 = 0 \tag{26}$$

and

$$p_{i,j}^{n+1} = \frac{p_{i,j}^n + \tau(\nabla(\operatorname{div}(p^n) - f/\lambda))_{i,j}}{1 + \tau|(\nabla(\operatorname{div}(p^n) - f/\lambda))_{i,j}|} \tag{27}$$

In [7] is given a sufficient condition ensuring the convergence of the algorithm:

Theorem 1 (Thm 1 [7]). *Assume that the parameter τ in (27) verifies $\tau \le 1/8$. Then $\lambda\operatorname{div}(p^n)$ converges to $P_{G_\lambda^d}(f)$ as $n \to +\infty$.*

2.4   Application to Problem (20)

Since J^* is the indicator function of G_1^d (see (16,22)), we can rewrite (19) as

$$F(u,v) = \frac{1}{2\lambda}\|f - u - v\|_X^2 + J(u) + J^*\left(\frac{v}{\mu}\right) \tag{28}$$

With this formulation, we see the symmetric roles played by u and v. And the problem we want to solve is:

$$\inf_{(u,v)\in X\times X} F(u,v) \tag{29}$$

To solve (29), we consider the two following problems:

- v being fixed, we search for u as a solution of:

$$\inf_{u\in X}\left(J(u) + \frac{1}{2\lambda}\|f - u - v\|_X^2\right) \tag{30}$$

- u being fixed, we search for v as a solution of:

$$\inf_{v\in G_\mu^d}\|f - u - v\|_X^2 \tag{31}$$

From Proposition 2, we know that the solution of (30) is given by: $\hat{u} = f - v - P_{G_\lambda^d}(f-v)$. And the solution of (31) is simply given by: $\hat{v} = P_{G_\mu^d}(f-u)$.

2.5   Algorithm

1. Initialization:

$$u_0 = v_0 = 0 \tag{32}$$

2. Iterations:

$$v_{n+1} = P_{G_\mu^d}(f - u_n) \tag{33}$$

$$u_{n+1} = f - v_{n+1} - P_{G_\lambda^d}(f - v_{n+1}) \tag{34}$$

3. Stopping test: we stop if

$$\max(|u_{n+1} - u_n|, |v_{n+1} - v_n|) \le \epsilon \tag{35}$$

3   Mathematical Results

In this section we carry out the mathematical study of the algorithm (32)–(35). We first show its convergence when λ is fixed. We then state more precisely the link of the limit of our model (when λ goes to 0) with Meyer's one.

3.1   Existence and Uniqueness of a Solution for (20)

Lemma 1. *There exists a unique couple* $(\hat{u}, \hat{v}) \in X \times G_\mu^d$ *minimizing* F *on* $X \times X$.

Proof: We split the proof into two steps.
Step 1: Existence

1. We first remark that the set $X \times G_\mu^d$ is convex, and then that F is convex on $X \times G_\mu^d$. We thus deduce that F is convex on $X \times X$.
2. It is immediate to see that F is continuous on $X \times G_\mu^d$. We then deduce that F is lower semi-continuous on $X \times X$.
3. Let $(u, v) \in X \times G_\mu^d$. We have $\|v\|_{G^d} \leq \mu$. Moreover, since X is of finite dimension, there exists $g \in X$ such that $v = \mathrm{div}\,(g)$ and $\|g\|_{L^\infty} = \|v\|_{G^d} \leq \mu$. We deduce from (12) that (N^2 is the size of the image):

$$\|v\|_X =\leq 4\mu N^2 \tag{36}$$

We recall that $X \times X$ is endowed with the Euclidean norm.

$$\|(u, v)\|_{X \times X} = \sqrt{\|u\|_X^2 + \|v\|_X^2} \tag{37}$$

Thus, if $\|(u, v)\|_{X \times X} \to +\infty$, then we get from (36) that $\|u\|_X \to +\infty$. We therefore deduce, since f is fixed, and since (36) holds, that $\|f - u - v\|_X^2 \to +\infty$. And since $F(u, v) \geq \frac{1}{2\lambda}\|f - u - v\|_2^2$, we get $F(u, v) \to +\infty$. Hence we deduce that F is coercive on $X \times G_\mu^d$. We therefore conclude that F is coercive on $X \times X$.

We deduce the existence of a minimizer $(\hat{u}, \hat{v})$.

Step 2: Uniqueness
To get the uniqueness, we first remark that F is strictly convex on $X \times G_\mu^d$, as the sum of a convex function and of a strictly convex function, except in the direction $(u, -u)$. Hence it suffices to check that if $(\hat{u}, \hat{v})$ is a minimizer of F then for $t \neq 0$, $(\hat{u} + t\hat{u}, \hat{v} - t\hat{u})$ is not a minimizer of F. The result is obvious if $\hat{v} - t\hat{u} \in X \backslash G_\mu^d$. Let us show that if $\hat{v} - t\hat{u} \in G_\mu^d$ then the result is still true. Indeed, if $\hat{v} - t\hat{u} \in G_\mu^d$, we have:

$$F(\hat{u} + t\hat{u}, \hat{v} - t\hat{u}) = F(\hat{u}, \hat{v}) + (|1 + t| - 1)J(\hat{u}) \tag{38}$$

By contradiction, let us assume that there exists $\hat{t} \neq \{-2, 0\}$ such that $\hat{v} - \hat{t}\hat{u} \in G_\mu^d$ and

$$F(\hat{u} + \hat{t}\hat{u}, \hat{v} - \hat{t}\hat{u}) \leq F(\hat{u}, \hat{v}) \tag{39}$$

As $(\hat{u}, \hat{v})$ minimizes F, (39) is an equality. From (38), we deduce that $(|1 + \hat{t}| - 1)J(\hat{u}) = 0$. And as $\hat{t} \neq \{-2, 0\}$, we get that $J(\hat{u}) = 0$. There exists therefore $\gamma \in \mathbb{R}$ such that for all (i, j), $\hat{u}_{i,j} = \gamma$.

1. If $\gamma = 0$, then $\hat{u} = 0$. Thus $(\hat{u} + \hat{t}\hat{u}, \hat{v} - \hat{t}\hat{u}) = (\hat{u}, \hat{v})$.
2. If $\gamma \neq 0$, then $\hat{v} - \hat{t}\hat{u}$ cannot belong to G_μ^d since its mean is not 0 (see Proposition 1). This contradicts our assumption.

There remains to check what happens in the case when $\hat{t} = -2$. In this case, we have: $F(-\hat{u}, \hat{v} + 2\hat{u}) \leq F(\hat{u}, \hat{v})$, i.e. $(-\hat{u}, \hat{v} + 2\hat{u})$ is also a minimizer of F. As we assume $\hat{v} + 2\hat{u} \in G_\mu^d$, and as F convex (and as G_μ^d convex), we get:

$$F(0, \hat{u} + \hat{v}) \leq \frac{1}{2}F(\hat{u}, \hat{v}) + \frac{1}{2}F(-\hat{u}, \hat{v} + 2\hat{u}) \tag{40}$$

And we deduce that $(0, \hat{u} + \hat{v})$ is also a minimizer of F. But $F(0, \hat{u} + \hat{v}) = F(\hat{u}, \hat{v})$, i.e. $\frac{1}{2\lambda}\|f - \hat{u} - \hat{v}\|_X^2 = J(\hat{u}) + \frac{1}{2\lambda}\|f - \hat{u} - \hat{v}\|_X^2$. We thus get that $J(\hat{u}) = 0$, and we conclude as before. Hence there exists a unique couple $(\hat{u}, \hat{v}) \in X \times G_\mu^d$ minimizing F on $X \times X$.

■

3.2   Convergence of the Algorithm

We show here that our algorithm gives asymptotically the solution of the discrete problem associated to (29).

Proposition 3. *The sequence $F(u_n, v_n)$ built in Section 2.5 converges to the minimum of F on $X \times X$.*

Proof: We first remark that, as we solve successive minimization problems, we have:

$$F(u_n, v_n) \geq F(u_n, v_{n+1}) \geq F(u_{n+1}, v_{n+1}) \tag{41}$$

In particular, the sequence $F(u_n, v_n)$ is nonincreasing. As it is bounded from below by 0, it thus converges in $\mathbb{R}$. We denote by m its limit. We want to show that

$$m = \inf_{(u,v) \in X \times X} F(u, v) \tag{42}$$

Without any restriction, we can assume that, $\forall n$, $(u_n, v_n) \in X \times G_\mu^d$. As F is coercive and as the sequence $F(u_n, v_n)$ converges, we deduce that the sequence (u_n, v_n) is bounded in $X \times G_\mu^d$. We can thus extract a subsequence (u_{n_k}, v_{n_k}) which converges to $(\hat{u}, \hat{v})$ as $n_k \to +\infty$, with $(\hat{u}, \hat{v}) \in X \times G_\mu^d$. Moreover, we have, for all $n_k \in \mathbb{N}$ and all v in X:

$$F(u_{n_k}, v_{n_k+1}) \leq F(u_{n_k}, v) \tag{43}$$

and for all $n_k \in \mathbb{N}$ and all u in X:

$$F(u_{n_k}, v_{n_k}) \leq F(u, v_{n_k}) \tag{44}$$

Let us denote by $\bar{v}$ a cluster point of (v_{n_k+1}). Considering (41), we get (since F is continuous on $X \times G_\mu^d$):

$$m = F(\hat{u}, \hat{v}) = F(\hat{u}, \bar{v}) \tag{45}$$

By passing to the limit in (33), we get: $\bar{v} = P_{G_\mu^d}(f - \hat{u})$. But from (45), we know that: $\|f - \hat{u} - \hat{v}\| = \|f - \hat{u} - \bar{v}\|$. By uniqueness of the projection, we conclude that $\bar{v} = \hat{v}$. Hence $v_{n_k+1} \to \hat{v}$. By passing to the limit in (43) (F is continuous on $X \times G_\mu^d$), we therefore have for all v:

$$F(\hat{u}, \hat{v}) \leq F(\hat{u}, v) \tag{46}$$

And by passing to the limit in (44), for all u:

$$F(\hat{u}, \hat{v}) \leq F(u, \hat{v}) \tag{47}$$

(46) and (47) can respectively be rewritten:

$$F(\hat{u}, \hat{v}) = \inf_{v \in X} F(\hat{u}, v) \tag{48}$$

$$F(\hat{u}, \hat{v}) = \inf_{u \in X} F(u, \hat{v}) \tag{49}$$

But, from the definition of $F(u, v)$ (see (28)), (49) is equivalent to (see [8]):

$$0 \in -f + \hat{u} + \hat{v} + \lambda \partial J(\hat{u}) \tag{50}$$

and (48) to:

$$0 \in -f + \hat{u} + \hat{v} + \lambda \partial J^* \left(\frac{\hat{v}}{\mu} \right) \tag{51}$$

The subdifferential ∂F of F at $(\hat{u}, \hat{v})$ is given by:

$$\partial F(\hat{u}, \hat{v}) = \frac{1}{\lambda} \begin{pmatrix} -f + \hat{u} + \hat{v} + \lambda \partial J(\hat{u}) \\ -f + \hat{u} + \hat{v} + \lambda \partial J^* \left(\frac{\hat{v}}{\mu} \right) \end{pmatrix} \tag{52}$$

And thus, according to (50) and (51), we have:

$$\begin{pmatrix} 0 \\ 0 \end{pmatrix} \in \partial F(\hat{u}, \hat{v}) \tag{53}$$

which is equivalent to: $F(\hat{u}, \hat{v}) = \inf_{(u,v) \in X^2} F(u, v) = m$. Hence the whole sequence $F(u_n, v_n)$ converges towards m, the unique minimum of F on $X \times G_\mu^d$. We deduce that the sequence (u_n, v_n) converges to $(\hat{u}, \hat{v})$, the minimizer of F, when n tends to $+\infty$.

■

3.3   Link with Meyer's Model

We examine here the link between the discrete model (29) and Meyer's problem. We first recall the discrete version of Meyer's problem:

$$\inf_{(u,v)\in X\times G^d/f=u+v} H_\alpha(u,v) \qquad (54)$$

with

$$H_\alpha(u,v) = (J(u) + \alpha\|v\|_{G^d}) \qquad (55)$$

The following result is straightforward:

Lemma 2. *There exists a solution $(\hat{u}, \hat{v}) \in X \times G^d$ of problem (54).*

Remark: We do not know if a uniqueness result holds for problem (54). We then recall problem (29):

$$\inf_{(u,v)\in X\times X} F_{\lambda,\mu}(u,v) \qquad (56)$$

with

$$F_{\lambda,\mu}(u,v) = \frac{1}{2\lambda}\|f - u - v\|^2 + J(u) + J^*\left(\frac{v}{\mu}\right) \qquad (57)$$

Let us consider the problem

$$\inf_{(u,v)\in X\times X/f=u+v} J(u) + J^*\left(\frac{v}{\mu}\right) \qquad (58)$$

One easily shows the next result:

Lemma 3. *There exists a solution $(\bar{u}, \bar{v}) \in X \times X$ of problem (58).*

Proposition 4. *Let us fix $\alpha > 0$ in problem (54). Let $(\hat{u}, \hat{v})$ a solution of problem (54). We fix $\mu = \|\hat{v}\|_{G^d}$ in (58). Then:*

- *$(\hat{u}, \hat{v})$ is also a solution of problem (58).*
- *Conversely, any solution $(\bar{u}, \bar{v})$ of (58) (with $\mu = \|\hat{v}\|_{G^d}$) is a solution of (54).*

Proof: We split the proof into two steps.
Step 1:
We first want to show that $(\hat{u}, \hat{v})$ is a solution of (58) (with $\mu = \|\hat{v}\|_{G^d}$). As $(\hat{u}, \hat{v})$ is a solution of (54) (the existence of $(\hat{u}, \hat{v})$ is given by Lemma 2) and as $\|\hat{v}\|_{G^d} = \mu$, then $\hat{u}$ is solution of

$$\inf_{u\in X/u=f-v,\|v\|_{G^d}=\mu} J(u) + \alpha\mu \qquad (59)$$

i.e. $\hat{u}$ is solution of

$$\inf_{u\in X/u=f-v,\|v\|_{G^d}=\mu} J(u) \qquad (60)$$

Since the set $\{u \in X/u = f - v, \|v\|_{G^d} = \mu\}$ is contained in $\{u \in X/u = f - v, \|v\|_{G^d} \leq \mu\}$, we have:

$$\inf_{u \in X/u=f-v, \|v\|_{G^d}=\mu} J(u) \geq \inf_{u \in X/u=f-v, \|v\|_{G^d} \leq \mu} J(u) \tag{61}$$

By contradiction, let us assume that

$$\inf_{u \in X/u=f-v, \|v\|_{G^d}=\mu} J(u) > \inf_{u \in X/u=f-v, \|v\|_{G^d} \leq \mu} J(u) \tag{62}$$

Thus, there exists $v' \in X$ such that $\|v'\|_{G^d} < \mu$ and

$$J(f - v') < \inf_{u \in X/u=f-v, \|v\|_{G^d}=\mu} J(u) \tag{63}$$

Denoting by $u' = f - v'$, we have: $J(u') + \alpha\|v'\|_{G^d} < J(u') + \alpha\mu$. But since $(\hat{u}, \hat{v})$ is a solution of (54):

$$J(\hat{u}) + \alpha\|\hat{v}\|_{G^d} \leq J(u') + \alpha\|v'\|_{G^d} < J(u') + \alpha\mu \tag{64}$$

Hence (we recall that $\|\hat{v}\|_{G^d} = \mu$), we get from (64) that $J(\hat{u}) < J(u')$. This contradicts (63). We conclude that (62) cannot hold. Hence:

$$\inf_{u \in X/u=f-v, \|v\|_{G^d}=\mu} J(u) = \inf_{u \in X/u=f-v, \|v\|_{G^d} \leq \mu} J(u) \tag{65}$$

From (60), we see that $\hat{u}$ is solution of $\inf_{u \in X/u=f-v, \|v\|_{G^d} \leq \mu} J(u)$, i.e. $\hat{u}$ is solution of

$$\inf_{u \in X/u=f-v} J(u) + J^*\left(\frac{v}{\mu}\right) \tag{66}$$

Hence $(\hat{u}, \hat{v})$ is also a solution of (58).

Step 2:

Let us now consider $(\bar{u}, \bar{v})$ a solution of (58) (the existence of $(\bar{u}, \bar{v})$ is given by Lemma 3). We can repeat the computations we made in Step 1. We get that $\bar{u}$ is a solution of:

$$\inf_{u \in X/u=f-v, \|v\|_{G^d}=\mu} J(u) + \alpha\mu \tag{67}$$

We therefore have: $J(\bar{u}) + \alpha\mu = J(\hat{u}) + \alpha\|\hat{v}\|_{G^d}$. But as $(\bar{u}, \bar{v})$ is a solution of (58), we have $\|\bar{v}\|_{G^d} \leq \mu$. Hence $J(\bar{u}) + \alpha\|\bar{v}\|_{G^d} \leq J(\hat{u}) + \alpha\|\hat{v}\|_{G^d}$. And since $(\hat{u}, \hat{v})$ is a solution of (54), we get that:

$$J(\bar{u}) + \alpha\|\bar{v}\|_{G^d} = J(\hat{u}) + \alpha\|\hat{v}\|_{G^d} \tag{68}$$

We thus conclude that $(\bar{u}, \bar{v})$ is a solution of (54).

■

In particular, we have thus shown that, when μ is correctly tuned, a solution of the limit problem (58) is in fact a solution of Meyer's problem (54).

3.4   Role of λ

We show here that problem (58) is obtained by passing to the limit λ goes to 0^+ in (56).

Proposition 5. *Let us fix $\alpha > 0$ in (54). Let us assume that problem (54) has a unique solution $(\hat{u}, \hat{v})$. Set $\mu = \|\hat{v}\|_{G^d}$ in (56) and (58). Let us denote (u_λ, v_λ) the solution of problem (56). Then (u_λ, v_λ) converges to $(u_0, v_0) \in X \times X$ as λ goes to 0. Moreover, $(u_0, v_0) = (\hat{u}, \hat{v})$ is the solution of problem (58).*

Remark: In the case when the solution of problem (54) is not unique, the result of Proposition 5 does not hold. We can just show that any cluster point of $(u_{\lambda_n}, v_{\lambda_n})$ is a solution of problem (58) and thus of (54)

Proof of Proposition 5: The existence of $(\hat{u}, \hat{v})$ is given by Lemma 3. The existence and uniqueness of (u_λ, v_λ) is given by Lemma 1.

Since (u_λ, v_λ) is the solution of problem (56), we have $v_\lambda \in G^d_\mu$, i.e. $\|v_\lambda\|_{G^d} \le \mu$. As we saw in the proof of Lemma 1, this inequality implies:

$$\|v_\lambda\|_X \le 4\mu N^2 \tag{69}$$

Since (u_λ, v_λ) is the solution of problem (56), we have:

$$F_{\lambda,\mu}(u_\lambda, v_\lambda) \le F_{\lambda,\mu}(f, 0) \tag{70}$$

which means

$$F_{\lambda,\mu}(u_\lambda, v_\lambda) \le J(f) \tag{71}$$

And the left hand-side of (71) is given by:

$$F_{\lambda,\mu}(u_\lambda, v_\lambda) = J(u_\lambda) + \frac{1}{2\lambda}\|f - u_\lambda - v_\lambda\|_X^2 + J^*\left(\frac{v_\lambda}{\mu}\right) = J(u_\lambda) + \frac{1}{2\lambda}\|f - u_\lambda - v_\lambda\|_X^2 \tag{72}$$

Hence $J(u_\lambda) + \frac{1}{2\lambda}\|f - u_\lambda - v_\lambda\|_X^2 \le J(f)$, and

$$\|f - u_\lambda - v_\lambda\|^2 \le 2\lambda J(f) \tag{73}$$

As $\|v_\lambda\|_X$ is bounded (from (69)), we conclude that if $\lambda \in [0;1]$, u_λ is bounded by a constant $C > 0$ which does not depend on λ.

Consider a sequence (λ_n) which goes to 0^+ as $n \to +\infty$. Then, up to an extraction (since $(u_{\lambda_n}, v_{\lambda_n})$ is bounded in $X \times X$), there exists $(u_0, v_0) \in X \times X$ such that $(u_{\lambda_n}, v_{\lambda_n})$ converges to (u_0, v_0). By passing to the limit in (73), we get: $\|f - u_0 - v_0\|_X = 0$, i.e. $f = u_0 + v_0$.

To conclude the proof of the proposition, there remains to show that (u_0, v_0) is a solution of problem (58). We first notice that as $\lambda > 0$, and since $\|v_\lambda\|_{G^d} \le \mu$, we get: $\|v_0\|_{G^d} \le \mu$. Let $(u, v) \in X \times X$ such that $f = u + v$. We have:

$$J(u) + J^*\left(\frac{v}{\mu}\right) + \frac{1}{2\lambda}\underbrace{\|f - u - v\|^2}_{=0}$$

$$\ge J(u_{\lambda_n}) + J^*\left(\frac{v_{\lambda_n}}{\mu}\right) + \frac{1}{2\lambda_n}\|f - u_{\lambda_n} - v_{\lambda_n}\|^2$$

$$\ge \underbrace{J(u_{\lambda_n}) + J^*\left(\frac{v_{\lambda_n}}{\mu}\right)}_{\to J(u_0) + J^*\left(\frac{v_0}{\mu}\right)}$$

Hence (u_0, v_0) is a solution of problem (58). And as we have assumed that problem (58) has a unique solution, we deduce that $(u_0, v_0) = (\hat{u}, \hat{v})$, i.e. (u_0, v_0) is the solution of problem (58).

∎

4 SAR Images Restoration

4.1 Introduction

Synthetic Aperture Radar (SAR) images are strongly corrupted by a noise called speckle. A radar sends a coherent wave which is reflected on the ground, and then registered by the radar sensor [9]. When one cares with the reflection of a coherent wave on a coarse surface, then one can see that the observed image is degraded by a noise of large amplitude. This gives a speckled aspect to the image. That is why such a noise is called speckle.

Link with our approach: Contrary to the usual modeling in SAR, the noise in our model is considered to be additive: the image f is decomposed into a component u belonging to BV, and a component v in G. But it is to be noticed that our model is completely different from the classical additive models: in these, v is often considered to be a Gaussian white noise, and therefore has a constant variance all over the image. Here, v belongs to G, a space in which signals can have large oscillations but small norm. Moreover the variance of the oscillations of v may not be uniform on the whole image. Note that by considering u as the restored image (without speckle) we assume that there is no texture in the SAR image.

4.2 Results on Sythetic Images

Restoration: Figure 1 shows why for a SAR image the decomposition proposed by Meyer is very interesting. Indeed, one checks that the v component contains the speckle, and the u component can be regarded as a restoration of the original image (if it does not contain textures). It is difficult to make comparisons with other methods [10], since the main criterion remains the visual interpretation.

Nevertheless, the results we achieve appear promising in comparison with existing methods. And above all, our approach being a variational one, computation time are very short. With a processor of 800 MHz and 128 kByte of RAM, it takes less than one minute to deal with an image of size 256*256.

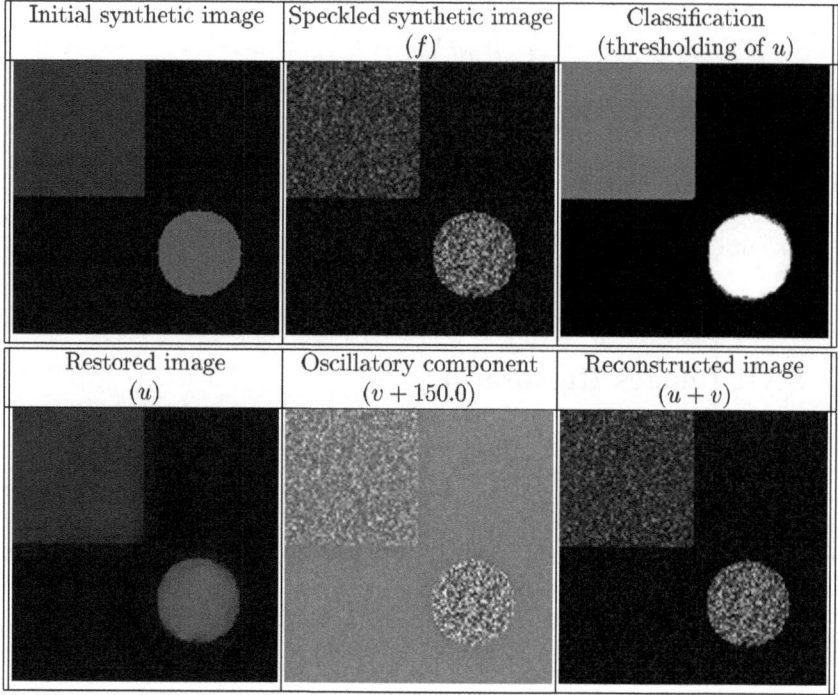

Fig. 1. Simple synthetic image ($\lambda = 0.01$ and $\mu = 80$)

4.3 Results on Real Images

We use SAR images of Bourges' area provided by the CNES. The reference image (also furnished by the CNES) has been obtained by amplitude summation. Image 2 shows the effect of parameter μ on the restoration process. The larger μ is, the more v contains information, and therefore the more u is averaged. According to the value of μ, we can thus get a more or less restored image, and also more or less of a smoother image.

5 Conclusion

In this article, we present a new algorithm to decompose a given image f into a component u belonging to BV and a component v containing the noise and

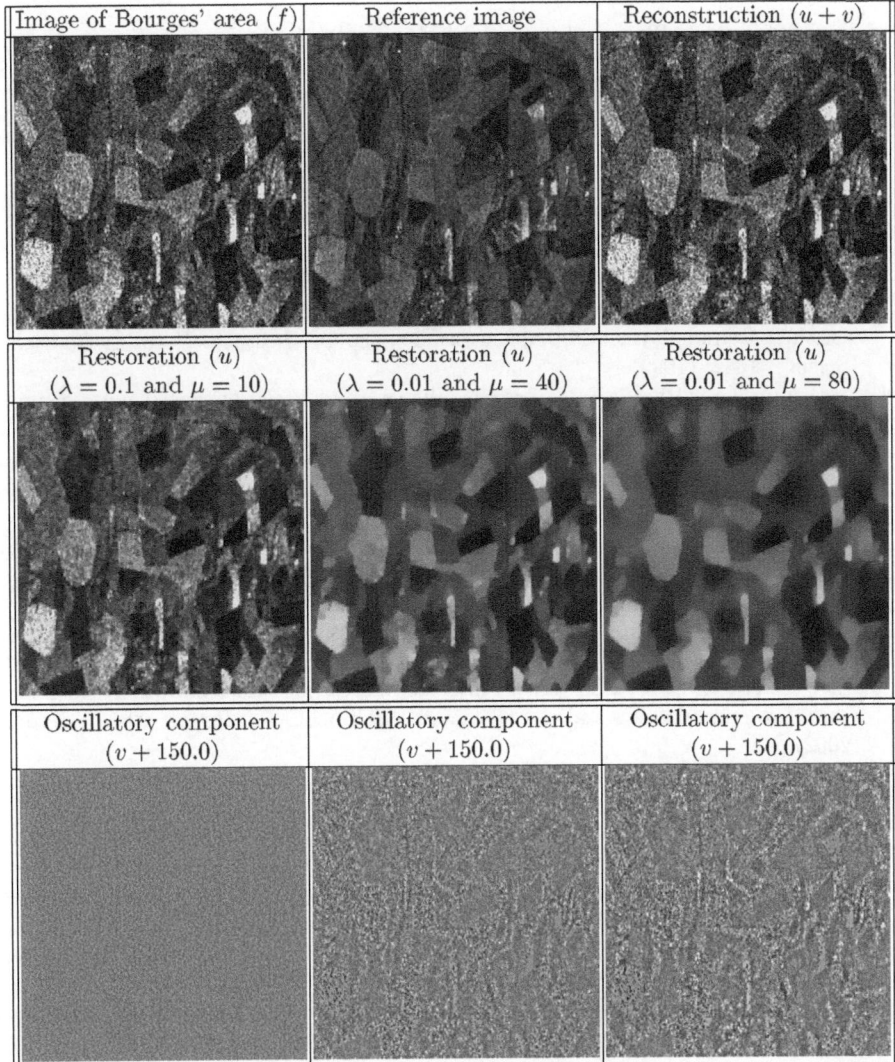

Fig. 2. Image of Bourges' area

the textures of the initial image. Our algorithm performs Meyer's program [3] when μ is suitably tuned. Moreover, we carry out the mathematical study of our model. We also show how the u component can be used for SAR image restoration. Further details about this work as well as comparisons with the standard BV filtering and with the Vese-Osher model [5] can be found in [11].

Acknowledgement: The authors would like to thank the French Space Agency CNES (Centre National d'Etudes Spatiales) and the French research center CES-BIO (Centre d'Etudes Spatiales de la Biosphère) for providing real SAR data extracted from the CD-ROM *Filtrage d'images SAR* (1999). Part of this work has been funded by GdR-PrC ISIS through the young researcher program.

References

[1] L. Alvarez, Y. Gousseau, J.M. Morel: Scales in Natural Images and a Consequence on their Bounded Variation Norm. Scale-Space '99, Lectures Notes in Computer Science, 1682, (1999).

[2] L. Rudin , S. Osher , E. Fatemi: Nonlinear total variation based noise removal algorithms. Physica D, 60, 259–268, (1992)

[3] Yves Meyer: Oscillating patterns in image processing and in some nonlinear evolution equations. The Fifteenth Dean Jacquelines B. Lewis Memorial Lectures, (March 2001)

[4] A. Chambolle , P.L. Lions: Image recovery via total variation minimization and related problems. Numerische Mathematik, 76, (3), 167–188, (1997)

[5] Luminita A. Vese , Stanley J. Osher: Modeling textures with total variation minimization and oscillating patterns in image processing. UCLA C.A.M. Report 02-19 (May 2002)

[6] S.J. Osher , A. Sole , L.A. Vese: Image decomposition and restoration using total variation minimization and the H^{-1} norm. UCLA C.A.M. Report 02-57 (October 2002)

[7] A. Chambolle: An algorithm for total variation minimization and applications. To appear in JMIV (2003).

[8] I. Ekeland , R. Temam: Analyse convexe et problèmes variationnels. Dunod, Grundlehren der mathematischen Wissenschaften, second edition, 224, (1983)

[9] Henderson Lewis: Principle and applications of imaging radar. J.Wiley and Sons, Manual of Remote Sensing, third edition, 2 (1998)

[10] G. Franceschetti , R. Lanari: Synthetic aperture radar processing CRC press, Electronic engineering systems series, (1999)

[11] J.F. Aujol , G. Aubert , L. Blanc-Féraud, A. Chambolle: Decomposing an image: Application to textured images and SAR images. Preprint (submitted)

Basic Morphological Operations, Band-Limited Images and Sampling

Cris L. Luengo Hendriks and Lucas J. van Vliet

Pattern Recognition Group, Delft University of Technology, The Netherlands

Abstract. Morphological operations are simple mathematical constructs, which have led to effective solution for many problems in image processing and computer vision. These solutions employ discrete operators and are applied to digitized images. The mathematics behind the morphological operators also exists in the continuous domain, the domain where the images came from. We observed that the discrete operators cannot reproduce the results obtained by the continuous operators. The reason for this is that neither the operator (the structuring element) nor the result of the operation are band-limited, and thus cannot be represented by equidistant samples without loss of information. The differences between continuous-domain and discrete-domain morphology are best shown by the dependency of the discrete morphology on sub-pixel translations and rotations of the images before digitization.

This article describes an algorithm that applies continuous-domain morphology to properly sampled images. We implemented the dilation for one-dimensional images (signals), and with it constructed the erosion, the closing and the opening. We provide a discussion on a possible extension to higher-dimensional images.

1 Introduction

1.1 Band-Limited Signals and Uniform Sampling

A large class of signals can be represented by an infinite set of equidistant samples without loss of information; that is, we can reconstruct the original signal from these samples. This class is composed of all band-limited signals in which the highest frequency that is needed to construct the signal (the cut-off frequency) is less than half the sampling frequency. This condition is called after Nyquist and/or Shannon [1,2].

A linear filter applied to a band-limited signal produces another band-limited signal with an equal or lower cut-off frequency. This implies that such an operation can be performed on the set of uniform samples, producing the sampled version of the continuous result [3]. However, almost all non-linear filters produce non-band-limited outputs, which cannot be represented correctly by equidistant samples. Therefore, the discrete implementations of these operators do not represent their continuous counterparts. This is certainly the case for morphological operations.

L.D. Griffin and M. Lillholm (Eds.): Scale-Space 2003, LNCS 2695, pp. 313–324, 2003.

1.2 Sampling Morphological Operations

The basic morphological operations, dilation and erosion with a flat structuring element (SE), are equivalent to a local maximum and local minimum filter respectively (assuming the SE is a closed set, and the function it is applied to is continuous) [4,5]. The output at each point is defined by the maximum (or minimum) over the neighborhood defined by the SE. When applied to a band-limited image, this produces an image that is not band-limited, and therefore cannot be represented correctly by equidistant samples. This can be seen by the fact that discontinuities in the first derivative are introduced.

For certain analysis operations, e.g. a granulometry, this is not important because the output of the morphological operation must be integrated to obtain a single value. That is, we are not interested in sampling the (continuous) result. When the result of a discrete operation that produces a non-band-limited result is integrated, an error is made. To reduce this error certain tricks can be used (see [6]). For example, it is possible to interpolate the input image. This causes the result of a discrete morphological operation to produce a better approximation to the sampled version of the corresponding continuous operation. Therefore a smaller error is made when integrating the result.

Because the morphological operations are local maximum or minimum filters, the result is heavily influenced by resampling the discrete image (either for interpolation, translation or rotation). This is because it is not expected that a sample exactly hits a local maximum or minimum of a function. Resampling will cause different values of the image to be sampled, thus changing the result of the local maximum or minimum filters. This dependency on the sampling grid can also be shown by translating it with respect to the original, continuous image; see Figure 1. We will call any difference between the continuous-domain and discrete-domain results *sampling error*.

In this article we propose an algorithm that implements continuous-domain morphological operators. It works for dilations as well as erosions with flat structuring elements. Openings and closings can be constructed using these two basic operations. This method is explicitly defined for 1D images (signals). It is theoretically possible to extend this method to higher dimensions, albeit with a constraint (Section 4). Some other operations, such as the watershed transform or the morphological reconstruction, might be implemented in this framework as well, but are beyond the scope of this paper.

2 Sampling-Free Dilations

To reduce the sampling error of morphological operations, we need a continuous representation of the signal, a function $f : \mathbb{R} \to \mathbb{R}$ defined on an interval $[x_0, x_N]$. We must be able to

- represent band-limited signals accurately,
- represent signals with discontinuities in the first and higher derivatives, and
- obtain such a representation from a set of given samples.

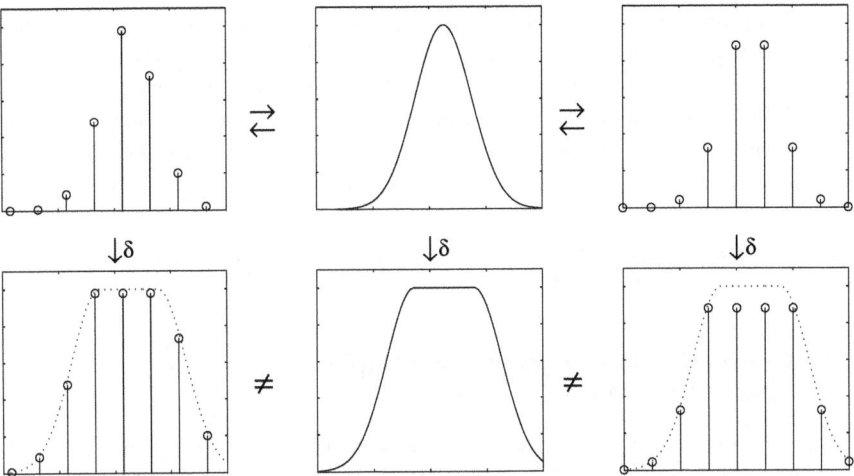

Fig. 1. The discrete dilation is not translation-invariant, as this example shows. In the middle of the top row is a continuous signal. We sample this signal twice, using uniform sampling, but with a different offset of the sampling grid. We are still able to recreate the original signal from both these instances, but the results of the dilation are different. Neither result is the same as a sampled version of the result of the continuous dilation.

We propose to use a piece-wise polynomial function, which is easy to work with. We limit ourselves to third-order polynomials, for which zero-crossings, maxima and minima can be found analytically. Also, it is possible to construct a good approximation of a band-limited function with third-order polynomials [7].

2.1   Representing a 1D Signal as a Piece-Wise Polynomial

To represent a continuous one-dimensional function as a set of third-order polynomial segments, the following information is required:

- Starting point of each polynomial (x_i)
- Polynomial coefficients (a_i, b_i, c_i, d_i)
- Length of each polynomial (l_i)

Since the function we are representing is defined everywhere in the signal domain, the end point of a polynomial is equal to the starting point of the next one. Thus the length is redundant, and we only need to store the starting points of each polynomial and the end point of the last polynomial. The function is then written as a collection of segments $S_i(x)$

$$S_i(x) = a_i + b_i(x - x_i) + c_i(x - x_i)^2 + d_i(x - x_i)^3 \quad , \tag{1}$$

$i \in [0, 1, 2, ...N - 1]$, plus a right bound x_N.

Certain operations on such a representation are trivial. For example, shifting the whole function just requires incrementing or decrementing the starting points x_i, and inverting the function is accomplished by negating all the polynomial coefficients. Other operations we apply to the polynomial function are sampling (evaluating the function at chosen locations) and integration. The integral over the function is the sum of the integral over each segment, determined by

$$\int_{x_0}^{x_N} f(x) = \sum_{i=0}^{N-1} \tfrac{1}{4}d_i x_{i+1}^4 + \tfrac{1}{3}c_i x_{i+1}^3 + \tfrac{1}{2}b_i x_{i+1}^2 + a_i x_{i+1} \quad . \tag{2}$$

2.2 Converting the Sequence of Samples into a Piece-Wise Polynomial

To create the piece-wise polynomial representation $f(x)$ from the given samples $f[n]$, we require an interpolation function that has certain characteristics:

- The resulting function must have as many continuous derivatives as possible (since the original band-limited signal is infinitely differentiable). We use third-order polynomials, thus we require that the second-order derivative be continuous.
- It must be a local representation. That is, the zone of influence of a single pixel must be limited, because only a limited number of samples is available.
- It must be capable of producing a polynomial representation.

An interpolator that satisfies these constraints is the cubic spline interpolator [8,9,10]. It produces polynomial segments in between each two sample points. Although its impulse response decays quite quickly, it requires a filter with an infinite impulse response to determine polynomial coefficients. This filter can be implemented recursively [11]. Note that a spline of infinite order equals the ideal interpolator (the sinc function) [7]. Thus, a cubic spline is an approximation of the ideal interpolator.

In the case of noisy input samples, it might be better to use a least squares spline [8]. In this case, the reconstructed function does not need to be equal to the samples at the sample locations, and thus can be smoother. Furthermore, by computing the piece-wise polynomial in this way the number of pieces is reduced, which makes further processing faster as well.

2.3 Dilations

Examining the 1D dilation operation with a flat, compact SE B, one can readily see that the result is composed of plateaus (constant sections) as well as slopes with the exact same shape as can be found in the input signal (see Fig. 2). Let us define the set B as

$$B = \{x | x \in [-r, r]\} \quad . \tag{3}$$

The plateaus are formed when, at a point x, the maximum value over the neighborhood B comes from a local maximum (see Fig. 2a). At points near x, the

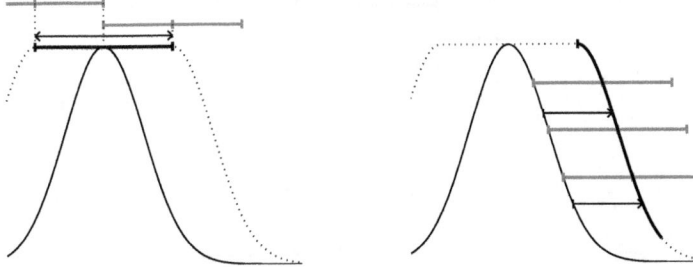

Fig. 2. Construction of the dilated function. **a:** maxima create plateaus in the output. **b:** slopes are shifted by a fixed distance, dictated by the size of the structuring element. In these graphs, the thin black line is the input signal, the dotted black line is the output signal, the thick black line shows how the output signal is constructed, and the thick grey lines give the size of the structuring element ($2r$).

maximum over the neighborhood will also come from the same local maximum, and will therefore receive the same value. These plateaus will have a width of at most $2r$, centered on the local maximum.

The sloped regions are produced when the maximum over B does not come from a local maximum. In this case, it must come from the border of the structuring element (see Fig. 2b). At nearby points, the resulting value also comes from the same edge of the neighborhood. Therefore, a slope is created that is the exact copy of a slope from the input signal, shifted by r or $-r$. This is known as the slope transform [12,13].

Thus, for a one-dimensional signal, the output of the dilation with a flat, compact SE is the point-wise (or, in our case, the segment-wise) maximum of three functions:

- the input signal translated by r: $f(x - r)$,
- the input signal translated by $-r$: $f(x + r)$, and
- a signal composed of plateaus centered around each of the local maxima.

The above analysis is valid for flat, compact and symmetric structuring elements. Any non-symmetric flat, compact SE C, defined by

$$C = \{x | x \in [-r - d, r - d]\} = \{x | x + d \in [-r, r]\} \quad , \tag{4}$$

can be converted into a symmetric SE B by translating the input or the output signal (δ denotes the dilation operator):

$$\delta_C f(x) = \delta_B f(x - d) = [\delta_B f](x - d) \quad . \tag{5}$$

A non-compact structuring element can be constructed with the union of compact structuring elements:

$$\delta_{[\bigcup_i B_i]} f(x) = \bigvee_i \delta_{B_i} f(x) \quad . \tag{6}$$

Thus, the above analysis suffices for any flat SE.

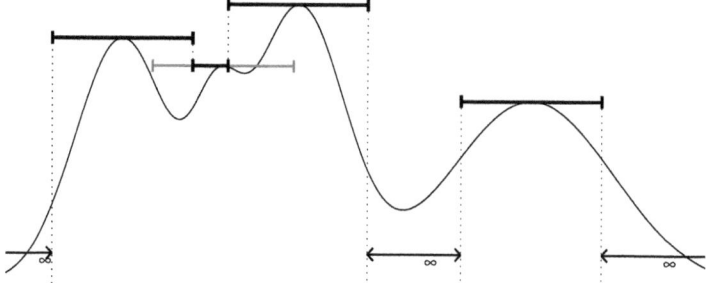

Fig. 3. Construction of the plateau function. At each local maximum a plateau (0th order polynomial) with the size of the structuring element is set. In the case of overlapping plateaus, the one with the highest value is kept intact; the other one must be cropped. Empty regions are filled with segments of value $-\infty$ so that the function is defined everywhere in the signal domain.

2.4   Implementation

The Plateau Function. Creating the function composed of the plateaus is not very complicated. First, all local maxima must be found. This is easily accomplished by examining the first and second order derivatives of each of the polynomials:

$$S_i'(x) = 0 \quad \wedge \quad S_i''(x) < 0 \quad \Leftrightarrow \quad x \text{ is a local maximum.} \tag{7}$$

Note that finding these derivatives is trivial, and finding the zero crossings of the first derivative is accomplished by solving a quadratic equation. Additionally, in the result of a previous morphological operation there can be maxima in the form of cusps and plateaus. These will be found only on knots (boundary points between polynomial segments), and are identified by comparing the derivatives of both polynomials at those points:

$$S_i'(x_{i+1}) \geq 0 \quad \wedge \quad S_{i+1}'(x_{i+1}) \leq 0 \quad \Leftrightarrow \quad x \text{ is a local maximum.} \tag{8}$$

These maxima must be sorted according to their value, largest first. Then the plateau image is created by adding a 0th order polynomial segment, ranging from $x-r$ to $x+r$, and with value $f(x)$, for each maximum at x (see Fig. 3). Each segment added must not overlap with any of the polynomials already present in the function, so it must be cropped to the available space (actually, we are taking the maximum over these segments implicitly). At the end of this process, eventual 'holes' must be filled with segments of value $-\infty$, so that the generated function is defined everywhere in the signal domain, and can be stored in the same manner as the input signal.

Maximum over the Segment Functions. The last step is to find the function that is the maximum of the three functions. This is a two-step process in which first two functions are compared, and then the result is compared to the third. To avoid complicated exceptions in the algorithm, we pad the three functions with zero-order polynomials so that they span the same interval (from $x_0 - r$ to $x_N + r$). The translated versions of the input signal are extended with the edge value (so as to keep them continuous). The function containing the plateaus is extended with $-\infty$.

This comparison is very simple, but potentially generates quite a lot of segments. For each (portion of a) segment $S_i^1(x)$ in one function that spans the same region as another (portion of a) segment $S_i^2(x)$ in the other function, the intersection points $S_i^1(x) = S_i^2(x)$ must be found (this is a cubic equation, the solution can be found in Bronstein [9]). There are up to three intersection points, and thus up to four sub-segments. For each of these, the polynomial with the larger value is used to construct the output signal.

2.5   Erosions, Closings and Openings

Since the erosion ε is the dual operation of the dilation [14], it is implemented by inverting the signal, applying the dilation, and inverting the result again:

$$\varepsilon_B f(x) = -\delta_B[-f(x)] \quad . \tag{9}$$

As stated above, inverting the piecewise polynomial function is easily accomplished by negating all the polynomial coefficients.

The closing ϕ is created by applying an erosion to the result of the dilation,

$$\phi_B f(x) = \varepsilon_{\check{B}}[\delta_B f(x)] \quad , \tag{10}$$

and the opening γ is constructed the other way around,

$$\gamma_B f(x) = \delta_{\check{B}}[\varepsilon_B f(x)] \quad . \tag{11}$$

The algorithm as described above can be applied on its own result, so that implementing closings and openings becomes trivial[1].

3   Results

3.1   A First Examination of the Algorithm

We extracted a line out of an image to apply our methods to. Figure 4 shows two portions of this line, along with the reconstructed continuous function, the result of a discrete dilation (i.e. one applied to the samples directly) and that of the sampling-free dilation proposed here. Figure 5 shows the results of the discrete and sampling-free closings on the same signal.

In these figures we can see that the sampling-free dilation reaches higher values than the discrete variant at some points, especially on plateaus. The value

---

[1] The source code for the sampling-free morphology is available through the author's web site at http://www.ph.tn.tudelft.nl/~cris/sfm.html.

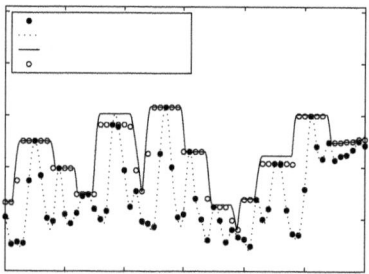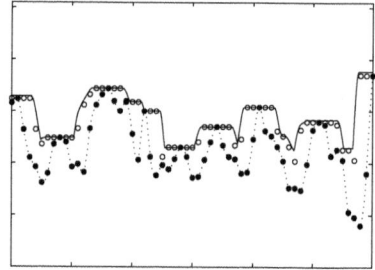

Fig. 4. Two interesting portions of a 1D signal, together with its sampling-free dilation. The open dots give the values of the discrete dilation for comparison. The SE has a length of 5 pixels.

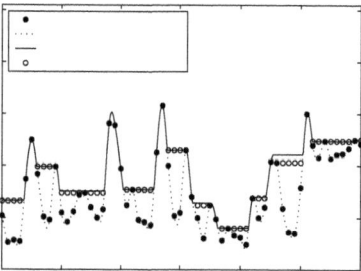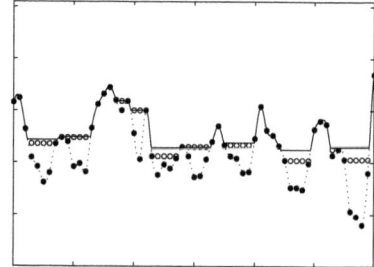

Fig. 5. Two interesting portions of a 1D signal, together with its sampling-free closing. The open dots give the values of the discrete closing for comparison. The SE has a length of 5 pixels.

of this signal at these points is equal to the value of the true local maximum of the input signal (or rather of the cubic spline approximation). Likewise, the closing has higher values at the plateaus (the continuous version is equal only in exceptional cases).

3.2 Granulometry

A granulometry is a multi-scale closing (or opening) [14]. The result of the operation at each scale is integrated to obtain a graph comparable to a cumulative size distribution. See Soille [5] for more information on granulometries. We normalize the measured granulometric curve so that the first value is 0 and the last value is 1.

We created a signal of which we know the function that represents the granulometric curve. To the samples of this signal we applied a granulometry with both the sampling-free and discrete closings, and compared the results to the theoretical granulometric curve.

The signal we used is a sine,

$$f(x) = sin\left(\frac{2\pi x}{T}\right) \quad , \tag{12}$$

with T the period. The sampling distance is 1, meaning that T must be larger than 2 for error-free sampling. The theoretical granulometric curve is described by

$$h(r) = \begin{cases} \frac{1}{\pi}sin\left(\frac{r\pi}{T}\right) - \frac{r}{T}cos\left(\frac{r\pi}{T}\right) & \text{for } r < T, \\ 1 & \text{for } r \geq T, \end{cases} \tag{13}$$

with r the size of the structuring element. We used two periods: $T_1 = \frac{200}{9}$ and $T_2 = \pi$. We chose these values because each period of the sine starts at a different offset with respect to the sampling grid. Both signals can be correctly sampled at a rate of 1. The first one can be interpolated very accurately using cubic splines, whereas the second will produce larger errors due to the inability of the spline to correctly reconstruct high-frequency signals (see Fig. 6). The frequency characteristic of the cardinal cubic spline can be found in Fig. 2 of [8]. The spline interpolation on the second signal produces a result that is obviously not an exact reproduction of the input signal. Therefore, the result of the granulometry is inaccurate as well. However, it lies much closer to the theoretical curve than the result of the discrete granulometry. Another obvious drawback of the discrete granulometry is the discreteness of the structuring element, which can only be constructed with integer lengths. Because of this, the granulometry with the sampling-free closing could be sampled much more densely.

We repeated the above experiments after adding noise to the input samples (see Fig. 7). The results show more or less the same characteristics, except that the granulometric curves deviate a bit more from the theoretical (noiseless) values. These results might improve when using regularized splines as mentioned above.

4   Extension to Multi-dimensional Images

Morphological operations can be defined for images of any dimensionality. Therefore, we would like to extend our algorithm to multi-dimensional images as well. This is, however, not an easy task.

Obviously, extension to multi-dimensional images by processing each dimension separately will not work. In this case, maxima lying in between raster lines will be missed. This shows that it is necessary to create a patch representation of the image (using multi-dimensional cubic splines), and work on that.

However, using multi-dimensional structuring elements with this representation also introduces a problem: we would need to create a translated version of the input image for each point along the boundary of this structuring element. Since there are an infinite number of these points, this is an impossible task. If we would simplify the structuring element by taking only a limited number of points along the contour, we would again miss some of the local maxima.

Thus, we should limit ourselves to one-dimensional structuring elements working on a patch representation of a multi-dimensional image. Now we can

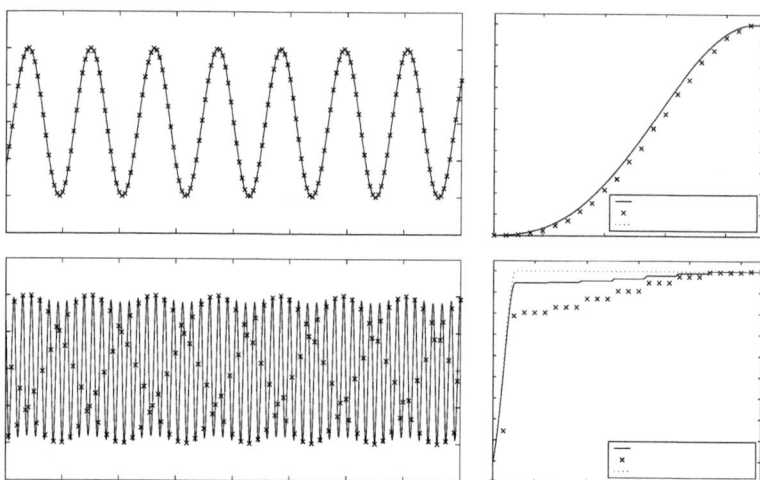

Fig. 6. Granulometry of a sine function sampled at different rates. On the left are the samples and the continuous function created with cubic splines. On the right is the result of the granulometry, computed with both discrete and continuous-domain morphology, compared to the theoretical granulometric function.

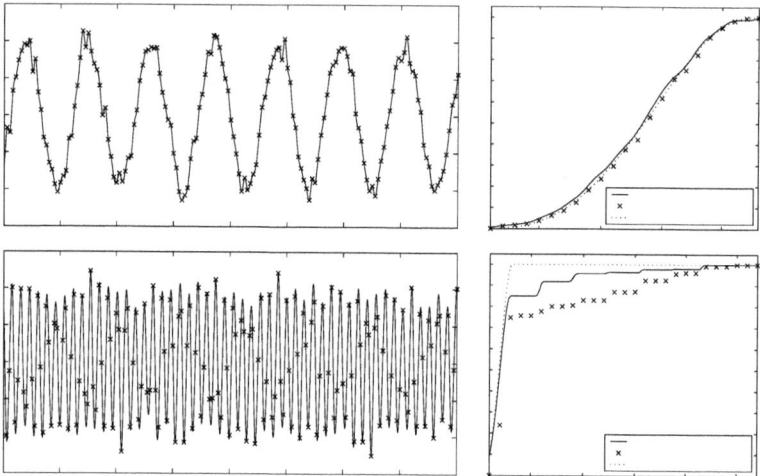

Fig. 7. Granulometry of a sine function sampled at different rates. Noise was added to the samples prior to the analysis. On the left are the samples and the continuous function created with cubic splines. On the right is the result of the granulometry, computed with both discrete and continuous-domain morphology, compared to the theoretical granulometric function of the noiseless signal.

implement the same operations we suggested above: create two translated versions of the input image (one for each end of the structuring element), and an image consisting of plateaus centered around each of the local maxima in the image; then take the maximum over these three images. In this case, however, local maxima are all points for which there is a maximum in the direction of the structuring element. These points form lines (in a two dimensional image; in an N-D image this is a $(N-1)$-D plane). The plateaus we create are therefore patches with a zero-order polynomial in one direction, and some third-order polynomial in all the orthogonal directions. These polynomials are taken from the input patches along the local maxima line.

The only problem with this approach is that the patches produced by a dilation have very complex boundaries (given by third-order polynomials). This makes an implementation difficult, although the theoretical formulation is trivial.

Using linear structuring elements it is possible to create more complex multidimensional structuring elements such as the rectangle, the hexagon, the octagon, etc. [14,5]. These shapes are increasingly better approximations of a disk. Thus, it is possible to create an arbitrarily accurate approximation to the isotropic SE.

5   Conclusion

In this paper we have shown that it is possible to apply continuous-domain morphology to one-dimensional images (signals). On a sampled signal, discrete morphology produces results that are not the same as the results produced by continuous morphology on the signal before sampling. By representing the continuous signal as a piece-wise polynomial function, we were able to compute dilations and erosions with flat structuring elements that produce the exact same results as their continuous-domain counterparts would produce. Some error is introduced when converting the sampled signal into a piece-wise polynomial function, but the morphological operations themselves do not introduce any further errors. It is possible to cascade dilations and erosions to obtain openings and closings and other, more complex morphological filters.

We applied a granulometry (multi-scale closings) to some sampled test signals using the sampling-free closings constructed with our algorithms. The resulting granulometric curve is compared to the theoretical result. The differences found were due to the difficulty in converting the samples of a high-frequency signal into a piece-wise polynomial function. This process is further complicated by the addition of noise to the input samples. These results should improve when using regularized splines instead of interpolating splines.

To apply these methods to higher-dimensional images, the decomposition principle must be used. Some structuring elements can be decomposed into one-dimensional operations. These can be applied to the polynomial patch representation of the multi-dimensional image. Even though it is trivial to state theoretically, the cubic boundary between these polynomial patches is complicated to use in an actual implementation.

References

1. Nyquist, H.: Certain topics in telegraph transmission theory. Transactions of the AIEE (1928) 617–644 [reprinted in: Proceedings of the IEEE 90 (2002) 280–305].
2. Shannon, C.: Communication in the presence of noise. Proceedings of the IRE 37 (1949) 10–21 [reprinted in: Proceedings of the IEEE 86 (1998) 447–457].
3. van Vliet, L.J.: Grey-Scale Measurements in Multi-Dimensional Digitized Images. PhD thesis, Pattern Recognition Group, Faculty of Applied Physics, Delft University of Technology, Delft (1993)
4. Serra, J.: Image Analysis and Mathematical Morphology. Academic Press, London (1982)
5. Soille, P.: Morphological Image Analysis. Springer-Verlag, Berlin (1999)
6. Luengo Hendriks, C.L., van Vliet, L.J., van Kempen, G.M.P., Bouwens, E.C.M.: (Using morphological sieves to detect minute differences in pore sizes) Submitted to Journal of Microscopy.
7. Unser, M., Aldroubi, A., Eden, M.: Polynomial spline signal approximations: Filter design and asymptotic equivalence with shannon's sampling theorem. IEEE Transactions on Information Theory 38 (1992) 95–103
8. Unser, M., Aldroubi, A., Eden, M.: B-spline signal processing: Part I - theory. IEEE Transactions on Signal Processing 41 (1993) 821–833
9. Bronstein, I.N., Semendjaev, K.A., Musiol, G., Mühlig, H.: Taschenbuch der Mathematik. 4th edn. Verlag Harri Deutsch, Thun, Frankfurt am Main (1999)
10. de Boor, C.: A Practical Guide to Splines. Springer, New York (2001)
11. Unser, M., Aldroubi, A., Eden, M.: Fast B-spline transforms for continuous image representation and interpolation. IEEE Transactions on Pattern Analysis and Machine Intelligence 13 (1991) 277–285
12. van den Boomgaard, R., Smeulders, A.: The morphological structure of images: The differential equations of morphological scale-space. IEEE Transactions on Pattern Analysis and Machine Intelligence 16 (1994) 1101–1113
13. Dorst, L., van den Boomgaard, R.: Morphological signal processing and the slope transform. Signal Processing 38 (1994) 79–98
14. Matheron, G.: Random Sets and Integral Geometry. Wiley, New York (1975)

An Explanation for the Logarithmic Connection between Linear and Morphological Systems

Bernhard Burgeth and Joachim Weickert

Mathematical Image Analysis Group
Faculty of Mathematics and Computer Science, Building 27
Saarland University, 66041 Saarbrücken, Germany
{burgeth,weickert}@mia.uni-saarland.de
http://www.mia.uni-saarland.de

Abstract. Since the introduction of the slope transform by Dorst/van den Boomgaard and Maragos as the morphological equivalent of the Fourier transform, people have been surprised about the almost logarithmic relation between linear and morphological system theory.
This article gives an explanation by revealing that morphology in essence is linear system theory in a specific algebra. While classical linear system theory uses the standard $(+, \times)$-algebra, the morphological system theory is based on the idempotent $(\max, +)$-algebra and the $(\min, +)$-algebra. We identify the nonlinear operations of erosion and dilation as linear convolutions $*_e$ and $*_d$ induced by these idempotent algebras. The slope transform in the $(\max, +)$-algebra, however, corresponds to the logarithmic multivariate Laplace transform in the $(+, \times)$-algebra. We study relevant properties of this transform and its links to convex analysis. This leads to the definition of the so-called Cramer transform as the Legendre-Fenchel transform of the logarithmic Laplace transform. Originally known from the theory of large deviations in stochastics, the Cramer transform maps standard convolution to $*_e$-convolution, and it maps Gaussians to quadratic functions.
The article is a step towards the unification of linear and morphological system theories on the basis of a general linear system theory in an appropriate algebra.

Keywords: linear system theory, morphology, convex analysis, MAX-PLUS algebra, MINPLUS algebra, slope transform, Cramer transform.

1 Introduction

Linear system theory is a successful and well established field in signal and image processing [6,14,15,29]. In the n-dimensional case, shift invariant linear filters can be described as convolutions of some signal $f : \mathbb{R}^n \to \mathbb{R}$ with a kernel function $b : \mathbb{R}^n \to \mathbb{R}$:

$$f * b(x) := \int_{\mathbb{R}^n} f(x - y)\, b(y)\, dy.$$

By means of the Fourier transform

$$\hat{f}(u) := \mathcal{F}[f](u) := \int_{\mathbb{R}^n} f(x)\, e^{-i2\pi u^\top x}\, dx$$

L.D. Griffin and M. Lillholm (Eds.): Scale-Space 2003, LNCS 2695, pp. 325–339, 2003.

and its backtransformation

$$\mathcal{F}^{-1}[g](x) := \int_{\mathbb{R}^n} g(u) \, e^{i2\pi u^\top x} \, du$$

one may conveniently compute a convolution in the spatial domain via a simple product in the Fourier domain:

$$\mathcal{F}[f * b] \;=\; \mathcal{F}[f] \cdot \mathcal{F}[b] \,.$$

In this context, Gaussians

$$K_\sigma(x) \;:=\; \frac{1}{(2\pi\sigma^2)^{n/2}} \, e^{-\frac{x^\top x}{2\sigma^2}}$$

play an important role as convolution kernels: They are the only separable and rotationally invariant function that preserve their shape under the Fourier transform. Convolutions of a signal f with the family $\{K_\sigma \mid \sigma > 0\}$ of Gaussians create the Gaussian scale-space [20,37,38], a multiscale representation that is useful in pattern recognition, image processing and computer vision [11,21,24,35]. Figure 1(a) shows an example.

Mathematical morphology is an interesting nonlinear alternative to linear systems theory [16,27,32,33,34]. It has been applied successfully to a large number of fields including cell biology, computer-aided quality control, mineralogy, remote sensing and medical imaging. Morphology is based on two fundamental processes: dilation and erosion. In the case of nonflat morphology, the dilation resp. erosion of some function $f : \mathbb{R}^n \to \mathbb{R}$ with a structuring function $b : \mathbb{R}^n \to \mathbb{R}$ can be defined as follows (see e.g. [25,36]):

$$(f \oplus b)(x) := \sup \, \{f(y) + b(x-y) \mid y \in \mathbb{R}^n\},$$
$$(f \ominus b)(x) := \inf \, \{f(y) - b(y-x) \mid y \in \mathbb{R}^n\}.$$

Dorst and van den Boomgaard [9] and Maragos [25] developed independently and simultaneously a morphological system theory that closely resembles linear system theory. Following [9], one may generalise the dilation to the tangential dilation via

$$(f \,\dot\oplus\, b)(x) \;:=\; \operatorname*{stat}_y \Big(f(y) + b(x-y)\Big)$$

with $\operatorname*{stat}_y f(y) := \{f(z) \mid \nabla f(z) = 0\}$. Then the morphological equivalent to the Fourier transform is given by the slope transform

$$\mathcal{S}[f](u) := \operatorname*{stat}_x (f(x) - u^\top x),$$

a transformation that is closely related to the Legendre transform and the Young–Fenchel conjugate in convex analysis. Its backtransformation is given by

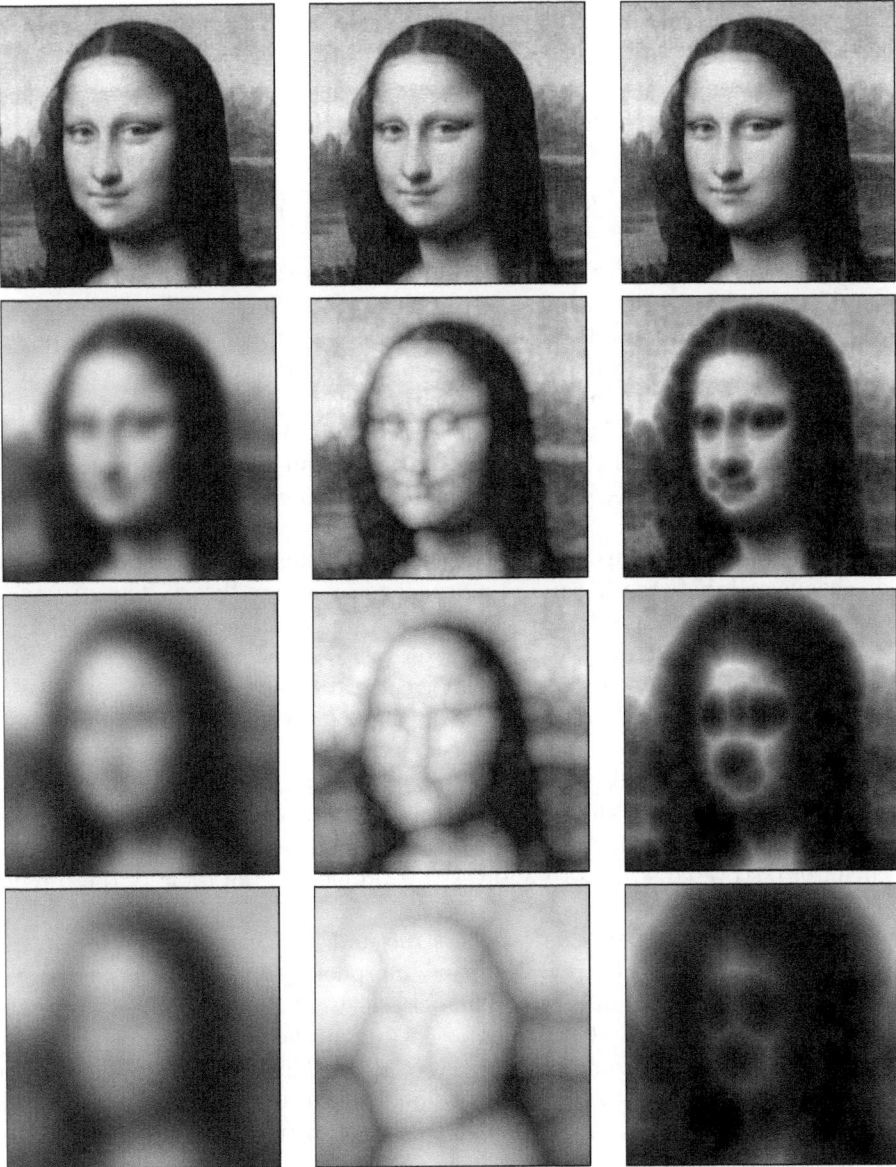

Fig. 1. Linear and morphological scale-spaces. **Top:** Mona Lisa painting by Leonardo da Vinci, 256×256 pixels. **(a) Left Column:** Gaussian scale-space, top to bottom: $\sigma = 0, 5, 10, 15$. **(b) Middle Column:** Dilation scale-space with quadratic structuring function, $t = 0, 0.25, 1, 4$. **(c) Right Column:** Erosion scale-space with quadratic structuring function, $t = 0, 0.25, 1, 4$.

$$\mathcal{S}^{-1}[g](x) \;=\; \operatorname*{stat}_{u}(g(u) + u^{\top}x).$$

The slope transform allows to replace the tangential dilation by simple addition in the slope domain:

$$\mathcal{S}[f \,\check{\oplus}\, b] \;=\; \mathcal{S}[f] + \mathcal{S}[b]\,.$$

Paraboloids

$$b(x,t) \;=\; -\frac{x^\top x}{4t} \qquad (t > 0)$$

are those structuring functions in morphological system theory that play a comparable role as Gaussians in linear system theory [36]: They are the only rotationally invariant and separable structuring functions that maintain their shape under the slope transformation. The corresponding dilation and erosion scale-spaces are depicted in Figure 1(b) and (c). For a detailed analysis of their scale-space properties, we refer to Jackway and Deriche [23]. Morphological scale-spaces with paraboloids as structuring functions are useful for computing Euclidean distance transformations [36], for image enhancement [31] and for multiscale segmentation [22].

From these discussions we observe that there seems to be an almost logarithmic connection between linear and morphological system theory. The structural similarities between linear and morphological processes have triggered Florack *et al.* [12,13] to construct a one-parameter process that incorporates Gaussian scale-space, and both types of morphological scale-spaces as limiting processes. Heijmans and van den Boomgaard [17,18] have investigated unifying algebraic definitions of scale-space concepts that include a number of linear and morphological approaches (cf. also [2]).

However, in spite of these very interesting contributions, the reason for the almost logarithmic connection between linear and morphological systems has not been discovered so far. To address this problem is the topic of the present paper.

We provide an explanation for the structural analogies between linear and morphological systems by revealing that morphology in essence is linear system theory in a specific algebra. While classical linear system theory uses the standard $(+, \times)$-algebra, the morphological system theory is based on the idempotent $(\max, +)$-algebra and the $(\min, +)$-algebra. This allows us to identify the nonlinear operations of erosion and dilation as linear convolutions $*_e$ and $*_d$ induced by these idempotent algebras. In this sense, morphology may be regarded as linear system theory in disguise.

These algebraic structures have already numerous interesting applications [4]: so-called discrete event dynamic systems (DEDS) can be modeled as *linear* systems with respect to these algebras. Discrete event dynamic systems in this algebraic formulation are used to find shortest paths in networks or to solve scheduling and communication problems in abstract project management, for instance. They are also employed to analyse queuing systems, traffic flow and the performance of special array processors. To the best of our knowledge, however, no attempt has been made so far to tackle problems from image analysis with this special algebraic approach.

A large part of our paper is devoted to the analysis of the role of the canonical integral transformations in the before mentioned algebras. First we show that the slope transform in the (max, +)-algebra corresponds to the logarithmic multivariate Laplace transform in the (+, ×)-algebra. We study relevant properties of this transform and point out links to convex analysis. This leads to the definition of the so-called Cramer transform as the Legendre-Fenchel transform of the logarithmic Laplace transform. The Cramer transform is well-known in the theory of large deviations in stochastics. In image analysis, it maps standard convolution to $*_e$-convolution, and Gaussians to quadratic functions. This explains why quadratic structuring functions are the morphological equivalent of Gaussian convolution kernels.

Our paper is organised as follows. In Section 2 we introduce the (max, +) and (min, +) algebras that will play a fundamental role for the analysis of morphological systems. In Section 3 we show that dilation and erosion are convolutions in these algebras. Connections to convex analysis are explained in Section 4, and the relations between the logarithmic Laplace transform and the Young-Fenchel conjugate are investigated in Section 5. Section 6 is devoted to the Cramer transform which constitutes the explanation for the logarithmic connection between linear and morphological systems. Finally we conclude our paper with a summary in Section 7.

2   The (max, +)- and the (min, +)-Algebra

In the theory of linear systems two algebraic structures play an important role: the (max, +)-algebra $\mathbb{R}_{max}$ and the (min, +)-algebra $\mathbb{R}_{min}$. Formally they emerge from the standard $(+, \cdot)$-algebra $(\mathbb{R}, +, \cdot)$ first by an extension of the real line with either the element $-\infty$ or $+\infty$, second by replacing the addition by a max- or min-operation, and the multiplication by $+$. We have the following table:

name	set	addition	multiplication
standard algebra $\mathbb{R}$	$\mathbb{R}$	$+$	$\times$
(max, +)-algebra $\mathbb{R}_{max}$	$\mathbb{R} \cup \{-\infty\}$	max	$+$
(min, +)-algebra $\mathbb{R}_{min}$	$\mathbb{R} \cup \{+\infty\}$	min	$+$

The algebraic structures $\mathbb{R}_{max}$ and $\mathbb{R}_{min}$ are examples of idempotent semifields. The idempotency has to serve as a substitute for the non-existing inverse w.r.t. the max- or min- operations. For a rather exhaustive amount of details, see [4]. The structural importance of these algebraic structures will become clear in the next section.

3   Convolution Induced by an Algebra

We equip the range of a scalar-valued function with the algebraic structure introduced above, that is, we consider functions

$$f : \mathbb{R}^n \longrightarrow \mathbb{R}_{max} \quad \text{or} \quad f : \mathbb{R}^n \longrightarrow \mathbb{R}_{min}.$$

This gives rise to two analogs to the well-known convolution $*$ stemming from the standard-algebra $(\mathbb{R}, +, \times)$,

$$(f * g)(x) := \int_{\mathbb{R}^n} f(x - y) \cdot g(y) \, dy,$$

for all $x \in \mathbb{R}^n$. The transition from the standard algebra to the other algebras

$$(\mathbb{R}, +, \times) \quad \rightleftharpoons \quad (\mathbb{R} \cup \{+\infty\}, \min, +) \quad \text{or} \quad (\mathbb{R} \cup \{-\infty\}, \max, +)$$

amounts to the replacement of integration (=summation) by taking the infimum or the supremum, and the replacement of multiplication by addition. This leads to the definitions

$$(f *_d g)(x) := \sup_{y \in \mathbb{R}^n} \left(f(x - y) + g(y) \right) = \sup_{y \in \mathbb{R}^n} \left(f(y) + g(x - y) \right),$$

$$(f *_e g)(x) := \inf_{y \in \mathbb{R}^n} \left(f(x - y) + g(y) \right) = \inf_{y \in \mathbb{R}^n} \left(f(y) + g(x - y) \right).$$

Hence the morphological operations of dilation $\oplus$ and erosion $\ominus$ as given in [9] or [26] appear as convolutions w.r.t. these algebras:

$$(f \oplus g)(x) = \sup_{y \in \mathbb{R}^n} \left(f(y) + g(x - y) \right) = f *_d g(x), \tag{1}$$

$$(f \ominus g)(x) = \inf_{y \in \mathbb{R}^n} \left(f(y) - g(y - x) \right) = f *_e \overline{g}(x), \tag{2}$$

with $\overline{g}(x) := -g(-x)$. This explains the notations $*_e$ and $*_d$. Furthermore, the operation $*_e$ coincides exactly with the so-called inf-convolution or epigraphic addition in convex analysis [19,30], denoted sometimes by $\overset{+}{\vee}$ or $\square$. Of vital importance in this field is the Legendre–Fenchel transform, which is intimately connected to the slope transform known from morphology. In order to explore this connection, it is worthwhile to pursue a short excursion into convex analysis.

4   Elements of Convex Analysis

Let $\overline{\mathrm{Conv}}\mathbb{R}^n$ be the set of functions $f : \mathbb{R}^n \longrightarrow \mathbb{R} \cup \{+\infty\}$ which are closed convex, that is, convex, lower semicontinuous and finite in at least one point. Let $\langle \cdot, \cdot \rangle$ denote the standard scalar product in $\mathbb{R}^n$. f has an affine minorant iff $f \geq \langle t', \cdot \rangle - c$ for some $(t', c) \in \mathbb{R}^n \times \mathbb{R}$. The convolution $f *_e g$ of two convex functions f, g that have a common affine minorant is again convex. The inf-convolution is an asociative, commutative, order-preserving binary operation. Defining for any subset $A \in \mathbb{R}^n$

$$i_A(x) := \begin{cases} 0 & x \in A, \\ +\infty & \text{otherwise}, \end{cases} \tag{3}$$

$i_{\{0\}}$ is recognised as the neutral element. This corresponds to the structural element in [9,23]. In general, closedness is not preserved under inf-convolution.

Definition 1. *The* Legendre-Fenchel transform *or* conjugacy operation *associates with each f with an affine minorant the function f^* defined by*

$$f^*(x) := \sup_{t \in \mathbb{R}^n} [\langle t, x \rangle - f(t)].$$

Remarkably, $f^* \in \overline{\mathrm{Conv}}\,\mathbb{R}^n$ as soon as f is affinely minorised, regardless of its convexity or closedness [19]. In morphology this operation is a variant of the slope transform [9,25].

The next theorem states that the function cone $\overline{\mathrm{Conv}}\,\mathbb{R}^n$ is indeed very suitable for this transform: the Legendre-Fenchel transform leaves $\overline{\mathrm{Conv}}\,\mathbb{R}^n$ invariant and is even an involution on this cone. Also of importance are the algebraic properties of this transform with respect to $*_e$:

Theorem 1. (Properties of the Legendre-Fenchel Transform)

If $f, g \in \overline{\mathrm{Conv}}\,\mathbb{R}^n$ then

1. $f^* \in \overline{\mathrm{Conv}}\,\mathbb{R}^n$.
2. *The Legendre-Fenchel transform is its own inverse:* $(f^*)^* = f.$
3. *It maps sums into erosions:* $(f + g)^* = f^* *_e g^*.$
4. *It also maps erosions into sums:* $(f *_e g)^* = f^* + g^*.$

For proofs of these assertions and more detailed results on the properties of conjugation, the reader is referred to [19]. Item 4 deserves a little remark: the conjugacy operation transforms convolution $*_e$ into the sum of the conjugates. It is not an incident that property 4 resembles very much the behaviour of the Laplace transform with respect to the standard convolution $*$ in the $(\mathbb{R}, +, \times)$-algebra. For any function $f : \mathbb{R}^n \longrightarrow [0, +\infty]$, we define the *multivariate Laplace transform* by

$$L[f] : x \longmapsto L[f](x) := \int_{\mathbb{R}^n} e^{\langle x, y \rangle} f(y)\, dy \qquad \text{with } x \in \mathbb{R}^n.$$

Indeed, $*$-convolution of functions is transformed into a multiplication of the Laplace transforms:

$$\int_{\mathbb{R}^n} e^{\langle x, y \rangle} \int_{\mathbb{R}^n} f(y - z) g(z)\, dz\, dy = \int_{\mathbb{R}^n} \int_{\mathbb{R}^n} e^{\langle x, y - z \rangle} f(y - z) e^{\langle x, z \rangle} g(z)\, dz\, dy$$

$$= \int_{\mathbb{R}^n} e^{\langle x, t \rangle} f(t)\, dt \cdot \int_{\mathbb{R}^n} e^{\langle x, z \rangle} g(z)\, dz.$$

To avoid confusion it should be mentioned that usually the (one-sided) Laplace transform is defined by

$$L_I[f](p) := \int_0^{+\infty} f(t)\, e^{-pt}\, dt$$

with $t \in \mathbb{R}$ and a complex number p. Hence the multivariate integral transform above is an n-dimensional generalisation of the so-called two-sided Laplace transform

$$L_{II}[f](p) := \int_{-\infty}^{+\infty} f(t)\, e^{-pt}\, dt.$$

More details and the close connection between the last two variants of the transform are discussed in [8].

5   A Link between Laplace Transform and Conjugation

Starting from the definition of the conjugacy operation the transition ($\rightleftharpoons$) from the $(\max, +)$-algebra to the $(+, \times)$-algebra entails

$$f^*(x) = \sup_{y \in \mathbb{R}^n} (\langle y, x \rangle - f(y)) = \log \sup_{y \in \mathbb{R}^n} \left(e^{\langle y, x \rangle - f(y)} \right)$$

$$\rightleftharpoons \log \int_{\mathbb{R}^n} e^{\langle y, x \rangle - f(y)} \, dy = \log \int_{\mathbb{R}^n} e^{\langle y, x \rangle} \cdot e^{-f(y)} \, dy = \log L[e^{-f}](x) \,.$$

In other words: the conjugate of f interpreted in the context of the $(\max, +)$-algebra corresponds to this logarithmic Laplace transform of e^{-f} in the standard algebra. A logarithmic relation between the two transforms becomes obvious: essentially it traces back to the homomorphism provided by the logarithm:

$$\log(a \cdot b) = \log a + \log b \,.$$

When compared to Theorem 1 (items 1 and 4), the following proposition emphasises the correspondence between conjugation and logarithmic Laplace transform.

Proposition 1. (Properties of the Logarithmic Laplace Transform)
For any functions $f, g : \mathbb{R}^n \longrightarrow [0, +\infty]$ with $f, g \not\equiv 0$ one has:

1. *The logarithmic Laplace transform is always convex and lower semicontinuous for non-negative functions:*

$$\log L[f] \in \overline{Conv} \, \mathbb{R}^n \,.$$

2. *Convolutions are maped into sums:*

$$\log L[f * g] = \log L[f] + \log L[g] \,.$$

PROOF: 1. Suppose $0 < \alpha < 1$, then

$$L[f](\alpha x_1 + (1 - \alpha)x_2) = \int_{\mathbb{R}^n} e^{\langle \alpha x_1 + (1-\alpha)x_2, y \rangle} f(y) \, dy$$

$$= \int_{\mathbb{R}^n} \left(e^{\langle x_1, y \rangle} \right)^\alpha \cdot \left(e^{\langle x_2, y \rangle} \right)^{(1-\alpha)} f(y) \, dy$$

$$\leq \left(L[f](x_1) \right)^\alpha \cdot \left(L[f](x_2) \right)^{1-\alpha}$$

by Hölder's Inequality with exponents $p = \frac{1}{\alpha}$ and $p' = \frac{1}{1-\alpha}$. Taking the logarithm proves the claimed convexity. The lower-semicontinuity follows directly from Fatou's lemma [5], since

$$\lim_{n \longrightarrow +\infty} x_n = x \quad \text{implies} \quad L[f](x) \leq \liminf_{n \longrightarrow +\infty} L[f](x_n),$$

for non-negative f and the fact that the logarithm is increasing and continuous. Property 2 follows directly from the properties of the Laplace transform and the logarithm.   □

In this context it is also worth mentioning that there is a continuous transition from the standard $*$-convolution of two positive functions f, g to their $*_e$-convolution, and again the logarithm makes its natural appearance. With reference to the usual Lebesgue norms $\| \cdot \|_p$ with $1 \leq p \leq +\infty$, we define for (strictly) positive functions f, g:

$$(f *_p g)(x) = \|f \cdot g(x - \cdot)\|_p \quad \text{for} \quad 1 \leq p \leq +\infty.$$

On the one hand, for $p = 1$ we regain the well-known convolution: $* = *_1$.
On the other hand, we infer directly from the definitions of the operations $*_d$ that

$$
\begin{aligned}
(f *_p g)(x) &= \|f \cdot g(x - \cdot)\|_p \\
&\xrightarrow{p \to +\infty} \|f \cdot g(x - \cdot)\|_\infty \\
&= \exp\left[\log\left[\sup_y (f(y) \cdot g(x - y))\right]\right] \\
&= \exp((\log f) *_d (\log g)(x)).
\end{aligned}
$$

In a similar fashion we obtain for not necessarily positive functions f, g:

$$\log((e^{-f} *_p e^{-g})(x)) \xrightarrow{p \to 1} \log((e^{-f} * e^{-g})(x))$$

as well as

$$
\begin{aligned}
\log((e^{-f} *_p e^{-g})(x)) &\xrightarrow{p \to \infty} \log\left[\sup_y (e^{-f(y)} \cdot e^{-g(x-y)})\right] \\
&= \log e^{-\inf_y (f(y) + g(x-y))} \\
&= -(f *_e g)(x).
\end{aligned}
$$

6   The Cramer Transform

The Cramer transform plays a key role in statistics, especially in the theory of large deviations [7,10]. From a functional point of view, it will allow us to make a connection between the usual convolution $*$, that appears in linear scale-space theory, and the morphological operations $\oplus$ and $\ominus$. This connection makes use of the convolutions $*_d$ and $*_e$. According to its appearance in statistics we will define the Cramer transform for non-negative functions only.

Definition 2. *For functions $f : \mathbb{R}^n \longrightarrow [0, +\infty]$, the transform*

$$C[f] := (\log L[f])^*$$

is called Cramer transform.

The reason why this transform is of importance in morphology is illuminated by the following theorem which is a direct consequence of the properties of the Laplace and Legendre-Fenchel transforms.

Theorem 2. (Convolution Theorem for the Cramer Transform)
If f and g are non-negative functions on $\mathbb{R}^n$, then

$$C[f * g] = C[f] *_e C[g].$$

In view of equations (1) and (2) this entails for nonnegative functions $f, g \neq 0$ the relations

$$-C[f * g] = (-C[f]) \oplus (-\overline{C[g]})$$

and

$$C[f * g] = C[f] \ominus \overline{C[g]}.$$

Let us now discuss some properties of the Cramer transform.
First we observe that, according to Proposition 1, the Cramer transform maps any non-negative function into $\overline{\text{Conv}} \, \mathbb{R}^n$. Hence it follows from Theorem 1 (2) that the conjugate of the Cramer transform is the logarithmic Laplace transform:

$$C^*[f] = \log \, L[f].$$

Examples of Cramer Transforms.

1. Let δ_a denote the Dirac measure in $a \in \mathbb{R}^n$. Then

$$C[\delta_a] = i_a$$

with i_a being defined in (3).

2. The Cramer transform is not additive:

$$C[(1 - p) \, \delta_0 + p \, \delta_1](x) = x \cdot \log\left(\frac{x}{p}\right) + (1 - x) \cdot \log\left(\frac{1 - x}{1 - p}\right) + i_{[0,1]}(x).$$

3. The Gauss distributions correlate to quadratic functions with reciprocal "variance": As mentioned in [1,3] this means for the one dimensional Gaussian with mean μ and variance σ that

$$C\left[\frac{1}{\sqrt{2\pi\sigma^2}} e^{-\frac{1}{2}\left(\frac{\cdot - m}{\sigma}\right)^2}\right](p) = \frac{1}{2} \left|\frac{p - m}{\sigma}\right|^2.$$

This can be extended to the n-variate case of a Gaussian with diagonal covariance matrix. We give a proof of both assertions by first calculating the Laplace transform of a one-dimensional Gaussian with mean $\mu = 0$ and $\sigma^2 > 0$:

$$\int_{-\infty}^{\infty} \frac{1}{\sqrt{2\pi\sigma^2}} e^{-\frac{t^2}{2\sigma^2}} e^{p \cdot t} \, dt$$

$$= \frac{1}{\sqrt{2\pi\sigma^2}} \int_{-\infty}^{\infty} e^{-\frac{t^2}{2\sigma^2}} e^{-p \cdot t} \, dt$$

$$= \frac{1}{\sqrt{2\pi\sigma^2}} \left(\int_{0}^{\infty} e^{-\frac{t^2}{2\sigma^2}} e^{-p \cdot t} \, dt + \int_{0}^{\infty} e^{-\frac{t^2}{2\sigma^2}} e^{-(-p) \cdot t} \, dt \right)$$

$$= \frac{1}{2} e^{\frac{1}{2}\sigma^2 p^2} \left(\mathrm{Erfc}(\frac{1}{2} p \sqrt{2\sigma^2}) + \mathrm{Erfc}(-\frac{1}{2} p \sqrt{2\sigma^2}) \right)$$

$$= e^{\frac{1}{2}\sigma^2 p^2}$$

where the complementary error function

$$\mathrm{Erfc}(x) := \frac{2}{\sqrt{\pi}} \int_{x}^{+\infty} e^{-t^2} \, dt$$

is used according to formula 5.41 in [28].
For a Gaussian with mean μ it follows immediately by a simple change of variables that

$$\int_{-\infty}^{\infty} \frac{1}{\sqrt{2\pi\sigma^2}} e^{\frac{(t-\mu)^2}{2\sigma^2}} e^{p \cdot t} \, dt = e^{p\mu} e^{\frac{1}{2}\sigma^2 p^2}.$$

An n-variate Gaussian distribution with mean vector $\mu \in \mathbb{R}^n$ and diagonal covariance matrix $D = diag(\sigma_1^2, \ldots, \sigma_n^2)$ has the separable density

$$g(y) = \frac{1}{(\sqrt{2\pi})^n \cdot \prod_{i=1}^{n} \sigma_i} e^{-\frac{1}{2} \sum_{i=1}^{n} \frac{(t_i - \mu_i)^2}{\sigma_i^2}}.$$

Making use of the results above, Fubini's theorem immediately gives

$$\int_{\mathbb{R}^n} g(y) e^{\langle p, y \rangle} \, dy = \int_{\mathbb{R}^n} \prod_{i=1}^{n} \frac{1}{\sqrt{2\pi\sigma^2}} e^{-\frac{1}{2} \frac{(t_i - \mu_i)^2}{\sigma_i^2}} \, dy$$

$$= \prod_{i=1}^{n} \int_{\mathbb{R}} \frac{1}{\sqrt{2\pi\sigma^2}} e^{-\frac{1}{2} \frac{(t_i - \mu_i)^2}{\sigma_i^2}} \, dy_i$$

$$= \prod_{i=1}^{n} e^{p_i \mu_i} \cdot e^{\frac{1}{2} p_i^2 \sigma_i^2}$$

$$= e^{\langle p, \mu \rangle + \frac{1}{2} p^\top D p}.$$

Hence we have

$$\log L[g](p) = \langle p, \mu \rangle + \frac{1}{2} p^\top D p.$$

Furthermore a straightforward calculation gives an optimal $p = D^{-1}(s - \mu)$ in $\sup_{p \in \mathbb{R}}\{\langle s, p\rangle - \log L[g](p)\}$ which results in

$$(\log L[g])^*(s) = \left(p \mapsto \langle p, \mu \rangle + \frac{1}{2}p^\top Dp\right)^*(s) \tag{4}$$

$$= \frac{1}{2}\langle s - \mu, D^{-1}(s - \mu)\rangle \tag{5}$$

for $x \in \mathbb{R}^n$. This result is in complete accordance with the findings in [36].

4. We conclude this set of examples with the numerical evaluation of Cramer transforms of positive, piecewise consant one-dimensional signals f sampled at equidistant points over the interval $[0, 1[$. Defining the indicator function $\mathbf{1}_A$ as $\mathbf{1}_A(x) = 1$ if $x \in A$, and 0 otherwise, these signals are of the form

$$f(x) = \sum_{i=1}^n \alpha_i \mathbf{1}_{[\frac{i-1}{n}, \frac{i}{n}[}.$$

Their logarithmic Laplace transforms read as

$$\log L[f](s) = \log\left(\frac{1}{s}\sum_{i=1}^n \alpha_i \left(e^{s\frac{i}{n}} - e^{s\frac{i-1}{n}}\right)\right)$$

but the corresponding conjugates, that means their Cramer transforms, cannot be computed explicitly. Therefore we depict the graphs of some signals together with their Cramer transforms in Figure 2 below. These results bring to light the very strong smoothing property of the Cramer transform.

7   Conclusions

In this paper we have given an explanation for the almost logarithmic connection between linear and morphological systems. This has been achieved by regarding morphological systems as linear systems in appropriate algebras. The link between these algebras and the standard algebra in linear system theory has been established by means of the Cramer transform.

The present article can be regarded as a step towards the unification of linear and morphological scale–space theory on the basis of a general linear system theory in an appropriate algebra. Taking full advantage of this connection may allow to translate results directly from one area to the other. This may trigger a more fruitful interaction of both paradigms that have evolved independently to powerful image processing tools. Finally, a unification within a more general algebraic framework may also help to identify novel image processing approaches that are based on other algebras. These points will be addressed in our future publications.

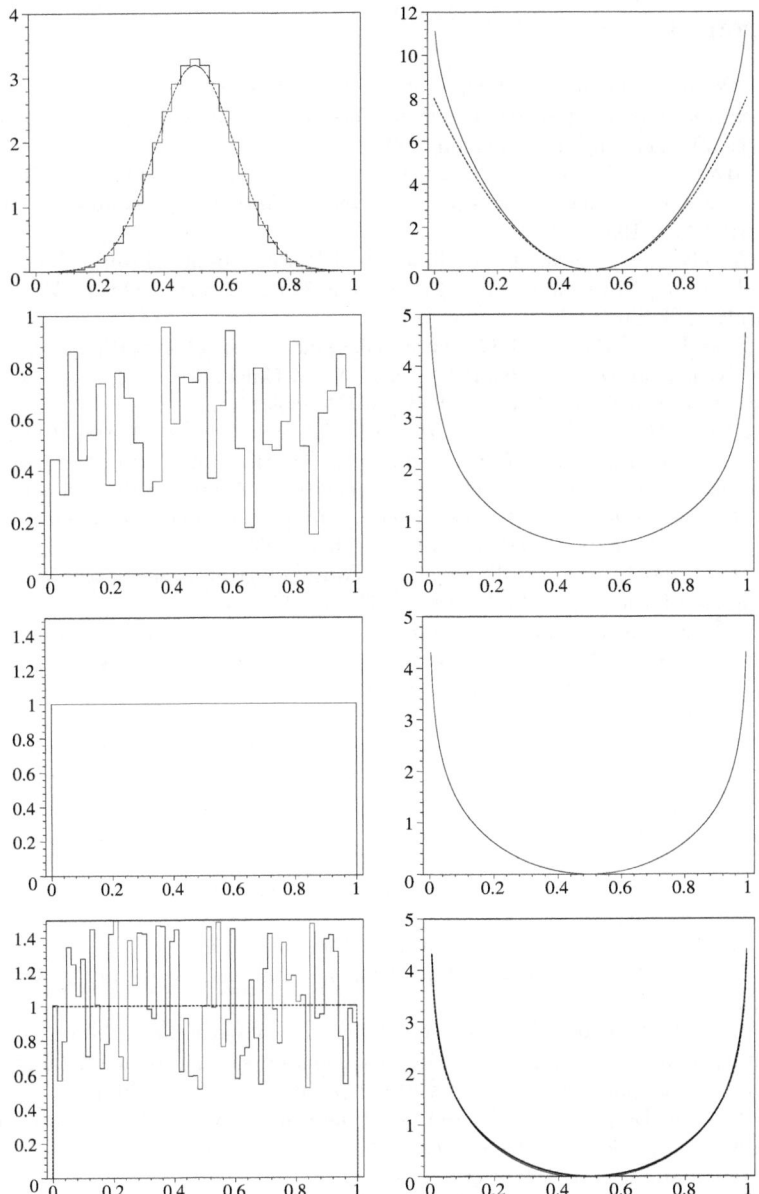

Fig. 2. The smoothing property of the Cramer transform (CT). **Top Row:** A Gaussian and its piecewise constant approximation on 33 subintervals (left) and their CTs (right). **2nd Row:** A random signal, piecewise constant on 33 subintervals (left) and its CT. **3rd Row:** 0-1 signal on the interval [0,1[(left) and its CT. **4th Row:** 0-1 signal on the interval [0,1[with 100% additive uniform noise (left) and its CT vs. the results of the 3rd row.

References

1. M. Akian, J. Quadrat, and M. Viot. Bellman processes. In *ICAOS '94: Discrete Event Systems*, volume 199 of *Lecture Notes in Control and Information Sciences*, pages 302–311. Springer, London, 1994.
2. L. Alvarez, F. Guichard, P.-L. Lions, and J.-M. Morel. Axioms and fundamental equations in image processing. *Archive for Rational Mechanics and Analysis*, 123:199–257, 1993.
3. R. Azencott, Y. Guivarc'h, R. Gundy, and P. Hennequin, editors. *Ecole d'Ete de Probalites de Saint-Flour VIII-1978*, volume 774 of *Lecture Notes in Mathematics*. Springer, Berlin, 1980.
4. F. Baccelli, G. Cohen, G. J. Olsder, and J. Quadrat. *Synchronization and Linearity: An Algebra for Discrete Event Systems*. Wiley, Chichester, 1992. www-rocq.inria.fr/scilab/cohen/SED/SED1-book.html.
5. H. Bauer. *Maß- und Integrationstheorie*. Walter de Gruyter, Berlin, 1990.
6. K. R. Castleman. *Digital Image Processing*. Prentice Hall, Englewood Cliffs, 1996.
7. J. Deuschel and D. W. Stroock. *Large Deviations*. Academic Press, Boston, 1989.
8. G. Doetsch. *Anleitung zum praktischen Gebrauch der Laplace-Transformation und der Z-Transformation*. Oldenbourg, München, 1967.
9. L. Dorst and R. van den Boomgaard. Morphological signal processing and the slope transform. *Signal Processing*, 38:79–98, 1994.
10. R. S. Ellis. *Entropy, Large Deviations, and Statistical Mechanics*, volume 271 of *Grundlehren der Mathematischen Wissenschaften*. Springer, New York, 1985.
11. L. Florack. *Image Structure*, volume 10 of *Computational Imaging and Vision*. Kluwer, Dordrecht, 1997.
12. L. Florack. Non-linear scale-spaces isomorphic to the linear case with applications to scalar, vector and multispectral images. *Journal of Mathematical Imaging and Vision*, 15(1–2):39–53, 2001.
13. L. M. J. Florack, R. Maas, and W. J. Niessen. Pseudo-linear scale-space theory. *International Journal of Computer Vision*, 31(2/3):247–259, Apr. 1999.
14. R. C. Gonzalez and R. E. Woods. *Digital Image Processing*. Addison–Wesley, Reading, second edition, 2002.
15. R. W. Hamming. *Digital Filters*. Dover, New York, 1998.
16. H. J. A. M. Heijmans. *Morphological Image Operators*. Academic Press, Boston, 1994.
17. H. J. A. M. Heijmans. Scale-spaces, PDEs and scale-invariance. In M. Kerckhove, editor, *Scale-Space and Morphology in Computer Vision*, volume 2106 of *Lecture Notes in Computer Science*, pages 215–226. Springer, Berlin, 2001.
18. H. J. A. M. Heijmans and R. van den Boomgaard. Algebraic framework for linear and morphological scale-spaces. *Journal of Visual Communication and Image Representation*, 13(1/2):269–301, 2001.
19. J.-B. Hiriart-Urruty and C. Lemarechal. *Fundamentals of Convex Analysis*. Springer, Heidelberg, 2001.
20. T. Iijima. Basic theory of pattern observation. In *Papers of Technical Group on Automata and Automatic Control*. IECE, Japan, Dec. 1959. In Japanese.
21. T. Iijima. *Pattern Recognition*. Corona Publishing, Tokyo, 1973. In Japanese.
22. P. T. Jackway. Gradient watersheds in morphological scale-space. *IEEE Transactions on Image Processing*, 5:913–921, 1996.
23. P. T. Jackway and M. Deriche. Scale-space properties of the multiscale morphological dilation–erosion. *IEEE Transactions on Pattern Analysis and Machine Intelligence*, 18:38–51, 1996.

24. T. Lindeberg. *Scale-Space Theory in Computer Vision*. Kluwer, Boston, 1994.
25. P. Maragos. Morphological systems: Slope transforms and max-min difference and differential equations. *Signal Processing*, 38(1):57–77, 1994.
26. P. Maragos. Differential morphology and image processing. *IEEE Transactions on Image Processing*, 5(6):922–937, 1996.
27. P. Maragos and R. W. Schafer. Morphological systems for multidimensional signal processing. *Proceedings of the IEEE*, 78(4):690–710, Apr. 1990.
28. F. Oberhettinger and L. Badii. *Tables of Laplace Transforms*. Springer, Berlin, 1973.
29. A. V. Oppenheim, R. W. Schafer, and J. R. Buck. *Discrete-Time Signal Processing*. Prentice Hall, Englewood Cliffs, second edition, 1999.
30. R. T. Rockafellar. *Convex Analysis*. Princeton University Press, Princeton, 1970.
31. J. G. M. Schavemaker, M. J. T. Reinders, J. J. Gerbrands, and E. Backer. Image sharpening by morphological filtering. *Pattern Recognition*, 33:997–1012, 2000.
32. J. Serra. *Image Analysis and Mathematical Morphology*, volume 1. Academic Press, London, 1982.
33. J. Serra. *Image Analysis and Mathematical Morphology*, volume 2. Academic Press, London, 1988.
34. P. Soille. *Morphological Image Analysis*. Springer, Berlin, 1999.
35. J. Sporring, M. Nielsen, L. Florack, and P. Johansen, editors. *Gaussian Scale-Space Theory*, volume 8 of *Computational Imaging and Vision*. Kluwer, Dordrecht, 1997.
36. R. van den Boomgaard. The morphological equivalent of the Gauss convolution. *Nieuw Archief Voor Wiskunde*, 10(3):219–236, Nov. 1992.
37. J. Weickert, S. Ishikawa, and A. Imiya. Linear scale-space has first been proposed in Japan. *Journal of Mathematical Imaging and Vision*, 10(3):237–252, May 1999.
38. A. P. Witkin. Scale-space filtering. In *Proc. Eighth International Joint Conference on Artificial Intelligence*, volume 2, pages 945–951, Karlsruhe, West Germany, August 1983.

Temporal Scale Spaces

Daniel Fagerström

Computational Vision and Active Perception Laboratory (CVAP)
Department of Numerical Analysis and Computing Science
KTH (Royal Institute of Technology) SE-100 44 Stockholm, Sweden
danielf@nada.kth.se

Abstract. In this paper we discuss how to define a scale space suitable for temporal measurements. We argue that such a temporal scale space should possess the properties of: temporal causality, linearity, continuity, positivity, recursitivity as well as translational and scaling covariance. It is shown that these requirements imply a one parameter family of convolution kernels. Furthermore it is shown that these measurements can be realized in a time recursive way, with the current data as input and the temporal scale space as state, i.e. there is no need for storing earlier input. This family of measurement processes contains the diffusion equation on the half line (that represents the temporal scale) with the input signal as boundary condition on the temporal axis. The diffusion equation is unique among the measurement processes in the sense that it is preserves positivity (in the scale domain) and is infinitesimally generated. A numerical scheme is developed and relations to other approaches are discussed.

1 Introduction

An important difference between spatial and temporal observation is that while all spatial directions basically can be treated in the same way there is certainly a difference between moving forward and backward in time. We have no access to future observations while we could have memorized earlier observations. A temporal observer must respect *causality*.

Furthermore, if we consider a causal scale space as an idealized device for performing real time measurement of a temporal signal *the observer cannot access the past*, only its memory of the past. An important property of a theory about real time temporal measurements is therefore that it has a time recursive formulation.

There have been several proposals on how to define a *causal scale space*. Koenderink [5] and Florack [3] map the half-line between current moment and the infinite past to a line by a logarithmic transformation. They then apply Gaussian scale space on the transformed signal. Lindeberg and Fagerström [7] base their axiomatization on the non-creation of local extrema. The axiomatization leads to scale spaces where either time or scale must be discrete and their scale spaces are not scale covariant. Salden et al. [11] derives the diffusion equation from conservation principles. They adapt their theory to the temporal

L.D. Griffin and M. Lillholm (Eds.): Scale-Space 2003, LNCS 2695, pp. 340–355, 2003.

domain by applying the diffusion equation on the past half-line, and by imposing reflecting boundary conditions.

The causal scale space theories of Koenderink [5], Florack [3] and Salden et al. [11] are formulated in terms of convolution against or diffusion on the past signal and have to our knowledge, no time recursive realization. The causal scale space theory of Lindeberg and Fagerström [7], has a time recursive formulation, but its lack of continuous formulation and scale covariance makes it less convenient to use as a basic theory about temporal observation.

In this article we will require a causal scale space to be temporally causal, linear, continuous, positive, having translational and scaling covariance and having a semigroup property. These requirements are the same as Pauwels et al. [8] used for defining scale spaces appropriate for spatial measurements, except for that we require temporal causality instead of reflectional symmetry. Each scale in the causal scale space is generated by convolving the input signal with a causal scale space kernel of a certain dilation. We show that there is a one parameter family of convolution kernels, know from probability theory as *extremal stable density functions*, that fulfills our requirements. We continue by showing that there is a time recursive realization of the causal scale spaces, using only the scale space it self as memory of earlier input. For only one of the parameter values, the recursive realization is infinitesimally generated. For this parameter the temporal scale space is given by the diffusion equation on the half line, a numerical scheme is given for this special case. We conclude by a comparison with earlier works.

2   Temporal Measurement

An temporal signal is usually thought about as a function from time coordinates to scalar values, $u(t)$. However, measurement of the instantaneous value of a signal is physically impossible, each measurement must take a non vanishing amount of time. Following the approach advocated by Florack [3] we instead start by designing our measurement apparatus, our space of *test functions* $\phi \in \Delta$, and try to make them as point like as possible . We then consider the temporal signal as a "black box", $u \in \Sigma$, that we can probe with our test functions, $u(\phi)$. Even if we cannot measure the instantaneous value of the signal we still need to associate each measurement with a certain moment, so that the measurement can be said to be performed at time t. To accomplish this a *measurement time*:

$$\pi(\phi) = t, \qquad \pi : \Delta \to \mathbb{R},$$

is defined for each test function, and the notation $\phi_t \in \pi^{-1}(t)$, is used for denoting a test function applied at time t.

For a seeing system with a wide range of visual competences it seems reasonable that the representation of the visual measurement as far as possible should avoid semantic interpretation of the environment, it should be *uncommitted* [6]. The representation of the input should just *embody* general structure in the environment.

2.1   Linearity

The measurement process is supposed to be linear,

$$u(\alpha\phi + \beta\psi) = \alpha u(\phi) + \beta u(\psi),$$

with $\alpha, \beta \in \mathbb{R}$. We also require that the sensor functions are well behaved in the following sense:

$$\Delta \subset L_1 \cap C^{\infty} \cap \{\phi | \phi(-\infty) = \phi(\infty) = 0\}.$$

The signal space Σ is defined as the topologically dual space to the space of test functions, $\Sigma = \Delta'$ [14]. In many cases a signal $u \in \Sigma$ can be represented, using the Riesz representation theorem, as $u(\phi) = \int u'(t)\phi(t)\,dt$, for some function u'. Differentiation in the signal space can be defined by

$$(\partial_t u)(\phi) = -u(\partial_t \phi). \tag{1}$$

We don't have to worry about the differentiability of the signals, it suffices to have differentiable test functions. The action of a diffeomorphism f on a signal can be defined as:

$$fu(\phi) = u(|J_{f^{-1}}|\phi \circ f^{-1}). \tag{2}$$

   Our main task is to define a *minimal* space of test functions, that suits the needs of the observation task.

2.2   Causality

A temporal measurement should not involve any future information.

$$\phi_t(t') = 0, \forall t' > t \tag{3}$$

This formalizes the differences between time and space that was discussed above.

2.3   Covariance

The structure of the sensor system should correspond to regularities in the environment. There is no a priori reason to believe that a certain moment of time, or a certain span of time should posses properties different from all the rest. This should be reflected in the measurement process: it should be *translation covariant*, the measurement process should be the same for each moment of time, and it should be *scaling covariant*, all length of time spans should be treated in the same way. To obtain translation and scaling covariance, a family of test functions $\phi'_{t,\tau} : \mathbb{R}\mathbb{R}_+ \to \Delta$, indexed by time t and scale τ is needed. Measurement at scale τ, is denoted by:

$$\Phi_\tau u(t) = u(\phi'_{t,\tau}), \qquad \Phi_\tau : \Sigma \to C^{\infty}. \tag{4}$$

Measurement is translation covariant if

$$\Phi_\tau T_a = T_a \Phi_\tau, \tag{5}$$

where $T_a f(x) = f(x - a)$. From the left hand side we get,

$$\Phi_\tau T_a u(t) = T_a u(\phi'_{t,\tau}) = u(T_{-a}\phi'_{t,\tau}),$$

by using equation (2), and from the right hand side we get,

$$T_a \Phi_\tau u(t) = \Phi_\tau u(t - a) = u(\phi'_{t-a,\tau})$$

As this holds for all u we get,

$$T_a \phi'_{t,\tau} = \phi'_{t+a,\tau},$$

and

$$\phi'_{t,\tau}(t') = T_t \phi'_{0,\tau}(t') = \phi'_{0,\tau}(t' - t) = \phi_{0,\tau}(t - t'),$$

where $\phi_{t,\tau}(t) = \phi'_{t,\tau}(-t)$. Using this in equation (4), we get,

$$\Phi_\tau u(t) = u(\phi_{0,\tau}(t - \cdot)) = u * \phi_{0,\tau}(t) = u * \phi_\tau(t), \tag{6}$$

where the convolution is in distribution sense and $\phi_\tau = \phi_{0,\tau}$. We also note that while convolution kernels are reflected compared to measurement functions, the temporal causality requirement for convolution kernels becomes:

$$\phi_\tau(t) = 0, \forall t < 0. \tag{7}$$

Scaling covariance has a somewhat more complicated form:

$$\Phi_\tau S_\gamma = S_\gamma \Phi_{\psi(\gamma)\tau}, \tag{8}$$

where $S_\gamma f(x) = f(\gamma x)$ and ψ is an invertible function such that $\psi(0) = 0$. Scaling up a signal and then applying measurement devices of a certain size corresponds to using smaller measurement devices at the original signal and performing the scaling afterward. Why there is a need for a coordinate change ψ on the rescaling parameter γ, will be clear later. From the left hand side of equation (8), and equation (2) we get

$$\Phi_\tau S_\gamma u(t) = S_\gamma u * \phi_\tau(t) = u(D_\gamma \phi_\tau(t - \cdot)),$$

where $D_\gamma f(x) = (1/\gamma)f(x/\gamma)$, and from the right hand side we get,

$$S_\gamma \Phi_{\psi(\gamma)\tau} u(t) = \Phi_{\psi(\gamma)\tau} u(\gamma t) = u(\phi_{\psi(\gamma)\tau}(\gamma t - \cdot)),$$

and by combining these two expressions and setting $t = 0$ we have,

$$D_\gamma \phi_\tau = \phi_{\psi(\gamma)\tau}. \tag{9}$$

2.4   Cascade Property

The result of a measurement of a signal can be considered to be a signal in turn. It then seems reasonable that a measurement of a measurement should correspond to single measurement. This can be formalized as:

$$\Phi_\tau \Phi_{t'} = \Phi_{t+t'}, \tag{10}$$

that is measurements form a semigroup. The semigroup property of measurement means that the measurement kernels form a convolution algebra. From the left hand side,

$$\Phi_\tau \Phi_{t'} u = (\Phi_{t'} u) * \phi_\tau = (u * \phi_{t'}) * \phi_\tau = u * (\phi_{t'} * \phi_\tau),$$

and from the right hand side, $\Phi_{t+t'} u = u * \phi_{t+t'}$, and by combining these we get,

$$\phi_\tau * \phi_{\tau'} = \phi_{t+t'}. \tag{11}$$

2.5   Extended Point

As mentioned earlier a measurement should approach a pointwise value of the signal. This could be described as:

$$\lim_{\tau \to 0} \phi_\tau = \delta, \tag{12}$$

the measurement kernel approaches a Dirac pulse. We also want the measurement kernel to have *unit area*:

$$\int \phi_\tau = 1, \tag{13}$$

and to be *positive*,

$$u \geq 0 \Rightarrow u(\phi_\tau) \geq 0. \tag{14}$$

3   Characterization of Causal Scale Space Kernels

We can now summarize the requirements on measurement kernels for a causal scale space.

Definition 1. $\phi_\tau(t)$ *is called a causal scale space kernel if it possesses the properties of: Continuity, Positivity (Eq. 14), Unit area (Eq. 13), Temporal causality (Eq. 7), Dilation covariance (Eq. 9), Convolution semigroup (Eq. 11)*

Pauwels et al [8] used the same axioms, with temporal causality replaced by reflection symmetry, to characterize spatial scale spaces (see Weickert et al. [15], for comparison between different scale space axiomatizations).

We now derive the form of causal scale space kernels in a sequence of lemmas.

Lemma 1 (Convolution semigroup). *Let $\phi_\tau(t)$ be an absolutely integrable causal function, then it is a convolution semigroup iff its Laplace transform $\tilde{\phi}_\tau(s) = e^{-g(s)\tau}$, where $g(s) \in \mathbb{R}$ for $s \in \mathbb{R}$.*

Proof. From absolute integrability it can be conclude that the Laplace transform of ϕ_τ exists and is given by

$$\tilde{\phi}_\tau(s) = \mathcal{L}[\phi_\tau(\cdot)] = \int_0^\infty e^{-st}\phi_\tau(t)\, dt.$$

In the Laplace transform domain the convolution semigroup property becomes $\tilde{\phi}_{\tau_1}\tilde{\phi}_{\tau_2} = \tilde{\phi}_{\tau_1+\tau_2}$. This is an instance of Cauchy's functional equation $f(x + y) = f(x)f(y)$. For continuous real functions it has the unique solution $f(x) = e^{-cx}, c \in \mathbb{R}$, (see e.g. [1]) and hence

$$\tilde{\phi}_\tau(s) = e^{-g(s)\tau}, \tag{15}$$

where $g(s) \in \mathbb{R}$.

Now we will use the dilation covariance to further restrict the form of ϕ_τ.

Lemma 2 (Dilatation covariant convolution semigroup). *Let $\phi_\tau(t)$ be a absolute integrable causal function, then it is a dilation covariant convolution semigroup iff its Laplace transform $\tilde{\phi}_\tau(s) = \tilde{\phi}_{\alpha,\tau}(s) = e^{-s^\alpha\tau}$, for $s \geq 0$ and $\alpha > 0$.*

Proof. The Laplace transform of the dilation covariant equation is

$$\tilde{\phi}_\tau(\gamma s) = \tilde{\phi}_{\psi(\gamma)\tau}(s). \tag{16}$$

Substituting (15) in (16) one obtains

$$e^{-g(\gamma s)\tau} = e^{-\psi(\gamma)g(s)\tau} \tag{17}$$

and, since the exponential function is invertible,

$$g(\gamma s) = \psi(\gamma)g(s) \tag{18}$$

must hold. Without loss of generality one can assume that $g(1) = 1$, and by inserting $s = 1$ in (18), we get $\psi(\gamma) = g(\gamma)$, and (18) becomes

$$g(\gamma s) = g(\gamma)g(s). \tag{19}$$

By setting $g(t) = f(\log(t))$, we can see that (19) is another form of Cauchy's functional equation, and that g must have the form $g(s) = s^\alpha$, $\alpha \in \mathbb{R}$. Substituted into (15), we get

$$\tilde{\phi}_{\alpha,\tau}(s) = e^{-s^\alpha\tau}. \tag{20}$$

From the proof above we also get the form for the function $\psi(\gamma) = \gamma^\alpha$, for the dilation covariance property (9)

Corollary 1. *Let $\{\phi_\tau | \tau \geq 0\}$ be a convolution semigroup, where ϕ_τ are absolutely integrable causal functions, then the functions are dilation covariant iff*

$$D_\gamma \phi_\tau = \phi_{\gamma^\alpha \tau}. \tag{21}$$

We still need to determine for what values of the parameter α that $\tilde{\phi}_\tau$ is a Laplace transform of a normalizing non negative function. To be able to do this we need a few facts from Laplace transform theory.

Definition 2. *A function f on $\mathbb{R}_+$ is called completely monotone (see e.g. [2]), if it has derivatives of all orders and fulfill*

$$(-1)^n f^{(n)}(s) \geq 0, \quad s > 0. \tag{22}$$

Theorem 1 (Bernstein). *A function f on $\mathbb{R}_+$ is the Laplace transform of a non negative normalizing function, iff it is completely monotone and $f(0) = 1$, (see e.g. [2]).*

More specifically a function e^{-f} is completely monotone if f is a positive function with a completely monotone derivative.

From this we can prove our main theorem:

Theorem 2 (The form of causal scale space kernels). *$\phi_{\alpha,\tau}(t)$ is a causal scale space kernel iff*

$$\phi_{\alpha,\tau}(t) = \mathcal{L}^{-1}[e^{-s^\alpha \tau}] \tag{23}$$

for a fixed $0 < \alpha < 1$.

Proof. s^α is a positive function if $\alpha \geq 0$ and it is completely monotone if $0 \leq \alpha \leq 1$. $\phi_{0,\tau} = \delta(t)$ and $\phi_{1,\tau} = \delta(t - \tau)$ respectively, and thus continuity implies that $\alpha \neq 0$ and $\alpha \neq 1$. For $0 < \alpha < 1$ $\phi_{\alpha,\tau}(t)$ is continuous for $t \geq 0$.

There does not seem to be any known closed form of $\phi_{\alpha,\tau}$, but a series expansion is possible.

$$\phi_{\alpha,\tau}(t) = \mathcal{L}^{-1}[e^{-s^\alpha \tau}] \tag{24}$$

$$= \sum_{k=0}^{\infty} \mathcal{L}^{-1}[\frac{(-s^\alpha \tau)^k}{k!}] \tag{25}$$

$$= \frac{1}{t} \sum_{k=0}^{\infty} \frac{(-\tau)^k}{k! \Gamma(-k\alpha)} t^{-k\alpha}. \tag{26}$$

Remark 1. For the particular case $\alpha = 1/2$ an explicit form of the causal scale space kernel is known to be

$$\phi_{1/2,\tau}(t) = \frac{\tau}{\sqrt{4\pi}} \frac{\exp(-\tau^2/4t)}{t^{3/2}} = -2\partial_\tau k_t(\tau), \quad t \geq 0 \tag{27}$$

where k_σ is the Gaussian function

$$k_\sigma(x) = \frac{e^{-x^2/4\sigma}}{\sqrt{4\pi\sigma}} \tag{28}$$

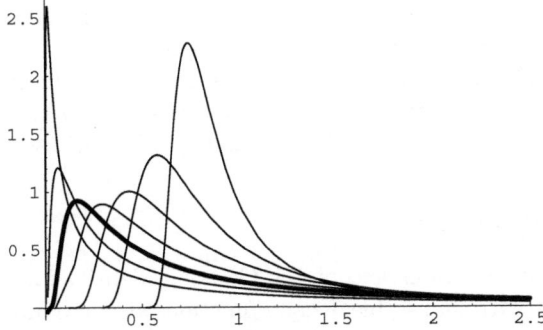

Fig. 1. $h_{\alpha,1}(t)$ for $\alpha = \{0.3, 0.4, \mathbf{0.5}, 0.6, 0.7, 0.8, 0.9\}$, where larger α corresponds to functions peaking further to the right and the thicker line corresponds to $\alpha = 0.5$.

4   Stable Density Functions

The one parameter family of causal scale space kernels derived above, is well known in the field of probability theory as *extremal stable density functions* (the results from probability theory reviewed here can be found in e.g. [2]). These functions were one of the findings from the various attempts to generalize the central limit theorem during the first half of the 20:th century. The central limit theorem basically states that if $X_1, X_2, \ldots$ are mutually independent one dimensional random variables with zero mean and finite variance, the distributions of the normalized sums

$$S_n = F_n(X_1 + \ldots + X_n),$$

where F_n is a normalization function with appropriate properties, tend to a normal distribution as $n \to \infty$.

P. Levy generalized the central limit theorem by removing the requirements on zero mean and finite variance. For this generalization, the densities of the limit sums of stochastic variables are denoted stable density functions, which is a two parameter family of functions. One sided (causal) stable density functions forms a one parameter subfamily called extremal stable density functions, the same family of functions that we derived from the causal scale space kernel axioms in definition 1. Symmetric stable density functions also form a one parameter subfamily of stable density functions, with the Cauchy and the Gaussian densities as notable members. The one parameter family of symmetric scale space kernel found by Pauwels et al. [8] is identical to the symmetric stable density functions.

Some of the relation between the generalized limit theorem and the causal scale space axioms (Definition 1) can roughly be sketched as follows: Density functions have unit area and are positive. Addition of stochastic variables correspond to convolution of their density functions. The dilation covariance requirement corresponds to properties of the normalization function in the limit theorems.

The stable density functions that are neither extremal nor symmetric can be said to lie between these two families and are more or less skew. If we were to define scale spaces (from the axioms in Definition 1 or the axioms in [8]), but without requiring causality or symmetry, the scale space kernels of those scale spaces would precisely correspond to the family of stable density functions.

4.1   Scaling Properties

All stable density functions except the Gaussian density function have infinite variance and the extremal stable density functions also have infinite mean. Stable density functions are also known to be unimodal. From causality and unimodality we can draw the conclusion that a temporal measurement give the highest weight to values of the input signals that occurred a while ago, (which can also be seen in Figure 1. This can be described as that there is a certain delay for each measurements, and this delay will be larger for measurement at larger scales. The delay cannot be described in terms of the mean as the mean is infinite for causal scale space kernels. The median i.e. the maximum of the density function can be used instead. For $\phi_{1/2,\tau}$ the median $t_m(\tau)$ easily can be shown to be

$$t_m(\tau) = \frac{\tau^2}{6},$$

by solving $(d/dt)\phi_{1/2,\tau}(t) = 0$, for t. For general causal scale space kernels it seems harder to find the median, but we can at least derive how the median is a function of scale modulo a constant.

Proposition 1. Let $T_{\alpha,m} = t_{\alpha,m}(1)$ then

$$t_{\alpha,m}(\tau) = T_{\alpha,m}\tau^{1/\alpha}. \tag{29}$$

Proof. By setting $\gamma^\alpha \tau = 1$ that is $\gamma = \tau^{-1/\alpha}$ in Equation 21 we get

$$\phi_{\alpha,1}(T_{\alpha,m}) = \frac{1}{\gamma}\phi_{\alpha,\tau}(\frac{T_{\alpha,m}}{\gamma}) = \frac{1}{\gamma}\phi_{\alpha,\tau}(t_{\alpha,m}(\tau))$$

and therefore

$$t_{\alpha,m}(\tau) = \frac{T_{\alpha,m}}{\gamma} = T_{\alpha,m}\tau^{1/\alpha}.$$

5   Recursive Formulation

A theory about temporal measurement must respect causality, the observer cannot access the future. Furthermore, *the observer cannot access the past*. The only way the observer can use information from its past is through its memory of past measurements, the observer must embody its past. That is, some kind of memory must be involved in the model and an important property for a temporal measurement theory is its memory model. From the result above, where the temporal measurement is described in terms of a convolution of a kernel with the past signal, it might seem like the memory must contain the whole past signal. However, we will show that, all information about the past that is needed is contained in the temporal scale space for the current moment. The scale space can be described in terms of an integrodifferential equation, where the temporal scale space evolves over time with the input signal as boundary condition.

5.1 Fractional Derivatives

The evolution equation involves fractional derivatives so we start by defining them.

Definition 3. $D^p_{x,+}$, is called the p-order left sided Riemann-Liouville fractional derivative [12] in x and is defined as,

$$D^p_{x,+} f(x) = \frac{1}{\Gamma(k-p)} (\partial_x)^k \int_0^x (x-y)^{k-p-1} f(y) \, dy, \quad (k-1 \le p < k) \quad (30)$$

$$= (\partial_x)^k \left(\frac{x^{k-p-1}_+}{\Gamma(k-p)} * f(x) \right), \tag{31}$$

where x_+ is equal to x for $x > 0$ and zero for $x \le 0$.

Fractional order derivatives are generalizations of ordinary derivatives. They are linear operators, satisfies a generalization of the Leibniz rule and integer order fractional derivatives are equivalent to ordinary derivatives.

Other families of fractional derivatives can be created by integrating over other intervals than in equation (30) [12]. We will also use left sided fractional derivatives in the sequel.

Definition 4. $D^p_{x,-}$, is called the p-order right sided Riemann-Liouville fractional derivative [12] in x and is defined as,

$$D^p_{x,-} f(x) = \frac{1}{\Gamma(k-p)} (-\partial_x)^k \int_x^\infty (y-x)^{k-p-1} f(y) \, dy, \qquad (k-1 \le p < k). \tag{32}$$

5.2 Infinitesimal Generator

Now we can state a partial integrodifferential equation that generates the causal scale spaces.

Theorem 3. The temporal scale space $u(t,x) = (\phi_{\alpha,t}(\cdot, x) * f)(t), 0 < \alpha < 1$, is the unique solution to the partial integrodifferential equation

$$\begin{cases} \partial_x u = -D^\alpha_{t,+} u \\ \lim_{x \to 0} u(t,x) = f(t) \end{cases} \tag{33}$$

Proof. A semigroup T, that fulfills a certain continuity requirement, *strong continuity*, is the unique solution to the *abstract Cauchy problem*, (see e.g. [4])

$$\begin{cases} \partial_t u = Au \\ \lim_{t \to 0} u(t) = f \end{cases} \tag{34}$$

A is denoted the *infinitesimal generator* of the semigroup, and is defined as $A = \lim_{h \to 0+} A_h$, where $A_h = (T(h) - I)/h$. A semigroup is strongly continuous if, for all f, $T(t)f$ is continuous in t on $\mathbb{R}_+$.

$\phi_{\alpha,\tau}$ is strongly continuous as a consequence of the continuity axiom (Definition 1). Its infinitesimal generator is readily found in the Laplace domain,

$$\mathcal{L}[A_h u] = \mathcal{L}[(\Phi_{\alpha,h} u - u)/h] = \frac{1}{h}(e^{-s^\alpha h} - 1)\mathcal{L}[u], \tag{35}$$

and

$$\mathcal{L}[A] = \lim_{h \to 0+} \frac{1}{h}(e^{-s^\alpha h} - 1) = -s^\alpha. \tag{36}$$

This is the Laplace transform of the left sided Riemann-Liouville fractional derivative of order α:

$$\mathcal{L}[D_{t,+}^\alpha u] = \mathcal{L}[(\partial_x)(\frac{x_+^{-\alpha}}{\Gamma(1-\alpha)} * u(x))] \tag{37}$$

$$= s\mathcal{L}[\frac{x_+^{-\alpha}}{\Gamma(1-\alpha)} * u(x)] - \left(\frac{x_+^{-\alpha}}{\Gamma(1-\alpha)} * u(x)\right)_{x=0} \tag{38}$$

$$= s\mathcal{L}[\frac{x_+^{-\alpha}}{\Gamma(1-\alpha)}]\mathcal{L}[u(x)] \tag{39}$$

$$= ss^{-(1-\alpha)}\mathcal{L}[u(x)] \tag{40}$$

$$= s^{-\alpha}\mathcal{L}[u(x)] \tag{41}$$

$$\tag{42}$$

It should be noted that we still lack a time recursive formulation, as the fractional operator $D_{t,+}^\alpha$ for $0 < \alpha < 1$ applied on the temporal signal is non-local, (it has in fact support on the whole half axis).

5.3   Evolution Equation

Interestingly enough, the partial integrodifferential equation from the theorem 3 above can be transformed to an partial integrodifferential equation that only applies a first derivative on the temporal signal.

Theorem 4. *The temporal scale space* $u(t,x) = (\phi_\alpha(\cdot,x) * f)(t), 0 < \alpha < 1$, *is the unique solution to the partial integrodifferential equation*

$$\begin{cases} \partial_t u = D_{x,-}^{1/\alpha} u \\ \lim_{x \to 0} u(t,x) = f(t) \\ u(0,x) = 0. \end{cases} \tag{43}$$

Proof. We need a linear operator A_x in x that satisfies:

$$\partial_t u = A_x u. \tag{44}$$

In the Laplace transform domain this becomes:

$$\mathcal{L}[\partial_t u] = se^{-s^\alpha x}\mathcal{L}[f] = \mathcal{L}[A_x u] = A_x \mathcal{L}[u] = A_x[e^{-s^\alpha x}]\mathcal{L}[f], \tag{45}$$

and thus A_x must satisfy,

$$A_x e^{-s^\alpha x} = se^{-s^\alpha x}. \tag{46}$$

A linear operator with this property is the right sided Riemann-Liouville fractional derivative defined above, we have [12]:

$$D_{x,-}^\beta e^{-\lambda x} = \lambda^\beta e^{-\lambda x}, \qquad \Re\lambda > 0, \tag{47}$$

setting $\beta = 1/\alpha$ and $\lambda = s^\alpha$, we can see that $A_x = D_{x,-}^{1/\alpha}$ satisfies (46). It is also known that equations of the type (43) have a unique solution.

Theorem 4 shows that the temporal scale space can be described in terms of an evolution equation on the half line where the position corresponds to temporal scale and to older information, (larger scale gives higher weight to older information). The input signal is fed to $x = 0$ where the evolution equation only applies a local operation (the temporal first derivative) on the signal in the present moment and as a consequence we have found a time recursive formulation of the causal scale spaces. For this realization of causal scale spaces the temporal scale space is the only needed memory of earlier input, and the content of the memory diffuses over time.

For $\alpha = 1/2$ we have $D_{x,-}^{1/\alpha} = D_{x,-}^2 = \partial_x^2$ and the evolution equation specializes to the *signaling equation*

$$\begin{cases} \partial_t u = \partial_x^2 u \\ \lim_{x\to 0} u(t,x) = f(t) \\ u(0,x) = 0. \end{cases} \tag{48}$$

The signaling equation describes how current is distributed in a semi infinite conductor when an temporally modulated electrical signal is applied at its end. It describes how heat is diffused in a semi infinite rod when a temporally modulated heat source is applied at its end, as well.

Theorem 5. *The signaling equation is unique among the evolution equations for causal scale spaces in the sense that it possesses booth locality and positivity in the scale domain.*

Proof. As already noted, equation (33) always uses a non local operator in the temporal direction. The evolution equation (43), becomes a partial differential equation for when $1/\alpha$, $0 < \alpha < 1$ is an integer i.e. for $\alpha = 1/k$, $k \geq 2$, where k is an integer. The locally generated evolution equations therefore has the form $\partial_t u = \partial_x^k u$, $k \geq 2$. And this equation has only a positive Greens function for $k = 2$, i.e. for $\alpha = 1/2$.

6   Discretization of Causal Scale Spaces

Booth the partial integrodifferential equations (33) and (43) can be numerically implemented by using discretization of fractional derivatives from e.g. [9]. Discretizations of fractional derivatives need to be computed for quite a large number of grid points to achieve a reasonable low numerical error. The locally generated signaling equation (48) can be much more efficiently implemented and we will focus on finding a numerical implementation for it.

An important step in finding a numerical implementation of the temporal scale space is to find a suitable discretization of the problem. Florack [3] states that the natural discretization of a space, is such that the grid steps are constant in the natural parametrization of the Lie group that generates the space. As we have decided that time is translation covariant we obtain constant intervals in the temporal direction. The scale is considered to be scale covariant which leads to a geometric progression of grid points in the scale direction.

6.1   Discrete Second Derivative on Log Spaced Grid

We need to derive a discretization of the second derivative operation for a grid with geometrical progression to be able to compute the signaling equation on such a grid.

We denote grid points by x_i and use the following notation.

$$u_i = u(x_i)$$
$$\Delta_{i+1} = x_{i+1} - x_i$$

Theorem 6 (Saulyev [13]). *For general non uniform grids the second derivative becomes:*

$$\frac{\partial^2 u_i}{\partial x^2} = \frac{2u_{i+1}}{\Delta_{i+1}(\Delta_{i+1} + \Delta_i)} + \frac{2u_{i-1}}{\Delta_i(\Delta_{i+1} + \Delta_i)} - \frac{2u_i}{\Delta_i \Delta_{i+1}} + o(\Delta_i), \qquad (49)$$

where the error term in general is of linear order in Δ_i.

If we specialize the above theorem for a grid with geometric progression:

$$x_i = x_0 h^i$$

where $h > 1$ we get an error term of order $o((h-1)^2)$.

Theorem 7. *For grids with geometric progression the second derivative becomes:*

$$\frac{\partial^2 u_i}{\partial x^2} = \frac{2}{x_i^2(h-1)(h-\frac{1}{h})}(u_{i+1} - (h+1)u_i + hu_{i-1}) + o((h-1)^2), \qquad (50)$$

where the error term is of quadratic order in $(h-1)$, if $(h-1)$ is small enough.

Proof. For a grid with geometric progression we have that:

$$\Delta_{i+1} = x_i h - x_i = x_i(h-1)$$
$$\Delta_i = x_i - x_i/h = x_i(1 - \frac{1}{h}).$$

Inserting this in equation (49), we get:

$$\delta_i^2 u = \frac{2u_{i+1}}{\Delta_{i+1}(\Delta_{i+1} + \Delta_i)} + \frac{2u_{i-1}}{\Delta_i(\Delta_{i+1} + \Delta_i)} - \frac{2u_i}{\Delta_i \Delta_{i+1}}$$

$$= \frac{2}{\Delta_{i+1}(\Delta_{i+1} + \Delta_i)} \left(u_{i+1} + \frac{\Delta_{i+1}}{\Delta_i} u_{i-1} - \frac{\Delta_{i+1} + \Delta_i}{\Delta_i} u_i \right)$$

$$= \frac{2}{x_i^2(h-1)(h-1/h)} \left(u_{i+1} + \frac{h-1}{1-1/h} u_{i-1} - \frac{h-1/h}{1-1/h} u_i \right)$$

$$= \frac{2}{x_i^2(h-1)(h-1/h)} \left(u_{i+1} + h u_{i-1} - (h+1)u_i \right).$$

For the error term we have:

$$e(h-1) = \sum_{j=3}^{\infty} \frac{2}{j!} \frac{\Delta_{i+1}^{j-1} + (-1)^j \Delta_i^{j-1}}{\Delta_{i+1} + \Delta_i} \frac{\partial^j u_i}{\partial x^j}$$

$$= \sum_{j=3}^{\infty} \frac{2}{j!} \frac{(h-1)^{j-1} + (-1)^j(1-1/h)^{j-1}}{(h-1) + (1-1/h)} x_i^{j-2} \frac{\partial^j u_i}{\partial x^j}$$

$$= \sum_{j=3}^{\infty} \frac{2(h-1)^{j-2}}{j!} \frac{h^{j-1} + (-1)^j}{h+1} x_0 \frac{\partial^j u_i}{\partial x^j}$$

$$= \sum_{j=3,5,7,\dots} \frac{2(h-1)^{j-1} \sum_{k=0}^{j-2} h^k}{j!} \frac{}{h+1} x_0 \frac{\partial^j u_i}{\partial x^j} +$$

$$= \sum_{j=4,6,8,\dots} \frac{2(h-1)^{j-2}}{j!} \frac{h^{j-1} + 1}{h+1} x_0 \frac{\partial^j u_i}{\partial x^j}$$

$$= o((h-1)^2).$$

7   Numerical Scheme for the Signaling Equation

Now we have what is needed for formulating a numerical scheme for the signaling equation. First we recall that for an explicit solution of the diffusion equation, $\Delta_t/\Delta_x^2 < 1/2$ must hold for the solution to be numerically stable (see e.g. [10]). This is fairly unattractive for the causal scale space as it means that for a given temporal sampling of the input signal there is a lower limit for how fine sampling we can choose in the scale domain. It therefore seems to be better to use an implicit numerical solution as it always is stable (see e.g. [10] for details about implicit solutions of the heat equation) .

For the temporal derivation the discretization

$$\delta_t u(t,x) = u(t,x) + \frac{3}{2}u(t - \Delta_t, x) - \frac{1}{2}u(t - 2\Delta_t, x),$$

is a good choice as it has quadratic stability $e = o(\Delta_t^2)$ and only uses past grid points. The proposed numerical scheme has second order stability booth in time and scale and can be implemented with four additions, four multiplications and two divisions per grid point.

8   Discussion

If we require the measurement kernels to both respect temporal causality and be scale covariant, the maximum (or the mean if it exists) will move backwards in time with increasing scale. Therefore we will never be able to measure what happens in the current moment, a measurement on a fine scale will reflect an event that happened just a short while ago while a measurement on a coarser scale will describe something that happened further back in time. A temporal measurement thus involves *two* different points in time: the one the measurement is performed at, t_0 and the one that has largest influence on the measurement t_m. The distance between these point is a function of scale $t_0 - t_m = f(\tau)$, the *influence curve*. From scale covariance considerations f typically should be on the form $f(\tau) = \beta\tau^\alpha$ for some $\alpha, \beta > 0$. Compare this with the situation for spatial measurements: reflectional or rotational symmetry means that the point where the measurement is performed also is the point that has maximal influence on the measurement.

Some earlier axiomatization of temporal scale spaces [5,7] have required non-creation of structure along the scale dimension. In Koenderink's axiomatization there were no solutions fulfilling temporal causality. He solved this by doing a remapping of the temporal dimension before applying ordinary scale space. In Lindeberg's and Fagerström's axiomatization there were solutions but they lacked scale covariance and had to be discrete in either time or space. From the above considerations about how larger scale leads to the measurement of events earlier in time the requirement of non-creation of structure along the scale dimension seems to be an unnecessarily strong. It might be more fruitful to require non-creation of structure along the influence curve $f(x)$ instead.

Koenderink motivates the logarithmic re-mapping of the time axis by analogy to our memories: we have finer temporal resolution on events taking place seconds ago than on events years ago. As already indicated, if we want the temporal scale space theory to be about measurement, we have to make careful distinction between the actual measurements and the memory of them. While a logarithmic mapping of time in the *memory* domain seem to be good first approximation of an uncommitted memory, we believe that the actual measurement process should be the same in every instant of time, i.e. it should be translationally invariant in time.

References

1. Aczél, J., Dhombres., J.: Functional Equations in Several Variables. Encyclopaedia of Mathematics and its Applications. Cambridge University Press (1989)
2. Feller, W.: An introduction to probability theory and its application, volume 2. John Willey & Sons, Inc. (1966)
3. Florack, L.M.J.: Image Structure. Series in Mathematical Imaging and Vision. Kluwer Academic Publishers, Dordrecht, Netherlands (1997)
4. Hille, E., Phillips, R.S.: Functional analysis and semi-groups. American Mathematical Society (1957)
5. Koenderink, J.J.: Scale-time. Biological Cybernetics **58** (1988) 169–162
6. Koenderink, J.J., Kappers, A., van Doorn, A.J.: Local operations: The embodiment of geometry. Orban, G.A., Nagel, H.H. (Eds), Artificial and Biological Vision Systems, Basic Research Series, Springer Verlag (1992) 1–23
7. Lindeberg, T., Fagerström, D.: Scale-space with causal time direction. Proc. 4th European Conference on Computer Vision, volume **1064**, Cambridge, UK, Springer Verlag, Berlin (1996) 229–240
8. Pauwels, E.J., VanGool, L.J., Fiddelaers, P., Moons, T.: An extended class of scale-invariant and recursive scale space filters. PAMI, **17**(7) (1995) 691–701
9. Podlubny, I.: Fractional Differential Equations. Academic Press (1999)
10. Richtmyer, R.D., Morton, K.W.: Difference Methods for Initial-Value Problems. Interscience Publishers, 2 ed. (1967)
11. Salden, A.H., Haar Romeny, B.M. ter, Viergever, M.A.: Linear scale-space theory from physical principles. JMIV, **9**(2) (1998) 103–139
12. Samko, S.G., Kilbas, A.A., Marichev, O.I.: Fractional integrals and derivatives : theory and applications. Gordon and Breach Science Publishers, cop., Yverdon (1992)
13. Saulyev, V.K.: Integration of Equatioons of Parabolic Type by the Method of Nets. Pergamon Press (1964)
14. Treves, F.:. Topological Vector Spaces, Distributions and Kernels. Academic Press (1967)
15. Weickert, J., Ishikawa, S., Imiya, A.: On the history of gaussian scale-space axiomatics. Gaussian Scale-Space Theory (1997) 45–59

Temporal Structure Tree
in Digital Linear Scale Space

Atsushi Imiya[1,2], Tateshi Sugiura[3], Tomoya Sakai[2], and Yuichiro Kato[4]

[1] National Institute of Informatics, Japan
[2] Institute of Media and Information Technology, Chiba University, Japan
[3] School of Science and Technology, Chiba University, Japan
[4] Department of Information and Image Sciences, Chiba University, Japan
imiya@{media.imit.chiba-u.ac.jp,nii.ac.jp}

Abstract. This paper focuses on the computation of stationary curves, which are sometimes called fingerprints for one dimensional real signals in the linear scale space. Images for the analysis in the linear scale space are expressed as digital images for each quantized scale. Therefore, we develop a discrete version of the linear scale space analysis, employing the results of digital image analysis. For the application of linear scale space analysis to the time-varying images and objects, our method has advantages, because our method is based on the digital geometry on a plane which is suitable for the computation in digital computers.

1 Introduction

This paper focuses on the computation of stationary curves [2,5,6], which are sometimes called fingerprints for one dimensional real signals, in the linear scale space [1,7], and its application for the description of the topological feature of temporal images and objects. We develop a method for the topological analysis of digital topographical maps defined by gray-valued images for each scale using digital geometry. We deal with stationary-curves of two- and three-dimensional real functions.

Zhao and Iijima [2,3] proposed a unique hierarchical expression of a gray-valued image using stationary points on the stationary-curves in the linear scale space. For the achievement of their method and the application of scale space analysis to temporal images and objects, we are required to develop accurate and fast methods for the computation of the stationary-curves from images and objects. Therefore, we design a discrete method for the computation of singular points on the topography of an image for each scale in linear scale space, since images in the linear scale space are expressed as digital images for each quantized scale. From this expression of images in the scale space, digital image analysis achieves the scale space analysis.

We first introduce digital versions of the linear scale space. Second, we define stationary points and stationary-curves for the digital linear scale space. Finally, we develop a method for the detection of stationary points in the digital linear scale space. As an application, we derive a sequence of time varying structure

L.D. Griffin and M. Lillholm (Eds.): Scale-Space 2003, LNCS 2695, pp. 356–371, 2003.

tree, the stationary version of which is defined by Zhao and Iijima as topological feature extracted from images and objects in the scale space.

2 Stationary-Curves and Tree in Linear Scale Space

2.1 Linear Scale Space Analysis

In the two-dimensional Euclidean space $\mathbf{R}^2$, for an orthogonal coordinate system x-y defined in $\mathbf{R}^2$, a vector in $\mathbf{R}^2$ is expressed by $\boldsymbol{x} = (x, y)^\top$ where $\cdot^\top$ is the transpose of a vector. Setting $|\boldsymbol{x}|$ to be the length of $\boldsymbol{x}$, the linear scale-space transform for function $f(\boldsymbol{x})$, such that

$$f(\boldsymbol{x}, \tau) = \frac{1}{(\sqrt{4\pi\tau})^2} \int_{\mathbf{R}^2} f(\boldsymbol{y}) \exp(-\frac{|\boldsymbol{x} - \boldsymbol{y}|^2}{\tau}) d\boldsymbol{y}, \tag{1}$$

defines the general image of function $f(\boldsymbol{x})$. Therefore, function $f(\boldsymbol{x}, \tau)$ is defined in $\mathbf{R}^2 \times \mathbf{R}_+$ [1]. The function $f(\boldsymbol{x}, \tau)$ is the solution of the linear diffusion equation

$$\frac{df(\boldsymbol{x}, \tau)}{d\tau} = \Delta f(\boldsymbol{x}, \tau), \ \tau > 0, \ f(\boldsymbol{x}, 0) = f(\boldsymbol{x}). \tag{2}$$

The solution of eq. (2) is formally expressed

$$f(\boldsymbol{x}, \tau) = \exp(\Delta\tau) f(\boldsymbol{x}) \tag{3}$$

using the theory of Lie group [9].

Stationary points for the topographical maps in the scale space [1,2] are defined as the solutions of the equation $\nabla f(\boldsymbol{x}, \tau) = 0$. Using the second derivations of $f(\boldsymbol{x}, \tau)$, we classify the topological properties of the stationary points on the topographical maps. Since the second directional derivation of $f(\boldsymbol{x}, \tau)$ for point $\boldsymbol{x}$ is defined as

$$D_{\boldsymbol{x}}^2(\theta) = \frac{d^2}{d\boldsymbol{n}(\theta)} f(\boldsymbol{x}, \tau), \tag{4}$$

where $\boldsymbol{n}(\theta) = \boldsymbol{\omega} - \boldsymbol{x}$ for $\boldsymbol{\omega} = (\cos\theta, \sin\theta)^\top$, $0 \le \theta \le 2\pi$. Equation (4) is rewritten as

$$D_{\boldsymbol{x}}^2(\theta) = f_{xx}(\boldsymbol{x}, \tau) \cos^2\theta + 2f_{xy}(\boldsymbol{x}, \tau) \cos\theta \sin\theta + f_{yy}(\boldsymbol{x}, \tau) \sin^2\theta. \tag{5}$$

$D_{\boldsymbol{x}}^2(\theta)$ satisfies the periodic relation $D_{\boldsymbol{x}}^2(\theta + \pi) = D_{\boldsymbol{x}}^2(\theta)$. The topological properties of point $\boldsymbol{x}$ for each τ is classified into four classes: the local maximum, the local minimum, the saddle point, and the singular point, based on the relations

$$
\begin{aligned}
&f(\boldsymbol{x}, \tau) \text{ is a local maximum} && \text{if } D_{\boldsymbol{x}}^2(\theta) < 0, \\
&f(\boldsymbol{x}, \tau) \text{ is a local minimum,} && \text{if } D_{\boldsymbol{x}}^2(\theta) > 0, \\
&f(\boldsymbol{x}, \tau) \text{ is on a saddle point,} && \text{if } \max_{0 \le \theta < \pi} D_{\boldsymbol{x}}^2(\theta) > 0 \text{ and} \\
& && \quad \min_{0 \le \theta < \pi} D_{\boldsymbol{x}}^2(\theta) < 0, \\
&f(\boldsymbol{x}, \tau) \text{ is on a singular point, if } \max_{0 \le \theta < \pi} D_{\boldsymbol{x}}^2(\theta) = 0 \text{ or} \\
& && \quad \min_{0 \le \theta < \pi} D_{\boldsymbol{x}}^2(\theta) = 0.
\end{aligned}
\tag{6}
$$

The stationary-curves in the scale space are the collections of the stationary points. We denote the trajectories of the stationary points as $\boldsymbol{x}(\tau)$. Setting $\boldsymbol{H}$ to be the Hessian matrix of $f(\boldsymbol{x}, \tau)$, Zhao and Iijima [2] showed that the stationary-curves for a two-dimensional image are the solution of,

$$\boldsymbol{H}\frac{d\boldsymbol{x}(\tau)}{d\tau} = -\nabla\varDelta f(\boldsymbol{x}(\tau), \tau) \tag{7}$$

and clarified topological properties of the stationary-curves for two-dimensional patterns. Since the Hessian matrix is always singular for singular points, this equation is valid for nonsingular points. Their definitions are formally valid to functions defined in $\mathbf{R}^n$ for $n \geq 3$. In section 5, we apply their ideas to the analysis of objects in a space.

According to the second directional derivation, we can define three types of stationary-curves: maximum curves, minimum curves, and saddle curves. Furthermore, since the stationary-curves consist of many curves for $\tau > 0$, we call each curve a branch curve. The point $\boldsymbol{x}_\infty$ for $\lim_{\tau \to \infty} \boldsymbol{x}(\tau) = \boldsymbol{x}_\infty$ is uniquely determined for any image. We call a curve on which point $\boldsymbol{x}_\infty$ lies and a curve which is open to the direction of $-\tau$ the trunk and branch, respectively. On the top of each branch, a singular point exists. For the construction of unique hierarchical expression of stationary points, Zhao and Iijima [2] considered that the subroot of a branch is the stationary point of the top of the branch curve and a subroot and the trunk are connected by a line segment parallel to x-y plane.

Zhao and Iijima [2,3] defined the stationary points on the stationary-curves which satisfy $S(\boldsymbol{r}, \tau) = 0$ or isolated points with the conditions

$$\frac{dS(\boldsymbol{r}, \tau)}{d\tau} = 0, \quad \frac{d^2 S(\boldsymbol{r}, \tau)}{d^2 \tau} = 0, \tag{8}$$

for $S(\boldsymbol{r}, \tau) = |\frac{d\boldsymbol{r}(\tau)}{d\tau}|$. They also developed an algorithm to define a unique tree whose nodes are the stationary points on the stationary-curves, and introduced a unique hierarchical expression of an image using this tree. From the stationary points on the stationary-curves, the tree is constructed according to the order of the stationary points on the stationary-curves.

Denoting a stationary point on the stationary-curve as $(\boldsymbol{x}_i, \tau_i)$, $\boldsymbol{x}_i$ and τ_i are called the stable view-point and the field of vision, and that

$$f(\boldsymbol{x}, \boldsymbol{x}_i, \tau_i) = \exp(-\frac{|\boldsymbol{x} - \boldsymbol{x}_i|^2}{\tau_i})f(\boldsymbol{x}) \tag{9}$$

is called a view-controlled image of the original image, since $f(\boldsymbol{x}, \boldsymbol{x}_i, \tau_i)$ approximates an image in the region of interest $\boldsymbol{R}(\boldsymbol{x}_i, \tau_i)$,

$$\boldsymbol{R}(\boldsymbol{x}_i, \tau_i) = \{\boldsymbol{x} \,|\, |\boldsymbol{x} - \boldsymbol{x}_i| < \tau_i\}, \tag{10}$$

observed by a vision system which has mechanisms similar to those of the view-controlling system of human beings [1].

2.2 Structure Tree and View-Field

Using the radii of the fields of views for the stationary points, in this paper, we reformulate the order of points along the stationary-curves. On the trunk, if $\tau > \tau'$, we define the order of the stationary points as $\boldsymbol{x}(\tau) \succ \boldsymbol{x}(\tau')$. On each branch $\boldsymbol{x}_i(\tau)$, we write the stationary point $\boldsymbol{x}_i(\tau_{i(j)})$. We assume that the maximum scale parameter on this branch is $\tau_{i(0)}$. We also set $\boldsymbol{x}_{i(j)} = (x_{ij}, y_{ij})^{\top}$ for point $\boldsymbol{x}_i(\tau_j)$ in the scale space. We define the order of the stationary points on each branch using the fields of views. On each branch curve, for $\tau_{i(m)} > \tau_{i(n)}$, if the relation

$$|\boldsymbol{x}_{i(m)} - \boldsymbol{x}_{i(n)}| \leq \sqrt{2\tau_{i(m)}} \tag{11}$$

is satisfied, then we define $\boldsymbol{x}_{i(m)} \succ \boldsymbol{x}_{i(n)}$. This definition of the order on each branch means that on the plane $\tau = 0$, vector $\boldsymbol{x}_{i(n)}$ lies in the field of view of vector $\boldsymbol{x}_{i(m)}$. For large scale parameters, we cannot detect stationary point $\boldsymbol{x}_{i(n)}$. Conversely we can detect stationary point $\boldsymbol{x}_{i(n)}$ on the planes for the small scale parameters. Therefore, this order describes the order of topological structures of images in the scale-space.

This order based on the field of view also permits to merge stationary points among branches and the trunk. For a pair of branch curves $\boldsymbol{x}_i(\tau)$ and $\boldsymbol{x}_j(\tau)$ and a pair of fixed scales $\tau_{i(m)}$ and $\tau_{j(0)}$, if the relation

$$|\boldsymbol{x}_{i(m)} - \boldsymbol{x}_{j(0)}| \leq \sqrt{2\tau_{i(m)}} \tag{12}$$

is satisfied, then we define $\boldsymbol{x}_{i(m)} \succ \boldsymbol{x}_{j(0)}$.

These definitions for the order of the stationary points along the stationary-curves defines the hierarchal tree of the stationary points for an image. We call this tree the structure tree of an image. For example, if the orders of the stationary points are

$$\boldsymbol{x}_{\infty} \succ \boldsymbol{x}(\tau_1), \boldsymbol{x}(\tau_1) \succ \boldsymbol{x}_2(\tau_{2(0)}),$$
$$\boldsymbol{x}_2(\tau_{2(0)}) \succ \boldsymbol{x}_2(\tau_{2(1)}), \quad \boldsymbol{x}_2(\tau_{2(0)}) \succ \boldsymbol{x}_2(\tau_{2(2)}), \tag{13}$$

we obtain the tree for the stationary points as

$$T = \langle \boldsymbol{x}_{\infty} \langle \boldsymbol{x}(\tau_1), \langle \boldsymbol{x}_2(\tau_{2(0)}), \langle \boldsymbol{x}_2(\tau_{2(1)}), \boldsymbol{x}_2(\tau_{2(2)}) \rangle \rangle \rangle \rangle, \tag{14}$$

where $T = \langle r, \langle T_1, T_2 \rangle \rangle$ means that the root of tree T is r and T_1 and T_2 are both subtrees whose roots are r.

3 Digital Scale-Space Analysis

In this section, we derive the complete digital version of the linear scale-space transformation. For the relations between our method and methods in signal processing see appendix. We assume that functions are defined on one-, two-, and three-dimensional lattice points $\mathbf{Z}$, $\mathbf{Z}^2$, and $\mathbf{Z}^3$, and the scale parameter τ is nonzero, that is $\tau \in \mathbf{Z}_0$, where $\mathbf{Z}$ and $\mathbf{Z}_0$ are the set of integers and the set of non-negative integers. We set $f_n(\tau) = f(n, \tau)$, $f_{mn}(\tau) = f(m, n, \tau)$ and $f_{kmn}(\tau) = f(k, m, n, \tau)$ for one-, two-, and three-dimensional functions.

3.1   Digital Linear Scale Space

First, we deal with the digital diffusion for one-dimensional digital functions. The digital diffusion equation is defined as

$$f_{m+1}(\tau+1) - f_m(\tau) = \alpha \left(f_{m+1}(\tau) - \frac{1}{2} f_m(\tau) + f_{m-1}(\tau) \right) \tag{15}$$

with an appropriate boundary condition.

Since on digital computers, we are required to deal with functions with finite support, we assume that functions satisfy the cyclic boundary condition $f(m + M, \tau) = f(m, \tau)$ for an appropriate large integer M. This cyclic condition derives the matrix notation of the diffusion equation

$$\boldsymbol{f}(\tau+1) - \boldsymbol{f}(\tau) = \alpha \frac{1}{2} \boldsymbol{F} \boldsymbol{f}(\tau), \quad \boldsymbol{F} = \begin{pmatrix} -2 & 1 & 0 & \cdots & 0 & 1 \\ 1 & -2 & 1 & \cdots & 0 & 0 \\ \vdots & \vdots & \vdots & \ddots & \vdots & \vdots \\ 1 & 0 & 0 & \cdots & 1 & -2 \end{pmatrix}, \tag{16}$$

for $\boldsymbol{f}(\tau) = (f_1(\tau), f_2(\tau), \cdots, f_M(\tau))^\top$ Since matrix $\boldsymbol{F}$ is a circlant matrix, the discrete Fourier transform matrix (DFT Matrix)

$$\boldsymbol{U} = \left(\left(\exp(2\pi i \frac{mn}{M}) \right) \right), \quad m, n = 1, 2, \cdots, M \tag{17}$$

is the eigenmatrix of matrix $\boldsymbol{F}$ which orthogonalizes matrix $\boldsymbol{F}$ as $\boldsymbol{F} = \boldsymbol{U}\boldsymbol{D}\boldsymbol{U}^*$.

If we assume the boundary condition of the first type, that is, $f(0) = f(M + 1) = 0$, the linear diffusion equation is expressed as

$$\boldsymbol{f}(\tau+1) - \boldsymbol{f}(\tau) = \alpha \frac{1}{2} \boldsymbol{S}, \quad \boldsymbol{S} = \begin{pmatrix} -2 & 1 & 0 & \cdots & 0 & 0 \\ 1 & -2 & 1 & \cdots & 0 & 0 \\ \vdots & \vdots & \vdots & \ddots & \vdots & \vdots \\ 0 & 0 & 0 & \cdots & 1 & -2 \end{pmatrix}. \tag{18}$$

Furthermore, if we assume the boundary condition of the second type, that is, $f(1) - f(0) = f(M + 1) - f(M) = 0$, the linear diffusion equation is expressed as

$$\boldsymbol{f}(\tau+1) - \boldsymbol{f}(\tau) = \alpha \frac{1}{2} \boldsymbol{C}, \quad \boldsymbol{C} = \begin{pmatrix} -1 & 1 & 0 & \cdots & 0 & 0 \\ 1 & -2 & 1 & \cdots & 0 & 0 \\ \vdots & \vdots & \vdots & \ddots & \vdots & \vdots \\ 0 & 0 & 0 & \cdots & 1 & -1 \end{pmatrix}. \tag{19}$$

The eigenmatrix of $\boldsymbol{S}$ and $\boldsymbol{C}$ are the digital sine transform matrix (DST matrix) and the digital cosine transform matrix (DCT matrix), respectively. Since the natural boundary conditions for the discrete diffusion equation determine the DFT, DST, and DCT matrices, the digital Fourier analysis plays an important role in the linear scale-space analysis for the digital signals.

For these three conditions, the image for each scale is computed as

$$f(\tau) = T(\tau)f = WS(\tau)W^*f, \tag{20}$$

where $S(\tau)$ and W are a diagonal matrix whose parameter is the scale argument τ and the unitary matrix which diagonalizes $T(\tau)$, respectively. Since all diagonal elements of diagonal matrix $S(\tau)$ is in the form γ^τ for an appropriate real constant γ, the matrix $T(\tau)$ satisfies the relation

$$T(\tau + \tau') = T(\tau)T(\tau'), \ T(0) = I. \tag{21}$$

Furthermore, if all elements of $S(\tau)$ are positive. Then $f(\tau) \geq 0$ for $f \geq 0$.

Setting $f = f(0)$, the equation

$$f(\tau) = U\Lambda^\tau U^*f, \ \Lambda = I + \frac{1}{2}\alpha D \tag{22}$$

produces the functions in the scale space, where D is the diagonal matrix whose elements are eigenvalues of F, S, and C [1]. Furthermore, eq. (22) is a discrete version of the Lie-group-based expression of the linear diffusion equation defined in eq. (3).

For two-dimensional digital images, the diffusion equations with cyclic, the first type, and the second type boundary conditions, are expressed as

$$f(\tau + 1) - f(\tau) = \alpha^2(F \otimes F)f(\tau), \tag{23}$$

$$f(\tau + 1) - f(\tau) = \alpha^2\frac{1}{4}(S \otimes S)f(\tau), \tag{24}$$

$$f(\tau + 1) - f(\tau) = \alpha^2\frac{1}{4}(C \otimes C)f(\tau), \tag{25}$$

respectively, where $A \otimes B$ is the Kronecker product [2] of two matrices A and B.

For the scale space analysis of objects and images in a space, the discrete linear scale-equations are expressed as

$$f(\tau + 1) - f(\tau) = \alpha^3(F \otimes F \otimes F)f(\tau), \tag{26}$$

$$f(\tau + 1) - f(\tau) = \alpha^3\frac{1}{4}(S \otimes S \otimes S)f(\tau), \tag{27}$$

$$f(\tau + 1) - f(\tau) = \alpha^3\frac{1}{4}(C \otimes C \otimes C)f(\tau). \tag{28}$$

---

[1] The eigenvalues of F, S, and C are

$$2\cos 2\frac{k\pi}{M}, \ 2\cos\frac{(k+1)\pi}{M+1}, \ 2\cos\frac{k\pi}{2N}.$$

[2] If $A = U\Lambda U^*$ and $B = V\Sigma V^*$ for diagonal matrixes Λ and Σ, and unitary matrices U and V, the Kronecker product satisfies

$$A \otimes B = (U \otimes V)(\Lambda \otimes \Sigma)(U^* \otimes V^*).$$

3.2   Stationary Points on Digital Topographical Maps

For the computation of $\nabla f(x, y) = 0$ for each scale τ, we adopt points $\boldsymbol{x} = (m, n)^\top$ which satisfies the system of inequalities

$$\operatorname{sign}(f_{mn} - f_{m-1\,n}) \neq \operatorname{sign}(f_{m+1\,n} - f_{m,n}),$$
$$\operatorname{sign}(f_{mn} - f_{m\,n-1}) \neq \operatorname{sign}(f_{m\,n+1} - f_{mn}), \tag{29}$$

or the system of inequalities

$$\operatorname{sign}(f_{mn} - f_{m-1\,n-1}) \neq \operatorname{sign}(f_{m+1\,n+1} - f_{mn}),$$
$$\operatorname{sign}(f_{mn} - f_{m+1\,n-1}) \neq \operatorname{sign}(f_{m-1\,n+1} - f_{mn}). \tag{30}$$

The first inequality in the first system is the change of derivation along a line such that the tangent is 0, and the second inequality in first system of is the change of derivation along a line such that the tangent is infinity. Furthermore, the first inequality in second system is the change of derivation along a line such that the tangent is 1, and the second inequality in the second system is the change of derivation along a line such that the tangent is -1. Therefore, for point $(m, n)^\top$, the sign of first derivation changes along the vertical line $x = m$ and the horizontal line $y = n$ or along the two diagonal lines $y + x = m + n$ and $y - x = n - m$; we conclude that $\nabla f(m, n) = 0$. If eqs. (29) and (30) are satisfied for a small region $\mathbf{A}$ on $\mathbf{Z}^2$, we adopt the point $([\bar{x}], [\bar{y}])^\top$ for the average centroid $(\bar{x}, \bar{y})^\top$ in region $\mathbf{A}$.

For the classification of topological properties of stationary points on a discrete topographical map, we define the directional derivation on the discrete plane. In the 5×5 neighborhood of point $\boldsymbol{x} = (m, n)^\top$, we define the second derivation of $f(m, n)$ at point $\boldsymbol{x}$ as

$$f_{mn}^{(2)}(0) = \frac{1}{2}(f_{m+1\,n} - 2f_{mn} + f_{m-1\,n})$$

$$f_{mn}^{(2)}\left(\frac{\pi}{4}\right) = \frac{1}{2}(f_{m+1\,n+1} - 2f_{mn} + f_{m-1\,n-1})$$

$$f_{mn}^{(2)}\left(\frac{\pi}{2}\right) = \frac{1}{2}(f_{m\,n+1} - 2f_{mn} + f_{m\,n-1})$$

$$f_{mn}^{(2)}\left(\frac{3\pi}{4}\right) = \frac{1}{2}(f_{m-1\,n+1} - 2f_{mn} + f_{m+1\,n-1})$$

$$f_{mn}^{(2)}(\alpha) = \frac{1}{3}(f_{m+2\,n+1} - 2f_{mn} + f_{m-2\,n-1})$$

$$f_{mn}^{(2)}\left(-\alpha + \frac{\pi}{2}\right) = \frac{1}{3}(f_{m+1\,n+2} - 2f_{mn} + f_{m-1\,n-2})$$

$$f_{mn}^{(2)}\left(\alpha + \frac{\pi}{2}\right) = \frac{1}{3}(f_{m-1\,n+2} - 2f_{mn} + f_{m+1\,n-2})$$

$$f_{mn}^{(2)}(-\alpha) = \frac{1}{3}(f_{m+2\,n-1} - 2f_{mn} + f_{m-2\,n+1})$$

for $\alpha = \tan^{-1}\frac{1}{2}$. These definitions lead to the discrete approximation of $D_{\boldsymbol{x}}^2$ as

$$\Delta_{\boldsymbol{x}}^2(\theta) = f_{mn}^{(2)}(\theta) \tag{31}$$

for $\theta = 0$, α, $\frac{\pi}{4}$, $-\alpha + \frac{\pi}{2}$, $\frac{\pi}{2}$, $\alpha + \frac{\pi}{2}$, $\alpha + \frac{\pi}{2}$, $\frac{3\pi}{4}$, and $-\alpha$. Using the signs of $\Delta_{\boldsymbol{x}}^2$, we classify the topological properties of point $(m,n)^\top$. For the computation of $\Delta_{\boldsymbol{x}}^2$, we require four copies of function $f(m,n)$, that is, $f(m+2,n)$, $f(m+1,n)$, $f(m,n+1)$, and $f(m,n+2)$.

3.3   Digital Stationary-Curves

In the digital scale space $\mathbf{Z}^2 \times \mathbf{Z}_0$, setting $\boldsymbol{x}(\tau) = (m,n,\tau)^\top$ to be a singular point, we define the neighborhood of $\boldsymbol{x}(\tau)$ as

$$\mathbf{N}(\boldsymbol{x}(\tau)) = \{(m',n',\tau')^\top \,||m - m'| + |n - n'| \le 1, |\tau - \tau'| \le T(\tau)\} \qquad (32)$$

for an appropriate constant T which depends on τ. Furthermore, we set

$$\mathbf{N}^k(\boldsymbol{x}(\tau)) = \{(m',n',\tau')^\top \,||m - m'| + |n - n'| \le k, |\tau - \tau'| \le T(\tau)\}. \qquad (33)$$

For a collection of singular points in the digital linear scale space, using the following definitions, we construct the stationary-curves.

Definition 1. *For a singular point $\boldsymbol{x} = (m,n,\tau)^\top$ and $\boldsymbol{y} = (m',n',\tau')^\top$ in the scale space such that $\tau' > \tau$, if $\boldsymbol{y}$ is a unique point which lies in the neighborhood of $\mathbf{N}^k$, we say $\boldsymbol{x}$ and $\boldsymbol{y}$ are connected.*

Definition 2. *The stationary-curves in the digital scale space are collections of all paths which connect singular points.*

As an example, we computed the stationary points and the structure trees for a regular triangle and a Kaniza triangle. As shown in Figure 1, the topological properties of the stationary curves for a triangle and a Kaniza triangle are same. These numerical results show that there is no topological difference between a regular triangle and Kaniza triangle. The geometric and topological properties of these two triangles suggest that the equivalence of features in the scale space might cause the illusion that human being can detect a triangle in a Kaniza triangle which contains no practical edges and vertices of a triangle.

3.4   Combinatorial Property of Structure Tree

Setting τ_α to be an arbitral large scale, for $\tau > \tau_\alpha$, it is possible to approximate the topographical maps in the scale space by the function

$$\hat{f}(\boldsymbol{x}) = \sum_{n=1}^{n(\tau_\alpha)} f_n \exp(-\frac{|\boldsymbol{x} - \boldsymbol{p}_n|^2}{\sigma_n}). \qquad (34)$$

Setting $w_n = \ln f_n$, from eq. (34), we have the generalized Voronoi distance

$$d(\boldsymbol{x}, \boldsymbol{p}_n) = \exp(-\frac{|\boldsymbol{x} - \boldsymbol{p}_n|^2}{\sigma_n} + w_n) \qquad (35)$$

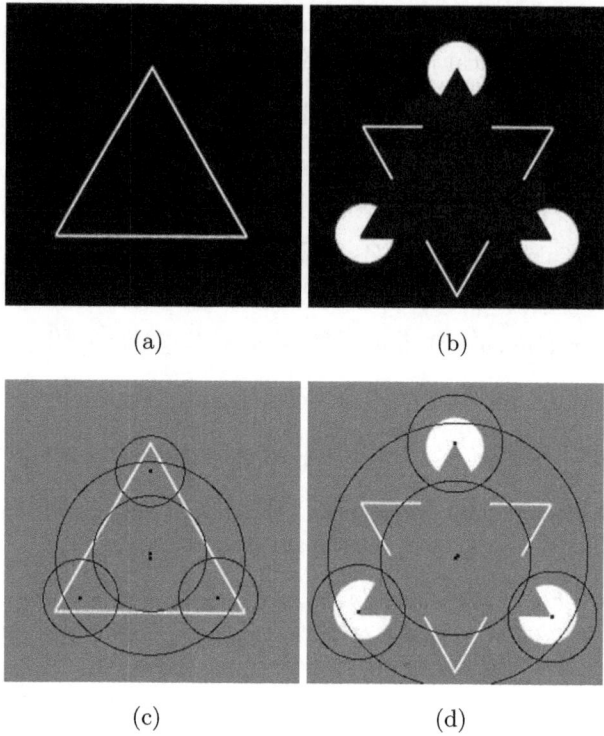

(a)   (b)

(c)   (d)

Fig. 1. Scale space analysis for the regular triangle and the Kaniza's triangle.

with respect to the generators $\{p_n\}_{n=1}^{n(\tau_\alpha)}$ [10]. The generators, Voronoi vertices, and Voronoi edges correspond to the local maximum, saddle, and local minimum points, respectively. Therefore, if we have function $\hat{f}(x,\tau)$ such that

$$\int_{\mathbf{R}^2} |\hat{f}(x,\tau) - f(x,\tau)|dx \leq \epsilon, \tag{36}$$

for a small positive constant ϵ, we can estimate the topological configuration of stationary points using combinatorial optimization.

For both usual triangles and Kaniza triangles, to a very large scale $\hat{f}(x,\tau)$ is a single Gaussian, and to a middle scale $\hat{f}(x,\tau)$ is approximated by the sum of three Gaussians whose peaks are at the vertices of a triangle. Therefore, we have the same configuration of the stationary points for both triangles.

Denoting the signs of the eigenvalues of the minus of the Hessian matrix as $(+,+)$, $(+,-)$ and $(-,-)$ in the linear scale space, these labels of points correspond to the local maximum points, the saddle points, and the local minimum points, respectively. The structure tree of triangles is

$$T_{triangle} = \langle\langle(+,+) \succ \langle(-,-),(+,+),(+,+),(+,+)\rangle\rangle. \tag{37}$$

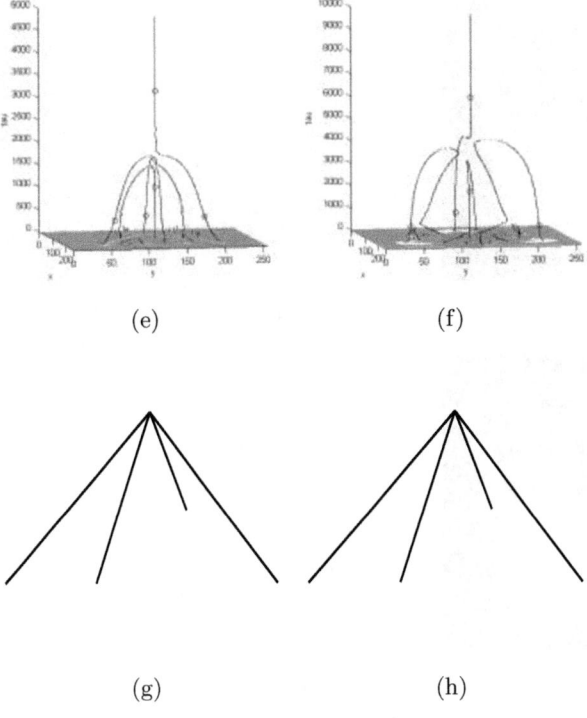

(e)   (f)

(g)   (h)

Fig. 1. (Continued).

The structure tree of Kaniza triangle

$$T_{Kaniza} = \langle(+,+) \succ \langle(-,-),(+,+),(+,+),(+,+)\rangle\rangle \qquad (38)$$

is equivalent to $T_{triangle}$ Moreover, if we consider the saddle curves, the tree of triangle is expressed as

$$\langle(+,+) \succ \langle(-,-),(+,+),(+,-),(+,+),(+,-),(+,+),(+,-)\rangle\rangle. \qquad (39)$$

Three saddle points appear if the branch of curves corresponding to the label $(-,-)$ appears.

4   Temporal Structure Tree

In this section, we define the structure forest for a time-varying image $f(x,y,t)$. For the sampled sequence $f(x,y,1)$, $f(x,y,2)$, $\cdots$, $f(x,y,t)$, $f(x,y,t+1)$, $\cdots$, we construct the structure tree $T(t)$ for each image in this sequence.

If each of pair of successive trees $T(t)$ and $T(t+1)$ are topologically different we affix new labels for nodes, except the root. Furthermore, for topologically equivalent trees $T(t)$ and $T(t+1)$, if stationary points of $T(t+1)$ do not remain in the field of view of each node, we consider these two trees to be different and $T(t+1)$ produces new nodes. We eliminate old symbols of nodes in $T(t)$ and affix

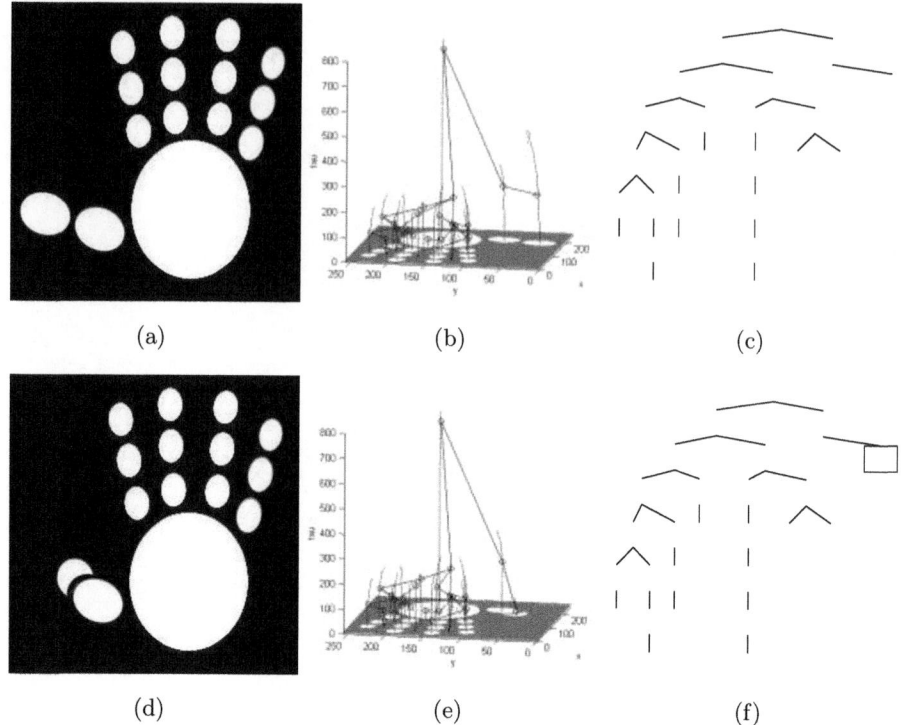

(a)   (b)   (c)

(d)   (e)   (f)

Fig. 2. Scale-space analysis of an image sequence. Nodes in the squares are the novel nodes.

new symbols to new nodes in $T(t + 1)$. Using these operations, we can extract the motion of stationary points and the change of the field of views on the image plane $\tau = 0$. This process detects moving parts in a sequence of images. For an example, we extracted the structure trees from a sequence of images shown in Figure 2. This synthetic image sequence expresses the motion of the first finger. In Figure, nodes in the squares are the novel nodes.

5   Scale Space Analysis for 3D Objects

In the digital scale space $\mathbf{Z}^3 \times \mathbf{Z}_0$, setting $\boldsymbol{x}(\tau) = (k'm, n, \tau)^\top$ to be a singular point, we define the neighborhood of $\boldsymbol{x}(\tau)$ as

$$\mathbf{N}(\boldsymbol{x}(\tau)) = \{(k'm', n', \tau')^\top \, | \, |k-k'|+|m-m'|+|n-n'| \leq 1, |\tau-\tau'| \leq T(\tau)\} \quad (40)$$

for an appropriate constant T which depends on τ. Furthermore, we set

$$\mathbf{N}^k(\boldsymbol{x}(\tau)) = $$
$$\{(k', m', n', \tau')^\top \, | \, |k - k'| + |m - m'| + |n - n'| \leq k, |\tau - \tau'| \leq T(\tau)\}. \quad (41)$$

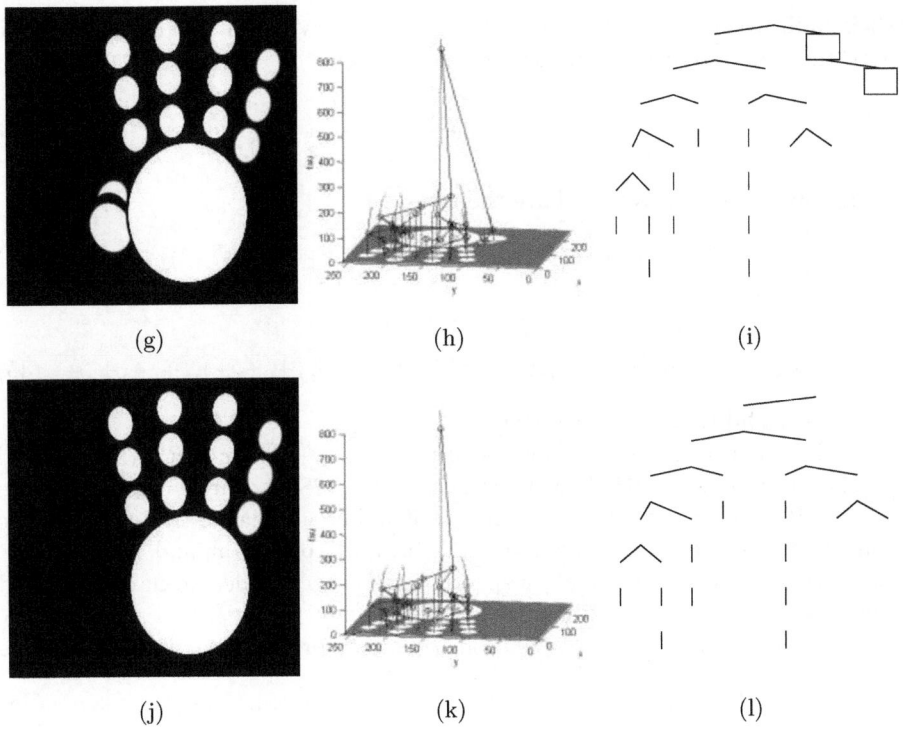

(g)   (h)   (i)

(j)   (k)   (l)

Fig. 2. (Continued).

Furthermore, as an extent ion of eq. (31), we define

$$f_{kmn}^{(2)}(\theta) = \frac{1}{\delta}(f_{k'm'n'} - 2f_{kmn} + f_{k''m''n''}),\tag{42}$$

for $k' - k = k - k''$, $m' - m = m - m''$, and $n' - n = n - n''$, and $\delta = |k'' - k'| + |m'' - m'| + |n'' - n'|$.

For a collection of singular points in the digital linear scale space, using the following definitions, we construct the stationary-curves for objects in three-dimensional space.

Definition 3. *For a singular point $x = (k, m, n, \tau)^{\top}$ and $y = (k', m', n', \tau')^{\top}$ in the scale space such that $\tau' > \tau$, if y is a unique point which lies in the neighborhood of $\mathbf{N}^k$, we say x and y are connected.*

As a simple example, we analysis the structure tree of moving balls. We assume that at time t_1 and time t_2, we have three balls and four balls whose centers are at the vertices of a tetrahedron, respectively. Figures 3 (a) and (b) show configurations of balls at time t_1 and time t_2, respectively. Furthermore, the stationary-curves in (c) and (d) are superimposed to the space $\tau = 0$. This simple sequence of 3D objects means that a ball appears and three balls change their radii between time t_1 and t_2. Then using the Voronoi tessellation of space with the generalized Voronoi distance

$$d(\boldsymbol{x}, \boldsymbol{p}_n) = \exp(-\frac{|\boldsymbol{x} - \boldsymbol{p}_n|^2}{\sigma_n} + w_n) \tag{43}$$

for $\boldsymbol{x} \in \mathbf{R}^3$ and $\boldsymbol{p}_n \in \mathbf{R}^3$, we have the trees

$$T_1 = \langle(+,+,+) \succ \langle(+,+,+),(+,+,+),(+,+,+)\rangle\rangle \tag{44}$$

and

$$T_2 = \langle(+,+,+) \succ \langle(-,-,-),(+,+,+),(+,+,+),(+,+,+),(+,+,+)\rangle\rangle \tag{45}$$

for time t_1 and time t_2, respectively.

In Figure 4, (a), (b), (c) show the slices of the $f(\boldsymbol{x}, \tau)$ $\boldsymbol{x} \in \mathbf{R}^3$. (a), (b), and (c) show the slices $x = 64$ of an object in the $128 \times 128 \times 128$ area, the distribution for $\tau = 100$, and the distribution for $\tau = 400$. The original distribution is three spheres whose radii and centers are 18, 16, and 16, and at and $(64, 64, 80)^\top$, $(64, 32, 64)^\top$, and $(64, 64, 32)^\top$, respectively. For each scale, we have evaluated the local minima, local maximal, and saddle points.

If we apply the Voronoi tessellation technique for this slice, we have the tree

$$T_{triangle} = \langle(+,+) \succ \langle(-,-),(+,+),(+,+),(+,+)\rangle\rangle. \tag{46}$$

Furthermore, if we consider the saddle points with the label $(+,-,-)$ for an object, we have the tree

$$T_{threeballs} = \langle(+,+,+) \succ \langle(+,-,-),(+,+,+,),(+,+,+)(+,+,+)\rangle\rangle. \tag{47}$$

Then, eliminating the first elements of labels, we have $T_{triangle}$ from $T_{threeballs}$. These two trees imply that for the scale space analysis of objects, we are require to consider the saddle points for the construction of structure tree of the two-dimensional projection of objects.

6   Conclusions

As a numerical example of the static image analysis, we computed the stationary curves and stationary points of a Kaniza triangle. The result shows that the topological configurations of the stationary curves and stationary points for a Kaniza triangle in the scale space is the same with these configurations for practical triangles, that is, topologically there is no difference in the scale-space between the practical tangles and a Kaniza tangle. This mathematical property suggests that we cannot distinguish these figures in the scale space. This unseparatability of a class of figures in the scale space might cause missrecognition of figures.

As an application of the scale-space analysis for time-varying images, we developed a method for the extraction of moving parts from a sequences of images.

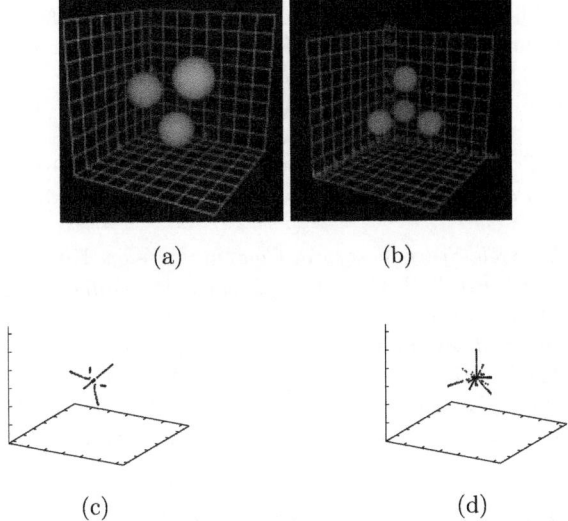

(a)   (b)

(c)   (d)

Fig. 3. A sequence of objects (a) and (b) and there fingerprints (c) and (d).

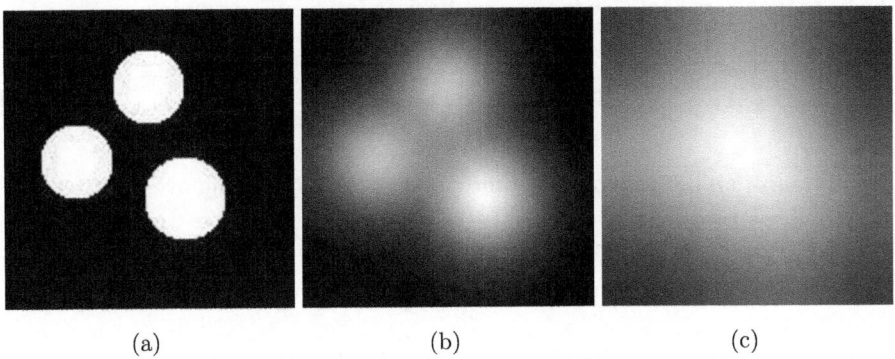

(a)   (b)   (c)

Fig. 4. (a), (b), and (c) show the slices $x = 64$ of an object in the $128 \times 128 \times 128$ area, the distribution for $\tau = 100$, and the distribution $\tau = 400$.

The method does not assume the scale-space transformation with respect to the time variable. Using this method, we can describe a spatiotemporal configurations and motions of fields of view in a sequence of images for moving objects. This method enables us to transform the sequences of images measured from an articulated object to a series of symbols.

References

1. Iijima, T., *Pattern Recognition*, Corona-sha, Tokyo, 1974 (in Japanese).
2. Zhao, N.-Y., Iijima, T., Theory on the method of determination of view-point and field of vision during observation and measurement of figure IECE Japan, Trans. D., **J68-D**, 508-514, 1985 (in Japanese).

3. Zhao, N.-Y., Iijima, T., A theory of feature extraction by the tree of stable viewpoints. IECE Japan, Trans. D., **J68-D**, 1125-1135, 1985 (in Japanese).
4. Zhao, N.-Y., *A Study of Feature Extraction by the Tree of Stable View-Points*, Dissertation to Doctor of Engineering, Tokyo Institute of Technology, 1985 (in Japanese).
5. Witkin, A.P., Scale space filtering, Pros. of 8th IJCAI, 1019-1022, 1993.
6. Yuille, A. L., Poggio, T., Scale space theory for zero crossings, IEEE PAMI, **8**, 15-25, 1986.
7. Lindeberg, T., *Scale-Space Theory in Computer Vision*, Kluwer, Boston 1994.
8. Grossmann, Ch., Roos, H.-G., *Numerik partieller Differentialgleichungen*, B.G. Teubner, Stuttgart 1994.
9. Otsu, N., *Mathematical Studies on Feature Extraction in Pattern Recognition*, Researches of The Electrotechnical Laboratory, **818**, 1981 (in Japanese).
10. Okabe, A., Boots, B., Sugihara, K., *Spatial Tessellations: Concepts and Applications of Voronoi Diagrams*, John Wiley & Sons, Chichester, 1992.

Appendix: Discritization of Convolution

For a sequence w_α $|\alpha| \leq k$ for an appropriate integer k, assuming

$$\sum_{\alpha=-k}^{k} w_\alpha = g, \ w_\alpha = w_{-\alpha}, \ |w_{|\alpha|}| > |w_{|\alpha'|}|, \tag{48}$$

for $|\alpha| > |\alpha'|$ and a positive constant g, the weighted moving average of $f_i(\tau)$,

$$f_m(\tau) = \sum_{\alpha=m-k}^{m+k} w_\alpha f_\alpha(\tau) \tag{49}$$

approximates the second derivative of sampled functions. [8]. For $w_{-1} = w_1 = 1/2$, $w_0 = 1$ and $g = 0$, we obtain the usual second derivation with samples on three successive points.

Setting

$$\boldsymbol{\Phi} = \sum_{\alpha=-k}^{k} w_k \boldsymbol{P}^k, \ \boldsymbol{P} = \begin{pmatrix} 0\,0\,0\cdots 0\,1 \\ 0\,1\,0\cdots 0\,0 \\ \vdots\,\vdots\,\vdots\,\ddots\,\vdots\,\vdots \\ 0\,0\,0\cdots 1\,0 \end{pmatrix}, \tag{50}$$

the equation

$$\boldsymbol{f}(\tau+1) - \boldsymbol{f}(\tau) = \alpha\boldsymbol{\Phi}\boldsymbol{f}(\tau) \tag{51}$$

approximates the diffusion equation for the cyclic condition. The eigenvalues of $\boldsymbol{\Phi}$ are

$$\lambda_m = \sum_{n=-k}^{k} w(n) \exp\left(2\pi i \frac{mn}{M}\right), \ m = 1, 2, \cdots, M. \tag{52}$$

A digital approximation of the convolution for the scale-space transformation is given as

$$f(m, n, \tau) = \sum_{m'=-\infty}^{\infty} \sum_{n'=-\infty}^{\infty} g(m', n', \tau) f(m - m', n - n'), \qquad (53)$$

where m and n are integers if $f(x, y)$ is sampled on lattice points $\mathbf{Z}^2$. For the achievement of the numerical integration of eq. (53) in digital computers, the equation is truncated as the finite sum,

$$\overline{f}(m, n, \tau) = \sum_{m'=-M}^{M} \sum_{n'=-N}^{N} g(m', n', \tau) f(m - m', n - n'), \qquad (54)$$

for a pair of appropriate large integers M and N. Zhao computed this convolution approximating the Gaussian kernel with the Normal distribution [4].

With the cyclic condition the one-dimensional version of eq. (54) is described as

$$f(\tau) = G(\tau)f, \; G(\tau) = \begin{pmatrix} g_1 & g_2 & g_3 & \cdots & g_{M-1} & g_M \\ g_M & g_1 & g_2 & \cdots & g_{M-2} & g_{M-1} \\ \vdots & \vdots & \vdots & \ddots & \vdots & \vdots \\ g_2 & g_3 & g_4 & \cdots & g_M & g_1 \end{pmatrix}, \qquad (55)$$

where

$$g_n = \begin{cases} \frac{1}{\sqrt{2\pi\tau}} \exp\left(\frac{n^2}{2\tau}\right) & \text{for } 1 \le n \le \frac{M-1}{2}, \\ \frac{1}{\sqrt{2\pi\tau}} \exp\left(\frac{(M-n)^2}{2\tau}\right) & \text{for } \frac{M-1}{2} < n \le M, \end{cases} \qquad (56)$$

if M is a positive odd integer. Therefore, we can rewrite the sample integration as

$$f(\tau) = UD(\tau)U^*f, \qquad (57)$$

where U is the DFT matrix. Furthermore, from eq. (52), the diagonal elements of matrix $D(\tau)$ is computed as the DFT of $\{g_i\}_{i=1}^{M}$. Since the Fourier transform of Gaussian $\frac{1}{\sqrt{2\pi\tau}} \exp\left(-\frac{x^2}{2\tau}\right)$ is $\sqrt{2\pi\tau} \exp\left(-\frac{\tau}{2}y^2\right)$, we approximate the diagonal elements of matrix $D(\tau)$ as

$$\lambda_n = \begin{cases} \sqrt{2\pi\tau} \exp\left(\frac{\tau n^2}{2}\right) & \text{for } 1 \le n \le \frac{M-1}{2}, \\ \sqrt{2\pi\tau} \exp\left(\frac{\tau(M-n)^2}{2}\right) & \text{for } \frac{M-1}{2} < n \le M. \end{cases} \qquad (58)$$

These analysis conclude that a usual discritization of convolution for the linear-scale-space transform is also described as a discrete transform for sampled signals and images.

Interest Point Detection and Scale Selection in Space-Time[*]

Ivan Laptev and Tony Lindeberg

Computational Vision and Active Perception Laboratory (CVAP)
Dept. of Numerical Analysis and Computing Science
KTH, S-100 44 Stockholm, Sweden

Abstract. Several types of interest point detectors have been proposed for spatial images. This paper investigates how this notion can be generalised to the detection of interesting events in space-time data. Moreover, we develop a mechanism for spatio-temporal scale selection and detect events at scales corresponding to their extent in both space and time.

To detect spatio-temporal events, we build on the idea of the Harris and Förstner interest point operators and detect regions in space-time where the image structures have significant local variations in both space and time. In this way, events that correspond to curved space-time structures are emphasised, while structures with locally constant motion are disregarded.

To construct this operator, we start from a multi-scale windowed second moment matrix in space-time, and combine the determinant and the trace in a similar way as for the spatial Harris operator. All space-time maxima of this operator are then adapted to characteristic scales by maximising a scale-normalised space-time Laplacian operator over both spatial scales and temporal scales. The motivation for performing temporal scale selection as a complement to previous approaches of spatial scale selection is to be able to robustly capture spatio-temporal events of different temporal extent. It is shown that the resulting approach is truly scale invariant with respect to both spatial scales and temporal scales.

The proposed concept is tested on synthetic and real image sequences. It is shown that the operator responds to distinct and stable points in space-time that often correspond to interesting events. The potential applications of the method are discussed.

1 Introduction

Analysing and interpreting video is a growing topic in computer vision and its applications. Video data contains information about changes in the environment and is highly important for many visual tasks including navigation, surveillance and video indexing.

Traditional approaches for motion analysis mainly involve the computation of optic flow (Barron, Fleet and Beauchemin 1994) and feature tracking (Smith

[*] The support from the Swedish Research Council and from the Royal Swedish Academy of Sciences as well as the Knut and Alice Wallenberg Foundation is gratefully acknowledged.

L.D. Griffin and M. Lillholm (Eds.): Scale-Space 2003, LNCS 2695, pp. 372–387, 2003.

and Brady 1995, Blake and Isard 1998). Although very effective for many tasks, both of these techniques have limitations. Optic flow approaches mostly capture first-order motion and often fail when the motion has sudden changes. Interesting solutions to this problem have been proposed by (Niyogi 1995, Fleet, Black and Jepson 1998). Feature trackers often assume the constant appearance of image patches over time and, hence, may fail when this appearance changes for example in situations when two objects in the image merge or split. Model-based solutions for this problem have been presented by (Black and Jepson 1998).

Image structures in video are not restricted to constant velocity and/or constant appearance over time. On the contrary, many interesting events in video are characterised by strong variations of the data in both the spatial and the temporal directions. For example, consider scenes with a person entering a room, applauding hand gestures, a car crash or a water splash. Moreover, it can be argued that changes of image velocity, i.e. accelerations of image structures are of particular interest since they may indicate the work of forces that act in the environment and change its structure.

In the spatial domain, points with a significant local variation of image intensities have been extensively investigated previously (Förstner and Gülch 1987, Harris and Stephens 1988, Schmid, Mohr and Bauckhage 2000). Such image points are frequently denoted as "interest points" and are attractive due to their high information contents. Highly successful applications of interest point detectors have been presented for image indexing (Schmid and Mohr 1997), stereo matching (Tuytelaars and Van Gool 2000, Mikolajczyk and Schmid 2002, Tell and Carlsson 2002), optic flow estimation and tracking (Smith and Brady 1995), and recognition (Lowe 1999, Hall, de Verdiere and Crowley 2000).

The purpose of this paper is to extend the notion of interest points into the spatio-temporal domain and to show that the resulting space-time features often correspond to interesting events in video. In particular we aim at the direct scheme for event detection that does not require feature tracking nor optic flow computation. As events often have characteristic extents in both space and time (Koenderink 1988, Lindeberg and Fagerström 1996, Florack 1997, Chomat, Martin and Crowley 2000b, Zelnik-Manor and Irani 2001), we investigate the behaviour of space-time interest points in spatio-temporal scale-space and adapt both the spatial and the temporal scales of the detected features to their characteristic extents in space-time. The idea of spatio-temporal interest points is illustrated in figure 1 where the result of a standard interest point detector applied to still images in a video is compared to the proposed spatio-temporal interest point detector. As can be seen, the spatio-temporal detector is more selective than the spatial one and detects specific events in the space-time cycle of a gait pattern.

To detect spatio-temporal events, we build on the idea of the Harris and Förstner interest point operators (Harris and Stephens 1988, Förstner and Gülch 1987) and derive the spatio-temporal event detector in section 2. We analyse its behaviour on synthetic image sequences and motivate the need for automatic temporal scale selection. In section 3 we investigate a mechanism for simulta-

Spatial interest points

Spatio-temporal interest points

Fig. 1. Detection of spatial and spatio-temporal interest points in a video sequence. Compared to a spatial detector that selects points with high variations of image values in space, the spatio-temporal detector selects areas corresponding to distinct *events* with high variations of image values in both space and time.

neous spatio-temporal scale selection based on the normalised spatio-temporal Laplace operator. In section 4 we propose an algorithm that adapts the detection of space-time interest points to their characteristic scales of observations by combining the theory from sections 2 and 3. The performance of the resulting detector on real image sequences is investigated in section 5. Finally, section 6 concludes the paper with the discussion of the method and its potential applications.

2   Interest Point Detection

2.1   Interest Points in Spatial Domain

In the spatial domain, we can model an image $f^s : \mathbb{R}^2 \mapsto \mathbb{R}$ by its linear scale-space representation (Witkin 1983, Koenderink and van Doorn 1992, Lindeberg 1994, Florack 1997) $L^s : \mathbb{R}^2 \times \mathbb{R}_+ \mapsto \mathbb{R}$

$$L^s(x, y; \sigma_l^2) = g^s(x, y; \sigma_l^2) * f^s(x, y), \tag{1}$$

defined by the convolution of f^s with Gaussian kernels of variance σ_l^2

$$g^s(x, y; \sigma_l^2) = \frac{1}{2\pi\sigma_l^2} \exp(-(x^2 + y^2)/2\sigma_l^2). \tag{2}$$

The idea of the Harris interest point detector is to find spatial locations where f^s has significant changes in both directions. For a given scale of observation σ_l^2, such points can be found using a second moment matrix integrated over a Gaussian window with the variance σ_i^2 (Förstner and Gülch 1987, Bigün, Granlund and Wiklund 1991, Garding and Lindeberg 1996):

$$\mu^s(\cdot; \sigma_l^2, \sigma_i^2) = g^s(\cdot; \sigma_i^2) * \left((\nabla L(\cdot; \sigma_l^2))(\nabla L(\cdot; \sigma_l^2))^T \right)$$

$$= g^s(\cdot; \sigma_i^2) * \begin{pmatrix} (L_x^s)^2 & L_x^s L_y^s \\ L_x^s L_y^s & (L_y^s)^2 \end{pmatrix} \tag{3}$$

where $'*'$ denotes convolution operator, and L_x^s and L_y^s are Gaussian derivatives computed at the local scale σ_l^2 and defined as $L_x^s = \partial_x(g^s(\cdot; \sigma_l^2) * f^s(\cdot))$, $L_y^s = \partial_y(g^s(\cdot; \sigma_l^2) * f^s(\cdot))$. The second moment descriptor can be thought of as the covariance matrix of a two-dimensional distribution of image orientations in the local neighbourhood of a point. Hence, the eigenvalues $\lambda_1, \lambda_2, (\lambda_1 \le \lambda_2)$ of μ^s represent characteristic variations of f^s in the both image directions while two significant values of λ_1, λ_2 indicate the presence of an interest point. To detect such points, Harris and Stephens (1988) proposed to detect positive maxima of the corner function

$$H^s = \det(\mu^s) - k\,\mathrm{trace}^2(\mu^s) = \lambda_1\lambda_2 - k(\lambda_1 + \lambda_2)^2. \tag{4}$$

The ratio of the eigenvalues $\alpha = \lambda_2/\lambda_1$ has to be high at the positions of the interest points. From (4) it follows that for positive local maxima of H^s the ratio α has to satisfy $k \le \alpha/(1 + \alpha)^2$. Hence, if we set $k = 0.25$, the positive maxima of H will only correspond to "ideal" interest points with $\alpha = 1$, i.e. $\lambda_1 = \lambda_2$. Lower values of k allow us to detect interest points with more elongated shape, corresponding to higher values of α. The commonly used value of k in the literature is $k = 0.04$ corresponding to the detection of points with $\alpha < 23$.

The result of detecting Harris interest points in an outdoor image sequence of a walking person is presented in the top row of figure 1.

2.2 Interest Points in the Spatio-Temporal Domain

In this section, we develop an operator that responds to events in temporal image sequences with specific positions and extents in space-time. The idea of interest points in the spatial domain can be extended into the spatio-temporal domain by requiring image values in space-time to have large variations in both the spatial and the temporal directions. Points with such properties will be spatial interest points with a distinct location in time corresponding to a local spatio-temporal neighbourhoods with non-constant motion.

To model a spatio-temporal image sequence we use a function $f\colon \mathbb{R}^2 \times \mathbb{R} \to \mathbb{R}$ and construct its linear scale-space representation $L\colon \mathbb{R}^2 \times \mathbb{R} \times \mathbb{R}_+^2 \mapsto \mathbb{R}$ by convolution of f with an anisotropic Gaussian kernel[1] with distinct spatial variance σ_l^2 and temporal variance τ_l^2

---

[1] In general, convolution with a Gaussian kernel in the temporal domain violates causality constraints since the temporal image data is available only for the past. For real-time implementation, time-causal scale-space filters thus have to be used (Koenderink 1988, Lindeberg and Fagerström 1996, Florack 1997). In this paper, however, we simplify the investigation and assume that the data is available for a sufficiently long period of time and the image sequence can be convolved with a Gaussian kernel in both space and time.

$$L(\cdot;\, \sigma_l^2, \tau_l^2) = g(\cdot;\, \sigma_l^2, \tau_l^2) * f(\cdot), \tag{5}$$

where the spatio-temporal separable Gaussian kernel is defined as

$$g(x, y, t;\, \sigma_l^2, \tau_l^2) = \frac{1}{\sqrt{(2\pi)^3 \sigma_l^4 \tau_l^2}} \exp(-(x^2 + y^2)/2\sigma_l^2 - t^2/2\tau_l^2). \tag{6}$$

The introduction of a separate scale parameter for the temporal domain is essential since the spatial and the temporal extents of events are in general independent. Moreover, as will be illustrated in section 2.3, events detected using our interest point operator depend on both spatial and temporal scales of observation and, hence, require separate treatment of the scale parameters σ_l^2 and τ_l^2.

Similar to the spatial domain, we consider the spatio-temporal second-moment matrix which is a 3-by-3 matrix composed of first order spatial and temporal derivatives averaged with a Gaussian weighting function $g(\cdot;\, \sigma_i^2, \tau_i^2)$

$$\mu = g(\cdot;\, \sigma_i^2, \tau_i^2) * \begin{pmatrix} L_x^2 & L_x L_y & L_x L_t \\ L_x L_y & L_y^2 & L_y L_t \\ L_x L_t & L_y L_t & L_t^2 \end{pmatrix}, \tag{7}$$

where the integration scales are $\sigma_i^2 = s\sigma_l^2$ and $\tau_i^2 = s\tau_l^2$ while the first-order derivatives are defined as

$$L_\xi(\cdot;\, \sigma_l^2, \tau_l^2) = \partial_\xi(g * f).$$

To detect interest points, we search for regions in f having significant eigenvalues $\lambda_1, \lambda_2, \lambda_3$ of μ. Among different approaches to find such regions we propose here to extend the Harris corner function (4) defined for the spatial domain into the spatio-temporal domain by combining the determinant and the trace of μ as follows

$$H = \det(\mu) - k\operatorname{trace}^3(\mu) = \lambda_1 \lambda_2 \lambda_3 - k(\lambda_1 + \lambda_2 + \lambda_3)^3. \tag{8}$$

To show that positive local maxima of H correspond to points with high values of $\lambda_1, \lambda_2, \lambda_3$ ($\lambda_1 \leq \lambda_2 \leq \lambda_3$), we define the ratios $\alpha = \lambda_2/\lambda_1$ and $\beta = \lambda_3/\lambda_1$ and re-write H as

$$H = \lambda_1^3(\alpha\beta - k(1 + \alpha + \beta)^3).$$

From the requirement $H \geq 0$ we get $k \leq \alpha\beta/(1 + \alpha + \beta)^3$ and it follows that k assumes its maximum possible value $k = 1/27$ when $\alpha = \beta = 1$. For sufficiently large values of k, positive local maxima of H correspond to points with high variation of image gray-values in both the spatial and the temporal directions. In particular, if we set the maximal value of α, β to 23 as in the spatial domain, the value of k to be used in H (8) will then be $k = 0.005$. Thus, spatio-temporal interest points of f can be found by detecting local positive spatio-temporal maxima in H.

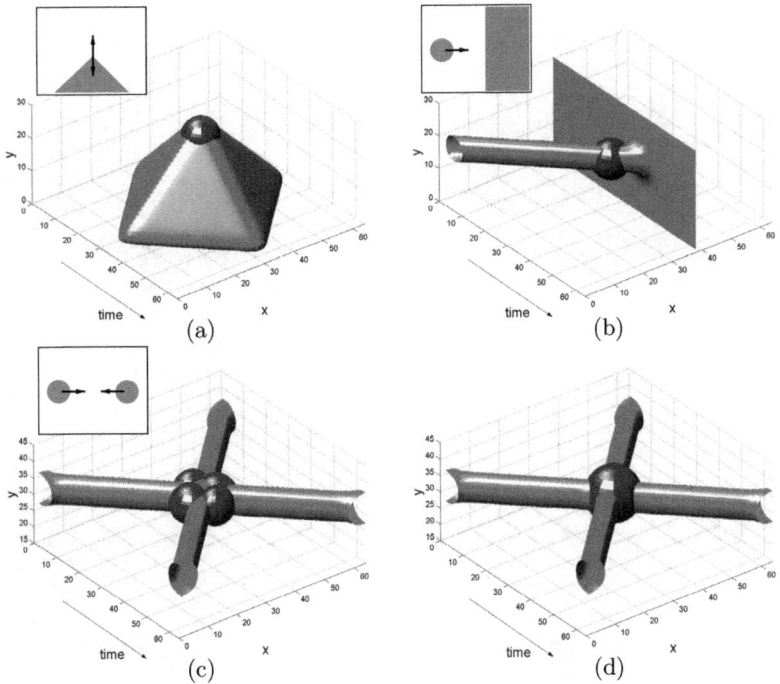

Fig. 2. Results of detecting spatio-temporal interest points on synthetic image sequences. (a): Moving corner; (b) A merge of a ball and a wall; (c): Collision of two balls with interest points detected at scales $\sigma_l^2 = 8$ and $\tau_l^2 = 8$; (d): the same as in (c) but with interest points detected at scales $\sigma_l^2 = 16$ and $\tau_l^2 = 16$.

2.3   Experimental Results on Synthetic Data

In this section, we illustrate the detection of spatio-temporal interest points on synthetic image sequences. For clarity of presentation, we show the spatio-temporal data as 3-D space-time plots where the original signal is represented by a threshold surface while the detected interest points are presented by ellipsoids with positions corresponding to the space-time location of interest points and the length of the semi-axes proportional to the local scale parameters σ_l and τ_l used in the computation of H.

Figure 2a illustrates a sequence with a moving corner. The interest point is detected at the moment in time when the motion of the corner changes direction. This type of event occurs frequently in natural sequences such as sequences of articulated motion. Note that image structures with constant motion do not give rise to the detection of interest points. Other typical types of events detected by the proposed method are splits and unifications of image structures. In figure 2b the interest point is detected at the moment and the position corresponding to the collision of a ball and a wall. Similarly, interest points are detected at the moment of collision and bouncing of two balls as shown in figure 2c-d. Note, that different types of events are detected depending on the scale of observation.

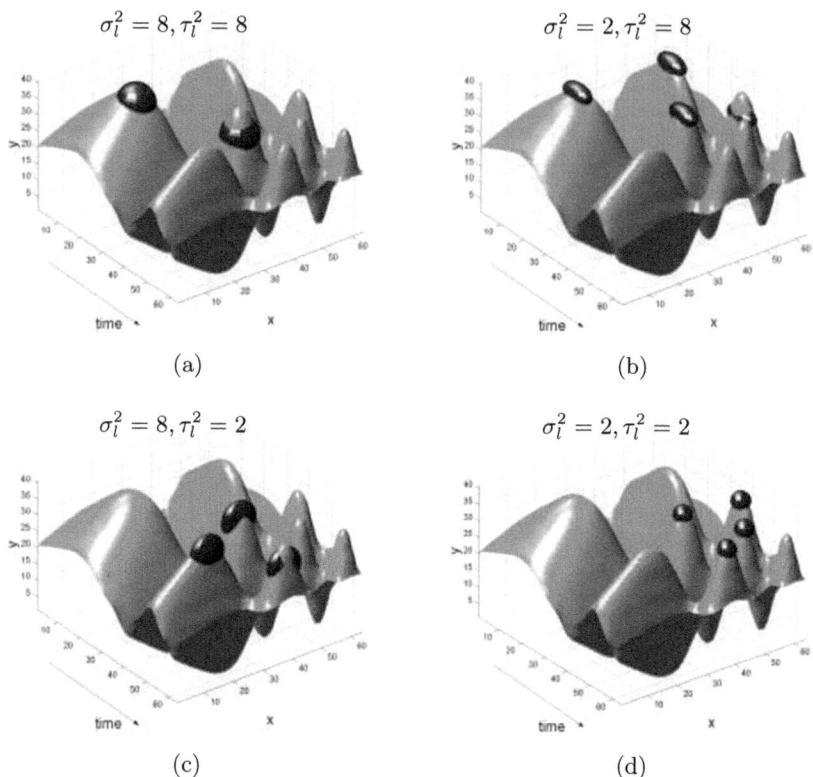

Fig. 3. Results of interest point detection at different spatial and temporal scales for a synthetic sequence with impulses having varying extents in space and time. The extents of the detected events roughly corresponds to the scale parameters σ_l^2 and τ_l^2 used in the computation of H.

To further emphasise the importance of the spatial and the temporal scales of observation, let us consider an oscillating signal with different spatial and temporal frequencies defined by the threshold surface $y = \sin(x^4) * \sin(t^4)$ (see figure 3). As can be seen, the result of detecting the strongest interest points highly depends on the scale parameters σ_l^2 and τ_l^2. We observe that space-time structures with long temporal extents are detected for large values of τ_l^2 while short events are preferred by the detector with small values of τ_l^2. Similarly, the spatial extent of events is related to the value of the spatial scale parameter σ_l^2.

From the presented examples it follows that a correct selection of temporal and spatial scales is crucial when capturing the events with different spatial and temporal extents. Moreover, estimation of the spatio-temporal extents of events can be interesting for their further interpretation. In the next section, we propose a mechanism for simultaneous estimation of spatio-temporal scales. This mechanism is combined with the interest point detector in section 4.

3   Scale Selection in Space-Time

During recent years, the problem of automatic scale selection has been addressed in several different ways, based on the maximisation of normalised derivative expressions over scale, or the behaviour of entropy measures or error measures over scales (see the companion paper by Lindeberg and Bretzner (2003) for a review). To estimate the spatio-temporal extent of an event in space-time we follow works on local scale selection proposed in the spatial domain by Lindeberg (1998) as well as in the temporal domain (Lindeberg 1997). The idea is to define a differential operator that assumes simultaneous extrema over the spatial and the temporal scales that are characteristic for an event with a particular spatio-temporal location.

For the purpose of analysis we study a prototype event represented by a spatio-temporal Gaussian blob $f = g(x, y, t; \sigma_0^2, \tau_0^2)$ with spatial variance σ_0^2 and temporal variance τ_0^2 (see figure 4a). Using the semi-group property of the Gaussian kernel, it follows that the scale-space representation of f is $L(x, y, t; \sigma^2, \tau^2) = g(x, y, t; \sigma_0^2 + \sigma^2, \tau_0^2 + \tau^2)$.

To recover the spatio-temporal extent (σ_0, τ_0) of f we consider second-order derivatives of L normalised by the scale parameters as follows

$$L_{xx,norm} = \sigma^{2a}\tau^{2b}L_{xx}, \quad L_{yy,norm} = \sigma^{2a}\tau^{2b}L_{yy}, \quad L_{tt,norm} = \sigma^{2c}\tau^{2d}L_{tt}. \quad (9)$$

All of these entities assume extrema over space and time at the centre of the blob f. Moreover, depending on the parameters a, b and c, d, they also assume extrema at certain spatial and temporal scales $\tilde{\sigma}^2$ and $\tilde{\tau}^2$.

The idea of scale selection we follow here is to determine the parameters a, b, c, d such that $L_{xx,norm}$, $L_{yy,norm}$ and $L_{tt,norm}$ assume extrema at scales $\tilde{\sigma}^2 = \sigma_0^2$ and $\tilde{\tau}^2 = \tau_0^2$. To find such extrema, we differentiate the expressions in (9) with respect to the spatial and the temporal scale parameters. For the spatial derivatives we obtain the following expressions at the centre of the blob

$$(L_{xx,norm})'_{\sigma^2} = -\frac{a\sigma^2 - 2\sigma^2 + a\sigma_0^2}{\sqrt{(2\pi)^3(\sigma_0^2 + \sigma^2)^6(\tau_0^2 + \tau^2)}}\sigma^{2(a-1)}\tau^{2b} \quad (10)$$

$$(L_{xx,norm})'_{\tau^2} = -\frac{2b\tau_0^2 + 2b\tau^2 - \tau^2}{\sqrt{2^5\pi^3(\sigma_0^2 + \sigma^2)^4(\tau_0^2 + \tau^2)^3}}\tau^{2(b-1)}\sigma^{2a}. \quad (11)$$

By setting these expressions to zero we obtain simple relations for a and b

$$a\sigma^2 - 2\sigma^2 + a\sigma_0^2 = 0, \quad 2b\tau_0^2 + 2b\tau^2 - \tau^2 = 0$$

that after substituting $\sigma^2 = \sigma_0^2$ and $\tau^2 = \tau_0^2$ lead to the values $a = 1$ and $b = 1/4$. Similarly, differentiating the second-order temporal derivative

$$(L_{tt,norm})'_{\sigma^2} = -\frac{c\sigma^2 - \sigma^2 + c\sigma_0^2}{\sqrt{(2\pi)^3(\sigma_0^2 + \sigma^2)^4(\tau_0^2 + \tau^2)^3}}\sigma^{2(c-1)}\tau^{2d} \quad (12)$$

$$(L_{tt,norm})'_{\tau^2} = -\frac{2d\tau_0^2 + 2d\tau^2 - 3\tau^2}{\sqrt{2^5\pi^3(\sigma_0^2 + \sigma^2)^2(\tau_0^2 + \tau^2)^5}}\tau^{2(d-1)}\sigma^{2c} \quad (13)$$

leads to the expressions

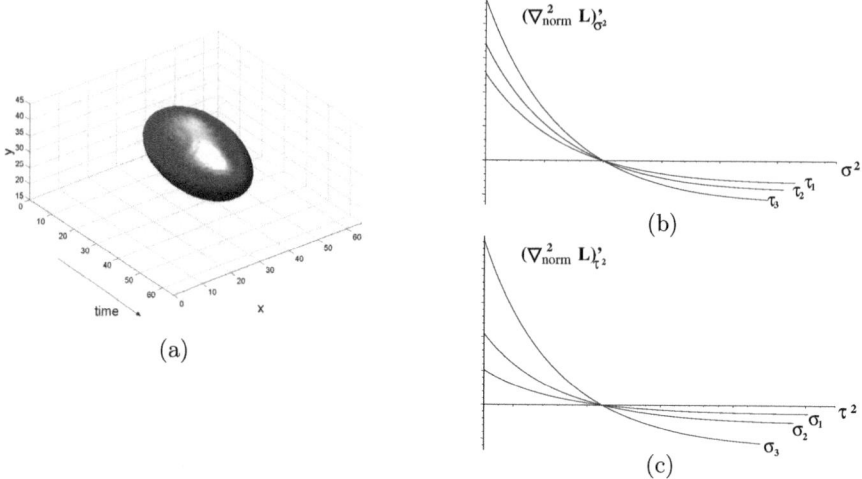

Fig. 4. (a): Spatio-temporal Gaussian blob with spatial variance $\sigma_0^2 = 4$ and temporal variance $\tau_0^2 = 16$; (b)-(c) derivatives of $\nabla^2_{norm}L$ with respect to scales. The zero-crossings of $(\nabla^2_{norm}L)'_{\sigma^2}$ and $(\nabla^2_{norm}L)'_{\tau^2}$ indicate extrema of $\nabla^2_{norm}L$ at scales corresponding to the spatial and the temporal extents of the blob.

$$c\sigma^2 - 2\sigma^2 + c\sigma_0^2 = 0, \quad 2d\tau_0^2 + 2d\tau^2 - \tau^2 = 0$$

that after substituting $\sigma^2 = \sigma_0^2$ and $\tau^2 = \tau_0^2$ result in $c = 1/2$ and $d = 3/4$.

The derived normalisation of derivatives in (9) guarantees that all of them assume space-time-scale extrema at the centre of the blob f and at scales corresponding to the spatial and the temporal extents of f, i.e. $\sigma = \sigma_0$ and $\tau = \tau_0$. The sum of these derivatives defines the normalised spatio-temporal Laplace operator

$$\nabla^2_{norm}L = L_{xx,norm} + L_{yy,norm} + L_{tt,norm}$$
$$= \sigma^2\tau^{1/2}(L_{xx} + L_{yy}) + \sigma\tau^{3/2}L_{tt}. \tag{14}$$

Figures 4b-c show derivatives of this operator with respect to the scale parameters evaluated at the centre of a spatio-temporal blob with spatial variance $\sigma_0^2 = 4$ and temporal variance $\tau_0^2 = 16$. The zero-crossings of the curves verify that $\nabla^2_{norm}L$ assumes extrema at the scales $\sigma^2 = \sigma_0^2$ and $\tau^2 = \tau_0^2$. Hence, the spatio-temporal extent of the blob can be estimated by finding the extrema of $\nabla^2_{norm}L$ over both spatial and temporal scales.

4   Scale-Adapted Space-Time Interest Points

Local scale estimation using the normalised Laplace operator has shown to be very useful in the spatial domain (Lindeberg 1998, Almansa and Lindeberg 2000, Chomat, de Verdiere, Hall and Crowley 2000a). In particular, Mikolajczyk and Schmid (2001) combined the Harris interest point operator with the normalised Laplace operator and derived a scale-invariant Harris-Laplace interest point detector. The idea is to find points in scale-space that are both spatial maxima of the Harris function H^s (4) and extrema over scale of the scale-normalised Laplace operator in space.

Here, we extend this idea and detect interest points that are simultaneous maxima of the spatio-temporal corner function H (8) over space and time (x, y, t) as well as extrema of the normalised spatio-temporal Laplace operator $\nabla^2_{norm}L$ (14) over scales (σ^2, τ^2). One way of detecting such points is to compute space-time maxima of H for each spatio-temporal scale level and then to select points that maximise $(\nabla^2_{norm}L)^2$ at the corresponding scale. This approach, however, requires dense sampling over the scale parameters and is therefore computationally expensive.

An alternative we follow here is to detect interest points for a set of sparsely distributed scale values and then to track these points in the spatio-temporal scale-time-space towards the extrema of $\nabla^2_{norm}L$. We do this by iteratively updating the scale and the position of the interest points by (i) selecting the neighbouring spatio-temporal scale that maximises $(\nabla^2_{norm}L)^2$ and (ii) re-detecting the space-time location of the interest point at a new scale. The corresponding algorithm is presented in figure 5.

The result of scale-adaptation of interest points for the spatio-temporal pattern in figure 3 is shown in figure 6. As can be seen, the chosen scales of the adapted interest points match the spatio-temporal extents of the corresponding structures in the pattern.

It should be noted, however, that the presented algorithm has been developed for processing pre-recorded video sequences. In real-time situations, when using causal scale-space representation based on recursive temporal filters (Lindeberg

---

1. Detect interest points $p_j = (x_j, y_j, t_j, \sigma^2_{l,j}, \tau^2_{l,j})$, $j = 1..N$ as maxima of H (8) over space and time using combinations of initial spatial scales $\sigma^2_l = \sigma^2_{l,1}, .., \sigma^2_{l,n}$ and temporal scales $\tau^2_l = \tau^2_{l,1}, .., \tau^2_{l,m}$ as well as integration scales $\sigma^2_i = s\sigma^2_l$ and $\tau^2_i = s\tau^2_l$.

2. **for** each interest point p_j **do**

3.   Compute $\nabla^2_{norm}L$ at position (x_j, y_j, t_j) and combinations of neighbouring scales $(\tilde{\sigma}^2_{i,j}, \tilde{\tau}^2_{i,j})$ where $\tilde{\sigma}^2_{i,j} = 2^\delta \sigma^2_{i,j}$, $\tilde{\tau}^2_{i,j} = 2^\delta \tau^2_{i,j}$, and $\delta = -0.25, 0, 0.25$

5.   Choose combination of integration scales $(\tilde{\sigma}^2_{i,j}, \tilde{\tau}^2_{i,j})$ that maximises $(\nabla^2_{norm}L)^2$

6.   **if** $\tilde{\sigma}^2_{i,j} \neq \sigma^2_{i,j}$ or $\tilde{\tau}^2_{i,j} \neq \tau^2_{i,j}$
       Re-detect interest point $\tilde{p}_j = (\tilde{x}_j, \tilde{y}_j, \tilde{t}_j, \tilde{\sigma}^2_{i,j}, \tilde{\tau}^2_{i,j})$ using integration scales $\tilde{\sigma}^2_{i,j} = \tilde{\sigma}^2_{i,j}$, $\tilde{\tau}^2_{i,j} = \tilde{\tau}^2_{i,j}$, local scales $\tilde{\sigma}^2_{l,j} = \frac{1}{s}\tilde{\sigma}^2_{i,j}$, $\tilde{\tau}^2_{l,j} = \frac{1}{s}\tilde{\tau}^2_{i,j}$ and position $(\tilde{x}_j, \tilde{y}_j, \tilde{t}_j)$ that is closest to (x_j, y_j, t_j);
       set $p_j := \tilde{p}_j$ and **goto 3**

7. **end**

Fig. 5. Algorithm for scale adaption of spatio-temporal interest points.

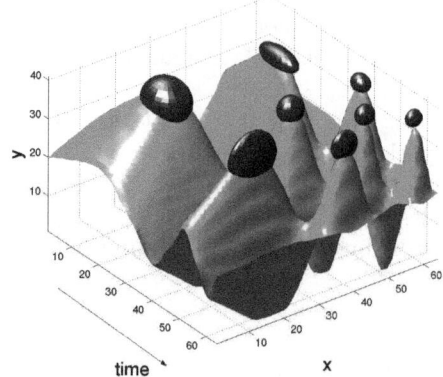

Fig. 6. The result of scale-adaptation of spatio-temporal interest points computed from a space-time pattern of the form $y = \sin(x^4) * \sin(t^4)$. The interest points are illustrated as ellipsoids showing the selected spatio-temporal scales overlayed on a surface plot of the intensity landscape.

and Fagerström 1996), only a fixed set of discrete temporal scales is available at any moment. In that case an approximate estimate of temporal scale can still be found by choosing interest points that maximise $(\nabla^2_{norm} L)^2$ in a local neighbourhood of the spatio-temporal scale-space.

5   Experiments

In this section we investigate the performance of the proposed scale-adapted spatio-temporal interest point detector applied to real image sequences. In the first example we consider a sequence of a walking person with non-constant image velocities due to the oscillating motion of the legs. As can be seen in figure 7, the pattern gives rise to stable interest points. Note that the detected interest points reflect well-localised events in both space and time, corresponding to space-time structures such as the starting and the stopping feet. From the space-time plot in figure 7(a) we can also observe how the selected spatial and temporal scales of the detected features roughly match the spatio-temporal extents of the corresponding image structures.

Figure 8 illustrates interest points detected in an outdoor sequence with a walking person and a zooming camera. The changing values of the selected spatial scales (illustrated by the size of the circles) illustrate the invariance of the method with respect to spatial scale changes of the image structures. Note that beside events in the leg pattern, the detector finds spurious points due to the non-constant motion of a coat and arms. However, image structures with constant motion in the background do not result in the response of the detector.

The third example explicitly illustrates how the proposed method is able to estimate the temporal extent of detected events. Figure 9 shows a person making hand-waving gestures with high frequency on the left and low frequency on the right. The distinct interest points are detected at the moments and at spatial

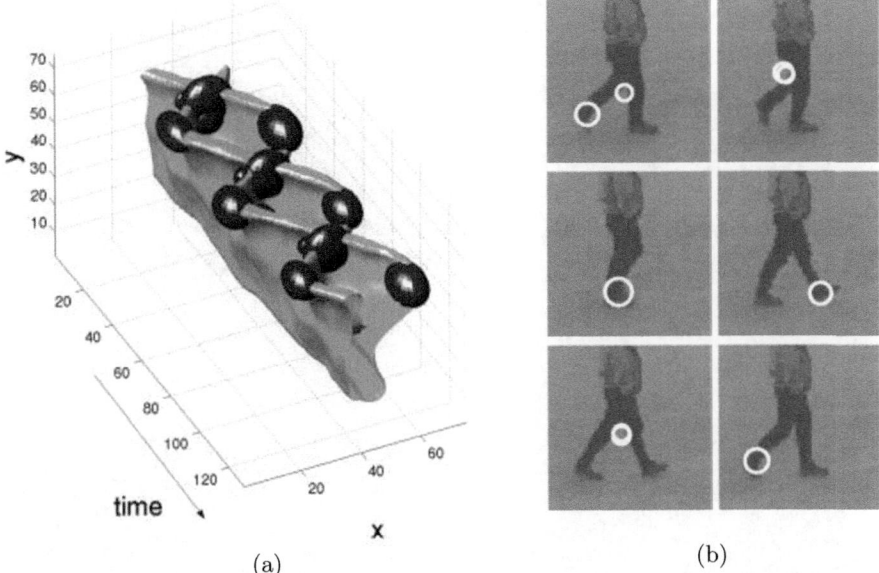

(a)   (b)

Fig. 7. Results of detecting spatio-temporal interest points for the motion of the legs of a walking person. (a): 3-D plot with a threshold surface of a leg pattern (up side down) and detected interest points; (b): interest points overlayed on single frames of a sequence.

Fig. 8. Results of interest point detection for a zoom-in sequence of a walking person. The spatial scale of the detected points (corresponding to the size of circles) matches the increasing spatial extent of image structures and verifies the invariance of the interest points with respect to changes in spatial scale.

positions where the palm of a hand changes its direction of motion. Whereas the spatial scale of the detected interest points remains constant, the selected temporal scale depends on the frequency of the wave pattern. The high frequency pattern results in short events and gives rise to interest points with small tem-

Hand waves with high frequency   *Hand waves with low frequency*

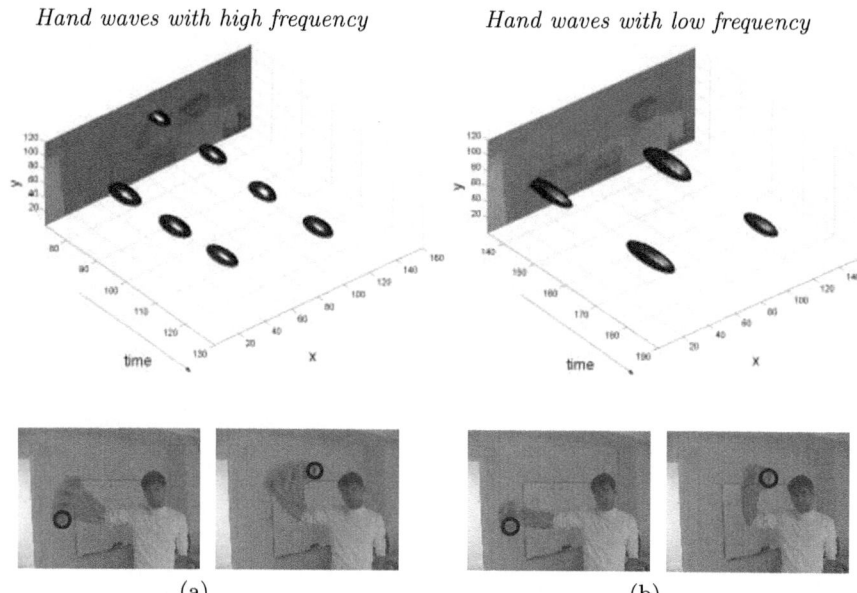

(a)   (b)

Fig. 9. Result of interest point detection for a sequence with waving hand gestures. (a) Interest points for hand gestures with high frequency; (b) Interest points for hand gestures with low frequency.

Fig. 10. Detection of the strongest interest point in a football sequence with a player heading the ball.

poral extent (see figure 9a). On the contrary, hand waves with low frequency result in interest points with long temporal extent as shown in figure 9b.

Figure 10 illustrates a football sequence with a player heading the ball. The sequence has multiple motions due to camera zooming and motion of objects in the scene and is probably hard to analyse using standard methods for motion estimation and tracking. However, the strongest output of the proposed detector (the interest point with the highest maxima of H in (8)) corresponds to the position and the moment of the most significant event in the sequence, i.e. the heading of the ball.

Fig. 11. Result of alignment of the model sequence (see figure 7) to the data sequence using classified spatio-temporal interest points. The details of the method are leaved out due to space limitations and will be presented elsewhere.

6   Summary and Discussion

We have proposed an interest point detector that finds events in space-time with high variations of the image values in space and non-constant motion in time. From the experimental results that have been presented in previous sections, it can be seen that many of the points detected in this way correspond to space-time structures that we would intuitively regard as meaningful events. For example, for the sequences with walking people (figures 7-8) we obtain responses at the beginning and the end of the gait cycle and at spatial locations corresponding to distinct body parts.

As temporal events exist over finite periods of time, the notion of temporal scale is incorporated in the detector and the method for automatic scale selection is used for estimating temporal as well as spatial extents of detected events.

The current implementation of this interest point detector is based on separable space-time filters and is therefore not invariant to Galilean transformations over time, e.g. caused by relative motions of the camera. To aim at Galilean invariance, one could either perform local stabilisation as done by Zelnik-Manor and Irani (2001) or consider (possibly ensembles of) spatio-temporal receptive fields that have been adapted to local directions in space-time (Laptev and Lindeberg 2002).

Regarding potential applications of the presented techniques, one area of interest concerns sparse representation of video data. A representation in terms of interest points could be used for matching between image sequences, or for matching an articulated model over time to a given video sequence. Furthermore, the spatio-temporal interest points could be attributed with higher order spatio-temporal derivatives or other types of image descriptors evaluated at positions, moments and scales estimated by the proposed detector. Combinations of classified spatio-temporal interest points could then be used for describing and analysing image sequences based on similar techniques as have been proposed for interest points in the spatial domain.

Figure 11 shows an example of using such an approach for matching a template gait pattern derived from one walking person to the gait pattern of another person using classified spatio-temporal interest points that have been grouped based on similar spatio-temporal receptive field responses and K-means clustering. Notably this result was obtained in relation to a complex cluttered background with multiple motions and using neither manual initialisation nor explicit tracking.

References

Almansa, A. and Lindeberg, T. (2000). Fingerprint enhancement by shape adaptation of scale-space operators with automatic scale-selection, *IEEE Transactions on Image Processing* **9**(12): 2027–2042.

Barron, J., Fleet, D. and Beauchemin, S. (1994). Performance of optical flow techniques, *International Journal of Computer Vision* **12**(1): 43–77.

Bigün, J., Granlund, G. and Wiklund, J. (1991). Multidimensional orientation estimation with applications to texture analysis and optical flow, *IEEE Transactions on Pattern Analysis and Machine Intelligence* **13**(8): 775–790.

Black, M. and Jepson, A. (1998). Eigen tracking: Robust matching and tracking of articulated objects using view-based representation, *International Journal of Computer Vision* **26**(1): 63–84.

Blake, A. and Isard, M. (1998). Condensation – conditional density propagation for visual tracking, *IJCV* **29**(1): 5–28.

Chomat, O., de Verdiere, V., Hall, D. and Crowley, J. (2000a). Local scale selection for Gaussian based description techniques, *Proc. Sixth European Conference on Computer Vision*, Vol. 1842 of *Lecture Notes in Computer Science*, Springer Verlag, Berlin, Dublin, Ireland, pp. 117–133.

Chomat, O., Martin, J. and Crowley, J. (2000b). A probabilistic sensor for the perception and recognition of activities, *Proc. Sixth European Conference on Computer Vision*, Dublin, Ireland, pp. I:487–503.

Fleet, D., Black, M. and Jepson, A. (1998). Motion feature detection using steerable flow fields, *Proc. Computer Vision and Pattern Recognition*, Santa Barbara, CA, pp. 274–281.

Florack, L. M. J. (1997). *Image Structure*, Kluwer Academic Publishers, Dordrecht, Netherlands.

Förstner, W. A. and Gülch, E. (1987). A fast operator for detection and precise location of distinct points, corners and centers of circular features, *Proc. Intercommission Workshop of the Int. Soc. for Photogrammetry and Remote Sensing*, Interlaken, Switzerland.

Garding, J. and Lindeberg, T. (1996). Direct computation of shape cues using scale-adapted spatial derivative operators, *International Journal of Computer Vision* **17**(2): 163–191.

Hall, D., de Verdiere, V. and Crowley, J. (2000). Object recognition using coloured receptive fields, *Proc. Sixth European Conference on Computer Vision*, Vol. 1842 of *Lecture Notes in Computer Science*, Springer Verlag, Berlin, Dublin, Ireland, pp. 164–177.

Harris, C. and Stephens, M. (1988). A combined corner and edge detector, *Alvey Vision Conference*, pp. 147–152.

Koenderink, J. J. (1988). Scale-time, *Biological Cybernetics* **58**: 159–162.

Koenderink, J. J. and van Doorn, A. J. (1992). Generic neighborhood operators, *IEEE Transactions on Pattern Analysis and Machine Intelligence* **14**(6): 597–605.

Laptev, I. and Lindeberg, T. (2002). Velocity-adaptation of spatio-temporal receptive fields for direct recognition of activities: An experimental study, *in* D. Suter (ed.), *Proc. ECCV'02 workshop on Statistical Methods in Video Processing*, Copenhagen, Denmark, pp. 61–66.

Lindeberg, T. (1994). *Scale-Space Theory in Computer Vision*, Kluwer Academic Publishers, Boston.

Lindeberg, T. (1997). On automatic selection of temporal scales in time-causal scale-space, *AFPAC'97: Algebraic Frames for the Perception-Action Cycle*, Vol. 1315 of *Lecture Notes in Computer Science*, Springer Verlag, Berlin, pp. 94–113.

Lindeberg, T. (1998). Feature detection with automatic scale selection, *International Journal of Computer Vision* **30**(2): 77–116.

Lindeberg, T. and Bretzner, L. (2003). Real-time scale selection in hybrid multi-scale representations, *Proc. Scale-Space'03*, LNCS, Springer Verlag, these proceedings.

Lindeberg, T. and Fagerström, D. (1996). Scale-space with causal time direction, *Proc. Fourth European Conference on Computer Vision*, Vol. 1064 of *Lecture Notes in Computer Science*, Springer Verlag, Berlin, Cambridge, UK, pp. I:229–240.

Lowe, D. (1999). Object recognition from local scale-invariant features, *Proc. Seventh International Conference on Computer Vision*, Corfu, Greece, pp. 1150–1157.

Mikolajczyk, K. and Schmid, C. (2001). Indexing based on scale invariant interest points, *Proc. Eighth International Conference on Computer Vision*, Vancouver, Canada, pp. I:525–531.

Mikolajczyk, K. and Schmid, C. (2002). An affine invariant interest point detector, *Proc. Seventh European Conference on Computer Vision*, Vol. 2350 of *Lecture Notes in Computer Science*, Springer Verlag, Berlin, Copenhagen, Denmark, pp. I:128–142.

Niyogi, S. A. (1995). Detecting kinetic occlusion, *Proc. Fifth International Conference on Computer Vision*, Cambridge, MA, pp. 1044–1049.

Schmid, C. and Mohr, R. (1997). Local grayvalue invariants for image retrieval, *IEEE Transactions on Pattern Analysis and Machine Intelligence* **19**(5): 530–535.

Schmid, C., Mohr, R. and Bauckhage, C. (2000). Evaluation of interest point detectors, *International Journal of Computer Vision* **37**(2): 151–172.

Smith, S. and Brady, J. (1995). ASSET-2: Real-time motion segmentation and shape tracking, *IEEE Transactions on Pattern Analysis and Machine Intelligence* **17**(8): 814–820.

Tell, D. and Carlsson, S. (2002). Combining topology and appearance for wide baseline matching, *Proc. Seventh European Conference on Computer Vision*, Vol. 2350 of *Lecture Notes in Computer Science*, Springer Verlag, Berlin, Copenhagen, Denmark, pp. I:68–83.

Tuytelaars, T. and Van Gool, L. (2000). Wide baseline stereo matching based on local, affinely invariant regions, *British Machine Vision Conference*, pp. 412–425.

Witkin, A. P. (1983). Scale-space filtering, *Proc. 8th Int. Joint Conf. Art. Intell.*, Karlsruhe, Germany, pp. 1019–1022.

Zelnik-Manor, L. and Irani, M. (2001). Event-based analysis of video, *Proc. Computer Vision and Pattern Recognition*, Kauai Marriott, Hawaii, pp. II:123–130.

Towards Recognition-Based Variational Segmentation Using Shape Priors and Dynamic Labeling

Daniel Cremers[1], Nir Sochen[2], and Christoph Schnörr[3]

[1] Department of Computer Science
University of California at Los Angeles, USA
http://www.cs.ucla.edu/~cremers
[2] Department of Applied Mathematics
Tel Aviv University, Israel
http://www.math.tau.ac.il/~sochen
[3] Department of Mathematics and Computer Science
University of Mannheim, Germany
http://www.cvgpr.uni-mannheim.de

Abstract. We propose a novel variational approach based on a level set formulation of the Mumford-Shah functional and shape priors. We extend the functional by a labeling function which indicates image regions in which the shape prior is enforced. By minimizing the proposed functional with respect to both the level set function and the labeling function, the algorithm selects image regions where it is favorable to enforce the shape prior. By this, the approach permits to segment multiple independent objects in an image, and to discriminate familiar objects from unfamiliar ones by means of the labeling function. Numerical results demonstrate the performance of our approach.

1   Introduction

The problem of segmenting an image into its semantically significant components has been eluding researchers in computer vision for over 30 years. In the early days it was believed to be merely a technical problem. The traditional bottom-up approach, which is still very common in the computer vision community, suggests to start with a denoising process and to apply some sort of threshold afterward. Once the image is successfully segmented one can go to higher levels such as 3D reconstruction, motion analysis, classification and recognition.

It was realized by Mumford and Shah in the mid 80's, that denoising and segmentation are two different aspects of the same problem. Indeed a good denoising process should distinguish between a set of significant regions and the border between them. Such a denoising process assumes, implicitly, that the segmentation is known. On the other hand, segmentation approaches based on a threshold process assume that the image is denoised such that jumps in gray value can be attributed to boundaries between objects and are not the result of noise. This line of reasoning culminated in the Mumford-Shah functional [14],

L.D. Griffin and M. Lillholm (Eds.): Scale-Space 2003, LNCS 2695, pp. 388–400, 2003.

which puts the denoised image and the boundary contours on the same footing. Minimizing the functional simultaneously with respects to the dynamic variables results in a denoised image AND the boundaries. The minimization procedure involves the solution of coupled equations for the image and its boundaries.

Despite the success of the Mumford-Shah functional, its many variants [13] and the extensive mathematical analysis that followed its introduction, the problem of segmentation still eludes us. There is no segmentation algorithm that comes near the performance of a 3 year old child.

This failure indicates that a pure bottom-up approach is inappropriate. In fact we believe that higher-level processes related to recognition should participate in the segmentation process. This idea is reflected in the works of several researches. For a recent non-PDE approach starting from such viewpoint see [1].

In this paper, we combine both data-driven and recognition-driven processing in an unbiased way by introducing a dynamic labeling function into a variational segmentation approach with shape priors. In analogy to the reasoning of Mumford and Shah, minimization of the proposed functional with respect to the dynamic variables results in denoising, reconstruction of boundaries *and* the evolution of a decision function which models a higher-level recognition process.

The integration of shape priors into PDE based segmentation methods has been a focus of research in past years (c.f. [10,21,17,6,12,20,8,7,16]). Commonly, such approaches introduce a shape prior into the contour evolution in such a way that (ideally) the object of interest is reconstructed and all unfamiliar image structures are suppressed.

More recently, implicit level set based representations of a contour [15] have become a popular framework for image segmentation (cf. [2,11,4]). Since the topology of the evolving boundary is not constrained, one can elegantly model topological changes such as splitting and merging. This permits to segment multiply connected objects or several independent objects in a given image.

The question of how to introduce higher-level prior shape knowledge into level set based contour evolutions has been addressed by a number of people in recent years (cf. [12,20,5,16]). In many of these approaches, the shape prior acts on the embedding level set function. As shown e.g. in [12], this approach permits to segment a multiply connected object with a statistical shape prior.

All of these approaches introduce the shape prior in such a way that only familiar structures of *one* given object can be recovered. They do not permit the simultaneous segmentation of several *independent* objects, comprising both familiar and unfamiliar ones. As an example, Figure 1, left side, shows a table scene containing two objects — the cover of a teapot, which is assumed to be familiar, and a pen assumed to be unfamiliar. The same scene is shown on the right, but the familiar object is corrupted: some parts are missing while others are occluded.

The aim of the present work is to introduce a shape prior into the Mumford-Shah functional, in a way which permits the simultaneous segmentation of several objects in one image (each of which may consist of several components). In particular, we will show that this approach permits to reconstruct the familiar

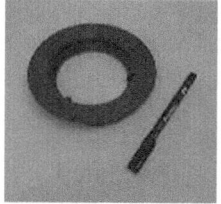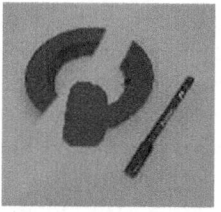

Fig. 1. Left: A table scene showing two objects — the cover of a tea pot which is assumed to be familiar, and a pen which is assumed to be unfamiliar.
Right: Corrupted version of the same image obtained by removing and occluding parts of the familiar object.

object in Figure 1, right side, while leaving the correct segmentation of the unfamiliar object unaffected. To this end, we extend a level set formulation of the Mumford-Shah functional by a labeling function which indicates the regions of the image plane in which a given shape prior is enforced. By simultaneously minimizing the proposed energy functional with respect to the level set function and the labeling function, the labeling function dynamically separates regions of familiar objects from regions of unfamiliar ones.

The organization of this paper is as follows: In Section 2, we briefly review a level set formulation of the piecewise constant Mumford-Shah functional, as proposed in [3]. In Section 3, we augment this variational framework by a shape prior which affects the evolution of the level set function globally. As a consequence, such commonly proposed global shape priors suppress all unfamiliar image structures in the resulting segmentation process. In Section 4, we introduce a static labeling function to explicitly restrict the effect of the shape prior to designated areas of the image plane. This permits to both reconstruct a corrupted version of a known object and segment a novel unknown object in the same input image. In Section 5, we finally propose a variational framework with a dynamic labeling function. Compared to the case of a static labeling, the dynamic one evolves in an unsupervised manner during energy minimization. Numerical results show that this approach permits to segment new objects and reconstruct known ones without specifying the respective image areas beforehand. In Sections 7 and 8, we discuss some limitations and future work and end with a conclusion.

2   Region-Based Segmentation with Level Sets

In this section, we will detail a level set method for image segmentation which aims at maximizing the gray value homogeneity in a set of disjoint regions. It is based on a level set formulation of the Mumford-Shah functional proposed by Chan and Vese [3]. This framework will then be extended by shape priors of increasing complexity in the subsequent sections.

2.1   A Level Set Framework for the Mumford-Shah Functional

Mumford and Shah [14] proposed to segment an input image $f : \Omega \to \mathbb{R}$ by minimizing the functional

$$E(u, C) = \frac{1}{2} \int_{\Omega} (f - u)^2 \, dx \; + \; \lambda^2 \frac{1}{2} \int_{\Omega - C} |\nabla u|^2 \, dx \; + \; \nu \, |C| \qquad (1)$$

simultaneously with respect to the segmenting boundary C and the piecewise smooth approximation u. If the smoothness constraint is further stressed, one obtains for $\lambda \to \infty$ the cartoon limit (or minimal partition problem) in which the input image f is approximated by a piecewise constant segmentation $u = \{u_i\}$:

$$E(u, C) = \frac{1}{2} \sum_i \int_{\Omega_i} (f - u_i)^2 \, dx \; + \; \nu \, |C|. \qquad (2)$$

During minimization of (2), the constants $\{u_i\}$ will take on the mean value of f over the set of disjoint regions $\{\Omega_i\}$ which partition the image plane ($\Omega = \bigcup \Omega_i$) and which are separated by the boundary C.

In several papers, Chan and Vese detailed a level set implementation of the Mumford-Shah functional (cf. [3,4]), which is based on the use of the Heaviside function as an indicator function for the separate phases. The focus of the present work is the modeling of selective shape priors. Therefore, we will restrict ourselves to the case of the piecewise constant Mumford-Shah model and a single level set function $\phi : \Omega \to \mathbb{R}$ to embed the contour: $C = \{x \in \Omega \, | \, \phi(x) = 0\}$.

A piecewise constant segmentation of an input image f with two gray values u_+ and u_- can be obtained by minimizing the functional [3]:

$$E_{CV}(u_+, u_-, \phi) = \int_{\Omega} (f - u_+)^2 H(\phi) + (f - u_-)^2 (1 - H(\phi)) \, dx + \nu \int_{\Omega} |\nabla H(\phi)|, \quad (3)$$

with respect to the scalar variables u_+ and u_- and the embedding level set function ϕ. Here $H(\phi)$ denotes the Heaviside function:

$$H(\phi) = \begin{cases} 1, & \phi \geq 0 \\ 0, & \text{else} \end{cases} \qquad (4)$$

The Euler-Lagrange equation for this functional can be implemented by the following gradient descent:

$$\frac{\partial \phi}{\partial t} = \delta_\epsilon(\phi) \left[\nu \, \mathrm{div} \left(\frac{\nabla \phi}{|\nabla \phi|} \right) - (f - u_+)^2 + (f - u_-)^2 \right], \qquad (5)$$

where the scalars u_+ and u_- are updated in alternation with the level set evolution to take on the mean gray value of the input image f in the regions with $\phi > 0$ and $\phi < 0$, respectively:

$$u_+ = \frac{\int f(x) H(\phi) dx}{\int H(\phi) dx}, \qquad u_- = \frac{\int f(x)(1 - H(\phi)) dx}{\int (1 - H(\phi)) dx}. \qquad (6)$$

The implementation in [3] is based on a smooth approximation of the delta function $\delta_\epsilon(s) = H'_\epsilon(s)$, which is chosen to have an infinite support:

$$\delta_\epsilon(s) = \frac{1}{\pi} \frac{\epsilon}{\epsilon^2 + s^2}. \tag{7}$$

In particular, a discretization with a support larger than zero permits the detection of interior contours – for example if one wants to segment a ring-like structure, starting from an initial contour located outside the ring.

2.2 Redistancing

During its evolution according to equation (5), the level set function ϕ generally grows to very large positive values in dark areas of the input image and very large negative values in bright areas of the image (or vice versa). At the zero crossings, it rises steeply, the gradient can become arbitrarily large. In numerical implementations, we found that a very steep slope of the level set function eventually inhibits the flexibility of the boundary to displace.

Many people have advocated the use of a redistancing procedure in the evolution of level set functions to constrain the slope of ϕ to $|\nabla\phi| = 1$, c.f. [9]. In order to reproject the evolving level set function to the space of distance functions, we intermittently iterate several steps of the redistancing equation [19]:

$$\frac{\partial\phi}{\partial t} = \mathrm{sign}(\hat{\phi}) \left(1 - |\nabla\phi|\right), \tag{8}$$

where $\hat{\phi}$ denotes the level set function before redistancing. As pointed out in [3], such a regularization is optional for the above level set model. Yet, we found this to have two favorable properties in our application. Firstly, it improves the convergence of the boundary evolution. And secondly, this normalization facilitates the introduction of shape priors which are encoded in terms of signed distance functions. We found this simple redistancing process to work well for our purpose, therefore we did not revert to more elaborate iterative redistancing schemes such as the one presented in [18].

2.3 Numerical Results

Minimization of the functional (3) is done by alternating the three fractional steps of iterating the gradient descent for the level set function ϕ, as given by equation (3), iterating the redistancing procedure given by equation (8) and updating the mean gray values for the two phases, as given in equation (6).

Figure 2 shows several steps of the evolution of the boundary C obtained by minimizing the functional (3) for the corrupted input image introduced in Figure 1. Due to the implicit representation, the boundary is free to perform splitting and merging. Due to the region-based formulation of the functional, the contour converges to the final segmentation over fairly large distances, while local edge and corner information is well preserved.

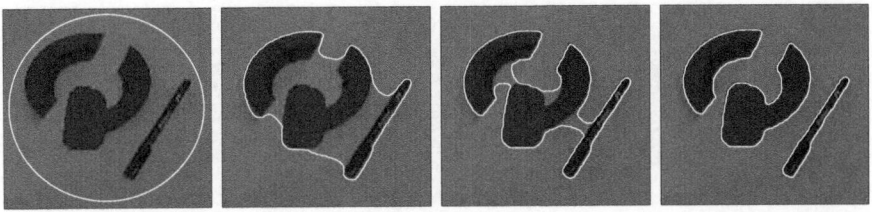

Fig. 2. Segmentation without shape prior. Evolution of the boundary for the Chan-Vese level set formulation of the piecewise constant Mumford-Shah functional (with a single level set function). The contour evolves so as to separate bright and dark areas. Due to the implicit level set representation, the topology is not constrained, which allows for splitting and merging of the boundary.

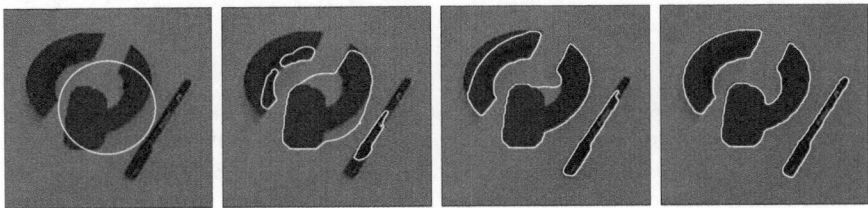

Fig. 3. Segmentation without shape prior. Evolution of the boundary for the Chan-Vese level set formulation with a different initialization as in Figure 2. There is no bias in form of a balloon term, therefore the contour can both expand and shrink for the same parameter value.

Yet, compared to many alternative approaches to level set segmentation, the above approach does not contain a balloon term which induces a bias favoring either contraction or expansion and therefore assumes prior knowledge about whether the objects of interest are inside or outside the initial contour. Figure 3 shows the corresponding contour evolution for a different initial contour. In this case, the contour expands from its initialization to converge to a similar segmentation of the given image.

3   Global Shape Prior in the Level Set Segmentation

In many applications of image segmentation, some knowledge about the shape of expected objects of interest is available. This prior shape information can be introduced into the level set functional in the following way. A number of train-ing shapes is embedded by the signed distance function. The set of associated distance functions is aligned and a statistical model for the level set function is inferred from the training set. This prior is then added either to the evolution equation (5) or directly as a shape energy to the functional (3). Invariance of the prior with respect to similarity transformations of the level set function can be incorporated into these approaches (cf. [12,16]). Since the focus of this work is selectivity of shape priors, we will not consider such invariances here. Moreover,

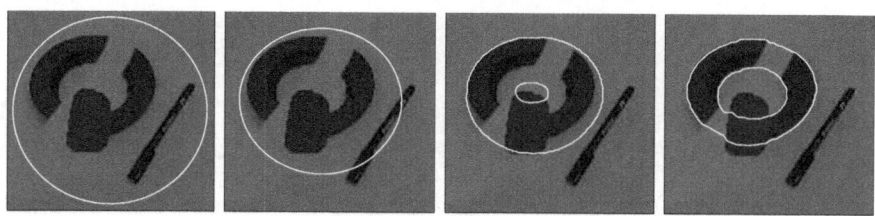

Fig. 4. Segmentation with a global shape prior. Evolution of the boundary for the Chan-Vese level set formulation with a shape prior which favors a ring. Compared to the corresponding segmentation without prior shown in Figure 2, the introduction of the global shape prior has several effects: The ring is reconstructed according to the shape prior, i.e. the partial occlusion is removed and the missing parts are filled in. At the same time, all other image structures – such as the pen on the right side – are removed.

we will only consider a single training shape, but we allow for more than one object in the image. Nevertheless, our approach can in principle be extended to more involved statistical shape priors [7].

A straight-forward extension of the functional (3) with an isotropic Gaussian shape prior is the following:

$$E(u_+, u_-, \phi) = E_{CV}(u_+, u_-, \phi) + \alpha\, E_{shape}(\phi), \tag{9}$$

with

$$E_{shape}(\phi) = \int_{\Omega} \big(\phi(x) - \phi_0(x)\big)^2 dx, \tag{10}$$

where ϕ_0 is the level set function embedding a given training shape (or the mean of a set of training shapes) and $\alpha \geq 0$ determines the weight of the prior.

Minimizing this functional with respect to ϕ results in an evolution equation of the form:

$$\frac{\partial \phi}{\partial t} = \delta_\epsilon(\phi) \left[\nu \, \mathrm{div} \left(\frac{\nabla \phi}{|\nabla \phi|} \right) - (f - u_+)^2 + (f - u_-)^2 \right]$$
$$- 2\,\alpha \left(\phi - \phi_0 \right). \tag{11}$$

Compared to the purely data-driven evolution in (5), we obtain an additional relaxation towards the learned shape ϕ_0.

Applied to the same image as in Figure 2, this generates the contour evolution shown in Figure 4. For sufficiently large weight of the shape prior, all image structures which are *not familiar* will be suppressed from the segmentation. In our example, the training shape consisted of the ring from a tea can. Due to the shape prior, missing parts of this object are recovered and occlusions removed. Moreover, the pen next to the ring is also removed.

4   Selective Shape Prior by Static Labeling

In the example of Figure 4, the shape prior permitted to reconstruct the learned object. In general, however, this object may not be present in a given image. Moreover, a given view of a scene may contain corrupted versions of a known object — the ring in our example — and other unfamiliar objects — in our example the pen next to it. In such cases, it may be desirable to have a *selective* shape prior, which permits to reconstruct the corrupted version of the known object, but which will not affect the segmentation of the unknown objects.

In this paper, we will model such a selective shape prior by a labeling function $L : \Omega \to \mathbb{R}$ which indicates the areas of the image plane in which a given prior should be enforced. This labeling function is to take on the values $+1$ and -1 depending on whether the prior should be enforced or not. For the beginning, we will assume this labeling to be known. This assumption will be removed in the following section.

We propose to segment an input image f by minimizing the functional (9) with a shape prior of the form:

$$E_{shape}(\phi) = \int_{\Omega} (\phi(x) - \phi_0(x))^2 (L + 1)^2 \, dx, \tag{12}$$

where a labeling L defines the parts of the image plane Ω where the shape prior should be active. The gradient descent equation for ϕ is given by:

$$\frac{\partial \phi}{\partial t} = \delta_\epsilon(\phi) \left[\nu \, \mathrm{div} \left(\frac{\nabla \phi}{|\nabla \phi|} \right) - (f - u_+)^2 + (f - u_-)^2 \right]$$

$$- 2\,\alpha\,(L + 1)^2\,(\phi - \phi_0). \tag{13}$$

Compared to the evolution equation with the global prior in (11), the additional relaxation towards the learned shape ϕ_0 is now restricted to the image areas where $L \neq -1$.

Figure 6 shows the result of minimizing this functional with respect to ϕ for the static labeling shown in Figure 5.

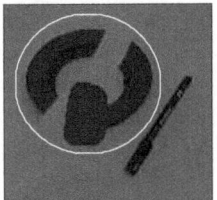

Fig. 5. Left: The labeling function L, shown in a 3D plot, indicates the area around the familiar object in which the prior should be applied ($L = 1$) and the remainder of the image plane, which should be segmented according to the gray value information only ($L = -1$). **Right:** Zero crossing of L superimposed on the input image.

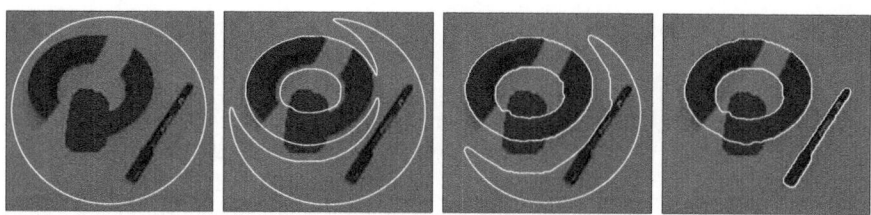

Fig. 6. Segmentation with static labeling function and shape prior.
Evolution of the boundary upon minimizing the functional (12) with the static labeling function shown in Figure 5. According to the labeling function, the prior is enforced in an area around the ring only and masked out in the areas away from the ring. Therefore — in contrast to the result in Figure 4 — the ring is reconstructed according to the shape prior while the pen is also segmented as an independent object.

Due to the definition of the labeling, the shape prior is only enforced in a region around the ring. In the remainder of the image plane the labeling function masks out the shape prior. The consequence is that the known object is reconstructed according to the prior, while the correct segmentation of the pen is unaffected by the prior. It remains to be shown how this labeling function L can be determined *from the image data* as well.

5   Selective Shape Prior by Dynamic Labeling

In the previous section, we introduced a labeling function L to indicate regions of the image plane where a given shape prior should be enforced. Yet, this labeling was specified beforehand.

In the present section, we will overcome this limitation by an approach with a *dynamic* labeling function. To this end, we propose to minimize the functional

$$E(u_+, u_-, \phi, L) = E_{CV}(u_+, u_-, \phi) + \alpha\, E_{shape}(\phi), \qquad (14)$$

with the shape prior:

$$E_{shape}(\phi, L) = \int (\phi - \phi_0)^2 \,(L{+}1)^2 \, dx + \int \lambda^2 \,(L{-}1)^2 \, dx + \gamma \int |\nabla H(L)| \, dx. \quad (15)$$

Compared to the previous approach of a static labeling function, we now assume the labeling L to be unknown. Instead of specifying the labeling, we simultaneously optimize the above cost functional with respect to both the segmenting level set function ϕ and the labeling function L.

According to the proposed cost functional, the labeling will evolve in an unsupervised manner driven by two criteria:

– The labeling should enforce the shape prior in those areas of the image where the level set function is similar to the prior. In particular, for fixed ϕ, minimizing the first two terms in (15) will generate the following qualitative behavior of the labeling:

$$L \to 1, \qquad \text{if } |\phi - \phi_0| < \lambda$$
$$L \to -1, \qquad \text{if } |\phi - \phi_0| > \lambda$$

- The boundary separating regions with shape prior from regions without shape prior should have minimal length. This regularizing constraint on the zero crossing of the labeling function – given by the last term in equation (15) — induces a topological "compactness" of the regions with and without shape prior.

6   Evolution of Labeling Function and Level Set Function

We simultaneously minimize the functional (14) with respect to both the labeling function L and the level set function ϕ by iterating the associated gradient descent equations in alternation with updates of the mean gray values u_+ and u_- according to (6).

For fixed ϕ, the gradient descent equation for the labeling function is given by:

$$\frac{\partial L}{\partial t} = -\frac{\partial E}{\partial L} = \alpha \left[2\lambda^2 (1 - L) - 2(\phi - \phi_0)^2 (1 + L) + \gamma \delta_\epsilon(L) \operatorname{div}\left(\frac{\nabla L}{|\nabla L|} \right) \right]. \quad (16)$$

The first two terms drive the labeling toward -1 or 1, depending on whether $|\phi - \phi_0|$ is larger or smaller than λ. And the last term in (16) minimizes the length of the zero-crossing of L, thereby enforcing decision regions with minimal boundary.

Conversely, for fixed labeling, the gradient descent equation for the level set function ϕ is given by:

$$\frac{\partial \phi}{\partial t} = -\frac{\partial E}{\partial \phi}$$
$$= \delta_\epsilon(\phi) \left[\nu \operatorname{div}\left(\frac{\nabla \phi}{|\nabla \phi|} \right) - (f - u_+)^2 + (f - u_-)^2 \right]$$
$$- 2\alpha (1 + L)^2 (\phi - \phi_0). \quad (17)$$

Compared to the purely data-driven evolution in equation (5), we have an additional relaxation toward the learned shape ϕ_0 in all areas of the image where $L \neq -1$.

The resulting evolution of the labeling function L, obtained by minimizing the total energy (14) for the same input image as before, is shown in Figure 7, bottom row: The area around the ring arises in an unsupervised manner as the region where the prior should be applied.

The corresponding evolution of the segmenting boundary given by the zero level set of the function ϕ is shown in Figure 7, top row. These results demonstrate the following favorable properties of our approach:

- Compared to the segmentation without shape prior shown in Figure 2, the ring is reconstructed according to the shape prior, i.e. the missing parts are filled in and the occlusion is removed.

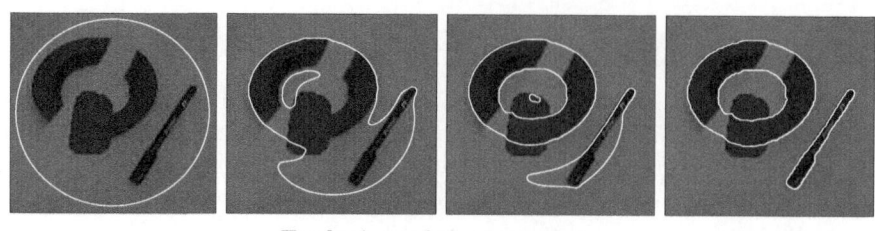

Evolution of the boundary

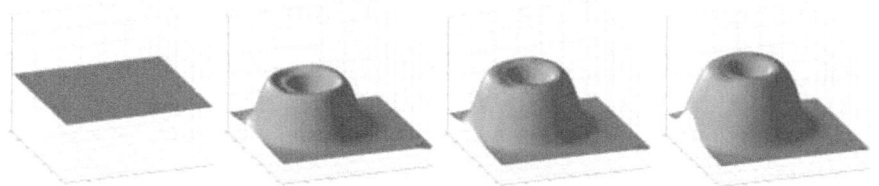

Evolution of the labeling

Fig. 7. Segmentation with dynamic labeling and shape prior. Top row: Evolution of the segmenting boundary obtained by minimizing the functional (14) for the same input image as in Figure 2. The ring object is reconstructed according to the shape prior, whereas the correct segmentation of the pen is unaffected by the prior. The prior affects the embedding level set function ϕ only in the regions indicated by the simultaneously evolving dynamic labeling function.
Bottom row: Evolution of the dynamic labeling L during energy minimization. The labeling function dynamically selects the region around the ring object to apply the shape prior. Compared to the static labeling shown in Figure 6, the dynamic labeling evolves in an unsupervised manner.

- Compared to the segmentation with a *global* shape prior shown in Figure 4, the correct segmentation of the pen is unaffected by the shape prior. The effect of the shape prior is restricted to the area around the familiar object by the labeling function.

- Compared to the case of a *static* labeling function shown in Figures 5 and 6, one no longer needs to specify beforehand the regions where the prior should be enforced. Instead, the labeling evolves in an unsupervised manner during the minimization of the functional (14). It dynamically selects image regions in which the prior is to be applied.

7   Limitations and Future Work

Segmentation results obtained with the dynamic labeling approach depend on an appropriate choice of the parameters λ and γ which affect the evolution of the labeling. We are currently working on a more rigorous probabilistic formulation of the dynamic labeling approach which, we believe, should result in automatic estimates of these parameters from the data. Moreover, we intend to generalize the proposed approach to *multiple* selective shape priors, and to more elaborate statistical shape priors such as the one introduced in [7]. We will also investigate methods to introduce pose invariance, such as the one presented e.g. in [12,16], into the proposed approach.

8   Conclusion

We presented a novel variational approach to integrate higher-level shape priors into level set based segmentation methods. In particular, we addressed the problem of applying shape priors selectively, such that a given prior permits the reconstruction of corrupted versions of a familiar object while not affecting the correct segmentation of independent unknown objects.

To this end, we extended the level set approach of Chan and Vese by three shape priors of increasing complexity. The first one is a simple globally active prior which permits the reconstruction of a known object but removes all unfamiliar objects. The second one is a shape prior with a static labeling, which allows to define areas of the input image in which the prior should be applied. This prior permits to reconstruct the known object in the selected area, but does not affect the correct segmentation of independent objects in the remainder of the image plane. Finally, the third shape prior is based on a dynamic labeling function. The latter is not specified beforehand, it rather evolves in an unsupervised manner in order to automatically select the image regions to which the prior should be applied. As a consequence, the familiar object is reconstructed, yet independent unfamiliar objects are correctly segmented. And the decision *which* areas correspond to familiar objects simultaneously evolves with the segmentation during minimization of the proposed functional.

We believe that the results presented in this work demonstrate the capacity of the dynamic labeling approach to model an *unsupervised decision process*. Such a decision process fundamentally extends the applicability of statistical shape priors in segmentation. It permits to combine both data-driven and recognition-driven processing on equal footings in a variational segmentation approach.

Acknowledgements

DC was supported by ONR N00014-02-1-0720 and AFOSR F49620-03-1-0095.

References

1. E. Borenstein and S. Ullman. Class-specific, top-down segmentation. In A. Heyden et al., editors, *Proc. of the Europ. Conf. on Comp. Vis.*, volume 2351 of *LNCS*, pages 109–122, Copenhagen, May 2002. Springer, Berlin.
2. V. Caselles, R. Kimmel, and G. Sapiro. Geodesic active contours. In *Proc. IEEE Internat. Conf. on Comp. Vis.*, pages 694–699, Boston, USA, 1995.
3. T. Chan and L. Vese. Active contours without edges. *IEEE Trans. Image Processing*, 10(2):266–277, 2001.

4. T. Chan and L. Vese. A level set algorithm for minimizing the Mumford–Shah functional in image processing. In *IEEE Workshop on Variational and Level Set Methods*, pages 161–168, Vancouver, CA, 2001.
5. Y. Chen, S. Thiruvenkadam, H. Tagare, F. Huang, D. Wilson, and E. Geiser. On the incorporation of shape priors into geometric active contours. In *IEEE Workshop on Variational and Level Set Methods*, pages 145–152, Vancouver, CA, 2001.
6. T. F. Cootes, A. Hill, C. J. Taylor, and J. Haslam. Use of active shape models for locating structures in medical images. *Image and Vision Computing*, 12(6):355–365, 1994.
7. D. Cremers, T. Kohlberger, and C. Schnörr. Nonlinear shape statistics in Mumford–Shah based segmentation. In A. Heyden et al., editors, *Proc. of the Europ. Conf. on Comp. Vis.*, volume 2351 of *LNCS*, pages 93–108, Copenhagen, May 2002. Springer, Berlin.
8. D. Cremers, F. Tischhäuser, J. Weickert, and C. Schnörr. Diffusion Snakes: Introducing statistical shape knowledge into the Mumford–Shah functional. *Int. J. of Comp. Vis.*, 50(3):295–313, 2002.
9. J. Gomes and O. D. Faugeras. Level sets and distance functions. In D. Vernon, editor, *Proc. of the Europ. Conf. on Comp. Vis.*, volume 1842 of *LNCS*, pages 588–602, Dublin, Ireland, 2000. Springer.
10. U. Grenander, Y. Chow, and D.M. Keenan. *Hands: A Pattern Theoretic Study of Biological Shapes*. Springer, New York, 1991.
11. S. Kichenassamy, A. Kumar, P. J. Olver, A. Tannenbaum, and A. J. Yezzi. Gradient flows and geometric active contour models. In *Proc. IEEE Internat. Conf. on Comp. Vis.*, pages 810–815, Boston, USA, 1995.
12. M. E. Leventon, W. E. L. Grimson, and O. Faugeras. Statistical shape influence in geodesic active contours. In *Proc. Conf. Computer Vis. and Pattern Recog.*, volume 1, pages 316–323, Hilton Head Island, SC, June 13–15, 2000.
13. J.-M. Morel and S. Solimini. *Variational Methods in Image Segmentation*. Birkhäuser, Boston, 1995.
14. D. Mumford and J. Shah. Optimal approximations by piecewise smooth functions and associated variational problems. *Comm. Pure Appl. Math.*, 42:577–685, 1989.
15. S. J. Osher and J. A. Sethian. Fronts propagation with curvature dependent speed: Algorithms based on Hamilton–Jacobi formulations. *J. of Comp. Phys.*, 79:12–49, 1988.
16. M. Rousson and N. Paragios. Shape priors for level set representations. In A. Heyden et al., editors, *Proc. of the Europ. Conf. on Comp. Vis.*, volume 2351 of *LNCS*, pages 78–92, Copenhagen, May 2002. Springer, Berlin.
17. L. H. Staib and J. S. Duncan. Boundary finding with parametrically deformable models. *IEEE Trans. on Patt. Anal. and Mach. Intell.*, 14(11):1061–1075, 1992.
18. M. Sussman and E. Fatemi. An efficient, interface-preserving level set redistancing algorithm and its application to interfacial incompressible fluid flow. *SIAM J. Sci. Comput.*, 20(4):1165–1191, 1999.
19. M. Sussman, Smereka P., and S. J. Osher. A level set approach for computing solutions to incompressible twophase flow. *J. of Comp. Phys.*, 94:146–159, 1994.
20. A. Tsai, A. Yezzi, W. Wells, C. Tempany, D. Tucker, A. Fan, E. Grimson, and A.. Willsky. Model–based curve evolution technique for image segmentation. In *Conf. on Comp. Vision Patt. Recog.*, pages 463–468, Kauai, Hawaii, 2001.
21. A. Yuille and P. Hallinan. Deformable templates. In A. Blake and A. Yuille, editors, *Active Vision*, pages 21–38. MIT Press, 1992.

PDE Based Shape from Specularities

Jan Erik Solem[1], Henrik Aanæs[2], and Anders Heyden[1]

[1] School of Technology and Society
Malmö University, Sweden
jes@ts.mah.se
[2] Informatics and Mathematical Modelling
Technical University of Denmark, Denmark

Abstract. When reconstructing surfaces from image data, reflections on specular surfaces are usually viewed as a nuisance that should be avoided. In this paper a different view is taken. Noting that such reflections contain information about the surface, this information could and should be used when estimating the shape of the surface. Specifically, assuming that the position of the light source and the cameras (i.e. the motion) are known, the reflection from a specular surface in a given image constrain the surface normal with respect to the corresponding camera.

Here the constraints on the normals, given by the reflections, are used to formulate a partial differential equation (PDE) for the surface. A smoothness term is added to this PDE and it is solved using a level set framework, thus giving a "shape from specularity" approach. The structure of the PDE also allows other properties to be included, e.g. the constraints from PDE based stereo.

The proposed PDE does not fit naturally into a level set framework. To address this issue it is proposed to couple a force field to the level set grid. To demonstrate the viability of the proposed method it has been applied successfully to synthetic data.

Keywords: Shape, Level Sets, Specularities, BRDF, Surface Estimation, Structure from Motion.

1 Introduction

Structure and motion, the reconstruction of an object and the motion of the camera from a sequence of images, is one of the most widely studied fields in computer vision. The final stage of a structure and motion system is estimating the shape of the object based on estimates of cameras and a sparse point cloud representing the structure. One assumption that is frequently made is that the object one tries to model is Lambertian, i.e. that light is reflected equally in all directions and therefore features have constant brightness from all viewpoints. Several methods for surface estimation have been proposed that uses the Lambertian surface assumption [4,7,9,13,16,24].

Non-Lambertian features are usually considered as outliers and removed. This means that the parts of the resulting 3D-model corresponding to specular

L.D. Griffin and M. Lillholm (Eds.): Scale-Space 2003, LNCS 2695, pp. 401–415, 2003.

Fig. 1. Estimating specular surfaces. The specular reflection on the car window gives information on the orientation of the surface normal. An image sequence showing the motion of the specular reflection can be used to estimate the shape of the window. It does not matter that the window is semi-transparent.

regions will usually be poor. The method proposed in this paper addresses this problem by estimating surfaces using information from the specular reflections.

The information contained in the specular reflections is that for a given camera position and light source the surface normal is constrained. This information can then be used to estimate the shape of the surface in specular regions to obtain a better 3D-model.

Consider the scenario of creating a 3D-model of a car. Cars contain smooth specular surfaces and are very hard to model with standard structure from motion techniques. Typically, the only features that can be extracted are sharp corners of the body or at edges of doors or windows. This is not enough to make a satisfactory model. Using the technique proposed in this paper it could be possible to estimate the surface where specular reflections are observed. In fact it is should even be possible to reconstruct the shape of semi-transparent specular surfaces such as windows, which is not possible with other methods, see Figure 1.

The problem setting can be summarized as follows. We wish to estimate a specular surface using the information contained in the specular reflections. This is done in order to complement the models obtained through structure from motion techniques. Our method is not a "stand alone" method since it requires the use of a structure from motion algorithm to determine the camera motion, which is necessary to determine surface normals. We see the shape from specularities scheme proposed in this paper as an integral part of a larger surface estimating scheme.

We use a level set representation of the surface which makes it easy to represent complex surfaces that can change topology as they evolve. We also derive constraints for the surface and in order to evolve the surface according to these constraints we introduce a coupled force field.

The paper is organized as follows: In Section 2 the preliminaries are explained and Section 3 describes the formalism. Section 4 shows how this is implemented and finally Section 5 shows experimental results.

1.1 Previous Work

The problem of recovering a surface from speculatities is related to the area of shape from shading [1,22]. Shape from shading deals with reconstructing the shape of smooth Lambertian surfaces from gray level images.

Some previous work has been done in the area of reconstructing or estimating surfaces from specularities. An early paper examining the information available from the motion of specularities under known camera motion is [25]. These methods all require some form of laboratory setup. Work has been done on recovering surfaces by illuminating them with circular light sources [23] or extended light sources [11]. Some work has also been done on reconstructing perfect mirror surfaces by studying reflections of a calibration object containing lines passing through certain points [17]. The method proposed in this paper is valid for general camera motion and general smooth specular surfaces as opposed to previous attempts where extended light sources, calibrated scenes, controlled camera motion or other constraints were used.

Our proposed method of adapting the level set framework is similar to the local operators proposed in [10]. But our work differs in that the local operators are different and are adapted to data fitting instead of 3D sculpturing.

Other methods have been proposed for fitting a surface to data using a variational approach, e.g. [2,3,8]

1.2 Contribution of Paper

The contribution of this paper is to propose a new method that makes it possible to estimate specular surfaces from image sequences taken with ordinary uncalibrated hand-held video cameras. Furthermore, the proposed method is valid for general camera motions. The only assumptions made are that the surface is smooth, i.e. that it has continuous derivatives, and that the light source is distant and point-shaped. The estimation of the surface is done using a level set approach where a force field is coupled to the level set grid.

2 Background

As a courtesy to the reader and to introduce notation a brief introduction of background material will be presented.

2.1 Camera Model

The following notations will be used: $\mathbf{X}$ denotes object points in homogeneous coordinates and $\mathbf{x}$ denotes image points in homogeneous coordinates. The focal point will be denoted $\mathbf{c}$ and the camera matrix P. We use the standard pin-hole camera model, cf. [5]. The object points are then related to the image points by a matrix multiplication,

$$\mathbf{x} \sim P\mathbf{X} , \tag{1}$$

where $\sim$ denotes equality up to scale. Given the camera matrix P, the line of sight corresponding to the image point $\mathbf{x}$ is given by

$$\mathbf{r}(\mathbf{x}) = \mathbf{c} + \lambda P^{+}\mathbf{x} , \tag{2}$$

where $\mathbf{c} = \mathcal{N}(P)$ denote the focal point, λ the depth parameter and P^{+} denote the pseudo-inverse of P. Hence if a specularity is observed at point $\mathbf{x}$ in a given camera, (2) denotes the possible locations of the surface reflecting the light.

2.2   Structure and Motion Estimation

To determine the shape of a surface with the proposed method it is necessary to know the motion of the camera. This is obtained using structure from motion techniques. Throughout this paper we assume that there are enough features in the scene to determine the motion of the camera and the structure of a limited number of feature points. This is done by extracting and tracking feature points through the image sequence. The fundamental matrix, F, and the trifocal tensor, T are estimated and an initial affine reconstruction is obtained from the cheriality constraints, cf. [12]. Finally an initial Euclidean reconstruction is obtained [6]. The Euclidean structure and camera motion is then refined using bundle adjustment, cf. [5,19,21].

2.3   Level Set Methods

In this paper we want to evolve a surface to find an estimate for a smooth specular surface using the level set method [15,14,18]. The time dependent surface $S(t)$ is implicitly represented as a level set of a scalar-valued function $\phi(\mathbf{x}, t)$ in $\mathbb{R}^3$ as

$$S(t) = \{\mathbf{x}(t) \; ; \; \phi(\mathbf{x}(t), t) = k\} , \tag{3}$$

where the value of $k \in \mathbb{R}$ is usually taken to be 0, making the surface a zero set of $\phi(\mathbf{x}, t)$. One of the advantages of this representation is that the topology of the surface is allowed to change as the surface evolves, thus making it easy to represent complex surfaces that can merge or split and also surfaces that contain holes.

Differentiating the expression $\phi(\mathbf{x}(t), t) = k$ in (3) using the chain-rule gives

$$\phi_t + \nabla\phi(\mathbf{x}(t), t) \cdot \mathbf{x}'(t) = 0 . \tag{4}$$

This is the fundamental equation of motion for the level set. The normal of the level set surface is given by,

$$\mathbf{n} = \frac{\nabla\phi}{|\nabla\phi|} . \tag{5}$$

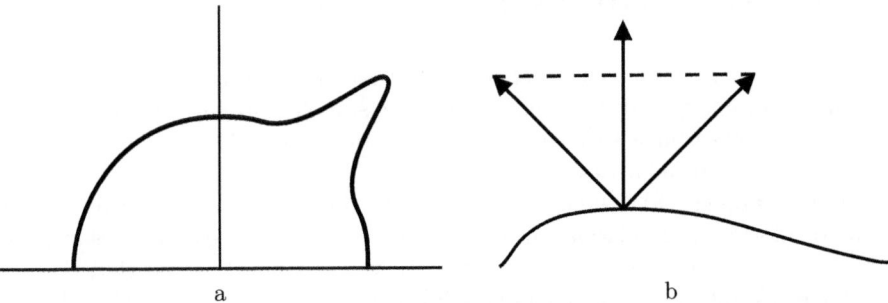

Fig. 2. Specular reflection. a) An example of a BRDF. The specular lobe shows the increased reflected intensity for certain viewing directions. b) The condition for specular reflection.

The surface evolves under the influence of a speed function F in the direction of the normal. The function F can be assumed to be normal to the surface such that $F = \mathbf{x}'(t) \cdot \mathbf{n}$, since motion in other directions can be considered as a reparametrization of the surface. The equation of evolution can then be written as

$$\phi_t + F|\nabla\phi| = 0 \ . \tag{6}$$

The surface is evolved by solving this PDE on a discrete grid. For a more thorough treatment of level set surfaces cf. [14,18].

2.4 Specular Reflection

A non-Lambertian, specular, surface reflects light according to some distribution function, called Bi-directional Reflectance Distribution Function (BRDF). For a specular surface the BRDF is not uniform as in the Lambertian case. An example of a BRDF for a specular surface is shown in Figure 2a. The specular component can be seen when the surface normal $\mathbf{N}$ bisects the viewing direction $\mathbf{R}$ and the light direction $\mathbf{L}$ as shown in Figure 2b. A Lambertian surface would have a symmetric reflectance function without a specular lobe. The condition for specular reflection shown in Figure 2b is valid for the limiting case when the specular lobe has very small width and can be considered a delta-function. This would give a hard constraint on the viewing direction. However, if the specular lobe is not a delta-function then there is some uncertainty in the viewing direction. This gives a soft constraint on the direction for observing specularities. If the BRDF of the surface is known this information can be used. This is however, beyond the scope of this paper. To sum up, a specular reflection gives directional information and if the light source direction is known, the surface normal is also known.

3   Surface Constraints from Specularites

The geometric conditions for specular reflection and the relation between a specularity in an image and the orientation of the surface normals leads us to formulate constraints that a surface has to satisfy in order to be consistent with the observation of specularities.

It is assumed, that there exist enough features in the scene to recover the camera motion and camera parameters, see Section 2.2. We also assume that the surface S is a smooth surface with observed specularities and that the light source is distant, point-shaped and its direction known. This means that the light source direction $\mathbf{L}$ is a constant vector. By smooth we mean that all components of S have continuous partial derivatives. These assumptions are reasonable since enough background can be included in the images by the person operating the camera and the distant light source assumption is valid for many scenarios, e.g. outdoor scenes with sunlight.

3.1   Specular Constrains

We use the following notation: S is a smooth surface in $\mathbb{R}^3$, $\mathbf{x}_i$ are the image coordinates for specularity i, c_i is the focal point of the corresponding camera and $\mathbf{r}_i$ is the ray from c_i through $\mathbf{x}_i$, see Figure 3. It is possible to have more than one specularity in each image so with image i we mean the image corresponding to specularity i. The total number of specularities in the sequence is denoted n.

The condition for observing a specular reflection is that the surface normal bisects the viewing direction and the incident light direction, see Figure 2. For a point on a surface S with normal $\mathbf{N}$ and light source direction $\mathbf{L}$, this relation is

$$\mathbf{R} + \mathbf{L} = (2\mathbf{N} \cdot \mathbf{L})\mathbf{N} \ , \tag{7}$$

and the specular reflection direction $\mathbf{R}$ can be determined as

$$\mathbf{R} = (2\mathbf{N} \cdot \mathbf{L})\mathbf{N} - \mathbf{L} \ . \tag{8}$$

Since we have computed the orientation and position of the camera for the whole sequence and the light source direction can easily be determined (e.g. by having one image where the shadow of the camera is visible) we get a series of constraints on the surface. The surface normal at the specular reflection fulfil the relation in (8) above. This means that at the intersection of the ray $\mathbf{r}_i$, given by (2), from the focal point c_i through the specularity $\mathbf{x}_i$ in image i, and the surface, the normal $\mathbf{N}_i$ is known. This relation is shown in Figure 3. Solving for $\mathbf{N}_i$ we get

$$\mathbf{N}_i = \frac{\mathbf{L} - \tilde{\mathbf{r}}_i}{|\mathbf{L} - \tilde{\mathbf{r}}_i|} \ , \tag{9}$$

where $\tilde{\mathbf{r}}_i$ is the directional vector for each ray, normalized so that $|\tilde{\mathbf{r}}_i| = 1$.

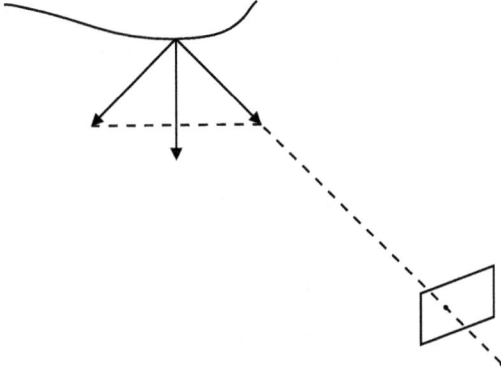

Fig. 3. The relation between a specularity in an image and the surface normal.

3.2   Regularization

The problem is that the depths λ_i from (2) cannot be determined. A distant
light source means that the condition for specular reflection will be fulfilled at
all points on the ray $\mathbf{r}_i$. Hence we get a whole family of surfaces that satisfy the
normal constraints (9). There is then an inherent ambiguity in the solutions since
there are many smooth surfaces at different depths λ_i that satisfy the conditions.
Note also that the ordering of the rays $\mathbf{r}_i$ is depth-dependent. This is illustrated
in Figure 4. To solve this ambiguity and to fix the surface in space we require
that one or more features corresponding to 3D-points $\mathbf{X}_j \in \mathbb{R}^3$ can be found on
the surface or the surface boundary.

Unfortunately, the constraints arising from known depths of a limited number
of points on the surface boundary and the constraints on the normal direction
to the surface arising from the detected reflections are not sufficient to uniquely
determine the shape of the surface. Thus we have to add additional constraints.
The most natural constraint to add is some kind of smoothness or regularity
constraint on the surface.

To find a surface estimate we then have three different constraints, *point
constraints* to position the surface in $\mathbb{R}^3$ and *normal constraints* to find the
shape of the surface. These are due to local properties of the surface. A global
smoothness constraint is also needed in order to obtain a reasonable surface
shape since there are many surfaces that fulfill the specular conditions even
after the depth of one or more points are fixed.

The point constraints are obtained from the structure from motion estimation, where the structure is represented as a cloud of points $\mathbf{X}_j$. The constraints
are then, that the surface S should pass through these points, or more formally

$$\forall \, \mathbf{X}_j \ \exists \, \mathbf{p}_j \in S \quad s.t. \quad |X_j - \mathbf{p}_j| = 0 \ . \tag{10}$$

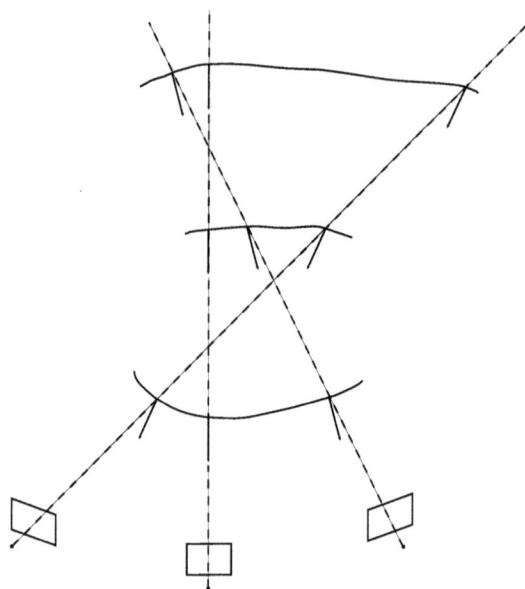

Fig. 4. The surface ambiguity due to unknown depth (note that the ordering of the rays $\mathbf{r}_i$ is not constant).

3.3   Induced Objective Function

We propose to apply the induced constraints, as expressed in (9) and (10), in the form of soft constraints as explained in Section 2.4. That is, instead of requiring that they hold exactly, an objective function will be used were deviations will be punished. As can be seen from the discussion above, such an objective function needs three terms corresponding to the normal constraints, the point constraints and a smoothing term.

We then obtain an expression for the objective function that looks like

$$\sum_i d_n(\mathbf{N}_i, \mathbf{n}_i) + \sum_j d_p(\mathbf{X}_j, S) + area \ , \tag{11}$$

where d_n and d_p are metrics for the deviation of the surface normal $\mathbf{n}_i$ from the desired normal $\mathbf{N}_i$ and for the point $\mathbf{X}_j$ from the surface. The last term is a mean curvature flow smoothness term [18]. Note that the additive structure of (11) makes it straight forward to incorporate other observed properties of the surface, e.g. those presented in [4].

Hence, the proposed scheme can be summarized as follows:

1. Compute camera motion from background features using structure from motion techniques, see Section 2.2.
2. Identify specularities in the images and determine the rays $\mathbf{r}_i$.

3. Calculate the light source direction **L** (from camera shadow or other method).
4. Determine the constraints for the surface normals $\mathbf{N}_i$ using (9) and the constraints for the points on the surface.
5. Find an estimate of the surface using (11) with the level set method as described in Section 4.

4   Level Set Implementation

The solution we are looking for is a smooth surface satisfying the constraints given in Section 3. We propose to do this by optimizing (11) using level sets. The third term is a standard mean curvature flow. We then propose to use a force field, derived from the normal constraints, when evolving the surface. The speed function F for the evolution equation then becomes

$$F = \alpha \mathcal{C} + \mathcal{F} \ , \tag{12}$$

where α is a real-valued constant that determines the amount of smoothing imposed, $\mathcal{C}$ is the mean curvature flow term and $\mathcal{F}$ is a scalar valued force field incorporating the two first constraints.

4.1   Force Field Method

To minimize the objective function (11), the normal constraints and the point constraints will change the surface locally. To do this a force field $\mathcal{F}$ is derived as the sum of the local forces contributed by each constraint

$$\mathcal{F} = \sum_i f_n(\mathbf{N}_i) + \sum_j f_p(\mathbf{X}_j) \ . \tag{13}$$

Here $f_n(\mathbf{N}_i)$ and $f_p(\mathbf{X}_j)$ are the force contributions for normal $\mathbf{N}_i$ and point $\mathbf{X}_j$. Since we are only interested in the force in the direction of the surface normal the direction of $\mathcal{F}$ is given. To obtain a scalar valued force field, we then only need to specify the amplitude.

A distance based function $\mathcal{D}$ is introduced to limit the volume affected by each constraint. We propose to use a symmetric three-dimensional gaussian with zero mean and width depending on the error of the corresponding constraint. The function $\mathcal{D}$ around point $\mathbf{p}$ is then

$$\mathcal{D}_p(\mathbf{d}) = \frac{1}{\sqrt{2\pi}\sigma} \ e^{-\mathbf{d}^T\mathbf{d}/2\sigma^2} \ , \tag{14}$$

where $\mathbf{d}$ is the distance from $\mathbf{p}$ and σ is the standard deviation.

This force field method is related to the approach taken in [20] where methods for surface processing are developed by operating on the surface normals in a two step iterative algorithm.

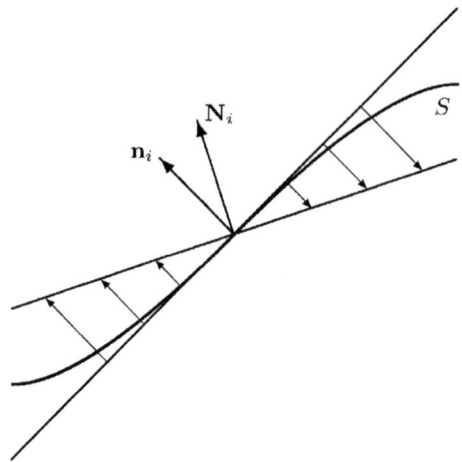

Fig. 5. The local force field $\tilde{f}_n$ induced from a normal constraint.

4.2   Normal Constraints

Deviation of the surface normal $\mathbf{n}$ from the desired value $\mathbf{N}$ will result in an error ϵ_n according to

$$\epsilon_n = |\mathbf{r}| \; , \tag{15}$$

where

$$\mathbf{r} = \mathbf{N} - \mathbf{n} \; . \tag{16}$$

To change the surface in order to minimize the error we propose to use a local linear force field $\tilde{f}_n$ induced by each constraint as

$$\tilde{f}_n = (\mathbf{p} - \mathbf{X}) \cdot \mathbf{R} \; , \tag{17}$$

Where $\mathbf{p}$ is the point at the intersection of $\mathbf{r}$, given by (2), and S, $\mathbf{X}$ is a point on the surface, $\cdot$ denotes the inner product and $\mathbf{R}$ is defined as

$$\mathbf{R} = \mathbf{r} - \frac{\mathbf{r} \cdot \mathbf{n}}{|\mathbf{n}|^2} \mathbf{n} \; . \tag{18}$$

This field is shown in Figure 5. This field is then convolved with the volume limiting function $\mathcal{D}$ centered at the point $\mathbf{p}$ where the normal constraint is applied. This gives the final force as

$$f_n(\mathbf{N}_i) = \tilde{f}_n(\mathbf{N}_i) * \mathcal{D} \; . \tag{19}$$

As the surface evolves, the value of λ in (2) needs to be updated since the intersection of $\mathbf{r}$ and S changes. This is done by finding the value that minimizes

$$\min_{\lambda} |\phi(\mathbf{c} + \lambda P^+ \mathbf{x})| \; . \tag{20}$$

for each ray $\mathbf{r}$, since the surface is defined as the zero set of ϕ.

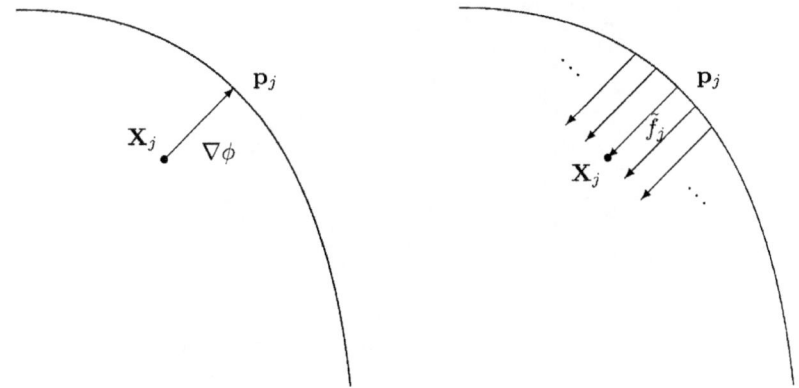

Fig. 6. The local force field $\tilde{f}_p$ induced from a point constraint.

4.3 Point Constraints

The effect of the point constraints is similar to the normal constraints. The constraint of (10) in effect imply that the surface S should pass through $\mathbf{X}$. Hence the force needed to achieve this moves S onto $\mathbf{X}$. This field is shown in Figure 6. Due to the nature of the level set method, the force needed is:

$$\tilde{f}_p(\mathbf{X}_j) = -\phi(\mathbf{X}_j) \ , \tag{21}$$

in that this will make the zero set of ϕ go through $\mathbf{X}_j$. Enforcing locality gives

$$f_p(\mathbf{X}_j) = \tilde{f}_p(\mathbf{X}_j) * \mathcal{D} \ . \tag{22}$$

This approach is similar to the point attractors used in [10] to locally move a surface closer to specified points.

4.4 Range Adaptation

A given constraint only specifies the properties of a single point on the surface. However, since S is a smooth surface the constraints will influence a region around this point. Hence the force associated with a constraint needs to have local effect. The influence of a constraint is determined by the width of $\mathcal{D}_p(\mathbf{d})$. If the error in a constraint is large, a larger part of the surface should be affected by the resulting force and closer to the desired value the effects should be more local. This means that the standard deviation σ in (14) should be different across constraints and iterations.

For the normal constraints we set the width of $\mathcal{D}_p(\mathbf{d})$ so that the value at ϵ_n is 0.99. And for the point constraints we set the value at ϵ_p to be 0.9.

Fig. 7. The original simulated data set.

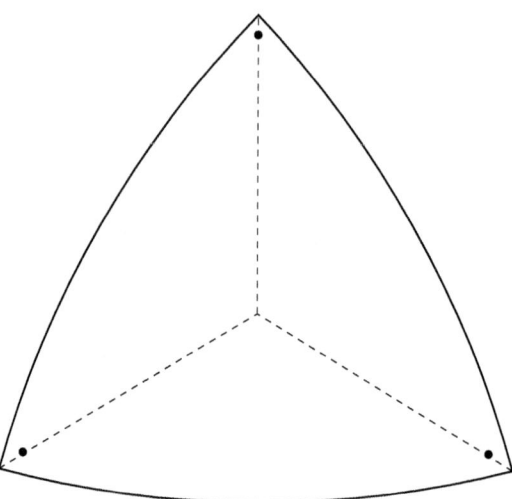

Fig. 8. A schematic overview of the test set up. The points illustrate the 3 point constraints and the 31 normal constraints are distributed uniformly along the dashed lines.

5   Experiments

To evaluate the proposed method, it was tested on a synthetic generated data set depicted in Figure 7, consisting of $\frac{1}{8}$ of a sphere. There are 3 point constraints, one in each corner and 31 normal constraints, as illustrated in Figure 8. The α value of (12) is 0.1.

The resulting surface is illustrated in Figure 9, where it can be seen, that the surface is reconstructed acceptably near the constraints, but not everywhere. The latter illustrates, that the surface chosen, i.e. the sphere, is not obtained by enforcing the prior of the curvature flow. For a more quantitative evaluation, the

Table 1. Mean differences between realized and specified properties of the surface.

Mean Deviation from:	
Normal Constraints	6.891 degrees
Point Contstraints	0.276 voxels

Fig. 9. The reconstructed surface. Note that the reconstruction is acceptable along the constraints specified in Figure 8.

mean deviation of the resulting normals from the specified is shown in Table 1. The same data are shown for the point constraints. Table 1 shows that the proposed method for incorporating the information from specularities preforms well. This should be seen in the light, of numerical noise, and that the area minimizing smoothness constraint induces a force on the surface away from the specified constraints.

As mentioned, the proposed method is not primarily thought of as a "stand alone" method, but is applied as such here to test its possibilities. In this respect it is noted that the smoothness constraint in form of the curvature flow is not a particulary good regularizer for interpolation. So if surface estimation from sparse specular constraints were to be attempted, it could be preferable to use another smoothness constraint. A likely candidate is a fourth order flow, which unfortunately, is hard to get numerically stable. Recently though, a feasible method has been proposed in [20].

6 Summary and Conclusions

We have proposed a method for reconstructing the shape of specular surfaces using a level set implementation to which a force field is coupled. This force field is derived from constraints on the surface that determines the shape of the surface, positions it is space and keeps the surface smooth. Experiments with synthetic data show very promising results.

The proposed method is not primarily intended to be used for surface estimation alone, and in future work it should be integrated with such methods, e.g.

[4]. It could also be interesting to try other smoothing schemes, e.g. fourth order flow, if the proposed method should be used alone. Lastly, it could be interesting to formalize the proposed framework more, e.g. by formulating it as minimizing a functional.

Acknowledgement

We would like to thank J. Andreas Bærentzen, for interesting discussions and advice regarding the level set implementation.

References

1. M. J. Brooks and B. K. P. Horn. *Shape and Source from Shading.* MIT Press, Cambridge, MA, 1989.
2. V. Caselles, R. Kimmel, and G. Sapiro. Geodesic active contours. *Int. Journal of Computer Vision*, 1997.
3. L. D. Cohen and R. Kimmel. Global minimum for active contour models: A minimal path approach. In *Proc. Conf. Computer Vision and Pattern Recognition*, pages 666–673, 1996.
4. O. Faugeras and R. Keriven. Variational principles, surface evolution, pdes, level set methods, and the stereo problem. *Image Processing, IEEE Transactions on*, 7(3):336 –344, 1998.
5. R. Hartley and A. Zisserman. *Multiple View Geometry.* Cambridge University Press, The Edinburgh Building, Cambridge CB2 2RU, UK, 2000.
6. A. Heyden and K. Åström. Euclidean reconstruction from image sequences with varying and unknown focal length and principal point. In *Proc. Conf. Computer Vision and Pattern Recognition*, pages 438–443, 1997.
7. H. Jin, A. Yezzi, and S. Soatto. Variational multiframe stereo in the presence of specular reflections. Technical Report TR01-0017, UCLA, 2001.
8. R. Kimmel and A. M. Bruckstein. Global shape from shading. In *Computer Vision and Image Understanding*, pages 120–125, 1995.
9. D. Morris and T. Kanade. Image-consistent surface triangulation. In *IEEE Conf. Computer Vision and Pattern Recognition'2000*, pages 332–338, 2000.
10. K. Museth, D.E. Breen, R.T. Whitaker, and A.H. Barr. Level set surface editing operators. *ACM Transactions on Graphics*, 21(3):330–8, 2002.
11. S. K. Nayar, K. Ikeuchi, and T. Kanade. Determining shape and reflectance of lambertian, specular, and hybrid surfaces using extended light sources. In *IEEE Workshop on Industrial Applications of Machine Intelligence and Vision*, 1989.
12. D. Nister. *Automatic Dense Reconstruction from Uncalibrated Video Sequences.* PhD thesis, Royal Institute of Technology, KTH, 2001.
13. M. Okutomi and T. Kanade. A multiple-baseline stereo. *Pattern Analysis and Machine Intelligence, IEEE Transactions on*, 15(4):353–363, 1993.
14. S. J. Osher and R. P. Fedkiw. *Level Set Methods and Dynamic Implicit Surfaces.* Springer Verlag, 1st edition, November 2002.
15. S. Osher and J. A. Sethian. Fronts propagating with curvature-dependent speed: Algorithms based on Hamilton-Jacobi formulations. *Journal of Computational Physics*, 79:12–49, 1988.

16. M. Pollefeys. *Self-calibration and metric 3D reconstruction from uncalibrated image sequences.* PhD thesis, K.U.Leuven, 1999.
17. S. Savarese and P. Perona. Local analysis for 3d reconstruction of specular surfaces. ii. *Computer Vision - ECCV 2002. 7th European Conference on Computer Vision. Proceedings (Lecture Notes in Computer Science Vol.2351)*, pages 759–74, 2002.
18. J.A. Sethian. *Level Set Methods and Fast Marching Methods Evolving Interfaces in Computational Geometry, Fluid Mechanics, Computer Vision, and Materials Science.* Cambridge University Press, 1999.
19. C.C. Slama. *Manual of Photogrammetry.* American Society of Photogrammetry, Falls Church, VA, 4:th edition, 1984.
20. T. Tasdizen, R. Whitaker, P. Burchard, and S. Osher. Geometric surface processing via normal maps. Technical Report UUCS-02-02, School of Computing, University of Utah, January 2002.
21. B Triggs, P F McLauchlan, R I Hartley, and A W Fitzgibbon. Special sessions - bundle adjustment - a modern synthesis. *Lecture Notes in Computer Science*, 1883:298–372, 2000.
22. Ruo Zhang, Ping-Sing Tsai, J.E. Cryer, and M. Shah. Shape-from-shading: a survey. *Pattern Analysis and Machine Intelligence, IEEE Transactions on*, 21(8):690–706, 1999.
23. J. Y. Zheng and A. Murata. Acquiring a complete 3d model from specular motion under the illumination of circular-shaped light sources. *Pattern Analysis and Machine Intelligence, IEEE Transactions on*, 22(8):913–920, 2000.
24. M. Ziegler, L. Falkenhagen, R. Horst, and D. Kalivas. Evolution of stereoscopic and three-dimensional video. *Image Communication*, 14, 1998.
25. A. Zisserman, P. Giblin, and A. Blakey. The information available to a moving observer from specularities. *Image and vision computing*, 1989.

A Markov Random Field Approach to Multi-scale Shape Analysis

Conglin Lu, Stephen M. Pizer, and Sarang Joshi

Medical Image Display and Analysis Group
University of North Carolina, Chapel Hill, USA
lu@cs.unc.edu

Abstract. With a mind towards achieving means of image comprehension by computer, we intend to convey the benefits of (1) characterizing the geometry of object complexes in the real world as contrasted with the geometric conformation of their images, and (2) describing populations of object complexes probabilistically. We show how a multi-scale description of inter-scale residues of geometric features provides a set of efficiently trainable probability distributions via a Markov random field approach, and specifics the location and scale of geometric differences between populations. These ideas and methods are illustrated using medial representations for 3D objects, depending on their properties (1) that local descriptors have an associated coordinate frame and distance metric, and (2) that continuous geometric random variables can be used to describe all members of a population of object complexes with a common structure and the variation among those members. We demonstrate with respect to the following object-complex-relative discrete scale levels: a whole object complex, individual objects, various object parts and sections, and fine boundary details. Using this illustrative framework, we show how to build Markov random field (MRF) models on the geometry scale space based on the statistics of shape residues across scales and between neighboring geometric entities at the level of locality given by its scale. In this paper, we present how to design and estimate MRF models on two scale levels, namely boundary displacement and object sections.

1   Introduction

Analysis of the geometry of one or a group of objects plays an important role in computer vision and image comprehension. In many applications, the focus is not on a single object instance, but rather on a population of objects in the real world. For instance, geometric information can be incorporated as priors to guide image segmentation. By imposing geometric constraints on the interpretation of data, one obtains more reliable results, as opposed to making decisions based solely on image intensity information. As another example, object discrimination usually involves characterizing and comparing classes that differ in their geometric conformation, e.g., their volume or their shape. In both of these cases, one needs to describe the geometry of populations of objects.

L.D. Griffin and M. Lillholm (Eds.): Scale-Space 2003, LNCS 2695, pp. 416–431, 2003.
© Springer-Verlag Berlin Heidelberg 2003

Objects or groups of objects are usually represented by certain geometric features. These features are most intuitive and provide locality if each is at a restricted range of scale. To describe an object complex at a coarse level, the relative poses of each object may be most informative. On the other hand, to characterize an individual object, one needs to go to finer scale levels to describe different sections and the boundary details of each section. A complete representation should be able to provide geometric information on all relevant scales. As such, each geometric entity can be regarded as an element of the *geometry scale-space*.

The classical scale-space theory is concerned with analysis of image intensity structure across scales. The most basic scale-space is generated by local Gaussian diffusions [1,2]. When applied to an image, this diffusion process provides descriptions of the image at different scale levels. The procedure is equivalent to solving a linear partial differential equation (PDE) with initial values. More general scale-spaces have been proposed, including those generated by various non-linear PDE's and morphological operations [3,4], as well as spatio-temporal scale-spaces [5]. The idea has also been applied to shape analysis, where objects are deformed in a geometrically consistent manner such that multi-scale descriptions are attained. The generic behaviors of surface evolution were studied using singularity theory [6,7,8]. In [9], a general "scale-based geometry" framework is proposed. Examples of multi-scale shape representation include description by singularities [10,11] and by geometric invariants [12].

The non-linear scale spaces have been used in an attempt to reflect a notion of locality, i.e., locally relevant size and distance, to the spatial summaries computed by diffusion. In our study of geometric entities such as objects and boundaries, locality must be taken relative to the components of which an entity is formed. The relevant size and distance of an object complex are determined by various objects in it, whereas those for an object are determined by the natural sections making up the object, and so on.

Based on this view, we focus on discrete object-relevant scale levels, rather than treating scale as a continuous parameter as in the powerful PDE theory. We follow the practice of making scale spaces from discrete levels and *residues* between levels as in the wavelet approaches [13,14]. Thus we describe the changes in geometry across scales rather than geometric features prominent at all scales. To do so, we choose geometric primitives at each level to describe the residue from the geometric information provided at the next larger scale level. Each primitive summarizes the geometric information contained in certain spatial domain of size relevant to that scale. The differences in description between scale levels reflect different levels of detail. Moreover, the relevant distance within a scale level induces a notion of neighbors, i.e., nearby geometric primitives at that scale: nearby objects, nearby object sections, etc. Neighbors at a larger scale level are typically more distant than neighbors at a smaller scale level. This neighbor relation, together with the spatial extent of primitives and levels of detail in description, realize the notion of locality. With this hierarchical, multi-scale description, a primitive at one scale level is seen relative both to the next larger level and to its neighbors at that level. As we will show, this approach enables one to efficiently describe shape models, and in turn efficiently train and apply these models to shape analysis applications. Furthermore, the study of multi-scale geometry also suggests a paradigm in which image intensity scale spaces can be analyzed in object-relevant coordinate systems, rather than in their native Euclidean spaces (see, e.g., [15].)

Since we are interested in shape class attributes, we consider objects or object complexes as members of a population of examples in the real world. A population of entities with similar geometric conformation can be effectively described by probabilistic models. For instance, 2D curve models were proposed in [16,17]. For 3D objects, principal component analysis on surface points [18,19] and spherical harmonic descriptors [20] have been studied extensively. In these models, a probability measure is put on the space of all possible deformations. The parameters of the measure can be estimated from a training data set.

In a large variety of applications, the population of interest can be described with a fixed topology. Take the shape of liver as an example. Globally all livers have the same general geometric conformation; yet on finer scales the geometry varies significantly from one to another. In deciding the topology of the representation of liver, i.e., whether to model it as a single blob or as a blob with branches, we need to take into account the common geometry as well as the variations among different instances, so that both of these pieces of information can be effectively described. Once a fixed topology is chosen, fixed correspondences can be established across the population between entities at a scale level and between scale levels. As a result, what differs between members of the population is the quantitative, geometric parameters and not qualitative properties of structure or topology. This provides a basis for building probabilistic shape models. Of course, there exist situations where one has to allow topological changes in representation, but that is not the focus of this paper.

A natural probabilistic framework to describe the inter-scale and intra-scale neighbor relationships is the Markov random field (MRF) approach. Given the configuration at each scale level, the geometric residue information at the next smaller scale level is summarized by describing the local interactions among the corresponding geometric primitives via an MRF model. In this approach there are a small number of parameters to be estimated at each location within each scale level, so accurate parameter estimation can be achieved with a limited number of training cases.

In this paper, we illustrate how to build such a multi-scale probabilistic framework for describing populations of objects based on a particular geometric representation called *m-reps* [21,22], which is medial-based and has a boundary displacement component. In m-reps, geometric entities are explicitly represented at discrete locations at the following discrete scale levels: a whole object complex, individual objects, different object parts and sections, and boundary points. Computing the fixed topology from a population in this framework is achieved using the method in [23], not discussed further here. At each scale level the geometry is represented by the configuration of certain geometric primitives that are most descriptive at that level. For example, an object as a whole is described by a global similarity transformation, whereas different sections of the object are characterized by more local transformations. This differs from the traditional scale-space in that different primitives are used at different scales in the representation. Using m-reps, we illustrate how MRF models can be designed on two scale levels, namely the boundary displacement level and the object section level.

In what follows, we first briefly describe the m-reps representation mechanism in section 2. The details of the Markov random field models are presented in sections 3 - 5. Section 6 shows some statistics of hippocampi. We finish with some concluding remarks in section 7.

2   Multi-scale Shape Representation by M-reps

We represent 3D entities by m-reps, which is a medial-based multi-scale representation with a boundary displacement component. It provides stable medial and boundary structures [21,22] as well as object-intrinsic local coordinate systems [24]. For its application in image segmentation and shape discrimination, refer to [21,25].

In m-reps, at all but the boundary scale level, an object is described by a set of continuous medial manifolds, which are sampled to yield discrete representations. Each sample point is called a medial atom, which describes a through section of the object (see Fig. ??). It is a 4-tuple $\mathbf{m} = (\mathbf{x}, \mathbf{R}, r, \theta)$ consisting of

- a translation $\mathbf{x} \in \mathbb{R}^3$, specifying the position of the medial point; we can consider this translation in units of the medial width r (defined below);
- a rotation $\mathbf{R} \in \mathrm{SO}(3)$, describing a local orthonormal frame $(\mathbf{n}, \mathbf{b}, \mathbf{b}^\perp)$, where $\mathbf{n}$ is the normal to the medial manifold, $\mathbf{b}$ is along the direction of the fastest narrowing of the implied boundary sections, and $\mathbf{b}^\perp = \mathbf{n} \times \mathbf{b}$;
- a magnification scalar $r \in \mathbb{R}^+$, describing the local width, defined as the distance from the medial point to the implied boundary points;
- a 2D rotation angle $\theta \in \mathrm{SO}(2)$, called the object angle, which determines the angulation of the implied boundary sections relative to $\mathbf{b}$.

Each of these 4 elements is a member of an algebraic group, a property that will be of importance later in the paper. $(\mathbf{x}, \mathbf{R}, r)$ defines a local coordinate system, in which the origin is given by $\mathbf{x}$, the coordinate axes are specified by $\mathbf{R}$, and the distance is measured in units of r. The two implied boundary points are specified as $\mathbf{y}_i = \mathbf{x} + r\mathbf{n}_i, i = 0, 1$, where $\mathbf{n}_0$ and $\mathbf{n}_1$ are the two respective surface normals given by $\mathbf{n}_{0,1} = \cos(\theta)\mathbf{b} \pm \sin(\theta)\mathbf{n}$.

An m-rep figure is a quadrilateral mesh of medial atoms, with spacing determined through the analysis of the training population [23]. It describes a slab-like object or object part. The 4-adjacency in the quad-mesh determines the atom neighbor relationship. Given an m-rep figure, a smooth boundary surface is generated by a subdivision surface algorithm [26] that approximates the boundary positions and normals implied by each atom. Objects are generally represented

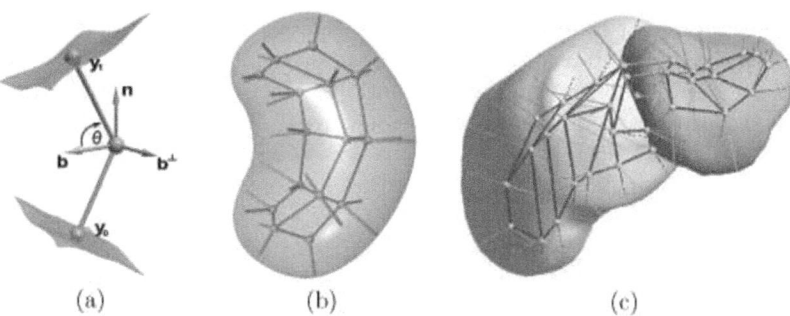

(a)   (b)   (c)

Fig. 1. M-reps. (a) A medial atom with a cross-section of the boundary surface it implies; (b) A single figure m-rep model of a kidney and its boundary surface; (c) A two-figure m-rep liver model.

Table 1. The scale levels of m-reps, with primitives, inter-scale residue transformations, and neighbor relations at each scale.

Scale level	Primitive	Inter-scale residue transformation	Intra-scale neighbors
Object complex	Object complex pose	Similarity	—
Object	Object pose	Similarity, relative to object complex level	Adjacent objects
Figure	Main figure and subfigure poses	Main figure and subfigure transformations, relative to object level	Adjacent figures
Medial atom	Atom configuration	Atom transformation, relative to figure level	Adjacent medial atoms
Boundary	Boundary vertex position	Displacement from medially implied position	Adjacent boundary vertices

by a linked figural model, together with boundary displacements. A main figure describes the main section of an object; various subfigures, each of which a single medial sheet, represent different branches, protrusions or indentations. Finally, an object complex is described by the configurations of individual objects.

In this multi-scale framework, residual geometric information between scale levels is specified by the appropriate residue transformations of the corresponding geometric primitives. Within a scale each primitive has neighbors of the same type. The residue of a primitive is expressible relative to those of its neighbors. For example, the residue of a medial atom can be described in terms of translation, rotation, etc. relative to its neighboring atoms. Similarly, inter-figural and inter-object relations can also be effectively described, since solid 3D regions and their boundaries are represented simultaneously. See Table 1 for a brief summary. A successively refined boundary representation can be derived as one goes from coarse to fine scale levels. At the finest level, each boundary point moves from its medially implied position along the medially implied normal direction to "fine tune" the description. There is a boundary tolerance associated with each scale level. At larger scales, the tolerance is higher, so details are ignored and global features are revealed; at smaller scales, the description focuses on refinements of the larger scales to describe more local geometric features.

To summarize, the m-rep framework defines a geometry scale-space with a discrete scale parameter, allowing local geometric features at different scale levels to be explicitly described. Furthermore, the medial structure, determined by a training population, provides a multi-scale object intrinsic coordinate system which is extremely well suited for statistical analysis of shapes, because correspondence among a population can be established systematically.

3   Probabilistic M-reps Models

3.1   Markov Random Fields

Given that any model has a fixed topology, we require a probability distribution on the space of geometric variables, i.e., real-valued random variables characterizing the primitives. Each of these variables specify an element of the algebraic group of operations applicable to that primitive. The total number of such random variables is usually very large. With an MRF approach, we can handle this problem by characterizing geometric information through local interactions among geometric primitives at various scale levels.

An MRF model is defined with respect to a simply-connected dependency graph $G = (V, E)$. Each vertex $v \in V$ corresponds to a random variable X_v in the model. The set of vertices that are connected to v via an edge in E is called the neighborhood of v, and is denoted by $\mathcal{N}(v)$. Thus $u \in \mathcal{N}(v) \Leftrightarrow v \in \mathcal{N}(u)$ and $v \notin \mathcal{N}(v)$. The completely connected subgraphs of G (including singletons) are called the cliques of G. A model P is an MRF with respect to G if

$$\text{Prob}\big(X_v \big| \text{ all other random variables }\big) = \text{Prob}\big(X_v \big| \{X_u : u \in \mathcal{N}(v)\}\big),$$

where $\text{Prob}(\cdot|\cdot)$ denotes conditional probability. P is said to be a Gibbs distribution with respect to G if the joint probability density of $\{X_v\}$ has the form

$$p_\Theta(\{X_v : v \in V\}) = \frac{1}{Z(\Theta)} \exp\{-\textstyle\sum_{C \in \mathcal{C}} A_C(X_C; \Theta)\}, \tag{1}$$

where $\mathcal{C}$ is the set of cliques of G, $X_C = \{X_v : v \in C\}$, Θ is a set of parameters, Z is a normalizing constant. Each $A_C \geq 0$ is called a potential function and depends only on those random variables whose indices are in C. The Hammersley-Clifford Theorem [27] establishes the equivalence between MRF's and Gibbs distributions with respect to the same dependency graph G. This allows one to specify an MRF by specifying the potentials in the corresponding Gibbs form.

The main advantage of the MRF approach is that the joint probability density to be estimated is specified by a relatively small number of parameters, which can

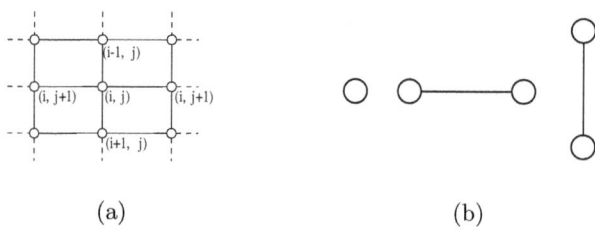

(a)   (b)

Fig. 2. The 4-neighbor structure for quad-mesh. (a) A typical node has 4 neighbors. (b) The cliques of the quad-mesh.

thus be efficiently learned from a training data set. One way of estimating them is to seek the parameter values that maximize the likelihood of the training data. These estimates are called the maximum likelihood (ML) estimates. In general this is done via stochastic sampling algorithms, such as Markov Chain Monte Carlo (MCMC) methods [27].

3.2   Markov Random Field M-reps Models

In the Markov random field approach for our framework, the neighbor relations are between scale levels and among neighbors at any scale level. Suppose there are l scale levels indexed by $1, 2, \ldots, l$, with scale 1 being the coarsest. Let $\mathbf{z}^k$ denote the collection of geometric primitives at scale level k. For a fixed k, every primitive $\mathbf{z}_j^k$ has a value implied by a corresponding primitive at the previous larger scale, which is called the parent primitive of $\mathbf{z}_j^k$ and denoted by $\mathcal{P}(\mathbf{z}_j^k)$. For example, at the figure level, each figure implies the medial atom primitives that make it up, with position, orientations, etc. relative to the figural geometry as in the mean of the training population. Let $\Delta\mathbf{z}_j^k$ denote the inter-scale residual giving the *difference* between $\mathbf{z}_j^k$ and $\mathcal{P}(\mathbf{z}_j^k)$, where differences are taken with respect to the group operations defining the primitive, e.g., translation, rotation, magnification, and object angulation for a medial atom. Let $\Delta\mathbf{z}^k = \{\Delta\mathbf{z}_j^k\}$. We describe each scale level by the residues $\Delta\mathbf{z}^k$ with the Markov assumption

$$\text{Prob}\big(\mathbf{z}^k|\{\mathbf{z}^1, \ldots, \mathbf{z}^{k-1}\}\big) = \text{Prob}\big(\mathbf{z}^k|\mathbf{z}^{k-1}\big) = P_k(\Delta\mathbf{z}^k), \quad \text{for } k > 1.$$

In doing so, we are describing the inter-scale-level relationship via the residuals and assuming that residuals at one scale are independent of those at other scales.

Each residual probability distribution $P_k(\Delta\mathbf{z}^k)$ is defined as an MRF model, with respect to the canonical neighborhood structure induced by the natural spatial relationship among primitives. For example, at the object level, where the primitives are objects, the neighbors of an object are its adjacent objects. At the atom level, the canonical neighborhood structure for medial atoms is the 4-adjacency graph induced by the quad-mesh structure, as shown in Fig. 2(a), since we sample the medial manifold by a quadrilateral array of atoms. The cliques of this dependency graph are single vertices and pairs of vertices that are connected by a horizontal or vertical edge in the quad-mesh (see Fig. 2(b)). If another sampling mesh, e.g. triangular mesh, is used, then appropriate canonical neighborhood structure can be induced similarly.

By the Hammersley-Clifford Theorem, the density of the MRF model P_k can be written in Gibbs form

$$p_k(\{\varDelta \mathbf{z}^k\}) \propto \exp\{-\textstyle\sum_{C \in \mathcal{C}} A_C(\varDelta \mathbf{z}^k_C)\},$$

where $\mathcal{C}$ is the set of cliques, and $\varDelta \mathbf{z}^k_C = \{\varDelta \mathbf{z}^k_j : j \in C\}$. Our goal is to design these MRF models so that they can be specified by a relatively small number of parameters. Two sets of details need to be given:

- in defining $\varDelta \mathbf{z}^k_j$, i.e., representing residual geometric information by primitives;
- in defining the potentials A_C, i.e., the form of the probability distribution and the means of estimating the parameters.

In the next two sections, we discuss these issues on two scale levels within the m-reps framework, namely the boundary level and the medial atom level.

4   MRF Models for Boundary Displacement

4.1   Model Description

For any m-rep figure, the medial manifold implies a 3D surface, which is represented by a dense set of boundary points $\{\mathbf{x}_i\}$. These are the geometric primitives at the boundary level. The medial manifold is parameterized by an object-intrinsic coordinate system, which provides correspondence between boundary points on different objects. Moreover, the description of each point $\mathbf{x}_i$ includes a local object width r_i, which is the distance between the point and the corresponding medial point, and a surface normal vector at that point.

At the boundary level, each medially implied boundary point is displaced along its normal vector to obtain a fine scale description. If the amount of movement of the i-th point is d_i, then by describing distance in multiples of object width r_i, so as to maintain magnification invariance, we define the displacement of that point to be the dimensionless variable $w_i = d_i/r_i$. This variable describes displacement as a member of the 1-dimensional additive group $\mathbb{R}$. The displacement field $\mathbf{w} = \{w_i\}$ on the medially implied boundary points is the residual geometric information at this scale level. The boundary surface is determined by the medial representation together with the displacement field.

Currently we use a quad-mesh to sample the boundary, thus the canonical neighborhood structure is the 4-adjacency structure. With respect to this graph, we define Markov random field model at the boundary level on the displacement field $\mathbf{w}$. The density has Gibbs form

$$p(\mathbf{w}) = \frac{1}{Z} \exp\left\{ -\textstyle\sum_i A_i(w_i) - \sum_{<i,j>} B_{ij}(w_i, w_j) \right\}, \tag{2}$$

with respect to Lebesgue measure, where $< i, j >$ denotes that points with indices i and j are neighbors. We assume that the distribution on $\mathbf{w}$ is a zero-mean Gaussian distribution, with the particular density form

$$p_q(\mathbf{w}) = \frac{1}{Z(q_1, q_2)} \exp\left\{ -\frac{q_1}{2} \sum_{i=1}^{n} s_i w_i^2 - \frac{q_2}{2} \sum_{<i,j>} b_{ij}(w_i - w_j)^2 \right\}, \quad (3)$$

where q_1, q_2 are positive parameters, $Z(q_1, q_2)$ is a constant depending on q_1 and q_2, n is the number of points, $\{s_i, b_{ij}\}$ are chosen so that the exponent above is a discrete approximation of the energy function

$$-\frac{q_1}{2} \int_S \frac{d^2(\mathbf{x})}{r^2(\mathbf{x})} d\mathbf{x} - \frac{q_2}{2} \int_S \|\nabla d(\mathbf{x})\|^2 d\mathbf{x},$$

where $d(\mathbf{x}), r(\mathbf{x})$ are the amount of movement and local radius at point $\mathbf{x}$, respectively. Notice that

$$p_q(w_i | \{w_j, j \neq i\}) \propto \exp\left\{ -\frac{q_1}{2} s_i w_i^2 - \frac{q_2 \sum_{<i,j>} b_{ij}}{2}\left(w_i - \sum_{<i,j>} \frac{b_{ij}}{\sum_{<i,j>} b_{ij}} w_j\right)^2 \right\}$$

This can be interpreted as putting a penalty on the amount of w_i as well as the difference between the displacement of point $\mathbf{x}_i$ and a weighted average of those of the neighboring points.

Different sections of the boundary can be modelled by the same MRF model (3) with different parameter values, which reflect the variation of boundary geometry in various sections.

4.2   Parameter Estimation

We now discuss how to estimate the parameters q_1 and q_2 in (3). Rewrite (3) as

$$p_q(\mathbf{w}) = \frac{1}{Z(q_1, q_2)} \exp\left\{ -\frac{1}{2} \mathbf{w}^T \mathbf{\Sigma}^{-1} \mathbf{w} \right\}. \quad (4)$$

Here $\mathbf{\Sigma}$ is the covariance matrix, and $\mathbf{\Sigma}^{-1} = q_1 \mathbf{D} + q_2 \mathbf{B}$, where $\mathbf{D}$ is an $n \times n$ diagonal matrix and $\mathbf{B}$ is an $n \times n$ symmetric, sparse matrix. The entries of $\mathbf{D}$ and $\mathbf{B}$ are determined by $\{s_i\}$ and $\{b_{ij}\}$. The normalizing constant is

$$Z(q_1, q_2) = (2\pi)^{\frac{n}{2}} |\det(q_1 \mathbf{D} + q_2 \mathbf{B})|^{-\frac{1}{2}}.$$

Since

$$|\det(q_1 \mathbf{D} + q_2 \mathbf{B})| = |\det(\mathbf{D})| \cdot |\det(q_1 \mathbf{I} + q_2 \mathbf{D}^{-\frac{1}{2}} \mathbf{B} \mathbf{D}^{-\frac{1}{2}})| = \prod_{i=1}^{n} s_i \prod_{i=1}^{n} (q_1 + q_2 \lambda_i),$$

where $\lambda_i, i = 1, \ldots, n$ are the eigenvalues of $\mathbf{D}^{-\frac{1}{2}} \mathbf{B} \mathbf{D}^{-\frac{1}{2}}$, we have

$$\log Z(q_1, q_2) = \frac{n}{2} \log(2\pi) - \frac{1}{2} |\det(q_1 \mathbf{D} + q_2 \mathbf{B})|$$

$$= \frac{n}{2} \log(2\pi) - \frac{1}{2} \sum_{i=1}^{n} \log(s_i) - \frac{1}{2} \sum_{i=1}^{n} \log(q_1 + q_2 \lambda_i).$$

Given a training data set, we seek the maximum likelihood estimates of q_1, q_2. Suppose there are M independent samples $\{\widehat{\mathbf{w}}_1, \widehat{\mathbf{w}}_2, \ldots, \widehat{\mathbf{w}}_M\}$, with

$$\widehat{\mathbf{w}}_i \sim \mathcal{N}\left(\mathbf{0}, (q_1 \mathbf{D}_i + q_2 \mathbf{B}_i)^{-1}\right), \qquad i = 1, 2, \ldots, M,$$

where $\{\mathbf{D}_i, \mathbf{B}_i\}$ have the same structure as $\mathbf{D}, \mathbf{B}$. The likelihood function is

$$L(q_1, q_2) = \sum_{i=1}^{M} \log\left(p_q^{(i)}(\widehat{\mathbf{w}}_i)\right) = -\sum_{i=1}^{M}\left(\log Z^{(i)} + \frac{1}{2}\widehat{\mathbf{w}}_i^T (q_1 \mathbf{D}_i + q_2 \mathbf{B}_i)\widehat{\mathbf{w}}_i\right).$$

Using the Cauchy-Schwartz inequality, we can show that the Hessian matrix $\nabla^2 L$ is negative semi-definite. Therefore, the maximum of L occurs at (q_1^*, q_2^*) such that $\nabla L(q_1^*, q_2^*) = \mathbf{0}$, which yields

$$\sum_{i=1}^{M}\sum_{j=1}^{n} \frac{1}{q_1^* + q_2^* \lambda_j^{(i)}} = \sum_{i=1}^{M} \widehat{\mathbf{w}}_i^T \mathbf{D}_i \widehat{\mathbf{w}}_i, \tag{5a}$$

$$\sum_{i=1}^{M}\sum_{j=1}^{n} \frac{\lambda_j^{(i)}}{q_1^* + q_2^* \lambda_j^{(i)}} = \sum_{i=1}^{M} \widehat{\mathbf{w}}_i^T \mathbf{B}_i \widehat{\mathbf{w}}_i, \tag{5b}$$

where $\{\lambda_j^{(i)}\}$ are the eigenvalues of $\mathbf{D}_i^{-\frac{1}{2}} \mathbf{B}_i \mathbf{D}_i^{-\frac{1}{2}}$. The above equations are solved numerically.

5   Markov Random Field Models for Object Sections

5.1   The MRF Model

The primitives at the atom level are medial atoms, which describe sections of an object. For simplicity we only consider single-figure objects. Let $\{\mathcal{A}_i\}$ denote the set of atoms of a figure at the atom scale level. At the previous larger scale, the figural scale, the object is described by another set of medial atoms $\{\mathcal{M}_i\}$, which are the parent primitives of $\{\mathcal{A}_i\}$. Thus each $\mathcal{A}_i$ is a modification of $\mathcal{M}_i$ at the atom level. The residue geometric information at the atom level is described by the differences between $\{\mathcal{A}_i\}$ and $\{\mathcal{M}_i\}$.

As discussed in section 2, each medial atom $\mathcal{A}_i$ is characterized by a 4-tuple $(\mathbf{x}_i, \mathbf{R}_i, r_i, \theta_i)$. Suppose that $\mathcal{M}_i = (\widetilde{\mathbf{x}}_i, \widetilde{\mathbf{R}}_i, \widetilde{r}_i, \widetilde{\theta}_i)$ is the corresponding parent atom at the previous scale level. We define the atom residue to be

$$\Delta\mathcal{A}_i = (\frac{\mathbf{x}_i - \widetilde{\mathbf{x}}_i}{\widetilde{r}_i}, \ \widetilde{\mathbf{R}}_i^{-1}\mathbf{R}_i, \ \frac{r_i}{\widetilde{r}_i}, \ \theta_i - \widetilde{\theta}_i) = (\Delta\mathbf{x}_i, \Delta\mathbf{R}_i, \Delta r_i, \Delta\theta_i).$$

It is an element of the product space $G = \mathbb{R}^3 \times SO(3) \times \mathbb{R}^+ \times SO(2)$. Each component of $\Delta\mathcal{A}_i$ can be regarded as an element of the corresponding group. Let $d_R(\cdot, \cdot)$ and $d_2(\cdot, \cdot)$ be the Riemannian distance on $SO(3)$ and $SO(2)$, respectively, with the corresponding norms denoted by $\|\cdot\|_R$ and $\|\cdot\|_2$. The distance d_G between two atom residues is defined to be

$$d_G(\Delta\mathcal{A}_i, \Delta\mathcal{A}_j) =$$
$$\sqrt{\|\Delta\mathbf{x}_i - \Delta\mathbf{x}_j\|^2 + d_R^2(\Delta\mathbf{R}_i, \Delta\mathbf{R}_j) + |\ln(\Delta r_i/\Delta r_j)|^2 + d_2^2(\Delta\theta_i, \Delta\theta_j)}. \quad (6)$$

The corresponding norm is denoted by $\|\cdot\|_G$.

We now describe the MRF model for atom residues with respect to the canonical 4-neighbor structure. The joint probability on the atom residues has a density of the Gibbs form

$$p(\{\Delta\mathcal{A}_i\}) \propto \exp\left\{-\sum_i f_i(\Delta\mathcal{A}_i) - \sum_{<i,j>} g_{ij}(\Delta\mathcal{A}_i, \Delta\mathcal{A}_j)\right\}. \quad (7)$$

This density is with respect to the Haar measure on G. We choose the potentials f_i and g_{ij} to be quadratic functions. In particular,

$$f_i(\Delta\mathcal{A}_i) = \frac{\sigma_i}{2} d_G^2(\Delta\mathcal{A}_i, \mathcal{E}) = \frac{\sigma_i}{2}\|\Delta\mathcal{A}_i\|_G^2,$$

where $\mathcal{E} = (\mathbf{0}, \mathbf{I}, 1, 0)$, and

$$g_{ij}(\Delta\mathcal{A}_i, \Delta\mathcal{A}_j) = \frac{\tau_{ij}}{2} d_G^2(\Delta\mathcal{A}_i, \Delta\mathcal{A}_j),$$

for neighboring pairs of medial atoms. The full probability density is

$$p(\{\Delta\mathcal{A}_i\}) \propto \exp\left\{-\sum_i \frac{\sigma_i}{2}\left(\|\Delta\mathbf{x}_i\|^2 + \|\Delta\mathbf{R}_i\|_R^2 + |\ln(\Delta r_i)|^2 + \|\Delta\theta_i\|_2^2\right)\right.$$
$$-\sum_{<i,j>} \frac{\tau_{ij}}{2}\left(\|\Delta\mathbf{x}_i - \Delta\mathbf{x}_j\|^2 + \|(\Delta\mathbf{R}_i)^{-1}\Delta\mathbf{R}_j\|_R^2 \right. \quad (8)$$
$$\left.\left. + |\ln(\Delta r_i) - \ln(\Delta r_j)|^2 + \|\Delta\theta_i - \Delta\theta_j\|_2^2\right)\right\}.$$

The conditional distribution of $\Delta\mathcal{A}_i$ given the rest of the residues has density

$$p(\Delta\mathcal{A}_i|\Delta\mathcal{A}_{\{j\neq i\}}) \propto \exp\left\{-\frac{\sigma_i}{2}\|\Delta\mathcal{A}_i\|_G^2 - \sum_{<i,j>} \frac{\tau_{ij}}{2} d_G^2(\Delta\mathcal{A}_i, \Delta\mathcal{A}_j)\right\}.$$

This model has an intuitive interpretation which is similar to the boundary displacement model (3). More precisely, the first term in the exponent penalizes the difference between $\mathcal{A}_i$ and its parent $\mathcal{M}_i$, whereas the second term penalizes $\Delta\mathcal{A}_i$ from being different to a weighted average of residues of the neighboring atoms, given the configurations of $\{\Delta\mathcal{A}_j : j \neq i\}$.

5.2   Discussion

Given a training data set, the parameters $\{\sigma_i, \tau_{ij}\}$ of the probability model (8) can be estimated using the maximum likelihood method. Since the space

G of atom residues is not Euclidean, even though the potentials are quadratic, the distribution (8) is not Gaussian. The maximum likelihood estimates of the parameters in this case are obtained by Markov Chain Monte Carlo methods. However, this is a computationally expensive procedure.

Here we present an alternative model whose parameters are easier to estimate. We start by noticing that the Riemannian distance between two rotations $\Delta \mathbf{R}_1$ and $\Delta \mathbf{R}_2$ is given by $\|\mathrm{Log}(\Delta \mathbf{R}_1^{-1} \Delta \mathbf{R}_2)\|_F$, where $\| \cdot \|_F$ is the Frobenius matrix norm, and for $\mathbf{R} \in \mathrm{SO}(3)$,

$$\mathrm{Log}(\mathbf{R}) = \begin{cases} \mathbf{0}, & \text{if } \theta = 0; \\ \dfrac{\theta}{2\sin\theta}(\mathbf{R} - \mathbf{R}^T), & \text{if } \theta \neq 0. \end{cases}$$

Here θ satisfies $\mathrm{tr}(\mathbf{R}) = 1 + 2\cos\theta$ and $|\theta| < \pi$. When $\Delta \mathbf{R}_1, \Delta \mathbf{R}_2$ are close to identity, as in the case for atom residues, their Riemannian distance can be approximated by $\|\mathrm{Log}(\Delta \mathbf{R}_2) - \mathrm{Log}(\Delta \mathbf{R}_1)\|_F$. Similarly, the distance between $\Delta\theta_1$ and $\Delta\theta_2$ in $\mathrm{SO}(2)$ is $|\Delta\theta_1 - \Delta\theta_2|$ when they are both close to 0. Now, define an invertible map L by

$$L : \Delta\mathcal{A} = \Big(\Delta\mathbf{x}, \Delta\mathbf{R}, \Delta r, \Delta\theta\Big) \in G \mapsto \Delta\mathcal{L} = \Big(\Delta\mathbf{x}, \mathrm{Log}(\Delta\mathbf{R}), \ln(\Delta r), \Delta\theta\Big) \in \mathfrak{g}.$$

Then we can approximate the distance d_G defined in (6) on G, the space of atom residues, by the distance $d_\mathfrak{g}$ on the linear space $\mathfrak{g}$:

$$\begin{aligned} d_G^2(\Delta\mathcal{A}_1, \Delta\mathcal{A}_2) &\approx d_\mathfrak{g}^2(\Delta\mathcal{L}_1, \Delta\mathcal{L}_2) \\ &= \|\Delta\mathbf{x}_1 - \Delta\mathbf{x}_2\|^2 + \|\mathrm{Log}(\Delta\mathbf{R}_1) - \mathrm{Log}(\Delta\mathbf{R}_2)\|_F^2 \\ &\quad + |\ln(\Delta r_1) - \ln(\Delta r_2)|^2 + |\Delta\theta_1 - \Delta\theta_2|^2. \end{aligned}$$

Instead of (8), we define a probability distribution on $\{\Delta\mathcal{L}_i\}$ with density

$$p(\{\Delta\mathcal{L}_i\}) = \frac{1}{Z} \exp\Big\{ - \sum_i \frac{\sigma_i}{2}\|\Delta\mathcal{L}_i\|_\mathfrak{g}^2 - \sum_{<i,j>} \frac{\tau_{ij}}{2} d_\mathfrak{g}^2(\Delta\mathcal{L}_i, \Delta\mathcal{L}_j) \Big\}, \qquad (9)$$

where $\| \cdot \|_\mathfrak{g}$ is the metric corresponding to $d_\mathfrak{g}$. (9) induces a probability distribution on G via L^{-1}, which takes each $\Delta\mathcal{L}$ back to $\Delta\mathcal{A}$.

Notice that since two of the components of $\Delta\mathcal{L}$, namely $\mathrm{Log}(\Delta\mathbf{R})$ and $\Delta\theta$, are defined on bounded domains, the model (9) is not a Gaussian distribution. However, it is closely approximated by the Gaussian model of the same form on the (unbounded) linear space $\mathfrak{g}$, because typically the values of the components of $\Delta\mathcal{L}$ are close to zero. With the Gaussian approximation, we can estimate the parameters $\{\sigma_i, \tau_{ij}\}$ from a training data set by the maximum likelihood principle, without using MCMC methods.

Remarks. In defining atom residues, the translation component is scaled by the local radius to make the units commensurate. We can further add explicit weights on different components if necessary. The idea of approximating distances on the group G by distances on a corresponding linear space $\mathfrak{g}$ can be formalized with Lie group theory, but is beyond the scope of this paper. For a detailed description of Lie groups and their application in statistical shape analysis, see [28,29] and the references therein.

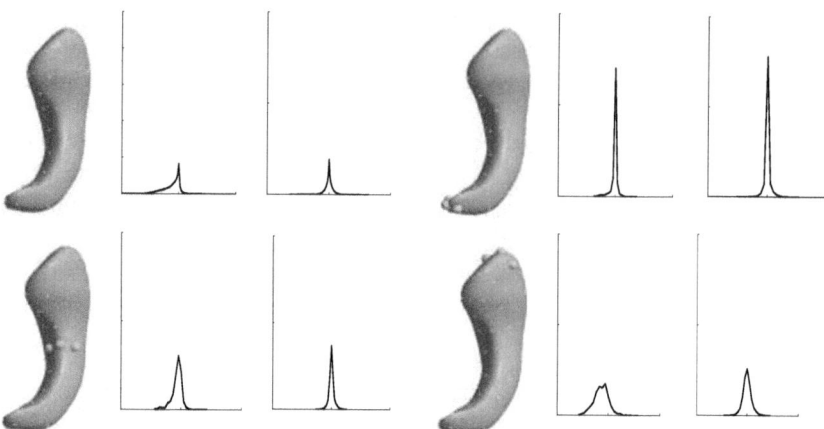

Fig. 3. Boundary displacement statistics of hippocampi. In each group, the left plot is the histogram of boundary displacement, and the right one is the histogram of the difference of displacements between neighbors.

6 Residue Statistics of Hippocampi

In this section we present some residue statistics of a sample population of hippocampi. The data set contains 86 left hippocampi, each being represented as a single m-rep figure by a 3×8 array of medial atoms.

Fig. 3 shows some boundary statistics of the samples. The top left group of figures show histograms of all boundary points, and the other three groups show histograms of different sections of the boundary. Each boundary section corresponds to a particular medial atom. The sections whose statistics are shown are indicated by the big dots. In each case, the left plot is the histogram of boundary displacements, and the right one shows the histogram of difference in displacements between neighboring boundary points. Notice how the histograms vary according to position changes. In each case the neighbor differences are more peaked around 0.

Fig. 4 illustrates some statistics on atom residues $\Delta \mathcal{A} = (\Delta \mathbf{x}, \Delta \mathbf{R}, \Delta r, \Delta \theta)$. The three rows of figures correspond to three different atoms. In each row, the left two plot are the histograms of $\ln(\Delta r)$ and $\|\Delta \mathbf{x}\|$. The right two show histograms of $\ln(\Delta r) - \overline{\ln(\Delta r)}$ and $\|\Delta \mathbf{x} - \overline{\Delta \mathbf{x}}\|$, where $\overline{\ln(\Delta r)}$ and $\overline{\Delta \mathbf{x}}$ are weighted averages of $\ln(\Delta r)$ and $\Delta \mathbf{x}$ of the atom's neighbors, respectively. Again, the histogram varies significantly with position, indicating that different sections of hippocampus vary in different ways among the sampling population.

These residue statistics are used to estimate the parameters of the corresponding MRF model. More results of these estimates can be found in [30].

7 Conclusions

Analysis of geometric shape space is essential to characterizing populations of object complexes in the real world, and, we suggest, will benefit the scale space

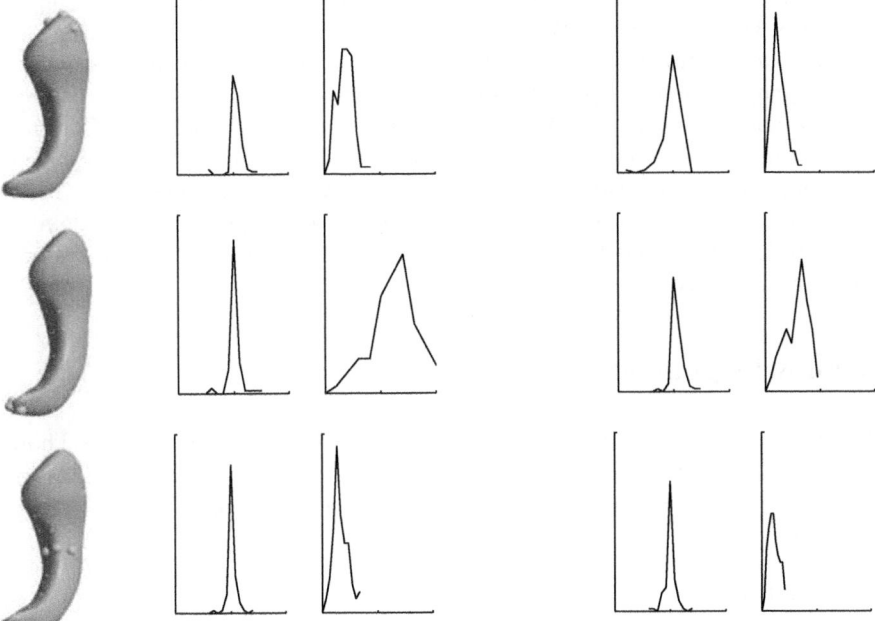

Fig. 4. Atom residue statistics of hippocampi. The big dots on the leftmost surface indicates the boundary section determined by the atom whose statistics is shown. In each row, the left two plots show histograms of $\ln(\Delta r)$ and $\|\Delta \mathbf{x}\|$, the right two show histograms of $\ln(\Delta r) - \overline{\ln(\Delta r)}$ and $\|\Delta \mathbf{x} - \overline{\Delta \mathbf{x}}\|$, where $\overline{\ln(\Delta r)}$ and $\overline{\Delta \mathbf{x}}$ are weighed averages of $\ln(\Delta r)$ and $\Delta \mathbf{x}$ of the atom's neighbors, respectively.

descriptions of images of those entities. We have shown that by describing the geometry at multiple scale levels using inter-scale residues and intra-scale neighbor relations, one can obtain descriptions that are localized in both scale and position. Furthermore, the geometric conformation of a population of object complexes is represented most effectively by probabilistic models, which incorporate both the common geometry and the deformations that occur among the population. We have seen that the inter-scale and intra-scale relations can be effectively modelled by a set of Markov random field models, whose neighborhood structures provide a mechanism for describing local features. Each of these MRF models has a relatively small number of parameters, yet is rich in geometric information. The models can also be tuned based on the statistics of a training data set, thus the same type of model can be used to describe different classes of geometric entities.

In this paper we have illustrated the above ideas based on m-reps, with which a geometric entity is represented at intuitive, object-intrinsic scale levels. We have shown how probabilistic descriptions of relations between adjacent scale levels and between neighboring primitives within a scale level can yield an intuitive and efficiently trainable overall description of populations of objects or object complexes.

We have given specifics for the MRF modelling and parameter estimation for two of the scale levels. We are going to apply the same idea to other scale levels, so that complete multi-scale probabilistic shape models can be established. Notice that although we find m-reps the appropriate geometric representation to work with, it is possible to build multi-scale MRF models based on other representation methods, as long as these representations can effectively describe residual geometric information both across and within scale levels.

We suggest that the framework discussed in this paper may provide a general paradigm for geometry and image description, with any effective system of geometric primitives.

Acknowledgements

We are grateful to contributions and suggestions of various aspects from Thomas Fletcher, Jeffrey Townsend, Martin Styner, and Michael Kerckhove. This work was supported by National Cancer Institute grant P01 CA47982.

References

1. Witkin, A.: Scale-space filtering. In: Proc. Intl. Joint Conf. on Artificial Intelligence, Kalsruhe, Germany (1983) 1019–1023
2. Koenderink, J.: The structure of images. Bio. Cybern. **50** (1984) 363–370
3. Lindeberg, T.: Scale-space theory in computer vision. Kluwer Academic Publishers (1994)
4. Cao, F.: Morphological scale space and mathematical morphology. In: Scale-Space '99. Volume 1682 of LNCS. Springer-Verlag (1999) 164–174
5. Lindeberg, T.: Linear spatio-temporal scale-space. In: Proc. 1st Int. Conf. Scale-Space Theory in Computer Vision. Volume 1252 of LNCS. Springer-Verlag (1997) 113–127
6. Bruce, J., Giblin, P., Tari, F.: Families of surfaces: height functions and projections to planes. Math. Scand. **82** (1998) 165–185
7. Bruce, J., Giblin, P., Tari, F.: Families of surfaces: focal sets, ridges and umbilics. Math. Proc. Cambridge Philos. Soc. **125** (1999) 243–268
8. Lu, C., Cao, Y., Mumford, D.: Surface evolution under curvature flows. J. Visual Communication and Image Representation **13** (2002) 65–81
9. Damon, J.: Scale-based geometry for nondifferentiable functions, measures, and distributions. Parts I-III. (Preprint)
10. Kimia, B.B., Tannenbaum, A.R., Zucker, S.W.: Shapes, shocks, and deformations I: the components of two-dimensional shape and the reaction-diffusion space. Int. J. Comput. Vision **15** (1995) 189–224
11. Siddiqi, K., Bouix, S., Tannenbaum, A.R., Zuker, S.W.: Hamilton-Jacobi skeletons. Int. J. Computer Vision **48** (2002) 215–231
12. Olver, P., Sapiro, G., Tannenbaum, A.: Invariant geometric evolutions of surfaces and volumetric smoothing. SIAM J. Appl. Math. **57** (1997)
13. Mallat, S.G.: Multifrequency channel decompositions of images and wavelet models. IEEE Trans. Acoust. Speech, Signal Processing **37** (1989) 2091–2110

14. Unser, M.: A review of wavelets in biomedical applications. Proceedings of the IEEE **84** (1996) 626–638
15. Ho, S., Gerig, G.: Scale-space on image profiles about an object boundary. In: Scale Space '03. (2003)
16. Zhu, S.C.: Embedding Gestalt laws in the Markov random fields – a theory for shape modeling and perceptual organization. IEEE T-PAMI **21** (1999)
17. Lu, C.: Curvature-based multi-scale shape analysis and stochastic shape modeling. PhD thesis, Brown University (2002)
18. Cootes, T.F., Edwards, G.J., Taylor, C.J.: Active appearance models. In: Fifth European Conference on Computer Vision. (1998) 484–498
19. Cootes, T.F., Taylor, C.J., Cooper, D.H., Graham, J.: Active shape models - their training and application. Computer Vision and Image Understanding **61** (1995) 38–59
20. Kelemen, A., Szekely, G., Gerig, G.: Three-dimensional model-based segmentation. IEEE-TMI **18** (1999) 828–839
21. Pizer, S.M., Chen, J.Z., Fletcher, P.T., Fridman, Y., Fritch, D.S., Gash, A.G., Glotzer, J.M., Jiroutek, M.R., Joshi, S., Lu, C., Muller, K.E., Thall, A., Tracton, G., Yushkevich, P., Chaney, E.L.: Deformable m-reps for 3D medical image segmentation. Int. J. Computer Vision (To appear)
22. Joshi, S., Pizer, S., Fletcher, P.T., Yushkevich, P., Thall, A., Marron, J.S.: Multi-scale deformable model segmentation and statistical shape analysis using medial descriptions. IEEE-TMI **21** (2002)
23. Styner, M., Gerig, G.: Medial models incorporating object variability for 3D shape analysis. In: IPMI '01. Volume 2082 of LNCS. Springer (2001) 502–516
24. Pizer, S.M., Fletcher, P.T., Thall, A., Styner, M., Gerig, G., Joshi, S.: Object models in multi-scale intrinsic coordinates via m-reps. Image and Vision Computing (To appear)
25. Gerig, G., Styner, M., Shenton, M., Lieberman, J.: Shape versus size: improved understanding of the morphology of brain structures. In: Proc. MICCAI 2001. Volume 2208 of LNCS. Springer (2001) 24–32
26. Thall, A.: Fast C^2 interpolating subdivision surfaces using iterative inversion of stationary subdivision rules. Technical report, Dept. of Computer Science, Univ. of North Carolina, `http://midag.cs.unc.edu/pubs/papers/Thall_TR02-001.pdf` (2002)
27. Geman, S., Geman, D.: Stochastic relaxation, Gibbs distributions, and the Bayesian restoration of images. IEEE T-PAMI **6** (1984) 721–741
28. Fletcher, P.T., Lu, C., Joshi, S.: Statistics of shape via principal component analysis on Lie groups. In: CVPR '03. (To appear)
29. Fletcher, P.T., Joshi, S., Lu, C., Pizer, S.M.: Gaussian distributions on Lie groups and their application to statistical shape analysis. (Submitted to IPMI '03)
30. Lu, C., Pizer, S.M., Joshi, S.: Multi-scale shape modeling by Markov random fields. (Submitted to IPMI '03)

Variational Dense Motion Estimation Using the Helmholtz Decomposition

Timo Kohlberger[1], Étienne Mémin[2], and Christoph Schnörr[3]

[1] Computer Vision, Graphics and Pattern Recognition Group
Department of Mathematics and Computer Science
University of Mannheim, 68131 Mannheim, Germany
{tiko,schnoerr}@uni-mannheim.de
[2] IRISA/Université de Rennes I
Campus universitaire de Beaulieu
35042 Rennes, Cedex, France
Etienne.Memin@irisa.fr

Abstract. We present a novel variational approach to dense motion estimation of highly non-rigid structures in image sequences. Our representation of the motion vector field is based on the extended Helmholtz Decomposition into its principal constituents: The laminar flow and two potential functions related to the solenoidal and irrotational flow, respectively. The potential functions, which are of primary interest for flow pattern analysis in numerous application fields like remote sensing or fluid mechanics, are *directly* estimated from image sequences with a variational approach. We use regularizers with derivatives up to third order to obtain unbiased high–quality solutions. Computationally, the approach is made tractable by means of auxiliary variables. The performance of the approach is demonstrated with ground-truth experiments and real-world data.

1 Introduction

In a number of domains affecting our everyday life, the analysis of image sequences involving fluid phenomena is of importance. This includes for instance domains such as visualization in experimental fluid mechanics [1,12], environmental sciences (meteorology [3,4,11,13,21], oceanography [6]) or medical imaging [2,7,18]. For all these domains it is of primary interest to extract reliable velocity fields and to the observed fluid flow. With respect to that goal, image sensors have considerable advantages compared to dedicated probes. Compared to these probes, image sensors provide a huge amount of almost continuous spatio-temporal data in a fast, tireless, reproducible and contact-free way. However, the sought motion information has then to be extracted from the luminance function which is not an easy task.

But in such a fluid context the extraction of a velocity field is far from being the ultimate goal of the analysis. Differential or integrated information from the velocity field is indeed far more valuable for concerned experts. For example, it

L.D. Griffin and M. Lillholm (Eds.): Scale-Space 2003, LNCS 2695, pp. 432–448, 2003.

is essential to characterize fluid flows to extract the vorticity fields, the stream-lines, or the singular points of the flows. All these features may be estimated indirectly from the velocity field by differentiation or by integration. Among all these information, the two potential functions called the *velocity potential* and the *stream function* are of great interest: (*i*) their gradients provide a description of the *irrotational* and the *solenoidal* components of the velocity fields; (*ii*) their Laplacians give access to the *vorticity* and the *divergence* of the velocity fields; (*iii*) their level lines allow us to extract directly the streamlines and the equipotential curves of the velocity potentials; (*iv*) their extrema provide the location of the singular points of major interest [5] (namely *sources, sinks and vortexes*).

Knowing the curl and the divergence of the flow, the extraction of such potential functions can be done by solving two Dirichlet problems. Such an estimation is particularly difficult for sparse velocity fields such as those obtained by the usual correlation methods [12] since an additional interpolation step is needed [16,17]. Dense motion estimation, on the other hand, allows to recover these potential functions more accurately. However, such an estimation as proposed in [5] is not "direct". It requires a process of three steps: First, the motion field has to be extracted, next the motion field is separated into its irrotational and solenoidal components, and finally the potential functions are estimated from these two vector fields by integration.

In this paper, we propose to estimate directly the *velocity potential* and the *stream function* from the luminance data. Instead of expressing the velocity field explicitly, it is represented by its *irrotational, solenoidal* and *laminar* components using the Helmholtz decomposition with a non-zero border condition. In addition, the first two components are expressed explicitly by the *velocity potential* and the *stream function*. This representation is applied to the Brightness Constancy Constraint Equation (BCCE) and embedded into an energy functional using a quadratic penalizer function. In a second step a structure preserving regularization based on the divergence of the *velocity potential* and the curl of the *stream function* is introduced. Since the original regularization contains high-order derivatives causing numerical difficulties it is then modified by introducing auxiliary variables. Finally the energy is integrated in a multi-resolution framework to minimize linearization errors.

The second section of this paper describes the representation of the decomposed velocity field and its integration into the BCCE. Furthermore, the regularization is introduced and the embedding into a multi-resolution framework as well as the energy minimization is described. The third section is dedicated to three experiments. The first one illustrates the influence of regularization parameters on the approximation quality. The second experiment compares the new approach to the approach of Corpetti et al. [5] by means of two dedicated cases. In the third and last experiment the approach is applied to real data. Finally, we conclude with a discussion in section four.

2   Description of the Approach

2.1   Helmholtz Decomposition and BCCE

Let us consider a smooth vector field $\boldsymbol{w} = (u(x,y), v(x,y))^\top : \mathbb{R}^2 \to \mathbb{R}^2$ defined over a section Ω of the image plane. Without loss of generality we can extend $\boldsymbol{w}$ to the whole plane and assume that it vanishes at infinity. As a consequence, it can be decomposed into a sum of a divergence free component (marked as *solenoidal*) and a curl free component (marked as *irrotational*). This decomposition is known as the *Helmholtz Decomposition* of a vector field

$$\boldsymbol{w} = \boldsymbol{w}_{so} + \boldsymbol{w}_{ir} \tag{1}$$

with

$$\mathrm{div}\boldsymbol{w}_{so} = \frac{\partial u}{\partial x} + \frac{\partial v}{\partial y} = 0 \quad \text{and} \quad \mathrm{curl}\boldsymbol{w}_{ir} = -\frac{\partial v}{\partial x} + \frac{\partial u}{\partial y} = 0. \tag{2}$$

In case of a non-zero border condition, the decomposition also includes a *laminar* component which is both irrotational and solenoidal:

$$\boldsymbol{w} = \boldsymbol{w}_{so} + \boldsymbol{w}_{ir} + \boldsymbol{w}_{lam}. \tag{3}$$

Furthermore, it is well known that both $\boldsymbol{w}_{so}$ and $\boldsymbol{w}_{ir}$ derive from potential functions ϕ and $\psi : \Omega \to \mathbb{R}$ denoted as *stream potential* and *velocity potential*, respectively:

$$\boldsymbol{w}_{ir} = \nabla\phi = \left(\frac{\partial\phi}{\partial x}, \frac{\partial\phi}{\partial y}\right)^\top \tag{4}$$

$$\boldsymbol{w}_{so} = \nabla\psi^\perp = \left(-\frac{\partial\psi}{\partial y}, \frac{\partial\psi}{\partial x}\right)^\top. \tag{5}$$

Let $\boldsymbol{w}$ denote the motion field of an image sequence $I : \Omega \times [0, T] \to \mathbb{R}$. Assuming brightness constancy, i.e. changes in intensity are due to motion only, leads to

$$I(\boldsymbol{x} + \boldsymbol{w}(\boldsymbol{x}, t), t + \Delta t) - I(\boldsymbol{x}, t) = 0 \tag{6}$$

with

$$\boldsymbol{w} = \nabla\phi + \nabla\psi^\perp + \boldsymbol{w}_{lam}. \tag{7}$$

The goal of the approach presented in the following paragraphs will be the *direct* approximation of those two potential fields ϕ and ψ from the brightness constancy constraint equation (6) in contrast to estimating the flow field first and followed by a path integration as proposed in [5].

The laminar component will not be of interest here. It is assumed that the motion in the given image sequence has no laminar component, which can be reached by approximating it separately in advance and removing it from the image sequence by defining $I(\boldsymbol{x}, t) := \bar{I}(\boldsymbol{x} + \boldsymbol{w}_{lam}(\boldsymbol{x}, t), t)$, with $\bar{I}$ representing the original sequence. In most applications $\boldsymbol{w}_{lam}$ can be estimated roughly by a multi-scale Horn and Schunck estimator using a strong regularization.

Embedding (6) into an energy framework leads to the functional

$$J(\phi, \psi) := \int_{\Omega} \left[I(\boldsymbol{x} + \nabla\phi(\boldsymbol{x}, t) + \nabla\psi^{\perp}(\boldsymbol{x}, t), t + \Delta t) - I(\boldsymbol{x}, t) \right]^2 d\boldsymbol{x}, \qquad (8)$$

for which optimal estimates $\hat{\phi}$ and $\hat{\psi}$ of both potentials are given by

$$(\hat{\phi}, \hat{\psi}) = \arg\min_{\phi, \psi} J(\phi, \psi) \qquad (9)$$

and the corresponding flow by $\hat{\boldsymbol{w}} = \nabla\phi + \nabla\psi^{\perp}$, respectively. Separate linearizations for $\nabla\psi^{\perp}$ and $\nabla\phi$

$$J_1(\phi, \psi) := \int_{\Omega} \left[\nabla I(\boldsymbol{x} + \nabla\phi, t + \Delta t)^{\top} \nabla\psi^{\perp} \right.$$

$$\left. + I(\boldsymbol{x} + \nabla\phi, t + \Delta t) - I(\boldsymbol{x}, t) \right]^2 d\boldsymbol{x} \qquad (10)$$

$$J_2(\phi, \psi) := \int_{\Omega} \left[\nabla I(\boldsymbol{x} + \nabla\psi^{\perp}, t + \Delta t)^{\top} \nabla\phi \right.$$

$$\left. + I(\boldsymbol{x} + \nabla\psi^{\perp}, t + \Delta t) - I(\boldsymbol{x}, t) \right]^2 d\boldsymbol{x} \qquad (11)$$

lead to two coupled minimization problems

$$\begin{cases} \arg\min_{\psi} J_1(\phi, \psi) \\[2mm] \arg\min_{\phi} J_2(\phi, \psi) \end{cases}. \qquad (12)$$

By defining abbreviations

$$\nabla I_{\nu}(\boldsymbol{x}) := \nabla I(\boldsymbol{x} + \nabla\nu, t + \Delta t) \qquad (13)$$
$$\partial I_{\nu} := I(\boldsymbol{x} + \nabla\nu, t + \Delta t) - I(\boldsymbol{x}, t) \qquad (14)$$

(12) can be written as

$$\begin{cases} \arg\min_{\psi} \int_{\Omega} \left[\nabla I_{\phi}^{\top} \nabla\psi^{\perp} + \partial I_{\phi} \right]^2 d\boldsymbol{x} \\[3mm] \arg\min_{\phi} \int_{\Omega} \left[\nabla I_{\psi}^{\top} \nabla\phi + \partial I_{\psi} \right]^2 d\boldsymbol{x}. \end{cases} \qquad (15)$$

Minimizing each energy J_1, J_2 requires the first variations to vanish, i.e.

$$\left. \frac{dJ_1(\phi, \psi + \tau\tilde{\psi})}{dt} \right|_{\tau=0} = 0 \quad \text{and} \quad \left. \frac{dJ_2(\phi + \tau\tilde{\phi}, \psi)}{dt} \right|_{\tau=0} = 0, \qquad (16)$$

for arbitrary test functions $\tilde{\psi}$ and $\tilde{\phi}$. This results in the system of equations:

$$
\begin{cases}
\int_{\Omega} \left[(\nabla I_\phi \nabla I_\phi^\top) \nabla \psi^\perp + \partial I_\phi \nabla I_\phi \right] \nabla \tilde{\psi}^\perp \, d\boldsymbol{x} = 0 \\
\int_{\Omega} \left[(\nabla I_\psi \nabla I_\psi^\top) \nabla \phi + \partial I_\psi \nabla I_\psi \right] \nabla \tilde{\phi} \, d\boldsymbol{x} = 0
\end{cases}
\tag{17}
$$

Since the matrices $\nabla I_\phi \nabla I_\phi^\top$ and $\nabla I_\psi \nabla I_\psi^\top$ are singular, terms of the form ϵI for small ϵ can be added to make the system well-posed. Numerical experiments revealed, however, that this regularization is too weak and does not yield satisfying results. Accordingly, we investigated additional regularizing smoothness terms which preserve the flow field structure as much as possible (see next section).

2.2   Structure-Preserving Regularization

Since the components of the motion fields (sources, sinks and vortexes) which we want to estimate explicitly in terms of potential functions ϕ and ψ contain discontinuities by their nature, first-order smoothness terms like those of the Horn and Schunck approach [9] or non-quadratic extensions [14,15,20] are inadequate. Instead we suggest using a second-order div-curl regularizer as proposed by Suter [19]:

$$
\int_{\Omega} ||\nabla \mathrm{div} \boldsymbol{w}||^2 + ||\nabla \mathrm{curl} \boldsymbol{w}||^2 d\boldsymbol{x}
\tag{18}
$$

which enforces a smoothing constraint not on the motion field $\boldsymbol{w}$ itself but only on its structural components we are interested in. Hence discontinuities in the motion field resulting from sources, sinks and vortexes are not penalized with this regularization term, but only abrupt changes in strength and direction of those components. Replacing $\boldsymbol{w}$ by its definition (7) (and neglecting the laminar component), this regularization term takes the form

$$
\int_{\Omega} ||\nabla \mathrm{div} \nabla \phi||^2 + ||\nabla \mathrm{curl} \nabla \psi^\perp||^2 d\boldsymbol{x}
\tag{19}
$$

since $\mathrm{div} \nabla \psi^\perp = 0$ and $\mathrm{curl} \nabla \phi = 0$ by definition (see (2), (4-5)). Unfortunately, the high-order derivatives render a direct numerical approach quite involved. To overcome this problem we follow the approach of Corpetti et al. [4] and introduce auxiliary variables ξ_1 and ξ_2, enforce them to approximate $\mathrm{curl} \nabla \psi^\perp$ and $\mathrm{div} \nabla \phi$ by additional (soft) constraints and impose the original regularization constraint on them, i.e.

$$
\int_{\Omega} \gamma \left([\mathrm{div} \nabla \phi - \xi_2]^2 + [\mathrm{curl} \nabla \psi^\perp - \xi_1]^2 \right) + \lambda \left(||\nabla \xi_2||^2 + ||\nabla \xi_1||^2 \right) d\boldsymbol{x}, \tag{20}
$$

with some regularization parameters $\gamma, \lambda \in \mathbb{R}^+$. As a consequence, the degree of derivation is lowered at the cost of slightly weakening the regularization strength.

The final form of the energy thus becomes

$$J(\phi, \psi, \xi_1, \xi_2) := \int_\Omega \left[I(\boldsymbol{x}, t) - I(\boldsymbol{x} + \nabla\phi + \nabla\psi^\perp, t + \Delta t) \right]^2 \tag{21}$$

$$+ \gamma \left([\operatorname{div}\nabla\phi - \xi_2]^2 + [\operatorname{curl}\nabla\psi^\perp - \xi_1]^2 \right)$$

$$+ \lambda \left(||\nabla\xi_2||^2 + ||\nabla\xi_1||^2 \right) d\boldsymbol{x},$$

leading to the minimization problem

$$\arg\min_{\phi,\psi,\xi_1,\xi_2} J(\phi, \psi, \xi_1, \xi_2). \tag{22}$$

As before, the data term can be approximated by separate linearizations with respect to $\nabla\psi^\perp$ and $\nabla\phi$ keeping only the involved regularization terms:

$$J_1(\phi, \psi, \xi_1) := \int_\Omega \left[\nabla I_\phi^\top \nabla\psi^\perp + \partial I_\phi \right]^2 + \gamma \left[\operatorname{curl}(\nabla\psi^\perp) - \xi_1 \right]^2$$

$$+ \lambda ||\nabla\xi_1||^2 d\boldsymbol{x} \tag{23}$$

$$J_2(\phi, \psi, \xi_2) := \int_\Omega \left[\nabla I_\psi^\top \nabla\phi + \partial I_\psi \right]^2 + \gamma \left[\operatorname{div}(\nabla\phi) - \xi_2 \right]^2 + \lambda ||\nabla\xi_2||^2 d\boldsymbol{x}. \tag{24}$$

As a result, we arrive at the regularized version of (22):

$$\begin{cases} \arg\min_{\psi,\xi_1} J_1(\phi, \psi, \xi_1) \\ \arg\min_{\phi,\xi_2} J_2(\phi, \psi, \xi_2) \end{cases} . \tag{25}$$

2.3 Extensions to Multiple Resolutions

The minimization furthermore can be integrated into a coarse-to-fine framework in order to keep the magnitudes of $\nabla\psi^\perp$ and $\nabla\phi$ on each resolution level preferably small, which lowers the error resulting from the data term linearization.

Given the solutions ϕ^{l-1} and ψ^{l-1} from resolution level $l-1$ (starting with $\phi^0 \equiv 0$ and $\psi^0 \equiv 0$) ϕ^l and ψ^l are calculated by

$$(\psi^l, \xi_1^l) = (P\psi^{l-1}, 0) + \arg\min_{\psi,\xi_1} J_1(\phi, P\psi^{l-1}, \psi, \xi_1, \gamma_l, \lambda_l) \tag{26}$$

$$(\phi^l, \xi_2^l) = (P\phi^{l-1}, 0) + \arg\min_{\phi,\xi_2} J_2(P\phi^{l-1}, \phi, \psi, \xi_2, \gamma_l, \lambda_l) \tag{27}$$

with

$$J_1(\phi, \psi_0, \psi, \xi_1, \gamma, \lambda) := \tag{28}$$

$$\int_\Omega \left[\nabla \bar{I}_\phi^\top \nabla \psi^\perp + \partial \bar{I}_\phi\right]^2 + \gamma \left[\mathrm{curl}\nabla(\psi_0 + \psi)^\perp - \xi_1\right]^2 + \lambda ||\nabla \xi_1||^2 d\boldsymbol{x}$$

$$J_2(\phi_0, \phi, \psi, \xi_2, \gamma, \lambda) := \tag{29}$$

$$\int_\Omega \left[\nabla \bar{I}_\psi^\top \nabla \phi + \partial \bar{I}_\psi\right]^2 + \gamma \left[\mathrm{div}\nabla(\phi_0 + \phi) - \xi_2\right]^2 + \lambda ||\nabla \xi_2||^2 d\boldsymbol{x}$$

and

$$\bar{I}(\boldsymbol{x}, t) := I(\boldsymbol{x}, t)$$
$$\bar{I}(\boldsymbol{x}, t + \Delta t) := I(\boldsymbol{x} + \nabla \phi_0 + \nabla \psi_0^\perp, t + \Delta t). \tag{30}$$

Let P be a prolongation operator mapping from resolution level $l-1$ to l. Except for the first resolution level ϕ and ψ are incremental refinements of the complete solution. Note that the linearization of the data terms is carried out only for the increments, in order to keep the linearization error small, since solutions from coarser resolutions ($P\phi^{l-1}$ and $P\psi^{l-1}$) are included in the data itself (cf. (30)). (30) is equivalent to an incremental back-mapping of a each frame to its predecessor.

In contrast to that the regularization terms apply to the complete potential functions ($P\phi^{l-1} + \phi$ and $P\psi^{l-1} + \psi$), not only to the increments. For the auxiliary variables the coarse-to-fine method is not applied here, since they are not involved in the linearization. Hence, ξ_1 and ξ_2 are non-increments and calculated independently at each resolution.

2.4 Minimization and Discretization

In order to evaluate (26) and (27) , we compute the Euler-Lagrange equations for J_1 w.r.t. ψ, ξ_1 and for J_2 w.r.t. ϕ and ξ_2:

$$\begin{cases} \mathrm{curl}\left((\nabla \bar{I}_\phi^\top \nabla \psi^\perp + \partial \bar{I}_\phi)^\top \nabla \bar{I}_\phi\right) + \gamma \left[\Delta^2(\psi_0 + \psi) + \Delta \xi_1\right] = 0 \\[2mm] \gamma \left[\Delta(\psi_0 + \psi) + \xi_1\right] - \lambda \Delta \xi_1 = 0 \\[2mm] \mathrm{div}\left((\nabla \bar{I}_\psi^\top \nabla \phi + \partial \bar{I}_\psi)^\top \nabla \bar{I}_\psi\right) + \gamma \left[\Delta^2(\phi_0 + \phi) - \Delta \xi_2)\right] = 0 \\[2mm] \gamma \left[-\Delta(\phi_0 + \phi) + \xi_2\right] - \lambda \Delta \xi_2 = 0 \end{cases} . \tag{31}$$

The Biharmonic Operator

$$\Delta^2 = \partial_{x^4} + 2\partial_{x^2 y^2} + \partial_{y^4} \tag{32}$$

occurring in two equations can be discretized by a standard 13-points-stencil [8]. In addition, we impose the boundary conditions

$$\phi(\boldsymbol{x}) = \sigma_\phi(\boldsymbol{x}) \quad \wedge \quad \nabla_n \phi(\boldsymbol{x}) = \rho_\phi(\boldsymbol{x}), \quad \forall \boldsymbol{x} \in \partial\Omega \tag{33}$$
$$\psi(\boldsymbol{x}) = \sigma_\psi(\boldsymbol{x}) \quad \wedge \quad \nabla_n \psi(\boldsymbol{x}) = \rho_\psi(\boldsymbol{x}), \quad \forall \boldsymbol{x} \in \partial\Omega \tag{34}$$

with boundary functions $\sigma_\phi, \sigma_\psi, \rho_\phi$ and ρ_ψ. ∇_n denotes the outer normal gradient with respect to Ω.

3   Experimental Results

This section is organized into three parts. First, the quantitative and qualitative influence of the parameters γ and λ is investigated in Section 3.1. Second, the proposed method is compared with the method of Corpetti et al. [5] on artificial motion fields, i.e. with ground truth, in Section 3.2. Finally, results for a real image sequence are presented.

In all experiments, the Euler-Lagrange equations (31) were solved sequentially in 3000 iterations using an incomplete CLG solver iterating 50 times in each (outer) iteration. Two resolution levels (including the original one) have been used for the experiments in Sections 3.1 and 3.2 and three for those in Section 3.3.

Furthermore, due to lack of data, the boundary values of ϕ and ψ have been set to zero, i.e. $\sigma_{\phi,\psi} \equiv 0$. The same has been done for the normal derivatives $\nabla_n\phi, \nabla_n\psi$, i.e. $\rho_{\phi,\psi} \equiv 0$. Since it is not adequate for the vector field to vanish at the border $\partial\Omega$, Ω was artificially enlarged by 30% of the width/height of the original image section and also set to zero[1].

As error measure the *average squared L_2 norm error* and the *average angular error* (mean and first standard derivation) proposed by [10] have been taken. Since the potentials ϕ and ψ are defined up to a constant it was difficult to find an appropriate error measure for the velocity potential and the stream functional directly instead for the derived flow components.

The intensities of all input images have been normalized to $[0, 1]$.

3.1   Parameter Studies

In first experimental studies the influence of the parameters γ and λ were investigated on synthetic potentials in order to have a ground truth (cf. Figure 1). The associated synthetic flow field was applied to a real image, i.e. the real image was mapped using the velocity field to obtain a second image resulting in the input image pair for the current experiments. Note that the vector field consist of an exact spatial overlap of the true components we wish to determine and distinguish.

Figure 3 shows the results for estimation the potential functions with parameter values $\gamma \in \{0.1, 0.25, 0.5, 1, 2, 3, 4\}$ and $\lambda = 10^3$ fixed. These results clearly illustrate the positive regularizing effect of the high-order smoothness terms in (23) and (24) which were made computationally tractable by means of auxiliary functions. It's remarkable that both vector field components can be distinguished — despite a complete spatial overlap and only partially given image structures — by subsequent linearizations of a single data term (cf. Section 2.1).

---

[1] Note that this is (approximately) consistent with the initial assumption (w.r.t. the Helmholtz Decomposition) that the velocity field vanishes at infinity.

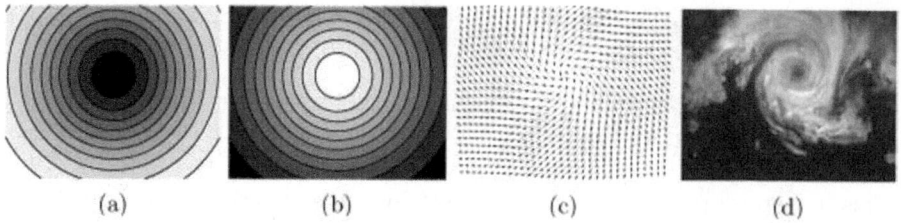

Fig. 1. Setting for Parameters Studies (a) Ground truth velocity potential ϕ **(b)** Ground truth stream function ψ **(c)** Associated synthetic velocity field $\boldsymbol{w} = \nabla\phi + \nabla\psi^{\perp}$ **(d)** The real image the velocity field $\boldsymbol{w}$ was applied to in order to generate the input image pair for the experiments (size: 128×100 pixels). Note that the vector field consist of an exact spatial overlap of the true components we wish to determine and distinguish.

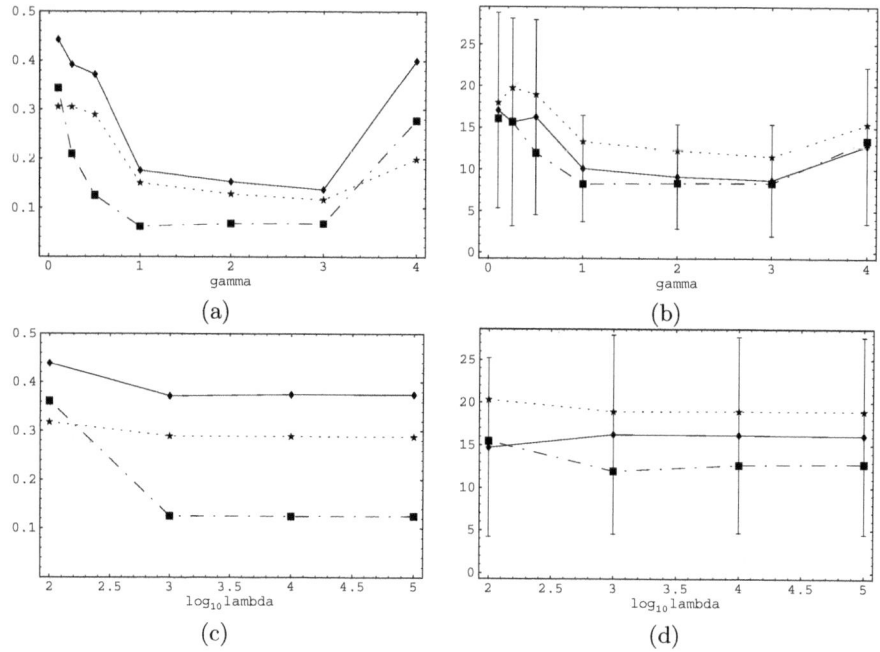

Fig. 2. Influence of parameters λ and γ on error measurements. (a,c) Average squared L_2 norm error of $\hat{\boldsymbol{w}}_{ir} + \hat{\boldsymbol{w}}_{so}$ (solid), $\hat{\boldsymbol{w}}_{ir}$ (pointed) and $\hat{\boldsymbol{w}}_{so}$ (dashed) depending on γ (a) and λ (c) **(b,d)** Average angular error of $\hat{\boldsymbol{w}}_{ir} + \hat{\boldsymbol{w}}_{so}$ (solid), $\hat{\boldsymbol{w}}_{ir}$ (pointed) and $\hat{\boldsymbol{w}}_{so}$ (dashed) depend-ending on γ (b) and λ (d). Direct estimation of potential functions as the objects of primary interest leads to a small global error of the corresponding gradient velocity fields (a,c) but to local angular errors from $10° - 20°$.

Whereas Figs. 2(a,c) and 3(d,k) show that the deformation pattern has been computed well, Figs. 2(b,d) reveal a relatively large angular error of $10° - 20°$ of the respective velocity fields. This is plausible since the velocity fields are related

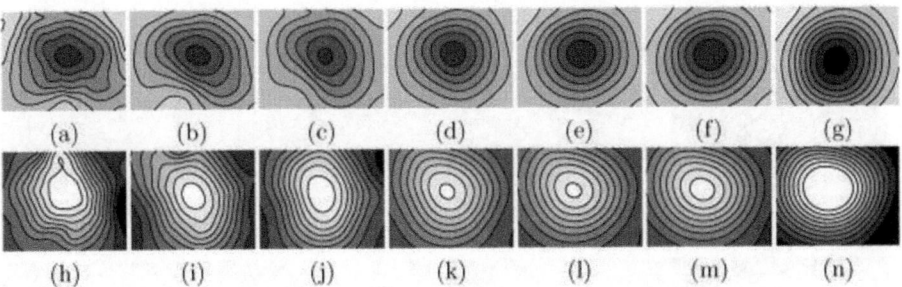

Fig. 3. Qualitative influence of parameter γ. Approximated velocity potentials ϕ **(a–g)** and stream functions ψ **(h–n)** for $\gamma \in \{0.1, 0.25, 0.5, 1, 2, 3, 4\}$ and $\lambda = 10^3$. Both vector field components can be distinguished, despite a complete spatial overlap. The positive effect of the high–order regularization implemented by means of auxiliary functions is clearly visible.

to the *derivatives* of the quantities we estimate directly (potential functions). As a consequence, inaccuracies are amplified. However, it should be noted again that the potential functions are the quantities of primary interest for flow pattern analysis. A comparison with an approach which uses *indirect* computation of the potential functions by integrating velocity fields along stream lines is the objective of the next section.

Figure 4 shows that for low values of the regularizing parameter γ the corresponding auxiliary function must not be smoothed too much (too large values of λ). This finding reveals a dependency between γ and λ, the closer investigation of which is left for future work.

3.2   Comparison with the Approaches of Corpetti et al. and Horn and Schunck

In [5] an approach was presented in which the potential functions were approximated *indirectly* by first estimating a motion field using a regularization similar to (21) and a subsequent integration along the stream lines in order to obtain the velocity potential and the stream function. Both approaches were compared in two experiments here based on given synthetic potential functions (cf. Figure 5). In order to have a reference both experiments have been carried out with a Horn and Schunck estimator also. The parameters of all methods have been optimized manually. Note that the setting of Comparison Experiment 1 is the same as for the parameters studies.

The results for Experiment 1 (Figure 6 and Table 1) and Experiment 2 (Figure 7 and Table 2) both show that the new approach yield similar good results as the approach of Corpetti et al., despite the higher order of differential equations involved in the minimization. Furthermore, they show that the standard regularization of Horn and Schunck is insufficient to preserve the desired image structures, since this regularization penalizes strong discontinuities like those in the center of the velocity field in Experiment 1 resulting from a vortex and a

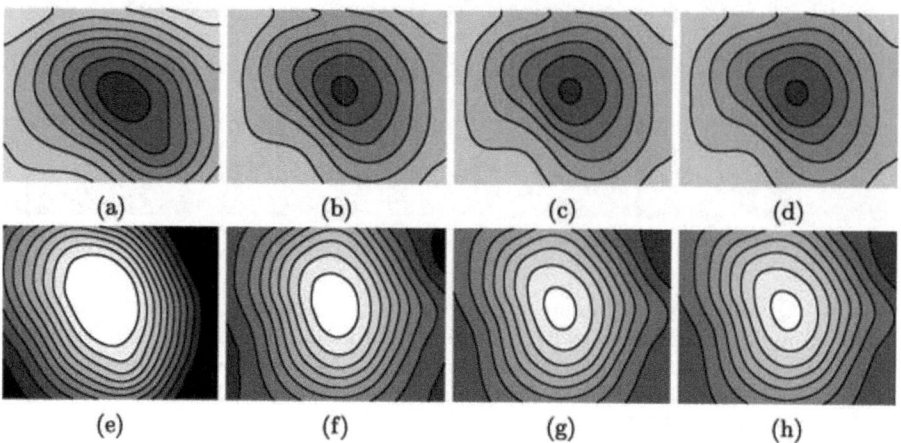

Fig. 4. Qualitative influence of parameter λ. Approximated velocity potentials ϕ **(a-d)** and stream functions ψ **(e-h)** for $\lambda \in \{10^2, 10^3, 10^4, 10^5\}$ and $\gamma = 0.5$. Parameters for enforcing regularization γ and smoothing the auxiliary functions λ are not independent. For low values of γ, the auxiliary functions must not be smoothed too much.

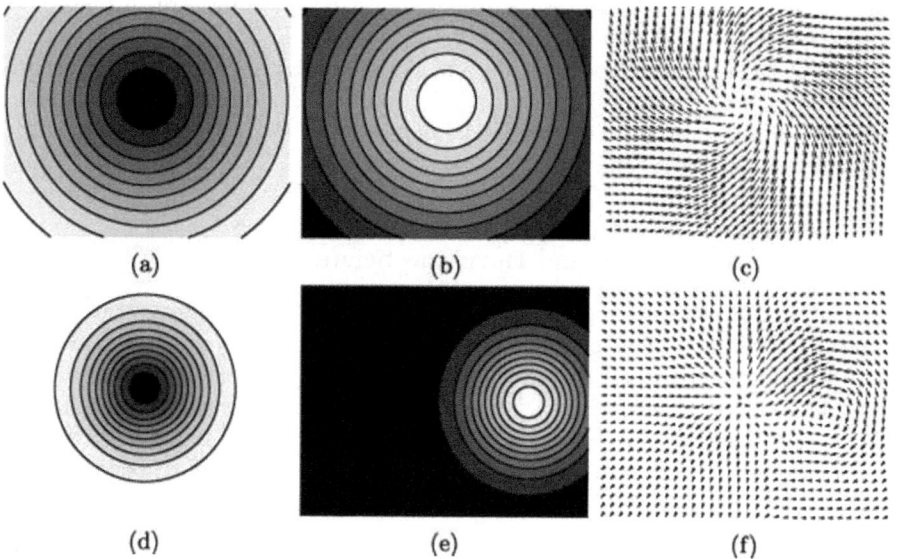

Fig. 5. Setting for Comparison Experiments 1–2. (a,b) Ground truth velocity potential ϕ and stream function ψ for Comparison Experiment 1 **(c)** Associated velocity field $w = \nabla\phi + \nabla\psi^\perp$, $\max \|w\| = 1.86$ **(d,e)** Ground truth velocity potential ϕ and stream function ψ for Comparison Experiment 2 **(f)** Associated velocity field $w = \nabla\phi + \nabla\psi^\perp$, $\max \|w\| = 1.89$. Both synthetic velocity fields have been used to map the real image in Figure 1(d) in order to generate the second images of the input image pairs.

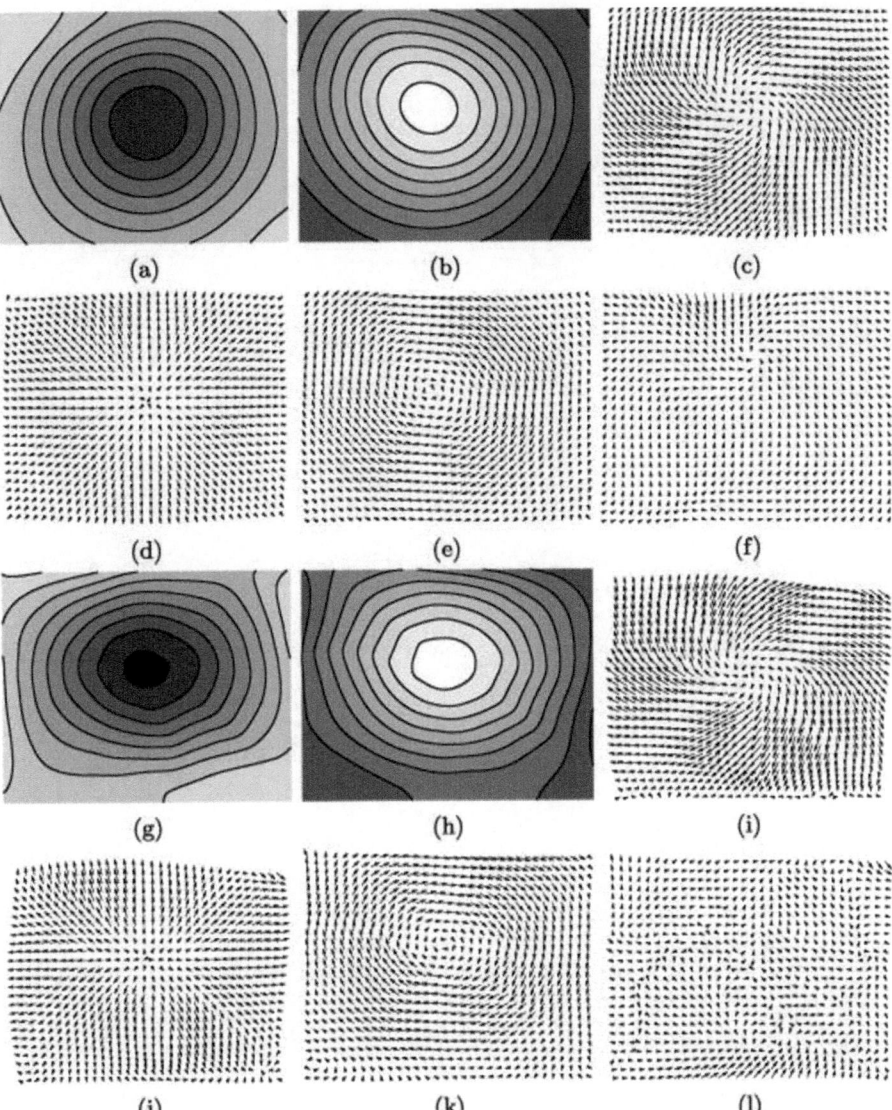

Fig. 6. Results of Comparison Experiment 1. Results of the direct approach ($\gamma = 3.0, \lambda = 1000$): **(a)** Velocity potential $\hat{\phi}$ **(b)** Stream function $\hat{\psi}$ **(c)** Associated velocity field $\hat{w}$ **(d)** Irrotational part of the vel. field $\hat{w}_{ir}$ **(e)** Solenoidal part of the vel. field $\hat{w}_{so}$ **(f)** Difference between approximated and true vel. field $\hat{w} - w$. Result of the Corpetti et al. approach ($\alpha = 300, \lambda = 250$): **(g)** Velocity potential $\hat{\phi}$ **(h)** Stream function $\hat{\psi}$ **(i)** Associated velocity field $\hat{w}$ **(j)** Irrotational part of the vel. field $\hat{w}_{ir}$ **(k)** Solenoidal part of the vel. field $\hat{w}_{so}$ **(l)** Difference between approximated and true vel. field $\hat{w} - w$. Both methods lead to very similar results in this case, despite significant differences in the approach.

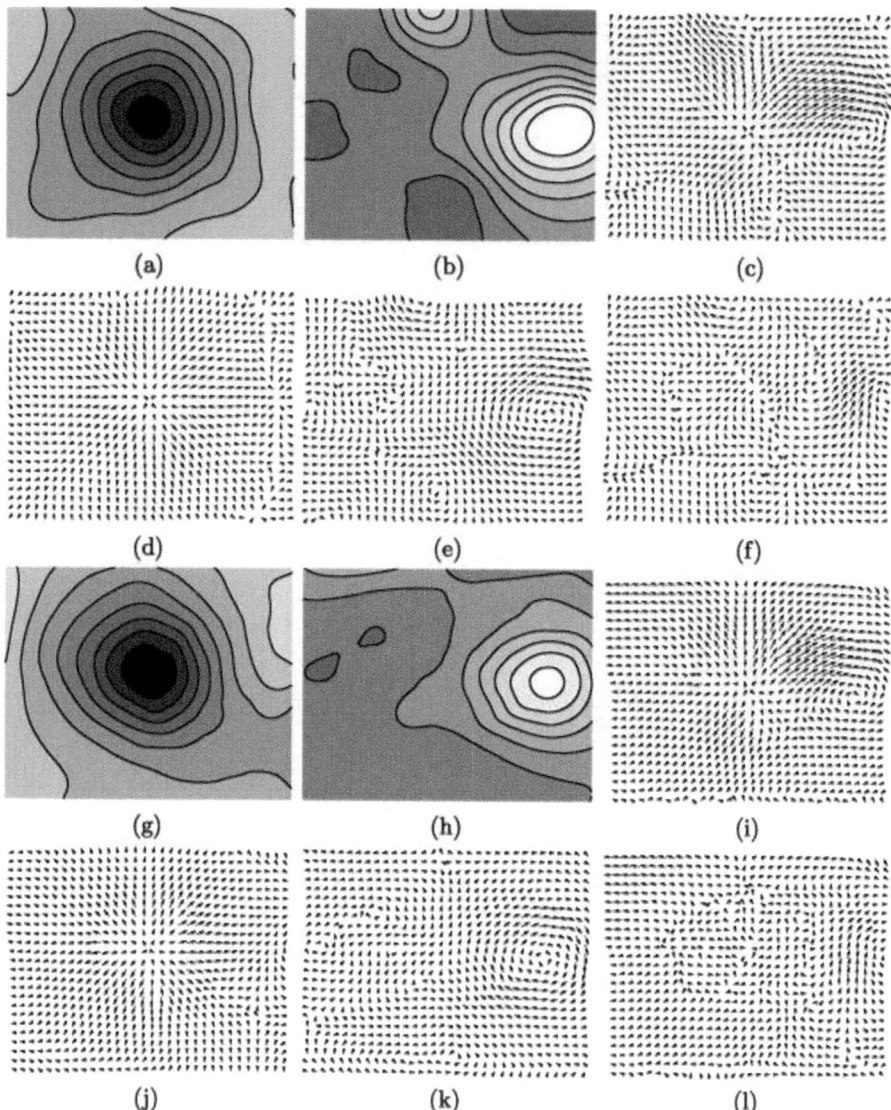

Fig. 7. Results of Comparison Experiment 2. Results of the direct approach ($\gamma = 0.5, \lambda = 100$): (a) Velocity potential $\hat{\phi}$ (b) Stream function $\hat{\psi}$ (c) Associated velocity field $\hat{w}$ (d) Irrotational part of the vel. field $\hat{w}_{ir}$ (e) Solenoidal part of the vel. field $\hat{w}_{so}$ (f) Difference between approximated and true vel. field $\hat{w} - w$. Results of the Corpetti et al. approach ($\alpha = 300, \lambda = 250$): (g) Velocity potential $\hat{\phi}$ (h) Stream function $\hat{\psi}$ (i) Associated velocity field $\hat{w}$ (j) Irrotational part of the vel. field $\hat{w}_{ir}$ (k) Solenoidal part of the vel. field $\hat{w}_{so}$ (l) Difference between approximated and true vel. field $\hat{w} - w$. Also here the results are quite similar, although the false second maximum of the stream function is more distinct with the direct approach.

Table 1. Approximation errors of Comparison Experiment 1. Both the *average squared L_2 norm error* and the *average angular error* of the complete velocity field ($\boldsymbol{w}_{ir} + \boldsymbol{w}_{so}$) as well as for the irrotational and solenoidal components ($\boldsymbol{w}_{ir}$, $\boldsymbol{w}_{so}$) are depicted. Approach dependent parameters have been optimized manually. The errors of the Horn and Schunck approach show clearly that, despite the flow directions are estimated similarly well, there is a significant higher error in the magnitudes which results from the penalization of the discontinuity in the velocity field (cf. Fig. 5(c)).

Approach/ Error Measure		Direct Approach $\gamma = 3.0, \lambda = 1000$	Corpetti et al. $\alpha = 300, \lambda = 250$	Horn&Schunck $\alpha^2 = 0.17$
Aver. Sq. L_2 Norm Error	$\boldsymbol{w}_{ir} + \boldsymbol{w}_{so}$	137.11	175.81	208.979
$\times 10^3$	$\boldsymbol{w}_{ir}$	116.38	152.89	
	$\boldsymbol{w}_{so}$	67.51	102.24	
Aver. Angular. Error	$\boldsymbol{w}_{ir} + \boldsymbol{w}_{so}$	$8.69° \pm 6.72°$	$10.12° \pm 7.95°$	$8.93° \pm 6.39°$
(Mean/1. Stand. Dev.)	$\boldsymbol{w}_{ir}$	$11.52° \pm 5.58°$	$12.89° \pm 7.71°$	
	$\boldsymbol{w}_{so}$	$8.35° \pm 5.39°$	$10.36° \pm 6.12°$	

Table 2. Approximation errors of Comparison Experiment 2. While the approximation quality compared to the indirect approach of Corpetti et al. is more equal here, the difference in approximation quality to the approach of Horn and Schunck is not as distinct as in Experiment 1 since the discontinuities of the velocity field are smaller in this case see Fig. 5(f)).

Approach/ Error Measure		Direct Approach $\gamma = 0.5, \lambda = 100$	Corpetti et al. $\alpha = 300, \lambda = 250$	Horn&Schunck $\alpha^2 = 0.07$
Aver. Sq. L_2 Norm Error	$\boldsymbol{w}_{ir} + \boldsymbol{w}_{so}$	162.18	168.7	170.514
$\times 10^3$	$\boldsymbol{w}_{ir}$	60.69	70.59	
	$\boldsymbol{w}_{so}$	129.56	65.76	
Aver. Angular. Error	$\boldsymbol{w}_{ir} + \boldsymbol{w}_{so}$	$14.68° \pm 9.59°$	$14.65° \pm 11.51°$	$16.19° \pm 10.26°$
(Mean/1. Stand. Dev.)	$\boldsymbol{w}_{ir}$	$10.56° \pm 5.93°$	$11.15° \pm 7.52°$	
	$\boldsymbol{w}_{so}$	$14.11° \pm 9.72°$	$10.29° \pm 7.36°$	

source, which we want to preserve. Even for weak regularizations the results of the Horn and Schunck estimator is less accurate in both experiments.

3.3   Reconstructing the Vortexes of a Landing Air Plane

Finally, the new approach has been applied on an image pair coming from a real image sequence. The sequence is a recording of the motion of smoke behind a landing passenger air plane. It contains a strong vortex in the center and a weaker but larger one in the other direction with its center laying outside the image plane, approximately 50% from the right border. In addition a weak source is present in the right half, centered vertically (cf. Figure 8(a,b)).

In order to eliminate the laminar component of the velocity field, the laminar flow has been approximated roughly by a Horn and Schunck estimator with a strong regularization $\alpha^2 = 10^6$ and used to map the second image of the input image pair back (cf. Section 2.1).

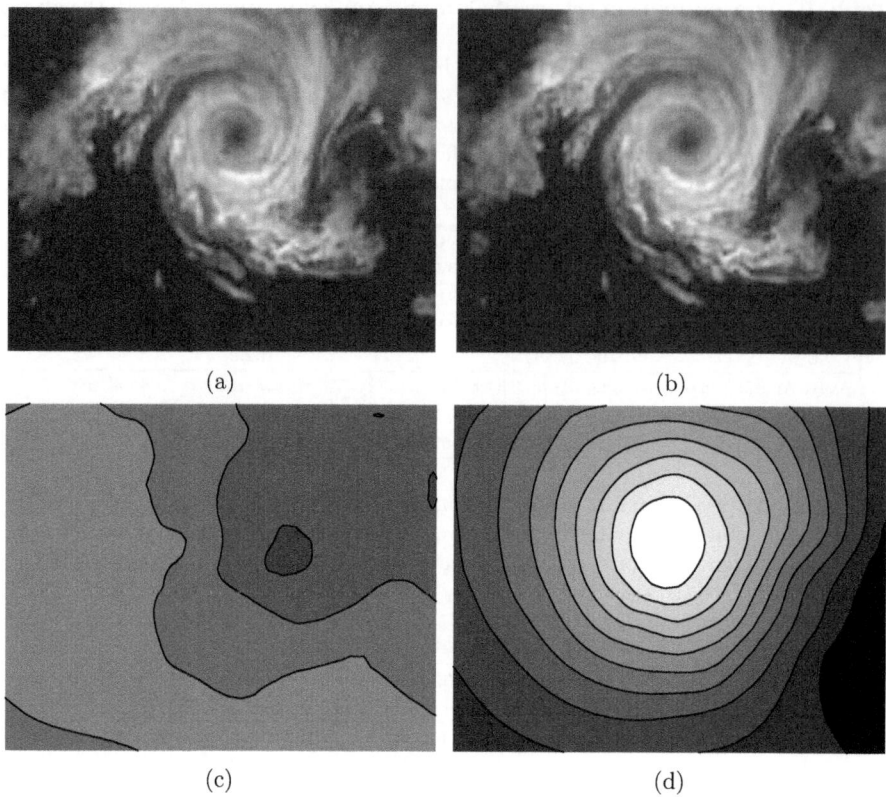

(a)   (b)

(c)   (d)

Fig. 8. Result of the Air Plane Sequence Experiment. (a,b) Two successive
images from the sequence, gray values, 128×100 pixels **(c)** Approximated velocity potential $\hat{\phi}$ **(d)** Approximated stream function $\hat{\psi}$ **(e)** Associated irrotational component
$\hat{\boldsymbol{w}}_{ir}$ **(f)** Associated solenoidal component $\hat{\boldsymbol{w}}_{so}$. Parameters were: $\gamma = 0.5$, $\lambda = 10^3$,
Maximum approximated displacement: $\max \|\hat{\boldsymbol{w}}\| = 2.75$. The image sequence contains
a strong vortex in the center and a weaker but larger one in the other direction laying
outside, approximately 50% from the right border. In addition a weak source is present
in the right half, vertically centered. Both the main vortex and the weak source are well
reconstructed despite the lack of image structures in the lower half of the sequence.
But the weaker counter-vortex is detected only in outlines. This is plausible since the
latter one has its maximum outside the image plane and the velocity potential ψ is set
to be zero on the enlarged border.

The results in Figure 8 show that both the main vortex and the weak source
are well reconstructed, despite the lack of image structures in the lower half of
the sequence. But the weaker counter-vortex is detected only in outlines. This
is plausible since the latter one has its maximum outside the image plane and
the stream function ψ is set to be zero on the large border (see beginning of this
section).

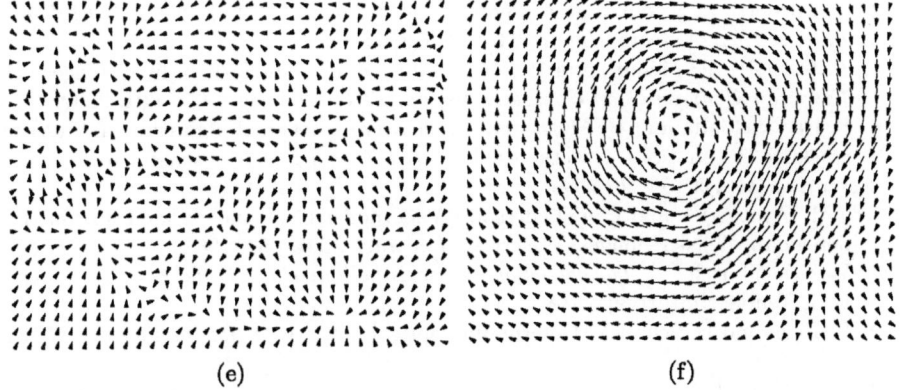

(e)   (f)

Fig. 8. (Continued).

4   Conclusion

Many important application areas pose the problem of computing highly non-rigid fluid flow from image sequences. In contrast to traditional variational approaches for optical flow computation which are not appropriate in this context, we dealt with this problem by using higher-order regularization terms which merely penalize changes of the principal flow constituents. The approach was made computationally tractable by the use of auxiliary functions. A significant feature of our approach is that the associated potential functions are directly computed. This is a favorable property regarding the recognition and analysis of flow patterns. Numerical experiments confirmed that both components can be estimated separately by subsequent linearizations of a single data term. A comparison with an indirect approach revealed no loss in performance, despite the higher order of differential equations to be solved.

Our further work will focus on the computation of the laminar flow in the same step as well, and on the problem of unknown boundary conditions at the image border (recall that, at present, we use standard boundary conditions for the biharmonic operator at an artificially enlarged image border). Moreover, we will investigate in more detail the dependency on the auxiliary functions and corresponding smoothing terms.

References

1. R. Adrian. Particle imaging techniques for experimental fluid mechanics. *Annal Rev. Fluid Mech.*, 23:261–304, 1991.
2. A. Amini. A scalar function formulation for optical flow. In *Proc. Europ. Conf. Computer Vision*, pages 125–131, 1994.
3. L. Bannehr, R. Rohn, and G. Warnecke. A functionnal analytic method to derive displacement vector fields from satellite image sequences. *Int. Journ. of Remote Sensing*, 17(2):383–392, 1996.

4. T. Corpetti, E. Mémin, and P. Pérez. Dense estimation of fluid flows. *IEEE Trans. Pattern Anal. Machine Intell.*, 24(3):365–380, 2002.
5. T. Corpetti, E. Mémin, and P. Pérez. Dense motion analysis in fluid imagery. In *European Conference on Computer Vision, ECCV'02*, pages 676–691, 2002.
6. S. Das Peddada and R. McDevitt. Least average residual algorithm (LARA) for tracking the motion of arctic sea ice. *IEEE trans. on Geoscience and Remote sensing*, 34(4):915–926, 1996.
7. J.M. Fitzpatrick and C.A. Pederson. A method for calculating fluid flow in time dependant density images. *Electronic Imaging*, 1:347–352, 1988.
8. W. Hackbusch. *Theorie und Numerik elliptischer Differentialgleichungen.* B.G. Teubner, Stuttgart, 1986.
9. B. Horn and B. Schunck. Determining optical flow. *Artificial Intelligence*, 17:185–203, 1981.
10. D. J. Fleet J. L. Barron and S. S. Beauchemin. Perfomance of optical flow techniques. *Int. J. Computer Vision*, 1994.
11. R. Larsen, K. Conradsen, and B.K. Ersboll. Estimation of dense image flow fields in fluids. *IEEE trans. on Geoscience and Remote sensing*, 36(1):256–264, 1998.
12. S.P. McKenna and W.R. McGillis. Performance of digital image velocimetry processing techniques. *Experiments in Fluids*, 32:106–115, 2002.
13. A. Ottenbacher, M. Tomasini, K. Holmlund, and J. Schmetz. Low-level cloud motion winds from Meteosat high-resolution visible imagery. *Weather and Forecasting*, 12(1):175–184, 1997.
14. C. Schnörr. Segmentation of visual motion by minimizing convex non-quadratic functionals. In *12th Int. Conf. on Pattern Recognition*, Jerusalem, Israel, Oct 9-13 1994.
15. C. Schnörr, R. Sprengel, and B. Neumann. A variational approach to the design of early vision algorithms. *Computing Suppl.*, 11:149–165, 1996.
16. J. Shukla and R. Saha. Computation of non-divergent streamfunction and irrotational velocity potential from the observed winds. *Monthly weather review*, 102:419–425, 1974.
17. J. Simpson and J. Gobat. Robust velocity estimates, stream functions, and simulated Lagrangian drifters from sequential spacecraft data. *IEEE trans. on Geosciences and Remote sensing*, 32(3):479–492, 1994.
18. S.M. Song and R.M. Leahy. Computation of 3D velocity fields from 3D cine and CT images of human heart. *IEEE trans. on medical imaging*, 10(3):295–306, 1991.
19. D. Suter. Motion estimation and vector splines. In *Proc. Conf. Comp. Vision Pattern Rec.*, 1994.
20. J. Weickert and C. Schnörr. A theoretical framework for convex regularizers in pde–based computation of image motion. *Int. J. Computer Vision*, 45(3):245–264, 2001.
21. L. Zhou, C. Kambhamettu, and D. Goldgof. Fluid structure and motion analysis from multi-spectrum 2D cloud images sequences. In *Proc. Conf. Comp. Vision Pattern Rec.*, volume 2, pages 744–751, Hilton Head Island, South Carolina, USA, 2000.

Regularizing a Set of Unstructured 3D Points from a Sequence of Stereo Images

Luis Álvarez-León, Carmelo Cuenca, and Javier Sánchez

Departamento de Informática y Sistemas
Universidad de Las Palmas de G.C.
Campus Universitario de Tafira
35017, Las Palmas
{lalvarez,ccuenca,jsanchez}@dis.ulpgc.es
http://serdis.dis.ulpgc.es/~{lalvarez,jsanchez}

Abstract. In this paper we present a method for the regularization of a set of unstructured 3D points obtained from a sequence of stereo images. This method takes into account the information supplied by the disparity maps computed between pairs of images to constraint the regularization of the set of 3D points. We propose a model based on an energy which is composed of two terms: an attachment term that minimizes the distance from 3D points to the projective lines of camera points, and a second term that allows for the regularization of the set of 3D points by preserving discontinuities presented on the disparity maps. We embed this energy in a 2D finite element method. After minimizing, this method results in a large system of equations that can be optimized for fast computations. We derive an efficient implicit numerical scheme which reduces the number of calculations and memory allocations.

1 Introduction

This paper deals with the problem of 3D geometry reconstruction from multiple 2D views. Recently, a new accurate technique based on a variational approach has been proposed in [7],[8]. Using a level set approach, this technique optimizes a 3D surface by minimizing an energy that takes into account the surface regularity as well as the projection of the surface on different images.

In this paper we propose a different approach which is also based on a variational formulation but only using a disparity estimation between images and without defining explicitly any 3D surface. We will assume that the cameras are calibrated in the strong sense (see [6], [9] or [11] for more details). In the last years, very accurate techniques to estimate the disparity map in a stereo pair of images have been proposed. To extend these techniques to the case of multiple views is not a trivial problem. Roughly speaking the 3D geometry estimation that we propose can be divided in the following steps:

- For each pair of consecutive images, we estimate a dense disparity map using the accurate technique developed in [1]. We estimate such disparity

L.D. Griffin and M. Lillholm (Eds.): Scale-Space 2003, LNCS 2695, pp. 449–463, 2003.

map forward and backward, that is, from one image to the next one and in the opposite direction.

- We estimate sequences of corresponding points across the multiple view image sequence. Basically, we try to connect points between images following the disparity map estimations. We select sequences of correspondent points for which the forward and backward disparity estimations are coherent.
- From each selected corresponding points sequence we recover a 3D point by intersecting the projection lines of the points in the sequence. By collecting the 3D points obtained from each sequence we recover an unstructured set of 3D points.
- Typically, the recovered set of 3D points is noisy, because of errors in the camera calibration process, errors in the disparity estimations, errors in the corresponding point sequences computations, etc., so some kind of regularization is needed. In this paper, we propose a new variational model to smooth the unstructured set of 3D points. This regularization model is based on the 2D image information and does not require to define any kind of geometric relation between the 3D points.

The proposed technique provides a smooth set of unstructured 3D points. In this paper we do not address the problem of defining one or several surfaces fitting the set of points. Such surfaces could be recovered using some standard methods like Alpha shapes [5], Ball pivot [4] or Voronoi filtering [3]. We notice that most of such techniques require the collection of points to be smooth enough to recover the surface. So the regularization step we propose is necessary to improve the results of such techniques.

The regularization model we propose, which is the main contribution of the paper is based on a variational approach. This model is designed in order to maintain the final 3D regularized surface next to the original surface and also to enable a regularization by preserving discontinuities on the disparity maps. The regularization is carried out by means of an operator which is similar to the Nagel–Enkelmann's operator [12]. This operator has already proven its efficiency in other fields like stereoscopic reconstruction [1], optical flow estimation [2], etc. We have modified this operator to include the information given by the forward and backward optical flows of every camera. In our case it is convenient to preserve these discontinuities because disparity maps represent the depth variation of the set of 3D points.

For every camera we will have a set of 2D projection points (extracted from the selected sequences of points described above) forming a grid. These grids are not necessary square and equally distributed as we would expect for regular images. On the contrary we will have in most cases some distributed points on the cameras with float precision. For this, we embed the energy in a finite element method and express the set of 3D points in terms of some basis functions. We will use a Delaunay triangulation for every camera and every point on a mesh will have a correspondent 3D point.

Deriving this energy and searching for the minimum yields a system of as many equations as 3D points are there in the set. The system matrix is a sparse

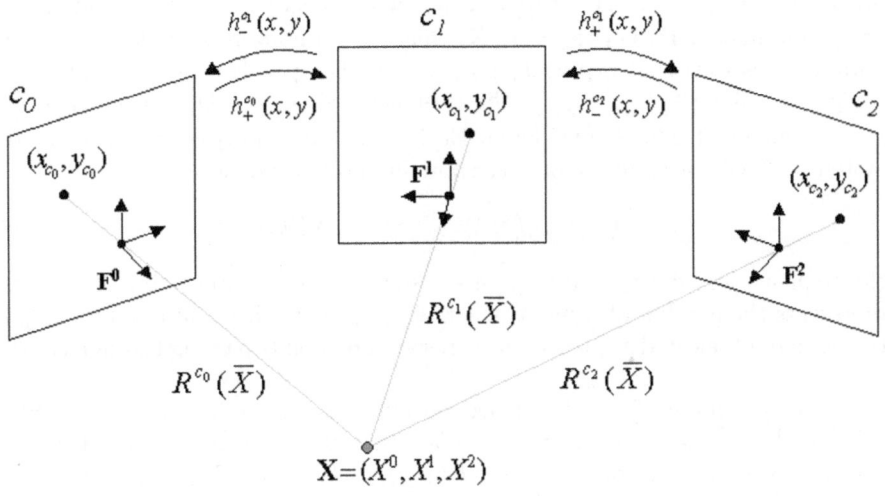

Fig. 1. Notation

matrix that will only have no-null values in some special locations. To solve the system we first arrange the matrices to optimize their size in memory and then implement a Gauss-Seidel numerical scheme that is very efficient in the number of calculations.

Finally we present two experimental results to evaluate the method. In the first experiment we use a single stereo pair of a face. We show the stereo pair and the reconstruction obtained from the method explained in [1]. Then we show the Delaunay triangulations of both cameras and finally some results of applying our regularizing method with different values for the parameters. In the other experiment we use larger sequences of stereo images. We show different results from several points of view.

In Sect. 2 the notation is explained. In Sect. 3 we study the model proposing an energy with the attachment and regularizing term and explaining the operator derived from the Nagel–Enkelmann operator. In Sect. 4 we explain how to embed the method into the 2D finite element approach. In Sect. 5 we present the numerical experiments.

2   Notation

As we will see later this method involves complex notation that is mainly affected by three different aspects: we use some notation for the camera configuration, notation for the stereo matching process and notation for the representation of the 3D points. We will focus on Fig. (1) to explain all these aspects.

We start explaining the camera notation. We suppose there is a set of projective cameras, c_i $(i = 1, .., N_c)$, N_c denoting the number of cameras. The projection model assumed for the cameras is the projective one. This model projects a

3D point $\mathbf{X} = [(X^0, X^1, X^2)]^t$ to a 2D image point $\mathbf{x_c} = [x_c, y_c]^t$ through a 3×4 projection matrix $\mathbf{P_c}$ via $s\hat{\mathbf{x}}_{\mathbf{c}} = \mathbf{P_c}\hat{\mathbf{X}}$, where s is a nonzero scale factor and the notation $\hat{\mathbf{p}}$ is such that if $\mathbf{p} = [x_1, x_2, \ldots, x_n]^t$ then $\hat{\mathbf{p}} = [x_1, x_2, \ldots, x_n, 1]^t$.

We note by $\{\mathbf{X_j}\}_{j=0,\ldots,N_p-1}$ the unstructured set of 3D points, where N_p is the number of points considered. Each point $\mathbf{X_j}$ is projected in the camera system following a sequence of corresponding points. We note by

$$(x_{c_i}(\mathbf{X_j}), y_{c_i}(\mathbf{X_j}))   i = 0, \ldots, N(\mathbf{X_j}) - 1 \tag{1}$$

the sequence of corresponding points associated to the 3D point X_j. $N(\mathbf{X_j})$ represents the number of cameras where X_j is projected. We also note by $N_p(c)$ the number of selected sequences of corresponding points passing through camera c.

We may compute from the projection matrices $\mathbf{P_c}$ (see [6]) the focuses $\mathbf{F}^c = (F^{c,0}, F^{c,1}, F^{c,2})$ for every camera c. For every 2D projection $\mathbf{x_c}$ and for every camera c we have an unitary vector $\bar{l}^c(x, y) = (l^{c,0}(x,y), l^{c,1}(x,y), l^{c,2}(x,y))$ in the direction of focus $\mathbf{F}^c$. We denote by $\bar{R}^c(x,y)$ the line passing through focus $\mathbf{F}^c$ and having $\bar{l}^c(x,y)$ as directional vector.

Let us now talk about the notation for the stereo aspects. For every pair of consecutive cameras on the sequence we have the estimated optical flow from camera c into the previous camera $c - 1$, $\bar{h}^c_-(x,y) = (u^c_-(x,y), v^c_-(x,y))$, and the corresponding flow from camera c into the following, $\bar{h}^c_+(x,y) = (u^c_+(x,y), v^c_+(x,y))$.

For every camera c we define a function $\bar{X}^c : \mathbb{R}^2 \longrightarrow \mathbb{R}^3$ that puts in correspondence bi-dimensional points (x, y) on the projection plane of camera c into 3D points $(X^{c,0}(x,y), X^{c,1}(x,y), X^{c,2}(x,y))$. Let us call $\mathbf{X} = (X^0, X^1, X^2)$ the 3D point obtained by the minimum distance to the set of lines $\bar{R}^c(x,y)$ that passes through focus $\mathbf{F}^c$ and has $\bar{l}^c(x,y)$ as directional vector for the cameras in where there is a correspondent projection. Suppose that every 3D point is visible from $N(\mathbf{X})$ cameras. These cameras are denoted by $c_i(\mathbf{X})$ for $i = 0, \ldots, N(\mathbf{X}) - 1$. The pixel coordinates for $\mathbf{X}$ on camera c_i is denoted by $(x_{c_i}(\mathbf{X}), y_{c_i}(\mathbf{X}))$. Therefore we have

$$\mathbf{X} = \bar{X}^c(x_{c_i}(\mathbf{X}), y_{c_i}(\mathbf{X}))   i = 0, \ldots, N(\mathbf{X}) - 1 . \tag{2}$$

Finally, the derivatives of a function f are represented as $\frac{df}{dx} = f_x$ and $\frac{df}{dy} = f_y$ and ∇X is the gradient of function X.

3   The 3D Regularizing Model

We tackle the problem of 3D regularizing by means of a variational approach in where the solution for the minimization of a global energy is the regularized set of 3D points.

Our method regularizes a set of 3D points by constraining the process with the information given by the optical flows computed for every camera.

The energy to be minimized for the regularization of the set of 3D point is:

$$E\left(\bar{X}^0, .., \bar{X}^{N_c-1}\right) =$$

$$\sum_{c=0}^{N_c-1} \left(\int_{\Omega} dist(\bar{X}^c, \bar{R}^c)^2 + \alpha \sum_{i=0}^{2} \int_{\Omega} \left(\nabla X^{c,i}\right)^T D(\bar{h}^c) \nabla X^{c,i} \right) \qquad (3)$$

where α is a parameter that states the balance between the two terms and $dist(\bar{X}^c, \bar{R}^c)$ denotes the distance from point $\bar{X}^c$ to the straight line $\bar{R}^c$ and is given by formula

$$dist(\bar{X}^c, \bar{R}^c)^2 = \sum_{i=0}^{2} \left(X^{c,i} - F^{c,i}\right)^2 - \left(\sum_{i=0}^{2} l^{c,i} \left(X^{c,i} - F^{c,i}\right)\right)^2 . \qquad (4)$$

Our unknowns are $\left(\bar{X}^0, ..., \bar{X}^{N_c-1}\right)$ for cameras $0, ..., N_c - 1$ respectively. We suppose every camera has its own set of 3D points but these sets are referred to a common global set of 3D points. For consecutive cameras a large part of their set of points are going to be coincident. In particular the functions X^c are related by (2). We will see later on Sect. 4 that formulating the energy in this way allows us to create a numerical scheme by summing the contribution of every camera.

The first term of (3) minimizes the distance from the 3D point $\bar{X}^c$ to the straight lines generated by the corresponding camera points passing through the camera focus. The second is a smoothness term that minimizes the variation of the 3D points according to the information given by the gradient of $\bar{h}_+^c$ and $\bar{h}_-^c$. In this case we minimize the variation of the 3D surface by using an operator which is very similar to the Nagel–Enkelmann operator. This operator allows the method to regularize isotropically when the 3D points set varies smoothly and anisotropically when there is a strong variation of the disparity map, thus, respecting regions of different depths.

Matrix $D(\bar{h}^c)$ is given by the following expression:

$$D(\bar{h}^c) = \frac{M(\bar{h}_+^c) + M(\bar{h}_-^c)}{\left\|\nabla \bar{h}_+^c\right\|^2 + \left\|\nabla \bar{h}_-^c\right\|^2} \qquad (5)$$

where

$$\left\|\nabla \bar{h}\right\|^2 = \left\|\nabla u\right\|^2 + \left\|\nabla v\right\|^2$$

and

$$M(\bar{h}) = \begin{cases} \left(\begin{array}{cc} \frac{\partial u}{\partial y}^2 + \frac{\partial v}{\partial y}^2 & -\left(\frac{\partial u}{\partial y}\frac{\partial u}{\partial x} + \frac{\partial v}{\partial y}\frac{\partial v}{\partial x}\right) \\ -\left(\frac{\partial u}{\partial y}\frac{\partial u}{\partial x} + \frac{\partial v}{\partial y}\frac{\partial v}{\partial x}\right) & \frac{\partial u}{\partial x}^2 + \frac{\partial v}{\partial x}^2 \end{array} \right) & if \ \lambda^2 \geq \left\|\nabla \bar{h}\right\|^2 \\ \left(\begin{array}{cc} \frac{1}{2} & 0 \\ 0 & \frac{1}{2} \end{array} \right) & if \ \lambda^2 < \left\|\nabla \bar{h}\right\|^2 \end{cases} \qquad (6)$$

This is a projection matrix on the orthogonal space to the vector field $h_{+/-}^c$. This matrix is similar to the Nagel-Enkelmann operator [12] but instead of using λ within the matrix we use it as a real threshold in a non-continuous function.

We also consider the optical flow as the projection space and take advantage of the information in both directions. λ is a threshold that states the contour value from where we obtain an anisotropic behavior. This parameter is obtained through an isotropy fraction, s, as is explained in [1]. When $s \to 0$ the method becomes anisotropical and when $s \to 1$ the method becomes isotropical.

The first problem that comes out when we discretize this energy is that in every camera, the point distribution $(x_c(\bar{X}), y_c(\bar{X}))$ is given in float precision and, therefore, it does not consist of a simple pixel grid where we may discretize the equations using the surrounding neighbors. To solve this problem we use a finite element method.

4   Finite Elements Method

We are going to discretize the function X^c using the discrete set of points $\{\mathbf{X_j}\}_{j=0,..,N_p-1}$ obtained from the sequence of 2D corresponding image points $(x_{c_i}(\mathbf{X_j}), y_{c_i}(\mathbf{X_j}))$, $i = 0, ..., N(\mathbf{X_j})-1$. Since $(x_{c_i}(\mathbf{X_j}), y_{c_i}(\mathbf{X_j}))$ are given in float precision and they are distributed in the images in an heterogeneous way, it is not convenient to use a classical finite difference scheme. So we propose a finite element scheme that takes into account the geometry of the set $(x_{c_i}(\mathbf{X_j}), y_{c_i}(\mathbf{X_j}))$ in every image.

Following the notation introduced in Sect. 2 let us call $N_p(c)$ the number of $3D$ points visible from camera c. For any of these points $(l = 0, .., N_p(c)-1)$, $j(l)$ denote the associated 3D point index in a global list of $3D$ points. We generate, using a Delaunay triangulation (see [10]), a grid of K_n^c triangles in every camera c with $(x_c(\bar{X}_{j(l)}), y_c(\bar{X}_{j(l)}))_{l=0,...,N_p(c)-1}$ points. We will call $N_t(c)$ the number of triangles generated on camera c.

In the finite element approach the $X^{c,i}(x, y)$ functions are approximated as

$$X^{c,i}(x, y) = \sum_{l=0}^{N_p(c)-1} X_{j(l)}^{c,i} \phi_l^c(x, y) \tag{7}$$

where $X_{j(l)}^{c,i}$ are the unknowns of our system and $\phi_{j(l)}^c(x, y)$ is a set of basis functions centered on the nodes $(x_c(\bar{X}_{j(l)}), y_c(\bar{X}_{j(l)}))$. These basis functions are defined over each triangle K_n^c of vertexes $(j(l), j(m), j(n))$ in the following way:

$$Dx(\tau, \upsilon) = (x_c(\bar{X}_{j(\tau)}) - x_c(\bar{X}_{j(\upsilon)}))$$
$$Dy(\tau, \upsilon) = (y_c(\bar{X}_{j(\tau)}) - y_c(\bar{X}_{j(\upsilon)}))$$

$$\phi_l^c(x, y) = \frac{(y - y_c(\bar{X}_{j(m)}))Dx(n, m) - (x - x_c(\bar{X}_{j(m)}))Dy(n, m)}{Dy(l, m)Dx(n, m) - Dx(l, m)Dy(n, m)}. \tag{8}$$

Note that the gradient of ϕ on the triangle is constant and is given by

$$\nabla\phi_l^c(x, y) = \begin{pmatrix} \frac{-Dy(n,m)}{Dy(l,m)Dx(n,m)-Dx(l,m)Dy(n,m)} \\ \frac{Dx(n,m)}{Dy(l,m)Dx(n,m)-Dx(l,m)Dy(n,m)} \end{pmatrix}. \tag{9}$$

First, we note that $\bar{X}^c_{j(l)}$ corresponds to a 3D point visible from camera c and that such a point references a 3D point, $\bar{X}_j$, of the global 3D point list, that is, the true unknowns are the $\bar{X}_j$ points. Replacing $X^{c,i}(x,y) = \sum_{l=0}^{N_c-1} X^{c,i}_{j(l)} \phi_{j(l)}(x,y)$ in (3) we obtain:

$$E\left(\{\bar{X}_j\}_{j=0,..,N_p-1}\right) = \sum_{c=0}^{N_c-1} \sum_{l=0}^{N_p(c)-1} dist(\bar{X}^c_{j(l)}, \bar{R}^c)^2 \tag{10}$$

$$+ \alpha \sum_{c=0}^{N_c-1} \sum_{i=0}^{2} \sum_{n=0}^{N_t(c)-1} \sum_{l,l'=0}^{N_p(c)-1} X^{c,i}_{j(l)} X^{c,i}_{j(l')} \int_{K^c_n} \left(\nabla \phi^c_{j(l)}\right)^T D(u^c, v^c) \nabla \phi^c_{j(l')} \, .$$

Deriving this energy by respect to X^i_j we obtain

$$\frac{\partial E}{\partial X^i_j} = \sum_{m=0}^{N(\bar{X}_j)-1} 2\left(X^{c_m(\bar{X}_j),i} - F^{c_m(\bar{X}_j),i}\right. \tag{11}$$

$$\left. - l^{c_m(\bar{X}_j),i} \sum_{k=0}^{2} l^{c_m(\bar{X}_j),k} \left(X^{c_m(\bar{X}_j),k} - F^{c_m(\bar{X}_j),k}\right)\right)$$

$$+ \alpha \sum_{m=0}^{N(\bar{X}_j)-1} \sum_{n=0}^{N_t(c_m(\bar{X}_j))-1} \sum_{l,l'=0,j(l')=j}^{N_p(c_m(\bar{X}_j))-1} 2X^{c,i}_{j(l)} \int_{K^c_n} (\nabla \phi^c_l)^T D(u^c, v^c) \nabla \phi^c_{l'}$$

for every $i = 0, 1, 2$. In order to find the minimum of this energy, these equations are equaled to 0. A system of $3N_p$ equations and unknowns is thus generated.

4.1   Solving the System

We are going to see how to build up the system and solve it numerically. We express the system as $Au = b$ where A is a square matrix, b is the independent constant vector and u is the vector of $3N_p$ unknowns to be ordered in the following way

$$u = (X^0_0, X^0_1, ..., X^0_{N_p-1}, X^1_0, ..., X^1_{N_p-1}, X^2_0, ..., X^2_{N_p-1}) \, . \tag{12}$$

We will concentrate on (11) to derive both matrix A and vector b. To compute vector b we realize that constant values generated from (11) are only affected by the first term in the following manner

$$b_{j+iN_p} = \sum_{m=0}^{N(\bar{X}_j)-1} \left(F^{c_m(\bar{X}_j),i} - l^{c_m(\bar{X}_j),i}\left(\sum_{k=0}^{2} l^{c_m(\bar{X}_j),k} F^{c_m(\bar{X}_j),k}\right)\right) \, . \tag{13}$$

Both terms of (11) supply information for matrix $A = (a_{j,j'})$. The first term of the energy supply the following information to the diagonal part of matrix A. For every $j = 0, ..., N_p - 1$ and $i = 0, 1, 2$ we have

$$a_{j+iN_p, j+iN_p} + = \sum_{m=0}^{N(\bar{X}_j)-1} \left(1 - \left(l^{c_m(\bar{X}_j),i}\right)^2\right) \, . \tag{14}$$

Note that += in this formulation means accumulation of values. Matrix A is initialized to null values. In the process of constructing this matrix, values are accumulated in different steps to facilitate the task of dividing the algorithm. Later on we will see that the diagonal part of this matrix is affected by the second term of the energy.

For the non-diagonal part of A such a term provides the following:

$$a_{j+N_p,j} = a_{j,j+N_p} + = - \sum_{m=0}^{N(\bar{X}_j)-1} l^{c_m(\bar{X}_j),0} l^{c_m(\bar{X}_j),1}$$

$$a_{j+2N_p,j} = a_{j,j+2N_p} + = - \sum_{m=0}^{N(\bar{X}_j)-1} l^{c_m(\bar{X}_j),0} l^{c_m(\bar{X}_j),2}$$

$$a_{j+2N_p,j+N_p} = a_{j+N_p,j+2N_p} + = - \sum_{m=0}^{N(\bar{X}_j)-1} l^{c_m(\bar{X}_j),1} l^{c_m(\bar{X}_j),2} . \tag{15}$$

If we now concentrate on the second term of the energy we realize that the contribution to matrix A is the same for the 3 coordinates $i = 0, 1, 2$. We examine every camera c and for every camera we go through every triangle K_n^c. Each triangle is composed of 3 vertexes $(x_c(\bar{X}_j), y_c(\bar{X}_j)), (x_c(\bar{X}_{j'}), y_c(\bar{X}_{j'})), (x_c(\bar{X}_{j''}), y_c(\bar{X}_{j''}))$. From every triangle we compute the following contributions to the matrix:

$$a_{j,j} + = \int_{K_n^c} \left(\nabla \phi_j^c\right)^T D(u^c, v^c) \nabla \phi_j^c$$

$$a_{j',j'} + = \int_{K_n^c} \left(\nabla \phi_{j'}^c\right)^T D(u^c, v^c) \nabla \phi_{j'}^c$$

$$a_{j'',j''} + = \int_{K_n^c} \left(\nabla \phi_{j''}^c\right)^T D(u^c, v^c) \nabla \phi_{j''}^c$$

$$a_{j',j} = a_{j,j'} + = \int_{K_n^c} \left(\nabla \phi_j^c\right)^T D(u^c, v^c) \nabla \phi_{j'}^c$$

$$a_{j'',j} = a_{j,j''} + = \int_{K_n^c} \left(\nabla \phi_j^c\right)^T D(u^c, v^c) \nabla \phi_{j''}^c$$

$$a_{j'',j'} = a_{j',j''} + = \int_{K_n^c} \left(\nabla \phi_{j'}^c\right)^T D(u^c, v^c) \nabla \phi_{j''}^c \tag{16}$$

these contributions are the same for the 3 coordinates.

4.2   Numerical Scheme

If we study matrix A we will realize that most of its values are null, therefore solving the previous system will induce a lot of unnecessary calculations and, therefore, a very slow method.

We have implemented an efficient numerical scheme for the resolution of the previous system. Studying the way the system matrix A is constructed we realize

Fig. 2. Stereoscopic pair for Herve's face

that it is a sparse matrix and most of its values are situated in symmetrical positions. Also we realize that in most cases, a single point will be surrounded by a small number of triangles, inducing, for every point on the mesh, a small number of values for the second term of the energy.

To take advantage of this we express the system $Au = b$ as $\alpha(D+H)u+Cu = b$ where C is a matrix with the contribution of the first term of the energy, D is the diagonal matrix with the contribution of the second term of the energy and H is the contribution of the non-diagonal part of the second term.

To solve this system we use a fixed point equation:

$$u = (\alpha D + C)^{-1}(b - \alpha Hu) \,. \tag{17}$$

5   Experimental Results

In this section we present some experiments. For the first experiment we propose a single stereo pair (Hervé's face). The second is a bust for which we have a complete round sequence of 47 calibrated images.

5.1   Single Stereo Pair

In Fig. (2) he have a stereo pair of a face. The purpose of this experiment is to show the validity of the method when there is only a single disparity map estimated for each camera. In Fig. (3) we show the profile and front of the Herve's reconstruction obtained through the method explained in [1]. We have only reconstructed the region belonging to the face.

During the process we have to construct two triangulations to be used within the finite element method. In Fig. (4) we show the Delaunay triangulations for both cameras. The reference system is situated on the left camera. We may appreciate that for the left image we have a regular mesh. This is because the points in the triangulation are coincident with the pixel distribution on the image. For the right image the points are a little warped in correspondence with the displacement given by the disparity map. The shape of the face could be guessed there. In Fig. (5) we have several 3D regularizations using different

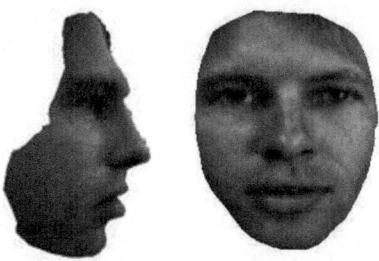

Fig. 3. Profile and front views of the Herve's 3D reconstruction

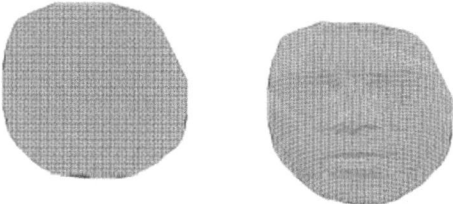

Fig. 4. Camera triangulations for the Herve's stereoscopic pair

values for α and s. The values of α increase by rows and the values of s increase by columns. As we can see the models become more regular for the last rows and more isotropically regularized for the last columns.

From these examples it is easy to see that s plays the role of an anisotropic/isotropic diffusion parameter as it does for other image diffusion related applications. The bigger s is the more the image is isotropically regularized. Concentrating on the first row: The first image on the left is computed for $\alpha = 0.1$ and $s = 0.1$ so the model is little regularized and its geometry (given by the disparity map) is preserved. On the other hand (third row and the right image) $\alpha = 1.0$ and $s = 1.0$ the model is regularized in all directions and the structures are also smoothed.

5.2   Stereo Sequence

Now we consider a more complex sequence. In this case the sequence is composed of 47 images taken around a face bust. Figure (6) shows the configuration of this sequence with the projection planes of the cameras. This is a closed sequence in where the first and last images are correlatives.

In Fig. (7) the 3D reconstruction for the bust sequence using the technique explained in [1] is shown. In Fig. (8) we show the triangulation and image for camera 6 for some set of 3D points. In this figure we may guess the shape of the faces and the set of triangles inside it.

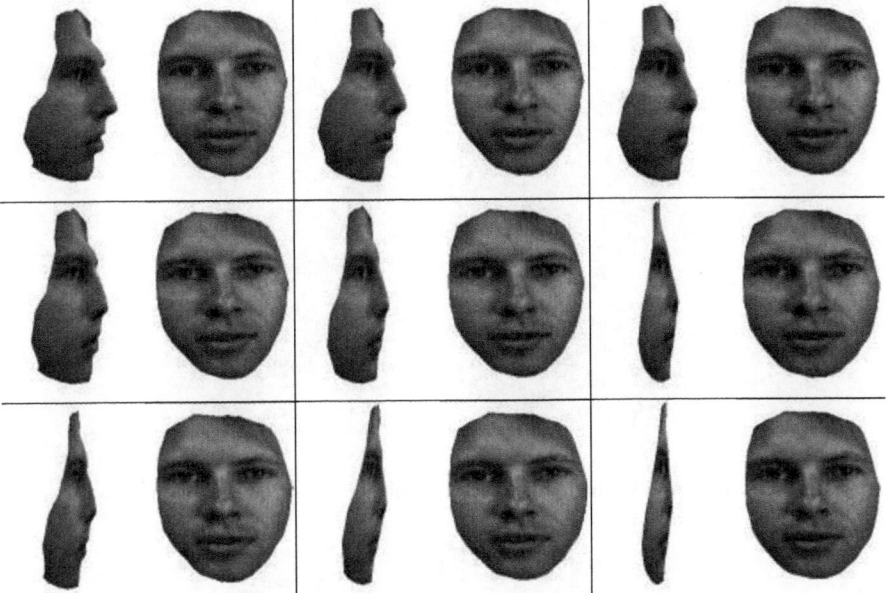

Fig. 5. Profile and front views of different 3D regularizations for Herve's sequence. First row: Left pair of images, $\alpha = 0.1$ and $s = 0.1$; center images, $\alpha = 0.1$ and $s = 0.5$ and; right images, $\alpha = 0.1$ and $s = 1.0$. Second row: Left images, $\alpha = 0.5$ and $s = 0.1$; center images, $\alpha = 0.5$ and $s = 0.5$; and right images, $\alpha = 0.5$ and $s = 1.0$. Third row: Left images, $\alpha = 1.0$ and $s = 0.1$; center images, $\alpha = 1.0$ and $s = 0.5$; and right images, $\alpha = 1.0$ and $s = 1.0$

Fig. 6. Bust configuration: This figure shows the 3D reconstructed bust and the distribution of the projection planes corresponding to the 47 cameras

Figures (9) and (10) contain several regularizations using a range of values for α and s. We may appreciate that when α is increased the set of points becomes more regular and when s tends to 1 the regularization becomes more isotropic.

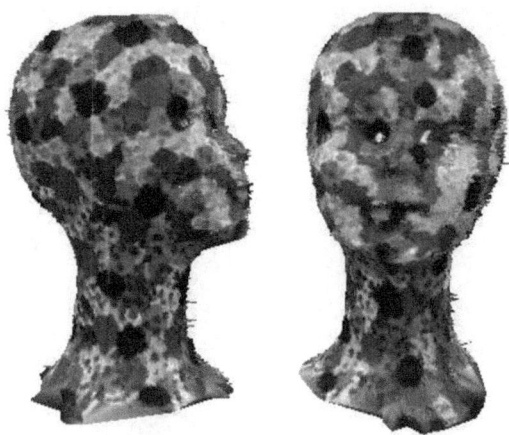

Fig. 7. Original 3D reconstruction for the Bust sequence

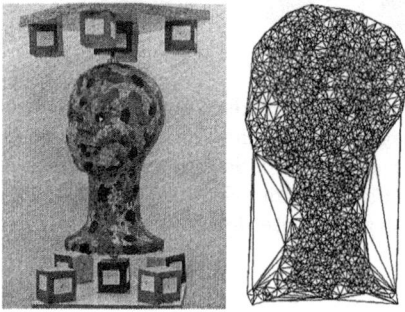

Fig. 8. Triangulation generated by the set of 3D points on the 6th projection plane

Some of the experimental results for the Bust sequence may be accessed at the web site `http://serdis.dis.ulpgc.es/~jsanchez/research/demos/`.

6   Conclusions

In this paper we have presented a novel method for the regularization of a set of 3D points. We have established an energy in a traditional attachment–regularizing couple of terms. In the regularizing term we have made use of an operator similar to the Nagel–Enkelmann operator for 3D regularizations. This new method has been embedded into a 2D finite element approach to take advantage of the underlying precision of data. Then we have managed to derive this energy and propose a very efficient and efficient numerical scheme that allows us to speed up the process and reduce the memory needs.

We have shown in the experiments that varying the α parameter results in a more regular set of points and varying the λ parameter implies a more regular

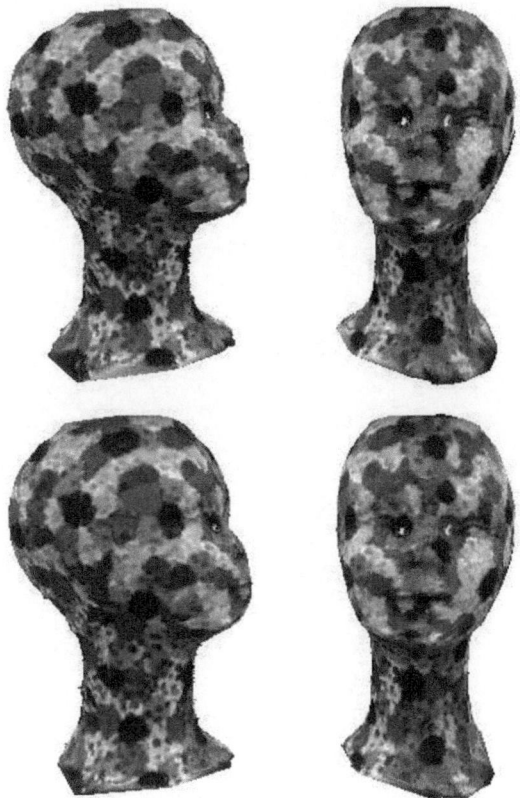

Fig. 9. Profile and front views of two different 3D regularizations for the Bust sequence. First row: Two views for $\alpha = 0.1$ and $s = 0.01$. Second row: Two views for $\alpha = 2.0$ and $s = 0.01$

set of points by preserving disparity map discontinuities as we have expected from the results obtained in other fields.

One of the main advantages of the method is that it regularizes sets of unstructured 3D points without using any geometric relation in 3D. In particular, this method could be used as a preprocessing step before the construction of 3D surfaces fitting the 3D points. Most of the techniques for such surface reconstruction are very sensitive to the noise in the 3D points representation, and they require the set of 3D points to be regular enough to work properly.

Acknowledgments

This work has been partially supported by the Spanish research project TIC 2000-0585 founded by the Ministerio de Ciencia y Tecnología and by the research project PI2002/193 founded by the Canary Islands Government.

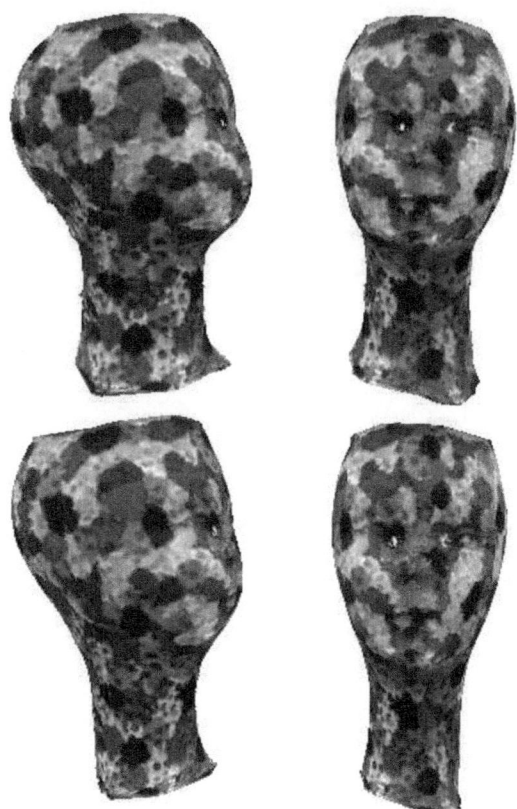

Fig. 10. Profile and front views of two different 3D regularizations for the Bust sequence. First row: Two views for $\alpha = 16.0$ and $s = 0.01$. Second row: Two views for $\alpha = 16.0$ and $s = 1.0$

References

1. L. Alvarez, R. Deriche, J. Sánchez and J. Weickert. Dense disparity map estimation respecting image derivatives: a PDE and scale-space based approach. *Journal of Visual Communication and Image Representation*, 13:3–21, January 2002. Also published as Inria Research Report n° 3874
2. L. Alvarez, J. Weickert and J. Sánchez. Reliable Estimation of Dense Optical Flow Fields with Large Displacements. *International Journal of Computer Vision*, 39(1):41–56, 2000.An extended version maybe be found at Technical Report n°2 del Instituto Universitario de Ciencias y Tecnologías Cibernéticas of the Universidad de Las Palmas de Gran Canaria, Spain
3. N. Amenta and M. Bern. Surface Reconstruction by Voronoi Filtering. *14th ACM Symposium on computational Geometry*, pages 248–253, June 1998
4. F. Bernardini, J. Mittleman and H. Rushmeier and C. Silva. The Ball Pivoting Algorithm for Surface Reconstruction. *IEEE Transactions on Visualization and Computer Graphics*, 5(4):349–359, Oct.–Dec. 1999

5. H. Edelsbrunner and E. P. Mucke. Three-dimensional Alpha Shapes. *ACM Transactions on Graphics*, 13(1):43–72, January 1994
6. O. Faugeras. Three-Dimensional Computer Vision: A Geometric Viewpoint. MIT Press, 1993
7. O. Faugeras and R. Keriven. Complete Dense Stereovision Using Level Set Methods. *Proceedings of Fifth European Conference on Computer Vision*, 1998
8. O. Faugeras and R. Keriven. Variational principles, surface evolution, PDEs, level set methods and the stereo problem. *IEEE Transactions on Image Processing. Special issue on Geometry driven diffusion and PDEs in image processing*, 7(3):336–344, 1998
9. O. Faugeras, and Q. Luong and T. Papadopoulo. The Geometry of Multiple Images. Mit Press, 2001
10. P. George and H. Borouchaki. Triangulation de Delaunay et Maillage. Hermes, 1997
11. R. Hartley and A. Zisserman. Multiple View Geometry in Computer Vision. Cambridge University Press, 2000
12. H.H. Nagel and W. Enkelmann. An Investigation of Smoothness Constraints for the Estimation of Displacement Vector Fields from Images Sequences. *IEEE Transactions on Pattern Analysis and Machine Intelligence*, 8(5):565–593, 1986

Image Reconstruction
from Multiscale Critical Points

Frans Kanters, Luc Florack, Bram Platel, and Bart M. ter Haar Romeny

Eindhoven University of Technology
Den Dolech 2, Postbus 513
5600 MB Eindhoven, the Netherlands
{F.M.W.Kanters,L.M.J.Florack,B.Platel,B.M.terHaarRomeny}@tue.nl
http://www.bmi2.bmt.tue.nl/image-analysis/

Abstract. A minimal variance reconstruction scheme is derived using derivatives of the Gaussian as filters. A closed form mixed correlation matrix for reconstructions from multiscale points and their local derivatives up to the second order is presented. With the inverse of this mixed correlation matrix, a reconstruction of the image can be easily calculated. Some interesting results of reconstructions from multiscale critical points are presented. The influence of limited calculation precision is considered, using the condition number of the mixed correlation matrix.

1   Introduction

There are still many open questions about the deep structure of images, and critical points in particular. One of the questions is how much information is contained in these critical points. Do these points contain sufficient information to compare images [6], to find substructures? Much research about reconstruction algorithms is done to get a hold on features which contain crucial image information. For example image reconstruction from sign information [12], reconstruction from zero crossings of a wavelet transform [11,13] and reconstruction from zero crossings in scale space [4]. Nielsen and Lillholm also look at the image information of different features [10]. In this paper, we look at the information contained in scale space critical points by proposing an algorithm for image reconstruction from multiscale points. It is based on the work of Nielsen and Lillholm [10]. We will also address some problems that occur due to limited machine precision.

2   Minimal Variance Reconstruction

2.1   Definitions

- $i = 1, ..., N$: enumeration index for scale space points
- $\phi(x, y, t)$: standard Gaussian at scale t centered at the origin, thus

$$\phi(x, y, t) = \frac{1}{4\pi t} e^{-\frac{x^2+y^2}{4t}}$$

L.D. Griffin and M. Lillholm (Eds.): Scale-Space 2003, LNCS 2695, pp. 464–478, 2003.
© Springer-Verlag Berlin Heidelberg 2003

- $f(x, y)$: arbitrary high resolution image
- $\hat{f}(x, y)$: approximation of $f(x, y)$
- $\langle f | g \rangle$: scalar product of $f(x, y)$ and $g(x, y)$,

$$\langle f | g \rangle = \int \int f(x, y) g(x, y) \, dx \, dy$$

2.2   Theory

The goal of the proposed algorithm is to make a reconstruction of an image from points in scale space. This reconstructed image should have the same local derivatives up to order N in the reconstruction points and the variance must be minimal. Note that the first constraint ensures that the reconstruction has the same critical points (points where the spatial gradient is zero) if the order N is at least 1 and has the same top points (points where the spatial gradient and Laplacian is zero) if the order N is at least 2. The second constraint makes the image as smooth as possible. The minimal variance reconstruction algorithm is based on [10]. First we derive a general minimal variance scheme. Given a set of filters Ψ_i, we have to minimize:

$$S[\hat{f}] \stackrel{def}{=} \frac{1}{2} || \hat{f} ||_{L^2}^2 + \sum_i \lambda_i \langle f - \hat{f} | \Psi_i \rangle \tag{1}$$

Where f is the original image, $\hat{f}$ is the reconstructed image and λ_i are Lagrange multipliers. The first part satisfies the minimal variance constraint, the second part makes sure that the features are preserved. Using the functional derivative we obtain:

$$\frac{\delta S[\hat{f}]}{\delta \hat{f}} \stackrel{def}{=} \hat{f} - \sum_i \lambda_i \Psi_i \tag{2}$$

we can determine the unique solution of $\frac{\delta S[\hat{f}]}{\delta \hat{f}} = 0$:

$$\hat{f} = \sum_i \lambda_i \Psi_i \tag{3}$$

which is the $\hat{f}$ that minimizes the variance if the coefficients λ_i are calculated by substitution of (3) in:

$$\langle f - \hat{f} | \Psi_j \rangle = 0 \tag{4}$$

Apparently the optimal solution lies in the span of the filters used to extract the linear features of interest. Now consider the Gaussian $\phi(x, y, t)$. Derivatives of this function can be defined as:

$$\phi_{,\nu_1 \ldots \nu_k}(x, y) = \nabla_{\nu_1 \ldots \nu_k} \phi \tag{5}$$

We can define the basic function:

$$\phi_{i,\nu_1...\nu_k}(x,y) = \sqrt{2t_i}^k \phi_{,\nu_1...\nu_k}(x-x_i, y-y_i, t_i) \tag{6}$$

as a normalized Gaussian of scale t_i, centered at x_i, y_i and differentiated to the k-th order with respect to $x^{\nu_1}, \ldots, x^{\nu_k}$ in which we identify $x^1 \equiv x$ and $x^2 \equiv y$. We may call (x_i, y_i) the spatial base point of $\phi_{i,\nu_1...\nu_k}(x,y)$ and (x_i, y_i, t_i) the scale space base point.

Furthermore, define $L_{i,\nu_1...\nu_k}$ as the features obtained by taking the scalar product of the original image $f(x,y)$ with the basic function $\phi_{i,\nu_1...\nu_k}(x,y)$:

$$L_{i,\nu_1...\nu_k} = \langle f \,|\, \phi_{i,\nu_1...\nu_k} \rangle \tag{7}$$

$L_{i,\nu_1...\nu_k}$ is called the Gaussian blurred derivative of the image with respect to $x^{\nu_1}, \ldots, x^{\nu_k}$ at point i. The indices $\nu_1, \ldots, \nu_k$ are referred to as spatial indices. A spatial index (in two dimensions) can take only two possible values, interchangeably denoted as "x" and "y", or as "1" resp. "2". The label i is sometimes referred to as an enumeration index. An enumeration index can take arbitrarily many values in principle, say $i = 1, \ldots, N$.

Second Order Case

Let us consider a number of scale space base points (x_i, y_i, t_i) with $i = 1, ..., N$ in scale space. For every point i, the features $L_{i,\nu_1...\nu_k}$ with $k = 0, 1, 2$ can be calculated. Given this feature space, a second order reconstruction $\hat{f}$ can be proposed, as follows:

$$\hat{f}(x,y) = \sum_{i=1}^{N} a_i\phi_i(x,y) + b_i^x\phi_{i,x}(x,y) + b_i^y\phi_{i,y}(x,y) +$$
$$c_i^{xx}\phi_{i,xx}(x,y) + c_i^{xy}\phi_{i,xy}(x,y) + c_i^{yy}\phi_{i,yy}(x,y) \tag{8}$$

which can be shortened using summation convention for the repeated spatial indices to:

$$\hat{f}(x,y) = \sum_{i=1}^{N} a_i\phi_i(x,y) + b_i^\mu\phi_{i,\mu}(x,y) + c_i^{\mu\rho}\phi_{i,\mu\rho}(x,y) \tag{9}$$

Which is cf. (3). As a constraint on the reconstruction, all features in every point $i = 1, ..., N$ of the reconstruction must be the same as those in the original image, thus in case we adopt the full set of second order constraints,

$$\langle f - \hat{f} \,|\, \phi_i \rangle = 0, \quad \langle f - \hat{f} \,|\, \phi_{i,\mu} \rangle = 0 \quad \text{and} \quad \langle f - \hat{f} \,|\, \phi_{i,\mu\rho} \rangle = 0 \tag{10}$$

for all $i = 1, ..., N$ and for $\mu = x, y$ and $\rho = x, y$, which is cf. (4). The features can be written as:

$$\langle f \,|\, \phi_i \rangle = L_i, \quad \langle f \,|\, \phi_{i,\mu} \rangle = L_{i,\mu} \quad \text{and} \quad \langle f \,|\, \phi_{i,\mu\rho} \rangle = L_{i,\mu\rho} \tag{11}$$

for all $i = 1...N$ and for $\mu = x, y$ and $\rho = x, y$.

If (9) is substituted in (10) using (11), the missing coefficients can be calculated from the following linear system of equations:

$$\langle \sum_{i=1}^{N} a_i\phi_i + b_i^{\mu}\phi_{i,\mu} + c_i^{\mu\rho}\phi_{i,\mu\rho} \,|\, \phi_j \rangle = L_j \tag{12}$$

$$\langle \sum_{i=1}^{N} a_i\phi_i + b_i^{\mu}\phi_{i,\mu} + c_i^{\mu\rho}\phi_{i,\mu\rho} \,|\, \phi_{j,\nu} \rangle = L_{j,\nu} \tag{13}$$

$$\langle \sum_{i=1}^{N} a_i\phi_i + b_i^{\mu}\phi_{i,\mu} + c_i^{\mu\rho}\phi_{i,\mu\rho} \,|\, \phi_{j,\nu\eta} \rangle = L_{j,\nu\eta} \tag{14}$$

with $\mu = x, y$, $\rho = x, y$, $\nu = x, y$ and $\eta = x, y$, which can be rewritten as:

$$\sum_{i=1}^{N} a_i\langle \phi_i \,|\, \phi_j \rangle + b_i^{\mu}\langle \phi_{i,\mu} \,|\, \phi_j \rangle + c_i^{\mu\rho}\langle \phi_{i,\mu\rho} \,|\, \phi_j \rangle = L_j \tag{15}$$

$$\sum_{i=1}^{N} -a_i\langle \phi_{i,\nu} \,|\, \phi_j \rangle - b_i^{\mu}\langle \phi_{i,\mu\nu} \,|\, \phi_j \rangle - c_i^{\mu\rho}\langle \phi_{i,\mu\rho\nu} \,|\, \phi_j \rangle = L_{j,\nu} \tag{16}$$

$$\sum_{i=1}^{N} a_i\langle \phi_{i,\nu\eta} \,|\, \phi_j \rangle + b_i^{\mu}\langle \phi_{i,\mu\nu\eta} \,|\, \phi_j \rangle + c_i^{\mu\rho}\langle \phi_{i,\mu\rho\nu\eta} \,|\, \phi_j \rangle = L_{j,\nu\eta} \tag{17}$$

2.3   Mixed Correlation Matrix

The linear system of equations of (15-17) can be solved using simple linear algebra. To simplify this we define a generalized correlation matrix:

Definition 1. *For each combination of spatial indices $\mu_1, ..., \mu_k$ the generalized correlation matrix $\boldsymbol{\Phi}_{\mu_1...\mu_k}$ is the $N \times N$-matrix with components $\Phi_{ij,\mu_1...\mu_k} = \langle \phi_{i,\mu_1...\mu_k} \,|\, \phi_j \rangle$*

Result 1. *The components of the generalized correlation matrix result in:*

$$\Phi_{ij,\mu_1...\mu_k} = \phi_{,\mu_1...\mu_k}(\bar{x}, t)|_{\bar{x}=\bar{x}_{ij}, t=t_{ij}} \quad \text{with } \bar{x}_{ij} = \bar{x}_i - \bar{x}_j \quad \text{and } t_{ij} = t_i + t_j$$

With Definition 1, (15–17) can be written in matrix form, using a mixed correlation matrix:

$$\overbrace{}^{\text{coefficient vector}} \quad \overbrace{}^{\text{feature vector}}$$

$$
\overbrace{
\begin{pmatrix}
\mathbf{\Phi} & \mathbf{\Phi}_x & \mathbf{\Phi}_y & \mathbf{\Phi}_{xx} & \mathbf{\Phi}_{xy} & \mathbf{\Phi}_{yy} \\
-\mathbf{\Phi}_x & -\mathbf{\Phi}_{xx} & -\mathbf{\Phi}_{xy} & -\mathbf{\Phi}_{xxx} & -\mathbf{\Phi}_{xxy} & -\mathbf{\Phi}_{xyy} \\
-\mathbf{\Phi}_y & -\mathbf{\Phi}_{xy} & -\mathbf{\Phi}_{yy} & -\mathbf{\Phi}_{xxy} & -\mathbf{\Phi}_{xyy} & -\mathbf{\Phi}_{yyy} \\
\mathbf{\Phi}_{xx} & \mathbf{\Phi}_{xxx} & \mathbf{\Phi}_{xxy} & \mathbf{\Phi}_{xxxx} & \mathbf{\Phi}_{xxxy} & \mathbf{\Phi}_{xxyy} \\
\mathbf{\Phi}_{xy} & \mathbf{\Phi}_{xxy} & \mathbf{\Phi}_{xyy} & \mathbf{\Phi}_{xxxy} & \mathbf{\Phi}_{xxyy} & \mathbf{\Phi}_{xyyy} \\
\mathbf{\Phi}_{yy} & \mathbf{\Phi}_{xyy} & \mathbf{\Phi}_{yyy} & \mathbf{\Phi}_{xxyy} & \mathbf{\Phi}_{xyyy} & \mathbf{\Phi}_{yyyy}
\end{pmatrix}
}^{\text{Mixed correlation matrix M}}
\times
\begin{pmatrix}
a_1 \\ \dots \\ a_N \\ b_1^x \\ \dots \\ b_N^x \\ b_1^y \\ \dots \\ b_N^y \\ c_1^{xx} \\ \dots \\ c_N^{xx} \\ c_1^{xy} \\ \dots \\ c_N^{xy} \\ c_1^{yy} \\ \dots \\ c_N^{yy}
\end{pmatrix}
=
\begin{pmatrix}
L_1 \\ \dots \\ L_N \\ L_1^x \\ \dots \\ L_N^x \\ L_1^y \\ \dots \\ L_N^y \\ L_1^{xx} \\ \dots \\ L_N^{xx} \\ L_1^{xy} \\ \dots \\ L_N^{xy} \\ L_1^{yy} \\ \dots \\ L_N^{yy}
\end{pmatrix}
\quad (18)
$$

By solving (18), all necessary coefficients of (8) can be calculated to make the reconstruction $\hat{f}$. Note that the full system has $6N$ equations and $6N$ unknowns. If not all features of order $0 \le k \le 2$ are needed, just remove the corresponding row and column in the matrix M, as well as the corresponding entries in the coefficient vector on the l.h.s. and the feature vector on the r.h.s. For example: If only the second order features are used, remove row 1-3 and column 1-3 from M and the first $3N$ elements of the coefficient vector and feature vector. Extension to higher order is also straight forward.

Note that in case of reconstruction from critical points (spatial gradient zero) L_i^μ in the feature vector is zero for all i and μ. In case of reconstruction from top points, the second order derivative in the direction where the Hessian degenerates is also zero. We can linearize this such that (9) becomes:

$$\hat{f}(x,y) = \sum_{i=1}^{N} a_i \phi_i(x,y) + b_i^\mu \phi_{i,\mu}(x,y) + c_i \left(\xi_i^\mu \xi_i^\rho \phi_{i,\mu\rho}(x,y) \right) \qquad (19)$$

subject to the constraints:

$$\langle f - \hat{f} \,|\, \phi_i \rangle = 0, \quad \langle f - \hat{f} \,|\, \phi_{i,\mu} \rangle = 0 \quad \text{and} \quad \langle f - \hat{f} \,|\, \xi_i^\mu \xi_i^\rho \phi_{i,\mu\rho} \rangle = 0 \qquad (20)$$

with features:

$$\langle f \,|\, \phi_i \rangle = L_i, \quad \langle f \,|\, \phi_{i,\mu} \rangle = 0 \quad \text{and} \quad \langle f \,|\, \xi_i^\mu \xi_i^\rho \phi_{i,\mu\rho} \rangle = 0 \qquad (21)$$

for all $i = 1...N$ and for $\mu = x, y$ and $\rho = x, y$. Here is ξ_i a unit vector based at the singular point that indicates the singular direction (the direction in which the Hessian degenerates). Note that the ξ_i vectors have to be calculated in advance for every point i, after which the vectors ξ_i can be considered known constants.

These vectors can be seen as an extra "feature" of the image at a certain point i. According to [1] the vector ξ_i can be calculated using third order derivatives:

$$\begin{bmatrix} \xi_i^x \\ \xi_i^y \end{bmatrix} = \begin{bmatrix} (L_{i,xxy} + L_{i,yyy})L_{i,xy} - (L_{i,xxx} + L_{i,xyy})L_{i,yy} \\ (L_{i,xxx} + L_{i,xyy})L_{i,xy} - (L_{i,xxy} + L_{i,yyy})L_{i,xx} \end{bmatrix} \tag{22}$$

3   Experimental Results

3.1   Random Points

Using (18) we can make a reconstruction from a number of points in scale space. For our first experiment the points are selected with random spatial location according to a uniform distribution and with decreasing probability as scale increases, such that:

$$N(\tau) = N_0\, e^{-n\tau} \tag{23}$$

with N_0 the number of points for $\tau = 0$ and n the dimension. Here τ is used as the "natural" scale parameter, instead of t, with $t = \frac{1}{2}e^{2\tau}$. Figure 1 shows reconstructions of a part of the famous Lena image (64×64 pixels), using 400 random points at a fixed scale $\tau = 0.0$, with different sets of features. Note the "holes" in the image, especially at reconstructions from points with few features. Figure 2 shows reconstructions of the same image, this time only using L as a feature, again using 400 points, but varying the maximum scale τ_{max}. Here $\tau_{max} = 3$ corresponds with $\sigma \approx 40$ pixels. Note that the difference between the reconstructions with $\tau_{max} = 3$ and $\tau_{max} = 2$ is small. From Fig. 1 and Fig. 2 one can conclude that image information increases if the number of features increases and that if little information is available, higher scales have to be used to get a visually correct reconstruction. Increasing the maximum scale even more does not make the reconstruction visually any better.

Using the previous results, another image is reconstructed. This time, 800 points are used, again with all combinations of features. The maximum scale τ_{max} is chosen in such a way that there are no more "holes" in the reconstructed image. This is done by choosing a τ_{max} and visually check the reconstruction for holes. A high value for τ_{max} will reduce the risk of holes, so this is a reason to choose τ_{max} as high as possible. One reason to choose τ_{max} as low as possible is the fact that for a constant number of points, N_0 is maximal if τ_{max} is minimal, which is important for small details in the image. The optimal τ_{max} will thus be the lowest value with no holes present in the reconstruction. Figure 3 shows the result of reconstructions from random points, where for every feature set, the optimal τ_{max} is visually determined. The original image is one slice of a MR brain scan (128×128 pixels). Note that in the MR image, part of the skull is artificially removed, for 3D visualization (ray-tracing) of the brains under the skull.

It is surprising that there is no clear relation between the optimal τ_{max} and the features used for reconstruction. However, there might be a relation between the amount of information contained in the used features and the optimal τ_{max}.

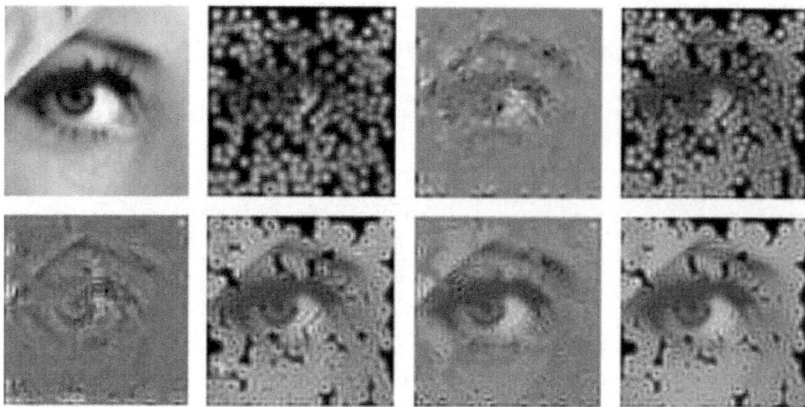

Fig. 1. Reconstructions of Lena's eye (64×64 pixels) from 400 random points at scale $\tau = 0.0$. From left to right, top to bottom: original image, reconstructions using $\{L\}$, $\{L_x, L_y\}$, $\{L, L_x, L_y\}$, $\{L_{xx}, L_{xy}, L_{yy}\}$, $\{L, L_{xx}, L_{xy}, L_{yy}\}$, $\{L_x, L_y, L_{xx}, L_{xy}, L_{yy}\}$ and $\{L, L_x, L_y, L_{xx}, L_{xy}, L_{yy}\}$ as features.

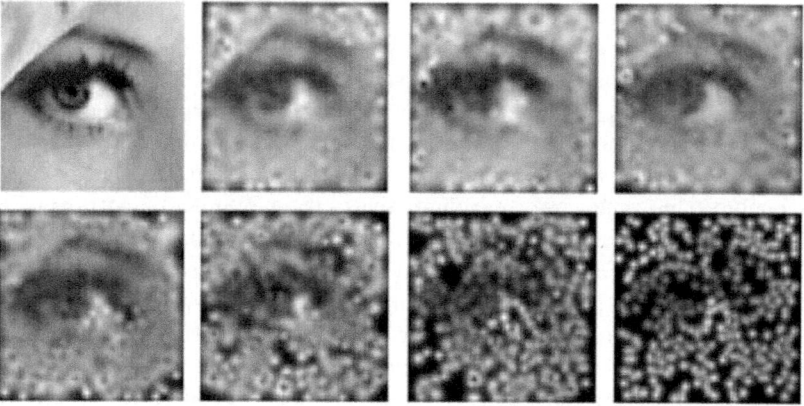

Fig. 2. Reconstructions of Lena's eye (64×64 pixels) from 400 random points using only L as feature. From left to right, top to bottom: original image, reconstructions using $\tau_{max} = 3.0$, $\tau_{max} = 2.5$, $\tau_{max} = 2.0$, $\tau_{max} = 1.5$, $\tau_{max} = 1.0$, $\tau_{max} = 0.5$, $\tau_{max} = 0.0$.

3.2   Critical Points

The random points of the previous section are usable for a reconstruction, but information can easily be lost, because of missing points in dense areas. According to Nielsen and Lillholm [10], there are points in scale space which contain more image information than others, for example edge points and blobs. They also examine which features are best suited for different points.

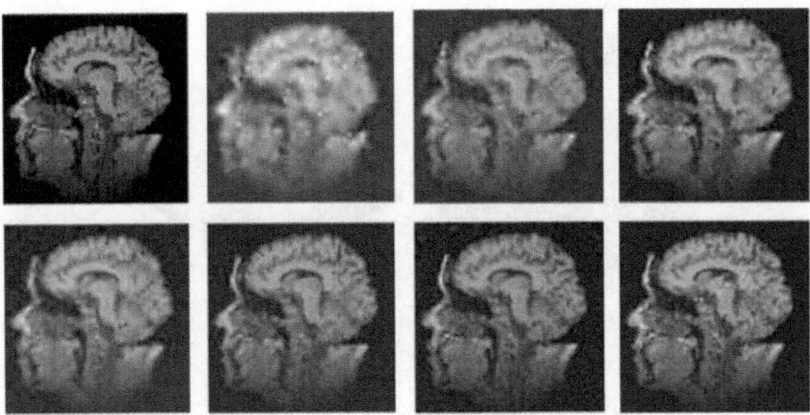

Fig. 3. Reconstructions of MR brain scan (128×128 pixels) from 800 random points. From left to right, top to bottom: original image, reconstructions using $\{L, \tau_{max} = 3.0, N_0 = 80\}$, $\{L_x, L_y, \tau_{max} = 3.0, N_0 = 80\}$, $\{L, L_x, L_y, \tau_{max} = 2.0, N_0 = 87\}$, $\{L_{xx}, L_{xy}, L_{yy}, \tau_{max} = 3.0, N_0 = 80\}$, $\{L, L_{xx}, L_{xy}, L_{yy}, \tau_{max} = 1.5, N_0 = 96\}$, $\{L_x, L_y, L_{xx}, L_{xy}, L_{yy}, \tau_{max} = 1.8, N_0 = 89\}$ and $\{L, L_x, L_y, L_{xx}, L_{xy}, L_{yy}, \tau_{max} = 1.1, N_0 = 110\}$.

In this section, an experiment[1] is done with reconstructions from multiscale top points as described by Florack et al. [2]. The approach is the same as the reconstructions by Nielsen and Lillholm [10], but with different points and features. First we define (spatial) critical points and top points:

Definition 2. *Spatial critical points are points where the spatial gradient is zero. For 2D images these points are maxima, minima or saddles.*

Definition 3. *Top points are critical points where the Hessian degenerates (det H=0). For generic 2D images, these points are annihilations or creations of saddles with maxima or minima.*

where the Hessian of a 2D image f is given by:

$$Hf = \triangledown \triangledown f = \begin{pmatrix} \partial_x^2 f & \partial_x \partial_y f \\ \partial_y \partial_x f & \partial_y^2 f \end{pmatrix} \qquad (24)$$

Given an image, top points can be found by tracking all maxima, minima and saddles in scale, and finding those points where pairs of saddles annihilate with maxima or minima and points where pairs of saddles and maxima or minima are created. Figure 4 shows a simple example of top points of an image. Note that according to Loog et al. [9], under certain weak conditions always one critical path from a maximum or minimum will be left at the highest scale. More about critical points and top points can be found in [1],[3],[5] and [8].

---

[1] Note that this is only a first try to reconstruct images from top points and much further research is needed to get a proper overview of the possibilities and limitations.

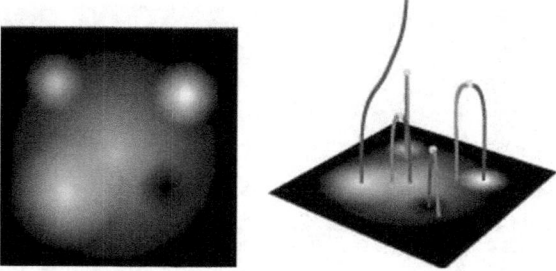

Fig. 4. Simple example top points. Left: original image, right: 3D view of critical paths and top points.

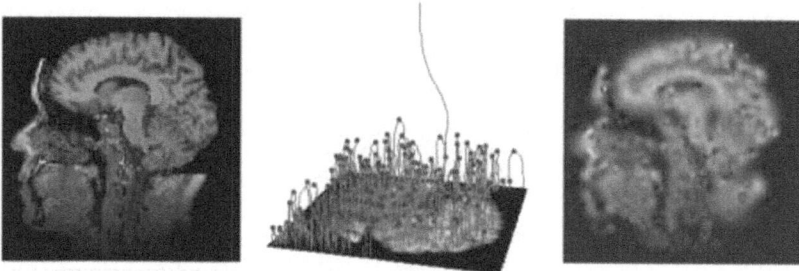

Fig. 5. Reconstruction from top points of mrbrain.tif. Left: original image, Center: 3D view of critical paths and top points, Right: Reconstruction from top points, using all features with order $k \leq 2$.

In our experiment, we use these top points for image reconstruction. Again we use the mrbrain.tif image. Figure 5 shows the original image, with the 3D view of the critical paths and top points. It also shows a reconstruction from all top points, using the full feature set with order $k \leq 2$. The number of top points for this image is 211. This seems to be insufficient, at least for the reconstruction order $k \leq 2$ actually used. Critical points are tracked in scale from $\tau = 0.0$ to $\tau = 4.0$. As can be seen in Fig. 5, the reconstruction does not contain much detail and resembles only very coarse the original. This is partly due to the fact that no top points can be found at scales $\tau \leq 0.0$ (which is equal to $t \leq \frac{1}{2}$), because we start tracking at that scale.

A better reconstruction can be made if all critical points at scale $\tau = 0$ are added to the top points. To prevent calculation problems (see also next chapter), points closer than a certain distance D_{opt} are deleted. The number of points increases to 779 if these points are added. In Fig. 6, the result of the reconstruction with top points combined with critical points at scale $\tau = 0$ is shown.

The reconstruction using top points and critical points at scale $\tau = 0$ looks reasonable, but compared to the reconstruction from random points shown in

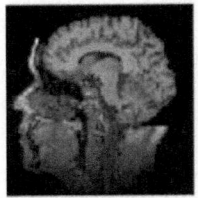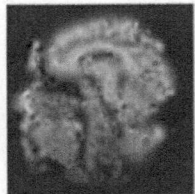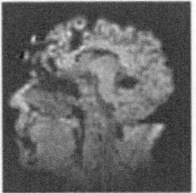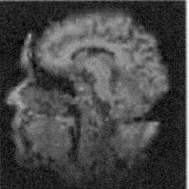

Fig. 6. Reconstruction from top points and critical points at scale $\tau = 0$ of mrbrain.tif. From left to right: original image, reconstructions from: top points, critical points at scale $\tau = 0$, top points and critical points at scale $\tau = 0$. All reconstructions use all features of order $k \leq 2$.

Fig. 3 it is not much better, considering the fact that in the latter case no image information is used to obtain the points. Note that for reconstruction from critical points, features L_x and L_y are known (gradient zero) and therefore the total amount of information needed for the reconstruction is reduced.

Probably better results can be achieved by using top points at lower scales, or by using more information from the critical paths, e.g. scale space saddles (points on the critical paths where the Laplacian is also zero).

4   Influence of Limited Calculation Precision on Reconstructions

As can be seen in Fig. 1 and Fig. 3, the visual quality of the reconstruction depends not only on the features used, but also on the interaction between the separate blobs. Therefore, the distance between blobs plays an important role in the quality of the reconstruction. In this section we use equidistant points for the reconstruction to easily measure the influence of the distance between blobs for the reconstruction quality. For measurement of the reconstruction quality the Root Mean Square error is often used. Although it is proven not to be a very good measurement for the visual quality, it is still good enough for our purpose of quantitative reconstruction. The RMS error is given by:

$$||f - \hat{f}||_{L^2} = \sqrt{\frac{1}{NM} \sum_{i=1}^{N} \sum_{j=1}^{M} \epsilon_{ij}^2} \tag{25}$$

with ϵ_{ij} the pixel-wise difference between the reconstruction and original and M and N the dimensions of the image.

The RMS error is caused by the lack of completeness of the feature set and by the error made due to the limited precision of the system. Especially the calculation of the inverse of the matrix in (18) can introduce errors due to limited machine precision. In our implementation, we used the Intel Math Kernel Library to calculate the inverse, which has the following properties:

Definition 4. *For a matrix* $\mathbf{A}$ *with dimensions* $n \times n$ *the norm* $\|\mathbf{A}\|_\infty$ *is defined by:*

$$\|\mathbf{A}\|_\infty \hat{=} \max_i \sum_{j=1}^n |a_{ij}|$$

Data perturbations. If $\bar{x}$ is the exact solution of $\mathbf{A}\bar{x} = \bar{b}$, and $\bar{x} + \delta\bar{x}$ is the exact solution of a perturbated problem $(\mathbf{A} + \delta\mathbf{A})\bar{x} = (\bar{b} + \delta\bar{b})$, then

$$\frac{\|\delta\bar{x}\|}{\|\bar{x}\|} \leq \kappa_\infty(\mathbf{A})(\frac{\|\delta\mathbf{A}\|}{\|\mathbf{A}\|} + \frac{\|\delta\bar{b}\|}{\|\bar{b}\|}), \text{ where } \kappa_\infty(\mathbf{A}) = \|\mathbf{A}\|_\infty \|\mathbf{A}^{-1}\|_\infty \qquad (26)$$

The amplification factor $\kappa_\infty(\mathbf{A})$ is called the condition number of matrix A. Note that the norm $\|.\|$ is the standard quadratic norm, while $\|\mathbf{A}\|_\infty$ is as in Def. 4.

Rounding errors. If ϵ is the *machine precision*, and $c(n)$ is a modest function of the matrix order n, then

$$\frac{\|\delta\bar{x}\|}{\|\bar{x}\|} \leq c(n)\,\kappa_\infty(\mathbf{A})\,\epsilon \qquad (27)$$

So if the condition number is very large, the error in the inverse due to rounding errors is also very large. In practice, $c(n) = \mathcal{O}(n)$.

Now let us consider a reconstruction from a number of points at a fixed scale τ, with $t = \frac{1}{2}e^{2\tau}$, which are equally distributed over the spatial domain. The points lie on a grid with distance D. Note that if D decreases, the number of points N increases. Figure 7 shows the RMS error and the condition number κ versus the distance D for different scales for the full feature set with order $k \leq 2$. The original image is a white rectangle in a black background. Some results of the reconstruction using different D can be seen in Fig. 8. Note that the optimal fit yields $D \approx 0.9\sigma$ with $\sigma = \sqrt{2t}$, in agreement with the expected linear scaling behavior.

The same measurement of the RMS error versus the distance D is done for the MR image of the brain, for some higher values of τ. Figure 9 shows the results, and some of the reconstructions. Note the odd result for $D = 4.8$, where the error is much higher than for $D = 4.4$, probably due to some special interaction between two pixels, such as very high derivative values in opposite directions, which causes the condition number to explode. More research about this problem is needed. Note that again the optimal $D \approx 0.9\sigma$ with $\sigma = \sqrt{2t}$, except for higher values of σ, where it is closer to $D \approx 0.8\sigma$.

The remarkable thing about the RMS graphs in Fig. 7 and Fig. 9 is the asymptotic behavior towards $D \rightarrow 0$. This indicates a numerical problem due to increasing mutual dependencies of features. This is in agreement with the fact that if less features are taken into account, for example if only L_{xx}, L_{xy} and L_{yy} or if only L is taken as a feature, the same results are found, but the curves are shifted to the left. Less features at a fixed distance D, means less pixel interaction, so the condition number will be lower. The conclusion is that points too close to each other will give problems due to machine limitations.

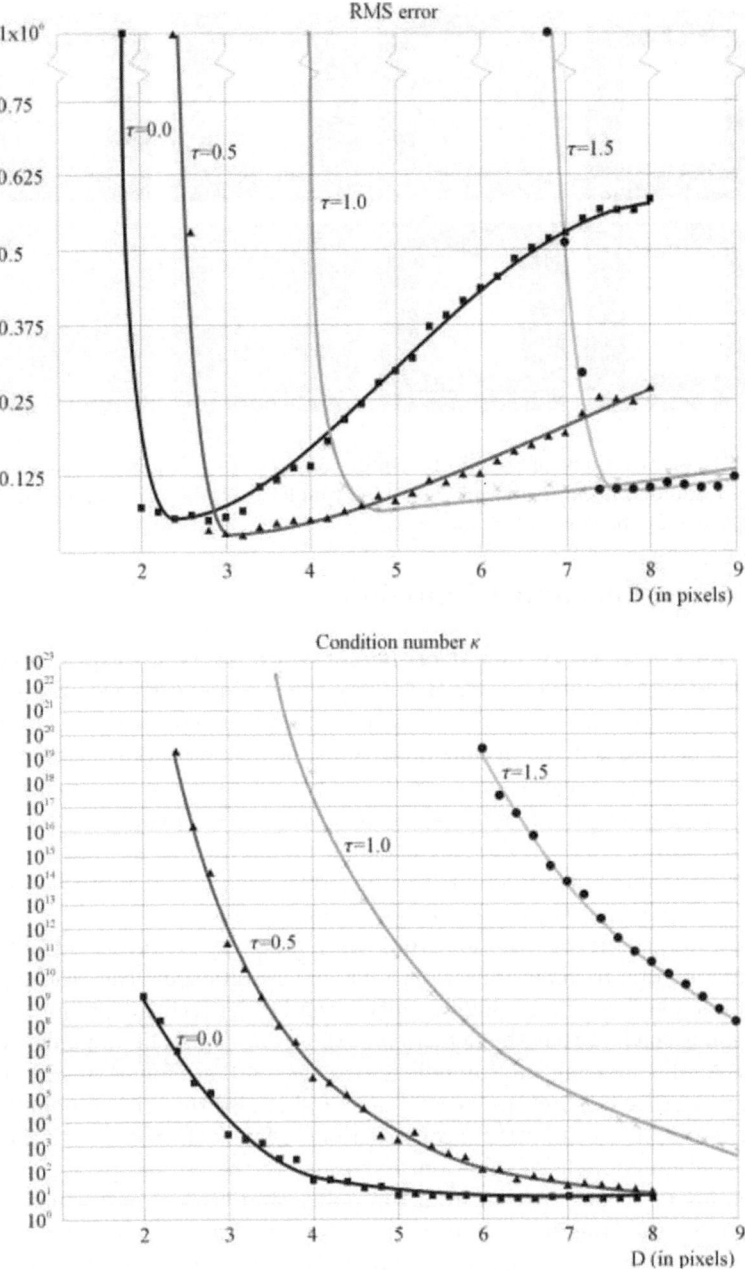

Fig. 7. Root Mean Square error and Condition number κ versus pixel distance D of image block.tif, using $L, L_x, L_y, L_{xx}, L_{xy}$ and L_{yy} as features. Scale parameter τ is used instead of t ($t = \frac{1}{2}e^{2\tau}$).

Fig. 8. Reconstructions of block.tif, using $L, L_x, L_y, L_{xx}, L_{xy}$ and L_{yy} as features. Points are equidistant at fixed scale $\tau = 0.5$. From left to right: original, reconstructions with distance D=6, 5, 4, 3, 2.8, 2.6, 2.4.

5   Conclusions and Discussion

In this paper, a reconstruction algorithm for multiscale points is presented, based on work by Nielsen and Lillholm [10]. Only the minimal variance prior described in [10] is used for simplicity. We introduce an explicit representation up to order 2 using a generalized co-variance matrix. This representation is used to make reconstructions of images using a number of points in scale space, such as random points, equidistant points and scale space top points (in contrast to [10] where feature points are used). The conditioning of the algorithm is analyzed using equidistant points with different features. With this paper, more insight about the reliability of the reconstruction algorithm is gained.

Regarding the reconstruction from top points, only a rather ad hoc experiment is done due to time limitation. Our first results show that the top points alone are not sufficient for a high quality second order reconstruction. One possible reason is the fact that the detection of those top points in an image is not yet perfect. At this moment, no top points at scale $\tau \leq 0$ are found. It also might be that higher order features are needed here. Using also the critical points at scale $\tau = 0$ resulted in much better reconstructions. Maybe the top points themselves do not contain enough information, but possibly the critical paths do. Further research about reconstruction from top points and critical paths must be done to determine the true image information contained in these points.

Reconstruction from equidistant points is useful for testing the possibilities and limitations of the algorithm. Problems due to limited machine precision can be pointed out clearly using equidistant points, which gives us an optimal distance between points. For the reconstruction, a large matrix has to be inverted, which can introduce errors if points are close together. This can be seen in the condition number of the matrix. A pseudo-inverse algorithm can be the solution, which has to be investigated. The experiments regarding equidistant points are

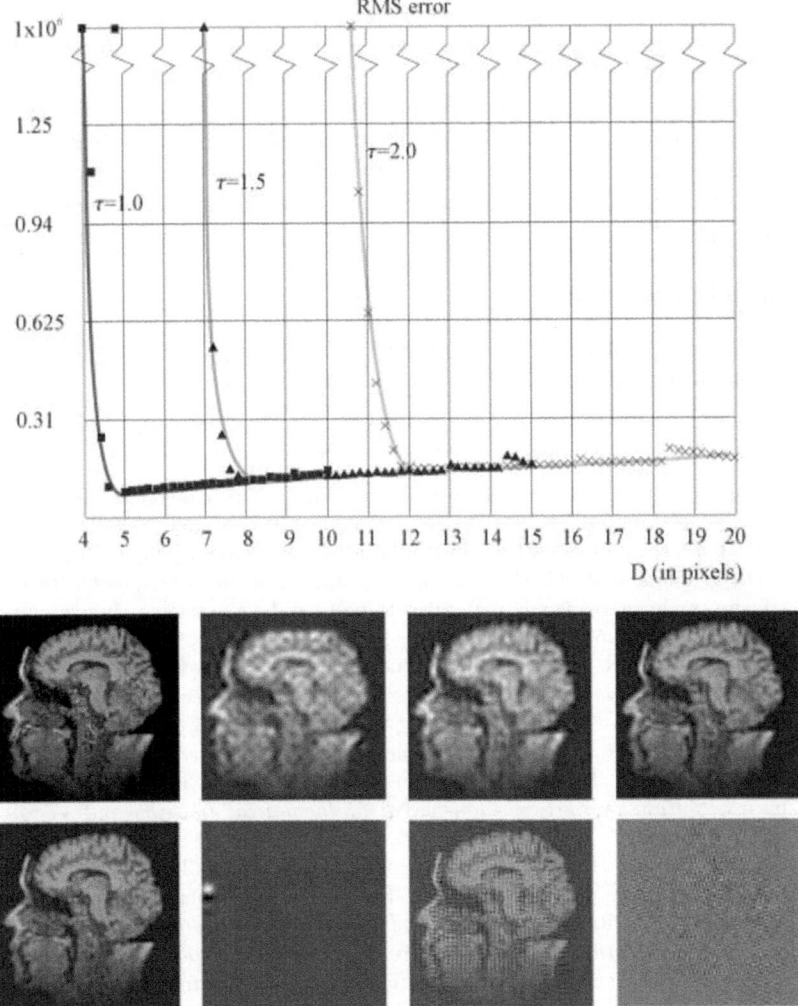

Fig. 9. Top: Root Mean Square error versus pixel distance D of image mrbrain.tif, using $L, L_x, L_y, L_{xx}, L_{xy}$ and L_{yy} as features. Scale parameter τ is used instead of t ($t = \frac{1}{2}e^{2\tau}$) Bottom: Reconstructions of mrbrain.tif, using $L, L_x, L_y, L_{xx}, L_{xy}$ and L_{yy} as features. Points are equidistant at fixed scale $\tau = 1.0$. From left to right: original, reconstructions with distance D=10, 8, 6, 5, 4.8, 4.4, 4.0.

only at one scale. Points at different scales, which are close to each other, sometimes lead to a high condition number, but sometimes they do not. The reason why is still an open question, which also has to be investigated.

Nielsen and Lillholm used in [10] an extra constraint to retain total image energy. The effect of this extra constraint must be explored. Nielsen and Lillholm also describe reconstructions using other priors, which should be investigated using top points. The reconstruction as described in this paper uses only features of order $k \leq 2$. The effect of using higher order derivatives as features is yet unknown (but it is unlikely that orders higher than 4 are of much influence, since the correlation between Gaussian derivatives of higher orders is high). The possibility to use different orders of features at different scales has to be implemented and examined, at this moment only one set of features for all points can be used. A coarse to fine approach of the reconstruction algorithm should also be explored. If it exists it may solve memory problems as well as ill-conditioning.

Acknowledgements

This work is part of the DSSCV project supported by the IST Programme of the European Union (IST-2001-35443).

References

1. L. Florack and A. Kuijper. The topological structure of scale-space images. *Journal of Mathematical Imaging and Vision*, 12(1):65–79, February 2000.
2. L. M. J. Florack. Reconstruction from scale space critical points. Internal report.
3. L. D. Griffin and A. C. F. Colchester. Superficial and deep structure in linear diffusion scale space: Isophotes, critical points and separatrices. *Image and Vision Computing*, 13(7):543–557, September 1995.
4. R. Hummel and R. Moniot. Reconstruction from zero crossings in scale space. In *IEEE Transactions on Acoustics, Speech, and Signal Processing*, volume 37, pages 2111–2130, 1989.
5. S. N. Kalitzin, B. M. ter Haar Romeny, A. H. Salden, P. F. M. Nacken, and M. A. Viergever. Topological numbers and singularities in scalar images: Scale-space evolution properties. *Journal of Mathematical Imaging and Vision*, 9(3), November 1998.
6. F.M.W. Kanters, B. Platel, L.M.J. Florack and B.M. ter Haar Romeny. Content based image retrieval using multiscale top points. Elsewhere in these proceedings.
7. M. Kerckhove, editor. *Scale-Space and Morphology in Computer Vision: Proceedings of the Third International Conference, Scale-Space 2001, Vancouver, Canada*, volume 2106 of *Lecture Notes in Computer Science*. Springer-Verlag, Berlin, July 2001.
8. A. Kuijper and L.M.J. Florack. The application of catastrophe theory to image analysis. Submitted to Image and Vision Computing.
9. M. Loog, J. J. Duistermaat, and L. M. J. Florack. On the behavior of spatial critical points under Gaussian blurring. a folklore theorem and scale-space constraints. In Kerckhove [7], pages 183–192.
10. M. Nielsen and M. Lillholm. What do features tell about images? In Kerckhove [7], pages 39–50.
11. J.L. Sanz. Multidimensional signal representation by zero crossings: an algebraic study. *Society for Industrial and Applied Mathematics, Philadelphia, PA, USA*, 49:281–295, 1989.
12. J.L.C. Sanz and T.T. Huang. Image representation by sign information. *IEEE Transactions on Pattern Analysis and Machine Intelligence*, 1(7):729–738, July 1989.
13. M. Shmouely and Y.Y. Zeevi. Image representation by level crossings of the wavelet transform. *IEEE Transactions on Image Processing*, 2:565–568, 1996.

Texture Classification through Multiscale Orientation Histogram Analysis

Miguel Alemán-Flores and Luis Álvarez-León

Departamento de Informática y Sistemas
Universidad de Las Palmas de Gran Canaria
35017 Las Palmas, Spain
{maleman,lalvarez}@dis.ulpgc.es

Abstract. This work presents a new approach to texture classification, in which orientation histograms and multiscale analysis have been combined to achieve a reliable method. From the outputs of a set of filters, the orientation and magnitude of the gradient in every point of a texture are estimated. By combining the orientations and relative magnitudes of the gradient, we build an orientation histogram for each texture. We have used Fourier analysis to measure the similarity between the histograms of different textures, considering the effects of a change in the size or orientation of the image to make our method invariant under these phenomena. Since different textures may generate very similar histograms, we have analyzed the evolution of these histograms at different scales, extracting a scale factor for each couple of compared textures to adjust the filters which are applied to them when the multiscale analysis is carried out.

1 Introduction

The visual identification of an object is not only provided by its shape. The texture in the inner region may be helpful to a large extent when we try to characterize materials, components, agglomerations, etc. Sonka et al. [1] define a texture as something consisting of mutually related elements. A texture consists of texture primitives or texture elements, sometimes called *texels* and, due to its wide variability, it is not simple to give a precise definition. An important problem when dealing with textures is that texture description is scale dependent. We may deal with the problem of texture classification from many different points of view, but we must take into account that the scale of the textures we are comparing is a crucial factor when measuring their similarity. The distribution of orientations has been previuosly used in [2] for the discrimination between city and suburb photos according to the presence or absence of dominant orientations. Other works have shown the results of different filters used in texture classification when applied to a certain texture benchmark set [3] and the evaluation of dissimilarity measures for color and texture [4]. A complete analysis on texture-related problems and applications, considering aspects like texture classification, segmentation or synthesis, is presented in [5].

L.D. Griffin and M. Lillholm (Eds.): Scale-Space 2003, LNCS 2695, pp. 479–493, 2003.
© Springer-Verlag Berlin Heidelberg 2003

Table 1. Modified Newton filters and corresponding orientation

$F_0 : 0$			$F_1 : \pi/4$			$F_2 : \pi/2$			$F_3 : 3\pi/4$		
1	1	−2	1	−2	−4	−2	−4	−2	−4	−2	1
2	2	−4	1	2	−2	1	2	1	−2	2	1
1	1	−2	2	1	1	1	2	1	1	1	2

$F_4 : \pi$			$F_5 : 5\pi/4$			$F_6 : 3\pi/2$			$F_7 : 7\pi/4$		
−2	1	1	1	1	2	1	2	1	2	1	1
−4	2	2	−2	2	1	1	2	1	1	2	−2
−2	1	1	−4	−2	1	−2	−4	−2	1	−2	−4

The purpose of this work is presenting an approach to texture classification based on the description given by the estimation of the orientation in the points within a textured region. Therefore, we must first extract a value for the orientation and magnitude of the gradient in every point within the region we are examining. From these estimations, we build an orientation histogram for each texture, which represents the distribution of the orientations.

Since an orientation histogram does not univocally characterize a texture, the same representation could have been extracted from different patterns. However, the study of texture histograms at many different scales allows comparing the evolution and interaction of the gradients, so that textures which are initially considered as very similar start to evolve in a quite different manner.

The paper has been structured as follows: Section 2 shows how the orientation of the edges can be estimated from the outputs of a set of filters. In Sect. 3, these estimations are used to build orientation histograms which describe the textures in terms of quantitative edge orientation distribution and the representation of the textures through these orientation histograms allows classifying them. The multiscale analysis of the textures is introduced in Sect. 4 to generate better texture classifications by comparing them at different scales. Section 5 shows how darkening, lightening and inverting the images affect the classification, which proves robust under these transformations. Finally, Sect. 6 shows some conclusions about this work.

2 Edge Orientation Estimation

From Newton filters [6][7], we have developed a set of filters which preserve their convenient properties, but which also avoid some of the undesirable phenomena by providing them with rotational invariance and non-null weights in all positions. The weights of the eight filters and the orientations they react to are shown in Table 1.

The output of these filters is independent of the particular gray value of the image border, i.e. F_k is invariant under a gray level translation as $I \rightarrow I + C$, where I contains the gray values of the image and C is any constant. This property is very important because the relevant information is provided by

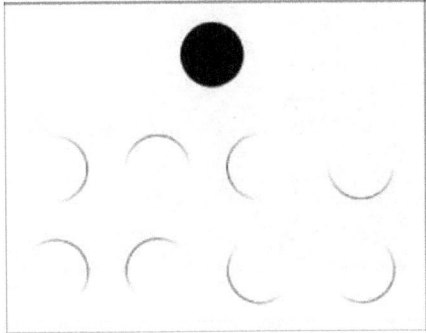

Fig. 1. Positive outputs of the modified Newton filters for the circle above (the higher the output value, the darker its representation)

the difference between neighbors, rather than the magnitude of the image gray values.

Instead of considering the maximum of the eight filters to assign a value to the orientation, a more accurate estimation can be obtained by considering the whole pattern provided by the eight filters. The output of these eight filters for a $\pi/2$ oriented edge of magnitude 1 is $(0, 5, 8, 5, 0, -4, -4, -4)$. If we increase or decrease the orientation in a multiple of $\pi/4$, the output is only cyclically shifted, but the values and their order are not altered. Figure 1 shows the outputs of the eight filters for a circle. When the real orientation does not correspond to one of these directions, we can estimate it by interpolating the higher value with its two neighbors, which provides an accurate estimation of edge orientation. A quadratic function y is built to interpolate these three values and its maximum is used as the estimated orientation, as shown in (1) and (2), where i is the index of the filter with the highest output, F_i, positions $i-1$ and $i+1$ are calculated modulo 8 and $x_{\max}$ is the estimation of the orientation for the current point:

$$y = \frac{8F_{i+1} - 16F_i + 8F_{i-1}}{\pi^2}\left(x - \frac{\pi(i-1)}{4}\right)^2 \tag{1}$$

$$+\frac{8F_i - 2F_{i+1} - 6F_{i-1})}{\pi}\left(x - \frac{\pi(i-1)}{4}\right) + F_{i-1} \; .$$

$$x_{\max} = \left(\frac{4F_i - 3F_{i-1} - F_{i+1}}{2\left[2F_i - F_{i-1} - F_{i+1}\right]} + i - 1\right)\frac{\pi}{4} \; . \tag{2}$$

By correlating the ideal pattern with the real one, we can determine how perfect the border is. As the change from one side of the border to the other one may be larger than 1, it is necessary to normalize the output. Thus, the main advantage of this kind of filters is not the location of edges, but their classification according to their orientation and the invariance under rotations and illumination changes.

Similar filters have been proposed by Prewitt, Sobel, Robinson or Kirsch [1], but they cause the duplication of edges, are independent of the central value or

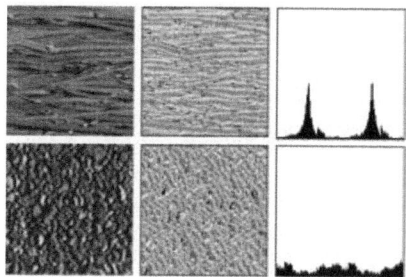

Fig. 2. Texture examples and corresponding orientation maps and orientation histograms

may produce the maximum output for imperfect edges. Furthermore, they do not constitute a set of linearly independent filters, as modified Newton filters do, and the information they provide is not complete.

3   Orientation-Based Texture Classification

With the modified Newton filters we are able to estimate accurately the orientation of the gradient in a certain pixel. Similarly, the values for the eight main orientations can be used to estimate the magnitude of this gradient. The interpolation of the outputs of the filters will provide us with a value for the direction of the gradient in every point as well as an estimation of its magnitude. With these estimations, we build a histogram of the orientations by adding the magnitude of the gradient in the points where the edges present the same orientation. This histogram describes how the orientations are distributed in that region, allowing us to determine the most significant ones, the proportion they represent and their relation in terms of orientation distance and concentration. Figure 2 shows an example of two textures and their corresponding orientation histograms, which can be used to compare these textures with others.

Using the interpolation polynomial in (1), the maximum value $o_{j,k}$, which estimates the orientation in point (j, k), is given by:

$$o_{j,k} = round\left(\frac{\left(\frac{4F_i - 3F_{i-1} - F_{i+1}}{2[2F_i - F_{i-1} - F_{i+1}]}\frac{\pi}{4} + \frac{\pi(i-1)}{4}\right)L}{2\pi}\right). \tag{3}$$

It has been rounded to adjust it to a discrete signal consisting of L equidistant values. For the magnitude, we use this orientation and substitute it in the polynomial:

$$(\nabla I)_{j,k} = \frac{8F_{i+1} - 16F_i + 8F_{i-1})}{\pi^2}\left(o_{j,k} - \frac{\pi(i-1)}{4}\right)^2 \tag{4}$$
$$+ \frac{8F_i - 2F_{i+1} - 6F_{i-1}}{\pi}\left(o_{j,k} - \frac{\pi(i-1)}{4}\right) + F_{i-1}.$$

The values of the histogram h_i are given by the following expression, where $(\nabla I)_{j,k}$ and $o_{j,k}$ are the magnitude and the orientation extracted for point (j,k). For normalization purposes, the global weight of all positions in the histogram is set to 1, thus dividing each resulting component of the histogram by the sum of all of them:

$$h_i = \sum_{\substack{j,k \\ o_{j,k}=i}} (\nabla I)_{j,k} \qquad \text{and} \qquad h'_i = \frac{h_i}{\sum_{j=0}^{L-1} h_j} \ . \tag{5}$$

In order to relate two textures, an energy function is built, in which the Fourier coefficients of both histograms are compared. We must achieve rotational invariance, in the sense that the result must not be affected if the textures are rotated. A change in the orientation of a texture will only cause a cyclical shift in the histogram. For this reason, the Fourier coefficients are modified as follows: let f_n and g_n be the orientation histograms of length L corresponding to the same texture but shifted a positions, i.e. the texture has been rotated an angle $\theta = 2\pi a/L$, and let f_k and g_k be the kth Fourier coefficients of these histograms, then $f_k = g_k e^{-i\frac{2\pi ka}{L}}$. Thus, a measure of how similar the coefficients of both textures are is given by:

$$E(a) = \sum_{k=1}^{\frac{L}{2}} \left(f_k - g_k e^{-i\frac{2\pi ka}{L}} \right) \left(f_k - g_k e^{-i\frac{2\pi ka}{L}} \right)^* \ . \tag{6}$$

In addition, the fact that the number of discrete orientations used for the histograms is constant and the normalization of the weights make the lengths of the signals and the total weight equal in both textures. Consequently, a change in the size of the region where the texture is analyzed will not cause the generation of a different distribution. Due to the fact that the higher frequencies are more affected by noise than the lower ones, a monotonic decreasing weighting function $w(.)$ can be used to emphasize the discrimination, thus obtaining the following expression, in which the first terms have a more important contribution than the last ones. The minimization of this function will provide the shift for which both histograms present the best matching, i.e. the rotation which makes both textures as similar as possible. The energy for that value is a measure of how similar they are:

$$E(a) = \sum_{k=1}^{\frac{L}{2}} w\left(\frac{2k}{L}\right) \left(f_k - g_k e^{-i\frac{2\pi ka}{L}} \right) \left(f_k - g_k e^{-i\frac{2\pi ka}{L}} \right)^* \ . \tag{7}$$

We have used different linear, quadratic and exponential weighting functions and the best results were obtained when $w(x) = e^{-10x}$. To test this technique, we have applied it to a set of textures contained in a database, which is shown in Fig. 3. This database has been made publicly available for research purposes by Columbia and Utrecht Universities, Columbia-Utrecht Reflectance and Texture Database [8]. We work with grayscale images and thus, a single histogram is used to represent the orientations of the edges in light intensity. Using the techniques

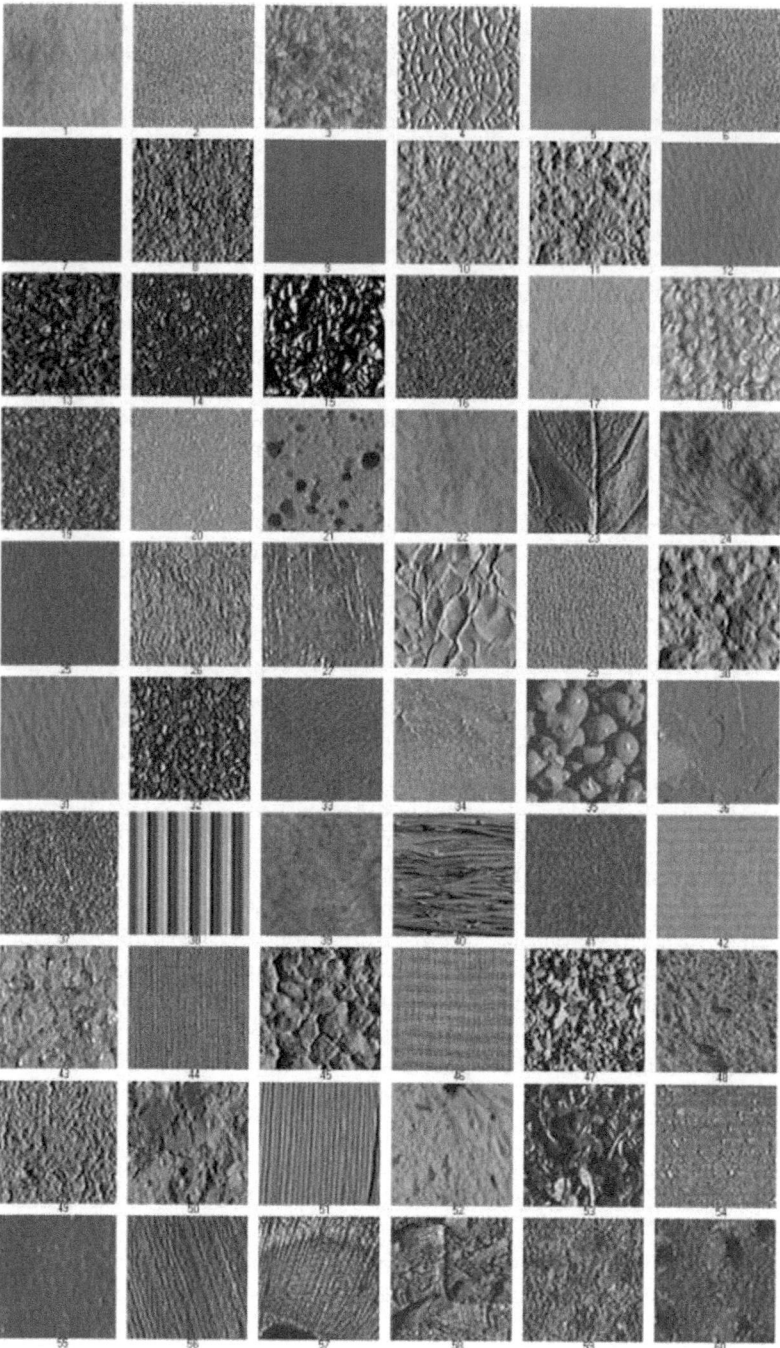

Fig. 3. Database textures

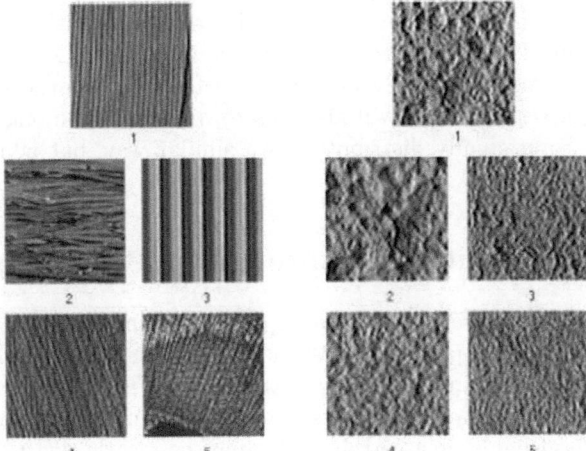

Fig. 4. Results of searching for similar textures for textures 51 and 11 using (7)

Table 2. Lowest energy values for textures 51 and 11

order	txt. number	wtd. energy	order	txt. number	wtd. energy
1	51	0.00	1	11	0.00
2	40	16.03	2	30	0.59
3	38	49.21	3	49	0.61
4	56	118.39	4	10	1.20
5	57	157.10	5	26	1.62

described in the previous sections, a certain texture is compared with all those in the database and the most similar ones are selected. The similarity between two textures is given by the energy obtained when comparing their orientation histograms.

In Fig. 4 and Table 2, we show some results of the application of the technique explained above. From the image database containing 60 textures of different natures, but visually difficult to classify, one is selected, and the 5 best comparisons are shown. Of course, as the selected image belongs to the set, the best match corresponds to itself, and the energy factor is 0.

As mentioned above, the orientation histograms extracted from the textures describe how the different orientations are quantitatively distributed across the region which is studied, but they do not provide any information about the spatial neighborhood of the pixels with a certain orientation. Thus, a completely noisy image, in which all orientations are equally but disorderly present in the image would generate a similar histogram than a circle, where the orientation is gradually increased along its outline. This forces us to search for a certain technique which complements the information provided by this kind of histograms in order to enhance the recognition capabilities.

A multiscale analysis of the images will provide us with a series of images which represent the evolution of each texture at different scales. In this evolution, the orientations will be differently affected by the others, depending on their spatial proximity. This will allow us to distinguish among textures where orientations are originally distributed in a similar way, but which are actually different.

4   Texture Classification through Multiscale Analysis

The interpretation of the information we perceive from the environment depends on the scale we use to process it. At the same time, the information provided by each scale is useful and the study of the same scene at different scales makes it possible to perceive a wider range of realities. Furthermore, elements which are not distinguishable at a certain scale may be clearly distinct at a different one and the rough and detailed information extracted from an image may help us decide when comparing textures. The multiscale analysis approach has been successfully used in the literature for texture enhancement and segmentation (see [9] and [10] for more details).

A multiscale analysis can be determined by a set of transformations $\{T_t\}_{t \geq 0}$, where t represents the scale. Let I be an image, i.e. $I : \Omega \longrightarrow \Re$, where Ω is the domain where the image is defined. In what follows, we will consider for simplicity in the exposition that $\Omega = \Re^n$ and $I \in H^2(\Omega)$ (Sobolev space, see [11] for more details). That is, I and ∇I have finite L^2 norm. $I_t = T_t(I)$ is a new image which corresponds to I at a scale t. For a given image I, to which the multiscale analysis is applied, we can extract a histogram $\{h_i^t\}_{i=0,..,L-1}$ which determines the distribution of the orientations of I at the scale t. In this case, the normalization of the values within a histogram is performed with respect to the initial addition, and not with respect to the addition at that scale. In order to compare the histograms of two images, the scale must be first adjusted.

4.1   Gaussian Multiscale Analysis

As said before, a multiscale analysis generates, for a given image, a series of images which show the evolution of the input signal when a certain process is applied. We will use a Gaussian filter, whose properties are described in [12] and [13]. In one dimension, we use the following Gaussian kernel, where the scale t is related to the standard deviation σ according to the expression $2t = \sigma^2$:

$$K_t(x) = \frac{1}{\sqrt{4\pi t}} e^{-\frac{x^2}{4t}} \implies T_t(f)(x) = \int_{\Re} \frac{1}{\sqrt{4\pi t}} e^{-\frac{(x-y)^2}{4t}} f(y) dy \ . \tag{8}$$

Afterwards, we quantize it as follows:

$$(K_t)_n = \frac{1}{\sqrt{4\pi t}} e^{-\frac{n^2}{4t}} \implies (x * K_t)_m = \sum_{n=-\infty}^{\infty} x_n \frac{1}{\sqrt{4\pi t}} e^{-\frac{(m-n)^2}{4t}} \ . \tag{9}$$

At this point, it is important to consider the relationship between the Gaussian filtering and the heat equation, given by $\partial u/\partial t = \partial^2 u/\partial x^2$, where $u(t,x)$ is the solution of the equation. Given a signal f, the result of convolving f with the Gaussian filter K_t is equivalent to the solution of the heat equation using f as the initial data $u(t, s) = K_t * f(x)$.

Considering this relationship, a discrete version of the heat equation can be used to accelerate the approximation of the Gaussian filtering (see [14] for more details), which results in a recursive scheme in three steps for each direction, as shown below, where I_0 is the original image:

$$
\begin{aligned}
I_j^{n+\frac{1}{3}} &= I_j^n + v I_{j-1}^{n+\frac{1}{3}} & \forall j \in Z \ . \\
I_j^{n+\frac{2}{3}} &= I_j^{n+\frac{1}{3}} + v I_{j+1}^{n+\frac{2}{3}} & \forall j \in Z \ . \\
I_j^{n+1} &= \tfrac{v}{\lambda} I_j^{n+\frac{2}{3}} & \forall j \in Z \ .
\end{aligned}
\tag{10}
$$

This process will be performed by rows and by columns in order to obtain a discrete expression for a two-dimensional Gaussian filtering. Making use of the features of the Gaussian kernels, the result of applying a Gaussian filter with an initial scale t can be used to obtain a Gaussian filtering of the initial image for a different scale without needing to start again from the input. We will discretize the scale considering $\sigma_n = n\sigma_0$ for a given σ_0. Taking into account the relation $\sigma^2 = 2t$, the step size Δt to go from σ_n to σ_{n+1} is given by:

$$
\Delta t = \frac{((n+1)\sigma_0)^2}{2} - \frac{(n\sigma_0)^2}{2} = \left(n + \frac{1}{2} \right) (\sigma_0)^2 \ .
\tag{11}
$$

If we use $niter$ iterations of the recursive scheme in (10) to compute $I_{\sigma_{n+1}}$ from I_{σ_n}, the discretization scheme for the heat equation is given by:

$$
\delta_t^{n+1} = \frac{\left(n + \frac{1}{2} \right) (\sigma_0)^2}{niter} \ .
\tag{12}
$$

4.2   Scale Estimation

We must take into account that, for a certain texture, the use of different resolutions forces us to apply Gaussian functions with different standard deviations, thus requiring an adaptation stage. To do that, we first extract the evolution of the addition of the squares of the gradients at different scales, and then we use these factors to compare the textures. Even if the quantitative distribution of the orientations may be alike for different textures, the spatial distribution will cause a divergence in the evolution and interaction, so the factors will differ.

One of the properties of the Gaussian filtering is the relationship between the resolution of two images and the effects of this kind of filters. In fact, the result of applying a Gaussian filter with standard deviation σ to an image with resolution factor x is equivalent to applying a Gaussian filter with standard deviation $k\sigma$ to the same image acquired with a resolution factor kx.

Lemma 1. *Let $I_0(x, y)$, $I_0'(x, y)$ be such that there exists a constant k satisfying that $I_0'(x, y) = I_0(kx, ky) \ \forall (x, y) \in \Omega$, then $I_t'(x, y) = I_{k^2 t}(kx, ky)$.*

Proof. The result follows from the uniqueness of the solution of the heat equation taking into account that the function $I_{k^2t}(kx, ky)$ is a solution of the heat equation for the initial datum $I'_0(x, y)$.

Given two textures, I_0 and I'_0, we will estimate the scale factor k using the normalized evolution of the norm of the gradient, that is, we will use:

$$\phi(I_0, \Omega, t) = \frac{\sqrt{\int_\Omega |\nabla I_t|^2}}{\sqrt{\int_\Omega |\nabla I_0|^2}} \; . \tag{13}$$

It is well known (see for instance [12]) that $\phi(I_0, \Omega, t)$ is a decreasing function with respect to t and $Lim_{t\to\infty}\phi(I_0, \Omega, t) = 0$. On the other hand, from the previous lemma, we deduce that if $I'_0(x, y) = I_0(kx, ky) \; \forall (x, y) \in \Omega$ then:

$$\phi(I_0, \Omega, t) = \phi(I'_0, k\Omega, k^2t) = \phi(I'_0, \Omega, k^2t) \; . \tag{14}$$

So in order to estimate a scale factor k between two textures I_0 and I'_0, we will compare the functions $\phi(I_0, \Omega, t)$ and $\phi(I'_0, \Omega, t)$. Let $r_n^1 = \phi(I_0, \Omega, (\sigma_n)^2/2)$ and $r_n^2 = \phi(I'_0, \Omega, (\sigma_n)^2/2)$ be the ratios obtained for two textures at the scale $\sigma_n = n\sigma_0$, the best adjusting coefficient k to fit the series of r_n^2 to that of r_n^1, both consisting of N terms, can be obtained as follows: First, we fit a value $0 < h < 1$ and we interpolate the values in the series r_n^1 and r_n^2 to obtain two new series σ_n^1 and σ_n^2 which estimate the scales for which the ratios $(1, 1 - h, 1 - 2h, 1 - 3h, ..., 1 - (N - 1)h)$ are obtained. In other words, we estimate the scale where $\phi(I, \Omega, (\sigma_n^1)^2/2) = 1 - nh$. We point out that if $nh < 1$, σ_n^1 and σ_n^2 are well-defined because $\phi(I, \Omega, t)$ is a decreasing function with respect to t and $Lim_{t\to\infty}\phi(I_0, \Omega, t) = 0$. With these values, we minimize the following error to obtain the scale factor k:

$$e(k) = \frac{1}{N} \sum_{i=0}^{N-1} \left(\sigma_i^1 - k\sigma_i^2\right)^2 \; . \tag{15}$$

$$\frac{de(k)}{dk} = 0 \implies \sum_{i=0}^{N-1} \left(\sigma_i^1 - k\sigma_i^2\right)\sigma_i^2 = 0 \implies k = \frac{\sum_{i=0}^{N-1}\left(\sigma_i^1\sigma_i^2\right)}{\sum_{i=0}^{N-1}\left(\sigma_i^2\right)^2} \; . \tag{16}$$

4.3   Multiscale Texture Orientation Histogram Comparison

We can study how the energy obtained when comparing the orientation histograms evolves as we apply a Gaussian filtering to the textures. We use the adjusting factor k, as in (16), to relate the scales to be compared. In practice, to estimate k, we take $h = 0.1$ and $N = 8$, $n = 0, 1, 2, ...7$. Finally, we obtain the energies for the comparison of the histograms at N different scales, given by (17), where $n = \{0, .., N - 1\}$ and σ_N is the minimum of σ_N^1 and $k\sigma_N^2$.

$$\sigma_n = \tfrac{n}{N}\sigma_N \qquad \text{and} \qquad \sigma'_n = \tfrac{nk}{N}\sigma_N \; . \tag{17}$$

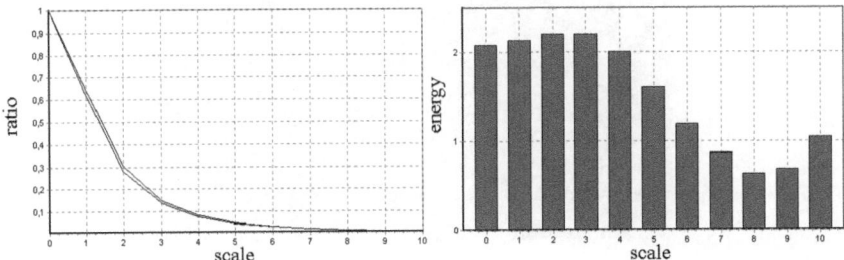

Fig. 5. Ratios (left) and energies (right) obtained when applying a Gaussian filter to textures 32 and 14 using (13) and (7) (horizontal axis represents the evolution of the scale)

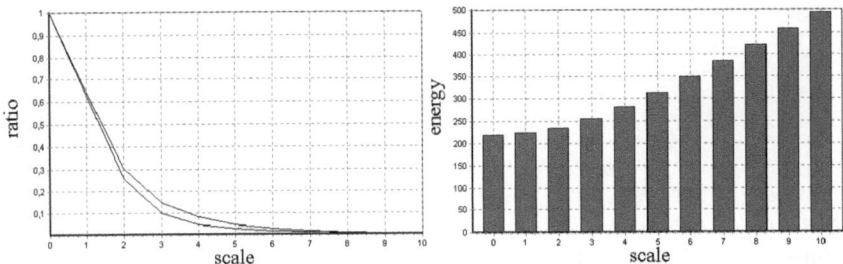

Fig. 6. Ratios (left) and energies (right) obtained when applying a Gaussian filter to textures 32 and 51 using (13) and (7) (horizontal axis represents the evolution of the scale)

Figure 5 shows the results of comparing two images of the database corresponding to the same texture, acquired with a different resolution. As observed, not only the initial energy is low, but also the subsequent energies, obtained when comparing the images at the corresponding scales, decrease as we increase the scale. On the other hand, Fig. 6 shows the comparison of two images of different textures and the energies, far from decreasing, increase from the initial value. Finally, Fig. 7 shows two images of the same texture acquired at different distances, and Fig. 8 shows the corresponding ratios and energies.

5   Robustness of Texture Classification under Darkening, Lightening and Inversion

The following examples show how darkening, lightening and inverting a pattern affect the results when calculating the energy which measures the similarity between two textures. This will allow us to test the robustness of our method when some kinds of transformations are performed in the input signal.

Fig. 7. Example of two images of the same texture acquired at different distances

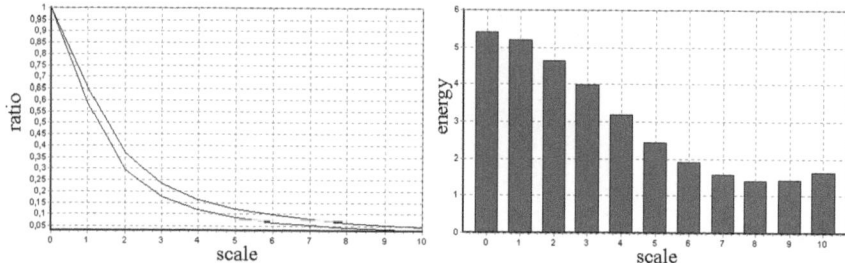

Fig. 8. Ratios (left) and energies (right) obtained when applying a Gaussian filter to textures in Fig. 7 using (13) and (7) (horizontal axis represents the evolution of the scale)

As observed in Fig. 9 and Table 3 (left), when a texture is darkened, the resulting energy is very low. These low results indicate that the textures are in fact almost identical. Only small differences in the new light intensity due to the representation limitations cause a negligible value. The use of integer values in intensity representation forces us to round the values once they have been reduced, generating small differences in gradient values.

Figure 10 and Table 3 (middle) show similar consequences when the textures are lightened. In this case, the overflow in light intensity values for the most bright points due to the increase they undergo, which forces us to truncate the values which exceed the maximum, causes a higher difference. However, it is still much lower than those observed when similar but different textures are compared, and thus, they can be neglected.

Finally, Fig. 11 and Table 3 (right) show the results when a texture has been inverted. In most cases, the values obtained are very low and the patterns can be considered as the same texture. Nevertheless, some cases present certain problems due to the asymmetry of the filters used for edge orientation estimation. As the values which result are not low enough to clarify the similarity of the textures, thus presenting a certain ambiguity, the multiscale analysis described in the previous sections is applied and the results dispel the doubts, since they are very low when a texture is compared with its inverted version.

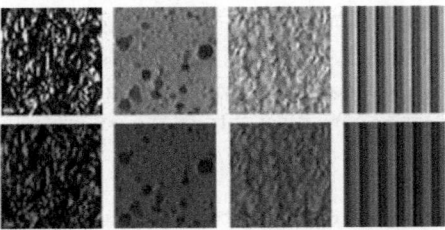

Fig. 9. Textures 15, 18, 21 and 38 before and after darkening

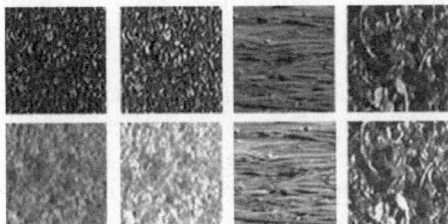

Fig. 10. Textures 3, 14, 40 and 53 before and after lightening

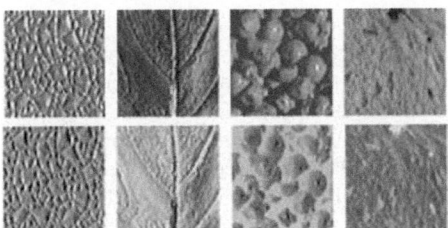

Fig. 11. Textures 4, 23, 35 and 52 before and after inversion

Table 3. Comparison with darkened, lightened and inverted textures

comparison	wtd. energy	comparison	wtd. energy	comparison	wtd. energy
15-drk.15	0.0083	03-lgt.03	0.0374	04-inv.04	0.2567
18-drk.18	0.0050	14-lgt.14	0.0117	23-inv.23	0.4859
21-drk.21	0.0129	40-lgt.40	0.0052	35-inv.35	0.6009
38-drk.38	0.0145	53-lgt.53	0.0757	52-inv.52	0.4927

6   Conclusion

In this work, we have presented a new approach to texture classification. By using
the modified Newton filters, we have obtained an estimation of the orientation of
the edges in every point of the textures. The extraction of orientation histograms
to describe the distribution of the orientations across a textured region permits
us to perform an initial comparison of the textures according to the quantitative
and relative distribution of the different orientations.

The comparison of the Fourier coefficients, certain normalization processes and the use of weighting functions allow a satisfactory classification in many cases, including size and rotational invariance. However, due to the ambiguities that are generated by the non-injectivity of the generation of these histograms, a further study has been carried out, by comparing the evolution of the histograms at different scales.

This multiscale analysis of the histograms has produced quite good results, since the visual similarity or difference between two textures is much more reliably detected by the evolution of the energies resulting when comparing histograms at different scales, which have been previously adjusted. We have extracted the scale factor which must be used when comparing two textures to perform the comparison appropriately.

The quite promising numerical results obtained in the tests which have been implemented confirm the usefulness of the multiple comparison of the images, since they endow us with a much more robust discrimination criterion. Furthermore, we have tested how our method reacts when the textures are darkened, lightened or when their grayscale levels are inverted, obtaining very satisfactory results.

References

1. Sonka, M., Hlavac, V., Boyle, R.: Image processing, analysis, and machine vision. PWS-ITP (1999)
2. Gorkani M.M., Picard, R. W.: Texture orientation for sorting photos "at a glance". In the proceedings of IEEE conference on pattern recognition **I** (1995) 459-464
3. Randen, T., Husøy, J. H.: Filtering for texture classification: a comparative study. IEEE Transactions on Pattern Analysis and Machine Intelligence **21:4** (1999) 291-310
4. Puzicha, J., Buhmann, J. M., Rubner, Y., Tomasi, C.: Empirical evaluation of dissimilarity measures for color and texture. In proceedings of the IEEE International Conference on Computer Vision (ICCV'99) (1999) 1165-1173
5. Tuceryan, M., Jain, A. K.: Texture analysis. In C.H. Chen, L.F. Pau, P.S.P. Wang, editors, The Handbook of Pattern Recognition and Computer Vision (2nd edition), World Scientific Publishing Co. (1998) 207-248
6. Alemán-Flores, M., Álvarez-León, L., Moreno-Díaz jr., R.: Modified Newton filters for edge orientation estimation, shape representation and motion analysis. Cuadernos del Instituto Universitario de Ciencias y Tecnologías Cibernéticas, Universidad de Las Palmas de Gran Canaria **17** (2001) 1-30
7. Moreno-Díaz jr., R.: Computación paralela y distribuida: relación estructura-función en retinas. Tesis Doctoral (1993)
8. Columbia University and Utrecht University. Columbia-Utrecht Reflectance and Texture Database. http://www.cs.columbia.edu/CAVE/curet/.index.html
9. Paragios, N., Deriche, R.: Geodesic active regions and level set methods for supervised texture segmentation. International Journal of Computer Vision **46:3** (2002) 223
10. Weickert, J.: Multiscale texture enhancement. R. Sara (Eds.), Computer analysis of images and patterns, Lecture Notes in Computer Science Springer Berlin **970** (1995) 230-237

11. Brezis, H.: Analyse fonctionelle. Masson (1983)
12. Evans, L.: Partial Differential Equations. American Mathematical Society (1998)
13. Lindeberg, T.: Scale space theory in computer vision. Kluwer Academic Publishers (1994)
14. Alvarez, L., Mazorra, L.: Signal and image restoration using shock filters and anisotropic diffusion. SIAM J. on Numerical Analysis **31:2** (1994) 590-605

α Scale Spaces on a Bounded Domain

Remco Duits, Michael Felsberg, Luc Florack, and Bram Platel

Eindhoven University of Technology, Department of Biomedical Engineering
P.O. box 513, NL-5600 MB Eindhoven, The Netherlands,
{R.Duits,L.M.J.Florack,B.Platel}@tue.nl
http://www.bmi2.bmt.tue.nl/image-analysis/
Linköping University, Computer Vision Laboratory
S-58183 Linköping, Sweden
mfe@isy.liu.se
http://www.isy.liu.se/~mfe

Abstract. We consider α scale spaces, a parameterized class ($\alpha \in (0,1]$) of scale space representations beyond the well-established Gaussian scale space, which are generated by the α-th power of the minus Laplace operator on a bounded domain using the Neumann boundary condition. The Neumann boundary condition ensures that there is no grey-value flux through the boundary. Thereby no artificial grey-values from outside the image affect the evolution proces, which is the case for the α scale spaces on an unbounded domain. Moreover, the connection between the α scale spaces which is not trivial in the unbounded domain case, becomes straightforward: The generator of the Gaussian semigroup extends to a compact, self-adjoint operator on the Hilbert space $\mathbb{L}_2(\Omega)$ and therefore it has a complete countable set of eigen functions. Taking the α-th power of the Gaussian generator simply boils down to taking the α-th power of the corresponding eigenvalues. Consequently, all α scale spaces have exactly the same eigen-modes and can be implemented simultaneously as scale dependent Fourier series. The only difference between them is the (relative) contribution of each eigen-mode to the evolution proces. By introducing the notion of (non-dimensional) relative scale in each α scale space, we are able to compare the various α scale spaces. The case $\alpha = 0.5$, where the generator equals the square root of the minus Laplace operator leads to Poisson scale space, which is at least as interesting as Gaussian scale space and can be extended to a (Clifford) analytic scale space.

1 Introduction

It is commonly taken for granted that the Gaussian scale space paradigm is the unique solution to a set of reasonable axioms if one disregards minor modifications, such as spatial inhomogeneities [7], diffeomorphisms [8], and anisotropies [17], which can be easily accounted for. That this is in fact not true has first been pointed out by Pauwels [13], who proposed a one-parameter class of scale space filters in Fourier space. In a more recent and extensive study [2] *all* reasonable scale space axioms are summarized and it is shown that they lead to a whole

L.D. Griffin and M. Lillholm (Eds.): Scale-Space 2003, LNCS 2695, pp. 494–510, 2003.

parameterized ($\alpha \in (0, 1]$) class of scale spaces, the so-called α scale spaces. They are shown to be related by taking a fractional power of the Gaussian semigroup generator, i.e. the generator of the α scale space is given by $-(-\Delta)^\alpha$. Before we focus on the (relatively easy) bounded domain case we summary the theory of α scale spaces on the unbounded domain $\{(\mathbf{x}, s) \mid \mathbf{x} \in \mathbb{R}^d, s > 0\}$ since it is essential background information with respect to the bounded domain case and it is not widely known in the computer vision community.

2   α Scale Spaces on the Unbounded Domain

The α scale spaces on the unbounded domain $u^{(\alpha)}$ are subject to the following pseudo partial differential system:

$$\begin{cases} \frac{\partial}{\partial s} u = -(-\Delta)^\alpha u \\ \lim_{s \downarrow 0} u(\cdot, s) = f(\cdot) & \text{in } \mathbb{L}_2(\mathbb{R}^d)\text{-sense} \end{cases} \quad \alpha \in (0, 1], s > 0, \quad , \tag{1}$$

where $-(-\Delta)^\alpha : \mathbb{H}_{2\alpha}(\mathbb{R}^d) \to \mathbb{L}_2(\mathbb{R}^d)$ is densely defined by

$$(-\Delta)^\alpha f = \frac{\sin \alpha \pi}{\pi} \int\limits_0^\infty \lambda^{\alpha-1} (\lambda I - \Delta)^{-1} (-\Delta f) \, d\lambda \qquad \text{for } f \in \mathrm{C}^2(\mathbb{R}^d) .$$

The corresponding kernels $K^{(\alpha)}$ are given by the somewhat awkward expression:

$$K^{(\alpha)}(\mathbf{x}, s) = \int\limits_0^\infty q_{s,\alpha}(t) \, G_t(\mathbf{x}) \, dt , \tag{2}$$

where $q_{s,\alpha}$ is the inverse Laplace transform of $\mu \mapsto e^{-s\mu^\alpha}$ and G_t is the usual Gaussian kernel $G_t(\mathbf{x}) = \frac{1}{(4\pi t)^{d/2}} e^{-\frac{\|\mathbf{x}\|^2}{4t}}$, but their Fourier transforms are quite simple:

$$\widehat{K^{(\alpha)}}(\boldsymbol{\omega}, s) = e^{-\|\boldsymbol{\omega}\|^{2\alpha} s} . \tag{3}$$

Although that many qualitative results about $(-\mathcal{A})^\alpha$ such as $(-\mathcal{A})^\alpha(-\mathcal{A})^\beta = (-\mathcal{A})^{\alpha+\beta}$ can be obtained, this expression is (probably) not an operational definition. There exist (more) operational expressions for $-(-\Delta)^\alpha$, see [15]p.216-217, but even they can not cope with the extremely simple and operational form of $-(-\Delta)^\alpha$ when working on a finite domain with (Neumann-)boundary conditions. The special cases $\alpha = 1/2$ and $\alpha = 1$ lead to respectively the Poisson and Gaussian scale space. Although Gaussian scale space is well established in the computer vision community, Poisson scale space is not. Nevertheless, Poisson scale space seems the most natural choice of all α-scale spaces, since this is the only scale space where the scale $s > 0$ has the same physical dimension as the spatial variables x_i, $i = 1 \ldots d$, allowing an Euclidean norm within *scale space.* Notice to this end that in recent work, concerning content-based image retrieval, cf.[11], Hausdorff-distances (constructed from *Euclidean* distances) on point-clouds within Gaussian scale spaces are used. This does not seem appropriate in Gaussian scale space, but is allowed within Poisson scale space.

2.1 Poisson Scale Space on the Unbounded Domain

In this subsection we focus on the Poisson scale space case ($\alpha = 1/2$). For more extensive information the reader is referred to earlier work of the authors who used to work independently, cf. [3], [2], [5].

The $d + 1$ dimensional Laplace operator (with respect to both $s > 0$ and $\mathbf{x} \in \mathbb{R}^d$) can be factorized:

$$u_{ss} + \Delta u = (\partial_s - \sqrt{-\Delta})(\partial_s + \sqrt{-\Delta})u = 0$$

and since the nil-space of the linear operator in the first factor of the factorization is zero, this equation is equivalent to

$$(\partial_s + \sqrt{-\Delta})u = 0 \Leftrightarrow u_s = -\sqrt{-\Delta}\,u \ .$$

As a result the pseudo differential system corresponding to the α-scale space (1) for $\alpha = 1/2$ is equivalent to a Dirichlet problem on the upper plane:

$$\begin{cases} \frac{\partial^2}{\partial s^2}u + \Delta u = 0 & s > 0, \mathbf{x} \in \mathbb{R}^d \\ \lim_{s \downarrow 0} u(\cdot, s) = f(\cdot) & \text{in } L_2(\mathbb{R}^d)\text{-sense} \end{cases} \tag{4}$$

The solution of this problem is given by a convolution with the Poisson kernel

$$u^{(1/2)}(\mathbf{x}, s) = (K_s^{1/2} * f)(\mathbf{x}) = \frac{2s}{\sigma_{d+1}} \int_{\mathbb{R}^d} \frac{f(\mathbf{y})}{(s^2 + \|\mathbf{x} - \mathbf{y}\|^2)^{\frac{d+1}{2}}} \, d\mathbf{y} \ ,$$

where $\sigma_{d+1} = \frac{2\pi^{\frac{d+1}{2}}}{\Gamma((d+1)/2)}$ equals the surface area of the unit sphere in $\mathbb{R}^{d+1}$.

The Poisson scale space generator satisfies the following relation

$$-\sqrt{-\Delta} = -\sum_{j=1}^{d} R_j \frac{\partial}{\partial x^j} f = -\mathbf{R} \cdot \nabla f \qquad , f \in \mathbb{H}_1(\mathbb{R}^d), \tag{5}$$

where the jth component R_j of the Riesz transform $\mathbf{R} = \sum \mathbf{e}_j R_j$ is given by the principle value integral

$$(R_j g)(x) = \frac{2}{\sigma_{d+1}} \int_{\mathbb{R}^d} \frac{x_j - u_j}{\|\mathbf{x} - \mathbf{u}\|^{d+1}} g(\mathbf{u}) \, d\mathbf{u}, \tag{6}$$

which boils down to a multiplication with $-i\frac{\omega^j}{\|\omega\|}$ in the Fourier domain. For $d = 1$ there exists only one component which is known as the Hilbert transform.

2.2 Clifford Analytic Extension of the Poisson Scale Space on the Unbounded Domain

In case of 1D-signals ($d = 1$) it is possible to extend the Poisson scale space to an analytic scale space $\tilde{u}(x + is) = u_A(x, s) = u(x, s) + iv(x, s)$, simply by adding i times the harmonic conjugate v which is determined (up to a constant) by Cauchy-Riemann ($u_x = v_s$, $u_s = -v_x$). The harmonic conjugate is given by $v = (Q_s * f)(x, s)$, where Q_s denotes the conjugate Poisson kernel which is given by the Hilbert transform of the Poisson kernel:

$$Q_s(\mathbf{x}) = (HK_s^{(1/2)})(x) = \frac{1}{\pi}\frac{x}{s^2 + x^2} \cdot$$

This follows directly by Cauchy's integral formula for analytic functions:

$$\tilde{u}(z) = \frac{1}{2\pi i}\oint_C \frac{\tilde{u}(w)}{w - z}\,dw \qquad z = x + is\ , \tag{7}$$

where C is any positively oriented simple curve around z, since

$$K_s^{(1/2)}(x) = \Re\left(\frac{1}{(2\pi i)(z)}\right)$$
$$Q_s(x) = \Im\left(\frac{1}{(2\pi i)(z)}\right) \cdot$$

In particular by taking $C = C_0 \cup C_R \cup C_\delta$, with $C_0 = [-R, R]$ and $C_R = \{z \in \mathbb{C}_+ \mid |z| = R\}$, $C_\delta = \{z \in \mathbb{C}_+ \mid |z| = \delta\}$ in (7) and letting $\delta \to 0$, $R \to \infty$ we obtain the Cauchy operator $C : \mathbb{L}_2 \to H^2(\mathbb{C}_+)$ which is given by

$$(Cf)(x, s) = \frac{1}{2\pi i}\int_{\mathbb{R}} \frac{f(t)}{t - z}\mathrm{dt} = \frac{1}{2}((K_s^{1/2} * f)(x) + i\,(Q_s * f)(x)) \qquad z = x + is \in \mathbb{C}_+ \tag{8}$$

where the space $H^2(\mathbb{C}_+)$ consists of all analytic functions F on $\mathbb{C}_+$ such that $\sup\limits_{t>0}\int_{-\infty}^{\infty}|F(x + it)|^2\,\mathrm{dx} < \infty$. Any signal can be split uniquely and orthogonally into an analytic and a non-analytic part:

$$f = f_{AN} + f_{NAN} = \frac{f + iHf}{2} + \frac{f - iHf}{2}.$$
$$\mathbb{L}_2(\mathbb{R}^d) = H^2(\partial\mathbb{C}_+) \bigoplus (H^2(\partial\mathbb{C}_+))^\perp\ ,$$

where the subspace of analytic signals is given by $H^2(\partial\mathbb{C}_+) = \{f \in \mathbb{L}_2(\mathbb{R}) \mid \mathrm{supp}(\hat{f}) \subset [0, \infty)\}$. To this end we notice $[\mathcal{F}(Hf)](\omega) = -i\mathrm{sgn}(\omega)[\mathcal{F}(f)](\omega)$ so $\hat{f}_{AN}(\omega) = 0$ for $\omega < 0$. Further we notice[1] $Cf = C(f_{AN}) + C(f_{NAN}) = C(f_{AN}) + 0$ and $\lim\limits_{s\downarrow 0} Cf(\cdot, s) = f_{AN}$. In practice f is real valued, so then $f = 2\Re(f_{AN})$ and consequently $u^{(1/2)}(x, s) = \Re\tilde{u}(x, s) = 2\Re(Cf_{AN})(x, s) = (K_s^{1/2} * f)(x)$.

---

[1] If restricted to the subspace of analytic signals the Cauchy operator is an isometric isomorphism such that the non tangential limit $\lim\limits_{s\downarrow 0} Cg(\cdot, s) = g(\cdot)$ for all $g \in H^2(\partial\mathbb{C}_+)$, cf.[9]p.113

Remarks:

- Physically, the Poisson scale space should be regarded as a potential problem rather than a heat problem. The isophotes within the Poisson scale space correspond to equi-potential curves and the isophotes within the conjugate Poisson scale space correspond to the flow-lines. By the Cauchy-Riemann equations these lines intersect each other orthogonal through each point (x, s):

$$(\partial_x, \partial_s)u \cdot (\partial_x, \partial_s)v = u_x v_s + u_s v_x = 0 .$$

For instance the isophotes of the Poisson kernel $K^{(1/2)}$ are the semi-circles $x^2 + (s - a)^2 = a^2$, $a, s > 0, x \in \mathbb{R}$ which intersect the flow lines $(x+a)^2 + s^2 = a^2$, $a, x \in \mathbb{R}, s > 0$ orthogonal. It might be tempting to regard f as charge density distribution, but this is not right: f is the potential at the boundary, due to some charge-distribution in the plane $s < 0$.

- The 2D Laplace operator can be split into two different ways:

$$\Delta_2 = (\partial_s + i\partial_x)(\partial_s - i\partial_x) = 4\partial_{\bar{z}}\partial_z$$
$$\Delta_2 = (\partial_s - \sqrt{-\partial_{xx}})(\partial_s + \sqrt{-\partial_{xx}}) .$$

The space of analytic signals $H_2(\partial\mathbb{C}^+)$ is very special since its elements are treated similarly by the operators $-\sqrt{-\Delta}$ and $i\partial_x$:

$$-\sqrt{-\partial_{xx}}f = i\partial_x f \qquad \text{for } f \in H_2(\partial\mathbb{C}^+),$$

which can be easily be verified in the Fourier domain. Consequently for sufficiently smooth f $(f \in \mathbb{H}_\infty)$:

$$u(x, s) = (K_s * f)(x) = (e^{-s\sqrt{-\partial_{xx}}}f)(x) = (e^{s\,i\partial_x}f)(x) = \tilde{u}(x + is) .$$

Complex analytic extension can only be done in the signal case $(d = 1)$. For images $d \geq 2$ an analogue recipe can be followed, using the more general notion of Clifford analytic functions. To this end some knowledge of Clifford algebra is necessary, cf.[3], [9] . Let $\{\mathbf{e}_i\}_{i=1}^n = \{\mathbf{e}_i\}_{i=1}^d \cup \{\mathbf{e}_{d+1}\}$, $n = d+1$, be an orthonormal base in $\mathbb{R}^n$ and let $\mathbb{R}_n$ and $\mathbb{R}_n^+$ be the Clifford algebra and its even subalgebra of $\mathbb{R}^n$. Let Ω be an open set in $\mathbb{R}^n$.

Definition 1. *A function $\tilde{u} \in C^\infty(\Omega, \mathbb{R}_n^+)$ is Clifford analytic on Ω if*

$$\nabla_n \tilde{u} = \sum_{j=1}^n \mathbf{e}_j \frac{\partial \tilde{u}}{\partial x^j} = 0 .$$

There again exists a (generalized) Cauchy integral theorem for these functions, cf.[9]p.103. Analogue to the $d = 1$ case we define the closed subspace of $\mathbb{L}_2(\mathbb{R}^d)$:

$$H^2(\partial\mathbb{R}_n^+) = \{f \in \mathbb{L}_2(\mathbb{R}^d) \mid (I - \mathbf{Re}_{d+1})f = 0\}.$$

Notice that $(R_j f, f) = (\mathcal{F}(R_j f), \mathcal{F}f) = 0$ for $j = 1 \ldots d$ and $(\mathbf{R})^2 = \sum R_j^2 = -I$, therefore we can split complex valued signals into a Clifford analytic and orthogonal to Clifford analytic part:

$$\mathbb{L}_2(\mathbb{R}^d) = H^2(\partial \mathbb{R}_n^+) \bigoplus (H^2(\partial \mathbb{R}_n^+))^{\perp}$$

$$f = \frac{f + \mathbf{Re}_{d+1} f}{2} + \frac{f - \mathbf{Re}_{d+1} f}{2} = f_{AN} + f_{NAN} \; ,$$

Notice that these two subspaces of $\mathbb{L}_2(\mathbb{R}^d)$ are precisely the irreducible subspaces of the semi-direct product of the dilation and translation group on $\mathbb{R}^d$.

We define the Cauchy operator $C : \mathbb{L}_2(\mathbb{R}^d) \to H^2(\mathbb{R}_n^+)$ by

$$(Cf)(\mathbf{x}, s) = \frac{1}{\sigma_{d+1}} \int_{\mathbb{R}^d} \frac{\mathbf{z} - \mathbf{u}}{\|\mathbf{z} - \mathbf{u}\|^{d+1}} e_{d+1} f(\mathbf{u}) \, d\mathbf{u} \qquad \mathbf{z} = \sum_{j=1}^{d} x^j \mathbf{e}_j + s \mathbf{e}_{d+1} \; , \quad (9)$$

which can again be expressed in the Poisson kernel and its harmonic conjugate:

$$\begin{aligned}
\mathbf{Q}_s(\mathbf{x}) &= \mathbf{R} K_s^{1/2}(\mathbf{x}) \\
&= \sum_j \mathbf{e}_j R_j K_s^{1/2}(\mathbf{x}) \\
&= \sum_j \mathbf{e}_j Q_s^{(j)}(\mathbf{x}) \\
&= \sum_j \frac{2}{\sigma_{d+1}} \frac{x^j \mathbf{e}_j}{(s^2 + \|\mathbf{x}\|^2)^{\frac{d+1}{2}}} \; ,
\end{aligned}$$

by

$$\begin{aligned}
(Cf)(\mathbf{x}, s) &= \tfrac{1}{2}(K_s^{(1/2)} * f)(\mathbf{x}) + \tfrac{1}{2} \sum_{j=1}^{d} \mathbf{e}_j e_{d+1}(Q_s^{(j)} * f)(\mathbf{x}) \\
&= (K_s^{(1/2)} * (\tfrac{1}{2}(I + \mathbf{Re}_{d+1}))f)(\mathbf{x}) \\
&= (K_s^{(1/2)} * f_{AN})(\mathbf{x}),
\end{aligned}$$

Remarks:

- The nil-space of C equals $(H^2(\partial \mathbb{R}_n^+))^{\perp}$, so $Cf = C(f_{AN} + f_{NAN}) = C(f_{AN})$.
- Let $d = 3$ and $\tilde{u}$ be Clifford analytic, then $\nabla_d \tilde{u} = 0$ and therefore $\nabla_d \tilde{u} e_{d+1} = \nabla_d \cdot (\tilde{u} e_{d+1}) + \nabla_d \wedge (\tilde{u} e_{d+1}) = 0 + \mathbf{0}$ so if we put $\mathbf{u} = \tilde{u} e_{d+1}$ we have $\mathrm{rot}\, \mathbf{u} = \mathbf{0}$ and $\mathrm{div}\, \mathbf{u} = 0$ from which it follows that $\mathbf{u}$ has a harmonic potential $\mathbf{u} = \nabla p$, with $\Delta p = 0$.
- The monogenic scale space $\mathbf{u}_M$ which is introduced by Felsberg and Sommer, cf.[3] (for $d = 2$) is given by

$$\begin{aligned}
\mathbf{u}_M(\mathbf{x}, s) &= \tilde{u}(\mathbf{x}, s) e_{d+1} = 2(Cf)(\mathbf{x}, s) e_{d+1} \\
&= (K_s^{(1/2)} + \sum_{j=1}^{d} \mathbf{e}_j e_{d+1} R_j K_s^{(1/2)}) * \mathbf{f}) \\
&= e_{d+1}(K_s^{(1/2)} * f)(\mathbf{x}) + \sum_{j=1}^{d} \mathbf{e}_j (Q_s^{(j)} * f)(\mathbf{x}) \qquad , \mathbf{f} = f e_{d+1} \; .
\end{aligned}$$

– By (5) we have $-\sqrt{-\Delta} = -\nabla \cdot \mathbf{R}$ with $\Delta = \nabla \cdot \nabla$ and therefore (by Gauss divergence Theorem) for all $\Omega' \subset \Omega$

$$\frac{\partial}{\partial s} \int_{\Omega'} u_\alpha(\mathbf{x}, s) \, d\mathbf{x} = \begin{cases} \int_{\partial\Omega'} \nabla u \cdot \mathbf{n} \, d\sigma & \text{for } \alpha = 1 \\ -\int_{\partial\Omega'} \mathbf{R}u \cdot \mathbf{n} \, d\sigma & \text{for } \alpha = 1/2 \end{cases}, \qquad (10)$$

so the other components in the monogenic scale space besides the Poisson scale space describe the Poisson image flow analogue to the fact that $-\nabla u$ describes the Gaussian image flow.

Some interesting local features can easily be obtained from the Monogenic/ Clifford analytic scale spaces, such as the local phase vector field, local amplitude (attenuation), local orientation. These concepts are again generalizations of the local phase analysis in signal analysis, cf. [3], [4], [6].

3 α Scale Spaces on the Bounded Domain with Neumann Boundary Conditions

In scale space theory the domain is usually taken $\{(\mathbf{x}, s) \mid s > 0, \mathbf{x} \in \mathbb{R}^d\}$, but this introduces some drawbacks. Real images are not defined on the whole $\mathbb{R}^d$ and we have to extend them somehow to the whole $\mathbb{R}^d$. Thereby, besides the original image f, external information will affect the scale space, especially at large scales. Some research has been going on concerning the topology (isophotes, critical paths, top-points, scale space saddles, multi scale graph representation) of Gaussian scale space, cf. [12], [14] . In this kind of research large scale information is essential. Moreover, since grey-values are allowed to flow out of the image, the average grey value is not maintained. To compensate this effect the image is often re-normalized at different scales with different normalization factors (depending on the external information and scale). As a result scale spaces in practice are quite different from the scale spaces in theory.

Another consequence of the unbounded domain is a continuous Fourier-spectrum, where a discrete Fourier spectrum and the corresponding Fourier series are easier to understand with respect to $\mathbb{L}_2$-considerations ($\mathbb{L}_2$ error estimation) rather than point-wise considerations. In practice Fourier transforms are always implemented by discrete Fourier transforms, which implicitly boils down to periodic extension of the image (so boundary conditions are used after all). A priori the behavior of an image at the boundary is not known and therefore it is impossible to solve *all* these problems perfectly. Rather than ignoring these problems, we choose an approach were this damage is reduced. Moreover, the connection between the α-scale spaces will become almost trivial.

The α scale space u_α on a bounded oriented domain Ω, with boundary $\partial\Omega$ and outward normal $\mathbf{n}$ with Neumann Boundary conditions is defined as the solution of the following boundary value problem (B.V.P.):

$$\begin{cases} \dfrac{\partial u}{\partial s} = -(-\varDelta)^{\alpha} u & \mathbf{x} \in \Omega, s > 0 \text{ (E.E.) Evolution Equation} \\ \dfrac{\partial u}{\partial n}\Big|_{\partial\Omega} = 0 & \text{(N.B.C.) Neumann Boundary} \\ & \qquad\qquad\quad \text{Condition} \\ \lim_{s\downarrow 0} u(\cdot, s) = f(\cdot) & \text{(I.C.) Initial Condition .} \end{cases} \quad (11)$$

By working on a finite domain two scale space axioms must be a adapted:

- Translation invariance ($\varPhi[T_\mathbf{a} f, s] = T_\mathbf{a}\varPhi[f, s]$) does not make sense unless the N.B.C. is translated together with f. As a result the solution of (11) is no longer a convolution.
- Monotonic increase of entropy on $\mathbb{R}^d$ must be replaced by increase of entropy on the bounded domain Ω, i.e. the Entropy function now becomes ($0 \leq u \leq 1$)

$$[\mathcal{E}(u)](s) = -\int_\Omega u(\mathbf{x}, s) \ln u(\mathbf{x}, s) \, d\mathbf{x} \qquad (s > 0) . \qquad (12)$$

- Rotation invariance ($\varPhi[\mathcal{P}_R f, s] = \mathcal{P}_R \varPhi[f, s]$) is only satisfied if Ω is a ball in $\mathbb{R}^d$. For 2D images this means that one must work on a disc to obtain rotational invariance.

With respect to the causality axiom we remark that similar to the unbounded case the α ($0 < \alpha < 1$) evolutions with N.B.C. do not satisfy Koenderink's principle of non-enhancement of local extrema: $(u_s \varDelta u)(x, y, s) \geq 0$ in (spacial) extrema $((x, y, s), u(x, y, s))$, but they do satisfy the weak causality constraint in the sense that every isophote is connected to the ground plane, following exactly the same argument as for the unbounded case, cf. [2] .

There are quite some fundamental reasons to impose N.B.C. . Before mentioning them we would like to remark that the right-hand side of Greens first and second identity vanishes

$$\int_\Omega g\varDelta f + \nabla g \cdot \nabla f \, d\mathbf{x} = \int_{\partial\Omega} g\frac{\partial f}{\partial n} \, d\sigma = 0 \qquad (13)$$

$$\int_\Omega g\varDelta f - f\varDelta g \, d\mathbf{x} = \int_{\partial\Omega} g\frac{\partial f}{\partial n} - f\frac{\partial g}{\partial n} \, d\sigma = 0 \qquad (14)$$

for functions $f, g \in C^2(\Omega) \cap C^1(\overline{\Omega})$, such that $\frac{\partial f}{\partial n} = \frac{\partial g}{\partial n} = 0$.

1. Average grey-value is maintained over scale, which immediately follows for the Gaussian case ($u_s = \varDelta u$) by the no-flux Neumann boundary condition:

$$\frac{\partial}{\partial s}(1, u)_{\mathbb{L}_2(\Omega)} = \frac{\partial}{\partial s}\int_\Omega u(\mathbf{x}, s) \, d\mathbf{x} = \int_\Omega \varDelta u(\mathbf{x}, s) \, d\mathbf{x} = \int_{\partial\Omega} \frac{\partial u}{\partial n} \, d\sigma = 0 . \quad (15)$$

As a result average grey-value is maintained for all α scale spaces, since all α scale spaces have a common complete orthonormal base of eigenfunctions including the constant function with corresponding eigenvalue $0 = -(0)^\alpha$, see the third item of this list.

2. It is essential for monotonically increase of entropy, which can be shown for the Poisson $\alpha = 1/2$ and Gaussian case $\alpha = 1$.
   – For Gaussian scale space, we have by Greens second identity(14):

$$\tfrac{\partial}{\partial s}[\mathcal{E}(u)](s) = -\int_\Omega \Delta u(\log u + 1)\,d\mathbf{x} = -\int_\Omega \Delta u(\log u)\,d\mathbf{x}$$

$$= -\int_\Omega u\,\mathrm{div}\left(\tfrac{\nabla u}{u}\right)\,d\mathbf{x} = \int_\Omega \left(\tfrac{\|\nabla u\|^2}{u}\right)\,d\mathbf{x} \geq 0 .$$

   – For Poisson scale space: Follow the proof of Theorem 5.3 in [2] use Ω in stead of $\mathbb{R}^d$ and notice that the right hand side of Greens first identity vanishes because of the N.B.C. .
3. The eigenvalue problem

$$\begin{cases} \Delta f = \lambda f \\ \tfrac{\partial f}{\partial n} = 0 \end{cases} \qquad f \in C^2(\Omega) \cap C^1(\overline{\Omega}) \tag{16}$$

is equivalent to the eigenvalue problem:

$$\mathcal{K}f = -\tfrac{1}{\lambda}f, \quad f \in \mathbb{L}_2(\mathbb{R}^d), \tag{17}$$

where the kernel operator $\mathcal{K} : \mathbb{L}_2(\Omega) \to \mathbb{L}_2(\Omega)$ is given by

$$\mathcal{K}(f)(\mathbf{x}) = (N_{\mathbf{x}}(\cdot), f(\cdot))_{\mathbb{L}_2(\Omega)} = \int_\Omega N(\mathbf{x},\mathbf{y})f(\mathbf{y})\,d\mathbf{y}$$

for almost every $\mathbf{x} \in \mathbb{R}^d$,

where N is the function of Neumann on Ω, i.e. the sum $N(\mathbf{x},\mathbf{y}) = S(\mathbf{x},\mathbf{y}) + n(\mathbf{x},\mathbf{y})$, of the fundamental solution S and a function h, which is the unique solution (up to a constant) of:

$$\begin{aligned} \Delta_{\mathbf{x}} h(\mathbf{x},\mathbf{y}) &= 0 & \mathbf{y} \in \Omega \\ \tfrac{\partial h}{\partial n} &= -\tfrac{\partial S}{\partial n} - 1/\mu(\partial\Omega)\ \mathbf{y} \in \Omega . \end{aligned}$$

If we would have worked with Dirichlet boundary conditions, such equivalence can be obtained using Greens function in stead of Neumanns function as a kernel for the kernel operator $\mathcal{K}$. Analogue to the Dirichlet problem, where the solution can be represented by $f(\mathbf{y}) = -\int_\Omega G(\mathbf{x},\mathbf{y})\Delta f\,d\mathbf{x}$, where G denotes Greens function, the solution of the Neumann problem is given by

$$f(\mathbf{y}) = -\int_\Omega N(\mathbf{x},\mathbf{y})(\Delta f)(\mathbf{x})\,d\mathbf{x} + \frac{1}{\mu(\partial\Omega)}\int_{\partial\Omega} f(\mathbf{x})\,d\sigma , \ \mathbf{y} \in \Omega$$

see [10]p.232-235 for formal treatment. Further we notice that the Neumann problem has a unique solution up to a constant, therefore the second term in the right hand side is not essential. By this result and the uniqueness of the Neumann problem (up to constants) we indeed find that system (17) follows from system (16). The reverse (in particular the twice continuously differentiable property of f) is more difficult to show since it is due to Weyls lemma, cf. [10]p.225-226, p.200. Operator $\mathcal{K}$ is positive and self-adjoint $\mathcal{K}^* = \mathcal{K}$ since $N(\mathbf{y}, \mathbf{x}) = N(\mathbf{x}, \mathbf{y})$. Moreover $\mathcal{K}$ is compact because it has a finite double norm $|||\mathcal{K}||| = \left[\int\limits_{\Omega \times \Omega} |N(\mathbf{x}, \mathbf{y})|^2 \, d\mathbf{x} \, d\mathbf{y} \right]^{1/2} < \infty$ for $d \leq 3$. For details, see [1]. From a fundamental result from functional analysis it now follows that there exists a complete (countable) set of eigen functions $\{f_n\}_{n \in \mathbb{N}}$, with positive eigenvalues μ_n which tend to zero as $n \to \infty$, cf. [18]p.167. By the equivalence between eigenvalue problems (16) and (17) these eigen functions are also the eigen functions of the generator of the Gaussian semi-group with corresponding eigenvalues $\lambda_n = -\frac{1}{\mu_n} < 0$ (tending to minus infinity as $n \to \infty$) and thereby these functions f_n are the eigen images of the Gaussian evolution operator $\Phi_1 : \mathbb{L}_2(\Omega) \times \mathbb{R}^+ \to \mathbb{L}_2(\Omega)$ with Neumann boundary conditions with eigenvalues $e^{\lambda_n s}$ (tending to 0 as $n \to \infty$):

$$u_1(\cdot, s) = \Phi_1[f, s] = \sum_{n \in \mathbb{N}} \Phi_1[f_n](f_n, f) = \sum_{n \in \mathbb{N}} f_n(f_n, f) e^{\lambda_n s} .$$

Taking the α-th power of the Gaussian generator now becomes fairly easy by taking the α-th power of the eigenvalues. As a result the α-scale space operator on a bounded domain with Neumann boundary conditions is given by:

$$u_\alpha(\cdot, s) = \Phi_\alpha[f, s] = \sum_{n \in \mathbb{N}} \Phi_\alpha f_n(f_n, f) = \sum_{n \in \mathbb{N}} f_n(f_n, f) e^{-(-\lambda_n)^\alpha s} .$$

In contrast to the unbounded domain case, it is now relatively easy to compare the α scale spaces. To this end we notice that the scale s in the α-scale space has a physical dimension that depends on α: $[s] = [\text{Length}]^{2\alpha}$. In the bounded domain we can define a dimensionless scale parameter which we will call relative scale $s_\alpha = \frac{s}{\rho^{2\alpha}}$, where ρ equals the radius of the smallest sphere around Ω.

In the next two subsections we will focus on the special cases where Ω is a rectangle respectively a disc. The first case is interesting purely from a pragmatic point of view, since in computer vision images are mostly rectangles and Euclidean coordinates are relatively easy to handle. Although the disc case is less pragmatic, it is still operational and from the theoretical point of view it is a better approach in the sense that rotational invariance axiom is satisfied. Moreover, with respect to human vision a disc seems more natural than the rectangle.

3.1   The Rectangle Case

By the method of separation $u_\alpha(x, y, s) = X(x)Y(y)G(s)$ we can find the orthonormal base of eigen images of all α scale spaces on a rectangle with Neumann boundary conditions, whose existence for the general domain case has been shown earlier in this section. We will only mention the result, for more details see [1]. For monogenic scale space implementation on rectangle with no flux boundary condition for its real component see [4].

Theorem 1. *The general solution of the α evolution problem with Neumann boundary condition on a rectangle $[0, a] \times [0, b]$ is given by:*

$$
u_\alpha^{(a,b)}(x, y, s) = \sum_{m=0}^{\infty} \sum_{n=0}^{\infty} \left(\int_0^a \int_0^b f(x', y') \, l_{mn}(x', y') \, \mathrm{dx}' \, \mathrm{dy}' \right)
$$
$$
\times l_{mn}(x, y) \, e^{-\left\{ \left(\frac{n\pi}{b} \right)^2 + \left(\frac{m\pi}{a} \right)^2 \right\}^\alpha s}
$$

$$\alpha \in (0, 1], a > 0, b > 0,$$

where $l_{mn} : \mathbb{R}^2 \to \mathbb{R}$ is given by

$$
l_{mn}(x, y) = \frac{2 \cos \left(\frac{m\pi x}{a} \right) \cos \left(\frac{n\pi y}{b} \right)}{\sqrt{ab}}.
$$

Notice that $\{l_{mn}\}_{\mathbb{L}_2(\Omega)}$ indeed is a complete orthonormal base for $\mathbb{L}_2([0, a] \times [0, b])$. Therefore the series in (18) converges in $\mathbb{L}_2$-sense. Under weak (Dirichlet) conditions it also converges pointwise and uniformly. The separation constants $\left\{ \left(\frac{n\pi}{b} \right)^2 + \left(\frac{m\pi}{a} \right)^2 \right\}$ are the eigenvalues of the Gaussian generator. The relative scale is given by $s_\alpha = \frac{2^{2\alpha} s}{(a^2 + b^2)^\alpha}$. It indeed follows by (18) that the only difference between the α scale spaces on a square is the relative contribution of each eigen mode to the evolution proces.

The Unbounded Domain Case as a Limit of the Rectangular Case

We will show that if $a, b \to \infty$ the discrete solution $u_\alpha^{(a,b)}(x, y, s)$ converges to the convolution $u^{(\alpha)}(x, y, s) = (K_s^{(\alpha)} * f)(x, y)$, which can be rewritten as

$$
(f * K_s^{(\alpha)})(x, y) = \mathcal{F}^{-1}[\mathcal{F}(f * K_s^{(\alpha)})](x, y)
$$
$$
= \mathcal{F}^{-1} \left[\boldsymbol{\omega} \mapsto e^{-\|\boldsymbol{\omega}\|^{2\alpha} s} [\mathcal{F}(f)](\omega_1, \omega_2) \right] (x, y)
$$
$$
\frac{1}{4\pi^2} \int_{\mathbb{R}} \int_{\mathbb{R}}
$$
$$
\times e^{-\|\boldsymbol{\omega}\|^{2\alpha} s} \left(\int_{\mathbb{R}} \int_{\mathbb{R}} f(x', y') \, e^{-i\omega_1 x' - i\omega_2 y'} \, \mathrm{dx}' \, \mathrm{dy}' \right)
$$
$$
\times e^{i\omega_1 x + i\omega_2 y} \, \mathrm{d}\omega_1 \, \mathrm{d}\omega_2
$$

(19)

Since $(f * K_s^{(\alpha)})$ is both even in x and y the last e-power can be replaced by a cosine. The first e-power is even ω, but this means that the second e-power can also be replaced by a cosine, i.e. (19) can be written

$$
\frac{1}{4\pi^2} \int_{\mathbb{R}^2} e^{-\|\boldsymbol{\omega}\|^{2\alpha} s} \left(\int_{\mathbb{R}^2} f(x', y') \, \cos(\omega_1 x') \cos(\omega_2 y') \, \mathrm{dx}' \, \mathrm{dy}' \right)
$$

$$\times \cos(\omega_1 x) \cos(\omega_2 y) \, d\omega_1 \, d\omega_2 \tag{20}$$

This integral equals the following Riemann sum, sampled equidistant in $\omega_{nm} = (\omega_{1m}, \omega_{2n}) = (\frac{m\pi}{a}, \frac{n\pi}{b})$, $m, n = 1, 2, \ldots$ and using the fact that $\text{supp}(f) \subseteq [0, a] \times [0, b]$ we indeed obtain

$$\lim_{a \to \infty} \lim_{b \to \infty} \sum_{m=0}^{\infty} \sum_{n=0}^{\infty} (f, l_{mn})_{\mathbb{L}_2([0,a] \times [0,b])} l_{mn}(x, y) \, e^{-\left\{ \left(\frac{n\pi}{b}\right)^2 + \left(\frac{m\pi}{a}\right)^2 \right\}^\alpha s}. \tag{21}$$

3.2 The Disc Case

The α evolution equation with Neumann boundary condition described in cylindrical coordinates on the disc is given by:

$$\begin{cases} \frac{\partial u}{\partial s} = -(-\Delta)^\alpha u & r = \|\mathbf{x}\| \leq a, s > 0 \text{ (E.E.) Evolution Equation} \\ \frac{\partial u}{\partial r}\big|_{r=a} = 0 & \text{(B.C.) Boundary Condition} \\ u(r, \phi, 0) = f(r, \phi) & \text{(I.C.) Initial Condition} \end{cases} \tag{22}$$

Obviously, We don not want u to explode to infinity if s tends to infinity and therefore we impose the additional constraint that u is finite at $r = 0$ or at least locally in $\mathbb{L}_2$. Again one can follow the general method of separation $u_\alpha(r, \phi, s) = R(r)\Phi(\phi)S(s)$ to obtain the ortonormal base of eigen functions of the diffusion problem ($\alpha = 1$), with corresponding eigenvalues. The other α scale spaces have the same eigenfunctions and their corresponding eigenvalues are obtain by taking the α-th power of the eigenvalues of the Gaussian case. Again we will only give the result, for detailed treatment and explicit proof see [1].

Define $h_{mn} : (0, a) \times (\pi, \pi] \to \mathbb{C}$, $m \in \mathbb{Z}, n \in \mathbb{N}$ by

$$h_{mn}(r, \phi) = \frac{1}{a\sqrt{\pi \left(1 - \left(\frac{m}{j'_{m,n}}\right)^2\right)}} \frac{J_m(j'_{m,n}r/a)}{J_m(j'_{m,n})} e^{im\phi}, \tag{23}$$

where the Bessel functions of the first kind are given by

$$J_m(z) = \sum_{k=0}^{\infty} \frac{(-1)^k}{k!\Gamma(m + k + 1)} \left(\frac{z}{2}\right)^{2k+m} \qquad m \in \mathbb{Z} \; z \in \mathbb{C}. \tag{24}$$

Note that $h_{mn} = \overline{h_{-mn}}$. Define $h_{00} : (0, a) \times (\pi, \pi] \to \mathbb{C}$, by $h_{00}(r, \phi) = \frac{1}{\sqrt{\pi a}}$.

Theorem 2. *The set $\{h_{mn}\}_{m \in \mathbb{Z}, n \in \mathbb{N} \cup \{0\}}$ is a complete orthonormal base of $\mathbb{L}_2(B_{0,a}) = \mathbb{L}_2(S_1) \otimes \mathbb{L}_2((0, a), rdr)$ and they are eigen functions of the infinitesimal generator of evolution system (22) with corresponding eigenvalue 0 respectively $-(\frac{j_{m,n}}{a})^{2\alpha}$.*

The general solution of (22) is given by:

$$u_\alpha(r,\phi,s) = \sum_{m=-\infty}^{\infty} \sum_{n=0}^{\infty} (f, h_{mn})_{\mathbb{L}_2(B_{0,a})} h_{mn}(r,\phi)\, e^{-\left(\frac{j'_{m,n}}{a}\right)^{2\alpha} s} =$$

$$f_{AV} + \sum_{m\in\mathbb{Z}} \sum_{n\in\mathbb{N}} \frac{\left[\int_0^a \int_{-\pi}^{\pi} f(\rho,\theta)\, J_m(j'_{m,n}\frac{\rho}{a})\, e^{i\,m\,\theta}\, \rho\,d\rho\,d\theta\right]}{a^2 J_m^2(j'_{m,n})\pi\left(1-(\frac{m}{j'_{m,n}})^2\right)} J_m(j'_{m,n}\frac{r}{a})\, e^{i\,m\,\phi}\, e^{-\left(\frac{j'_{m,n}}{a}\right)^{2\alpha} s}.$$

$$(25)$$

Remarks:

- We have shown in general, cf. (15), that average grey-value f_{AV} is maintained. If s tends to infinity u indeed tends to $\frac{h_{00}(r,\phi)}{\sqrt{\pi}a}(f,h_{00})_{\mathbb{L}_2(B_{0,a})} = f_{AV}$.
- Under small conditions the series also converges point-wise and uniformly, cf. [16]pp.591 sec 18-24.
- Note that

$$j'_{-m,n} = j'_{m,-n} = -j'_{-m,-n} = -j'_{m,n}.$$

$$(26)$$

- Using the fact that $J_m(\frac{j'_{mn}}{a}r)$ satisfies the second order differential equation

$$(\frac{j'_{mn}}{a}r)J_m''(\frac{j'_{mn}}{a}r) + J_m'(\frac{j'_{mn}}{a}r) + (kr - \frac{m^2}{kr})J_m(\frac{j'_{mn}}{a}r) = 0$$

and one partial integration step it directly follows that

$$(1, \gamma_{n\,m})_{\mathbb{L}_2(B_{0,a})} = \begin{cases} 0 & \text{if } m = 0 \\ \int_0^a \frac{a\,m^2}{j'_{mn}\,r} J_m(\frac{j'_{mn}}{a}r)\,dr & \text{if } m \neq 0 \end{cases}$$

$$(27)$$

- The physical dimension of the evolution parameter within the αth scale space s equals $[\text{Length}]^{2\alpha}$, so the exponent in (25) has no physical dimension. The *relative scale* in the disc case is given by $s_\alpha = \frac{s}{a^{2\alpha}}$.
- By (26) we have $\overline{h_{mn}} = h_{-mn}$. From this it follows that $\overline{c_{mn}} = c_{-mn}$ if f is real valued. Moreover we have $j'_{m,0} = 0 \Leftrightarrow m \neq 0$. So then we have

$$u_\alpha(r,\phi,s) = f_{AV} + 2 \sum_{m=1}^{\infty} \sum_{n=1}^{\infty} \text{Re}\left[(h_{mn}, f)_{\mathbb{L}_2(B_{0,a})} h_{mn}(r,\phi)\right] e^{-\left(\frac{j'_{m,n}}{a}\right)^{2\alpha} s}.$$

- It is shown in [1] that if the radius a of the disc tend to infinity one obtains the usual α-scale spaces and corresponding convolution kernels, see (2) (3). Although, this is straightforward in a conceptually sense, (the influence from the boundary will disappear), it is extremely hard to show this in a clean mathematical way analogue to the square case.

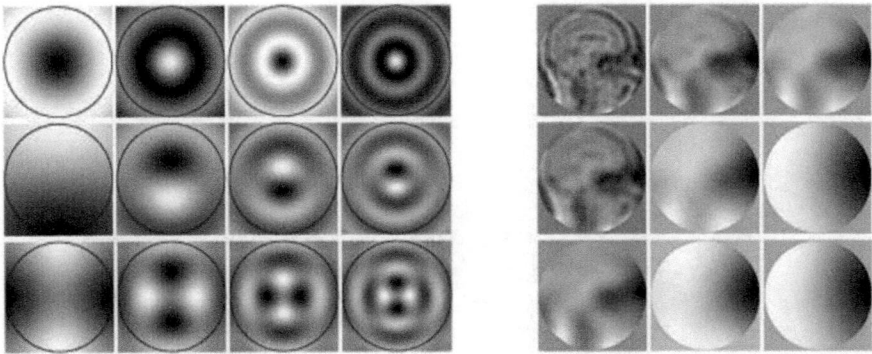

Fig. 1. Left: The real part of various evolution eigen modes h_{mn} of the α-scale spaces on a disc. Top row $m = 0$ and $n = 1, \ldots 4$, middle row $m = 1$ for $n = 1 \ldots 4$, bottom row $m = 2$ and $n = 1, \ldots 4$. The circle denotes the boundary of the disc.
Right: Various scale space representations of a 128×128 MR brain slice. Top row: $\alpha = \frac{1}{2}$ (Poisson scale space), middle row: $\alpha = \frac{3}{4}$, bottom row: $\alpha = 1$ (Gaussian scale space). On respective relative scales $s_\alpha = \frac{s}{a^{2\alpha}} = 0.01, 0.1, 0.2$.

4   Truncation of the Fourier Series Expansions

In practice f can not be expanded in the infinite(countable) base $\{f_{mn}\}$ of common eigen functions of the α scale space operators; the Fourier series must be truncated at say $m = M$ and $n = N$. The most natural norm for error estimation is the $\mathbb{L}_2(B_{0,a})$-norm:

$$
(\varepsilon_{MN}^{(\alpha)}(s))^2 = \|u_\alpha(\cdot, s) - \sum_{m=0}^{M} \sum_{n=0}^{N} (f_{mn}, f) f_{mn} e^{-(-\lambda_{mn})^\alpha s}\|_{\mathbb{L}_2(B_{0,a})}^2
$$
$$
= \left(\sum_{m=0}^{\infty} \sum_{n=N+1}^{\infty} + \sum_{n=0}^{N} \sum_{m=M+1}^{\infty} \right) \left| (f_{mn}, f)_{\mathbb{L}_2(\Omega)} \right|^2 e^{-2(-\lambda_{mn})^\alpha s} . \tag{28}
$$

The α scale spaces on a bounded domain have the same eigen functions and can thereby be implemented simultaneously. For such an implementation it is important to have a sharp error estimation which depends explicitly on M, N, s and α, since as α increases the series can be chopped sooner to maintain the same amount of accuracy. Moreover, as the relative scale $s_\alpha > 0$ increases, the higher frequency components vanish so the series can also be chopped sooner as s_α increases. Notice that in practice for sampled images frequencies above the Nyquist-frequencies M' and N' are omitted, so one can replace the infinite upper limits by these frequencies. In the disc case we found $f_{mn} = h_{mn}$, $\lambda_{mn} = -\frac{j'_{m,n}}{a}$ and in the square case we found $f_{mn} = l_{mn}$ and $\lambda_{mn} = -\left(\frac{n\pi}{b}\right)^2 - \left(\frac{m\pi}{a}\right)^2$. Here we will only estimate the error in the square case. For details with regard to disk case, see [1]. In order to have a sharp estimation we assume that

$$
|(l_{mn}, f)|^2 \leq C(\frac{1}{m^v n^w}) \text{ for } m > M \text{ and } n > N, \tag{29}
$$

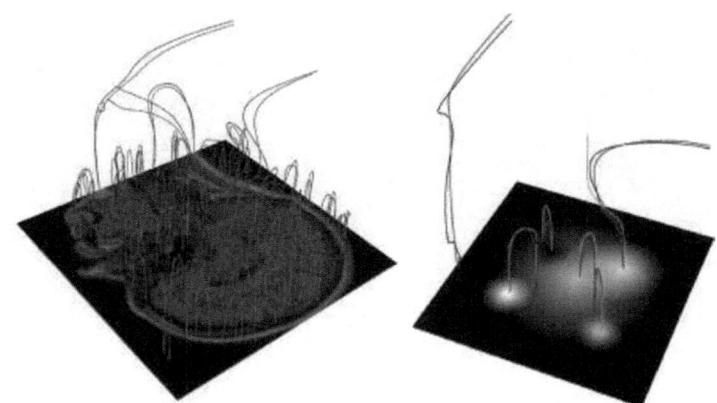

Fig. 2. Critical curves through scale space. Left: Critical curves in respective Poisson (red) and Gaussian (blue) in finite domain scale space of square MRI-image. Right: synthetic image, with additional in green crit. curves unbounded Gaussian scale space.

where $v, w > 1, C > 0$ are fixed and not difficult to estimate in practice. Due to Cauchy-Scwarz one can estmimate *roughly* with $C = \|f\|^2_{\mathbb{L}_2([0,a]\times[0,b])}$, $v = w = 0$. In estimating the $\mathbb{L}_2$ error (below) we use the Hölder inequality $(\mathbf{x}, \mathbf{y})_2 \leq \|\mathbf{x}\|_p\|\mathbf{y}\|_q$, for $1/p + 1/q = 1$, applied with $p = 1/\alpha$, $\mathbf{x} = (|\frac{m\pi}{a}|^{2\alpha}, |\frac{n\pi}{b}|^{2\alpha})^T$ and $\mathbf{y} = (1,1)^T$, yielding

$$\left(\left(\frac{n\pi}{b}\right)^2 + \left(\frac{m\pi}{a}\right)^2\right)^\alpha \geq 2^{\alpha-1}\left(\left(\frac{m\pi}{a}\right)^{2\alpha} + \left(\frac{n\pi}{b}\right)^{2\alpha}\right) \quad \text{and therefore,}$$

$$
\begin{aligned}
(\varepsilon_{MN}^{(\alpha)}(s))^2 &= \left(\sum_{m=0}^{\infty}\sum_{n=N+1}^{\infty} + \sum_{n=0}^{N}\sum_{m=M+1}^{\infty}\right)\left|(l_{mn}, f)_{\mathbb{L}_2([0,a]\times[0,b])}\right|^2 \\
&\quad \times e^{-2\left(\left(\frac{n\pi}{b}\right)^2 + \left(\frac{m\pi}{a}\right)^2\right)^\alpha s} \\
&\leq \left(\sum_{m=0}^{\infty}\sum_{n=N+1}^{\infty} + \sum_{n=0}^{N}\sum_{m=M+1}^{\infty}\right) C\, e^{-2^\alpha\left(\left(\frac{n\pi}{b}\right)^{2\alpha} + \left(\frac{m\pi}{a}\right)^2\right)^{2\alpha} s}\frac{1}{m^v\,n^w} \\
&= C\left(\sum_{m=0}^{\infty}\sum_{n=N+1}^{\infty} + \sum_{n=0}^{N}\sum_{m=M+1}^{\infty}\right)\frac{(\eta(a,\alpha,s))^{m2\alpha}}{m^v}\frac{(\eta(b,\alpha,s))^{n2\alpha}}{n^w} \\
&= C\left(\sum_{m=0}^{\infty}\frac{(\eta(a,\alpha,s))^{m2\alpha}}{m^v}\right)\left(\sum_{n=N+1}^{\infty}\frac{(\eta(b,\alpha,s))^{n2\alpha}}{n^w}\right) \\
&\quad + C\left(\sum_{n=0}^{N}\frac{(\eta(b,\alpha,s))^{m2\alpha}}{m^v}\right)\left(\sum_{m=M+1}^{\infty}\frac{(\eta(a,\alpha,s))^{n2\alpha}}{n^w}\right),
\end{aligned}
$$

$$(30)$$

where $\eta(a, \alpha, s) = e^{-2^\alpha\left(\frac{\pi}{a}\right)^{2\alpha}s} \ll 1$. The Poisson case $\alpha = 1/2$ is easiest to estimate, since standard series expansions can be used, cf. [1]. For instance estimate (30) in case $v = w = 0$ becomes:

$$(\varepsilon_{MN}^{(1/2)}(s))^2 \leq \|f\|^2 \left(\frac{1}{1 - \eta(a, 1/2, s)}\frac{(\eta(b, 1/2, s))^{N+1}}{1 - \eta(b, 1/2, s)}\right.$$

$$+\frac{1-(\eta(b,1/2,s))^{N+1}}{1-\eta(b,1/2,s)}\,\frac{(\eta(a,1/2,s))^{M+1}}{1-\eta(a,1/2,s)}\Bigg)\,.$$

But also in the general α-case one can find a sharp estimation, using the following inequality

$$\sum_{m=M+1}^{\infty}\frac{r^{-m^{2\alpha}}}{m^{v}}\leq\int_{M}^{\infty}\frac{r^{-x^{2\alpha}}}{x^{v}}\,dx=\frac{(\log r)^{\frac{m-1}{2\alpha}}\Gamma[-\frac{m-1}{2\alpha},M^{2\alpha}\log r]}{2\alpha}\qquad r=\frac{1}{\eta}>1,$$

which follows by monotonic decrease of the function $x\mapsto\frac{r^{-x^{2\alpha}}}{x^{v}}$ on $\mathbb{R}$ and where $\Gamma[a,z]=\int_{z}^{\infty}t^{a-1}e^{-t}dt$ denotes the incomplete Gamma function.

5   Conclusion

A unified framework to scale space theory is presented. First we summarized the theory of α scale spaces ($\alpha\in(0,1]$) on the unbounded domain, which satisfy all reasonable scale space axioms and which have analogue properties as the Gaussian scale space ($\alpha=1$). Particularly interesting is the Poisson scale space ($\alpha=1/2$) and its Clifford analytic extension. Since images are given on a bounded domain in practice and because it is not desirable that grey-vales from outside the image affect the scale space, we observed the α-scale spaces on a bounded domain with Neumann-boundary conditions. Moreover, the mathematical description and connection between the α scale spaces become relatively easy. First some general results are proven and then the explicit solutions expanded in a *common* complete orthonormal base of eigenfunctions are given for the (pragmatic) case of the rectangle and the (more difficult, but stil operational) case of the disc. Further we have shown that the α scale spaces on the unbounded domain are limits of these bounded domain cases. Finally, a sharp estimate of the $\mathbb{L}_2$-error as a function of α and scale, which is caused by cutting of the Fourier series of the α scale space representation, is given, which can be used for simultaneous implementation.

References

1. R. Duits, M. Felsberg, and L. Florack. α scale spaces on a bounded domain. Technical report, TUE, Eindhoven, March 2003. In Preparation.
2. R. Duits, L. M. J. Florack, J. De Graaf, and B. M. ter Haar Romeny. On the axioms of scale space theory. *Accepted for publication in Journal of Mathematical Imaging and Vision*, 2002.
3. M. Felsberg. *Low-Level Image Processing with the Structure Multivector*. PhD thesis, Institute of Computer Science and Applied Mathematics Christian-Albrechts-University of Kiel, 2002.
4. M. Felsberg, R. Duits, and L.M.J. Florack. The monogenic scale space on a bounded domain and its applications. Accepted for publication in proceedings Scale Space Conference 2003.

5. M. Felsberg and G. Sommer. Scale adaptive filtering derived from the Laplace equation. In B. Radig and S. Florczyk, editors, *23. DAGM Symposium Mustererkennung, München*, volume 2191 of *Lecture Notes in Computer Science*, pages 124–131. Springer, Heidelberg, 2001.

6. M. Felsberg and G. Sommer. The Poisson scale-space: A unified approach to phase-based image processing in scale-space. *Journal of Mathematical Imaging and Vision*, 2002. submitted.

7. L. M. J. Florack. A geometric model for cortical magnification. In S.-W. Lee, H. H. Bülthoff, and T. Poggio, editors, *Biologically Motivated Computer Vision: Proceedings of the First IEEE International Workshop, BMCV 2000 (Seoul, Korea, May 2000)*, volume 1811 of *Lecture Notes in Computer Science*, pages 574–583, Berlin, May 2000. Springer-Verlag.

8. L. M. J. Florack, R. Maas, and W. J. Niessen. Pseudo-linear scale-space theory. *International Journal of Computer Vision*, 31(2/3):247–259, April 1999.

9. J.E. Gilbert and M.A.M. Murray. *Clifford algebras and Dirac operators in harmonic analysis*. Cambridge University Press, Cambridge, 1991.

10. G. Hellwig. *Partial Differential Equations*. Blaisdell Publishing Company, 1964.

11. F.M.W. Kanters, B. Platel, L.M.J. Florack and B.M. ter Haar Romeny. Content based image retrieval using multiscale top points. In proceedings of the scale space conference, 2003.

12. A. Kuijper and L. M. J. Florack. Hierarchical pre-segmentation without prior knowledge. In *Proceedings of the 8th International Conference on Computer Vision (Vancouver, Canada, July 9–12, 2001)*, pages 487–493. IEEE Computer Society Press, 2001.

13. E. J. Pauwels, L. J. Van Gool, P. Fiddelaers, and T. Moons. An extended class of scale-invariant and recursive scale space filters. *IEEE Transactions on Pattern Analysis and Machine Intelligence*, 17(7):691–701, July 1995.

14. B. Platel, L.M.J. Florack, F. Kanters, and B. M. ter Haar Romeny. Multi-scale hierarchical segmentation. In Scale Space Conference 2003 submitted.

15. K. I. Sato. *Lévy processes and Infinitely Divisible Distributions*. Cambridge University Press, Cambridge, 1999.

16. J.N. Watson. *A Treatise on the Theory of Bessel Functions*. Cambridge University Press, Cambridge, 1996. Originally published by Cambridge University Press 1922.

17. J. A. Weickert. *Anisotropic Diffusion in Image Processing*. ECMI Series. Teubner, Stuttgart, January 1998.

18. J. Wloka. *Partial Differential Equations*. Cambridge University Press, Cambridge, 1987.

Efficient Beltrami Flow
Using a Short Time Kernel

Alon Spira[1], Ron Kimmel[1], and Nir Sochen[2]

[1] Department of Computer Science, Technion, Israel
{salon,ron}@cs.technion.ac.il
[2] Department of Applied Mathematics, University of Tel-Aviv, Israel
sochen@math.tau.ac.il

Abstract. We introduce a short time kernel for the Beltrami image enhancing flow. The flow is implemented by 'convolving' the image with a space dependent kernel in a similar fashion to the implementation of the heat equation by a convolution with a gaussian kernel. The expression for the kernel shows, yet again, the connection between the Beltrami flow and the Bilateral filter. The kernel is calculated by measuring distances on the image manifold by an efficient variation of the fast marching method. The kernel, thus obtained, can be used for arbitrary large time steps in order to produce adaptive smoothing and/or a new scale-space. We apply it to gray scale and color images to demonstrate its flow like behavior.

1 Introduction

The Beltrami flow [4,12] is a powerful tool for image enhancement. Its good visual effect results from de-noising the image while keeping the edges intact. The flow originates from minimizing the area of the 2-dimensional Riemannian image manifold embedded in $\mathbb{R}^N$, where $N = 3$ for gray scale images and $N = 5$ for color images.

A short time kernel has been presented for 1D non-linear diffusion in [10] and an approximation for the 2D Beltrami operator in [9]. These kernels enable the implementation of the flows by 'convolving' the signals with the kernels, similar to the implementation of the heat equation by a convolution with a gaussian kernel. This implementation replaces the conventional method of solving the first variation as a gradient descent PDE process by the appropriate numerical schemes. One of the main advantages of this approach is the ability to select an arbitrary time step for the kernel.

In order to compute the short time kernel we need to calculate distances on the image manifold. Measuring distances on manifolds has been done before for triangulated manifolds [5], graphs of functions [8], and implicit manifolds [6]. Here, we propose a new variation of the fast marching method for calculating the distances, especially suited for image manifolds.

This paper is organized as follows. The first section describes the Beltrami flow for gray scale and color images. In Section 2 the derivation of the short

L.D. Griffin and M. Lillholm (Eds.): Scale-Space 2003, LNCS 2695, pp. 511–522, 2003.
© Springer-Verlag Berlin Heidelberg 2003

time kernel is presented. Section 3 reviews our new contribution for calculating geodesic distances on the image manifold, which is required for the implementation of the short time kernel. The simulations and results are in Section 4 and the conclusions in Section 5.

2   The Beltrami Flow

In the Beltrami framework the image is regarded as the embedding $X : U \to \mathbb{R}^N$, with U the 2-dimensional image manifold and $\mathbb{R}^N$ the space-feature manifold. For gray scale images

$$X(u^1, u^2) = \{u^1, u^2, I(u^1, u^2)\}, \tag{1}$$

where u^1, u^2 are the space coordinates and I is the intensity component. The metric h_{ij} of the space-feature manifold is

$$H = (h_{ij}) = \begin{pmatrix} 1 & 0 & 0 \\ 0 & 1 & 0 \\ 0 & 0 & \beta^2 \end{pmatrix}, \tag{2}$$

where β is the relative scale between the space coordinates and the intensity component. This is an Euclidean space-feature manifold. Non-Euclidean manifolds were addressed in [11,13]. The metric elements g_{ij} of the image manifold U are derived from the metric elements h_{ij} and the embedding X by the pullback procedure

$$G = (g_{ij}) = \begin{pmatrix} 1 + \beta^2 I_1^2 & \beta^2 I_1 I_2 \\ \beta^2 I_1 I_2 & 1 + \beta^2 I_2^2 \end{pmatrix}, \tag{3}$$

where $I_i \triangleq \frac{\partial I}{\partial u^i}$.

For color images

$$X(u^1, u^2) = \{u^1, u^2, I^1(u^1, u^2), I^2(u^1, u^2), I^3(u^1, u^2)\}, \tag{4}$$

where I^1, I^2, I^3 are the three color components (for instance red, green and blue for the RGB color space). The metric h_{ij} of the space-feature manifold is

$$H = (h_{ij}) = \begin{pmatrix} 1 & 0 & 0 & 0 & 0 \\ 0 & 1 & 0 & 0 & 0 \\ 0 & 0 & \beta^2 & 0 & 0 \\ 0 & 0 & 0 & \beta^2 & 0 \\ 0 & 0 & 0 & 0 & \beta^2 \end{pmatrix}, \tag{5}$$

and the metric of the image manifold is

$$G = (g_{ij}) = \begin{pmatrix} 1 + \beta^2 \sum_a (I_1^a)^2 & \beta^2 \sum_a I_1^a I_2^a \\ \beta^2 \sum_a I_1^a I_2^a & 1 + \beta^2 \sum_a (I_2^a)^2 \end{pmatrix}. \tag{6}$$

The Beltrami flow can be obtained by minimizing the area of the image manifold

$$S = \iint \sqrt{g} du_1 du_2, \tag{7}$$

with respect to the embedding, where $g = \det(G) = g_{11}g_{22} - g_{12}^2$. The corresponding Euler-Lagrange equations as a gradient descent process are

$$X_t^a = -g^{-\frac{1}{2}} h^{ab} \frac{\delta S}{\delta X^b} = g^{-\frac{1}{2}} \partial_i (g^{\frac{1}{2}} g^{ij} \partial_j X^a), \tag{8}$$

with g^{ij} the contravariant metric of the image manifold (the inverse of the metric tensor g_{ij}) and using Einstein's summation convention. In a matricial form it reads

$$X_t^a = \underbrace{\frac{1}{\sqrt{g}} \mathrm{Div} \left(\sqrt{g} G^{-1} \nabla X^a \right)}_{\Delta_g X^a} \tag{9}$$

where Δ_g is the Laplace-Beltrami operator which is the extension of the Laplacian to manifolds.

For gray scale images we get

$$I_t = \Delta_g I. \tag{10}$$

For color images we get for each color component

$$I_t^i = \Delta_g I^i. \tag{11}$$

We introduce in the next sections the kernel method for solving these coupled and highly non-linear partial differential equations.

3   A Short Time Kernel for the Beltrami Flow

It can be shown that applying the heat equation

$$I_t = \Delta I \tag{12}$$

to the 2-dimensional data $I(u^1, u^2, t_0)$ for the duration t is equivalent to convolving the data with a Gaussian kernel

$$\begin{aligned}
I(u^1, u^2, t_0 + t) &= \int I(\tilde{u}^1, \tilde{u}^2, t_0) K(|u^1 - \tilde{u}^1|, |u^2 - \tilde{u}^2|; t) d\tilde{u}^1 d\tilde{u}^2 \\
&= I(u^1, u^2, t_0) * K(u^1, u^2; t) ,
\end{aligned} \tag{13}$$

where the kernel is given by

$$K(u^1, u^2; t) = \frac{1}{4\pi t} \exp \left(-\frac{(u^1)^2 + (u^2)^2}{4t} \right) . \tag{14}$$

An iterative implementation of the PDE is replaced, in this approach, by a one step filter.

In this section we extend this result to the Beltrami flow. Because of the non-linearity of this flow (the Beltrami operator depends on the data I), a global (in time) kernel is impossible. We therefore develop a short time kernel that if used iteratively, has an equivalent effect to that of the Beltrami flow. We replace Equation (13) with

$$I^i(u^1, u^2, t_0 + t) = \int I^i(\tilde{u}^1, \tilde{u}^2, t_0) K(u^1, u^2, \tilde{u}^1, \tilde{u}^2; t) d\tilde{u}^1 d\tilde{u}^2 , \qquad (15)$$

which we denote by

$$I^i(u^1, u^2, t_0 + t) = I^i(u^1, u^2, t_0) *_g K(u^1, u^2; t), \qquad (16)$$

This is not a convolution in the strict sense, because K does not depend on the differences $u^i - \tilde{u}^i$. In general, the coordinates u^i are arbitrary local coordinates on the manifold. These coordinates are not a geometric object and the difference between coordinates, therefore, has no intrinsic meaning. We will justify our definition of a convolution on a manifold after we develop the explicit form of the kernel. The general form of K is

$$K(u^1, u^2; t) = \frac{H(u^1, u^2; t)}{t} \exp\left(-\frac{\psi^2(u^1, u^2)}{t} \right), \qquad (17)$$

where we take, without lose of generality, $(\tilde{u}^1, \tilde{u}^2) = (0,0)$ and omit from K the notation of dependency on these coordinates. It will be re-instated later on by fixing the integration constants. Note, that ψ does not depend on t at all, while H is a regular function of t and can be expanded as a Taylor series $H(x, y, t) = \sum_{n=0}^{\infty} H_n(x, y) t^n$. In order to find K, we use the fact that it should satisfy Equation (11). Therefore,

$$K_t = \Delta_g K. \qquad (18)$$

The left hand side of the equation is

$$
\begin{aligned}
K_t &= \partial_t \left(\frac{H}{t} \exp\left(-\frac{\psi^2}{t} \right) \right) \\
&= \left(\frac{H_1}{t} - \frac{H_0}{t^2} - \frac{H_1}{t} + \frac{H_0 \psi^2}{t^3} + \frac{H_1 \psi^2}{t^2} + O(\frac{1}{t}) \right) \exp\left(-\frac{\psi^2}{t} \right) \\
&= \left(\frac{H_0 \psi^2}{t^3} + \frac{H_1 \psi^2 - H_0}{t^2} + O(\frac{1}{t}) \right) \exp\left(-\frac{\psi^2}{t} \right) \qquad (19)
\end{aligned}
$$

For the right hand side of Equation (18) we calculate

$$K_i = \frac{\partial K}{\partial u^i} = \left(\frac{H_{0i}}{t} - \frac{2H_0 \psi \psi_i}{t^2} - \frac{2H_1 \psi \psi_i}{t} + O(1) \right) \exp\left(-\frac{\psi^2}{t} \right) . \qquad (20)$$

The second derivative is calculated similarly. The first two leading terms that multiply the exponential are

$$\frac{4H_0\psi^2\psi_i\psi_j}{t^3} - \frac{2(H_{0i} - 2H_1\psi\psi_i)\psi\psi_j + \partial_j(2H_0\psi\psi_i)}{t^2} . \tag{21}$$

Putting everything together, we get for the leading order

$$\Delta_g K = \frac{4}{t^3}g^{ij}\left(\psi^2\psi_i\psi_j + O\left(t\right)\right)K. \tag{22}$$

Equating the leading terms in Equations (19) and (18), yields

$$g^{ij}\psi_i\psi_j = \|\nabla_g\psi\|^2 = \frac{1}{4}, \tag{23}$$

with ∇_g the extension of the gradient to the manifold. This is the Eikonal equation on the manifold, and its viscosity solution is a geodesic distance map ψ on the manifold. An efficient solution of the Eikonal equation for image manifolds is given in the next section. The H_n coefficients, which depend on the spatial variables, are solutions of the PDEs that are obtained by equating the coefficients of powers of t. It is not too difficult to be convinced that H_0 is a constant (see [10] for an example of such a computation).

The resulting short time kernel is thereby

$$K(u^1, u^2, \tilde{u}^1, \tilde{u}^2; t) = \frac{H_0}{t}\exp\left(-\frac{\left(\int_{(u^1,u^2)}^{(\tilde{u}^1,\tilde{u}^2)} ds\right)^2}{4t}\right)$$

$$= \frac{H_0}{t}\exp\left(-\frac{d_g^2\left((u^1, u^2), (\tilde{u}^1, \tilde{u}^2)\right)}{4t}\right), \tag{24}$$

where ds is an arc-length element on the manifold, and $d_g(p_1, p_2)$ is the geodesic distance between two points, p_1 and p_2, on the manifold. Note that in the Euclidean space with Cartesian coordinate system $d_E(p_1, p_2) = |p_1 - p_2|$. The geodesic distance on manifolds is therefore the natural generalization of the difference between coordinates in the Euclidean space. It is natural then to define the convolution on a manifold by

$$I^i(u^1, u^2) *_g K(u^1, u^2; t) = \int I^i(\tilde{u}^1, \tilde{u}^2)K\left(d_g\left((u^1, u^2), (\tilde{u}^1, \tilde{u}^2)\right)\right) d\tilde{u}^1 d\tilde{u}^2 \tag{25}$$

The update step for the image is

$$I^i(u^1, u^2, t_0+t) = \frac{H_0}{t}\iint_{(\tilde{u}^1,\tilde{u}^2)\in N(u^1,u^2)} I^i(\tilde{u}^1, \tilde{u}^2, t_0)\exp\left(-\frac{\left(\int_{(u^1,u^2)}^{(\tilde{u}^1,\tilde{u}^2)} ds\right)^2}{4t}\right) d\tilde{u}^1 d\tilde{u}^2,$$

$$\tag{26}$$

with $N(u^1, u^2)$ the neighborhood of the point (u^1, u^2), where the value of the kernel is above a certain threshold. Because of the monotone nature of the fast marching algorithm used in the next section for the solution of the Eikonal equation, once a point is reached, where the value of the kernel is smaller than the threshold, the algorithm can stop and thereby naturally bound the numerical support of the kernel. The value of the kernel for the remaining points of the manifold would be negligible. Therefore, the Eikonal equation is solved only in a small neighborhood of each image point. H_0 is taken such that integration over the kernel in the neighborhood $N(u^1, u^2)$ of the point equals one.

The short time Beltrami kernel in Equation (24) is very similar to the Bilateral filter kernel [15,3]. The difference between them is that the Beltrami kernel uses geodesic distances on the image manifold, while the Bilateral kernel uses Euclidean distances. The derivation of the Beltrami kernel shows that the Bilateral filter originates from image manifold area minimization. The Bilateral filter can actually be viewed as an Euclidean approximation of the Beltrami flow. Another connection between the Beltrami flow and the Bilateral filter appears in [2].

The Euclidean distance used in the Bilateral filter, while being easier to calculate, does not take into account the image intensity values between two image points. A point can have a relatively high kernel value, although it belongs to a different object than that of the filtered image point. The Beltrami kernel takes this effect into account and penalizes a point that belongs to a different 'connected component'. That is, it is not 'as blind' as the Bilateral filter to the spatial structure of the image.

4   Solving the Eikonal Equation on Image Manifolds

The image manifold is a parametric manifold, where the metric g_{ij} is given for every point. We present here an efficient solution for the Eikonal equation on parametric manifolds, based on the fast marching approach [7]. A more detailed description appears in [14].

The original fast marching algorithm [7] solves the Eikonal equation in an orthogonal coordinate system. In this case, the numerical support for the update of a grid point consists of one or two points out of its four neighbors. The first point is from the up/down pair and the second is from the left/right pair. The two selected grid points, together with the updated one, compose the vertices of a right triangle. See Figure 1.

This is not the case for image manifolds. There $g_{12} \neq 0$ and we get a non-orthogonal coordinate system on the manifold, see Figure 2. The resulting angles are not necessarily right angles. If a grid point is updated by a stencil which includes an obtuse angle, a problem may arise. The value of one of the points of the stencil might not be set in time and cannot be used. There is a similar problem with fast marching on triangulated domains which include obtuse angles [5].

Our solution is similar to that of [5]. We perform a pre-processing stage for the grid, in which we split every obtuse angle into two acute ones, see Figure 3.

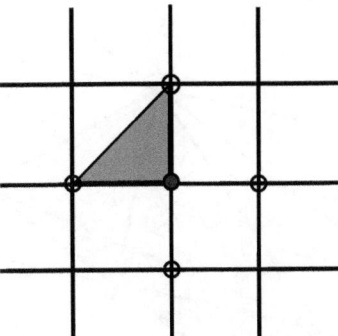

Fig. 1. The numerical support for the orthogonal fast marching algorithm.

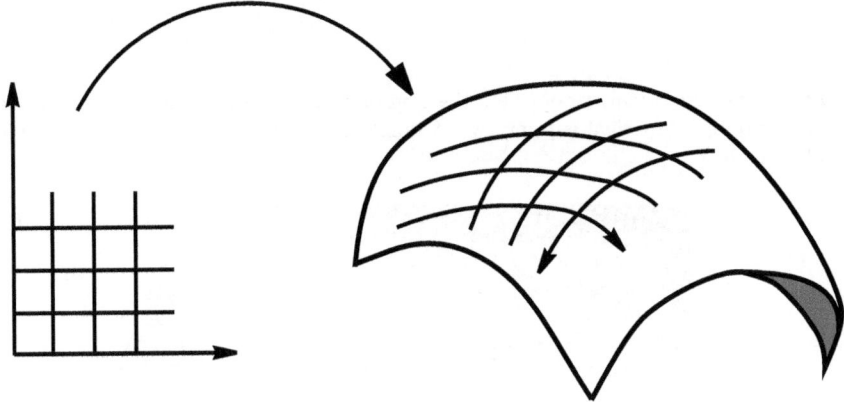

Fig. 2. The orthogonal grid on the parameterization plane is transformed into a non-orthogonal grid on the manifold.

The split is performed by adding an additional edge, connecting the updated grid point with a non-neighboring grid point. The distant grid point becomes part of the numerical stencil. The need for splitting is determined according to the angle between the non-orthogonal axes at the grid point. It is calculated by

$$\cos(\alpha) = \left(\frac{X_1 \cdot X_2}{\|X_1\|\|X_2\|} \right) = \frac{g_{12}}{\sqrt{g_{11}g_{22}}}. \tag{27}$$

If $\cos(\alpha) = 0$, the axes are perpendicular, and no splitting is required. If $\cos(\alpha) < 0$, the angle α is obtuse and should be split. The denominator of Equation (27) is always positive, so we need only check the sign of the numerator g_{12}.

In order to split an angle, we should connect the updated grid point with another point, located m grid points from the point in the direction of X_1 and n grid points in the direction of X_2 (m and n can be negative). The point is

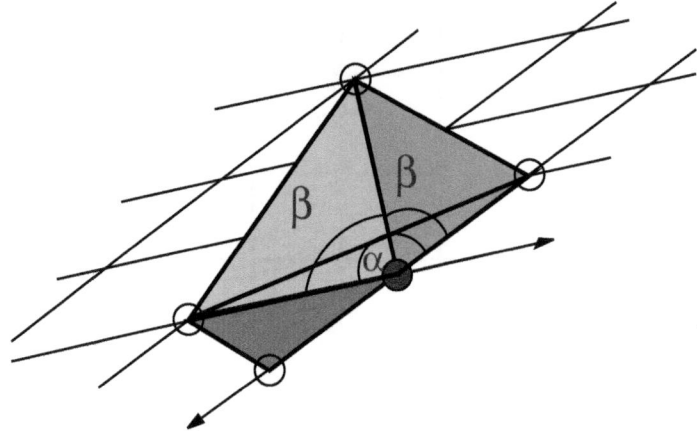

Fig. 3. The numerical support for the non-orthogonal coordinate system. Triangle 1 gives a proper numerical support, yet triangle 2 is obtuse. It is replaced by triangle 3 and triangle 4.

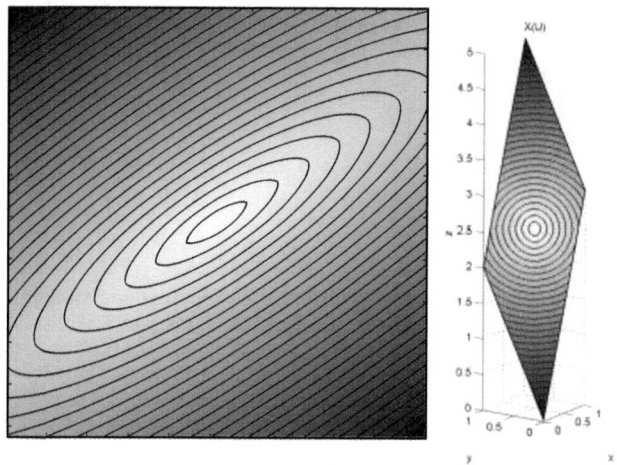

Fig. 4. Fast marching on the manifold $z = 3x + 2y$. Left: implemented on the parameterization plane. Right: projected on the manifold. Lower values are assigned brighter colors. The black curves are the level curves.

a proper supporting point, if the obtuse angle is split into two acute ones. For $\cos(\alpha) < 0$ this is the case if

$$\cos(\beta_1) = \left(\frac{X_1 \cdot (mX_1 + nX_2)}{\|X_1\|\|mX_1 + nX_2\|} \right) = \frac{mg_{11} + ng_{12}}{\sqrt{g_{11}(m^2g_{11} + 2mng_{12} + n^2g_{22})}} > 0,$$

$$(28)$$

and

Fig. 5. Beltrami flow using a short time kernel. The original image is in the top left. The order of the images is from top to bottom and left to right.

$$\cos\left(\beta_2\right) = \left(\frac{X_2 \cdot \left(mX_1 + nX_2\right)}{\|X_2\|\|mX_1 + nX_2\|}\right) = \frac{mg_{12} + ng_{22}}{\sqrt{g_{22}\left(m^2 g_{11} + 2mn g_{12} + n^2 g_{22}\right)}} > 0. \tag{29}$$

Here it is enough to check the sign of the numerator. For $\cos\left(\alpha\right) > 0$, $\cos\left(\beta_2\right)$ changes its sign and the constraints are

$$mg_{11} + ng_{12} > 0, \tag{30}$$

and

$$mg_{12} + ng_{22} < 0. \tag{31}$$

This process is done for all grid points. Once the pre-processing stage is done, we have a suitable numerical stencil for each grid point and we can solve the Eikonal equation numerically. The numerical scheme used is similar to that of solving the Eikonal equation on triangulated manifolds [5] with the exception that there is no need to perform the unfolding step. The supporting grid points

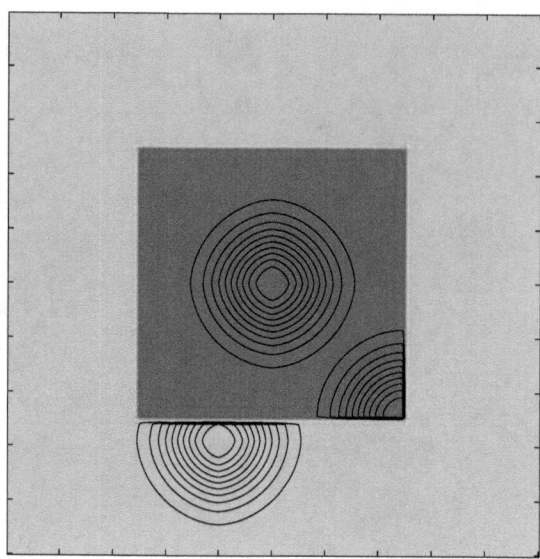

Fig. 6. Level curves of the kernel at various locations in a synthetic image.

that split the obtuse angles can be found more efficiently. The required triangle edge lengths and angles are calculated according to the image metric g_{ij} at the grid point, see [14].

5   Simulations and Results

First, we demonstrate the performance of our algorithm for the solution of the Eikonal equation. The algorithm is tested for a parametric manifold with a non-orthogonal coordinate system. In Figure 4 it is implemented on the tilted plane $z = 3x + 2y$. The correctness of the distance map is evident from the resulting level curves, which are concentric circles on the manifold.

Next, we use the solution of the Eikonal equation to create the short time kernel for the Beltrami flow. Figure 5 shows the implementation of the Beltrami flow for a gray scale image using a short time kernel. In this case $\beta = 3$, the time step taken was $t = 0.5$, and only grid points with a kernel value above 0.01 were used for the filtering. The time difference between the images is 1. Similar results for color images appear in [1].

The use of pixels with a weight larger than 0.01 resulted in an average of 25 neighboring pixels that take part in the filtering of each image pixel. When the threshold is reached, the fast marching algorithm is stopped, and the calculation of the distance to unnecessary points is avoided. In order to make the fast marching algorithm even faster, we can bound in advance the neighborhood in which the Eikonal equation is solved. This way, the pre-processing stage of the algorithm, including the splitting of obtuse angles, is done only for relevant pixels. In Figure 5 the size of this neighborhood is 7×7.

In order to demonstrate the spatial structure of the kernel, we tested it on the synthetic image in Figure 6. At isotropic areas of the image, the kernel is isotropic and its weights are determined solely by the spatial distance from the filtered pixel. Across edges the significant change in intensity is translated into a long geodesic distance, which results in negligent kernel weights on the other side of the edge. The filtered pixel is computed as an average of the pixels on the 'right' side of the edge.

6   Conclusions

A short time kernel was derived for the Beltrami image enhancing flow. Geodesic distances on the image manifold, which are required for the implementation of the kernel, were calculated in a new efficient way. From the theoretical stand point, a connection has been shown between the Beltrami flow and the Bilateral filter. The Bilateral filter is found to be a Euclidean approximation of the Beltrami flow. From a practical stand point, the kernel filter enables an arbitrary time step for a Beltrami-like adaptive smoothing, which is impossible for the explicit numerical schemes currently existing for color images.

References

1. http://www.cs.technion.ac.il/~ron/.
2. D. Barash. Fundamental relationship between bilateral filtering, adaptive smoothing, and the nonlinear diffusion equation. *IEEE Transactions on Pattern Analysis and Machine Intelligence*, 24(6):844–847, June 2002.
3. M. Elad. On the bilateral filter and ways to improve it. *IEEE Transactions on Image Processing*, 11(10):1141–1151, October 2002.
4. R. Kimmel, R. Malladi, and N. Sochen. Image processing via the beltrami operator. In *Proc. of 3-rd Asian Conf. on Computer Vision*, Hong Kong, January 1998.
5. R. Kimmel and J. Sethian. Computing geodesic paths on manifolds. *Proceedings of National Academy of Sciences*, 95(15):8431–8435, July 1998.
6. F. Mémoli and G. Sapiro. Fast computation of weighted distance functions and geodesics on implicit hyper-surfaces. *Journal of Computational Physics*, 173(2):730–764, 2001.
7. J. Sethian. A fast marching level set method for monotonically advancing fronts. *Proceedings of National Academy of Sciences*, 93(4):1591–1595, 1996.
8. J. Sethian and A. Vladimirsky. Ordered upwind methods for static hamilton-jacobi equations: theory and applications. Technical Report PAM-792(UCB), Center for Pure and Applied Mathematics, May 2001. submitted for publication to SIAM Journal on Numerical Analysis in July 2001.
9. N. Sochen. Stochastic processes in vision: From langevin to beltrami. In *Proc. of International Conference on Computer Vision*, Vancouver, Canada, July 2001.
10. N. Sochen, R. Kimmel, and A. Bruckstein. Diffusions and confusions in signal and image processing. *Journal of Mathematical Imaging and Vision*, 14(3):195–209, 2001.
11. N. Sochen, R. Kimmel, and R. Malladi. From high energy physics to low level vision. LBNL report LBNL-39243, UC Berkeley, August 1996.

12. N. Sochen, R. Kimmel, and R. Malladi. A general framework for low level vision. *IEEE Trans. on Image Processing*, 7(3):310–318, 1998.
13. N. Sochen and Y. Y. Zeevi. Representation of colored images by manifolds embedded in higher dimensional non-euclidean space. In *Proc. of ICIP98*, pages 166–170, Chicago, IL, January 1998.
14. A. Spira and R. Kimmel. An efficient solution to the eikonal equation on parametric manifolds. Submitted for publication, March 2003.
15. C. Tomasi and R. Manduchi. Bilateral filtering for gray and color images. In *Sixth International Conference on Computer Vision*, Bombay, India, January 1998.

Evolution of the Critical Points in the Curvature and Affine Morphological Scale Spaces

Marcos Craizer

Dep. Matemática, Pontifícia Universidade Católica do Rio de Janeiro
R. Marquês de São Vicente, 225, Rio de Janeiro, RJ, Brazil, 22453-900
craizer@mat.puc-rio.br

Abstract. In this work we analyse the evolution of the critical points of an image by the curvature and affine morphological scale spaces. We define the notions of circular and elliptic extremum and show that an extremum becomes circular by the curvature scale space and elliptic by the affine morphological scale space. The evolution of a saddle point by the curvature scale space is also described. And we show how these properties can lead to numerical methods for the simulation of the curvature scale space.

1 Introduction

In this paper we shall consider two important scale spaces that appear in computer vision: the scale space generated by the curvature equation, that we shall refer as Curvature Scale Space (CSS), and the Affine Morphological Scale Space (AMSS). In the CSS, the level curves evolve according to the curvature motion, while in the AMSS, the level curves evolve according to the affine curve evolution. In the last decade, these evolutions have been the subject of a lot of research, both empirical and theoretical. Besides the mathematical beauty, they have a lot applications in image processing and computer vision. Some of the tasks we can do with them are edge detection, image smoothing and image enhancement ([1],[10],[11],[13]).

The CSS and AMSS are described by second-order partial differential equations. The equation describing CSS is given by

$$\frac{\partial u}{\partial t} = \frac{\frac{\partial^2 u}{\partial x^2}\left(\frac{\partial u}{\partial y}\right)^2 - 2\frac{\partial^2 u}{\partial x \partial y}\frac{\partial u}{\partial x}\frac{\partial u}{\partial y} + \frac{\partial^2 u}{\partial y^2}\left(\frac{\partial u}{\partial x}\right)^2}{\left(\frac{\partial u}{\partial x}\right)^2 + \left(\frac{\partial u}{\partial y}\right)^2}, \tag{1}$$

where u is the image, (x,y) are the space variables and t the time variable. The equation describing AMSS is

$$\frac{\partial u}{\partial t} = \left[\frac{\partial^2 u}{\partial x^2}\left(\frac{\partial u}{\partial y}\right)^2 - 2\frac{\partial^2 u}{\partial x \partial y}\frac{\partial u}{\partial x}\frac{\partial u}{\partial y} + \frac{\partial^2 u}{\partial y^2}\left(\frac{\partial u}{\partial x}\right)^2\right]^{1/3}. \tag{2}$$

L.D. Griffin and M. Lillholm (Eds.): Scale-Space 2003, LNCS 2695, pp. 523–536, 2003.

The critical points of the image are the points where its gradient is zero. At these points, the second member of equation (1) is not defined, while the second member of equation (2) is zero. The behaviour of these equations near the critical points is very delicate, both in the theoretical and in the numerical point of view, and is the topic of the present paper.

In the case that the critical point is an extremum point, we can describe quite well the behaviour of the CSS and AMSS near it. The results can be summarized as follows: It is well known that the curvature motion evolves a curve until it disappears as a "circular" point ([8],[9]), while the affine motion evolves a curve until it disappears as an "elliptic" point ([2],[12]). The corresponding results for the scale spaces are that the CSS makes an extremum "circular" and the AMSS makes an extremum "elliptic" after any positive time. In this paper, we shall give precise statements of these results. For the CSS, we shall not provide mathematical proofs, since they can be found in [4]. In the AMSS case, we shall give some of the mathematical proofs. The original part of this paper are the results describing the elliptization of an isolated extremum by the AMSS.

In the case that the critical point is a saddle point, the behaviour of the CSS and AMSS near it is more difficult to describe. One interesting property is that the CSS develop a plateau near any saddle point ([6],[7]). In this paper, we shall describe this property. We also show how these properties suggested modifications in the usual numerical method used for simulation of the CSS ([5]).

This paper is organized as follows: In section 2 we fix the notation and show some preliminary facts. In section 3, we describe the behaviour of the critical points under the CSS, and in section 4, we describe the elliptization of an isolated extremum by the AMSS. The appendix contains the proof of some estimates of section 4.

2   Preliminaries and Notation

Let us introduce some notation. Consider a function $u(x, y)$ with a local extremum (x_m, y_m) at height $z_m = u(x_m, y_m)$. For the sake of simplicity, assume that the extremum (x_m, y_m) of u is a local minimum, the case of local maximum being analogous. For a fixed $\Delta > z_m$ and $z_m \leq z \leq \Delta$, denote by $D(z)$ the connected component of the level set $\{(x, y) : u(x, y) \leq z\}$ containing (x_m, y_m) and by $C(z)$ its boundary.

Denote by $\overline{D}(z)$ the region obtained from $D(z)$ by a similarity of rate $\sqrt{1/A(z)}$ and center (x_m, y_m), where $A(z)$ is the area of $D(z)$.

Definition 1. *We say that (x_m, y_m) is a circular minimum if $\overline{D}(z)$ converges to a disk in the Hausdorff metric, when $z \to z_m$ and that (x_m, y_m) is an elliptic minimum if $\overline{D}(z)$ converges to an ellipse in the Hausdorff metric, when $z \to z_m$.*

Example 1. Consider the function $u(x, y) = ax^{2n} + by^{2n}$, where $a > 0, b > 0$ and n is a positive integer. We have that, for any $z > 0$, $\overline{D}(z)$ is the region bounded by the curve $[A(1)]^n \left(ax^{2n} + by^{2n}\right) = 1$, where $A(1)$ denotes the area inside $ax^{2n} + by^{2n} = 1$. Hence the minimum $(0,0)$ is elliptic if and only if $n = 1$ and is circular if and only $n = 1$ and $a = b$.

We shall denote by $D(z,t)$ the evolution of $D(z)$ by the curvature motion or the affine curve evolution with $D(z,0) = D(z)$. The evolution $u(x, y, t)$ of $u(x,y)$ by the CSS or the AMSS is the function whose level sets are $D(z,t)$. Assuming that each $D(z)$ is convex and smooth, it is known that $D(z,t)$ is also convex and smooth until it disappears ([9],[12]). We shall denote by $T(z)$ the time that the region $D(z)$ takes to disappear and by $O(z)$ its final point.

In this paper, we shall always assume that $D(z)$ is convex and smooth for $z_m < z \leq \Delta$.

Lemma 1. *Considering the CSS or the AMSS evolution, the final time $T(z)$ is a continuous increasing function of z. And the final point $O(z)$ is a continuous function of z.*

Proof. Fix $z_m < z \leq \Delta$. Given $\varepsilon > 0$, we can choose $\delta > 0$ such that for $z - \delta \leq z_1 \leq z$, $D(z, \varepsilon) \subset D(z_1, 0)$ and for $z \leq z_2 \leq z + \delta$, $D(z_2, \varepsilon) \subset D(z, 0)$. This implies that for any $0 \leq t \leq T(z) - \varepsilon$,

$$D(z, \varepsilon + t) \subset D(z_1, t) \subset D(z, t),$$

and

$$D(z, \varepsilon + t) \subset D(z_2, \varepsilon + t) \subset D(z, t).$$

Therefore $T(z) - \varepsilon \leq T(z_1) \leq T(z)$ and $T(z) \leq T(z_2) \leq T(z) + \varepsilon$. This implies that the function $T(z)$ is continuous and increasing. Also, the inclusions above imply that $O(z_1)$ and $O(z_2)$ belong to $D(z, T(z) - \varepsilon)$. So the function $O(z)$ is continuous.

Denote by $z_m(t)$ the inverse function of $T(z)$. Then it is clear that $O(z_m(t)) = (x_m(t), y_m(t))$ is an isolated minimum of $u(x, y, t)$ at height $z_m(t)$. Moreover, $z_m(t)$ is also a continuous increasing function of t.

Denote by $A(z,t)$ the area of $D(z,t)$ and by $C(z,t)$ its boundary. When we supress the argument t, we understand that we have chosen $t = 0$. Since the curves $C(z,t)$ are convex, we can parameterize them by the angle θ that the tangent vectors make with the x-axis. For each (z,t), denote by $\varphi(z,t) : \mathbf{S}^1 \to \mathbf{R}^2$ this parameterization and by $k(z,t,\theta)$ the curvature of the curve $C(z,t)$ at the point corresponding to θ.

Denote by $\overline{D}(z,t)$ the region obtained from $D(z,t)$ by a similarity of rate $\sqrt{1/A(z,t)}$ and center $O(z)$ and by $\overline{C}(z,t)$ its boundary. In the following sections, we shall also consider the parametrization $\overline{\varphi}(z,t) : \mathbf{S}^1 \to \mathbf{R}^2$ of $\overline{C}(z,t)$ defined by

$$\overline{\varphi}(z,t,\theta) = \sqrt{\frac{1}{A(z,t)}}\left(\varphi(z,t,\theta) - O(z)\right),$$

and the corresponding curvature $\overline{k}(z,t,\theta) = k(z,t,\theta)A^{1/2}(z,t)$.

3   The Behaviour of the CSS Near Critical Points

In this section , we shall study the evolution of a critical point under the CSS. In subsection 3.1, we shall describe the circularization of the extrema. In subsection 3.2, we shall describe the formation of a plateau near a saddle point. And in subsection 3.3, we show how these facts were used to justify a modification in the usual numerical method for the CSS simulation.

3.1   Circularization of an Extremum by the CSS

We shall now describe in 3 different ways the circularization of extrema by the CSS. All results of this subsection are proved in [4]. We shall make the general hypothesis that the initial image is smooth in a neighborhood of the extremum, except possibly at the extremum.

The first result says that, for any fixed $t_0 > 0$, the level curves $C(z, t_0)$ converge to a circle in the isoperimetric sense, when z tends to $z_m(t_0)$. The corresponding result for a single curve was proved in [8].

Theorem 1. *Denote by $\mathcal{L}(z,t)$ the euclidean length of $C(z,t)$. Then, for any $t_0 > 0$, we have that*

$$\lim_{z \to z_m(t_0)} \frac{\mathcal{L}^2}{A}(z, t_0) = 4\pi.$$

Our second result is stronger than the first and states that the level curves $C(z, t_0)$ converge to a circle in the curvature sense. The corresponding result for a single curve was proved in [9].

Theorem 2. *For any $t_0 > 0$, we have that*

$$\lim_{z \to z_m(t_0)} \overline{k}(\theta, z, t_0) = 1,$$

uniformly in θ. Moreover, for any $1 \le j < \infty$,

$$\lim_{z \to z_m(t_0)} \frac{\partial^j}{\partial \theta^j} \overline{k}(\theta, z, t_0) = 0,$$

uniformly in θ.

Our third result says that, for any $t_0 > 0$, the solution has a quadratic expansion with equal "eigenvalues" at the extremum. This result, besides saying that the extremum is circular, also gives information about the relation between the "radius" of the level set at height z and the distance between z and $z_m(t_0)$. The hypothesis we need is that the initial area function $A(z)$ has a non-zero derivative with respect to z at the point $z_m(t_0)$.

Theorem 3. *Assume that the extremum is a minimum. For any $t > 0$ such that $\frac{\partial A}{\partial z}(z_m(t_0)) \neq 0$, we can write*

$$u(x, y, t_0) = z_m(t_0) + \frac{\lambda(t_0)}{2} \left((x - x_m(t_0))^2 + (y - y_m(t_0))^2 \right) + h(x, y),$$

with

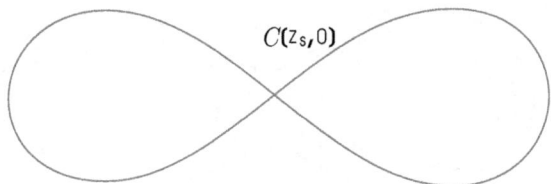

$C(z_s, 0)$

Fig. 1. The initial level curve.

$$\lim_{(x,y)\to(x_m(t_0),y_m(t_0))} \frac{h(x,y)}{(x-x_m(t_0))^2+(y-y_m(t_0))^2} = 0,$$

and

$$\lambda(t_0) = 2\pi\left[\frac{\partial A}{\partial z}(z_m(t_0))\right]^{-1}.$$

When the original image is smooth at the extremum and this extremum is non-degenerate, we can find explicitly the limit of $\lambda(t_0)$ of the above theorem, when t_0 goes to zero.

Proposition 1. *Denote by a and b are the eigenvalues of the Hessian matrix of the original image at the non-degenerate minimum. Then*

$$\lim_{t_0\to 0} 2\pi\left[\frac{\partial A}{\partial z}(z_m(t_0))\right]^{-1} = \sqrt{ab}.$$

3.2   Formation of a Plateau Near a Saddle Point

Consider a non-degenerate saddle point (x_s, y_s) and let $z_s = u(x_s, y_s, 0)$. Denote by $C(z_s, t)$ the connected components of the level set

$$\{(x,y) \in \mathbf{R}^2 |\ u(x,y,t) = z_s\}$$

that passes through a neighborhood of the saddle point, where $u(x,y,t)$ is the evolution of $u(x,y,0)$ by the CSS. The level line $C(z_s, 0)$ is a figure eight curve, as the one showed in figure 1.

One can show that $C(z_s, t_0)$ has positive area, for any $t_0 > 0$ ([5]). This fact implies that $C(z_s, t_0)$ is not a curve. In fact, it is shown in [7], page 138, that this set looks like the "fat eight" of figure 2. For more details related to the creation of plateaus by the CSS, see [6] and [7].

3.3   Numerical Method for Implementing CSS

In proposition 1, it is proved that the vertical speed of a non-degenerate minimum is given by the geometric mean of the eigenvalues of the second derivative. And the arguments above show that, since the initial saddle point is contained in the "fat eight", the vertical speed of this point must be zero. Taking into account

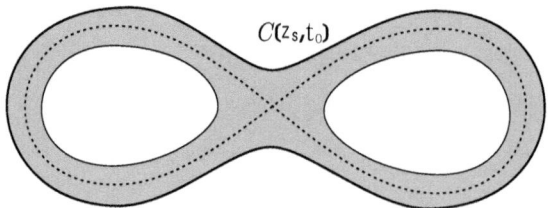

Fig. 2. Evolution of the eight by the CSS.

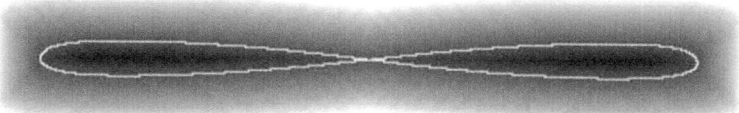

Fig. 3. Level curve through the saddle point in the original image.

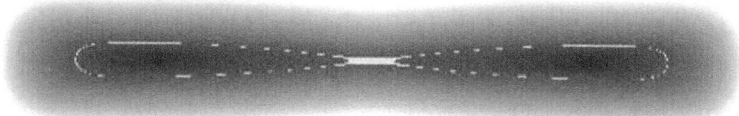

Fig. 4. Evolution of the level curve under the CSS using the modified algorithm.

these facts, a modification of the usual numerical method used to simulate the CSS was proposed in [5].

The usual numerical methods for implementing the CSS uses a discretization of the equation 1 outside the critical points and add some diffusion near them. The algorithm proposed in [5] also discretizes equation 1 outside the critical points, but uses the geometric mean of the eigenvalues of the second derivative as the vertical speed of the extrema and zero as the vertical speed of the saddle points. In figure 3, we see the level curve of an image containing a saddle point and in figure 4 we can see the same level curve after some time of evolution by the CSS. We observe that the plateau that appears in figure 4 does not appear so clearly using other algorithms.

4   Elliptization of an Extremum by the AMSS

In this section, we shall describe the evolution of the extrema under the AMSS. Let $\varphi(q)$, $0 \leq q \leq Q$, be a parameterized closed convex curve in the plane. Denote by

$$s = \int_0^q \left[\frac{\partial \varphi}{\partial q}, \frac{\partial^2 \varphi}{\partial q^2} \right]^{1/3} dq$$

the affine parameter of the curve, where $[X, Y]$ represents the determinant of the 2×2 matrix whose columns are X and Y. The affine curve evolution is defined by the intrinsic heat equation

$$\frac{\partial \varphi}{\partial t} = \frac{\partial^2 \varphi}{\partial s^2}.$$

In terms of curve evolution, it is equivalent to the evolution determined by the equation

$$\frac{\partial \varphi}{\partial t} = k^{1/3}\mathbf{n},$$

where $\mathbf{n}$ is the euclidean normal to the curve pointing inwards and k is the euclidean curvature. When the level sets of an image u evolves by the above equations, the image itself satisfies equation (2).

Consider an isolated extremum (x_m, y_m) of an image u that is evolving according to the AMSS. It it proved in [2] that, for any z fixed, $\overline{\varphi}(z, t, \theta)$ converges in the C^∞ topology to $\overline{\varphi}(z, T(z), \theta)$, which is a parametrization of an ellipse $E(z)$, when t goes to $T(z)$. In the following, we shall fix any time $t_0 > 0$ and prove that $\overline{\varphi}(z, t_0, \theta)$ converges in the C^∞ topology to $\overline{\varphi}(z_m(t_0), t_0, \theta)$, when z decreases to $z_m(t_0)$. In particular, we obtain that $\overline{D}(z, t_0)$ converges to $E(z_m(t_0))$ in the Hausdorff metric, and thus the extremum $(x_m(t_0), y_m(t_0))$ is elliptic.

4.1   Some Important Estimates

In order to prove the above mentioned convergence, we need estimates of the relevant quantities. These estimates were obtained in [2] for a single curve, but we need them to be uniform in z. The hypothesis that we consider is that the initial image is smooth in a neighborhood of the extremum, except possibly at the extremum.

Isoperimetric and Area Estimates. The affine isoperimetric ratio is defined as $\frac{L^3}{A}$, where L is the affine length given by

$$L = \int_0^Q \left[\frac{\partial \varphi}{\partial q}, \frac{\partial^2 \varphi}{\partial q^2}\right]^{1/3} dq.$$

For each $z_m < z \le \Delta$, it is well known that the isoperimetric ratio $\frac{L^3}{A}(z, t)$ converges to $8\pi^2$ when $t \to T(z)$ ([12]). For our purposes, we need to verify some kind of uniformity in this convergence. We shall write $(z, t) \to (z_m(t_0), t_0)$ to indicate that t is approaching t_0 and z is approaching $z_m(t_0)$ by above.

Proposition 2. *For any $t_0 > 0$,*

$$\lim_{(z,t) \to (z_m(t_0), t_0)} \frac{L^3}{A}(z, t) = 8\pi^2$$

The proof of this proposition can be found in the appendix.

Corollary 1. *We have that*

$$\lim_{(z,t) \to (z_m(t_0), t_0)} \frac{A^{2/3}(z, t)}{T(z) - t} = \frac{4}{3}\pi^{2/3}.$$

Proof. Since $\frac{\partial A}{\partial t} = -L$ (see [12]), proposition 2 implies that

$$\frac{\partial}{\partial t} A^{2/3}(z,t) = -\frac{2}{3}\frac{L(z,t)}{A^{1/3}(z,t)} = -\frac{4\pi^{2/3}}{3} + r(z,t),$$

where $\lim_{(z,t)\to(z_m(t_0),t_0)} r(z,t) = 0$. So

$$\frac{A^{2/3}(z,t)}{T(z)-t} = -\frac{1}{T(z)-t}\int_t^{T(z)}\frac{\partial}{\partial t}A^{2/3}(z,t)dt = \frac{4\pi^{2/3}}{3} - \frac{1}{T(z)-t}\int_t^{T(z)} r(z,t)dt.$$

But since $r(z,t)$ goes to 0, we conclude the proof of the corollary.

Affine and Euclidean Curvature Estimates. The affine curvature is defined by

$$\mu(s) = [\frac{\partial^2\varphi}{\partial s^2}, \frac{\partial^3\varphi}{\partial s^3}].$$

For a geometrical interpretation of this curvature, see [3]. Denote by $osc(f(\theta))$ the oscillation of a function $f : \mathbf{S}^1 \to \mathbf{R}$ and by $\overline{\mu}(z,t,\theta) = \mu(z,t,\theta)A^{2/3}(z,t)$ the affine curvature of $\overline{C}(z,t)$ at the point θ.

The following estimates are fundamental to prove that the extrema become elliptic.

Theorem 4. *Fix $t_0 > 0$. There exist constants C_l, $l = 0,1,2,...$ such that for for any $z_m(t_0) \leq z \leq \Delta$ and $0 \leq t < T(z)$, there exists an affine transformation $H(z,t)$ of $\mathbf{R}^2$ with $A(H(z,t) \circ D(z,t)) = \pi$ and*

$$\|H(z,t) \circ \varphi(z,t)\|_{C^l} \leq C_l.$$

Theorem 5. *Fix $t_0 > 0$. There exist positive constants α and C (depending only on t_0) such that*

$$osc(\overline{\mu}(z,t,\theta)) \leq C(T(z) - t)^\alpha,$$

for any $z_m(t_0) \leq z \leq \Delta$ and $0 \leq t \leq T(z)$.

Detailed proofs of theorems 4 and 5 will not be presented in this paper, since they would be very long. Theorem 4 can be proved in the same way as theorem 5.13 in [2], taking into account that in our case we have a family of level curves instead of a single curve. And theorem 5 can be proved in the same way as lemma 6.18 in [2], again taking into account that we have a family of level curves. A hypothesis that we need to prove these theorems is that the initial level curves have affine curvatures bounded below and uniformly bounded derivatives. These conditions are automatically fullfilled in the case of a image that is smooth outside the extremum.

Theorem 5 allows us to obtain an estimate of the normalized affine curvature, and this estimate has as consequence the following estimate of the normalized euclidean curvature:

Proposition 3. *Fix $t_0 > 0$. There are positive constants C_1 and α independent of z, t and θ such that*

$$\exp\left\{-C_1(T(z) - t)^\alpha\right\} \le \frac{\overline{k}(z, T(z), \theta)}{\overline{k}(z, t, \theta)} \le \exp\left\{C_1(T(z) - t)^\alpha\right\},$$

for any $z_m(t_0) \le z \le \Delta$, $0 \le t \le T(z)$ and $\theta \in \mathbf{S}^1$.

The proof of this proposition can be also obtained from [2]. In fact, the proof given in [2], page 224, is for a single curve, but a similar proof works also in our case.

Corollary 2. *Assume that the initial image is smooth outside the extremum and fix $t_0 > 0$. Then*

$$\lim_{z \to z_m(t_0)} \frac{\overline{k}(z, t_0, \theta)}{\overline{k}(z_m(t_0), t_0, \theta)} = 1,$$

uniformly in $\theta \in \mathbf{S}^1$.

The proof of this corollary is given in the appendix.

4.2   Elliptization of the Extrema

In this subsection, we show the elliptization of the extrema in three different ways, as we have done for the CSS case. At the end of the subsection we give two illustrative examples.

Elliptization in the Hausdorff Metric. We can now show that for any $t_0 > 0$, $(x_m(t_0), y_m(t_0))$ is an elliptic extremum. This is equivalent to show that $\overline{D}(z, t_0)$ converges to an ellipse in the Hausdorff metric, when z goes to $z_m(t_0)$.

Theorem 6. *Assume that the initial function is smooth except possibly at the extremum (x_m, y_m). Then, for any $t_0 > 0$, $\overline{D}(z, t_0)$ converges to the ellipse $E(z_m(t_0))$ in the Hausdorff topology, when z goes to $z_m(t_0)$.*

Proof. Direct consequence of corollary 2 and the fact that $O(z)$ converges to $O(z_m(t_0))$.

Elliptization in the C^∞-Topology. We now show that $\overline{D}(z, t_0)$ converges to $E(z_m(t_0))$ in the C^∞ topology, which is much stronger than the convergence in the Hausdorff metric.

Theorem 7. *Assume that the initial function is smooth except possibly at the extremum (x_m, y_m). Then, for any $t_0 > 0$, $\overline{D}(z, t_0)$ converges to $E(z_m(t_0))$ in the C^∞ topology.*

Proof. We can rewrite theorem 4 as

$$\|H(z,t) \circ \overline{\varphi}(z,t)\|_{C^l} \leq C_l,$$

$H(z,t)$ being an affine transformation of the plane with $\det(H) = 1$. Since $\overline{D}(z,t)$ converges to $E(z)$ in the Hausdorff metric, we conclude that $\|H(z,t)\|$ is bounded, for any (t,z) in a neighborhood of $(t_0, z_m(t_0))$. Therefore we can write

$$\|\overline{\varphi}(z,t)\|_{C^l} \leq B_l.$$

A standard argument shows now that $\overline{\varphi}(z,t_0)$ converges to $\overline{\varphi}(z_m(t_0),t_0)$ in the C^∞ topology.

Expansion Near the Extrema. We now show a normal form that describes the behaviour of the AMSS near the extrema, for almost every time. Next results, besides saying that the extremum is elliptic, also relates the length of the "axes of the ellipse" at height z with the distance between z and $z_m(t_0)$.

Proposition 4. *We have that, for almost all $0 < t_0 < T(\Delta)$,*

$$\lim_{z \to z_m(t_0)} \frac{A^{2/3}(z,t_0)}{z - z_m(t_0)} > 0$$

Proof. The function $t \to z_m(t)$ is strictly increasing. Therefore for almost every $t_0 > 0$, $\frac{dz_m}{dt}(t_0) > 0$. This is equivalent to $\frac{dT}{dz}(z_m(t_0)) > 0$. But

$$\frac{A^{2/3}(z,t_0)}{z - z_m(t_0)} = \frac{A^{2/3}(z,t_0)}{T(z) - t_0} \cdot \frac{T(z) - t_0}{z - z_m(t_0)}$$

and, by proposition 1,

$$\lim_{z \to z_m(t_0)} \frac{A^{2/3}(z,t_0)}{T(z) - t_0} = \frac{4}{3}\pi^{2/3}.$$

We conclude that

$$\lim_{z \to z_m(t_0)} \frac{A^{2/3}(z,t_0)}{z - z_m(t_0)} = \frac{4}{3}\pi^{2/3} \cdot \frac{dT}{dz}(z_m(t_0)),$$

thus proving the proposition.

Corollary 3. *For almost all $0 < t_0 < T(\Delta)$, $u(x,y,t_0)$ can be written, possibly after a rotation of the plane, as*

$$u(x,y,t_0) = z_m(t_0) + (1 + \gamma(x,y))$$
$$\times \left(a\,(x - x_m(t_0))^2 + b\,(y - y_m(t_0))^2 + h(x,y) \right)^{2/3},$$

where $a > 0, b > 0$, h and γ are continuous functions satisfying $\gamma(0,0) = 0$ and

$$\lim_{(x,y) \to (x_m(t_0), y_m(t_0))} \frac{h(x,y)}{(x - x_m(t_0))^2 + (y - y_m(t_0))^2} = 0.$$

Proof. The ellipticity of the minimum $(x_m(t_0), y_m(t_0))$ implies that, possibly after a rotation of the plane, we can write

$$A(z, t_0) = a_1 (x - x_m(t_0))^2 + b_1 (y - y_m(t_0))^2 + h(x, y), \qquad (3)$$

for some $a_1 > 0$, $b_1 > 0$ and $h(x, y)$ satisfying

$$\lim_{(x,y)\to(x_m(t_0), y_m(t_0))} \frac{h(x, y)}{(x - x_m(t_0))^2 + (y - y_m(t_0))^2} = 0.$$

Now, the above proposition says that, for almost every $0 < t_0 < T(\Delta)$,

$$A^{2/3}(z, t_0) = (c + \gamma(x, y)) (z - z_m(t_0)),$$

with $c > 0$ and $\gamma(0, 0) = 0$. If we use this equation in formula (3), we complete the proof of the corollary.

Examples. We complete the section with two illustrative examples:

Example 2. Consider the initial function

$$u(x, y) = \frac{1}{2} \left(x^2 + \sqrt{x^2 + 4y^2} \right),$$

which has an isolated minimum at $(0, 0)$. Denoting by $E(a, b)$ the ellipse centered at the origin and with axes $2a$ and $2b$, parallel to the coordinate axes, we have that the level curves of u are $E(\sqrt{z}, z)$. So the minimum is not elliptic. After time $t_0 > 0$, each level set has evolved to

$$E\left(\sqrt{z} \left(1 - \frac{4t_0}{3z} \right)^{3/4}, z \left(1 - \frac{4t_0}{3z} \right)^{3/4} \right)$$

(see for example [12]). Hence $z_m(t_0) = \frac{4t_0}{3}$ and the normalized level set $\overline{D}(z, t_0)$ converge to

$$E\left(\frac{1}{\sqrt{\pi}} \left(\frac{4t_0}{3} \right)^{-1/4}, \frac{1}{\sqrt{\pi}} \left(\frac{4t_0}{3} \right)^{1/4} \right),$$

when z goes to $z_m(t_0)$.

Example 3. Suppose that the initial function is given by $u(x, y) = f_0(r)$, where f is an increasing smooth function with $f(0) = 0$ and $r = \sqrt{x^2 + y^2}$. Then the level set at height z is a disk of radius $g_0(z)$, where g_0 is the inverse of f_0. Since

$$\frac{\partial r}{\partial t} = -\frac{1}{r^{1/3}},$$

we obtain

$$r^{4/3}(z, t_0) = g_0^{4/3}(z) - \frac{4t_0}{3},$$

which implies that

$$u(x, y, t_0) = f_0 \left[\left(r^{4/3} + \frac{4t_0}{3} \right)^{3/4} \right].$$

This formula can be rewritten as

$$u(x, y, t_0) = z_m(t_0) + (a + \alpha(r)) \, r^{4/3}$$

where $z_m(t_0) = f_0 \left[\left(\frac{4t_0}{3} \right)^{3/4} \right]$, $a = f_0' \left[\left(\frac{4t_0}{3} \right)^{3/4} \right] \frac{3}{4} \left(\frac{4t_0}{3} \right)^{-1/4}$ and

$$\lim_{r \to 0} \alpha(r) = 0,$$

which agrees with the formula of corollary 3.

5   Conclusion and Future Work

In this paper we have considered the behaviour of the CSS and the AMSS near critical points. The behaviour of both spaces near extrema are quite well understood: the CSS makes an extremum become circular and the AMSS makes an extremum become elliptic. We have also described how the CSS forms plateaus near saddle points.

An interesting problem for future work is the description of the AMSS near a saddle point. Another interesting theoretical question is to understand what happens to the critical points for scale spaces whose level curves evolve with a speed given by a power of the curvature different from 1 or 1/3.

References

1. L.Alvarez, P.L.Lions and J.M.Morel, Image selective smoothing and edge detection by nonlinear diffusion, SIAM J.Numer.Analysis 29, 845-866, 1992.
2. B.Andrews, Contraction of convex hypersurfaces by their affine normal. J.Diff.Geometry 43, n.2, 207-230, 1996.
3. S.Buchin, Affine differential geometry, Gordon & Breach, NY, 1983.
4. M.Craizer and R.Teixeira, Evolution of an extremum by curvature motion, pre-print.
5. M.Craizer, S.Pesco and R.Teixeira, A numerical scheme for the curvature equation near the singularities, pre-print.
6. L.C.Evans and J.Spruck, Motion of level sets by mean curvature II, Trans.Amer.Math.Soc. 330, n.1, 321-332, 1992.
7. L.C.Evans and J.Spruck, Motion of level sets by mean curvature III, J.Geom.Analysis 2, n.2, 121-150, 1992.
8. M.E.Gage, Curve shortening makes convex curves circular, Inv.Mathematicae 76, 357-364, 1984.
9. M.Gage and R.S.Hamilton, The heat equation shrinking convex plane curves, J.Diff.Geometry 23, 69-96, 1986.
10. B.B.Kimia, A.Tannenbaum and S.W.Zucker, Shapes, shocks and deformations, I, Int.J.Comput.Vision 15, 189-224, 1995.

11. P.Perona and J.Malik, Scale space and edge detection using anisotropic diffusion, Proc.IEEE Comp.Soc.Workshop Computer Vision, 1987.
12. G.Sapiro and A.Tannenbaum, On affine plane curve evolution. J.Funct.Analysis 119, 79-120, 1994.
13. G.Sapiro, Geometric partial differential equations and image analysis. Cambridge University Press, 2001.

Appendix: Proof of some Estimates of Section 5

Proof of Proposition 2.

In the proof of proposition 2, we shall use the functional F defined by

$$F(C) = L \left(\frac{L^2}{A} - 2 \oint_C \mu ds \right).$$

It is easy to see that F is absolutely affine invariant and that $F(C) = 0$ if and only if C is an ellipse. Moreover, it is proved in [12] that

$$\frac{\partial}{\partial t} \left(\frac{L^3}{A} \right) = \frac{L}{A} F(C).$$

We now proceed to the proof of the proposition. Given $\varepsilon > 0$, suppose that there exists a sequence $z_j \searrow z_m(t_0)$ and a sequence of times $t_j \to t_0$ such that

$$\frac{L^3}{A}(z_j, t_j) \leq 8\pi^2 - \varepsilon.$$

Then, since $\frac{L^3}{A}(z_j, t)$ is increasing with t,

$$\frac{L^3}{A}(z_j, t) \leq 8\pi^2 - \varepsilon,$$

for any $0 \leq t \leq t_j$. Therefore there exists $c = c(\varepsilon) > 0$ such that $F(z_j, t) > c$. Hence

$$\frac{\partial}{\partial t} \left(\frac{L^3}{A} \right)(z_j, t) \geq \frac{Lc}{A} = -c \frac{\partial}{\partial t} (\ln A)(z_j, t),$$

where we have used that $\frac{\partial A}{\partial t} = -L$. Integrating from $t = 0$ to $t = t_j$ we obtain

$$\frac{L^3}{A}(z_j, t_j) - \frac{L^3}{A}(z_j, 0) \geq c \ln(A(z_j, 0)) - c \ln(A(z_j, t_j)),$$

and so

$$\frac{L^3}{A}(z_j, t_j) \geq \frac{L^3}{A}(z_j, 0) + c \ln \left(\frac{A(z_j, 0)}{A(z_j, t_j)} \right).$$

Making $z_j \to z_m(t_0)$ and $t_j \to t_0$, we obtain that the second member goes to infinity, contradicting the fact that $\frac{L^3}{A}$ is bounded by $8\pi^2$.

Proof of Corollary 2.

Let us now prove corollary 2. Given $\varepsilon > 0$, take $\Delta > 0$ and $\delta > 0$ such that for any $z_m(t_0) \leq z \leq z_m(t_0) + \Delta$,

$$(1 + \varepsilon)^{-1} \leq \frac{\overline{k}(z_m(t_0), t_0 - \delta, \theta)}{\overline{k}(z_m(t_0), t_0, \theta)} \leq 1 + \varepsilon$$

and

$$(1 + \varepsilon)^{-1} \leq \frac{\overline{k}(z, t_0, \theta)}{\overline{k}(z, t_0 - \delta, \theta)} \leq 1 + \varepsilon.$$

Now, if Δ is sufficiently small, we can assume that

$$(1 + \varepsilon)^{-1} \leq \frac{\overline{k}(z, t_0 - \delta, \theta)}{\overline{k}(z_m(t_0), t_0 - \delta, \theta)} \leq 1 + \varepsilon.$$

Multiplying these inequalities we obtain

$$(1 + \varepsilon)^{-3} \leq \frac{\overline{k}(z, t_0, \theta)}{\overline{k}(z_m(t_0), t_0, \theta)} \leq (1 + \varepsilon)^3,$$

thus proving the corollary.

MAPS: Multiscale Attention-Based PreSegmentation of Color Images

Nabil Ouerhani and Heinz Hügli

Institute of Microtechnology, University of Neuchâtel
Rue A.-L. Breguet 2, CH-2000 Neuchâtel, Switzerland
{Nabil.Ouerhani,Heinz.Hugli}@unine.ch

Abstract. This paper reports a novel Multiscale Attention-based Pre-Segmentation method (MAPS) which is built around the multi-feature, multiscale, saliency-based model of visual attention. From the saliency map, provided by the attention algorithm, MAPS first derives the spatial locations of salient regions that will be considered further in the segmentation process. Then, the salient scale and the salient feature of each salient region is determined by exploring the scale and feature spaces computed by the model of attention. A first and rough multiscale segmentation of the salient regions is performed on the corresponding salient scale. This innovative presegmentation but yet uncomplete procedure is followed by some refined segmentation that operates in the salient feature at full resolution.

Keywords: color image segmentation, visual attention, attentive vision, salient regions, salient features, salient scales, multiscale processing.

1 Introduction

Image segmentation is an essential preprocessing step towards scene understanding in computer vision. The segmentation task aims at grouping together spatially connected pixels which fulfill certain homogeneity criteria. Image segmentation algorithms can be roughly classified into three categories:

(i) **Pixel-based segmentation** is the most local method to address the task of image segmentation. The property of single pixels is used to classify the image points into regions. Histogram thresholding [1] and data clustering [2] (among others techniques) belong to this category.

(ii) **Edge-based segmentation** relies on discontinuities of image data. The algorithms belonging to this category are generally composed of three main steps. Firstly, edges are extracted using edge detection techniques [3]. In the second step, non connected edges which belong to the same physical regions border are connected. Finally, regions are derived from the close edges.

(iii) **Region-based segmentation** is based on two main principles. First, the feature homogeneity, which means that pixels of the same region must fulfill certain homogeneity criteria. The other principle is the spatial connectivity of pixels of the same region. Split and merge [4], as well as region growing [5] are classical examples of this category.

L.D. Griffin and M. Lillholm (Eds.): Scale-Space 2003, LNCS 2695, pp. 537–549, 2003.
© Springer-Verlag Berlin Heidelberg 2003

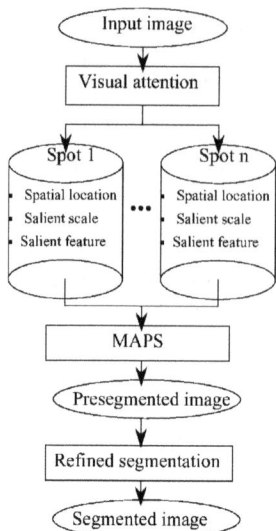

Fig. 1. MAPS: The saliency-based model of visual attention provides a set of data about visually salient regions, such as spatial location, the salient scale of regions and their salient features. These data are efficiently used by the segmentation module to optimally segment the visually relevant image parts.

Of course there exist hybrid segmentation methods that combine techniques from different categories to obtain better results [6,7].

Although built around different concepts, the segmentation techniques described above have at least two major properties in common. All of them try to partition the **entire** image into regions. Despite the fact that this approach is widely used in the computer vision field, it does not have its foundation from human vision. Human vision is principally attentive. Only a small subset of the sensory information, which is selected by our visual attention mechanism, is processed by our brain to perform scene understanding tasks. Thus, and in a computer vision context, focusing the segmentation task on significant regions speeds up not only the segmentation itself but also the subsequent tasks such as object recognition. A previous work that has dealt with attentive segmentation of color images was presented in [8].

The second property, which numerous segmentation techniques (especially the region-based ones) have in common is the use of the same homogeneity criteria to segment all scene regions. This tendency can be seen as a limitation since it neglects the feature-related specificities of single image segments. The idea is to adapt the homogeneity criteria according to the features that discriminate the region to be segmented from its surroundings: the salient feature.

In this paper we present a novel Multiscale Attention-based PreSegmentation (MAPS) method, which addresses the segmentation issues mentioned above. Inspired from psychophysical findings, our method is built around the multi-feature, multiscale, saliency-based model of visual attention. From the saliency

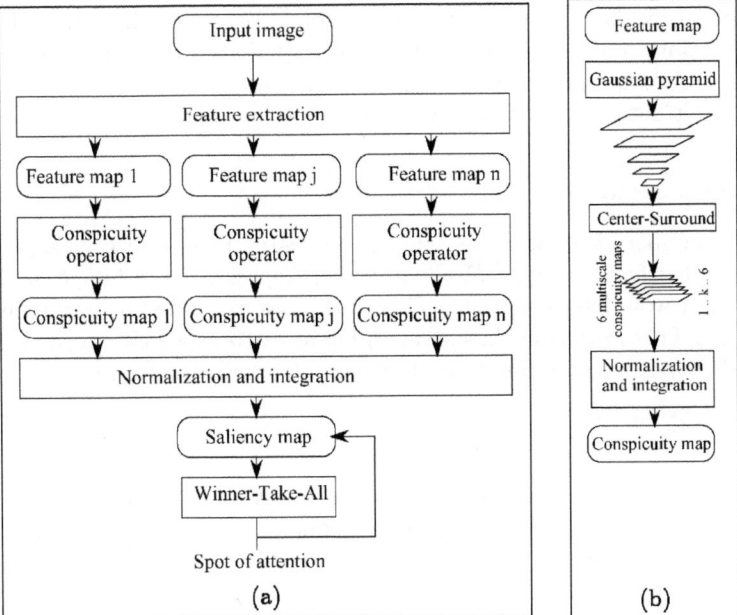

Fig. 2. Saliency-based model of visual attention. (a) represents the four main steps of the visual attention model. Feature extraction, conspicuity computation (for each feature), saliency map computation by integrating all conspicuity maps and finally the detection of spots of attention by means of a winner-take-all network. (b) illustrates, with more details, the conspicuity operator, which computes six intermediate multiscale conspicuity maps. Then, it normalizes and integrates them into the feature-related conspicuity map.

map, provided by the attention algorithm, MAPS derives the spatial locations of the visually salient regions, which will be considered in the segmentation process. Then, our segmentation method determines the salient scale of each visually salient region as well as its salient feature, that is the feature that distinguishes a region from its surrounding. A first and rough multiscale segmentation, which is performed in the salient scale, is used in a later step to achieve a refined segmentation at full resolution, where the homogeneity criterion is adapted to the region's salient feature.

The remainder of this paper is organized as follows. Section 2 reports the saliency-based model of visual attention that we used to develop our segmentation method. The presegmentation method MAPS is presented in Section 3. The usability of the relevant data, provided by MAPS, in an accurate segmentation task is showed in Section 4. Finally, the conclusions are stated in section 5.

2   Visual Attention Model

According to a generally admitted model of visual perception [9], visually salient regions are defined as those scene parts that differ, according to one or a combination of features, from their neighborhood. Based on this principle Koch *et al.* [10] reported a computational model of visual attention that gave rise to numerous software and hardware implementations [11,12,13]. The saliency-based model of attention consists of four main steps (see Fig. 2).

2.1   Feature Maps

First, a number of features $(1..j..n)$ are extracted from the scene by computing the so called feature maps F_j. Such a map represents the image of the scene, based on a well-defined feature. This leads to a multi-feature representation of the scene. This work considers seven different features which are computed from an RGB color image and can be classified in three main groups.

- Intensity feature

$$F_1 = (R + G + B)/3 \tag{1}$$

- Two chromatic features based on the two color opponency filters R^+G^- and B^+Y^- where the yellow signal is defined by $Y = \frac{R+G}{2}$. Such chromatic opponency exists in human visual cortex [14].

$$F_2 = R - G$$
$$F_3 = B - Y \tag{2}$$

Before computing these two features, the color components are first normalized by F_1 in order to decouple hue from intensity.
- Four local orientation features $F_{4..7}$ according to the angles $\theta \in \{0°, 45°, 90°, 135°\}$ [15].

2.2   Conspicuity Maps

In a second step, each feature map is transformed in its conspicuity map which highlights the parts of the scene that strongly differ, according to a specific feature, from their surrounding. In biologically plausible models, this is usually achieved by using a center-surround-mechanism. Practically, this mechanism can be implemented with a difference-of-Gaussians-filter, $\mathcal{D}o\mathcal{G}$, which can be applied on feature maps to extract local activities for each feature type. A visual attention task has to detect conspicuous regions, regardless of their sizes. Thus, a multi-scale conspicuity operator is required. It has been shown in [16], that applying variable size center-surround filter on fixed size images, has a high computational cost. An interesting alternative method to implement the center-surround-mechanism at lower computational cost has been presented in [17]. This method is based on a multi-resolution representation of images. For each feature j, a nine scale gaussian pyramid $\mathcal{P}_j$ is created by progressively lowpass filter and subsample the feature map F_j, using a gaussian filter G (see Eq. 3).

$$\mathcal{P}_j(0) = F_j$$
$$\mathcal{P}_j(i) = \mathcal{P}_j(i-1) * G \tag{3}$$

Where ($*$) refers to the spatial convolution operator.

Center-surround is then implemented as the difference between fine and coarse scales. For each feature j, six intermediate multiscale conspicuity maps $M_{j,k}$ $(1..k..6)$ are computed according to equation 4, giving rise to 42 maps for the considered seven features.

$$M_{j,1} = |\mathcal{P}_j(2) - \mathcal{P}_j(5)|, \quad M_{j,2} = |\mathcal{P}_j(2) - \mathcal{P}_j(6)|$$
$$M_{j,3} = |\mathcal{P}_j(3) - \mathcal{P}_j(6)|, \quad M_{j,4} = |\mathcal{P}_j(3) - \mathcal{P}_j(7)|$$
$$M_{j,5} = |\mathcal{P}_j(4) - \mathcal{P}_j(7)|, \quad M_{j,6} = |\mathcal{P}_j(4) - \mathcal{P}_j(8)| \tag{4}$$

The absolute value of the difference between the center and the surround allows the simultaneous computing of both sensitivities, dark center on bright surround and bright center on dark surround (red/green and green/red or blue/yellow and yellow/blue for color). For the orientation features, an oriented Gabor pyramid $\mathcal{O}(\theta)$ is used instead of the gaussian one. For each of the four preferred orientations, six maps are computed according to equation 4 ($\mathcal{P}_j$ is simply replaced by $\mathcal{O}(\theta)$).

Note that these intermediate multiscale conspicuity maps are sensitive to different spatial frequencies. Fine maps (e.g. $M_{j,1}$) detect high frequencies and thus small image regions, whereas coarse maps, such as $M_{j,6}$, detect low frequencies and thus large regions.

For each feature j, the six multiscale maps $M_{j,k}$ are then combined, in a competitive way into a unique feature-related conspicuity map C_j:

$$C_j = \sum_{k=1}^{6} w_k M_{j,k} \tag{5}$$

The weighting function w, which simulates the competition between the different scales, is described in Section 2.3.

Finally, the seven conspicuity maps C_j, are transformed into three cue conspicuity maps: $\hat{C}_1$ for the intensity cue, $\hat{C}_2$ for the color cue and $\hat{C}_3$ for the orientation cue, according to Equation 6.

$$\hat{C}_1 = C_1$$
$$\hat{C}_2 = \sum_{j=2}^{3} w_j C_j$$
$$\hat{C}_3 = \sum_{j=4}^{7} w_j C_j \tag{6}$$

2.3   Saliency Map

In the last stage of the attention model, the three conspicuity maps are integrated together, in a competitive manner, into a saliency map $\mathcal{S}$ in accordance with equation 7.

$$\mathcal{S} = \sum_{i=1}^{3} w_i \hat{C}_i \tag{7}$$

The competition between conspicuity maps is usually established by selecting weights w_i according to a weighting function w, like the one presented in [17]: $w = (M - \overline{m})^2$, where M is the maximum activity of the conspicuity map and $\overline{m}$ is the average of all its local maxima. w measures how the most active locations differ from the average of local maxima. Thus, this weighting function promotes conspicuity maps in which a small number of strong peaks of activity is present. Maps that contain numerous comparable peak responses are demoted. It is obvious that this competitive mechanism is purely data-driven and does not require any a priori knowledge about the analyzed scene.

2.4 Selection of Salient Locations

At any given time, the maximum of the saliency map defines the most salient location, which represents the actual spot of attention. A "winner-take-all" (WTA) mechanism [10] is used to detect, successively, the significant regions. Given a saliency map computed by the saliency-based model of visual attention, the WTA mechanism starts with selecting the location with the maximum value of the map. This selected region is considered as the most salient part of the image (winner). The spot of attention is then shifted to this location. Local inhibition is activated in the saliency map, in an area around the actual spot. This yields dynamical shifts of the spot of attention by allowing the next most salient location to subsequently become the winner. Besides, the inhibition mechanism prevents the spot of attention from returning to previously attended locations. The number of the detected locations can be either set by the user or determined automatically through the activities of the saliency map.

To summarize this section, the saliency-based model of visual attention provides the following data:

- 7 feature maps F_j computed from an RGB image.
- 42 normalized multiscale conspicuity maps $M_{j,k}$ which express the conspicuousness of each image location at different spatial scales and for different scene features.
- 7 feature-related conspicuity maps C_j.
- 3 cue conspicuity maps $\hat{C}_i$ related to intensity, color and orientation.
- A saliency map
- A set of spots of attention.

3   Multiscale Attention-Based PreSegmentation (MAPS)

MAPS is thought to take advantage of the data provided by the visual attention algorithm in order to guide the segmentation process. This section describes the segmentation-relevant information which can be derived from the visual attention model, and later on presents the presegmentation method.

3.1   Segmentation-Relevant Scene Data

First of all the visual attention model localizes the visually salient regions in the image by detecting a set of spots of attention. Instead of segmenting the entire color image, MAPS considers only the regions around the detected spots. The detected locations will serve in our method as a kind of seed points around which the segmentation is performed. Thus, the segmentation task can be achieved in an attentive manner.

Furthermore and for each detected spot, we determine the multiscale conspicuity map M_{j^*,k^*} (among the 42 maps) that mostly contributed to the saliency of that location. Since equation 7 can be rewritten as follows:

$$S = \sum_{j=1}^{7}\sum_{k=1}^{6} w_{jk}M_{j,k} \tag{8}$$

(j^*, k^*) can be computed according to equation 9.

$$(j^*, k^*) = argmax_{j,k}(M_{j,k}(\mathbf{x})) \tag{9}$$

Where $\mathbf{x}$ is the spatial location of the considered spot of attention.

M_{j^*,k^*} is of special interest because it contains two kinds of information about the detected image location:

– The salient feature of the detected location, namely j^*. This information is useful for the refined segmentation (Section 4).
– The salient scale k^*, which provides information about the spatial size of the region to which belongs the detected spot of attention.

To summarize, three kinds of segmentation-relevant information are now available. Spatial information (location of the region), feature-based information (j^*) and scale-based information (k^*).

Figure 3 and Figure 5 ((2b) and (3b)) illustrate these presegmentation information which will play an essential role in the segmentation task. Figure 3 depicts the salient scale and the map of the salient feature for the first spot of attention on a traffic scene image, whereas Figure 5 shows the salient scales and the salient features for the eight spots of attention computed on two traffic scene images.

3.2   Presegmentation of the Multiscale Map M_{j^*,k^*}

In this section we aim at finding an approximative segmentation (presegmentation) of each detected region, based on their conspicuousness or salience. This presegmentation is best achieved at the salient scale of the considered region. Therefore, we apply, for each detected spot, a seeded region growing (SRG) algorithm on the corresponding multiscale map M_{j^*,k^*}. The seed point of SRG is the location of the spot of attention and the homogeneity criteria is the conspicuousness.

First spot of attention

$$M_{j^*,k^*} \ ((j^*,k^*) = (3,3)) \qquad\qquad F_{j^*}$$

Fig. 3. The segmentation-relevant data provided by MAPS about the first spot of attention on a traffic scene image. $M_{3,3}$ is the multiscale conspicuity map with the highest conspicuity value around this spot. Thus, F_3 (opponent colors B/Y) is the map of the salient feature at this location.

This first step can not be seen as a final segmentation result. Further information collected through the attention model should be used to accurately refine the presegmented region.

Examples of the presegmentation results are illustrate in Figure 4 and Figure 5 ((2c) and (3c)). In Figure 4 only the first spot of attention is considered, whereas Figure 5 illustrates the presegmentation of the eight first detected regions on two natural images.

4   Refined Segmentation

The presegmentation step supplies rough segments that must be described more accurately in a refined segmentation step. The refined segmentation method should be applied to F_{j^*} of each presegmented region, since it is the feature map that contains the clearest discrimination of that segment from the rest of the image.

Although the refined segmentation is not the main issue of this work, this section describes a possible refined segmentation method and presents some segmentation results.

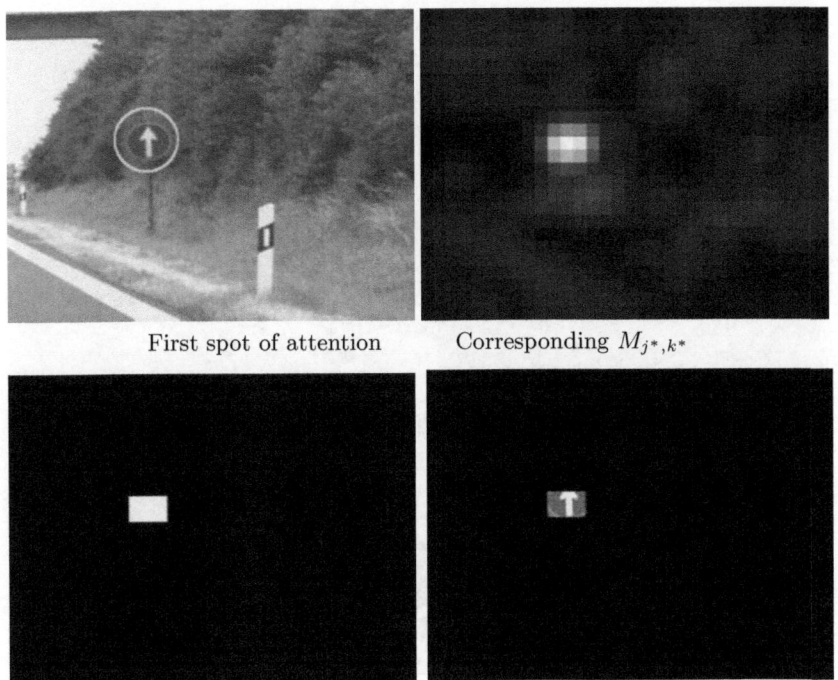

First spot of attention   Corresponding M_{j^*,k^*}

Presegmentation of M_{j^*,k^*} Result mapped on co lor image

Fig. 4. Presegmentation of M_{j^*,k^*} for the first spot of attention.

4.1   The Method

The refined segmentation method uses thresholding [1] and region expansion [5] to refine the rough segments. The thresholding step removes the pixels which are falsely included in the presegmented region. Thus, a threshold T, which clearly separate the region pixels and the outliers, must be computed automatically. We know that a detected region strongly differs, according to the corresponding feature j^*, from its neighbors. On F_{j^*}, this region R_i is either bright with dark background ($R_i(b/d)$) or dark with bright background ($R_i(d/b)$). In order to determine to which of the two categories the region R_i belongs, we use a statistical method. Two mean values μ_1 and μ_2 are computed on F_{j^*}. μ_1 is the mean value of presegmented region and μ_2 is computed within an enlarged (by factor 2) version of the same presegmented region. A decision about the nature of the region R_i is taken in accordance with equation 10.

$$R_i = \begin{cases} R_i(b/d) & if \; \mu_1 > \mu_2 \\ R_i(d/b) & otherwise \end{cases} \qquad (10)$$

The threshold T, which can be seen as the typical value of the detected region R_i, is then computed according to equation 11.

$$T = \begin{cases} argmax_{(i>\mu_1)}(h(i)) & if \; R_i = R_i(b/d) \\ argmax_{(i<\mu_1)}(h(i)) & if \; R_i = R_i(d/b) \end{cases} \qquad (11)$$

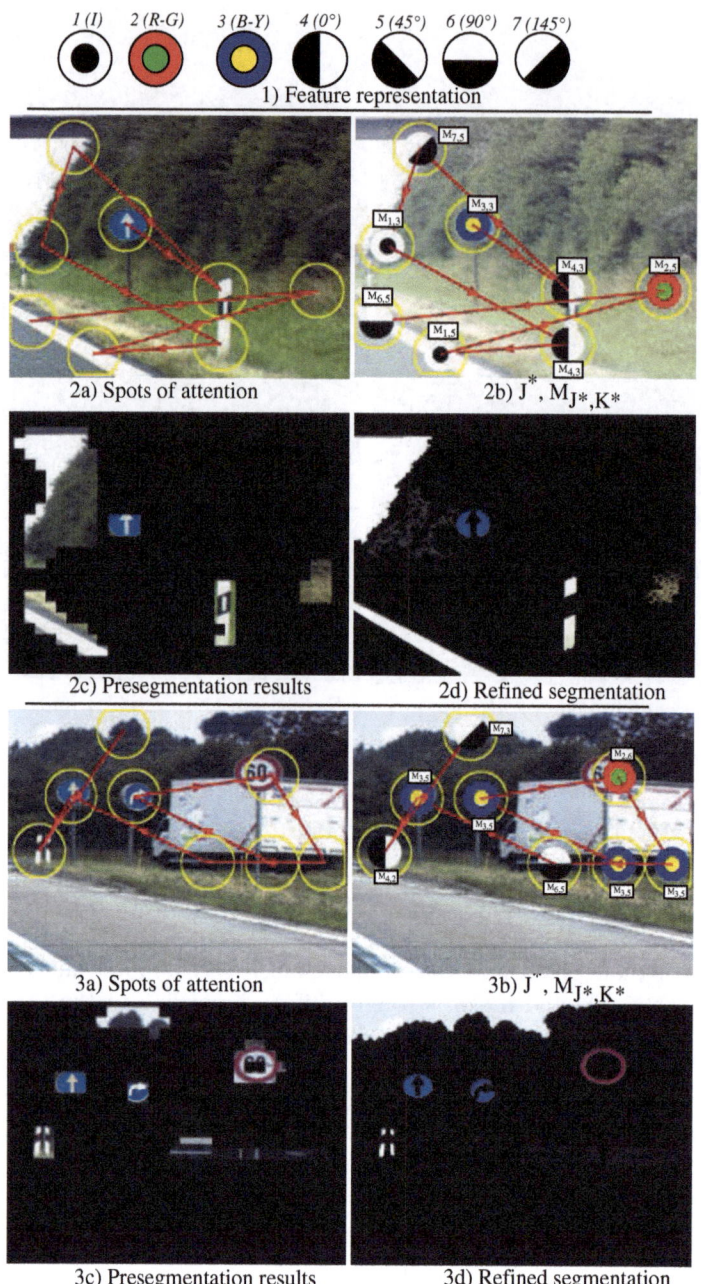

Fig. 5. Examples of MAPS results. 1): Colored representation of the seven scene features. 2) and 3): Results on two traffic scene images: (a) the spots of attention, (b) corresponding salient features and scales, (c) presegmentation results, and (d) refined segmentation results.

Spots of attention

Spot-based segmentation   MAPS segmentation

Fig. 6. Spot-based segmentation[8] Vs. MAPS segmentation.

Where $h(i)$ is the histogram of F_{j*} within the presegmented region. R_i is then segmented through applying a two level thresholding to F_{j*} within the presegmented region, using the thresholds $T - \epsilon$ and $T + \epsilon$.

Finally, the expansion step expands the region to those pixels which belong to the region R_i and which were not included into the presegmented region during the presegmentation step.

4.2 Segmentation Examples

Figure 5 ((2d) and (3d)) illustrates two examples of the refined segmentation on traffic scene color images.

Figure 6 depicts a comparison between results performed by MAPS and segmentation results of a spot-based segmentation algorithm developed also in our lab [8]. The spot-based algorithm uses visual attention only to determine the spatial locations of the salient regions. It performs segmentation at a single scale and uses color-based euclidian distance as homogeneity criterion for all regions. On one hand, the absence of a multiscale concept limits the spot-based algorithm to segment only fine structure of the image (the arrows of the blue traffic signs instead of the entire signs). On the other hand, the non-consideration of the salient features of regions to adapt the homogeneity criteria leads to the segmentation of unsignificant image parts like the segmentation of a part of the forest instead of the sky.

Note that in these examples the orientation feature maps ($F_{4..7}$) were not used for the refined segmentation. Since strong edges indicate the presence of region borders, an adequate use of this feature should improve the performance of the segmentation procedure. This extension will be considered in future work.

5   Conclusion

This work reports a novel Multiscale Attention-based PreSegmentation method (MAPS). Unlike classical segmentation methods, MAPS performs the segmentation task as an attentive process, during which only visually salient image regions are segmented. Furthermore and instead of using the same homogeneity criteria for all regions, MAPS uses the salient feature of each single region, which represents the clearest separation criteria between the region in question and the background. In addition, MAPS involves a multiscale concept allowing the segmentation of regions at the corresponding salient scale. This presegmentation is followed by a refined segmentation step that is best performed in the optimal feature map. Two extensions of MAPS will be considered in future work. First, multi-feature characterization of regions seems to be more promising than the single-feature characterization used in this paper. Second, presegmentation techniques which adequately handle the presence of strong gradients will be conceived and integrated into MAPS.

Acknowledgment

This work was partially supported by the CSEM-IMT Common Research Program.

References

1. J. Puzicha, T. Hofmann, and J. Buhmann. Histogram clustering for unsupervised image segmentation. *Proceedings of the IEEE International Conference on Computer Vision and Pattern Recognition (CVPR'99), pp. 602-608*, 1999.
2. D. Comaniciu and P. Meer. Robust analysis of feature spaces: Color image segmentation. *Computer Vision and Pattern Recognition 97. pp. 750-755*, 1997.
3. J. Canny. A computational approach to edge detection. *IEEE Transactions on Pattern Analysis and Machine Intelligence (PAMI), Vol.8, pp. 679-698*, 1986.
4. SY Chen, WC Lin, and CT Chen. Split and merge image segmentation based on localized feature analysis and statistical tests. *CVGIP, Vol. 53, pp. 457-475*, 1991.
5. R. Adams and L. Bischof. Seeded region growing. *IEEE Trans. on Pattern Analysis and Maschine Intelligence (PAMI), vol 16, no 6*, 1994.
6. A. Chakraborty and J.S. Duncan. Game-theoretic integration for image segmentation. *PAMI, Vol. 21(1), pp. 12-30*, Jan 1999.

7. J. Fan, D.K.Y. Yau, A.K. Elmagarmid, and W.G. Aref. Automatic image segmentation by integrating color edge extraction and seeded region growing. *IEEE Trans. On Image Processing, Vol. 10, No. 10, pp. 1454-1466*, October 2001.

8. N. Ouerhani, N. Archip, H. Hugli, and P. J. Erard. A color image segmentation method based on seeded region growing and visual attention. *Int. Journal of Image Processing and Communication, Vol. 8, Nr. 1, pp. 3-11*, 2002.

9. A.M. Treisman and G. Gelade. A feature-integration theory of attention. *Cognitive Psychology, pp. 97-136*, Dec. 1980.

10. Ch. Koch and S. Ullman. Shifts in selective visual attention: Towards the underlying neural circuitry. *Human Neurobiology (1985) 4, pp. 219-227*, 1985.

11. N. Ouerhani and H. Hugli. Computing visual attention from scene depth. *Proc. ICPR 2000, IEEE Computer Society Press, Vol. 1, pp. 375-378, Barcelona, Spain*, Sept. 2000.

12. N. Ouerhani, H. Hugli, P Y. Burgi, and P F. Ruedi. A real time implementation of visual attention on a simd architecture. *Proc. DAGM 2002, Springer Verlag, Lecture Notes in Computer Science (LNCS) 2449, pp. 282-289*, 2002.

13. L. Itti and Ch. Koch. Computational modeling of visual attention. *Nature Reviews Neuroscience, Vol. 2, No. 3, pp. 194-203*, March 2001.

14. S Engel, X. Zhang, and B. Wandell. Colour tuning in human visual cortex measured with functional magnetic resonance imaging. *Nature, Vol. 388, no. 6637, pp. 68-71*, Jul. 1997.

15. H. Greenspan, S. Belongie, R. Goodman, P. Perona, S. Rakshit, and C.H. Anderson. Overcomplete steerable pyramid filters and rotation invariance. *Proc. IEEE Computer Vision and Pattern Recognition (CVPR), Seattle, USA, pp. 222-228*, Jun. 1994.

16. R. Milanese. Detecting salient regions in an image: from biological evidence to computer implementation. *Ph.D. Thesis, Dept. of Computer Science, University of Geneva, Switzerland*, Dec. 1993.

17. L. Itti, Ch. Koch, and E. Niebur. A model of saliency-based visual attention for rapid scene analysis. *IEEE Transactions on Pattern Analysis and Machine Intelligence (PAMI), Vol. 20(11), pp. 1254-1259*, 1998.

Convex Colour Sieves

Stuart Gibson, Richard Harvey, and Graham Finlayson

University of East Anglia, Norwich, NR4 7TJ, UK
{s.e.gibson,r.w.harvey,g.finlayson}@uea.ac.uk
http://www.sys.uea.ac.uk

Abstract. Sieves and their variants are established processors for simplifying greyscale images. Because combined outputs of these filters satisfy the scale-space causality property they are often referred to as scale-
space filters although they have quite different characteristics compared
to systems based around diffusion. In this paper we implement several
possible extensions of sieves for colour images which include: applying
the processor on separate channels; and enforcing an ordering on the
colour vectors. We show that a new definition, based on convex hulls in
colour space, can lead to an effective algorithm. As with the greyscale
method, the colour sieve produces a tree-based representation of image
that form the first step to a meaningful hierarchical decomposition.

1 Background

Since the early work of Iijima [10] the idea of simplifying images through a
diffusion equation, perhaps where the conductivity varies, has received much
attention [27,14,21,13,16,18,26] and is generally known as *scale-space* processing.
However there are alternative processors, based in mathematical morphology,
that have some, or all, of the properties of diffusion-based systems. Some of
these alternatives have been compared to linear scale-spaces [8] and the variant
known as *sieves* has been shown to have useful properties. Sieves [4, 2, 5] are
a class of graph morphology algorithms that simplify images by scale. Their
properties are usually derived using a formal graph-theory approach [2] and
include the preservation of scale-space causality, computational efficiency and
some robustness to noise and occlusion [7].

There are several colour scale-spaces based on smoothed derivatives [20,12].
Likewise morphological operators can, with some care, be extended [6] to handle
colour images but, so far, sieves and their variants [19,9] have had only quite
restricted extensions into colour (by application to each channel for example).
This paper reviews the options for extending scalar sieves to vector quantities
with particular reference to colour images and shows how the right choice can
lead to a tree-based representation of the image that might be useful for later
stage processing.

L.D. Griffin and M. Lillholm (Eds.): Scale-Space 2003, LNCS 2695, pp. 550–563, 2003.

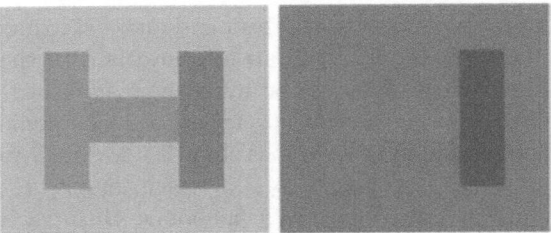

Fig. 1. An RGB image (left) when converted to greyscale (right) may not show the same features.

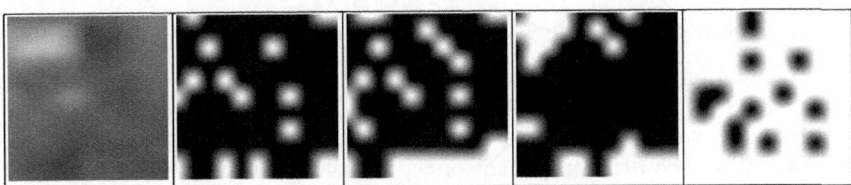

Fig. 2. A small part of an image (left) and starting extrema shown as white pixels. From left to right: extrema measured in greyscale; in the RGB channels separately; global convex hull; and local convex hull.

2   Colour Sieves

It is evident that there are likely to be many possible extensions of the scalar algorithm of which the most obvious is to apply the conventional greyscale algorithm on the individual colour channels. Although this can sometimes give aesthetically pleasing results, the method is *ad-hoc* because it allows only for extrema that align to certain preferred directions.

To define a more principled extension we first review how the greyscale sieve operates. The greyscale sieve can be defined in terms of graph morphology operators [2, 25, 23, 24, 22, 17, 11] but algorithmically there are two steps:

1. The image is scanned to identify which are extremal connected sets that are brighter or darker than their neighbours.
2. The smallest connected sets of scale (area) 1, are merged to their next most extreme neighbouring region. This process is then repeated for the area-two extremal connected sets and so on.

Step 2 contains two subtleties: it must be possible to define a distance measure so that the next-most extreme neighbouring region can be determined and also there must be a rule for over-writing the merge regions and creating new connected sets. To extend the conventional sieve to colour we need to find satisfactory extensions of these rules to colour.

Figure 2 shows several approaches to defining extrema in colour. It might be possible to start with the original greyscale extrema (second from left), but as Figure 1 shows, this approach will miss salient regions. Considering the colour channels (RGB say) separately generates many more starting points (centre of Figure 2) but the identified extrema are not invariant to the choice of colour axes. Ideally colour extrema should be invariant to typical transformations of the colour axes such as linear transformations or monotonic transformations such as brightness or gamma corrections. In other words, if we define a matrix $C_4 = [c_{m,n}, c_{m+1,n}, c_{m,n+1}, c_{m,n-1}, c_{m-1,n}]$ which consists of the (m, n)th pixel of colour $c_{m,n}$ and its four-connected neighbours[1] then we can also define an extremity operator

$$\mathcal{E}(C_4) = \begin{cases} 1 \text{ if } c_{m,n} \text{ is extreme compared to its neighbours} \\ 0 \text{ otherwise} \end{cases}$$

For a successful definition of colour extrema we demand that

$$\mathcal{E}(C_4) = \mathcal{E}(\tilde{C}_4) \tag{1}$$

where $\tilde{C}_4$ is a pixel and its neighbours in a transformed colour space. Specifically we allow the nth component of the transformed space to be:

$$\tilde{c}_{p,q}(n) = f_n \left(\sum_{m=1}^{N} \alpha_m g_m \left(c_{p,q}(m) \right) \right) \tag{2}$$

where $\tilde{c}_{p,q}$ and $c_{p,q}$ are N-dimensional. α_m are the coefficients of any invertible linear transformation and $f_n(\cdot)$ and $g_m(\cdot)$ are strictly increasing or strictly decreasing monotonic functions. Note that (1) is satisfied by a conventional greyscale sieve in which $N = 1$ and $c_{p,q}$ is the greyscale value so that $\tilde{c}_{p,q} = f(\alpha g(c_{p,q}))$.

3   Colour Sieve Algorithm

Figure 2 also shows (second from right) extrema defined through a convex hull [15]. The convex hull of a set S of points is the smallest polygon P for which each point of S is either on the boundary or in the interior of P. Fast divide and conquer algorithms exist that have complexity of roughly $N \log N$ where N is the number of points. Taking all the colour vectors in the image on the left of Figure 2 and fitting a convex hull to all theses points gives the extrema shown. Linear transformations of the axes and monotonic scalings will affect the geometry of the hull but not its topology – points on the exterior remain on the exterior. However the global hull is analogous to global extrema in the greyscale case and in practice, there are too few extrema to successfully initiate a colour sieve. A solution is to fit a convex hull to the points in C_4 (we call these local convex hulls since they are formed from pixels and their neighbours). Similarly any pixels that lie on the exterior of the local convex hull are defined as extreme. In practice we are working in three-space (RGB) so if we are considering a four-connected sieve then often we have to consider volumes with at most five vertices (for an eight-connected sieve, at most nine vertices). However often there are degenerate cases and these are shown in Figure 3.

---

[1] Any other connectivity follows *mutatis mutandis*.

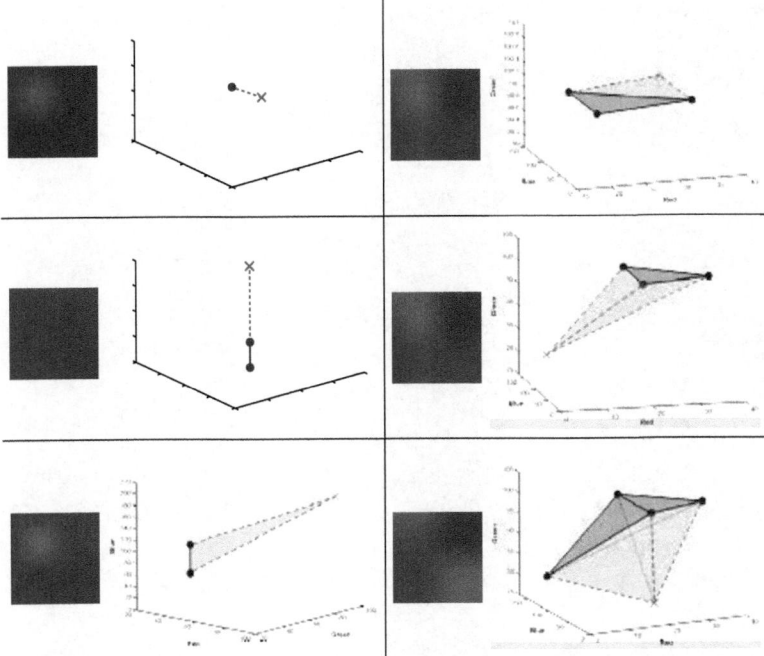

Fig. 3. Cases to be considered when constructing a convex hull. Each cell shows, on the left a pixel and its neighbours and, on the right, the convex hull plotted in RGB space. Blue dots on the hull represent the neighbours and the red cross is the pixel. In the degenerate cases the hull has reduced dimensionality and is either a line or a plane. A point is labelled as extreme if it is on the exterior of the reduced dimensionality convex hull.

If the region is extreme, it is added to a processing list. This list is firstly ordered by scale and secondly ordered by the region's top-left 2D image coordinate. The list is then processed, merging extremal regions to their nearest spatially connected neighbour. However to define merging in colour requires a merge rule.

Finding an exact analogy to the greyscale merge rule is more challenging. Ideally we wish to merge smaller regions into larger ones without introducing additional extrema. Achieving this in greyscale is tricky and requires a two pass algorithm: the M or N sieves. In these, maxima (resp. minima) are merged to their neighbour with the closest greyscale, then minima (resp. maxima) likewise. In the convex colour sieve we merge to the neighbour with the closest Euclidean distance, but because two different colours can have identical distances and there is no longer the possibility of handling maxima and minima separately, we make two modifications.

1. Neighbouring regions with identical colour distances are further ordered by computing the difference of their luminance $L = (r + g + b)/3$. If this does not resolve the tie then regions are ordered by their G,R and B values.

Fig. 4. Original image 384×256 pixels (top left) convex colour sieved in a RGB space to scales $2^n, n = 0 \ldots 14$ (shown left-to-right).

2. After performing a merge we check for the introduction of new extrema. If there are new extrema then they are merged up to the current scale. This step repeats until no new extrema are introduced.

Step 2, which is similar to the processing that occurs in the recursive median sieve or m-sieve, ensures idempotence.

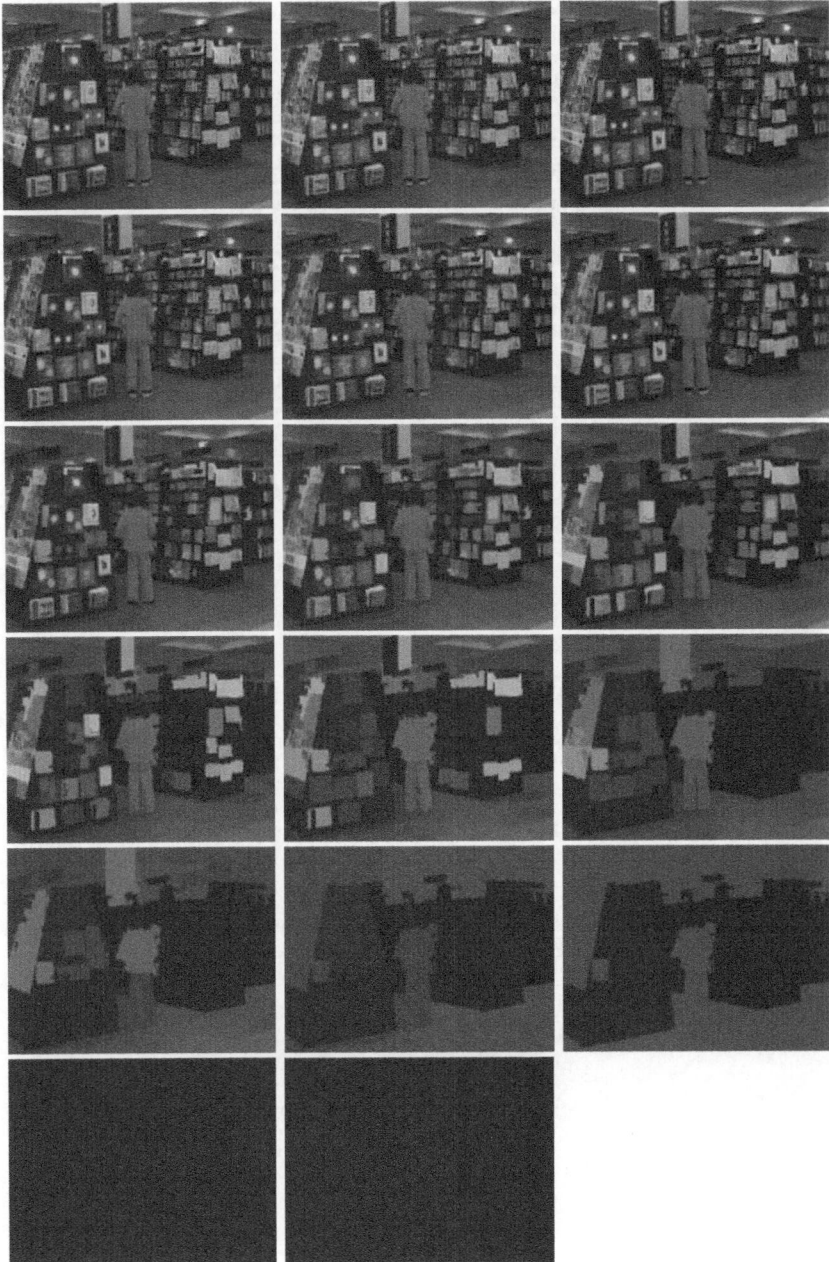

Fig. 5. Original image 320×240 pixels (top left) convex colour sieved in a RGB space to scales $2^n, n = 0 \ldots 15$ (shown left-to-right).

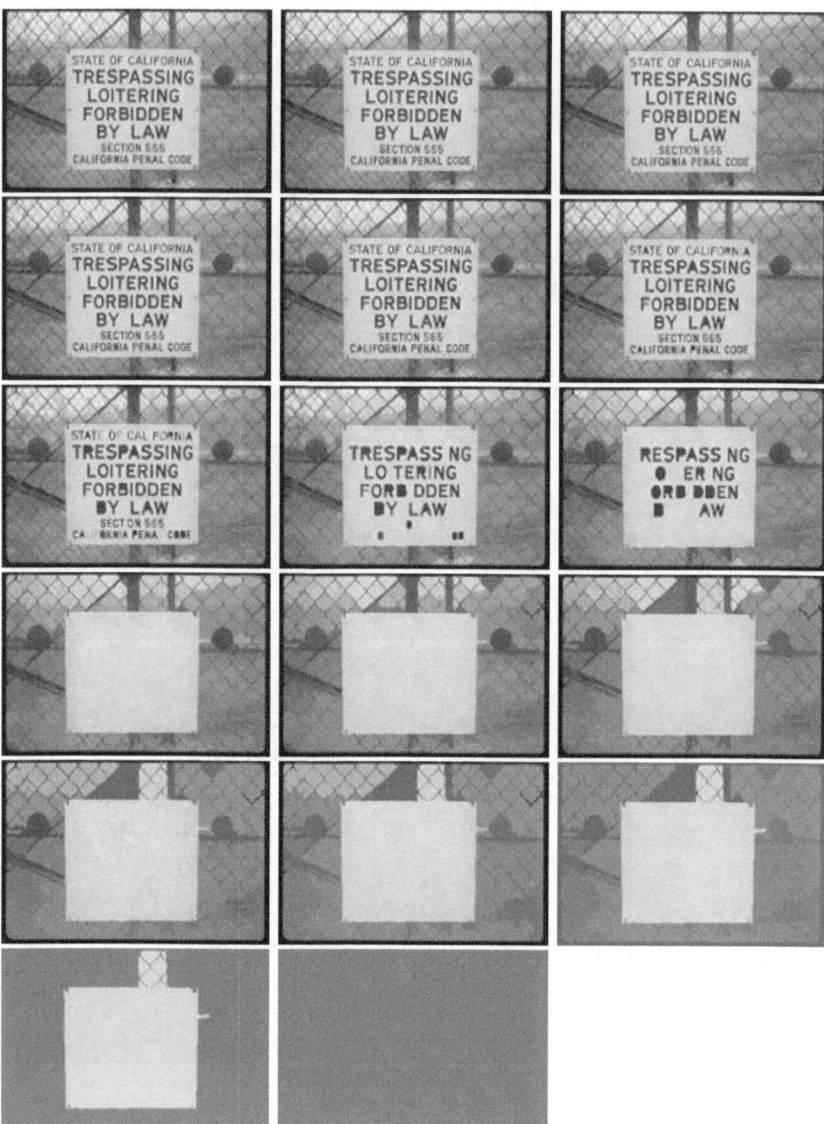

Fig. 6. Original image 384×256 pixels (top left) convex colour sieved in a RGB space to scales $2^n, n = 0 \ldots 15$ (shown left-to-right).

We write the image sieved to scale s, X_s,

$$X_s = C_s(X_{s-1}) = K_s(X) \tag{3}$$

where C_s is the recursive convex colour sieve operator that transforms an image sieved to scale $s - 1$ to scale s and K_s is the cascade of recursive operators that transforms the original image $X = X_0$ to one sieved to scale s. Thus

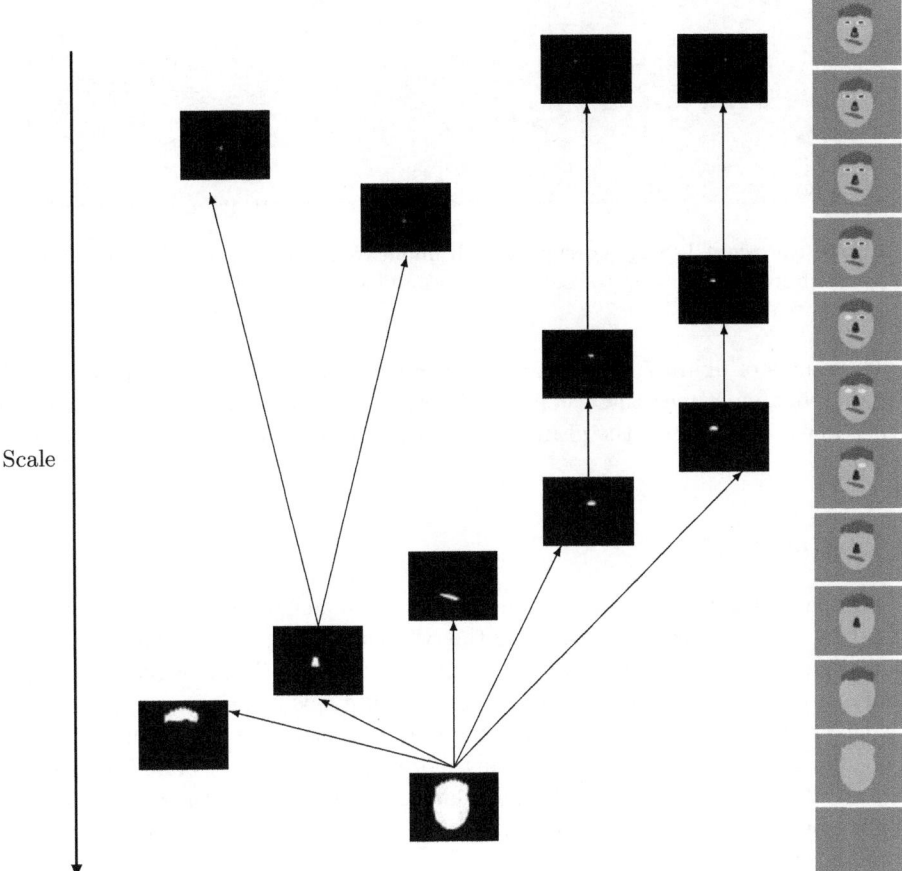

Fig. 7. A colour sieve decomposition of a 126 by 168 pixel image showing the original (top right) and simplifications at scales 4, 10, 11, 28, 32, 92, 98, 194, 203, 1219 and 5118. The differences between successive simplifications are called *granules* and are shown as a tree where the edges indicate containment.

$$K_s(\cdot) = C_s\left(C_{s-1}\left(\cdots C_1(\cdot)\right)\right) \tag{4}$$

Idempotence is the property that $X_s = C_s(X_s)$.

Example decompositions for images from the MPEG-7 Common Color Dataset [1] are shown in Figures 4, 5 and 6. The algorithm generates all scales from 0 to the size of the image but here we have shown a restricted range.

An important property of the greyscale system, not mentioned so far, is the implicit construction of a tree. The application of operator C_s removes extremal regions of size (area) s which are guaranteed to be contained within larger regions. This containment can be represented by an edge in a tree [19,3,9]. Figure 7 shows an example. On the right is shown, what is sometimes called, the low-pass decomposition in which the image gets progressively simpler from top to bottom.

Fig. 8. An original image corrupted with impulsive noise with $p = 0.1$ (left) and the result after colour sieving to scale 1 (centre) and scale 6 (right).

The centre of figure 7 shows the tree with connected sets that are progressively removed. Sometimes these connected sets are called *granules*. Note that, unlike linear scale-space, the granules represent sharp-edged regions that might be objects.

4   Discussion

This paper has introduced a colour variant of a scale-space processor known as a sieve. It was devised by considering the two parts of the greyscale sieve, extrema identification and merging, and suggesting alternatives. The new method of identifying extrema has considerable appeal because it is invariant to typical colour space transformations. It identifies more extrema than the greyscale equivalent so allows the possibility of segmenting more regions. In a test of 50 images taken from the MPEG-7 Common Colour Dataset we found that the probability of a pixel being a greyscale extrema was 0.17 whereas the probability of it being a colour extrema was 0.56.

The merge rule is less satisfactory for two reasons: it is not invariant to the colour space and it forces a recursive implementation in which we recheck for extrema. Nevertheless the algorithm executes in a few minutes for a 384x256 pixel image(24-bit pixels) on a domestic PC.

The processor appears to inherit the robustness to noise of the greyscale sieve. Figure 8 shows an original image corrupted with impulsive noise in which, with probability $p = 0.1$, a pixel is replaced with a random colour. In the sieved versions most of the noise is removed at small scale which is to be expected since the probability of a large noise region is small. In future work we hope to test the robustness of the new processor against other colour scale-space processors ([8] contains a methodology for such tests).

The effect of operating in a different colour space is shown in Figures 9, 10 and 11 where we compute a Karhunen-Loeve, Principal Component or Mahalanobis transform. If the image contains N pixels $c_n, n = 1...N$ then computing their mean and covariance over the image as

$$\mu = E\{c_n\}, \qquad R = E\{c_n c_n^T\} \tag{5}$$

allows the construction of a new space aligned along the eigenvalues in which

Fig. 9. Original image 384×256 pixels (top left) convex colour sieved in a Mahalanobis space to scales $2^n, n = 0 \ldots 14$ (shown left-to-right).

$$\tilde{c}_n = \lambda^{-1/2} \left(c_n - \mu\right)^T \left[e_1 e_2 e_3\right] \tag{6}$$

This is the well known principal component transformation in which the first component of $\tilde{c}_n$ is aligned with the axis of maximum variance of c_n. In the convex colour sieve operating in this Mahalanobis space the initial extrema are identical to the convex colour sieve operating in RGB. However the merges differ

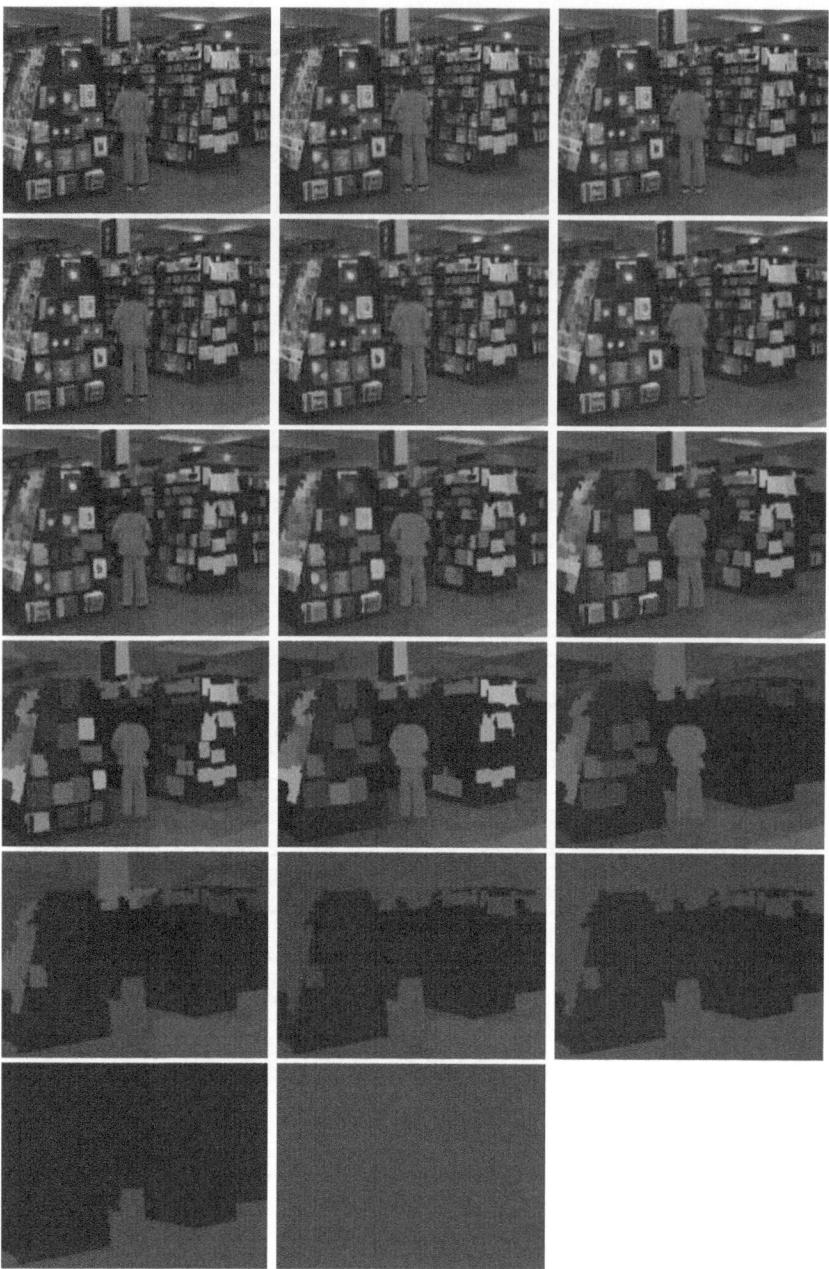

Fig. 10. Original image 320×240 pixels (top left) convex colour sieved in a Mahalanobis space to scales $2^n, n = 0 \dots 15$ (shown left-to-right).

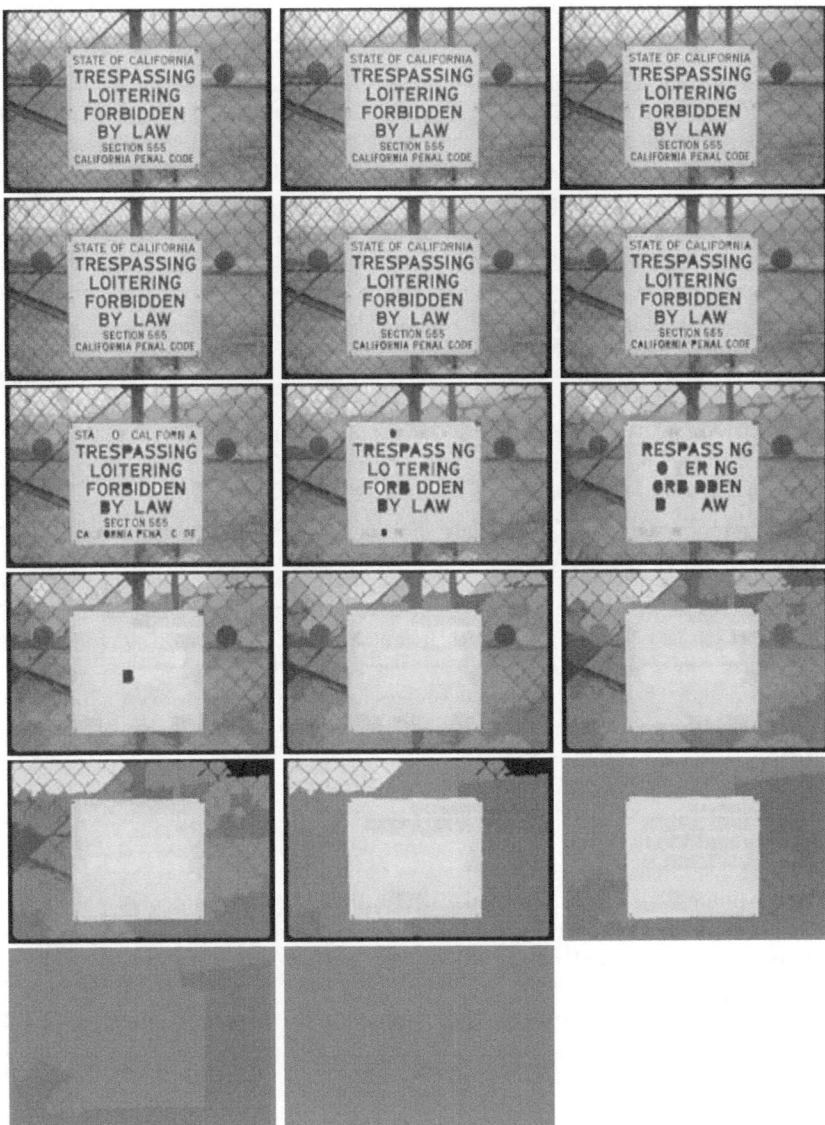

Fig. 11. Original image 384×256 pixels (top left) convex colour sieved in a Mahalanobis space to scales $2^n, n = 0 \ldots 15$ (shown left-to-right).

because the variance weighting introduced by the Mahalanobis distance alters the Euclidean distance between neighbours.

The Mahalanobis space appears to delay the merging of objects into the background so that, in Figure 11 for example, the sign is prevented from merging into the fence and a better approximation to the ideal object tree [9] results. However we note that the correct choice of colour space is a matter of current interest.

References

1. Common datasets and queries in MPEG-7 color core experiments. Technical Report ISO/IEC JTC1/SC29/WG11/MPEG99/M5060, MPEG-7 Standards Body, October 1999.
2. J. Andrew Bangham, Richard Harvey, and Paul D. Ling. Morphological scale-space preserving transforms in many dimensions. *Journal of ELectronic Imaging*, 5(3):283–299, July 1996.
3. J. Andrew Bangham, Javier Ruiz Hidalgo, and Richard Harvey. Robust morphological scale-trees. In *Proceedings of Noblesse Workshop on Non-linear Model-Based Image Analysis*, pages 133–139. Springer-Verlag, July 1998.
4. J. Andrew Bangham, Paul W. Ling, and Richard Harvey. Nonlinear scale-space causality preserving filters. *IEEE Transactions in Pattern Analysis and Machine Intelligence*, 18:520–528, 1996.
5. J.A. Bangham, P. Chardaire, C.J. Pye, and P.D. Ling. Multiscale nonlinear decomposition: The sieve decomposition theorem. *IEEE Transactions on Pattern Analysis and Machine Intelligence*, 18(5):529–539, May 1996.
6. M.L. Comer and E.J. Delp. Morphological operations for color image processing. *Journal of Electronic Imaging*, 8(3):279–289, 1999.
7. R. Harvey, J.A. Bangham, and A. Bosson. Scale-space filters and their robustness. In *Proc. First Int. Conf. on Scale-space theory*, pages 341–344. Springer, 1997.
8. R. W. Harvey, A. Bosson, and J. A. Bangham. The robustness of some scale-spaces. In *British machine vision Conference*, pages 11 – 20, 1997.
9. J.R. Hidalgo. The representation of images using scale trees. Master's thesis, School of Information Systems, University of East Anglia, Norwich, NR4 7TJ, October 1998.
10. T. Iijima. Basic theory of pattern normalization (for the case of a typical one-dimensional pattern. *Bulletin of the Electrotechnical Laboratory*, 26:368–388, 1962.
11. J.Serra and P.Salembier. Connected operators and pyramids. In *Proceedings SPIE*, volume 2030, pages 65–76, 1994.
12. Ron Kimmel, Nir A. Sochen, and Ravi Malladi. From high energy physics to low level vision. In *Proc. First Int. Conf. on Scale-space theory*, pages 236–247, 1997.
13. J. Koenderink. The structure of images. *Biological Cybernetics*, 50(5):363–370, August 1984.
14. Tony Lindeberg. *Scale-space theory in computer vision*. Kluwer, 1994. ISBN 9-7923-9418-6.
15. Joseph O'Rourke. *Computational Geometry in C*. Cambridge University Press, 1998.
16. P. Perona and J. Malik. Scale-space and edge detection using anisotropic diffusion. *IEEE Transactions on Pattern Analysis and Machine Intelligence*, 12(7):629–639, July 1990.
17. P.Salembier and J.Serra. Flat zones filtering, connected operators and filters by reconstruction. *IEEE Transactions on Image Processing*, 8(4):1153–1160, AUGUST 1995.
18. L.I. Rudin, S. Osher, and E. Fatemi. Nonlinear total variation based noise removal algorithms. *Physica D: Nonlinear Phenomena*, 60:259–268, November 1992.
19. P. Salembier and L. Garrido. Binary partition tree as an efficient representation for filtering, segmentation and information retrieval. In *International Conference on Image Processing (ICIP'98)*, volume II, pages 252–256, Chicago, USA, October 1998.

20. Bart. ter Haar Romeny, Jan-Mark Geusebroek, Peter van Osta, R. van den Boom-gaard, and Jan Koenderink. Color differential structure. In *Scale-Space 2001*, page 353 ff. Springer, 2001.

21. Bart M. ter Harr Romeny, editor. *Geometry-driven diffusion in Computer vision.* Kluwer Academic, Dordrecht, Netherlands, 1994. ISBN 0-7923-3087-0.

22. L. Vincent. Graphs and mathematical morphology. *Signal Processing*, 16:365–388, 1989.

23. L. Vincent. Morphological area openings and closings of greyscale images. In *Workshop on Shape in Picture*. NATO, 1992.

24. Luc Vincent. Grayscale area openings and closings, their efficent implementation and applications. In Jean Serra and Phillipe Salembier, editors, *Proceedings of the international workshop on mathematical morphology and its applications to signal processing*, pages 22–27, May 1993.

25. Luc Vincent. Morphological grayscale reconstruction in image analysis: applications and efficient algorithms. *IEEE Transactions on Image Processing*, 2(2):176–201, April 1993.

26. J. Weickert. *Anisotropic Diffusion in Image Processing*. Teubner-Verlag, Stuttgart, Germany, 1998.

27. A. P. Witkin. Scale-space filtering. In *8th Int. Joint Conf. Artificial Intelligence*, pages 1019–1022. IEEE, 1983.

Scale-Space on Image Profiles about an Object Boundary

Sean Ho* and Guido Gerig

Department of Computer Science
University of North Carolina, Chapel Hill, NC 27599, USA
seanho@cs.unc.edu

Abstract. Traditionally, image blurring by diffusion is done in Euclidean space, in an image-based coordinate system. This will blur edges at object boundaries, making segmentation difficult. Geometry-driven diffusion [1] uses a geometric model to steer the blurring, so as to blur along the boundary (to overcome noise) but edge-detect across the object boundary. In this paper, we present a scale-space on image profiles taken about the object boundary, in an object-intrinsic coordinate system. The profiles are sampled from the image in the fashion of Active Shape Models [2], and a scale-space is constructed on the profiles, where diffusion is run only in directions tangent to the boundary. Features from the scale-space are then used to build a statistical model of the image structure about the boundary, trained on a population of images with corresponding geometric models. This statistical image match model can be used in an image segmentation framework. Results are shown in 2D on synthetic and real-world objects; the methods can also be extended to 3D.

1   Introduction

Features in images exist at scale, hence scale-spaces of images are useful for characterizing them. In image segmentation, we wish to localize the boundary of an object within the image. The traditional linear scale space can blur the boundary, complicating segmentation. Early on, Canny's [3] edge-detection work motivated a scheme which blurs along the boundary, but edge-detects across the boundary. Variable-conductance diffusion [4] and geometry-driven diffusion [1] aim to achieve this, by using the geometry of the image intensity field to orient and modulate the diffusion locally.

In deformable model-based segmentation, the deformable model has geometry of its own. In this paper, we develop an implementation of a geometry-driven scale-space which uses the geometry of a deformable model instead of the geometry of the image intensity field to drive the diffusion. This scale-space in an object-intrinsic coordinate system is then used to construct a statistical model of image structure about the object boundary, as trained on a population of segmented images. The statistical model forms the image-match likelihood, which

---

* Supported by NIH-NCI P01 CA47982.

L.D. Griffin and M. Lillholm (Eds.): Scale-Space 2003, LNCS 2695, pp. 564–575, 2003.

Fig. 1. Sagittal slice of the hippocampus in MRI, showing complex boundary appearance. The hippocampus is outlined in red. The tail of the hippocampus is to the right; the head is to the left. The hippocampus is a 3D object; this image only shows a cross-section.

when coupled with a prior on the shape of the deformable model, forms a complete segmentation framework.

Objects in images, particularly medical images, often exhibit complex boundary profiles, which can not be represented by simple step edges. Hence the image gradient magnitude map alone is not sufficient to drive segmentation of these objects. Frequently an object in a medical image will have different image profiles at different locations around the boundary. For instance, the hippocampus in T1-weighted MRI shows a sharp step edge at the tail from grey inside the hippocampus to black outside (Figure 1). However, parts of the bottom of the hippocampus might be characterized by a bar-like edge, and at the head of the hippocampus, the transition zone to the adjacent amygdala is not always visible, resulting in no edge at all in the profile at that location. Successful segmentation of these real-world objects in real-world images that have noise and large variability necessitates the development of an image model which can represent these complex boundary profiles; one that is local, statistical, and multiscale.

2   Related Work

2.1   Geometry-Driven Diffusion

Viewing an image as an embedding map between Riemannian manifolds, e.g. viewing the intensity field as a height field, provides the framework to apply principles from Riemannian differential geometry to image analysis. Much work has been done on geometry-driven diffusion [1], where the geometry which drives the diffusion is taken from the image itself. Mean curvature flow and anisotropic diffusion [4] are well-known examples. The Beltrami flow [5] smooths the image while preserving edges by using the Laplace-Beltrami diffusion operator on the manifold given by the intensity field. The Laplace-Beltrami operator is the generalization of the Laplacian from Euclidean space to arbitrary manifolds, incorporating the Riemannian metric tensors of the manifold.

Sapiro et al [6] apply Laplace-Beltrami diffusion to directional data defined on a flat planar domain, although the theory they develop is applicable to maps between arbitrary manifolds.

Chung et al [7] use the Laplace-Beltrami operator to perform diffusion on a manifold given by a triangulation of the brain cortical surface, to perform diffusion smoothing of a scalar field (e.g. mean curvature) defined on the manifold of the cortex. With this method, the geometry which drives the diffusion is no longer an image intensity field, but rather externally provided geometry, in the form of a triangle mesh outlining the cortical surface. The "image" is a map from the manifold of the cortical surface to the 1D range of mean curvatures; i.e. a texture map on a triangulated mesh.

2.2 Profile Model

Cootes and Taylor's seminal Active Shape Model work [2] samples the image along 1D profiles around boundary points, normal to the boundary, using correspondence given by the Point Distribution Model of geometry. Profiles are extracted at corresponding locations in a population of segmented training images. At each location along the boundary, a probability distribution is trained on the image-derived features in the 1D profile, and these probability distributions form the image-match model for segmentation.

The "hedgehog" model in the 3D spherical harmonic segmentation framework [8] can be seen as a variant of ASMs, and uses a training population linked with correspondence from the geometric model. It can be extended with a coarse-to-fine sampling of the object boundary in order to improve robustness in the face of image noise and jitter in correspondence.

2.3 Active Appearance Models

Perhaps the most well-developed work on modeling intensity variation in objects is the Active Appearance Model [9], which uses the point correspondences given by an ASM to warp images into a common coordinate system, within a region of interest across the whole object, given by a triangulation of the PDM. A global Principal Components Analysis is performed on the intensities across the whole object (the size of the feature space is the number of pixels in the region of interest). The use of a global PCA is particularly well-suited for capturing global illumination changes, for instance in Cootes and Taylor's face recognition applications.

It should be noted that both ASMs and AAMs have multiresolution extensions, which help to speed the optimization process of segmentation, and avoid local minima. However, the image pyramids used in the multiresolution extensions are created using isotropic diffusion in the original Euclidean image space, hence the object boundary will also be smoothed.

3 Method: Profile Scale-Space Model

We propose sampling intensities around the object boundary in an intrinsic coordinate system derived from the geometric model, which preserves the local

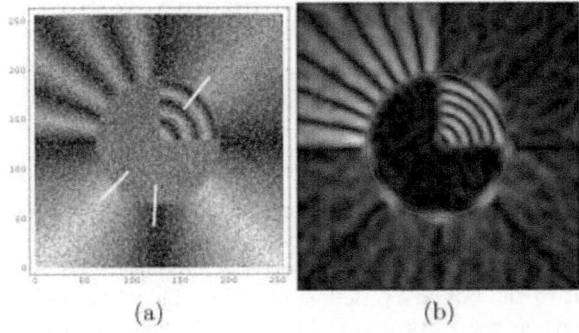

(a)   (b)

Fig. 2. (a) A sample from the toy example training population. (b) The corresponding gradient magnitude image at $\sigma = 4$ pixels, for reference. The contour from which profiles will be taken is shown in red. Each of the 30 images in the population have different additive noise, as well as multiplicative shading (linear ramp) in different directions.

orientation relative to the object boundary: i.e. which direction is across (normal to) the boundary and which directions are along (tangential to) the boundary. We want to look for edges across the boundary, but smooth away noise along the boundary. To accomplish this, we construct a scale-space in a non-Euclidean object-intrinsic coordinate system which preserves the along-boundary versus across-boundary distinction.

3.1   Toy Example

Throughout the discussion of our method, a toy example population of images will be used for ease of explanation, where the object shape is very simple (a circle in 2D), but the greyscale intensities are complex and have variability. Section 4 shows some results on real-world objects with shape variability. We use 2D Fourier coefficients [10] as a parameterized shape representation; the parameter space is the unit circle. The toy example images all have the same underlying structure, with sine waves of varying frequency at different sectors of the boundary, and boundary profiles more complex than a simple step edge. In addition, each image in the population has some global multiplicative shading in a different direction, and some uniform additive noise. The images are 256x256 pixels, and 30 sample images were created.

Figure 2 shows a sample from this toy population, as well as its gradient magnitude image at scale $\sigma = 4$ pixels. Note that a simple gradient magnitude search would not yield an appropriate image match model in this case, e.g. in the top right quadrant of the circle.

To illustrate the different edge structure at different locations around the boundary, profiles at three locations are shown in Figure 3. The locations from which the profiles are taken are shown overlaid on Figure 2(a) in yellow. The profile from the lower left quadrant of the circle shows a descending step edge

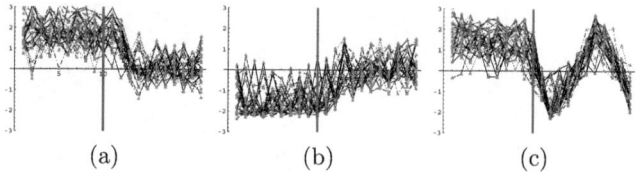

Fig. 3. Profiles from the training population, at the three boundary locations indicated in yellow in Figure 2 (counter-clockwise from lower left), showing: (a) negative step edge, (b) positive step edge, and (c) complex line-like edge. The x-axis indicates distance normal to the boundary (left is outside, right is inside the object). Object boundary is shown as red vertical line. The y-axis is image intensity.

from outside to inside the boundary; the profile at the bottom of the circle shows an ascending step edge, and the profile at the top right of the circle shows a complex structure with a line-like edge at the boundary and additional step edges further inside the object. Also shown in Figure 3 is the large variability in the training population, such that the step edge at places is overwhelmed by the image noise and variability.

3.2   Scale-Space in Object-Intrinsic Coordinates

Our object shape representation (2D Fourier harmonics) provides a mapping from a standardized parameter space to the object boundary in each image in the population. In the case of our toy example, the unit circle parameter space always maps to the same circle in each of the training images, but in a real-world population, the object shapes will differ within the training population. If we were to build a linear scale space in the original (x, y) image coordinates, edge structure across the boundary would also be blurred away. In the spirit of geometry-driven diffusion, we wish to blur along the boundary (to deal with noise), but detect edges across the boundary.

An elegant solution for diffusion along the object boundary is to map the blurring kernel from the parameter space onto the object. The image intensities sampled at the object boundary form a texture map on the surface of the object. The Laplace-Beltrami operator can then be used to blur this texture map. Extending the blurring to a collar of thickness about the boundary can be done by using the original object boundary to define a family of surfaces at different distances away from the object, e.g. a wavefront propagation using the distance transform on the boundary. Our implementation is an approximation to the Laplace-Beltrami diffusion.

In our implementation, we sample the image along 1D profiles normal to the boundary, unrolling the profiles into a flat rectangular grid. The "across-boundary" and "along-boundary" directions then become the x and y axes of the unrolled profile image, and we can construct a standard linear scale-space in one direction of the unrolled profiles, leaving the across-boundary direction unfiltered. The convolution in the along-boundary direction is cyclic, since the

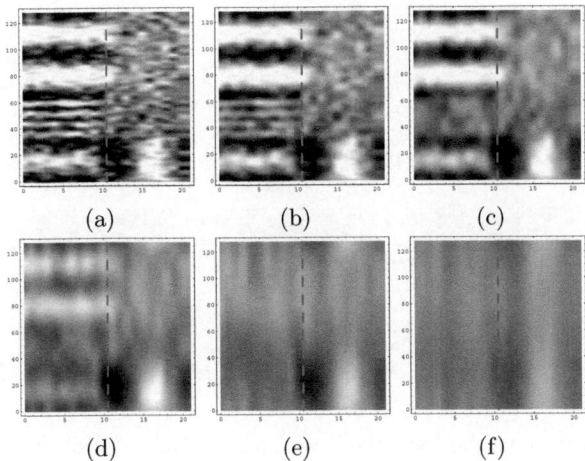

(a) (b) (c)

(d) (e) (f)

Fig. 4. The Gaussian scale-space constructed from one case in the toy example training population. Levels of the scale-space are (a)–(f), fine-scale to coarse-scale. Each row within a plot represents a boundary profile at that scale. The object boundary runs vertically through the center of each plot; left is exterior of the object and right is interior.

parameter space is the unit circle. We sample scale-space at $\sigma = 2, 4, 8, 16, 32, 64$. As a global brightness/contrast normalization pre-processing step, the profile images are scaled to have zero mean and unit variance, before the scale-space is constructed.

We build the Laplacian scale-space ($\partial/\partial\sigma$) on the profile image, representing each level of the scale-space as a residual from the next coarser level (Figure 5). This sets up a Markov chain in scale-space, as will be discussed in Section 3.3. One could also calculate local differences in space between neighboring profiles ($\partial/\partial\sigma \cdot \partial/\partial u$) to set up slightly different Markov neighborhood relations. So that the features are complete (we can reconstruct the original profiles from the features), we leave the coarsest level of the Gaussian scale-space (Figure 4(f)) alone in the Laplacian scale-space.

3.3   Local Statistical Model

The Laplacian scale-space is constructed on profiles taken from all images in the training population. We treat each profile at every location and scale level in the Laplacian scale-space as an independent feature vector, with its own PCA and (assumed Gaussian) distribution. Let $I(\sigma, u, \tau, i)$ be the normalized image intensity (a scalar), where:

- σ is the scale level in the Laplacian scale-space,
- u is the location of the profile around the boundary,
- τ is the position along the profile (in/out from the boundary),
- i is the index of the image in the training population.

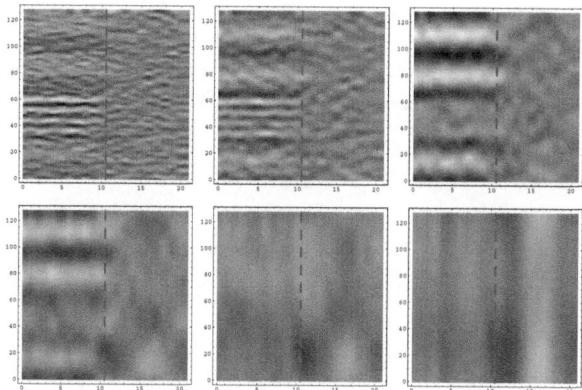

Fig. 5. The corresponding Laplacian scale-space. Levels of scale-space are shown fine-to-coarse, as in Figure 4. The coarsest level is duplicated from the Gaussian scale-space.

With our toy example, each profile is 13 samples long, there are six scale levels, with 128 profiles around the boundary, and 30 images in the training population. The 13 samples along each profile are spaced in units of two pixels. At each combination of (σ, u) in the scale-space, we have a 13-dimensional feature space and a cluster of 30 points in it.

The statistical distributions are local to each scale and location, so that we can model local variability, such as the high-frequency sine waves in the northwest quadrant of our toy example. Because we use the Laplacian scale-space, the local distributions are tied together into one joint distribution through a Markov model, where we model only how each scale level differs from the next coarser scale level. The joint distribution over the whole profile image can be complex and non-Gaussian (Gibbs distribution), but each local distribution can be assumed to be Gaussian and hence easy to model. Non-parametric estimation of the local distributions (e.g. Parzen windowing), as in [11], could also be done instead of the Gaussian estimation.

By treating each profile in the scale-space as a feature vector, we avoid a model which is tied to a particular type of edge (step edge, bar-like edge, etc.), and this approach is different from feature selection from an n-jet [12] . The training population "tells us" what kind of edge to look for locally, including expected variability.

Figure 6 shows the mean profiles at all locations and scale levels in the scale-space. What is not shown is the local covariance structure which is also part of the statistical model. For instance, at the very coarse levels of the scale-space, the noise has been smoothed away, so that even though the edge structure is very weak, the variability across the training population is very small. At the finest scale level, some locations show a clear edge structure in the mean profile, however the variability due to noise at that scale overwhelms the step edge; compare with the scale-space of a single case in Figure 5.

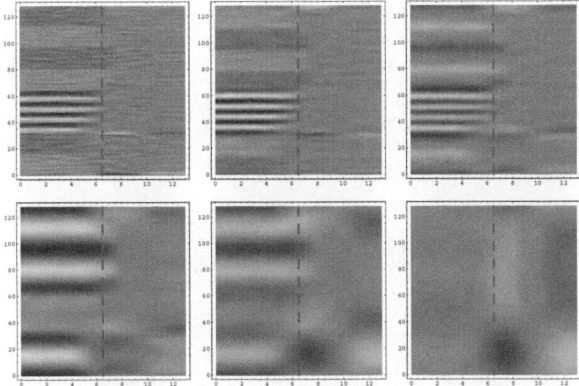

Fig. 6. Training population mean, Laplacian scale-space. Levels of the scale-space are shown fine-to-coarse, as in Figure 4.

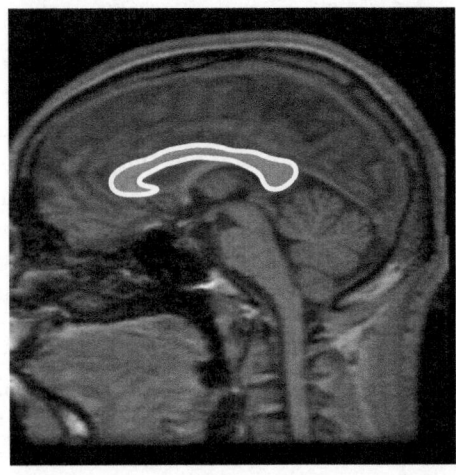

Fig. 7. Midsagittal MRI slice of the brain, showing a corpus callosum from the training dataset. Overlaid in red is the manual segmentation, represented as a 2D Fourier contour.

4   Application to the Corpus Callosum

We have built a statistical profile scale-space model on 71 segmented 2D corpus callosum images, for use in a segmentation framework with a shape representation of 2D Fourier harmonics [13]. Each of the 71 corpora callosa have been hand-segmented by experts, and the Fourier harmonics constructed. An example from the training population is shown in Figure 7, together with its Fourier boundary.

Profiles were extracted from the images (128 profiles around the boundary), and the Laplacian scale-space built for each set of profiles. The large scale edge

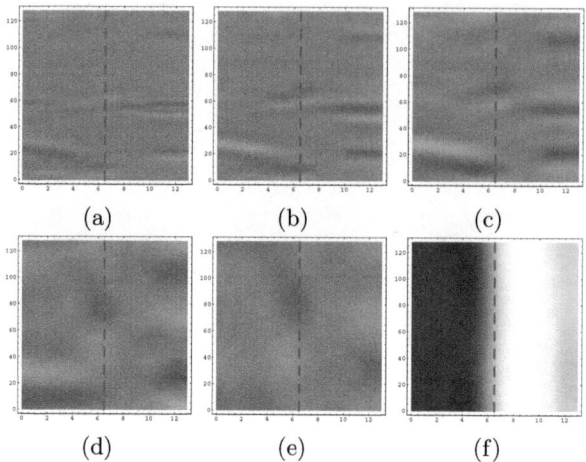

Fig. 8. Population mean of profiles around the corpus callosum; Laplacian scale-space. Levels of scale-space are shown fine-to-coarse, as in Figure 4.

structure of the corpus callosum can indeed be represented by a single step edge, which is in the very top level of the Gaussian scale-space, as shown in Figure 8(f).

Figure 8 shows the mean profiles of the Laplacian scale-space. As before there is also covariance structure behind each profile at each scale level that is not shown. The lower levels of the Laplacian scale-space show the local variability where the local edge structure differs from a simple step edge. The bright area of the fornix can be seen in levels 3 and 4 of the Laplacian scale-space, near the bottom of the graphs on the left hand side of each graph (exterior to the boundary). The fornix is a thin white-matter structure extending from the lower edge of the corpus callosum and not considered part of the corpus callosum; it can be seen in the MRI slice in Figure 7. Where the fornix joins the corpus callosum, there is no greylevel step edge at the boundary. The fornix is represented in the fine scales of the Laplacian pyramid as an *inverse* step edge (white outside, grey inside), which is the local residual needed to "cancel out" the global-scale step edge (dark outside, grey inside) shown in Figure 8(f).

5   Discussion / Conclusion

5.1   Application to Analysis and Discrimination

Our method constructs a statistical scale-space model of image profiles in a collar about the boundary of a deformable shape model placed in the image. Given a segmented training population, where each image in the population has a corresponding deformable shape model placed in it, our profile scale-space model allows one to understand how the image can be expected to vary along the object boundary. For example, the profile scale-space model built on the disk toy example clearly shows the sine waves of different frequency that are expected at different locations about the boundary. As mentioned in the introduction, the hippocampus exhibits different edge profiles at different parts of the boundary. Our profile scale-space model can be used to quantify these different edge profiles locally, including natural variability in the population.

Another application of the profile scale-space model is in discrimination. Discrimination is often done on shape features to understand differences between subpopulations, e.g. a diseased patient group vs. a control group. Using local profile scale-space features, one could, instead of performing PCA, use discriminant analyses like Independent Component Analysis [14] to understand how the patient group differs from the control group in image features. For example, a clinical hypothesis might be that one portion of the hippocampus darkens slightly relative to the rest of the hippocampus in diseased patients. This (made-up) hypothesis is an example of what can be tested using discriminant analysis on the profile scale-space model.

In both analysis and discrmination, the profile scale-space model looks at image features after the shape variability has been factored away, by use of the correspondence given by the set of deformed shape models. As mentioned in Section 2.3, this is similar in spirit to Active Appearance Models [9]. PCA and discriminant analysis are often performed on shape features; similar analysis can be done on image features in a shape-normalized space. The image feature analysis is in complement to the shape analysis.

5.2   Application to Segmentation

Bayesian image segmentation by deformable models optimizes two competing terms, the image-match likelihood and the shape prior (shape typicality measure). The statistical model derived from the profile scale-space provides the image-match likelihood. Integrated with an appropriate shape prior and an optimization strategy, this can provide a complete segmentation framework. The deformable model is initialized in the target image, and the image profiles are sampled relative to the deformable model. A scale-space is built on the profiles, and features derived from the scale-space are compared to the scale-space features to the statistical model to get a goodness-of-fit of the deformable model into the target image. In addition to the global goodness-of-fit, local feature selection from the scale-space can allow a local confidence measure in the goodness-of-fit to be constructed. This goodness-of-fit is balanced against the shape prior, and the contour is deformed to optimize the combination. We have such a segmentation framework in 2D using Fourier harmonic shape descriptors; future work could extend this to 3D using analogous spherical harmonic shape descriptors.

5.3   Extension to 3D/nD Images

Geometry-driven diffusion literature has described generalized images as embedding maps between two manifolds. Chung et al [7] blur what amounts to a 2D texture map on a triangle mesh in 3D. If we let M be the triangle mesh (a 2D manifold embedded in 3D), then their image can be seen as a function

$I : M \rightarrow \mathbb{R}^+$, where $\mathbb{R}^+$ represents the positive reals, the range of intensity values in the image. Alternately, we can think of the image as a manifold embedded in the space $M \times \mathbb{R}^+$.

Now, to extend this to a collar of thickness about an object boundary (a volume of the image), the domain of the image can be extended to a 3D space about the boundary, instead of just the 2D manifold M. If we use the formalism of image profiles normal to the boundary, we can think of the image as a function $I : M \times [-1, 1] \rightarrow \mathbb{R}^+$, where $[-1, 1]$ indicates normalized distance from the object boundary M. Hence, for a fixed $d \in [-1, 1]$, the domain $\{(u, d) : u \in M\}$ represents a "shell" of distance d from the original boundary, according to a distance transform. For a fixed $u \in M$ on the surface, the set $\{I(u, d) : d \in [-1, 1]\}$ is a single image profile across the boundary at that point. Diffusion using the Laplace-Beltrami operator can then be run on each shell, using the metric tensor of the manifold M, to construct a scale-space. A non-Euclidean distance could certainly be used for d, for instance the intrinsic figural coordinate system in M-reps [15] [16].

5.4   Conclusion

We present a statistically trained scale-space model on image profiles around the boundary of an object embedded within images. There is a sampling of the image in a coordinate system relative to the object boundary (image profiles). A scale-space is constructed on these profiles which preserves the across-boundary features, but looks at along-boundary features in a multiscale fashion. The process is done for a population of images, with correspondence given by the shape models which have been deformed to fit each image. A statistical model is then built on the profile scale-space, which can then be incorporated into a full Bayesian segmentation framework.

References

1. Bart M. ter Haar Romeny, Ed., *Geometry-Driven Diffusion in Computer Vision*, Computational Imaging and Vision. Kluwer Academic Publisher, 1994.
2. Timothy F. Cootes, A. Hill, Christopher J. Taylor, and J. Haslam, "The use of active shape models for locating structures in medical images," in *IPMI*, 1993, pp. 33–47.
3. J.F. Canny, "A computational approach to edge detection," *IEEE Trans on Pattern Analysis and Machine Intelligence (PAMI)*, vol. 8, no. 6, pp. 679–697, November 1986.
4. P. Perona and J. Malik, "Scale-space and edge detection using anisotropic diffusion," *IEEE Trans on Pattern Analysis and Machine Intelligence (PAMI)*, vol. 12, no. 7, pp. 629–639, July 1990.
5. N. Sochen, R. Kimmel, and R. Malladi, "A general framework for low level vision," *IEEE Transactions on Image Processing*, vol. 7, no. 3, pp. 310–318, 1998.
6. Bei Tang, Guillermo Sapiro, and Vicent Caselles, "Diffusion of general data on non-flat manifolds via harmonic maps theory: The direction diffusion case," *International Journal of Computer Vision (IJCV)*, vol. 36, no. 2, pp. 149–161, 2000.

7. M.K. Chung, K.J. Worsley, J. Taylor, J.O. Ramsay, S. Robbins, and A.C. Evans, "Diffusion smoothing on the cortical surface," *NeuroImage*, vol. 13S, no. 95, 2001.
8. András Kelemen, Gábor Székely, and Guido Gerig, "Elastic model-based segmentation of 3d neuroradiological data sets," *IEEE Transactions on Medical Imaging (TMI)*, vol. 18, pp. 828–839, October 1999.
9. Timothy F. Cootes, Gareth J. Edwards, and Christopher J. Taylor, "Active appearance models," *IEEE Transactions on Pattern Analysis and Machine Intelligence (PAMI)*, vol. 23, no. 6, pp. 681–685, 2001.
10. L.H. Staib and J.S. Duncan, "Boundary finding with parametrically deformable contour models," *IEEE Transactions on Pattern Analysis and Machine Intelligence (PAMI)*, vol. 14, no. 11, pp. 1061–1075, Nov 1992.
11. M. Leventon, O. Faugeraus, and W. Grimson, "Level set based segmentation with intensity and curvature priors," in *Workshop on Mathematical Methods in Biomedical Image Analysis Proceedings (MMBIA)*, June 2000, pp. 4–11.
12. L. M. J. Florack, B. M. ter Haar Romeny, J. J. Koenderink, and M. A. Viergever, "The Gaussian scale-space paradigm and the multiscale local jet," *International Journal of Computer Vision (IJCV)*, vol. 18, no. 1, pp. 61–75, April 1996.
13. G. Székely, A. Kelemen, Ch. Brechbühler, and G. Gerig, "Segmentation of 2-D and 3-D objects from MRI volume data using constrained elastic deformations of flexible Fourier contour and surface models," *Medical Image Analysis*, vol. 1, no. 1, pp. 19–34, 1996.
14. A. Hyvärinen, J. Karhunen, and E. Oja, *Independent Component Analysis*, John Wiley & Sons, New York, 2001.
15. SM Pizer, T Fletcher, A Thall, M Styner, G Gerig, and S Joshi, "Object models in multiscale intrinsic coordinates via m-reps," in *Generative-Model-Based Vision (GMBV)*, 2002.
16. Conglin Lu, Stephen M. Pizer, and Sarang Joshi, "A markov random field approach to multi-scale shape analysis," in *ScaleSpace*, 2003, vol. this volume.

Iris Feature Extraction and Matching Based on Multiscale and Directional Image Representation

Chul-Hyun Park[1], Joon-Jae Lee[2], Sang-Keun Oh[1], Young-Chul Song[1],
Doo-Hyun Choi[1], and Kil-Houm Park[1]

[1] School of Electrical Engineering and Computer Science
Kyungpook National University, Daegu, Korea
{nagne,taesa}@palgong.knu.ac.kr, {songyc03,dhc,khpark}@ee.knu.ac.kr
[2] Division of Internet Engineering, Dongseo University, Busan, Korea
jjlee@dongseo.ac.kr

Abstract. This paper presents a new filterbank-based iris recognition method that effectively extracts the spatial and directional features of iris patterns on multiple scales, then performs matching. First, the proposed method localizes the iris area from an input image and establishes a region of interest (ROI) for feature extraction. Second, the iris features are extracted on multiple scales from the ROI and a feature vector generated using a band pass filter and directional filter bank (DFB), which decomposes the image into several directional subband outputs. Finally, iris pattern matching robust to various rotations of the input is performed based on finding the Hamming distance between the corresponding feature vectors. Experimental results demonstrate that the proposed method is both effective in extracting directional and multiresolutional features from iris patterns and robust to input image rotation due to head tilt.

1 Introduction

Iris recognition refers to a biometric technology that identifies an individual using the annular iris pattern that surrounds the pupil of the eye. A human iris pattern is unique to each person and nearly immutable through life, plus it includes rich discriminatory information, such as collagenous fibers, contraction furrows, coronas, crypts, colors, freckles, rifts, and pits (See Fig. 1(c)) [1]. Therefore, iris recognition can be a highly reliable solution for personal identification or verification.

The use of iris patterns as a biometric method has several other advantages besides its high fidelity. One is that it is non-invasive, unlike fingerprint recognition that requires physical contact with a sensor. Consequently, the user does not experience any discomfort or uneasiness during the authentication process. In addition, since an iris is circular and much darker than the neighboring (white) sclera, the iris region can be easily detected in the input image. Accordingly, this facilitates the development of a recognition method that is robust to both rotation due to head tilt and changes in the position and size of the iris. Although iris patterns are rich in information, exploiting their geometric features is not

L.D. Griffin and M. Lillholm (Eds.): Scale-Space 2003, LNCS 2695, pp. 576–583, 2003.

as easy as in fingerprints or faces. Thus, most existing iris recognition methods extract either the multiresolutional or the directional features of an iris pattern using Gabor filters or wavelet transforms [1], [2], [3], [4], and in most cases are unable to well extract both, plus the rotations of an input iris image due to head tilt are not considered.

Therefore, this paper presents a new filterbank-based iris recognition method that is effective in extracting spatial and directional features on a multiple scale, and also robust to various rotations of an input iris image. The proposed method extracts various directional features using a directional filter bank (DFB), which accurately decomposes the image into several directional subband outputs [5], [6]. The multiscale or multiresolutional characteristics are also reflected in the feature values by performing band pass filtering before the DFB. In the proposed method, the subband regions used to extract the feature values can be adjusted by changing the cut-off frequencies of the band pass filter in front of the DFB. Matching is then carried out based on finding the minimum Hamming distance between a set of input feature vectors in which various rotations are considered and the enrolled template feature vector, thereby achieving rotational alignment between the input and template feature vectors. Section 2 explains the DFB used in the proposed method, while Sections 3 and 4 describe the feature extraction and subsequent matching procedures. Section 5 presents the experimental results and some final conclusions are given in Section 6.

2 Directional Filter Bank (DFB)

The DFB divides the two-dimensional spectrum of an image into wedge-like directional subbands, as shown in Fig. 1(a) [5], [6]. Eight directional subband outputs can be obtained using the 8-band DFB, as shown in Fig. 1(b). Figure 1(d) shows an example of the directional subband images decomposed by the 8-band DFB, where each directional component is captured in its subband image.

The DFB efficiently and accurately divides an image into directional subband outputs in the spatial domain using quincunx matrices, which rotate and downsample the image, and a low pass filter, which has a diamond-shape filtering characteristic. Each direction subband output has a rectangular shape where the width and height differ due to the back sampling matrices used to remove frequency scrambling [6]. For an $N \times N$ image, the first half of the 2^n subband outputs is $N/2^{n-1} \times N/2$ in size, while the other half is $N/2 \times N/2^{n-1}$.

3 Feature Extraction

The proposed method first locates the iris region from an input image, then establishes the ROI for feature extraction from the iris region. The iris features are extracted from the ROI using a band pass filter and DFB.

3.1 Iris Localization

Since iris images mostly contain a lot of noise, for example, reflections due to the illuminator and pattern occlusion due to eyelashes or eyelids, the proposed

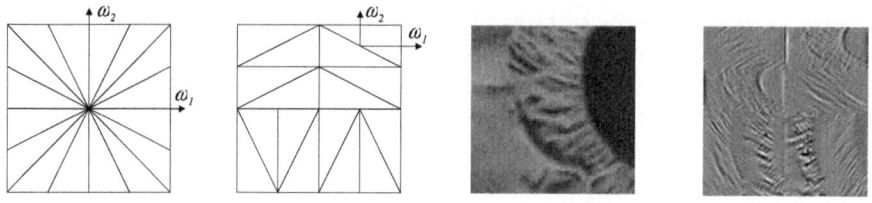

Fig. 1. Frequency partition map of (a) input, (b) 8 subband outputs, (c) sample of iris pattern, and (d) decomposed subband outputs of (c) when using 8-band DFB

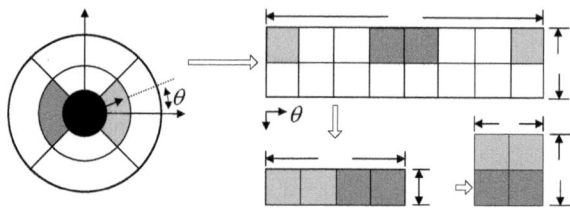

Fig. 2. Conversion of iris region coordinate system into polar coordinate system, and region used for matching. (a) Iris region, (b) iris region converted to polar coordinate system, (c) extracted ROI, and (d) ROI converted to square form

method conducts preprocessing to identify the iris region and exclude the region with most noise. After acquiring an iris image, the inner and outer boundaries of the iris are found using a circular edge detector [1]. The detected iris region is then converted from a Cartesian coordinate system into a polar coordinate system to facilitate the subsequent processing. Since the center of the pupil and the center of the iris are not generally the same, the detected region is normalized into one with a fixed size of $N \times 4N$, as shown in Fig. 2(b), using information on the center coordinates of the pupil and the iris so that the same portion of the iris is matched. The regions corresponding to the lower and upper $90°$ cones and the outer half of the iris region are excluded, since occlusions by the eyelids commonly occur or glint appears due to the illuminator (see Fig. 2(c)). The remaining region with a size of $N/2 \times 2N$ is then converted again into a square form with a size of $N \times N$ by repositioning the right half of the region below the left half, as shown in Fig. 2(d), as the DFB is only optimized to filter a square image where the width and height are equal. In this work, N was set at 144.

3.2   Generation of Feature Vector

Iris patterns include many linear features that can be considered as a combination of directional linear pattern components. As such, the unique characteristics of an iris pattern can be effectively represented by a feature vector constructed by extracting the linear components of an iris according to directionality. In

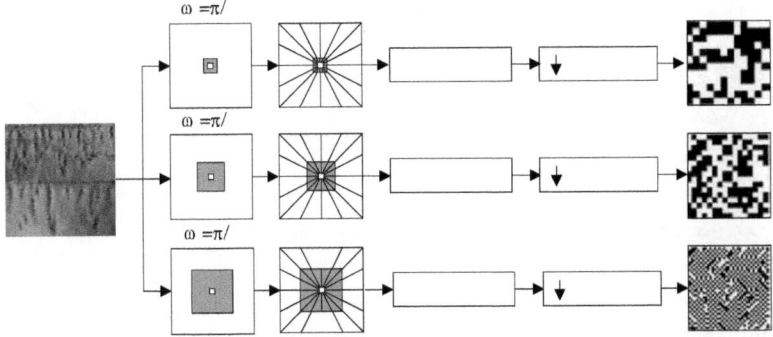

Fig. 3. Procedure for generating feature vector

this sense, a DFB is suitable for extracting iris features, since it can accurately decompose an image into directional subband outputs. In a previous application using a DFB, Rosiles et al. [7] classified textural images using the variances of directional subband images as the feature vector. However, if the variances of an entire subband image are used as the feature vector for iris recognition, this does not produce a good performance, as such variances do not effectively represent the spatial information in the image. Alternatively, if sub-block variances are considered as the feature vector, the performance becomes sensitive to the brightness or contrast in an image. Therefore, the proposed method binarizes the subband outputs and performs sampling at regular intervals to capture both the spatial and directional information and generate a feature vector robust to changes in illumination or brightness. Since each decomposed subband output value has an average value of almost 0, those values thresholded by 0 preserve the directional linear features and are robust to changes in illumination or brightness.

The proposed method uses an additional band-pass filter to extract the iris features on multiple scales [7]. The lower cut-off frequency is determined to remove the DC energy. The extracted ROI is band-pass filtered and decomposed by an 8-band DFB. The resultant subband outputs are then thresholded at either 1 or 0 according to their signs and sampled at regular intervals. For the subband outputs of an image filtered by a band-pass filter with a higher cut-off frequency of π/n, sampling is performed every $n/2$ pixels. The overall iris feature vector is then completed by combining the feature values obtained from three different scales. Fig. 3 shows the procedures used to generate a feature vector.

The main difference between the proposed method and the Gabor filter bank-based method in [1] is the use of different subband ranges for feature extraction, as illustrated in Fig. 4. With a Gabor filter bank, there are always some overlapping or missing subband regions, whereas a DFB has a directionally accurate subband separation characteristic, as shown in Fig. 4(b). Accordingly, a DFB can represent linear patterns, as found in iris patterns, more effectively than a Gabor filter bank (See Fig. 1(c)).

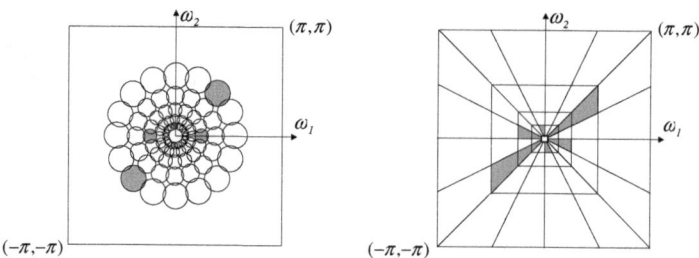

Fig. 4. Subband ranges used for generating feature vector by (a) Gabor filter bank and (b) multiresolution DFB

4   Matching

Matching is performed based on the Hamming distance between the input feature vector and the template feature vector enrolled in the database. To make the matching robust to various rotations of the iris image, the proposed method generates a set of input feature vectors where diverse rotations are considered by shifting the thresholded subband outputs and resampling the feature values. Thereafter, the method matches these input feature vectors with the template feature vector enrolled in the database to find the minimum Hamming distance. As a result, a rotational alignment between the input and template feature vectors is achieved. When the rotation of an input iris is taken into account, the subband regions corresponding to the extent of the considered rotation can not be used for feature extraction, as there will be a discontinuity between the upper and lower halves of the ROI.

The minimum unit of horizontal shift in the subband images is a one-pixel shift in 4 to 7 direction subband images, and this one-pixel shift means a four-pixel shift in the image domain. For an $N \times N$ image, a four-pixel shift in the ROI is equivalent to a rotation of $90 \times (4/N)$ degrees in the original image. For a 144×144 image, the minimum unit of rotation that can be effectively compensated for is $2.5°$, as such, the feature vector generated by the proposed method is invariant to small perturbations within $1.25°$. Let A_j^R denote the jth feature value of the input feature vector in which $R \times 90 \times (4/N)$ degree rotation is considered and let B_j denote the jth feature value of the template feature vector, then the Hamming distance between input and template feature vectors, HD, is given by

$$HD = \min_R \frac{1}{N} \sum_{j=1}^{N} A_j^R \oplus B_j \qquad (1)$$

where $R \in \{-10, -9, \ldots, -2, -1, 0, 1, 2, \ldots, 9, 10\}$, N is the size of the feature vector, and $\oplus$ is an exclusive-OR operator that yields one if A_j^R is not equal to B_j, and zero otherwise. The range for R is established on the assumption that the rotation of the input iris image is between $-25°$ and $+25°$.

5   Experimental Results

To acquire iris images for experiments, a digital movie camera and 50 W halogen lamp were used. Close-up shots were taken from a distance of about 15 cm to obtain iris images large enough for recognition. The light was located below the camera so that the glint due to the illuminator only appeared in the lower 90° cone of the iris.

The original iris images acquired by the camera were 640×480 color images, however, they were converted into an 8-bit grayscale for the experiments. First, the test images were divided into two groups. One image set was used as the enrolled database, while the other was used as the input images. The iris images were acquired over a period of more than six months, as such, there were many differences in the brightness and iris size. A total of 200 images were acquired from 10 persons for the enrolled database and 234 images from 10 people for the input image set.

To evaluate the proposed method, its performance was compared with that of the Gabor filter bank-based method in [1] and the Haar wavelet transform-based method in [4]. The performance of each method was evaluated for both recognition (or identification) and verification (or authentication). To evaluate the recognition performance, the feature vector generated by each method was compared with the feature vectors of the database images, and recognition conducted by identifying the class of the iris image with the minimum Hamming distance in the database.

The recognition rate was found to increase in proportion to the number of scales used to generate a feature vector, indicating that a method using features on multiple scale levels can provide a better recognition performance than a method using features on a single scale. As regards the size of the feature vector, since the Gabor filter bank-based method used 24 Gabor filters with 3 frequencies ($2^{-3}\pi$, $2^{-5/2}\pi$, $2^{-4}\pi$), 8 directions in the experiment, and the feature values are two bit quantized ones, such as 00, 01, 10, and 11, it required $2,128(2\times8\times(144/16\times144/16 + 144/24\times144/24 + 144/36\times144/36))$ bits of storage for the feature vector. For the proposed method, to obtain three different resolutions of thresholded subband outputs, the band pass filters with the higher cut-off frequencies of $\pi/8, \pi/12, \pi/16$ were used. Thus the proposed method required $2,196(144/4\times144/4 + 144/6\times144/6 + 144/8\times144/8)$ bits to store the feature vector. Meanwhile, the method using a Haar wavelet transform generated a feature vector with $84(144/16\times144/16 + 3)$ components, representing a more compact feature vector than either of the other two methods.

Table 1 shows the recognition results for each method with and without compensation for input image rotation. The matching results were significantly influenced by the input image rotation. For the proposed method, rotational alignment between the input and enrolled template data was automatically performed, as described in Section 4. For the other two methods, manually compensated images were used to obtain recognition results that included compensation for rotation. When the proposed method was applied to the manually compensated dataset, the recognition rate was 99.97%, confirming that the proposed

Table 1. Recognition rate for each method

Method	Recognition rate	
	Without CR	With CR
Haar wavelet-based	34.18%	77.78%
Gabor filter bank-based	61.11%	98.29%
Proposed	68.37%	98.71%

CR : Compensation for Rotation

method could effectively compensate for rotation. The proposed method exhibited a similar recognition performance to the Gabor filter bank-based method, yet a better recognition rate by about 20% than the Haar wavelet transform-based method. Despite the advantage of a compact feature vector with the Haar wavelet transform-based method, its ineffective use of multiresolutional and directional features resulted in a lower recognition rate compared to the other two methods.

To evaluate the performances as a personal verification method, genuine and imposter distributions of the Hamming distances were examined for each method. A genuine distribution represents the distribution of the Hamming distances between all possible intra-class image pairs in the database, while an imposter distribution represents the distribution of the Hamming distances between all possible inter-class image pairs in the database. The more separated the two distributions and the smaller the standard deviation for each distribution, the more advantageous for a personal verification method. A decidability index is also a good measure of how well the two distributions are separated [1]. Let μ_1 and μ_2 be the means of the two distributions, respectively, and σ_1 and σ_2 the standard deviations, then the decidability index can be defined as $d' = |\mu_1 - \mu_2|/\sqrt{(\sigma_1^2 + \sigma_2^2)/2}$. The decidability index for the proposed method was 3.2892, while that for the Gabor filter bank-based method and Haar wavelet transform method was 1.8901 and 3.0498, respectively. This result shows that the distributions for the proposed method were better separated than those for the other two methods.

Another measure that can be used to estimate the verification performance is the EER (Equal Error Rate), when the FRR (False Reject Rate) is equal to the FAR (False Acceptance Rate). The EER and GAR (Genuine Acceptance Rate) for each method are shown in Table 2. Therefore, for verification, the proposed method exhibited a better performance than the other two methods. For the proposed method and Gabor filter bank-based method, the recognition and verification performances changed according to the subband ranges used for feature extraction, and properly determined subband ranges produced the best performance for each method. Although the sub-band ranges for feature extraction can be adjusted with the Gabor filter bank-based method by changing the values of the orientation and frequency parameters, such changes are restricted due to the characteristics of the Gabor filter itself. However, with the proposed method, the subband ranges can be adjusted by simply changing the cut-off frequency of the band pass filter and the number of directional bands decomposed by the

Table 2. Decidability index, EER, and GAR at EER for each method

Method	d' index	EER	GAR
Haar wavelet transform-based	1.8901	16.84%	82.79%
Gabor filter bank-based	3.0498	4.25%	95.74%
Proposed	3.2892	3.13%	96.90%

DFB. Consequently, the proposed feature extraction method is also an effective way of extracting features from patterns containing various kinds of frequency and directional components.

6   Conclusion

A new iris feature extraction and matching method was presented based on a mutiresolution DFB. The proposed method is effective in extracting directional features from iris patterns at various resolution levels and robust to input image rotation due to head tilt. In addition, since the subband ranges used for feature extraction can be adjusted by simply changing the cut-off frequency values of the band pass filter attached in front of the DFB, the proposed method has advantages when designing an optimal feature extractor for object patterns that include various kinds of frequency and directional components. To further enhance the verification and recognition accuracy, more study is needed on expanding the ROI for feature extraction.

References

1. Daugman, J. G.: High confidence visual recognition of persons by a test of statistical independence. IEEE Trans. Pattern Anal. Machine Intell. **15** (1992) 1148–1161
2. Wildes, R. P.: Iris recognition: An emerging biometric technology. Proc. IEEE **85** (1997) 1348–1363
3. Boles, W. W., Boashash, B.: A Human Identification Technique Using Images of the Iris and Wavelet Transform. IEEE Trans. Signal Processing **46** (1998) 1185–1188
4. Lim, S., Lee, K., Byeon, O., Kim, T.: Efficient iris recognition through improvement of feature vector and classifier. ETRI Journal **23** (2001) 61–70
5. Bamberger, R. H., Smith, M. J. T.: A filter bank for the directional decomposition of images: Theory and design. IEEE Trans. Signal Processing **40** (1992) 882–893
6. Park, S., Smith, M. J. T., Mersereau, R. M.: A new directional filter bank for image analysis and classification. Proc. IEEE Intl. Conf. on Acoustics, Speech, and Signal Processing **3** (1999) 1417–1420
7. Rosiles, J. G., Smith, M. J. T.: Texture Classification with a Biorthogonal Directional FilteBank. Proc. IEEE Intl. Conf. on Acoustics, Speech, and Signal Processing **3** (2001) 1549–1552

Fast Computation of Scale Normalised Gaussian Receptive Fields

James L. Crowley and Olivier Riff

Laboratoire GRAVIR, INRIA Rhône Alpes
655 Ave de l'Europe, F-38330 Montbonnot, France
{Crowley, Riff}@inrialpes.fr
http://www-prima.imag.fr

Abstract. The characteristic (or intrinsic) scale of a local image pattern is the scale parameter at which the Laplacian provides a local maximum. Nearly every position in an image will exhibit a small number of such characteristic scales. Computing a vector of Gaussian derivatives (a Gaussian jet) at a characteristic scale provides a scale invariant feature vector for tracking, matching, indexing and recognition. However, the computational cost of directly searching the scale axis for the characteristic scale at each image position can be prohibitively expensive. We describe a fast method for computing a vector of Gaussian derivatives that are normalised to the characteristic scale at each pixel. This method is based on a scale equivariant half-octave binomial pyramid. The characteristic scale for each pixel is determined by an interpolated maximum in the Difference of Gaussian as a function of scale. We show that interpolation between pixels across scales can be used to provide an accurate estimate of the intrinsic scale at each image point. We present an experimental evaluation that compares the scale invariance of this method to direct computation using FIR filters, and to an implementation using recursive filters. With this method we obtain a scale normalised Gaussian Jet at video rate for a 1/4 size PAL image on a standard 1.5 Ghz Pentium workstation.

1   Introduction

The visual appearance of a neighborhood can be described by a local Taylor series [1]. The coefficients of this series constitute a feature vector that compactly represents the neighborhood appearance for indexing[2] and matching[3]. The set of possible local image neighborhoods that project to the same feature vector are referred to as the "Local Jet". A key problem in computing the local jet is determining the scale at which to evaluate the image derivatives.

Lindeberg [4] has described scale invariant features based on profiles of Gaussian derivatives across scales. In particular, the profile of the Laplacian, evaluated over a range of scales at an image point, provides a local description that is "equivariant" to changes in scale. Equivariance means that the feature vector translates exactly with scale and can thus be used to track, index and recognize structures in the presence of changes in scale.

L.D. Griffin and M. Lillholm (Eds.): Scale-Space 2003, LNCS 2695, pp. 584–598, 2003.

The problem with this approach is that a direct computation of the characteristic scale at each image position appears to make real-time implementation unfeasible. This paper presents a method to obtain the characteristic scale by interpolating the samples of a half-octave Laplacian Pyramid along both the image and the scale axes. The Laplacian for any image position is obtained by bi-linear interpolation between adjacent sample pixels. Local maxima over scale are determined by a fitting a parabolic function to samples in the scale direction at a pixel. However, not just any multi-resolution pyramid can be used for such calculations. Scale-invariant image description requires that the sampled impulse response be the same at every level of the pyramid.

2   Fast Computation of Chromatic Receptive Fields

Multi-resolution methods have been used in computer vision since the 1970's. Early work in multi-resolution image description was primarily motivated by a desire to reduce the computational cost of methods for image description and image matching. One of the earliest uses was a technique referred to as "planning", in which image resolution was reduced by summing pixels in non-overlapping 8x8 blocks [5]. The results of edge detection at low resolution were used to select regions for edge detection at high resolution.

Multi-resolution processing was soon generalized to computing multiple copies of an image by repeatedly summing non-overlapping blocks of pixels and re-sampling until the image reduced to a small number of pixels. Such a structure became known as a multi-resolution pyramid [6]. In a typical early pyramid algorithm, non-overlapping blocks of 4x4 pixels were summed at each level to produce the next reduced resolution level. Such pyramid structures were used to construct fast algorithms for image segmentation, edge detection, and to accelerate correlation for stereo matching. Unfortunately, computing a pyramid by averaging non-overlapping windows resulted in substantial aliasing. Such aliasing is most noticeable as a large component of additive random noise generated by image translation. Such noise can render most image analysis algorithms unreliable.

The problem of segmentation and classification of textures led a number of researchers to look for general-purpose multi-resolution representations. Burt proposed a multi-resolution pyramid algorithm using smoothing with overlapping windows [7]. Weights for the smoothing filters were obtained by postulating a set of four principles. These principles resulted in the use of a mask that serves as a smoothing filter for repeated re-sampling. While smoothing with these masks did reduce noise, significant aliasing effects still remained. Moreover, Burt's pyramid was not scale invariant.

During this period, a half-octave scale-invariant pyramid algorithm was proposed based on considerations from signal processing [8]. This algorithm was explicitly designed to maintain the same sampled impulse response at each level. Images were smoothed by a Gaussian filter designed to avoid aliasing effects. Unfortunately, the use of large FIR Gaussian filters led to computing times on the order of an hour for a single image.

By the mid-1980's, the multi-resolution pyramid had become a standard structure for use in stereo matching and motion analysis [9]. The use of techniques from digital signal processing provided mathematical tools to understand the effects of repeated smoothing and sampling. By the late 1980's, pyramids were generally computed using Gaussian filters of sufficiently large size so as to minimize the random noise dues to aliasing. However, generally little attention was paid to the scale-invariant properties.

3   A Scale Invariant Half Octave Pyramid

A scale-equivariant space can be constructed using any kernel function. Let $x(t)$ be a signal defined over a continuous variable t. A kernel function, $k(t)$, can be scaled to any scale factor, s, by dividing t by s. Thus for continuous variables, a scale-equivariant "scale-space" representation of a signal is easily defined, as

$$p(t,s) = x(t) * k(\frac{t}{s})$$

Computing a sampled digital representation of such a space requires choosing the appropriate sample rates for t and for s. The sample rate, T_o , for the t variable is determined by the frequency content of the signal that should be preserved in the sampled representation. For a scale-invariant representation, the variable s should be sampled using an exponential series

$$S_k = s_0^k$$

This is easily shown by taking the logarithm of t/s. The logarithm converts the $1/s$ term into translation along the scale axis. Thus changes in scale are expressed as translation in a logarithmic scale space.

The set of possible scales range from 1 to the number of samples. The desired sample rate in scale will often depend on the smoothness of the kernel. The cost of brute force sampling of such a space is the number of signal samples, N, times the number of scale samples, LogN. Thus computational cost of such a space is, in principle, O(N Log N) . Unless the bandwidth of $x(t)$ is limited and the kernel is properly chosen, the actual constants required for such a space are computationally prohibitive.

A multi-resolution pyramid algorithm produces a sampled scale-space representation of a signal, p(t,s), with a computational complexity of O(N). The reduction in complexity is achieved by re-using each scale-sampled representation of the signal as an intermediate result for producing the next. Strict scale equivariance requires that convolution of a kernel filter with itself produce a scaled copy of the kernel filter:

$$k(\frac{t}{S_0}) = k(t) * k(t)$$

The Gaussian function:

$$g(t, \sigma) = e^{-\frac{t^2}{2\sigma^2}}$$

obeys this property, with a scale factor of $s_0 = \sqrt{2}$. More generally, the Gaussian functions are closed under convolution. That is, the convolution of two Gaussians of variance σ_1^2 and σ_2^2 results in a Gaussian of variance $\sigma_3^2 = \sigma_1^2 + \sigma_2^2$. As a result, a scale-invariant pyramid can be defined by cascaded convolution with a Gaussian kernel.

The Gaussian function has a number of other properties that make it ideally suited for use as a kernel filter for computing a scale-invariant pyramid. Among these is the fact that a circularly symmetric Gaussian is separable into a product of 1-D components. This property allows us to compute the convolution of an NxN Gaussian by a series of two 1-D convolutions. Thus the convolution with a Gaussian remains O(N), even when applied to a 2-D NxN signal.

3.1   The O(N) Scale-Invariant Pyramid

A multi-resolution pyramid is an O(N) method for computing a sampled scale space. The reduction in computation is achieved by reusing each level as an intermediate result to compute the next level. This pyramid algorithm is scale equivariant. Each level is resampled at a step size that exactly equals the increase in scale. Thus the ratio of scale to sample rate is constant. Scaling a signal translates its response in the scale axis.

The scale-equivariant pyramid algorithm shown in figure 4 is composed of an initial convolution with the kernel filter followed by a series of processing stages, k=0 to K. For each stage, k, the pyramid is composed of three signals $p_0(n,k)$, $p_1(n,k)$ and $p_2(n,k)$. The output of each stage is resampled to produce the input for the next stage. Because of resampling, each stage is composed of $N_k = N/2^k$ samples (in the case of a 1-D signal).

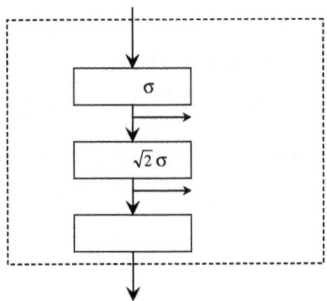

Fig. 4. First stage of the scale invariant pyramid algorithm.

The signal $p_0(n,k)$ serves as the input to the kth stage. This signal is convolved with the kernel filter, $g(n, \sigma)$, to provide $p_1(n,k)$:

$$p_1(n,k) = p_0(n,k) * g(n,\sigma)$$

The second stage is computed by convolution with a scaled copy of the kernel filter:

$$p_2(n,k) = p_1(n,k) * g(n,\sqrt{2}\sigma)$$

This scaled copy can be obtained by cascaded convolution with the kernel filter:

$$p_2(n,k) = p_1(n,k) * g(n,\sigma) * g(n,\sigma)$$

To demonstrate the scale equivariance, consider the impulse response for a scale-invariant pyramid with a Gaussian kernel $g(n,0)$ using a typical value of $\sigma=1$. Thus the kernel filter is:

$$g(n,1) = e^{-\frac{n^2}{2}}$$

To have an impulse as input, assume an N-sample input signal $s(n) = \delta(n - \frac{N}{2})$ composed of zero values, except at position N/2 where the value is set to 1. The initialization step convolves the impulse with the kernel filter:

$$p_0(n,0) = g(n,1)$$

Thus the variance and standard deviation at $p_0(n,k)$ are both 1.0. The next step is

$$p_1(n,0) = p_0(n,0) * g(n,1)$$

Thus the variance at $p_1(n,k)$ is $\sigma_{01}^2 = 2$ and the scale factor is $\sigma_{01} = \sqrt{2}$. Continuing,

$$p_2(n,0) = p_1(n,0) * g(n,1) * g(n,1)$$

The variance of $p_2(n,k)$ is $\sigma_{02}^2 = 4$, and thus $\sigma_{02} = 2$. The result is resampled at $T_1 = 2$ to provide stage k=1. To show the effects of sampling, consider a change in variables, m=2n, to obtain

$$p_1(m,1) = p_2(2m,0)$$

Expressed in the original variable, n, resampling does not effect the variance or σ of the signal. Thus $\sigma_{01}^2 = 2$ and $\sigma_{01} = 2$. However, convolution with a resampled signal is the same as scaling the kernel filter. Thus,

$$p_1(m,1) = p_0(m,1) * g(m,1) = p_2(n,0) * g(2n,1)$$

By virtue of resampling, the Gaussian kernel has effectively been rescaled by a factor of $\sigma=2$. This is equivalent to rescaling the variance of the Gaussian by 4. Thus

$$\sigma_{11}^2 = 8, \text{ and } \sigma_{11} = 2\sqrt{2}.$$

Continuing the stage,

$$p_2(m,1) = p_1(m,1) * g(m,1) * g(m,1)$$

which gives $\sigma_{12}^2 = 16$, and $\sigma_{12} = 4$. The result is resampled to provide the input to the next stage and the process is repeated:

$$p_0(m,3) = p_2(2m,2)$$

The result is a sequence of signals in which both the sample rate and the scale factor grow in powers of 2. At each stage, an intermediate result for $p_1(n,k)$ provides a $\sqrt{2}$ scaling of the impulse response.

The 1-D algorithm defined above is easily generalized to 2-D by replacing the variable n with x, y. This input signal is changed from p(n) of size N sample to p(x,y) of size N^2. However, the Gaussian kernel is separable:

$$g(x,y,\sigma) = e^{-\frac{x^2+y^2}{2\sigma^2}} = e^{-\frac{x^2}{2\sigma^2}} * e^{-\frac{y^2}{2\sigma^2}}$$

Thus, convolution with the kernel with an NxN image can be computed as a series of two O(N) 1-D convolutions. Thus the cost of convolution with a Gaussian remains O(N) and the resulting pyramid is an O(N) algorithm.

4   Experimental Comparison of Fast Gaussian Filters

4.1   Fast Gaussian Filters

Digital filters can be designed using either a direct (FIR) or recursive (IIR) form. The direct form is obtained as a finite number of samples of the desired impulse response. The recursive form is designed as a ratio of polynomials in the z domain. Closure under convolution provides a third method for designing Gaussian filters by cascade convolution. The following section compares these three implementation methods for a 1-D Gaussian filter.

4.1.1   The FIR Implementation of a Gaussian

The simplest means to implement a digital Gaussian filter is to sample the Gaussian function at integer multiples of T_0. For $\sigma=1$, a reasonably good approximation is obtained using a kernel width of 9 pixels. This gives

$$G(x) = e^{-\frac{x^2}{2}}$$

for integer values of x in the range $x \in [-4, 4]$.

4.2   Binomial Filters

Binomial filters are obtained with cascaded convolution of a kernel filter composed of [1, 1]. The coefficients for the nth filter in the series, $b_n(m)$, are defined by:

$$b_n(m) = [1,1]^{*n}$$

where the exponent *n denotes n auto-convolutions. The set of filter coefficients is well known as the binomial series, often computed using Pascal's triangle. This series provide the best (least sum of squares error) approximation to a Gaussian function by an integer coefficient sequence of finite duration. The properties of the binomial filters are particularly easy to compute. For example, for the nth binomial $b_n(m)$, there are n coefficients, whose sum is 2^n. The midpoint (or center of gravity) is the coefficients at $m = \frac{n}{2}$ and the variance is $\sigma^2 = \frac{n}{4}$.

The binomial filters $b_2(m)$ (with coefficients [1, 2, 1]) and $b_4(m)$ (with coefficients [1, 4, 6, 4, 1]) are of special interest. The Fourier transform of $b_2(m)$ is a single period of a cosine on platform and thus is a monotonic low-pass filter with no ripples in the stop band:

$$B_2(\omega) = 2 + 2\cos(\omega)$$

Since the even-order binomials are auto-convolutions of this filter, their Fourier transforms are powers of $B_2(\omega)$ and thus have no ripples in the stop band. The filters $b_2(m)$ and $b_4(m)$ have variances of 0.5 and 1, respectively. The filter is $b_4(m)$ equivalent to $b_2(m) * b_2(m)$. Thus, a $\sigma=1$ Gaussian filter can be computed by two convolutions with the kernel [1, 2, 1] at a cost of two multiplications and 4 additions per pixel.

4.2.1   Recursive Filters

Recursive implementations of Gaussian filters have been proposed by Deriche [10] and by Vliet, Young and Verbeek [11]. To maintain shift invariance (or zero phase), the filter is implemented as a cascade of forward and backward difference equations with real-valued coefficients b.

$$\textit{Backward:} \qquad v[n] = \alpha x[n] - \sum_{i=1}^{N} b_i v[n-i]$$

$$\textit{Forward:} \qquad y[n] = \alpha v[n] - \sum_{i=1}^{N} b_i y[n+i]$$

with. $\alpha = 1 + \sum_{i=1}^{N} b_i$

An interesting property of recursive filters is that the number of operations is independent of the variance of the filter. In the following we consider recursive filters of size N=5.

4.3   Laplacian as a Difference of Gaussians

A difference of Gaussians (DoG) is widely used as an approximation for the Laplacian of a Gaussian. A Gaussian low-pass pyramid is thus easily used to compute a Laplacian pyramid. However, the precision of this approximation is rarely studied. In radial form, the normalized Laplacian is a second derivative, given by:

$$\nabla^2 G(r,\sigma) = \frac{r^2 - \sigma^2}{\sigma^4 \sqrt{2\pi}} e^{-\frac{1}{2}\frac{r^2}{\sigma^2}}$$

The difference of Gaussians is:

$$DOG(r,\sigma_{dog}) = \frac{1}{\sigma_1 \sqrt{2\pi}} e^{-\frac{1}{2}\frac{r^2}{\sigma^2}} - \frac{1}{\sigma_{dog} \sqrt{2\pi}} e^{-\frac{1}{2}\frac{r^2}{\sigma_{dog}^2}}$$

Approximating the Laplacian with a difference of Gaussians requires the specification of the two parameters σ_1 and σ_{dog}. Our Gaussian pyramid provides Gaussians in scale step sizes of $\sqrt{2}$ so that $\sigma_1 = \sqrt{2}\sigma_{dog}$. To determine the σ of the corresponding Laplacian, we wrote a simple script search for the value of σ for which the sum of squares of the difference is minimized. The minimum error energy was obtained when $\sigma_{lap} = 1.18\sigma_{dog}$. Figure 5 shows the difference between a Laplacian in radial form and a difference of Gaussians.

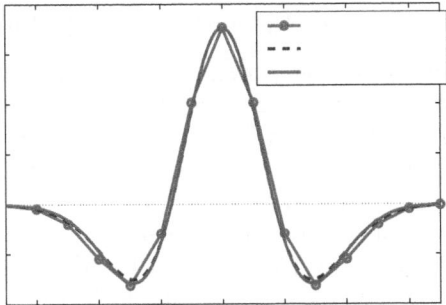

Fig. 5. Comparisons of real Laplacian versus real DoG and binomial DoG for $\sigma_{dog} = \sqrt{2}$ and $\sigma_1 = \sqrt{2}\sigma_{dog}$ and $\sigma_{lap} = 1.7$

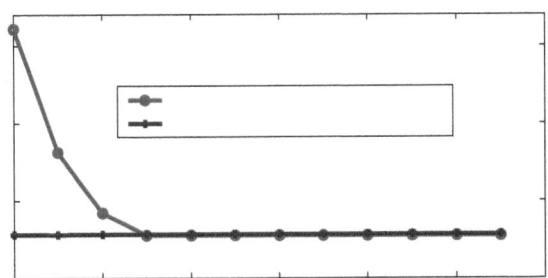

Fig. 6. Evaluation of accuracy of approximation of a Laplacian with a binomial DoG

In Figure 6, a DoG computed with binomial coefficients and a DoG computed using an FIR Gaussian are compared to a true Laplacian. The FIR DoG demonstrates a constant error of approximately 3.6% at all scales. The Binomial DoG starts with an error of 16% but rapidly descends to match the 3.6% error of the FIR implementation by the third image of the pyramid.

The binomial pyramid based on the Kernel filter [1, 4, 6, 4, 1] provides the fastest implementation of the methods tested. The experiments indicate that this method provides sufficiently accurate approximation for a Laplacian.

5   Comparison of Scale Invariance

The scale invariance of the impulse response for a pyramid with $\sigma_0=1$ was evaluated on an image where the central pixel has a value of 100 and all others pixels are set to zero. Gaussian Pyramids with $\sigma_0=1$ were computed using the three filter methods: FIR (N=9), Recursive (N=5) and Binomial. Two DoG images were computed at each level:

$$d_{01}(i,j,k) = p_1(i,j,k) - p_0(i,j,k)$$
$$d_{12}(i,j,k) = p_2(i,j,k) - p_1(i,j,k)$$

All three filters exhibited rapid convergence to a scale-invariant impulse response. For example, the percentage of change for the center pixel at levels k=1,2,3,4 are shown for $d_{01}(i,j,k)$ and $d_{12}(i,j,k)$ in Figure 7. These are representative of the errors observed at other pixel positions. One can note that the invariance error for d_{01} is within 3%. The binomial, the recursive and the FIR filter implementations rapidly converged to extremely small errors (less than 0.0001%).

The percentage error for $d_{12}(i,j,k)$ are within 1% with the same rapid convergence. The improvement in error rates is primarily due to the extra smoothing provided by a larger ratio of σ to sample rate, which results in less error due to sampling. The experiments also validate our choice of $\sigma_0=1.0$ for our pyramid by showing that such pyramid provides reasonably accurate scale invariance.

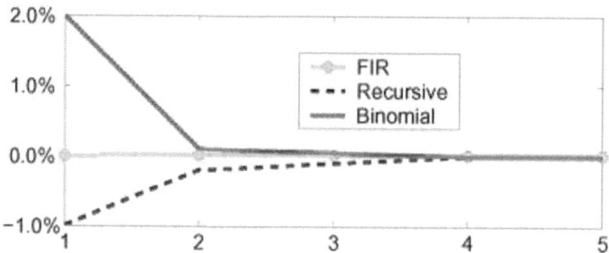

Fig. 7a. Scale invariance of $d_{01}(i,j,k)$ for FIR, Recursive and Binomial Laplacians

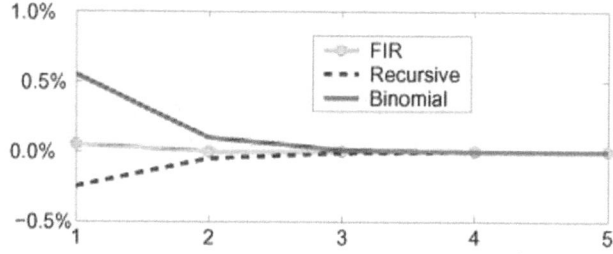

Fig. 7b. Scale invariance of $d_{12}(i,j,k)$ for FIR, Recursive and Binomial Laplacians

Table 1 recapitulates the previous results in operations per pixel for filters g_0 and g_1 with the FIR (N=9), the binomial [1, 2, 1] and 2 recursive filters (N=3 and N=5). This shows that a pyramid computed using the binomial filter has a lower cost than either the recursive filter or the direct FIR filter.

Table 1. Computational cost (Standard Ops) per pixel for different filter types

Filter	FIR N=9	Binomial	IIR N=3	IIR N=5
$g_0(n)$	36	16	28	44
$g_1(n)$	72	32	28	44
$g_0(n)*g_1(n)$	108	48	56	88

6   Determining Intrinsic Scale

Determining characteristic scale requires comparison of Laplacian values along the scale axis. However, because the pyramid is computed on resampled images, Laplacian values are not directly available at most pixels. These samples were eliminated with minimal loss of information due to smoothing. Thus they can be recovered through bi-linear interpolation.

Suppose that we seek the value at pixel i,j at level k, and that this pixel falls between pixels (i_0, j_0) and (i_1, j_1). Note that $T_k = 2^k$ is the sample rate at level k. Given

$$a = \frac{p(i_1, j_0, k) - p(i_0, j_0, k)}{T_k}$$

$$b = \frac{p(i_0, j_1, k) - p(i_0, j_0, k)}{T_k}$$

$$c = a + \frac{p(i_0, j_1, k) - p(i_1, j_1, k)}{T_k}$$

the interpolated value at pixel i, j is

$$p(i, j, k) = a(i - i_0) + b(j - j_0) + c(i - i_0)(j - j_0) + p(i_1, j_1, k)$$

6.1   Computing Characteristic Scale

Let us refer to the difference of Gaussian images at each level k as l=0 for d_{01} and l=1 for d_{12}. We can define an integer scale index n=2k+l. For a typical 6-level pyramid, n runs from 0 to 11. Using this index as a free variable, the Laplacian profile, at pixel (i,j) is the series of interpolated Laplacian values, the d(n) determined for each pixel i,j. The peak in this profile is equivariant with scale. We refer to the scale of this peak as the characteristic scale of the signal at that image position.

The precision of the characteristic scale can be improved by interpolation using a parabola for the three samples closest to the peak. Let $d(n_0)$ be a local peak in d(n). The interpolated extremum is

$$\sigma_{max} = n_0 + 1 + \frac{d(n_0 - 1) - d(n_0 + 1)}{2(d(n_0 - 1) + d(n_0 + 1) - 2d(n_0))}$$

Multiple characteristic scales correspond to concentric patterns in an image. The half-octave pyramid limits discrimination of such patterns to concentric scale changes of powers of 2. This is a fundamental limitation due to sampling scale at multiples of $\sqrt{2}$. Fortunately denser concentric scales tend to be rare in real images.

The following graph (Figure 8) shows an example of Laplacian values as a function of the characteristic scale on a 12-level pyramid (i.e., 6 stages). The extremum of the curve in figure 9 is located around a characteristic scale of 10 pixels. The interpolated curve is shown as a dashed line on this figure.

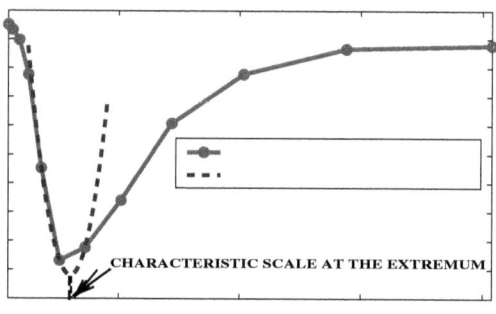

Fig. 8. Interpolation of the Laplacian profile

6.2   Estimating Size from Intrinsic Scale

To evaluate the ability of intrinsic scale to recover size, we constructed an image set containing uniform disks of radius from 1 to 100 pixels. Each image was processed with a binomial pyramid, and the profile of Laplacian values was computed at the center of the circle. This profile was interpolated using parabolic interpolation. The interpolated values of the Laplacian at each extremum are compared in Figure 9 to an ideal straight line. The constancy of these curves further confirms the scale invariance of the pyramid.

7   Invariance to Rotation

Figure 10 demonstrates the invariance to rotation of the characteristic scales. In this experiment, the characteristic scale was computed at every pixel of an image containing a Dirac impulse. The resulting image of characteristic scales, encoded as gray levels, is displayed together with a set of level curves. Note the slight deviations from perfect radial symmetry.

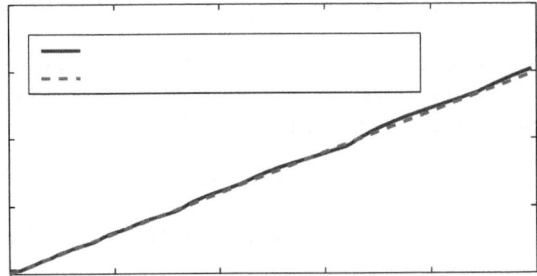

Fig. 9. Scale invariance: The characteristic scale was estimated at the center pixel for 100 images containing each containing disks of radius from 1 to 100 pixels

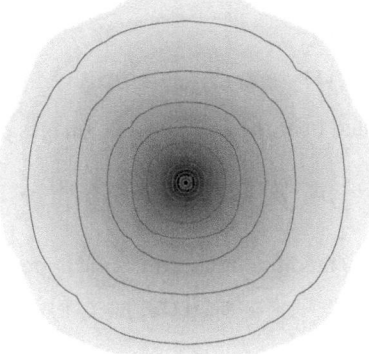

Fig. 10. Rotational invariance of Intrinsic Scale

8   Synthesis of Normalized Receptive Fields

Receptive fields at canonical (row and column) directions can be directly computed from differences of adjacent pixels at level of the binomial pyramid. Such derivative filters are close approximations to Gaussian derivatives at the same scale. In this section we explain how to calculate such receptive fields from the binomial pyramid, how to steer the receptive fields to the intrinsic orientation, and demonstrate that the impulse response has a scale invariance that is similar to the value demonstrated above for the binomial filters.

For each image point in a logo or ROI, a local feature vector can be produced by using the binomial pyramid to compute a vector of 9 chromatic receptive fields [12]. This vector can be computed in a manner that is normalized to the intrinsic scale and orientation at that point. Such normalization provides a vector for robust matching invariant to transformations scale and orientation.

The receptive field vector is based on computing the product with a vector of image differences in the row and column directions of the luminance and chrominance images. The luminance (L) and chrominance (C_1, C_2) images can be obtained by

$$L = R+G+B, \quad C_1 = \frac{1}{2}(R-B) \quad C_2 = \frac{1}{3}(R+B-G)$$

A first derivative along the row or column direction is obtained by convolution with the mask [1, 0, −1] in that direction. For position (x, y) at sampled pyramid image k, this is equivalent to computing the differences of adjacent pixels

$$P_x(x, y, k) = P(x-1, y, k) - P(x+1, y, k)$$

$$P_y(x, y, k) = P(x, y-1, k) - P(x, y+1, k)$$

The second difference is computed by a convolution of the mask [1, −2, 1]. For the rows and columns at pyramid level k, this is equivalent to.

$$P_{xx}(x, y, k) = P(x-1, y, k) + P(x+1, y, k) - 2 \cdot P(x, y, k)$$

$$P_{yy}(x, y, k) = P(x, y-1, k) + P(x, y+1, k) - 2 \cdot P(x, y, k)$$

The mixed derivative is

$$P_{xy}(x,y,k) = P(x-1,yk) + P(x+1,y,k) + P(x,y-1,k) + P(x,y+1,k) - 4 \cdot P(x,y,k)$$

Image differences computed over the L, C_1, and C_2 images compose the vector.

$$\vec{P} = (P^L, P_x^L, P_y^L, P_{xx}^L, P_{xy}^L, P_{yy}^L, P^{C_1}, P_x^{C_1}, P_y^{C_1}, P^{C_2}, P_x^{C_2}, P_y^{C_2})$$

This vector gives an un-oriented feature vector at each point equivalent to

$$\vec{P} = < \vec{G}(\sigma), p(x, y) >$$

where

$$\vec{G} = (G^L, G_x^L, G_y^L, G_{xx}^L, G_{xy}^L, G_{yy}^L, G^{C_1}, G_x^{C_1}, G_y^{C_1}, G^{C_2}, G_x^{C_2}, G_y^{C_2})$$

is the vector of Gaussian derivatives at the intrinsic scale.

The differences in row and column directions can be steered to the intrinsic orientation, θ, at pixel (x, y) using the steerable filter formulas of Freeman and Adelsen [13]:

$$V_1 = P_x \cos(\theta) + P_y \sin(\theta)$$

$$V_2 = P_{xx} \cos(\theta)^2 + P_{xy} \cos(\theta) \sin(\theta) + P_y \sin(\theta)^2$$

where the intrinsic orientation for each pixel is provided by :

$$\theta = \tan^{-1}\left(\frac{P_y}{P_x}\right)$$

The steered local feature vector for luminance and chrominance at the intrinsic scale and orientation can be written as

$$\vec{V}(x, y) = (V^L, V_1^L, V_2^L, V^{C_1}, V_1^{C_1}, V_2^{C_1}, V^{C_2}, V_1^{C_2}, V_2^{C_2})$$

where the subscript 1 represents a first derivative and the subscript 2 represents a second derivative.

Figure 11a and 11b show the impulse response obtained from such a calculation. White pixels represent positive values and purple negative values. A second derivative can be obtained from a convolution of [1, –2, 1]. Figure 11c and 11d show examples of the resulting impulse response. The mixed impulse responses can be obtained by convolving the row directions with [1, 0, –1] followed by convolving the column directions with this filter. Figure 11e shows an example of the resulting impulse response. Synthetic filters at any desired angle can be computed from these filters using a weighted sum of the derivatives.

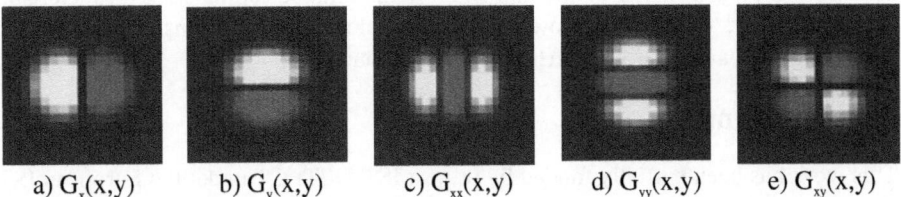

a) $G_x(x,y)$ b) $G_y(x,y)$ c) $G_{xx}(x,y)$ d) $G_{yy}(x,y)$ e) $G_{xy}(x,y)$

Fig. 11. The impulse responses for receptive fields computed from the pyramid at level 2.

Because the sampled impulse response of the pyramid is the same at every level (beyond the first), the impulse responses of the derivatives are also equivalent at all levels. As with the Laplacian, these derivative impulse responses are "equivariant" with scale. As a demonstration of the invariance of the impulse response, Figure 12 shows the impulse responses from filter $G_x(i, j, k)$ and $G_{xx}(i, j, k)$ computed at the first image in each of the stages of the binomial pyramid. That is, these impulse responses are

$$G_x(i, j, k) = [-1,0,1] * p_0(i, j, k)$$

$$G_{xx}(i, j, k) = [1, -2, 1] * p_l(i, j, k)$$

for k = 0, 1, 2, 3, 4, 5. It can be seen from figure 12 that after the first level of the pyramid, the impulse response is invariant to scale.

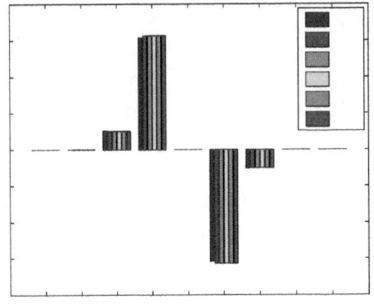   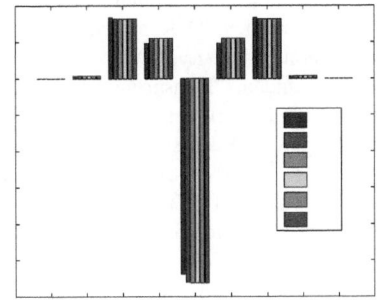

a) Scale invariance for $G_{x0}(x,y,k)$ b) Scale invariance of $G_{xx0}(x,y,k)$

Fig. 12. Comparison of impulse response for first and second Gaussian derivatives for pyramid stages k=0,1,2,3,4,5.

9   Summary and Conclusion

The binomial pyramid gives a simple and fast method to evaluate characteristic scales at any pixel in an image. This method is based on the computation of differences of Gaussians obtained by binomial filtering in a pyramid. The experiments described above demonstrate that a scale-invariant half-octave pyramid computed with a binomial kernel can provide an efficient and precise means to compute characteristic scales. At first glance, it may seem surprising that a relatively crude Gaussian approximation such as a 1-2-1 binomial filter yields reliable estimates of characteristic scale. However, this kernel allows video rate calculation of intrinsic scale for 1/4 PAL images on a standard 1.5 GHz personal computer.

Acknowledgment

This work has been partially funded by project IST DETECT under the European IST Cognitive Vision Program.

References

1. J. J. Koenderink and A. J. van Doorn, "Representation of local geometry in the visual system", Biological Cybernetics, 55:367-375, 1987.
2. D. G. Lowe, "Object Recognition from local scale-invariant features", in 1999 International Conference on Computer Vision (ICCV-99), Corfu Greece, pp 1150-1157, Sept. 1999.
3. C. Schmid and R. Mohr. "Local greyvalue invariants for image retrieval", IEEE Transactions on PAMI, PAMI Vol 19, No. 5, pages 530-534, 1997.
4. T. Lindeberg, "Feature detection with automatic scale selection", International Journal of Computer Vision, IJCV 30(2):77-116, 1998.
5. M. D. Kelly, "Edge detection by computer in pictures using planning", Machine Intelligence, 6:379–409, 1971.
6. S. L. Tanimoto and T. Pavlidis, "A hierarchical data structure for picture processing", Computer Graphics and Image Processing, 4:104-119, 1975.
7. P. J. Burt and E. H. Adelson, "The Laplacian pyramid as a compact image code", IEEE Transactions on Communications, 31:532-540, 1983.
8. J. L. Crowley, "A Representation for Visual Information", Doctoral Dissertation, Carnegie-Mellon University, 1981.
9. P. Anandan, "Measuring Visual Motion from Image Sequences", PhD thesis, Computer Science Department, Doctoral Thesis, University of Massachusetts, 1987.
10. R. Deriche. Recursively implementing the Gaussian and its derivatives. Rapport de Recherche 1893, INRIA, Sophia Antipolis, France, Apr. 1993.
11. L. J. van Vliet, I. T. Young, and P. W. Verbeek. Recursive Gaussian derivative filters. In Proc. 14th International Conference on Pattern Recognition (ICPR'98), volume 1, pages 509-514. IEEE Computer Society Press, Aug. 1998.
12. D. Hall, V. Colin de Verdiere and J. L. Crowley, "Object Recognition using Coloured Receptive Field", 6th European Conference on Computer Vision, Springer Verlag, Dublin, pp 164-178, June 2000.
13. W.T. Freeman, E.H. Adelson, "The Design and Use of Steerable Filters", Transactions on Pattern Analysis and Machine Intelligence, (PAMI), Vol 13, No. 9, pp 891-906, September 1991.

A Multiphase Level Set Framework for Motion Segmentation

Daniel Cremers

Department of Computer Science
University of California at Los Angeles
http://www.cs.ucla.edu/~cremers

Abstract. We present a novel variational approach for segmenting the image plane into a set of regions of piecewise constant motion on the basis of only two consecutive frames from an image sequence.

To this end, we formulate the problem of estimating a motion field in the framework of Bayesian inference. Our model is based on a conditional probability for the spatio-temporal image gradient, given a particular velocity vector, and on a prior on the estimated motion field favoring motion boundaries of minimal length. The corresponding negative log likelihood is a functional which depends on motion vectors for a set of regions and on the boundary separating these regions. It can be considered an extension of the Mumford-Shah functional from intensity segmentation to motion segmentation.

We propose an implementation of this functional by a multiphase level set framework. Minimizing the functional with respect to its dynamic variables results in an evolution equation for a vector-valued level set function and in an eigenvalue problem for the motion vectors. Compared to most alternative approaches, we jointly solve the problems of segmentation and motion estimation by minimizing a *single* functional. Numerical results both for simulated ground truth experiments and for real-world sequences demonstrate the capacity of our approach to segment several – possibly multiply connected – objects based on their relative motion.

1 Introduction

Motion estimation from image sequences has a long tradition in computer vision. Two seminal variational methods were proposed by Horn and Schunck [11] and by Lucas and Kanade [15]. Both of these methods are based on a least-squares criterion for the optic flow constraint and some global or local smoothness assumption on the flow field.

In practice, flow fields are usually not smooth. The boundaries of moving objects will correspond to discontinuities in the motion field. Such motion discontinuities have been modeled implicitly by non-quadratic robust estimators [2,17,14,28]. Other approaches tackled the problem of segmenting the motion field by treating the problems of motion estimation in disjoint sets and optimization of the motion boundaries separately [25,3,21,9]. Some approaches are based

L.D. Griffin and M. Lillholm (Eds.): Scale-Space 2003, LNCS 2695, pp. 599–614, 2003.

on Markov Random Field formulations and the EM algorithm (cf. [12,1,29]). Yet, as pointed out in [29], exact solutions to the EM algorithm are computationally expensive and therefore suboptimal approximations are employed. For certain tracking applications, it may also be sufficient to perform segmentation on the basis of temporal change detection [23]. Yet this approach does not extend to the cases of moving background and multiple motion considered here.

In [8], we presented a variational approach to motion segmentation with an explicit contour where both the motion estimation and the boundary optimization are derived from minimizing a *single* energy functional. Yet, this approach had two drawbacks: Firstly, satisfactory results were only obtained upon applying two posterior normalizations to the terms driving the evolution of the motion boundary. And secondly, due to the explicit representation of this motion boundary, the segmentation of multiple moving objects is not straight-forward.

In [6], we addressed these drawbacks by proposing a novel geometric interpretation of the optic flow constraint and by reverting to a two-phase level set representation of the motion boundary. Due to this geometric interpretation of the optic flow constraint, all normalizations of the boundary evolution forces are derived in a consistent manner by minimizing the proposed cost functional. And the level set formulation permits a segmentation of the image plane into multiple regions, each of which is associated with one of two motion models.

The present paper extends this work in several ways: Firstly, we formulate the problem of motion estimation in the framework of Bayesian inference to derive an energy functional which extends the Mumford-Shah functional [19] from gray value segmentation to motion segmentation. Secondly, we detail a multiphase level set implementation of this functional, which is based on the corresponding gray value model of Chan and Vese [5]. The multiphase formulation permits to segment an arbitrary number of differently moving regions. We show that minimization leads to an eigenvalue problem for the motion parameters, and to a gradient descent evolution for the level set functions embedding the motion discontinuities. Numerical results are demonstrated on simulated ground-truth experiments and on real world problems.

A related approach to motion segmentation was proposed in [16]. Firstly, our cost functional differs from theirs in that it includes normalizations of the residuals which we believe are important for comparing differently moving regions. As a consequence, most of our purely motion-based segmentations are highly accurate. Secondly, we use an efficient multiphase model which does not suffer from the formation of vacuum or overlap regions. Thirdly, the motion models in [16] are estimated in a separate process on the basis of "feature points".

This paper is organized as follows. In Section 2, we formulate motion estimation as a problem of Bayesian Inference. In Section 3, we consistently derive a variational framework for motion segmentation. In Section 4, we propose a multiphase level set formulation of the motion segmentation functional. In particular, we detail the special cases of a two-phase and a four-phase model. In Sections 5 and 6, we show numerical results obtained on simulated ground truth data and on real world image sequences. In Section 7, we end with a conclusion.

2   Motion Estimation as Bayesian Inference

Let $\Omega \subset \mathbb{R}^2$ denote the image plane and let $f : \Omega \times \mathbb{R} \to \mathbb{R}$ be a gray value image sequence. Denote the spatio-temporal image gradient of $f(x, t)$ by

$$\nabla_3 f = \left(\frac{\partial f}{\partial x_1}, \frac{\partial f}{\partial x_2}, \frac{\partial f}{\partial t} \right)^t. \tag{1}$$

Let

$$v : \Omega \to \mathbb{R}^3, \qquad v(x) = (u(x), w(x), 1)^t, \tag{2}$$

be the velocity vector at a point x in homogeneous coordinates[1].

With these definitions, the problem of motion estimation now consists in maximizing the conditional probability

$$P(v \,|\, \nabla_3 f) = \frac{P(\nabla_3 f \,|\, v) \; P(v)}{P(\nabla_3 f)}, \tag{3}$$

with respect to the motion field v.

To this end, we make the following assumptions:

- We assume that the intensity of a moving point remains constant throughout time. Expressed in differential form, this gives us a relation between the spatio-temporal image gradient and the homogeneous velocity vector, known as *optic flow constraint*:

$$\frac{df}{dt} = \frac{\partial f}{\partial t} + \frac{\partial f}{\partial x_1} \frac{dx_1}{dt} + \frac{\partial f}{\partial x_2} \frac{dx_2}{dt} = v^t \, \nabla_3 f = 0. \tag{4}$$

Except for locations where the spatio-temporal gradient vanishes, this constraint states that the homogeneous velocity vector must be orthogonal to the spatio-temporal image gradient. Therefore we propose to use a measure of this orthogonality as a conditional probability on the spatio-temporal image gradient. Let β be the angle between the two vectors, then:

$$P(\nabla_3 f(x) \,|\, v(x)) \propto \exp\left(-\cos^2(\beta)\right) = \exp\left(-\frac{(v(x)^t \nabla_3 f(x))^2}{|v(x)|^2 \,|\, \nabla_3 f(x)|^2}\right). \tag{5}$$

By construction, this probability is independent of the length of the two vectors and monotonically increases the more orthogonal the two vectors are. We regularize this expression by replacing

$$|\nabla_3 f(x)| \;\longrightarrow\; |\nabla_3 f(x)| + \epsilon \tag{6}$$

in the denominator. This guarantees that the probability is maximal if the gradient vanishes, while not affecting the result for gradients much larger than ϵ. As long as ϵ is chosen sufficiently small, we did not find a noticeable influence of its precise value in numerical implementations.

---

[1] Since we are only concerned with two consecutive frames from a sequence, we will drop the time coordinate in the notation of the velocity field.

- We discretize the velocity field v by a set of disjoint regions $R_i \subset \Omega$ with constant velocity v_i:

$$v(x) = \{v_i, \text{ if } x \in R_i\} \qquad (7)$$

Note that such a discretization in itself does not restrict the class of permissible motion fields, since each image pixel could be considered a separate region. We now assume the prior probability on the velocity field to only depend on the length of the boundary C separating these regions:

$$\mathcal{P}(v) \propto \exp(-\nu |C|) \qquad (8)$$

In particular, this means that we do not make any prior assumptions on the velocity vectors v_i. Such a term would necessarily introduce a bias favoring certain velocities.

3   A Variational Framework for Motion Segmentation

With the above assumptions, we can use the framework of Bayesian inference to derive a variational method for motion segmentation. The first term in the numerator of equation (3) can be written as:

$$\mathcal{P}(\nabla_3 f \mid v) = \prod_{x \in \Omega} \mathcal{P}(\nabla_3 f(x) \mid v(x))^h = \prod_{i=1}^{n} \prod_{x \in R_i} \mathcal{P}(\nabla_3 f(x) \mid v_i)^h, \qquad (9)$$

where h denotes the grid size of the discretization of Ω. The first step is based on the assumption that the velocity affects the spatio-temporal gradient only locally. And the second step is based on the discretization of the velocity field given in (7).

With the formulas (5), (8) and (9), maximizing the conditional probability (3) is equivalent to minimizing its negative logarithm, which is given (up to a constant) by the energy functional:

$$E(C, \{v_i\}) = \sum_{i=1}^{n} \int_{R_i} \frac{(v_i^t \nabla_3 f(x))^2}{|v_i|^2 \, |\nabla_3 f(x)|^2} \, dx + \nu |C|. \qquad (10)$$

Let us make the following remarks about this functional:

- The functional (10) can be considered an extension of the piecewise constant Mumford-Shah functional [19] from the case of gray value segmentation to the case of motion segmentation. Rather than having a mean gray values f_i for each region R_i, we now have a homogeneous velocity vector v_i for each region R_i.

- Minimizing the functional (10) with respect to the boundary C and the set of motion vectors $\{v_i\}$, jointly solves the problems of segmentation and motion estimation. In our view, this aspect is crucial since we believe that these two problems are tightly coupled. Many alternative approaches to motion segmentation tend to instead treat the two problems separately by first (globally) estimating the motion and then trying to segment the estimated motion into a set of sensible regions.

- Note that the integrand in the data term differs from the one commonly used in the optic flow community for motion estimation: Rather than minimizing the deviation from the optic flow constraint in a least-squares manner, as done e.g. in the seminal work of Horn and Schunck [11], our measure (5) of orthogonality introduces an additional normalization with respect to the length of the two vectors. We found this to be essential in the case of motion *segmentation*, where one needs to compare differently moving regions.

- The functional (10) contains only one free parameter ν, which determines the relative weight of the length constraint. Larger values of ν will induce a segmentation of the image motion on a coarser scale. As argued by Morel and Solimini [18], such a scale parameter is fundamental in all segmentation approaches.

4   A Multiphase Level Set Implementation

In order to minimize the functional (10), we need to specify an appropriate representation for the boundary C. In this paper, we choose an implicit level set representation of the boundary [22]. Level set based contour representations have become a popular framework in image segmentation (cf. [4,13,5]), because they do not depend on a particular choice of parameterization, and because they do not restrict the topology of the evolving interface. This permits splitting and merging of the contour during evolution and therefore makes level set representations well suited for the segmentation of several objects or multiply connected objects.

Based on the work of Chan and Vese [5], we will first present a two-phase level set model for the functional (10) with a single level set function ϕ. This model is subsequently extended to a multi-phase model with a vector-valued level set function.

4.1   The Two Phase Model

In this subsection, we restrict the class of permissible motion segmentations to two-phase solutions, i.e. to segmentations of the image plane for which each point can be ascribed to one of two velocities v_1 and v_2. The general case of several velocities $\{v_i\}_{i=1,\ldots,n}$ will be treated in the next subsection.

Let the boundary C in the functional (10) be represented as the zero level set of a function $\phi : \Omega \to \mathbb{R}$:

$$C = \{x \in \Omega \mid \phi(x) = 0\}. \tag{11}$$

Using the Heaviside step function

$$H(\phi) = \begin{cases} 1 \text{ if } \phi \geq 0 \\ 0 \text{ if } \phi < 0 \end{cases}, \tag{12}$$

and, for notational simplification, the matrix

$$T(x) = \frac{\nabla_3 f \, \nabla_3 f^t}{|\nabla_3 f|^2}, \tag{13}$$

again with the regularization (6) in numerical implementations, we can embed the motion energy (10) by the following *two-phase functional*:

$$E(v_1, v_2, \phi) = \int_\Omega \frac{v_1^t T v_1}{|v_1|^2} H(\phi) dx + \int_\Omega \frac{v_2^t T v_2}{|v_2|^2} (1 - H(\phi)) dx + \nu \int_\Omega |\nabla H(\phi)| dx. \tag{14}$$

This functional is now simultaneously minimized with respect to the velocity vectors v_1 and v_2, and with respect to the embedding level set function ϕ defining the motion boundaries. To this end, we alternate the two fractional steps:

(a) **An Eigenvalue Problem for the Motion Vectors.**

For fixed ϕ, minimization of the functional (14) with respect to the motion vectors v_1 and v_2 results in the eigenvalue problem:

$$v_i = \arg\min_v \frac{v^t M_i v}{v^t v}, \tag{15}$$

for the 3×3-matrices

$$M_1 = \int_\Omega T(x) H(\phi) \, dx \quad \text{and} \quad M_2 = \int_\Omega T(x) (1 - H(\phi)) \, dx. \tag{16}$$

The solution of (15) is given by the eigenvectors corresponding to the smallest eigenvalues of M_1 and M_2.

(b) **Evolution of the Level Set Function.**

Conversely, for fixed motion vectors, the gradient descent on the functional (14) for the level set function ϕ is given by:

$$\frac{\partial \phi}{\partial t} = \delta(\phi) \left[\nu \, \mathrm{div} \left(\frac{\nabla \phi}{|\nabla \phi|} \right) + e_2 - e_1 \right], \tag{17}$$

with the energy densities e_i given by

$$e_i(x) = \frac{v_i^t T(x) v_i}{v_i^t v_i}. \tag{18}$$

As suggested in [5], we implement the Delta function $\delta(\phi) = \frac{d}{d\phi} H(\phi)$ by a smooth approximation of finite width τ:

$$\delta_\tau(s) = \frac{1}{\pi} \frac{\tau}{\tau^2 + s^2}. \tag{19}$$

Depending on the size of τ, this permits to detect interior motion boundaries.

4.2   The General Multiphase Model

The above approach to represent the motion boundary with a single level set function ϕ permits to model motion fields with only two phases (i.e. it permits only two different velocity vectors). Moreover, one cannot represent certain geometrical features of the boundary, such as triple junctions, by the zero level set of a single function ϕ. There are various ways to overcome these limitations by using multiple level set functions.

One approach, investigated e.g. in [30,24,16], is to represent each phase $R_i \subset \Omega$ by a different level set function ϕ_i: $R_i = \{x \in \Omega \,|\, \phi_i(x) \geq 0\}$. Although this approach permits to overcome the above limitations, it has two disadvantages: Firstly, it is computationally expensive to represent a large number of phases by a separate level set function for each phase. And secondly, one needs to suppress the formation of vacuum and overlap regions by introducing additional energy terms.

An alternative more elegant approach to model multiple phases was proposed by Chan and Vese in [5]. They introduce a more compact representation of up to n phases which needs only $m = \log_2(n)$ level set functions. Moreover, by definition, it generates a partition of the image plane and therefore does not suffer from overlap or vacuum formation. We will adopt this representation which shall be detailed in the following.

Let $\Phi = (\phi_1, \ldots, \phi_m)$ be a vector level set function, with $\phi_i : \Omega \to \mathbb{R}$. Let $H(\Phi(x)) = (H(\phi_1(x)), \ldots, H(\phi_m(x)))$ be the associated vector Heaviside function. This function maps each point $x \in \Omega$ to a binary vector and therefore permits to encode a set of $n = 2^m$ phases R_i defined by:

$$R = \{x \in \Omega \,|\, H(\Phi(x)) = \text{constant}\}. \tag{20}$$

In analogy to the case of the Mumford-Shah functional treated in [5], we propose to replace the two-phase functional (14) by the *multiphase functional*:

$$E(\{v_i\}, \Phi) = \sum_{i=1}^{n} \int_{\Omega} \frac{v_i^t \, T \, v_i}{|v_i|^2} \, \chi_i(\Phi) \, dx \; + \; \nu \sum_{i=1}^{n} \int_{\Omega} |\nabla H(\phi_i)| \, dx, \tag{21}$$

where χ_i denotes the indicator function for the region R_i. Note, that for $n = 2$, this is equivalent to the two-phase model introduced in (14).

For the purpose of illustration, we explicitly give the functional for the case of $n = 4$ phases:

$$E(\{v_i\}, \Phi) = \int_{\Omega} \frac{v_{11}^t T v_{11}}{|v_{11}|^2} H(\phi_1) H(\phi_2) \, dx \; + \int_{\Omega} \frac{v_{10}^t T v_{10}}{|v_{10}|^2} H(\phi_1) \big(1 - H(\phi_2)\big) \, dx$$

$$+ \int_{\Omega} \frac{v_{01}^t T v_{01}}{|v_{01}|^2} \big(1 - H(\phi_1)\big) H(\phi_2) \, dx + \int_{\Omega} \frac{v_{00}^t T v_{00}}{|v_{00}|^2} \big(1 - H(\phi_1)\big)\big(1 - H(\phi_2)\big) \, dx$$

$$+ \nu \int_{\Omega} |\nabla H(\phi_1)| \, dx + \nu \int_{\Omega} |\nabla H(\phi_2)| \, dx. \tag{22}$$

Minimization of this functional with respect to the motion vectors $\{v_i\}$ for fixed Φ results in the eigenvalue problems:

$$v_i = \arg\min_v \frac{v^t M_i v}{v^t v}, \tag{23}$$

with four 3×3-matrices M_i given by

$$\begin{cases} M_{11} = \text{mean}(T) \text{ in } \{\phi_1 \geq 0, \, \phi_2 \geq 0\} \\ M_{10} = \text{mean}(T) \text{ in } \{\phi_1 \geq 0, \, \phi_2 < 0\} \\ M_{01} = \text{mean}(T) \text{ in } \{\phi_1 < 0, \, \phi_2 \geq 0\} \\ M_{00} = \text{mean}(T) \text{ in } \{\phi_1 < 0, \, \phi_2 < 0\} \end{cases} \tag{24}$$

Conversely, for fixed velocity vectors, the evolution equations for the two level set functions are given by:

$$\frac{\partial \phi_1}{\partial t} = \delta(\phi_1) \left[\nu \, \text{div} \left(\frac{\nabla \phi_1}{|\nabla \phi_1|} \right) + (e_{01} - e_{11}) \, H(\phi_2) + (e_{00} - e_{10}) \left(1 - H(\phi_2)\right) \right],$$

$$\tag{25}$$

$$\frac{\partial \phi_2}{\partial t} = \delta(\phi_2) \left[\nu \, \text{div} \left(\frac{\nabla \phi_2}{|\nabla \phi_2|} \right) + (e_{10} - e_{11}) \, H(\phi_1) + (e_{00} - e_{01}) \left(1 - H(\phi_1)\right) \right],$$

with the energy densities e_i defined in (18).

4.3   Redistancing

During their evolution according to equations (17) or (25), the level set functions ϕ_i generally grow to very large positive or negative values in the respective areas of the input image corresponding to a particular motion hypothesis. At the zero crossings, they rise steeply, the gradient can become arbitrarily large. In numerical implementations, we found that a very steep slope of the level set functions eventually inhibits the flexibility of the boundary to displace.

Many people have advocated the use of a redistancing procedure to constrain the slope of ϕ to $|\nabla \phi| = 1$, c.f. [10]. In order to reproject the evolving level set function to the space of distance functions, we intermittently iterate several steps of the redistancing equation [27]:

$$\frac{\partial \phi}{\partial t} = \text{sign}(\hat{\phi}) \left(1 - |\nabla \phi|\right), \tag{26}$$

where $\hat{\phi}$ denotes the level set function before redistancing. Although this regularization is optional in the proposed level set model — see also [5] — it improves the convergence of the boundary evolution. Since the data term given by the image motion information dominates the evolution of the boundary, we found this simple redistancing process to be sufficiently accurate for our application. Therefore we did not revert to more elaborate iterative redistancing schemes such as the one presented in [26].

 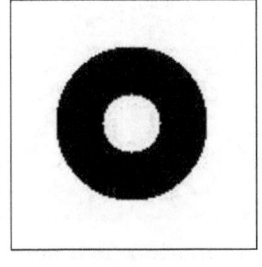 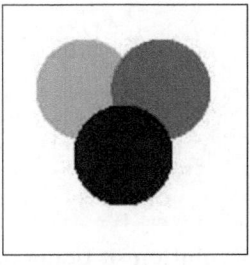

Wallpaper Input Image   Ring-shaped region   Multiple regions

Fig. 1. Data for ground truth experiments. Left: Specific image regions of the wallpaper shot are artificially translated to generate input data. **Middle and Right:** Chosen image regions indicated by different gray values.

5   Ground Truth Experiments

In order to verify the precision of our segmentation approach, we performed a number of ground truth experiments in the following way. We took a snapshot of homogeneously structured wallpaper, which is shown in Figure 1, left side. Then we artificially translated certain image regions according to a particular motion. The respective image regions are highlighted in various shades of gray in Figure 1, middle and right side.

We determined the spatio-temporal image gradient from two consecutive images and specified a particular initialization of the boundary. Then we minimized the functional (21) by alternating the three fractional steps of:

- iterating the gradient descent (17) or (25) for the level set functions,
- iterating the redistancing procedure (26) for the level set functions,
- and updating the motion vectors for all phases by solving the corresponding eigenvalue problem (23).

For all experiments, we show the evolving motion boundaries and corresponding motion estimates superimposed onto the ground truth region information. Yet, it should be noted that in these experiments the objects cannot be distinguished from the background based on their appearance, as they correspond to homogeneously textured parts of the wallpaper. Therefore, all results are obtained exclusively on the basis of the motion information. For the purpose of illustration, we also show in the first example the evolution of the level set function, the zero level set of which represents the motion boundary.

The three ground truth experiments are chosen so as to highlight different properties of the proposed approach: The first example shows a result obtained with the two-phase model, in which a multiply connected moving object is segmented on a differently moving background. The second and third experiment show an application of the four-phase model in which three differently moving regions are segmented, once on a static and once on a moving background.

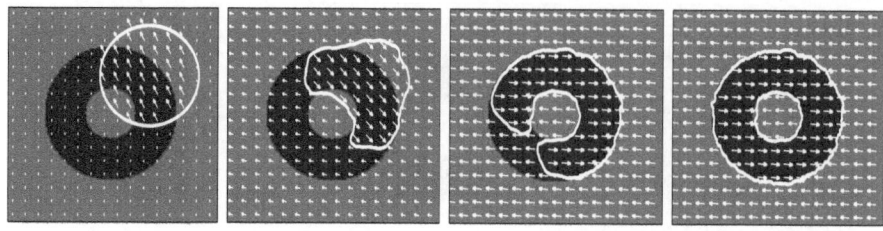

Evolution of boundary and motion field superimposed on true region.

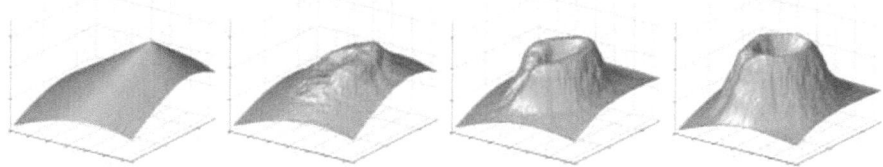

Corresponding evolution of the embedding level set function.

Fig. 2. Segmenting a multiply connected moving object. The two input images show the wallpaper of Figure 1, left side, with a ring-shaped region translating to the right and the remaining region translating to the left. Both the motion estimates and the evolution of the boundary between the two phases are obtained by minimizing the two-phase motion functional (14) simultaneously with respect to both the level set function ϕ and the motion vectors v_1 and v_2. Thus, the minimization of the *single* energy functional generates both the precise object location and the motion information for object and background. Due to the level set representation, the evolving motion discontinuity set can change topology. Note also that in the input data the region of interest is not perceivable based on its appearance.

5.1   Segmenting Multiply Connected Moving Objects

In this experiment, we demonstrate the capacity of the proposed approach to segment multiply connected moving objects. The two input images consisted of the wallpaper shown in Figure 1, left side, and the same image with the ring-shaped area indicated in Figure 1, center, translated to the right, and the remaining image area translated to the left.

Figure 2, bottom row, shows four steps in the evolution of the level set function ϕ, generated by minimizing the two-phase model (14). The top row shows the evolution of the corresponding motion boundary, given by the zero level set of ϕ, and the estimated motion field, superimposed on the ground truth information about the ring region. The results show that one can obtain both a precise information about the location of a multiply connected object and about the motion of object and background by minimizing a *single* energy functional.

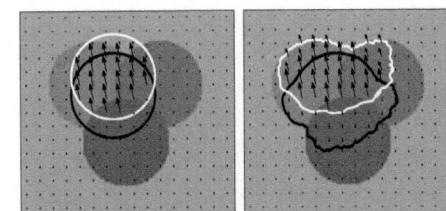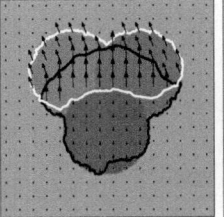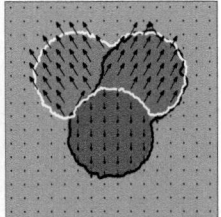

Fig. 3. Segmenting multiple moving regions. The two input images show the wallpaper of Figure 1, left side, with three circular regions moving away from the center. The magnitude of the velocity of the upper two regions is 1.4 times larger than that of the bottom region. Superimposed on the true region information are the evolving zero level sets of ϕ_1 (black contour) and ϕ_2 (white contour), which define four different phases. The simultaneously evolving piecewise constant motion field is represented by the black arrows. Both the phase boundaries and the motion field are obtained by minimizing the four-phase model (22) with respect to the level set functions and the motion vectors. Note that in the final solution, the two boundaries clearly separate the four phases corresponding to the three moving regions and the static background.

5.2 Segmenting Several Differently Moving Regions

In this experiment, we demonstrate an application of the four-phase model (22) to the segmentation of up to four different regions based on their motion information. The input data consists of two images showing the wallpaper from Figure 1, left side, with three regions (shown in Figure 1, right side) moving away from the center. The upper two regions move by a factor 1.4 faster than the lower region.

Figure 3 shows several steps in the minimization of the functional (22). Superimposed onto the ground truth region information are the evolution of the zero level sets of the two embedding functions ϕ_1 (black contour) and ϕ_2 (white contour), and the estimated piecewise constant motion field indicated by the black arrows.

Note that the two contours represent a set of four different phases:

$$R_1 = \{x \in \Omega \mid \phi_1 \geq 0, \phi_2 \geq 0\}, \qquad R_2 = \{x \in \Omega \mid \phi_1 \geq 0, \phi_2 < 0\},$$

$$R_3 = \{x \in \Omega \mid \phi_1 < 0, \phi_2 \geq 0\}, \qquad R_4 = \{x \in \Omega \mid \phi_1 < 0, \phi_2 < 0\}.$$

Upon convergence, these four phases clearly separate the three moving regions and the static background. The resulting final segmentation of the image, which is not explicitly shown here, is essentially identical to the ground truth region information. Again, we stress that the segmentation is obtained purely on the basis of the *motion information*: In the input images, the different regions cannot be distinguished from the background on the basis of their *appearance*.

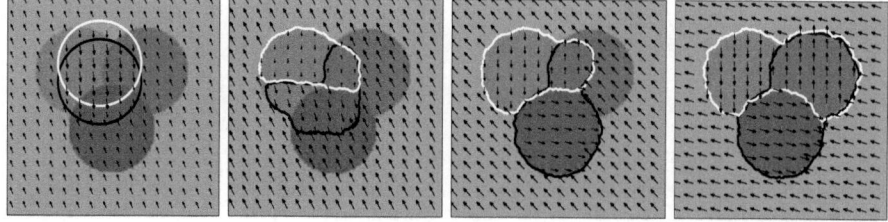

Fig. 4. Segmenting multiple moving regions on moving background.
The two input images show the wallpaper of Figure 1, left side, with the three circular regions and the background moving in different directions. The true motion directions for all regions are down (top left), up (top right), right (bottom) and left (background). Superimposed on the true region information are the evolving zero level sets of ϕ_1 (black contour) and ϕ_2 (white contour), defining four different phases. The simultaneously evolving piecewise constant motion field is represented by the black arrows. Both the phase boundaries and the piecewise constant motion field are obtained by minimizing the four-phase model (22). Upon convergence, the two boundaries clearly separate the four motion phases corresponding to the three regions and the background.

5.3   Multiple Moving Objects and Moving Background

In the previous example, the three regions were moving, while the background was static. In many real-world applications of motion estimation and motion segmentation, the background may also undergo a certain motion — for example in a motion sequence filmed by a moving camera. This problem has been addressed by a number of researchers. In particular, it has been proposed to estimate the *dominant* motion in a robust estimator framework (cf. [20,2]). Although this may permit to compensate for the background motion, it strongly relies on the assumption that the background forms the dominant part of the image plane[2].

Our approach does not rely on any assumptions about the relative size of the different moving regions. Figure 4 shows the segmentation of a sequence containing three moving regions on a moving background, obtained by minimizing the four-phase model (22). During energy minimization, both the motion estimates and the motion boundary are progressively improved. The boundaries of the four motion phases converge over a fairly large distance, yet the region boundaries are precisely reconstructed in the final segmentation. As in the previous example, the zero level sets of ϕ_1 and ϕ_2 define a segmentation which is essentially identical with the ground truth. The directions of the motion estimated for the lower region and the background deviate slightly from the ground truth. It is unclear where this small discrepancy stems from.

---

[2] In [2], for example, it is stated that the robust estimation of the background motion works well on an artificial sequence (involving translatory motion only) if the background motion takes up at least 60% of the image plane.

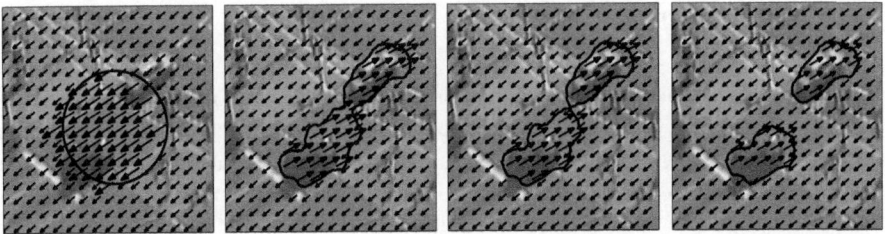

Evolution of boundary and motion field superimposed on the first frame.

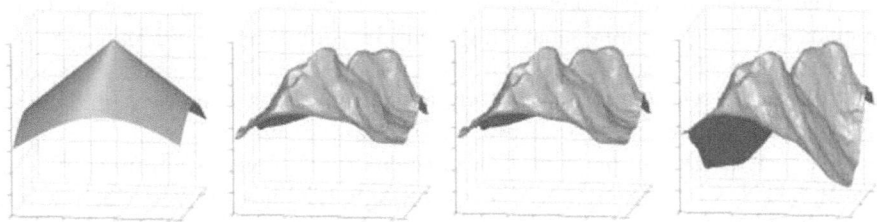

Corresponding evolution of embedding level set function.

Fig. 5. Segmenting moving cars captured by a moving camera.
The image sequence shows two cars moving to the top right, and the background moving to the bottom left. Due to the level set representation, the topology of the motion boundary is not constrained, such that splitting and merging is possible. By minimizing the two-phase model (14), one obtains a fairly accurate segmentation of the two cars and an estimate of the motion of cars and background.

6   Real-World Application: Segmentation of Moving Cars

As a final demonstration of the proposed level set based motion segmentation approach, we now present an application to a real-world traffic scene showing two moving cars on a differently moving background. We used two consecutive images from a sequence recorded by D. Koller and H.-H. Nagel[3]. The sequence shows several cars moving in the same direction, filmed by a static camera. In order to increase the complexity of the sequence, we artificially induced a background motion by shifting one of the two frames, thereby simulating the case of a moving camera.

The images in figure 5, top row, show the contour evolution with the corresponding motion estimates superimposed on one of the two frames. These were generated by minimizing the two-phase model (14). The bottom row shows the evolution of the underlying level set function. Due to the level set representation of the boundary, the zero level set can undergo topological changes such as the splitting and merging from the third to the fourth frame.

---

[3] KOGS/IAKS, Univ. of Karlsruhe, http://i21www.ira.uka.de/image_sequences/

7 Conclusion

We approached the problems of motion estimation and motion segmentation from the viewpoint of Bayesian inference. Based on a few very basic assumptions, we rigorously derived a variational framework for segmenting the image plane into a set of regions of homogeneous motion. Our model is based on a conditional probability for the spatio-temporal image gradient, given a particular velocity vector, and on a prior on the estimated piecewise constant motion field favoring motion boundaries of minimal length.

The proposed functional depends on velocity vectors for a set of disjoint regions and the boundary separating them. It can be considered an extension of the Mumford-Shah functional from the case of gray value segmentation to the case of motion segmentation. The only free parameter is given by a fundamental scale parameter intrinsic to all segmentation approaches.

We proposed an implementation of the motion segmentation functional in a multiphase level set framework. The resulting model has the following favorable properties:

- The minimization of a *single* functional with respect to its dynamic variables jointly solves the problems of motion estimation and motion segmentation. It generates a segmentation of the image plane into a set of disjoint regions of homogeneous velocity.

- The *implicit* representation of the motion discontinuity set does not depend on a particular choice of parameterization. Moreover, it allows for topological changes of the boundary such as splitting or merging.

- The *multiphase* level set formulation permits a segmentation of the image plane into several (possibly multiply-connected) motion phases.

- Minimizing the proposed functional is straight-forward. It results in an eigenvalue problem for the motion vectors, and a gradient descent evolution for the level set functions embedding the motion boundary.

- Due to the region-based homogeneity criterion rather than an edge-based formulation, the functional is robust to noise and the motion boundaries tend to converge over fairly large spatial distances.

- The segmentation and motion estimates are obtained on the basis of the spatio-temporal image gradient calculated from only two consecutive frames of an image sequence. Therefore the presented approach is in principle amenable to real-time implementations and tracking.

We demonstrated these properties by a number of experimental results which were obtained both on simulated ground truth data and on real-world sequence data. Present work focuses on extending the proposed approach to the simultaneous segmentation of multiple frames in a sequence and to estimating piecewise parametric motion fields (cf. [7]).

Acknowledgements

The author would like to thank P. Favaro, S. Soatto, A. Yuille and S.-C. Zhu for fruitful discussions. This research was supported by ONR N00014-02-1-0720 and AFOSR F49620-03-1-0095.

References

1. S. Ayer and H.S. Sawhney. Layered representation of motion video using robust maximum likelihood estimation of mixture models and MDL encoding. In *Proc. of the Int. Conf. on Comp. Vis.*, pages 777–784, Boston, USA, 1995.
2. M. J. Black and P. Anandan. The robust estimation of multiple motions: Parametric and piecewise–smooth flow fields. *Comp. Vis. Graph. Image Proc.: IU*, 63(1):75–104, 1996.
3. V. Caselles and B. Coll. Snakes in movement. *SIAM J. Numer. Anal.*, 33:2445–2456, 1996.
4. V. Caselles, R. Kimmel, and G. Sapiro. Geodesic active contours. In *Proc. IEEE Internat. Conf. on Comp. Vis.*, pages 694–699, Boston, USA, 1995.
5. T. Chan and L. Vese. Active contours without edges. *IEEE Trans. Image Processing*, 10(2):266–277, 2001.
6. D. Cremers. A variational framework for image segmentation combining motion estimation and shape regularization. In C. Dyer and P. Perona, editors, *IEEE Int. Conf. on Comp. Vis. and Patt. Recog.*, Madison, Wisconsin, June 2003. To appear.
7. D. Cremers and C. Schnörr. Motion Competition: Variational integration of motion segmentation and shape regularization. In L. van Gool, editor, *Pattern Recognition*, volume 2449 of *LNCS*, pages 472–480, Zürich, Sept. 2002. Springer.
8. D. Cremers and C. Schnörr. Statistical shape knowledge in variational motion segmentation. *Image and Vision Computing*, 21(1):77–86, 2003.
9. G. Farnebäck. Very high accuracy velocity estimation using orientation tensors, parametric motion, and segmentation of the motion field. In *Proc. 8th ICCV*, volume 1, pages 171–177, 2001.
10. J. Gomes and O. D. Faugeras. Level sets and distance functions. In D. Vernon, editor, *Proc. of the Europ. Conf. on Comp. Vis.*, volume 1842 of *LNCS*, pages 588–602, Dublin, Ireland, 2000. Springer.
11. B.K.P. Horn and B.G. Schunck. Determining optical flow. *Artif. Intell.*, 17:185–203, 1981.
12. A. Jepson and M.J. Black. Mixture models for optic flow computation. In *Proc. IEEE Conf. on Comp. Vision Patt. Recog.*, pages 760–761, New York, 1993.
13. S. Kichenassamy, A. Kumar, P. J. Olver, A. Tannenbaum, and A. J. Yezzi. Gradient flows and geometric active contour models. In *Proc. IEEE Internat. Conf. on Comp. Vis.*, pages 810–815, Boston, USA, 1995.
14. P. Kornprobst, R. Deriche, and G. Aubert. Image sequence analysis via partial differential equations. *J. Math. Im. Vis.*, 11(1):5–26, 1999.
15. B. D. Lucas and T. Kanade. An iterative image registration technique with an application to stereo vision. In *Proc. 7th International Joint Conference on Artificial Intelligence*, pages 674–679, Vancouver, 1981.
16. A. Mansouri, B. Sirivong, and J. Konrad. Multiple motion segmentation with level set. In *Proc. SPIE Conf. on Image and Video Communications and Processing*, pages 584–595, Santa Fe, 2000.

17. E. Memin and P. Perez. Dense estimation and object-based segmentation of the optical flow with robust techniques. *IEEE Trans. on Im. Proc.*, 7(5):703–719, 1998.

18. J.-M. Morel and S. Solimini. *Variational Methods in Image Segmentation.* Birkhäuser, Boston, 1995.

19. D. Mumford and J. Shah. Optimal approximations by piecewise smooth functions and associated variational problems. *Comm. Pure Appl. Math.*, 42:577–685, 1989.

20. J.-M. Odobez and P. Bouthemy. Robust multiresolution estimation of parametric motion models. *J. of Visual Commun. and Image Repr.*, 6(4):348–365, 1995.

21. J.-M. Odobez and P. Bouthemy. Direct incremental model-based image motion segmentation for video analysis. *Signal Proc.*, 66:143–155, 1998.

22. S. J. Osher and J. A. Sethian. Fronts propagation with curvature dependent speed: Algorithms based on Hamilton–Jacobi formulations. *J. of Comp. Phys.*, 79:12–49, 1988.

23. N. Paragios and R. Deriche. Geodesic active contours and level sets for the detection and tracking of moving objects. *IEEE Trans. on Patt. Anal. and Mach. Intell.*, 22(3):266–280, 2000.

24. C. Samson, L. Blanc-Féraud, G. Aubert, and J. Zerubia. A level set model for image classification. *Int. J. of Comp. Vis.*, 40(3):187–197, 2000.

25. C. Schnörr. Computation of discontinuous optical flow by domain decomposition and shape optimization. *Int. J. of Comp. Vis.*, 8(2):153–165, 1992.

26. M. Sussman and E. Fatemi. An efficient, interface-preserving level set redistancing algorithm and its application to interfacial incompressible fluid flow. *SIAM J. Sci. Comput.*, 20(4):1165–1191, 1999.

27. M. Sussman, Smereka P., and S. J. Osher. A level set approach for computing solutions to incompressible twophase flow. *J. of Comp. Phys.*, 94:146–159, 1994.

28. J. Weickert and C. Schnörr. A theoretical framework for convex regularizers in PDE–based computation of image motion. *Int. J. of Comp. Vis.*, 45(3):245–264, 2001.

29. Y. Weiss. Smoothness in layers: Motion segmentation using nonparametric mixture estimation. In *Proc. IEEE Conf. on Comp. Vision Patt. Recog.*, pages 520–527, Puerto Rico, 1997.

30. H.-K. Zhao, T. Chan, B. Merriman, and S. Osher. A variational level set approach to multiphase motion. *J. of Comp. Phys.*, 127:179–195, 1996.

Segmentation of Coarse and Fine Scale Features Using Multi-scale Diffusion and Mumford-Shah

Jeremy D. Jackson[1], Anthony Yezzi, Jr.[1,*],
Wes Wallace[2], and Mark F. Bear[2,**]

[1] School of Electrical and Computer Engineering, Georgia Institute of Technology
Atlanta, GA 30332
gtg120d@prism.gatech.edu
[2] Department of Neuroscience, Howard Hughes Medical Institute/Brown University
Providence, RI 02912
wwallace@brown.edu

Abstract. Here we present a segmentation algorithm that uses multi-scale diffusion with the Mumford-Shah model. The image data inside and outside a surface is smoothed by minimizing an energy functional using a partial differential equation that results in a trade-off between smoothing and data fidelity. We propose a scale-space approach that uses a good deal of diffusion as its coarse scale space and that gradually reduces the diffusion to get a fine scale space. So our algorithm continually moves to a particular diffusion level rather than just using a set diffusion coefficient with the Mumford-Shah model. Each time the smoothing is decreased, the data fidelity term increases and the surface is moved to a steady state. This method is useful in segmenting biomedical images acquired using high-resolution confocal fluorescence microscopy. Here we tested the method on images of individual dendrites of neurons in rat visual cortex. These dendrites are studded with dendritic spines, which have very small heads and faint necks. The coarse scale segments out the dendrite and the brighter spine heads, while avoiding noise. Backing off the diffusion to a medium scale fills in more of the structure, which gets some of the brighter spine necks. The finest scale fills in the small and detailed features of the spines that are missed in the initial segmentation. Because of the thin, faint structure of the spine necks, we incorporate into our level set framework a topology preservation method for the surface which aids in segmentation and keeps a simple topology.

1 Introduction

Global segmentation algorithms have the benefit of being able to extract an object and its prominent features from an image or image volume. They have this capability because they segment an image based on properties such as average pixel intensity of a region or differing textures of regions. Some of these methods

---

* Supported by NSF grant CCR-0133736 and NIH grant R01-HLS0004-01A1.
** This work is supported in part by a grant from the National Eye Institute.

L.D. Griffin and M. Lillholm (Eds.): Scale-Space 2003, LNCS 2695, pp. 615–624, 2003.
© Springer-Verlag Berlin Heidelberg 2003

are detailed in [2], [5],[7], [8],[9]. While a global perspective avoids the noise that more edge-based detectors would get caught up on, it can lose the fine scale features of the object in capturing a coarse estimate of the object. It would be useful to have a method with which to include some of these finer scale features after this coarse segmentation has been done. In our work, a method of this type was necessary to solve the problem of segmenting a topographically complex biological structure from a three dimensional image volume.

The structures in question are the dendrites of pyramidal neurons in rat visual cortex. These dendrites are studded with individual tiny branchlets called spines. The spines are sites of synaptic contact between neurons, and their 3-d morphology is thought to be a marker of the functional state of individual synapses. The fine structure of spines has been extensively investigated at the electron-microscope level – they are known to be bulbous in shape and always connected to the dendrite by very thin necks (with diameter on the order of 0.1 micron). [10]

We obtained 3D image volumes of spiny dendrites as follows: pyramidal neurons in fixed tissue slices of rat visual cortex were intracellularly injected with the fluorescent dye Alexa-488 (Molecular Probes Inc., Eugene, OR; emission peak = 517 nm). Individual dendritic segments were imaged in 3D using an olympus fluoview confocal microscope, at zoom factor 8, with a 63x NA 1.2 water-immersion lens. The voxel size of these images was 0.09 x 0.09 x 0.15 microns (actually slightly above the diffraction limit of this imaging system). 3D images were preprocessed using simple operations to improve contrast and reduce noise. Images were then deconvolved using an adaptive blind deconvolution algorithm (Autoquant Imaging, Watervliet, NY).

In these images, the dendrite is more brightly fluorescent than the spines, due to the greater volume of fluorescent dye it contains. The spine necks in particular can be very faint both because of their very small volume, and because their size is at the limit of resolution of the confocal microscope. Some of the spine heads are dim as well. This is apparent by looking at a full 2D slice of the 3D images in Fig. 1 and a close up of a section of the dendrite and its spines in Fig. 2.

The regional methods only capture the dendrite and some of the spine heads. The first step to solve this problem is to set a smoothing parameter in the Mumford-Shah segmentation method so that it becomes a regional algorithm that gives a coarse segmentation of the dendrite. Then this smoothing term is gradually reduced to capture some fine scale features. It is this stepping down of the diffusion term that gradually gets a correct segmentation of the spine heads and the necks that connect them to the dendrite.

2   The Mumford Shah Model

Here in this section we present the variational formulation of the main segmentation algorithm (a multi-scale version of Mumford-Shah) that was used in this project. This algorithm was implemented in a level set formulation according to [11]. Other level set implementations of Mumford-Shah are in [1], [6] and the

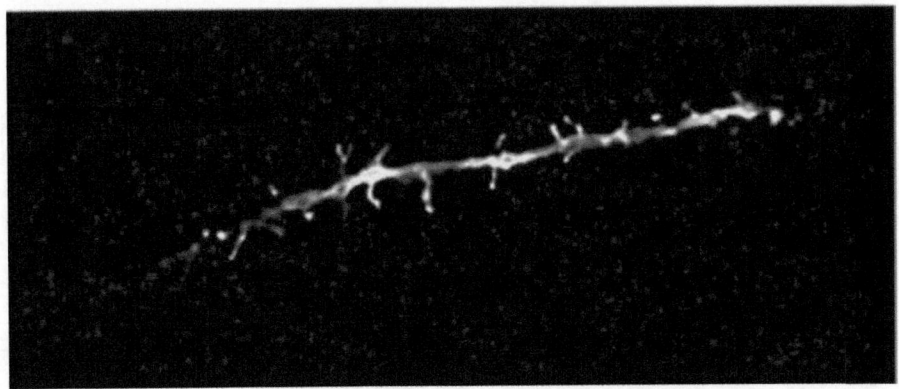

Fig. 1. 2D image plane from the middle of a 3D volume, showing the dendrite with spines branching off

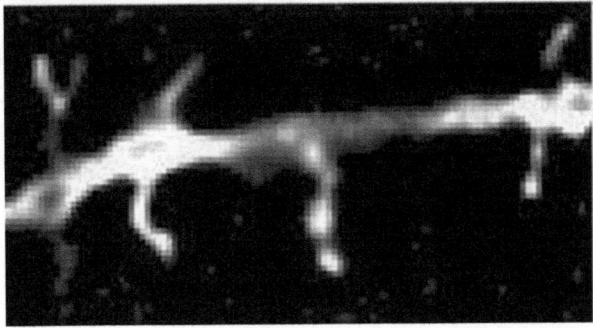

Fig. 2. Closeup of Fig. 1, showing the fine structure of spines. Pixel size = 0.09 x 0.09 microns

model itself is in [4]. The level set is evolved using a PDE that minimizes a given energy functional. More about level set theory can be found in [12].

To implement the Mumford and Shah model, a smooth estimate of the foreground and one of the background is needed so there can exist a piecewise smooth estimate of the image data with the surface being the discontinuity between the two estimates. Based on these smooth estimates, the level set which contains the surface (in this case, a three dimensional surface) is evolved to minimize the following energy functional:

$$E = \alpha \iiint_R (I - f)^2 dV + \alpha \iiint_{R^c} (I - g)^2 dV + \beta \iiint_R |\nabla f|^2 dV$$
$$+ \beta \iiint_{R^c} |\nabla g|^2 dV + \gamma \iint_S d\sigma \tag{1}$$

where I is the image volume, f is the smooth estimate of the image in the foreground R, g is the smooth estimate of the image in the background R^c, and S is the surface. The first two terms in the energy functional are data fidelity terms that make sure that the smooth estimates of the foreground and background match the image data as much as possible. The next two terms keep the norm squared gradients of the smooth estimates f and g as small as possible which results in a smoother f and g. The last term of the energy functional is used to penalize surface area. The parameters $\alpha, \beta, \gamma \in [0, 1] \subset \Re$ should all add up to 1 so they can be used as weights to either increase data fidelity or smoothness or penalizing of surface area. So then the level set is evolved according to the flow

$$\phi_t = -\alpha((I - g)^2 - (I - f)^2)N + \beta(|\nabla g|^2 - |\nabla f|^2)N + \gamma \kappa N \qquad (2)$$

where N is the inward normal of the surface S. The derivation of this can be found in [1] and [4]. With each evolution of the level set ϕ, we need to get the new smooth estimates f and g. This is done using the same energy functional as above but minimization is done with respect to f when evolving the smooth estimate f. Using the Calculus of Variations, the first variation is used to get the Euler-Lagrange equations necessary to evolve the smooth function to a steady state based on the the energy functional. The resulting equation to evolve the smooth function f is

$$f_t = 2(\alpha(I - f) + \beta \Delta f) \qquad (3)$$

where Δf is the laplacian of f:

$$\Delta f = f_{xx} + f_{yy} + f_{zz}. \qquad (4)$$

Evolving g is similar.

A piecewise constant version of this is given by Chan and Vese in [5]. The energy functional is given by:

$$E = \alpha \iiint_R (I - u)^2 dV + \alpha \iiint_{R^c} (I - v)^2 dV + \gamma \iint_S d\sigma. \qquad (5)$$

where u and v are the means inside and outside the surface respectively. The evolution of the the level set is given by

$$\phi_t = -\alpha(u - v)(I - u + I - v)N + \gamma \kappa N. \qquad (6)$$

The Chan-Vese flow can also be looked at as the $(\beta = \infty)$ case (total smoothing) of Mumford-Shah.

3   Multi-scale Diffusion with Mumford-Shah

So in our algorithm, the coarse Mumford-Shah segmentation that we begin with is the $(\beta = \infty)$ case which is equivalent to the Chan-Vese piecewise constant model. We evolve the Mumford-Shah flow to steady state, decrease the smoothing parameter and increase the data fidelity term.

First let us see why the ($\beta = \infty$) case is our coarse scale space which will segment prominent features of the image only. The update of the level set can be rearranged as such:

$$\phi_t = -2\alpha(u - v)(I - \frac{u + v}{2})\boldsymbol{N} + \gamma\kappa\boldsymbol{N}. \tag{7}$$

If the surface is initialized so that it is outside of the object we want to segment, then the term $-2\alpha(u - v)$ should not change sign while the surface is evolving. The term $I - \frac{u+v}{2}$ shows us that the flow will move the surface according to u (the mean of the image data inside the surface) and v (the mean of the image data outside the surface) so that the energy

$$E = \alpha \iiint_{R} (I - u)^2 dV + \alpha \iiint_{R^c} (I - v)^2 dV \tag{8}$$

is as small as possible. So what happens is with each iteration the means are computed and the surface moves past a pixel in I if it is less than $\frac{u+v}{2}$. This is the case if we ignore the surface area penalty which gets rid of bright pieces of noise because they have high curvature. The value $\frac{u+v}{2}$ in this case can be looked at as a threshold that gets larger as the surface segments a bright object. This flow gives a segmentation of all of the very prominent features of the object. The problem with this is the single value $\frac{u+v}{2}$ that is used to move the surface at all points in the image. This tends to skip over fine detail that might be fainter than most of the rest of the object. In the case of dendrites, the main dendrite and the head of the spines are segmented very well, but the dimmer spine necks are totally skipped over.

To fix this we need the Mumford-Shah flow (with $\beta \neq \infty$) which uses a value $\frac{f+g}{2}$ to decide whether to pass by a pixel or not. Since f and g are smooth functions, there is a more adaptive threshold that passes by pixels depending on a value that is more local to the pixel since f or g at each pixel is smoothed out by its neighboring pixels. This is preferred over a global smoothing ($\beta = \infty$) which results in $f = u$ and $g = v$. This allows Mumford-Shah to capture some of the fine detail. So the premise of our algorithm is to keep backing off the smoothing to acquire more and more detail of the object from a very nice, but rough initial estimate. This gradual aquiring of features in a multi-step fashion allows the flow to accurately capture more detail than a Mumford-Shah flow with a set diffusion. The set Mumford-Shah flow does not get these details as well as the multi-step version because it has no good coarse segmentation to build upon.

Also an assumption that we made in the segmentation of dendrites is that the background is constant (fairly close to zero) which turns out to be true for all the data we have worked on. This allows us to use v or zero as the estimate for the background which speeds up the process since it is not necessary to use a PDE to find the smooth function g each time the surface needs to be evolved.

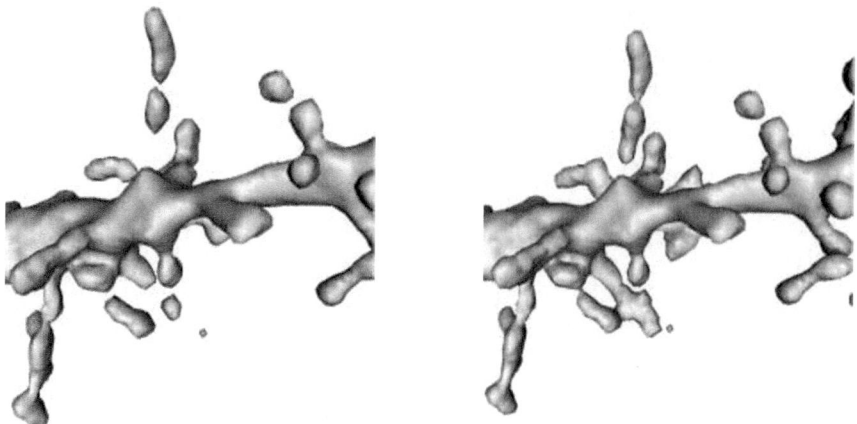

620   Jeremy D. Jackson et al.

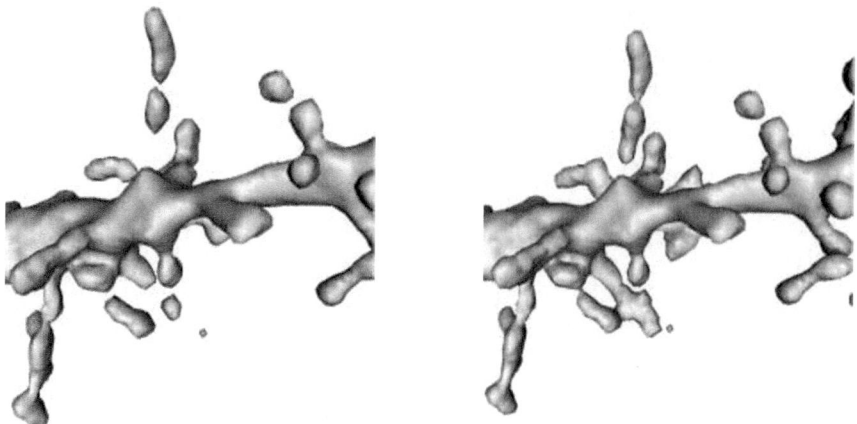

Fig. 3. Set Mumford-Shah vs.Multi-Scale Mumford-Shah

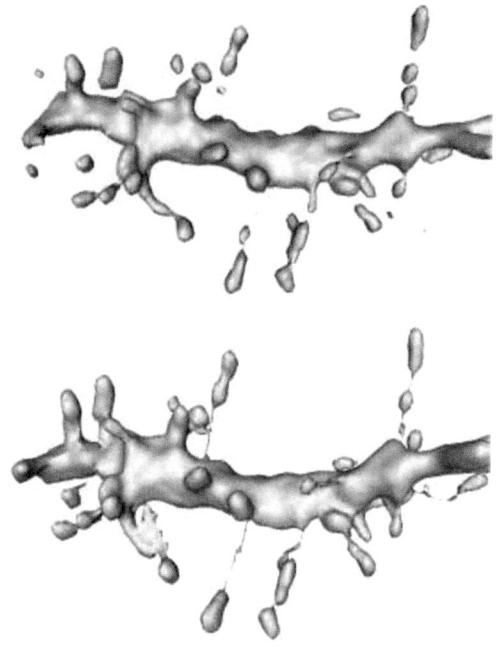

Fig. 4. Mumford-Shah ($\beta = \infty$) case: no topology preservation vs. topology preservation

4   Topology Preservation with Mumford Shah

It would be nice to keep objects that we segment to be as realistic as possible. In the case with dendrites there are no holes of any kind; so a dendrite should be

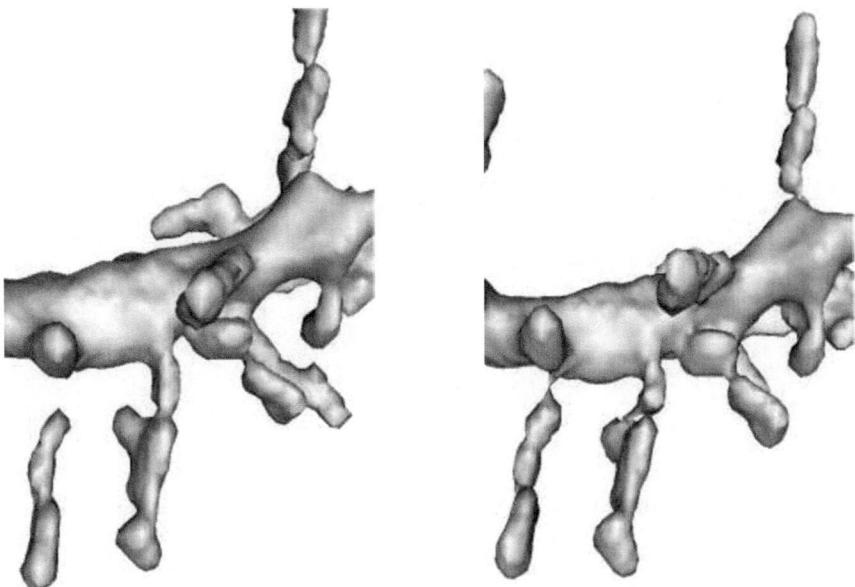

Fig. 5. Connection example: Multi-Scale Mumford-Shah without and with topology preservation

topologically equivalent to a sphere. To keep this realism in our surface we need a flow that preserves the topology of the surface. This will also preserve fine-scale features (i.e. the necks of the dendritic spines). We use the method in [14] which preserves the topology of a surface in a level-set methodology. This method looks for *simple points* as described in [15], [16], and [17]. If this preservation is not done, the surface will pinch off the necks and just segment the dendrite and the spine heads and will not have a simple topology.

Our level set function uses values below zero to denote the inside of the surface (the zero level set) and values above zero to denote pixels that are outside the surface. When a value of our level set ϕ wants to change sign, i.e. a pixel wants to change from foreground to the background or vice versa, it is possible that the change will cause a change in topology. To keep this from happening we look at a point in the level set when it is going to change sign. If this will cause a break in topology (the point is not a *simple point*), we just set the value of the level set at that point to be some small number ϵ that has the same sign as the point had before.

This topology preservation helps at each step in the evolution of our surface. The initial ($\beta = \infty$) Mumford-Shah flow needs to have this preservation so that it will not break topology so our initial coarse estimate is still topologically equivalent to a sphere. If this topology preservation is not in place, the necks of the dendritic spines would get pinched off as shown in Fig. 4.

It is possible to get these necks back without doing topology preservation and just running the multi-step Mumford-Shah. The advantage of having topology

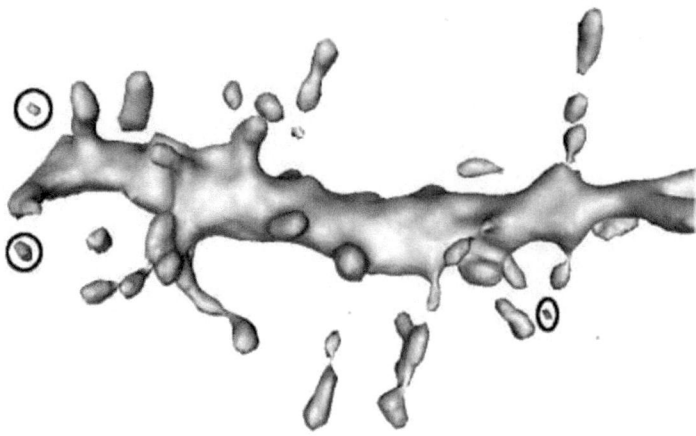

Fig. 6. Noise example: Mumford-Shah ($\beta = \infty$) case without topology preservation

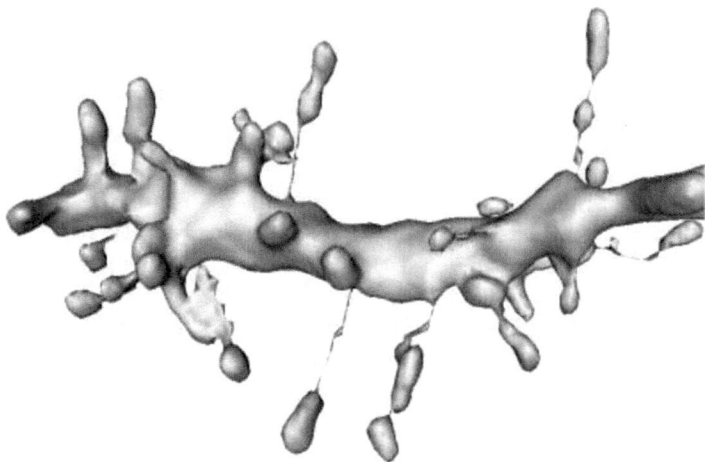

Fig. 7. Noise example: Mumford-Shah ($\beta = \infty$) case with topology preservation

preservation is that there is a piece of surface that is already connecting the spine head and the dendrite where the neck should be. This makes it easier for the multi-step Mumford-Shah to expand out over that neck. Whereas without the neck surface there, the neck does get found, but in the case of a totally missing neck or extremely faint data the multi-scale Mumford-Shah will not fill in the neck completely and so it will not totally connect the spine head to the dendrite.

Another benefit of having the topology preservation is that it helps get rid of pieces of noise. With topology preservation, there is some surface that connects the noise to the dendrite. Without the surface connecting the noise to the dendrite, the noise has its own local smooth function and the areas near it have a

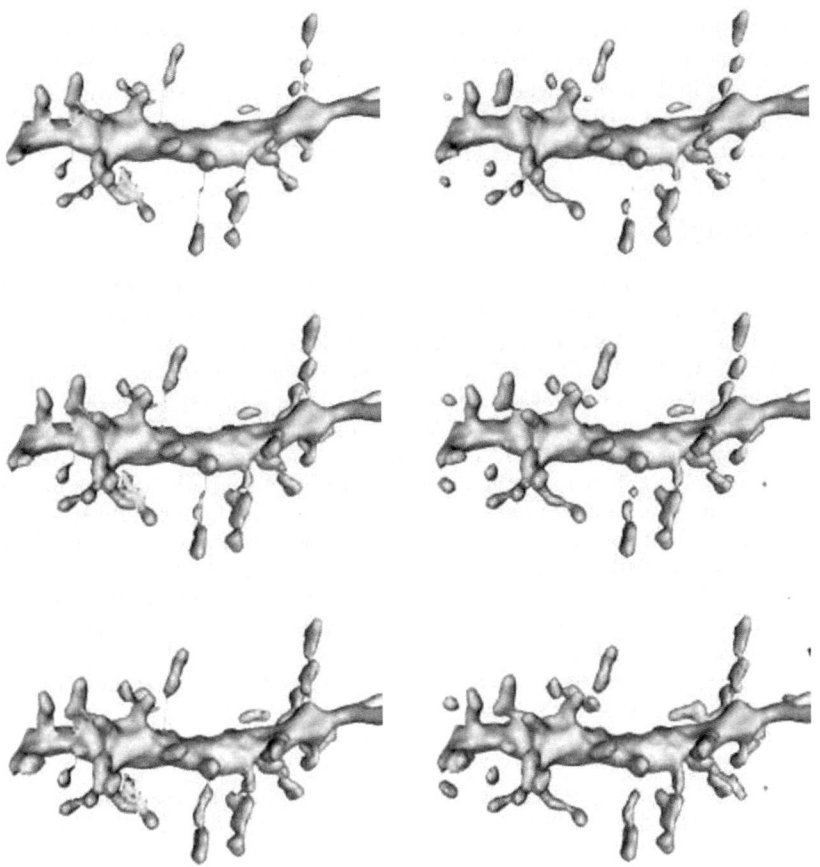

Fig. 8. Progression of Multi-Scale Mumford-Shah with and without topology preservation

smooth function close to or equal to zero because they are background. Therefore the noise has a greater chance of staying in the segmentation as shown in Figure 6. With a surface connecting the noise to the dendrite, the smooth function close to that region of noise will be higher causing the noise to disappear as in Figure 7.

5   Conclusion

Here we have shown a method to segment fine scale features of a biological object. This scale-space approach of a multi-scale Mumford-Shah is very good for capturing coarse and then fine-scale features. Also the preservation of topology allows for a more realistic segmentation with no breaks in topology. In the case of dendritic spines, we have prior knowledge of their topology and therefore we can require that spine heads remain connected to the dendrite by a neck, even when there is no data for a neck. Topology preservation also improves segmentation of spine necks in cases where the data for the neck exists but is very faint. This is evident in Figure 8 where the progression of Multi-scale Mumford-Shah is shown with and without topology preservation. The Multi-scale Mumford-Shah with topology preservation captures the dendrite quite well.

References

1. Tsai, A., Yezzi A., Wilsky A.: Curve Evolution Implementation of the Mumford-Shah Functional for Image Segmentation, Denoising, Interpolation, and Magnification. IEEE Trans. on Image Processing (2001) 1169–1184
2. Yezzi, A., Tsai, A., Wilsky, A.: A statistical approach to snakes for bimodal and trimodal imagery. Int. Conf. on Computer Vision. (1999) **2** 898–903
3. Yezzi, A.: Modified curvature motion for image smoothing and enhancement. IEEE Trans. Image Processing **7** (1998) 345–352
4. Mumford D., Shah J.: Optimal approximations by piecewise smooth functions and associated variational problems. Commun. Pure Appl. Math (1989)
5. Chan, T.F, Vese, L.A.: Active contours without edges. IEEE Trans. Image Processing **10** (2001) 266-277
6. Chan, T.F, Vcsc, L.A.: A level set algorithm for minimizing the Mumford-Shah functional in image processing. IEEE Proc. on Variational and Level Set Meth. in Comp. Vision (2001) 161–168
7. Paragios, N., Deriche, R.: Geodesic active regions for texture segmentation. INRIA, France, Res. Rep. 3440 (1998)
8. Ronfard, R.: Region-based strategies for active contour models. Int. J. Comput. Vis. **13** (1994) 229–251
9. Zhu, S., Yuille, A.: Region competition: Unifying snakes, region growing, and Bayes/MDL for multiband image segmentation. IEEE Trans. Pattern Anal. Machine Intell. **18** (1996) 884–900
10. Nimchinsky E. A., Sabatini B.L. and Svoboda K. (2002). "Structure and function of dendritic spines." Annu. Rev. Physiol. 64: 313-353.
11. Osher, S., Sethian, J.: Fronts propagation with curvature dependent speed: Algorithms based on Hamilton-Jacobi formulations. J. Comput. Physics **79** (1988) 12–49
12. J.A. Sethian: Level Set Methods and Fast Marching Methods: Evolving Interfaces in Geometry, Fluid Mechanics, Computer Vision, and Material Science. Cambridge, U.K.:Cambridge University Press (1999)
13. Kass, M., Witkin, A., Terzopoulos, D.: Snakes: Active contour models. Int. J. Comput. Vis **1** (1987) 321-331
14. Han, X., Xu, C., Tosun, D., Prince, J.L.: Corical Surface Reconstruction Using a Topology Preserving Geometric Model. IEEE Trans. on Medical Imaging (2002) 109-121
15. Malandain, G., Bertrand G.: Fast Characterization of 3D Simple Points. IEEE Pattern Recognition (1992) 232–235
16. Malandain, G., Bertrand G.: A new characterization of three-dimensional simple points. Pattern Recognition Letters **15** (1994) 169–175
17. Bertrand, G.: Simple points, toplogical numbers and geodesic negihborhoods in cubic grids. Pattern Recognition Letters **15** (1994) 1003–1011
18. Weinstock, R.: Calculus of Variations: With Applications to Physics and Engineering. New York:Dover Pub. Inc. (1974)

On the Number of Modes of a Gaussian Mixture

Miguel Á. Carreira-Perpiñán[1] and Christopher K.I. Williams[2]

[1] Dept. of Computer Science, University of Toronto
miguel@cs.toronto.edu
[2] School of Informatics, University of Edinburgh
c.k.i.williams@ed.ac.uk

Abstract. We consider a problem intimately related to the creation of maxima under Gaussian blurring: the number of modes of a Gaussian mixture in D dimensions. To our knowledge, a general answer to this question is not known. We conjecture that if the components of the mixture have the same covariance matrix (or the same covariance matrix up to a scaling factor), then the number of modes cannot exceed the number of components. We demonstrate that the number of modes can exceed the number of components when the components are allowed to have arbitrary and different covariance matrices.

We will review related results from scale-space theory, statistics and machine learning, including a proof of the conjecture in 1D. We present a convergent, EM-like algorithm for mode finding and compare results of searching for all modes starting from the centers of the mixture components with a brute-force search. We also discuss applications to data reconstruction and clustering.

1   Introduction

We propose a mathematical conjecture about Gaussian mixtures (GMs): that, under certain conditions, the number of modes cannot exceed the number of components. Although we originally came across this conjecture in a pattern recognition problem (sequential data reconstruction), it is intimately related to scale-space theory (since some GMs are the convolution of a delta mixture with a Gaussian kernel) and statistical smoothing (since Gaussian kernel density estimates are GMs). Bounding the number of modes and the region where they lie, and finding all these modes, is of interest in these areas. The widespread use of GMs makes the conjecture relevant not only theoretically but also in applications of these areas, such as data reconstruction, image segmentation or clustering.

We state formally the conjecture and prove part of it in Sect. 2, and review related proof approaches in Sect. 3. We show the convergence of an algorithm that tries to find all modes in Sect. 4 and discuss applications in Sect. 5. An extended version of this paper appears as [1].

L.D. Griffin and M. Lillholm (Eds.): Scale-Space 2003, LNCS 2695, pp. 625–640, 2003.
© Springer-Verlag Berlin Heidelberg 2003

2   The Conjecture

Consider a GM density of $M > 1$ components in $\mathbb{R}^D$ for $D \geq 1$, with mixture proportions $\{\pi_m\}_{m=1}^M \subset (0, 1)$ satisfying $\sum_{m=1}^M \pi_m = 1$, component means $\{\boldsymbol{\mu}_m\}_{m=1}^M \subset \mathbb{R}^D$ and positive definite covariance matrices $\{\boldsymbol{\Sigma}_m\}_{m=1}^M$:

$$p(\mathbf{x}) \overset{\text{def}}{=} \sum_{m=1}^M p(m)p(\mathbf{x}|m) \overset{\text{def}}{=} \sum_{m=1}^M \pi_m p(\mathbf{x}|m) \quad \forall \mathbf{x} \in \mathbb{R}^D \quad \mathbf{x}|m \sim \mathcal{N}_D(\boldsymbol{\mu}_m, \boldsymbol{\Sigma}_m).$$

In general, there is no explicit expression for the modes of p, i.e., no analytic solution for the stationary points in eq. (1); we do not even know how many modes p has. Intuitively, it seems reasonable that the number of modes of p will not exceed the number M of components in the GM: the more the different components interact (depending on their mutual separation and on their covariance matrices), the more they will coalesce and the fewer modes will exist. Besides, modes should always appear inside the region enclosed by the component centroids $\{\boldsymbol{\mu}_m\}_{m=1}^M$—more precisely, in their convex hull[1]. Based on this reasoning, Carreira-Perpiñán [2] (see also [3]) proposed the following conjecture.

Conjecture. Let $p(\mathbf{x}) = \sum_{m=1}^M p(m)p(\mathbf{x}|m)$, where $\mathbf{x}|m \sim \mathcal{N}_D(\boldsymbol{\mu}_m, \boldsymbol{\Sigma}_m)$, be a mixture of M D-variate normal distributions. Then $p(\mathbf{x})$ has M modes at most, all of which are in the convex hull of $\{\boldsymbol{\mu}_m\}_{m=1}^M$, if one of the following conditions holds:

1. $D = 1$ *(one-dimensional mixture).*
2. $D \geq 1$ and the covariance matrices are arbitrary but equal: $\boldsymbol{\Sigma}_m = \boldsymbol{\Sigma} \; \forall m = 1, \ldots, M$ *(homoscedastic mixture).*
3. $D \geq 1$ and the covariance matrices are isotropic: $\boldsymbol{\Sigma}_m = \sigma_m^2 \mathbf{I}_D$ *(isotropic mixture).*

Several parts of this conjecture hold, namely the modes (and all other stationary points) lie in the convex hull, and for $D = 1$ the number of modes does not exceed M. We will prove this below. Besides, the conditions of the conjecture are necessary. Figure 1 gives examples of a GM with nonisotropic, different component covariance matrices that has more modes than components and the modes lie outside the convex hull of the centroids. Also, if the kernel $p(\mathbf{x}|m)$ is not Gaussian one can construct examples where the conjecture does not hold. This may seem counterintuitive, since one may expect that localised, tapering kernels would behave like the Gaussian. However, small modes can typically arise where kernels interact—although it may occur only rarely. The necessity that the kernel be Gaussian has been established in scale-space theory (see Sect. 3.2).

The Modes Lie in the Convex Hull for Any Dimension D. All modes lie in the convex hull of the centroids for the case of isotropic GMs. One proof is given by the stationary-point eq. (3), which also shows that in generic cases the modes must lie strictly in the interior of the convex hull and not on its boundary. An alternative proof is given in [2, p. 218].

---

[1] Defined as the set $\left\{ \mathbf{x} : \mathbf{x} = \sum_{m=1}^M \lambda_m \boldsymbol{\mu}_m \text{ with } \{\lambda_m\}_{m=1}^M \subset [0, 1] \text{ and } \sum_{m=1}^M \lambda_m = 1 \right\}$.

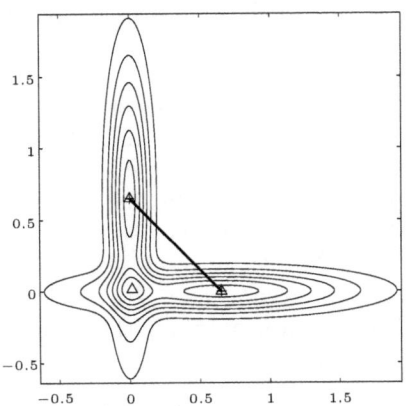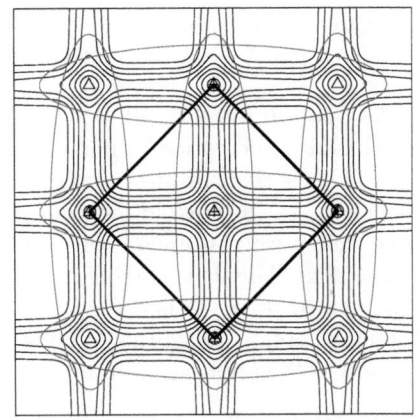

Fig. 1. GMs in dimension $D \geq 2$ that have different, non-isotropic covariances do not generally verify conjecture 1. The left graph shows a contour plot for a bicomponent GM with $\pi_1 = \pi_2 = \frac{1}{2}$, $\boldsymbol{\mu}_1 = \left(\begin{smallmatrix} 0.6 \\ 0 \end{smallmatrix}\right)$, $\boldsymbol{\mu}_2 = \left(\begin{smallmatrix} 0 \\ 0.6 \end{smallmatrix}\right)$, $\boldsymbol{\Sigma}_1 = \left(\begin{smallmatrix} 0.65 & 0 \\ 0 & 0.1 \end{smallmatrix}\right)$ and $\boldsymbol{\Sigma}_2 = \left(\begin{smallmatrix} 0.1 & 0 \\ 0 & 0.65 \end{smallmatrix}\right)$. This GM has three modes (marked "$\triangle$"): two nearly coincident with the centroids $\boldsymbol{\mu}_m$ (marked "+") and a third one near the meeting point of the components' principal axes. All the modes are outside the convex hull of the centroids (marked by the thick line). More complicated arrangements can result in a multiplicity of modes, as shown in the right graph (inspired by Fig. 2 of [4]).

The Homoscedastic Case Is Equivalent to the Homoscedastic Isotropic One. The following theorem shows that the modes problem for a homoscedastic GM with a given arbitrary covariance $\boldsymbol{\Sigma}$ is equivalent to that of another homoscedastic GM with isotropic covariance $\sigma^2 \mathbf{I}$ (for a certain σ). Thus, one can try to prove a result for the simple case of isotropic covariances and then the result will also hold for $\boldsymbol{\Sigma}_m = \boldsymbol{\Sigma}$ arbitrary. The reason is that, by rotating and rescaling the coordinate axes, we can spherise each component.

Theorem 1. *The mixtures* $p(\mathbf{x}) = \sum_{m=1}^{M} \pi_m |2\pi \boldsymbol{\Sigma}|^{-\frac{1}{2}} e^{-\frac{1}{2}(\mathbf{x}-\boldsymbol{\mu}_m)^T \boldsymbol{\Sigma}^{-1}(\mathbf{x}-\boldsymbol{\mu}_m)}$ *(arbitrary but equal covariances) and* $p(\mathbf{u}) = \sum_{m=1}^{M} \pi_m (2\pi)^{-\frac{D}{2}} e^{-\frac{1}{2}\|\mathbf{u}-\boldsymbol{\nu}_m\|^2}$ *(unit covariances), related by a rotation and scaling, have the same number of modes, which lie in the respective centroid convex hulls.*

Proof. Let $\boldsymbol{\Sigma}^{-1} = \mathbf{U}\boldsymbol{\Lambda}\mathbf{U}^T$ be the spectral decomposition of $\boldsymbol{\Sigma}^{-1}$, with $\mathbf{U}$ orthogonal and $\boldsymbol{\Lambda}$ diagonal and positive definite. Consider the coordinate transformation $\mathbf{u} \overset{\text{def}}{=} \boldsymbol{\Lambda}^{\frac{1}{2}} \mathbf{U}^T \mathbf{x}$ (orthogonal rotation followed by scaling), so that $p(\mathbf{u}) = \sum_{m=1}^{M} \pi_m (2\pi)^{-\frac{D}{2}} e^{-\frac{1}{2}\|\mathbf{u}-\boldsymbol{\nu}_m\|^2}$ and $\nabla_{\mathbf{x}} p = \mathbf{U}\boldsymbol{\Lambda}^{\frac{1}{2}} \nabla_{\mathbf{u}} p$, and define $\boldsymbol{\nu}_m \overset{\text{def}}{=} \boldsymbol{\Lambda}^{\frac{1}{2}} \mathbf{U}^T \boldsymbol{\mu}_m$. Since $\mathbf{U}\boldsymbol{\Lambda}^{\frac{1}{2}}$ is nonsingular, $\nabla_{\mathbf{x}} p = \mathbf{0} \Leftrightarrow \nabla_{\mathbf{u}} p = \mathbf{0}$ and so the stationary points are preserved by the transformation.

Now, if $\mathbf{x}$ is a point in the convex hull of $\{\boldsymbol{\mu}_m\}_{m=1}^M$ then $\mathbf{x} = \sum_{m=1}^{M} \lambda_m \boldsymbol{\mu}_m$ where $\{\lambda_m\}_{m=1}^M \subset [0,1]$ and $\sum_{m=1}^{M} \lambda_m = 1$. So $\mathbf{u} = \boldsymbol{\Lambda}^{\frac{1}{2}} \mathbf{U}^T \mathbf{x} = \sum_{m=1}^{M} \lambda_m \boldsymbol{\Lambda}^{\frac{1}{2}} \mathbf{U}^T \boldsymbol{\mu}_m = \sum_{m=1}^{M} \lambda_m \boldsymbol{\nu}_m$ which is in the convex hull of $\{\boldsymbol{\nu}_m\}_{m=1}^M$. $\square$

Theorem 1 shows that case 2 of conjecture 1 is a particular case of case 3 (case 1 is also a particular case of case 3, obviously).

The Conjecture Holds for $D = 1$. We can prove this using the scale-space theory proofs of non-creation of maxima with Gaussian blurring (Sect. 3.2). The intuitive idea is that, by alternating the operations of "planting" a delta function (of value π_m) at a centroid location $\boldsymbol{\mu}_m$ and applying Gaussian blurring (to fatten the delta) we can create any isotropic GM. If planting a delta adds a single mode and Gaussian blurring never creates modes (this latter result given by the mentioned proofs), then the number of modes will never exceed the number of components M. Our proof is by induction. Note that the only step that requires $D = 1$ is the application of the scale-space theorem.

Theorem 2. *In 1D, any Gaussian mixture with M components has at most M modes.*

Proof. By induction on M. The statement holds trivially for $M = 1$. Assume it holds for $M - 1$ components and consider an arbitrary GM p with $M > 1$ components. Consider the component with narrowest variance and call this σ_M^2, perhaps by reordering the components, so that $\sigma_M < \sigma_m \;\forall m < M$ (in the nongeneric case of ties, simply choose any of the narrowest ones and the argument holds likewise). Now apply Gaussian deblurring of variance σ_M^2, recalling that the convolution of two isotropic Gaussians of variances σ_a^2 and σ_b^2 is a Gaussian of variance $\sigma_a^2 + \sigma_b^2$ (the semigroup structure). We obtain a mixture density p' where each component for $m = 1, \ldots, M - 1$ is a Gaussian of mixing proportion π_m and variance $\sigma_m^2 - \sigma_M^2$, and component M is a delta function of mixing proportion π_M. Thus, p' is a mixture of a delta and a GM with $M - 1$ components. By the induction hypothesis the latter has $M - 1$ modes at most, so p' has M modes at most. Now apply Gaussian blurring to p'. By the scale-space theorems of Sect. 3.2, no new maxima can appear, and so the original GM p has M modes at most. $\square$

The following corollary results from the fact that all marginal and conditional distributions of a GM (of arbitrary covariances) are also GMs.

Corollary 1. *Any 1D projection (marginal or conditional distribution) of any Gaussian mixture in D dimensions with M components has at most M modes.*

3   Approaches to Proving the Conjecture

We review results from different fields that concern the conjecture. Additional results applicable only in particular cases are given in [2].

3.1   System of Equations for the Stationary Points of the Density

In [2] the problem was approached by trying to determine the stationary (or critical) points of the GM density p as follows. Consider the case with $\boldsymbol{\Sigma}_m =$

Σ, $m = 1, \ldots, M$ (homoscedastic GM) and assume $\mathbf{x}$ is a stationary point of p. Then

$$\nabla p(\mathbf{x}) = \sum_{m=1}^{M} p(\mathbf{x}, m) \Sigma^{-1} (\boldsymbol{\mu}_m - \mathbf{x}) = \mathbf{0} \Longrightarrow \mathbf{x} = \sum_{m=1}^{M} p(m|\mathbf{x}) \boldsymbol{\mu}_m. \quad (1)$$

This is a nonlinear system of D equations and D unknowns $x_1, \ldots, x_D \subset \mathbb{R}$. Since $p(m|\mathbf{x}) \in (0, 1)$ for all m and $\sum_{m=1}^{M} p(m|\mathbf{x}) = 1$, $\mathbf{x}$ is a convex linear combination of the centroids and so all stationary points lie in the convex hull of the centroids. Instead, write $\mathbf{x} = \sum_{m=1}^{M} \lambda_m \boldsymbol{\mu}_m$ with $\lambda_m \in (0, 1)$ and $\sum_{m=1}^{M} \lambda_m = 1$. Then we can consider ($m = 1, \ldots, M$):

$$\lambda_m = p(m|\mathbf{x}) = \frac{\pi_m e^{-\frac{1}{2} \mathbf{u}_m^T \Sigma^{-1} \mathbf{u}_m}}{\sum_{m'=1}^{M} \pi_{m'} e^{-\frac{1}{2} \mathbf{u}_{m'}^T \Sigma^{-1} \mathbf{u}_{m'}}}$$

$$\mathbf{u}_m \stackrel{\text{def}}{=} \mathbf{x} - \boldsymbol{\mu}_m = \sum_{m'=1}^{M} \lambda_{m'} \boldsymbol{\mu}_{m'} - \boldsymbol{\mu}_m \quad (2)$$

as a nonlinear system of M equations and M unknowns $\lambda_1, \ldots, \lambda_M \in (0, 1)$ subject to $\sum_{m=1}^{M} \lambda_m = 1$.

For $M = 2$ with $\lambda \stackrel{\text{def}}{=} \lambda_1$, $\lambda_2 = 1 - \lambda$, and $\pi \stackrel{\text{def}}{=} p(1)$, $p(2) = 1 - \pi$, eq. (2) reduces to the transcendental equation $\lambda = \frac{1}{1 + e^{-\alpha(\lambda - \lambda_0)}}$ with $\alpha = (\boldsymbol{\mu}_1 - \boldsymbol{\mu}_2)^T \Sigma^{-1} (\boldsymbol{\mu}_1 - \boldsymbol{\mu}_2) \in (0, \infty)$ and $\lambda_0 = \frac{1}{2} + \frac{1}{\alpha} \log \frac{1-\pi}{\pi} \in (-\infty, \infty)$. This can have at most 3 roots in $(0, 1)$, as can be easily seen geometrically in Fig. 2, and so at most 2 can be maxima.

Unfortunately, for higher M the system becomes very difficult to study. Besides, if a counterexample to the conjecture does exist, it is likely to require a nontrivial number of components M in $D \geq 2$, which makes very difficult to look for such a counterexample in terms of the λ_m's.

In case $\Sigma_m = \sigma_m^2 \mathbf{I}$, $m = 1, \ldots, M$ (isotropic components), we get the system

$$\lambda_m = q(m|\mathbf{x}) \stackrel{\text{def}}{=} \frac{p(m|\mathbf{x}) \sigma_m^{-2}}{\sum_{m'=1}^{M} p(m'|\mathbf{x}) \sigma_{m'}^{-2}} = \frac{\pi_m \sigma_m^{-(D+2)} e^{-\frac{1}{2} \left\| \frac{\mathbf{u}_m}{\sigma_m} \right\|^2}}{\sum_{m'=1}^{M} \pi_{m'} \sigma_{m'}^{-(D+2)} e^{-\frac{1}{2} \left\| \frac{\mathbf{u}_{m'}}{\sigma_{m'}} \right\|^2}} \quad (3)$$

again with $\mathbf{x} = \sum_{m=1}^{M} \lambda_m \boldsymbol{\mu}_m$, where $\lambda_m \in (0, 1)$ and $\sum_{m=1}^{M} \lambda_m = 1$, and $\mathbf{u}_m$ as in eq. (2). In effect, λ_m equals the responsibility $p(m|\mathbf{x})$ but reweighted by the inverse variance and renormalised. An analogous analysis shows that there are 3 stationary points at most for the case $M = 2$ but becomes difficult in general.

A further problem with this approach is that the equations apply to all stationary points (maxima, minima and saddles) rather than to maxima only.

Note that the modes must lie strictly in the interior of the convex hull and not on its boundary, since $p(m|\mathbf{x})$ and $q(m|\mathbf{x}) < 1$ (except in non-generic cases such as when all the centroids are equal or some σ_m is zero).

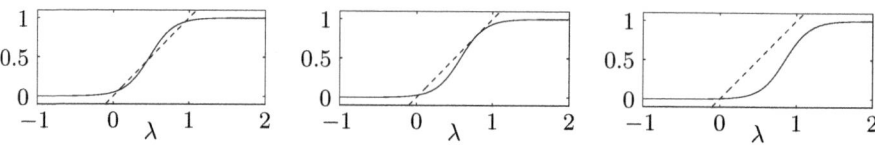

Fig. 2. Three possible cases for the solutions of the equation $\lambda = \frac{1}{1+e^{-\alpha(\lambda-\lambda_0)}}$.

3.2   Scale-Space Theory

We give here a short summary of the creation of maxima with Gaussian blurring in the scale space framework. The central issue of linear Gaussian scale space [5] is the generation of a family of functions $I(\mathbf{x}; s)$ by convolution or blurring of the original D-dimensional function $I(\mathbf{x})$ (the "greyscale image") with a Gaussian kernel of scale $s = \sigma^2$:

$$I(\mathbf{x}; s) \stackrel{\text{def}}{=} (G_s * I)(\mathbf{x}) = \int (2\pi s)^{-\frac{D}{2}} e^{-\frac{1}{2s}\|\mathbf{y}\|^2} I(\mathbf{x} - \mathbf{y}) \, d\mathbf{y} \qquad \mathbf{x} \in \mathbb{R}^D$$

with $I(\mathbf{x}) \equiv I(\mathbf{x}; 0)$. As the scale increases, $I(\mathbf{x}; s)$ represents coarser structure. Several researchers (among others [6,7,8,9]) proved that the Gaussian kernel never creates new maxima in 1D and, further, is the only kernel to do so. Their proofs are typically based on the following points. (1) Causality principle: since the Gaussian kernel is the Green's function of the diffusion equation, the family $I(\mathbf{x}; s)$ is the solution of the diffusion equation (where the time is given by the scale s) with initial condition $I(\mathbf{x}; 0) = I(\mathbf{x})$: $\frac{\partial I}{\partial s} = \frac{1}{2}\nabla^2_\mathbf{x} I$. (2) Particular properties of the blurring process and the Gaussian kernel, such as semigroup structure, homogeneity or isotropy. (3) The implicit function theorem applied to the variables $\mathbf{x}$ and s guarantees that the maxima trajectories $\mathbf{x} = \mathbf{x}(s)$ (along which the gradient $\nabla_\mathbf{x} I$ is zero) are continuously differentiable except at bifurcation points[2] where the Hessian of I with respect to $\mathbf{x}$ becomes singular and the topology changes. A concave level surface corresponds to the annihilation of a pair (a maximum with a minimum or saddle-point), while a convex one corresponds to the creation of a pair. The fact that the family satisfies the diffusion equation forbids the latter.

However, this does not hold in 2D (counterintuitive as it may seem, and though some of the mentioned proofs claimed it did) as originally evidenced by a counterexample proposed by Lifshitz and Pizer [10]: the original image is unimodal, made up by a low hill from whose summit a narrow ramp ascends over a deep valley towards a high hill (which contains the maximum). Gaussian blurring produces a dip in the ramp, creating a new maximum on the low hill, that later annihilates with the dip. This and further examples are analysed by Kuijper and Florack [11,12], who also suggest that such created maxima are rare (being associated with elongated structures) and short-lived in scale space.

The definitive explanation of the creation of maxima was given by Damon [13] using Morse theory and catastrophe theory. Thom's theorem classifies the

---

[2] Also called degenerate critical points, top-points or catastrophes.

behaviour at bifurcation points of a family of functions dependent on parameters (such as the scale). This cannot be applied directly because the family is not unconstrained, but must obey the diffusion equation. Damon showed that maxima creations are associated with an umbilic catastrophe that occurs generically, i.e., does not disappear by perturbing the function.

In summary, in 2D or higher, there exist functions upon which Gaussian blurring results in occasional, but generic, creations of maxima as the scale increases. In 1D no such functions exist: Gaussian blurring never creates maxima, and is the only kernel to do so—for any other kernel, there exist functions on which it creates maxima.

How does this apply to the GM case? Our original "image" is a delta mixture $I(\mathbf{x}) = \sum_{m=1}^{M} \pi_m \delta(\mathbf{x} - \boldsymbol{\mu}_m)$, which by convolution with a Gaussian of variance $s = \sigma^2$ results in a homoscedastic isotropic GM with component covariances $\boldsymbol{\Sigma}_m = \sigma^2 \mathbf{I}_D$. At zero scale the mixture has M modes, one on each centroid $\boldsymbol{\mu}_m$. Therefore, in 1D the scale-space theorems state that no new modes appear as σ increases, which proves the conjecture for the homoscedastic case; and our Theorem 2 extends the proof to the *isotropic* case. In 2D or higher, the possibility that the Gaussian blurring may create modes does not necessarily disprove the conjecture. Firstly, it could be that for mixtures of Gaussians or deltas new modes can never appear; we have never succeeded to replicate a sequence of events such as that of [10]. Perhaps an approach based on catastrophe theory but restricted to initial images which are delta mixtures would resolve this question. Secondly, even if new modes can appear when blurring a GM, the total number of modes may still never exceed M. In other words, a situation of mode creation may require a large number of Gaussian components that interact to result in a GM with only a few modes before and after the creation. A brute-force search has failed to find counterexamples of the conjecture (see Sect. 4.3).

Note also that all catastrophes, being stationary points, must lie in the interior of the convex hull of the centroids (or, nongenerically, on its boundary), as mentioned in Sect. 2.

3.3   Kernel Density Estimation in 1D

Given a data sample $\{x_n\}_{n=1}^{N} \subset \mathbb{R}$, Silverman [14] considers the 1D Gaussian kernel density estimate $p(x; h)$ (Sect. 5.2). This is, of course, a homoscedastic isotropic GM of centroids $\{x_n\}_{n=1}^{N}$, variance h^2 and and equal mixing proportions $\pi_n = \frac{1}{N}$. In his proof, which we believe is not known to the scale-space community, Silverman shows that the number of maxima of $p(x; h)$ (or generally of $\partial^m p / \partial x^m$ for integer $m \geq 0$) is a right continuous decreasing function of h. His proof is based on the total positivity and the semigroup structure of the Gaussian kernel and the variation diminishing property of functions generated by convolutions with totally positive kernels. However, the proof uses the counts of sign changes of the mixture derivative and so it seems difficult to extend it to dimensions higher than 1.

4   Algorithms for Finding All the Modes

We now turn to the practically important question of finding all the modes of
a GM. No direct solution exists, so we need to use numerical iterative methods.
Carreira-Perpiñán [15] suggested starting a mode-seeking algorithm from every
centroid to locate all the modes. He gave two hill-climbing algorithms applicable
to GMs with components of arbitrary covariance: a gradient-quadratic one and
a fixed-point iteration one. Here we deal only with the latter because it allows to
define in a unique way a basin of attraction for each mode, which is relevant both
for the conjecture and for mean-shift algorithms. We also prove its convergence
by deriving it as an EM algorithm.

4.1   The Fixed-Point Iteration Algorithm as an EM Algorithm

By equating the gradient of the GM density to zero, using Bayes' theorem and
rearranging we obtain a fixed-point iterative scheme [15]:

$$\mathbf{x}^{(\tau+1)} = \mathbf{f}(\mathbf{x}^{(\tau)}) \text{ with } \mathbf{f}(\mathbf{x}) \stackrel{\text{def}}{=} \left(\sum_{m=1}^{M} p(m|\mathbf{x}) \boldsymbol{\Sigma}_m^{-1} \right)^{-1} \sum_{m=1}^{M} p(m|\mathbf{x}) \boldsymbol{\Sigma}_m^{-1} \boldsymbol{\mu}_m. \quad (4)$$

Following a suggestion from the second author, Carreira-Perpiñán [2] showed
that this algorithm can also be derived as an *expectation-maximisation (EM)
algorithm* [16,17] as follows[3]. Consider the following density model with param-
eters $\mathbf{v} = (v_1, \dots, v_D)^T$ and fixed $\{\pi_m, \boldsymbol{\mu}_m, \boldsymbol{\Sigma}_m\}_{m=1}^{M}$:

$$p(\mathbf{x}|\mathbf{v}) = \sum_{m=1}^{M} \pi_m \left| 2\pi \boldsymbol{\Sigma}_m \right|^{-\frac{1}{2}} e^{-\frac{1}{2}(\mathbf{x}-(\boldsymbol{\mu}_m-\mathbf{v}))^T \boldsymbol{\Sigma}_m^{-1}(\mathbf{x}-(\boldsymbol{\mu}_m-\mathbf{v}))}.$$

That is, $\mathbf{x}|\mathbf{v}$ is a D-dimensional GM where component m has mixing proportion
π_m (fixed), mean vector $\boldsymbol{\mu}_m - \mathbf{v}$ ($\boldsymbol{\mu}_m$ fixed) and covariance matrix $\boldsymbol{\Sigma}_m$ (fixed).
Varying $\mathbf{v}$ results in a rigid translation of the whole GM as a block rather than
the individual components varying separately. Now consider fitting this model
by maximum likelihood to a data set $\{\mathbf{x}_n\}_{n=1}^{N}$ and let us derive an EM algorithm
to estimate the parameters $\mathbf{v}$. Call $z_n \in \{1, \dots, M\}$ the (unknown) index of the
mixture component that generated data point $\mathbf{x}_n$. Then:

E step The complete-data log-likelihood, as if all $\{z_n\}_{n=1}^{N}$ were known, and
assuming iid data, is $\sum_{n=1}^{N} \mathcal{L}_{n,\text{complete}}(\mathbf{v}) = \sum_{n=1}^{N} \log p(\mathbf{x}_n, z_n|\mathbf{v})$ and so its
expectation with respect to the current posterior distribution is

---

[3] We recently learned of an independent derivation by Y. Weiss (unpubl. manuscript).

$$Q(\mathbf{v}|\mathbf{v}^{(\tau)}) \overset{\text{def}}{=} \sum_{n=1}^{N} E_{p(z_n|\mathbf{x}_n,\mathbf{v}^{(\tau)})}\{\mathcal{L}_{n,\text{complete}}(\mathbf{v})\}$$

$$= \sum_{n=1}^{N} \sum_{z_n=1}^{M} p(z_n|\mathbf{x}_n, \mathbf{v}^{(\tau)}) \log\{p(z_n|\mathbf{v})p(\mathbf{x}_n|z_n, \mathbf{v})\}$$

$$= \sum_{n=1}^{N} \sum_{z_n=1}^{M} p(z_n|\mathbf{x}_n, \mathbf{v}^{(\tau)}) \log p(\mathbf{x}_n|z_n, \mathbf{v}) + K$$

where $K \overset{\text{def}}{=} \sum_{n=1}^{N} \sum_{z_n=1}^{M} p(z_n|\mathbf{x}_n, \mathbf{v}^{(\tau)}) \log \pi_{z_n}$ is independent of $\mathbf{v}$.

M step The new parameter estimates $\mathbf{v}^{(\tau+1)}$ are obtained from the old ones $\mathbf{v}^{(\tau)}$ as $\mathbf{v}^{(\tau+1)} = \arg\max_{\mathbf{v}} Q(\mathbf{v}|\mathbf{v}^{(\tau)})$. To perform this maximisation, we equate the gradient of Q with respect to $\mathbf{v}$ to zero:

$$\frac{\partial Q}{\partial \mathbf{v}} = \sum_{n=1}^{N} \sum_{z_n=1}^{M} p(z_n|\mathbf{x}_n, \mathbf{v}^{(\tau)}) \frac{1}{p(\mathbf{x}_n|z_n, \mathbf{v})} \frac{\partial p(\mathbf{x}_n|z_n, \mathbf{v})}{\partial \mathbf{v}} = \mathbf{0}. \tag{5}$$

Solving for $\mathbf{v}$ in eq. (5) results in

$$\mathbf{v}^{(\tau+1)} = \left(\sum_{n=1}^{N} \sum_{z_n=1}^{M} p(z_n|\mathbf{x}_n, \mathbf{v}^{(\tau)}) \mathbf{\Sigma}_{z_n}^{-1}\right)^{-1}$$

$$\times \sum_{n=1}^{N} \sum_{z_n=1}^{M} p(z_n|\mathbf{x}_n, \mathbf{v}^{(\tau)}) \mathbf{\Sigma}_{z_n}^{-1}(\boldsymbol{\mu}_{z_n} - \mathbf{x}_n).$$

If now we choose the data set as simply containing the origin, $\{\mathbf{x}_n\}_{n=1}^{N} = \{\mathbf{0}\}$, rename $z_1 = m$ and omit $\mathbf{x}_1 = \mathbf{0}$ for clarity, we obtain the M step as:

$$\mathbf{v}^{(\tau+1)} = \left(\sum_{m=1}^{M} p(m|\mathbf{v}^{(\tau)}) \mathbf{\Sigma}_m^{-1}\right)^{-1} \sum_{m=1}^{M} p(m|\mathbf{v}^{(\tau)}) \mathbf{\Sigma}_m^{-1} \boldsymbol{\mu}_m \tag{6}$$

which is formally identical to the iterative scheme of eq. (4).

General properties of the EM algorithm for GMs [16,18,17] show that the convergence of (6) is global and linear. Firstly, at every iteration τ, the iterative scheme (6) will either increase or leave unchanged the log-likelihood $\sum_{n=1}^{N} \log p(\mathbf{x}_n|\mathbf{v}) = \log p(\mathbf{0}|\mathbf{v})$ so, correspondingly, the iterative scheme (4) will monotonically increase the density value $p(\mathbf{x})$ or leave it unchanged. Thus, (4) converges from any initial value of $\mathbf{x}$ to a local stationary point of $p(\mathbf{x})$ [17, Th. 3.2]. Although convergence can occur to a saddle point or to a minimum as well as to a mode. Since both saddle points and minima are unstable for maximisation, a small random perturbation will cause the EM algorithm to diverge from them. Thus, practical convergence will almost always be to a mode. Secondly, its convergence rate is linear (first-order) and so is very slow except when the mixture components are very separated, in which case the convergence becomes superlinear.

Note in Fig. 3 the slow crawl along ridges of the density and how the iterates may be attracted to saddle points, to then deviate towards a mode.

The EM view of the fixed-point algorithm should also be applicable to mixtures of other kernels.

4.2   Particular Cases

In the case of isotropic GMs the fixed-point scheme reduces to:

$$\mathbf{x}^{(\tau+1)} = \sum_{m=1}^{M} q(m|\mathbf{x}^{(\tau)})\boldsymbol{\mu}_m \qquad q(m|\mathbf{x}) = \frac{p(m|\mathbf{x})\sigma_m^{-2}}{\sum_{m'=1}^{M} p(m'|\mathbf{x})\sigma_{m'}^{-2}} \qquad (7)$$

where $p(m|\mathbf{x})$ is the posterior probability or responsibility of component m given point $\mathbf{x}$ and the $q(m|\mathbf{x})$ values are the responsibilities $p(m|\mathbf{x})$ reweighted by the inverse variance and renormalised. For homoscedastic GMs, this simplifies even more with $q(m|\mathbf{x}) = p(m|\mathbf{x})$ so that the new point $\mathbf{x}^{(\tau+1)}$ is the conditional mean of the GM under the current point $\mathbf{x}^{(\tau)}$. This is formally akin to clustering by deterministic annealing [19], to algorithms for finding pre-images in kernel-based methods [20] and to mean-shift algorithms (Sect. 5.2).

In both cases, each iterate is a convex linear combination of the centroids, as are the stationary points, and so the sequence lies in the interior of the convex hull of the centroids. In general for finite mixtures of densities from the exponential family, the EM algorithm always stays in the convex hull of a certain set of parameters [18, eq. (5.3)].

4.3   Brute-Force Search for Counterexamples

Whether starting the algorithm from each centroid can indeed find all modes depends on the conjecture. It certainly does not hold in the general case where the covariance matrices are not isotropic and different, since then we can have more modes than centroids (although we may expect the algorithm to find many of the modes). Deriving an efficient algorithm to find all modes for this case is difficult, because we do not even know where to look for the modes: they need not lie inside the convex hull of the centroids, and may lie far away from them.

What happens in the cases where the conjecture may hold? Even if the number of modes is fewer than or equal to the number of components, some modes might conceivably not be reachable from any centroid. Since we can associate almost every point $\mathbf{x} \in \mathbb{R}^D$ with a unique mode (except for saddles, minima and points converging to them), we can define the *basin of attraction* of each mode as the region of $\mathbb{R}^D$ of all points that converge to that mode. The claim that the algorithm finds all modes if started from every centroid is equivalent to the claim that the basin of attraction of every mode contains at least one centroid.

Theoretically, this question seems as difficult as the modes conjecture, so we decided to run a brute-force search to look for counterexamples (we thank Geoff Hinton for suggesting us this idea). We uniformly randomly generated $M = 30$ centroids in the rectangle $[0,1] \times [0, 0.7]$, mixing proportions $\pi_m \in$

$(0, 1)$ and isotropic covariance matrices with $\sigma_m \in [0.05, 0.15]$. Then we run the algorithm starting (a) from every centroid and (b) from every point in a grid of 100×70 of the rectangle. Call π_a and π_b the number of modes found in each case, respectively. We repeated the process 1500 times and considered only those cases where $\pi_a \neq \pi_b$; cases where $\pi_a = \pi_b$ cannot disprove the modes conjecture since by construction $\pi_a \leq M$. A difference $\pi_a \neq \pi_b$ was considered a false alarm if due to a single mode appearing as two or more with a small numerical difference[4]. We found 3 differences in the homoscedastic GM case (all false alarms) and 10 in the isotropic case (7 genuine, 3 false alarms). One of the genuine differences is shown in Fig. 3(right column): note how the green basin of attraction at the top right contains no centroids. The associated mode lies in a very flat area of the density, as indicated by the lack of contours[5]; the same happened in all other cases. We conclude that, in the isotropic case, it is possible (but rare) that a mode may not be reachable from any centroid.

In all our experiments the number of modes found by brute-force search π_b was $\leq M$. This reinforces our belief that the modes conjecture holds, or that if it does not, then it may fail only rarely. The results also show that the algorithm almost always finds all the modes when the component covariances are isotropic, perhaps always when they are equal.

Figure 4 shows that, unlike a Voronoi tessellation, the basins of attraction need be neither convex (left plot) nor connected sets (right plot). The points on the basin boundaries are either saddle points or minima, or converge to a saddle point. Note the following: (1) the basins often have very thin streaks extending for long distances, sandwiched between other basins; (2) one basin can be completely included in another; and (3) some points (typically minima) lie in the boundary of several basins simultaneously. Also, the sharper a mode is (e.g. for high π_m and low σ_m), the smaller its basin is. However, such small-basin modes will not be missed since they will lie near a centroid.

The brute-force search extension to 3D is computationally prohibitive.

5   Applications

The conjecture and mode-finding algorithms are relevant in statistical and machine learning applications such as function approximation, data visualisation, data reconstruction, clustering or image processing. The basic idea is that modes can be associated with important structure in an empirical distribution. We discuss the problems of regression and clustering.

5.1   Multivalued Regression and Data Reconstruction

In traditional nonlinear regression, one wants to derive a (parametric or nonparametric) mapping $\mathbf{y} = \mathbf{f}(\mathbf{x})$ given data pairs $\{(\mathbf{x}_n, \mathbf{y}_n)\}_{n=1}^{N} \subset \mathbb{R}^D \times \mathbb{R}^E$. The

---

[4] The implementation of the algorithm considers that two modes are the same if their distance is less than a user parameter `min_diff` that has a very small value [15]. This helps to remove duplicated modes, but can occasionally fail.

[5] It might be argued that perhaps such modes are not really modes, but lie in the limit of numerical accuracy.

Homoscedastic GM:
$\boldsymbol{\Sigma}_m = \sigma^2 \mathbf{I}$, $m = 1, \ldots, M$

Isotropic GM:
$\boldsymbol{\Sigma}_m = \sigma_m^2 \mathbf{I}$, $m = 1, \ldots, M$

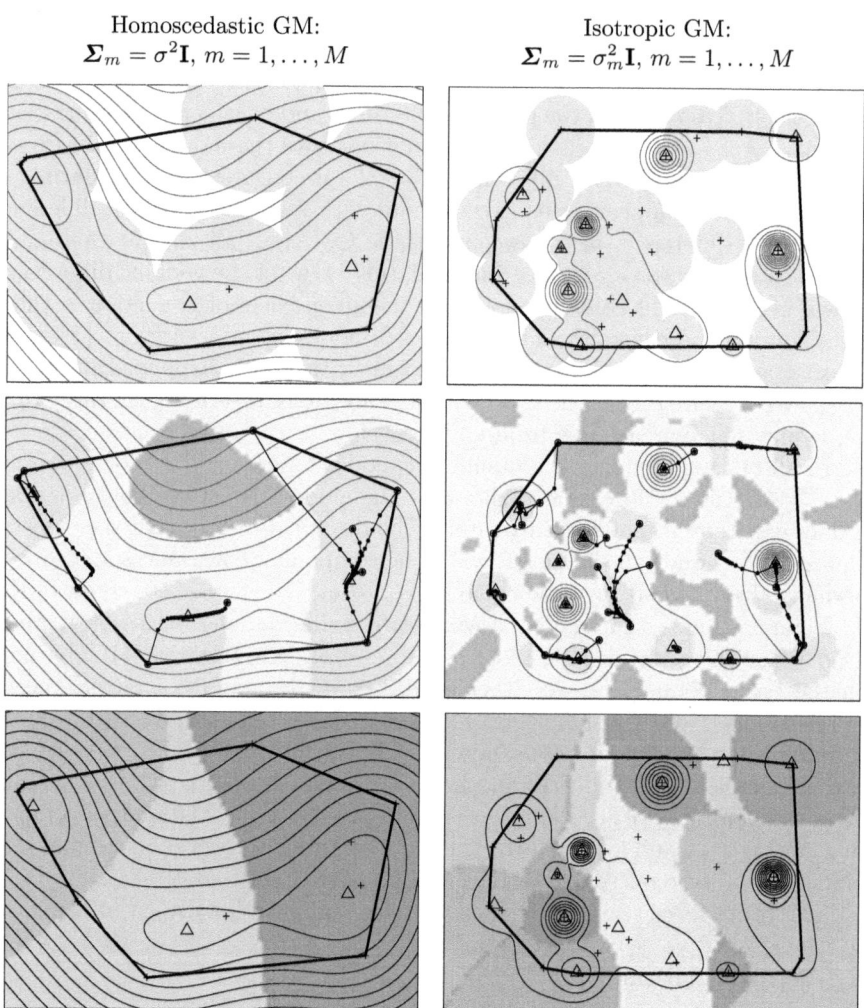

Fig. 3. The fixed-point iterative algorithm for exhaustive mode finding in 2D. The left column shows an example of homoscedastic GM ($\boldsymbol{\Sigma}_m = \sigma^2 \mathbf{I}$) and the right one an example of isotropic GM ($\boldsymbol{\Sigma}_m = \sigma_m^2 \mathbf{I}$). The latter is a very rare case where the algorithm did not find all modes (compare the top and bottom rows: in the top-row plot, a mode is missing at the top right). All parameters $\boldsymbol{\mu}_m$, π_m, σ_m and σ were drawn randomly. The GM modes are marked "$\triangle$" and the GM centroids "+". The thick-line polygon is the convex hull of the centroids. *Top row*: contour plot of the GM density $p(\mathbf{x})$. Each original component is indicated by a grey disk of radius σ or σ_m centred on the corresponding mean vector $\boldsymbol{\mu}_m$ (marked "+"). *Middle row*: plot of the Hessian character (dark colour: positive definite; white: indefinite; light colour: negative definite). The search paths from the centroids are given. *Bottom row*: plot of the basins of attraction of each mode (i.e., the geometric locus of points that converge to each mode). Figures 3 and 4 may require to be viewed in colour to appreciate the different basins.

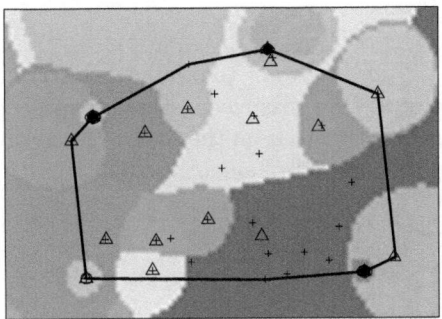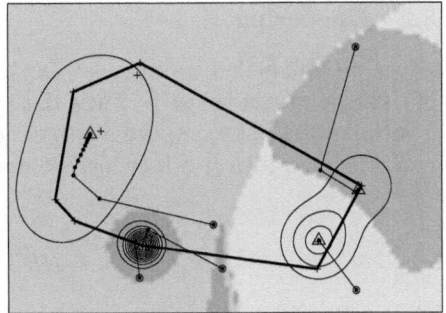

Fig. 4. Basins of attraction of each mode for examples that are heteroscedastic GMs with isotropic components. *Left*: basins may not be convex sets. *Right*: basins may not be connected sets (note the sample search paths).

mapping $\mathbf{f}$ assigns a unique value $\mathbf{y}$ to every input $\mathbf{x}$; often some unimodal noise model is also assumed (e.g. Gaussian noise for a sum-squared error function). However, this is an unreasonable model if $p(\mathbf{y}|\mathbf{x})$ can be multimodal; typically this occurs when inverting a non-injective forward mapping such as $g(x) = x^2$. Representing $p(\mathbf{y}|\mathbf{x})$ as a mixture model has been proposed in a number of contexts. For example it arises with the mixture of experts model [21] (see also the mixture density networks [22, §6.4]), where the number of mixtures is chosen in some fashion. Also, the Nadaraya-Watson estimator [23] gives rise to an N-component mixture for $p(\mathbf{y}|\mathbf{x})$. Carreira-Perpiñán [24] proposed a flexible way to represent multivalued mappings by first estimating a probability density function $p(\mathbf{x}, \mathbf{y})$ for the joint variables from the training data, and then defining a multivalued mapping $\mathbf{y} = \mathbf{f}(\mathbf{x})$ as the collection of modes of the conditional distribution $p(\mathbf{y}|\mathbf{x})$. It is computationally convenient to model $p(\mathbf{x}, \mathbf{y})$ as a homoscedastic GM[6], since then computing $p(\mathbf{y}|\mathbf{x})$ or any other conditional distribution is trivial, and we can use the algorithms of [15] (such as that of Sect. 4.1) to find the modes. Since (ideally at least) *every mode corresponds to a branch of the multivalued mapping and vice versa* it is of interest to locate *all* the modes of the conditional distribution, which leads us to the conjecture.

These ideas can be used to reconstruct missing data in a sequence of vectors $\mathbf{t}_1, \ldots, \mathbf{t}_N$ in a two-step procedure. First, at each vector in the sequence, one finds all the modes of the conditional distribution $p(\mathbf{t}_{n,\mathcal{M}}|\mathbf{t}_{n,\mathcal{P}})$, where $\mathbf{t}_{n,\mathcal{M}}$ (resp. $\mathbf{t}_{n,\mathcal{P}}$) means the missing variables (resp. present) at vector $\mathbf{t}_n$. This gives several candidate reconstructions for each vector. Second, a unique candidate at each n is selected by minimising a continuity constraint (such as the trajectory length) over the whole sequence. This results in a unique reconstruction of the whole sequence. This method was applied [2] to inverse mappings in speech (the acoustic-to-articulatory mapping) and robotics (the inverse kinematics).

---

[6] Or any other model that results in it, such as the generative topographic mapping [25] or kernel density estimation.

5.2   Clustering

Given an unlabelled training set $\{\mathbf{x}_n\}_{n=1}^N \subset \mathbb{R}^D$, we want to obtain a clustering of these points and classify a new data point $\mathbf{x}$. One possible clustering approach is as follows. First, compute a kernel density estimate from the data of kernel K and window width $h > 0$ (which controls the amount of smoothing [23]):

$$p(\mathbf{x}; h) = \frac{1}{Nh^D} \sum_{n=1}^N K\left(\frac{\mathbf{x} - \mathbf{x}_n}{h}\right). \tag{8}$$

Then associate each mode of it with a cluster. If we define an iterative mode seeking algorithm such as gradient ascent or our EM algorithm, then we can assign a new point $\mathbf{x}$ to the mode to which the algorithm converges if started from $\mathbf{x}$. It is of interest to know how many modes exist at a given width h, which brings us to the conjecture when Gaussian kernels are used.

Perhaps the earliest proposal of this approach was the *mean-shift algorithm* of Fukunaga and Hostetler [26], recently extended in [27] and [28]. The mean-shift algorithm was defined as

$$\mathbf{x} \leftarrow \mathbf{m}(\mathbf{x}) = \frac{\sum_{n=1}^N K\left(\frac{\mathbf{x}-\mathbf{x}_n}{h}\right) \mathbf{x}_n}{\sum_{n=1}^N K\left(\frac{\mathbf{x}-\mathbf{x}_n}{h}\right)}$$

where $\mathbf{m}(\mathbf{x}) - \mathbf{x}$ is called the mean shift. The algorithm was derived for the Epanechnikov kernel for computational convenience (since it has finite support) as gradient ascent on $\log p(\mathbf{x})$ with a variable step size; no convergence proof was given. For the Gaussian kernel it coincides with our algorithm for homoscedastic GMs of eq. (7) with $q(m|\mathbf{x}) = p(m|\mathbf{x})$—thus, *the mean-shift algorithm with the Gaussian kernel (and probably other kernels) is an EM algorithm*, which proves it has first-order convergence from any starting point. Comaniciu and Meer [28], in an image segmentation application, gave a different convergence proof for the mean-shift algorithm for certain isotropic kernels (including the Gaussian and Epanechnikov) and noted empirically its slow convergence for the Gaussian kernel. Note that the fact that the clusters defined by mean-shift may not be connected sets (Fig. 4) could be undesirable for some applications. Related clustering methods have been proposed [29,30,9,31]. The mode trajectories in the scale space of h have also been used as a tool for data visualisation [32].

In scale-space clustering the mode-finding algorithms of [15] can also be used in a fast incremental way, where the modes at scale s_1 are found from the modes at scale $s_0 < s_1$ (rather than starting from every centroid). If the number of modes decreases with the scale, this will not miss any mode.

6   Conclusion

We have presented theoretical and experimental evidence for the conjecture that, in any dimension, the number of modes of a Gaussian mixture where all components are isotropic or equal cannot exceed the number of components (and proven it in 1D). It may hold even if Gaussian blurring of a delta mixture can create modes. A possible approach to (dis)prove the conjecture is to particularise Morse theory to Gaussian blurring of delta mixtures. Practically, it seems that the conjecture will hold for almost all isotropic Gaussian mixtures and that hill-climbing algorithms started from each centroid of the mixture will usually find all modes. The conjecture may also typically hold for mixtures of certain non-gaussian kernels even though these are known to create modes upon blurring. Our derivation of the fixed-point iterative algorithm (which can also be seen as a mean-shift algorithm) as an EM algorithm guarantees it has first-order convergence from any starting point.

References

1. Carreira-Perpiñán, M.Á., Williams, C.K.I.:   On the number of modes of a Gaussian mixture.   Technical Report EDI–INF–RR–0159, School of Informatics, University of Edinburgh, UK (2003). Available online at http://www.informatics.ed.ac.uk/publications/report/0159.html.
2. Carreira-Perpiñán, M.Á.: Continuous Latent Variable Models for Dimensionality Reduction and Sequential Data Reconstruction. PhD thesis, Dept. of Computer Science, University of Sheffield, UK (2001)
3. Carreira-Perpiñán, M.Á.:   Mode-finding for mixtures of Gaussian distributions.   Technical Report CS–99–03, Dept. of Computer Science, University of Sheffield, UK (1999), revised August 4, 2000. Available online at http://www.dcs.shef.ac.uk/~miguel/papers/cs-99-03.html.
4. Hinton, G.E.: Training products of experts by minimizing contrastive divergence. Neural Computation **14** (2002) 1771–1800
5. Lindeberg, T.: Scale-Space Theory in Computer Vision. Kluwer Academic Publishers Group, Dordrecht, The Netherlands (1994)
6. Koenderink, J.J.: The structure of images. Biol. Cybern. **50** (1984) 363–370
7. Babaud, J., Witkin, A.P., Baudin, M., Duda, R.O.: Uniqueness of the Gaussian kernel for scale-space filtering. IEEE Trans. on Pattern Anal. and Machine Intel. **8** (1986) 26–33
8. Yuille, A.L., Poggio, T.A.: Scaling theorems for zero crossings. IEEE Trans. on Pattern Anal. and Machine Intel. **8** (1986) 15–25
9. Roberts, S.J.: Parametric and non-parametric unsupervised cluster analysis. Pattern Recognition **30** (1997) 261–272
10. Lifshitz, L.M., Pizer, S.M.: A multiresolution hierarchical approach to image segmentation based on intensity extrema. IEEE Trans. on Pattern Anal. and Machine Intel. **12** (1990) 529–540
11. Kuijper, A., Florack, L.M.J.:   The application of catastrophe theory to image analysis.   Technical Report UU–CS–2001–23, Dept. of Computer Science, Utrecht University (2001). Available online at ftp://ftp.cs.uu.nl/pub/RUU/CS/techreps/CS-2001/2001-23.pdf.
12. Kuijper, A., Florack, L.M.J.: The relevance of non-generic events in scale space models. In Heyden, A., Sparr, G., Nielsen, M., Johansen, P., eds.: Proc. 7th European Conf. Computer Vision (ECCV'02), Copenhagen, Denmark (2002)
13. Damon, J.: Local Morse theory for solutions to the heat equation and Gaussian blurring. J. Diff. Equations **115** (1995) 368–401

14. Silverman, B.W.: Using kernel density estimates to investigate multimodality. Journal of the Royal Statistical Society, B **43** (1981) 97–99
15. Carreira-Perpiñán, M.Á.: Mode-finding for mixtures of Gaussian distributions. IEEE Trans. on Pattern Anal. and Machine Intel. **22** (2000) 1318–1323
16. Dempster, A.P., Laird, N.M., Rubin, D.B.: Maximum likelihood from incomplete data via the *EM* algorithm. Journal of the Royal Statistical Society, B **39** (1977) 1–38
17. McLachlan, G.J., Krishnan, T.: The EM Algorithm and Extensions. Wiley Series in Probability and Mathematical Statistics. John Wiley & Sons (1997)
18. Redner, R.A., Walker, H.F.: Mixture densities, maximum likelihood and the EM algorithm. SIAM Review **26** (1984) 195–239
19. Rose, K.: Deterministic annealing for clustering, compression, classification, regression, and related optimization problems. Proc. IEEE **86** (1998) 2210–2239
20. Schölkopf, B., Mika, S., Burges, C.J.C., Knirsch, P., Müller, K.R., Rätsch, G., Smola, A.: Input space vs. feature space in kernel-based methods. IEEE Trans. Neural Networks **10** (1999) 1000–1017
21. Jacobs, R.A., Jordan, M.I., Nowlan, S.J., Hinton, G.E.: Adaptive mixtures of local experts. Neural Computation **3** (1991) 79–87
22. Bishop, C.M.: Neural Networks for Pattern Recognition. Oxford University Press, New York, Oxford (1995)
23. Silverman, B.W.: Density Estimation for Statistics and Data Analysis. Number 26 in Monographs on Statistics and Applied Probability. Chapman & Hall, London, New York (1986)
24. Carreira-Perpiñán, M.Á.: Reconstruction of sequential data with probabilistic models and continuity constraints. In Solla, S.A., Leen, T.K., Müller, K.R., eds.: Advances in Neural Information Processing Systems. Volume 12, MIT Press, Cambridge, MA (2000) 414–420
25. Bishop, C.M., Svensén, M., Williams, C.K.I.: GTM: The generative topographic mapping. Neural Computation **10** (1998) 215–234
26. Fukunaga, K., Hostetler, L.D.: The estimation of the gradient of a density function, with application in pattern recognition. IEEE Trans. Inf. Theory **IT–21** (1975) 32–40
27. Cheng, Y.: Mean shift, mode seeking, and clustering. IEEE Trans. on Pattern Anal. and Machine Intel. **17** (1995) 790–799
28. Comaniciu, D., Meer, P.: Mean shift: A robust approach toward feature space analysis. IEEE Trans. on Pattern Anal. and Machine Intel. **24** (2002) 603–619
29. Wong, Y.: Clustering data by melting. Neural Computation **5** (1993) 89–104
30. Chakravarthy, S.V., Ghosh, J.: Scale-based clustering using the radial basis function network. IEEE Trans. Neural Networks **7** (1996) 1250–1261
31. Leung, Y., Zhang, J.S., Xu, Z.B.: Clustering by scale-space filtering. IEEE Trans. on Pattern Anal. and Machine Intel. **22** (2000) 1396–1410
32. Minnotte, M.C., Scott, D.W.: The mode tree: A tool for visualization of nonparametric density features. Journal of Computational and Graphical Statistics **2** (1993) 51–68

Fully Automatic Segmentation of MRI Brain Images Using Probabilistic Anisotropic Diffusion and Multi-scale Watersheds

Carl Undeman[1,2] and Tony Lindeberg[1]

[1] Computational Vision and Active Perception Laboratory (CVAP)
Department of Numerical Analysis and Computer Science
KTH, SE-100 44 Stockholm, Sweden
[2] Department of Neuroscience, Division of Human Brain Research
Karolinska Institutet, SE-171 77 Solna, Sweden

Abstract. This article presents a fully automatic method for segmenting the brain from other tissue in a 3-D MR image of the human head. The method is a an extension and combination of previous techniques, and consists of the following processing steps: (i) After an initial intensity normalization, an affine alignment is performed to a standard anatomical space, where the unsegmented image can be compared to a segmented standard brain. (ii) Probabilistic diffusion, guided by probability measures between white matter, grey matter and cerebrospinal fluid, is performed in order to suppress the influence of extra-cerebral tissue. (iii) A multi-scale watershed segmentation step creates a slightly over-segmented image, where the brain contour constitutes a subset of the watershed boundaries. (iv) A segmentation of the over-segmented brain is then selected by using spatial information from the pre-segmented standard brain in combination with additional stages of probabilistic diffusion, morphological operations and thresholding.

The composed algorithm has been evaluated on 50 T1-weighted MR volumes, by visual inspection and by computing quantitative measures of (i) the similarity between the segmented brain and a manual segmentation of the same brain, and (ii) the ratio of the volumetric difference between automatically and manually segmented brains relative to the volume of the manually segmented brain. The mean value of the similarity index was 0.9961 with standard deviation 0.0034 (worst value 0.9813, best 0.9998). The mean percentage volume error was 0.77 % with standard deviation 0.69 % (maximum percentage error 3.81 %, minimum percentage error 0.05 %).

1   Introduction

Segmenting the brain from other tissue in a 3-D MR image of the human head is an important pre-processing stage for many tasks, for example:

L.D. Griffin and M. Lillholm (Eds.): Scale-Space 2003, LNCS 2695, pp. 641–656, 2003.

- when aligning individual brains to a standard anatomical format,
- when delimiting a volume in the brain where statistical measurements of brain activation are to be performed, and
- when computing morphological measures of the shape of the brain.

Until recently, high-quality segmentation was carried out semi-automatically at our laboratory, which implied a large amount of manual intervention. The purpose of this article is to develop a procedure for carrying out this task in a fully automatic manner. This automation step is also important for the development of brain databases containing functional brain activation images as well as procedures for analyzing such data in a fully automated manner.

Despite the fact that the MR imaging technique provides images with high spatial resolution and good soft-tissue contrast, computing a high-quality and fully automatic segmentation of the brain from other tissue is a non-trivial problem. Some reasons why the problem is hard, are imperfections in the data due to electrical and thermal noise, errors in the scanner due to inhomogeneities in the magnetic field, partial volume effects and biological variations between subjects.

To address this problem, we propose a multi-stage solution that combines and extends earlier image processing methods, essentially based on:

- an approximate normalization to standard anatomical format, so that anatomical information from a pre-segmented standard brain can be exploited,
- an anisotropic diffusion algorithm, guided by approximate probabilities for different types of tissue, in order to improve the performance of subsequent multi-scale watershed segmentation and morphological processing.

The integrated algorithm has been evaluated on 50 T1-weighted MR images of the human head, which were also segmented manually (see section 4). In addition, the algorithm has been applied to at least 200 more brain images, with highly successful results.

2   Approaches to Brain Segmentation

To address the brain segmentation problem, several different types of approaches have been considered in the literature. Since the brain consists of a known set of tissue types (mainly white matter, grey matter, bone and cerebrospinal fluid) a natural first approach consists of capturing the distributions of these tissues, and aiming at a *classification from the intensity values*, see for example (Atkins & Mackiewich 1997), who fitted a mixture of Rayleigh distributions to the histogram of the original MR image. The main strength of this method is that it allows for fast and easy implementation. The main weakness is that the distributions of the different tissue types in general overlap, which means that it will not be possible to find thresholds that separate brain and non-brain tissue.

Another approach is to align a pre-segmented template brain to the brain of any individual subject, and thereby propagate the segmentation to this subject.

Such an approach has been used by (Collins et al. 1994), (Dawant et al. 1999) and (Holden et al. 2001) for identifying anatomical substructures, and provides an efficient way to incorporate anatomical information into segmentation algorithms. These works, however, depend on high quality registration between the template brain and the unsegmented brain, which may be hard to obtain.

To detect tissue boundaries from local information, *edge detection* is a natural approach to consider. However, since the contrast may vary substantially in an MR image of a brain, it will in general not be possible to find thresholds and scale levels that lead to connected edges that separate different tissues into connected regions. A more refined approach has been considered by (Zeng et al. 1999), who developed an edge detector aimed at detecting only edges that correspond to boundaries between pre-defined tissue types, e.g. between cerebrospinal fluid and white matter. The method is based on approximating the distribution of tissue types A and B by Gaussian distributions, and then computing the probability $p_A(x)$ that a voxel x belongs to tissue A, and the probability $p_B(N(x))$ that a neighbour $N(x)$ of x belongs to tissue type B. The product $p_A(x)\,p_B(N(x))$ will only assume high values when both the factors are high, and in this way the probability of a border between any two tissue types can be estimated.

Another limitation of edge detection preceded by Gaussian smoothing is that edges with low contrast may be lost during the smoothing operation and that highly curved edges may be rounded. To address these issues, several *anisotropic diffusion schemes* have been developed, where the Gaussian smoothing operator is replaced by an edge preserving anisotropic diffusion step, see e.g. (ter Haar Romeny 1994, Weickert 1998) for overviews and (Gerig et al. 1992) for an application to brain segmentation.

More specific *multi-scale methods* have also been developed. By extending the ideas of multi-scale extrema linking (Koenderink 1984, Lifshitz & Pizer 1990, Lindeberg 1994), (Olsen 1997) developed a *multi-scale watershed approach* based on the definition of sinks in the gradient magnitude map at all scales in a Gaussian scale-space representation of the original image. Provided that an appropriate scale level can be selected, this approach can be expected to give rise to an over-segmentation of the anatomical MR image, where the boundaries of the brain constitute a subset of the boundaries of the watersheds computed from the gradient magnitude image. By linking the watershed segments between different scales, a coarse scale segmentation can be propagated to finer scales by following links over scales. In continued work by (Dam & Nielsen 2000), non-linear diffusion was added as a complement. As far as we know, however, these methods have not been fully automated, and require a certain degree of operator assistance. An interesting alternative in the area of automated watershed approaches has been presented by (Hahn & Peitgen 2000), based on a modified watershed transform.

Linking of image structures over scales has been considered also (Vincken et al. 1997) by establishing *probabilistic links between image structures over scales*, and then propagating a coarse-scale segmentation to finer scales in a probabilistic manner. A combination of probabilistic relations and anisotropic diffusion referred to as *probabilistic diffusion* has been presented by (Arridge & Simmons 1997), based on the idea of using probabilities of different classes to control the conductance in a diffusion equation, as opposed to intensity differences. In close relation to these notions, three-dimensional *Markov random field models* have been developed by e.g. (Rajapakse et al. 1997) and (Held et al. 1997). Moreover, *morphological processing* is used in several works (Atkins & Mackiewich 1997, Lemieux et al. 1999). For further overviews of brain segmentation, see e.g. (Atkins & Mackiewich 1997) and (Hahn & Peitgen 2000).

3   Proposed Method

Based on the abovementioned literature survey, we propose to address the problem of segmenting MR images of human brains using a combination of the following types of methods:

- Empirically, we have found that MR images often contain spurious high values. To reduce the possibly negative influence of these, an *intensity normalization* will be performed as a first step in the proposed algorithm.
- A *spatial normalization* is a natural pre-processing step, since using a pre-segmented standard brain, prior knowledge about the shape of the brain can be effectively represented by transforming the given image into a standard anatomical space.
- We will make use of *probabilistic anisotropic diffusion* guided by edges between different types of tissue as an important pre-processing stage to multi-scale watershed segmentation. Probabilistic anisotropic diffusion may also be used for reducing the negative influence of inhomogeneities in the unsegmented image.
- *Multi-scale watershed segmentation* transforms the image into a collection of volumetric elements. Identifying the volumetric elements that constitute the brain can be accomplished by using prior anatomical knowledge obtained via the spatial normalization step.
- *Post-processing* using a combination of probabilistic anisotropic diffusion, morphological operations and thresholding for computing the final result.

In the following we will describe each method and show that a highly useful method for brain segmentation can be obtained by combining these modules in the proposed novel way.

3.1   Intensity Normalization

To avoid the possibly negative influence of spurious high values in the original MR image, an intensity normalization was performed prior to the affine normalization. This intensity normalization was carried out by estimating a central region of the brain, by computing the weighted center of gravity in the MR image, and computing mean and standard deviation of the intensity values in the central region. A high threshold was chosen two standard deviations above the mean value in the central region, and each voxel having an intensity value above the high threshold was set to this value.

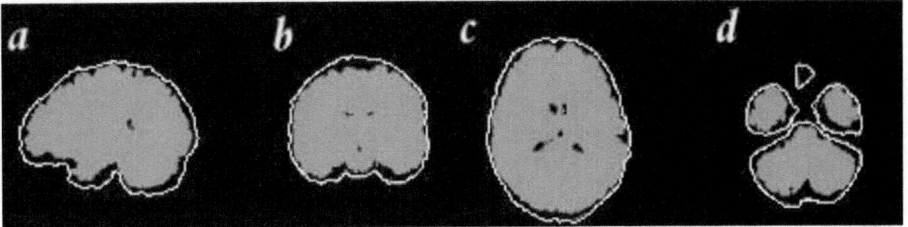

Fig. 1. The contour of the intersection of 50 stripped brains, linearly transformed to the standard brain, compared to the binary standard brain mask. (a) a sagittal slice, (b) a coronal slice (b), (c)–(d) two horizontal views.

3.2   Spatial Normalization by Affine Warping

To transform the brain images to a standard anatomical space, we apply an affine transformation in 3-D which in practice is computed using the AIR package (Woods et al. 1998) with a 12 parameter model and a 6 mm Gaussian FWHM (FWHM $= 2\sigma\sqrt{2\ln 2}$) pre-filtering of both the target and the input image. Figure 1 shows a summary of an evaluation of this processing stage, by computing the intersection of 50 manually segmented brains that were transformed to a standard anatomical space in this way. As can be seen, this processing stage effectively reduces the main variations in brain anatomy between different subjects. Certain individual variations, however, remain, which leads to the need for more refined processing stages that will be described next.

3.3   Capturing the Intensity Distributions of Different Tissues

Due to the alignment of the input image to a standard anatomical brain, we can use a manual pre-segmentation of the standard brain into different tissue types (white matter, grey matter, cerebrospinal fluid and bone) as an initial estimate of the segmentation that is to be computed. In particular, since the volume of the transformed brain ideally will be equal to the volume of the standard brain, we can initially assume that the volume of white matter, grey matter and cerebrospinal fluid is equal in these two brains. Thus, threshold values for each tissue type can be easily computed from the idea of sorting the voxels in the transformed brain image, and assuming that the intensities are ordered $L_{CSF} < L_{GM} < L_{WM} < L_{BONE}$ (for simplicity we denote both bone, eye muscles and vascular tissue by "bone"). If the volume of cerebrospinal fluid is V_{CSF}, L_{CSF} can be determined from the lowest V_{CSF} voxels, etc.

Due to the fact that some background voxels may be (and often are) included in the support region of the pre-segmented brain mask, the lower threshold L_{CSF} can sometimes be close to the intensity of the background. This is however not a problem since we want to treat background and CSF voxels the same way when segmenting the brain from surrounding tissue. Moreover, the threshold between white matter and bone, can sometimes be in the bone intensity region, due to

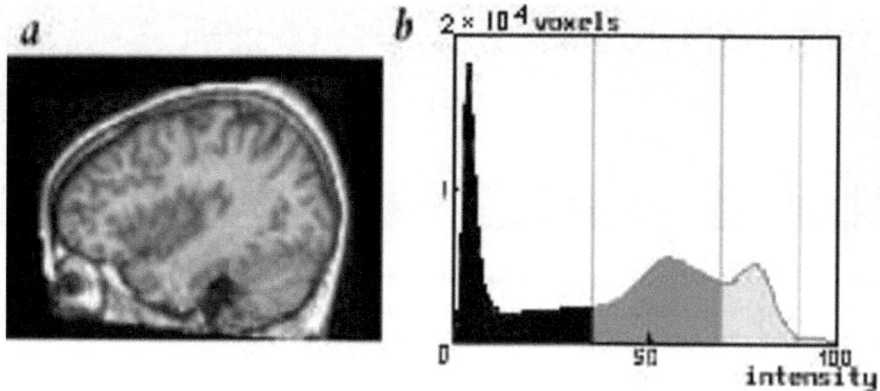

Fig. 2. (a) An MR image of a brain after intensity normalization and scaled to values between 0.0 and 100.0. (b) The computed intensity intervals are marked in the histogram: CSF [0.0-36.2], GM [36.2-69.5], WM [69.5-89.6] and BONE [89.6-100.0].

the fact that some bone voxels are likely to be included in the support region of the pre-segmented standard brain. Thus, the following modification is used in practice: Every voxel *outside* the standard brain mask in the transformed image with an intensity above the mean value of the intensity for white matter is taken to be a bone voxel. The low threshold value for the bone voxels is then computed as the average of the mean value for white matter and the mean value for the bone voxels outside the standard brain mask.

Figure 2 shows a histogram of an MR image with associated intensity intervals for CSF, GM, WM and $BONE$ determined in this way.

3.4   Multi-scale Watershed Segmentation

For segmenting an MR image using edge information, it is natural to compute gradient magnitude maps and to associate a watershed with each local minimum. If the watersheds are computed at a too fine scale, however, there will be a large over-segmentation, while if we choose a too coarse scale the boundaries between the watershed regions will not constitute a superset of the boundaries between the different tissue types. Hence, selecting a proper scale level for computing the gradient magnitude is of crucial importance. The selection of reasonable subsets may be simplified by considering coarse-to-fine propagations of watershed support regions, using criteria based on overlap of support regions, intensity levels and inclusion of local minima. Figure 3 shows a few examples of watersheds, computed according to the method in (Vincent & Soille 1991), propagated in this manner. In general, however, it may still be non-trivial to automatically select which watershed regions should be included in the final segmentation. For this reason, (Olsen 1997) and (Dam & Nielsen 2000) considered semi-automatic watershed segmentation method complemented by operator assistance to obtain reliable segmentations.

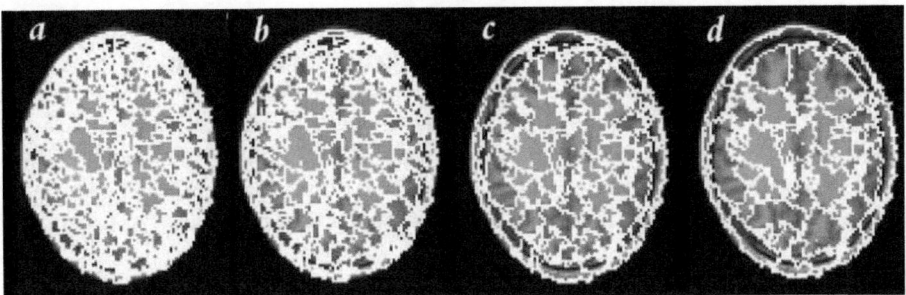

Fig. 3. (a)-(h) Watersheds computed from the gradient magnitude of a brain image smoothed with Gaussian filters of widths: (a) 7 mm, (b) 11 mm, (c) 15 mm, (d) 19 mm propagated down to the 1 mm scale using criteria based on overlap.

During our experiments with multi-scale watershed segmentation some characteristics of the method became apparent:

- Sharp edges in the input image are likely to be present even at coarser scales.
- Thin structures close to the brain surface will be lost at coarser scales. This has the effect that the watersheds sometimes coincide with the brain surface and sometimes with structures just outside the brain.
- Large volumes with slowly varying intensities give rise to large catchment basins.

These observations suggested that a pre-processing of the MR image that transformed the image into an image with slowly varying intensity inside the brain and sharp edges between brain and non-brain tissue, with the fine structures close to the brain surface suppressed would be beneficial to the performance of the multi-scale watershed method.

3.5   Probabilistic Diffusion Using Prior Knowledge

Edge preserving smoothing is a common pre-processing step in applications for brain segmentation. A common approach in this context is to make use of local gradient information, to construct either anisotropic diffusion schemes based on either a scalar conductivity $c(x, y, z, t)$

$$\partial_t L = \frac{1}{2} \nabla^T (c(x, y, z, t) \nabla L) \tag{1}$$

or a conductivity matrix $C(x, y, z, t)$ that allows for different conductivities in different directions

$$\partial_t L = \frac{1}{2} \nabla^T (C(x, y, z, t) \nabla L) \tag{2}$$

In computer vision applications, the motivation for setting the conductivities from local gradient information (Perona & Malik 1990, ter Haar Romeny 1994,

Weickert 1998) originates from the fact that in intensity images the absolute intensity information rarely has meaningful semantic interpretation. In our case, however, the situation is different, since we can associate probability distributions of the different tissue types to the image intensities. To formulate this notion, let us discretize the evolution equations (1) and (2). For both of these types of equations, the result of a spatial semi-discretization only, can be expressed in the form (with slightly different interpretation of $w_{(u,v,w)}(x,y,z)$)

$$\partial_t L(x,y,z) = \sum_{(u,v,w) \in N(x,y,z)} w_{(u,v,w)}(x,y,z)\, (L(u,v,w) - L(x,y,z)) \quad (3)$$

where (u,v,w) represents any neighbour $N(x,y,z)$ of (x,y,z) on the grid that is used, and the weights $w_{(u,v,w)}(x,y,z)$ can be interpreted as local conductivities between any neighbouring points (u,v,w) and (x,y,z). If a boundary preserving operation is desired, the weights should be high at connections where it is regarded as likely that (u,v,w) and (x,y,z) belong to the same type of tissue, while the weight should be low at connections where is is regarded as likely that (u,v,w) and (x,y,z) belong to different types of tissue. In this way, we can aim at a natural unification between the ideas of probabilistic edge detection (Zeng et al. 1999), tensor-based anisotropic diffusion (Nitzberg & Shiota 1992, Lindeberg 1994, Weickert 1998, Almansa & Lindeberg 2000) and a previously proposed notion of probabilistic diffusion (Arridge & Simmons 1997).

Determination of Weights in Probabilistic Anisotropic Diffusion. A problem that remains concerns how to choose the weights $w_{(u,v,w)}(x,y,z)$. One natural approach could be to fit parameterized models to the distributions of the different tissue types as described in section 3.3, and using these distributions to express an iterative scheme within a Bayesian setting. Such an approach would have close similarities to a Markov random field model (Rajapakse et al. 1997, Held et al. 1997). When initiating this work, however, we started by experimenting with empirically determined weights, which turned out to give highly useful results, based on the following ideas:

For the purpose of segmenting the outer boundary of the brain, the boundary between grey matter and cerebrospinal fluid is of major interest. Hence, the intensity ranges corresponding to these types of tissues constitute a main source of information (see also figure 2). With reference to the method described in section 3.3 for classifying voxels into different types of tissue, this method classifies background and cerebrospinal fluid as non-brain tissue (not white matter and not grey matter) with high accuracy. The voxels classified as CSF can therefore be regarded as certain non-brain voxels. Thus, an edge detector for the outer boundary of the brain should give a maximum response when a CSF voxel has a certain GM voxel (or a voxel with a higher value) as a neighbour. Sometimes, however, the intensity interval for GM includes some cerebrospinal fluid. The mean value GM_{mean} of the set of GM voxels is, however, always well inside the true intensity interval for grey matter. Therefore, the probability for an edge between CSF and GM should be low for a voxel with intensity close to the lower threshold GM_{low} of the GM intensity interval, and increase as the voxel intensity approaches the mean value GM_{mean} of grey matter. In practice, we have chosen a linear function to approximate this transition of edge probabilities. Thus, the probability $p_{edge}(L(x,y,z))$ of an edge between cerebrospinal fluid CSF and grey matter GM at a voxel (x,y,z) is approximated by

$$p_{edge} = \begin{cases} 1 & \text{if } L(x,y,z) \in CSF \text{ and } \exists L(N(x,y,z)) \geq GM_{mean} \\ \frac{GM_{mean}-L(x,y,z)}{GM_{mean}-GM_{low}} & \text{if } GM_{low} \leq L(x,y,z) \leq GM_{mean} \\ 0 & \text{otherwise} \end{cases}$$

(4)

Then, in turn the conductivity weights $\omega_{(u,v,w),(x,y,z)}$ as function of the edge probabilities $p_{edge}(u,v,w)$ and $p_{edge}(x,y,z)$ of adjacent image points (u,v,w) and (x,y,z) are determined according to

$$\omega_{(u,v,w),(x,y,z)} = 1 - \mid p_{edge}(u,v,w) - p_{edge}(x,y,z) \mid^{\frac{1}{\alpha}} \tag{5}$$

which gives high conduction only between voxels with similar edge probability.

The motivation for $1/\alpha$ in the expression above is that we want to control the influence of edges of intermediate magnitudes. Empirically, we have found that $\alpha = 4$ to give a probabilistic diffusion scheme with desirable properties.

3.6 Combining Probabilistic Diffusion with Watershed Segmentation

Let us recapitulate from section 3.4 that an ideal input for the multi-scale watershed algorithm would be a brain image with slowly varying intensities inside the brain, sharp edges between brain and non-brain tissue and structures close to the brain surface suppressed. We propose to use probabilistic diffusion according to section 3.5 to achieve these properties. Applied on the unsegmented MR image (Figure 4), probabilistic diffusion guided by the probabilities of the brain edge produces an image with slowly varying intensities inside the brain and sharp edges between brain and non-brain tissue. However, the thin structures just outside the brain surface are still present, making a correct segmentation difficult to obtain. A way to suppress these structures is to not use the unsegmented MR image as initial condition for the probabilistic diffusion, but something that is well inside the brain, typically the white matter. If only the white matter voxels are used as initial condition and all other voxels are set to zero, the intensity of the white matter will flow towards the brain boundaries as new states of the diffusion equation (heat equation) are computed. The high edge probabilities at the brain contour will hinder voxels outside the brain to be reached by the white matter intensity. The voxels outside the brain will therefore in general have lower values than voxels inside the brain, where the white matter intensity is distributed.

Using white matter as initial condition has, however, its problems: (i) Local intensity variations in the MR image may make it difficult to segment the white matter (ii) The presence of white matter is very small in some parts of the

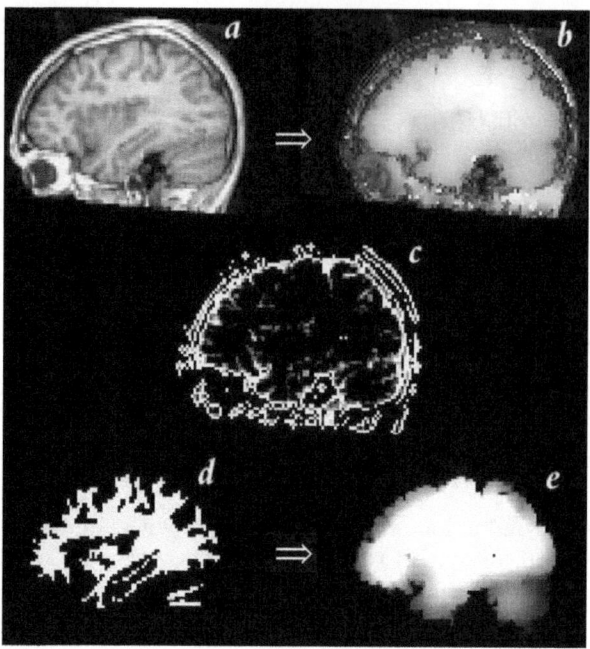

Fig. 4. Original MR image (a) diffused (50 iterations) with edge probabilities (c) produces (b). If instead the white matter (d) is used as initial condition the fine structures outside the brain are suppressed (e).

brain, e.g. the temporal lobes and the cerebellum which has the effect that it takes many time steps for the intensity to reach the lower parts of the temporal lobes and cerebellum. A side effect of many iterations is that intensity will begin to leak out from other parts of the brain where the amount of white matter is large and the estimated edge probabilities at the brain contour is less than one. Thus, a better initial condition than just the white matter is desirable.

In short, such an initial condition can be computed by using probabilistic diffusion in an iterative manner. First, white matter is used as an initial condition to the diffusion equation iterated 50 time steps. In the resulting image, the mean value is computed from all the voxel sites occupied by GM voxels in the original MR image (transformed to standard space). All voxels above this mean value in the diffused white matter image are then saved as a binary mask that will be a slightly undersized approximation of the brain. This mask can in turn be used as initial condition to the probabilistic diffusion under the restriction that the heat transportation only is allowed from "hotter" to "colder" voxels. After 50 time steps the voxels with an intensity larger than 10 percent of the intensity of the voxels in the input mask will be an excellent initial condition to the probabilistic diffusion. (This procedure is derived in an empirical way and is described in full detail in (Undeman 2001). See also (Atkins & Mackiewich 1997) for another way of segmenting white matter.)

After 50 time steps the initial condition computed as described above will produce an image with slowly varying intensity inside the brain and sharp edges between brain and non-brain tissue, with the fine structures close to the brain surface suppressed, i.e ideal input for the multi-scale watershed segmentation.

The diffused image is then filtered with Gaussian FWHM-filters of sizes ranging from 1 mm to 19 mm with a 2 mm increment between each scale. At each scale, catchment basins are computed from the gradient magnitude image of the filtered image. The 19 mm scale catchment basins are then propagated down to the 1 mm scale using a linking scheme based on maximum overlap of support regions. The selection of catchment basins that constitutes the brain is then done by multiplying the percentage overlap of each basin with the pre-segmented standard brain and the percentage of the voxels in the basin that is classified as GM or WM. If the product is more than 0.25 (that is 0.5 * 0.5), the catchment basin is regarded as a part of the brain.

When inspecting the result from the multi-scale watershed segmentation it turns out that a good estimate of the true brain mask is obtained. The algorithm works particularly well where the cerebrospinal fluid is clearly present just outside the brain contour (almost everywhere). In some parts, however, where the partial volume effect causes the border between CSF and GM to be diffuse, the computed brain mask is sometimes a bit oversized due to leakage (caused by low edge probabilities) in the probabilistic diffusion step. This problem is most common close to the sagittal sinus. To reduce these minor errors the following post-processing algorithms were implemented.

3.7   Post-processing

Morphological Operations in Combination with Probabilistic Diffusion and Thresholding. When looking at the results from the multi-scale watershed segmentation we observed that where the algorithm has included thin structures just outside the brain, there is often a weak edge coinciding with the correct brain contour very close to the incorrect edge suggested by the watershed segmentation (Fig 5.c). If the contour computed by the watershed segmentation (Fig 5.b) is used as input to the probabilistic diffusion guided by the edge image, with the restriction that no heat transfer is allowed outside the mask computed by the watershed segmentation (Fig 5.a), the transportation of heat away from the regions where the watershed segmentation has made mistakes will be lower than in the correct regions due to the presence of the correct edge (although weak). This will result in an image where the misclassified thin structures close to the brain surface will have higher intensity than the regions where the watershed segmentation has made no error (Fig 5.d). These incorrect regions can be thresholded away. In the first post-processing step of the suggested algorithm, all voxels on the border of the brain mask computed by the watershed segmentation are set to one. These voxels are used as input to the probabilistic diffusion equation, applied 60 steps. Then, a threshold of 0.33 removes the incorrect voxels (Fig 5.e) from the brain mask.

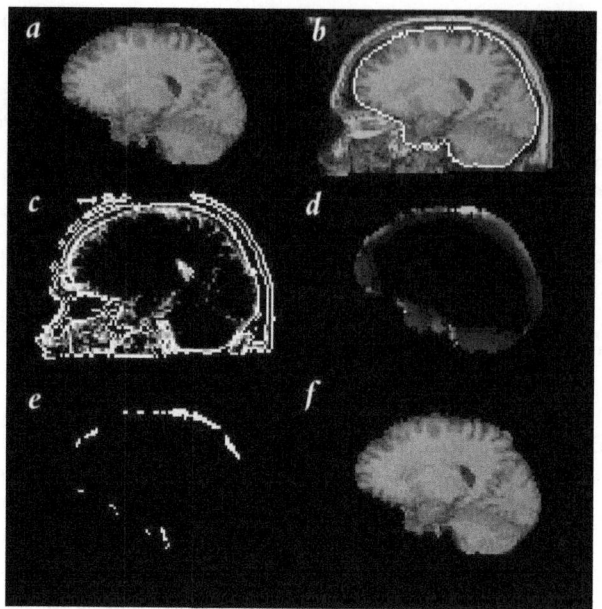

Fig. 5. Sometimes the watershed segmentation leaves non-brain tissue (a) in the mask. By using the outline (b) of the mask computed by the watershed segmentation as input to probabilistic diffusion guided by (c), an image where the misclassified voxels have high intensity (d) can be generated. The image in (d) is thresholded, leaving the voxels shown in (e), which are used for removing some misclassified voxels from (a), producing (f).

Morphological Operations in Combination with Region Growing. A T1-weighted brain image in general has lower intensities at the brain contour than inside the brain. This observation can be used for reducing the amount of misclassified voxels using the boundary of the computed brain mask in combination with a simple region growing technique. All voxels on the border of the brain mask obtained from the watershed segmentation are used as a seed to a region growing step. The seed is allowed to grow in every direction where there is a neighbouring voxel with lower intensity. When the growing is finished all voxels included in the growing are removed from the mask obtained from the watershed segmentation.

4   Experimental Results

To evaluate the proposed method, we will in this section present the results of applying it to 50 T1-weighted MR images of the human head, acquired from a GE Signa 1.5 T scanner. An illustration of one volume in this data set is shown in figure 6, where the difference between manual and automatic segmentation can be seen. For display purposes, a set of representative slices have been selected.

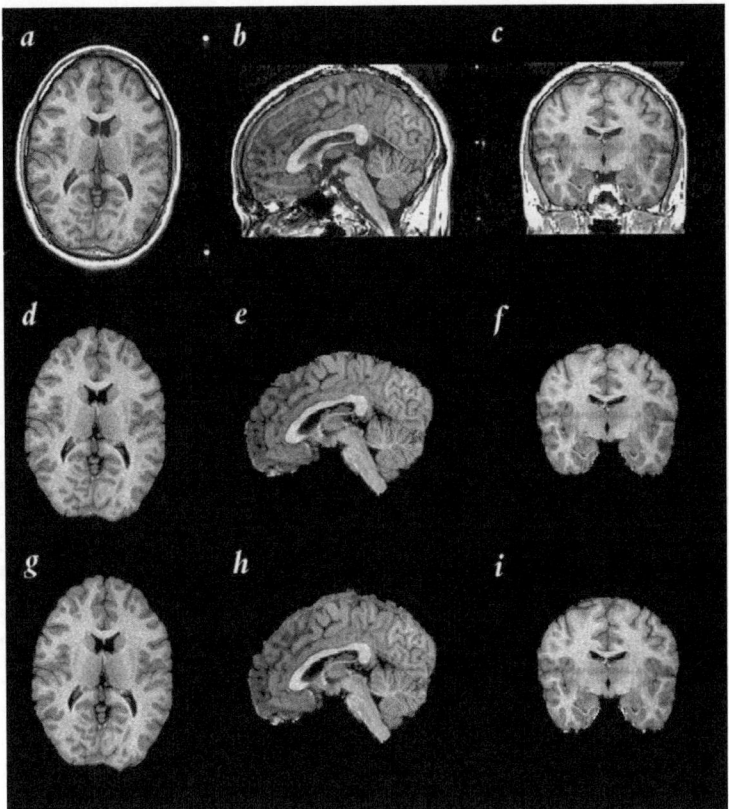

Fig. 6. Illustration of an average result: (a)-(c) The unsegmented brain. (d)-(f) The manually segmented brain. (g)-(i) The automatically segmented brain.

4.1   Qualitative Evaluation

Criteria for a successful segmentation are that (i) no brain tissue should be removed and that (ii) the amount of non-brain tissue left should be so small that further analysis will not be influenced in a negative manner by remaining non-brain tissue. To give a preview of the results, the method performs well in about 99 % of the volume. In some brains, however, the method has problems. These problems usually occur in regions close to the pituary gland, basilar artery, the sagittal sinus and the internal carotid artery. The most common mistake made by the proposed algorithm is to include small parts of the sagittal sinus.

4.2   Quantitative Performance Measures

To measure the quality of the computed segmentation in a quantitative way, an approach similar to the procedure in (Atkins & Mackiewich 1997) was applied:

- A set of manually segmented brains was generated by correcting any errors in the automatic segmentations.
- A similarity index was computed between the manually corrected segmentations and the corresponding automatically segmented brains.
- For each brain, the difference between the volume of the automatically segmented brain and the manually corrected segmentation was computed.

The following quantitative performance measures were used:

- *Similarity index:* Consider a binary segmentation as a set A containing the voxels that are considered to belong to the segmentation. The similarity index of two segmentations A_1 and A_2 is given by

$$S = 2 * \frac{|A_1 \cap A_2|}{|A_1| + |A_2|} \tag{6}$$

where $A_1 \cap A_2$ denotes the intersection of two sets A_1 and A_2, and $|A|$ denotes the volume of any set A. The algorithm described in (Atkins & Mackiewich 1997), which is claimed to compare very favorably with other methods, gives similarity indices that in general are above 0.95 and at 0.99 at its best.
- *Percentage volume change error:* This entity measures the relative size of the difference between the two segmentations in relation to the manually segmented brain:

$$P = 100 * \frac{|A_1 - A_2|}{|A_2|} \tag{7}$$

where $|A_1 - A_2|$ denotes the absolute value of the volume difference between the sets A_1 and A_2, and $|A|$ is the volume of any set A. In (Atkins & Mackiewich 1997) this measure is within 4% in most cases.

Our method gave a mean similarity index of 0.9961 (standard deviation 0.0034) for the 50 automatically segmented brains. The worst brain had a similarity index of 0.9813, the best 0.9998. Our method gave at worst a percentage volume change error of 3,81 % and 0.049 % at its best. 47 of the 50 segmented brains had a percentage error of less than 2.0 %, and 38 of the 50 segmented brains had a percentage error of less than 1.0 %. The mean percentage error was 0.77 % (standard deviation 0.69 %).

While our performance measures are, in general, better than those reported by (Atkins & Mackiewich 1997), we cannot jump to the conclusion that the proposed method by necessity is better, since the results are based on different data sets. The combination of qualitative and quantitative results, however, strongly suggests that the method satisfactorily solves the task it was designed for – to automatically segment brains in functional brain databases.

5   Summary and Discussion

We have presented an integrated method for brain segmentation, based on the primary components of (i) anatomical normalization, (ii) probabilistic diffusion, (iii) multi-scale watershed analysis and (iv) morphological processing.

Anatomical standardization as a pre-processing stage allows us to explore anatomical prior knowledge and to estimate the distributions of the different types of tissue in the skull. Probabilistic diffusion, in turn, allows us to express a tensor-like diffusion scheme, that instead of local gradient information allows us to make use of probabilities of different tissue types to control an anisotropic diffusion schemes. We propose that these two mechanisms are natural components to consider for future developers of brain segmentation systems and that probabilistic diffusion *per se* warrants further study due to its efficiency in specific situations.

Using these components as primary tools, an integrated brain segmentation system has been developed and evaluated. The qualitative properties of the results and the quantitative performance values are highly satisfactory. To state firm conclusions of the performance relative to other works, however, full availability of the original image data and the original software is needed.

Concerning suggestions for further work, a natural extension consists of expressing a Bayesian derivation of the conduction coefficients $w_{(u,v,w)}(x,y,z)$ in the discrete diffusion equation (3) and to estimate the probabilities more accurately from the image data. Another natural extension is to refine the watershed module, e.g. based on the ideas presented by (Hahn & Peitgen 2000). Thus, a natural next step is to investigate if the composed scheme can be simplified by such extensions. Nonwithstanding these possibilities for additional improvements, it should be emphasized that besides the detailed quantitative evaluation presented in section 4, the method has been applied to at least 200 more brain images, with highly successful results.

Acknowledgements

We gratefully acknowledge the support from the EU project NeuroGenerator, the Swedish Research Council for Engineering Sciences, the Royal Swedish Academy of Sciences and the Knut and Alice Wallenberg Foundation.

References

Almansa, A. & Lindeberg, T. (2000), 'Fingerprint enhancement by shape adaptation of scale-space operators with automatic scale-selection', *IEEE Transactions on Image Processing* **9**(12), 2027–2042.

Arridge, S. R. & Simmons, A. (1997), Multi-spectral probabilistic diffusion using Bayesian classification, *Proc. Scale-Space'97*, Springer LNCS vol 1252, 224–235.

Atkins, M. S. & Mackiewich, B. T. (1997), 'Fully automatic segmentation of the brain in MRI', *IEEE Transactions on Medical Imaging* **16**, 41–54.

Collins D.L., Peters T.M., Dai W. & Evans A.cC (1994), 'Model based Segmentation of Individual Brain Structures from MRI Data', *SPIE Visualization in Biomedical Computing* **vol. 1808**, 10–19.

Dam, E. & Nielsen, M. (2000), Non-linear diffusion for interactive multi-scale watershed segmentation, *Proc. MICCAI'00*, Springer LNCS vol 1935, pp. 216–225.

Dawant B. M., Hartman S. L., Thirion J.-P., Maes F., Vandermeulen D. & Demaerel P. (1999), 'Automatic 3-D segmentation of internal structures of the head in MR images using a combination of similarity and free-form transformations', *IEEE Transactions on Medical Imaging* **18**(10), 909–916.

Gerig, G., Kübler, O., Kikinis, R. & Jolesz, F. A. (1992), 'Nonlinear anisotropic filtering of MRI data', *IEEE Transactions on Medical Imaging* **11**(2), 221–232.

Hahn, H. K. & Peitgen, H.-O. (2000), The skulle stripping problem in MRI solved by a single 3D watershed transform, *MICCAI'00*, Springer LNCS vol 1935, 134–144.

Holden, M., Schnabel J.A. & Hill D. (2001), Quantifying Small Changes in Brain Ventricular Volume Using Non-rigid Registration, *MICCAI'01*, 49–56.

Held, K., Kops, E. R., Krause, B. J., Wells, W. M., Kikinis, R. & Mueller-Gaertner, H.-W. (1997), 'Markov random field segmentation of brain MR images', *IEEE Transactions on Medical Imaging* **16**(6), 878–886.

Koenderink, J. J. (1984), 'The structure of images', *Biological Cybernetics* **50**, 363–370.

Lemieux, L., Hagemann, G., Krakow, K. & Woermann, F. (1999), 'Fast, accurate, and reproducible automatic segmentation of the brain in T1-weighted volume MRI data', *Magnetic Resonance in Medicine* **42**, 127–135.

Lifshitz, L. & Pizer, S. (1990), 'A multiresolution hierarchical approach to image segmentation based on intensity extrema', *IEEE-PAMI* **12**(6), 529–541.

Lindeberg, T. (1994), *Scale-Space Theory in Computer Vision*, Kluwer.

Nitzberg, M. & Shiota, T. (1992), 'Non-linear image filtering with edge and corner enhancement', *IEEE Trans. Pattern Analysis and Machine Intell.* **14**(8), 826–833.

Olsen, O. F. (1997), Multi-scale watershed segmentation, *in* J. Sporring et al, eds, 'Gaussian Scale-Space Theory', Kluwers, 191–200.

Perona, P. & Malik, J. (1990), 'Scale-space and edge detection using anisotropic diffusion', *IEEE Trans. Pattern Analysis and Machine Intell.* **12**(7), 629–639.

Rajapakse, J., Giedd, J. & Rapoport, J. (1997), 'Statistical approach to segmentation of single-channel cerebral MR images', *IEEE Tran. Med. Im.* **16**(2), 176–186.

ter Haar Romeny, B., ed. (1994), *Geometry-Driven Diffusion in Computer Vision*.

Undeman, C. (2001), Fully automatic segmentation of MRI brain images using probabilistic diffusion and a watershed scale-space approach, MSc thesis TRITA-NA-E0161, KTH, Stockholm, Sweden.

Vincent & Soille (1991), 'Watersheds in Digital Spaces: An Efficient Algorithm Based on Immersion Simulations', *IEEE Trans. Pat. Anal. Mac. Intell.* **13**(6), 589–598.

Vincken, K., Koster, A. & Viergever, M. (1997), 'Probabilistic multiscale image segmentation', *IEEE Trans. Pattern Analysis and Machine Intell.* **19**(2), 109–120.

Weickert, J. (1998), *Anisotropic Diffusion in Image Processing*, Teubner-Verlag.

Woods, R. P., Grafton, S. T., Holmes, C. J., Cherry, S. R. & Mazziotta, J. (1998), 'Automated image registration: I. General methods and intrasubject, intramodality validation', *Journal of Computer Assisted Tomography* **22**, 141–154.

Zeng, X., Staib, L. H., Schultz, R. T. & S.Duncan, J. (1999), 'Segmentation and measurement of the cortex from 3-D MR images using coupled-surfaces propagation', *IEEE Transactions on Medical Imaging* **18**(10), 927–937.

Error-Bounds on Curvature Estimation

Sven Utcke

Universität Hamburg, Fachbereich Informatik, Arbeitsbereich Kognitive Systeme
Vogt-Kölln-Str. 30, 22527 Hamburg, Germany
utcke@informatik.uni-hamburg.de
http://kogs-www.informatik.uni-hamburg.de/~utcke

Abstract. Estimation of a digital curve's curvature at any given point is needed for many tasks in computer vision, be it differential invariants or curvature scale space. However, curvature estimation is known to be very susceptible to noise on the contour. We shall show how noise on the contour affects the relative accuracy of the curvature computation. One interesting result is that, contrary to intuition, the accurate calculation of the curvature for low-curvature regions is in fact impossible for common image-sizes, while reasonable results may under favourable conditions be obtained for higher-curvature regions.

1 Introduction

In the early years of computer vision, the need to calculate a digital curve's curvature used to arise quite frequently. However, this did turn out to be a rather harder task than one originally appreciated (see [1] which compares several different approaches), and today the knowledge about the difficulty of the task has become so ingrained in the computer vision community that curvature is hardly ever used outside the scale space community, where much of the work is reduced to locating the zero crossings of curvature [2].

We believe, however, that only little attention has so far been given to the question *why* the calculation of curvature is hard. Worring's approach [1] is only relevant for edgels on a pixel-grid; Kovalevsky [3] recently extended this to edgels with sub-pixel accuracy, using his own edge-finder. However, both authors basically follow the assumption that curvature for low-curvature regions can be calculated more accurately than for high-curvature regions, since for the former smoothing will be less invasive for the latter, and as a consequence concentrate on high-curvature regions with a radius of curvature below 40 pxl.

This paper strives to demonstrate that for real images the curvature of low-curvature regions is in fact just as difficult to compute. This is due to the limited size of outlines in real images, and we shall see that neither very low nor very high curvature can be estimated reliably under real life conditions.

This paper is organised as follows: Section 2 gives the analytic derivation of the expected relative error for a very simple case, this serves mainly to give a feeling for the mathematics involved. Section 3 gives results for a more involved, but also more realistic, case, but without the analytic derivation. Based on these,

L.D. Griffin and M. Lillholm (Eds.): Scale-Space 2003, LNCS 2695, pp. 657–666, 2003.

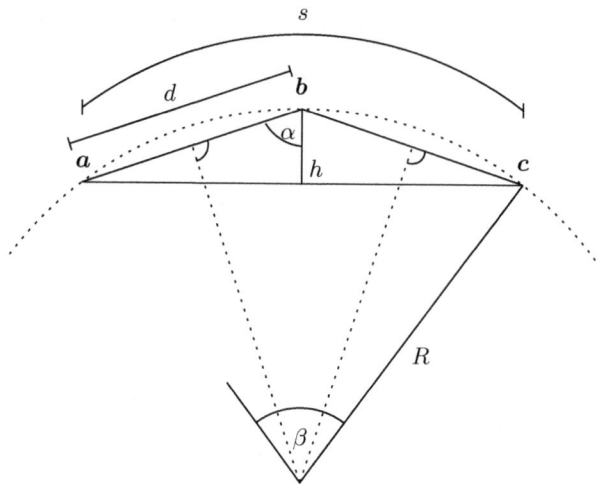

Fig. 1. Simple model for the calculation of curvature

Sec. 4 gives numeric examples of what can be expected from locally operating approaches for the calculation of the curvature, Sec. 5 discusses some additional aspects of curvature calculation, and Sec. 6 summarises our findings.

2   Analytic Derivation of a Simple Case

We will first analyse a very simple case. Given 3 points $\{a, b, c\}$ on a circle, we can calculate the circle's midpoint as the intersection of the two perpendicular bisectors of the 2 lines connecting point a with b and b with c; the circle's radius is then straightforwardly given as the distance from the circle's midpoint to any of the three points on the circle. Figure 1 gives an overview of the construction.

The main source of error using this setup is the uncertainty in the position of the three points. For a digital contour given as edgel-chains, only the edgels' uncertainty perpendicular to the curve is of interest, while the exact location of the edgels along the curve is of no significance.

In order to simplify the example, we will assume that the position of the middle edgel varies with standard deviation σ along the vertical-direction, and that the two other edgels vary with standard deviation σ orthogonal to the line-segments ab and bc (so that d is constant – this only introduces a small error, which will later be ignored anyway). In addition, we will introduce a further simplification by assuming symmetry between the two sides of the construction (this introduces a much bigger error, but gives us a lower bound for the error and a much simpler model). The above simplifications allow us to replace the individual uncertainties in the positions of the edgels by a single uncertainty of the height h of the triangle connecting the three edgels, it is

$$\sigma_h = \sigma\sqrt{1 + \sin^2(\alpha)} \ . \tag{1}$$

In doing so we are projecting the edgels' uncertainty onto the vertical axis.

Using similarity triangles, we can calculate the radius R as

$$R = \frac{d^2}{2h} \ .$$ (2)

So how does the uncertainty in h propagate to R? Using simple linear error propagation we can approximate the standard deviation of the radius σ_R by multiplying the covariance-matrix of all independent variables – in this case this is simply the scalar σ_h^2 – on both sides with the Jacobian of the dependent variables with respect to all independent variables – in this case this is simply the derivative of R with respect to h,

$$\frac{\partial R}{\partial h} = -\frac{d^2}{2h^2} \ .$$ (3)

We therefore get the approximation

$$\sigma_R \approx \sqrt{\frac{d^2}{2h^2} \cdot \sigma_h^2 \cdot \frac{d^2}{2h^2}} = \sigma_h \frac{d^2}{2h^2} \ ,$$ (4)

for the standard deviation of the radius.

We are, however, interested in the relative accuracy of the radius as a function of the arc-length s between the outer edgels. From Fig. 1 we see that

$$\alpha = \frac{\pi}{2} - \frac{\beta}{4}$$ (5)

$$\beta = \frac{s}{R}$$ (6)

$$h = d\cos(\alpha)$$ (7)

and substituting (1) as well as (5), (6), and (7) in (4) leads to

$$\sigma_R = \frac{\sigma}{2} \frac{\sqrt{1 + \sin^2(\alpha)}}{\cos^2(\alpha)} = \frac{\sigma}{2} \frac{\sqrt{1 + \cos^2\left(\frac{s}{4R}\right)}}{\sin^2\left(\frac{s}{4R}\right)} \ .$$ (8)

In order for a result R to be usable we would require its relative error to remain sufficiently small, i. e.

$$\frac{\sigma_R}{R} = \rho \ll 1 \ .$$ (9)

Note that, as we only analyse the *relative* error ρ, it does of course not make any difference whether we calculate the radius R or the curvature $\kappa = 1/R$.

We will now calculate the minimum arc-length s necessary to satisfy (9), and from there, assuming a known average distance from edgel to edgel (set to 1 pxl in the following for simplicity), the number of edgels necessary to achieve a desired relative accuracy $\rho = \sigma_R/R$. As

$$0 \le \cos^2\left(\frac{s}{4R}\right) \le 1$$ (10)

we get from (8)

$$4R\arcsin\left(\sqrt{\frac{\sigma}{2R\rho}}\right) \leq s \leq 4R\arcsin\left(\sqrt{\frac{\sigma}{\sqrt{2}R\rho}}\right) \tag{11}$$

and for $\sigma \ll R\rho = \sigma_R$, which is almost always the case

$$\frac{4}{\sqrt{2}}\sqrt{\frac{\sigma}{\rho}R} \leq s \leq \frac{4}{2^{1/4}}\sqrt{\frac{\sigma}{\rho}R} \ . \tag{12}$$

Note that the left and right side of (12) differ only by a factor of $2^{1/4} \approx 1.19$; in both cases the minimum required arc-length s for a required relative accuracy $\rho = \sigma_R/R$ is proportional to the same term and approximately

$$s \approx 3\left(\frac{\sigma}{\rho}R\right)^{0.5} \ . \tag{13}$$

The entire model is of course an oversimplification: we only use 3 out of N edgels, the end-edgels' covariance should be taken in radial direction (i. e. orthogonal to the contour), we assume that the distributions of the two end-edgels are not independent of each other (which they are). We also assume that curvature is constant over the entire section of the curve under consideration (which most likely it isn't). However, we do feel that the above model is well suited to give an initial idea of the behaviour of curvature calculation.

In the next section we will analyse a more realistic approach which does away with all of the above simplifications except that of constant curvature, in particular using all edgels along s.

3   Using N Edgels

When looking for a description of a curve-segment, instead of calculating this description directly from the edgels it is not uncommon to fit an algebraic curve to the edgels and use the parameters of the curve with the smallest error of fit to calculate the description. These curves are often polynomials in x and y, and in our particular application, where we are trying to find the radius of curvature of a given curve, it is a natural choice to fit part of a circle to the edgels[1].

Different methods exist to calculate the error of fit between a curve and N edgels. Probably the most intuitive method is to minimise the sum of orthographic, squared distances, which for a circle with midpoint (x_0, y_0) and radius R would be

$$C = \frac{1}{N}\sum_{i=1}^{N}\left(\sqrt{(x_i - x_0)^2 + (y_i - y_0)^2} - R\right)^2 \ . \tag{14}$$

However, in general no closed form solution exists for (14), and it has therefore become customary to use the so called algebraic distance

$$C = \frac{1}{N}\sum_{i=1}^{N}\left((x_i - x_0)^2 + (y_i - y_0)^2 - R^2\right)^2 \ , \tag{15}$$

---

[1] This is especially the case as we assume a curve with constant curvature – in general more complicated models will yield better results, although one should be wary of too complex models [4].

for which a closed form solution exists [5]. It is worth noting that minimising either (14) or (15) in their above form will introduce an additional bias into the solution, which we will ignore in the following.

We now once again want to calculate the uncertainty of the radius calculated by minimising (14) or (15) and from there the number of edgels N (or the length s) of the segment to which we need to fit the circle in order to achieve a given relative error $\rho = \sigma_R/R$. The uncertainty σ_R could, in theory, be calculated in direct analogy with the last section as the matrix-product of the edgels' covariance matrix, multiplied on both sides with the Jacobian of R with respect to all edgels, $\boldsymbol{J}_R$. However, since no explicit solution for R exists, at least not for (14), we need a way to calculate the Jacobian without explicit knowledge about R as a function of the edgels $\boldsymbol{e} = (x_1, \ldots, x_N, y_1, \ldots, y_N)^T$. The tool to do so is the implicit function theorem, it is

$$
\boldsymbol{J}_R = -\left(\frac{\partial^2 C}{\partial R^2}\right)^{-1} \left(\frac{\partial^2 C}{\partial R \partial e}\right)^T \Bigg|_{R_{\min}}
\tag{16}
$$

provided that the Hessian $\frac{\partial^2 C}{\partial R^2}$ is indeed invertible at the point of the solution $R_{\min}$. If we further assume that all edgels $(x_i, y_i)^T$ share the same covariance $\sigma \boldsymbol{I}_2$, where $\boldsymbol{I}_2$ is the 2×2 identity matrix – this is a reasonable model for variance σ perpendicular to the curve – we can calculate the variance of R as

$$
\boldsymbol{J}_R\; \sigma \boldsymbol{I}_{2N}\; \boldsymbol{J}_R^T = \sigma \left(\frac{\partial^2 C}{\partial R^2}\right)^{-1} \left(\frac{\partial^2 C}{\partial R \partial e}\right)^T \left(\frac{\partial^2 C}{\partial R \partial e}\right) \left(\frac{\partial^2 C}{\partial R^2}\right)^{-1^T} \Bigg|_{R_{\min}} .
\tag{17}
$$

Evaluating (17) analytically, a purely mechanical process, leads to rather lengthy (although for (15) still quite manageable) equations outside the scope of this paper. A numerical evaluation is, however, perfectly possible, and we can then fit a function to the numerical results; using the squared orthographic distance (14) the output of these calculations is well approximated by a function

$$
N \approx 2\left(\frac{\sigma}{\rho} R\right)^{0.4},
\tag{18}
$$

while for the algebraic distance (15) a good approximation is

$$
N \approx 3.7\left(\frac{\sigma}{\rho} R\right)^{0.4}.
\tag{19}
$$

This is very similar in structure to (13), but with an exponent of 0.4 rather than 0.5. We will see in the next section what this means for practical applications.

4   Numeric Examples

In Sec. 2 and 3 we have calculated the minimum number of edgels required to compute a radius of curvature R with given relative accuracy $\rho = \sigma_R/R$,

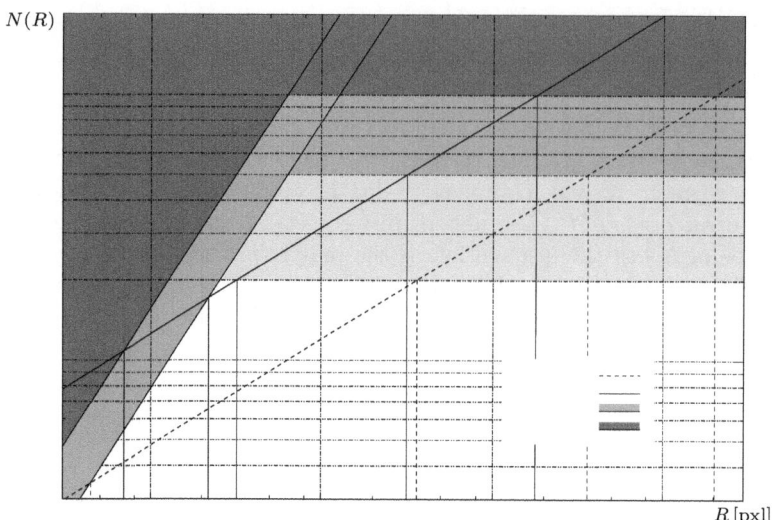

Fig. 2. Number of edgels $N(R, \rho, \sigma = 0.3\,\mathrm{pxl})$ needed to calculate the radius of curvature R with a relative error of less than $\rho = 33\%$ (*dashed line*) or $\rho = 3\%$ (*solid line*). Inside the *grayed out* regions R can not even be determined with this accuracy if we assume a maximum arc $\beta = \pi/4$ ($\pi/2$) and a maximum number of edgels $N = 20$ (50, 100)

compare (13), (18), and (19). It is now instructive to see what actual values for s or N are required when dealing with real images. For this it is important to know with what subpixel-precision we can expect to locate edges in real images. As cited in [6] a basic implementation of the Canny algorithm [7] will typically produce edgels with a standard derivation perpendicular to the edge from somewhere about $\sigma \approx 0.1\,\mathrm{pxl}$ to $\sigma \approx 0.3\,\mathrm{pxl}$ depending on the actual CCD (3-chip RGB versus BW versus Bayer RGB) and lens used as well as the contrast between regions. As most affordable colour cameras use only one CCD with a Bayer-filter and often rather cheap lenses, $\sigma = 0.3\,\mathrm{pxl}$ is usually all we can hope for. Together with NTSC or, at best, PAL sized images as they are still predominant in computer vision, we are therefore hampered both by low image quality and small images.

Figure 2 shows how many edgels $N(R)$ are needed to compute the curvature (radius) with a relative accuracy of both $\rho = \sigma_R/R = 3\%$ (*solid line*) and $\rho = 33\%$ (*dashed line*), using (18). It is worth remembering that the latter corresponds to a worst-case error of about $3\rho \approx 100\%$, which means that not even the sign of curvature can be calculated reliably! But even accepting such an error we see from Fig. 2 that for a radius of approx. $R = 400\,\mathrm{pxl}$ we already need to fit a circle to $N = 20$ edgels, by many authors considered the maximum region size which is still sensible to use [3]. The much more realistic (though still big) relative error of $\rho = 3\%$ would, for a radius of about $R = 400\,\mathrm{pxl}$ already require a fit to $N > 50$ edgels, and even with $N \approx 100$ edgels could we only calculate radii up to $R \approx 2000$ pxl. Similar limits exist for high-curvature regions, in order to calculate a radius of $R = 8$ pxl with a relative error of $\rho = 3\%$ we need $N \approx 12$ edgels, that's nearly one quarter of the full circle.

So why is this need for $N > 20$ (or even $N > 100$) a problem? After all, a PAL sized image comes with 768 pxl × 576 pxl, so 100 pxl isn't really that big. Or is it? In order to understand the problem it is important to realise that so far we have calculated the curvature of something we *knew* to be a circle, i. e. a contour with constant curvature. In real-life, however, we will usually encounter curves with non-constant curvature, and it is important to realise that for such a contour, what we really calculate is not the curvature at the particular point under consideration, but some sort of average curvature. Assuming noise-free data, and as long as the real curvature in the interval under consideration is a monotonic function of the arc-length, we can at least guarantee that the curvature thus calculated would be correct for some point inside the interval – although almost certainly not for the point under consideration. This is called localisation uncertainty in [3]. However, as soon as the curvature over arc-length ceases to be a monotonic function (i. e. contains extrema), arbitrary results are possible even in the case of noise-free data.

From the above it is immediately clear that the region of N edgels from which curvature is calculated should not cross extrema of curvature (whose position we only know after the fact); to keep the localisation error small it is also desirable to keep N small enough so that curvature inside the interval doesn't vary too much. And the fact that the entire object, usually containing several extrema of curvature, has to fit into a PAL-sized (768 pxl × 576 pxl) image, severely limits the maximum usable number of edgels, [3] cites values around 4 pxl–20 pxl. This is the factor we are most interested in. By contrast, most authors so far have been interested in high-curvature regions [1,3], where visible arc-length is the criterion; as a rule of thumb most authors would limit the arc under consideration to somewhere between $\beta \leq \pi/4$ and $\beta \leq \pi/2$.

In Fig. 2 regions with an arc $\beta > \pi/4$ and $\beta > \pi/2$ as well as $N > 20$, $N > 50$, and $N > 100$ have been grayed-out; only the remaining region (with white background) is suitable for curvature calculation, which clearly limits both the maximum, but also the minimum curvature which can be calculated with a given relative error ρ.

Figure 3 shows an exemplary calculation of curvature for a contour with $\sigma \approx 0.3$ and $N \in \{20, 50, 100\}$. We see that with $N = 20$ we get reasonable (although still quite noisy) results for 20 pxl $< R <$ 80 pxl. For smaller radii the calculated results are biased (due to extrema of curvature within the fitting-window) and for bigger radii results are so noisy as to be useless. Using more edgels clearly removes some of that noise, but at the same time introduces a bias into the result which can easily reach 100% of the true result or more (note that the scale is logarithmic). Choosing a different scale for different regions on the contour is an obvious way to improve the results, but note Fig. 3 (*bottom left*), where for radii of around $R = 20$ pxl and $N = 50$ we already observe a noticeable bias, while curvature estimates are still quite noisy.

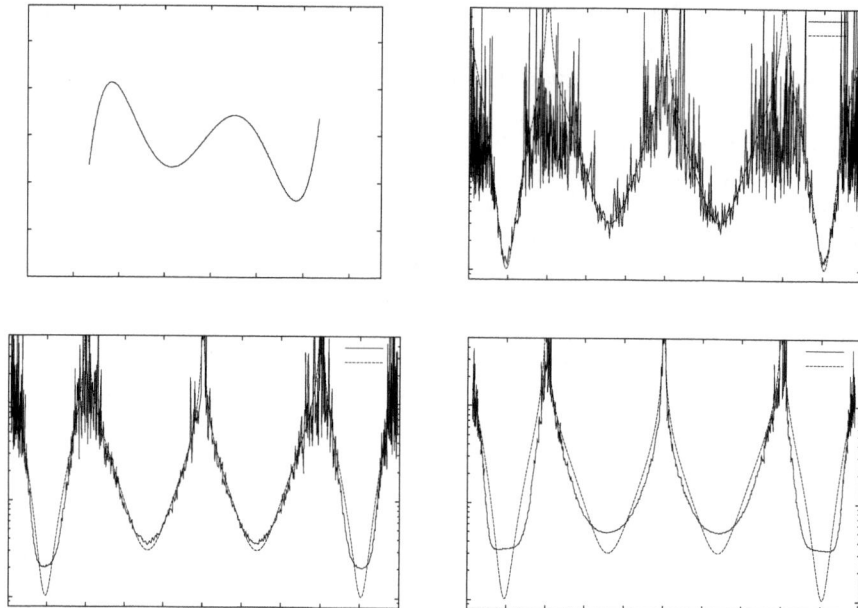

Fig. 3. Gradient calculated for a contour with $\sigma \approx 0.3$ (*top left*) and $N = 20$ (*top right*), $N = 50$ (*bottom left*), and $N = 100$ (*bottom right*)

5   Discussion

In this paper we presented an estimation of the minimum number of edgels required in order to calculate the curvature (-radius) with a given relative error $\rho = \sigma_R/R$. In our calculations we assumed the most simple case possible, that of a contour with constant curvature (essentially part of a circle) and except for the standard deviation of the edgel-positions ignored all additional error-sources (such as the systematic error inherent in (14) and (15)). Even under those optimal conditions did we find that curvature for low curvature regions is essentially impossible to calculate accurately – for real-life, non-constant curve segments the error will increase, due to location inaccuracy for contour segments with non-monotonic curvature (which introduces another systematic error) and the possible presence of curvature extrema within the segment, in which case arbitrary (and arbitrarily wrong) results are possible.

We can of course increase the location accuracy (i. e. the likelihood that the curvature estimated belongs to a point near the one for which we estimated it) by fitting a sufficiently complicated model to the curve; even just fitting a general conic cross-section rather than a circle would improve localisation. It should however be clear that the curves in Fig. 2 already constitute a best-case estimate – curvature calculated from real, non-circular data will always be less correct unless an *exact* model of the curve is known.

So is there no hope at all? An intuitive reaction would be to try and somehow "smooth" the contour – however, there is no better smoothing algorithm available than to try and fit a circle to data which we *know* to be circular. Any other smoothing algorithm can, with the same number of edgels N, only produce inferior results to such a fit. As long as we are calculating curvature based solely on edgel position, we can not get better than (13), (18), and (19)[2]. This means that in order to increase the accuracy of our calculations, we can either decrease σ, or increase the resolution of the image, thus increasing the number of edgels N on any given curve-segment, which would increase the maximum radius of curvature.

Decreasing σ, i. e. increasing the accuracy with which edgels can be located, unfortunately only offers limited room for improvements. Canny's algorithm is already quite good at this, and even algorithms which were specifically tuned to improved location accuracy[8] promise little better than $\sigma = 0.1$ pxl. This basically leaves higher resolution as the only way out, and thankfully modern digital cameras offer just that. Using a 6 Mega-pixel camera (i. e. increasing the resolution by a factor of approx. 3.7) means that only about $3.7^{-0.6} \approx 45\,\%$ of the same curve still needs to be visible when compared to PAL, as the number of edgels on any given piece of curve increases, while at the same time their variance decreases.

Finally, I would like to comment on the use of discreet edgels (but with arbitrary coordinates) throughout this paper (rather than pretending that we are dealing with a continuous contour). The reason for this is of course that with real images discreet edgels are all we know how to compute. Using something like the Canny algorithm [7] these edgels are, however, approximately uniformly distributed along the contour (this is different from [3]), and since we are fitting a circle rather than calculating curvature directly from the edgel positions any irregularities would not have much of an effect. However, if you like to pretend that you are dealing with a continuous contour the same arguments still apply; in this case the location of the entire contour is uncertain, you can visualise the contour's error-distribution as that of a 1D-Gaussian distribution swept along the curve in such a way that it is orthogonal to the curve in each point. The radii of the set of circles fully contained within the 1-σ-band around the contour are approximately bounded by a 1-σ_R region around the median value of R, and the same goes for the k-σ-band. This way it might actually be easier to visualise what is happening.

6   Conclusion

In this paper, we have shown that using locally operating methods the curvature of both high-curvature as well as low-curvature regions can not be calculated reliably from standard PAL-sized images even under optimal conditions and

---

[2] Strictly speaking, this is only true for sufficiently small values of ρ, since we used linear error propagation throughout this paper. For larger values only qualitative statements are possible.

extending the meaning of "locally operating" to rather large regions; only for a possibly quite small range of medium-small curvatures is it at all possible to calculate a meaningful result. Increasing the resolution of the image by a factor u (linearly) will increase the maximum radius of curvature which can be calculated reliably (by $u^{0.6}$), but this does not eliminate the underlying problem.

As did [1,3], this paper only considered local algorithms for the calculation of curvature. It is possible that more global methods (gray-level- or gradient-based, spline-fit, or scale-space based methods, where the biased curvature-estimates in Fig. 3 are somehow propagated back to their true values) could provide additional accuracy. However, this has not been analysed.

Acknowledgements

I wish to thank the anonymous reviewers, especially reviewer two, and in particular Lewis D. Griffin for thankful comments and suggestions.

References

1. Worring, M., Smeulders, A.W.M.: Digital curvature estimation. Computer Vision, Graphics and Image Processing: Image Understanding **58** (1993) 366–382
2. Mokhtarian, F., Mackworth, A.K.: A theory of multiscale, curvature-based shape representation for planar curves. IEEE Transactions on Pattern Analysis and Machine Intelligence **14** (1992) 789–805
3. Kovalevsky, V.: Curvature in digital 2d images. International Journal of Pattern Recognition and Artificial Intelligence **15** (2001) 1183–1200
4. Kanatani, K.: Geometric information criterion for model selection. International Journal of Computer Vision **26** (1998) 171–189
5. Pratt, V.: Direct least-squares fitting of algebraic surfaces. Computer Graphics **21** (1987) 145–152
6. Utcke, S.: Grouping based on projective geometry constraints and uncertainty. In: Proceedings of the Sixth International Conference on Computer Vision, Bombay, IEEE Computer Society, Narosa Publishing House, New Delhi (1998) 739–746
7. Canny, J.F.: A computational approach to edge detection. IEEE Transactions on Pattern Analysis and Machine Intelligence **8** (1986) 679–698
8. Overington, I.: Computer Vision: A unified, biologically-inspired approach. Elsevier, Amsterdam (1992)

Multiresolution Approach
to Biomedical Image Segmentation
with Statistical Models of Appearance

Špela Iveković[1] and Aleš Leonardis[2]

[1] Dep. of Electrical, Electronic and Computer Eng., Heriot-Watt University
Edinburgh EH14 4AS, United Kingdom
si1@hw.ac.uk

[2] Faculty of Computer and Information Science, University of Ljubljana
SI-1001 Ljubljana, Slovenia
ales.leonardis@fri.uni-lj.si

Abstract. Structural variability present in biomedical images is known to aggravate the segmentation process. Statistical models of appearance proved successful in exploiting the structural variability information in the learning set to segment a previously unseen medical image more reliably. In this paper we show that biomedical image segmentation with statistical models of appearance can be improved in terms of accuracy and efficiency by a multiresolution approach. We outline two different multiresolution approaches. The first demonstrates a straightforward extension of the original statistical model and uses a pyramid of statistical models to segment the input image on various resolution levels. The second applies the idea of direct coefficient propagation through the Gaussian image pyramid and uses only one statistical model to perform the multiresolution segmentation in a much simpler manner. Experimental results illustrate the scale of improvement achieved by using the multiresolution approaches described. Possible further improvements are discussed at the end.

1   Introduction

Biomedical images typically contain information about human body and its health condition. Aging, illnesses, injuries, and similar factors induce the *biological variability* of human beings. In the process of biomedical image acquisition there are several variable factors such as patient position and equipment properties which result in so called *technical variability*. Both biological and technical variability reflect in the acquired biomedical image and their volatile nature aggravates the segmentation process.

Statistical model of appearance [5,6] successfully models the structural variability present in biomedical images. The model's training process is based on a training set of images, representing the texture of the structure in question, and the training set of structure's shapes, provided by medical experts' manual annotation of training images. The variability information is then extracted from

L.D. Griffin and M. Lillholm (Eds.): Scale-Space 2003, LNCS 2695, pp. 667–682, 2003.

both training sets by means of PCA analysis. After the training phase, statistical model has the ability to segment a structure which potentially belongs to the family of structures seen in the training set but exercises variability different from any of the structures actually present in the training set.

The statistical model on its own successfully deals with variability of objects in the images. However, by applying a multiresolution approach to the basic statistical model image segmentation, computational complexity can be reduced and quality of the results improved.

We applied two different multiresolution approaches to the problem of medical image segmentation. The first approach uses a Gaussian pyramid of statistical models. An independent model is trained for each of the pyramid levels. During the segmentation process the resulting shape and pose estimate of the structure in question are passed down the pyramid with each next model improving the result's accuracy. The second approach uses only one statistical model of appearance. The model is trained with the original training set. After the training phase a Gaussian pyramid of the model's eigenvectors is constructed. Consequently, the same model can perform the segmentation of the input image on various resolution levels.

The paper is organized as follows: Section 2 describes the related work. In section 3, we outline the basic concepts of image segmentation with statistical models of appearance. A detailed description of each of the applied multiresolution approaches is presented in section 4. Section 5 elaborates on our experimental results and section 6 concludes with a summary and discussion.

2   Related Work

Medical image segmentation [7] presents a demanding task for segmentation methods. Structures present in medical images exercise large variability in shape and texture. Consequently, the most popular methods used in medical image segmentation area are the *semiautomatic* methods, also called *user-steered* methods [8,9,12,13], where the expert-operator initializes the segmentation process and then directs the segmentation algorithm towards the correct solution.

Although the semiautomatic methods are satisfyingly accurate, the drawback is their speed and the fact that an expert should be present during the segmentation process to guide it towards the correct solution. To avoid the need for operator-guided segmentation and instead fully automatize the segmentation process, medical image segmentation with deformable models [4,11] has been introduced. Deformable models maintain the essential characteristics of a class of structures they represent but can also deform to fit a range of examples. A particular representative of the deformable model class is the statistical model of appearance [5,6]. It is possible to reliably segment medical images with the statistical model of appearance. However, speed and accuracy of the segmentation performance can be further improved by applying the multiresolution approach [1,10]. Yoshimura and Kanade [14] apply the idea of multiresolution approach to eigenimages and suggest the eigenspace construction on each resolution level.

Leonardis and Bischof [3] explore the possibility of building the eigenspace on the highest resolution only and afterwards adjusting it to the necessary resolution levels. In this paper, we further explore the idea of one eigenspace for all resolution levels and apply it to the segmentation with statistical models of appearance [6].

3   Image Segmentation with Statistical Models of Appearance

In order to synthesize a complete appearance of an image structure, we must model both its shape and texture [6]. For this purpose, statistical model of appearance consists of two submodels - statistical model of shape maintaining the structure's shape variation behaviour, and statistical model of texture maintaining the structure's texture variation behaviour. Both submodels cooperate with each other during the segmentation process. In the sequel, we provide a general description of the statistical model, describe its training process, and present the application of the model to image segmentation. Description applies to both, statistical model of shape and statistical model of texture, providing that the training set consists of shape vectors and texture vectors respectively.

3.1   Statistical Model – General Description

Statistical model is a deformable model which in a compact way describes the information contained in the training set data. During the training process modes of variability are extracted from the training set data and presented as model's parameter space base vectors. When performing the segmentation of a previously unseen image, the mode of variation of the new image structure is expected to comply to a great extent with the modes described by the base vectors. Consequently, if the new image structure belongs to the class of structures present in the training set data, it can be accurately described as a linear combination of model's parameter space base vectors. In the opposite case the structure is assumed not to belong to the same class as those in the learning set.

3.2   Training the Statistical Model

We start with a training set of data vectors $x_1, x_2, ..., x_s$ which exercise as many different modes of structure variability as possible. We first calculate the mean data vector

$$\overline{x} = \frac{1}{s} \sum_{i=1}^{s} x_i, \tag{1}$$

then we compute the data covariance matrix C

$$C = \frac{1}{s-1} \sum_{i=1}^{s} (x_i - \overline{x})(x_i - \overline{x})^T. \tag{2}$$

In the next step we use the SVD to compute the eigenvectors e_i and corresponding eigenvalues λ_i of the matrix C (sorted so that $\lambda_i \geq \lambda_{i-1}$). When we sort the eigenvectors according to their corresponding eigenvalues we get an ordered orthogonal base of the model's parametric space. The directions of eigenvectors represent the directions of training data variability. The larger the corresponding eigenvalue, the larger the data variability along that eigenvector direction. It is sufficient to keep only the first t, $t < s$, eigenvectors to accurately describe training data variability. Number t can be calculated as

$$\sum_{i=1}^{t} \lambda_i \geq f_v V_T, \tag{3}$$

where $V_T = \sum_{i=1}^{s} \lambda_i$ is the total variance of the training data and f_v is the proportion of variance we wish to represent with t eigenvectors (e.g. $f_v = 0.98$). If the columns of the matrix Θ contain the t eigenvectors, corresponding to the t largest eigenvalues, $\Theta = (e_1|e_2|...|e_t)$, we can then approximate any data vector x of the training set using

$$x \approx \overline{x} + \Theta p, \tag{4}$$

where p is a t-dimensional vector given by

$$p = \Theta^T (x - \overline{x}). \tag{5}$$

The approximation sign $\approx$ in the equation (4) is due to the fact that we are only using the first t eigenvectors instead of all s eigenvectors available as a result of SVD. The elements of the vector p are called *parameter space coefficients*. If we use the equation (5) for a specific input data vector x, we transform the input data vector from the reference space, where it exists as a texture, shape, etc., into a point in the statistical model's t-dimensional parameter space. The elements of the vector p are the coordinate values of this point along the model's parameter space base vectors. Parameter space representation of the input vector x is useful when trying out different configurations of the input data vector. With the help of equation (4) varying the elements of p results in varying the data vector x in the reference space.

3.3   Image Segmentation with the Statistical Model

Segmentation of the input image with the statistical model of appearance is in effect an optimization procedure. Segmentation result is obtained by *minimizing* the energy function for the values of the parameter space coefficients p_i of the model. The energy function is defined as a weighted sum of *internal* and *external* energy

$$E = \alpha E_{int} + (1 - \alpha) E_{ext}. \tag{6}$$

The *internal* energy measures the deformation of the model and is defined as a weighted sum of model's parameter space coefficient values:

$$E_{int}(p) = \sum_{i=1}^{t} u(p_i). \tag{7}$$

As mentioned before, variation of parameter space coefficient values results in the change of the reference space data vector. We refer to this change as *deformation*. The larger the absolute values of coefficients, the more expressed the deformation of data vector. The weight function $u(x)$ imposes a penalty on configurations of coefficient values which result in a large deformation and in this way limits the extent of deformation tried out during the optimization procedure.

The *external* energy measures the congruity of the texture of the currently segmented image area with the model's ability to reconstruct this same texture. It is defined as the sum of squared differences between the corresponding pixels of segmented texture and of reconstructed texture:

$$E_{ext}(M(\boldsymbol{p}), I) = \sum_{i=1}^{n} (x_i^{seg} - x_i^{rec})^2, \tag{8}$$

where n stands for the length of the vectors representing both textures. The bigger the similarity between segmented and reconstructed texture, the lower the external energy.

The idea behind the definition of the external energy is the following: statistical submodel of texture represents typical texture variation modes of the texture training set. If the texture of the currently segmented area belongs to the same family of structures as those in the training set, it should be possible to accurately represent this texture with the linear combination of the statistical submodel of texture's base vectors. In other words, if the modes of variability of the currently segmented texture comply with those of the training set, we can sequentially apply the equations (4) and (5) to the currently segmented texture input vector and the recovered texture which results from the equation (5) applied after (4) should match closely with the input texture used in equation (4). A mismatch of the two textures is an obvious sign for the segmentation procedure that it is not segmenting the right area of the image and should thus either move elsewhere around the image or change the shape of the captured area and so capture a slightly different texture.

The importance of both energies, internal and external, can be influenced through the weight parameter α. By giving a higher significance to the internal energy, the segmentation process will avoid large deformations and instead put more effort in discovering the right position in the image (capture the right texture area). On the other hand, given a higher significance to the external energy, the model will tend to look for a better solution by trying out different deformations rather than looking for a better position in the image.

4   Multiresolution Approach

The general idea behind the multiresolution approach [1,10] is to construct a Gaussian input image pyramid in order to perform coarse to fine segmentation of the input image. The original definition of the statistical model of appearance deals only with segmentation of the input image on the original resolution level. In order to perform multiresolution segmentation, some adjustments of the original model have to be made. In the next two sections we describe two different ways in which we adjusted the original statistical model of appearance to perform the segmentation on different resolution levels of the input image.

4.1 First Multiresolution Approach – Gaussian Pyramid of Statistical Models

The first multiresolution approach is a straightforward extension of the original statistical model. Original statistical model is only trained on the original resolution training set and therefore performs the input image segmentation on the original resolution level only. In the multiresolution extension of this approach, we construct a Gaussian pyramid of the learning set vectors and train an independent statistical model on every level of the pyramid. The Gaussian input image pyramid also has to be constructed with the input image pyramid resolution levels corresponding to those in the learning set pyramid.

The statistical model trained on the lowest resolution training set (i.e. the highest level of the Gaussian pyramid) begins the segmentation. Since the resolution is expected to be rather low at the highest level of the pyramid, the task of the first model merely concentrates on finding the best initial guess about the possible position of the image structure that the segmentation is trying to extract from the image.

Once the first model is finished it passes the result down the Gaussian pyramid to the second model, which accepts this result as its initial estimate. The second model is performing the segmentation on the same input image but on a higher resolution level, consequently having more image details at its disposal. The greater level of detail enables the second model to improve the initial estimate received from the first model. Since the second model already has the information about the possible initial position of the image structure in question, it can concentrate on improving the quality of segmentation and doesn't spend a lot of time searching through the image for possible structure positions. This fact results in significant reductions of time needed to perform a successful segmentation.

The result obtained by the second model is then passed to the third model and so forth, eventually arriving down to the lowest level of the pyramid (i.e. the highest resolution level). The last model in the pyramid performs the segmentation of the original resolution input image and, already having a significant amount of information available about the pose and shape of the segmented structure, merely performs the possible improvements in defining the shape of the structure.

There is one more issue that we have to consider when performing the multiresolution segmentation by using a pyramid of statistical models. As we mentioned before, the models are trained independently, each on its own training set. Consequently a transformation is needed to pass the result obtained by one model as an initial estimate to the other model. By training two statistical models independently of each other with two different training sets, each of the models will have its own eigenvectors defining the base of its parameter space. Therefore, the result obtained by the first model in form of parameter space coefficients cannot be passed directly into the parameter space of the second model but must instead undergo the transformation described in the sequel.

Transformation. The transformation description depends on the actual nature of coefficients (i.e. the result of segmentation) that are being passed down the pyramid from one model to the other. In our case the coefficients passed down the pyramid are shape coefficients complemented with the information about the position of the shape in the image. Texture coefficients are obtained during the energy evaluation part of the optimization and don't need to be explicitly passed down the pyramid. Adjustment of the position information to a higher resolution level is straightforward. We will describe the shape coefficient transformation in more detail. Shape is defined as a collection of n two-dimensional image points:

$$(x_1, y_1)(x_2, y_2)(x_3, y_3)...(x_n, y_n) \tag{9}$$

In the reference space shape can be represented as vector of $2n$ image point coordinates:

$$\boldsymbol{x}_{sh} = [x_1, x_2, ..., x_n, y_1, y_2, ..., y_n] \tag{10}$$

As we mentioned, the following equation is used to transform the data vector from the reference space into the corresponding coefficients in the parameter space:

$$\boldsymbol{p}_{sh} = \Theta^T(\boldsymbol{x}_{sh} - \overline{\boldsymbol{x}}_{sh}), \tag{11}$$

and to transform the coefficients back into the reference space data vector:

$$\boldsymbol{x}_{sh} \approx \overline{\boldsymbol{x}}_{sh} + \Theta\boldsymbol{p}_{sh}, \tag{12}$$

where Θ stands for the matrix of eigenvectors, $\overline{\boldsymbol{x}}_{sh}$ for the average shape vector, and $\boldsymbol{p}_{sh}$ for the vector of shape coefficients.

As mentioned above, the consequence of the independently trained models is the incompatibility of their parameter spaces. However, both models interpret the reference space information in the same way, the only difference being the resolution on which the model was initially trained. We base our transformation on this reference space compatibility of the models.

Let's call the eigenvector matrices of the first and the second model Ψ and Φ, respectively, and assume that the first model obtained the resulting shape coefficients $\boldsymbol{p}_{sh}^1$. In order to pass this result to the second model down the pyramid, we first transform the coefficients into the shape vector in the reference space of the first model:

$$\boldsymbol{x}_{sh}^1 \approx \overline{\boldsymbol{x}}_{sh}^1 + \Psi\boldsymbol{p}_{sh}^1. \tag{13}$$

The next step is to adjust the shape vector to the reference space of the second model by means of upsampling. If the resolution of the first model is n-times smaller than the resolution of the second model, we get:

$$x_{sh_i}^2 = nx_{sh_i}^1, \tag{14}$$

$$y_{sh_i}^2 = ny_{sh_i}^1. \tag{15}$$

The remaining bit of the transformation is now to transform the new shape vector

$$\boldsymbol{x}_{sh}^2 = [x_{sh_1}^2, x_{sh_2}^2, ..., x_{sh_n}^2, y_{sh_1}^2, y_{sh_2}^2, ..., y_{sh_n}^2] \tag{16}$$

into the parameter space of the second statistical model using the equation:

$$\boldsymbol{p}_{sh}^2 = \Phi^T (\boldsymbol{x}_{sh}^2 - \overline{\boldsymbol{x}}_{sh}^2). \tag{17}$$

Vector of shape coefficient values $\boldsymbol{p}_{sh}^2$ can now be used as an initial estimate for the second statistical model.

4.2 Second Multiresolution Approach – Gaussian Pyramid within the Statistical Model

The first approach in a straightforward manner upgrades the basic statistical model. However, the straightforward upgrade involves at least two redundant steps. First redundancy is the separate training of an independent statistical model for each resolution level. Consequently, the second redundancy is the transformation needed to pass the result down the pyramid due to incompatibility of the models' parameter spaces.

Our goal in the second multiresolution approach was to replace the training of several independent statistical models with only one model handling all resolution levels of the input image and being trained only once on the original resolution training set. Training only one single statistical model simplifies the application of the model to different learning sets extensively. The training process in the second multiresolution approach is hardly any more demanding than that one of the basic statistical model. As a result, time needed to train the model is reduced and, since we are still using the multiresolution approach, the quality of the results is preserved, as we will see later.

Our second multiresolution approach is based on the work of Bischof and Leonardis [3]. In this work Bischof and Leonardis discuss among other things the multiresolution coefficient estimation. They point out that one can use the same coefficients to reconstruct a filtered and subsampled image from the filtered and subsampled eigenimages as well as to reconstruct the original resolution image from the original resolution eigenimages. This idea constitutes the essence of our second multiresolution approach that is why the following section describes it into some more detail.

Multiresolution Coefficient Estimation. The idea we are about to present stems from the efforts to find a method of eigenimage coefficient estimation which would be robust against occlusion, varying background, and other types of non-Gaussian noise [3]. It is based on the following observation: If we take into account all s eigenvectors and if there is no noise in the data, then, in order to calculate the coefficients p_i, we only need s points $\boldsymbol{r} = (r_1, r_2, ..., r_s)$ where $\boldsymbol{r}$ is a vector of point coordinates. It is sufficient to compute the coefficients p_i by simply solving the following system of linear equations:

$$x(r_j) = \sum_{i=1}^{s} p_i e_i(r_j) \tag{18}$$

where $x(r_j)$ denotes the recovered value of x at position r_j. This notation emphasizes the fact that the equation is valid pointwise.

In a more common case we assume that we have eigenvectors e_i ordered in descending order with respect to the corresponding eigenvalues λ_i. Then, depending on the correlation among the data vectors in the learning set, only t, $t < s$, eigenvectors are needed to represent the data vector x to a sufficient degree of accuracy:

$$\tilde{x}(r_j) = \sum_{i=1}^{t} p_i e_i(r_j). \tag{19}$$

We can now derive the following property which holds due to the linearity of the equation:

$$(f * \tilde{x})(r_j) = \sum_{i=1}^{t} p_i (f * e_i)(r_j), \tag{20}$$

where f denotes a filter and $*$ the convolution.

The next property holds because Eq.(19) is valid for each point r_j, therefore

$$\tilde{x}_{\downarrow}(r_j) = \sum_{i=1}^{t} p_i e_{i\downarrow}(r_j), \tag{21}$$

where $\downarrow$ denotes the subsampling operation. Combining (20) and (21), we obtain

$$(f * \tilde{x})_{\downarrow}(r_j) = \sum_{i=1}^{t} p_i (f * e_i)_{\downarrow}(r_j), \tag{22}$$

which states that using the same coefficients we can reconstruct a filtered and subsampled image from the filtered and subsampled eigenimages.

Application in the Second Multiresolution Approach. The multiresolution coefficient estimation presents some significant advantages for the statistical model segmentation. By incorporating the results from the previous section, we can consequently perform the multiresolution segmentation of the input image using only one single statistical model trained on the original training set.

Training the statistical model essentially means extracting the data variability information from the learning set and representing it in terms of eigenvector directions. Once we have trained the original statistical model using the original training set, we can adjust the model to different resolution levels by adjusting its eigenvectors. As it follows from the previous section, the coefficients representing the input vector in the parameter space can be recovered from the original resolution image and corresponding original resolution eigenvectors as well as from subsampled input vector and corresponding subsampled eigenvectors. There is a direct relationship between the coefficients recovered using the first or the second approach - they are essentially the same, their accuracy being the only difference.

Applying this idea to the multiresolution statistical model segmentation is very straightforward. We first build the Gaussian pyramid of the input image. According to the resolution levels of this pyramid we then build the Gaussian pyramid of model's eigenvectors. The segmentation process starts on the highest level of the pyramid and performs the segmentation of the lowest resolution input image with the lowest resolution eigenvectors. The resulting coefficients are now passed straight down the pyramid to the next resolution level. The transformation from the first approach is not necessary anymore, since what we had on the previous level was only a filtered and subsampled version of the eigenvectors which will be used on the following level. According to the robust coefficient estimation idea [3] in this case coefficients are interpreted in the same way on both pyramid levels. However, the accuracy of coefficients improves as they are passed down the pyramid since every next level contains more details and thus enables the segmentation process to more reliably determine the coefficient values.

The fact that we are using the filtered and subsampled eigenvectors raises another issue. In the first multiresolution approach we had an independent model trained on each resolution level. As a consequence each model had its own orthogonal set of parameter space base vectors. As a result we could use the projection equations (4) and (5) to transform the data from the parameter space representation into the reference space representation and vice versa. However, in the second multiresolution approach this is not anymore the case. In the training process of the original model we do get an orthogonal set of eigenvectors composing the parameter space base. As soon as we filter and subsample these eigenvectors, however, we lose the orthogonal property. Consequently, the projection equations (4) and (5) are no longer valid. Following from equation (19), we solve a predetermined system of m linear equations with t unknowns instead. To illustrate the approach let Ψ be a matrix of eigenvector values at m points

$$
\Psi = \begin{pmatrix}
e_1(r_1) & e_2(r_1) & \dots & e_t(r_1) \\
e_1(r_2) & e_2(r_2) & \dots & e_t(r_2) \\
\cdot & \cdot & & \cdot \\
\cdot & \cdot & \cdot & \cdot \\
\cdot & \cdot & & \cdot \\
e_1(r_m) & e_2(r_m) & \dots & e_t(r_m)
\end{pmatrix} , \tag{23}
$$

x_m be the input data vector of length m (in principle t points from the input data vector suffice to calculate the value of the coefficients, however, for the purpose of simplicity we rather take all m elements of the input vector and solve an overdetermined linear system of equations instead), and p the vector of t parameter space coefficients

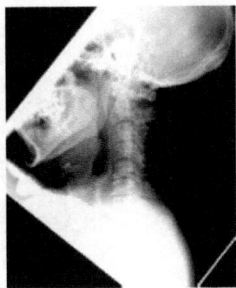

Fig. 1. Example of a training set image.

$$x_m = \begin{pmatrix} x(r_1) \\ x(r_2) \\ \cdot \\ \cdot \\ \cdot \\ x(r_m) \end{pmatrix}, \ p = \begin{pmatrix} p_1 \\ p_2 \\ \cdot \\ \cdot \\ \cdot \\ p_t \end{pmatrix}. \tag{24}$$

Now we can, for the purpose of the second multiresolution approach, replace the equation (4) by

$$x_m = \Psi p \tag{25}$$

and the equation (5) by

$$p = \Psi^+ x_m. \tag{26}$$

Note that Ψ represents the matrix of an overdetermined system of linear equations and thus has to be inverted using pseudoinverse.

In this section we described the idea behind the second multiresolution approach and explained the mathematical background to it. In the following section we proceed to tests and test results which show the scale and mode of improvement possible when using the described multiresolution approaches.

5 Experimental Results

The statistical model was trained on the training set consisting of 36 x-ray images of 366×499 original resolution. The images were x-ray scans of the human cervical vertebrae, courtesy of Jesenice hospital, Slovenia. An example of the training set image is given in Figure 1. Medical experts annotated every image in the training set with 7 points describing the shape of a single vertebra. These points were copied to shape vectors and assembled in the shape training set.

Database size is a distinctive problem when it comes to medical image processing. It is difficult to obtain a large database of quality medical images, especially when dealing with x-ray images. The primary aim of a doctor taking an x-ray scan of a patient is not to expose the patient to the radiation for longer than necessary. It often happens that the image which is incomplete from computer

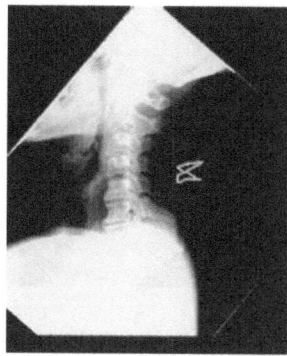

Fig. 2. The original model tends to get lost when initialized too far away from the structure's expected position.

processing point of view still has enough information for doctor's interpretation. There are severe health reasons against additional x-ray scans so one has to compromise and use the database available.

Because of the relatively small database size we conducted the leave-one-out test on the database of 36 images. The criteria used as a quality measure of the segmentation result was the Euclidean distance in pixels between the shape estimate given by the model and the shape annotated by the medical expert.

We first conducted the test with the original statistical model. The results were used as a reference in terms of segmentation quality and number of iterations. When segmenting the images with the original model, on average 35 iterations were necessary for the model to converge to a solution. In many cases the solutions were in the right position but not properly shaped.

When we forced the original model to initialize the segmentation process far from the actual solution, it regularly got lost and terminated the optimization due to exceeded maximum number of iterations. Example of a case where the model got lost is shown in Figure 2. The position of the "butterfly" in the image shows where the model terminated the segmentation. The reason behind the poor performance on the shifted structures or bad initialization was on one hand the restriction of the model to only segment the area which was covered in the training set and on the other hand the large amount of detail present in the test image which got the model confused before it arrived at the correct solution. It is always possible to perform several segmentations of the same image with the original model, every time initializing the model in a different image area and therefore increasing the chances of finding the right solution even when it is shifted from its usual position, but it is computationally prohibitive. Our goal was to explore how much the segmentation performance could improve by using the multiresolution approach.

The multiresolution approach turned out to be significantly more successful than its original model counterpart. The pyramid consisted of 5 resolution levels. While the original model was forced to work with the global optimization method

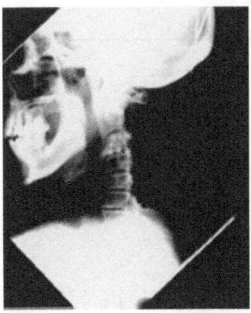

Fig. 3. The multiresolution approach is far more successful when dealing with bad initialization.

due to the multimodality of the problem, the multiresolution approach arrived at the same or even better quality solution using only the local optimization technique. Due to the multiresolution approach the number of iterations necessary to converge to a solution got much smaller. On average two iterations per resolution level were necessary. By contrast with the original model the multiresolution approach never segmented a shape that would be too small or wrongly oriented, which was often the case with the original model.

Another advantage of the multiresolution approach is the segmentation of shifted structures. It is relatively easy to perform an exhaustive search through the lowest resolution level image in order to determine the location of the shifted structure when using the multiresolution approach (see Figure 3).

This can be a problem, however, when the image contains several instances of the structure that is being segmented, such as in the case of cervical vertebrae. However, to arrive at the correct solution in this case, instead of performing the exhaustive search of the entire image on the lowest resolution level, only the appropriate surroundings of the expected solution position are examined. In a case like ours where there are several vertebra structures exercising very similar modes of variability, it is helpful to exploit the spatial relationship between the individual structures in order to segment the correct one [2]. Speaking about automatizing the segmentation process, it is also good to keep in mind that when exploiting the spatial relationship and trying to locate a shifted structure, we should not strive to segment the structure no matter what its position in the image. If the structure is significantly shifted from its expected position and therefore the segmentation process fails to locate it, this could well be the indicator that there's something wrong with the structure's position in the first place and that it thus needs, for example, doctor's special attention.

Both the first and the second approach produced better results than the original approach. Figure 4 shows the comparison of the original model results and results obtained by the second multiresolution approach.

The correct solutions in Figure 4 are those which are no more than 25-30 pixels high. One would expect the correct solution to be close to 0. The reason

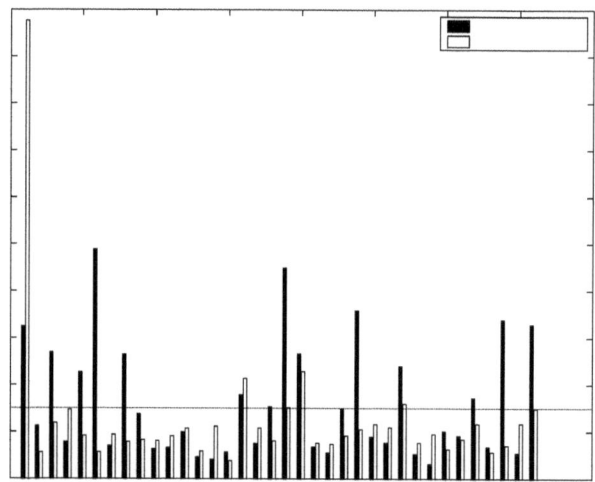

Fig. 4. Comparison of original model and second multiresolution approach test results.

for such a large number lies in the distance criteria. The quality of the solution is defined as a distance between the expert annotated shape and the segmented shape estimated by the statistical model. The expert annotated shape cannot be taken as 100% correct. It is the expert's decision where the distinctive anatomic points lie in the image. When the model suggests its own solution, this can differ up to as much as 30 pixels from the expert annotated solution and still make sense. The upper boundary for the correct solutions was estimated by manually examining all 36 segmentation results.

6   Summary and Discussion

We suggested the upgrade of the original statistical model with the multiresolution approach and described two different multiresolution approaches. We conducted segmentation tests on the medical image database available and showed that the complexity and quality of segmentation can be improved by using the multiresolution approach.

The computational complexity got reduced due to the multiresolutional approach since the segmentation process was reliably performed with the local optimization technique. In more demanding cases, e.g. shifted structures, the lowest resolution level segmentation process could be initialized in several different areas of image in order to enhance the segmentation robustness. In spite of several repetitions of the segmentation on the lowest resolution level, the multiresolution approach is still far less computationally prohibitive than the original statistical model and at the same time producing better quality solutions.

The multiresolution approach also proved to be much more flexible when segmenting shifted structures. The only drawback was the quality of the images which often contained shadows and missing information. These images were already very unclear on the original resolution level. After downsampling they lost the important distinctive information about the structure since through the downsampling process the shadows mixed with the actual vertebrae. In such cases, in order to reliably segment the images with the multiresolution approach, some extra image preprocessing, e.g. removing the shadows, enhancing the edges, and similar transformations are required that would preserve their effects all the way down to the lowest resolution level and so make it easier for the model to discover the initial structure position. However, the multiresolution approach has proved to be very reliable when segmenting the images which didn't contain strong shadow problems and large portions of missing information.

When comparing both multiresolution approaches applied, the second proves to be particularly useful in spite of the results being pretty similar for both described approaches. Apart from subsampling the input image, the second approach appears to be the same on the outside as the original statistical model of appearance. The end user does not feel the difference when using the second approach adjusted model, since it does not require the downsampling of the training set. The second approach also allows for arbitrary choice of number of resolution levels actually used during the segmentation process. The coefficients obtained on the penultimate level of resolution can in most cases be directly propagated onto the last resolution level without actually performing the last level segmentation. This is due to the same interpretation of the coefficients on all resolution levels.

The elegant implementation of the second multiresolution approach and its simple application to various training sets is a significant advantage. Using this idea together with the information about the spatial relationship between the structures in the image (called *topology model* and described in [2]) could significantly improve the medical image segmentation's reliability as well as speed and remains an area to be explored.

References

1. R. Bajcsy, S. Kovacic, "Multiresolution Elastic Matching", *Computer Vision, Graphics and Image Processing*, 46, 1-21, April 1989.
2. R. Bernard, B. Likar, F. Pernus, "Segmenting Articulated Structures by Hierarchical Statistical Modeling of Shape, Appearance, and Topology", *Proceedings of the 4th International Conference on Medical Image Computing and Computer-Assisted Intervention–MICCAI 2001*, Utrecht, The Netherlands, October 2001, 499-506, Springer 2001.
3. H. Bischof, A. Leonardis, "Robust Recognition of Scaled Eigenimages Through a Hierarchical Approach", *Proceedings of Conference on Computer Vision and Pattern Recognition - CVPR'98*, 664-670, IEEE 1998.
4. A. Blake, M. Isard, *Active Contours*, Springer, 1998

5. T. F. Cootes, G. J. Edwards, C. J. Taylor "Active Appearance Models" *IEEE Transactions on Pattern Analysis and Machine Intelligence*, Vol. 23, No. 6, June 2001.
6. T. F. Cootes, C. J. Taylor, "Statistical Models of Appearance for Medical Image Analysis and Computer Vision", *Proceedings of SPIE Medical Imaging 2001*, Vol. 4322.
7. J. Duncan, N. Ayache, "Medical Image Analysis: Progress Over Two Decades and the Challenges Ahead", *IEEE Trans. PAMI 22(1)*, 85-106, 2000
8. A. X. Falcao, K. Jayaram, J. K. Udupa, F. K. Miyazava "An ultra-fast user-steered image segmentation paradigm: Live-wire-on-the-fly", *SPIE on Medical Imaging*, vol. 3661, 184-191, 1999
9. M. Kass, A. Witkin, D. Terzopoulus, "Snakes: active contour models", *International Journal of Computer Vision* 1(4), 321-331, 1988
10. T. Lindeberg, *Scale-Space Theory in Computer Vision*, Kluwer Academic Publishers, Dordrecht, Netherlands, 1994
11. T. McInerney, D. Terzopoulos, "Deformable models in medical image analysis: a survey", *Medical Image Analysis*, 1 (2) 91-108, 1996.
12. A. Schenk, G. P . M. Prause, H. -O. Peitgen, "Efficient semiautomatic segmentation of 3D objects in medical images", *Lecture Notes in Computer Science*, Vol. 1935, 186-195, Springer, New York, 2000.
13. D. Stalling, H. -C. Hege, "Intelligent scissors for medical image segmentation", *Digitale Bildverarbeitung fuer die Medizin*, Freiburg, 32-36, 1996
14. S. Yoshimura, T. Kanade, "Fast template matching based on the normalized correlation by using multiresolution eigenimages", *Proceedings of IROS '94*, 2086-2093, 1994

A Common Viewpoint on Broad Kernel Filtering and Nonlinear Diffusion

Danny Barash[1] and Dorin Comaniciu[2]

[1] Department of Chemistry and Courant Institute of Mathematical Sciences
New York University and Howard Hughes Medical Institute
31 Washington Place, Main 1021, New York, NY 10003
barash@biomath.nyu.edu
[2] Real-Time Vision and Modeling Department
Siemens Corporate Research
755 College Road East, Princeton, NJ 08540
comanici@scr.siemens.com

Abstract. Using a consistent adaptive smoothing formulation we show that both nonlinear diffusion and adaptive smoothing can be extended to an arbitrary window, a process called broad kernel filtering. Based on this idea, this paper presents a unified treatment of a number of well known nonlinear techniques for filtering. We show that bilateral filtering represents a particular choice of weights in the extended diffusion process, that is obtained from geometrical considerations. We then show that kernel density estimation applied in the joint spatial-range domain yields a powerful processing paradigm - the mean shift procedure, related to bilateral filtering but having additional flexibility. This establishes an attractive relationship between the theory of statistics and that of diffusion and energy minimization. We experimentally compare the discussed methods and give insights on their performance.

1 Introduction

Nonlinear operations are becoming increasingly important in visual processing applications. Since they are substantially more difficult to analyze, formulate and predict compared to linear operations, various innovative approaches have been proposed independently for low-level computer vision tasks. The integration of several approaches that rely on different mathematical tools (e.g., functional minimization, nonlinear PDEs, statistics and data analysis) is essential for obtaining high-quality results in real-life applications.

This paper concentrates on edge-preserving smoothing. It extends previous work [1], [2] on the relationship between nonlinear diffusion [25], [29], [34], adaptive smoothing [30], and bilateral filtering [33] to establish a connection to the mean shift procedure [9,10] in the joint spatial-range domain. Both nonlinear diffusion and adaptive smoothing are generalized to encompass large neighborhoods, while the bilateral filtering serves as a link between the *extended nonlinear diffusion* (i.e., nonlinear diffusion on extended neighborhoods) and mean shift filtering.

L.D. Griffin and M. Lillholm (Eds.): Scale-Space 2003, LNCS 2695, pp. 683–698, 2003.

The paper is divided as follows. Section 2 emphasizes the importance of extended neighborhoods in edge-preserving smoothing by analyzing smoothing on 1D 3-neighborhood, smoothing on 1D 5-neighborhood, and adaptive smoothing on 1D 5-neighborhood. This leads to the formulation in Section 3 of the *extended nonlinear diffusion* on 2D $(2S + 1 \times 2S + 1)$-neighborhood. In Section 4, it is shown that a specific choice of weights in the *extended nonlinear diffusion*, that is based on geometrical considerations, leads to bilateral filtering. By defining kernel density estimation in the spatial-range domain, we derive in Section 5 the mean shift procedure for filtering and show its extended flexibility over bilateral filtering. In Section 6 experiments and comparisons are presented, while in Section 7, conclusions are drawn based on the common framework that unifies several fundamental approaches for low-level vision.

2   Importance of Extended Neighborhood

The extension of gradient based, edge-preserving smoothing to include information from non-nearest neighboring pixels is natural and has been considered before in various contexts (e.g., [5,33]). Here, we start from Saint-Marc–Chen–Medioni's adaptive smoothing [30] that was reformulated in [2] for consistency with the diffusion equation, and extend the approach from the original 3×3 window to a window of arbitrary size.

The adaptive smoothing approach is fundamental and intuitive. Given an image $I^{(t)}(\boldsymbol{x})$, where $\boldsymbol{x} = (x_1, x_2)$ denotes space coordinates, an iteration of adaptive smoothing yields:

$$I^{(t+1)}(\boldsymbol{x}) = \frac{\sum_{i=-1}^{+1} \sum_{j=-1}^{+1} I^{(t)}(x_1 + i, x_2 + j) w^{(t)}}{\sum_{i=-1}^{+1} \sum_{j=-1}^{+1} w^{(t)}} \tag{1}$$

where the convolution mask $w^{(t)}$ is defined as:

$$w^{(t)}(x_1, x_2) = \exp\left(-\frac{\left|d^{(t)}(x_1, x_2)\right|^2}{2k^2}\right) \tag{2}$$

where k is the variance of the Gaussian mask. In [30], $d^{(t)}(x_1, x_2)$ is chosen to depend on the magnitude of the gradient computed in a 3×3 window:

$$d^{(t)}(x_1, x_2) = \sqrt{G_{x_1}^2 + G_{x_2}^2} \tag{3}$$

where,

$$(G_{x_1}, G_{x_2}) = \left(\frac{\partial I^{(t)}(x_1, x_2)}{\partial x_1}, \frac{\partial I^{(t)}(x_1, x_2)}{\partial x_2}\right) \tag{4}$$

noting the similarity (see also [4] for further analogies) of the convolution mask with the diffusion coefficient in anisotropic diffusion [25], [34], or more specifically, the total variation in Rudin–Osher–Fatemi's original work [29] that demonstrated how edge-preserving smoothing can be achieved from energy minimization.

2.1   Smoothing on 1D 3-Neighborhood

It was suggested in [30] that equation (1) is an implementation of anisotropic diffusion. Briefly sketched, lets consider the case of a one-dimensional signal $I^t(x)$ and reformulate the averaging process as follows:

$$I^{t+1}(x) = c_1 I^t(x-1) + c_2 I^t(x) + c_3 I^t(x+1), \tag{5}$$

with

$$c_1 + c_2 + c_3 = 1. \tag{6}$$

Therefore, it is possible to write the above iteration scheme as:

$$I^{t+1}(x) - I^t(x) = c_1(I^t(x-1) - I^t(x)) + c_3(I^t(x+1) - I^t(x)) \tag{7}$$

Taking $c_1 = c_3 = c$, this reduces to:

$$I^{t+1}(x) - I^t(x) = c(I^t(x-1) - 2I^t(x) + I^t(x+1)) \tag{8}$$

which is a discrete approximation of the linear diffusion equation:

$$\frac{\partial I}{\partial t} = c\nabla^2 I. \tag{9}$$

2.2   Smoothing on 1D 5-Neighborhood

The averaging process can be extended to include second-neighbors:

$$I^{t+1}(x) = c_1 I^t(x-2) + c_2 I^t(x-1) + c_3 I^t(x) + c_4 I^t(x+1) + c_5 I^t(x+2), \tag{10}$$

with

$$c_1 + c_2 + c_3 + c_4 + c_5 = 1. \tag{11}$$

Taking $c_1 = c_5 = w_2$, $c_2 = c_4 = w_1$, and $c_3 = 1 - 2w_2 - 2w_1$ this reduces to:

$$I^{t+1}(x) = w_2(I^t(x-2) + I^t(x+2)) + (1 - 2w_2 - 2w_1)I^t(x) \\ + w_1(I^t(x-1) + I^t(x+1)) \tag{12}$$

rearrangement of terms leads to:

$$I^{t+1}(x) - I^t(x) = w_2(I^t(x-2) - 2I^t(x) + I^t(x+2)) \\ + w_1(I^t(x-1) - 2I^t(x) + I^t(x+1)) \tag{13}$$

which is a discrete approximation of the linear diffusion equation:

$$\frac{\partial I}{\partial t} = w_1 \nabla_1^2 I + w_2 \nabla_2^2 I, \tag{14}$$

where ∇_1 denotes ∇ over a grid containing only the nearest-neighbors, and ∇_2 denotes ∇ over a grid containing only the second-neighbors. Typically $w_1 > w_2$ since nearest-neighbors have more influence than second-neighbors.

2.3 Adaptive Smoothing on 1D 5-Neighborhood

When the weights are space-dependent, one should write the weighted averaging scheme (see [2] for adaptive smoothing on 1D 3-Neighborhood that results in consistency to the diffusion equation) as follows:

$$I^{t+1}(x) = \left(\frac{c^t(x-2)+c^t(x)}{2}\right)I^t(x-2) + \left(\frac{c^t(x-1)+c^t(x)}{2}\right)I^t(x-1) \quad (15)$$

$$+c^t(x)I^t(x) + \left(\frac{c^t(x+1)+c^t(x)}{2}\right)I^t(x+1) + \left(\frac{c^t(x+2)+c^t(x)}{2}\right)I^t(x+2)$$

with

$$\frac{c^t(x-2)+c^t(x)}{2} + \frac{c^t(x-1)+c^t(x)}{2} + c^t(x) \quad (16)$$

$$+\frac{c^t(x+1)+c^t(x)}{2} + \frac{c^t(x+2)+c^t(x)}{2} = 1$$

Plugging (16) into (15) and rearranging finally leads to:

$$I^{t+1}(x) - I^t(x) = \frac{c^t(x+2)+c^t(x)}{2}\left[I^t(x+2)-I^t(x)\right]$$

$$-\frac{c^t(x-2)+c^t(x)}{2}\left[I^t(x)-I^t(x-2)\right]$$

$$+\frac{c^t(x+1)+c^t(x)}{2}\left[I^t(x+1)-I^t(x)\right]$$

$$-\frac{c^t(x-1)+c^t(x)}{2}\left[I^t(x)-I^t(x-1)\right] \quad (17)$$

which is a consistent implementation of the nonlinear diffusion equation:

$$\frac{\partial I}{\partial t} = \nabla_1(w_1(x_1,x_2)\nabla_1 I) + \nabla_2(w_2(x_1,x_2)\nabla_2 I), \quad (18)$$

where we have used w instead of c, since the variable w was adopted in (14) instead of c in (9). Thus, the weights $w_1(x_1,x_2)$, $w_2(x_1,x_2)$ are the nonlinear diffusion coefficients in the nearest-neighbors grid or second-neighbor grid, respectively, typically taken as:

$$w_{1,2}(x_1,x_2) = g(\|\nabla_{1,2}I(x_1,x_2)\|), \quad (19)$$

where $\|\nabla_{1,2}I\|$ is the gradient magnitude on either the nearest-neighbors grid or the second-neighbors grid, respectively, and $g(\|\nabla_{1,2}I\|)$ is an "edge-stopping" function. This function is chosen to satisfy $g(x) \to 0$ when $x \to \infty$ so that the diffusion is stopped across edges. Thus, a fundamental link between the nonlinear diffusion equation and edge-preserving smoothing filters is noticed. This link will be extended in the next Section to $(2S+1 \times 2S+1)$-neighborhood, and will lead to constructing the bilateral filter as a basic mechanism for the mean shift procedure.

3 Generalized Adaptive Smoothing and Nonlinear Diffusion on Extended Neighborhood

Adaptive smoothing was introduced in [30] as a local process applying a 3×3 window at each iteration, as defined in (1). However, it is natural to extend this definition to an arbitrary, $(2S + 1 \times 2S + 1)$ window

$$I^{(t+1)}(\boldsymbol{x}) = \frac{\sum_{i=-S}^{+S} \sum_{j=-S}^{+S} I^{(t)}(x_1 + i, x_2 + j) w^{(t)}}{\sum_{i=-S}^{+S} \sum_{j=-S}^{+S} w^{(t)}}, \tag{20}$$

where I is a three-element vector that describes color images. In the rest of this Section, the *extended nonlinear diffusion* is derived using the generalization of adaptive smoothing outlined in (20).

First, it is instructive to derive the *extended nonlinear diffusion* in one-dimension. Considering adaptive smoothing on 1D $(2S + 1)$-neighborhood, it is possible to generalize (15) by using vector notation to:

$$I^{t+1}(x) = \left(\frac{\boldsymbol{c}_{(-)} + c(x)\hat{\boldsymbol{1}}}{2} \right) \cdot \boldsymbol{I}^t_{(-)} + c(x) I^t(x) + \left(\frac{\boldsymbol{c}_{(+)} + c(x)\hat{\boldsymbol{1}}}{2} \right) \cdot \boldsymbol{I}^t_{(+)}, \tag{21}$$

where $\hat{\boldsymbol{1}} = [1, 1, .., 1]$ is the unity vector, "$\cdot$" denotes the dot product, $\boldsymbol{c}_{(-)} = [c(x - S), c(x - S + 1), ..., c(1)]$, $\boldsymbol{c}_{(+)} = [c(1), c(2), .., c(x + S)]$, $\boldsymbol{I}_{(-)} = [I(x - S), I(x - S + 1), ..., I(1)]$, $\boldsymbol{I}_{(+)} = [I(1), I(2), .., I(x + S)]$, $c(x)$ and $I(x)$ are scalars, $I(x)$ being the gray-level intensity at the point of interest x. In color images, I becomes a three-element vector, as in (20), but this extension is avoided in (21) for the purpose of clarity and is deferred until the final expression for the two-dimensional case. By analogy to (16), or adaptive smoothing in 1D 3-neighborhood outlined in [2], normalization of the weights can be written in vector notation as:

$$\left(\frac{\boldsymbol{c}_{(-)} + c(x)\hat{\boldsymbol{1}}}{2} \right) \cdot \hat{\boldsymbol{1}} + c(x) + \left(\frac{\boldsymbol{c}_{(+)} + c(x)\hat{\boldsymbol{1}}}{2} \right) \cdot \hat{\boldsymbol{1}} = 1, \tag{22}$$

and by analogy to (17), or (13) of [2], we obtain:

$$I^{t+1}(x) - I^t(x) = \left(\frac{\boldsymbol{c}_{(+)} + c(x)\hat{\boldsymbol{1}}}{2} \right) \cdot \left[\boldsymbol{I}^t_{(+)} - I^t(x)\hat{\boldsymbol{1}} \right]$$

$$- \left(\frac{\boldsymbol{c}_{(-)} + c(x)\hat{\boldsymbol{1}}}{2} \right) \cdot \left[I^t(x)\hat{\boldsymbol{1}} - \boldsymbol{I}^t_{(-)} \right], \tag{23}$$

which is an implementation of the nonlinear diffusion equation:

$$\frac{\partial I(x)}{\partial t} = \boldsymbol{\nabla} \cdot (\boldsymbol{w}(x) \boldsymbol{\nabla} I(x)), \tag{24}$$

where $\boldsymbol{\nabla} = [\nabla_{-S}, \nabla_{-S+1}, ..., \nabla_S]$ is a vector containing gradients taken at different neighboring configurations (i.e., nearest-neighbors, second-neighbors, etc.)

and $w = [w_{-S}, w_{-S+1}, ..., w_S]$ are the nonlinear diffusion coefficients. It is also possible to write (24) as:

$$\frac{\partial I(x)}{\partial t} = \sum_{j=-S}^{S} \nabla_j(w_j(x)\nabla_j I(x)), \tag{25}$$

expanding the vector notation used in (24).

Second, the generalization of adaptive smoothing in two-dimensions written in (20) leads to the *extended nonlinear diffusion* in two-dimensions by simple analogy to (24). Taking matrices instead of vectors for ∇, w, and using a three-element vector I instead of a scalar I to represent color images leads to the *extended nonlinear diffusion*:

$$\frac{\partial I(x_1, x_2)}{\partial t} = \tilde{\nabla} \cdot (\tilde{w}(x_1, x_2)\tilde{\nabla} I(x_1, x_2)), \tag{26}$$

where "$\cdot$" denotes the scalar product between two matrices, and $\tilde{\nabla}$, $\tilde{w}$ are $(2S+1) \times (2S+1)$ matrices that correspond to different neighbor combinations with respect to the center pixel of interest. The generalized adaptive smoothing (20) is a discrete approximation of the *extended nonlinear diffusion* (26). It is noted that in practice, S need not be taken too large (i.e., $S \leq 3$), otherwise the generalized adaptive smoothing becomes an inaccurate representation of the extended diffusion equation. The extended nonlinear diffusion (26) can be further generalized to digital TV filtering [7,24], and an idea from computational biology to use spectral information for targeted filtering is suggested in the Appendix based on scales analogy between the digital TV filter and RNA folding.

4   Bilateral Filtering

The idea of combining space and color for computer vision tasks has been explored in several works (e.g. [32], [31], [20], [6]) and consequently, related digital filters have been proposed in [33], [7]. In this Section, the Kimmel–Malladi–Sochen approach [32] of using the geometry of spatial-color space to perform edge-preserving smoothing is used to systematically choose the weights of the extended nonlinear diffusion that yields bilateral filtering. Properties of bilateral filtering can be found in the references [33], [1], [2], [14], [12], [31]. In [31], it was shown that the bilateral filter is closely related to the short time kernel of the Beltrami [32].

Bilateral filtering was introduced [33] as a nonlinear filter which combines domain and range filtering. Given an input image $I(x)$, using a continuous representation notation as in [33], the output image $h(x)$ is obtained by:

$$h(x) = \frac{\int_{-\infty}^{\infty} \int_{-\infty}^{\infty} I(\xi)c(\xi, x)s(I(\xi), I(x))d\xi}{\int_{-\infty}^{\infty} \int_{-\infty}^{\infty} c(\xi, x)s(I(\xi), I(x))d\xi}, \tag{27}$$

where $\boldsymbol{x} = (x_1, x_2), \boldsymbol{\xi} = (\xi_1, \xi_2)$ are space variables and $\boldsymbol{I} = (I_R, I_G, I_B)$ is the intensity. The convolution mask is the product of the functions c and s, which represent 'closeness' (in the domain) and 'similarity' (in the range), respectively.

It was demonstrated in [2] that a discrete version of Gaussian bilateral filtering can be written as follows:

$$I^{(t+1)}(\boldsymbol{x}) = \frac{\sum_{i=-S}^{+S} \sum_{j=-S}^{+S} \boldsymbol{I}^{(t)}(x_1 + i, x_2 + j) w^{(t)}}{\sum_{i=-S}^{+S} \sum_{j=-S}^{+S} w^{(t)}}, \tag{28}$$

with the weights given by:

$$w^{(t)}(\boldsymbol{x}, \boldsymbol{\xi}) = \exp(\frac{-(\boldsymbol{\xi} - \boldsymbol{x})^2}{2\sigma_D^2}) \exp(\frac{-(I(\boldsymbol{\xi}) - I(\boldsymbol{x}))^2}{2\sigma_R^2}), \tag{29}$$

where S is the window size of the filter. Since (28) and the generalized adaptive smoothing (20) are equivalent, what remains to be shown is an explanation for the origin of the weights given in (29).

In color images, it was demonstrated in [32] that the image can be represented as a $2D$ surface embedded in the $5D$ spatial-color space and denoising can be achieved by using the Beltrami flow. For representing the geometry of the $5D$ (x, y, R, G, B) space, it is simple and logical to define the local measure as

$$ds^2 = dx^2 + dy^2 + \beta^2(dR^2 + dG^2 + dB^2), \tag{30}$$

which is the geometric arclength in the hybrid spatial-color space. Thus, the distance measure in (2) is given by:

$$\left| d^{(t)}(x_1, x_2) \right|^2 = \Delta x_1^2 + \Delta x_2^2 + \beta^2(\Delta R^2 + \Delta G^2 + \Delta B^2). \tag{31}$$

In [1], [2] a rigorous analysis was worked out, by defining the *generalized intensity* in the $5D$ spatio-color space, for what can be intuitively conjectured; namely, plugging (31) into (2) yields:

$$w^{(t)}(\boldsymbol{x}, \boldsymbol{\xi}) = \exp\left(\frac{-(\boldsymbol{\xi} - \boldsymbol{x})^2}{2\sigma_D^2}\right) \exp\left(\frac{-(I(\boldsymbol{\xi}) - I(\boldsymbol{x}))^2}{2\sigma_R^2}\right), \tag{32}$$

where $\beta = \sigma_D/\sigma_R$, $\boldsymbol{x}$ is the location of the pixel of interest, $\boldsymbol{\xi}$ is the location of a pixel in its vicinity inside the window, and

$$(I(\boldsymbol{x}) - I(\boldsymbol{\xi}))^2 = (\Delta R)^2 + (\Delta G)^2 + (\Delta B)^2, \tag{33}$$

noting that in (30), the RGB color space was chosen in defining the geometric arclength for illustrative purposes, but different color spaces of interest such as the CIE Luv or CIE Lab can be chosen for I, depending on the application.

The extended adaptive smoothing or equivalently, the *extended nonlinear diffusion* (26), along with the choice of weights according to (32) yields precisely the Gaussian bilateral filter (28), (29) proposed in [33]. Thus, bilateral filtering is obtained from the extended nonlinear diffusion, with the weights (i.e., the nonlinear diffusivities) chosen according to the geometric arclength in the $5D$ spatio-color space defined by (30).

5   Mean Shift-Based Filtering

This section introduces the mean shift procedure as an iterative algorithm for local mode detection in the joint spatial-range domain. The mean shift-based filtering is defined in the sequel followed by a discussion on the relation to bilateral filtering.

Let us denote by $\mathbf{x}_i$, $i = 1 \ldots n$ a set of n data points in the d-dimensional space R^d. The multivariate kernel density estimator with normal kernel and a symmetric positive definite $d \times d$ bandwidth matrix $\mathbf{H}$, computed at the point $\mathbf{x}$ is given by

$$\hat{f}(\mathbf{x}) = \frac{1}{n \mid 2\pi\mathbf{H} \mid^{1/2}} \sum_{i=1}^{n} \exp\left(-\frac{1}{2} d^2(\mathbf{x}, \mathbf{x}_i, \mathbf{H})\right) \tag{34}$$

where

$$d^2(\mathbf{x}, \mathbf{x}_i, \mathbf{H}) \equiv (\mathbf{x} - \mathbf{x}_i)^\top \mathbf{H}^{-1}(\mathbf{x} - \mathbf{x}_i) \tag{35}$$

is the Mahalanobis distance from $\mathbf{x}$ to $\mathbf{x}_i$. By computing the gradient of $\hat{f}(\mathbf{x})$

$$\nabla \hat{f}(\mathbf{x}) = \frac{\mathbf{H}^{-1}}{n \mid 2\pi\mathbf{H} \mid^{1/2}} \sum_{i=1}^{n} (\mathbf{x}_i - \mathbf{x})\exp\left(-\frac{1}{2} d^2(\mathbf{x}, \mathbf{x}_i, \mathbf{H})\right) \tag{36}$$

after some algebra we have

$$\mathbf{m}(\mathbf{x}) = \mathbf{H}\frac{\nabla \hat{f}(\mathbf{x})}{\hat{f}(\mathbf{x})} \tag{37}$$

where

$$\mathbf{m}(\mathbf{x}) \equiv \frac{\sum_{i=1}^{n} \mathbf{x}_i \exp\left(-\frac{1}{2} d^2(\mathbf{x}, \mathbf{x}_i, \mathbf{H})\right)}{\sum_{i=1}^{n} \exp\left(-\frac{1}{2} d^2(\mathbf{x}, \mathbf{x}_i, \mathbf{H})\right)} - \mathbf{x} \tag{38}$$

is the mean shift vector. Observe that $\mathbf{m}(\mathbf{x})$ is an estimator of the normalized gradient of the underlying density. The repetitive computation of (38) followed by the translation of the kernel according to the mean shift vector defines a procedure which leads to a local mode of the density [9,10].

Assume now that the data points $\mathbf{x}_i$ are the *generalized pixels* of the input image. This means the vector components of $\mathbf{x}$ contain both the spatial lattice information $\mathbf{z} = (z_1, z_2)^\top$ and range information $\mathbf{c}$, i.e.,

$$\mathbf{x}_i = (\mathbf{z}_i^\top, \mathbf{c}_i^\top)^\top \tag{39}$$

with $i = 1 \ldots n$. The dimension of vector $\mathbf{c}$ is $r = 1$ when only the intensity values are considered, $r = 3$ for color images, or $r > 3$ in the multispectral case. Although a more complex form of the bandwidth can be useful in certain applications, we will assume henceforth that the bandwidth matrix $\mathbf{H}$ is diagonal having the diagonal terms equal to σ_D^2 for the spatial part and σ_R^2 for the range part. Using these notations, the mean shift vector (38) can be expressed as

$$\mathbf{m}(\mathbf{x}) = \frac{\sum_{i=1}^{n} \mathbf{x}_i \exp\left(-\frac{\|\mathbf{z}-\mathbf{z}_i\|^2}{2\sigma_D^2}\right) \exp\left(-\frac{\|\mathbf{c}-\mathbf{c}_i\|^2}{2\sigma_R^2}\right)}{\sum_{i=1}^{n} \exp\left(-\frac{\|\mathbf{z}-\mathbf{z}_i\|^2}{2\sigma_D^2}\right) \exp\left(-\frac{\|\mathbf{c}-\mathbf{c}_i\|^2}{2\sigma_R^2}\right)} - \mathbf{x} \tag{40}$$

Denote by $\mathbf{x}_{j,conv} = (\mathbf{z}_{j,conv}^\top, \mathbf{c}_{j,conv}^\top)^\top$ the convergence point of the iterative mean shift procedure initialized in $\mathbf{x}_j = (\mathbf{z}_j^\top, \mathbf{c}_j^\top)^\top$. By running the procedure for all $j = 1 \ldots n$, each data point is associated to a local mode in the joint spatial-range domain. The *mean shift-based filtered* image $\mathbf{y}_j$, $j = 1 \ldots n$ is defined by the range information carried by the point of convergence

$$\mathbf{y}_j = (\mathbf{z}_j^\top, \mathbf{c}_{j,conv}^\top)^\top \tag{41}$$

The algorithm has been first published in [9] and achieves a high quality discontinuity preserving filtering, by identifying local modes in the joint domain. The novelty is that the kernel moves iteratively in both spatial and range domains, in contrast to the methods discussed up to now, which maintain a fixed spatial component.

A particular case of mean shift filtering related to bilateral filtering can be obtained by fixing the spatial component of the vectors during iterations. The algorithm will again search for the local mode, but only by evolving in the range domain. We call this variant, *restricted* mean shift filtering.

An important feature of mean shift filtering is that the image structure does not change during iterations. The algorithm evolution is driven by the initial image structure. By contrast, both nonlinear diffusion and bilateral filtering change the initial image structure and will converge to a flat image, if run until convergence (although, this can be remedied in nonlinear diffusion by introducing an extra term to the diffusion equation as in [29]). We note that the principles of mean shift filtering were recently rediscovered in [36], where the algorithm is called local mode filtering.

6   Experiments

We compare in this section the performance of a simple nonlinear diffusion (as in [25]), bilateral filtering, and mean shift filtering. The comparison is performed on a the B/W squirrel image (Figure 1) in order to observe the main features of the underlying mechanisms.

Nonlinear diffusion results with $\Delta t = 1.0$ are presented in Figure 2 after 10, 20, 50 and 100 iterations. In Figure 3 and Figure 4 we show bilateral filtering with $\sigma_D = 3.0$, $\sigma_R = 25.0$ and $\sigma_D = 3.0$, $\sigma_R = 35.0$, respectively, after 1, 2, 5 and 10 iterations. Restricted mean shift results (kernel moving in range domain) are shown in Figure 5 for $\sigma_D = 3.0$ and 5.0, $\sigma_R = 25.0$ and 35.0. The same parameters are used to generate the unrestricted mean shift results (kernel moving in both domains), presented in Figure 6.

The following observations can be derived:

– Nonlinear diffusion obtains a pleasant result after 50 iterations, although many regions around the squirrel tail are excessively smoothed, while the borders are not very well defined. The amount of excessive smoothing increases at 100 iterations.

Fig. 1. Original squirrel image.

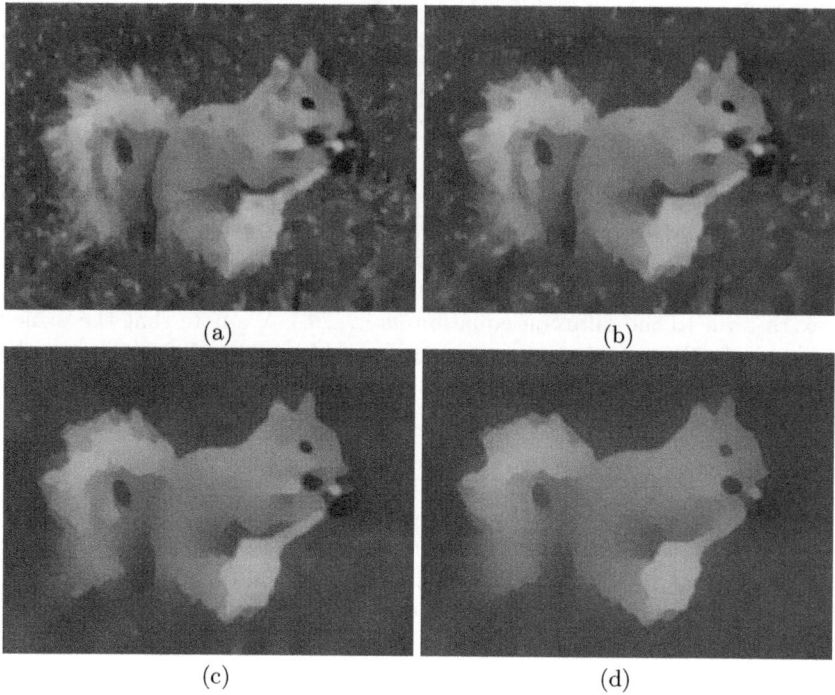

Fig. 2. Nonlinear Diffusion with $\Delta t = 1.0$. (a) 10 iterations. (b) 20 iterations. (c) 50 iterations. (d) 100 iterations.

- Similar comments are valid for bilateral filtering. After 2 iterations a good compromise between the amount of smoothing and preserved edges is reached. Nevertheless, excessive smoothing is present around the tail regions. The gradual collapse of the processed data to a flat image is noticeable after five iterations.
- Restricted mean shift filtering bears resemblance to the first two techniques in terms of quality of the preserved edges.

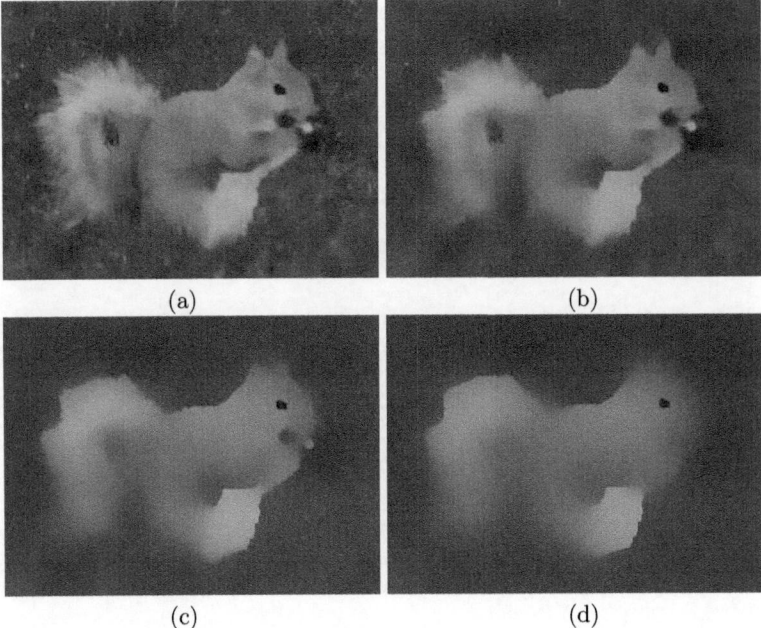

Fig. 3. Bilateral filtering with $\sigma_D = 3.0$ and $\sigma_R = 25.0$. (a) 1 iteration. (b) 2 iterations. (c) 5 iterations. (d) 10 iterations.

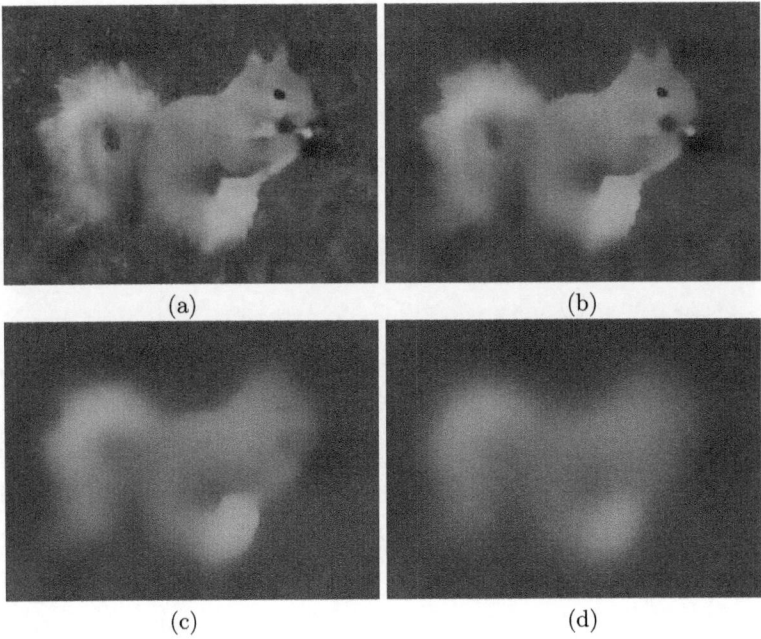

Fig. 4. Bilateral filtering with $\sigma_D = 3.0$ and $\sigma_R = 35.0$. (a) 1 iteration. (b) 2 iterations. (c) 5 iterations. (d) 10 iterations.

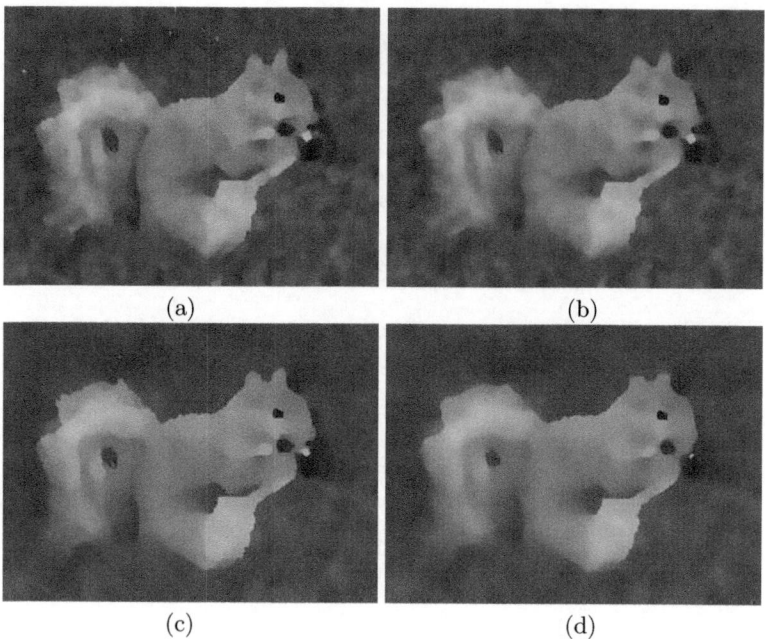

Fig. 5. Restricted mean shift filtering. (a) $\sigma_D = 3.0$, $\sigma_R = 25.0$. (b) $\sigma_D = 3.0$, $\sigma_R = 35.0$. (c) $\sigma_D = 5.0$, $\sigma_R = 25.0$. (d) $\sigma_D = 5.0$, $\sigma_R = 35.0$.

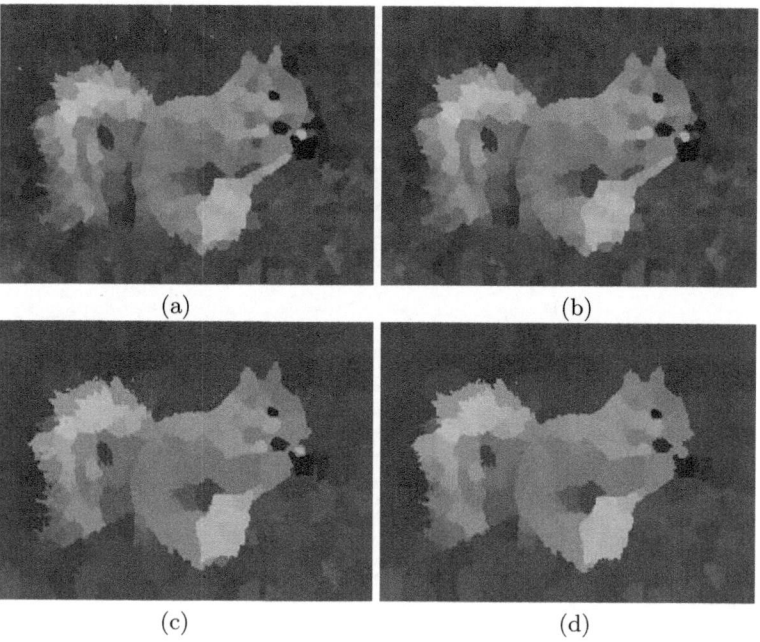

Fig. 6. Mean shift filtering (unrestricted). (a) $\sigma_D = 3.0$, $\sigma_R = 25.0$. (b) $\sigma_D = 3.0$, $\sigma_R = 35.0$. (c) $\sigma_D = 5.0$, $\sigma_R = 25.0$. (d) $\sigma_D = 5.0$, $\sigma_R = 35.0$.

– Unrestricted mean-shift is successful in achieving the sharpest boundaries among all the various approaches examined (see the quality of results in Figure 6c and d). The reason is that the local structure is better exploited by letting the kernel to simultaneously move in both spatial and range domains.

7   Conclusion

A common framework has been formulated for nonlinear diffusion [25], [29], [34], adaptive smoothing [30], bilateral filtering [33], and the mean shift paradigm [10]. Emphasizing the importance of extended neighborhoods, both nonlinear diffusion and adaptive smoothing can be generalized and unified to a single approach that accomplishes edge-preserving smoothing by using $(2S + 1 \times 2S + 1)$ instead of 3×3 window at each iteration (i.e., an *extended nonlinear diffusion*). The extended nonlinear diffusion process can then be casted into bilateral filtering by a specific choice of weights, based on geometrical considerations [32]. The bilateral mechanism is in turn related to a robust iterative procedure (i.e., the mean shift) which achieves edge-preserving filtering by searching for local modes in the joint spatial-range domain. We have thus established a noteworthy link between the nonlinear diffusion and the kernel methods from statistics. As a result, various tools derived with statistical motivations such as bandwidth selection [11] could be interpreted and exploited for parameter selection in the diffusion process.

8   Appendix: Scales in RNAs and Digital TV Filtering

In Section 3, it was mentioned that the extended nonlinear diffusion (26) can be further generalized based on graph connectivities to encompass unstructured grids, resulting in the digital TV filter [7,24]. Here, an interesting analogy between the digital TV filter and RNA folding, based on scales, is described.

An RNA is an important biomolecule that consists of a sequence of nucleic acids (A,C,G,U) and folds in three dimensional space to form a unique structure. In order to relate between structure and function, computational structure prediction of RNA has been a biologically important area of active research over the past 20 years. It is almost impossible to approach the RNA folding problem without the exploitation of scales. The different scales are illustrated in Figure 7. The *tertiary structure* can be reduced to a *secondary structure* [13], for which sophisticated computational structure prediction methods based on energy minimization exist (e.g., Zuker's *mfold* [38,23,37], the Vienna RNA package [19], and some other folding algorithms [28,8,17,18]). The secondary structure can be further downscaled to a graph [21] (see also [3,27]) and other graphs [16]. Furthermore, for predicting clever nucleotide mutations that will perturb a given secondary structure, the first author proposed an idea that he encountered in domain decomposition to represent these graphs by a Laplacian matrix, and seek the second eigenvalue of the Laplacian matrix (often denoted as the Fiedler eigenvalue [15] that represents the connectivity of the graph). This complements previous detailed work [22] that address mutations at the tree-graph scale, by

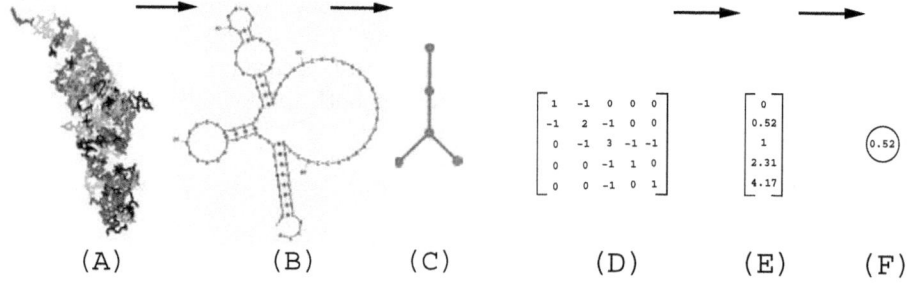

Fig. 7. Scales in the RNA biomolecule: (A) tertiary structure of yeast phenylalanine tRNA. (B) secondary structure. (C) graph representation. (D) Laplacian matrix. (E) spectra of the Laplacian matrix. (F) second eigenvalue of the Laplacian matrix.

further downscaling to capture essential structural information. Thus, at the coarsest scale, a single real positive number (the second eigenvalue of the Laplacian) exists that roughly indicates how the folding occurred as opposed to other folding possibilities. This reduced spatial information coexists with the energy of the folded structure, both being real positive numbers that are calculated independently. The two numbers can then be used as signatures in searching for interesting nucleotide mutations by examining the spatio-energy landscape. In addition, spectral graph partitioning [26] may be used to chop large RNAs into smaller fragments, as will be reported in future work.

In image processing and computer vision, such mathematical concepts based on graph theory have been proposed in the formulation of the Chan-Shen-Osher digital TV filter [7] that is based on a graph representation for performing nonlinear diffusion filtering. In [24], the idea of obtaining further information by the spectra of a Laplacian matrix is mentioned and may perhaps be brought to use during the filtering for performing an extended nonlinear diffusion targeted filtering, based on a reduced structural information that has been obtained by a decomposition to scales.

References

1. D. Barash, "Bilateral Filtering and Anisotropic Diffusion: Towards a Unified Viewpoint," *Hewlett-Packard Laboratories Technical Report*, HPL-2000-18(R.1), 2000.
2. D. Barash, "A Fundamental Relationship between Bilateral Filtering, Adaptive Smoothing, and the Nonlinear Diffusion Equation," *IEEE Transactions on Pattern Analysis and Machine Intelligence*, Vol. 24, No. 6, p.844, 2002.
3. G. Benedetti, S. Morosetti, "A Graph-Topological Approach to Recognition of Pattern and Similarity in RNA Secondary Structure," *Biophysical Chemistry*, Vol. 59, p.179, 1996.
4. M.J. Black, G. Sapiro, D. Marimont, and D. Heeger, "Robust Anisotropic Diffusion," *IEEE Transactions on Image Processing*, Vol. 7, No. 3, p.421, 1998.
5. T. Boult, R.A. Melter, F. Skorina, and I. Stojmenovic, "G-neighbors," *Proceedings of the SPIE, Vision Geometry II*, Vol.2060, p.96, 1993.

6. T.F. Chan and J. Shen, "Variational Restoration of Non-flat Image Features: Models and Algorithms," *SIAM Journal of Applied Mathematics*, Vol. 61(4), p.1338, 2000.

7. T.F. Chan, S. Osher, and J. Shen, "The Digital TV Filter and Nonlinear Denoising," *IEEE Transactions on Image Processing*, Vol. 10, No. 2, p.231, 2001.

8. J.H. Chen, S.Y. Le, J. Maizel, "Prediction of Common Secondary Structures of RNAs: A genetic Algorithm Approach," *Nucleic Acids Research*, Vol. 28, No. 4, p. 991, 2000.

9. D. Comaniciu and P. Meer, "Mean Shift Analysis and Applications," *Proceedings of the 1999 IEEE International Conference on Computer Vision*, Kerkyra, Greece, p. 1197, 1999.

10. D. Comaniciu and P. Meer, "Mean Shift: A Robust Approach towards Feature Space," *IEEE Transactions on Pattern Analysis and Machine Intelligence*, Vol. 24, No. 5, p.603, 2002.

11. D. Comaniciu, "An Algorithm for Data-Driven Bandwidth Selection," *IEEE Transactions on Pattern Analysis and Machine Intelligence*, Vol. 25, No.2, 2003.

12. F. Durand and J. Dorsey, "Fast Bilateral Filtering for the Display of High-Dynamic Range Image," *Proceedings of ACM SIGGRAPH 2002*, in Computer Graphics Proceedings, San Antonio, TX, 2002.

13. R. Durbin, S. Eddy, A. Krogh, G. Mitchison, "Biological Sequence Analysis: Probabilistic Models of Proteins and Nucleic Acids", Cambridge University Press, 1998.

14. M. Elad, "On the Bilateral Filter and Ways to Improve It," *IEEE Transactions on Image Processing*, Vol. 11, No. 10, p.1141, 2002.

15. M. Fiedler, "Algebraic Connectivity of Graphs," *Czechoslovak Mathematical Journal*, Vol.23, p.298, 1973.

16. H.H. Gan, S. Pasquali, T. Schlick, "Exploring The Repertoire of RNA Secondary Motifs Using Graph Theory with Implications for RNA Design," *Nucleic Acid Research*, 2002, Submitted.

17. A.P. Gultyaev, F.H.D. van Batenburg, C.W.A. Pleij, "The Computer Simulation of RNA Folding Pathways Using a Genetic Algorithm," *Journal of Molecular Biology*, Vol.250, p.37, 1995.

18. D.A.M. Konings and R.R. Gutell, "A Comparison of Thermodynamic Foldings with Comparatively Derived Structures of 16S and 16S-like rRNAs," *RNA*, Vol.1, No.6, p.559.

19. I.L. Hofacker, W. Fontana, P.F. Stadler, L.S. Bonhoeffer, M. Tacker, P. Schuster, "Fast Folding and Comparison of RNA Secondary Structures," *Monatshefte fur Chemie*, Vol. 125, p. 167, 1994.

20. J.J. Koenderink and A.J. Van Doorn, "The Structure of Locally Orderless Images," *International Journal of Computer Vision*, 21(2/3), p.159, 1999.

21. S. Y. Le, R. Nussinov, J.V. Maizel, "Tree Graphs of RNA Secondary Structures and Their Comparisons", *Computers and Biomedical Research*, Vol. 22, p.461, 1989.

22. H. Margalit, B.A. Shapiro, A.B. Oppenheim, J.V. Maizel, "Detection of Common Motifs in RNA Secondary Structure," *Nucleic Acids Research*, Vol. 17, No. 12, p. 4829, 1989.

23. D.H. Mathews, J. Sabina, M. Zuker, D.H. Turner, "Expanded Sequence Dependence of Thermodynamic Parameters Improves Prediction of RNA Secondary Structure," *Journal of Molecular Biology*, Vol. 288, p.911-940, 1999.

24. S. Osher, J. Shen, "Digitized PDE Method for Data Restoration," *Analytic-Computational Methods in Applied Mathematics*, G.A. Anastassiou, Ed., 2000.

25. P. Perona and J. Malik, "Scale-Space and Edge Detection Using Anisotropic Diffusion," *IEEE Transactions on Pattern Analysis and Machine Intelligence*, Vol. 12, No. 7, p.629, 1990.
26. A. Pothen, H. Simon, K.P. Liou, "Partitioning Sparse Matrices with Eigenvectors of Graphs," *SIAM Journal on Matrix Analysis and Applications*, Vol. 11, p.430, 1990.
27. C. M. Reidys, P. F. Stadler, P. Schuster, "Generic Properties of Combinatory Maps: Neural Networks of RNA Secondary Structures," *Bulletin of Mathematical Biology*, Vol. 59, No. 2, p.339, 1997.
28. E. Rivas, S.E. Eddy, "A Dynamic Programming Algorithm for RNA Structure Prediction Including Pseudoknots," *Journal of Molecular BIology*, Vol. 185, No. 5, p.2053, 1999.
29. L.I. Rudin, S. Osher, and F. Fatemi, "Nonlinear Total Variation Based Noise Removal Algorithms," *Physica D.*, Vol. 60, p.259, 1992.
30. P. Saint-Marc, J.S. Chen, and G. Medioni, "Adaptive Smoothing: A General Tool for Early Vision," *IEEE Transactions on Pattern Analysis and Machine Intelligence*, Vol. 13, No. 6, p.514, 1991.
31. N. Sochen, R. Kimmel, and A.M. Bruckstein, "Diffusions and Confusions in Signal and Image Processing," *Journal of Mathematical Imaging and Vision*, 14(3), p.195, 2001.
32. N. Sochen, R. Kimmel, and R. Malladi, "A Geometrical Framework for Low Level Vision," *IEEE Transactions on Image Processing*, Vol. 7, No. 3, p.310, 1998.
33. C. Tomasi and R. Manduchi, "Bilateral Filtering for Gray and Color Images," *Proceedings of the 1998 IEEE International Conference on Computer Vision*, Bombay, India, 1998.
34. J. Weickert, *Anisotropic Diffusion in Image Processing*, Tuebner Stuttgart, 1998. ISBN 3-519-02606-6.
35. J. Weickert, B.M. ter Haar Romeny, and M. Viergever, "Efficient and Reliable Schemes for Nonlinear Diffusion Filtering," *IEEE Transactions on Image Processing*, Vol. 7, No. 3, p.398, 1998.
36. J. van de Weijer and R. van den Boomgaard, "Local Mode Filtering," *Proceedings of the 2001 IEEE Conference on Computer Vision and Pattern Recognition*, Hawaii, Vol. 2, p. 428, 2001.
37. M. Zuker, "On Finding All Suboptimal Foldings of an RNA Molecule," *Science*, Vol. 244, p.48, 1989.
38. M. Zuker, D.H. Mathews, D.H. Turner, "Algorithms and Thermodynamics for RNA Secondary Structure Prediction: A Practical Guide In RNA Biochemistry and Biotechnology", *NATO ASI Series*, J.J. Barciszewski and B.F.C. Clark, eds., p.11-43, Kluwer Academic Publishers, 1999.

Efficient and Consistent Recursive Filtering of Images with Reflective Extension

Ben Appleton[1] and Hugues Talbot[2]

[1] Intelligent Real-Time Imaging and Sensing Group, ITEE
The University of Queensland, Brisbane, QLD 4072, Australia
appleton@itee.uq.edu.au
[2] CSIRO Mathematical and Information Sciences
Locked Bag 17, North Ryde, NSW 1670, Australia
hugues.talbot@csiro.au

Abstract. Recursive filters are commonly used in scale space construction for their efficiency and simple implementation. However these filters have an initialisation problem which either produces unusable results near the image boundaries or requires costly approximate solutions such as extending the boundary manually.

In this paper, we describe a method for the recursive filtering of reflectively extended images for filters with symmetric denominator. We begin with an analysis of reflective extensions and their effect on non-recursive filtering operators. Based on the non-recursive case, we derive a formulation of recursive filtering on reflective domains as a linear but time-varying implicit operator. We then give an efficient method for decomposing and solving the linear implicit system. This decomposition needs to be performed only once for each dimension of the image. This yields a filtering which is both stable and consistent with the ideal infinite extension. The filter is efficient, requiring the same order of computation as the standard recursive filtering.

We give experimental evidence to verify these claims.

1 Introduction

Recursive filters such as those described in [1] are commonly used in scale space construction for their efficiency and simple implementation. However these filters have an initialisation problem which produces unusable results near the image boundaries or requires costly approximations such as extending the boundary manually. These problems are exacerbated by the high scales encountered in linear scale space construction.

Martucci [2] investigated the relationships between symmetric convolution and the various Discrete Trigonometric Transforms. He demonstrated that DTTs could be used for an efficient implementation of the linear filtering of finite signals with reflective boundary conditions. However as Deriche observed in [3], methods relating to the Fast Fourier Transform may take many orders of magnitude greater computational effort than a direct implementation for recursive filters of low order.

L.D. Griffin and M. Lillholm (Eds.): Scale-Space 2003, LNCS 2695, pp. 699–712, 2003.
© Springer-Verlag Berlin Heidelberg 2003

A reflectively extended signal of length N may be treated as a periodic signal of length $2N$. Cunha [4] considered the problem of computing the initial values for a recursive filter on a periodically extended domain. He noted that the solution required more effort than the actual filtering. Smith and Eddins [5] suggest explicitly computing the impulse response of the recursive filter and using this to compute the initial values. As given, their solution is restricted to filters that have only single poles. Both methods suffer from the additional disadvantage that they must be repeated for each row of the image being filtered.

In [6] Weickert, ter Haar Romeny and Viergever give a method for recursive Gaussian filtering on a reflective domain. They derive their method from the relationship between linear diffusion filtering and Gaussian convolution, using a first order semi-implicit approximation to a linear diffusion PDE on a domain with Neumann boundary conditions. Their method requires a number of iterations proportional to the variance of the Gaussian. Their implementation of a single iteration is similar to the general method developed here.

In this paper, we describe a method for the recursive filtering of reflectively extended images. In Section 2 we introduce discrete filtering on infinite and finite domains. Section 3 develops a theory of filtering on finite domains with reflective extension. This theory forms the basis of the method for recursive filtering presented in Section 4. Here we analyse the existence and numerical stability of the scheme and propose an efficient implementation. Section 5 gives results on the accuracy and timings.

2 Discrete Filtering

2.1 Filtering on Infinite Domains

Linear filtering is widely used in image analysis for feature extraction, coding, enhancement and transformation. Here we briefly introduce non-recursive and recursive filtering, focussing in particular on the formulation of linear filtering as a system of explicit or implicit linear equations.

The simplest form of linear filter is the non-recursive filter, also known as the moving average or finite impulse response filter. Consider a signal x an element of the real vector space $V(\mathbb{Z})$, and a filter h. Then the filtering of x by h is defined by the convolution

$$(h \star x)[i] = \sum_{j=-\infty}^{\infty} h[j]x[i-j]$$

where h is known as the *kernel* of the filter. Here we assume h is stable in the sense that a bounded input produces a bounded output.

For a filter $h = (\ldots, h_{-2}, h_{-1}, h_0, h_1, h_2, \ldots)$ we may express this relationship as the (infinite) matrix-vector product

$$y = Hx \tag{1}$$

where y is the result of the convolution. Here $H_{ij} = h[i - j]$. For a filter h of length b, H is a banded Toeplitz matrix with total bandwidth b.

Recursive filters, also known as autoregressive or infinite impulse response filters, are expressed implicitly via a convolution. Let x be the input to a recursive filter with kernel h and let y be the output. We solve the implicit system

$$x = h \star y$$

to obtain y. Unfortunately the term 'recursive filter' is also used to describe the combination of a non-recursive filter and a recursive filter. Their sequential separability allows us to focus on the implementation of purely recursive filters with the natural extension to more general filters implicit throughout.

As with non-recursive filtering, we may implicitly define recursive filtering in matrix notation

$$x = Hy \tag{2}$$

where we solve the implicit system to obtain y. Typically a recursive filter will require only low order to approximate a desired impulse response, producing a narrowly banded matrix H [7].

2.2   Filtering on Finite Domains

The definitions given above apply only to signals on infinite discrete domains. When presented with a finite signal defined on the discrete interval $[1, N]$ we must define the action of a filter on that signal. To do so we define an extension of the signal to the domain $V(\mathbb{Z})$ and induce the definition of filtering from the choice of extension.

Zero Extension. The simplest extension is the zero extension. This extension assigns the signal zero value outside of $[1, N]$. For the filter

$$h = (\ldots, h_{-2}, h_{-1}, h_0, h_1, h_2, \ldots)$$

we obtain the filter matrix

$$H_0 = \begin{bmatrix} h_0 & h_{-1} & h_{-2} & \ldots & h_{2-N} & h_{1-N} \\ h_1 & h_0 & h_{-1} & \ldots & h_{3-N} & h_{2-N} \\ h_2 & h_1 & h_0 & \ldots & h_{4-N} & h_{3-N} \\ \vdots & \vdots & \vdots & \ddots & \vdots & \vdots \\ h_{N-2} & h_{N-3} & h_{N-4} & \ldots & h_0 & h_{-1} \\ h_{N-1} & h_{N-2} & h_{N-3} & \ldots & h_1 & h_0 \end{bmatrix}$$

Periodic Extension. Another possibility is to periodically extend the signal beyond the original finite domain. The resulting filter operation is a circular convolution. It may be expressed in matrix notation using Equation 1 where

$$H_P = \begin{bmatrix} h_0 & h_{-1} & h_{-2} & \dots & h_2 & h_1 \\ h_1 & h_0 & h_{-1} & \dots & h_3 & h_2 \\ h_2 & h_1 & h_0 & \dots & h_4 & h_3 \\ \vdots & \vdots & \vdots & \ddots & \vdots & \vdots \\ h_{-2} & h_{-3} & h_{-4} & \dots & h_0 & h_{-1} \\ h_{-1} & h_{-2} & h_{-3} & \dots & h_1 & h_0 \end{bmatrix}$$

is a circulant matrix. Circulant matrices have a close connection to the Discrete Fourier Transform which may be used for fast multiplication in Equation 1 or fast inversion in Equation 2 [8].

3   Reflective Domains

3.1   Indexing

Both zero extension and periodic extension have undesirable effects on the image border. Zero extension introduces large discontinuities on the order of the signal magnitude at the image borders. Periodic extension introduces a large discontinuitiy when connecting the two endpoints of the signal.

As a more appropriate alternative, the reflective extension of a signal can be described as placing a mirror at each end of the finite domain. In the case of Gaussian blurring this is equivalent to imposing an adiabatic boundary on the equivalent linear diffusion problem. This maintains the sum of intensities in the image, a desirable property in scale space construction. In the discrete case this is equivalent to reflective extension with repetition of the endpoints, known in the context of the Discrete Cosine Transforms as type-2 symmetric extension [7].

It is instructive in the following to consider the algebra of reflective indexing. For a signal defined on a finite discrete interval $I = \{1, 2, \dots, N\}$ we define the following reflectively extended domain:

$$R = I \times \{+, -\}$$

with the group $\mathbb{Z}_{2N}$ acting on it by addition. For $j \in \mathbb{Z}_{2N}$, $i \in I$, we have

$$j + (i, +) = (i + j, +)$$
$$j + (i, -) = (i - j, -)$$
$$(i, +) = (2N + 1 - i, -)$$

Observe that the action of $\mathbb{Z}_{2N}$ on R is isomorphic to the additive group $\mathbb{Z}_{2N}$. The following diagram depicts the increment action on R:

$$(1, +) \xrightarrow{+1} (2, +) \xrightarrow{+1} \dots \xrightarrow{+1} (N - 1, +) \xrightarrow{+1} (N, +)$$
$$\uparrow{+1} \qquad\qquad\qquad\qquad\qquad\qquad\qquad\qquad \downarrow{+1}$$
$$(1, -) \xleftarrow{+1} (2, -) \xleftarrow{+1} \dots \xleftarrow{+1} (N - 1, -) \xleftarrow{+1} (N, -)$$

3.2   Convolution

The definition of the algebra of indices in R allows us to define a convolution in this space. For a signal $g \in V(R)$ and a convolution kernel $h \in V(\mathbb{Z}_{2N})$ we define the convolution as

$$(h \star g)\,[r] = \sum_{j=1}^{2N} g[r]h[j - r]$$

the standard periodic convolution on a domain of size $2N$. As noted earlier, the operator $(h\star)$ may also be expressed as the circulant matrix $H_{ij} = h[i - j]$.

We seek to obtain a natural definition for the reflective convolution $h\odot$ in $V(I)$. To do so, we define an epimorphism ϕ from $V(I)$ to $V(R)$ and a projection π from $V(R)$ to $V(I)$. The definition for the reflective convolution is then induced by the following diagram

$$\begin{array}{ccc} V(I) & \xrightarrow{\phi} & V(R) \\ \downarrow{h\odot} & & \downarrow{h\star} \\ V(I) & \xleftarrow{\pi} & V(R) \end{array}$$

Let $\phi : V(I) \to V(R)$ be defined as follows. If $f \in V(I)$ then

$$\phi\,(f)\,(i, +) = \phi\,(f)\,(i, -) = f\,[i]$$

Then ϕ is the canonical *reflective extension* epimorphism from $V(I)$ into $V(R)$. We may express ϕ as the $2N \times N$ block matrix:

$$\Phi = \begin{bmatrix} I_N \\ J_N \end{bmatrix}$$

where I_N is the $N \times N$ identity matrix and J_N is the $N \times N$ reflection matrix

$$J_N = \begin{bmatrix} & & 1 \\ & \cdot^{\cdot^{\cdot}} & \\ 1 & & \end{bmatrix}$$

Then $\phi\,(f) = \Phi f$.

Let $\pi : V(R) \to V(I)$ be defined as follows: if $g \in V(R)$ then

$$\pi(g)[i] = \frac{1}{2}\,(g\,(i, +) + g\,(i, -))$$

Then π is a projection from $V(R)$ to $V(I)$. Here we have chosen π such that $\phi\pi$ is the identity mapping and π acts symmetrically in the indices of R. We may express π in matrix form as

$$\Pi = \frac{1}{2}\Phi^T = \frac{1}{2}\begin{bmatrix} I_N & J_N \end{bmatrix}$$

We then obtain the definition for reflective convolution:

$$h \odot f = \pi\,(h \star \phi\,(f))$$

Observe that, as the composition of linear operators, $(h \odot)$ is linear. We express it in matrix form as

$$H_\odot = \Pi H \Phi = \frac{1}{2} \Phi^T H \Phi \qquad (3)$$

3.3 Equivalence to Symmetric Filtering

We now analyse the structure of $H_\odot$. Recall that H is the circulant $2N \times 2N$ matrix corresponding to periodic convolution in $V(R)$ by the kernel h. We partition it into a 2×2 block matrix as follows:

$$H = \begin{bmatrix} H_{11} & H_{12} \\ H_{21} & H_{22} \end{bmatrix}$$

As H is cyclic it is also block cyclic, so $H_{11} = H_{22}$ and $H_{12} = H_{21}$. We can now compute $H_\odot$ as follows:

$$
\begin{aligned}
H_\odot &= \frac{1}{2} \begin{bmatrix} I_N & J_N \end{bmatrix} \begin{bmatrix} H_{11} & H_{12} \\ H_{12} & H_{11} \end{bmatrix} \begin{bmatrix} I_N \\ J_N \end{bmatrix} \\
&= \frac{1}{2} \left(H_{11} + H_{12} J_N + J_N H_{12} + J_N H_{11} J_N \right) \\
&= \frac{1}{2} \left(H_{11} + H_{11}^T \right) + \frac{1}{2} \left(H_{12} + H_{12}^T \right) J_N
\end{aligned}
$$

We observe that $H_\odot$ is symmetric. In fact, consider the decomposition of our filter $h \in V(\mathbb{Z})$ into the sum of an odd and an even function $h = h^{\mathrm{even}} + h^{\mathrm{odd}}$. The equivalent circulant matrix is $H = H^{\mathrm{even}} + H^{\mathrm{odd}}$, where $(H^{\mathrm{even}})^T = H^{\mathrm{even}}$ and $(H^{\mathrm{odd}})^T = -H^{\mathrm{odd}}$. Computing H gives

$$
\begin{aligned}
H &= \frac{1}{2} \left(\left(H_{11}^{\mathrm{even}} + (H_{11}^{\mathrm{even}})^T + H_{11}^{\mathrm{odd}} + (H_{11}^{\mathrm{odd}})^T \right) \right) \\
&\quad + \frac{1}{2} \left(\left(H_{12}^{\mathrm{even}} + (H_{12}^{\mathrm{even}})^T + H_{12}^{\mathrm{odd}} + (H_{12}^{\mathrm{odd}})^T \right) J_N \right) \\
&= H_{11}^{\mathrm{even}} + H_{12}^{\mathrm{even}} J_N
\end{aligned}
$$

We see that the anti-symmetric component of our filter h does not contribute to the reflective convolution in $V(I)$. By our choice of projection π all reflective recursive filters are equivalent to an odd-length symmetric filter. Consequently, in the following we assume without loss of generality that h is symmetric about the index $0 \in \mathbb{Z}$ and hence that H is symmetric. Practically, *this constraint applies only to the recursive component of the filter.*

4 Recursive Filtering in a Reflective Domain

4.1 Existence

We now demonstrate that recursive filtering in $V(I)$, defined as the solution of Equation 2 for the filter matrix $H_\odot$, is always possible for an invertible kernel h.

If $(h\star)$ is invertible in $V(R)$ then H is invertible as a matrix. Observe that if H is invertible then so is $H_\odot$, with its inverse given by:

$$H_\odot^{-1} = \Phi^T H^{-1} \Pi^T$$

as

$$
\begin{aligned}
H_\odot^{-1} H_\odot &= \frac{1}{4} \begin{bmatrix} I_N & J_N \end{bmatrix} H^{-1} \begin{bmatrix} I_N \\ J_N \end{bmatrix} \begin{bmatrix} I_N & J_N \end{bmatrix} H \begin{bmatrix} I_N \\ J_N \end{bmatrix} \\
&= \frac{1}{4} \begin{bmatrix} I_N & J_N \end{bmatrix} H^{-1} \left(I_{2N} + J_{2N} \right) H \begin{bmatrix} I_N \\ J_N \end{bmatrix} \\
&= \frac{1}{4} \begin{bmatrix} I_N & J_N \end{bmatrix} H^{-1} H \begin{bmatrix} I_N \\ J_N \end{bmatrix} + \frac{1}{4} \begin{bmatrix} I_N & J_N \end{bmatrix} H^{-1} J_{2N} H \begin{bmatrix} I_N \\ J_N \end{bmatrix} \\
&= \frac{1}{2} I_N + \frac{1}{4} \begin{bmatrix} I_N & J_N \end{bmatrix} H^{-1} H^T J_{2N} \begin{bmatrix} I_N \\ J_N \end{bmatrix} \\
&= \frac{1}{2} I_N + \frac{1}{4} \begin{bmatrix} I_N & J_N \end{bmatrix} H^{-1} H \begin{bmatrix} I_N \\ J_N \end{bmatrix} \\
&= I_N
\end{aligned}
$$

4.2   Stability

From this we may bound the condition number κ_∞ of $H_\odot$, a measure of the numerical stability of the corresponding implicit system [8].

$$
\begin{aligned}
\kappa_\infty \left(H_\odot \right) &= \| H_\odot \|_\infty \| H_\odot^{-1} \|_\infty \\
&= \| \Pi H \Phi \|_\infty \| \Phi^T H^{-1} \Pi^T \|_\infty \\
&\leq \frac{1}{4} \| \Phi^T \|_\infty \| H \|_\infty \| \Phi \|_\infty \| \Phi^T \|_\infty \| H^{-1} \|_\infty \| \Phi \|_\infty \\
&\leq \| H \|_\infty \| H^{-1} \|_\infty \\
&\leq \kappa_\infty \left(H \right)
\end{aligned}
$$

Here we have noted that $\| \Phi \|_\infty = 2$ and $\| \Phi^T \|_\infty = 1$. So we see that the inversion of the reflective convolution operator $(h\odot)$ on $V(I)$ is numerically at least as stable as the inversion of the periodic convolution operator $(h\star)$ on $V(R)$.

4.3 LDL^T Decomposition and Solution

Let $A \in I\!\!R^{n \times n}$ be a symmetric, positive definite matrix with real coefficients. Then A may be decomposed into the product $A = LDL^T$ where L is a unit lower triangular matrix and D a diagonal matrix. An algorithm for the numerical implementation of this decomposition is described in [8]. For an $N \times N$ matrix with total bandwidth b it requires $O(Nb^2)$ computation and $O(Nb)$ storage.

We will not always be dealing with positive definite matrices. The condition of positive definiteness may be relaxed to a more general existence criterion for the LDL^T decomposition of a symmetric matrix A: All principal submatrices of A must have non-zero determinant. So

$$\det A(1:s,1:s) \neq 0 \quad 1 \leq s < n$$

where $A(1:s,1:s)$ denotes the intersection of the first s rows and columns of A. To show this consider the following: $A(1:s,1:s) = L(1:s,1:s)D(1:s,1:s)L(1:s,1:s)^T$ as L is lower triangular, so

$$\det A(1:s,1:s) = \det L(1:s,1:s)\det D(1:s,1:s)\det L^T(1:s,1:s)$$
$$= \det D(1:s,1:s)$$
$$= \prod_{i=1}^{s} D_{ii}$$

where we have observed that $\det L(1:s,1:s) = 1$ as L is unit lower triangular. So

$$D_{ii} = \frac{\det A(1:i,1:i)}{\det A(1:i-1,1:i-1)}$$

where we take $D_{11} = A_{11}$. As we will be solving the system in Equation 2 using the LDL^T decomposition we require that $\det A(1:s,1:s) \neq 0$ for $1 \leq s \leq n$.

Once the LDL^T decomposition has been computed it is simple to solve Equation 2 to recursively filter a signal x to obtain the filtered signal y. In sequence solve:

$$x = Lu \tag{4}$$
$$u = Dv \tag{5}$$
$$v = L^T y \tag{6}$$

The solution to Equations 4, 5, and 6 may be considered as the sequential application of a causal filtering, a point-scaling, and an anti-causal filtering. This requires $O(Nb)$ computation for a kernel h of length b and may be solved in place. In fact, it requires the same number of arithmetic operations per pixel as the standard recursive filtering on an infinite domain.

When implementing the LDL^T decomposition in finite numerical precision the stability criterion is very simple to verify. As each D_{ii} is computed simply check that it is sufficiently far from zero:

$$|D_{ii}| > \eta$$

where η is a user-defined threshold for numerical stability, eg. $\eta = 10^{-7}$.

We have observed that quite surprisingly the LDL^T decomposition of $H_{\odot}$ appears to always exist for an invertible filter h. All filters which produce a non-decomposable matrix $H_{\odot}$ are not invertible in $V(Z)$. While the authors have proved this for low filter orders the general proof remains an open problem.

4.4   A Recursive Filtering Algorithm for Reflective Domains

Here we give an algorithm for the recursive filtering of images with reflective extension.

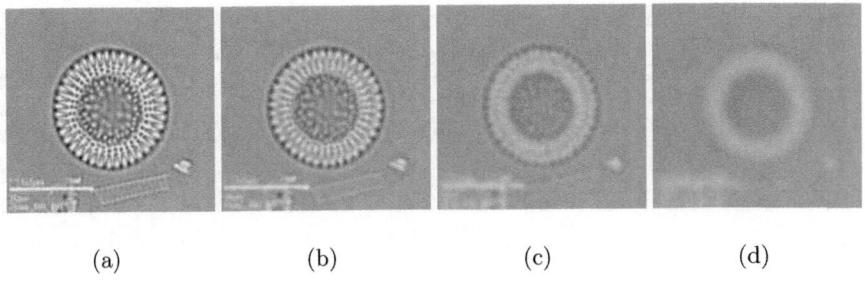

(a)   (b)   (c)   (d)

Fig. 1. Symmetric filtering: Gaussian blurring of *Cyclostephanos Dubius* (329 × 303). (a) The original image. (b) $\sigma = 2$. (c) $\sigma = 4$. (d) $\sigma = 8$.

Algorithm. For each image axis:

1. Form the $H_\odot$ matrix along the current axis according to Equation 3
2. Compute the decomposition $LDL^T = H_\odot$
3. For each row of data x, solve in place Equations 4, 5, 6.

The decomposition is only computed once per image axis. The total amount of computation on a d-dimensional image of sidelength N is $O(N^d b + N b^2)$. When applied to images of dimension $d \geq 2$ with a typical recursive filter of low order $b \ll N^{d-1}$, this reduces to $O(N^d b)$. This is the same order of computation as the standard implementation of a recursive filter. The algorithm requires auxiliary storage of $O(Nb)$ which is trivial when compared to the size of the image.

5 Application to Deriche's Recursive Gaussian Approximations

Here we consider Deriche's approximations to filtering by Gaussians and their derivatives, which require scale-invariant computation in theory [3]. However on a finite domain the standard implementation manually extends the boundary by an amount proportional to the scale of the Gaussian in order to minimise border effects. As a result the application of these filters to images is not truly invariant to scale and is doubly approximate, approximating both the Gaussian's impulse response and its effect on a finite domain.

We apply the new filtering scheme developed here to both symmetric and anti-symmetric filters. Figure 1 demonstrates the application of Deriche's 4th order Gaussian approximation to a microscope image of a diatom, while Figure 2 demonstrates the application of Deriche's 4th order derivative of Gaussian approximation to computing its spatial gradient. The application of the new scheme to other filters is straightforward.

Deriche's 4th order approximation to a Gaussian of scale σ for $x \geq 0$ is:

$$e^{-\frac{x^2}{2\sigma^2}} \approx \left(1.68\cos\left(0.6318\frac{x}{\sigma}\right) + 3.735\sin\left(0.6318\frac{x}{\sigma}\right)\right) e^{-1.783\frac{x}{\sigma}} - \left(0.6803\cos\left(1.997\frac{x}{\sigma}\right) + 0.2598\sin\left(1.997\frac{x}{\sigma}\right)\right) e^{-1.723\frac{x}{\sigma}}$$

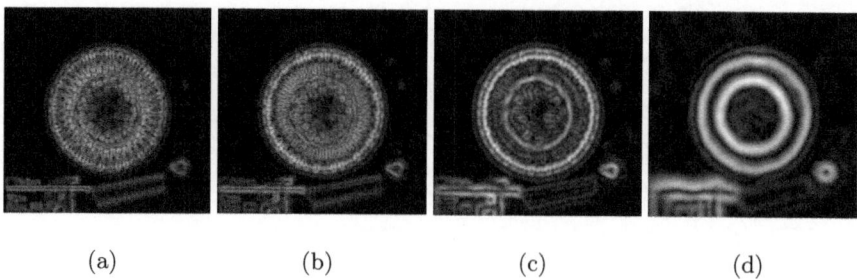

(a)   (b)   (c)   (d)

Fig. 2. Anti-symmetric filtering: the gradient magnitudes of *Cyclostephanos Dubius*. (a) $\sigma = 1$. (b) $\sigma = 2$. (c) $\sigma = 4$. (d) $\sigma = 8$.

His 4th order approximation to the first derivative of a Gaussian is:

$$xe^{-\frac{x^2}{2\sigma^2}} \approx \left(-0.6472\cos\left(0.6719\frac{x}{\sigma}\right) - 4.531\sin\left(0.6719\frac{x}{\sigma}\right)\right)e^{-1.527\frac{x}{\sigma}} +$$
$$\left(0.6494\cos\left(2.072\frac{x}{\sigma}\right) + 0.9557\sin\left(2.072\frac{x}{\sigma}\right)\right)e^{-1.516\frac{x}{\sigma}}$$

Deriche shows how these one-sided approximations may be used to compute a causal and anti-causal filter whose sum approximates a Gaussian of the desired scale. The causal filter is of the form:

$$H_+(z^{-1}) = \frac{n_{00}^+ + n_{11}^+ z^{-1} + n_{22}^+ z^{-2} + n_{33}^+ z^{-3}}{1 + d_{11}z^{-1} + d_{22}z^{-2} + d_{33}z^{-3} + d_{44}z^{-4}}$$

while the anti-causal filter is of the form:

$$H_-(z) = \frac{n_{11}^- z + n_{22}^- z^2 + n_{33}^- z^3 + n_{44}^- z^4}{1 + d_{11}z + d_{22}z^2 + d_{33}z^3 + d_{44}z^4}$$

These may be combined to produce a symmetric filter

$$H(z^{-1}) = H_+(z^{-1}) + H_-(z)$$
$$= \frac{n_0 + n_1(z^{-1} + z) + n_2(z^{-2} + z^2) + n_3(z^{-3} + z^3)}{d_0 + d_1(z^{-1} + z) + d_2(z^{-2} + z^2) + d_3(z^{-3} + z^3) + d_4(z^{-4} + z^4)}$$

or an anti-symmetric filter

$$H(z^{-1}) = H_+(z^{-1}) - H_+(z)$$
$$= \frac{n_1(z^{-1} - z) + n_2(z^{-2} - z^2) + n_3(z^{-3} - z^3) + n_4(z^{-4} - z^4)}{d_0 + d_1(z^{-1} + z) + d_2(z^{-2} + z^2) + d_3(z^{-3} + z^3) + d_4(z^{-4} + z^4)}$$

Observe that the denominators are symmetric in both cases. For any anti-symmetric filter realisable by a stable recursive filter we may always manipulate the filter into a form with symmetric denominator. Anti-symmetric denominators are not considered because they are unstable, having a pole at $z = 1$.

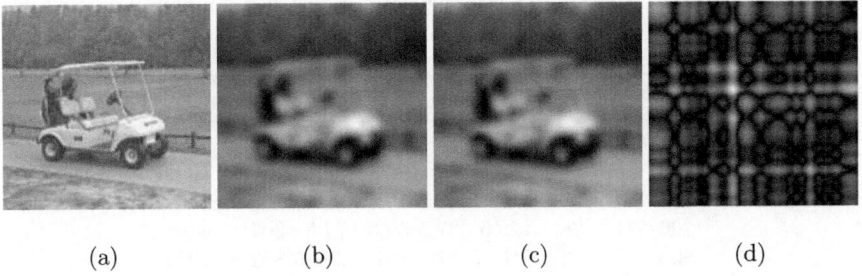

(a)   (b)   (c)   (d)

Fig. 3. Gaussian blurring of a golf cart image (548 × 509). (a) The original image. (b) Ground truth, $\sigma = 10$. (c) Proposed algorithm, $\sigma = 10$. (d) Difference between (b) and (c), scaled by 1.2×10^9 in order to be visible.

Due to Deriche's construction of the anti-causal filter from the causal filter to ensure symmetry, 4th order terms in the numerator of $H(z^{-1})$ cancel to reduce the order of the symmetric filtering. Likewise in the anti-symmetric case the constant term in the numerator cancels. These improve the efficiency of the filter and may be considered a benefit of solving an inherently symmetric problem in a symmetric manner.

The non-recursive and recursive components of Deriche's filtering are performed in sequence. The non-recursive component is performed by manually extending the image by the highest power in the numerator. The recursive component uses the method proposed in this paper.

6   Results

All tests were performed on a 700MHz Toshiba P-III laptop with 192MB of RAM under the Linux operating system. The algorithm presented here has been implemented in double precision floating point arithmetic in C and has not been optimised significantly.

6.1   Accuracy

We apply Deriche's 4th order recursive Gaussian approximation to blur the image of Figure 3 and compare this to a ground truth result. The ground truth result is obtained by symmetrically extending the image before filtering with a zero extension implementation of Deriche's 4th order recursive Gaussian. The border is extended by 20σ so that border artifacts do not contribute measureably to the error. For a Gaussian of scale $\sigma = 10$ the error has root mean square magnitude 4.8×10^{-8} with peak magnitude 2.1×10^{-7}. This error is trivially small compared both to the amplitude of the image and to the error of Deriche's 4th order Gaussian approximation with relative root mean squared magnitude of 2.93×10^{-4}.

Table 1. Running times (ms) for Deriche's 4th order Gaussian approximation on 2D images, implemented with manual extension by 4σ.

$\sigma =$	10	20	50	100	200	500	1000
Sidelength							
50	2.5	4	5.5	9	17	74	109.5
100	9.5	10	12.5	19.5	32.5	148.5	216.5
200	28	28.5	37.5	50.5	74.5	307	449
500	151.5	155	173	207	267.5	852.5	1214
1000	647.5	644.5	689	753.5	876.5	2058.5	2942.5

Table 2. Running times (ms) for the algorithm presented here on 2D images.

$\sigma =$	10	20	50	100	200	500	1000
Sidelength							
50	3.5	2	3	2	1.5	3	2.5
100	7.5	7.5	7.5	6.5	8	5.5	8
200	31	30	30.5	31	30.5	30	29.5
500	199	194.5	194	193.5	195	196.5	195
1000	875.5	859.5	857.5	857.5	856	867.5	858.5

6.2 Timing

Here we compare the running time of the algorithm proposed in this paper with the standard implementation via border extension. Borders are manually extended by 4σ as a reasonable tradeoff between additional computation and border effects. We consider a range of image sizes and scales for 2D and 3D images. Although we have chosen here for simplicity to test square and cubic images the filter decomposition has been repeated for each image axis. Results for 2D images are given in Tables 1 and 2 while results for 3D images are given in Tables 3 and 4. Over the range of image sizes considered we observe that in 2D this method is faster for scales greater than or equal to one fifth of the image size, while in 3D it is faster for all scales. Finally we observe that our method has a constant computing time irrespective of scale σ.

7   Conclusion

In this paper we have described a method for the recursive filtering of reflectively extended images by filters with symmetric denominator. This method is efficient and consistent with the infinite reflectively extended case. It is based on a time-varying formulation of convolution derived by considering the algebra of reflective extensions. In the recursive case this leads to an implicit linear system whose solution for invertible filters was shown to exist and have the same numerical stability as the corresponding recursive filter in the symmetrically extended domain. An efficient algorithm was given for the decomposition and

Table 3. Running times (ms) for Deriche's 4th order Gaussian approximation on 3D images, implemented with manual extension by 4σ.

$\sigma =$	2	5	10	20	50	100
Sidelength						
10	3.5	3	3.5	7.5	13	22
20	12	13	17	26.5	50	89
50	130	143	168.5	218.5	368	618.5
100	998.5	1063.5	1160	1351.5	1973.5	2952

Table 4. Running times (ms) for the algorithm presented here on 3D images.

$\sigma =$	2	5	10	20	50	100
Sidelength						
10	2	2.5	2.5	1.5	3.5	4
20	7.5	8.5	8	6	9	7
50	109	108.5	107	109	111	109
100	919	922.5	919	918	929.5	926

solution of this implicit system. The solution requires the same order of computation as a standard recursive filtering while the matrix decomposition needs to be performed only once along each axis.

The method was applied to both symmetric and anti-symmetric filters. It has been demonstrated on Deriche's recursive approximations to filtering by a Gaussian and its first derivative. Results demonstrate that the proposed method has excellent numerical accuracy. In contrast to standard implementations it requires a constant amount of computation irrespective of scale. Its speed is similar to a standard implementation of recursive filtering on 2D images and is faster on 3D images.

Acknowledgements

The diatom image in Figures 1 and 2 was taken from the ADIAC public data web page: http://www.ualg.pt/adiac/pubdat/pubdat.html   (CEC   contract MAS3-CT97-0122). The author would like to thank Peter Kootsookos of the University of Queensland and David Chan and Carolyn Evans of CSIRO Mathematical and Information Sciences for interesting discussions and assistance in proof-reading this paper.

References

1. Deriche, R.: Fast algorithms for low-level vision. IEEE Tr. on Pattern Analysis and Machine Intelligence **12** (1990) 78–87
2. Martucci, S.A.: Symmetric convolution and the discrete sine and cosine transforms. IEEE Transactions on Signal Processing **42** (1994) 1038–1051

3. Deriche, R.: Recursively implementing the gaussian and its derivatives. Technical Report 1893, Programme 4 - Robotique, Image et Vision, INRIA - Institut National en Informatique et en Automatique (1993)
4. da Cunha, A.M.: Espaços de escala e detecção de arestas. Master's thesis, IMPA, Rio de Janeiro (2000) http://www.visgraf.impa.br/escala.html.
5. Smith, M.J.T., Eddins, S.L.: Analysis/synthesis techniques for subband image coding. IEEE Transactions on Acoustics, Speech, and Signal Processing **38** (1990) 1446–1456
6. Weickert, J., ter Haar Romeny, B.M., Viergever, M.A.: Efficient and reliable schemes for nonlinear diffusion filtering. IEEE Transactions on Image Processing **7** (1998) 398–410
7. Oppenheim, A.V., with John R. Buck, R.W.S.: Discrete-Time Signal Processing. second edn. Prentice-Hall (1999)
8. Golub, G.H., Loan, C.F.V.: Matrix computations. Third edn. Johns Hopkins University Press (1996)

Shape Description
Using Gradient Vector Field Histograms

Wooi-Boon Goh and Kai-Yun Chan

School of Computer Engineering, Nanyang Technological University,
Nanyang Avenue, Singapore 639798
{aswbgoh,askychan}@ntu.edu.sg

Abstract. We present a novel approach to shape representation that describes a shape using a set of histograms derived at salient points within the shape. A computationally efficient multiresolution pyramidal framework is used to generate a dense gradient vector field whose characteristics can be altered through the use of a scale parameter α. This parameter regulates the proportion of low and high spatial frequency components used in creating the vector field and can be set such that minor boundary distortions do not significantly change the representation of the shape. Local maximas of the directional disparity measure in the vector field are used for locating shape axes, from where polar sampling of the vector field is then used to build scale and rotational invariant histograms that describes subparts of the shape. A saliency measure based on the size of a part is introduced to provide appropriate weighting to each part during the shape matching process. Experimental results involving silhouettes images are presented to demonstrate the effectiveness of the proposed gradient vector field histograms for similarity-based shape retrieval.

1 Introduction

There have been many criteria used to classify the different approaches to shape description and analysis. A common classification divides the various approaches to either one that is boundary-based, region-based or a hybrid of the two. Boundary-based techniques represent shapes using their outline. Outline curves typically do not describe the interior of the shape well and as such, it has several drawbacks such difficulty in handling rotation and scale invariance, sensitivity to articulation, deformation and spatial rearrangement of parts [20]. Despite these limitations, it has been popular and has been effectively deployed in several applications [9], [13].

Region-based techniques such as those that use moment invariants [18] are very robust to noise and have the ability to perform shape recognition under affine and projective transformation. However, because of its more global nature, it is sensitive to occlusion and articulation.

The hybrid techniques seek to combine the strengths of both the boundary and region-based approaches. Medial axis representation and its variants describe both the interior and outline of the shape [4], [14], [21], [24]. A particularly rich medial axis

L.D. Griffin and M. Lillholm (Eds.): Scale-Space 2003, LNCS 2695, pp. 713–728, 2003.
© Springer-Verlag Berlin Heidelberg 2003

shape descriptor is the shock graph as it encapsulates the shape geometry in the dynamics of the shock formation during the boundary curve evolution process [12]. Shock graph matching has been used in object recognition and image indexing applications [21], [24]. The shape descriptor proposed in this paper is similar in principle to the medial axis representation approach but unlike [14], [21], [24]; we do not represent shapes by graphs or trees. In our approach, the shape axes provide a means to partition the shape into parts. Each part is then described by histograms that contain information about the shape of the part and its geometric relationship to other parts in its vicinity. Shape similarity is then computed by comparing the histograms.

One of the weaknesses of medial axes extraction algorithms such as Blum's [4] is its sensitivity to boundary perturbations. Most of the approaches proposed to address this problem adopt some form of multiscale or multiresolution framework [17]. The Hierarchic Voronoi skeleton approach of Ogniewicz and Kübler [15] adopts an axis-pruning process that reduces the insignificant medial branches extracted and organizes the more relevant ones into a hierarchical framework (skeleton-space) based of their measure of prominence. Pizer et al. [16] proposed a scale-space approach that describes the medialness function of a shape over a width-proportional scale. The cores describing the interior medialness of shapes are extracted in the wider scales, whilst the cores of minor boundary protrusions are extracted at finer scales. The cores in the interior of the shape and those near the boundary are not necessarily connected. We argue that this is an important feature of boundary noise robust medial axis representation, which our proposed multiresolution gradient vector field histogram approach shares with cores [16] and other techniques like [19].

In section 2, we detail the multiresolution technique used to generate the gradient vector field and vector field disparity map. We also describe the effect of varying the scale parameter α. In section 3, we describe how the proposed the shape descriptors are derived from the gradient vector field. Section 4 describes how the proposed shape descriptors are used to compute a shape similarity measure. Section 5 presents experimental results to illustrate the effectiveness of the proposed shape descriptor in handling invariance to similarity transform and boundary distortions. We also present some results from shape classification experiments.

2   The Multiresolution Pyramidal Framework

In our implementation, a computationally efficient multiresolution pyramidal framework is used to derive the gradient vector field and vector field disparity map. We first review several useful pyramid operations used in this work. The process which generates a lower resolution image from its predecessor will be called a *REDUCE* operation [5]. If G_0 is the original image $I(x, y)$ and G_N is the top level of the pyramid, then for $0 < l \leq N$, $G_l = REDUCE [G_{l-1}]$ is define as

$$G_l(x, y) = \sum_{m,n=-k}^{k} w(m, n) G_{l-1}(2x + m, 2y + n) \qquad (1)$$

where the generating kernel w of size $(2k+1)$ performs smoothing before the sub-sampling process. For computational efficiency reasons [5], a separable, normalized and symmetric Gaussian kernel of $k = 2$ is used throughout. Another pyramid operation, *EXPAND* is used to expand an image of size $M+1$ to $2M+1$ by interpolating sample values from a low resolution image. If G_l is derived by expanding it low resolution image G_{l+1}, then for $0 \leq l < N$, $G_l = EXPAND [G_{l+1}]$ is define as

$$G_l(x, y) = 4 \sum_{m,n=-k}^{k} w(m,n)G_{l+1}(\frac{x+m}{2}, \frac{y+n}{2}) \qquad (2)$$

where summation is only carried out when $x+m$ and $y+n$ are even numbers.

2.1 The Gradient Vector Field Pyramid

It is assumed that the input image is a 2D binary image $I(x, y)$. The procedure for creating a Gradient Vector Field pyramid is illustrated in Fig. 1.

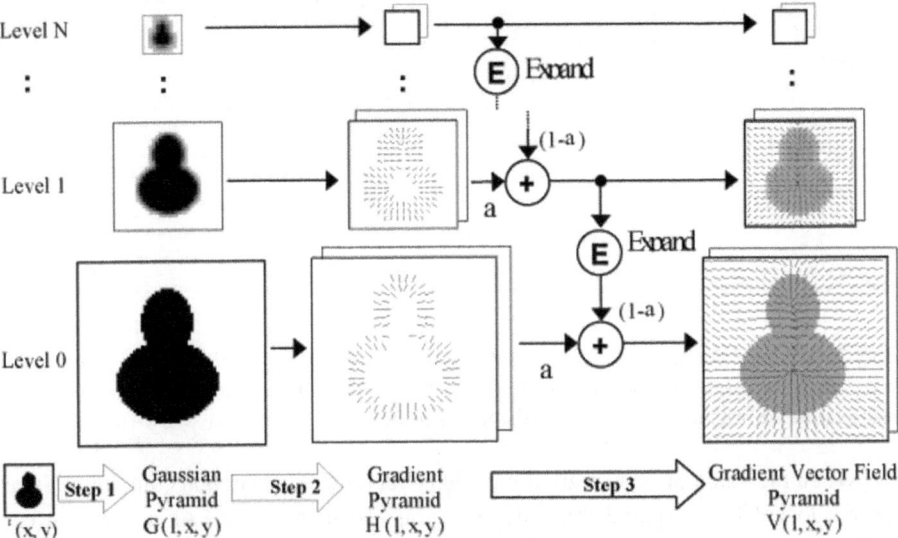

Fig. 1. A summary of the steps involved in constructing the Gradient Vector Field pyramid from a 2D binary image $I(x, y)$.

Firstly, a Gaussian pyramid $G(l, x, y)$ of $N+1$ levels is created by iteratively applying the *REDUCE* operation N times on each consecutive output image, starting with $I(x, y)$. From the scalar Gaussian pyramid, we then derived the vectorial Gradient pyramid $H(l, x, y)$, which consist of two pyramids $H^x(l, x, y)$ and $H^y(l, x, y)$ given by

$$H_l^x(x, y) = g_{\sigma_H}^x(x, y) * G_l(x, y) \text{ and } H_l^y(x, y) = g_{\sigma_H}^y(x, y) * G_l(x, y) \text{ for } 0 \leq l \leq N \qquad (3)$$

The convolution kernels $g_{\sigma_H}^x$ and $g_{\sigma_H}^y$ are Gaussian derivatives of the first-order in the x and y directions, respectively and are given by

$$g_{\sigma_H}^x (x, y) = -\frac{x}{\sigma_H^2}\exp(-\frac{x^2 + y^2}{2\sigma_H^2}) \quad \text{and} \quad g_{\sigma_H}^y (x, y) = -\frac{y}{\sigma_H^2}\exp(-\frac{x^2 + y^2}{2\sigma_H^2}) \quad (4)$$

The Gradient pyramid essentially describes the gradient of the image $I(x, y)$ at octave scales as the pyramid is traversed from one level to another. Finally, starting at $l = N$-1, each level H_l of the Gradient pyramid is expanded and then combined to derived the Gradient Vector Field pyramid $V(l, x, y)$, where each level V_l is defines as

$$V_l = \alpha\, H_l + (1 - \alpha)\, EXPAND\,[V_{l+1}] \qquad \text{for } 0 \leq l < N \qquad (5)$$

At the top, we have $V_N = H_N$ and the scale parameter $\alpha \in [0, 1]$ determines the smoothness of the gradient vector field. Therefore, by varying α, we can control the characteristic of the extracted shape axes as illustrated by the examples in Fig. 2.

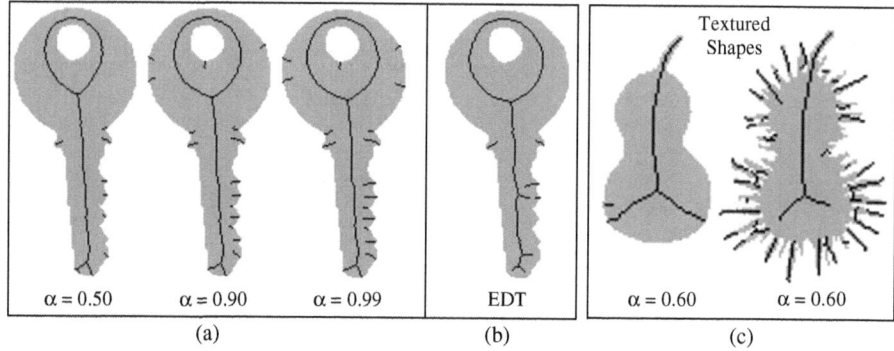

| $\alpha = 0.50$ | $\alpha = 0.90$ | $\alpha = 0.99$ | EDT | $\alpha = 0.60$ | $\alpha = 0.60$ |
| (a) | | | (b) | (c) | |

Fig. 2. (a) Shape axes extracted from a key shape at three different values of α. (b) The medial axis extracted from a Euclidean Distance Transform image of the same key shape. (c) Shape axes extracted from textured pear-like shapes taken from [12].

When the value of α is small, the gradient vector field generated contains a higher proportion of lower spatial frequency components of the shape. As a result, at $\alpha = 0.50$, boundary artifacts such as the serrations on the key shape in Fig. 2a are deemphasied but they become more prominent when α is increased to 0.90. Fig 2b shows the medial axes extracted from the Euclidean Distance Transform (EDT) image of the key shape, which is a good approximation to a subset of the medial locus defined by Blum's medial axes [4], where the locus is defined by the centre of all maximal disks in the interior of the shape that contacts the boundary in two or more separate positions. As can be observed in Fig. 2a when $\alpha = 0.50$ and $\alpha = 0.90$, the shape axes in the interior of the key do not conform to the medial locus as define by Blum. This is due to the non-trivial combination of high and low spatial frequency gradient information at the smaller inner boundary (hole in the key) and the larger outer boundary at various discrete spatial scales. The conformity improves as $\alpha \rightarrow 1$

but complete conformity is not possible as a $\alpha = 1$ means using only the gradient information of a full resolution image. This may not produce significant gradient information within the interior of a large-sized shape due to the limited spatial extent of the Gaussian derivative kernels (see Gradient pyramid in Fig. 1). Notice at $\alpha = 0.99$, the major shape axes in the interior of the shape recovered from the gradient vector field are almost identical to that recovered using EDT.

Fig. 2c shows how a low value of $\alpha = 0.60$ can result in a gradient vector field that provides relatively stable major shape axes that describe the gross visual form of textured shapes. Unconnected axes that remain at the boundary also capture boundary texture information.

2.2 The Vector Field Disparity Map

In order to describe a complex shape by decomposing it into suitable parts, the shape axes must first be extracted from the gradient vector field. Given the gradient vector field $V(x, y)$, the shape axes can be located by detecting locations in the vector field where the local gradient vectors exhibit high directional disparity. This is similar in principle to the technique proposed in [25], where Siddiqi et al. used the measure of average outward flux over a very small neighborhood to detect medial points. Extending the idea Ben-Arie and Wang in [2], we propose an alternative vector disparity operator $\mathcal{D}$ given by

$$\mathcal{D}(V(x), \sigma_D) = \frac{g_{\sigma_D}(x) * \|V(x)\| - \|g_{\sigma_D}(x) * V(x)\|}{g_{\sigma_D}(x) * \|V(x)\|} \tag{6}$$

where $x = (x, y)$ and the local disparity measure is computed within a weighted locality defined by the Gaussian convolution kernel $g_{\sigma_D}(x)$ given by

$$g_{\sigma_D}(x, y) = \frac{1}{\sqrt{2\pi}\sigma_D} \exp(-\frac{x^2 + y^2}{2\sigma_D^2}) \tag{7}$$

The Vector Field Disparity pyramid $D(l, x, y)$ is derived from Gradient Vector Field pyramid $V(l, x, y)$, where each level D_l is define as

$$D_l(x, y) = \mathcal{D}(V_l(x, y), \sigma_D) \quad \text{for } 0 \leq l \leq N \tag{8}$$

The normalized vector field disparity measure $D_l(x, y) \in [0, 1]$ gives a value close to 1 in localities of high disparity such as at the centre of a circle. In order to detect consistent shape axes over different scales of a shape, a full resolution vector field disparity map $M(x, y)$ is obtained by iteratively applying EXPAND to the sum of D_l and EXPAND $[D_{l+1}]$, starting at $l = N-1$ to $l = 0$. This multi-level integration process is illustrated in Fig. 3a. The final summed output is divided by $N+1$ to re-normalise the disparity map such that $M(x, y) \in [0, 1]$.

Pizer et al. [16] proposed extracting relevant cores of shapes and its subparts of varying width by tracking optimal scale ridges over a 3D scale-space volume. This approach is computationally costly and requires large amount of memory if fine scale resolutions are required. The proposed integration of all levels of in Vector Field Disparity pyramid $D(l, x, y)$ into one full resolution disparity map $M(x, y)$ allows us to simultaneously extract all relevant shape axes from a single 2D spatial vector, as can be seen in Fig. 3b. The integration of what are essentially multiscale features spread over the discrete pyramidal scale-space into a single 2D spatial image is possible here because the medial features we are attempting to extract, though relevant at different scales, are not spatially co-located. This is true of the symmetry set defined by the medial locus of Blum [4]. The shape axes extracted are also consistent over a wide scale variation of a given shape, as demonstrated by the three rabbits in Fig. 3b.

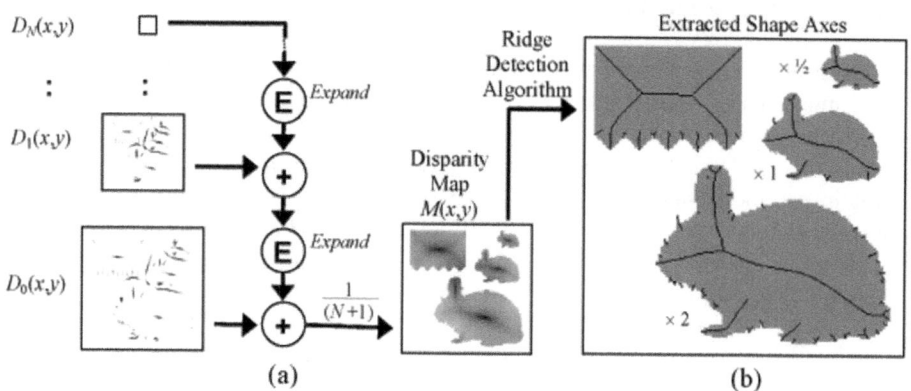

Fig. 3. (a) A multi-level integration technique for generating a vector field disparity map $M(x, y)$. (b) Shape axes extracted from an image containing a test shape taken from [16] and three similar rabbit shapes spanning scales of two octaves.

2.3   Shape Axes Extraction

Given a full resolution disparity map $M(x, y)$, the shape axes are extracted by locating the local maximas in the disparity measure. However, as seen in Fig. 4a, the ridges in the interior of the shape has been significantly blurred due to the multiple *EXPAND* operations used to derive $M(x, y)$. This poses difficulty in the ridge extraction stage. We propose resolving this problem by first obtaining the derivative of $M(x, y)$ given by

$$M_V(x, y) = \nabla g_{\sigma_V} * M(x, y) \tag{9}$$

where ∇g_{σ_V} produces first-order Gaussian derivative convolution kernels with a standard deviation of σ_V, as defined in (4) and the resulting disparity gradient map M_V is a vector consisting of the x and y derivatives of the disparity map given by M_V^x and M_V^y respectively. We then apply the vector disparity operator $\mathcal{D}$ in (6) to M_V to get an enhanced disparity map $M_E(x, y)$ shown in Fig. 4b, which is given by

$$M_E(x, y) = \mathcal{D}(M_V, \sigma_E) \tag{10}$$

where σ_E determines the size of the effective neighbourhood in which the vector disparity is computed. Unfortunately, the vector disparity operator $\mathcal{D}$ does not differentiate between ridges and valleys. As a result, both are accentuated. Ridge and valley points in the disparity map $M(x, y)$ can be readily differentiated by observing the characteristics of its gradient vectors define in M_V. Vectors around ridge points are outward flowing and those around valleys are inward flowing. As such, a landscape map $M_L(x, y)$ that highlights 'ridgeness' and 'valleyness' in the disparity map can be constructed by finding the sum of the resulting two convolutions of each x and y gradient component of the disparity map $M(x, y)$, with their respective x and y first-order Gaussian derivative kernels defined in (4), and is given by

$$M_L(x, y) = \sum_{i=x,y} g^i_{\sigma_R}(x, y) * M^i_V(x, y) \tag{11}$$

$M_L(x, y)$ gives large positive values at ridge points and large negative values at valleys points. A ridge-enhanced disparity map $M_R(x, y)$ can now be created using

$$M_R(x, y) = M_E(x, y) \, M_L(x, y) \tag{12}$$

The local maximas on the ridge-enhanced map $M_R(x, y)$ are detected using a simple local ridge detection algorithm, which flags out a pixel as a ridge pixel when a specified number of its neighbouring pixels have disparity values less than its own. In our implementation, a 5×5 square neighbourhood is used and at least 14 pixels must have disparity values less than the centre pixel to classify it as a ridge point. The result of the local ridge detection algorithm is shown in Fig. 4c. Standard disparity-weighted morphological thinning can be further applied to thin the extracted ridges to obtain unit-width shape axes as shown in Fig. 4d. In this process, connectedness is maintained while ridge pixels are progressively eroded from outside-in. Removal preference is given to pixels with lower disparity values, thus ensuring the single-width shape axes reside as close to the axes of maximum vector field disparity.

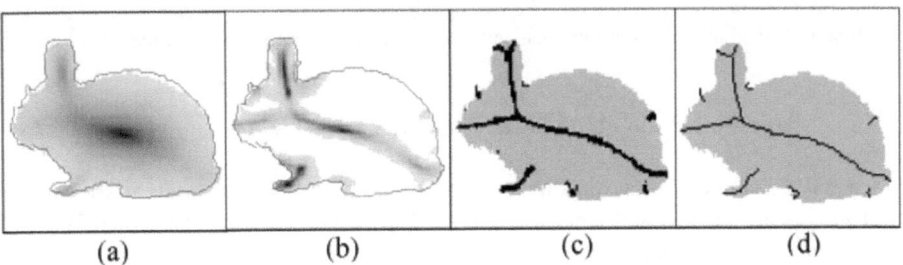

(a)   (b)   (c)   (d)

Fig. 4. (a) Disparity map $M(x, y)$. (b) Enhanced disparity map $M_E(x, y)$. (c) Ridge extracted using a local ridge detection algorithm. (d) Shape axes thinned to unit-width.

3   Shape Description

3.1   Description by Parts

There is much evidence that human perception of shapes involves some form of part-based representation [11], [13], [23]. It is not difficult to see why as this more local form of shape description allows for recognition that is more robust to occlusion and articulation of limbs [23]. Furthermore, from the image database retrieval perspective, a part-based description simplifies the model database since a complex shape can be described by a collection of part descriptors that are lower in dimension and more amenable to simple indexing and matching schemes. In our work, a shape is decomposed into parts based on the extracted shape axes and each part is statistically described by two histograms derived from the gradient vector field within the locality of the segmented shape axis, as shown in Fig. 5.

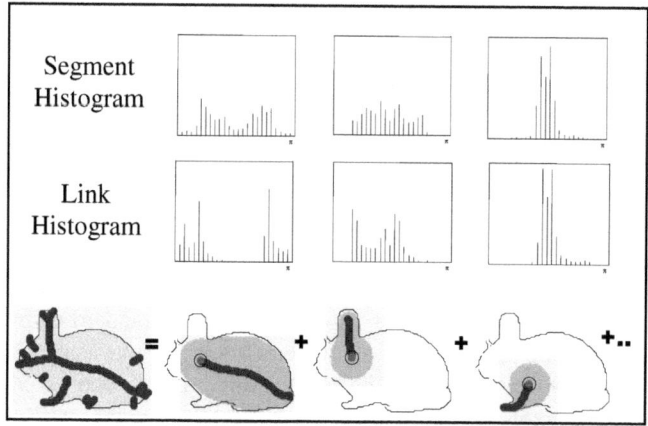

Fig. 5. Decomposition of a rabbit shape into subparts based of its extracted shape axes. The gray-shaded area of each part shows the locality U_k, defined in (13), from where the segment and link histograms are computed. The size of this locality gives a good indication of the part's saliency. The small circle marking one end of each of the part axis is a link node.

3.2   Shape Description Using the Gradient Vector Field

Several researchers in the past have proposed the use of vector fields in the analysis of gray-scale images and shapes [2], [8], [22]. More recently, Shroff and Ben-Arie [22] modeled the gradient of a shape as magnetic dipoles and extracted smooth shape axes at point where the magnetic field interaction resulted in a local minima. Cross and Hancock [8] proposed a multiscale framework where a vector field is obtained by computing the curl of vector potential found through volume averaging of the tangential edge gradient vectors. The additional z-dimension provides a scale-space representation of the extracted edge and symmetry lines. However, both these works were only limited to the task of extracting symmetry axes of shapes, and in the case of the later, the detection of edges as well. The work of Ben-Arie and Wang [2] is closest in scope and concept to the work described in this paper. They proposed extracting hierarchical shape descriptors at location of high vectorial disparity (based on a measure they termed Cancellation Energy) when the image gradients at shape boundaries are radially propagated. The issue of scale was addressed by their use of a Vectorial Gradual Lattice Pyramid and their normalized shape feature tokens were shown to be invariant to scaling and rotation. However, the characteristics of the vector field and the axes of symmetry embedded within it were not fully exploited in order to extract a richer description of the shape and the subtle geometric relationship between shape parts.

Different shapes produce gradient vector fields with differing statistical properties as can be seen in the histograms of the vector orientations shown in Fig. 6. The vector field essentially describes the regional interaction of the shape boundary and is thus very descriptive of the general visual form and geometric relationship between parts of the shape. A statistical descriptors such as a histogram of the vector orientation is relatively robust to minor perturbation on the shape boundary (see Fig. 6d), especially if the deformation of the boundary is relatively small compared to the overall size of the shape. This correlates well to the perceptual significance of such distortions.

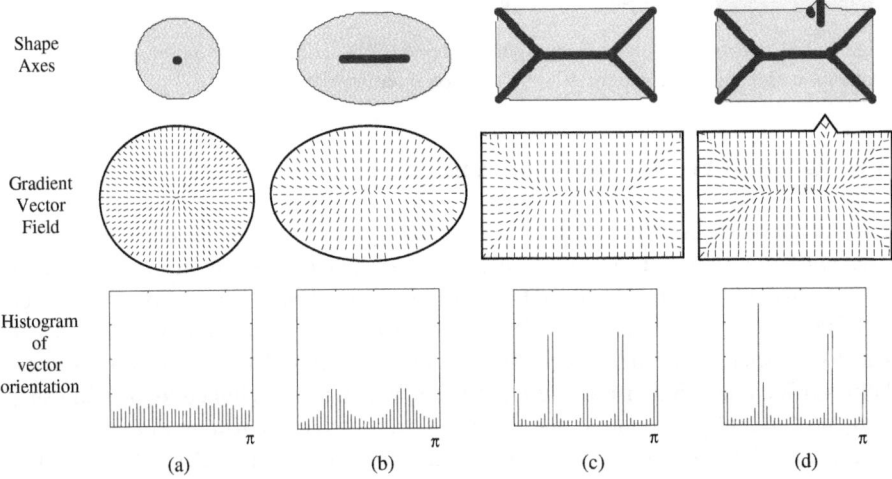

Fig. 6. (a), (b) and (c) show the extracted shape axes, gradient vector field and orientation histograms for various simple shapes. (d) A minor distortion in the shape boundary results in a proportionally small change in the vector field histogram.

3.3 Part Axes

In keeping with a part-based representation, we decompose a shape into subparts defined by a continuous segment of the shape axes. Using skeleton analysis algorithms, the shape axes are segmented at every intersections and junctions into Q continuous segments called *part axes*. With reference to Fig. 7, let the kth part axis $P_k = \{p_k^i\}_{i=1}^M$ be an ordered set of M discrete points starting at one end of a continuous segment and ending at the other. Let r_1 and r_M be the respective radial distances of points p_k^1 and p_k^M to their nearest edge points on the shape boundary. If the ordering of the points $\{p_k^i\}_{i=1}^M$ on P_k is such that $r_1 > r_M$, then the starting point p_k^1 is termed the *link node* of the part axis P_k. For computational and memory storage efficiency reasons, our shape descriptor adopts a single link node (as oppose to link nodes at both ends). The tradeoff is a slight degradation in discriminative power of the shape descriptors.

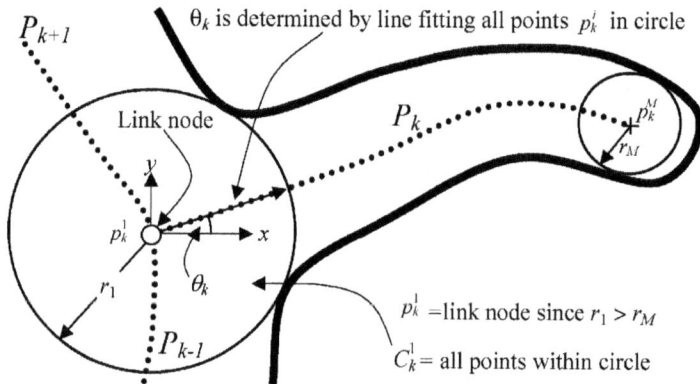

θ_k is determined by line fitting all points p_k^i in circle

P_{k+1}

Link node

P_k

p_k^M

r_M

y

p_k^1

x

r_1

θ_k

P_{k-1}

p_k^1 =link node since $r_1 > r_M$

C_k^1 = all points within circle

Fig. 7. The part axis P_k and relevant parameters for link node determination, histogram normalization and angular compensation for rotation invariance (see text).

3.4   Gradient Vector Field Histograms

Each part axis is associated with two normalized gradient vector field histograms, the *segment histogram* and the *link histogram*. The segment histogram S_k describes the general shape of the part, such as its length-to-width ratio, its convexity, its taper, etc. The link histogram L_k^1 contains information pertaining to the common space that the part axis k shares with other part axes within its vicinity. This relational information helps differentiate similar looking protrusions, which may be linked to the main shape body at different places and in varying configurations. Link histogram L_k^1 is first constructed by determining the vector orientation associated with all C_k^1 discrete image pixels within the circle of radius r_1, centered about the start point p_k^1. The histogram L_k^1 with n bins representing the value range $[0, 2\pi)$ cumulates the quantised orientation of all C_k^1 gradient vectors. The histogram is made rotationally invariant by adding a value θ_k^1 to all orientation values before cumulation. The angle θ_k^1 is derived from the orientation of a straight line fitted along all part axis points $\{p_k^i\}_{i=1}^M \cap C_k^1$ that lie within the circle of radius r_1. By using the link node as an orientation reference for the part, the start point of the straight line is the end closest to p_k^1. The link histogram L_k^1 is normalized with the value C_k^1. The segment histogram S_k is obtained by repeating the procedure described for extracting histogram L_k^1 but this time summing all the computed histograms L_k^i for $1 \leq i \leq M$ over the entire length of the part axis P_k. More formally, for a part axis P_k, whose locality is define by a region U_k, its segment histogram S_k and its associated normalisation value W_k are given by

$$S_k = \sum_{i=1}^{M} L_k^i \quad \text{and} \quad W_k = \sum_{i=1}^{M} C_k^i \quad \text{and} \quad U_k = \bigcup_{i=1}^{M} C_k^i \tag{13}$$

3.5   Saliency of Part

For effective part-based shape similarity computation, it is essential that a mismatch of a more salient part should carry a higher penalty then one that is small and not perceptually significant compared to the overall shape. An example is the boundary kink on the rectangle in Fig. 6d. In our work, a salient part is defined as one that covers a large region of the overall shape (see Fig. 5). A useful measure that relates to this definition of saliency is the value W_k given in (13). Therefore, if a shape A has Q_A part axes, then the relative saliency $R_{A,k} \in (0, 1]$ of the part axes $P_{A,k}$ is given by

$$R_{A,k} = W_k \left/ \sum_{i=1}^{Q_A} W_i \right. \tag{14}$$

The shape axes extracted by our proposed multiresolution technique differ from the shocks of [12] and medial axes of [4] because minor distortions in the shape boundary results in disconnected shape axes that remain mainly at the boundary (see Fig. 2). This characteristic fits neatly into our definition of part saliency given in (14), as minor boundary details like small protrusions and texture remain disconnected and close to the boundary, thus resulting in low saliency features. This saliency measure can be used to speed up image retrieval operations. An appropriate scale-invariant saliency threshold can be set to remove detailed shape features during the first stage of shape comparison. Subsequent accurate shape comparison using all available histograms can then be done on the smaller subset of shapes retrieved earlier.

4   Matching Parts and Shapes

The triangular inequality does not generally hold in shape similarity comparison [3], especially in situation where parts of the shape are occluded or a shape is a composite of two or more simpler shapes. As a consequence, the distance between two shapes should not be evaluated based on just a metric distance. But from a computational viewpoint, some form of metric distance must be used to allow part-based descriptors to be organized into a database and retrieved through multi-dimensional index structures such as R-trees or M-trees [6], [10], for example. Similar to the work of [3], we propose reconciling the two contrasting requirements with the use of two distinct measures called *part distance* and *shape distance*. The part distance is a metric distance used to measure similarity between two parts. On the other hand, the shape distance is a non-metric distance defined as the optimal combination of part distances between two shapes and it yields a global measure of shape similarity that is more consistent with human visual perception.

4.1   Part and Shape Distances

The part distance $d_p(P_{A,k}, P_{B,k})$ between two part axes $P_{A,k}$ and $P_{B,k}$ extracted from shapes A and B is given by the combined χ^2 distance between their respective n-bin normalised link and segment histograms and is given by

$$d_p(P_{A,k}, P_{B,k}) = \frac{\beta}{2} \sum_{i=1}^{n} \frac{[L_{A,k}^1(i) - L_{B,k}^1(i)]^2}{L_{A,k}^1(i) + L_{B,k}^1(i)} + \frac{(1-\beta)}{2} \sum_{i=1}^{n} \frac{[S_{A,k}(i) - S_{B,k}(i)]^2}{S_{A,k}(i) + S_{B,k}(i)} \tag{15}$$

where $d_p(P_{A,k}, P_{B,k}) \in [0,1]$ and the parameter $\beta \in [0,1]$, determines the relative importance attached to the matching of the link and segment histograms. A value of $\beta = 0.5$ will give equal emphasis to both when computing the matching cost. Using (15), the part distance matrix PM_{AB} of size $Q_A \times Q_B$ is obtained by matching all part axes in shape A to those in shape B and we assume $Q_A \leq Q_B$ (if not, shapes A and B are reversed). The matching costs in matrix PM_{AB} is then weighted by the part saliency of shape A as defined in (14), to yield a weighted part distance matrix WM_{AB} given by

$$WM_{AB}(i,j) = PM_{AB}(i,j) R_{A,i} \quad \text{for } 1 \leq i \leq Q_A \text{ and } 1 \leq j \leq Q_B \tag{16}$$

The shape distance between two shapes A and B, where $Q_A \leq Q_B$, is the optimal saliency weighted match between all parts in A and a subset of parts in B that results in the lowest total matching cost. Given the weighted part distance matrix WM_{AB}, we can determine the optimal match by minimizing the total cost of matching subjected to the constraint that a one-to-one match exist [1]. Since $Q_A \leq Q_B$, we can realize a one-to-one assignment by adding unit dummy costs to WM_{AB} to turn it into a square matrix. With a square WM_{AB} of size $Q_B \times Q_B$, we have essentially reduced this task to a matching problem for bipartite graphs and this can be solved in $O(Q_B^3)$ time by the Hungarian method [1], [7]. The resulting output from the Hungarian method is a permutation of $(P_{A,i}, P_{B,j})$ pairs that results in the lowest overall matching cost $d_h(A, B)$. The shape distance $d_s(A, B)$ between shapes A and B is then obtained by removing the added unit dummy costs and is given by $d_s(A, B) = d_h(A, B) - (Q_B - Q_A)$.

In order to handle reflection, a mirrored version of shape A must also be matched to shape B to see which yields a lower shape distance. To obtain a mirrored version of shape A, the link and segment histograms of all its part axes $P_{A,k}$ are reversed. This means $L_{A,k}^1(i)$ and $S_{A,k}(i)$ becomes $L_{A,k}^1(n-i+1)$ and $S_{A,k}(n-i+1)$ respectively, where n is the number of bins in the histograms. Choosing the lower of the two shape distances will ensure invariance to the mirror transform.

5   Experimental Results

All results presented were obtained with standard deviation values of $\sigma_H = 1.0$, $\sigma_D = 1.5$, $\sigma_V = 1.0$, $\sigma_E = 1.5$, $\sigma_R = 1.0$. The top pyramid level for image sizes 257×257 and 129×129 (default size) pixels are $N = 5$ and $N = 4$, respectively. All histograms' bin sizes are $n = 24$. The values of $\alpha = 0.9$ and $\beta = 0.5$ unless otherwise stated.

5.1   Geometric Invariance

The first experiment demonstrates the proposed shape descriptor's invariance to the similarity transform and mirror reflection. Twelve 257×257 pixel-sized images comprising of three basic shapes (i.e. hare, rabbit and turtle) in various orientations and spanning scales of two octaves were matched each to every other. The closest three matches to each image are shown in Fig. 8. Except for the one image marked (×), all shapes were correctly matches to their respective shape type. Errors in the gradient vector field histograms are more significant when the shape is small due to the reduced sample points making up the histograms. It is not surprising that within the best 3 matches, the largest-scaled rabbit was not matched to the smallest-scaled rabbit but to a scale (×1) hare as the shape distances between hares and rabbits are small.

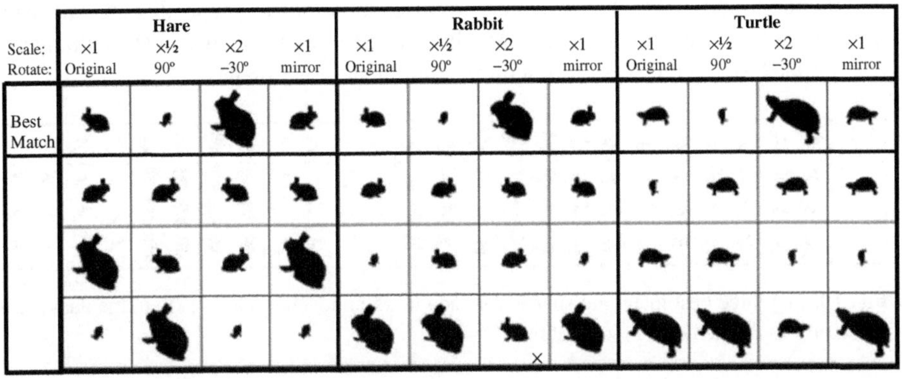

	Hare				Rabbit				Turtle			
Scale:	×1	×½	×2	×1	×1	×½	×2	×1	×1	×½	×2	×1
Rotate:	Original	90°	−30°	mirror	Original	90°	−30°	mirror	Original	90°	−30°	mirror
Best Match												
										×		

Fig. 8. Results demonstrating invariance to scaling, rotation and mirror reflection.

5.2   Robustness to Boundary Distortions

The results in Fig. 9 demonstrate the advantage of our disconnected medial representation and the incorporation of a region-based saliency weighting in the shape similarity computation. Short unconnected part axes at the boundary that describe textural information have a lower saliency measure compared to those in the interior of the shape, which describes the gross shape form. As a result, all 15 textured shapes in Fig. 9 were able to find their other two 'compatriot' shapes within the first two best matches. A scale parameter $\alpha = 0.6$ was used to obtain the results shown.

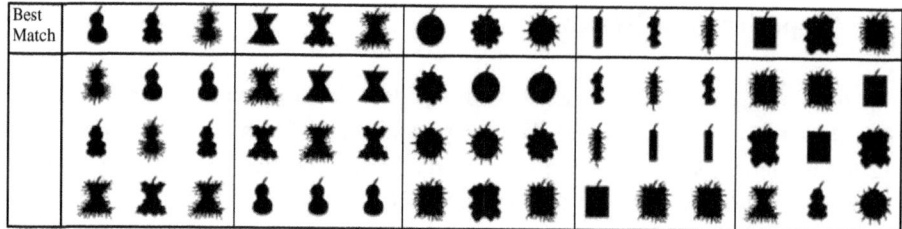

Fig. 9. The three best matching shapes for each textured shape. 5 general shape forms, each with 3 different boundary textures (smooth, bumpy and spiky) were used.

5.3   Silhouettes Classification

A database of 25 silhouettes images forming 6 different classes were compared one to another. Fig. 10 displays the three smallest shape distances for each of the 25 shapes, with the smallest shape distance label best match #1. Other authors presented similar tests using the same image data set and they present results in the form of the 1st, 2nd and 3rd best matches that fall into the correct category. Measured this way; the result presented in Fig. 10 is 25/25, 21/25, 21/25. This is comparable to results present in other works. The result reported in [1] is 25/25, 24/25, 23/25, and in [21] it is 23/25, 21/25, 20/25, and in [9] it is 25/25, 21/25, 19/25. Notice most errors occur between the spanners and fishes because our proposed non-metric shape distance measure produces a good match when a simpler shape like the spanner, finds a subset match to the similar-looking tail section of a more complicated fish shape.

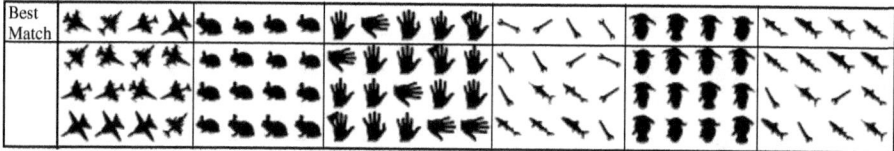

Fig. 10. The three best matching shapes for each of 25 silhouette images [21] in the data set. A value of β as defined in (15) was set to 0.5.

In order to test the effectiveness of incorporating the link histogram into our shape descriptor, the same test was re-run with $\beta = 0$. This means only the segment histogram is used in computing the shape distances. Fig. 11a shows an increase in classification errors, from 8 to 12. Without the influence of the link histogram, it is even more difficult to distinguish the spanner from the tail section of the fish.

We also tested the improved discriminative power when the link histograms at both ends of a part axes are available. The part distance $d_p(P_{A,k}, P_{B,k})$ in (15) is modified such that $L^1_{B,k}(i)$ is now $0.5[L^1_{B,k}(i)|L^M_{B,k}(i)]$, where $|$ means an ordered concatenation of histograms. $L^1_{A,k}(i)$ becomes either $0.5[L^1_{A,k}(i)|L^M_{A,k}(i)]$ or $0.5[L^M_{A,k}(i)|L^1_{A,k}(i)]$ and we select the order that gives the lowest part distance. Using $\beta = 0.5$, shape matching using two links histograms has reduced the classification error from 8 in

Fig. 10 to 4 in Fig. 11b. This comes at the expense of increased computation during shape matching and about 50% increase in shape descriptor vector size.

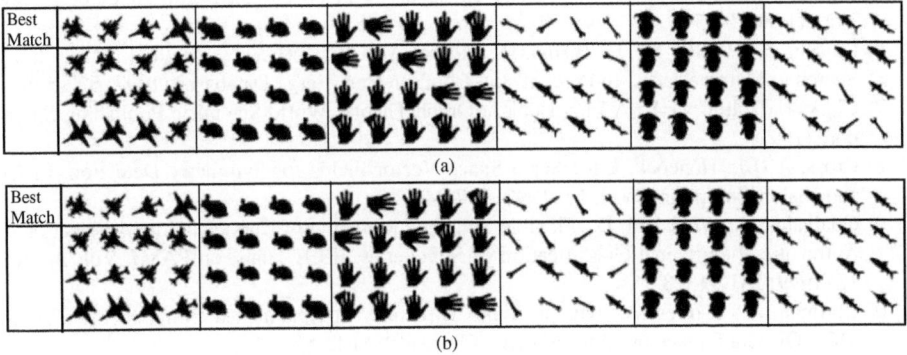

Fig. 11. (a) Results from using only the segment histogram ($\beta = 0$). (b) Improved results when histograms from two link nodes (one at each end) are available.

6   Conclusions

We have presented a novel shape descriptor that is derived from the gradient vector field of shapes in binary images. These part-based shape descriptors encapsulates both the form of the part as well as it geometric relationship to other parts within it vicinity. They are also invariant to scaling, rotation and mirror reflection. Experimental results show that the use of semi-local statistics in the form of segment and link orientation histograms are sufficient in providing reliable shape matching and classification without any need for graph and tree matching. A novel computationally efficient multiresolution pyramidal framework was proposed for computing the gradient vector field and vector field disparity map. We demonstrated how the gradient vector field could be regulated via a scale parameter α such that the extracted shape descriptors within the interior of the shape describes the gross shape form, whilst those at the boundary describe boundary details. Our work suggest that this characteristic is important for effective shape representation as we have demonstrated how this representation naturally results in a saliency measure that facilitates robust part-based shape comparison in the presence of significant amount of boundary noise and artifacts.

References

1. Belongie, S., Malik, J., Puzicha, J.: Shape Matching and Object Recognition using Shape Contexts. IEEE Trans. on PAMI, Vol. 24, No. 24 (2002) 509-522
2. Ben-Arie, J., Wang, W.: Shape Description and Invariant Recognition Employing Connectionist Approach. Int. J. of Pattern Recognition and AI, Vol. 16, No.1 (2002) 69-83
3. Berretti, S., Bimbo, A.D., Pala, P.: Retrieval by Shape Similarity with Perceptual Distance and Effective Indexing. IEEE Trans. on Multimedia, Vol. 2, No. 4 (2000) 225-239

4. Blum, H.: A Transformation for Extracting New Descriptors of Shape. In Proc. Symp. Models for the Perception of Speech and Visual Form, Cambridge, MA: MIT Press (1964)
5. Burt, P.J.: The Pyramid as a Structure for Efficient Computation. In: Rosenfeld, A. (ed.): Multiresolution Image Processing and Analysis. Springer-Verlag, Berlin Heidelberg New York (1984) 6-35
6. Ciaccia, P., Patella, M., F, Zezula, P.: M-tree: An Efficient Access Method for similarity Search in Metric Spaces. In Proc. of Int. Conf. on Very Large Databases (1997) 522-525
7. Clark, J., Holton, D.A.: A First Look at Graph Theory. World Scientific Publishing Singapore (1991)
8. Cross, A.D.J., Hancock, E.R.: Scale Space Vector Fields for Symmetry Detection. Image and Vision Computing, Vol. 17 (1999) 337-345
9. Gdalyahu, Y., Weinshall, D.: Flexible Syntactic Matching of Curves and its Application to Automatic Hierarchical Classification of Silhouettes. IEEE Trans. on PAMI, Vol. 21, No. 12 (1999) 1312-1328
10. Guttman, A.: R-trees: A Dynamic Index Structure for Spatial Searching. In Proc. ACM SIGMOD Conference on Management of Data (1984) 47-57
11. Hoffman, D.D., Richards, W.A.: Parts of Recognition. Cognition, Vol. 18 (1985) 65-96
12. Kimia, B.B. Tannenbaum, A., Zucker, S.W.: Shape, Shocks and Deformation, I: The Components of 2-D Shape and the Reaction-diffusion Space, International Journal of Computer Vision, Vol. 15 (1995) 189-224
13. Latecki, L.J., Lakamper, R.: Convexity Rule for Shape Decomposition based on Discrete Contour Evolution. Comp. Vision and Image Understanding, Vol. 73, No.3 (1999) 441-454
14. Liu, T., Geiger, D.: Approximate Tree Matching and Shape Similarity. In Proc. International Conference on Computer Vision (1999) 456-462
15. Ogniewicz, R.L., Kübler, O.: Hierarchic Voronoi Skeletons, Pattern Recognition, Vol. 28, No. 3 (1995) 342-359
16. Pizer, S.M., Eberly, D., Fritsch, D.S., Morse, B.S.: Zoom-invariant Vision of Figural Shape: The Mathematics of Cores, Computer Vision & Image Understanding, Vol. 69, No. 1 (1998) 55-71
17. Pizer, S.M., Siddiqi, K., Székely, G., Damon, J.N., Zucker, S.W.: Multiscale Medial Loci and Their Properties, to appear in UNC-MIDAG issue of Int. Journal of Computer Vision, 2002, available at http://midag.cs.unc.edu/pubs/papers/IJCV01-Pizer-medloci.pdf.
18. Prokop, R.J., Reeves, A.P.: A Survey of Moment-based Techniques for Unoccluded Object Representation and Recognition. CVGIP: Graphical Models Image Processing, Vol. 54 (1992) 438-460
19. Rom, H., Medioni, G.: Hierarchical Decomposition and Axial Shape Description, IEEE Trans. on PAMI, Vol. 15, No. 10 (1993) 973-981
20. Sebastian, T.B., Kimia, B.B.: Curves vs Skeletons in Object Recognition. In Proc. International Conference on Image Processing, Vol. 3 (2001) 22-25
21. Sharvit, D., Chan, J., Tak, H., Kimia, B.: Symmetry-based Indexing of Image Databases. J. Visual Communication and Image Representation (1998)
22. Shroff, H., Ben-Arie, J.: Finding Shape Axes using Magnetic Field. IEEE Trans. on Image Processing. Vol. 8, No. 10 (1995) 1388-1394
23. Siddiqi, K., Kimia, B.B.: Parts of Visual Form: Computational Aspects. IEEE Trans. on PAMI, Vol. 17, No. 3 (1995) 239-251
24. Siddiqi, K., Shokoufandeh, A., Dickinson, S., Zucker, S.: Shock Graphs and Shape Matching. International Journal of Computer Vision, Vol. 35, No. 1 (1999) 13-32
25. Siddiqi, K.S., Bouix, S., Tannenbaum, A., & Zucker, S.W.: Hamilton-Jacobi Skeletons, International Journal of Computer Vision, Vol. 48, No. 3 (2002) 215-231

Comparing Objective and Subjective Quality Results for Compression Pre-processing with Non-linear Diffusion

Ivan Kopilovic[1] and Tamás Szirányi[2]

[1] University of Konstanz, Department of Computer & Information Science, Fach M 697
D-78457 Konstanz, Germany
kopilovi@inf.uni-konstanz.de
[2] Analogical and Neural Computing Laboratory, Comp. & Automation Inst.
Hungarian Academy of Sciences
H-1111 Budapest, Kende u. 13-17., Hungary
and
University of Veszprém, Department of Image Processing and Neurocomputing
sziranyi@sztaki.hu

Abstract. Compression systems like JPEG include optional pre-processing with filtering to avoid compression artefacts. At higher compression ratios a stronger filtering is needed that impacts the large scale image content. To preserve the large scale information we have previously proposed to use non-linear diffusion as a pre-processing for filtering out small scale details irrelevant at a given compression ratio and acting as noise. Now we compare typical diffusion processes applied before the blockwise DCT compression using the peak signal to noise ration (PSNR) as an objective quality measure. We give a simple measure of artefact reduction in terms of PSNR, and show that a considerable artefact reduction is achieved by pre-processing at the same bit rate as and with no greater error than the original compression. We did tests to see if the above artefact reduction implies a better subjective impression of quality. The images processed with the PSNR-based algorithm had nearly the same but greater PSNR value as the original compression. Subjects preferred noisy image content to the lack of small scale details, so the subjective preference of the images with reduced artefact is worse that of the original compression. Results suggest however that non-linear diffusion is more efficient for artefact reduction than non-adaptive smoothing like Gaussian filtering in terms of the subjective preference.

1   Introduction

Lossy image and video compression yield typical error patterns on the decompressed images or video sequences due to the quantisation error. Depending on the compression scheme and the bit rate, these can be ringing patterns around the edges, false or blurred texture, visible block-boundaries in block-partitioning schemes [20]. These phenomena are called compression artefacts. Compression artefacts not only deteriorate the visual quality of images but also perturb image processing and machine vision algorithms like edge detection and motion estimation.

A possible solution is to do a post-processing on the decompressed image, which removes compression artefacts. Some post-processing methods improve the visual

L.D. Griffin and M. Lillholm (Eds.): Scale-Space 2003, LNCS 2695, pp. 729–743, 2003.

quality of the compressed image by various adaptive filters that alleviate the artefacts and enhance the edges [3,4,9,19,20]. The blocking artefact for the blockwise transform compression can also be reduced by prediction methods for the quantised or missing transform coefficients [14]. Post-processing can also be formulated as a reconstruction problem for the original image and solved by optimisation with respect to an objective error measure like MSE, to probability considerations, to regularity constraints, or to a combination of these [11,17, 24].

The quality of the compressed images is affected not only by the bit rate and the compression algorithm, but also by the data source. In particular, images with more details usually degrade more than those with fewer details when compressed at the same bit rate [20]. It is therefore sensible to do some pre-processing on the image before compression, which alleviates the components of the image susceptible to artefacts. Pre-filtering methods were proposed for video coding [15]. Although pre-filtering options for still images exist in compression software products (e.g. in [25]), little research has been done in this area. Pre-processing has several advantages. It does not require a change in the decompression, and the compression standard remains unaffected. Moreover, pre-processing is done only once, while any post-processing must be done each time the image is decompressed, involving additional computational complexity. As a practical application, digital cameras use motion-JPEG for storing images and involve some kind of pre processing compensating for the different optical and/or compression artefacts.

In this paper, we consider the idea in [21] in a systematic way. We define what we mean by artefact reduction, and use a simple way of measuring and expressing it through PSNR. Relying on the theory of anisotropic diffusion filtering, we claim that it is possible to achieve artefact reduction, while preserving the main structure of the image. To check whether artefact reduction results in images that are more preferred by human observers, subjective tests were done to rate the different methods. Results on test images suggest that the block boundaries and false textures are alleviated in the pre-processed images. The diffusion strength is controlled by PSNR measurements. The results of the subjective tests show however that the maximal artefact reduction obtained this way does not increase the subjective preference of the compressed image. This follows from the over-smoothing of some textured regions. The subjects prefer compressed images with noisy texture to images with reduced noise but less small-scale details. The tests show that there are significant differences between the diffusion methods. In particular, the more edge-adaptive the method is in the mathematical sense, the more it will be preferred by the observers.

2   Non-linear Diffusion and Adaptive Filtering

We consider diffusions that are potentially useful for artefact reduction having different levels of edge-adaptability. We begin by analysing the *linear diffusion* (LD) process.

We treat images as positively valued smooth functions defined at the points $\mathbf{x} \in \Re^2$. Let us take the smoothing of an image f with Gaussian kernels

$$G_t(\mathbf{x}) = \frac{1}{4\pi t} \exp\left(-\frac{|\mathbf{x}|^2}{4t} \right),$$

$\mathbf{x} \in \Re^2$ having various parameters $t > 0$. We obtain a family of images $(u_t)_{t \geq 0}$ with $u_0 = f$ and $u_t = G_t * f$. Each element u_t of this family can also be obtained [6,10] by an LD process done on the image up to time t, i.e., $u_t(\mathbf{x}) = u(t,\mathbf{x})$ for all $t > 0$ and $\mathbf{x} \in \Re^2$, where u is the solution of the LD equation

$$\partial_t u = \Delta u, \tag{1}$$

with the initial condition $u_0 = f$. The LD generates a multiscale representations of images [1,6], where t is called the scale of the diffusion.

To understand the usefulness of this filtering for artefact reduction, the Laplacian operator Δ is written as a sum of two orthogonal components

$$\Delta u = u_\| + u_\perp, \tag{2}$$

where $u_\|$ denotes the second spatial derivative in the direction orthogonal to the gradient ∇u, and $u_\perp$ is the second spatial derivative in the direction parallel with the gradient ∇u. The term $u_\|$ can be interpreted as an "infinitesimal" Gaussian filtering along the edge, and $u_\perp$ as an "infinitesimal" Gaussian filtering across the edge. This low-pass filtering can contribute to the reduction of the ringing artefact, which is the result of the sharp frequency cut-off caused by quantisation at lower bit rates. In Eq. 2, the edge-parallel and edge-normal directions have equal weights, and both diffusion terms depend only on the direction of ∇u and not on $|\nabla u|$, where the latter indicates the local contrast difference. The smoothing obtained by the LD respects neither the direction nor the contrast of the edges.

To add contrast and directional sensitivity we extend Eq. 2 as

$$\partial_t u = p\left(|\nabla G_\sigma * u|\right) u_\| + n\left(|\nabla G_\sigma * u|\right) u_\perp, \tag{3}$$

where p and n are weighting functions controlling the diffusion along and across the edges, respectively, and $\sigma > 0$ is a fixed parameter. The purpose of the pre-smoothing with G_σ is to obtain a reliable estimate on edges, and to make the equation robust against noise. We allow full diffusion at uniform regions where the value of $|\nabla G_\sigma * u|$ is small and to inhibit the diffusion at edge locations where $|\nabla G_\sigma * u|$ is large. One possibility to control the diffusion in this way is to use the weighting function

$$w_K(x) = \begin{cases} 2\exp\left(-\dfrac{|x|^2}{K}\right) & ,0 < K < \infty \\ 2 & ,K = \infty \end{cases}, \tag{4}$$

where $x \in \Re$ and $K \in (0,+\infty]$ is a fixed parameter [16]. With the special choice $p = (1-\alpha)w$ and $n = \alpha w$, where $0 \leq \alpha \leq 0.5$, we obtain the diffusions examined in this paper

Table 1. Diffusions with different degree of adaptability.

	Directionally insensitive (isotropic) $\alpha = 0.5$	Directionally adaptive (anisotropic) $\alpha = 0$
Contrast insensitive $K = \infty$	Linear Diffusion (LD)	Mean Curvature Motion Diffusion (MCMD) [1]
Contrast adaptive $0 < K < \infty$	Non-linear Isotropic Diffusion (NLID) [18]	Pure Anisotropic Diffusion (PAD) [2,5]

$$\partial_t u = w_K \left(\left| \nabla G_\sigma * u \right| \right) \left((1-\alpha) u_\parallel + \alpha u_\perp \right), \tag{5}$$

The parameter α controls the directional adaptability and the parameter K controls the contrast sensitivity. The diffusions obtained for particular choices of the parameters are listed in Table 1.

Whatever the parameters α and K are, the diffusion can contribute to the suppression of the ringing artefact to some degree, as explained above. Moreover, since the minima and the maxima of the intensity values get closer (the contrast decreases), the DC values of the neighbouring DCT blocks of a flat area are more likely to fall into the same quantisation bin after the pre-processing, thus decreasing the blocking artefact.

We used forward numerical schemes for the above equations in all of our experiments with the fixed step size $\lambda = 0.1$, and parameters $K=0.05$, $\sigma = 0.4$, by recasting $(1-\alpha) u_\parallel + \alpha u_\perp = \alpha \Delta u + (1-2\alpha) u_\parallel$ and computing $u_\parallel$ as in [2]. The smoothing with G_σ for the gradient computation was done with LD. The scale values whenever indicated were computed as $t = \lambda m$, where m is the number of iterations done by the numerical scheme.

3   Artefact Reduction

To explain artefact reduction, we observe the details of the different compressed versions of the test image "Goldhill" shown in Fig. 3. The JPEG compressed image clearly suffers from ringing and blocking artefacts. By diffusion pre-filtering the artefacts will be reduced. It can be seen in Figs. 3b-d, if we compare the images along the high-contrast edges, or surfaces where block-boundaries become less visible.

Apart from visual quality, artefacts can also impair the stability of a computer vision algorithm. As an example, we give the edge detection results for the compressed versions of the image "Boat" from Fig. 4. The contours were extracted with a Canny edge detector [12] using fixed parameters. We compared the edges with the edges of the original image, and separated the true edge points from the falsely detected edge points. The true edge points are shown in Figs. 5 and the false detections are shown Fig. 6. Clearly, though the compressed image without pre-processing lets more original edge points detected (mainly texture), it will also give rise to more false structures. The ration of the number of true and false edge points in this example is 1.38 for the compressed image and 1.68 for the compressed image with pre-processing.

We try to define and measure artefact reduction first. Let P_t denote a diffusion pre-processing method done up to scale t and C the compression method (JPEG in our case). We assume for now that the bit rate is fixed. We say that an image f can be compressed at better quality than an image g at the given rate if $PSNR(C(f), f) \geq PSNR(C(g), g)$, where $C(f)$ and $C(g)$ denote the compressed images, and $PSNR(f, g) = -10\log(255^2 / MSE(f, g))$, with N denoting the image size, and MSE the mean square error. This quality measure is used usually to evaluate compression results. Our goal is to transform f into an image f' by pre-processing so that we can compress f' at a better quality than f.

Suppose that we pre-process the image f with a diffusion method P up to scale t. The reconstruction quality of the pre-processed image Pf at the given fixed bit rate will be

$$Q_{PP}(t) = PSNR(C(P_t f), P_t f).$$

The quality of the compression with pre-processing relative to the original image f is

$$Q_P(t) = PSNR(C(P_t f), f).$$

The reconstruction quality of the compression without pre-processing is

$$Q_0 = Q_{PP}(0) = Q_P(0) = PSNR(C(f), f).$$

A diffusion processing method is said to reduce artefact if there is a scale $t \geq 0$ such that Pf can be compressed at a better quality than f, i.e., if $Q_{PP}(t) > Q_0$, under the constraint $Q_P(t) \geq Q_0$. We impose the latter constraint, since any processing is meaningful only if the original quality does not decrease.

The tendencies of the curves Q_P and Q_{PP} as a function of t are shown for "Goldhill" in Fig. 1a and 1b for the pre-processing with NLID at bit rate $c = 0.25$ *bits/pixel*. The combined plot of $Q_P(t)$ and $Q_{PP}(t)$ as a function of the increasing scale $t \geq 0$ is show in Fig. 1c. Important to note is that the value $Q_{PP}(t)$ increases monotonously with t, and that $Q_P(t)$ changes around the value Q_o, reaches a maximum, and drops for larger scales. We can use $Q_{PP}(t)$ to measure the artefact reduction.

The latter observations and our definition of the artefact reduction lead us to the definition of the following two characteristic scales for each particular pre-processing method P:

1. The scale corresponding to the largest artefact reduction with maximal quality improvement,

$$t_1 = \max \; \arg\max\{Q_P(t) \,|\, t \geq 0\}. \tag{6}$$

2. The scale corresponding to the maximal artefact reduction,

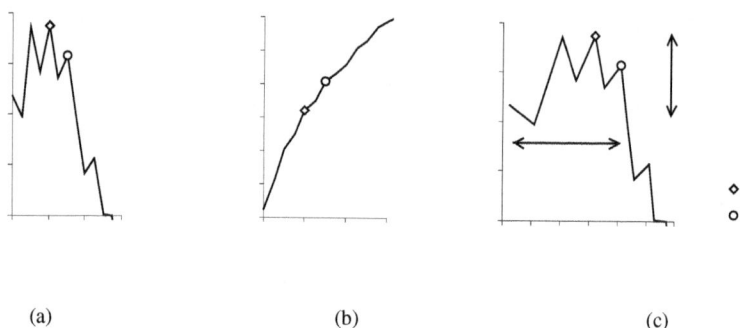

(a)   (b)   (c)

Fig. 1. Pre-processing for "Goldhill" with NLID (c = 0.25 bits/pixel). (a) the PSNR values v. the original image $Q_P(t) = PSNR(C(P_t f), f)$, (b) the PSNR values v. the processed image $Q_{PP}(t) = PSNR(C(P_t f), P_t f)$, (c) the curve $t \mapsto (Q_{PP}(t), Q_P(t))$.

$$t_2 = \max\left\{ t \geq 0 \,\big|\, Q_P(t) \geq Q_0 \right\}. \tag{7}$$

These scale values are indicated in Fig. 1, t_1 with a diamond and t_2 with a circle. Note that the pre-processing up to scale t_2 will redistribute the original compression error, so that the maximum portion of the quality will be devoted to the main structure of the image, as defined by the underlying multiscale representation, and the smallest portion of quality will be allotted to the small-scale details and noise. Different diffusion methods will do this redistribution in different ways. According to the theory concerning the filtering with non-linear diffusion, the more adaptive diffusion is, the better this redistribution will be. The results of the subjective tests support this claim.

4   Results

We compare the pre-processing with three diffusion processes (LD, NLID, PAD) for three typical test images at two selected bit rates. The results are summarized in Table 2. The table contains the original compression quality Q_0 without pre-processing , the maximal improvement in quality $Q_P(t_1)$ with the corresponding value $Q_{PP}(t_1)$ indicating the artefact reduction, and the quality values $Q_P(t_2)$ and $Q_{PP}(t_2)$ corresponding to the maximal artefact reduction (compare Fig. 1). Note that the value $Q_P(t_2)$ is very close, but in general it is not equal to Q_0. This is because we cannot change the scale parameter and the quantisation intervals continuously.

These values show that it is possible to improve the compression quality in PSNR up to 1.5%, depending on the bit rate. As regards the artefact reduction, the pre-processed images can be compressed at 5 - 10% better quality than the original image at the pre-processing scale t_1, and at up to 20% better quality at the pre-processing scale t_2, without loss in the quality as compared to the original image (Q_0). The relative improvement in any sense is better for the lower bit rate.

Table 2. The PSNR values Q_P (compression quality) and Q_{PP} (artefact reduction) for the maximal quality improvement (scale t_1) and the maximal artefact reduction (scale t_2). Q_0 is the original JPEG compression quality in PSNR (dB).

		0.25 bits/pixel			0.4 bits/pixel		
		Goldhill	Boat	Bridge	Goldhill	Boat	Bridge
Q_0		29.23	30.11	24.09	30.87	32.40	25.39
$Q_P(t_1)$	LD	29.36	30.26	24.37	30.91	32.43	25.67
	NLID	29.37	30.32	24.41	30.97	32.46	25.61
	PAD	29.41	30.37	24.42	31.02	32.56	25.67
$Q_{PP}(t_1)$	LD	31.15	31.77	26.32	33.18	34.53	28.12
	NLID	32.19	32.28	25.74	34.00	33.28	27.90
	PAD	30.46	31.00	25.31	32.33	33.43	27.20
$Q_P(t_2)$	LD	29.32	30.12	24.24	30.91	32.43	25.48
	NLID	29.32	30.13	24.12	30.97	32.43	25.44
	PAD	29.24	30.14	24.09	30.87	32.40	25.43
$Q_{PP}(t_2)$	LD	32.52	32.99	29.98	33.18	34.53	29.86
	NLID	33.10	33.13	29.21	34.00	33.98	29.74
	PAD	32.70	32.15	28.70	33.23	34.01	28.51

If we look at the images in Fig. 3, we can see that the pre-processed images have less artefact. The LD-pre-processed image has got more blurred than those pre-processed with non-linear diffusion.

Though we have shown that it is possible to obtain a maximal improvement in the PSNR compression quality by pre-processing up to scale t_1, this will not yield a large perceivable artefact reduction in general. With the maximal artefact reduction, there is a larger perceivable artefact reduction, but in spite of the PSNR results, which suggest improvement, we should test whether the subjective quality is deteriorated by the blur involved in these diffusion processes.

For the above mentioned reasons, we were interested in comparing the different pre-processing methods on a subjective scale for the maximal artefact reduction (scale t_2). We did a so-called Thurstone-scaling experiment [22,23]. The subjective scale is constructed based on a larger number of pairwise comparisons of the different stimuli. Thurstone-scaling is typically used in situations where the stimuli are very similar (in our case, we have nearly equal MSE for the images) or where the measured quantity is hard to describe exactly (image quality is hard to define in general).

In the scaling experiment, a test image is taken, and the test person has to compare each version of the image which each other version of the image. The versions constitute the compressed image without pre-processing, and the compressed images with the different pre-processing methods (only one pair is shown on the display at a time). We made the test for three different images. The scaling method gives useful results if the stimuli are very similar and do not contain outliers. For these reasons, the original images were excluded from the comparisons. *In this way, we did not measure the similarity (distance) to the original image, but a subjective impression of quality manifesting itself in the preference decisions of the human observers.*

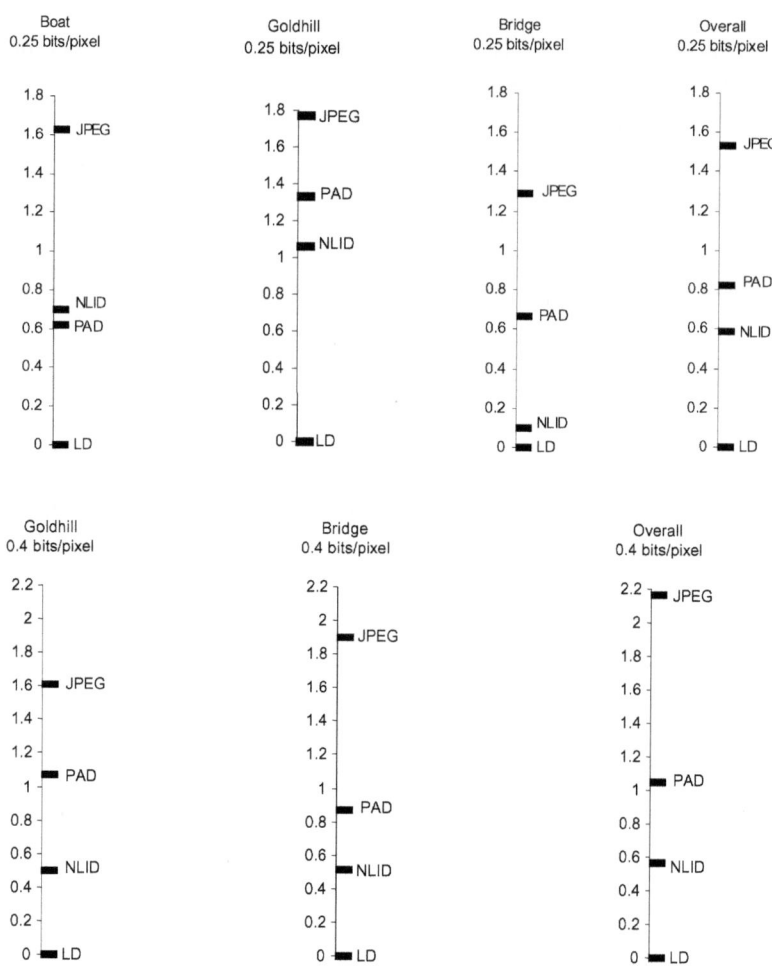

Fig. 2. Scaling results. The scales show the expected values of a model observers quality assessment with respect to a chosen reference point, e.g., the position of LD. Positive difference on the scale means "better in expectance". The unit is the uncertainty of the decision. Large differences mean a better discrimination.

We show the results in Fig. 2 obtained for a subjective test with 21 test persons. We constructed scales for the two compression rates by collecting the scores for the different methods across the different test images. The scales for these overall scores are also shown in Fig. 2. The scale is interpreted roughly as follows: the guesses of subjects for each item (JPEG, LD, etc.) are modelled as normally distributed random variables having the same variance, which is equal to the unit of the scale. The scale value for each item is the expected value of the corresponding random variable. The unit of the scale is interpreted as the uncertainty of the guesses of subjects. We emphasize that the scales in Fig. 2 are not directly comparable, since the units are obtained implicitly through the construction of each particular scale, and depend on the image.

Table 3. Results of the χ^2-test of Mosteller for the subjective scaling experiment. The hypothesis is that the obtained scale values are correct. The sample size is n, the degree of freedom of the test is df=3. The probability that the scales are not random is $P(\chi^2)$. The regression coefficient $0\leq r \leq 1$ gives the goodness of fit of the preference probabilities reconstructed from the model to those obtained by counting the scores.

		n	$P(\chi^2)$	r
0.25 bits/pixel	1. Boat	21	0.8611	0.9790
	2. Goldhill	21	0.72271	0.9627
	3. Bridge	21	0.79388	0.9627
	Overall (1 & 2 & 3)	63	0.9985272	0.9965
0.4 bits/pixel	1. Boat	21	0.033233	0.9665
	2. Goldhill	21	0.70797	0.9693
	3. Bridge	21	0.96986	0.9952
	Overall (2 & 3)	42	0.99432	0.9986

The parameters of the above Gaussian distributions are computed based on the preference probabilities for each pair of images (the probability that one image is preferred to the other). These probabilities are in turn obtained by counting the scores for each pair. There are more possibilities for measuring the reliability of the scales. The χ^2-test of Mosteller [13] is typically used for these purposes. Another way to test reliability is to compute the regression coefficient between the set of preference probabilities and the preference probabilities recomputed with the parameters of the Gaussian model. The results of the reliability test are summarized in Table 3. Because the scale for "Boat" at 0.4 *bits/pixels* is not very reliable, the scores for this case were not considered in the construction of the overall scale for 0.4 *bits/pixels*.

Although the compared images have almost the same PSNR error (quality values $Q_P(t_2)$ in Table 2), their preference by the subjects is quite different. The subjective preference of the non-linear diffusion methods is significantly larger than that of the LD. The adaptive diffusions NLID and PAD are quite similar (compare Figs. 3 and 3) suggesting that the contrast adaptability is more important than the directional sensitivity, though since PAD is almost always better than NLID, directional adaptability is also important. We conclude that better adaptability leads to a better redistribution of the original bit/quality-rate by devoting larger portions of these resources on perceptually important information.

The compressed images *without* pre-processing show a superior subjective preference over the pre-processed ones, though they are actually worse in PSNR. This can be due to the fact that many subjects preferred sharp though falsely textured regions rather than smoothed areas, where the undersampled textures and compression-artifacts were removed by the pre-processing . It may follow the well-known effects of psycho visual illusions (like as Kanizsa figures [8]), which can be derived from the description of brain stimuli responses of neurosciences (e.g. [7]). It says that the brain, based on its high adaptability, may detect anomalous contours from the partially degraded details, like from artefacts.

(a)   (b)

(c)   (d)

Fig. 3. Maximal artifact reduction with pre-processing for "Goldhill" (c=0.25 bits/pixel, scale t_2). Details of the (a) JPEG compressed image, (b) pre-processing with LD, (c) NLID, (d) PAD.

5   Conclusion

We have considered the application of different diffusion methods as a pre-processing step for artefact reduction for the blockwise DCT compression, which we previously proposed in [21].

The compression quality improvement achievable by diffusion pre-processing is only up to 1.5% in PSNR for our test images and for low bit rates. For some of these

(a)

(b)

Fig. 4. Boat at 0.25 bits/pixel. (a) JPEG, (b) JPEG with PAD preprocessing.

(a)

(b)

Fig. 5. Edge detection for Boat 0.25 bits/pixel. (a) True edge points for JPEG, (b) true edge points for JPEG with PAD preprocessing.

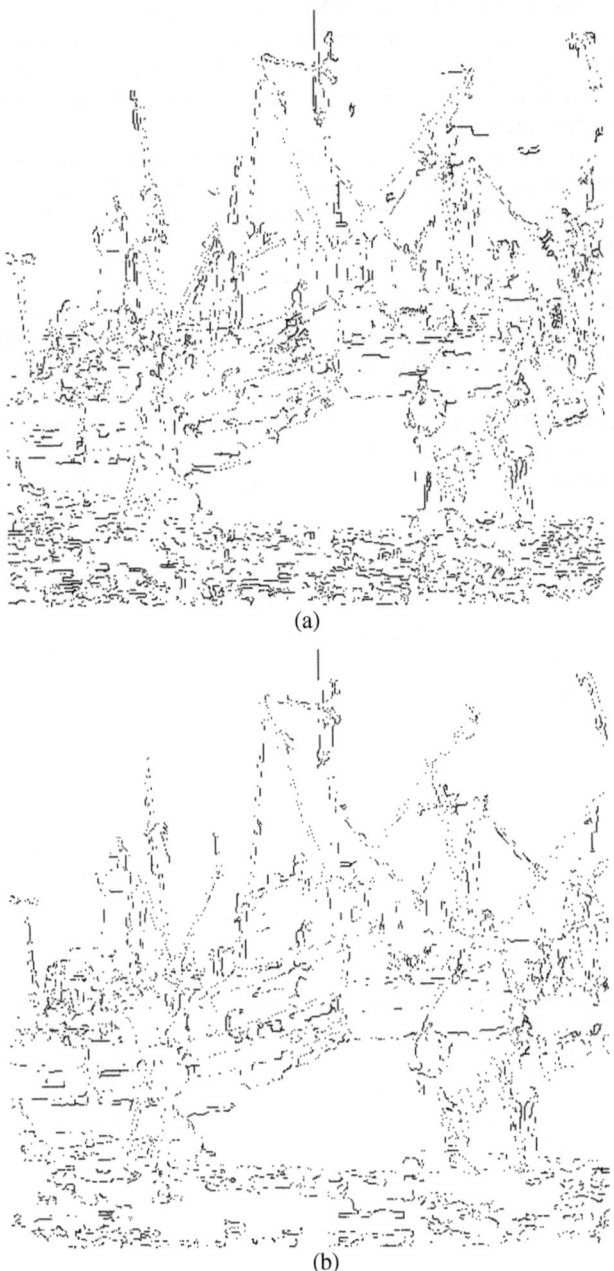

(a)

(b)

Fig. 6. Edge detection for Boat. (a) False edges for JPEG, (d) false edges for JPEG with PAD preprocessing.

test images, 2-3% improvement in PSNR was reported for post-processing methods [20] for the same bit rates. Note that diffusion pre-processing is done only once in the compression phase, and it consists of local iterative local operations. We have also seen that such small differences in the PSNR can yield surprisingly different results in the subjective evaluation.

The pre-processed images can be compressed with a significantly better PSNR than the original image, where the quality of all these compressed images have a better or equal value as compared to the original image. This means that these images are less susceptible to artefacts.

The subjective tests have shown that for a fixed bit rate and error-rate, adaptive diffusion captures better the visually important information in images. However, since the diffusion pre-processing algorithm is tuned by using PSNR to estimate the image quality and the artefact reduction, and since the blur involved in the artefact-removal filtering is less tolerated by the human observers than the artefacts, the perceived subjective quality of the images will generally be decreased by pre-processing when done up to the maximal artefact reduction. We have to remark, that to our knowledge, no such subjective tests were done for other quality enhancement methods like post-processing.

Currently we are trying to find methods for establishing better diffusion pre-processing parameters in order to improve the subjective performance. We also do a modified subjective experiment, where the original image is displayed along with the two images to be compared. Preliminary results show that there are better parameter choices and that the two experiments, the one in the presence of the original image and the one without the original displayed, can yield quite different results.

References

1. L. Alvarez, F. Guichard, P. L. Lions, and J. M. Morel, "Axioms and fundamental equations of image processing", *Arch. Rational Mech. Anal.*, vol. 123, pp. 199-257, 1993.
2. L. Alvarez, P. L. Lions, and J. M. Morel, "Image selective smoothing and edge detection by non-linear diffusion II", *SIAM J. of Num. Anal.*, vol. 29, no. 3, pp. 845-866, 1992.
3. J. G. Apostolopoulos and N. S. Jayant, "Postprocessing for very low bit rate video compression", *IEEE Trans. Image Processing, vol.* 8, no.8, pp.1125-1129, 1999.
4. R. Castagno, S. Marsi, and G. Ramponi, "A simple algorithm for the reduction of blocking artifacts in images and its implementation*", IEEE Trans. on Consumer Electronics*, vol. 44, no.3, pp. 1062-1070, 1998.
5. F. Catté, T. Coll, P. L. Lions, and J. M. Morel, "Image selective smoothing and edge detection by non-linear diffusion", *SIAM J. Numerical Anal.*, vol. 29, pp.182-193, 1992.
6. L. Florack:, "Image Structure", *Kluwer Academic Publishers*, 1997.
7. R. Heydt, E. Peterhans, "Mechanism of contours in monkey visual cortex. I. Lines of pattern discontinuity", *Journal of Neuroscience*, vol.9, pp.1731-1748, 1989.
8. G. Kanizsa, "Subjective Contours", *Scientific American*, vol.234, pp.48-52, 1976.
9. Y. L. Lee, H. C. Kim, and H. W. Park, "Blocking effect reduction of JPEG images by signal adaptive filtering", *IEEE Trans. on Image Processing*, vol.7, no.2, pp.229-234, 1998.
10. T. Lindeberg and B. M. Haar Romeny, "Linear scale-space I-II*". In: Geometry-Driven Diffusion In Computer Vision*, Kluwer Academic Publishers, pp.1-72, 1992.
11. J. Luo, C. W. Chen, K. J. Parker, and T. S. Huang, "Artifact Reduction in Low Bit rate DCT-Based Image Compression", *IEEE Trans. on Image Proc.*, vol.5, no.9, pp. 1363-1370, 1996.

12. J O. Monga, R. Deriche, G. Malandain and J.-P. Cocquerez, "Recursive filtering and edge tracking: two primary tools for 3-D edge detection", *Image and Vision Computing* vol. 4, no. 9, pp 203-214, 1991.
13. F. Mosteller: "Remarks on the method of paired comparisons III", *Psychometrika,* 16, pp. 207-218, 1951.
14. W. B. Pennebaker and J. L. Mitchell, "JPEG Still Image Data Compression Standard", *Van Nostrand Reinhold*, 1993.
15. M. K. Ozkan, M. I. Sezan, and A. M. Tekalp, "Adaptive motion-compensated filtering of noisy image sequences", *IEEE Trans. on Circuits and Systems for Video Tech.*, vol. 3, no. 4, pp. 277-290, 1993.
16. P. Perona and J. Malik, "Scale-space and edge detection using anisotropic diffusion", *IEEE Trans. Pattern Anal. Machine Intel.*, vol. 12, no. 7, pp. 629-639, 1990.
17. R. Prost and A. Baskurt, "JPEG dequatisation array for regularized decompression", *IEEE Trans. on Image Proc.*, vol.6., no.6., pp. 883-888, 1997.
18. T. Roska and T. Szirányi, "Classes of analogic CNN algorithms and their practical use in complex processing", In: *Proc. IEEE Non-linear Signal and Image Processing*, pp.767-770, June, 1995.
19. K. Sauer, "Enhancement of low bit rate images using edge detection and estimation", *CVGIP*: Graphical Models and Image Processing, vol.53., no.1, pp. 52-65, 1991.
20. M. Y. Shen and C. C. J. Kuo, "Review of postprocessing techniques for compression artifact removal", *J. of Visual Comm. And Image Rep.*, vol. 9, no. 1, pp. 2-14, 1998.
21. T. Szirányi, I. Kopilovi•, and B. P. Tóth, "Anisotropic diffusion as a pre-processing step for efficient image compression", In: *Proc. ICPR,* pp. 1565-1567, Brisbane, Australia, August, 1998.
22. L. L. Thurstone, "A law of comparative judgment", Psychol. Rew., 34, pp. 273-286, 1927.
23. W. S. Torgerson, "Theory and Methods of Scaling", *John Wiley and Sons, Inc.*, 1958.
24. Y. Yang, N. Galatsanos, and A. K. Katsaggelos, "Projection-based spatially adaptive reconstruction of block-transform compressed images", *IEEE Trans. on Circuits and Systems for Video Tech.*, vol. 3, no. 6, pp. 421-432, 1993.
25. Independent JPEG Group's CJPEG, DJPEG, version 6a, 7-Feb-96, Copyright (C) 1996, Thomas G. Lane (http://ww.ijp.com).

Computation of Generic Features
for Object Classification

Daniela Hall* and James L. Crowley

Projet PRIMA – Lab. GRAVIR–IMAG
INRIA Rhônes–Alpes
655, avenue de l'Europe
38330 – Montbonnot Saint Martin, France

Abstract. In this article we learn significant local appearance features
for visual classes. Generic feature detectors are obtained by unsupervised learning using clustering. The resulting clusters, referred to as
"classtons", identify the significant class characteristics from a small set
of sample images. The classton channels mark these characteristics reliably using a probabilistic cluster representation. The classtons demonstrate good generalisation with respect to viewpoint changes and previously unseen objects. In all experiments, the classton channels of similar
images have the same spatial relations. Learning of these relations allows to generate a classification model that combines the generalisation
ability from the classtons and the discriminative power from the spatial
relations.

Keywords: local image features, classification, clustering

1 Introduction

Structural matching is a classical approach for object recognition. Gaussian
derivatives measure the basic geometries of the appearance of local features.
In such a feature space, similarity of features can be measured by the distance
between their vectorial representation. This feature matching principle is widely
used for image indexing, and object identification [8,14].

Classification is a task that requires the assignment of previously unseen
objects to the corresponding class of visually similar objects. Classical feature
matching fails in many cases due to large feature variations among objects a
class. For this reason vision systems have difficulties to generalize from a small
set of images to other images of the same class. This makes classification a much
harder problem than identification of previously seen objects.

Successful classification relies on the extraction of significant class features
that should be robust to changes in viewpoint, object identity, position, scale
and lighting conditions. This article adresses the problem of the extraction of
such significant features. Generic feature detectors have the property that they
mark the most characteristic features with respect to a learned class. In our

* This research is funded by IST CAVIAR 2001 37540

L.D. Griffin and M. Lillholm (Eds.): Scale-Space 2003, LNCS 2695, pp. 744–756, 2003.

method, the generic features are computed automatically by unsupervised clustering. We propose a measure for the selection of the most significant clusters and several experiments show that the selected clusters detect those significant features robust to changes in viewpoint and object identity.

2   Composition of Generic Features (Classtons)

The idea of vector quantization or clustering of the outputs of linear filter sets has been applied by Leung and Malik for texture recognition and image segmentation [6,9]. They define texture as entity with spatially repeating properties. Zhu and his collaborators obtain clusters robust to rotation and scale changes by applying a transform component analysis to image patches before clustering [15]. The obtained textons that represent the texture clusters allow the efficient modeling of textures. Schmid has applied the same k-means clustering scheme to compose generic features for image indexing [13]. We want to extend this idea and use exclusively clusters in feature space for image description, recognition and classification.

A visual object class consists of visually similar images with spatially repeating properties over these images. Under these constraints the clustering of vector representations of local features is able to detect automatically the repeating features and learn their variations. Clustering is therefore a means for the computation of the desired generic features.

3   Clustering Approaches

The success of classification depends on the generic features (the classton vocabulary). The choice of an appropriate clustering algorithm is crucial. In this section we evaluate k-means, k-means with pruning and DBScan. The methods are compared on several test databases.

The choice of the comparison of those three methods is motivated by the work of Leung, Malik, Schmid, and Zhu, who all use k-means. Leung [6] uses k-means with pruning. This method is less sensitive to cluster center shifts due to outliers than the original k-means algorithm. We compare these standard methods to a new clustering algorithm from the data mining community. Ester [1,2] developed DBScan for the expansion of density clusters of arbitrary shape with a minimum of domain knowledge. The definition of DBScan allows to find natural boundaries between clusters. This property has the effect that the number and the shape of the significant feature clusters is automatically adapted to the data.

3.1   K-Means Clustering

K-means is an agglomerative clustering method with a specific objective function. Assuming that there are k clusters and each cluster is represented by its center of gravity, an objective function is obtained by evaluating the distances of image points, $x_j \in D$, to their respective cluster center, c_i:

$$E(C, D) = \sum_{i=1}^{k} \left(\sum_{j \in C_i} (x_j - c_i)^T (x_j - c_i) \right), C = \{C_1, \ldots, C_k\} \qquad (1)$$

The algorithm assigns each point to the closest cluster center and updates the centers. These steps are iterated until the objective function reaches a minimum. K-means has linear complexity $O(n)$ with n number of points. The simplicity and the efficiency of the algorithm explains its popularity.

In most applications the optimal number of clusters is unknown. The objective function is proven to converge to a local minimum, not the global minimum. The problem of convergence to a suboptimal solution can be overcome by running k-means many times and retaining the best solution. This multiplies the computation time and there is no guarantee of the quality of the solution [3].

Standard k-means has the disadvantage that data points are assigned to the currently closest cluster center. Outliers as a result to noise are always present in the data. In extreme cases, the assignment of outliers shifts significantly the center of gravity of the cluster and decrease the overall quality of the solution. K-means with pruning is less sensitive to outliers.

3.2   K-Means with Pruning

In the first step, standard k-means is applied with a large number of clusters (in the range of 500 to 8000 clusters). Then these clusters are reduced subsequently. Close clusters are merged and clusters with few elements are suppressed. This pruning step takes as parameters the distance ε between clusters for merging and a number of required elements $MinPts$. The pruning step is repeated with increasing ε until the desired number of clusters is reached.

This algorithm is computationally more expensive, because the data is represented by a larger number of initial clusters requiring more iterations. Subsequent merging and discarding of small clusters allow to assign fewer outliers to the clusters. The remaining clusters are more representative for the image characteristics.

3.3   DBScan Clustering

In this section we describe an alternative clustering algorithm that is based on expanding clusters from a seed. This algorithm, referred to as DBScan, has been proposed by Ester [1,2] for the organisation of spatial databases with minimal requirements of domain knowledge. The algorithm is density based and can discover density clusters of arbitrary shape. This algorithm is interesting for the computation of classtons, because the number of clusters is determined automatically.

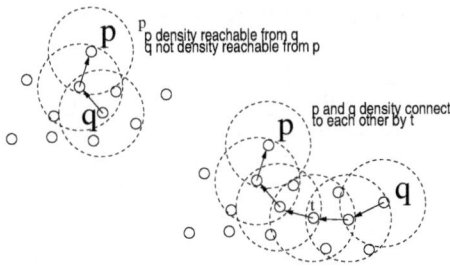

Fig. 1. Density reachability and density connectivity

The key idea of DBScan is that the neighborhood of cluster points has to contain a minimum number of data points $MinPts$. In other words, the cardinality of a sphere with radius ε has to exceed the threshold $MinPts$. Such points can serve as seed for cluster expansion. The algorithm is formalised by following definitions.

Definition 1 (directly density reachable) A point p is directly density reachable from point q with respect to ε and $MinPts$ in the point set D if

- $p \in S_\varepsilon(q)$
- $card(S_\varepsilon(q)) > MinPts$

Definition 2 (density reachable) A point p is density reachable from q with respect to ε and $MinPts$ in D, denoted as $p >_D q$, if there is a chain $q = p_1, \ldots, p_n = p$ such that $p_i \in D$ and p_{i+1} is directly density reachable from p_i (see Figure 1).

Definition 3 (density connectivity) p is density connected to q with respect to ε and $MinPts$ in D if there is a point $t \in D$ such that both p and q are density reachable from t (see Figure 1).

$$card(S_\varepsilon(q)) <= MinPts \qquad (2)$$
$$\exists p : q >_D p \text{ and } card(S_\varepsilon(p)) > MinPts \qquad (3)$$

A cluster is defined as a set of density connected points which is maximal with respect to density reachability. Noise is the set of points that are not contained in any cluster. A cluster contains core and boundary points. Core points are those points that fulfill $card(S_\varepsilon(q)) > MinPts$. Boundary points are points that fulfill

To find a cluster, DBScan starts with an arbitrary point p and retrieves all density reachable points. If p is a core object, this yields a new cluster. If p is a non-core object, no points are density reachable from p and it is assigned to noise. In the first case, the density reachable points are used to expand the cluster until maximality. The algorithm continues until all points are labelled.

Table 1. Processing time of image features from (A) unsegmented and (B) segmented images

	k-means	k-means with pruning	DBScan
A (faces)	27 s, 6 iterations	1442 s, 27 iterations	7698 s, no iteration
B (toy cars, segmented)	12 s, 13 iterations	1712 s, 28 iterations	143 s, no iteration

DBScan has the advantage that every point is treated only once, by computing $card(S_\varepsilon(p))$ and assigning the point to the current cluster or to noise. A sophisticated implementation of DBScan has a computational complexity of $O(n\log(n))$. The number of clusters is determined automatically. Standard k-means does not reject outliers which decreases the overall quality of the clustering. DBScan detects outliers automatically and assigns them to noise.

3.4 Evaluation

K-means with pruning has a computational complexity of $O(nk)$, with n number of points and k number of clusters. For standard k-means, the number of clusters is small which results in a linear complexity of $O(n)$. DBScan needs to compute the nearest neighbors for every point. In the current straight forward implementation the nearest neighbor search requires $O(n)$. By using for example binary search trees, the complexity can be reduced to $O(\log(n))$. The overall complexity of the current implementation is $O(n^2)$.

The computational difference becomes clear in following experiment. The results of the proposed clustering algorithms are evaluated on two data sets. Data set A has 63000 local image features extracted according to section 4 from four frontal face images from the AR face database [10]. Data set B consists of 9600 local image features extracted from four toy car pictures. The toy cars are segmented from the background, the faces images are unsegmented.

Table 1 displays processing times (on a 600MHz Pentium III). Standard k-means is fast, but the optimal number of clusters is unknown. To ensure that a good solution is found, k-means should be run several times with changing k. This multiplies the computation costs. K-means with pruning requires more iterations in order to minimise the overall error. Pruning is then called subsequently with parameters ε and $MinPts$ until the desired number of clusters is reached. For segmented images, DBScan is faster than k-means with pruning.

4   Feature Prototype Generation

In this section we describe the feature space used for feature extraction. In section 4.2 different cluster representations are evaluated. In order to select those clusters that correspond to the desired signficant features we need to be able to evaluate the quality of a particular cluster. Appropriate measures are proposed in section 4.3.

4.1   Feature Description

Gaussian derivative receptive fields are used by many researchers for the description of local feature appearance [4,8,11,12,14]. Low order derivatives measure the basic geometries of features [5]. Local features are represented by the response to a bank of Gaussian derivative receptive fields centered on the image position. The receptive fields are scale invariant due to normalization for intrinsic scale. We compute the intrinsic scale as an extremum in the normalised Laplacian over scale as proposed by Lindeberg [7].

We experiment with following feature spaces: first and second order derivatives and first, second and third order derivatives. The suppression of the derivative of order zero makes the feature less sensitive to illumination variations. The features are extracted either at a fixed scale, or at the specific intrinsic scale. In the second case, only those features are considered that actually display a maximum over scale within the predefined range (in our experiments $\sigma \in [1.34, 8.00]$). The data is normalised to compensate for the dynamic of receptive fields of different orders such that the distribution has 0 mean and 1.0 standard deviation.

Scale normalized features are clustered without taking into account their local scale. This has the advantage that features that occur at different scales due to perspective transformation are assigned to the same cluster. On the other hand, the relative scale between features of the same objects is lost. The scale relations between characteristic features of an object is discriminant and worth preserving. Instead of using the feature space $(L_x, L_y, L_{xx}, L_{xy}, L_{yy})$, we propose to use $(L_x, L_y, L_{xx}, L_{xy}, L_{yy}, \sigma)$. This adds the local scale to the feature space. A relatively scale invariant object representation is obtained that preserves the internal scale relations.

Figure 2 shows examples of clusters obtained from the toy car example using k-means. The linear combination of the cluster prototypes are shown. By comparing the left figures with the right figures, it can be observed that the classtons on the left and the classtons on the right display the same basic geometries. This means that the extension of the feature space to third order derivatives does not increase significantly the ability to describe the present geometries. A feature space up to second order covers sufficiently the geometry of the local features in the experiments.

4.2   Cluster Representation

To enable classification we need to compute classton channels. In a classton channel those points that belong to the classton are marked. The generation of the classton channel requires an assignment algorithm that decides if a particular feature belongs to the classton. The quality of the assignment is closely related to the cluster representation.

The assignment can be computed by several algorithms. Many researchers use minimum distance to prototype [6,9,13]. This is a fast measure, but it does not take into account the distribution of the cluster points. This measure is acceptable when the clusters have close spherical shape or clusters are sufficiently

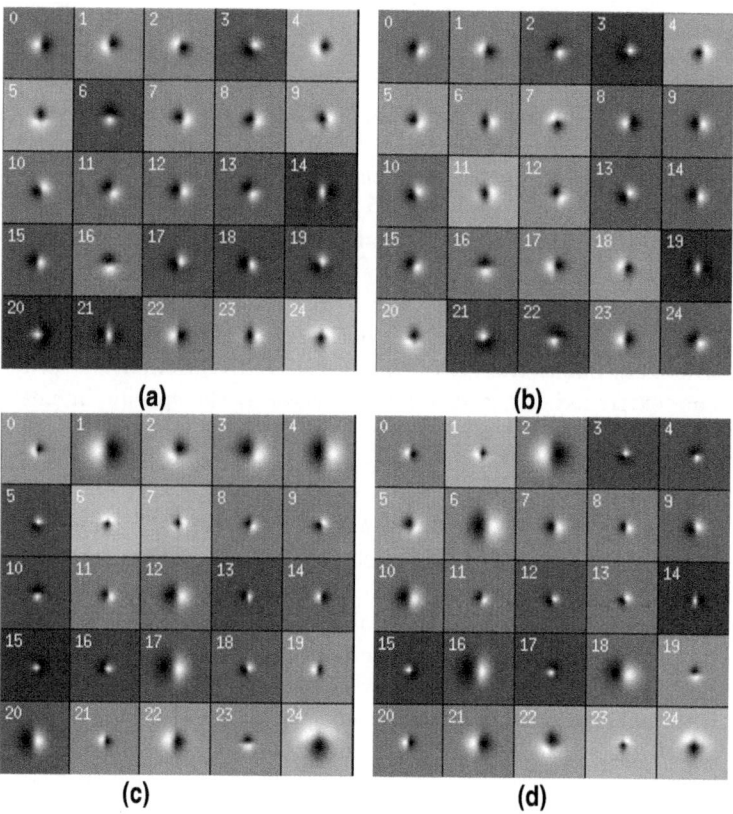

Fig. 2. Cluster examples. (a) feature space up to order 2. (b) feature space up to order 3. (c) feature space up to order 2 with scale. (d) feature space up to order 3 with scale

far from each other. In the typical case where clusters are elliptic point clouds, a better representation is obtained by using a probabilistic measure based on the Mahalanobis distance.

$$p(C_i|x) = \exp\left(-(x - \mu_i)^T \Sigma_i^{-1}(x - \mu_i)\right) \tag{4}$$

where Σ_i is the covariance matrix of cluster C_i. Clusters of arbitrary shape can be represented by a set of elliptic point clouds.

Figure 3 illustrates the effect of the different representation methods. The labels of the assigned cluster are coded as grey values. We observe connected regions in the channel images. This is due to the effect that spatially close features have similar appearance and are assigned to the same cluster. The connected regions of the probabilistic method are more stable than the regions obtained by distance to prototype. Although, in this example the differences are not significant, the probabilistic measure should be used to reduce incorrect assignments.

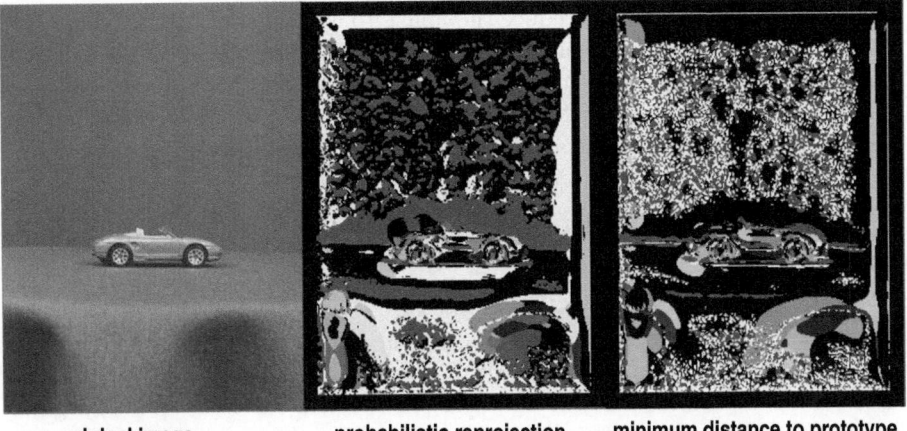

original image   **probabilistic reprojection**   **minimum distance to prototype**

Fig. 3. Effect on feature assignment using different cluster representations (probabilistic or center of gravity)

4.3   Quality of Clustering Results

Clusters are by definition dense collections of data points. They are useful for classification because they represent a collection of highly similar features. Under the condition that the training images are visually similar, those dense clusters represent the most significant features for the trained image class.

Several parameters can be used to judge the quality of a cluster, such as the density and the compactness in feature space and the connectedness of the regions in the classton channel. The density of clusters depends on the total number of feature points. For this reason, a threshold for reliable detection of dense clusters can not be found. Compactness has the advantage that it is independent from the number of features, under the condition that the learned features represent sufficiently the true feature distribution. A generic feature with good generalisation ability produces large connected regions. Figure 5 shows an example of compact and connected clusters that specify forehead, hair, eyes, nose, and lips as significant features of faces.

Connectedness of regions can be measured as the average number of pixels per region. Compactness is defined as the ratio of a volume and the enclosing sphere. In order to compute the compactness of a point cloud, we modify the geometrical definition of compactness as follows:

$$\text{Compact}(C_k) = \frac{\prod_{i=1}^N \sigma_i}{\max_i(\sigma_i)^N} \tag{5}$$

The volume of a cluster C_k is approximated by the product of standard deviation of its members in each dimension $i = 1, \dots, N$. The volume of the enclosing sphere is computed as the maximum standard deviation to the power of N. Density can be computed as average number of points per volume unit.

Connectedness allows to reduce the image to a number of regions. This gives the required tolerance to region positioning that enables classification robust to viewpoint changes and object identity. Figure 5 and Figure 6 show the classton channels of the most compact and the most connected classtons computed from 15 frontal faces and 13 toy cars from 2 viewpoints respectively.

5 Experiments and Observations

5.1 The Test Database

We use three different test databases. 15 frontal faces of size 256×192 from the AR face database (men with and without glasses). Toy cars and toy animals of size 341×256 from the ETH 80 database (13 cars and 5 horses). We use segmentation maps to focus on the object features. Examples are shown in Figures 4 and 5.

5.2 Experiments

This section shows the results for the databases. We only display classton channels that have high compactness and connectedness. The face example in Figure 5 demonstrates the robustness to scale changes and occlusions caused by facial hair or glasses. The example on the ETH 80 database shows that the classton channels are stable for visually similar views. In all our experiments, the classton channels of visually similar images produce the same spatial relations. These relations can be learned and the resulting model can then be used for classification. Such a model inherits the generalisation ability from the classtons and obtains discriminative power from the relative spatial relations.

The feature space used in the experiments is not normalised for orientation. The computed classtons are therefore orientation dependent. Orientation is an important feature for discrimination and in our examples orientation is needed to discriminate horizontal features from vertical features. If rotation invariance is required, than the invariance should be introduced by choosing a rotation invariant feature space as in [13].

Robustness to Object Identity. We compute classtons from 15 frontal face images. Out of the 37 k-means clusters we choose the 4 classtons that mark the most significant features over several individuals (see Figure 5, corresponding clusters are marked with same grey level). Note, that nose, forehead, cheeks, eyes, chin, upper lips are marked by the same classton invariant of scale changes. The overall structure of the face is recovered even in the case of occlusions by facial hair, or glasses.

Robustness to Viewpoint Changes. This section demonstrates the robustness to viewpoint changes on two examples. Figure 6 shows the 5 most compact and most connected classtons superposed on the image. We observe that the

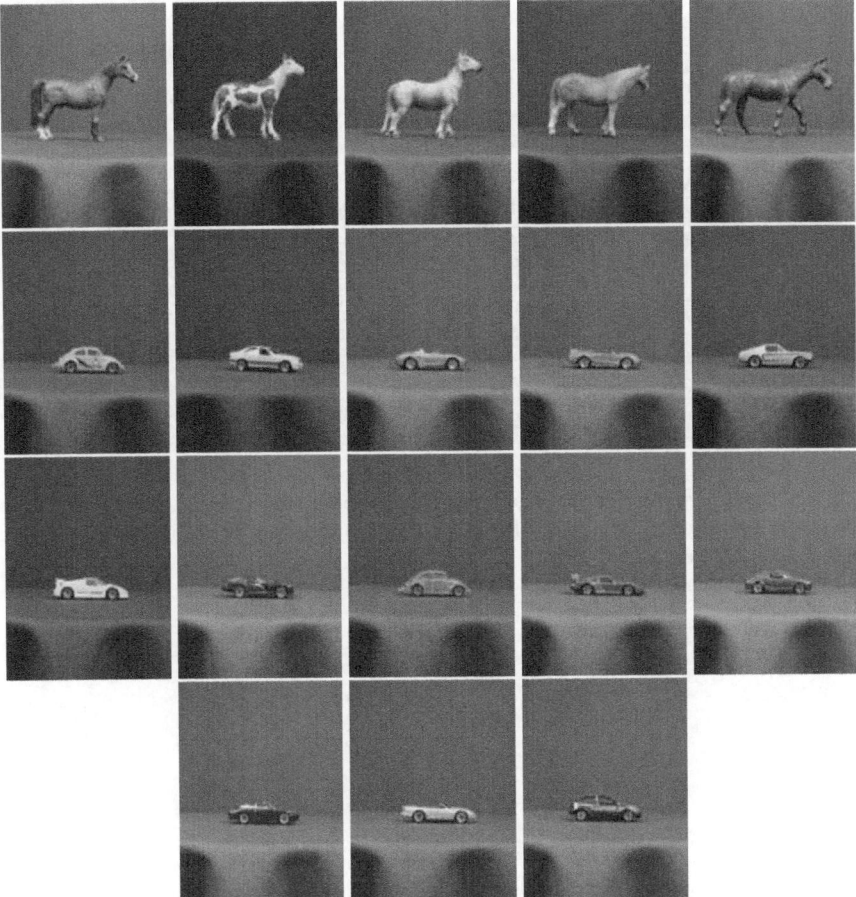

Fig. 4. Example of the ETH 80 database

classtons determine the wheels as significant features among the set of 13 toy cars shown in Figure 4. Due to the very different car shapes, the wheels are the only common feature. Figure 7 shows an example of the robustness to viewpoint changes. Significant features are stable for visually similar views (the three left-most figures). For views that are visually not similar (example Figure 7 right), the appearance of the channels changes considerably and classification would be difficult.

6   Conclusion

We propose a method to detect significant parts of the learned object robust to object identity, viewpoint, lighting conditions, pose, and scale of observation. Local appearance features are described by an appropriate feature space. Generic

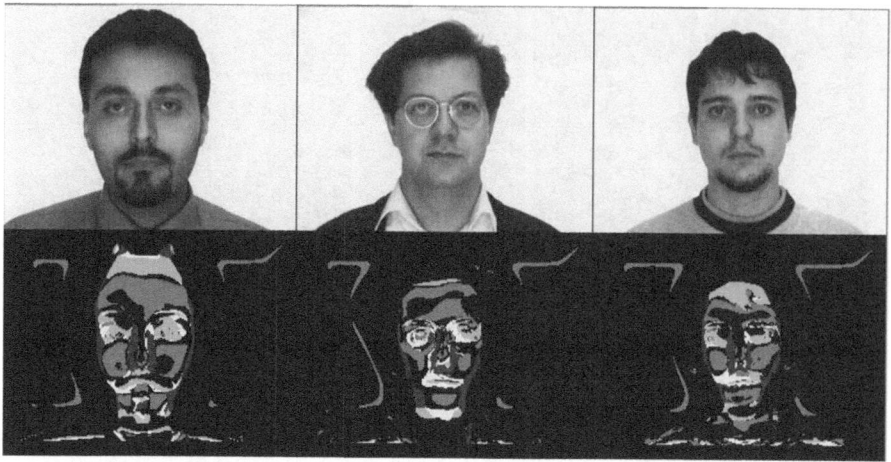

Fig. 5. Classton channel for frontal faces from the AR face database. Significant facial features are marked by the same classton channels, independent of identity, facial hair, glasses and scale change

lat 90 lon −22   lat 90 lon 0   lat 90 lon 22

Fig. 6. Classton channels of the most compact and the most connected clusters computed from 26 side views of 13 toy cars (displayed are 6 channels coded as different grey levels). The classton channels are superposed on the original images. We observe the robustness to viewpoint changes. The wheels are automatically identified as significant features

features are computed by unsupervised learning using clustering. The resulting classtons automatically identify significant class characteristics from a small set of examples. These significant characteristics are then reliably detected by means of the classton channels using a probabilistic cluster representation.

Local image features are often affected by noise. As a consequence, noise is present in the clusters computed from the features. We propose DBScan and k-means with pruning to reduce the sensitivity to noisy features. In order to

lat 90 lon 0
(Trained image)   lat 90 lon 22   lat 90 lon 45   lat 90 lon 68

Fig. 7. Classton channels learned from full side view of 5 different horses. The channels are stable for visually similar views (here 45 degrees). Classification is possible. The channels are unstable for views that are not similar (right)

select those classtons that display the best generalisation ability, we consider the density, the compactness of the clusters in feature space and the connectivity of the features in image space.

The reprojection of the clusters demonstrates generalisation ability with respect to previously unseen objects and robustness to viewpoint changes. The fusion of several such classton channels provides a powerful means for robust classification by preserving the internal scale relations of the class features.

Without the robust detection of significant class features any classification algorithm is going to fail. This is the motivation for this article and the presented clustering technique is a means to robustly detect and identify the significant class features which are essential for the composition of a model for classification.

The exact structure of the classification model is another complex problem and merits an article on its own. The classton approach that preserves internal scale relations of the features opens several interesting possibilities for the construction of classification models, among these scale invariant classification by shifting the class model in scale.

References

1. M. Ester, H.-P. Kriegel, J. Sander, and X. Xu. A density-based algorithm for discovering clusters in large spatial databases with noise. In *International Conference on Knowledge Discovery and Data Mining (KDD-96)*, 1996.
2. M. Ester, H.-P. Krieger, J. Sander, M. Wimmer, and X. Xu. Incremental clustering for mining in a data warehouse environment. In *24th VLDB Conference*, New York, USA, 1998.
3. D.A. Forsyth and J. Ponce. *Computer Vision a Modern Approach*. Prentice Hall, 2003.
4. D. Hall, V. Colin de Verdière, and J.L. Crowley. Object recognition using coloured receptive fields. In *ECCV00*, Dublin, Ireland, June 2000.
5. J.J. Koenderink and A.J. van Doorn. Generic neighborhood operators. *Pattern Analysis and Machine Intelligence*, 14(6):597–605, June 1992.
6. T. Leung and J. Malik. Recognizing surfaces using three-dimensional textons. In *ICCV*, Corfu, Greece, September 1999.
7. T. Lindeberg. Feature detection with automatic scale selection. *IJCV*, 30(2):79–116, 1998.
8. D.G. Lowe. Object recognition from local scale-invariant features. In *ICCV99*, pages 1150–1157, 1999.
9. J. Malik, S. Belongie, T. Leung, and J. Shi. Contour and texture analysis for image segmentation. *IJCV*, 43(1):7–27, June 2001.
10. A.M. Martinez and R. Benavente. The ar face database. Technical Report 24, CVC, June 1998.
11. R.P.N. Rao and D.H. Ballard. An active vision architecture based on iconic representations. *Artificial Intelligence*, 78(1–2):461–505, 1995.
12. B. Schiele and J.L. Crowley. Recognition without correspondence using multidimensional receptive field histograms. *IJCV*, 36(1):31–50, January 2000.
13. C. Schmid. Constructing models for content-based image retrieval. In *Computer Vision and Pattern Recognition*, Kauai, USA, December 2001.
14. C. Schmid and R. Mohr. Local greyvalue invariants for image retrieval. *TPAMI*, 1997.
15. S.-C. Zhu, C. Guo, Y. Wu, and Y. Wang. What are textons? In *ECCV02*, pages IV 793–807, 2002.

Gaussian Scale Space
from Insufficient Image Information

Marco Loog[1], Martin Lillholm[2], Mads Nielsen[2], and Max A. Viergever[1]

[1] Image Sciences Institute, University Medical Center Utrecht
Utrecht, The Netherlands
[2] Image Processing Group, IT University Copenhagen
Copenhagen, Denmark

Abstract. Gaussian scale space is properly defined and well-developed for images completely known and defined on the d dimensional Euclidean space $\mathbb{R}^d$. However, as soon as image information is only partly available, say, on a subset V of $\mathbb{R}^d$, the Gaussian scale space paradigm is not readily applicable and one has to resort to different approaches to come to a scale space on V. Examples are the theory dealing with scale space on $\mathbb{Z}^d \subset \mathbb{R}^d$, i.e., discrete scale space; the approach based on the heat equation satisfying certain boundary conditions; and the ad hoc approaches dealing with (hyper)rectangular images, e.g. zero-padding of the area outside of V, or periodic continuation of the image.

We propose to solve the foregoing problem for general V from a Bayesian viewpoint. Assuming that the observed image is obtained by linearly sampling a real underlying image that is actually defined on the complete d dimensional Euclidean space, we can infer this latter image and from that image build the scale space. Re-sampling this scale space then gives rise to the scale space on V. Necessary for inferring the underlying image is knowledge on the linear apertures (or receptive field) used for sampling this image, and information on the prior over the class of all images.

1   Introduction

The definition and construction of Gaussian scale space [3,5] for images—a function from $\mathbb{R}^d$ to $\mathbb{R}$—with insufficient data is considered, i.e., given a d-dimensional *observed* image L for which the gray values are merely known on a proper subset V of the Euclidean space $\mathbb{R}^d$, scale space is defined, some explicit formulations are given, and several scale spaces for a certain image are constructed and analyzed experimentally.

Several approaches solving this problem in part exist. Koenderink, in his original paper [5], gave a possible approach that is based on a solution of the heat equation fulfilling certain boundary conditions. However, this method is not applicable to the situation in which the subset V is, for example, discrete.

In case that V equals $\mathbb{Z}^d$, a solution was proposed by Lindeberg in [7]. Here the discrete version of the diffusion equation is considered and used for constructing the Gaussian scale space on $\mathbb{Z}^d$. This approach exploits the regular structure of the sampling of L and in this way comes to a satisfactory formulation of discrete scale space. Extending the suggested approach to, for example, irregularly sampled images, is however not straightforward.

L.D. Griffin and M. Lillholm (Eds.): Scale-Space 2003, LNCS 2695, pp. 757–769, 2003.

Finally, we mention the ad hoc techniques most often preferred in practice: Zero-padding of the area outside of V or, when V is a hyper-parallelepiped, assuming this image to be periodically continuable over the entire d-dimensional Euclidean space.

We cast our approach in a Bayesian framework and solve the problem in theory and illustrate its practical performance. Section 2 presents the theoretical formulation of the problem and our solution to it. It also discusses the assumptions made to come to a unique solution. Furthermore, Subsection 2.2 presents methods for solving the optimization problem associated with the solutions to certain specific cases of V. Section 3 elaborates on the findings in Section 2, and provides illustrative examples. Provisional conclusions and remarks, and a discussion of the experiments are in Section 4.

2   Scale Space from Insufficient Image Information

Consider a d-dimensional image L merely defined on a proper subset V of the Euclidean space $\mathbb{R}^d$, i.e., the gray value of the image is only known for vectors x in V. We look at the problem of defining Gaussian scale space for such an *observed* image from a Bayesian perspective. We call such scale space the i scale space, where i could stand for incomplete, insufficiency, inferred or induced.

Our formulation partially involves an optimization problem, which infers the *real underlaying* image Γ from the observed L, assuming that the observations on the subset V are obtained by linearly sampling this underlying underlying image. From that we build the Gaussian i scale space on V for the image L. Actually, Γ can be considered to be a de-blurring or super-sampling of L [1,9], as such even enabling us to construct scale space on the entire $\mathbb{R}^d$.

Necessary for inference of Γ are assumptions on the kind of linear apertures used for sampling this image, and the presumptive knowledge on the prior over the class of all images.

2.1   Bayesian Formulation

Assume that the gray values of L on V are obtained via linearly sampling some underlying image Γ with receptive fields (or apertures) f_x, $x \in V$:

$$L(x) := \int f_x(y)\, \Gamma(y)\, dy. \tag{1}$$

(N.B. all integrals of such inner products are over $\mathbb{R}^d$, unless stated otherwise.) If the system of linear constraints given in Equation (1) is underdetermined, there is a complete metameric class of images Γ that would lead to the same observed image L when using the receptive fields f_x [6], i.e., we have insufficient information for constructing the image in a unique way. Assuming in addition a certain log-convex prior Π over the space of all images, we can construct a unique representative from this metameric class. From Bayes theorem, it directly follows that

$$P(\Gamma|L) \propto P(L|\Gamma)\Pi(\Gamma). \tag{2}$$

A natural choice for the underlying image Γ we like to infer, is the maximum a posteriori estimate given by maximizing the conditional probability $P(\Gamma|L)$ in Equation (2). For fixed L, this can be done by maximizing $P(L|\Gamma)\Pi(\Gamma)$, which solution is given by the underlying image Γ that maximizes the prior Π under the constraint that the Equations in (1) are satisfied for all x in V.

$$
\begin{array}{ccc}
\Gamma_\sigma \text{ (scale space)} & \xrightarrow{\;f_x\;} & L_\sigma \text{ (i scale space)} \\[4pt]
{\scriptstyle .*g_\sigma}\Big\uparrow & & \Big\uparrow{\scriptstyle .*\tilde{g}_\sigma} \\[4pt]
\Gamma & \xleftarrow[\;\text{argmax}P(\Gamma|.)\;]{} & L
\end{array}
\tag{3}
$$

Diagram (3) gives a simple, schematic overview of the process of building i scale space. Here, the operator we are actually interested in, i.e., the one that 'blurs' the observed image L, is denoted by $*\tilde{g}_\sigma$.

As a final remark, if the linear constraints in (1) are overdetermined, not a single solution exists. Therefore, throughout the paper, we assume underdeterminacy of the system. We briefly return to this topic in the discussion.

2.2 Image Priors and Inference

This subsection introduces the two image priors used throughout the rest of the paper. Subsequently, a few specific instances of the Bayesian formulation are considered, and it is shown how for these cases (2) is maximized. For a general formulation for optimizing (2), from which the solutions below are derived, the reader is referred to [6].

Many image priors are proposed in the literature—e.g. the maximum entropy prior, the uniform intensity prior, or priors learned from training data (see for example [2,6,8,10], and references therein). We restrict ourselves to the following two priors [6].

The Gaussian Intensity Prior.

$$
\Pi(\Gamma) \propto \exp\left[-\int \frac{(\Gamma(x) - \mu)^2}{2\varsigma^2} dx\right] ,
\tag{4}
$$

where μ is the mean and ς the standard deviation of the underlying Gaussian intensity distribution. Maximizing this prior is equivalent to minimizing the intensity variance measure

$$
V(\Gamma) = \int (\Gamma(x) - \mu)^2 dx .
$$

For notational convenience, we merely consider the prior in which μ is set to 0 (which would be a reasonable assumption if the mean gray value over V approximates 0). Using different μs leads to similar procedures and solutions.

The Brownian Motion Intensity Prior.

$$\Pi(\Gamma) \propto \exp\left[-\int \frac{\|\nabla \Gamma(x)\|^2}{2\varsigma^2} dx\right],\tag{5}$$

which optimization comes down to minimizing the measure of local variation

$$B(\Gamma) = \int \|\nabla \Gamma(x)\|^2 dx,$$

and is in accordance with results on natural images from [2].

Inference of Γ under the Gaussian intensity prior can be done by solving a system of linear equations. For notational convenience, we merely consider the prior in which μ is set to 0. Then, following [6], using the technique of Lagrange multipliers, the underlying image Γ satisfying the linear constraints from Equation (1) can be expressed in terms of a linear combination of these receptive fields f_x

$$\Gamma(z) = \sum_{x,y \in V} f_x(z)(F^{-1})_{xy} L(y).\tag{6}$$

Here, F is the matrix with entries

$$F_{xy} = \int f_x(z) f_y(z) dz\tag{7}$$

over all pairs x and y from V. In the experiment in Section 3, the discrete formulation of the above is used. This formulation is based on the calculation of a certain pseudo-inverse, and actually solves a linear equation involving the measurements of L and the apertures f_x (see [6] for details). Note that in all of the foregoing, we used a notation as if the number of receptive fields is finite, however similar expression can be defined for an infinite number of receptive fields. In this case, where appropriate, one should turn the summation into integration.

For the Brownian motion intensity prior, we suggest to solve the optimization problem via a variational reconstruction. Again, following [6], using a specific instance of their observation-constrained evolution, we start with an initial metamere Γ_0—typically the minimum variance reconstruction—and iteratively update this estimate according to the following rule

$$\Gamma_{n+1} = \Gamma_n - \gamma(G[\Gamma_n] - G_\perp[\Gamma_n]),\tag{8}$$

in which $G[\Gamma] = -\triangle\Gamma$ (which can be derived using variational calculus [4]), and

$$G_\perp[\Gamma] = -\sum_{x,y \in V} f_x(F^{-1})_{xy} \int f_y \triangle\Gamma\, dx.\tag{9}$$

Because the Brownian motion intensity prior is log-convex, B is convex, and hence a unique solution exists, which can be approximated using the foregoing scheme.

Based on specific prior knowledge, the foregoing formulations may simplify.

Scale Space under the Gaussian Intensity Prior. Using the Gaussian prior, Equation (6) gives the solution to the inference problem right away. From that, i scale space L_σ for L can be expressed simply as follows

$$L_\sigma(z) = \int f_z(a) \Big(\sum_{x,y \in V} f_x(F^{-1})_{xy} L(y) * g_\sigma \Big)(a) \, da \,, \tag{10}$$

which thus gives an explicit form for i scale space on $z \in V$.

Orthogonal Receptive Fields. If we assume that the receptive fields f_x are orthogonal, i.e., $\int f_x(z) f_y(z) \, dz = 0$ if $x \neq y$, the matrix entries in (7) are by definition zero when $x \neq y$, and equal the 2-norm of f_x for $x = y$. Therefore, F becomes readily invertible and Equation (6), the minimum variance solution, reduces to $\Gamma(z) = \sum_{x \in V} \frac{L(x) f_x(z)}{\|f_x\|^2}$. Of course, i scale space can be constructed from Equation (10).

For the Brownian motion intensity prior, the update rule (8) becomes

$$\Gamma_{n+1}(z) = \Gamma_n(z) + \gamma \Big(\triangle \Gamma_n(z) - \sum_{x \in V} \frac{f_x(z)}{\|f_x\|^2} \int f_x(y) \triangle \Gamma_n(y) \, dy \Big) \tag{11}$$

We should note however that this update rule is only readily applicable when using smooth receptive fields f_x, a more general formulation is allowed for when using the formalism of distributional derivatives.

Note that for the minimum variance solution $\Gamma(z)$ equals 0, when z is not an element of $\bigcup_{x \in V} \operatorname{supp} f_x$, i.e., if z does not reside in at least one of the supports associated to the receptive fields f_x, it holds that $\Gamma(z) = 0$, i.e., the d-dimensional Euclidean space is *padded with zeros* outside of the support $\bigcup_{x \in V} \operatorname{supp} f_x$.

Orthonormal and Position Invariant Receptive Fields. In addition to the foregoing, we assume the receptive fields to be normalized with respect to the 2-norm, and that they are translation invariant in the sense that $f_x = h(. - x)$ for some fixed kernel function h for all $x \in V$.

Using this, the procedures can now be further simplified using convolutions. For the Gaussian intensity prior we have

$$\Gamma(z) = \sum_{x \in V} L(x) \, h(z - x) = \Big(\sum_{x \in V} L(x) \delta_x * h \Big)(z) \,,$$

and the iteration step for the Brownian motion prior equals

$$\Gamma_{n+1}(z) = \Gamma_n(z) + \gamma \Big(\triangle \Gamma_n(z) - \sum_{x \in V} h(z - x) \int h(y - x) \triangle \Gamma_n(y) \, dy \Big)$$

$$= \Gamma_n(z) + \gamma \Big(\triangle \Gamma_n(z) - \sum_{x \in V} h(z - x) \triangle (h(-.) * \Gamma_n)(x) \Big) \,,$$

resembling an iterative procedure for regularized de-convolution.

Again, under the Gaussian prior, i scale space can be constructed from Equation (10), which gives

$$L_\sigma(z) = \int h(a - z) \sum_{x \in V} L(x) \, (h * g_\sigma)(a - x) \, da \, .$$

3   Reconstruction Examples and Induced Scale Spaces: Some Experiments

This section presents a pictorial overview of some possible inferred images, under the two priors presented earlier, using different apertures (in form and size) and measurement domains. Furthermore, it illustrates the different i scale spaces obtained from them. Before doing so, we first discuss the settings and the data used for the experiments conducted.

3.1   The Observed Images

The two example images L_C and L_S, used to illustrate several aspects of constructing scale space on a particular subset V of $\mathbb{R}^2$, are as follows.

The observed image L_C is defined on a 'circular' subset V_C of $\mathbb{Z}^2 \subset \mathbb{R}^2$ of radius 5,

$$V_C = \{x \in \mathbb{Z}^2 | (x_1 - \tfrac{1}{2})^2 + (x_2 - \tfrac{1}{2}^2) < 5^2\} \, ,$$

on which only negative – and positive + measurements are made, say –1 and +1. See Figure 1 on the left for an illustration of V_C and the positions of the several observed values. For our results and subsequent discussion, the actual value of the positive and negative measurements are unimportant, however – is chosen to equal the negation of +. The observed image L_S is defined on a 'square' subset V_S of $\mathbb{Z}^2 \subset \mathbb{R}^2$ with sides 12,

$$V_S = \{x \in \mathbb{Z}^2 | x \in [-5, 6] \times [-5, 6]\} \, ,$$

on which negative –, positive +, as well as zero ◦ measurements are made. Again, say the positive observations are +1 and the negative are –1. The right half of Figure 1 depicts V_S and the positions of the several observed values. In fact, both images L_C and L_S are assumed to come from the same underlying image Γ. The observed image L_S, however, has more observations of this underlying image, which may result in a different inferred image, as we will see below.

3.2   The Apertures

In our experiments, we use two different kernels as aperture. The first one is a box kernel, the second a Gaussian kernel. Both come in two different scales. The small box kernel has sides of length $\tfrac{2}{5}$, while the large box has sides of $\tfrac{12}{5}$. The sizes of these boxes are chosen in such a way, that the small kernels do not overlap—and hence are orthogonal to each other, and the large kernels do overlap with neighboring kernels. The Gaussian kernels always overlap, however for the small kernel this overlap is less severe than for the large one. The scales of the Gaussian kernels are $\tfrac{3}{10}$ and $\tfrac{18}{10}$, respectively.

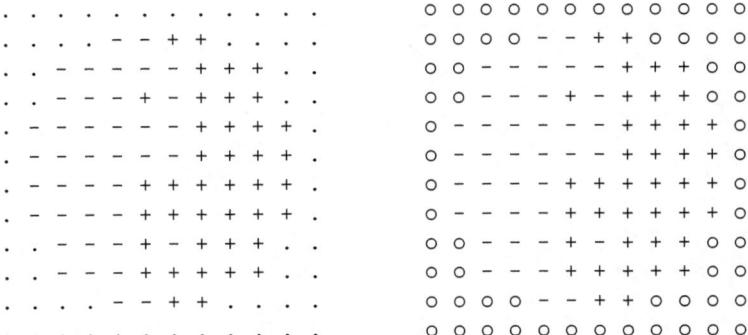

Fig. 1. On the left: Spatial arrangement of the circular subset V_C on which the observed image L_C is defined. The observed values are indicated by + and -, i.e., merely two different values are observed. On the right: Spatial arrangement of the square subset V_S on which the observed image L_S is defined. The observed values are indicated by o, +, and -.

3.3   Inferred Images and Their Induced Scale Space

Inferred Images. For the observed images L_C and L_S, Figures 2 and 3 give all eight possible inferences of Γ (per domain) using the two different priors, the two different apertures, and the two different scales (see the captions to these figures).

For the inferred images under the Gaussian prior with box apertures, it is clearly visible that these images are superpositions of box kernels. The compact support of these kernels results in images having a compact support also. Outside of this support the images are zero-padded. Comparing the inference, using the `large` box aperture, on the circular domain with the one on the square domain, we see differences in the surrounding of the *yin and yang* symbol. These differences are not there when using the `small` box apertures. Differences in inferences using different domains become even more apparent when comparing the several inferred images using the `large` Gaussian kernel.

A next observation we make is that the images inferred using the Brownian motion intensity prior look more diffuse than many of the Gaussian intensity prior inferred images. Furthermore, the `small` scale inferred all are visually similar to each other. On the other hand, `large` scale inference completely obscure the *yin and yang* symbol and can look very different. Especially for the four regular 'checkerboard-like' images inferred assuming the apertures to be `large` Gaussian kernels it seems implausible that applying the Gaussian apertures to these images could reconstruct the observed images L_C and L_S. However, simply applying the f_xs to these images does bring back the initial observed images.

Improbable Observations. What might be implausible, i.e. rather improbable, is that the observed image is obtained by such a large aperture as the `large` Gaussian. Figure 4 shows what happens when we perturb the receptive fields. In this figure, for the right image, all initial f_xs are shifted slightly to the right, resulting in a large change of visual appearance and therefore the observed image. This illustrates that some inferences can

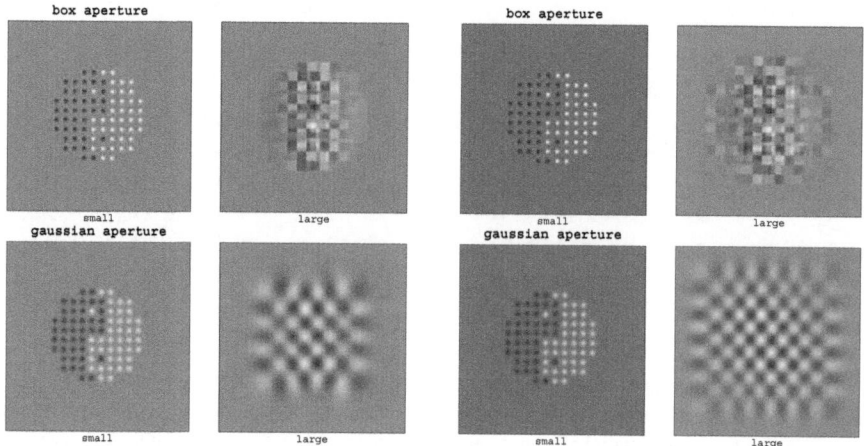

Fig. 2. Possible inferences under the Gaussian intensity prior on the circular (left four) and the box (right four) domain. The top row assumes the measurement to come from a box aperture, while the bottom row gives results for using a Gaussian aperture. The sub-captions indicate whether the small scale or large scale aperture is used.

be unstable in a certain sense and that we were rather fortunate obtaining the observed image under these prior assumptions.

This latter observation holds for many of the inferred images. Consider for example the images inferred using the Gaussian prior and the small apertures. A small spatial shift of the f_xs will result in a large change of observed values because the amount of overlap between the inferred image and the apertures diminished rapidly in this case.

More stable and more probable are the inferred images under the Brownian motion prior using small apertures. In these cases small perturbations of the f_xs lead to similar observed images.

i **Scale Spaces.** The inferred images can look rather different from each other. Our main interest, however, is the i scale spaces, on both the square and the circular domain, constructed from these underlying images. In Figure 5 on page 767 four different i scale spaces based on the Gaussian prior are shown: Two for the circular domain and two for the square domain. Figure 6 on page 768 gives two additional i scale spaces are shown are in row 1 and 2, and row 5 and 6, respectively. Per i scale space, ten scales are taken, ranging from $1.4^1 = 1.4$ to $1.4^{10} \approx 28.9$. In both Figure 5 and Figure 6, in the sub-captions to every scale, we show the maximum intensity value of the function on its domain V. The ten i scale spaces not displayed behave in the same manner and henceforth were omitted.

As one can see in Figure 5, the i scale spaces under these different prior assumptions may also differ considerably in visual appearance. However, the i scale spaces in 5(c), 6(a), 6(c), and even 5(a) are not very different from each other. One might note that the i scale space first depicted in 5(a) looks slightly less blurred for the smaller scales in comparison to the other three i, but their behavior as one goes up in scale is comparable

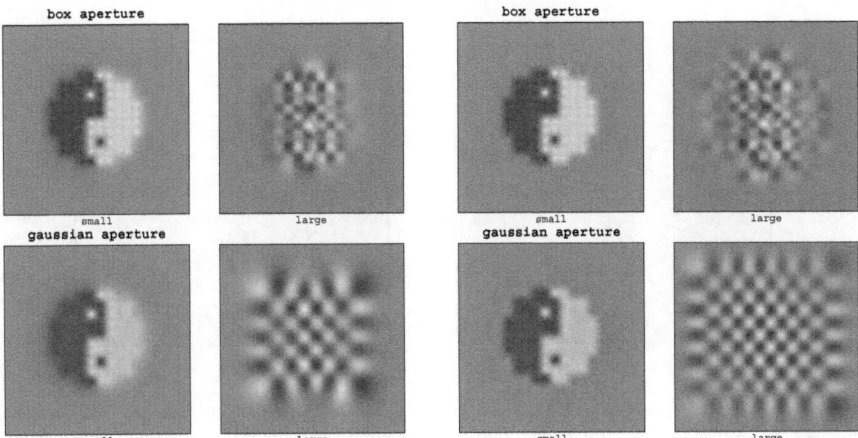

Fig. 3. Possible inferences under the Brownian motion intensity prior on the circular (left) and the square (right) domain. The top row assumes the measurement to come from a box aperture, while the bottom row gives results for using a Gaussian aperture. The sub-captions indicate whether the small scale or large scale aperture is used.

for all these i scale spaces even though 5(a) is restricted to the circular domain. In fact all i scale spaces not based on the large Gaussian aperture look rather similar. This is remarkable, especially considering the four inferences assuming large box apertures (see Figures 2 and 3), it is certainly not obvious that there are no larger differences present in the several i scale spaces.

Maximum Principle. The four i scale spaces based on the large Gaussian aperture (two of which are in Figures 5 (b) and (d)) are are quite different from the other ones. They are also relatively different in comparison to each other, which was less apparent for the other twelve i scale spaces. What is most noteworthy is the fact that the maximum intensities in all cases go up dramatically, at least for a couple of subsequent scales (see the sub-captions to Figures 5 (b) and (d)). Because the underlying image Γ will go to the zero function eventually if scale is large enough, this also holds for its induced scale spaces. Note that also in Figure 6 (c), the maximum intensity first increases.

For the underlying image the maximum principle will hold [5] when going up in scale, and therefore the maximum intensity over $\mathbb{R}^d$ will decrease starting from scale zero. On the contrary, when restricting our attention to the subsets on which our observed image is defined, it can happen that maximum intensity increases with increasing scale, i.e, these domain restricted i scale spaces may violate the maximum principle.

Comparison with 'Normal' Scale Space. Some additional experiments were conducted for which we used the two i scale space based on the square domain and the small Gaussian aperture.

Firstly, under these prior assumptions, we determined the i scale spaces and, in addition, we constructed the scale spaces in the 'normal' way, i.e., by interpreting every

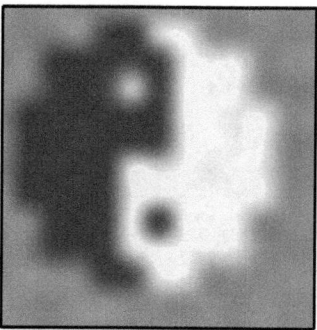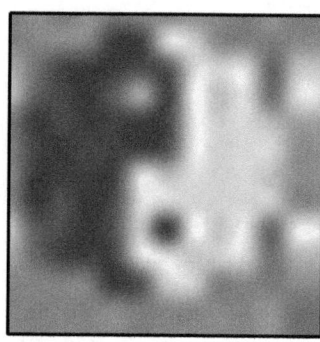

Fig. 4. The inferred image on the square domain under the Brownian motion intensity prior using the `large` Gaussian apertures is sub-sampled at scale 0, i.e., the initial observed image is reconstructed. On the left is the correctly sub-sampled version. The right perturbed image is obtained by shifting all receptive fields f_x slightly to the right, which has a large effect on the final result.

measurement as the gray value in a pixel and blurring this image using discretized Gaussian kernels and zero-padding. In Figure 6, (a) and (c) are the i scale spaces under the Gaussian and Brownian prior, respectively. Illustrations (b) and (d) in addition give the absolute difference of the i scale spaces and the 'normal' scale space on the square domains. In the sub-caption to these images we give the root mean squared error (left from the /) and the maximum absolute error (right from the /) measured on the respective difference images.

Keeping in mind that the maximum absolute pixel value of the initial observed image equal 1, we see that especially the root mean squared errors under the Gaussian prior get quite large; up to 0.372 over the whole image. For both examples, the maximum absolute error can become larger than 0.5. Both errors drop of more quickly in case of using the Brownian prior.

4   Discussion, Provisional Conclusions and Remarks

As we are aware of the exploratory nature of this paper, we realize that the following remarks and conclusions drawn from the experiments—and our few theoretical findings—may not be more than provisional. However, we think we provided a very general and interesting possibility to constructing scale space from insufficient image information on a general subset V of the d-dimensional Euclidean space.

In conclusion, we raise the following issues.

– Our experiments show that i scale spaces constructed via our Bayesian formulation may look very different compared to each other, but can in many other cases be visually similar. The differences, however, may only be appreciable when the number of observations in V is rather small, because when many constraint have to be fulfilled the metameric class becomes very small. So repeating the experiments for many more observations than the number considered here might reveal most i scale spaces to be indiscernible from each other.

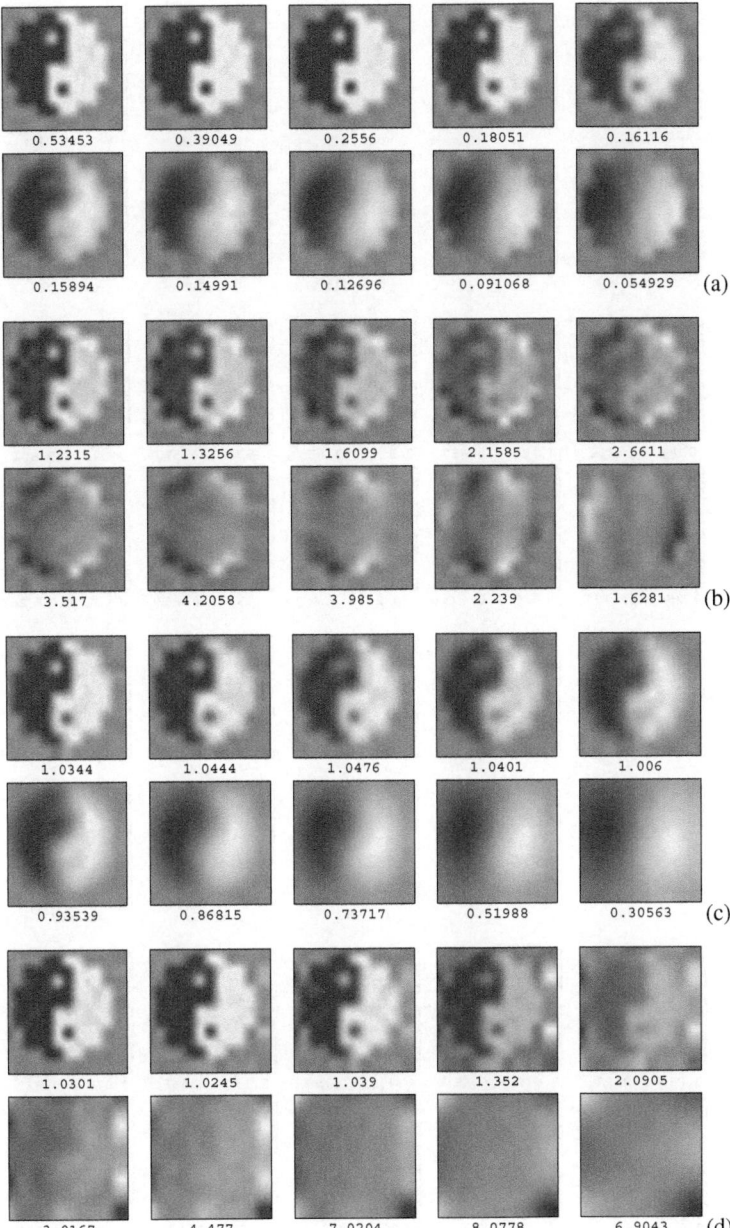

Fig. 5. Four i scale spaces induced from four different prior settings (see text Section 3.3). In the sub-caption the maximum intensity value at that scale on the domain is given. The pixel intensities are scaled to fully use the range from black to white. In (a) is the i scale space using the `small` box aperture, and the circular domain; (b) the `large` Gaussian, and the circular domain; (c) the `large` box aperture, and the square domain; (d) the `large` Gaussian, and the square domain.

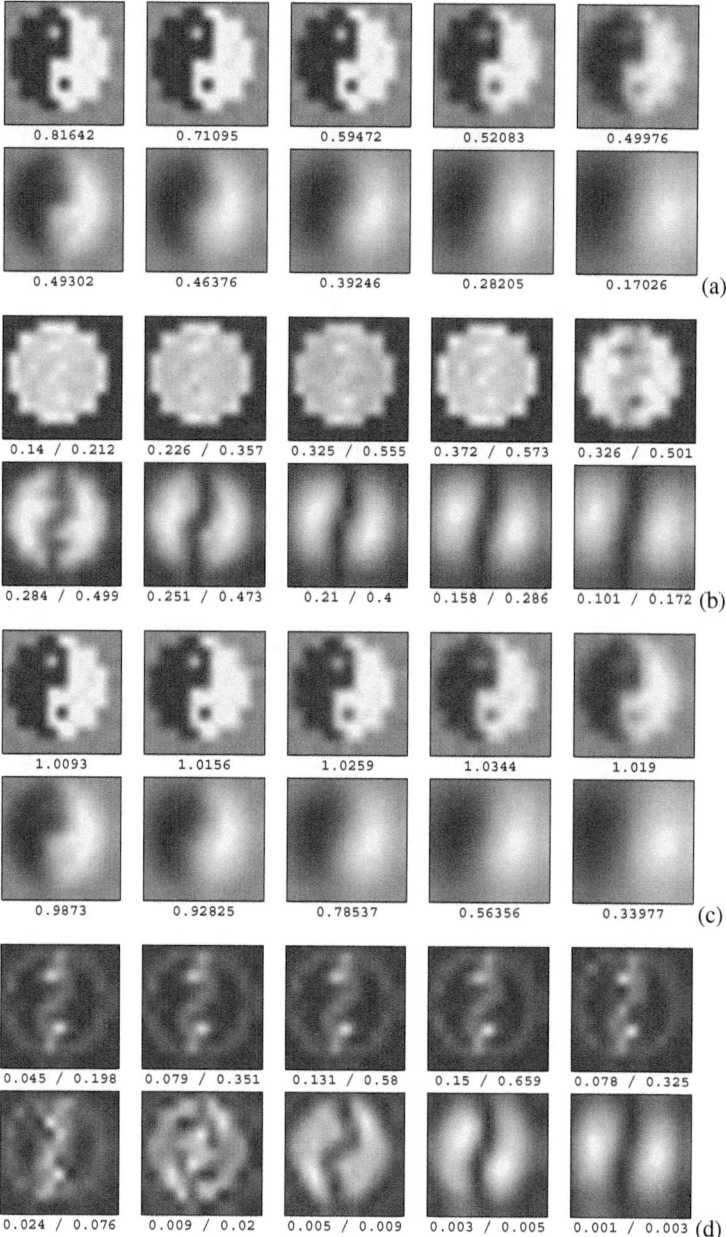

Fig. 6. The two i scale spaces in (a) and (c) on the square domain are inferred using the `small` Gaussian aperture under the Gaussian and Brownian prior, respectively. The absolute difference images of the i scale spaces with the 'normal' scale space are in (b) and (d) (see Section 3.3). Sub-caption to (a) and (c): maximum intensity value at that scale on the square domain. Sub-caption to (b) and (d): root mean squared / maximum absolute error.

- When using the Gaussian prior, $\mathbb{R}^d$ is always padded with zeros for those points that lie outside of the conjunction of the apertures' supports. We may take this as an argument for solving the insufficient information problem by padding the remaining space with zeros. However, the experiments described at the end of Subsection 3.3 show that the difference between this 'normal', pixel-based way of dealing with Gaussian blurring can give results rather different from our i scale space approach.
- With both theoretical and experimental analysis we shall try to gain additional insight in the inference process and the associated i scale spaces. In practice, especially i scale spaces for images observed on a compact (or even (hyper)rectangular) $V \subset \mathbb{Z}^d$ are interesting. From a more theoretical point of view this holds for V equal to $\mathbb{Z}^d$, or some compact and connected subset of $\mathbb{R}^d$. Analyzing these instances may lead to interesting links with the scale space of Lindeberg [7] or the PDE-based approach of Koenderink [5].
- Further analysis should also go out to the possible violation of the maximum principle noted at the end of Subsection 3.3. This violation is due to the sub-sampling of the real scale space to get to the i scale space on V and so could be avoided by working with the scale space based on the underlying image. In practice, we almost always work with some discrete subset V of $\mathbb{R}^n$, and hence we might consider designing prior assumptions for inference that do not lead to the violation of the maximum principle on the domain V.
- In the light of the foregoing, we can think of other conditions that restrict the possible prior assumptions. For this, looking at axiomatic approaches to defining scale space (see for example [3]) might be interesting. One example would be the condition that $(L * \tilde{g}_\sigma) * \tilde{g}_\tau$ equals $L * \tilde{g}_{\sqrt{\sigma^2+\tau^2}}$. For scale spaces on $\mathbb{R}^d$ this follows directly from the semigroup property [3].

References

1. S. Chaudhuri. *Super-Resolution Imaging*, volume 632 of *International Series in Engineering and Computer Science*. Kluwer Academic Publishers, Boston, 2001.
2. D. J. Field. Relations between the statistics of natural images and the response proporties of cortical cells. *Journal of the Optical Society of America. A, Optics and Image Science*, 4(12):2379–2394, 1987.
3. L. M. J. Florack. *Image Structure*, volume 10 of *Computational Imaging and Vision*. Kluwer, Dordrecht . Boston . London, 1997.
4. C. Fox. *An Introduction to the Calculus of Variations*. Dover Press, New York, 1987.
5. J. J. Koenderink. The structure of images. *Biological Cybernetics*, 50:363–370, 1984.
6. M. Lillholm, M. Nielsen, and L. D. Griffin. Feature-based image analysis. *International Journal of Computer Vision*, in press, 2003.
7. T. Lindeberg. Scale-space for discrete signals. *IEEE Transactions on Pattern Analysis and Machine Intelligence*, 12(3):234–245, 1990.
8. D. L. Ruderman and W. Bialek. Statistics of natural images: Scaling in the woods. *Physical Review Letters*, 73(6):100–105, 1994.
9. J. L. Starck, E. Pantin, and F. Murtagh. Deconvolution in astronomy: A review. *Publications of the Astronomical Society of the Pacific*, 114:1051–1069, 2002.
10. S. Zhu, Y. Wu, and D. Mumford. Minimax entropy principle and its application to texture modeling. *Neural Computation*, 9:1627–1660, 1997.

Families of Generalised
Morphological Scale Spaces

Martin Welk

Saarland University, Faculty of Mathematics and Computer Science
welk@mia.uni-saarland.de

Abstract. Morphological and linear scale spaces are well-established instruments in image analysis. They display interesting analogies which make a deeper insight into their mutual relation desirable. A contribution to the understanding of this relation is presented here.

We embed morphological dilation and erosion scale spaces with paraboloid structure functions into families of scale spaces which are found to include linear Gaussian scale space as limit cases. The scale-space families are obtained by deforming the algebraic operations underlying the morphological scale spaces within a family of algebraic operations related to l^p norms and generalised means. Alternatively, the deformation of the morphological scale spaces can be described in terms of grey-scale isomorphisms.

We discuss aspects of the newly constructed scale space families such as continuity, invariance, and separability, and the limiting procedure leading to linear scale space. This limiting procedure requires a suitable renormalisation of the scaling parameter. In this sense, our approach turns out to be complementary to that proposed by L. Florack et al. in 1999 which comprises a continuous deformation of linear scale space including morphological scale spaces as limit cases provided an appropriate renormalisation.

Keywords: morphological scale space, linear scale space, dilation, erosion, deformation

1   Introduction

A scale space [10,11,1,2,3,5,6,13,18,15] can be described as a family of filters which transform a given signal into a simplified signal. The family of operators is equipped with a linear ordering, from fine to coarse resolution, having the identity as minimal element, and is required to fulfil the causality condition, i.e. structural details of the signal such as extrema must not be enhanced under the filter action. For a scale space in strict sense it is also required that concatenation of filter operators is equivalent to one single filter operator of coarser resolution, such that the filter operators form a semi-group.

An outstanding example with multiple applications is the Gaussian linear scale space made up of convolutions of the original signal with Gaussians of increasing standard deviation [10,11,5]. Since a two-dimensional Gaussian can

L.D. Griffin and M. Lillholm (Eds.): Scale-Space 2003, LNCS 2695, pp. 770–784, 2003.
© Springer-Verlag Berlin Heidelberg 2003

be written as the product of two one-dimensional Gaussians, this scale space is separable. Here, separability means that the filtering in two dimensions can be performed by filtering in x and y directions subsequently, and it is a highly desirable property particularly from the computational point of view. The semigroup property holds because the convolution of Gaussians is again a Gaussian with the sum of variances. Last but not least, Gaussian filtering is rotationally invariant which is an important requirement particularly in image processing applications.

Morphological dilation and erosion, with structure functions of increasing size in the scaling parameter, define another class of scale spaces [4,16,17,12]. Provided that a rotation-symmetric quadratic structure function is used, one obtains again rotational invariance, separability and semigroup property.

Besides sharing many useful properties making them valuable for denoising and other image analysis applications, both before-mentioned classes of scale spaces display also similarities in their structure which have been noted e.g. in [16,17,8,7]. The defining formulas of dilation and erosion can formally be obtained from that of convolution by replacing addition with maximum or minimum, and multiplication with addition. Now the real numbers equipped with maximum and addition form a semi-ring, the so-called max-plus algebra, cf. e.g. [14]. Being only a semi-ring, the max-plus algebra is a weaker algebraic structure than the usual, plus-product algebra, but still has many parallels to the latter. The most important difference is the lacking of a neutral and inverse for maximum. Quite alike, there is also a min-plus algebra. Essentially, dilation and erosion are in the max-plus and min-plus algebras what convolution is in the plus-product algebra. Furthermore, convolution in plus-product algebra is in close relation to the Fourier transform which carries convolution to multiplication and vice versa. In max-plus algebra, there is the slope transform which stands in mostly the same relation to dilation: it carries over dilation to addition and vice versa [4]. We shall not pursue the latter analogy but concentrate on the scale spaces instead.

Starting from the observation that all algebraic operations involved – addition, multiplication, minimum/maximum – fit into one single parametrised family of operations which stands in close relation to generalised means and l^p norms, we describe a variety of scale spaces which in some sense interpolate between morphological and linear scale spaces.

Parametrised families of scale spaces that allow a continuous transition, in some sense, between different fundamental scale spaces have already been proposed in the literature, see [8,7,5,9]. While in [5] Poisson and Gaussian scale spaces are considered, the construction by Florack et al. from [8,7] is of particular interest for us since it is also concerned with linking morphological and Gaussian scale spaces. The scaling procedure used in [9] also includes the grey-value transformations via power functions that can be used to describe subsets of the filter families discussed here, see section 2.4.

The paper is organised as follows: In paragraph 2.1 we introduce the algebraic operations to be used, and we collect some basic facts about them. In paragraph 2.2 we define the family of scale spaces that are treated in this paper. These objects are studied in more detail then. While paragraph 2.3 contains limit statements securing the continuity of the family of scale spaces as a whole, the properties of the individual scale spaces are investigated in paragraph 2.4. A comparison to the family of scale spaces proposed by Florack et al. is given in paragraph 2.5. In section 3, the interpolation property of our scale space family is illustrated with an example picture.

2   Generalised Morphological Scale Spaces

2.1   A Family of Algebraic Operations

First we introduce algebraic operations and integrals for later use. Throughout the paper, $\mathbb{R}_0^+$ and $\mathbb{R}^+$ denote non-negative and positive real numbers, resp.

Definition 1. *Let $\varphi : R \to R'$ be a continuous, monotonic, one-to-one function where each of R and R' may stand for $\mathbb{R}$, $\mathbb{R}_0^+$ or $\mathbb{R}^+$. Then we define*

$$a +_{\varphi} b := \varphi^{-1}(\varphi(a) + \varphi(b)) \tag{1}$$

and call it φ-deformed addition. Analogously, we define the φ-deformed integral of a function f over a domain $D \subset \mathbb{R}^n$ by

$$\varphi\!\!\int_D f(x)\,\mathrm{d}x := \varphi^{-1}\left(\int_D \varphi(f(x))\,\mathrm{d}x\right). \tag{2}$$

Given a second continuous, monotonic, one-to-one function $\psi : R \to R''$, $R'' \in \{\mathbb{R}, \mathbb{R}_0^+, \mathbb{R}^+\}$, we call

$$(f *_{\varphi,\psi} g)(x) := \varphi\!\!\int_{\mathbb{R}} f(x-y) +_{\psi} g(y)\,\mathrm{d}x \tag{3}$$

(φ, ψ)-deformed convolution.

Keeping this in mind, we turn to have a – rather grazing – look at generalised means.

Definition 2. *Assume $p \in \mathbb{R} \setminus \{0\}$. For $a, b \in \mathbb{R}^+$, or even $a, b \in \mathbb{R}_0^+$ if $p > 0$, let*

$$M_p(a, b) := \left(\frac{a^p + b^p}{2}\right)^{1/p}. \tag{4}$$

Moreover, let for $a, b \in \mathbb{R}_0^+$

$$M_{-\infty}(a, b) := \min(a, b), \quad M_0(a, b) := (ab)^{1/2}, \quad M_{+\infty}(a, b) := \max(a, b). \tag{5}$$

For $p \in \mathbb{R} \cup \{\pm\infty\}$, $M_p : \mathbb{R}^+ \times \mathbb{R}^+ \to \mathbb{R}^+$ is called p-th generalised mean.

It is well-known that for $a, b \in \mathbb{R}^+$, $M_p(a, b)$ as a function in $p \in \mathbb{R} \cup \{\pm\infty\}$ is continuous and monotonically increasing everywhere. If $a \neq b$, monotony is strict. In this way the geometric mean fits smoothly into the series of power means, as do maximum and minimum as limit cases. This fact motivates us to interpolate between the three algebraic operations multiplication, addition and maximum in the following way.

Definition 3. *Let $p \in \mathbb{R} \setminus \{0\}$. For $a, b \in \mathbb{R}^+$, or even $a, b \in \mathbb{R}_0^+$ provided that p is positive, define*

$$a +_p b := (a^p + b^p)^{1/p}. \tag{6}$$

Further let for $a, b \in \mathbb{R}_0^+$

$$a +_{-\infty} b := \min(a, b), \quad a +_0 b := ab, \quad a +_{+\infty} b := \max(a, b). \tag{7}$$

Note that $+_p$, $p \in \mathbb{R}$ is exactly the φ-deformed addition in the sense of definition 1 if $\varphi(x) = x^p$ for $p \neq 0$, $\varphi(x) = \ln x$ for $p = 0$. Besides this, $+_p$ for $p \in [1, +\infty)$ is just the l^p-norm of the finite sequence (a, b).

It is obvious that all $+_p$, $p \in \mathbb{R} \cup \{\pm\infty\}$, are commutative and associative. Distributivity between two of these operations, $(a +_q b) +_p c = (a +_p c) +_q (b +_p c)$ for all admissible a, b, c, holds if and only if $q = \pm\infty$ or $p = 0$.

For non-negative real p or $p = +\infty$ it makes sense to define a partially inverse operation for $+_p$ in the following way:

Definition 4. *For $p \in [0, +\infty]$, $a, b \in \mathbb{R}_0^+$, we define $a -_p b := \inf\{c \in \mathbb{R}_0^+ \mid c +_p b \geq a\}$.*

For $p = 0$, $-_p$ coincides with division for all $a \geq 0$, $b > 0$. If $p > 0$, $(a -_p b) +_p b = a$ holds only for $a \geq b$. It is clear that in this case one can calculate $a -_p b = (a^p - b^p)^{1/p}$ while $a -_p b = a$ for $p = +\infty$.

Using the same deformation functions φ as for $+_p$, we can also introduce modified integrals.

Definition 5. *For continuous functions $f : \mathbb{R}^n \to \mathbb{R}_0^+$ and domains $D \subset \mathbb{R}^n$ we define the p-integral by*

$$_p\!\!\int_D f(x) \, dx := \left(\int_D (f(x))^p \, dx \right)^{1/p} \qquad \text{for } p \in \mathbb{R} \setminus \{0\}, \tag{8}$$

$$_0\!\!\int_D f(x) \, dx := \exp \int_D \ln f(x) \, dx, \tag{9}$$

$$_{+\infty}\!\!\int_D f(x) \, dx := \sup_{x \in D} f(x), \qquad _{-\infty}\!\!\int_D f(x) \, dx := \inf_{x \in D} f(x). \tag{10}$$

For $p \in [1, +\infty]$ these p-integrals coincide with the $L^p(D)$-norms of f.

As in definition 1, generalised addition and integral can be combined to form a (q, p)-convolution of two functions. We refrain from carrying this out in a formal expression at this point; we shall use the idea in a slightly modified manner when introducing (q, p)-dilation.

2.2   Definition of Generalised Dilation and Erosion Scale Spaces

We write a scale space as a family $\{F_t \mid t \in \mathrm{I\!R}_0^+\}$ of mappings of some function space $\mathcal{F}$ over $\mathrm{I\!R}^n$ into itself, with F_0 being the identity. The causality condition states that for any given function $u_0(x) = f(x)$ from this function space and any $t > 0$, the function $u_t(x) = F_t f(x)$ contains no details which are not contained in $u_{t'} = F_{t'} f(x)$ for all $0 \leq t' \leq t$.

Throughout the following, it is understood that $\mathcal{F}$ consists of the continuous, bounded functions over $\mathrm{I\!R}^n$ with compact support. Since $f \in \mathcal{F}$ is to represent a given image, we shall also assume that the range of f, representing grey values, is contained in $[0, 1]$.

Gaussian convolution linear scale space is given by

$$F_t f(x) = \int_{\mathrm{I\!R}^n} f(x - y)\phi_{\sqrt{t}}(y)\, dy, \quad \phi_\sigma(y) = \frac{1}{(\sigma\sqrt{2\pi})^n} \exp\left(-\frac{\|y\|}{2\sigma^2}\right). \quad (11)$$

The morphological scale spaces of dilation and erosion are defined by

$$F_t f(x) = (f \oplus b_t)(x), \quad t > 0, \quad (12)$$
$$F_t f(x) = (f \ominus b_t)(x), \quad t > 0, \quad (13)$$

resp., with families of quadratic structure functions, $b_t = \|x\|^2/(2t)$. Here, dilation $\oplus$ and erosion $\ominus$ are given by

$$(f \oplus b)(x) = \max_{y \in \mathrm{I\!R}^n} (f(x - y) - b(y)),$$
$$(f \ominus b)(x) = \min_{y \in \mathrm{I\!R}^n} (f(x - y) + b(y)).$$

Motivated by the analogies between these scale spaces we look for a more general class of scale spaces on the ground of the algebraic operations introduced in section 2.1. We start with the definition of generalised dilations. Not all parameter values will lead to scale spaces in strict sense, i.e. with the semi-group property; we shall deal with this issue in proposition 2.

Definition 6. *Let* $f : \mathrm{I\!R}^n \to [0, 1]$ *be a signal and* $b : \mathrm{I\!R}^n \to \mathrm{I\!R}_0^+$ *continuous such that* $\{x \in \mathrm{I\!R}^n \mid b(x) < B\}$ *is bounded for any* $B \geq 0$. *For* $q \in [1, +\infty]$, $p \in [0, 1]$ *we define*

$$(f \oplus_{q,p} b)(x) := {}^q\!\!\int_{\mathrm{I\!R}^n} f(x - y) -_p b(y)\, dy \quad (14)$$

and call $f \oplus_{q,p} b$ *the* (q, p)-*dilation of* f *w.r.t. the kernel* b.

Obviously, ordinary dilation is recovered for $q = +\infty$, $p = 1$. For $q = 1$, $p = 0$, we have convolution with the kernel $1/b$. As a third special case we mention the "multiplicative dilation" $\oplus_{+\infty,0}$ with

$$(f \oplus_{+\infty,0} b)(x) = \sup_{y \in \mathrm{I\!R}^n} f(x - y)/b(y).$$

It must be pointed out that our definition of $-_p$ implies that the range of $f(x-y) -_p b(y)$ is truncated from below at zero. Since f is assumed to have $[0,1]$ range, and $b_p(0) = 0$, this has no effect whatsoever on the result of the generalised dilation as long as the integral is in fact a maximum, i.e. for $q = +\infty$. The truncation at zero has also no effect in the case $p = 0$ since then the integrand is in fact $f(x-y)/b(y)$ which never becomes negative. However, for $q < +\infty$ and $p > 0$ the truncation at zero is in fact somehow arbitrary; we come back to this issue in section 2.4 where the properties of the family of (q,p)-dilations will be discussed in more detail.

Proposition 1. *The (q,p)-dilation with kernel b is rotationally invariant for all continuous $f : \mathbb{R}^n \to \mathbb{R}_0^+$ if and only if $b(x)$ depends only on $\|x\|$. The (q,p)-dilation with kernel b is rotationally invariant and separable if and only if either $q = +\infty$, $p > 0$, $b(x) = b_{p;\lambda}(x) = \lambda \|x\|^{2/p}$, or $p = 0$, $q \in [1,+\infty]$ arbitrary, and $b(x) = b_{0;k,\sigma}(x) = k^n \exp(\|x\|^2/(2\sigma^2))$, with λ, k and σ being arbitrary positive real numbers.*

Proof. Rotational invariance means that whatever f may be given, $(f \oplus_{q,p} b)_\varrho$ is identical with $f_\varrho \oplus_{q,p} b$ for any rotation $\varrho : \mathbb{R}^n \to \mathbb{R}^n$. Here, g_ϱ is defined as $g_\varrho(x) := g(\varrho x)$ for all $x \in \mathbb{R}^n$. Now we have

$$(f \oplus_{q,p} b)_\varrho(x) = q\!\!\int_{\mathbb{R}^n} f(\varrho x - \varrho y) +_p b(\varrho y)\, dy$$
$$= q\!\!\int_{\mathbb{R}^n} f_\varrho(x - y) +_p b_\varrho(y)\, dy$$

on one side and

$$(f_\varrho \oplus_{q,p} b)(x) = q\!\!\int_{\mathbb{R}^n} f_\varrho(x - y) +_p b(y)\, dy$$

on the other side. Identity for all f can hold only if $b_\varrho(x) = b(x)$ for all $x \in \mathbb{R}^n$ and all rotations ϱ which implies that $b(x)$ depends only on $\|x\|$ since rotations act transitive on each sphere $\|x\| = $ const.

To study separability, it is sufficient to consider the decomposition of two-dimensional (q,p)-dilation into two one-dimensional (q,p)-dilations. Requiring that for all admissible functions f the two-dimensional (q,p)-dilation

$$q\!\!\int_{\mathbb{R}^2} f(x - y) +_p b(y)\, dy,$$

$x = (x_1, x_2)^{\mathrm{T}}$, $y = (y_1, y_2)^{\mathrm{T}}$, be equal to the concatenation

$$q\!\!\int_{\mathbb{R}} q\!\!\int_{\mathbb{R}} f((x_1 - y_1, x_2 - y_2)^{\mathrm{T}}) +_p b_1(y_1)\, dy_1 +_p b_2(y_2)\, dy_2$$

implies that the p-addition of $b(y_2)$ commutes with the inner q-integration, and $b_1(x_1) +_p b_2(x_2) = b(x)$ holds for the kernels b, b_1, b_2. The first restriction boils down to the distributivity of the two operations $+_q$, $+_p$ and thus to the condition ($q = +\infty$ or $p = 0$). Evaluation of the second condition for $p > 0$ together with rotational invariance leads to $b_i(x_i) = \lambda|x_i|^{2/p}$, $i = 1, 2$, and $b(x) = \lambda\|x\|^{2/p}$. For $p = 0$ the second condition becomes $b_1(x_1)b_2(x_2) = b(x)$; again, combination with rotational invariance yields $b_i(x_i) = k\exp(x_i^2/(2\sigma^2))$, $b(x) = k^2\exp(\|x\|^2/(2\sigma^2))$ with constants k, σ.

Finally, one easily checks that with $q = +\infty$ or $p = 0$ and the described kernels one has indeed rotational invariance from the first part of the proposition and also the intended separability.

This result is in accordance with the known facts about dilation and convolution in cases $(q, p) = (1, 0), (+\infty, 1)$. Note that the exponent in b_0 has positive sign; this is just a side-effect of our choice of notation for the generalised dilation. The conventional Gauss kernel is $1/b_0$, with $k = \sigma\sqrt{2\pi}$.

With classical morphological operations, it can be observed that erosion and dilation are related via $1 - (f \ominus b) = (1 - f) \oplus b$ for all f, b. This allows us to introduce generalised erosion as follows.

Definition 7. *For $q \in [1, +\infty]$, $p \in [0, 1]$, $f : \mathbb{R}^n \to [0, 1]$ and $b : \mathbb{R}^n \to \mathbb{R}_0^+$ as in definition 6 let*

$$(f \ominus_{q,p} b)(x) := 1 - ((1 - f) \oplus_{q,p} b)(x). \tag{15}$$

We shall call this operation (q, p)-erosion.

From the above-mentioned relation between conventional dilation and erosion it is clear that ordinary erosion is recovered, again, for $(q, p) = (+\infty, 1)$. Note that the $(1, 0)$-erosion of f w.r.t. b is just the convolution of f and $1/b$, plus a constant which is zero for kernels of total weight 1.

Definition 8. *For each $(q, p) \in [1, +\infty] \times [0, 1]$ a family of dilation filters $F_t^{q,p} : \mathcal{F} \to \mathcal{F}$ can be defined by $F_0 := \mathrm{id}$ and, for $t > 0$, $F_t^{q,p}f := f \oplus b_{p;\lambda(t)}$ with $\lambda(t) \sim t^{-1/p}$ if $p > 0$, $F_t^{q,0}f := f \oplus b_{0;k,\sigma(t)}$ with $\sigma(t) \sim \sqrt{t}$. By replacing dilation with erosion in the definition of F_t, a family of erosion filters is obtained.*

These families of filters obey most properties of scale spaces – allow for rescaling to satisfy maximum-minimum principle – but still the question is open whether they are semi-groups. We answer this by the following proposition.

Proposition 2. *The (q, p)-dilation filters $F_t^{q,p}$, $t \geq 0$ from def. 8 form a semi-group if and only if $q = +\infty$ or $p = 0$. The same is true for (q, p)-erosion filters.*

Proof. By an easy computation it is seen that one has indeed $F_{t_2}^{q,p} \circ F_{t_1}^{q,p} = F_{t_1+t_2}^{q,p}$ if the condition on (q, p) is satisfied.

On the other hand, semi-group property requires that for given $t_1, t_2 > 0$ the concatenation $F_{t_2}^{q,p} \circ F_{t_1}^{q,p}$ can be represented by one single (q, p)-dilation $f \mapsto f \oplus_{q,p} b$ for all f, with b independent on f. Like in the proof of the separability statement of proposition 1 one concludes that to enable this, $+_q$ and $+_p$ have to fulfil a distributivity law. Thus, $q = +\infty$ or $p = 0$ is necessary. Transfer to erosions is obvious.

2.3 Limit Statements

We want to investigate now in which sense the families of generalised morphological operations as introduced in defs. 6 and 7 of the previous section are
continuous w.r.t. the parameters q and p. To this purpose, we assume that the
input image f is arbitrarily chosen but fixed. Then it is clear that we have continuity – even uniform continuity – in q at any $(q,p) \in [1,+\infty) \times [0,1]$ and also
in p at any $(q,p) \in [1,+\infty] \times (0,1]$ because of the continuity of the family of
power functions used. It remains to describe the continuity of the transitions
$q \to +\infty$, $p \to +0$. The following proposition deals with the limit in p. Note that
as p tends to zero, the kernel parameters have to be adjusted.

Proposition 3. *Define $b_{0;k,\sigma}$ and $b_{p;\lambda}(x)$ for $p \in (0,1]$ as in proposition 1, with
$\lambda = \lambda_p := (p/(2\sigma^2))^{1/p}$ for $p \in (0,1]$. Then we have for all $q \in [1,+\infty]$ and
for any continuous, bounded function $f \in \mathcal{F}$, $f : \mathbb{R}^n \to \mathbb{R}_0^+$ the pointwise limit
equations*

$$\lim_{p \to +0} (f \oplus_{q,p} b_{p;\lambda})(x) = (f \oplus_{q,0} b_{0;1,\sigma})(x), \tag{16}$$

$$\lim_{p \to +0} (f \ominus_{q,p} b_{p;\lambda})(x) = (f \ominus_{q,0} b_{0;1,\sigma})(x), \tag{17}$$

for all $x \in \mathbb{R}^n$.

Proof. First, we consider dilations with $q = +\infty$. We have

$$\lim_{p \to +0} (f \oplus_{+\infty,p} b_{p;\lambda})(x) = \lim_{p \to +0} \max_y (f(x-y)^p - p\|y\|^2/(2\sigma^2))^{1/p}$$

$$= \lim_{p \to +0} \max_y f(x-y)(1 - f(x-y)^{-p}p\|y\|^2/(2\sigma^2))^{1/p}$$

$$= \max_y f(x-y) \lim_{p \to +0} (1 - f(x-y)^{-p}p\|y\|^2/(2\sigma^2))^{1/p}$$

$$= \max_y f(x-y) \exp(-\|y\|^2/(2\sigma^2))$$

$$= (f \oplus_{+\infty,0} b_{0;1,\sigma})(x).$$

Let now $q \in [1,+\infty)$. Then

$$\lim_{p \to +0} (f \oplus_{q,p} b_{p;\lambda})(x) = \lim_{p \to +0} \left(\int_{D_{p,x}} (f(x-y)^p - p\|y\|^2/(2\sigma^2))^{1/p} \, dy \right)^{1/q}$$

$$= \left(\lim_{p \to +0} \int_{D_{p,x}} f(x-y)^q (1 - f(x-y)^{-p}p\|y\|^2/(2\sigma^2))^{q/p} \, dy \right)^{1/q}$$

$$= \left(\int_{\mathbb{R}^n} f(x-y)^q \lim_{p \to +0} (1 - f(x-y)^{-p}p\|y\|^2/(2\sigma^2))^{q/p} \, dy \right)^{1/q}$$

$$= \left(\int_{\mathbb{R}^n} f(x-y)^q \exp(-q\|y\|^2/(2\sigma^2)) \, dy \right)^{1/q}$$

$$= (f \oplus_{q,0} b_{0;1,\sigma})(x)$$

with $D_{p,x} := \{y \in \mathbb{R}^n \mid f(x - y) \geq b_{p;\lambda}(y)\}$, where we have made use of the monotonic convergence theorem.

Replacing f by $1 - f$ and subtracting the resulting equations from 1, both limit results are easily transferred to erosions.

We turn now to the limit case $q \to +\infty$.

Proposition 4. *Let $p \in [0, 1]$ fixed and $b : \mathbb{R}^n \to \mathbb{R}_0^+$ continuous and bounded. Then we have for any continuous, bounded $f \in \mathcal{F}$, $f : \mathbb{R}^n \to \mathbb{R}_0^+$ the pointwise limit equations*

$$\lim_{q \to +\infty} (f \oplus_{q,p} b)(x) = (f \oplus_{+\infty,p} b)(x), \tag{18}$$

$$\lim_{q \to +\infty} (f \ominus_{q,p} b)(x) = (f \ominus_{+\infty,p} b)(x). \tag{19}$$

Proof. For any f as required and $x \in \mathbb{R}^n$, one has

$$(f \oplus_{q,p} b)(x) = \left(\int_{D_x} g_x(y)^q \, dy \right)^{1/q}$$

where $g_x(y) := (f(x - y)^p - b(y)^p)^{1/p}$ if $p > 0$, $g_x(y) = f(x - y)b(y)$ if $p = 0$, and $D_x = \{y \in \mathbb{R}^n \mid f(x - y) \geq b(y)\}$ if $p > 0$, $D_x = \mathbb{R}^n$ if $p = 0$. In both cases, $g_x(y)$ is continuous, bounded and takes only non-negative values on D_x. Thus,

$$\lim_{q \to +\infty} \left(\int_{D_x} g_x(y)^q \, dy \right)^{1/q} = \sup_y g_x(y) = (f \oplus_{+\infty,p} b)(x).$$

Transfer to the erosion case is clear.

2.4   Properties of the (q, p)-Dilations

We discuss now in more detail several features of the (q, p)-dilations. Everything said here transfers to (q, p)-erosions in an obvious way.

Let us first look at invariance properties of the (q, p)-dilations. Both morphological and Gaussian scale spaces are invariant under grey-value shifts, $f \mapsto f + C$. Gaussian scale space also displays invariance under scalar multiplication of grey-values, $f \mapsto C \cdot f$. We ask therefore which (q, p)-dilations share one of these invariances.

It turns out that the grey-value shift invariance is restricted to the two parameter pairs $(q, p) = (+\infty, 1), (1, 0)$ corresponding to morphological and Gaussian scale space themselves. Invariance under scalar multiplication of grey-values holds for all $(q, 0)$-dilations. However, a closer look shows that the grey-shift invariance of ordinary dilation is not simply lost but turns into an invariance under the grey-value transform $f \mapsto f +_p C$ for $q = +\infty$, $p \in [0, 1]$. In the limit $p = 0$ it coincides with the scalar multiplication invariance of the $(q, 0)$ case. Thus, it is the grey-shift invariance in $(1, 0)$ case which is truly an additional symmetry.

There is another way how the (q, p)-dilations for $q = +\infty$ or $p = 0$ can be understood. It coincides with one of the scaling operations used by Heijmans and van den Boomgaard in [9], making clear that these particular generalised dilations are also included in their framework. In $(+\infty, p)$-dilation, only the "inner" operation differs from that in ordinary dilation by the action of $\varphi : z \mapsto z^p$: simple addition is replaced by mapping the arguments via φ, executing the original addition and transforming back the result. In particular, the kernels b_p transform to simple quadratic kernels of the type b_1 under φ. Since φ is strictly increasing, the inverse mapping φ^{-1} commutes with taking the maximum. We can therefore describe $(+\infty, p)$-dilation as ordinary dilation performed on a signal which is obtained from the original one by a strictly monotonic transformation of grey-values, $f \oplus_{+\infty, p} b_p = \varphi^{-1}(\varphi(f) \oplus b_1)$ or, in terms of the filtering operators from def. 8, $F_t^{+\infty, p} = \varphi^{-1} \circ F_t^{+\infty, 1} \circ \varphi$. An analogous argument applies to the $(q, 0)$-dilations with $q \in (1, +\infty)$. Again, the commutation of two operations is crucial – here, $\psi : z \mapsto z^q$ may be applied before, instead of after, the multiplication of $f(x - y)$ by $b_0(y)$, provided the Gaussian b_0 is replaced with $\tilde{b}_0 := b_0^q$ which is a Gaussian, too, just with different standard deviation. We have $f \oplus_{q, 0} b_0 = \psi^{-1}(\psi(f) * (1/\tilde{b}_0))$ and $F_t^{q, 0} = \psi^{-1} \circ F_{t'}^{1, 0} \circ \psi$. Unfortunately, the grey-value transformation picture does not allow to include the case $(q, p) = (+\infty, 0)$ from either side.

As can be seen from the preceding paragraphs, (q, p)-dilations make sense for $(q, p) \in [1, +\infty] \times [0, 1]$. The algebraic definition is clear, and we have studied the continuity properties. However, there are considerable drawbacks for the parameter values $(q, p) \in Y := [1, +\infty) \times (0, 1]$ which strongly suggest that the boundary cases with $q = +\infty$ or $p = 0$ are actually the interesting ones.

First, we have pointed out earlier that the definition of the $-_p$ operation contains a truncation at zero which constitutes no problem for $q = +\infty$ or $p = 0$ since it does not influence the result. The truncation itself can't be avoided in this construction since the $+_p$ operations can't be defined for negative numbers in a sensible way. But for $(q, p) \in Y$ this truncation introduces an arbitrariness into the definition of (q, p)-dilations.

Second, since no distributivity law between $+_q$ and $+_p$ applies for $(q, p) \in Y$, it is not easy at all to interpret the algebraic operation of q-integrating over p-sums. Qualitatively, the image f and kernel b reduce their true interaction as q and p approach to each other, and for $q = p = 1$ the whole operation degenerates into a summation of paraboloid hats. An even more severe consequence of the lack of distributivity is, third, the non-separability for $(q, p) \in Y$. This constitutes an obstacle to efficient numerical computation of (q, p)-dilations with $(q, p) \in Y$.

Finally, the filter family $F_t^{q, p}$ with $(q, p) \in Y$ has no semi-group structure and is, therefore, not a scale space in strict sense.

2.5   Comparison to the Construction of Florack et al.

We want now to compare our family of generalised morphological scale spaces to the family of pseudo-linear scale spaces introduced by Florack et al. in [8,7] which also links morphological and Gaussian scale spaces.

Pseudo-linear scale spaces are introduced as a one-parameter deformation of linear Gaussian scale space via the grey-value transformation

Fig. 1. The simple 128×128 image used to illustrate the parametrised dilations. A compilation of five simple geometric shapes is superposed by uncorrelated Gaussian noise with a standard deviation of 15 % of the highest grey value.

$$\gamma_\mu : x \mapsto [x]_\mu := \frac{\exp(\mu x) - 1}{\exp \mu - 1}, \ \mu \in \mathbb{R} \setminus \{0\}, \ \gamma_0 = \mathrm{id}. \tag{20}$$

Morphological dilation and erosion scale spaces with quadratic structure functions are recovered as limit cases $\mu \to \pm\infty$ of the pseudo-linear family.

As opposed to this, the approach presented here varies morphological scale spaces in a way that includes Gaussian scale space. More precisely, deformed versions of both scale space categories are given that share the $(+\infty, 0)$ limit case. While the algebraic operations used here are somewhat simpler and allow generalisations of the same type to be inserted for the "inner" and "outer" operations of the dilation, the proposal of Florack et al. has the clear advantage of being linked to a simple modification in the Laplace-Beltrami operator which is to be used when the pseudo-linear filter is to be described as a diffusion process – an aspect that could not be regarded in the present paper.

A crucial point in the limiting process on pseudo-linear scale spaces leading to the morphological scale spaces is that a rescaling of the standard deviation σ of the Gaussian kernel is used, such that $\sigma \sqrt{|\mu|}$ is kept constant while $|\mu|$ tends to infinity. It might be that this type of renormalisation during the transition process is principally inevitable – note that we had to use a quite analogous procedure in proposition 3.

3   Experiment

As an illustrating example we show the results of (q, p)-dilations with different values of q and p on a simple image showing a few geometrical figures contaminated with Gaussian noise. The original image is shown in fig. 1.

The dilated images (fig. 2) show the interpolation property of the family of (q, p)-dilations. Note how the granular structure typical for the ordinary dilation of noisy images at small t (left bottom) is gradually reduced as the parameters are changed towards those of ordinary Gaussian convolution (right top). Also, it is worth noting that those (q, p)-dilations having neither $q = +\infty$ nor $p = 0$, in spite of their theoretical shortcomings, do not turn out obviously disastrous

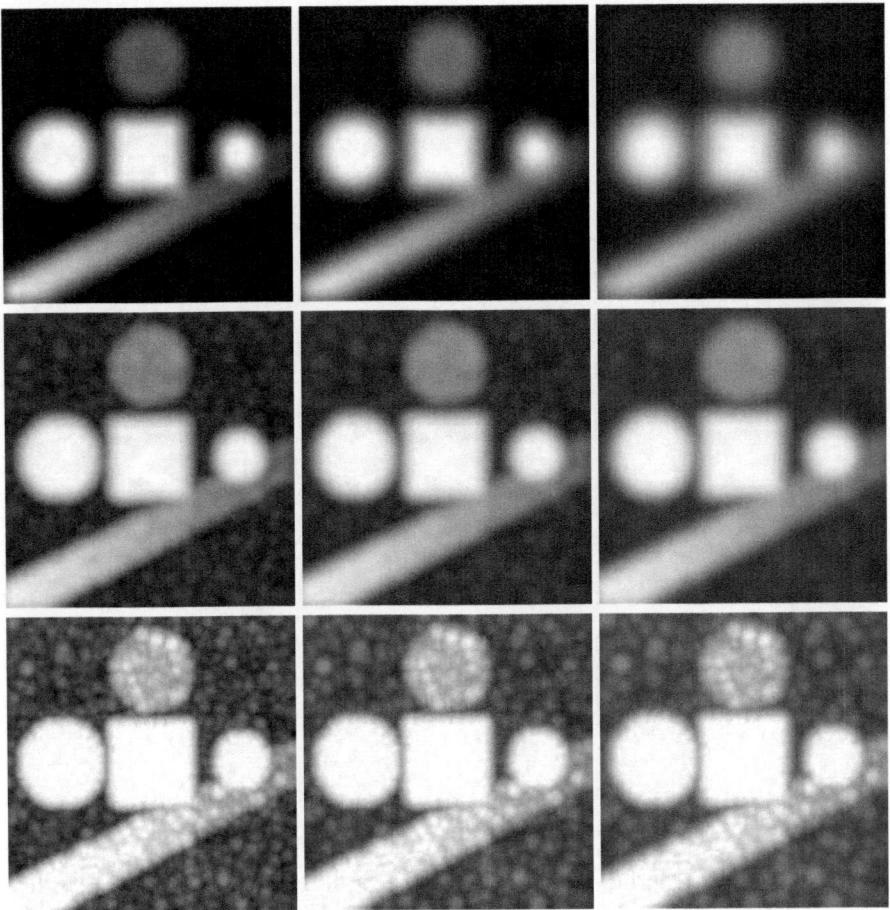

Fig. 2. Results of (q, p)-dilation of the simple image from fig. 1. Columns from left to right correspond to $p = 1, 0.5, 0$, rows from top to bottom correspond to $q = 1, 4, +\infty$. In all pictures, t^2 is set to 5. In the upper two rows the grey-values are linearly remapped to $[0, 1]$.

in the numerical experiment. Of course, for lack of separability, they consume considerably more computing time.

In fig. 3, the same dilated images are shown decorated with selected level-lines. The reduction of the granular structure becomes even more eye-catching, along with the changes in topology of the level-lines particularly in the transition zones between the geometrical elements. Finally, fig. 4 shows one-dimensional sections of the same images.

4   Conclusion

We have introduced two-parameter families of generalised scale spaces that connect the well-studied morphological scale spaces of dilation and erosion with

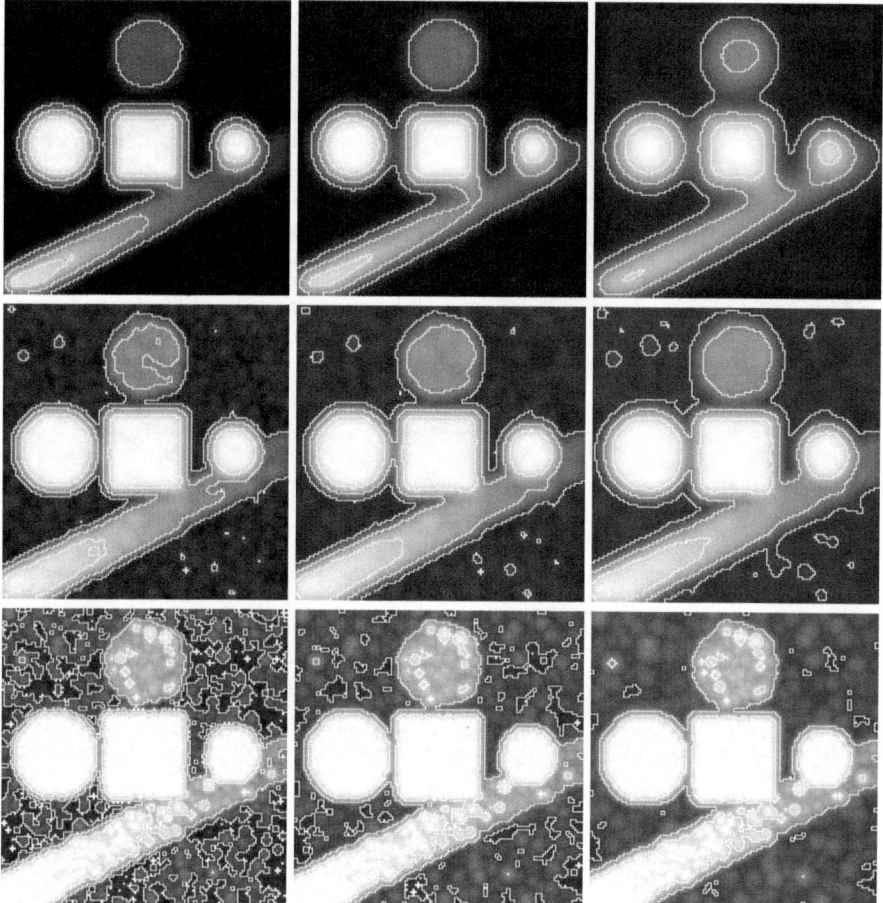

Fig. 3. The same dilated images as in fig. 2 but with level lines corresponding to 0.2, 0.5 and 0.8 times the highest grey-value.

the Gaussian convolution linear scale space. For distinguished sub-families, the semi-group property holds, making them into scale spaces in strict sense. The construction relies on a family of algebraic operations and integrals which correspond to l^p and L^p norms and generalised (power) means.

The results are primarily of theoretical interest in the theory of scale spaces since they hopefully will enrich the picture of structural analogies between the above-mentioned classes of scale spaces. In some sense, the construction presented here is complementary to that of pseudo-linear scale spaces by Florack et al. [8,7]. In particular, both approaches share the need for a renormalisation of the scale parameter in the transition between morphological and Gaussian scale spaces.

An interesting point for possible applications is the simplicity of the algebraic operations used in defining the family of scale spaces which still allows for good control over their algebraic properties.

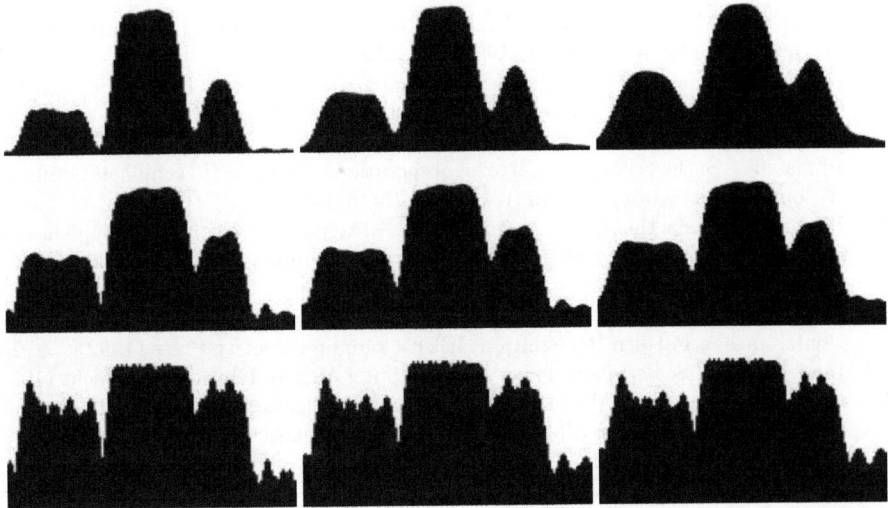

Fig. 4. One-dimensional sections of the images from fig. 2 along the vertical middle-axis.

Future work should also deal with the question how the pseudo-linear scale space approach of Florack et al. and the construction shown here could be integrated into a unified framework. Investigations should include as well possible relations to diffusion-like processes. An extension of the construction to Fourier and slope transforms would be desirable.

References

1. ter Haar Romeny, B. M., Florack, L. M. J., Koenderink, J. J., Viergever, M. A. (eds.): Scale Space Theory in Computer Vision. Proceedings of the First International Conference Scale Space '97, Utrecht, The Netherlands. LNCS vol. 1252. Springer, Berlin (1997)
2. Nielsen, M., Johansen, P., Olsen, O. F., Weickert, J. (eds.): Scale Space Theory in Computer Vision. Proceedings of the Second International Conference Scale Space '99, Corfu, Greece. LNCS vol. 1682. Springer, Berlin (1999)
3. Kerckhove, M. (ed.): Scale-Space and Morphology in Computer Vision. Third International Conference Scale-Space 2001, Vancouver, Canada, July 7–8, 2001, Proceedings. LNCS vol. 2106. Springer, Berlin (2001)
4. Dorst, L., van den Boomgaard, R.: Morphological signal processing and the slope transform. Signal Processing **38** (1994) 79–98
5. Duits, R., de Graaf, J., Florack, L. M. J., ter Haar Romeny, B. M.: Scale Space Axioms Critically Revisited. Signals and Image Processing (2002)
6. Florack, L.: Image Structure. Kluwer, Dordrecht (1997)
7. Florack, L.: Non-Linear Scale-Spaces Isomorphic to the Linear Case with Applications to Scalar, Vector and Multispectral Images. Journal of Mathematical Imaging and Vision **15** (2001) 39–53

8. Florack, L., Maas, R., Niessen, W.: Pseudo-Linear Scale-Space Theory. Int. Journal of Computer Vision **31** **(2/3)** (1999) 247–259
9. Heijmans, H. J. A. M., van den Boomgaard, R.: Algebraic Framework for Linear and Morphological Scale-Spaces. Journal of Visual Communication and Image Representation **13** (2002) 269–301
10. Iijima, T.: Basic theory of pattern observation. Papers of Technical Group on Automata and Automatic Control (1959) (in Japanese)
11. Iijima, T.: Basic theory on normalization of a pattern (in case of typical one-dimensional pattern). Bulletin of Electrical Laboratory, **26** (1962) 368–388 (in Japanese)
12. Jackway, P. T.: Morphological scale-space. Proceedings 11th IAPR International Conference on Pattern Recognition, IEEE Computer Society Press (1992) 252–255
13. Lindeberg, T.: Scale-Space Theory in Computer Vision. Kluwer, Dordrecht (1994)
14. Quadrat, J. P., et al.: Max-Plus Algebra and Applications to System Theory and Optimal Control. International Congress of Mathematicians, Zürich 1994.
15. Sporring, J., Nielsen, M., Florack, L., Johansen, P. (eds.): Gaussian Scale-Space Theory. Kluwer, Dordrecht (1997)
16. van den Boomgaard, R.: The morphological equivalent of the Gauss convolution. Nieuw Archief voor Wiskunde, **10**, 3 (1992) 219–236
17. van den Boomgaard, R., Smeulders, A. W. M.: The morphological structure of images, the differential equations of morphological scale-space. IEEE Transactions on Pattern Analysis and Machine Intelligence, **16, 11** (1994) 1101–1113, Nov 1994.
18. Weickert, J., Ishikawa, S., Imiya, A.: Linear Scale-Space has First been Proposed in Japan. Journal of Mathematical Imaging and Vision **10** (1999) 237–252

Detection and Localization of Random Signals[*]

Jon Sporring[1], Niels Holm Olsen[2], and Mads Nielsen[3]

[1] 3DLab, School of Dentistry
University of Copenhagen
Nørre Alle 20
DK–2200 Copenhagen
[2] Dept. of Computer Science
University of Copenhagen
Nørre Alle 1
DK–2100 Copenhagen
[3] IT-University
Glentevej 67
DK–2400 Copenhagen

Abstract. Object detection and localization are common tasks in image analysis.

Correlation based detection algorithms are known to work well, when dealing with objects with known geometry in Gaussianly distributed additive noise. In the Bayes' view, correlation is linearly related to the logarithm of the probability density, and optimal object detection is obtained by the integral of the exponentiated squared correlation under appropriate normalization.

Correlation with a model is linear in the input image, and can be computed effectively for all possible positions of the model using Fourier based linear filtering techniques. It is therefore interesting to extend the application to objects with many but small degrees of freedom in their geometry. These geometric variations deteriorate the linear correlation signal, both regarding its strength and localization with multiple peaks from a single object.

Localization is typically preferred over detection, and Bayesian localization may be obtained as local integration of the probability density. In this work, Gaussian kernels of the exponentiated correlation are studied, and the use of Linear Scale-Space allows us to extend the Bayes detection with a well-posed localization, to extend the usage of correlation to a larger class of shapes, and to argue for the use of mathematical morphology with quadratic structuring elements on correlation images.

1 Introduction

In this article we study the detection and localization of objects given by simplified examples. The background of the present work is the study of Hoffman

[*] This project is supported in part by the Danish Research Agency, project "Computing Natural Shape", no. 2051-01-0008 and in part by the DSSCV project under the IST Programme of the European Union (IST-2001-35443)

L.D. Griffin and M. Lillholm (Eds.): Scale-Space 2003, LNCS 2695, pp. 785–797, 2003.

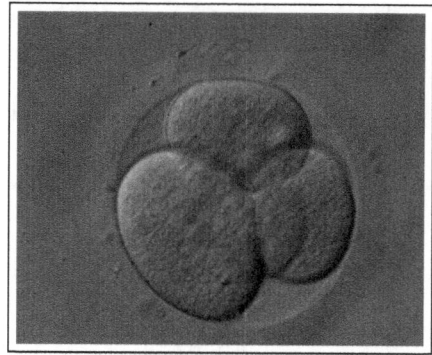

Fig. 1. Example of a HMC microscopy image of a human embryo. The microscope is focused at the so-called Equator plane, where most cells appear to be in focus.

Modulation Contrast (HMC) microscopy [1] images of human embryos in the Fertimorph project[1], an example of which is shown in Figure 1[2]. The cells are almost circular, and when focusing on the equatorial plane of each cell, the HMC technique causes their images to resemble the directional derivative of a (half)sphere thickness profile. In this setting there are essentially three approaches to detect the position, size, and shape of cells, which we have considered:

- Regularized directional integration essentially, inverting the apparent directional derivative, followed by a standard gradient based edge detection [2]. Unfortunately, the image modality is not linear, and out of focused cells are not easily compensated for.
- Data driven template matching [3], where the image model is built by examples based on manual segmentation and alignment. To obtain a useful model, a large number of annotations must be collected in order to cover the statistical variation in both shape an appearance, and this approach has therefore been considered outside the resource capabilities of the project.
- Ideal template matching with simplified geometry. This implies a two step analysis of the images. The idealized models are compared with an image followed by an analysis of the spatial correlation between models and data.

The topic of the present article is detection and localization of random signals using ideal templates. We will study the use of "circular model images" that mimic the appearance of idealized cells "with average geometry" under HMC microscopy and investigate a Bayesian analysis of their detection and localization. This analysis leads to "averaging of the detections" of idealized cells rather than detecting cells with average appearance.

---

[1] Industrial research project at IH-Medical, Image House A/S, DK, in collaboration with the Fertility Clinic at the University Hospital of Copenhagen, 3D-Lab at School of Dentistry, University of Copenhagen and the IT University of Copenhagen.

[2] Images from the Fertimorph project, acquired by Christina Hnida and Søren Ziebe at the Fertility Clinic, University Hospital of Copenhagen, Denmark.

Shape variation can be difficult to quantify. Empirical models of shape variation require a large number of samples unless based on strict model assumptions like the local linear models [4] which just cover simple shape transformations, or the slightly more flexible and globally defined models of shape variability by linear transformation groups [5]. In all cases such models must be inferred from annotated examples. Studying shape as a scale-space in the dual space of some over-complete basis appears to be an alternative route for object detections, and we consider points in the dual space for a given scale as equivalence classes of shapes.

The simplest shape variation we can imagine is the deformation of circles into ellipses. Elliptical templates have 5 parameters: location (2), orientation (1), and scaling along the two main axes (2), and it seems prohibitive to investigate all parameters. In contrast, circles have only 3 degrees of freedom, location (2) and radius (1), and we will use circles as idealized templates for matching. Another argument for using circles is, that they are approximations of curve patches up to second order in Euclidean geometry.

Detection, formulated as an optimal Bayesian decision, can be achieved by correlation of the image with a template, followed by an integration of the exponentiated correlations and comparing with a threshold as thoroughly discussed in [6]. Localization then appears to be the generalization of the integral with locally weighted integrals with kernels at different scales and positions. However, we will show in the following sections that the pseudo-linear scale-space with the exponential function as non-linearity function [7] is more suitable. A key property of the pseudo-linear scale-space is, that it generates a one-parameter family of scale-spaces with mathematical morphology as one limit and linear scale-space as the other. In terms of probabilistic detection, this corresponds to estimating the local maximum versus the mean probability, i.e. given a detection scale, we may balance the localization accuracy versus the number of hypothesized objects inside the detection area.

This paper is organized as follows: Detection and localization of random signals is discussed in the seminal paper on Bayesian detection [6] and briefly summarized in Section 2, and we will add how detection can be generalized to localization. Localization requires either local integration or application of the maximum operator on either energies or probabilities. In Section 2.6 we recognize that pseudo-linear scale-space [7] on energies generates a one parameter family of decision methods with Gaussian integration and maximum operator as the extremities and with a straight-forward probabilistic interpretation. In Section 3, we illustrate the use of the theory on simple images, and finally we conclude on this article in Section 4.

2   Bayesian Detection and Localization of Random Signals

In this section we shortly introduce our notation and review the theory of optimal (Bayesian minimal risk) decisions as applied to object detection in images. Accounts of Bayesian inference and decisions can be found in e.g. [8] (Bayesian inference in general) and [9,10] (in the setting of statistical pattern recognition).

2.1 Detection as Binary Decision

Barret and Abbey consider Bayesian detection of random signals on noisy additive random backgrounds in [6], in terms of a binary decision problem, between
the two hypotheses:

H₀ The null hypothesis of "no signal present".

H₁ The alternate hypothesis of "signal present".

The hypotheses are formulated as probabilistic models of the image formation
conditional on the two hypotheses, which we shall denote abstractly by $P(I|\mathbf{H}_0)$
and $P(I|\mathbf{H}_1)$. Optimal detection is then a matter of optimal decision, which can
be achieved by thresholding the likelihood ratio[6]:

$$\Lambda(I) = \frac{P(I|\mathbf{H}_1)}{P(I|\mathbf{H}_0)}. \tag{1}$$

That is, an optimal "signal present" estimate $\overline{\mathtt{sig}}_\theta$, is defined by

$$\overline{\mathtt{sig}}_\theta = \begin{cases} 1 \text{ if } \Lambda(I) > \theta \\ 0 \text{ otherwise} \end{cases} \tag{2}$$

The detection is optimal in the sense that it achieves maximal probability of
true positive detection (among all deterministic detection schemes) with a given
maximal probability of false positive detection.

2.2 Random Object Variability and Image Formation

In the abstract Bayesian inference model, the randomness of the image data I
is described as a two-fold randomness.

First, there is the random variation of the "signal", described by a prior
probabilistic model $P(M)$ of object variability. The object models $M \in \mathbf{M}$ serve
as "causes" or "model-explanations" of the observed image I. In terms of these
abstract model-explanations, the two alternate hypotheses, $\mathbf{H}_0$ and $\mathbf{H}_1$, are a
disjoint classification of $\mathbf{M}$ representing a binary model of explanations: $\mathbf{M} = \mathbf{H}_0 \cup \mathbf{H}_1$ with $\mathbf{H}_0 \cap \mathbf{H}_1 = \emptyset$.

Secondly, a probabilistic model $P(I|M)$ of the image formation $(M, B) \mapsto I(M, B)$, describes the additional randomness in images of the same "model
explanation", arising from random imaging conditions, like noise or a randomly
varying background here denoted by $B \in \mathbf{B}$. The image formation model $P(I|M)$
can be seen as the result of marginalizing a deterministic model $P(I|M, B) = \delta_{I(M,B)}$ ($\delta_{I(M,B)}$ being the Dirac delta distribution with probability mass 1 lo-
cated at $I(M, B)$) over the background variability $P(B|M)$, which here is as-
sumed to be independent of the object $P(B|M) = P(B)$:

$$P(I|M) = \frac{\int P(I,M,B)dB}{P(M)} \tag{3}$$

$$= \int \frac{P(I,M,B)}{P(M,B)} \frac{P(M,B)}{P(M)} dB \tag{4}$$

$$= \int P(I|M,B)P(B|M)dB \tag{5}$$

$$= \int \delta_{I(M,B)} P(B)dB \tag{6}$$

To simplify formulas, we shall restrict the probabilistic noise and background model of the image formation to the case of additive $(I(M,B) = I(M) + I(B))$ iid. Gaussian pixel noise $(I(B) = B, P(B) \propto \exp(-||B||^2/(2\sigma_{\text{noise}}^2)))$, for an $N \times N$ discretely sampled image $I \in \mathbf{I} = \mathbf{R}^{(N^2)}$:

$$P(I = I_{\text{obs}}|M) = \frac{1}{Z_{\sigma_{\text{noise}}}} \exp\left(-\frac{||I_{\text{obs}} - I(M)||^2}{2\sigma_{\text{noise}}^2}\right), \tag{7}$$

where $|| \cdot ||$ is the standard Euclidean norm on $\mathbf{R}^{(N^2)}$. Clearly having a simple noise and background model pushes the complexity of the overall imaging model on to the model of the "object variability" $P(M)$ and "noiseless image formation" $I(M)$ - and vice versa. In [6] they considered a more general additive Gaussian background model $P(B)$, corresponding to the use of a different norm $|| \cdot ||$, and discussed the validity and accuracy of such noise and background models for transillumination imaging modalities. Linearity of the HMC imaging modality is discussed in e.g. [11].

To discuss signal variability, we formally assume that the object variability in $\mathbf{H}_1$ can be abstractly factorized into variabilities of "position" ($\mathbf{M}_{\text{position}}$), "pose" ($\mathbf{M}_{\text{pose}}$), "size" ($\mathbf{M}_{\text{size}}$), and "morphology" ($\mathbf{M}_{\text{morphology}}$), where morphology is the remaining "shape" and "appearance" variabilities:

$$\mathbf{H}_1 = \mathbf{M}_{\text{position}} \times \mathbf{M}_{\text{pose}} \times \mathbf{M}_{\text{size}} \times \mathbf{M}_{\text{morphology}}. \tag{8}$$

Although it is often a good and simplifying approximation to further assume that these variabilities are all independent, we might as well for now keep the general case, making the above factorization a merely formal one:

$$P(M|\mathbf{H}_1) = P(M_{\text{position}}, M_{\text{pose}}, M_{\text{size}}, M_{\text{morphology}}).$$

In terms of this prior model of object variability the probabilistic model of the "signal present" hypothesis $\mathbf{H}_1$ is given by a marginalizing integration of the image formation $P(I|M)$ over $\mathbf{H}_1$:

$$P(I|\mathbf{H}_1) = \int_{M \in \mathbf{H}_1} P(I = I_{\text{obs}}|M)P(M|\mathbf{H}_1)dM \tag{9}$$

$$= \iiiint P(I = I_{\text{obs}}|M_{\text{position}}, M_{\text{pose}}, M_{\text{size}}, M_{\text{morphology}}) \tag{10}$$
$$P(M_{\text{position}}, M_{\text{pose}}, M_{\text{size}}, M_{\text{morphology}})$$
$$dM_{\text{position}}dM_{\text{pose}}dM_{\text{size}}dM_{\text{morphology}}.$$

With $I(\mathbf{H}_0) = 0$, the model of the "no signal present" hypothesis $\mathbf{H}_0$, is

$$P(I = I_{\text{obs}}|\mathbf{H}_0) = \frac{1}{Z_{\sigma_{\text{noise}}}} \exp\left(-\frac{\|I_{\text{obs}}\|^2}{2\sigma_{\text{noise}}^2}\right). \tag{11}$$

2.3 Detection under Random Position and Size

In addition to analyzing detection in different additive noise and background models $P(B)$, Barret and Abbey also analyzed detecting a known signal with unknown position ([6]) (and size ([12])) by marginalizing over a uniform prior distribution of M_{position}. This corresponds to assuming that $P(M_{\text{position}})$ and $P(M_{\text{size}})$ are uniform, while $P(M_{\text{pose}})$ and $P(M_{\text{morphology}})$ are degenerate with all the probability mass centered at a single parameter set which we shall denote $(M_{\text{pose}_0}, M_{\text{morphology}_0})$.

$$P(I|\text{signal with unknown position and size}) \tag{12}$$
$$= \iint P(I|M_{\text{sig}} = 1, M_{\text{pose}_0}, M_{\text{morphology}_0}, M_{\text{position}}, M_{\text{size}})$$
$$P(M_{\text{position}}, M_{\text{size}})dM_{\text{position}}dM_{\text{size}}$$

They found that optimal detection of a signal with only the position unknown is realized by integrating exponentiated cross correlations of the observed image I_{obs} with the model image $I(M)$, with respect to the prior $P(M_{\text{position}})$ on M_{position} and comparing to a threshold [6, p.164].

2.4 Detection and Localization under Random Shape

We now turn to the study of both detection and localization of objects which also exhibit morphology or shape variability (non-degenerate $P(M_{\text{morphology}})$) in addition to the position and size variability. However following the general theory by extending the integration to also cover non-degenerate morphology variability, is likely to be computational infeasible, because of the many integrations involved and the demand for explicit models of morphology variability.

Instead, we shall analyze the effect on the correlation values of small shape variabilities around a mean shape (see figure 2 and 3). The observation is, that a small shape deviation has the effect of both damping and smoothing the correlation values in a nontrivial way which resembles a generalized autocorrelation. The need to handle shape variability thus motivates considering a smooth scale-space. Detecting shape transformed objects can then be considered as detecting consequent deteriorated correlation peaks (see figure 3).

2.5 Bayesian Localization of Random Signals

In the following, we will extend the detection problem to a combined detection and localization problem by generalizing the binary family of hypotheses to a full scale-space of partial localization hypotheses. An obvious extension of optimal

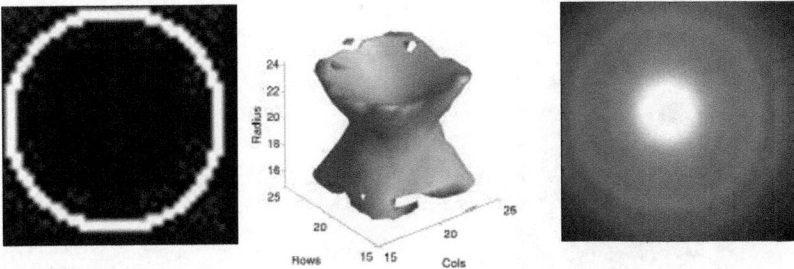

Fig. 2. Detection of a circle with circles. LEFT: The original image, and also an example of the images of circles we correlate with. MIDDLE: The 3.5 isosurface in energy space. RIGHT: The mean energy over radius.

Bayes' detection [6] is to subdivide the spatial domain into small non-overlapping regions and applying the detection theory independently on each.

First localization detection is considered in terms of a continuous family of hypotheses $\mathbf{H}_L(M_{\mathrm{mean}}, \tau) = \{M \in \mathbf{H}_1 | \|M - M_{\mathrm{mean}}\| < \tau$, parameterized by the possible object or mean signal models $M_{\mathrm{mean}} \in \mathbf{H}_1$. In terms of these, "localized detection" can formally be phrased in terms of the probabilistic models of this scale-space of hypotheses $P(I|\mathbf{H}_L(M_{\mathrm{mean}}, \tau))$.

In the standard least-committed manner of linear scale-space [13], we notice that the concept of regions should properly be generalized to soft Gaussian windows, i.e.

$$P(I|\mathbf{H}_L(M_{\mathrm{mean}}, \tau)) = P(I|M) * G_\tau(M)|_{M=M_{\mathrm{mean}}},$$

where $\cdot * \cdot$ is the convolution operator. Besides being the obvious extension of detection to localization, the implication for the corresponding energies is that point of maximum is preferred under the Gaussian window, when the image noise is low, and that the mean energy is preferred when the image noise is high. This will be discussed in Section 2.6.

The structure of our detection mechanism is that radius of the circle-templates is monotonically increased, and it should thus be expected, that detection structures similar to the evolutes should dominate. If we consider the simple images of circular and elliptical bands as shown in Figure 2 and 3 we notice that the energy isosurfaces have a strong evolute-like structure. The circle has, naturally, highest energy in a single point corresponding to the exact match of the original image with the model. Conversely, as can be seen by the the the integration over radius, the ellipse is much less compactly located than the circle. We conclude, that using a single maximum for detecting circular structures in an image is not at all likely to reflect the mass center of the structure. Optimally, the detection of ellipses with circles should be followed by a second correlation of a prototypical detection image, which recursively would not seem to lead to any decision. We have considered two alternative routes for exploiting this structure. One option would be to perform a three dimensional Gaussian scale-space

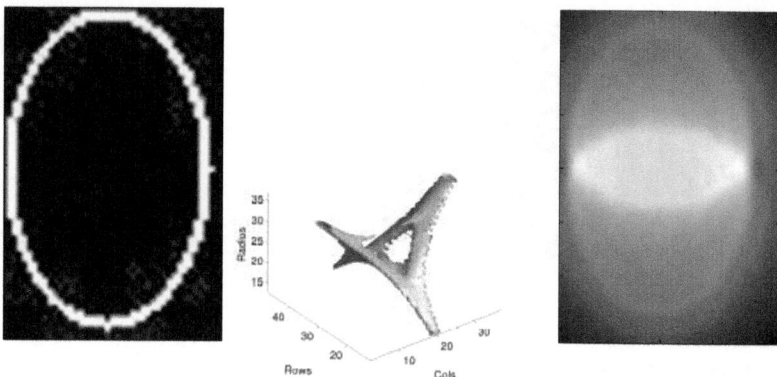

Fig. 3. Detection of an ellipse with circles. LEFT: The original image. An example of the model is shown in Figure 2(LEFT). MIDDLE: The 3.5 isosurface in energy space. RIGHT: The mean energy over radius.

in the position and radius space, however, since the radius increase is constant, we know that the ridges in the correlation images essentially have moved the identical distance, and we may therefore tailor the Gaussian integration along the radius accordingly as demonstrated in Section 3.

2.6   Pseudo-Linear Scale-Space on Energies and Probabilities

In images where there are several objects present, global detection has to be replaced by a trade-off between integration and maxima detection. The family of isotropic Gaussian kernels is a least-committed choice of localized distributions, and the stack of images under all sizes of the Gaussian is the linear scale-space [13]. I.e. detecting the existence of a model in an image is equivalent to calculating the linear scale-space of the exponential of the square correlation image and read any value at infinite scale. Conversely, mathematical morphology [14] is a natural method for studying maxima and minima under neighborhoods of varying size. We note that linear scale-space is equivariant with affine transformation of the intensities. Conversely, mathematical morphology is only invariant under addition of constants, and normalization factors cannot be ignored.

When deciding upon the detection and localization of several objects in an image, one may view the maximum operator and the linear scale-space as two opposing operations in a one parameter family of scale-spaces, the pseudo-linear scale-space [7]. The Pseudo-Linear Scale-Space $I(x, y, \tau, \sigma)$ on $I(x, y)$ and with non-linearity f is defined as,

$$I(x, y, \tau, \sigma) = f_\sigma^{-1}\left(f_\sigma\big(I(x, y)\big) * G_\tau(x, y)\right),$$

$$f_\sigma(v) = \exp\left(\frac{v}{\sigma}\right),$$

$$G_\tau(x,y) = \frac{1}{\tau^2 2\pi} \exp\left(-\frac{x^2 + y^2}{2\tau^2}\right),$$ (13)

and can be shown to converge to Mathematical Morphology with a quadratic structuring element on $I(x,y)$, when $\sigma \to 0$, and to converge to the linear scale-space, i.e. $I(x,y,\tau,\sigma) = I * G_\tau$ when $\sigma \to \infty$. In the Bayes' view, we note that σ corresponds to the noise of the observed image, while τ is the detection scale.

3   Experiments

We have performed a number of simple experiments to validate and visualize the use of pseudo-linear scale-space. The goal is to detect semi-circular band type structure in images, where the width of the HMC images seems to be a parameter of the microscopy independent on the size of the cell. We wish to detect such cells at their equatorial plane using stiff circular models in order to avoid a prohibitive amount of manual labor..

The algorithm we've build is given in pseudo-Matlab code in Figure 4. Two parameters must be specified: `sigma` and `tau`. The first is an estimate of the signal noise and could be estimated from the observed image, I. The second should be set according to the expected size of structures to be detected. If no prior knowledge is available on expected sizes, then all scales should be studied. Figure 5 illustrates a simplified example of the use of the algorithm. Please notice the high size and position precision obtained. The limits `sigma` $= \infty$, implies a Gaussian smoothing, while `sigma` $= 0$, implies a morphological operation on the energies at scale `tau` according to pseudo-linear scale-space theory. We conclude that the algorithm is able to detect the approximate center of the ellipse in very noisy images.

The HMC images have a directional structure, which may be exploited in the detection of cells. In an oversimplified view, the circular structures in Figures 2 and 3 are multiplied with a plane slanted in a particular direction defined by the microscope. This is illustrated in Figure 6. In the figure are two almost identical experiments presented, where the difference is the amount of iid. Gaussian noise. The two parameter σ has been set to sufficiently separate the four extrema of the evolute-like structure from the background, and τ has been set to integrate all four peaks. In these oversimplified images, the integration scale cannot be set too high, but we emphasize that all τ-scales should be studied in images, where several circular structures are present. From the experiments we find that the method is extremely robust to the noise. Another illustration of the advantage of having an integration scale is shown in Figure 7. Here we compare the maximum of the correlation squared and the smoothed probability images. Due to the elongated structure of the most prominent cell, the correlation squared peaks for the circle corresponding to the lower evolute cusp. On the other hand, the probability successfully integrates all the evolute-like structure of the cell.

```
Algorithm D = Detect(I,sigma,tau)
% Detect calculates the detection image D in R^(NxNxN/2)
%   I      - the original image in R^(NxN)
%   sigma  - the estimated noise level in the image
%   tau    - the detection scale

max_radii = size(I,1)/2;

% First we calculate the correlation, C in R^(NxNxN/2)
For r = 1:max_radii
  M = circularBand(r);
  C(r) = correlation(I,M);
end

% Then we calculate the probabilities up to noise level,
% P in R^(NxNxN/2)
P = exp(C.^2/sigma);

% Finally we detect at scale tau, D in R^(NxNxN/2)
for r = 1:max_radii
  D(r) = scalespace(P(r),tau);
  for s = 1:tau
    D(r) = D(r) + scalespace(P(r+s),tau+s)+scalespace(P(r-s),tau+s);
  end
end
```

Fig. 4. Algorithm Detect. The algorithm is designed to circular-like objects of unknown position and size using circles of varying radii. The function `scalespace(I,s)` smoothes image I with a isotropic Gaussian of standard deviation s.

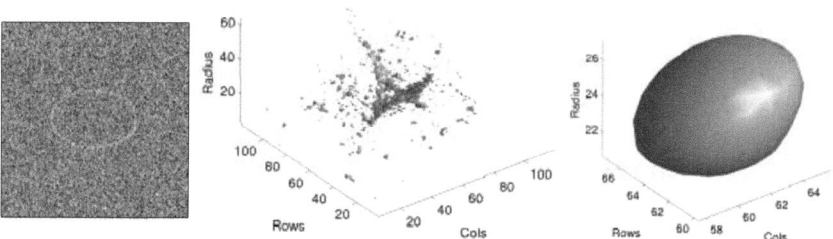

Fig. 5. Detecting ellipses using Algorithm Detect in noise images. Parameters sigma and tau are both set to 10. LEFT: The original synthetic image with i.i.d. additive Gaussian Noise. MIDDLE: Isosurface of the energy image showing the evolute like structure. RIGHT: The near maximum isosurface in the detection image.

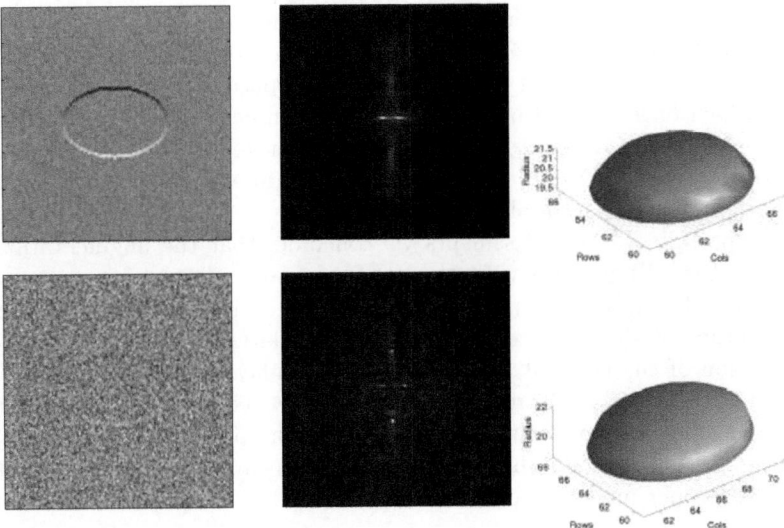

Fig. 6. Detecting synthetic crater ellipses. Top and bottom row is the same experiment with varying amount of added i.i.d. Gaussian noise. LEFT: The original images. MIDDLE: The mean energies. RIGHT: The detection blobs.

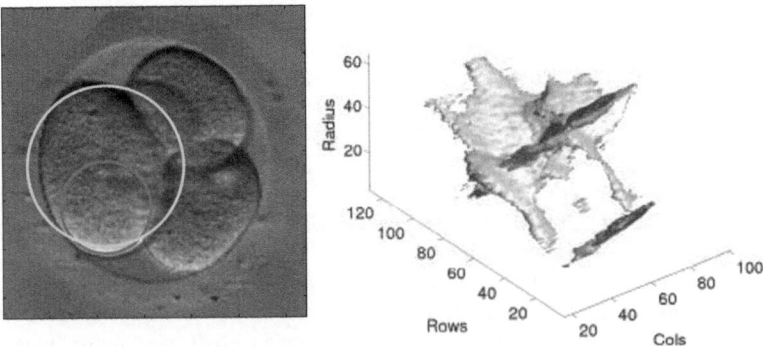

Fig. 7. Comparing the difference between maximum detection in the correlation squared and the scale-space of the probabilities. LEFT: The original image overlaid with the circle corresponding to maximum correlation squared (small and red), and the circle corresponding to the maximum smoothed probability. RIGHT. An isosurface from the correlation image.

4   Conclusion

In this paper we have discussed detecting and locating objects in images. The objects, which we have been concerned with, are semi-circular and have non-trivial appearances. We have investigated the use of simplified template models and optimal Bayes' detection followed by an integration using pseudo-linear scale-space. We have used pseudo-linear scale-space with the exponential function, which have allowed us to interpret the temperature and integration scale of pseudo-linear scale-space in a probabilistic sense, with the temperature being the noise of the observed image and the integration scale being the localization uncertainty for Bayesian object detection.

Supported by the experiments we conclude, that the augmentation of the optimal Bayes' detection with the pseudo-linear scale-space is very useful for detecting objects with simplified templates, essentially by grouping the shape-variabilities of the observed image after correlation with the templates. The estimation of the noise of the observed image plays a key role in the resulting detection, since it determines the trade-off between maximum detection and mean calculations, and while it may be difficult to estimate on small samples, our view is that the noise should be considered as part of the model and estimated through proper model selection at a super level.

References

1. Robert Hoffman and Leo Gros. The modulation contrast microscope. *Nature*, 254:568–588, April 1975.
2. N.H. Olsen, J. Sporring, and M. Nielsen. Reconstruction of optical thickness. In Peter Johansen, editor, *Den 11. Danske Konference om Mønstergenkendelse og Billedanalyse*. Departement of Computer Science, University of Copenhagen, 2002. Technical Report no. 2002/15.
3. T.F. Cootes, G.J. Edwards, and C.J. Taylor. Active appearance models. *PAMI*, 23(6):681–684, June 2001.
4. C. J. Taylor, T. F. Cootes, A. Hill, and J. Haslam. Medical image segmentation using active shape models. In L. Beolchi and M. Kuhn, editors, *Medical Imaging*. IOS Press, 1994.
5. Andrew Swann and Niels Holm Olsen. Linear transformation groups and shape space. *JMIV*, accepted 2002. revised.
6. H.H. Barrett and C.K. Abbey. Bayesian detection of random signals on random backgrounds. In J. Duncan and G. Gindi, editors, *Proceedings of the 15th International Conference on Information Processing in Medical Imaging*, pages 155–166. Springer-Verlag, 1997.
7. Luc Florack, Robert Maas, and Wiro Niessen. Pseudo-linear scale-space theory. *International Journal of Computer Vision*, 31(2/3):247–259, 1999.
8. Christian P. Robert. *The Bayesian Choice: a decision-theoretic motivation*. Springer, 1997.
9. Yurij Kharin. *Robustness in Statistical Pattern Recognition*, volume 380 of *Mathematics and Its Applications*. Kluwer Academic Publishers, 1996.
10. Vladimir N. Vapnik. *The Nature of Statistical Learning Theory*. Springer, 1995.
11. Niels Holm Olsen, Jon Sporring, Mads Nielsen, Christina Hnida, and Søren Ziebe. Reconstructing the optical thickness from hoffman modulation contrast images. In Søren Olsen, editor, *Den 11. Danske Konference i Mønstergenkendelse og Billedanalyze*, volume 15 of *DIKU Technical Reports*, pages 11–18. DIKU, August 2002. ISSN: 0107-8283.

12. Eric Clarkson and Harrison Barrett. Bayesian detection with amplitude, scale, orientation and position uncertainty. In James Duncan and Gene Gindi, editors, *Information Processing in Medical Imaging, 15 th Internanational Conference, IPMI'97*, volume 1230 of *Lecture Notes in Computer Science*, pages 549–554, Poultney, Vermont, USA, June 1997. Springer.

13. J. Weickert, S. Ishikawa, and A. Imiya. On the history of Gaussian scale-space axiomatics. In Sporring et al. [15], chapter 4, pages 45–59.

14. Rein van den Boomgaard and Leo Dorst. The morphological equivalent of gaussian scale-space. In Sporring et al. [15], chapter 15.

15. Jon Sporring, Mads Nielsen, Luc Florack, and Peter Johansen, editors. *Gaussian Scale-Space Theory*. Kluwer Academic Publishers, Dordrecht, The Netherlands, 1997.

Continuous Curve Matching with Scale-Space Curvature and Extrema-Based Scale Selection

Brian Avants and James Gee

University of Pennsylvania
Philadelphia, PA, USA 19104-2644
{avants,gee}@grasp.cis.upenn.edu

Abstract. We extend a symmetric parametric curve matching algorithm designed for recognition and morphometry by incorporating Gaussian smoothing and curvature scale-space. A general statement of the matching theory and the properties of the associated algorithm is given. Gaussian smoothing is used to assist in approximating the continuous solution from the discrete solution given by dynamic programming. The method is then investigated in a multi-scale framework, which has the advantage of reducing the effects of noise and occlusion. A novel scale-space derived energy functional that incorporates geometric information from many scales at once is proposed. The related issue of selecting a smoothing kernel for a given matching problem is also explored, resulting in a topologically based method of scale-selection. This application requires estimating the matching between the fine and coarse scale versions of the same curve. We provide a tool for finding this inter-scale, intra-curve correspondence, based on tracking curvature extrema through scales. These novel algorithms are demonstrated on both 2D and 3D data.

1 Introduction

Object boundaries, typically segmented from images or given by hand drawings, provide an efficient shape representation for use in computer vision. Contours may be used for shape recognition [1], in medical or biological analysis [2] or in a tracking algorithm [3]. Any of these applications may require finding correspondences between pairs of curves under varying conditions of noise. A solution to this problem involves treating the contours as pure geometric curves. One may then exploit a rich mathematical foundation for algorithm development and to understand one's curves with the use of invariant "signature functions" [4]. Here, the object is encoded as a list of points from which differential information may be approximated. Another approach involves computing the skeleton of the object [5–7] and using a graph representation for the object parts. While this description is also complete and particularly useful for capturing object articulation, and in the case of occlusion, medial representations are very sensitive to noise and require more expensive algorithms for matching [8]. A third approach is to use the contour as the boundary of an image region [3], where the contour separates an image into inside and outside. This regional information is then used for tracking the contour between frames.

L.D. Griffin and M. Lillholm (Eds.): Scale-Space 2003, LNCS 2695, pp. 798–813, 2003.

Each of these approaches, and many others, use the initially extracted contour as a starting point for more involved shape or image analysis. This paper will investigate scale-space diffusion as a method for using and selecting scales at which curve pairs may be represented when computing correspondences. An efficient dynamic programming (DP) matching algorithm that focuses on the intrinsic properties of the contour is investigated. Gaussian smoothing will also play a role in estimating a continuous solution from the discrete DP solution.

Dynamic programming provides a well-known solution to curve matching, [9,10,2,8,11] The methods, in general, attempt to find a re-sampling, or reparameterization, of one or both of the curves to be matched such that a variational energy or cost function is minimized. The advantage of this approach is that the solution is found in the domain of the parameter and therefore is exact. Methods that deform the range of the curves do not guarantee an exact correspondence and may be more computationally expensive because the match may have to be computed over the entire object, though the boundary completely specifies the solution.

Scale-space approaches to multi-scale parameteric DP matching have not yet been investigated. Previous researchers have focused more often on changing sampling rates [11] or on successive polygonal approximations [12]. Sclaroff [13] uses modal matching, which naturally allows a multi-scale description by changing the number of modes included in the shape description. Younes [14] approaches multi-scale matching by using coarse-to-fine polygonal approximations. This avoids having to deal with the reparameterization that is implicit in smoothing [15]. However, coarse polygonal approximations do not have the same noise-reduction properties that come with scale-space diffusion [16,17].

One of the main advances of scale-space research has been in providing methods for selecting characteristic scales [18,19]. Typically, these methods use the maximum response of a rescaled derivative operator over the image to find a characteristic scale. We will, in this paper, attempt to use the extrema of the shape description to choose scales when matching pairs of object contours. This also requires us to consider the problem of finding the correspondence of a curve with itself as it is smoothed. A related approach is given in [20], in which the authors evolve a curve until a fixed number of extrema exist. These extrema are subsequently tracked back to zero-scale to provide a boundary segmentation. Individual contour parts are then compared with Procrustes alignment.

Previous work on reparameterization-based curve matching is now reviewed. Tagare proposed an original reparameterization matching in [21,22] and also showed the solution space is isomorphic to a torus, for the case of closed planar curves. Sebastian and Kimia [1] focus on symmetric matching of curvature, but do not deal with situations in which noise is prevalent. Younes formally developed a group theory for diffeomorphically matching 1D signals before applying the classical DP solution in a multi-scale framework. Amini applies the DP solution to match with an elastic energy function [23]. Viterbi used a similar algorithm for recognition of speech patterns [24]. Serra's work [11,25] provides unique modifications of the basic discrete DP solution to allow for sub-pixel approximations, thus reducing the error caused by sampling and discretization of the solution curve. Mokhtarian used curvature zero-crossings through scale to create a scale-space signature of a curve that could be used directly for matching [26]. Curvature zero-crossings have the advantage of bounding salient features (as there is an extrema between every pair of zero-crossings) at coarse to fine scales. Mokhtarian uses these signatures with circular shift to match 2D planar curves, but does not use them explicity in a variational, non-rigid deformation-based approach. Geiger [8] uses variational energies to find curve self-similarities and to match the shape axis tree (similar to the medial axis) with an $O(N^4)$ algorithm.

The methods introduced below are designed to broaden the applicability of parametric curve matching. One such method improves the performance of the discrete algorithm so it better mimics a continuous solution. We will also show a matching tool similar to [15] but which uses the non-rigid variational context. Furthermore, we advance a novel approach to scale selection. The scale-selection ideas proposed in this paper may be useful in conjunction with a variety of other methods, including image analysis, although we focus on its use with our own DP curve matching algorithm. Many of the contours used in this paper are taken from the shape databases [27,28].

2   Curve Matching Theory and Algorithm

Our symmetric DP variational curve matching formulation is most similar to that given in [21,8,10]. However, the implementation is most similar to [10].

2.1   Curves and Reparameterization

The data is assumed to be given in the form of an ordered point set. We desire a continuous curve model for the discrete data so that $C : t \in \tau \to \mathbb{R}^d$, where τ is a real interval and d gives the dimension of the curve. In practice, a spline obtains this parameterization as well as derivative information [29]. In matrix form, we have $C(s) = BX$, where B is the $1 \times p$ vector of basis functions and X is the $p \times d$ matrix of control vectors with $p - 1$ the degree of the spline. This allows us to reparameterize the curves at any time and choose either B-spline (for approximation) or Hermite basis functions (for interpolation). If the points are unordered, methods exist for obtaining error minimizing ordered (sub)sets of the original points for use as control vectors [30] Note the constant arc-length parameterization guarantees a unique interpolation, given a set of data points and is always used here. Good approximations to the constant arc-length parameterization are available [29]. If they are inadequate, they can be refined relatively quickly with binary search.

2.2   Curve Matching Formulation

Given a pair of such curves, (C_1, C_2), we want to find a reparameterization such that the pair is better aligned with respect to some cost. The reparameterization

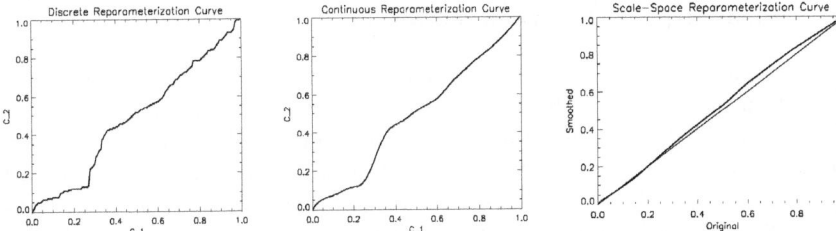

Fig. 1. A discrete reparameterization curve is shown on the left. The axes represent the curve parameters and vary from 0 to 1. A continuously corrected match curve is shown in the center. A coarse-to-fine reparameterization curve, for use with inter-scale, intra-curve matching, is at the right. The identity is also present to show the size of the reparameterization caused by Gaussian smoothing. More complex curves and more smoothing result in greater deviation from identity.

curve is denoted $g \in C_R^p : \xi \in \tau_g = [0,1] \to \tau_1 \times \tau_2$, where C_R^p denotes the p-continuous regular curves. Then, $g(\xi) = (t_1(\xi), t_2(\xi))$ which is a pairing of curve parameters for (C_1, C_2). The dependence of t on ξ is suppressed from here except where necessary. If $\tau_1 = [0, L_1]$, $\tau_2 = [0, L_2]$, where L is curve length, then each point on g defines a pairing of arc lengths s_1 and s_2 from C_1 and C_2, respectively. Given this definition, the endpoints for g are required to be $(0,0)$ and (L_1, L_2). This ensures that our matching is defined over both curves.

We also require that topology is preserved by reparameterization. That is, if a curve does not self intersect, no self intersections should be created (for example, by a reordering of the points.) For this reason, the space of permissible reparameterizations requires that g is monotonic; its first derivative does not change sign. The identity reparameterization simply pairs every value with itself so $g_I(\xi_I) = (s, s)$ with subscript I denoting the identity. It will be the shortest possible curve linking the points $(0,0)$ and (L_1, L_2) in $\tau_1 \times \tau_2$, if $L_1 = L_2$.

The approach given here addresses the issue of reparameterization curve symmetry dealt with previously by other authors [10,21]. Symmetry, here, is guaranteed due to the arbitrariness of labelling t_1 and t_2. Applying g polymorphically reparameterizes the curve pair such that,

$$g : (\ C_1(s_1), C_2(s_2) \ ) \to (\ C_1(t_1), C_2(t_2) \ ). \tag{1}$$

We will denote the reparameterization as $C(t)$ or $C \circ t$ for either curve. New reparameterizations can be found by composing sets of g_i so

$$g^n = g^{n-1} \circ \cdots \circ g^0, \tag{2}$$

where $g^i \circ g^j = (t_1^i \circ t_1^j, t_2^i \circ t_2^j)$. This property will be important when we match through scales. An example of a reparameterization is found in Figure 1.

2.3   Diffeomorphic Reparameterization

The diffeomorphisms $g_d\colon s_1 \to s_2$ and $g_d^{-1}\colon s_2 \to s_1$ can be gained from this formulation by projecting the curve, g, onto its axes. This is only true if $\dot{g}_d > 0$, otherwise the function g_d is not one-to-one and thus is not a diffeomorphism [31]. This can always be enforced if we approximate stationary points on g_d with regular points such that $\dot{g}_d > \epsilon$, for small ϵ. More generally, the reparameterization curve will imply diffeomorphisms if $\dot{t}_1 > 0 \wedge \dot{t}_2 > 0$.

2.4   Reparameterization Matching Energy Functions

Minimizing variational energies allows reparameterization of curve pairs with respect to geometric features. We first form the energy space, E, over which the variational problem is defined,

$$E\colon \tau_1 \times \tau_2 \to \mathbb{R}^+.$$

The Euclidean distance between a pair of real valued functions (f_1, f_2) defined over the arc lengths of (C_1, C_2) gives a valid energy space. Then, for any pairing (s_1, s_2) of points on the curve pair, we have some value for E, which depends on the choice of (f_1, f_2).

The reparameterization curve, g, will be a minimum energy path in E, such that,

$$g = \operatorname{argmin} E(f_1, f_2, g). \tag{3}$$

Application of the optimal g will then reparameterize the curves such that their points are aligned with respect to the features measured by E. We illustrate with an energy function that measures differences in curvature,

$$E(g) = \int ((\kappa_1 \circ t_1 - \kappa_2 \circ t_2)^2)^p d\xi. \tag{4}$$

Recall that curvature is invariant to rotation and translation and that affine-invariant curvatures also exist [37]. The power, p, defines the steepness of the energy landscape.

2.5   Matching Algorithm Implementation

Dynamic programming provides an efficient method for minimizing functionals similar to those given above. Given optimal substructure, DP algorithms yield a global minimum (maximum) cost path in the discrete domain. Because our energies are all based on the Euclidean distance, the triangle inequality guarantees that we have such substructure. We refer the reader to [32] for proof and for the algorithm's implementation. A brief summary of one iteration in the DP algorithm follows. For other implementations, see [33].

The admissible search region, N, respects monotonicity and is in the neighborhood of $\xi_{\min}$, the optimal point on some g. The total cost at each specific

---

Algorithm 1 *Dynamic Programming Curve Match*

$\xi_{\min} = \min(D) \notin K$.

$K \leftarrow \xi_{\min} \cup K$.

for all $\xi_i \in N(\xi_{\min}) = \{\xi_1, \cdots, \xi_n\}$ **do**

   $D(\xi_i) \leftarrow$ **if** $D(\xi_i) > D(\xi_{\min}) + d(\xi_{\min}, \xi_i)$ **then** $D(\xi_{\min}) + d(\xi_{\min}, \xi_i)$.

end for

---

match point, $D \colon \xi \to \mathbb{R}^+$, is initialized to ∞ for all except the starting points which are zero. The set $K = \{$ Known match costs $\}$ is initialized to $\emptyset$. This solution stops when the end of the curves are set as the new $\xi_{\min}$. The value of $d(\xi_{\min}, \xi_i)$ is derived from the correct application-specific formulation. One form for the local cost is,

$$d(\xi_{\min}, \xi) = \|\dot{g}\| \, M(\xi), \tag{5}$$

which gives a trapezoidal approximation to the integral (4). The matrix M is a discrete representation of the energy space, E, defined above. We access the matrix, M, with the 2-dimensional index ξ_i. We may set the neighborhood $N = \xi_{\min} + \{(1,0), (1,1), (0,1)\}$. Alternative choices can be made [10]. The speed of the algorithm depends upon this neighborhood choice as well as the method used to calculate the minimum cost at each iteration. We use a greedy approach which increases the efficiency of the DP solution from $O(N^2)$ to $O(N \log N)$. Methods for finding an optimal starting point for the DP solution may be found in [34,1].

The correspondence curve g is found by backtracking along the minimum path given by the DP solution. Note that this curve is constrained to lie in the discrete intervals provided by the original sampling of the curves. Dense samplings of the curve pair increase the resolution of the correspondence curve but also increase the execution time. Thus, a novel post-processing step that both allows a coarser sampling of the curves and approximates a continuous solution is used.

2.6 Continuous Approximation from Discrete Solution

The discrete solution often causes artificial many-to-one correspondences. This is undesirable from a theoretical point of view as well as for visualization. Serra's method [11,25] estimates sub-pixel costs during the DP solution. While possible for cost functions that depend only on position or first derivatives, it becomes problematic when using higher-order or vector-valued measurements. This correction also comes with a significant time cost.

An efficient second energy minimization step, based on Gaussian smoothing, is introduced to better approximate a continuous solution. Consider the energy, $E(G_\sigma \star g)$, where g is the 1D diffeomorphic projection of the discrete solution. We then perform a line search on the energy (e.g. the curvature difference) with σ as the variable over which to minimize. Note that if σ is 0, we get the discrete solution and that as σ becomes large, g approaches the identity mapping.

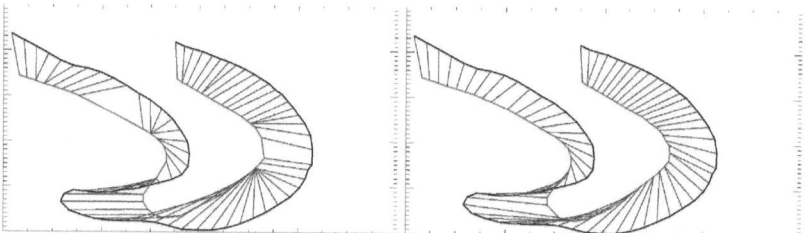

Fig. 2. A detail of the vector field arising from a discrete (left) and continuous (right) matching is shown.

Furthermore, if the discrete g is monotonically increasing, then $G_\sigma \star g$ will be as well (due to the diffusion properties of Gaussian filtering), thus preserving topology. This procedure is particularly important in morphometry, which relies on smooth, diffeomorphic vector fields to make statistical measurements of shape. Discretization error, as shown in Figure 1, reduces the confidence of the measurements. The continuous correction is also shown. The effect on the vector field is shown in Figure 2.

3   Matching Smoothed Curves

Parametric curve matching within a framework of linear diffusion provides robustness to noise and a multi-scale solution space, but also requires additional consideration. We first give an example that illustrates the usefulness of the diffusion space for noise reduction. Next, two alternatives for using multi-scale information are introduced. The first incorporates information from many scales simultaneously and the second selects the best scale for solving a given problem. Toward both of these ends, a modification of the curvature scale-space image, the curvature conservative scale-space image, is defined. This image is incoporated in our variational framework to define a robust-to-noise curvature matching function. We follow this with a tool for finding the coarse-to-fine reparameterization needed for matching a curve between scales. Finally, a scale-selection principle, based on global topology, is given for efficiently choosing the smoothing parameters for matching.

3.1   Motivation for Linear Diffusion

Curvature is a very useful description in computer vision due to its invariance properties. However, its computation is subject to noise due to its dependence on second order derivatives. In real applications, noise can almost always be expected to exist. For example, consider the case of matching a template to new data. The new data often originates from another image processing or curve extraction algorithm. We simulate this situation with an image of a Mig-29 and its corruption with normally distributed noise, as shown in Figure 3. The matching is done with curvature at both full-scale and at a coarser scale. The full-scale matching fails, but a modest amount of smoothing results in an adequate

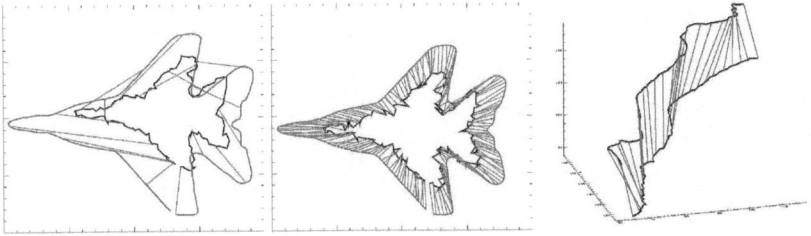

Fig. 3. This example shows the result of matching with curvature at full scale (left) and matching with the curvature scale-space image (center). An example 3D matching of real human-extracted sulcal curves with the same algorithm is at right. The lines connecting the contours show the correspondences.

match. However, this does not solve the problem of how to select the amount of smoothing to apply, nor does it address the issue of how smoothing affects the parameterization.

3.2   Intra-Curve Inter-Scale Matching

Mokhtarian noted in [26] that smoothing does not preserve the parameterization of a curve. That is, diffusion induces a reparameterization, which deviates from the identity as the smoothing parameter increases. To match at a coarse scale and apply this solution at a finer scale, we estimate the reparameterization caused by the smoothing.

We propose to approximate this reparameterization curve by tracking curvature extrema through scale-space. Curvature extrema are known to capture visually salient features [35] and are more stable than inflection points [20]. Zero-crossings only bracket curve features, whereas curvature extrema explicitly mark feature peaks or valleys, as shown in Figure 4. By following these features through scale space, we locate "landmark" points which tell us how the parameterization is changing. We use a simple version of the mean shift algorithm [36] to follow extrema. The mean shift kernel typically covers three discrete curve points and centers over curve points which provide maximum curvature. The kernel is initialized at a coarse scale extremum which is then followed to its fine-scale final position. The curve parameter of this final position is paired with the curve parameter at the coarse scale to provide a single point of the inter-scale, intra-curve reparameterization curve. This process is repeated for the desired number of coarse-scale extrema. Linear interpolation is used between these landmark points, along with a small amount of smoothing, to find the final reparameterization curve, g_σ, as in Figure 1. This curve should be composed with a coarse-scale correspondence curve, g_c, to find the fine-scale correspondence, s.t. $g_f = g_c \circ g_\sigma$. This procedure effectively tracks the lifetime of curvature extrema which can be useful for automatic landmarking and in computing characteristic lengths of features.

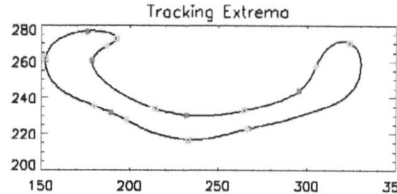 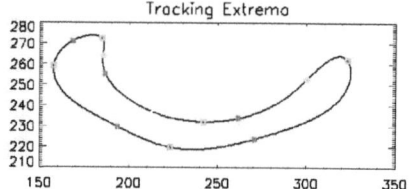

Fig. 4. A coarse version of a corpus callosum is shown on the left and a smooth version on the right. Curvature maxima are marked with an open green square. Curvature minima are marked with a small closed red square. Zero-crossings are marked with a blue diamond.

3.3 Curvature Conservative Scale-Space Image and Variational Energy

We now introduce a modification of the curvature scale-space image, the curvature conservative scale-space image, for use with variational matching.

The curvature scale-space image is defined as the curvature along the curve parameter versus the scale parameter. If the curve parameter is normalized to the interval $[0,1]$ at every scale, then this defines a multi-valued vector function of the curve parameter. This approach, however, weights the finer scales more than coarser scales. The *curvature conservative scale-space image* preserves the integral of the curvature across scales. A pair of these images is shown in Figure 5. Denote the non-smoothed curvature function as κ and its integral over the domain of the curve as $\bar{\kappa} = \int_0^1 \kappa(t)dt$. The curvature conservative scale space function is then defined as,

$$\Pi(t,\sigma) = \frac{\kappa(t,\sigma)}{\int \kappa(t,\sigma)}\bar{\kappa}, \tag{6}$$

where $\kappa(t,\sigma)$ denotes the curvature of $G_\sigma \star C$. Given this definition, Euclidean distance measures the difference of a pair (Π_1, Π_2) at parameter pair (t_1, t_2). The variational energy is,

$$E(g) = \int (\|\Pi_1 \circ t_1 - \Pi_2 \circ t_2\|)^p d\xi. \tag{7}$$

We have found that a good heuristic choice of p is $\frac{1}{2}$. This effectively makes the energy "valleys" less steep. An example of the improvement of using this metric over single-scale curvature is in Figure 6. The advantage of using this method on real data is shown in Figure 7.

Issues with this measure are the computational cost as well as the fact that fine scales may not be desirable at all, due to noise. Pruning the number of scales we use in the evaluation of the Euclidean distance,

$$D(\Pi_1(t_1), \Pi_2(t_2)),$$

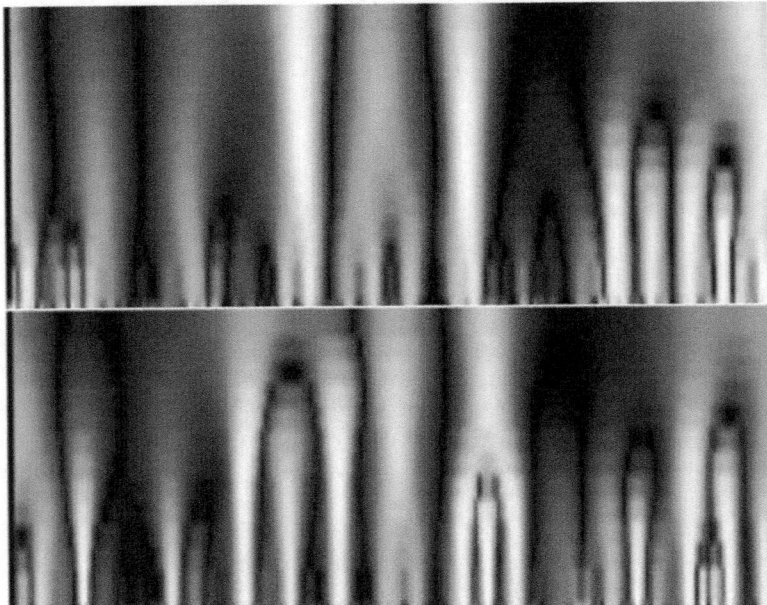

Fig. 5. Here we show a pair of curvature conservative scale-space images for a deer (top) and moose (bottom).

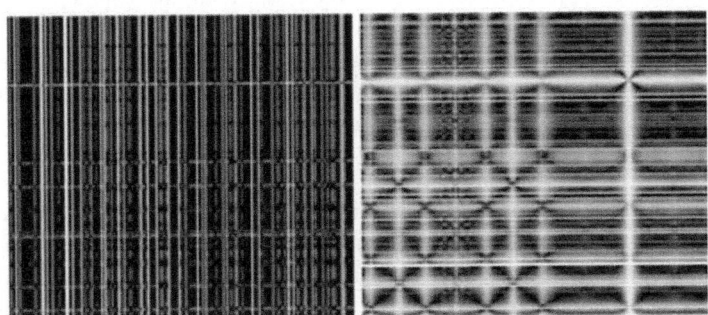

Fig. 6. This figure displays the improvement in the curvature matching energy space by using the curvature conservative scale-space image. The image on the left gives the Euclidean distance of the curvature functions at all possible pairings of Mig-29 and noisy Mig-29 arc lengths. No clear path is visible. The image on the right, however, uses the curvature conservative scale-space image. The dark line along the diagonal is the minimal path, as expected.

alleviates these drawbacks. Principled selection of scale, combined with an estimate of the reparameterization caused by Gaussian smoothing, will allow a coarse scale for finding the correspondence and then application at the fine scale.

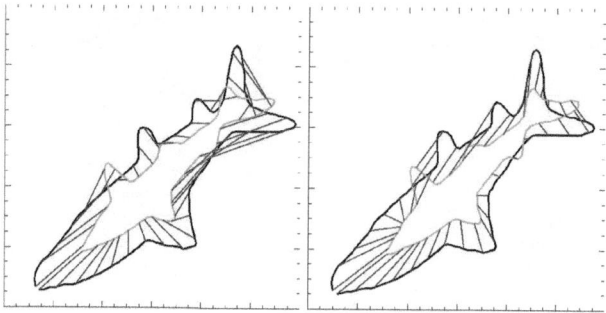

Fig. 7. The result of matching the shark and the bonefish with curvature is shown on the left, with a total curvature difference of 3.35. The matching is confused by the presence of features that exist on one contour and not the other. The result of matching via minimization of the curvature conservative variational energy is shown on the right. A natural solution is found although the full-scale curvature difference is larger, 5.4.

3.4 Cross-Product Space of Extrema Topology for Choosing Matching Scales

Lindeberg's scale-selection methods [18] focus on locating scales which expose the characteristic lengths of structures in images. Our approach differs in its basic hypothesis - we hope to locate scales at which the topology of our data pair, (C_1, C_2), is similar. Topology, here, refers to the number of local extrema in curvature (or some other function) over the global domain (which we consider as equivalent to "holes"). Note that these properties do not change under affine transformation, except in degenerate cases (e.g. the circle).

Scale-Comparison Function Via Number of Extrema. We define a curvature scale-space based descriptor similar to [15], except that we focus on extrema, rather than zero-crossings. Extrema are used due to their stability and because zero-crossings do not exist unless the curvature is signed. Furthermore, as a curve becomes convex, zero-crossings disappear although maxima and minima still exist. The function is defined as follows, $\forall \sigma \in \{\sigma_i \mid 1 \leq \sigma_i \leq L\}$ compute $C(s, \sigma) = C(s) \star G_\sigma$, denoting convolution with a σ-width Gaussian. Then $F \colon C(s, \sigma) \to n$ where n is an integer counting the number of curvature extrema. Note that this function, F, is weakly invariant to affine transformations at a single scale. Examples of this function for a pair of camels is shown in Figure 9. Changing the scale at which F is evaluated will not affect affine invariance if the correct scale-space is used [37].

We now define a distance image based on F. Define $E_F \colon \sigma_1 \times \sigma_2 \to \mathbb{R}^+$ as the Euclidean distance function in the cross-product space of (F_1, F_2). Thus, $E_F(\sigma_1, \sigma_2) = (F_1(\sigma_1) - F_2(\sigma_2))^2$. This comparison is motivated by the assumption that our curves are best matched when their curvature functions have similar topology. Note that one may follow the basic principle to construct functions

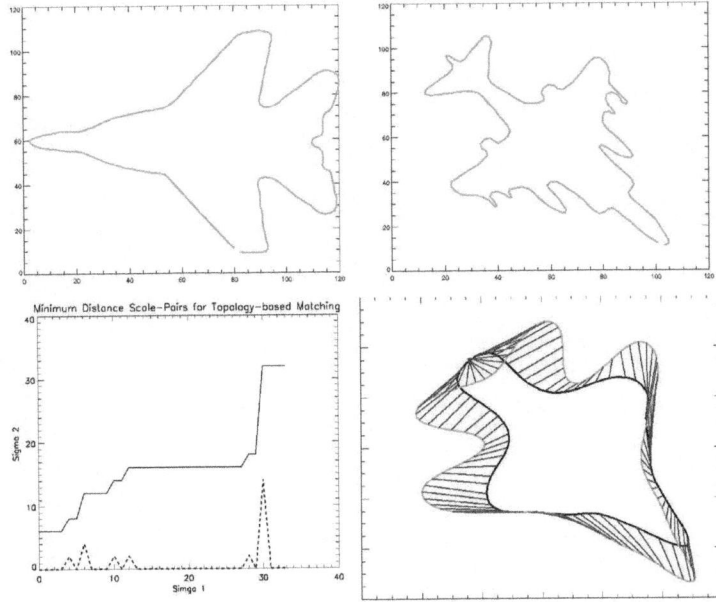

Fig. 8. The top left image shows the Mig-29 contour, C_1. The top right shows the harrier, C_2. These curves do not match well at full-scale. The bottom left shows the result of the extrema topology comparison. The dotted curve overlaid is the size of the derivative of the curve. The second largest of these was selected for use in the matching shown on the bottom right. The largest also provided a reasonable matching, but was oversmoothed.

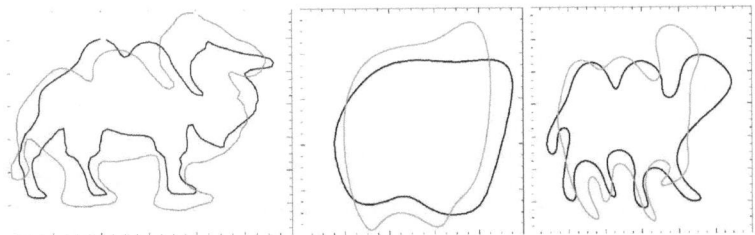

Fig. 9. A pair of camel contours are shown on the left. A coarse (center) and fine (right) scale selection for matching is also shown.

similar to E_F for scale-selection with either images (measuring a function of intensity) or curves.

One then chooses one of the two axes of E_F (suppose σ_1) and extracts the location, (here, along σ_2), of the minimum of E_F. This is done for each value of σ_1, constructing a plot as in Figure 8. An advantage for choosing F in this manner is that it produces a monotic function. The salient features on this curve (where the first derivative is large) are chosen to define scales for matching. This

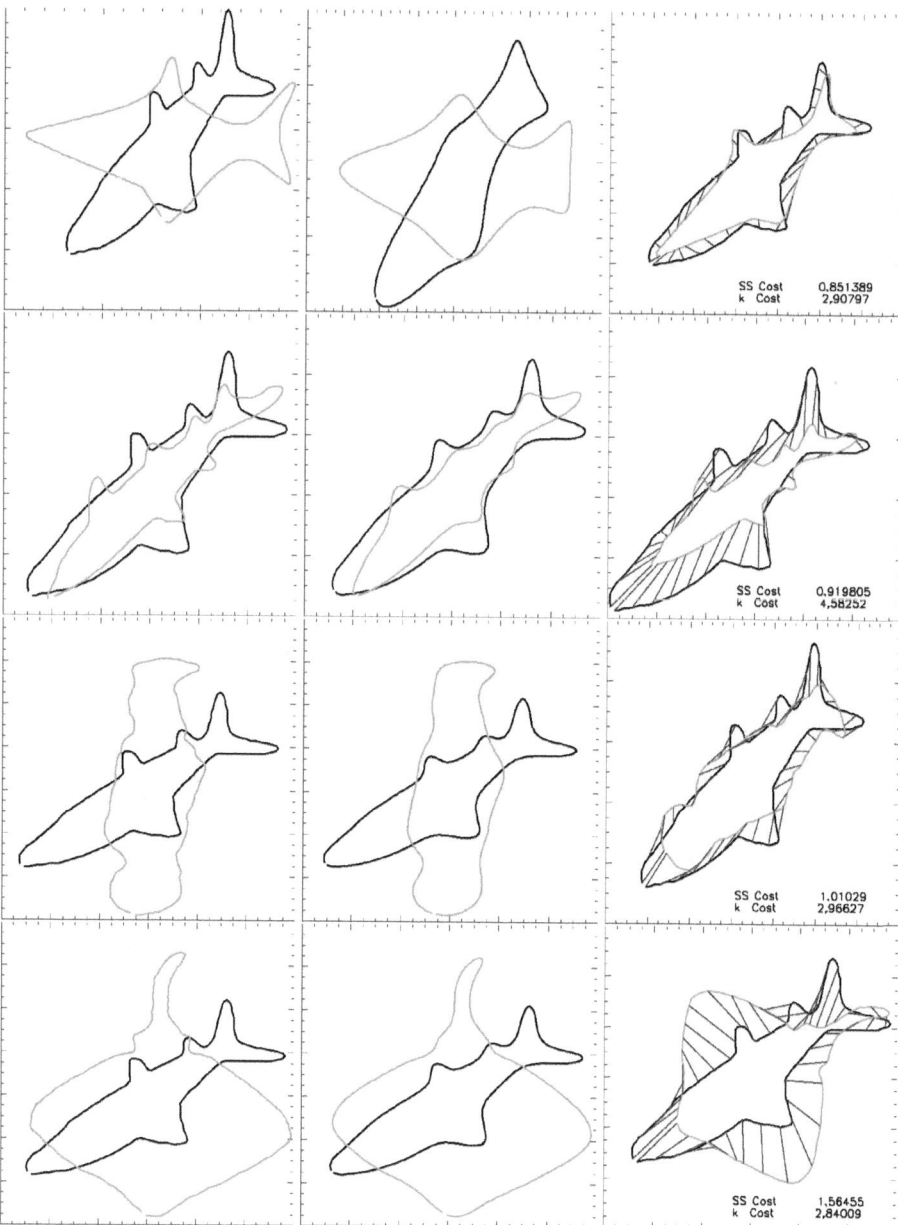

Fig. 10. We show representative results of a study that used scale-selection with the curvature conservative variational energy. The left column shows the initialization. The center column shows the scale selection. The last column shows the resulting matching as well as the curvature (k) and curvature conservative (SS) costs. Note that the latter cost better reflects visual intuition.

curve will always have slope $\geq$ zero because of the simplifying properties of the Gaussian kernel. These points are rank-ordered by the slope in a set of triplets, $(\sigma_1, \sigma_2, \frac{d\sigma_2}{d\sigma_1})$. Thus, matching occurs at scales that guarantee similar (not necessarily identical) topology. Typically, one need only select one or two points from this curve. It is often useful to choose the scales where $\sigma_1 = \sigma_2$ and $F_1 = F_2$, for very noisy but similarly shaped data. A result of the process is shown in Figure 8. Some representative results from an object recognition and comparison study are shown in Figure 10. Comparing the cost of the curvature and the curvature conservative scale-space variational energies, we conclude that the latter reflects visual intuition more closely, although this needs to be validated with human experiments.

4   Conclusion

A variational parametric curve matching algorithm as well as scale-space enhancements to its performance and robustness were introduced. Linear diffusion was applied for locating a continuous correspondence solution from the discrete initialization provided by the traditional dynamic programming geodesic path. We then investigated the use of multi-scale information within this variational framework. In the process, a tool for intra-curve, inter-scale matching is given. This extra level of matching is not always necessary in practice, but becomes more important as the complexity of one's curves increases. Curvature was used at multiple scales simultaneously, avoiding the need for scale-selection. However, a practical topological principle for selecting scales for curve matching was also developed. In summary, we provided a continuous improvement to DP curve matching as well as two approaches to using multi-scale curvature information for curvature-based matching. We hope to perform more extensive numerical tests of the methods given here in the future. We also intend to compare minimization of these variational energies with human perception.

References

1. T. Sebastian and B. Kimia, "On curve alignment," *IEEE Trans. Pattern Analysis and Machine Intelligence*, vol. 25, no. 1, pp. 116–124, 2003.
2. A. Amini, T. Weymouth, and R. Jain, "Using dynamic programming for solving variational problems in vision," *IEEE Trans. on Pattern Analysis and Machine Intelligence*, vol. 12, no. 9, pp. 855–867, 1990.
3. N. Paragios and R. Deriche, "Geodesic active contours for supervised texture segmentation," in *Computer Vision and Pattern Recognition*, 1999, pp. 422–427.
4. T. Moons, E. J. Pauwels, L. J. van Gool, and A. Oosterlinck, "Foundations of semi-differential invariants," *International Journal of Computer Vision*, vol. 14, pp. 25–47, 1995.
5. T. Liu and D. Geiger, "Approximate tree matching and shape similarity," *International Conference on Computer Vision*, pp. 456–462, 1999.
6. P. Yushkevich, S. Pizer, S. Joshi, and J. Marron, "Intuitive, localized analysis of shape variability," *Image Processing in Medical Imaging*, pp. 402–408, 2001.

7. K. Siddiqi, A. Shokoufandeh, S. Dickinson, and S. Zuker, "Shock graphs and shape matching," *Proc. of International Conference on Computer Vision*, pp. 222–229, 1998.

8. D. Geiger, T. Liu, and R. Kohn, "Representation and self-similarity of shapes," *IEEE Trans. Pattern Analysis and Machine Intelligence*, vol. 25, pp. 86–99, 2003.

9. L. Younes, "Computable elastic distance between shapes," *SIAM J. Appl. Math*, vol. 58, pp. 565–586, 1998.

10. T. Sebastian, P. Klein, B. Kimia, and J. Crisco, "Constructing 2D curve atlases," *Mathematical Methods in Biomedical Image Analysis*, pp. 70–77, 2000.

11. B. Serra and M. Berthold, "Subpixel contour matching using continuous dynamic programming," *Computer Vision and Pattern Recognition*, pp. 202–207, 1994.

12. H. Wolfson, "On curve matching," *IEEE Trans. Pattern Analysis and Machine Intelligence*, vol. 12, no. 5, pp. 483–489, 1990.

13. S. Sclaroff and A. Pentland, "Modal matching for correspondence and recognition," *IEEE Trans. Pattern Analysis and Machine Intelligence*, vol. 17, pp. 545–561, 1995.

14. A. Trouve and L. Younes, "On a class of diffeomorphic matching problems in one dimension," *SIAM Journal on Control and Optimization*, vol. 39, no. 4, pp. 1112–1135, 2000.

15. F. Mokhtarian and A. Mackworth, "Scale-based description and recognition of planar curves and two-dimensional shapes," *IEEE Trans. Pattern Analysis and Machine Intelligence*, vol. 8, pp. 34–44, 1986.

16. J. Babaud, A. Witkin, M. Baudin, and R. Duda, "Uniqueness of the gaussian kernel for scale space filtering," *IEEE Trans. Pattern Analysis and Machine Intelligence*, vol. 8, no. 1, pp. 26–33, 1986.

17. T. Lindeberg, "Scale-space for discrete signals," *IEEE Trans. Pattern Analysis and Machine Intelligence*, vol. 12, pp. 234–254, March 1990.

18. T. Lindeberg, "Feature detection with automatic scale selection," *International Journal of Computer Vision (IJCV)*, vol. 30, no. 2, pp. 79–116, 1998.

19. T. Lindeberg, "A scale selection principle for estimating image deformations," *Image and Vision Computing*, vol. 16, pp. 961–997, 1998.

20. J. Sporring, X. Zabulis, P.E. Trahanias, and S. Orphanoudakis, "Shape similarity by piecewise linear alignment," in *Proceedings of the Fourth Asian Conference on Computer Vision (ACCV'00)*, Taipei, Taiwan, January 2000, pp. 306–311.

21. H.D. Tagare, D. O'Shea, and A. Rangarajan, "A geometric criterion for shape based non-rigid correspondence," *Fifth Intl. Conf. on Computer Vision*, pp. 434–439, 1995.

22. H.D. Tagare, "Shape-based nonrigid correspondence with application to heart motion analysis," *IEEE Trans. on Medical Imaging*, vol. 18, no. 7, pp. 570–579, 1999.

23. A.A. Amini and J.S. Duncan, "Pointwise tracking of left-ventricular motion in 3D," *WVM*, vol. 91, pp. 294–299, 1995.

24. A. Viterbi, "Error bounds for convolutional codes and an asymptotically optimum decoding algorithm," *IEEE Trans. on Information Theory*, vol. 13, no. 2, pp. 260–269, 1967.

25. B. Serra and M. Berthod, "Optimal subpixel matching of contour chains and segments," *IEEE Proceedings of the International Conference on Computer Vision*, pp. 402–407, 1995.

26. F. Mokhtarian and A. Mackworth, "A theory of multiscale, curvature-based shape representation for planar curves," *IEEE Trans. Pattern Analysis and Machine Intelligence*, vol. 14, no. 8, pp. 789–805, 1992.

27. D. Sharvit, J. Chan, H. Tek, and B. Kimia, "Symmetry-based indexing of image databases," *Journal of Visual Communication and Image Representation*, vol. 9, no. 4, pp. 366–380, December 1998.

28. T. Sebastian, P. Klein, and B. Kimia, "Recognition of shapes by editing shock graphs," in *International Conference on Computer Vision*, 2001, pp. 755–762.

29. L. Piegl and W. Tiller, *The NURBS Book*, Springer-Verlag, 1995.

30. I. Lee, "Curve reconstruction from unorganized points," *Computer Aided Design*, vol. 17, pp. 161–177, 2000.

31. M. DoCarmo, *Differential Geometry of Curves and Surfaces*, Prentice-Hall, 1976.

32. R. Rivest, T. Cormen, and C. Leiserson, *Introduction to Algorithms*, MIT Press, Cambridge, MA, 1992.

33. B. Cherkassky, A. Goldberg, and T. Radzik, "Shortest path algorithms: Theory and experimental evaluation," in *Proceedings of 5th Annual ACM SIAM Symposium on Discrete Algorithms*, 1994, pp. 516–525.

34. J. Marques and A. Abrantes, "Shape alignment optimal initial point and pose estimation," *Pattern Recognition Letters*, vol. 18, no. 1, pp. 49–53, 1997.

35. H. Asada and M. Brady, "The curvature primal sketch," *IEEE Trans. on Pattern Analysis and Machine Intelligence*, vol. 8, pp. 2–14, 1986.

36. D. Comaniciu, V. Ramesh, and P. Meer, "Real-time tracking of non-rigid objects using mean shift," in *Computer Vision and Pattern Recognition*, 2000, pp. 142–151.

37. G. Sapiro and A. Tannenbaum, "Affine invariant scale-space," *International Journal of Computer Vision*, vol. 11, no. 1, pp. 25–44, 1993.

Author Index

Lecture Notes in Computer Science

For information about Vols. 1–2599

please contact your bookseller or Springer-Verlag

Vol. 2641: P.J. Nürnberg (Ed.), Metainformatics. Proceedings, 2002. VIII, 187 pages. 2003.

Vol. 2642: X. Zhou, Y. Zhang, M.E. Orlowska (Eds.), Web Technologies and Applications. Proceedings, 2003. XIII, 608 pages. 2003.

Vol. 2643: M. Fossorier, T. Høholdt, A. Poli (Eds.), Applied Algebra, Algebraic Algorithms and Error-Correcting Codes. Proceedings, 2003. X, 256 pages. 2003.

Vol. 2644: D. Hogrefe, A. Wiles (Eds.), Testing of Communicating Systems. Proceedings, 2003. XII, 311 pages. 2003.

Vol. 2645: M.A. Wimmer (Ed.), Knowledge Management in Electronic Government. Proceedings, 2003. XI, 320 pages. 2003. (Subseries LNAI).

Vol. 2646: H. Geuvers, F, Wiedijk (Eds.), Types for Proofs and Programs. Proceedings, 2002. VIII, 331 pages. 2003.

Vol. 2647: K.Jansen, M. Margraf, M. Mastrolli, J.D.P. Rolim (Eds.), Experimental and Efficient Algorithms. Proceedings, 2003. VIII, 267 pages. 2003.

Vol. 2648: T. Ball, S.K. Rajamani (Eds.), Model Checking Software. Proceedings, 2003. VIII, 241 pages. 2003.

Vol. 2649: B. Westfechtel, A. van der Hoek (Eds.), Software Configuration Management. Proceedings, 2003. VIII, 241 pages. 2003.

Vol. 2651: D. Bert, J.P. Bowen, S. King, M, Waldén (Eds.), ZB 2003: Formal Specification and Development in Z and B. Proceedings, 2003. XIII, 547 pages. 2003.

Vol. 2652: F.J. Perales, A.J.C. Campilho, N. Pérez de la Blanca, A. Sanfeliu (Eds.), Pattern Recognition and Image Analysis. Proceedings, 2003. XIX, 1142 pages. 2003.

Vol. 2653: R. Petreschi, Giuseppe Persiano, R. Silvestri (Eds.), Algorithms and Complexity. Proceedings, 2003. XI, 289 pages. 2003.

Vol. 2656: E. Biham (Ed.), Advances in Cryptology – EUROCRPYT 2003. Proceedings, 2003. XIV, 649 pages. 2003.

Vol. 2657: P.M.A. Sloot, D. Abramson, A.V. Bogdanov, J.J. Dongarra, A.Y. Zomaya, Y.E. Gorbachev (Eds.), Computational Science – ICCS 2003. Proceedings, Part I. 2003. LV, 1095 pages. 2003.

Vol. 2658: P.M.A. Sloot, D. Abramson, A.V. Bogdanov, J.J. Dongarra, A.Y. Zomaya, Y.E. Gorbachev (Eds.), Computational Science – ICCS 2003. Proceedings, Part II. 2003. LV, 1129 pages. 2003.

Vol. 2659: P.M.A. Sloot, D. Abramson, A.V. Bogdanov, J.J. Dongarra, A.Y. Zomaya, Y.E. Gorbachev (Eds.), Computational Science – ICCS 2003. Proceedings, Part III. 2003. LV, 1165 pages. 2003.

Vol. 2660: P.M.A. Sloot, D. Abramson, A.V. Bogdanov, J.J. Dongarra, A.Y. Zomaya, Y.E. Gorbachev (Eds.), Computational Science – ICCS 2003. Proceedings, Part IV. 2003. LVI, 1161 pages. 2003.

Vol. 2663: E. Menasalvas, J. Segovia, P.S. Szczepaniak (Eds.), Advances in Web Intelligence. Proceedings, 2003. XII, 350 pages. 2003. (Subseries LNAI).

Vol. 2665: H. Chen, R. Miranda, D.D. Zeng, C. Demchak, J. Schroeder, T. Madhusudan (Eds.), Intelligence and Security Informatics. Proceedings, 2003. XIV, 392 pages. 2003.

Vol. 2667: V. Kumar, M.L. Gavrilova, C.J.K. Tan, P. L'Ecuyer (Eds.), Computational Science and Its Applications – ICCSA 2003. Proceedings, Part I. 2003. XXXIV, 1060 pages. 2003.

Vol. 2668: V. Kumar, M.L. Gavrilova, C.J.K. Tan, P. L'Ecuyer (Eds.), Computational Science and Its Applications – ICCSA 2003. Proceedings, Part II. 2003. XXXIV, 942 pages. 2003.

Vol. 2669: V. Kumar, M.L. Gavrilova, C.J.K. Tan, P. L'Ecuyer (Eds.), Computational Science and Its Applications – ICCSA 2003. Proceedings, Part III. 2003. XXXIV, 948 pages. 2003.

Vol. 2670: R. Peña, T. Arts (Eds.), Implementation of Functional Languages. Proceedings, 2002. X, 249 pages. 2003.

Vol. 2671: Y. Xiang, B. Chaib-draa (Eds.), Advances in Artificial Intelligence. Proceedings, 2003. XIV, 642 pages. 2003. (Subseries LNAI).

Vol. 2674: I.E. Magnin, J. Montagnat, P. Clarysse, J. Nenonen, T. Katila (Eds.), Functional Imaging and Modeling of the Heart. Proceedings, 2003. XI, 308 pages. 2003.

Vol. 2675: M. Marchesi, G. Succi (Eds.), Extreme Programming and Agile Processes in Software Engineering. Proceedings, 2003. XV, 464 pages. 2003.

Vol. 2676: R. Baeza-Yates, E. Chávez, M. Crochemore (Eds.), Combinatorial Pattern Matching. Proceedings, 2003. XI, 403 pages. 2003.

Vol. 2678: W. van der Aalst, A. ter Hofstede, M. Weske (Eds.), Business Process Management. Proceedings, 2003. XI, 391 pages. 2003.

Vol. 2679: W. van der Aalst, E. Best (Eds.), Applications and Theory of Petri Nets 2003. Proceedings, 2003. XI, 508 pages. 2003.

Vol. 2686: J. Mira, J.R. Álvarez (Eds.), Computational Methods in Neural Modeling. Proceedings, Part I. 2003. XXVII, 764 pages. 2003.

Vol. 2687: J. Mira, J.R. Álvarez (Eds.), Artificial Neural Nets Problem Solving Methods. Proceedings, Part II. 2003. XXVII, 820 pages. 2003.

Vol. 2688: J. Kittler, M.S. Nixon (Eds.), Audio- and Video-Based Biometric Person Authentication. Proceedings, 2003. XVII, 978 pages. 2003.

Vol. 2692: P. Nixon, S. Terzis (Eds.), Trust Management. Proceedings, 2003. X, 349 pages. 2003.

Vol. 2694: R. Cousot (Ed.), Static Analysis. Proceedings, 2003. XIV, 505 pages. 2003.

Vol. 2695: L.D. Griffin, M. Lillholm (Eds.), Scale Space Methods in Computer Vision. Proceedings, 2003. XII, 816 pages. 2003.

Vol. 2701: M. Hofmann (Ed.), Typed Lambda Calculi and Applications. Proceedings, 2003. VIII, 317 pages. 2003.

Vol. 2706: R. Nieuwenhuis (Ed.), Rewriting Techniques and Applications. Proceedings, 2003. XI, 515 pages. 2003.

Vol. 2707: K. Jeffay, I. Stoica, K. Wehrle (Eds.), Quality of Service – IWQoS 2003. Proceedings, 2003. XI, 517 pages. 2003.